PHYSICS

For Scientists and Engineers

Fifth Edition

Solar System Data

Body	Mass (kg)	Mean Radius (m)	Period (s)	Distance from the Sun (m)
Mercury	3.18×10^{23}	2.43×10^6	7.60×10^6	5.79×10^{10}
Venus	4.88×10^{24}	6.06×10^6	1.94×10^7	1.08×10^{11}
Earth	5.98×10^{24}	6.37×10^6	3.156×10^7	1.496×10^{11}
Mars	6.42×10^{23}	3.37×10^6	5.94×10^7	2.28×10^{11}
Jupiter	1.90×10^{27}	6.99×10^7	3.74×10^8	7.78×10^{11}
Saturn	5.68×10^{26}	5.85×10^7	9.35×10^8	1.43×10^{12}
Uranus	8.68×10^{25}	2.33×10^7	2.64×10^9	2.87×10^{12}
Neptune	1.03×10^{26}	2.21×10^7	5.22×10^9	4.50×10^{12}
Pluto	$\approx 1.4 \times 10^{22}$	$\approx 1.5 \times 10^6$	7.82×10^9	5.91×10^{12}
Moon	7.36×10^{22}	1.74×10^6	—	—
Sun	1.991×10^{30}	6.96×10^8	—	—

Physical Data Often Used[a]

Average Earth–Moon distance	3.84×10^8 m
Average Earth–Sun distance	1.496×10^{11} m
Average radius of the Earth	6.37×10^6 m
Density of air (0°C and 1 atm)	1.29 kg/m^3
Density of water (20°C and 1 atm)	1.00×10^3 kg/m^3
Free-fall acceleration	9.80 m/s^2
Mass of the Earth	5.98×10^{24} kg
Mass of the Moon	7.36×10^{22} kg
Mass of the Sun	1.99×10^{30} kg
Standard atmospheric pressure	1.013×10^5 Pa

[a] These are the values of the constants as used in the text.

Some Prefixes for Powers of Ten

Power	Prefix	Abbreviation	Power	Prefix	Abbreviation
10^{-24}	yocto	y	10^1	deka	da
10^{-21}	zepto	z	10^2	hecto	h
10^{-18}	atto	a	10^3	kilo	k
10^{-15}	femto	f	10^6	mega	M
10^{-12}	pico	p	10^9	giga	G
10^{-9}	nano	n	10^{12}	tera	T
10^{-6}	micro	μ	10^{15}	peta	P
10^{-3}	milli	m	10^{18}	exa	E
10^{-2}	centi	c	10^{21}	zetta	Z
10^{-1}	deci	d	10^{24}	yotta	Y

PHYSICS
For Scientists and Engineers

Fifth Edition

Volume 1

Raymond A. Serway
North Carolina State University

Robert J. Beichner
North Carolina State University

John W. Jewett, Jr., Contributing Author
California State Polytechnic University—Pomona

SAUNDERS GOLDEN SUNBURST SERIES

SAUNDERS COLLEGE PUBLISHING

A Harcourt Higher Learning Company

FORT WORTH   PHILADELPHIA   SAN DIEGO   NEW YORK   ORLANDO   AUSTIN   SAN ANTONIO   TORONTO   MONTREAL   LONDON   SYDNEY   TOKYO

Publisher: Emily Barrosse
Publisher: John Vondeling
Marketing Manager: Pauline Mula
Developmental Editor: Susan Dust Pashos
Project Editor: Frank Messina
Production Manager: Charlene Catlett Squibb
Manager of Art and Design: Carol Bleistine
Text and Cover Designer: Ruth A. Hoover

Cover Image and Credit: Victoria Falls, Zimbabwe, at sunset (© *Schafer & Hill/Tony Stone Images*)
Frontmatter Images and Credits: Title page: ripples on water (© *Yagi Studio/Superstock*); water droplets on flower (© *Richard H. Johnson/FPG International*); p. v: View of Saturn from *Voyager 2 (Courtesy of NASA)*; p. vii: Crocus and bee *(E. Webber/Visuals Unlimited)* and speed skaters *(Bill Bachman/Photo Researchers, Inc.)*; p. viii: Sky surfer *(Jump Run Productions/The Image Bank)* and launch of space shuttle *Columbia (Courtesy of NASA)*; p. ix: Strings in a piano *(Charles D. Winters)* and lightning storm over a desert *(Richard Kaylin/Tony Stone Images)*; p. x: jogger *(Jim Cummins/FPG International)*; p. xi: Penny-farthing bicycle race (© *Steve Lovegrove/ Tasmanian Photo Library*); p. xiii: Long-jumper *(Chuck Muhlstock/FPG International)*; p.xiv: Satellite and astronaut *(Courtesy of NASA)*; p. xv: "Corkscrew" roller coaster *(Robin Smith/Tony Stone Images)*; p.xvi: Cyclists pedaling uphill *(David Madison/Tony Stone Images)*; p. xvii: Twin Falls on the island of Kauai, Hawaii *(Bruce Byers/FPG)*; p. xviii: Bowling ball striking a pin *(Ben Rose/The Image Bank)*; p. xix: Sprinters at staggered starting positions (© *Gerard Vandystadt/Photo Researchers, Inc.)*; p. xx: U.S. Marine Corps CH-53E helicopter in flight *(Ross Harrison Koty/Tony Stone Images)*; p. xxi: Cheering crowd *(Gregg Adams/Tony Stone Images)*; p. xxii: Wind chime *(Joseph L. Fontenot/Visuals Unlimited)*; p. xxiii: Bat in flight *(Joe McDonald/Visuals Unlimited)*; p. xxiv: Welder (© *The Telegraph Colour Library/FPG*); p. xxv: Basketball player dunking ball *(Ron Chapple/FPG International)*; p. xxvi: Athlete throwing discus *(Bruce Ayres/Tony Stone Images)*; p. xxvii: Chum salmon *(Daniel J. Cox/Tony Stone Images)*; p.xxviii: Hurricane Mitch *(Courtesy of NOAA)*; p. xxix: Bottle-nosed dolphin *(Stuart Westmorland/Tony Stone Images)*.

PHYSICS FOR SCIENTISTS AND ENGINEERS, Fifth Edition, Volume 1
ISBN 0-03-020968-4
Library of Congress Catalog Card Number: 99-61820

Address for domestic orders:
Saunders College Publishing, 6277 Sea Harbor Drive, Orlando, FL 32887-6777
1-800-782-4479

Address for international orders:
International Customer Service, Harcourt Brace & Company
6277 Sea Harbor Drive, Orlando FL 32887-6777
(407) 345-3800
Fax (407) 345-4060
e-mail hbintl@harcourtbrace.com

Address for editorial correspondence:
Saunders College Publishing, Public Ledger Building, Suite 1250
150 S. Independence Mall West, Philadelphia, PA 19106-3412

Web Site Address
http://www.hbcollege.com

Contents Overview

Table of Contents

Preface

*I*n writing this fifth edition of *Physics for Scientists and Engineers*, we have made a major effort to improve the clarity of presentation and to include new pedagogical features that help support the learning and teaching processes. Drawing on positive feedback from users of the fourth edition and reviewers' suggestions, we have made refinements in order to better meet the needs of students and teachers. We have also streamlined the supplements package, which now includes a CD-ROM containing student tutorials and interactive problem-solving software, as well as offerings on the World Wide Web.

This textbook is intended for a course in introductory physics for students majoring in science or engineering. The entire contents of the text could be covered in a three-semester course, but it is possible to use the material in shorter sequences with the omission of selected chapters and sections. The mathematical background of the student taking this course should ideally include one semester of calculus. If that is not possible, the student should be enrolled in a concurrent course in introductory calculus.

OBJECTIVES

This introductory physics textbook has two main objectives: to provide the student with a clear and logical presentation of the basic concepts and principles of physics, and to strengthen an understanding of the concepts and principles through a broad range of interesting applications to the real world. To meet these objectives, we have placed emphasis on sound physical arguments and problem-solving methodology. At the same time, we have attempted to motivate the student through practical examples that demonstrate the role of physics in other disciplines, including engineering, chemistry, and medicine.

CHANGES IN THE FIFTH EDITION

A large number of changes and improvements have been made in preparing the fifth edition of this text. Some of the new features are based on our experiences and on current trends in science education. Other changes have been incorporated in response to comments and suggestions offered by users of the fourth edition and by reviewers of the manuscript. The following represent the major changes in the fifth edition:

Improved Illustrations

- **Time-sequenced events** are represented by circled letters in selected mechanics illustrations. For example, Figure 2.1b (see page 25) shows such letters at the appropriate places on a position–time graph. This construction helps students "translate" the observed motion into its graphical representation.

- **Motion diagrams** are used early in the text to illustrate the difference between velocity and acceleration, concepts easily confused by beginning students. (For example, see Figure 2.9 on page 34, Figure 4.5 on page 81, and Figure 4.8 on page 84.) Students will benefit greatly from sketching their own motion diagrams, as they are asked to do in the Quick Quizzes found in Chapter 4.

- **Greater realism** is achieved with the superimposition of photographs and line art in selected figures (see pages 96 and 97). Also, the three-dimensional appearance of "blocks" in figures accompanying examples and problems in mechanics has been improved (see pages 142 and 143).

More Realistic Worked Examples   Readers familiar with the fourth edition may recall that Example 12.5 involved raising a cylinder of radius R onto a step of height h. In this idealized example, we calculated the minimum force **F** necessary to raise the cylinder, as well as the reaction force exerted by the step on the cylinder. In the fifth edition, we are pleased to present the revised Example 12.5 (see page 370), in which we calculate the force that a person must apply to a wheelchair's main wheel to roll it up over an uncut sidewalk curb. Although the revised Example 12.5 involves essentially the same calculation as its fourth-edition predecessor, we think that the increased realism makes the example more interesting and provides new motivation for studying physics.

Puzzlers   Every chapter begins with an interesting photograph and a caption that includes a Puzzler. Each Puzzler poses a thought-provoking question that is intended to motivate students' curiosity and enhance their interest in the chapter's subject matter. Part or all of the answer to each Puzzler is contained within the chapter text and is indicated by the icon.

Chapter Outlines   The opening page of each chapter now includes an outline of the chapter's major headings. This outline gives students and instructors a preview of the chapter's content.

QuickLabs   This new feature encourages students to perform simple experiments on their own, thereby engaging them actively in the learning process. Most QuickLab experiments can be performed with low-cost items such as string, rubber bands, tape, a ruler, drinking straws, and balloons. In most cases, students are asked to observe the outcome of an experiment and to explain their results in terms of what they have learned in the chapter. When appropriate, students are asked to record data and to graph their results.

Quick Quizzes   Several Quick Quiz questions are included in each chapter to provide students with opportunities to test their understanding of the physical concepts presented. Many questions are written in multiple-choice format and require students to make decisions and defend them on the basis of sound reasoning. Some of them have been written to help students overcome common misconceptions. (Instructors should look to the Instructor's Notes in the margins of the Instructor's Annotated Edition for tips regarding certain Quick Quizzes.) Answers to all Quick Quiz questions are found at the end of each chapter.

Marginal Comments and Icons   To provide students with further guidance, common misconceptions and pitfalls are pointed out in comments in the margin of the text. Often, references to the *Saunders Core Concepts in Physics CD-ROM* and useful World Wide Web site addresses are given in these comments to encourage students to expand their understanding of physical concepts. The 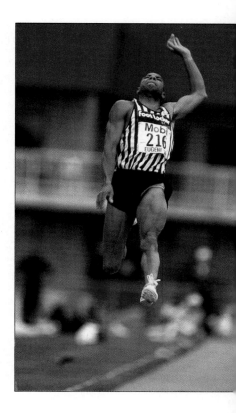 icon in the text margin refers students to the specific module and screen number(s) of the *Saunders Core Concepts in Physics CD-ROM* that deals with the topic under discussion. A text illustration, example, Quick Quiz, or problem marked with the 📷 icon indicates that it is accompanied by an Interactive Physics™ simulation that can be found on the *Student Tools CD-ROM*. See the Student Ancillaries section (page xvii) for descriptions of these two electronic learning packages.

Applications   Some chapters include Applications, which are about the same length as or slightly longer than worked examples. The Applications demonstrate to students how the physical principles covered in a chapter apply to practical problems of everyday life or engineering. For instance, Applications in Volume 1 discuss antilock brakes within the context of static and kinetic friction (see Chapter 5); analyze the tension and compressional forces on the struts of a truss bridge (see Chapter 12); and explore the power delivered in automobile and diesel engine cycles (see Chapter 22).

Problems   A substantial revision of the end-of-chapter problems was made in an effort to improve their clarity and quality. Approximately 20 percent of the problems (about 650) are new, and most of these new problems are at the intermediate level (as identified by blue problem numbers). Many of the new problems require students to make order-of-magnitude calculations. All problems have been carefully edited and reworded when necessary. Solutions to approximately 20 percent of the end-of-chapter problems are included in the *Student Solutions Manual and Study Guide*. These problems are identified in the text by boxes around their numbers. A smaller subset of solutions are posted on the World Wide Web **(http://www.saunderscollege.com/physics/)** and are accessible to students and instructors using *Physics for Scientists and Engineers*. These problems are identified in the text by the **WEB** icon. See the next section for a complete description of other features of the problem set.

Line-by-Line Revision   The entire text has been carefully edited to improve clarity of presentation and precision of language. We believe that the result is a book that is both accurate and enjoyable to read.

Typographical and Notation Changes   The Text Features section (see page xiv) mentions the use of **boldface** type and screens for emphasizing important statements and definitions. Boldfaced passages in the text of the fifth edition replace the less legible passages appearing in italics in the fourth edition. Similarly, the symbols for vectors stand out very clearly from the surrounding text owing to the strong boldface type used in the fifth edition. As a step toward making equations more transparent and therefore more easily understood, the use of the subscripts "i" and "f" for initial and final values replaces the fourth edition's older notation, which makes use of subscript 0 (usually pronounced "naught") for an initial value and no subscript for a final value. In equations describing motion or direction, variables carry the subscripts x, y, or z whenever added clarity is needed.

Content Changes   Examination of the full Table of Contents might lead one to the impression that the content and organization of the textbook are essentially the same as those of the fourth edition. However, a number of subtle yet significant improvements in content have been made. Following are some examples:

- Section 16.8 contains a more complete and careful derivation of the power or rate of energy transfer for sinusoidal waves on strings. A similar development occurs in Section 17.3, which deals with the intensity of periodic sound waves.

- Section 18.2 contains an improved discussion of the envelope function of a standing wave.

- Chapter 20 contains an updated discussion of the distinction between heat and internal energy. Both heat and work are described and clarified as means of changing the energy of a system.

- Chapter 22 contains a new description of microstates and macrostates of a system, beginning with Section 22.6 on entropy and continuing through the end of the chapter.

Many sections in these and other chapters have been streamlined, deleted, or combined with other sections to allow for a more balanced presentation. Looking ahead to Volume 2, the former Chapter 44 on superconductivity in the fourth edition of *Physics for Scientists and Engineers with Modern Physics* has been deleted, and an abridged section on this topic has been added to Chapter 43. Some of the sections deleted from the fourth edition may be found on the textbook's Web sites for both instructors and students.

Instructor's Notes　For the first time, tips and comments are offered to instructors in blue marginal Instructor's Notes, which appear only in the Instructor's Annotated Edition. These annotations expand on common student misconceptions; call attention to certain worked examples, QuickLabs, and Quick Quizzes; or cite key physics education research literature that bears on the topic at hand. In some chapters, Instructor's Notes appear as footnotes in the end-of-chapter problem sets; these notes point out related groups of problems found in other chapters of the textbook. The Instructor's Annotated Edition includes Chapters 1 to 39.

CONTENT

The material in this book covers fundamental topics in classical physics and provides an introduction to modern physics. The book is divided into six parts. Part I (Chapters 1 to 15) deals with the fundamentals of Newtonian mechanics and the physics of fluids, Part II (Chapters 16 to 18) covers wave motion and sound, Part III (Chapters 19 to 22) addresses heat and thermodynamics, Part IV (Chapters 23 to 34) treats electricity and magnetism, Part V (Chapters 35 to 38) covers light and optics, and Part VI (Chapters 39 to 46) deals with relativity and modern physics. Each part opener includes an overview of the subject matter covered in that part, as well as some historical perspectives.

TEXT FEATURES

Most instructors would agree that the textbook selected for a course should be the student's primary guide for understanding and learning the subject matter. Furthermore, the textbook should be easily accessible and should be styled and written to facilitate instruction and learning. With these points in mind, we have included many pedagogical features in the textbook that are intended to enhance its usefulness to both students and instructors. These features are as follows:

Style   To facilitate rapid comprehension, we have attempted to write the book in a style that is clear, logical, and engaging. We have chosen a writing style that is somewhat informal and relaxed so that students will find the text appealing and enjoyable to read. New terms are carefully defined, and we have tried to avoid the use of jargon.

Previews   Most chapters begin with a brief preview that includes a discussion of the chapters' objectives and content.

Important Statements and Equations   Most important statements and definitions are set in boldface type or are highlighted with a tan background screen for added emphasis and ease of review. Similarly, important equations are highlighted with a tan background screen to facilitate location.

Problem-Solving Hints   We have included general strategies for solving the types of problems featured both in the examples and in the end-of-chapter problems. This feature helps students to identify necessary steps in problem-solving and to eliminate any uncertainty they might have. Problem-Solving Hints are highlighted with a light blue-gray screen for emphasis and ease of location.

Marginal Notes   Comments and notes appearing in the margin can be used to locate important statements, equations, and concepts in the text.

Illustrations and Tables   The readability and effectiveness of the text material and worked examples are enhanced by the large number of figures, diagrams, photographs, and tables. Full color is used to add clarity to the artwork and to make illustrations as realistic as possible. For example, vectors are color-coded, and curves in *xy* plots are drawn in color. The three-dimensional appearance of many illustrations has been improved in this fifth edition. The color photographs have been carefully selected, and their accompanying captions have been written so that they serve as an added instructional tool.

Mathematical Level   We have introduced calculus gradually, keeping in mind that students often take introductory courses in calculus and physics concurrently. Most steps are shown when basic equations are developed, and reference is often made to mathematical appendices at the end of the textbook. Vector products are introduced later in the text, where they are needed in physical applications. The dot product is introduced in Chapter 7, which addresses work and energy; the cross product is introduced in Chapter 11, which deals with rotational dynamics.

Worked Examples   A large number of worked examples of varying difficulty are presented to promote students' understanding of concepts. In many cases, the examples serve as models for solving the end-of-chapter problems. The examples are set off in boxes, and the answers to examples with numerical solutions are highlighted with a light blue-gray screen.

Worked Example Exercises   Many of the worked examples are followed immediately by exercises with answers. These exercises are intended to promote interactivity between the student and the textbook and to immediately reinforce the student's understanding of concepts and problem-solving techniques. The exercises represent extensions of the worked examples.

Conceptual Examples   As in the fourth edition, we have made a concerted effort to emphasize critical thinking and the teaching of physical concepts. We have accomplished this by including Conceptual Examples (for instance, see page 41). These examples provide students with a means of reviewing and applying the concepts presented in a section. Some Conceptual Examples demonstrate the connection between concepts presented in a chapter and other disciplines. The Conceptual Examples can serve as models for students when they are asked to respond to end-of-chapter questions, which are largely conceptual in nature.

Questions   Questions requiring verbal responses are provided at the end of each chapter. Over 1,000 questions are included in this edition. Some questions provide students with a means of testing their mastery of the concepts presented in the chapter. Others could serve as a basis for initiating classroom discussions. Answers to selected questions are included in the *Student Solutions Manual and Study Guide.*

Significant Figures   Significant figures in both worked examples and end-of-chapter problems have been handled with care. Most numerical examples and problems are worked out to either two or three significant figures, depending on the accuracy of the data provided.

Problems   An extensive set of problems is included at the end of each chapter; in all, over 3,000 problems are given throughout the text. Answers to odd-numbered problems are provided at the end of the book in a section whose pages have colored edges for ease of location. For the convenience of both the student and the instructor, about two thirds of the problems are keyed to specific sections of the chapter. The remaining problems, labeled "Additional Problems," are not keyed to specific sections.

   Usually, the problems within a given section are presented so that the straightforward problems (those with black problem numbers) appear first; these straightforward problems are followed by problems of increasing difficulty. For ease of identification, the numbers of intermediate-level problems are printed in blue, and those of a small number of challenging problems are printed in magenta.

Review Problems   Many chapters include review problems that require the student to draw on numerous concepts covered in the chapter, as well as on those discussed in previous chapters. These problems could be used by students in preparing for tests and by instructors for special assignments and classroom discussions.

Paired Problems   Some end-of-chapter numerical problems are paired with the same problems in symbolic form. Two paired problems are identified by a common tan background screen.

Computer- and Calculator-Based Problems   Most chapters include one or more problems whose solution requires the use of a computer or graphing calculator. These problems are identified by the ▣ icon. Modeling of physical phenomena enables students to obtain graphical representations of variables and to perform numerical analyses. Student support for working computer-based problems is provided on the *Student Tools CD-ROM.*

Units   The international system of units (SI) is used throughout the text. The British engineering system of units (conventional system) is used only to a limited extent in the chapters on mechanics, heat, and thermodynamics.

Summaries   Each chapter contains a summary that reviews the important concepts and equations discussed in that chapter.

Appendices and Endpapers   Several appendices are provided at the end of the textbook. Most of the appendix material represents a review of mathematical concepts and techniques used in the text, including scientific notation, algebra, geometry, trigonometry, differential calculus, and integral calculus. Reference to these appendices is made throughout the text. Most mathematical review sections in the appendices include worked examples and exercises with answers. In addition to the mathematical reviews, the appendices contain tables of physical data, conversion factors, atomic masses, and the SI units of physical quantities, as well as a periodic table of the elements. Other useful information, including fundamental constants and physical data, planetary data, a list of standard prefixes, mathematical symbols, the Greek alphabet, and standard abbreviations of units of measure, appears on the endpapers.

ANCILLARIES

The ancillary package has been updated substantially and streamlined in response to suggestions from users of the fourth edition. The most essential changes in the student package are a *Student Solutions Manual and Study Guide* with a tighter focus on problem-solving, the *Student Tools CD-ROM,* and the *Saunders Core Concepts in Physics CD-ROM* developed by Archipelago Productions. Instructors will find increased support for their teaching efforts with new electronic materials.

Student Ancillaries

Student Solutions Manual and Study Guide by John R. Gordon, Ralph McGrew, and Raymond A. Serway, with contributions by Duane Deardorff. This two-volume manual features detailed solutions to 20 percent of the end-of-chapter problems from the textbook. Problems in the textbook whose complete solutions are found in the manual are identified by boxes around their numbers. The solutions to many problems follow the **GOAL** protocol described in the textbook (see page 47). The manual also features a list of important equations and concepts, as well as answers to selected end-of-chapter questions.

Pocket Guide by V. Gordon Lind. This $5'' \times 7''$ paperback is a section-by-section capsule of the textbook and serves as a handy guide for looking up important concepts, formulas, and problem-solving hints.

Student Tools CD-ROM   This CD-ROM contains tools that are designed to enhance the learning of physical concepts and train students to become better problem-solvers. It includes a textbook version of the highly acclaimed Interactive Physics™ software by Knowledge Revolution, more than 100 Interactive Physics™ simulations keyed to appropriate figures, worked examples, Quick Quizzes, and selected end-of-chapter problems (as identified by the 📕 icon), a graphing calculator, and support for working those end-of-chapter problems that require the use of a computer.

Saunders Core Concepts in Physics CD-ROM   This CD-ROM package developed by Archipelago Productions applies the power of multimedia to the introductory physics course, offering full-motion animation and video, engaging interactive graphics, clear and concise text, and guiding narration. *Saunders Core Concepts in Physics CD-ROM* focuses on those concepts students usually find most difficult in the course, drawing from topics in mechanics, thermodynamics, electric fields, magnetic fields, and optics. The animations and graphics are presented to aid the student in developing accurate conceptual models of difficult topics—topics often too complex to be explained in words or chalkboard illustrations. The CD-ROM also presents step-by-step explorations of problem-solving strategies and provides animations of problems in order to promote conceptual understanding and sharpen problem-solving skills. Topics in the textbook that are further explored on the CD-ROM are identified by marginal icons that give the appropriate module and screen number(s). Students should look to the CD-ROM to aid in their understanding of these topics.

Student Web Site   Students will have access to an abundance of material at **http://www.saunderscollege.com/physics/.** The Web Site features special topic essays by guest authors, practice problems with answers, and optional topics that accompany selected chapters of the textbook. Also included are selected solutions from the *Student Solutions Manual and Study Guide,* a sampling of the *Pocket Guide,* and a glossary that includes more than 300 physics terms. Students also can take practice quizzes in our Practice Exercises and Testing area.

Physics Laboratory Manual, Second Edition by David Loyd. Updated and redesigned, this manual supplements the learning of basic physical principles while introducing laboratory procedures and equipment. Each chapter includes a pre-laboratory assignment, objectives, an equipment list, the theory behind the experiment, step-by-step experimental procedures, and questions. A laboratory report form is provided for each experiment so that students can record data and make calculations. Students are encouraged to apply statistical analysis to their data so that they can develop their ability to judge the validity of their results.

So You Want to Learn Physics: A Preparatory Course with Calculus by Rodney Cole. This introductory-level book is useful to those students who need additional preparation before or during a calculus-based course in physics. The friendly, straightforward style makes it easier to understand how mathematics is used in the context of physics.

Life Science Applications for Physics by Jerry Faughn. This book provides examples, readings, and problems from the biological sciences as they relate to physics. Topics include "Friction in Human Joints," "Physics of the Human Circulatory System," "Physics of the Nervous System," and "Ultrasound and Its Applications." This supplement is useful in those courses taken by a significant number of pre-med students.

Instructor's Ancillaries

Instructor's Manual with Solutions by Ralph McGrew, Jeffery Saul, and Charles Teague, with contributions by Duane Deardorff. This manual consists of complete, worked solutions to all the problems in the textbook. The solutions to problems new to the fifth edition are marked for easy identification by the instructor. New to this edition of the manual are suggestions on how to teach difficult topics and how to help students overcome common misconceptions. These suggestions are based on recent research in physics education.

Instructor's Web Site   The instructor's area at **http://www.saunderscollege.com/physics/** includes a listing of overhead transparencies; a guide to relevant experiments in David Loyd's *Physics Laboratory Manual, Second Edition*; a correlation guide between sections in *Physics for Scientists and Engineers* and modules in the *Saunders Core Concepts in Physics CD-ROM*; supplemental problems with answers; optional topics to accompany selected chapters of the textbook; and a testing administration system.

Instructor's Resource CD-ROM   This CD-ROM accompanying the fifth edition of *Physics for Scientists and Engineers* has been created to provide instructors with an exciting new tool for classroom presentation. The CD-ROM contains a collection of graphics files of line art from the textbook. These files can be opened directly, can be imported into a variety of presentation packages, or can be used in the presentation package included on the CD-ROM. The labels for each piece of art have been enlarged and boldfaced to facilitate classroom viewing. The CD-ROM contains electronic files of the *Instructor's Manual, Test Bank,* and *Practice Problems with Solutions.*

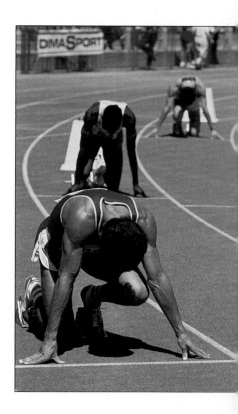

CAPA: A Computer-Assisted Personalized Approach   CAPA is a network system for learning, teaching, assessment, and administration. It provides students with personalized problem sets, quizzes, and examinations consisting of qualitative conceptual problems and quantitative problems, including problems from *Physics for Scientists and Engineers.* CAPA was developed through a collaborative effort of the Physics–Astronomy, Computer Science, and Chemistry Departments at Michigan State University. Students are given instant feedback and relevant hints via the Internet and may correct errors without penalty before an assignment's due date. The system records each student's participation and performance on assignments, quizzes, and examinations; and records are available on-line to both the individual student and to his or her instructor. For more information, visit the CAPA Web site at **http://www.pa.msu.edu/educ/CAPA/**

WebAssign: A Web-Based Homework System   WebAssign is a Web-based homework delivery, collection, grading, and recording service developed at North Carolina State University. Instructors who sign up for WebAssign can assign homework to their students, using questions and problems taken directly from *Physics for Scientists and Engineers.* WebAssign gives students immediate feedback on their homework that helps them to master information and skills, leading to greater competence and better grades. WebAssign can free instructors from the drudgery of grading homework and recording scores, allowing them to devote more time to meeting with students and preparing classroom presentations. Details about and a demonstration of WebAssign are available at **http://webassign.net/info**. For more information about ordering this service, contact WebAssign at **webassign@ncsu.edu**

Homework Service   With this service, instructor's can reduce their grading workload by assigning thought-provoking homework problems using the World Wide Web. Instructors browse problem banks that include problems from *Physics for Scientists and Engineers,* select those they wish to assign to their students, and then let the Homework Service take over the delivery and grading. This system was developed and is maintained by Fred Moore at the University of Texas (**moore@physics.utexas.edu**). Students download their unique problems, submit their answers, and obtain immediate feedback; if students' answers are incorrect, they can resubmit them. This rapid grading feature facilitates effective learning.

After the due date of their assignments, students can obtain the solutions to their problems. Minimal on-line connect time is required. The Homework Service uses algorithm-based problems: This means that each student solves sets of problems that are different from those given to other students. Details about and a demonstration of this service are available at **http://hw.ph.utexas.edu/hw.html**

Printed Test Bank   The *Printed Test Bank* contains approximately 2,300 multiple-choice questions. It is provided for the instructor who does not have access to a computer.

Computerized Test Bank   Available in Windows™ and Macintosh® formats, the *Computerized Test Bank* contains more than 2,300 multiple-choice questions, representing every chapter of the text. The *Computerized Test Bank* enables the instructor to create many unique tests by allowing the editing of questions and the addition of new questions. The software program solves all problems and prints each answer on a separate grading key. All questions have been reviewed for accuracy.

Overhead Transparency Acetates   This collection of transparencies consists of 300 full-color figures from the text and features large print for easy viewing in the classroom.

Instructor's Manual for Physics Laboratory Manual   by David Loyd. Each chapter contains a discussion of the experiment, teaching hints, answers to selected questions, and a post-laboratory quiz with short-answer and essay questions. It also includes a list of the suppliers of scientific equipment and a summary of the equipment needed for each of the laboratory experiments in the manual.

> Saunders College Publishing may provide complementary instructional aids and supplements or supplement packages to those adopters qualified under our adoption policy. Please contact your sales representative for more information. If as an adopter or potential user you receive supplements you do not need, please return them to your sales representative or send them to
>
> Attn: Returns Department
> Troy Warehouse
> 465 South Lincoln Drive
> Troy, MO 63379

TEACHING OPTIONS

The topics in this textbook are presented in the following sequence: classical mechanics, matter waves, and heat and thermodynamics (Volume 1); followed by electricity and magnetism, light waves, optics, relativity, and modern physics (Volume 2). This presentation represents a more traditional sequence, with the subject of matter waves being presented before the topics of electricity and magnetism. Some instructors may prefer to cover this material after completing electricity and magnetism (i.e., after Chapter 34). The chapter on relativity was placed near the end of the text because this topic often is treated as an introduction to the era of "modern physics." If time permits, instructors may choose to cover Chapter 39 in Volume 2 after completing Chapter 14, as it concludes the material on Newtonian mechanics.

Instructors teaching a two-semester sequence can delete sections and chapters without any loss of continuity. We have labeled these as "Optional" in the Table of Contents and in the appropriate sections of the text. For student enrichment, instructors can assign some of these sections or chapters as extra reading.

ACKNOWLEDGMENTS

The fifth edition of this textbook was prepared with the guidance and assistance of many professors who reviewed part or all of the manuscript, the pre-revision text, or both. We wish to acknowledge the following scholars and express our sincere appreciation for their suggestions, criticisms, and encouragement:

Edward Adelson, *Ohio State University*

Roger Bengtson, *University of Texas at Austin*

Joseph Biegen, *Broome Community College*

Ronald J. Bieniek, *University of Missouri at Rolla*

Ronald Brown, *California Polytechnic State University—San Luis Obispo*

Michael E. Browne, *University of Idaho*

Tim Burns, *Leeward Community College*

Randall Caton, *Christopher Newport University*

Sekhar Chivukula, *Boston University*

Alfonso Diáz-Jiménez, *ADJOIN Research Center*

N. John DiNardo, *Drexel University*

F. Eugene Dunnum, *University of Florida*

William Ellis, *Cornell University*

F. Paul Esposito, *University of Cincinnati*

Paul Fahey, *University of Scranton*

Arnold Feldman, *University of Hawaii at Manoa*

Alexander Firestone, *Iowa State University*

Robert Forsythe, *Broome Community College*

Philip Fraundorf, *University of Missouri at St. Louis*

John Gerty, *Broome Community College*

John B. Gruber, *San Jose State University*

John Hubisz, *North Carolina State University*

Joey Huston, *Michigan State University*

John W. Jewett, Jr., *California State Polytechnic State University—Pomona*

Calvin S. Kalman, *Concordia University*

Natalie Kerr, M.D., *University of Tennessee, Memphis*

Peter Killen, *University of Queensland (Australia)*

Earl Koller, *Stevens Institute of Technology*

David LaGraffe, *U.S. Military Academy*

Ying-Cheng Lai, *University of Kansas*

Donald Larson, *Drexel University*

Robert Lieberman, *Cornell University*

Ralph McGrew, *Broome Community College*

David Mills, *Monash University (Australia)*

Clement J. Moses, *Utica College*

Peter Parker, *Yale University*

John Parsons, *Columbia University*

Arnold Perlmutter, *University of Miami*

Henry Schriemer, *Queen's University (Canada)*

Paul Snow, *University of Bath (U.K.)*

Edward W. Thomas, *Georgia Institute of Technology*

Charles C. Vuille, *Embry-Riddle Aeronautical University*

Xiaojun Wang, *Georgia Southern University*

Gail Welsh, *Salisbury State University*

This book was carefully checked for accuracy by James H. Smith *(University of Illinois at Urbana-Champaign)*, Gregory Snow *(University of Nebraska—Lincoln)*, Edward Gibson *(California State University—Sacramento)*, Ronald Jodoin *(Rochester Institute of Technology)*, and Arnold Perlmutter *(University of Miami)*.

We thank the following people for their suggestions and assistance during the preparation of earlier editions of this textbook:

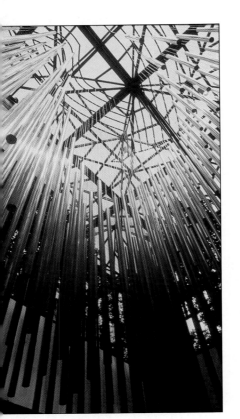

George Alexandrakis, *University of Miami*

Elmer E. Anderson, *University of Alabama*

Wallace Arthur, *Fairleigh Dickinson University*

Duane Aston, *California State University—Sacramento*

Stephen Baker, *Rice University*

Richard Barnes, *Iowa State University*

Stanley Bashkin, *University of Arizona*

Robert Bauman, *University of Alabama*

Marvin Blecher, *Virginia Polytechnic Institute and State University*

Jeffrey J. Braun, *University of Evansville*

Kenneth Brownstein, *University of Maine*

William A. Butler, *Eastern Illinois University*

Louis H. Cadwell, *Providence College*

Ron Canterna, *University of Wyoming*

Bo Casserberg, *University of Minnesota*

Soumya Chakravarti, *California Polytechnic State University*

C. H. Chan, *The University of Alabama at Huntsville*

Edward Chang, *University of Massachusetts at Amherst*

Don Chodrow, *James Madison University*

Clifton Bob Clark, *University of North Carolina at Greensboro*

Walter C. Connolly, *Appalachian State University*

Hans Courant, *University of Minnesota*

Lance E. De Long, *University of Kentucky*

James L. DuBard, *Binghamton-Southern College*

F. Paul Esposito, *University of Cincinnati*

Jerry S. Faughn, *Eastern Kentucky University*

Paul Feldker, *Florissant Valley Community College*

Joe L. Ferguson, *Mississippi State University*

R. H. Garstang, *University of Colorado at Boulder*

James B. Gerhart, *University of Washington*

John R. Gordon, *James Madison University*

Clark D. Hamilton, *National Bureau of Standards*

Mark Heald, *Swarthmore College*

Herb Helbig, *Rome Air Development Center*

Howard Herzog, *Broome Community College*

Paul Holoday, *Henry Ford Community College*

Jerome W. Hosken, *City College of San Francisco*

Larry Hmurcik, *University of Bridgeport*

William Ingham, *James Madison University*

Mario Iona, *University of Denver*

Karen L. Johnston, *North Carolina State University*

Brij M. Khorana, *Rose-Hulman Institute of Technology*

Larry Kirkpatrick, *Montana State University*

Carl Kocher, *Oregon State University*

Robert E. Kribel, *Jacksonville State University*

Barry Kunz, *Michigan Technological University*

Douglas A. Kurtze, *Clarkson University*

Fred Lipschultz, *University of Connecticut*

Francis A. Liuima, *Boston College*

Robert Long, *Worcester Polytechnic Institute*

Roger Ludin, *California Polytechnic State University*

Nolen G. Massey, *University of Texas at Arlington*

Charles E. McFarland, *University of Missouri at Rolla*

Ralph V. McGrew, *Broome Community College*

James Monroe, *The Pennsylvania State University, Beaver Campus*

Bruce Morgan, *United States Naval Academy*

Clement J. Moses, *Utica College*

Curt Moyer, *Clarkson University*

David Murdock, *Tennessee Technological University*

A. Wilson Nolle, *University of Texas at Austin*

Thomas L. O'Kuma, *San Jacinto College North*

Fred A. Otter, *University of Connecticut*

George Parker, *North Carolina State University*

William F. Parks, *University of Missouri at Rolla*

Philip B. Peters, *Virginia Military Institute*

Eric Peterson, *Highland Community College*

Richard Reimann, *Boise State University*

Joseph W. Rudmin, *James Madison University*

Jill Rugare, *DeVry Institute of Technology*

Charles Scherr, *University of Texas at Austin*

Eric Sheldon, *University of Massachusetts—Lowell*

John Shelton, *College of Lake County*

Stan Shepard, *The Pennsylvania State University*

James H. Smith, *University of Illinois at Urbana-Champaign*

Richard R. Sommerfield, *Foothill College*

Kervork Spartalian, *University of Vermont*

Robert W. Stewart, *University of Victoria*

James Stith, *American Institute of Physics*

Charles D. Teague, *Eastern Kentucky University*

Edward W. Thomas, *Georgia Institute of Technology*

Carl T. Tomizuka, *University of Arizona*

Herman Trivilino, *San Jacinto College North*

Som Tyagi, *Drexel University*

Steve Van Wyk, *Chapman College*

Joseph Veit, *Western Washington University*

T. S. Venkataraman, *Drexel University*

Noboru Wada, *Colorado School of Mines*

James Walker, *Washington State University*

Gary Williams, *University of California, Los Angeles*

George Williams, *University of Utah*

Edward Zimmerman, *University of Nebraska, Lincoln*

Earl Zwicker, *Illinois Institute of Technology*

We are grateful to Ralph McGrew for organizing the end-of-chapter problems, writing many new problems, and his suggestions for improving the content of the textbook. The new end-of-chapter problems were written by Rich Cohen, John DiNardo, Robert Forsythe, and Ralph McGrew, with suggestions by Liz McGrew, Alexandra Héder, and Richard McGrew. We thank Laurent Hodges for permission to use selected end-of-chapter problems. We are grateful to John R. Gordon, Ralph McGrew, and Duane Deardorff for writing the *Student Solutions Manual and Study Guide,* and we thank Michael Rudmin for its attractive layout. Ralph McGrew, Jeffery Saul, and Charles Teague have prepared an excellent *Instructor's Manual,* and we thank them. We thank Gloria Langer, Linda Miller, and Jennifer Serway for their excellent work in preparing the *Instructor's Manual* and the supplemental materials that appear on our Web site.

Special thanks and recognition go to the professional staff at Saunders College Publishing—in particular, Susan Pashos, Sally Kusch, Carol Bleistine, Frank Messina, Suzanne Hakanen, Ruth Hoover, Alexandra Buczek, Pauline Mula, and John Vondeling—for their fine work during the development and production of this textbook. We are most appreciative of the intelligent line editing by Irene Nunes, the final copy editing by Sue Nelson and Mary Patton, the excellent artwork produced by Rolin Graphics, and the dedicated photo research efforts of Dena Digilio-Betz.

Finally, we are deeply indebted to our wives and children for their love, support, and long-term sacrifices.

Raymond A. Serway
Chapel Hill, North Carolina

Robert J. Beichner
Raleigh, North Carolina

John W. Jewett, Jr.
Pomona, California

To the Student

*I*t is appropriate to offer some words of advice that should be of benefit to you, the student. Before doing so, we assume that you have read the Preface, which describes the various features of the text that will help you through the course.

HOW TO STUDY

Very often instructors are asked, "How should I study physics and prepare for examinations?" There is no simple answer to this question, but we would like to offer some suggestions that are based on our own experiences in learning and teaching over the years.

First and foremost, maintain a positive attitude toward the subject matter, keeping in mind that physics is the most fundamental of all natural sciences. Other science courses that follow will use the same physical principles, so it is important that you understand and are able to apply the various concepts and theories discussed in the text.

CONCEPTS AND PRINCIPLES

It is essential that you understand the basic concepts and principles before attempting to solve assigned problems. You can best accomplish this goal by carefully reading the textbook before you attend your lecture on the covered material. When reading the text, you should jot down those points that are not clear to you. We've purposely left wide margins in the text to give you space for doing this. Also be sure to make a diligent attempt at answering the questions in the Quick Quizzes as you come to them in your reading. We have worked hard to prepare questions that help you judge for yourself how well you understand the material. The QuickLabs provide an occasional break from your reading and will help you to experience some of the new concepts you are trying to learn. During class, take careful notes and ask questions about those ideas that are unclear to you. Keep in mind that few people are able to absorb the full meaning of scientific material after only one reading. Several readings of the text and your notes may be necessary. Your lectures and laboratory work supplement reading of the textbook and should clarify some of the more difficult material. You should minimize your memorization of material. Successful memorization of passages from the text, equations, and derivations does not necessarily indicate that you understand the material. Your understanding of the material will be enhanced through a combination of efficient study habits, discussions with other students and with instructors, and your ability to solve the problems presented in the textbook. Ask questions whenever you feel clarification of a concept is necessary.

STUDY SCHEDULE

It is important that you set up a regular study schedule, preferably one that is daily. Make sure that you to read the syllabus for the course and adhere to the schedule set by your instructor. The lectures will be much more meaningful if you read the corresponding textual material before attending them. As a general rule, you should devote about two hours of study time for every hour you are in class. If you are having trouble with the course, seek the advice of the instructor or other students who have taken the course. You may find it necessary to seek further instruction from experienced students. Very often, instructors offer review sessions in addition to regular class periods. It is important that you avoid the practice of delaying study until a day or two before an exam. More often than not, this approach has disastrous results. Rather than undertake an all-night study session, briefly review the basic concepts and equations and get a good night's rest. If you feel you need additional help in understanding the concepts, in preparing for exams, or in problem-solving, we suggest that you acquire a copy of the *Student Solutions Manual and Study Guide* that accompanies this textbook; this manual should be available at your college bookstore.

USE THE FEATURES

You should make full use of the various features of the text discussed in the preface. For example, marginal notes are useful for locating and describing important equations and concepts, and **boldfaced** type indicates important statements and definitions. Many useful tables are contained in the Appendices, but most are incorporated in the text where they are most often referenced. Appendix B is a convenient review of mathematical techniques.

Answers to odd-numbered problems are given at the end of the textbook, answers to Quick Quizzes are located at the end of each chapter, and answers to selected end-of-chapter questions are provided in the *Student Solutions Manual and Study Guide*. The exercises (with answers) that follow some worked examples represent extensions of those examples; in most of these exercises, you are expected to perform a simple calculation (see Example 4.7 on page 90). Their purpose is to test your problem-solving skills as you read through the text. Problem-Solving Hints are included in selected chapters throughout the text and give you additional information about how you should solve problems. The Table of Contents provides an overview of the entire text, while the Index enables you to locate specific material quickly. Footnotes sometimes are used to supplement the text or to cite other references on the subject discussed.

After reading a chapter, you should be able to define any new quantities introduced in that chapter and to discuss the principles and assumptions that were used to arrive at certain key relations. The chapter summaries and the review sections of the *Student Solutions Manual and Study Guide* should help you in this regard. In some cases, it may be necessary for you to refer to the index of the text to locate certain topics. You should be able to correctly associate with each physical quantity the symbol used to represent that quantity and the unit in which the quantity is specified. Furthermore, you should be able to express each important relation in a concise and accurate prose statement.

PROBLEM-SOLVING

R. P. Feynman, Nobel laureate in physics, once said, "You do not know anything until you have practiced." In keeping with this statement, we strongly advise that you develop the skills necessary to solve a wide range of problems. Your ability to solve problems will be one of the main tests of your knowledge of physics, and therefore you should try to solve as many problems as possible. It is essential that you understand basic concepts and principles before attempting to solve problems. It is good practice to try to find alternate solutions to the same problem. For example, you can solve problems in mechanics using Newton's laws, but very often an alternative method that draws on energy considerations is more direct. You should not deceive yourself into thinking that you understand a problem merely because you have seen it solved in class. You must be able to solve the problem and similar problems on your own.

The approach to solving problems should be carefully planned. A systematic plan is especially important when a problem involves several concepts. First, read the problem several times until you are confident you understand what is being asked. Look for any key words that will help you interpret the problem and perhaps allow you to make certain assumptions. Your ability to interpret a question properly is an integral part of problem-solving. Second, you should acquire the habit of writing down the information given in a problem and those quantities that need to be found; for example, you might construct a table listing both the quantities given and the quantities to be found. This procedure is sometimes used in the worked examples of the textbook. Finally, after you have decided on the method you feel is appropriate for a given problem, proceed with your solution. General problem-solving strategies of this type are included in the text and are highlighted with a light blue-gray screen. We have also developed the **GOAL** protocol (see page 47) to help guide you through complex problems. If you follow the steps of this procedure (**G**ather information, **O**rganize your approach, carry out your **A**nalysis, and finally **L**earn from your work), you will not only find it easier to come up with a solution, but you will also gain more from your efforts.

Often, students fail to recognize the limitations of certain formulas or physical laws in a particular situation. It is very important that you understand and remember the assumptions that underlie a particular theory or formalism. For example, certain equations in kinematics apply only to a particle moving with constant acceleration. These equations are not valid for describing motion whose acceleration is not constant, such as the motion of an object connected to a spring or the motion of an object through a fluid.

General Problem-Solving Strategy

Most courses in general physics require the student to learn the skills of problem-solving, and exams are largely composed of problems that test such skills. This brief section describes some useful ideas that will enable you to increase your accuracy in solving problems, enhance your understanding of physical concepts, eliminate initial panic or lack of direction in approaching a problem, and organize your work. One way to help accomplish these goals is to adopt a problem-solving strategy. Many chapters in this text include Problem-Solving Hints that should help you through the "rough spots."

In developing problem-solving strategies, five basic steps are commonly followed:

- Draw a suitable diagram with appropriate labels and coordinate axes (if needed).

- As you examine what is being asked in the problem, identify the basic physical principle (or principles) that are involved, listing the knowns and the unknowns.

- Select a basic relationship or derive an equation that can be used to find the unknown, and then solve the equation for the unknown symbolically.

- Substitute the given values along with the appropriate units into the equation.

- Obtain a numerical value for the unknown. The problem is verified and receives a check mark if the following questions can be properly answered: Do the units match? Is the answer reasonable? Is the plus or minus sign proper or meaningful?

One of the purposes of this strategy is to promote accuracy. Properly drawn diagrams can eliminate many sign errors. Diagrams also help to isolate the physical principles of the problem. Obtaining symbolic solutions and carefully labeling knowns and unknowns will help eliminate other careless errors. The use of symbolic solutions should help you think in terms of the physics of the problem. A check of units at the end of the problem can indicate a possible algebraic error. The physical layout and organization of your problem will make the final product more understandable and easier to follow. Once you have developed an organized system for examining problems and extracting relevant information, you will become a more confident problem-solver.

EXPERIMENTS

Physics is a science based on experimental observations. In view of this fact, we recommend that you try to supplement your reading of the text by performing various types of "hands-on" experiments, either at home or in the laboratory. Most chapters include one or two QuickLabs that describe simple experiments you can do on your own. These can be used to test ideas and models discussed in class or in the textbook. For example, the common Slinky™ toy is an excellent tool for studying traveling waves; a ball swinging on the end of a long string can be used to investigate pendulum motion; various masses attached to the end of a vertical spring or rubber band can be used to determine their elastic nature; an old pair of Polaroid sunglasses, some discarded lenses, and a magnifying glass are the components of various experiments in optics; and the approximate measure of the acceleration of gravity can be determined simply by measuring with a stopwatch the time it takes for a ball to drop from a known height. The list of such experiments is endless. When physical models are not available, be imaginative and try to develop models of your own.

NEW MEDIA

We strongly encourage you to use one or more of the following multimedia products that accompany this textbook. It is far easier to understand physics if you see it in action, and these new materials will enable you to become a part of that action.

Student Tools CD-ROM   You may purchase the the dual-platform (Windows™- and Macintosh®-compatible) *Student Tools CD-ROM*. In addition to a graphing calculator and support for working those end-of-chapter problems that require the use of a computer, this CD-ROM contains a textbook version of the Interactive Physics™ program by Knowledge Revolution. Interactive Physics™ simulations are keyed to the following figures, worked examples, Quick Quizzes, and end-of-chapter problems (identified in the text with the 🖱 icon).

Saunders Core Concepts in Physics CD-ROM   Alternatively or in addition, you may purchase the *Saunders Core Concepts in Physics CD-ROM* developed by Archipelago Productions. This CD-ROM provides a complete multimedia presentation of selected topics in mechanics, thermodynamics, electromagnetism, and optics. It contains more than 350 movies—both animated and live video—that bring to life laboratory demonstrations, "real-world" examples, graphic models, and step-by-step explanations of essential mathematics. Those CD-ROM modules that supplement the material in *Physics for Scientists and Engineers* are identified in the margin of the text by the 💿 icon.

AN INVITATION TO PHYSICS

It is our sincere hope that you too will find physics an exciting and enjoyable experience and that you will profit from this experience, regardless of your chosen profession. Welcome to the exciting world of physics!

The scientist does not study nature because it is useful; he studies it because he delights in it, and he delights in it because it is beautiful. If nature were not beautiful, it would not be worth knowing, and if nature were not worth knowing, life would not be worth living.

—Henri Poincaré

About the Authors

Raymond A. Serway received his doctorate at Illinois Institute of Technology and is Professor Emeritus at James Madison University. In 1990, he received the Madison Scholar award at James Madison University, where he taught for 17 years. Dr. Serway began his teaching career at Clarkson University, where he conducted research and taught from 1967 to 1980. He was the recipient of the Distinguished Teaching Award at Clarkson University in 1977 and of the Alumni Achievement Award from Utica College in 1985. As Guest Scientist at the IBM Research Laboratory in Zurich, Switzerland, he worked with K. Alex Müller, 1987 Nobel Prize recipient. Dr. Serway also was a visiting scientist at Argonne National Laboratory, where he collaborated with his mentor and friend, Sam Marshall. In addition to earlier editions of this textbook, Dr. Serway is the author of *Principles of Physics, Second Edition,* and co-author of *College Physics, Fifth Edition,* and *Modern Physics, Second Edition*; he also is the author of the high-school textbook *Physics,* published by Holt, Rinehart, & Winston. In addition, Dr. Serway has published more than 40 research papers in the field of condensed matter physics and has given more than 60 presentations at professional meetings. Dr. Serway and his wife Elizabeth enjoy traveling, golfing, and spending quality time with their four children and four grandchildren.

Robert J. Beichner received his doctorate at the State University of New York at Buffalo. Currently, he is Associate Professor of Physics at North Carolina State University, where he directs the Physics Education Research and Development Group. He has more than 20 years of teaching experience at the community-college, four year–college, and university levels. His research interests are centered on improving physics instruction: In his work, he has published studies of video-based laboratories, collaborative learning, technology-supplemented learning environments, and the assessment of student understanding of various physics topics. Dr. Beichner has held several leadership roles in the field of physics education research and has given numerous talks and colloquia on his work. In addition to being an author of this textbook, he is the co-author of two CD-ROMs, several commercially available software packages, and two books for preservice elementary school teachers. Dr. Beichner is an avid sea kayaker and enjoys spending time with his wife Mary and their two daughters, Sarah and Julie.

John W. Jewett, Jr. earned his doctorate at Ohio State University, specializing in optical and magnetic properties of condensed matter. He is currently Professor of Physics at California State Polytechnic University—Pomona. Throughout his teaching career, Dr. Jewett has been active in promoting science education. In addition to receiving four National Science Foundation grants, he helped found and direct the Southern California Area Modern Physics Institute (SCAMPI). He also is the director of Science IMPACT (Institute for Modern Pedagogy and Creative Teaching), which works with teachers and schools to develop effective science curricula. Both organizations operate in the United States and abroad. Dr. Jewett's honors include four Meritorious Performance and Professional Promise awards, selection as Outstanding Professor at California State Polytechnic University for 1991–1992, and the Excellence in Undergraduate Physics Teaching Award from the American Association of Physics Teachers (AAPT) in 1998. He has given many presentations both domestically and abroad, including multiple presentations at national meetings of the AAPT. He will co-author the third edition of *Principles of Physics* with Dr. Serway. Dr. Jewett enjoys playing piano, traveling, and collecting antique quack medical devices, as well as spending time with his wife Lisa and their children.

Pedagogical Color Chart

Part 1 (Chapters 1–15) : Mechanics

Displacement and position vectors

Linear (**v**) and angular (**ω**) velocity vectors

Velocity component vectors

Force vectors (**F**)

Force component vectors

Acceleration vectors (**a**)

Acceleration component vectors

Linear (**p**) and angular (**L**) momentum vectors

Torque vectors (**τ**)

Linear or rotational motion directions

Springs

Pulleys

Part 4 (Chapters 23–34) : Electricity and Magnetism

Electric fields

Magnetic fields

Positive charges

Negative charges

Resistors

Batteries and other dc power supplies

Switches

Capacitors

Inductors (coils)

Voltmeters

Ammeters

Galvanometers

ac Generators

Ground symbol

Part 5 (Chapters 35–38) : Light and Optics

Light rays

Lenses and prisms

Mirrors

Objects

Images

part 1

Mechanics

Physics, the most fundamental physical science, is concerned with the basic principles of the Universe. It is the foundation upon which the other sciences — astronomy, biology, chemistry, and geology — are based. The beauty of physics lies in the simplicity of the fundamental physical theories and in the manner in which just a small number of fundamental concepts, equations, and assumptions can alter and expand our view of the world around us.

All of physics can be divided into five main areas:

1. Classical mechanics, which is concerned with the motion of objects that are large relative to atoms and move at speeds much slower than the speed of light
2. Relativity, which is a theory describing objects moving at any speed, even speeds approaching the speed of light
3. Thermodynamics, which deals with heat, work, temperature, and the statistical behavior of a large number of particles
4. Electromagnetism, which is concerned with electricity, magnetism, and electromagnetic fields
5. Quantum mechanics, a collection of theories dealing with the behavior of matter at both the submicroscopic and macroscopic levels

The first part of this textbook deals with classical mechanics, sometimes referred to as *Newtonian mechanics* or simply *mechanics.* This is an appropriate place to begin an introductory text because many of the basic principles used to understand mechanical systems can later be used to describe such natural phenomena as waves and energy transfer. Furthermore, the laws of conservation of energy and momentum introduced in mechanics retain their importance in the fundamental theories of other areas of physics.

Today, classical mechanics is of vital importance to students from all disciplines. It is highly successful in describing the motions of different bodies, such as planets, rockets, and baseballs. In the first part of the text, we shall describe the laws of classical mechanics and examine a wide range of phenomena that can be understood with these fundamental ideas. We shall treat the various disciplines of classical physics in separate sections; however, we shall see that the disciplines of mechanics and electromagnetism are basic to all other branches of classical and modern physics.

c h a p t e r

Physics and Measurement

$\mathbf{L}$ike all other sciences, physics is based on experimental observations and quantitative measurements. The main objective of physics is to find the limited number of fundamental laws that govern natural phenomena and to use them to develop theories that can predict the results of future experiments. The fundamental laws used in developing theories are expressed in the language of mathematics, the tool that provides a bridge between theory and experiment.

When a discrepancy between theory and experiment arises, new theories must be formulated to remove the discrepancy. Many times a theory is satisfactory only under limited conditions; a more general theory might be satisfactory without such limitations. For example, the laws of motion discovered by Isaac Newton (1642–1727) in the 17th century accurately describe the motion of bodies at normal speeds but do not apply to objects moving at speeds comparable with the speed of light. In contrast, the special theory of relativity developed by Albert Einstein (1879–1955) in the early 1900s gives the same results as Newton's laws at low speeds but also correctly describes motion at speeds approaching the speed of light. Hence, Einstein's is a more general theory of motion.

Classical physics, which means all of the physics developed before 1900, includes the theories, concepts, laws, and experiments in classical mechanics, thermodynamics, and electromagnetism.

Important contributions to classical physics were provided by Newton, who developed classical mechanics as a systematic theory and was one of the originators of calculus as a mathematical tool. Major developments in mechanics continued in the 18th century, but the fields of thermodynamics and electricity and magnetism were not developed until the latter part of the 19th century, principally because before that time the apparatus for controlled experiments was either too crude or unavailable.

A new era in physics, usually referred to as *modern physics*, began near the end of the 19th century. Modern physics developed mainly because of the discovery that many physical phenomena could not be explained by classical physics. The two most important developments in modern physics were the theories of relativity and quantum mechanics. Einstein's theory of relativity revolutionized the traditional concepts of space, time, and energy; quantum mechanics, which applies to both the microscopic and macroscopic worlds, was originally formulated by a number of distinguished scientists to provide descriptions of physical phenomena at the atomic level.

Scientists constantly work at improving our understanding of phenomena and fundamental laws, and new discoveries are made every day. In many research areas, a great deal of overlap exists between physics, chemistry, geology, and biology, as well as engineering. Some of the most notable developments are (1) numerous space missions and the landing of astronauts on the Moon, (2) microcircuitry and high-speed computers, and (3) sophisticated imaging techniques used in scientific research and medicine. The impact such developments and discoveries have had on our society has indeed been great, and it is very likely that future discoveries and developments will be just as exciting and challenging and of great benefit to humanity.

 ## STANDARDS OF LENGTH, MASS, AND TIME

The laws of physics are expressed in terms of basic quantities that require a clear definition. In mechanics, the three basic quantities are length (L), mass (M), and time (T). All other quantities in mechanics can be expressed in terms of these three.

If we are to report the results of a measurement to someone who wishes to reproduce this measurement, a *standard* must be defined. It would be meaningless if a visitor from another planet were to talk to us about a length of 8 "glitches" if we do not know the meaning of the unit glitch. On the other hand, if someone familiar with our system of measurement reports that a wall is 2 meters high and our unit of length is defined to be 1 meter, we know that the height of the wall is twice our basic length unit. Likewise, if we are told that a person has a mass of 75 kilograms and our unit of mass is defined to be 1 kilogram, then that person is 75 times as massive as our basic unit.[1] Whatever is chosen as a standard must be readily accessible and possess some property that can be measured reliably—measurements taken by different people in different places must yield the same result.

In 1960, an international committee established a set of standards for length, mass, and other basic quantities. The system established is an adaptation of the metric system, and it is called the **SI system** of units. (The abbreviation SI comes from the system's French name "Système International.") In this system, the units of length, mass, and time are the meter, kilogram, and second, respectively. Other SI standards established by the committee are those for temperature (the *kelvin*), electric current (the *ampere*), luminous intensity (the *candela*), and the amount of substance (the *mole*). In our study of mechanics we shall be concerned only with the units of length, mass, and time.

Length

In A.D. 1120 the king of England decreed that the standard of length in his country would be named the *yard* and would be precisely equal to the distance from the tip of his nose to the end of his outstretched arm. Similarly, the original standard for the foot adopted by the French was the length of the royal foot of King Louis XIV. This standard prevailed until 1799, when the legal standard of length in France became the *meter*, defined as one ten-millionth the distance from the equator to the North Pole along one particular longitudinal line that passes through Paris.

Many other systems for measuring length have been developed over the years, but the advantages of the French system have caused it to prevail in almost all countries and in scientific circles everywhere. As recently as 1960, the length of the meter was defined as the distance between two lines on a specific platinum–iridium bar stored under controlled conditions in France. This standard was abandoned for several reasons, a principal one being that the limited accuracy with which the separation between the lines on the bar can be determined does not meet the current requirements of science and technology. In the 1960s and 1970s, the meter was defined as 1 650 763.73 wavelengths of orange-red light emitted from a krypton-86 lamp. However, in October 1983, the **meter (m) was redefined as the distance traveled by light in vacuum during a time of 1/299 792 458 second.** In effect, this latest definition establishes that the speed of light in vacuum is precisely 299 792 458 m per second.

Table 1.1 lists approximate values of some measured lengths.

---

[1] The need for assigning numerical values to various measured physical quantities was expressed by Lord Kelvin (William Thomson) as follows: "I often say that when you can measure what you are speaking about, and express it in numbers, you should know something about it, but when you cannot express it in numbers, your knowledge is of a meagre and unsatisfactory kind. It may be the beginning of knowledge but you have scarcely in your thoughts advanced to the state of science."

TABLE 1.1 Approximate Values of Some Measured Lengths

	Length (m)
Distance from the Earth to most remote known quasar	1.4×10^{26}
Distance from the Earth to most remote known normal galaxies	9×10^{25}
Distance from the Earth to nearest large galaxy	
(M 31, the Andromeda galaxy)	2×10^{22}
Distance from the Sun to nearest star (Proxima Centauri)	4×10^{16}
One lightyear	9.46×10^{15}
Mean orbit radius of the Earth about the Sun	1.50×10^{11}
Mean distance from the Earth to the Moon	3.84×10^{8}
Distance from the equator to the North Pole	1.00×10^{7}
Mean radius of the Earth	6.37×10^{6}
Typical altitude (above the surface) of a satellite orbiting the Earth	2×10^{5}
Length of a football field	9.1×10^{1}
Length of a housefly	5×10^{-3}
Size of smallest dust particles	$\sim 10^{-4}$
Size of cells of most living organisms	$\sim 10^{-5}$
Diameter of a hydrogen atom	$\sim 10^{-10}$
Diameter of an atomic nucleus	$\sim 10^{-14}$
Diameter of a proton	$\sim 10^{-15}$

Mass

The basic SI unit of mass, **the kilogram (kg), is defined as the mass of a specific platinum–iridium alloy cylinder kept at the International Bureau of Weights and Measures at Sèvres, France.** This mass standard was established in 1887 and has not been changed since that time because platinum–iridium is an unusually stable alloy (Fig. 1.1a). A duplicate of the Sèvres cylinder is kept at the National Institute of Standards and Technology (NIST) in Gaithersburg, Maryland.

Table 1.2 lists approximate values of the masses of various objects.

Time

Before 1960, the standard of time was defined in terms of the *mean solar day* for the year 1900.[2] The *mean solar second* was originally defined as $\left(\frac{1}{60}\right)\left(\frac{1}{60}\right)\left(\frac{1}{24}\right)$ of a mean solar day. The rotation of the Earth is now known to vary slightly with time, however, and therefore this motion is not a good one to use for defining a standard.

In 1967, consequently, the second was redefined to take advantage of the high precision obtainable in a device known as an *atomic clock* (Fig. 1.1b). In this device, the frequencies associated with certain atomic transitions can be measured to a precision of one part in 10^{12}. This is equivalent to an uncertainty of less than one second every 30 000 years. Thus, in 1967 the SI unit of time, the *second*, was redefined using the characteristic frequency of a particular kind of cesium atom as the "reference clock." The basic SI unit of time, **the second (s), is defined as 9 192 631 770 times the period of vibration of radiation from the cesium-133 atom.**[3] To keep these atomic clocks—and therefore all common clocks and

**TABLE 1.2
Masses of Various Bodies
(Approximate Values)**

Body	Mass (kg)
Visible Universe	$\sim 10^{52}$
Milky Way galaxy	7×10^{41}
Sun	1.99×10^{30}
Earth	5.98×10^{24}
Moon	7.36×10^{22}
Horse	$\sim 10^{3}$
Human	$\sim 10^{2}$
Frog	$\sim 10^{-1}$
Mosquito	$\sim 10^{-5}$
Bacterium	$\sim 10^{-15}$
Hydrogen atom	1.67×10^{-27}
Electron	9.11×10^{-31}

---

[2] One solar day is the time interval between successive appearances of the Sun at the highest point it reaches in the sky each day.

[3] *Period* is defined as the time interval needed for one complete vibration.

Figure 1.1   *(Left)* The National Standard Kilogram No. 20, an accurate copy of the International Standard Kilogram kept at Sèvres, France, is housed under a double bell jar in a vault at the National Institute of Standards and Technology (NIST). *(Right)* The primary frequency standard (an atomic clock) at the NIST. This device keeps time with an accuracy of about 3 millionths of a second per year. *(Courtesy of National Institute of Standards and Technology, U.S. Department of Commerce)*

watches that are set to them—synchronized, it has sometimes been necessary to add leap seconds to our clocks. This is not a new idea. In 46 B.C. Julius Caesar began the practice of adding extra days to the calendar during leap years so that the seasons occurred at about the same date each year.

Since Einstein's discovery of the linkage between space and time, precise measurement of time intervals requires that we know both the state of motion of the clock used to measure the interval and, in some cases, the location of the clock as well. Otherwise, for example, global positioning system satellites might be unable to pinpoint your location with sufficient accuracy, should you need rescuing.

Approximate values of time intervals are presented in Table 1.3.

In addition to SI, another system of units, the *British engineering system* (sometimes called the *conventional system*), is still used in the United States despite acceptance of SI by the rest of the world. In this system, the units of length, mass, and

TABLE 1.3   Approximate Values of Some Time Intervals

	Interval (s)
Age of the Universe	5×10^{17}
Age of the Earth	1.3×10^{17}
Average age of a college student	6.3×10^{8}
One year	3.16×10^{7}
One day (time for one rotation of the Earth about its axis)	8.64×10^{4}
Time between normal heartbeats	8×10^{-1}
Period of audible sound waves	$\sim 10^{-3}$
Period of typical radio waves	$\sim 10^{-6}$
Period of vibration of an atom in a solid	$\sim 10^{-13}$
Period of visible light waves	$\sim 10^{-15}$
Duration of a nuclear collision	$\sim 10^{-22}$
Time for light to cross a proton	$\sim 10^{-24}$

time are the foot (ft), slug, and second, respectively. In this text we shall use SI units because they are almost universally accepted in science and industry. We shall make some limited use of British engineering units in the study of classical mechanics.

In addition to the basic SI units of meter, kilogram, and second, we can also use other units, such as millimeters and nanoseconds, where the prefixes *milli-* and *nano-* denote various powers of ten. Some of the most frequently used prefixes for the various powers of ten and their abbreviations are listed in Table 1.4. For

TABLE 1.4   Prefixes for SI Units

Power	Prefix	Abbreviation
10^{-24}	yocto	y
10^{-21}	zepto	z
10^{-18}	atto	a
10^{-15}	femto	f
10^{-12}	pico	p
10^{-9}	nano	n
10^{-6}	micro	μ
10^{-3}	milli	m
10^{-2}	centi	c
10^{-1}	deci	d
10^{1}	deka	da
10^{3}	kilo	k
10^{6}	mega	M
10^{9}	giga	G
10^{12}	tera	T
10^{15}	peta	P
10^{18}	exa	E
10^{21}	zetta	Z
10^{24}	yotta	Y

example, 10^{-3} m is equivalent to 1 millimeter (mm), and 10^3 m corresponds to 1 kilometer (km). Likewise, 1 kg is 10^3 grams (g), and 1 megavolt (MV) is 10^6 volts (V).

Figure 1.2 Levels of organization in matter. Ordinary matter consists of atoms, and at the center of each atom is a compact nucleus consisting of protons and neutrons. Protons and neutrons are composed of quarks. The quark composition of a proton is shown.

1.2 THE BUILDING BLOCKS OF MATTER

A 1-kg cube of solid gold has a length of 3.73 cm on a side. Is this cube nothing but wall-to-wall gold, with no empty space? If the cube is cut in half, the two pieces still retain their chemical identity as solid gold. But what if the pieces are cut again and again, indefinitely? Will the smaller and smaller pieces always be gold? Questions such as these can be traced back to early Greek philosophers. Two of them—Leucippus and his student Democritus—could not accept the idea that such cuttings could go on forever. They speculated that the process ultimately must end when it produces a particle that can no longer be cut. In Greek, *atomos* means "not sliceable." From this comes our English word *atom*.

Let us review briefly what is known about the structure of matter. All ordinary matter consists of atoms, and each atom is made up of electrons surrounding a central nucleus. Following the discovery of the nucleus in 1911, the question arose: Does it have structure? That is, is the nucleus a single particle or a collection of particles? The exact composition of the nucleus is not known completely even today, but by the early 1930s a model evolved that helped us understand how the nucleus behaves. Specifically, scientists determined that occupying the nucleus are two basic entities, protons and neutrons. The *proton* carries a positive charge, and a specific element is identified by the number of protons in its nucleus. This number is called the **atomic number** of the element. For instance, the nucleus of a hydrogen atom contains one proton (and so the atomic number of hydrogen is 1), the nucleus of a helium atom contains two protons (atomic number 2), and the nucleus of a uranium atom contains 92 protons (atomic number 92). In addition to atomic number, there is a second number characterizing atoms—**mass number,** defined as the number of protons plus neutrons in a nucleus. As we shall see, the atomic number of an element never varies (i.e., the number of protons does not vary) but the mass number can vary (i.e., the number of neutrons varies). Two or more atoms of the same element having different mass numbers are **isotopes** of one another.

The existence of neutrons was verified conclusively in 1932. A *neutron* has no charge and a mass that is about equal to that of a proton. One of its primary purposes is to act as a "glue" that holds the nucleus together. If neutrons were not present in the nucleus, the repulsive force between the positively charged particles would cause the nucleus to come apart.

But is this where the breaking down stops? Protons, neutrons, and a host of other exotic particles are now known to be composed of six different varieties of particles called **quarks,** which have been given the names of *up, down, strange, charm, bottom,* and *top.* The up, charm, and top quarks have charges of $+\frac{2}{3}$ that of the proton, whereas the down, strange, and bottom quarks have charges of $-\frac{1}{3}$ that of the proton. The proton consists of two up quarks and one down quark (Fig. 1.2), which you can easily show leads to the correct charge for the proton. Likewise, the neutron consists of two down quarks and one up quark, giving a net charge of zero.

1.3 ▷ DENSITY

A property of any substance is its **density** ρ (Greek letter rho), defined as the amount of mass contained in a unit volume, which we usually express as *mass per unit volume:*

A table of the letters in the Greek alphabet is provided on the back endsheet of this textbook.

$$\rho \equiv \frac{m}{V} \qquad \textbf{(1.1)}$$

For example, aluminum has a density of 2.70 g/cm^3, and lead has a density of 11.3 g/cm^3. Therefore, a piece of aluminum of volume 10.0 cm^3 has a mass of 27.0 g, whereas an equivalent volume of lead has a mass of 113 g. A list of densities for various substances is given Table 1.5.

The difference in density between aluminum and lead is due, in part, to their different *atomic masses*. The **atomic mass** of an element is the average mass of one atom in a sample of the element that contains all the element's isotopes, where the relative amounts of isotopes are the same as the relative amounts found in nature. The unit for atomic mass is the *atomic mass unit* (u), where 1 u = 1.660 540 2 × 10^{-27} kg. The atomic mass of lead is 207 u, and that of aluminum is 27.0 u. However, the ratio of atomic masses, 207 u/27.0 u = 7.67, does not correspond to the ratio of densities, (11.3 g/cm^3)/(2.70 g/cm^3) = 4.19. The discrepancy is due to the difference in atomic separations and atomic arrangements in the crystal structure of these two substances.

The mass of a nucleus is measured relative to the mass of the nucleus of the carbon-12 isotope, often written as ^{12}C. (This isotope of carbon has six protons and six neutrons. Other carbon isotopes have six protons but different numbers of neutrons.) Practically all of the mass of an atom is contained within the nucleus. Because the atomic mass of ^{12}C is defined to be exactly 12 u, the proton and neutron each have a mass of about 1 u.

One mole (mol) of a substance is that amount of the substance that contains as many particles (atoms, molecules, or other particles) as there are atoms in 12 g of the carbon-12 isotope. One mole of substance A contains the same number of particles as there are in 1 mol of any other substance B. For example, 1 mol of aluminum contains the same number of atoms as 1 mol of lead.

TABLE 1.5	Densities of Various Substances
Substance	**Density ρ (10^3 kg/m³)**
Gold	19.3
Uranium	18.7
Lead	11.3
Copper	8.92
Iron	7.86
Aluminum	2.70
Magnesium	1.75
Water	1.00
Air	0.0012

Experiments have shown that this number, known as Avogadro's number, N_A, is

$$N_A = 6.022\ 137 \times 10^{23} \text{ particles/mol}$$

Avogadro's number is defined so that 1 mol of carbon-12 atoms has a mass of exactly 12 g. In general, the mass in 1 mol of any element is the element's atomic mass expressed in grams. For example, 1 mol of iron (atomic mass = 55.85 u) has a mass of 55.85 g (we say its *molar mass* is 55.85 g/mol), and 1 mol of lead (atomic mass = 207 u) has a mass of 207 g (its molar mass is 207 g/mol). Because there are 6.02×10^{23} particles in 1 mol of *any* element, the mass per atom for a given element is

$$m_{\text{atom}} = \frac{\text{molar mass}}{N_A} \qquad \textbf{(1.2)}$$

For example, the mass of an iron atom is

$$m_{\text{Fe}} = \frac{55.85 \text{ g/mol}}{6.02 \times 10^{23} \text{ atoms/mol}} = 9.28 \times 10^{-23} \text{ g/atom}$$

EXAMPLE 1.1   How Many Atoms in the Cube?

A solid cube of aluminum (density 2.7 g/cm^3) has a volume of 0.20 cm^3. How many aluminum atoms are contained in the cube?

Solution Since density equals mass per unit volume, the mass m of the cube is

$$m = \rho V = (2.7 \text{ g/cm}^3)(0.20 \text{ cm}^3) = 0.54 \text{ g}$$

To find the number of atoms N in this mass of aluminum, we can set up a proportion using the fact that one mole of alu-

minum (27 g) contains 6.02×10^{23} atoms:

$$\frac{N_A}{27 \text{ g}} = \frac{N}{0.54 \text{ g}}$$

$$\frac{6.02 \times 10^{23} \text{ atoms}}{27 \text{ g}} = \frac{N}{0.54 \text{ g}}$$

$$N = \frac{(0.54 \text{ g})(6.02 \times 10^{23} \text{ atoms})}{27 \text{ g}} = \boxed{1.2 \times 10^{22} \text{ atoms}}$$

1.4   DIMENSIONAL ANALYSIS

The word *dimension* has a special meaning in physics. It usually denotes the physical nature of a quantity. Whether a distance is measured in the length unit feet or the length unit meters, it is still a distance. We say the dimension—the physical nature—of distance is *length*.

The symbols we use in this book to specify length, mass, and time are L, M, and T, respectively. We shall often use brackets [ ] to denote the dimensions of a physical quantity. For example, the symbol we use for speed in this book is v, and in our notation the dimensions of speed are written $[v] = \text{L/T}$. As another example, the dimensions of area, for which we use the symbol A, are $[A] = \text{L}^2$. The dimensions of area, volume, speed, and acceleration are listed in Table 1.6.

In solving problems in physics, there is a useful and powerful procedure called *dimensional analysis*. This procedure, which should always be used, will help minimize the need for rote memorization of equations. Dimensional analysis makes use of the fact that **dimensions can be treated as algebraic quantities.** That is, quantities can be added or subtracted only if they have the same dimensions. Furthermore, the terms on both sides of an equation must have the same dimensions.

TABLE 1.6 **Dimensions and Common Units of Area, Volume, Speed, and Acceleration**

System	Area (L^2)	Volume (L^3)	Speed (L/T)	Acceleration (L/T^2)
SI	m^2	m^3	m/s	m/s^2
British engineering	ft^2	ft^3	ft/s	ft/s^2

By following these simple rules, you can use dimensional analysis to help determine whether an expression has the correct form. The relationship can be correct only if the dimensions are the same on both sides of the equation.

To illustrate this procedure, suppose you wish to derive a formula for the distance x traveled by a car in a time t if the car starts from rest and moves with constant acceleration a. In Chapter 2, we shall find that the correct expression is $x = \frac{1}{2}at^2$. Let us use dimensional analysis to check the validity of this expression. The quantity x on the left side has the dimension of length. For the equation to be dimensionally correct, the quantity on the right side must also have the dimension of length. We can perform a dimensional check by substituting the dimensions for acceleration, L/T^2, and time, T, into the equation. That is, the dimensional form of the equation $x = \frac{1}{2}at^2$ is

$$L = \frac{L}{T^2} \cdot T^2 = L$$

The units of time squared cancel as shown, leaving the unit of length.

A more general procedure using dimensional analysis is to set up an expression of the form

$$x \propto a^n t^m$$

where n and m are exponents that must be determined and the symbol $\propto$ indicates a proportionality. This relationship is correct only if the dimensions of both sides are the same. Because the dimension of the left side is length, the dimension of the right side must also be length. That is,

$$[a^n t^m] = L = LT^0$$

Because the dimensions of acceleration are L/T^2 and the dimension of time is T, we have

$$\left(\frac{L}{T^2}\right)^n T^m = L^1$$

$$L^n T^{m-2n} = L^1$$

Because the exponents of L and T must be the same on both sides, the dimensional equation is balanced under the conditions $m - 2n = 0$, $n = 1$, and $m = 2$. Returning to our original expression $x \propto a^n t^m$, we conclude that $x \propto at^2$. This result differs by a factor of 2 from the correct expression, which is $x = \frac{1}{2}at^2$. Because the factor $\frac{1}{2}$ is dimensionless, there is no way of determining it using dimensional analysis.

Quick Quiz 1.1

True or False: Dimensional analysis can give you the numerical value of constants of proportionality that may appear in an algebraic expression.

EXAMPLE 1.2   Analysis of an Equation

Show that the expression $v = at$ is dimensionally correct, where v represents speed, a acceleration, and t a time interval.

Solution   For the speed term, we have from Table 1.6

$$[v] = \frac{L}{T}$$

The same table gives us L/T^2 for the dimensions of acceleration, and so the dimensions of at are

$$[at] = \left(\frac{L}{T^2}\right)(T) = \frac{L}{T}$$

Therefore, the expression is dimensionally correct. (If the expression were given as $v = at^2$, it would be dimensionally *incorrect*. Try it and see!)

EXAMPLE 1.3   Analysis of a Power Law

Suppose we are told that the acceleration a of a particle moving with uniform speed v in a circle of radius r is proportional to some power of r, say r^n, and some power of v, say v^m. How can we determine the values of n and m?

Solution   Let us take a to be

$$a = kr^n v^m$$

where k is a dimensionless constant of proportionality. Knowing the dimensions of a, r, and v, we see that the dimensional equation must be

$$L/T^2 = L^n(L/T)^m = L^{n+m}/T^m$$

This dimensional equation is balanced under the conditions

$$n + m = 1 \quad \text{and} \quad m = 2$$

Therefore $n = -1$, and we can write the acceleration expression as

$$a = kr^{-1}v^2 = k\frac{v^2}{r}$$

When we discuss uniform circular motion later, we shall see that $k = 1$ if a consistent set of units is used. The constant k would not equal 1 if, for example, v were in km/h and you wanted a in m/s^2.

QuickLab

Estimate the weight (in pounds) of two large bottles of soda pop. Note that 1 L of water has a mass of about 1 kg. Use the fact that an object weighing 2.2 lb has a mass of 1 kg. Find some bathroom scales and check your estimate.

1.5   CONVERSION OF UNITS

Sometimes it is necessary to convert units from one system to another. Conversion factors between the SI units and conventional units of length are as follows:

$$1 \text{ mi} = 1\ 609 \text{ m} = 1.609 \text{ km} \qquad 1 \text{ ft} = 0.304\ 8 \text{ m} = 30.48 \text{ cm}$$

$$1 \text{ m} = 39.37 \text{ in.} = 3.281 \text{ ft} \qquad 1 \text{ in.} \equiv 0.025\ 4 \text{ m} = 2.54 \text{ cm (exactly)}$$

A more complete list of conversion factors can be found in Appendix A.

Units can be treated as algebraic quantities that can cancel each other. For example, suppose we wish to convert 15.0 in. to centimeters. Because 1 in. is defined as exactly 2.54 cm, we find that

$$15.0 \text{ in.} = (15.0 \text{ in.})(2.54 \text{ cm/in.}) = 38.1 \text{ cm}$$

This works because multiplying by $\left(\frac{2.54 \text{ cm}}{1 \text{ in.}}\right)$ is the same as multiplying by 1, because the numerator and denominator describe identical things.

(Left) This road sign near Raleigh, North Carolina, shows distances in miles and kilometers. How accurate are the conversions? *(Billy E. Barnes/Stock Boston).*

(Right) This vehicle's speedometer gives speed readings in miles per hour and in kilometers per hour. Try confirming the conversion between the two sets of units for a few readings of the dial. *(Paul Silverman/Fundamental Photographs)*

ΕXAMPLE **1.4**   The Density of a Cube

The mass of a solid cube is 856 g, and each edge has a length of 5.35 cm. Determine the density ρ of the cube in basic SI units.

Solution   Because $1 \text{ g} = 10^{-3} \text{ kg}$ and $1 \text{ cm} = 10^{-2} \text{ m}$, the mass m and volume V in basic SI units are

$$m = 856 \text{ g} \times 10^{-3} \text{ kg/g} = 0.856 \text{ kg}$$

$$V = L^3 = (5.35 \text{ cm} \times 10^{-2} \text{ m/cm})^3$$
$$= (5.35)^3 \times 10^{-6} \text{ m}^3 = 1.53 \times 10^{-4} \text{ m}^3$$

Therefore,

$$\rho = \frac{m}{V} = \frac{0.856 \text{ kg}}{1.53 \times 10^{-4} \text{ m}^3} = \boxed{5.59 \times 10^3 \text{ kg/m}^3}$$

ΕSTIMATES AND ORDER-OF-MAGNITUDE CALCULATIONS

It is often useful to compute an approximate answer to a physical problem even where little information is available. Such an approximate answer can then be used to determine whether a more accurate calculation is necessary. Approximations are usually based on certain assumptions, which must be modified if greater accuracy is needed. Thus, we shall sometimes refer to the order of magnitude of a certain quantity as the power of ten of the number that describes that quantity. If, for example, we say that a quantity increases in value by three orders of magnitude, this means that its value is increased by a factor of $10^3 = 1000$. Also, if a quantity is given as 3×10^3, we say that the order of magnitude of that quantity is 10^3 (or in symbolic form, $3 \times 10^3 \sim 10^3$). Likewise, the quantity $8 \times 10^7 \sim 10^8$.

The spirit of order-of-magnitude calculations, sometimes referred to as "guesstimates" or "ball-park figures," is given in the following quotation: "Make an estimate before every calculation, try a simple physical argument . . . before every derivation, guess the answer to every puzzle. Courage: no one else needs to know what the guess is."[4] Inaccuracies caused by guessing too low for one number are often canceled out by other guesses that are too high. You will find that with practice your guesstimates get better and better. Estimation problems can be fun to work as you freely drop digits, venture reasonable approximations for unknown numbers, make simplifying assumptions, and turn the question around into something you can answer in your head.

EXAMPLE 1.5   **Breaths in a Lifetime**

Estimate the number of breaths taken during an average life span.

Solution   We shall start by guessing that the typical life span is about 70 years. The only other estimate we must make in this example is the average number of breaths that a person takes in 1 min. This number varies, depending on whether the person is exercising, sleeping, angry, serene, and so forth. To the nearest order of magnitude, we shall choose 10 breaths per minute as our estimate of the average. (This is certainly closer to the true value than 1 breath per minute or 100 breaths per minute.) The number of minutes in a year is

approximately

$$1 \, \cancel{y} \times 400 \, \frac{\cancel{days}}{\cancel{y}} \times 25 \, \frac{\cancel{h}}{\cancel{day}} \times 60 \, \frac{min}{\cancel{h}} = 6 \times 10^5 \, min$$

Notice how much simpler it is to multiply 400×25 than it is to work with the more accurate 365×24. These approximate values for the number of days in a year and the number of hours in a day are close enough for our purposes. Thus, in 70 years there will be $(70 \, yr)(6 \times 10^5 \, min/yr) = 4 \times 10^7$ min. At a rate of 10 breaths/min, an individual would take

4×10^8 breaths in a lifetime.

EXAMPLE 1.6   **It's a Long Way to San Jose**

Estimate the number of steps a person would take walking from New York to Los Angeles.

Solution   Without looking up the distance between these two cities, you might remember from a geography class that they are about 3 000 mi apart. The next approximation we must make is the length of one step. Of course, this length depends on the person doing the walking, but we can estimate that each step covers about 2 ft. With our estimated step size, we can determine the number of steps in 1 mi. Because this is a rough calculation, we round 5 280 ft/mi to 5 000 ft/mi. (What percentage error does this introduce?) This conversion factor gives us

$$\frac{5\,000 \, ft/mi}{2 \, ft/step} = 2\,500 \, steps/mi$$

Now we switch to scientific notation so that we can do the calculation mentally:

$$(3 \times 10^3 \, \cancel{mi})(2.5 \times 10^3 \, steps/\cancel{mi}) = 7.5 \times 10^6 \, steps$$

$$\sim \quad 10^7 \, steps$$

So if we intend to walk across the United States, it will take us on the order of ten million steps. This estimate is almost certainly too small because we have not accounted for curving roads and going up and down hills and mountains. Nonetheless, it is probably within an order of magnitude of the correct answer.

EXAMPLE 1.7   **How Much Gas Do We Use?**

Estimate the number of gallons of gasoline used each year by all the cars in the United States.

Solution   There are about 270 million people in the United States, and so we estimate that the number of cars in the country is 100 million (guessing that there are between two and three people per car). We also estimate that the aver-

age distance each car travels per year is 10 000 mi. If we assume a gasoline consumption of 20 mi/gal or 0.05 gal/mi, then each car uses about 500 gal/yr. Multiplying this by the total number of cars in the United States gives an estimated

total consumption of 5×10^{10} gal $\sim$ 10^{11} gal.

---

[4] E. Taylor and J. A. Wheeler, *Spacetime Physics*, San Francisco, W. H. Freeman & Company, Publishers, 1966, p. 60.

1.7 SIGNIFICANT FIGURES

When physical quantities are measured, the measured values are known only to within the limits of the experimental uncertainty. The value of this uncertainty can depend on various factors, such as the quality of the apparatus, the skill of the experimenter, and the number of measurements performed.

Suppose that we are asked to measure the area of a computer disk label using a meter stick as a measuring instrument. Let us assume that the accuracy to which we can measure with this stick is ± 0.1 cm. If the length of the label is measured to be 5.5 cm, we can claim only that its length lies somewhere between 5.4 cm and 5.6 cm. In this case, we say that the measured value has two significant figures. Likewise, if the label's width is measured to be 6.4 cm, the actual value lies between 6.3 cm and 6.5 cm. Note that the significant figures include the first estimated digit. Thus we could write the measured values as (5.5 ± 0.1) cm and (6.4 ± 0.1) cm.

Now suppose we want to find the area of the label by multiplying the two measured values. If we were to claim the area is $(5.5 \text{ cm})(6.4 \text{ cm}) = 35.2 \text{ cm}^2$, our answer would be unjustifiable because it contains three significant figures, which is greater than the number of significant figures in either of the measured lengths. A good rule of thumb to use in determining the number of significant figures that can be claimed is as follows:

> When multiplying several quantities, the number of significant figures in the final answer is the same as the number of significant figures in the *least* accurate of the quantities being multiplied, where "least accurate" means "having the lowest number of significant figures." The same rule applies to division.

Applying this rule to the multiplication example above, we see that the answer for the area can have only two significant figures because our measured lengths have only two significant figures. Thus, all we can claim is that the area is 35 cm^2, realizing that the value can range between $(5.4 \text{ cm})(6.3 \text{ cm}) = 34 \text{ cm}^2$ and $(5.6 \text{ cm})(6.5 \text{ cm}) = 36 \text{ cm}^2$.

Zeros may or may not be significant figures. Those used to position the decimal point in such numbers as 0.03 and 0.007 5 are not significant. Thus, there are one and two significant figures, respectively, in these two values. When the zeros come after other digits, however, there is the possibility of misinterpretation. For example, suppose the mass of an object is given as 1 500 g. This value is ambiguous because we do not know whether the last two zeros are being used to locate the decimal point or whether they represent significant figures in the measurement. To remove this ambiguity, it is common to use scientific notation to indicate the number of significant figures. In this case, we would express the mass as 1.5×10^3 g if there are two significant figures in the measured value, 1.50×10^3 g if there are three significant figures, and 1.500×10^3 g if there are four. The same rule holds when the number is less than 1, so that 2.3×10^{-4} has two significant figures (and so could be written 0.000 23) and 2.30×10^{-4} has three significant figures (also written 0.000 230). In general, **a significant figure is a reliably known digit** (other than a zero used to locate the decimal point).

For addition and subtraction, you must consider the number of decimal places when you are determining how many significant figures to report.

QuickLab

Determine the thickness of a page from this book. (Note that numbers that have no measurement errors—like the count of a number of pages—do not affect the significant figures in a calculation.) In terms of significant figures, why is it better to measure the thickness of as many pages as possible and then divide by the number of sheets?

> When numbers are added or subtracted, the number of decimal places in the result should equal the smallest number of decimal places of any term in the sum.

For example, if we wish to compute $123 + 5.35$, the answer given to the correct number of significant figures is 128 and not 128.35. If we compute the sum $1.000\ 1 + 0.000\ 3 = 1.000\ 4$, the result has five significant figures, even though one of the terms in the sum, $0.000\ 3$, has only one significant figure. Likewise, if we perform the subtraction $1.002 - 0.998 = 0.004$, the result has only one significant figure even though one term has four significant figures and the other has three. In this book, **most of the numerical examples and end-of-chapter problems will yield answers having three significant figures.** When carrying out estimates we shall typically work with a single significant figure.

Quick Quiz 1.2

Suppose you measure the position of a chair with a meter stick and record that the center of the seat is 1.043 860 564 2 m from a wall. What would a reader conclude from this recorded measurement?

EXAMPLE 1.8   The Area of a Rectangle

A rectangular plate has a length of (21.3 ± 0.2) cm and a width of (9.80 ± 0.1) cm. Find the area of the plate and the uncertainty in the calculated area.

Solution

$$\text{Area} = \ell w = (21.3 \pm 0.2\ \text{cm}) \times (9.80 \pm 0.1\ \text{cm})$$

$$\approx (21.3 \times 9.80 \pm 21.3 \times 0.1 \pm 0.2 \times 9.80)\ \text{cm}^2$$

$$\approx (209 \pm 4)\ \text{cm}^2$$

Because the input data were given to only three significant figures, we cannot claim any more in our result. Do you see why we did not need to multiply the uncertainties 0.2 cm and 0.1 cm?

EXAMPLE 1.9   Installing a Carpet

A carpet is to be installed in a room whose length is measured to be 12.71 m and whose width is measured to be 3.46 m. Find the area of the room.

Solution   If you multiply 12.71 m by 3.46 m on your calculator, you will get an answer of $43.976\ 6\ \text{m}^2$. How many of these numbers should you claim? Our rule of thumb for multiplication tells us that you can claim only the number of significant figures in the least accurate of the quantities being measured. In this example, we have only three significant figures in our least accurate measurement, so we should express our final answer as $44.0\ \text{m}^2$.

Note that in reducing 43.976 6 to three significant figures for our answer, we used a general rule for rounding off numbers that states that the last digit retained (the 9 in this example) is increased by 1 if the first digit dropped (here, the 7) is 5 or greater. (A technique for avoiding error accumulation is to delay rounding of numbers in a long calculation until you have the final result. Wait until you are ready to copy the answer from your calculator before rounding to the correct number of significant figures.)

SUMMARY

The three physical quantities of mechanics are length, mass, and time, which in the SI system have the units meters (m), kilograms (kg), and seconds (s), respectively. Prefixes indicating various powers of ten are used with these three basic units. The **density** of a substance is defined as its *mass per unit volume*. Different substances have different densities mainly because of differences in their atomic masses and atomic arrangements.

The number of particles in one mole of any element or compound, called **Avogadro's number,** N_A, is 6.02×10^{23}.

The method of *dimensional analysis* is very powerful in solving physics problems. Dimensions can be treated as algebraic quantities. By making estimates and making order-of-magnitude calculations, you should be able to approximate the answer to a problem when there is not enough information available to completely specify an exact solution.

When you compute a result from several measured numbers, each of which has a certain accuracy, you should give the result with the correct number of significant figures.

QUESTIONS

1. In this chapter we described how the Earth's daily rotation on its axis was once used to define the standard unit of time. What other types of natural phenomena could serve as alternative time standards?

2. Suppose that the three fundamental standards of the metric system were length, density, and time rather than length, mass, and time. The standard of density in this system is to be defined as that of water. What considerations about water would you need to address to make sure that the standard of density is as accurate as possible?

3. A hand is defined as 4 in.; a foot is defined as 12 in. Why should the hand be any less acceptable as a unit than the foot, which we use all the time?

4. Express the following quantities using the prefixes given in Table 1.4: (a) 3×10^{-4} m (b) 5×10^{-5} s (c) 72×10^2 g.

5. Suppose that two quantities A and B have different dimensions. Determine which of the following arithmetic operations *could* be physically meaningful: (a) $A + B$ (b) A/B (c) $B - A$ (d) AB.

6. What level of accuracy is implied in an order-of-magnitude calculation?

7. Do an order-of-magnitude calculation for an everyday situation you might encounter. For example, how far do you walk or drive each day?

8. Estimate your age in seconds.

9. Estimate the mass of this textbook in kilograms. If a scale is available, check your estimate.

PROBLEMS

1, 2, 3 = straightforward, intermediate, challenging   ☐ = full solution available in the *Student Solutions Manual and Study Guide*
WEB = solution posted at **http://www.saunderscollege.com/physics/**   🖥 = Computer useful in solving problem   🖼 = Interactive Physics
☐ = paired numerical/symbolic problems

Section 1.3   Density

1. The standard kilogram is a platinum–iridium cylinder 39.0 mm in height and 39.0 mm in diameter. What is the density of the material?

2. The mass of the planet Saturn (Fig. P1.2) is 5.64×10^{26} kg, and its radius is 6.00×10^7 m. Calculate its density.

3. How many grams of copper are required to make a hollow spherical shell having an inner radius of 5.70 cm and an outer radius of 5.75 cm? The density of copper is 8.92 g/cm^3.

4. What mass of a material with density ρ is required to make a hollow spherical shell having inner radius r_1 and outer radius r_2?

5. Iron has molar mass 55.8 g/mol. (a) Find the volume of 1 mol of iron. (b) Use the value found in (a) to determine the volume of one iron atom. (c) Calculate the cube root of the atomic volume, to have an estimate for the distance between atoms in the solid. (d) Repeat the calculations for uranium, finding its molar mass in the periodic table of the elements in Appendix C.

Figure P1.2   A view of Saturn from *Voyager 2*. *(Courtesy of NASA)*

6. Two spheres are cut from a certain uniform rock. One has radius 4.50 cm. The mass of the other is five times greater. Find its radius.

WEB 7. Calculate the mass of an atom of (a) helium, (b) iron, and (c) lead. Give your answers in atomic mass units and in grams. The molar masses are 4.00, 55.9, and 207 g/mol, respectively, for the atoms given.

8. On your wedding day your lover gives you a gold ring of mass 3.80 g. Fifty years later its mass is 3.35 g. As an average, how many atoms were abraded from the ring during each second of your marriage? The molar mass of gold is 197 g/mol.

9. A small cube of iron is observed under a microscope. The edge of the cube is 5.00×10^{-6} cm long. Find (a) the mass of the cube and (b) the number of iron atoms in the cube. The molar mass of iron is 55.9 g/mol, and its density is 7.86 g/cm^3.

10. A structural I-beam is made of steel. A view of its cross-section and its dimensions are shown in Figure P1.10.

15.0 cm

1.00 cm

36.0 cm

1.00 cm

Figure P1.10

(a) What is the mass of a section 1.50 m long? (b) How many atoms are there in this section? The density of steel is 7.56×10^3 kg/m^3.

11. A child at the beach digs a hole in the sand and, using a pail, fills it with water having a mass of 1.20 kg. The molar mass of water is 18.0 g/mol. (a) Find the number of water molecules in this pail of water. (b) Suppose the quantity of water on the Earth is 1.32×10^{21} kg and remains constant. How many of the water molecules in this pail of water were likely to have been in an equal quantity of water that once filled a particular claw print left by a dinosaur?

Section 1.4   Dimensional Analysis

12. The radius r of a circle inscribed in any triangle whose sides are a, b, and c is given by

$$r = [(s - a)(s - b)(s - c)/s]^{1/2}$$

where s is an abbreviation for $(a + b + c)/2$. Check this formula for dimensional consistency.

13. The displacement of a particle moving under uniform acceleration is some function of the elapsed time and the acceleration. Suppose we write this displacement $s = ka^m t^n$, where k is a dimensionless constant. Show by dimensional analysis that this expression is satisfied if $m = 1$ and $n = 2$. Can this analysis give the value of k?

14. The period T of a simple pendulum is measured in time units and is described by

$$T = 2\pi \sqrt{\frac{\ell}{g}}$$

where ℓ is the length of the pendulum and g is the free-fall acceleration in units of length divided by the square of time. Show that this equation is dimensionally correct.

15. Which of the equations below are dimensionally correct?
    (a) $v = v_0 + ax$
    (b) $y = (2 \text{ m}) \cos(kx)$, where $k = 2 \text{ m}^{-1}$

16. Newton's law of universal gravitation is represented by

$$F = \frac{GMm}{r^2}$$

Here F is the gravitational force, M and m are masses, and r is a length. Force has the SI units kg·m/s^2. What are the SI units of the proportionality constant G?

WEB 17. The consumption of natural gas by a company satisfies the empirical equation $V = 1.50t + 0.008\,00t^2$, where V is the volume in millions of cubic feet and t the time in months. Express this equation in units of cubic feet and seconds. Put the proper units on the coefficients. Assume a month is 30.0 days.

Section 1.5   Conversion of Units

18. Suppose your hair grows at the rate 1/32 in. per day. Find the rate at which it grows in nanometers per second. Since the distance between atoms in a molecule is

on the order of 0.1 nm, your answer suggests how rapidly layers of atoms are assembled in this protein synthesis.

19. A rectangular building lot is 100 ft by 150 ft. Determine the area of this lot in m².

20. An auditorium measures 40.0 m × 20.0 m × 12.0 m. The density of air is 1.20 kg/m³. What are (a) the volume of the room in cubic feet and (b) the weight of air in the room in pounds?

21. Assume that it takes 7.00 min to fill a 30.0-gal gasoline tank. (a) Calculate the rate at which the tank is filled in gallons per second. (b) Calculate the rate at which the tank is filled in cubic meters per second. (c) Determine the time, in hours, required to fill a 1-cubic-meter volume at the same rate. (1 U.S. gal = 231 in.³)

22. A creature moves at a speed of 5.00 furlongs per fortnight (not a very common unit of speed). Given that 1 furlong = 220 yards and 1 fortnight = 14 days, determine the speed of the creature in meters per second. What kind of creature do you think it might be?

23. A section of land has an area of 1 mi² and contains 640 acres. Determine the number of square meters in 1 acre.

24. A quart container of ice cream is to be made in the form of a cube. What should be the length of each edge in centimeters? (Use the conversion 1 gal = 3.786 L.)

25. A solid piece of lead has a mass of 23.94 g and a volume of 2.10 cm³. From these data, calculate the density of lead in SI units (kg/m³).

26. An astronomical unit (AU) is defined as the average distance between the Earth and the Sun. (a) How many astronomical units are there in one lightyear? (b) Determine the distance from the Earth to the Andromeda galaxy in astronomical units.

27. The mass of the Sun is 1.99×10^{30} kg, and the mass of an atom of hydrogen, of which the Sun is mostly composed, is 1.67×10^{-27} kg. How many atoms are there in the Sun?

28. (a) Find a conversion factor to convert from miles per hour to kilometers per hour. (b) In the past, a federal law mandated that highway speed limits would be 55 mi/h. Use the conversion factor of part (a) to find this speed in kilometers per hour. (c) The maximum highway speed is now 65 mi/h in some places. In kilometers per hour, how much of an increase is this over the 55-mi/h limit?

29. At the time of this book's printing, the U. S. national debt is about $6 trillion. (a) If payments were made at the rate of $1 000/s, how many years would it take to pay off a $6-trillion debt, assuming no interest were charged? (b) A dollar bill is about 15.5 cm long. If six trillion dollar bills were laid end to end around the Earth's equator, how many times would they encircle the Earth? Take the radius of the Earth at the equator to be 6 378 km. (*Note:* Before doing any of these calculations, try to guess at the answers. You may be very surprised.)

30. (a) How many seconds are there in a year? (b) If one micrometeorite (a sphere with a diameter of 1.00×10^{-6} m) strikes each square meter of the Moon each second, how many years will it take to cover the Moon to a depth of 1.00 m? (*Hint:* Consider a cubic box on the Moon 1.00 m on a side, and find how long it will take to fill the box.)

WEB **31.** One gallon of paint (volume = 3.78×10^{-3} m³) covers an area of 25.0 m². What is the thickness of the paint on the wall?

32. A pyramid has a height of 481 ft, and its base covers an area of 13.0 acres (Fig. P1.32). If the volume of a pyramid is given by the expression $V = \frac{1}{3}Bh$, where B is the area of the base and h is the height, find the volume of this pyramid in cubic meters. (1 acre = 43 560 ft²)

Figure P1.32 Problems 32 and 33.

33. The pyramid described in Problem 32 contains approximately two million stone blocks that average 2.50 tons each. Find the weight of this pyramid in pounds.

34. Assuming that 70% of the Earth's surface is covered with water at an average depth of 2.3 mi, estimate the mass of the water on the Earth in kilograms.

35. The amount of water in reservoirs is often measured in acre-feet. One acre-foot is a volume that covers an area of 1 acre to a depth of 1 ft. An acre is an area of 43 560 ft². Find the volume in SI units of a reservoir containing 25.0 acre-ft of water.

36. A hydrogen atom has a diameter of approximately 1.06×10^{-10} m, as defined by the diameter of the spherical electron cloud around the nucleus. The hydrogen nucleus has a diameter of approximately 2.40×10^{-15} m. (a) For a scale model, represent the diameter of the hydrogen atom by the length of an American football field (100 yards = 300 ft), and determine the diameter of the nucleus in millimeters. (b) The atom is how many times larger in volume than its nucleus?

37. The diameter of our disk-shaped galaxy, the Milky Way, is about 1.0×10^5 lightyears. The distance to Messier 31—which is Andromeda, the spiral galaxy nearest to the Milky Way—is about 2.0 million lightyears. If a scale model represents the Milky Way and Andromeda galaxies as dinner plates 25 cm in diameter, determine the distance between the two plates.

38. The mean radius of the Earth is 6.37×10^6 m, and that of the Moon is 1.74×10^8 cm. From these data calculate (a) the ratio of the Earth's surface area to that of the Moon and (b) the ratio of the Earth's volume to that of the Moon. Recall that the surface area of a sphere is $4\pi r^2$ and that the volume of a sphere is $\frac{4}{3}\pi r^3$.

WEB 39. One cubic meter (1.00 m^3) of aluminum has a mass of 2.70×10^3 kg, and 1.00 m^3 of iron has a mass of 7.86×10^3 kg. Find the radius of a solid aluminum sphere that balances a solid iron sphere of radius 2.00 cm on an equal-arm balance.

40. Let ρ_{Al} represent the density of aluminum and ρ_{Fe} that of iron. Find the radius of a solid aluminum sphere that balances a solid iron sphere of radius r_{Fe} on an equal-arm balance.

Section 1.6   Estimates and Order-of-Magnitude Calculations

WEB 41. Estimate the number of Ping-Pong balls that would fit into an average-size room (without being crushed). In your solution state the quantities you measure or estimate and the values you take for them.

42. McDonald's sells about 250 million packages of French fries per year. If these fries were placed end to end, estimate how far they would reach.

43. An automobile tire is rated to last for 50 000 miles. Estimate the number of revolutions the tire will make in its lifetime.

44. Approximately how many raindrops fall on a 1.0-acre lot during a 1.0-in. rainfall?

45. Grass grows densely everywhere on a quarter-acre plot of land. What is the order of magnitude of the number of blades of grass on this plot of land? Explain your reasoning. (1 acre = 43 560 ft^2.)

46. Suppose that someone offers to give you $1 billion if you can finish counting it out using only one-dollar bills. Should you accept this offer? Assume you can count one bill every second, and be sure to note that you need about 8 hours a day for sleeping and eating and that right now you are probably at least 18 years old.

47. Compute the order of magnitude of the mass of a bathtub half full of water and of the mass of a bathtub half full of pennies. In your solution, list the quantities you take as data and the value you measure or estimate for each.

48. Soft drinks are commonly sold in aluminum containers. Estimate the number of such containers thrown away or recycled each year by U.S. consumers. Approximately how many tons of aluminum does this represent?

49. To an order of magnitude, how many piano tuners are there in New York City? The physicist Enrico Fermi was famous for asking questions like this on oral Ph.D. qual-

ifying examinations and for his own facility in making order-of-magnitude calculations.

Section 1.7   Significant Figures

50. Determine the number of significant figures in the following measured values: (a) 23 cm (b) 3.589 s (c) 4.67×10^3 m/s (d) 0.003 2 m.

51. The radius of a circle is measured to be 10.5 ± 0.2 m. Calculate the (a) area and (b) circumference of the circle and give the uncertainty in each value.

52. Carry out the following arithmetic operations: (a) the sum of the measured values 756, 37.2, 0.83, and 2.5; (b) the product $0.003\ 2 \times 356.3$; (c) the product $5.620 \times \pi$.

53. The radius of a solid sphere is measured to be (6.50 ± 0.20) cm, and its mass is measured to be (1.85 ± 0.02) kg. Determine the density of the sphere in kilograms per cubic meter and the uncertainty in the density.

54. How many significant figures are in the following numbers: (a) 78.9 ± 0.2, (b) 3.788×10^9, (c) 2.46×10^{-6}, and (d) 0.005 3?

55. A farmer measures the distance around a rectangular field. The length of the long sides of the rectangle is found to be 38.44 m, and the length of the short sides is found to be 19.5 m. What is the total distance around the field?

56. A sidewalk is to be constructed around a swimming pool that measures (10.0 ± 0.1) m by (17.0 ± 0.1) m. If the sidewalk is to measure (1.00 ± 0.01) m wide by (9.0 ± 0.1) cm thick, what volume of concrete is needed, and what is the approximate uncertainty of this volume?

ADDITIONAL PROBLEMS

57. In a situation where data are known to three significant digits, we write 6.379 m = 6.38 m and 6.374 m = 6.37 m. When a number ends in 5, we arbitrarily choose to write 6.375 m = 6.38 m. We could equally well write 6.375 m = 6.37 m, "rounding down" instead of "rounding up," since we would change the number 6.375 by equal increments in both cases. Now consider an order-of-magnitude estimate, in which we consider factors rather than increments. We write 500 m $\sim 10^3$ m because 500 differs from 100 by a factor of 5 whereas it differs from 1000 by only a factor of 2. We write 437 m $\sim 10^3$ m and 305 m $\sim 10^2$ m. What distance differs from 100 m and from 1000 m by equal factors, so that we could equally well choose to represent its order of magnitude either as $\sim 10^2$ m or as $\sim 10^3$ m?

58. When a droplet of oil spreads out on a smooth water surface, the resulting "oil slick" is approximately one molecule thick. An oil droplet of mass 9.00×10^{-7} kg and density 918 kg/m^3 spreads out into a circle of radius 41.8 cm on the water surface. What is the diameter of an oil molecule?

59. The basic function of the carburetor of an automobile is to "atomize" the gasoline and mix it with air to promote rapid combustion. As an example, assume that 30.0 cm^3 of gasoline is atomized into N spherical droplets, each with a radius of 2.00×10^{-5} m. What is the total surface area of these N spherical droplets?

60. In physics it is important to use mathematical approximations. Demonstrate for yourself that for small angles ($< 20°$)

$$\tan \alpha \approx \sin \alpha \approx \alpha = \pi \alpha' / 180°$$

where α is in radians and α' is in degrees. Use a calculator to find the largest angle for which $\tan \alpha$ may be approximated by $\sin \alpha$ if the error is to be less than 10.0%.

61. A high fountain of water is located at the center of a circular pool as in Figure P1.61. Not wishing to get his feet wet, a student walks around the pool and measures its circumference to be 15.0 m. Next, the student stands at the edge of the pool and uses a protractor to gauge the angle of elevation of the top of the fountain to be 55.0°. How high is the fountain?

Figure P1.61

62. A shape covers an area A and has a uniform height h. Its cross-sectional area is uniform over its height, and its volume is given by $V = Ah$. (a) Show that $V = Ah$ is dimensionally correct. (b) Show that the volumes of a cylinder and of a rectangular box can be written in the form $V = Ah$, identifying A in each case. (Note that A, sometimes called the "footprint" of the object, can have any shape and that the height can be replaced by average thickness in general.)

63. A useful fact is that there are about $\pi \times 10^7$ s in one year. Find the percentage error in this approximation, where "percentage error" is defined as

$$\frac{|\text{Assumed value} - \text{true value}|}{\text{True value}} \times 100\%$$

64. A crystalline solid consists of atoms stacked up in a repeating lattice structure. Consider a crystal as shown in Figure P1.64a. The atoms reside at the corners of cubes of side $L = 0.200$ nm. One piece of evidence for the regular arrangement of atoms comes from the flat surfaces along which a crystal separates, or "cleaves," when it is broken. Suppose this crystal cleaves along a face diagonal, as shown in Figure P1.64b. Calculate the spacing d between two adjacent atomic planes that separate when the crystal cleaves.

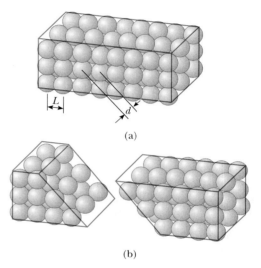

(a)

(b)

Figure P1.64

65. A child loves to watch as you fill a transparent plastic bottle with shampoo. Every horizontal cross-section of the bottle is a circle, but the diameters of the circles all have different values, so that the bottle is much wider in some places than in others. You pour in bright green shampoo with constant volume flow rate 16.5 cm^3/s. At what rate is its level in the bottle rising (a) at a point where the diameter of the bottle is 6.30 cm and (b) at a point where the diameter is 1.35 cm?

66. As a child, the educator and national leader Booker T. Washington was given a spoonful (about 12.0 cm^3) of molasses as a treat. He pretended that the quantity increased when he spread it out to cover uniformly all of a tin plate (with a diameter of about 23.0 cm). How thick a layer did it make?

67. Assume there are 100 million passenger cars in the United States and that the average fuel consumption is 20 mi/gal of gasoline. If the average distance traveled by each car is 10 000 mi/yr, how much gasoline would be saved per year if average fuel consumption could be increased to 25 mi/gal?

68. One cubic centimeter of water has a mass of 1.00×10^{-3} kg. (a) Determine the mass of 1.00 m^3 of water. (b) Assuming biological substances are 98% water, estimate the mass of a cell that has a diameter of 1.0 μm, a human kidney, and a fly. Assume that a kidney is roughly a sphere with a radius of 4.0 cm and that a fly is roughly a cylinder 4.0 mm long and 2.0 mm in diameter.

69. The distance from the Sun to the nearest star is 4×10^{16} m. The Milky Way galaxy is roughly a disk of diameter $\sim 10^{21}$ m and thickness $\sim 10^{19}$ m. Find the order of magnitude of the number of stars in the Milky Way. Assume the 4×10^{16}-m distance between the Sun and the nearest star is typical.

70. The data in the following table represent measurements of the masses and dimensions of solid cylinders of alu-

minum, copper, brass, tin, and iron. Use these data to calculate the densities of these substances. Compare your results for aluminum, copper, and iron with those given in Table 1.5.

Substance	Mass (g)	Diameter (cm)	Length (cm)
Aluminum	51.5	2.52	3.75
Copper	56.3	1.23	5.06
Brass	94.4	1.54	5.69
Tin	69.1	1.75	3.74
Iron	216.1	1.89	9.77

ANSWERS TO QUICK QUIZZES

1.1   False. Dimensional analysis gives the units of the proportionality constant but provides no information about its numerical value. For example, experiments show that doubling the radius of a solid sphere increases its mass 8-fold, and tripling the radius increases the mass 27-fold. Therefore, its mass is proportional to the cube of its radius. Because $m \propto r^3$, we can write $m = kr^3$. Dimensional analysis shows that the proportionality constant k must have units kg/m^3, but to determine its numerical value requires either experimental data or geometrical reasoning.

1.2   Reporting all these digits implies you have determined the location of the center of the chair's seat to the nearest ± 0.000 000 000 1 m. This roughly corresponds to being able to count the atoms in your meter stick because each of them is about that size! It would probably be better to record the measurement as 1.044 m: this indicates that you know the position to the nearest millimeter, assuming the meter stick has millimeter markings on its scale.

THE WIZARD OF ID   By Parker and Hart

By permission of John Hart and Field Enterprises, Inc.

c h a p t e r

2

Motion in One Dimension

Chapter Outline

As a first step in studying classical mechanics, we describe motion in terms of space and time while ignoring the agents that caused that motion. This portion of classical mechanics is called *kinematics*. (The word *kinematics* has the same root as *cinema*. Can you see why?) In this chapter we consider only motion in one dimension. We first define displacement, velocity, and acceleration. Then, using these concepts, we study the motion of objects traveling in one dimension with a constant acceleration.

From everyday experience we recognize that motion represents a continuous change in the position of an object. In physics we are concerned with three types of motion: translational, rotational, and vibrational. A car moving down a highway is an example of translational motion, the Earth's spin on its axis is an example of rotational motion, and the back-and-forth movement of a pendulum is an example of vibrational motion. In this and the next few chapters, we are concerned only with translational motion. (Later in the book we shall discuss rotational and vibrational motions.)

In our study of translational motion, we describe the moving object as a *particle* regardless of its size. In general, **a particle is a point-like mass having infinitesimal size.** For example, if we wish to describe the motion of the Earth around the Sun, we can treat the Earth as a particle and obtain reasonably accurate data about its orbit. This approximation is justified because the radius of the Earth's orbit is large compared with the dimensions of the Earth and the Sun. As an example on a much smaller scale, it is possible to explain the pressure exerted by a gas on the walls of a container by treating the gas molecules as particles.

2.1 DISPLACEMENT, VELOCITY, AND SPEED

The motion of a particle is completely known if the particle's position in space is known at all times. Consider a car moving back and forth along the x axis, as shown in Figure 2.1a. When we begin collecting position data, the car is 30 m to the right of a road sign. (Let us assume that all data in this example are known to two significant figures. To convey this information, we should report the initial position as 3.0×10^1 m. We have written this value in this simpler form to make the discussion easier to follow.) We start our clock and once every 10 s note the car's location relative to the sign. As you can see from Table 2.1, the car is moving to the right (which we have defined as the positive direction) during the first 10 s of motion, from position Ⓐ to position Ⓑ. The position values now begin to decrease, however, because the car is backing up from position Ⓑ through position Ⓕ. In fact, at Ⓓ, 30 s after we start measuring, the car is alongside the sign we are using as our origin of coordinates. It continues moving to the left and is more than 50 m to the left of the sign when we stop recording information after our sixth data point. A graph of this information is presented in Figure 2.1b. Such a plot is called a *position–time graph*.

If a particle is moving, we can easily determine its change in position. **The displacement of a particle is defined as its change in position.** As it moves from an initial position x_i to a final position x_f, its displacement is given by $x_f - x_i$. We use the Greek letter delta (Δ) to denote the *change* in a quantity. Therefore, we write the displacement, or change in position, of the particle as

$$\Delta x \equiv x_f - x_i \tag{2.1}$$

From this definition we see that Δx is positive if x_f is greater than x_i and negative if x_f is less than x_i.

TABLE 2.1
Position of the Car at Various Times

Position	$t(s)$	$x(m)$
Ⓐ	0	30
Ⓑ	10	52
Ⓒ	20	38
Ⓓ	30	0
Ⓔ	40	−37
Ⓕ	50	−53

Figure 2.1 (a) A car moves back and forth along a straight line taken to be the *x* axis. Because we are interested only in the car's translational motion, we can treat it as a particle. (b) Position–time graph for the motion of the "particle."

A very easy mistake to make is not to recognize the difference between displacement and distance traveled (Fig. 2.2). A baseball player hitting a home run travels a distance of 360 ft in the trip around the bases. However, the player's displacement is zero because his final and initial positions are identical.

Displacement is an example of a vector quantity. Many other physical quantities, including velocity and acceleration, also are vectors. In general, **a vector is a physical quantity that requires the specification of both direction and magnitude.** By contrast, **a scalar is a quantity that has magnitude and no direction.** In this chapter, we use plus and minus signs to indicate vector direction. We can do this because the chapter deals with one-dimensional motion only; this means that any object we study can be moving only along a straight line. For example, for horizontal motion, let us arbitrarily specify to the right as being the positive direction. It follows that any object always moving to the right undergoes a

Figure 2.2 Bird's-eye view of a baseball diamond. A batter who hits a home run travels 360 ft as he rounds the bases, but his displacement for the round trip is zero. *(Mark C. Burnett/Photo Researchers, Inc.)*

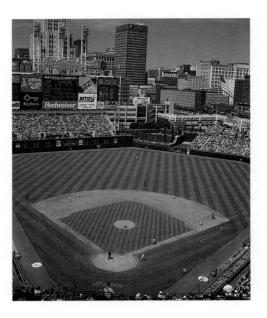

positive displacement $+\Delta x$, and any object moving to the left undergoes a negative displacement $-\Delta x$. We shall treat vectors in greater detail in Chapter 3.

There is one very important point that has not yet been mentioned. Note that the graph in Figure 2.1b does not consist of just six data points but is actually a smooth curve. The graph contains information about the entire 50-s interval during which we watched the car move. It is much easier to see changes in position from the graph than from a verbal description or even a table of numbers. For example, it is clear that the car was covering more ground during the middle of the 50-s interval than at the end. Between positions ⓒ and ⓓ, the car traveled almost 40 m, but during the last 10 s, between positions ⓔ and ⓕ, it moved less than half that far. A common way of comparing these different motions is to divide the displacement Δx that occurs between two clock readings by the length of that particular time interval Δt. This turns out to be a very useful ratio, one that we shall use many times. For convenience, the ratio has been given a special name—*average velocity*. **The average velocity $\bar{v}_x$ of a particle is defined as the particle's displacement Δx divided by the time interval Δt during which that displacement occurred:**

Average velocity

$$\bar{v}_x \equiv \frac{\Delta x}{\Delta t} \qquad (2.2)$$

where the subscript x indicates motion along the x axis. From this definition we see that average velocity has dimensions of length divided by time (L/T)—meters per second in SI units.

Although the distance traveled for any motion is always positive, the average velocity of a particle moving in one dimension can be positive or negative, depending on the sign of the displacement. (The time interval Δt is always positive.) If the coordinate of the particle increases in time (that is, if $x_f > x_i$), then Δx is positive and $\bar{v}_x = \Delta x/\Delta t$ is positive. This case corresponds to motion in the positive x direction. If the coordinate decreases in time (that is, if $x_f < x_i$), then Δx is negative and hence $\bar{v}_x$ is negative. This case corresponds to motion in the negative x direction.

We can interpret average velocity geometrically by drawing a straight line between any two points on the position–time graph in Figure 2.1b. This line forms the hypotenuse of a right triangle of height Δx and base Δt. The slope of this line is the ratio $\Delta x/\Delta t$. For example, the line between positions Ⓐ and Ⓑ has a slope equal to the average velocity of the car between those two times, $(52\text{ m} - 30\text{ m})/(10\text{ s} - 0) = 2.2\text{ m/s}$.

In everyday usage, the terms *speed* and *velocity* are interchangeable. In physics, however, there is a clear distinction between these two quantities. Consider a marathon runner who runs more than 40 km, yet ends up at his starting point. His average velocity is zero! Nonetheless, we need to be able to quantify how fast he was running. A slightly different ratio accomplishes this for us. **The average speed of a particle, a scalar quantity, is defined as the total distance traveled divided by the total time it takes to travel that distance:**

$$\text{Average speed} = \frac{\text{total distance}}{\text{total time}}$$

Average speed

The SI unit of average speed is the same as the unit of average velocity: meters per second. However, unlike average velocity, average speed has no direction and hence carries no algebraic sign.

Knowledge of the average speed of a particle tells us nothing about the details of the trip. For example, suppose it takes you 8.0 h to travel 280 km in your car. The average speed for your trip is 35 km/h. However, you most likely traveled at various speeds during the trip, and the average speed of 35 km/h could result from an infinite number of possible speed values.

EXAMPLE 2.1   Calculating the Variables of Motion

Find the displacement, average velocity, and average speed of the car in Figure 2.1a between positions Ⓐ and Ⓕ.

Solution The units of displacement must be meters, and the numerical result should be of the same order of magnitude as the given position data (which means probably not 10 or 100 times bigger or smaller). From the position–time graph given in Figure 2.1b, note that $x_A = 30$ m at $t_A = 0$ s and that $x_F = -53$ m at $t_F = 50$ s. Using these values along with the definition of displacement, Equation 2.1, we find that

$$\Delta x = x_F - x_A = -53\text{ m} - 30\text{ m} = \boxed{-83\text{ m}}$$

This result means that the car ends up 83 m in the negative direction (to the left, in this case) from where it started. This number has the correct units and is of the same order of magnitude as the supplied data. A quick look at Figure 2.1a indicates that this is the correct answer.

It is difficult to estimate the average velocity without completing the calculation, but we expect the units to be meters per second. Because the car ends up to the left of where we started taking data, we know the average velocity must be negative. From Equation 2.2,

$$\bar{v}_x = \frac{\Delta x}{\Delta t} = \frac{x_f - x_i}{t_f - t_i} = \frac{x_F - x_A}{t_F - t_A}$$

$$= \frac{-53\text{ m} - 30\text{ m}}{50\text{ s} - 0\text{ s}} = \frac{-83\text{ m}}{50\text{ s}} = \boxed{-1.7\text{ m/s}}$$

We find the car's average speed for this trip by adding the distances traveled and dividing by the total time:

$$\text{Average} = \frac{22\text{ m} + 52\text{ m} + 53\text{ m}}{50\text{ s}} = \boxed{2.5\text{ m/s}}$$

2.2   INSTANTANEOUS VELOCITY AND SPEED

Often we need to know the velocity of a particle at a particular instant in time, rather than over a finite time interval. For example, even though you might want to calculate your average velocity during a long automobile trip, you would be especially interested in knowing your velocity at the *instant* you noticed the police

(a)

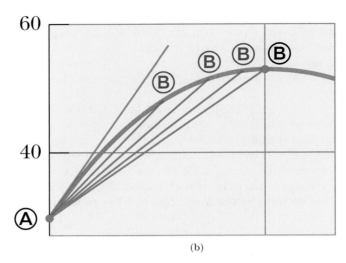

(b)

Figure 2.3 (a) Graph representing the motion of the car in Figure 2.1. An enlargement of the upper left-hand corner of the graph shows how the blue line between positions Ⓐ and Ⓑ approaches the green tangent line as point Ⓑ gets closer to point Ⓐ.

car parked alongside the road in front of you. In other words, you would like to be able to specify your velocity just as precisely as you can specify your position by noting what is happening at a specific clock reading—that is, at some specific instant. It may not be immediately obvious how to do this. What does it mean to talk about how fast something is moving if we "freeze time" and talk only about an individual instant? This is a subtle point not thoroughly understood until the late 1600s. At that time, with the invention of calculus, scientists began to understand how to describe an object's motion at any moment in time.

To see how this is done, consider Figure 2.3a. We have already discussed the average velocity for the interval during which the car moved from position Ⓐ to position Ⓑ (given by the slope of the dark blue line) and for the interval during which it moved from Ⓐ to Ⓕ (represented by the slope of the light blue line). Which of these two lines do you think is a closer approximation of the initial velocity of the car? The car starts out by moving to the right, which we defined to be the positive direction. Therefore, being positive, the value of the average velocity during the Ⓐ to Ⓑ interval is probably closer to the initial value than is the value of the average velocity during the Ⓐ to Ⓕ interval, which we determined to be negative in Example 2.1. Now imagine that we start with the dark blue line and slide point Ⓑ to the left along the curve, toward point Ⓐ, as in Figure 2.3b. The line between the points becomes steeper and steeper, and as the two points get extremely close together, the line becomes a tangent line to the curve, indicated by the green line on the graph. The slope of this tangent line represents the velocity of the car at the moment we started taking data, at point Ⓐ. What we have done is determine the *instantaneous velocity* at that moment. In other words, the **instantaneous velocity v_x equals the limiting value of the ratio $\Delta x/\Delta t$ as Δt approaches zero:**[1]

Definition of instantaneous velocity

$$v_x \equiv \lim_{\Delta t \to 0} \frac{\Delta x}{\Delta t} \qquad \textbf{(2.3)}$$

3.3

[1] Note that the displacement Δx also approaches zero as Δt approaches zero. As Δx and Δt become smaller and smaller, the ratio $\Delta x/\Delta t$ approaches a value equal to the slope of the line tangent to the *x*-versus-*t* curve.

In calculus notation, this limit is called the *derivative* of x with respect to t, written dx/dt:

$$v_x \equiv \lim_{\Delta t \to 0} \frac{\Delta x}{\Delta t} = \frac{dx}{dt} \qquad \textbf{(2.4)}$$

The instantaneous velocity can be positive, negative, or zero. When the slope of the position–time graph is positive, such as at any time during the first 10 s in Figure 2.3, v_x is positive. After point Ⓑ, v_x is negative because the slope is negative. At the peak, the slope and the instantaneous velocity are zero.

From here on, we use the word *velocity* to designate instantaneous velocity. When it is *average velocity* we are interested in, we always use the adjective *average*.

The instantaneous speed of a particle is defined as the magnitude of its velocity. As with average speed, instantaneous speed has no direction associated with it and hence carries no algebraic sign. For example, if one particle has a velocity of $+25$ m/s along a given line and another particle has a velocity of -25 m/s along the same line, both have a speed[2] of 25 m/s.

EXAMPLE 2.2   **Average and Instantaneous Velocity**

A particle moves along the x axis. Its x coordinate varies with time according to the expression $x = -4t + 2t^2$, where x is in meters and t is in seconds.[3] The position–time graph for this motion is shown in Figure 2.4. Note that the particle moves in the negative x direction for the first second of motion, is at rest at the moment $t = 1$ s, and moves in the positive x direction for $t > 1$ s. (a) Determine the displacement of the particle in the time intervals $t = 0$ to $t = 1$ s and $t = 1$ s to $t = 3$ s.

Solution   During the first time interval, we have a negative slope and hence a negative velocity. Thus, we know that the displacement between Ⓐ and Ⓑ must be a negative number having units of meters. Similarly, we expect the displacement between Ⓑ and Ⓓ to be positive.

In the first time interval, we set $t_i = t_A = 0$ and $t_f = t_B = 1$ s. Using Equation 2.1, with $x = -4t + 2t^2$, we obtain for the first displacement

$$\Delta x_{A \to B} = x_f - x_i = x_B - x_A$$
$$= [-4(1) + 2(1)^2] - [-4(0) + 2(0)^2]$$
$$= \boxed{-2 \text{ m}}$$

To calculate the displacement during the second time interval, we set $t_i = t_B = 1$ s and $t_f = t_D = 3$ s:

$$\Delta x_{B \to D} = x_f - x_i = x_D - x_B$$

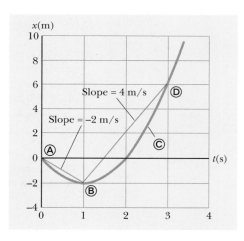

Figure 2.4   Position–time graph for a particle having an x coordinate that varies in time according to the expression $x = -4t + 2t^2$.

$$= [-4(3) + 2(3)^2] - [-4(1) + 2(1)^2]$$
$$= \boxed{+8 \text{ m}}$$

These displacements can also be read directly from the position–time graph.

---

[2] As with velocity, we drop the adjective for instantaneous speed: "Speed" means instantaneous speed.

[3] Simply to make it easier to read, we write the empirical equation as $x = -4t + 2t^2$ rather than as $x = (-4.00 \text{ m/s})t + (2.00 \text{ m/s}^2)t^{2.00}$. When an equation summarizes measurements, consider its coefficients to have as many significant digits as other data quoted in a problem. Consider its coefficients to have the units required for dimensional consistency. When we start our clocks at $t = 0$ s, we usually do not mean to limit the precision to a single digit. Consider any zero value in this book to have as many significant figures as you need.

(b) Calculate the average velocity during these two time intervals.

Solution In the first time interval, $\Delta t = t_f - t_i = t_B - t_A = 1$ s. Therefore, using Equation 2.2 and the displacement calculated in (a), we find that

$$\bar{v}_{x(A \to B)} = \frac{\Delta x_{A \to B}}{\Delta t} = \frac{-2 \text{ m}}{1 \text{ s}} = \boxed{-2 \text{ m/s}}$$

In the second time interval, $\Delta t = 2$ s; therefore,

$$\bar{v}_{x(B \to D)} = \frac{\Delta x_{B \to D}}{\Delta t} = \frac{8 \text{ m}}{2 \text{ s}} = \boxed{+4 \text{ m/s}}$$

These values agree with the slopes of the lines joining these points in Figure 2.4.

(c) Find the instantaneous velocity of the particle at $t = 2.5$ s.

Solution Certainly we can guess that this instantaneous velocity must be of the same order of magnitude as our previous results, that is, around 4 m/s. Examining the graph, we see that the slope of the tangent at position Ⓒ is greater than the slope of the blue line connecting points Ⓑ and Ⓓ. Thus, we expect the answer to be greater than 4 m/s. By measuring the slope of the position–time graph at $t = 2.5$ s, we find that

$$v_x = \boxed{+6 \text{ m/s}}$$

2.3 ACCELERATION

In the last example, we worked with a situation in which the velocity of a particle changed while the particle was moving. This is an extremely common occurrence. (How constant is your velocity as you ride a city bus?) It is easy to quantify changes in velocity as a function of time in exactly the same way we quantify changes in position as a function of time. When the velocity of a particle changes with time, the particle is said to be *accelerating*. For example, the velocity of a car increases when you step on the gas and decreases when you apply the brakes. However, we need a better definition of acceleration than this.

Suppose a particle moving along the x axis has a velocity v_{xi} at time t_i and a velocity v_{xf} at time t_f, as in Figure 2.5a.

The average acceleration of the particle is defined as the *change* in velocity Δv_x divided by the time interval Δt during which that change occurred:

Average acceleration

$$\bar{a}_x \equiv \frac{\Delta v_x}{\Delta t} = \frac{v_{xf} - v_{xi}}{t_f - t_i} \tag{2.5}$$

As with velocity, when the motion being analyzed is one-dimensional, we can use positive and negative signs to indicate the direction of the acceleration. Because the dimensions of velocity are L/T and the dimension of time is T, accelera-

Figure 2.5 (a) A "particle" moving along the x axis from Ⓐ to Ⓑ has velocity v_{xi} at $t = t_i$ and velocity v_{xf} at $t = t_f$. (b) Velocity–time graph for the particle moving in a straight line. The slope of the blue straight line connecting Ⓐ and Ⓑ is the average acceleration in the time interval $\Delta t = t_f - t_i$.

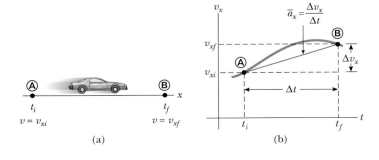

tion has dimensions of length divided by time squared, or L/T^2. The SI unit of acceleration is meters per second squared (m/s^2). It might be easier to interpret these units if you think of them as meters per second per second. For example, suppose an object has an acceleration of 2 m/s^2. You should form a mental image of the object having a velocity that is along a straight line and is increasing by 2 m/s during every 1-s interval. If the object starts from rest, you should be able to picture it moving at a velocity of +2 m/s after 1 s, at +4 m/s after 2 s, and so on.

In some situations, the value of the average acceleration may be different over different time intervals. It is therefore useful to define the *instantaneous acceleration* as the limit of the average acceleration as Δt approaches zero. This concept is analogous to the definition of instantaneous velocity discussed in the previous section. If we imagine that point Ⓑ is brought closer and closer to point Ⓐ in Figure 2.5a and take the limit of $\Delta v_x / \Delta t$ as Δt approaches zero, we obtain the instantaneous acceleration:

$$a_x \equiv \lim_{\Delta t \to 0} \frac{\Delta v_x}{\Delta t} = \frac{dv_x}{dt} \qquad \text{(2.6)}$$

Instantaneous acceleration

That is, **the instantaneous acceleration equals the derivative of the velocity with respect to time,** which by definition is the slope of the velocity–time graph (Fig. 2.5b). Thus, we see that just as the velocity of a moving particle is the slope of the particle's x-t graph, the acceleration of a particle is the slope of the particle's v_x-t graph. One can interpret the derivative of the velocity with respect to time as the time rate of change of velocity. If a_x is positive, then the acceleration is in the positive x direction; if a_x is negative, then the acceleration is in the negative x direction.

From now on we shall use the term *acceleration* to mean instantaneous acceleration. When we mean average acceleration, we shall always use the adjective *average.*

Because $v_x = dx/dt$, the acceleration can also be written

$$a_x = \frac{dv_x}{dt} = \frac{d}{dt}\left(\frac{dx}{dt}\right) = \frac{d^2x}{dt^2} \qquad \text{(2.7)}$$

That is, in one-dimensional motion, the acceleration equals the *second derivative* of x with respect to time.

Figure 2.6 illustrates how an acceleration–time graph is related to a velocity–time graph. The acceleration at any time is the slope of the velocity–time graph at that time. Positive values of acceleration correspond to those points in Figure 2.6a where the velocity is increasing in the positive x direction. The acceler-

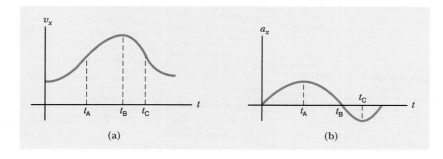

(a)   (b)

Figure 2.6 Instantaneous acceleration can be obtained from the v_x-t graph. (a) The velocity–time graph for some motion. (b) The acceleration–time graph for the same motion. The acceleration given by the a_x-t graph for any value of t equals the slope of the line tangent to the v_x-t graph at the same value of t.

ation reaches a maximum at time t_A, when the slope of the velocity–time graph is a maximum. The acceleration then goes to zero at time t_B, when the velocity is a maximum (that is, when the slope of the v_x-t graph is zero). The acceleration is negative when the velocity is decreasing in the positive x direction, and it reaches its most negative value at time t_C.

CONCEPTUAL EXAMPLE 2.3   Graphical Relationships Between x, v_x, and a_x

The position of an object moving along the x axis varies with time as in Figure 2.7a. Graph the velocity versus time and the acceleration versus time for the object.

Solution   The velocity at any instant is the slope of the tangent to the x-t graph at that instant. Between $t = 0$ and $t = t_A$, the slope of the x-t graph increases uniformly, and so the velocity increases linearly, as shown in Figure 2.7b. Between t_A and t_B, the slope of the x-t graph is constant, and so the velocity remains constant. At t_D, the slope of the x-t graph is zero, so the velocity is zero at that instant. Between t_D and t_E, the slope of the x-t graph and thus the velocity are negative and decrease uniformly in this interval. In the interval t_E to t_F, the slope of the x-t graph is still negative, and at t_F it goes to zero. Finally, after t_F, the slope of the x-t graph is zero, meaning that the object is at rest for $t > t_F$.

The acceleration at any instant is the slope of the tangent to the v_x-t graph at that instant. The graph of acceleration versus time for this object is shown in Figure 2.7c. The acceleration is constant and positive between 0 and t_A, where the slope of the v_x-t graph is positive. It is zero between t_A and t_B and for $t > t_F$ because the slope of the v_x-t graph is zero at these times. It is negative between t_B and t_E because the slope of the v_x-t graph is negative during this interval.

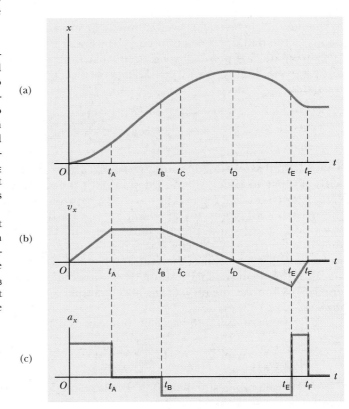

(a)

(b)

(c)

Figure 2.7   (a) Position–time graph for an object moving along the x axis. (b) The velocity–time graph for the object is obtained by measuring the slope of the position–time graph at each instant. (c) The acceleration–time graph for the object is obtained by measuring the slope of the velocity–time graph at each instant.

Quick Quiz 2.1

Make a velocity–time graph for the car in Figure 2.1a and use your graph to determine whether the car ever exceeds the speed limit posted on the road sign (30 km/h).

EXAMPLE 2.4   Average and Instantaneous Acceleration

The velocity of a particle moving along the x axis varies in time according to the expression $v_x = (40 - 5t^2)$ m/s, where t is in seconds. (a) Find the average acceleration in the time interval $t = 0$ to $t = 2.0$ s.

Solution   Figure 2.8 is a v_x-t graph that was created from the velocity versus time expression given in the problem statement. Because the slope of the entire v_x-t curve is negative, we expect the acceleration to be negative.

v_x(m/s)

Slope = –20 m/s²

Ⓐ Ⓑ

t(s)

Figure 2.8 The velocity–time graph for a particle moving along the x axis according to the expression $v_x = (40 - 5t^2)$ m/s. The acceleration at $t = 2$ s is equal to the slope of the blue tangent line at that time.

We find the velocities at $t_i = t_A = 0$ and $t_f = t_B = 2.0$ s by substituting these values of t into the expression for the velocity:

$$v_{xA} = (40 - 5t_A^2) \text{ m/s} = [40 - 5(0)^2] \text{ m/s} = +40 \text{ m/s}$$

$$v_{xB} = (40 - 5t_B^2) \text{ m/s} = [40 - 5(2.0)^2] \text{ m/s} = +20 \text{ m/s}$$

Therefore, the average acceleration in the specified time interval $\Delta t = t_B - t_A = 2.0$ s is

$$\bar{a}_x = \frac{v_{xf} - v_{xi}}{t_f - t_i} = \frac{v_{xB} - v_{xA}}{t_B - t_A} = \frac{(20 - 40) \text{ m/s}}{(2.0 - 0) \text{ s}}$$

$$= -10 \text{ m/s}^2$$

The negative sign is consistent with our expectations—namely, that the average acceleration, which is represented by the slope of the line (not shown) joining the initial and final points on the velocity–time graph, is negative.

(b) Determine the acceleration at $t = 2.0$ s.

Solution The velocity at any time t is $v_{xi} = (40 - 5t^2)$ m/s, and the velocity at any later time $t + \Delta t$ is

$$v_{xf} = 40 - 5(t + \Delta t)^2 = 40 - 5t^2 - 10t\,\Delta t - 5(\Delta t)^2$$

Therefore, the change in velocity over the time interval Δt is

$$\Delta v_x = v_{xf} - v_{xi} = [-10t\,\Delta t - 5(\Delta t)^2] \text{ m/s}$$

Dividing this expression by Δt and taking the limit of the result as Δt approaches zero gives the acceleration at *any* time t:

$$a_x = \lim_{\Delta t \to 0} \frac{\Delta v_x}{\Delta t} = \lim_{\Delta t \to 0} (-10t - 5\Delta t) = -10t \text{ m/s}^2$$

Therefore, at $t = 2.0$ s,

$$a_x = (-10)(2.0) \text{ m/s}^2 = \boxed{-20 \text{ m/s}^2}$$

What we have done by comparing the average acceleration during the interval between Ⓐ and Ⓑ (-10 m/s²) with the instantaneous value at Ⓑ (-20 m/s²) is compare the slope of the line (not shown) joining Ⓐ and Ⓑ with the slope of the tangent at Ⓑ.

Note that the acceleration is not constant in this example. Situations involving constant acceleration are treated in Section 2.5.

So far we have evaluated the derivatives of a function by starting with the definition of the function and then taking the limit of a specific ratio. Those of you familiar with calculus should recognize that there are specific rules for taking derivatives. These rules, which are listed in Appendix B.6, enable us to evaluate derivatives quickly. For instance, one rule tells us that the derivative of any constant is zero. As another example, suppose x is proportional to some power of t, such as in the expression

$$x = At^n$$

where A and n are constants. (This is a very common functional form.) The derivative of x with respect to t is

$$\frac{dx}{dt} = nAt^{n-1}$$

Applying this rule to Example 2.4, in which $v_x = 40 - 5t^2$, we find that $a_x = dv_x/dt = -10t$.

2.4 ▶ MOTION DIAGRAMS

The concepts of velocity and acceleration are often confused with each other, but in fact they are quite different quantities. It is instructive to use motion diagrams to describe the velocity and acceleration while an object is in motion. In order not to confuse these two vector quantities, for which both magnitude and direction are important, we use red for velocity vectors and violet for acceleration vectors, as shown in Figure 2.9. The vectors are sketched at several instants during the motion of the object, and the time intervals between adjacent positions are assumed to be equal. This illustration represents three sets of strobe photographs of a car moving from left to right along a straight roadway. The time intervals between flashes are equal in each diagram.

In Figure 2.9a, the images of the car are equally spaced, showing us that the car moves the same distance in each time interval. Thus, the car moves with *constant positive velocity* and has *zero acceleration.*

In Figure 2.9b, the images become farther apart as time progresses. In this case, the velocity vector increases in time because the car's displacement between adjacent positions increases in time. The car is moving with a *positive velocity* and a *positive acceleration.*

In Figure 2.9c, we can tell that the car slows as it moves to the right because its displacement between adjacent images decreases with time. In this case, the car moves to the right with a constant negative acceleration. The velocity vector decreases in time and eventually reaches zero. From this diagram we see that the acceleration and velocity vectors are *not* in the same direction. The car is moving with a *positive velocity* but with a *negative acceleration.*

You should be able to construct motion diagrams for a car that moves initially to the left with a constant positive or negative acceleration.

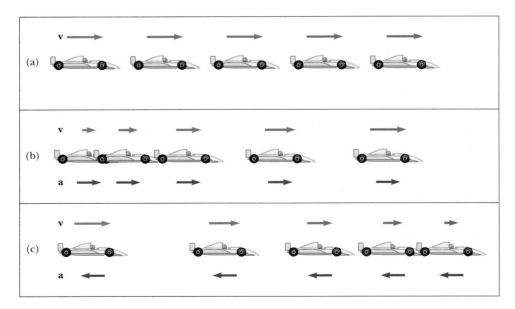

Figure 2.9   (a) Motion diagram for a car moving at constant velocity (zero acceleration). (b) Motion diagram for a car whose constant acceleration is in the direction of its velocity. The velocity vector at each instant is indicated by a red arrow, and the constant acceleration by a violet arrow. (c) Motion diagram for a car whose constant acceleration is in the direction *opposite* the velocity at each instant.

(a) If a car is traveling eastward, can its acceleration be westward? (b) If a car is slowing down, can its acceleration be positive?

2.5 ▸ ONE-DIMENSIONAL MOTION WITH CONSTANT ACCELERATION

If the acceleration of a particle varies in time, its motion can be complex and difficult to analyze. However, a very common and simple type of one-dimensional motion is that in which the acceleration is constant. When this is the case, the average acceleration over any time interval equals the instantaneous acceleration at any instant within the interval, and the velocity changes at the same rate throughout the motion.

If we replace $\bar{a}_x$ by a_x in Equation 2.5 and take $t_i = 0$ and t_f to be any later time t, we find that

$$a_x = \frac{v_{xf} - v_{xi}}{t}$$

or

$$v_{xf} = v_{xi} + a_x t \qquad \text{(for constant } a_x\text{)} \qquad \text{(2.8)}$$

| Velocity as a function of time |

This powerful expression enables us to determine an object's velocity at *any* time t if we know the object's initial velocity and its (constant) acceleration. A velocity–time graph for this constant-acceleration motion is shown in Figure 2.10a. The graph is a straight line, the (constant) slope of which is the acceleration a_x; this is consistent with the fact that $a_x = dv_x/dt$ is a constant. Note that the slope is positive; this indicates a positive acceleration. If the acceleration were negative, then the slope of the line in Figure 2.10a would be negative.

When the acceleration is constant, the graph of acceleration versus time (Fig. 2.10b) is a straight line having a slope of zero.

Describe the meaning of each term in Equation 2.8.

(a)

(b)

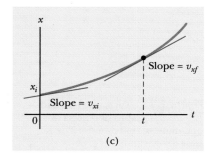
(c)

Figure 2.10   An object moving along the x axis with constant acceleration a_x. (a) The velocity–time graph. (b) The acceleration–time graph. (c) The position–time graph.

Because velocity at constant acceleration varies linearly in time according to Equation 2.8, we can express the average velocity in any time interval as the arithmetic mean of the initial velocity v_{xi} and the final velocity v_{xf}:

$$\bar{v}_x = \frac{v_{xi} + v_{xf}}{2} \qquad \text{(for constant } a_x\text{)} \qquad \textbf{(2.9)}$$

Note that this expression for average velocity applies *only* in situations in which the acceleration is constant.

We can now use Equations 2.1, 2.2, and 2.9 to obtain the displacement of any object as a function of time. Recalling that Δx in Equation 2.2 represents $x_f - x_i$, and now using t in place of Δt (because we take $t_i = 0$), we can say

Displacement as a function of velocity and time

$$x_f - x_i = \bar{v}_x t = \tfrac{1}{2}(v_{xi} + v_{xf})t \qquad \text{(for constant } a_x\text{)} \qquad \textbf{(2.10)}$$

We can obtain another useful expression for displacement at constant acceleration by substituting Equation 2.8 into Equation 2.10:

$$x_f - x_i = \tfrac{1}{2}(v_{xi} + v_{xi} + a_x t)t$$

$$x_f - x_i = v_{xi}t + \tfrac{1}{2}a_x t^2 \qquad \textbf{(2.11)}$$

The position–time graph for motion at constant (positive) acceleration shown in Figure 2.10c is obtained from Equation 2.11. Note that the curve is a parabola. The slope of the tangent line to this curve at $t = t_i = 0$ equals the initial velocity v_{xi}, and the slope of the tangent line at any later time t equals the velocity at that time, v_{xf}.

We can check the validity of Equation 2.11 by moving the x_i term to the right-hand side of the equation and differentiating the equation with respect to time:

$$v_{xf} = \frac{dx_f}{dt} = \frac{d}{dt}\left(x_i + v_{xi}t + \frac{1}{2}a_x t^2\right) = v_{xi} + a_x t$$

Finally, we can obtain an expression for the final velocity that does not contain a time interval by substituting the value of t from Equation 2.8 into Equation 2.10:

$$x_f - x_i = \frac{1}{2}(v_{xi} + v_{xf})\left(\frac{v_{xf} - v_{xi}}{a_x}\right) = \frac{v_{xf}^2 - v_{xi}^2}{2a_x}$$

$$v_{xf}^2 = v_{xi}^2 + 2a_x(x_f - x_i) \qquad \text{(for constant } a_x\text{)} \qquad \textbf{(2.12)}$$

For motion at *zero* acceleration, we see from Equations 2.8 and 2.11 that

$$\left.\begin{array}{c} v_{xf} = v_{xi} = v_x \\ x_f - x_i = v_x t \end{array}\right\} \quad \text{when } a_x = 0$$

That is, when acceleration is zero, velocity is constant and displacement changes linearly with time.

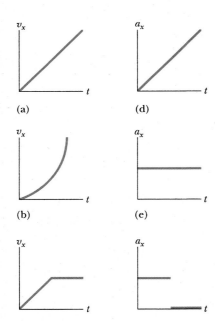

Figure 2.11 Parts (a), (b), and (c) are v_x-t graphs of objects in one-dimensional motion. The possible accelerations of each object as a function of time are shown in scrambled order in (d), (e), and (f).

Quick Quiz 2.4

In Figure 2.11, match each v_x-t graph with the a_x-t graph that best describes the motion.

Equations 2.8 through 2.12 are **kinematic expressions that may be used to solve any problem involving one-dimensional motion at constant accelera-**

TABLE 2.2   Kinematic Equations for Motion in a Straight Line Under Constant Acceleration

Equation	Information Given by Equation
$v_{xf} = v_{xi} + a_x t$	Velocity as a function of time
$x_f - x_i = \frac{1}{2}(v_{xi} + v_{xf})t$	Displacement as a function of velocity and time
$x_f - x_i = v_{xi}t + \frac{1}{2}a_x t^2$	Displacement as a function of time
$v_{xf}^2 = v_{xi}^2 + 2a_x(x_f - x_i)$	Velocity as a function of displacement

Note: Motion is along the x axis.

tion. Keep in mind that these relationships were derived from the definitions of velocity and acceleration, together with some simple algebraic manipulations and the requirement that the acceleration be constant.

The four kinematic equations used most often are listed in Table 2.2 for convenience. The choice of which equation you use in a given situation depends on what you know beforehand. Sometimes it is necessary to use two of these equations to solve for two unknowns. For example, suppose initial velocity v_{xi} and acceleration a_x are given. You can then find (1) the velocity after an interval t has elapsed, using $v_{xf} = v_{xi} + a_x t$, and (2) the displacement after an interval t has elapsed, using $x_f - x_i = v_{xi}t + \frac{1}{2}a_x t^2$. You should recognize that the quantities that vary during the motion are velocity, displacement, and time.

You will get a great deal of practice in the use of these equations by solving a number of exercises and problems. Many times you will discover that more than one method can be used to obtain a solution. Remember that these equations of kinematics *cannot* be used in a situation in which the acceleration varies with time. They can be used only when the acceleration is constant.

CONCEPTUAL EXAMPLE 2.5 ▶ The Velocity of Different Objects

Consider the following one-dimensional motions: (a) A ball thrown directly upward rises to a highest point and falls back into the thrower's hand. (b) A race car starts from rest and speeds up to 100 m/s. (c) A spacecraft drifts through space at constant velocity. Are there any points in the motion of these objects at which the instantaneous velocity is the same as the average velocity over the entire motion? If so, identify the point(s).

Solution   (a) The average velocity for the thrown ball is zero because the ball returns to the starting point; thus its displacement is zero. (Remember that average velocity is de-

fined as $\Delta x / \Delta t$.) There is one point at which the instantaneous velocity is zero—at the top of the motion.

(b) The car's average velocity cannot be evaluated unambiguously with the information given, but it must be some value between 0 and 100 m/s. Because the car will have every instantaneous velocity between 0 and 100 m/s at some time during the interval, there must be some instant at which the instantaneous velocity is equal to the average velocity.

(c) Because the spacecraft's instantaneous velocity is constant, its instantaneous velocity at *any* time and its average velocity over *any* time interval are the same.

EXAMPLE 2.6 ▶ Entering the Traffic Flow

(a) Estimate your average acceleration as you drive up the entrance ramp to an interstate highway.

Solution   This problem involves more than our usual amount of estimating! We are trying to come up with a value

of a_x, but that value is hard to guess directly. The other three variables involved in kinematics are position, velocity, and time. Velocity is probably the easiest one to approximate. Let us assume a final velocity of 100 km/h, so that you can merge with traffic. We multiply this value by 1 000 to convert kilometers to meters and then divide by 3 600 to convert hours to seconds. These two calculations together are roughly equivalent to dividing by 3. In fact, let us just say that the final velocity is $v_{xf} \approx 30$ m/s. (Remember, you can get away with this type of approximation and with dropping digits when performing mental calculations. If you were starting with British units, you could approximate 1 mi/h as roughly 0.5 m/s and continue from there.)

Now we assume that you started up the ramp at about one-third your final velocity, so that $v_{xi} \approx 10$ m/s. Finally, we assume that it takes about 10 s to get from v_{xi} to v_{xf}, basing this guess on our previous experience in automobiles. We can then find the acceleration, using Equation 2.8:

$$a_x = \frac{v_{xf} - v_{xi}}{t} \approx \frac{30 \text{ m/s} - 10 \text{ m/s}}{10 \text{ s}} = \boxed{2 \text{ m/s}^2}$$

Granted, we made many approximations along the way, but this type of mental effort can be surprisingly useful and often

yields results that are not too different from those derived from from careful measurements.

(b) How far did you go during the first half of the time interval during which you accelerated?

Solution We can calculate the distance traveled during the first 5 s from Equation 2.11:

$$x_f - x_i = v_{xi}t + \tfrac{1}{2}a_x t^2 \approx (10 \text{ m/s})(5 \text{ s}) + \tfrac{1}{2}(2 \text{ m/s}^2)(5 \text{ s})^2$$

$$= 50 \text{ m} + 25 \text{ m} = \boxed{75 \text{ m}}$$

This result indicates that if you had not accelerated, your initial velocity of 10 m/s would have resulted in a 50-m movement up the ramp during the first 5 s. The additional 25 m is the result of your increasing velocity during that interval.

Do not be afraid to attempt making educated guesses and doing some fairly drastic number rounding to simplify mental calculations. Physicists engage in this type of thought analysis all the time.

EXAMPLE 2.7   Carrier Landing

A jet lands on an aircraft carrier at 140 mi/h (≈ 63 m/s). (a) What is its acceleration if it stops in 2.0 s?

Solution We define our x axis as the direction of motion of the jet. A careful reading of the problem reveals that in addition to being given the initial speed of 63 m/s, we also know that the final speed is zero. We also note that we are not given the displacement of the jet while it is slowing down. Equation 2.8 is the only equation in Table 2.2 that does not involve displacement, and so we use it to find the acceleration:

$$a_x = \frac{v_{xf} - v_{xi}}{t} \approx \frac{0 - 63 \text{ m/s}}{2.0 \text{ s}} = \boxed{-31 \text{ m/s}^2}$$

(b) What is the displacement of the plane while it is stopping?

Solution We can now use any of the other three equations in Table 2.2 to solve for the displacement. Let us choose Equation 2.10:

$$x_f - x_i = \tfrac{1}{2}(v_{xi} + v_{xf})t = \tfrac{1}{2}(63 \text{ m/s} + 0)(2.0 \text{ s}) = \boxed{63 \text{ m}}$$

If the plane travels much farther than this, it might fall into the ocean. Although the idea of using arresting cables to enable planes to land safely on ships originated at about the time of the First World War, the cables are still a vital part of the operation of modern aircraft carriers.

EXAMPLE 2.8   Watch Out for the Speed Limit!

A car traveling at a constant speed of 45.0 m/s passes a trooper hidden behind a billboard. One second after the speeding car passes the billboard, the trooper sets out from the billboard to catch it, accelerating at a constant rate of 3.00 m/s². How long does it take her to overtake the car?

Solution A careful reading lets us categorize this as a constant-acceleration problem. We know that after the 1-s delay in starting, it will take the trooper 15 additional seconds to accelerate up to 45.0 m/s. Of course, she then has to continue to pick up speed (at a rate of 3.00 m/s per second) to

catch up to the car. While all this is going on, the car continues to move. We should therefore expect our result to be well over 15 s. A sketch (Fig. 2.12) helps clarify the sequence of events.

First, we write expressions for the position of each vehicle as a function of time. It is convenient to choose the position of the billboard as the origin and to set $t_B \equiv 0$ as the time the trooper begins moving. At that instant, the car has already traveled a distance of 45.0 m because it has traveled at a constant speed of $v_x = 45.0$ m/s for 1 s. Thus, the initial position of the speeding car is $x_B = 45.0$ m.

Because the car moves with constant speed, its accelera-

$v_{x\,car} = 45.0$ m/s
$a_{x\,car} = 0$
$a_{x\,trooper} = 3.00$ m/s^2

$t_A = -1.00$ s $t_B = 0$ $t_C = ?$
Ⓐ   Ⓑ   Ⓒ

Figure 2.12   A speeding car passes a hidden police officer.

tion is zero, and applying Equation 2.11 (with $a_x = 0$) gives for the car's position at any time t:

$$x_{car} = x_B + v_{x\,car}t = 45.0 \text{ m} + (45.0 \text{ m/s})t$$

A quick check shows that at $t = 0$, this expression gives the car's correct initial position when the trooper begins to move: $x_{car} = x_B = 45.0$ m. Looking at limiting cases to see whether they yield expected values is a very useful way to make sure that you are obtaining reasonable results.

The trooper starts from rest at $t = 0$ and accelerates at 3.00 m/s^2 away from the origin. Hence, her position after any time interval t can be found from Equation 2.11:

$$x_f = x_i + v_{xi}t + \tfrac{1}{2}a_x t^2$$

$$x_{trooper} = 0 + 0t + \tfrac{1}{2}a_x t^2 = \tfrac{1}{2}(3.00 \text{ m/s}^2)t^2$$

The trooper overtakes the car at the instant her position matches that of the car, which is position Ⓒ:

$$x_{trooper} = x_{car}$$

$$\tfrac{1}{2}(3.00 \text{ m/s}^2)t^2 = 45.0 \text{ m} + (45.0 \text{ m/s})t$$

This gives the quadratic equation

$$1.50t^2 - 45.0t - 45.0 = 0$$

The positive solution of this equation is $t = \boxed{31.0 \text{ s}}$.

(For help in solving quadratic equations, see Appendix B.2.) Note that in this 31.0-s time interval, the trooper travels a distance of about 1440 m. [This distance can be calculated from the car's constant speed: (45.0 m/s)(31 + 1) s = 1 440 m.]

Exercise   This problem can be solved graphically. On the same graph, plot position versus time for each vehicle, and from the intersection of the two curves determine the time at which the trooper overtakes the car.

2.6   FREELY FALLING OBJECTS

It is now well known that, in the absence of air resistance, all objects dropped near the Earth's surface fall toward the Earth with the same constant acceleration under the influence of the Earth's gravity. It was not until about 1600 that this conclusion was accepted. Before that time, the teachings of the great philosopher Aristotle (384–322 B.C.) had held that heavier objects fall faster than lighter ones.

It was the Italian Galileo Galilei (1564–1642) who originated our present-day ideas concerning falling objects. There is a legend that he demonstrated the law of falling objects by observing that two different weights dropped simultaneously from the Leaning Tower of Pisa hit the ground at approximately the same time. Although there is some doubt that he carried out this particular experiment, it is well established that Galileo performed many experiments on objects moving on inclined planes. In his experiments he rolled balls down a slight incline and measured the distances they covered in successive time intervals. The purpose of the incline was to reduce the acceleration; with the acceleration reduced, Galileo was able to make accurate measurements of the time intervals. By gradually increasing the slope of the incline, he was finally able to draw conclusions about freely falling objects because a freely falling ball is equivalent to a ball moving down a vertical incline.

Astronaut David Scott released a hammer and a feather simultaneously, and they fell in unison to the lunar surface. *(Courtesy of NASA)*

QuickLab ▷

Use a pencil to poke a hole in the bottom of a paper or polystyrene cup. Cover the hole with your finger and fill the cup with water. Hold the cup up in front of you and release it. Does water come out of the hole while the cup is falling? Why or why not?

You might want to try the following experiment. Simultaneously drop a coin and a crumpled-up piece of paper from the same height. If the effects of air resistance are negligible, both will have the same motion and will hit the floor at the same time. In the idealized case, in which air resistance is absent, such motion is referred to as *free fall*. If this same experiment could be conducted in a vacuum, in which air resistance is truly negligible, the paper and coin would fall with the same acceleration even when the paper is not crumpled. On August 2, 1971, such a demonstration was conducted on the Moon by astronaut David Scott. He simultaneously released a hammer and a feather, and in unison they fell to the lunar surface. This demonstration surely would have pleased Galileo!

Definition of free fall

When we use the expression *freely falling object,* we do not necessarily refer to an object dropped from rest. **A freely falling object is any object moving freely under the influence of gravity alone, regardless of its initial motion. Objects thrown upward or downward and those released from rest are all falling freely once they are released. Any freely falling object experiences an acceleration directed *downward*, regardless of its initial motion.**

Free-fall acceleration
$g = 9.80 \text{ m/s}^2$

We shall denote the magnitude of the *free-fall acceleration* by the symbol g. The value of g near the Earth's surface decreases with increasing altitude. Furthermore, slight variations in g occur with changes in latitude. It is common to define "up" as the $+y$ direction and to use y as the position variable in the kinematic equations. At the Earth's surface, the value of g is approximately 9.80 m/s^2. Unless stated otherwise, we shall use this value for g when performing calculations. For making quick estimates, use $g = 10 \text{ m/s}^2$.

If we neglect air resistance and assume that the free-fall acceleration does not vary with altitude over short vertical distances, then the motion of a freely falling object moving vertically is equivalent to motion in one dimension under constant acceleration. Therefore, the equations developed in Section 2.5 for objects moving with constant acceleration can be applied. The only modification that we need to make in these equations for freely falling objects is to note that the motion is in the vertical direction (the y direction) rather than in the horizontal (x) direction and that the acceleration is downward and has a magnitude of 9.80 m/s^2. Thus, we always take $a_y = -g = -9.80 \text{ m/s}^2$, where the minus sign means that the acceleration of a freely falling object is downward. In Chapter 14 we shall study how to deal with variations in g with altitude.

CONCEPTUAL EXAMPLE 2.9   The Daring Sky Divers

A sky diver jumps out of a hovering helicopter. A few seconds later, another sky diver jumps out, and they both fall along the same vertical line. Ignore air resistance, so that both sky divers fall with the same acceleration. Does the difference in their speeds stay the same throughout the fall? Does the vertical distance between them stay the same throughout the fall? If the two divers were connected by a long bungee cord, would the tension in the cord increase, lessen, or stay the same during the fall?

Solution   At any given instant, the speeds of the divers are different because one had a head start. In any time interval

Δt after this instant, however, the two divers increase their speeds by the same amount because they have the same acceleration. Thus, the difference in their speeds remains the same throughout the fall.

The first jumper always has a greater speed than the second. Thus, in a given time interval, the first diver covers a greater distance than the second. Thus, the separation distance between them increases.

Once the distance between the divers reaches the length of the bungee cord, the tension in the cord begins to increase. As the tension increases, the distance between the divers becomes greater and greater.

EXAMPLE 2.10 Describing the Motion of a Tossed Ball

A ball is tossed straight up at 25 m/s. Estimate its velocity at 1-s intervals.

Solution Let us choose the upward direction to be positive. Regardless of whether the ball is moving upward or downward, its vertical velocity changes by approximately −10 m/s for every second it remains in the air. It starts out at 25 m/s. After 1 s has elapsed, it is still moving upward but at 15 m/s because its acceleration is downward (downward acceleration causes its velocity to decrease). After another second, its upward velocity has dropped to 5 m/s. Now comes the tricky part—after another half second, its velocity is zero.

The ball has gone as high as it will go. After the last half of this 1-s interval, the ball is moving at −5 m/s. (The minus sign tells us that the ball is now moving in the negative direction, that is, *downward*. Its velocity has changed from +5 m/s to −5 m/s during that 1-s interval. The change in velocity is still −5 − [+5] = −10 m/s in that second.) It continues downward, and after another 1 s, has elapsed, it is falling at a velocity of −15 m/s. Finally, after another 1 s, it has reached its original starting point and is moving downward at −25 m/s. If the ball had been tossed vertically off a cliff so that it could continue downward, its velocity would continue to change by about −10 m/s every second.

CONCEPTUAL EXAMPLE 2.11 Follow the Bouncing Ball

A tennis ball is dropped from shoulder height (about 1.5 m) and bounces three times before it is caught. Sketch graphs of its position, velocity, and acceleration as functions of time, with the + y direction defined as upward.

Solution For our sketch let us stretch things out horizontally so that we can see what is going on. (Even if the ball were moving horizontally, this motion would not affect its vertical motion.)

From Figure 2.13 we see that the ball is in contact with the floor at points Ⓑ, Ⓓ, and Ⓕ. Because the velocity of the ball changes from negative to positive three times during these bounces, the slope of the position–time graph must change in the same way. Note that the time interval between bounces decreases. Why is that?

During the rest of the ball's motion, the slope of the velocity–time graph should be −9.80 m/s². The acceleration–time graph is a horizontal line at these times because the acceleration does not change when the ball is in free fall. When the ball is in contact with the floor, the velocity changes substantially during a very short time interval, and so the acceleration must be quite great. This corresponds to the very steep upward lines on the velocity–time graph and to the spikes on the acceleration–time graph.

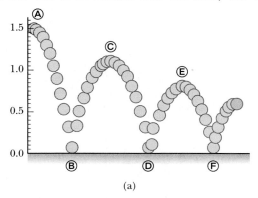

(a)

Figure 2.13 (a) A ball is dropped from a height of 1.5 m and bounces from the floor. (The horizontal motion is not considered here because it does not affect the vertical motion.) (b) Graphs of position, velocity, and acceleration versus time.

(b)

Quick Quiz 2.5

Which values represent the ball's velocity and acceleration at points Ⓐ, Ⓒ, and Ⓔ in Figure 2.13?

(a) $v_y = 0$, $a_y = 0$
(b) $v_y = 0$, $a_y = 9.80$ m/s^2
(c) $v_y = 0$, $a_y = -9.80$ m/s^2
(d) $v_y = -9.80$ m/s, $a_y = 0$

EXAMPLE 2.12   Not a Bad Throw for a Rookie!

A stone thrown from the top of a building is given an initial velocity of 20.0 m/s straight upward. The building is 50.0 m high, and the stone just misses the edge of the roof on its way down, as shown in Figure 2.14. Using $t_A = 0$ as the time the stone leaves the thrower's hand at position Ⓐ, determine (a) the time at which the stone reaches its maximum height, (b) the maximum height, (c) the time at which the stone returns to the height from which it was thrown, (d) the velocity of the stone at this instant, and (e) the velocity and position of the stone at $t = 5.00$ s.

Solution   (a) As the stone travels from Ⓐ to Ⓑ, its velocity must change by 20 m/s because it stops at Ⓑ. Because gravity causes vertical velocities to change by about 10 m/s for every second of free fall, it should take the stone about 2 s to go from Ⓐ to Ⓑ in our drawing. (In a problem like this, a sketch definitely helps you organize your thoughts.) To calculate the time t_B at which the stone reaches maximum height, we use Equation 2.8, $v_{yB} = v_{yA} + a_y t$, noting that $v_{yB} = 0$ and setting the start of our clock readings at $t_A \equiv 0$:

$$20.0 \text{ m/s} + (-9.80 \text{ m/s}^2)t = 0$$

$$t = t_B = \frac{20.0 \text{ m/s}}{9.80 \text{ m/s}^2} = \boxed{2.04 \text{ s}}$$

Our estimate was pretty close.

(b) Because the average velocity for this first interval is 10 m/s (the average of 20 m/s and 0 m/s) and because it travels for about 2 s, we expect the stone to travel about 20 m. By substituting our time interval into Equation 2.11, we can find the maximum height as measured from the position of the thrower, where we set $y_i = y_A = 0$:

$$y_{\max} = y_B = v_{yA} t + \tfrac{1}{2}a_y t^2$$

$$y_B = (20.0 \text{ m/s})(2.04 \text{ s}) + \tfrac{1}{2}(-9.80 \text{ m/s}^2)(2.04 \text{ s})^2$$

$$= \boxed{20.4 \text{ m}}$$

Our free-fall estimates are very accurate.

(c) There is no reason to believe that the stone's motion from Ⓑ to Ⓒ is anything other than the reverse of its motion

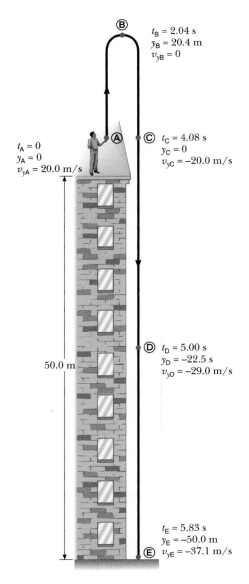

Ⓑ $t_B = 2.04$ s
$y_B = 20.4$ m
$v_{yB} = 0$

$t_A = 0$
$y_A = 0$
$v_{yA} = 20.0$ m/s   Ⓐ

Ⓒ $t_C = 4.08$ s
$y_C = 0$
$v_{yC} = -20.0$ m/s

Ⓓ $t_D = 5.00$ s
$y_D = -22.5$ s
$v_{yD} = -29.0$ m/s

50.0 m

Ⓔ $t_E = 5.83$ s
$y_E = -50.0$ m
$v_{yE} = -37.1$ m/s

Figure 2.14   Position and velocity versus time for a freely falling stone thrown initially upward with a velocity $v_{yi} = 20.0$ m/s.

from Ⓐ to Ⓑ. Thus, the time needed for it to go from Ⓐ to Ⓒ should be twice the time needed for it to go from Ⓐ to Ⓑ. When the stone is back at the height from which it was thrown (position Ⓒ), the y coordinate is again zero. Using Equation 2.11, with $y_f = y_C = 0$ and $y_i = y_A = 0$, we obtain

$$y_C - y_A = v_{yA}\, t + \tfrac{1}{2} a_y t^2$$

$$0 = 20.0t - 4.90t^2$$

This is a quadratic equation and so has two solutions for $t = t_C$. The equation can be factored to give

$$t(20.0 - 4.90t) = 0$$

One solution is $t = 0$, corresponding to the time the stone starts its motion. The other solution is $t = 4.08$ s, which is the solution we are after. Notice that it is double the value we calculated for t_B.

(d) Again, we expect everything at Ⓒ to be the same as it is at Ⓐ, except that the velocity is now in the opposite direction. The value for t found in (c) can be inserted into Equation 2.8 to give

$$v_{yC} = v_{yA} + a_y t = 20.0 \text{ m/s} + (-9.80 \text{ m/s}^2)(4.08 \text{ s})$$

$$= -20.0 \text{ m/s}$$

The velocity of the stone when it arrives back at its original height is equal in magnitude to its initial velocity but opposite in direction. This indicates that the motion is symmetric.

(e) For this part we consider what happens as the stone falls from position Ⓑ, where it had zero vertical velocity, to position Ⓓ. Because the elapsed time for this part of the motion is about 3 s, we estimate that the acceleration due to gravity will have changed the speed by about 30 m/s. We can calculate this from Equation 2.8, where we take $t = t_D - t_B$:

$$v_{yD} = v_{yB} + a_y t = 0 \text{ m/s} + (-9.80 \text{ m/s}^2)(5.00 \text{ s} - 2.04 \text{ s})$$

$$= -29.0 \text{ m/s}$$

We could just as easily have made our calculation between positions Ⓐ and Ⓓ by making sure we use the correct time interval, $t = t_D - t_A = 5.00$ s:

$$v_{yD} = v_{yA} + a_y t = 20.0 \text{ m/s} + (-9.80 \text{ m/s}^2)(5.00 \text{ s})$$

$$= -29.0 \text{ m/s}$$

To demonstrate the power of our kinematic equations, we can use Equation 2.11 to find the position of the stone at $t_D = 5.00$ s by considering the change in position between a different pair of positions, Ⓒ and Ⓓ. In this case, the time is $t_D - t_C$:

$$y_D = y_C + v_{yC}t + \tfrac{1}{2}a_y t^2$$

$$= 0 \text{ m} + (-20.0 \text{ m/s})(5.00 \text{ s} - 4.08 \text{ s})$$

$$+ \tfrac{1}{2}(-9.80 \text{ m/s}^2)(5.00 \text{ s} - 4.08 \text{ s})^2$$

$$= -22.5 \text{ m}$$

Exercise Find (a) the velocity of the stone just before it hits the ground at Ⓔ and (b) the total time the stone is in the air.

Answer (a) -37.1 m/s   (b) 5.83 s

Optional Section

2.7 KINEMATIC EQUATIONS DERIVED FROM CALCULUS

This is an optional section that assumes the reader is familiar with the techniques of integral calculus. If you have not yet studied integration in your calculus course, you should skip this section or cover it after you become familiar with integration.

The velocity of a particle moving in a straight line can be obtained if its position as a function of time is known. Mathematically, the velocity equals the derivative of the position coordinate with respect to time. It is also possible to find the displacement of a particle if its velocity is known as a function of time. In calculus, the procedure used to perform this task is referred to either as *integration* or as finding the *antiderivative*. Graphically, it is equivalent to finding the area under a curve.

Suppose the v_x-t graph for a particle moving along the x axis is as shown in Figure 2.15. Let us divide the time interval $t_f - t_i$ into many small intervals, each of duration Δt_n. From the definition of average velocity we see that the displacement during any small interval, such as the one shaded in Figure 2.15, is given by $\Delta x_n = \bar{v}_{xn} \Delta t_n$, where $\bar{v}_{xn}$ is the average velocity in that interval. Therefore, the displacement during this small interval is simply the area of the shaded rectangle.

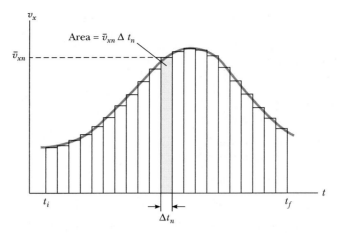

Figure 2.15   Velocity versus time for a particle moving along the *x* axis. The area of the shaded rectangle is equal to the displacement Δx in the time interval Δt_n, while the total area under the curve is the total displacement of the particle.

The total displacement for the interval $t_f - t_i$ is the sum of the areas of all the rectangles:

$$\Delta x = \sum_n \overline{v}_{xn} \, \Delta t_n$$

where the symbol Σ (upper case Greek sigma) signifies a sum over all terms. In this case, the sum is taken over all the rectangles from t_i to t_f. Now, as the intervals are made smaller and smaller, the number of terms in the sum increases and the sum approaches a value equal to the area under the velocity–time graph. Therefore, in the limit $n \rightarrow \infty$, or $\Delta t_n \rightarrow 0$, the displacement is

$$\Delta x = \lim_{\Delta t_n \to 0} \sum_n v_{xn} \, \Delta t_n \qquad \textbf{(2.13)}$$

or

  Displacement = area under the v_x-t graph

Note that we have replaced the average velocity $\overline{v}_{xn}$ with the instantaneous velocity v_{xn} in the sum. As you can see from Figure 2.15, this approximation is clearly valid in the limit of very small intervals. We conclude that if we know the v_x-t graph for motion along a straight line, we can obtain the displacement during any time interval by measuring the area under the curve corresponding to that time interval.

The limit of the sum shown in Equation 2.13 is called a **definite integral** and is written

Definite integral

$$\lim_{\Delta t_n \to 0} \sum_n v_{xn} \Delta t_n = \int_{t_i}^{t_f} v_x(t) \, dt \qquad \textbf{(2.14)}$$

where $v_x(t)$ denotes the velocity at any time t. If the explicit functional form of $v_x(t)$ is known and the limits are given, then the integral can be evaluated.

Sometimes the v_x-t graph for a moving particle has a shape much simpler than that shown in Figure 2.15. For example, suppose a particle moves at a constant ve-

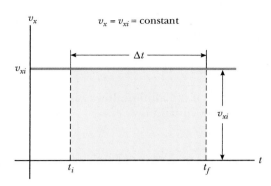

Figure 2.16 The velocity–time curve for a particle moving with constant velocity v_{xi}. The displacement of the particle during the time interval $t_f - t_i$ is equal to the area of the shaded rectangle.

locity v_{xi}. In this case, the v_x-t graph is a horizontal line, as shown in Figure 2.16, and its displacement during the time interval Δt is simply the area of the shaded rectangle:

$$\Delta x = v_{xi}\Delta t \qquad (\text{when } v_{xf} = v_{xi} = \text{constant})$$

As another example, consider a particle moving with a velocity that is proportional to t, as shown in Figure 2.17. Taking $v_x = a_x t$, where a_x is the constant of proportionality (the acceleration), we find that the displacement of the particle during the time interval $t = 0$ to $t = t_A$ is equal to the area of the shaded triangle in Figure 2.17:

$$\Delta x = \tfrac{1}{2}(t_A)(a_x t_A) = \tfrac{1}{2}a_x t_A^2$$

Kinematic Equations

We now use the defining equations for acceleration and velocity to derive two of our kinematic equations, Equations 2.8 and 2.11.

The defining equation for acceleration (Eq. 2.6),

$$a_x = \frac{dv_x}{dt}$$

may be written as $dv_x = a_x dt$ or, in terms of an integral (or antiderivative), as

$$v_x = \int a_x \, dt + C_1$$

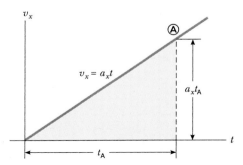

Figure 2.17 The velocity–time curve for a particle moving with a velocity that is proportional to the time.

where C_1 is a constant of integration. For the special case in which the acceleration is constant, the a_x can be removed from the integral to give

$$v_x = a_x \int dt + C_1 = a_x t + C_1 \qquad \textbf{(2.15)}$$

The value of C_1 depends on the initial conditions of the motion. If we take $v_x = v_{xi}$ when $t = 0$ and substitute these values into the last equation, we have

$$v_{xi} = a_x(0) + C_1$$

$$C_1 = v_{xi}$$

Calling $v_x = v_{xf}$ the velocity after the time interval t has passed and substituting this and the value just found for C_1 into Equation 2.15, we obtain kinematic Equation 2.8:

$$v_{xf} = v_{xi} + a_x t \qquad \text{(for constant } a_x\text{)}$$

Now let us consider the defining equation for velocity (Eq. 2.4):

$$v_x = \frac{dx}{dt}$$

We can write this as $dx = v_x dt$ or in integral form as

$$x = \int v_x \, dt + C_2$$

where C_2 is another constant of integration. Because $v_x = v_{xf} = v_{xi} + a_x t$, this expression becomes

$$x = \int (v_{xi} + a_x t) \, dt + C_2$$

$$x = \int v_{xi} \, dt + a_x \int t \, dt + C_2$$

$$x = v_{xi} t + \tfrac{1}{2} a_x t^2 + C_2$$

To find C_2, we make use of the initial condition that $x = x_i$ when $t = 0$. This gives $C_2 = x_i$. Therefore, after substituting x_f for x, we have

$$x_f = x_i + v_{xi} t + \tfrac{1}{2} a_x t^2 \qquad \text{(for constant } a_x\text{)}$$

Once we move x_i to the left side of the equation, we have kinematic Equation 2.11. Recall that $x_f - x_i$ is equal to the displacement of the object, where x_i is its initial position.

*B*esides what you might expect to learn about physics concepts, a very valuable skill you should hope to take away from your physics course is the ability to solve complicated problems. The way physicists approach complex situations and break them down into manageable pieces is extremely useful. We have developed a memory aid to help you easily recall the steps required for successful problem solving. When working on problems, the secret is to keep your GOAL in mind!

GOAL PROBLEM-SOLVING STEPS

*G*ather information

The first thing to do when approaching a problem is to understand the situation. Carefully read the problem statement, looking for key phrases like "at rest" or "freely falls." What information is given? Exactly what is the question asking? Don't forget to gather information from your own experiences and common sense. What should a reasonable answer look like? You wouldn't expect to calculate the speed of an automobile to be 5×10^6 m/s. Do you know what units to expect? Are there any limiting cases you can consider? What happens when an angle approaches 0° or 90° or when a mass becomes huge or goes to zero? Also make sure you carefully study any drawings that accompany the problem.

*O*rganize your approach

Once you have a really good idea of what the problem is about, you need to think about what to do next. Have you seen this type of question before? Being able to classify a problem can make it much easier to lay out a plan to solve it. You should almost always make a quick drawing of the situation. Label important events with circled letters. Indicate any known values, perhaps in a table or directly on your sketch.

*A*nalyze the problem

Because you have already categorized the problem, it should not be too difficult to select relevant equations that apply to this type of situation. Use algebra (and calculus, if necessary) to solve for the unknown variable in terms of what is given. Substitute in the appropriate numbers, calculate the result, and round it to the proper number of significant figures.

*L*earn from your efforts

This is the most important part. Examine your numerical answer. Does it meet your expectations from the first step? What about the algebraic form of the result—before you plugged in numbers? Does it make sense? (Try looking at the variables in it to see whether the answer would change in a physically meaningful way if they were drastically increased or decreased or even became zero.) Think about how this problem compares with others you have done. How was it similar? In what critical ways did it differ? Why was this problem assigned? You should have learned something by doing it. Can you figure out what?

When solving complex problems, you may need to identify a series of subproblems and apply the GOAL process to each. For very simple problems, you probably don't need GOAL at all. But when you are looking at a problem and you don't know what to do next, remember what the letters in GOAL stand for and use that as a guide.

SUMMARY

After a particle moves along the x axis from some initial position x_i to some final position x_f, its **displacement** is

$$\Delta x \equiv x_f - x_i \tag{2.1}$$

The **average velocity** of a particle during some time interval is the displacement Δx divided by the time interval Δt during which that displacement occurred:

$$\bar{v}_x \equiv \frac{\Delta x}{\Delta t} \tag{2.2}$$

The **average speed** of a particle is equal to the ratio of the total distance it travels to the total time it takes to travel that distance.

The **instantaneous velocity** of a particle is defined as the limit of the ratio $\Delta x / \Delta t$ as Δt approaches zero. By definition, this limit equals the derivative of x with respect to t, or the time rate of change of the position:

$$v_x \equiv \lim_{\Delta t \to 0} \frac{\Delta x}{\Delta t} = \frac{dx}{dt} \tag{2.4}$$

The **instantaneous speed** of a particle is equal to the magnitude of its velocity.

The **average acceleration** of a particle is defined as the ratio of the change in its velocity Δv_x divided by the time interval Δt during which that change occurred:

$$\bar{a}_x \equiv \frac{\Delta v_x}{\Delta t} = \frac{v_{xf} - v_{xi}}{t_f - t_i} \tag{2.5}$$

The **instantaneous acceleration** is equal to the limit of the ratio $\Delta v_x / \Delta t$ as Δt approaches 0. By definition, this limit equals the derivative of v_x with respect to t, or the time rate of change of the velocity:

$$a_x \equiv \lim_{\Delta t \to 0} \frac{\Delta v_x}{\Delta t} = \frac{dv_x}{dt} \tag{2.6}$$

The **equations of kinematics** for a particle moving along the x axis with uniform acceleration a_x (constant in magnitude and direction) are

$$v_{xf} = v_{xi} + a_x t \tag{2.8}$$

$$x_f - x_i = \bar{v}_x t = \tfrac{1}{2}(v_{xi} + v_{xf}) t \tag{2.10}$$

$$x_f - x_i = v_{xi} t + \tfrac{1}{2} a_x t^2 \tag{2.11}$$

$$v_{xf}^2 = v_{xi}^2 + 2 a_x (x_f - x_i) \tag{2.12}$$

You should be able to use these equations and the definitions in this chapter to analyze the motion of any object moving with constant acceleration.

An object falling freely in the presence of the Earth's gravity experiences a free-fall acceleration directed toward the center of the Earth. If air resistance is neglected, if the motion occurs near the surface of the Earth, and if the range of the motion is small compared with the Earth's radius, then the free-fall acceleration g is constant over the range of motion, where g is equal to 9.80 m/s^2.

Complicated problems are best approached in an organized manner. You should be able to recall and apply the steps of the GOAL strategy when you need them.

QUESTIONS

1. Average velocity and instantaneous velocity are generally different quantities. Can they ever be equal for a specific type of motion? Explain.

2. If the average velocity is nonzero for some time interval, does this mean that the instantaneous velocity is never zero during this interval? Explain.

3. If the average velocity equals zero for some time interval Δt and if $v_x(t)$ is a continuous function, show that the instantaneous velocity must go to zero at some time in this interval. (A sketch of x versus t might be useful in your proof.)

4. Is it possible to have a situation in which the velocity and acceleration have opposite signs? If so, sketch a velocity–time graph to prove your point.

5. If the velocity of a particle is nonzero, can its acceleration be zero? Explain.

6. If the velocity of a particle is zero, can its acceleration be nonzero? Explain.

7. Can an object having constant acceleration ever stop and stay stopped?

8. A stone is thrown vertically upward from the top of a building. Does the stone's displacement depend on the location of the origin of the coordinate system? Does the stone's velocity depend on the origin? (Assume that the coordinate system is stationary with respect to the building.) Explain.

9. A student at the top of a building of height h throws one ball upward with an initial speed v_{yi} and then throws a second ball downward with the same initial speed. How do the final speeds of the balls compare when they reach the ground?

10. Can the magnitude of the instantaneous velocity of an object ever be greater than the magnitude of its average velocity? Can it ever be less?

11. If the average velocity of an object is zero in some time interval, what can you say about the displacement of the object for that interval?

12. A rapidly growing plant doubles in height each week. At the end of the 25th day, the plant reaches the height of a building. At what time was the plant one-fourth the height of the building?

13. Two cars are moving in the same direction in parallel lanes along a highway. At some instant, the velocity of car A exceeds the velocity of car B. Does this mean that the acceleration of car A is greater than that of car B? Explain.

14. An apple is dropped from some height above the Earth's surface. Neglecting air resistance, how much does the apple's speed increase each second during its descent?

15. Consider the following combinations of signs and values for velocity and acceleration of a particle with respect to a one-dimensional x axis:

	Velocity	Acceleration
a.	Positive	Positive
b.	Positive	Negative
c.	Positive	Zero
d.	Negative	Positive
e.	Negative	Negative
f.	Negative	Zero
g.	Zero	Positive
h.	Zero	Negative

Describe what the particle is doing in each case, and give a real-life example for an automobile on an east-west one-dimensional axis, with east considered to be the positive direction.

16. A pebble is dropped into a water well, and the splash is heard 16 s later, as illustrated in Figure Q2.16. Estimate the distance from the rim of the well to the water's surface.

17. Average velocity is an entirely contrived quantity, and other combinations of data may prove useful in other contexts. For example, the ratio $\Delta t/\Delta x$, called the "slowness" of a moving object, is used by geophysicists when discussing the motion of continental plates. Explain what this quantity means.

B.C.   By John Hart

By permission of John Hart and Field Enterprises Inc.

Figure Q2.16

PROBLEMS

1, 2, 3 = straightforward, intermediate, challenging ▢ = full solution available in the *Student Solutions Manual and Study Guide*
WEB = solution posted at **http://www.saunderscollege.com/physics/** 🖥 = Computer useful in solving problem 📱 = Interactive Physics
▢ = paired numerical/symbolic problems

Section 2.1   Displacement, Velocity, and Speed

1. The position of a pinewood derby car was observed at various times; the results are summarized in the table at the top of page 50. Find the average velocity of the car for (a) the first second, (b) the last 3 s, and (c) the entire period of observation.

x (m)	0	2.3	9.2	20.7	36.8	57.5
t (s)	0	1.0	2.0	3.0	4.0	5.0

2. A motorist drives north for 35.0 min at 85.0 km/h and then stops for 15.0 min. He then continues north, traveling 130 km in 2.00 h. (a) What is his total displacement? (b) What is his average velocity?

3. The displacement versus time for a certain particle moving along the x axis is shown in Figure P2.3. Find the average velocity in the time intervals (a) 0 to 2 s, (b) 0 to 4 s, (c) 2 s to 4 s, (d) 4 s to 7 s, (e) 0 to 8 s.

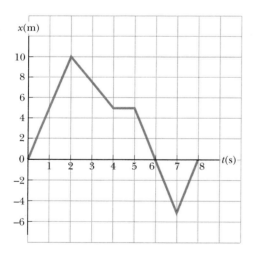

Figure P2.3   Problems 3 and 11.

4. A particle moves according to the equation $x = 10t^2$, where x is in meters and t is in seconds. (a) Find the average velocity for the time interval from 2.0 s to 3.0 s. (b) Find the average velocity for the time interval from 2.0 s to 2.1 s.

5. A person walks first at a constant speed of 5.00 m/s along a straight line from point A to point B and then back along the line from B to A at a constant speed of 3.00 m/s. What are (a) her average speed over the entire trip and (b) her average velocity over the entire trip?

6. A person first walks at a constant speed v_1 along a straight line from A to B and then back along the line from B to A at a constant speed v_2. What are (a) her average speed over the entire trip and (b) her average velocity over the entire trip?

Section 2.2   Instantaneous Velocity and Speed

7. At $t = 1.00$ s, a particle moving with constant velocity is located at $x = -3.00$ m, and at $t = 6.00$ s the particle is located at $x = 5.00$ m. (a) From this information, plot the position as a function of time. (b) Determine the velocity of the particle from the slope of this graph.

8. The position of a particle moving along the x axis varies in time according to the expression $x = 3t^2$, where x is in meters and t is in seconds. Evaluate its position (a) at $t = 3.00$ s and (b) at 3.00 s $+ \Delta t$. (c) Evaluate the limit of $\Delta x/\Delta t$ as Δt approaches zero to find the velocity at $t = 3.00$ s.

WEB 9. A position–time graph for a particle moving along the x axis is shown in Figure P2.9. (a) Find the average velocity in the time interval $t = 1.5$ s to $t = 4.0$ s. (b) Determine the instantaneous velocity at $t = 2.0$ s by measuring the slope of the tangent line shown in the graph. (c) At what value of t is the velocity zero?

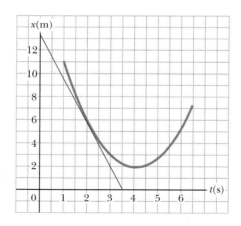

Figure P2.9

10. (a) Use the data in Problem 1 to construct a smooth graph of position versus time. (b) By constructing tangents to the $x(t)$ curve, find the instantaneous velocity of the car at several instants. (c) Plot the instantaneous velocity versus time and, from this, determine the average acceleration of the car. (d) What was the initial velocity of the car?

11. Find the instantaneous velocity of the particle described in Figure P2.3 at the following times: (a) $t = 1.0$ s, (b) $t = 3.0$ s, (c) $t = 4.5$ s, and (d) $t = 7.5$ s.

Section 2.3 Acceleration

12. A particle is moving with a velocity of 60.0 m/s in the positive x direction at $t = 0$. Between $t = 0$ and $t = 15.0$ s, the velocity decreases uniformly to zero. What was the acceleration during this 15.0-s interval? What is the significance of the sign of your answer?

13. A 50.0-g superball traveling at 25.0 m/s bounces off a brick wall and rebounds at 22.0 m/s. A high-speed camera records this event. If the ball is in contact with the wall for 3.50 ms, what is the magnitude of the average acceleration of the ball during this time interval? (*Note:* 1 ms = 10^{-3} s.)

14. A particle starts from rest and accelerates as shown in Figure P2.14. Determine: (a) the particle's speed at $t = 10$ s and at $t = 20$ s, and (b) the distance traveled in the first 20 s.

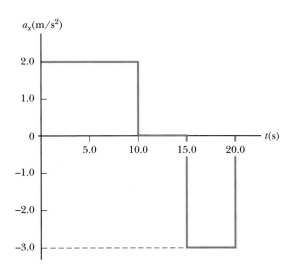

Figure P2.14

15. A velocity–time graph for an object moving along the x axis is shown in Figure P2.15. (a) Plot a graph of the acceleration versus time. (b) Determine the average acceleration of the object in the time intervals $t = 5.00$ s to $t = 15.0$ s and $t = 0$ to $t = 20.0$ s.

16. A student drives a moped along a straight road as described by the velocity–time graph in Figure P2.16. Sketch this graph in the middle of a sheet of graph paper. (a) Directly above your graph, sketch a graph of the position versus time, aligning the time coordinates of the two graphs. (b) Sketch a graph of the acceleration versus time directly below the v_x-t graph, again aligning the time coordinates. On each graph, show the

Figure P2.15

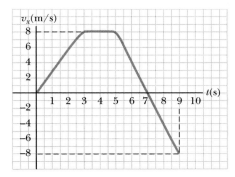

Figure P2.16

numerical values of x and a_x for all points of inflection. (c) What is the acceleration at $t = 6$ s? (d) Find the position (relative to the starting point) at $t = 6$ s. (e) What is the moped's final position at $t = 9$ s?

WEB 17. A particle moves along the x axis according to the equation $x = 2.00 + 3.00t - t^2$, where x is in meters and t is in seconds. At $t = 3.00$ s, find (a) the position of the particle, (b) its velocity, and (c) its acceleration.

18. An object moves along the x axis according to the equation $x = (3.00t^2 - 2.00t + 3.00)$ m. Determine (a) the average speed between $t = 2.00$ s and $t = 3.00$ s, (b) the instantaneous speed at $t = 2.00$ s and at $t = 3.00$ s, (c) the average acceleration between $t = 2.00$ s and $t = 3.00$ s, and (d) the instantaneous acceleration at $t = 2.00$ s and $t = 3.00$ s.

19. Figure P2.19 shows a graph of v_x versus t for the motion of a motorcyclist as he starts from rest and moves along the road in a straight line. (a) Find the average acceleration for the time interval $t = 0$ to $t = 6.00$ s. (b) Estimate the time at which the acceleration has its greatest positive value and the value of the acceleration at that instant. (c) When is the acceleration zero? (d) Estimate the maximum negative value of the acceleration and the time at which it occurs.

Figure P2.19

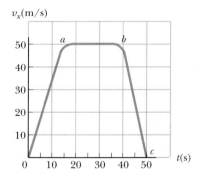

Figure P2.26

Section 2.4   Motion Diagrams

20. Draw motion diagrams for (a) an object moving to the right at constant speed, (b) an object moving to the right and speeding up at a constant rate, (c) an object moving to the right and slowing down at a constant rate, (d) an object moving to the left and speeding up at a constant rate, and (e) an object moving to the left and slowing down at a constant rate. (f) How would your drawings change if the changes in speed were not uniform; that is, if the speed were not changing at a constant rate?

Section 2.5   One-Dimensional Motion with Constant Acceleration

21. Jules Verne in 1865 proposed sending people to the Moon by firing a space capsule from a 220-m-long cannon with a final velocity of 10.97 km/s. What would have been the unrealistically large acceleration experienced by the space travelers during launch? Compare your answer with the free-fall acceleration, 9.80 m/s².

22. A certain automobile manufacturer claims that its super-deluxe sports car will accelerate from rest to a speed of 42.0 m/s in 8.00 s. Under the (improbable) assumption that the acceleration is constant, (a) determine the acceleration of the car. (b) Find the distance the car travels in the first 8.00 s. (c) What is the speed of the car 10.0 s after it begins its motion, assuming it continues to move with the same acceleration?

23. A truck covers 40.0 m in 8.50 s while smoothly slowing down to a final speed of 2.80 m/s. (a) Find its original speed. (b) Find its acceleration.

24. The minimum distance required to stop a car moving at 35.0 mi/h is 40.0 ft. What is the minimum stopping distance for the same car moving at 70.0 mi/h, assuming the same rate of acceleration?

WEB 25. A body moving with uniform acceleration has a velocity of 12.0 cm/s in the positive x direction when its x coordinate is 3.00 cm. If its x coordinate 2.00 s later is -5.00 cm, what is the magnitude of its acceleration?

26. Figure P2.26 represents part of the performance data of a car owned by a proud physics student. (a) Calculate from the graph the total distance traveled. (b) What distance does the car travel between the times $t = 10$ s and $t = 40$ s? (c) Draw a graph of its ac-celeration versus time between $t = 0$ and $t = 50$ s. (d) Write an equation for x as a function of time for each phase of the motion, represented by (i) $0a$, (ii) ab, (iii) bc. (e) What is the average velocity of the car between $t = 0$ and $t = 50$ s?

27. A particle moves along the x axis. Its position is given by the equation $x = 2.00 + 3.00t - 4.00t^2$ with x in meters and t in seconds. Determine (a) its position at the instant it changes direction and (b) its velocity when it returns to the position it had at $t = 0$.

28. The initial speed of a body is 5.20 m/s. What is its speed after 2.50 s (a) if it accelerates uniformly at 3.00 m/s² and (b) if it accelerates uniformly at -3.00 m/s²?

29. A drag racer starts her car from rest and accelerates at 10.0 m/s² for the entire distance of 400 m ($\frac{1}{4}$ mi). (a) How long did it take the race car to travel this distance? (b) What is the speed of the race car at the end of the run?

30. A car is approaching a hill at 30.0 m/s when its engine suddenly fails, just at the bottom of the hill. The car moves with a constant acceleration of -2.00 m/s² while coasting up the hill. (a) Write equations for the position along the slope and for the velocity as functions of time, taking $x = 0$ at the bottom of the hill, where $v_i = 30.0$ m/s. (b) Determine the maximum distance the car travels up the hill.

31. A jet plane lands with a speed of 100 m/s and can accelerate at a maximum rate of -5.00 m/s² as it comes to rest. (a) From the instant the plane touches the runway, what is the minimum time it needs before it can come to rest? (b) Can this plane land at a small tropical island airport where the runway is 0.800 km long?

32. The driver of a car slams on the brakes when he sees a tree blocking the road. The car slows uniformly with an acceleration of -5.60 m/s² for 4.20 s, making straight skid marks 62.4 m long ending at the tree. With what speed does the car then strike the tree?

33. *Help! One of our equations is missing!* We describe constant-acceleration motion with the variables and parameters v_{xi}, v_{xf}, a_x, t, and $x_f - x_i$. Of the equations in Table 2.2, the first does not involve $x_f - x_i$. The second does not contain a_x, the third omits v_{xf}, and the last

Figure P2.37 *(Left)* Col. John Stapp on rocket sled. *(Courtesy of the U.S. Air Force)*
(Right) Col. Stapp's face is contorted by the stress of rapid negative acceleration. *(Photri, Inc.)*

leaves out *t*. So to complete the set there should be an equation *not* involving v_{xi}. Derive it from the others. Use it to solve Problem 32 in one step.

34. An indestructible bullet 2.00 cm long is fired straight through a board that is 10.0 cm thick. The bullet strikes the board with a speed of 420 m/s and emerges with a speed of 280 m/s. (a) What is the average acceleration of the bullet as it passes through the board? (b) What is the total time that the bullet is in contact with the board? (c) What thickness of board (calculated to 0.1 cm) would it take to stop the bullet, assuming the bullet's acceleration through all boards is the same?

35. A truck on a straight road starts from rest, accelerating at 2.00 m/s² until it reaches a speed of 20.0 m/s. Then the truck travels for 20.0 s at constant speed until the brakes are applied, stopping the truck in a uniform manner in an additional 5.00 s. (a) How long is the truck in motion? (b) What is the average velocity of the truck for the motion described?

36. A train is traveling down a straight track at 20.0 m/s when the engineer applies the brakes. This results in an acceleration of − 1.00 m/s² as long as the train is in motion. How far does the train move during a 40.0-s time interval starting at the instant the brakes are applied?

37. For many years the world's land speed record was held by Colonel John P. Stapp, USAF (Fig. P2.37). On March 19, 1954, he rode a rocket-propelled sled that moved down the track at 632 mi/h. He and the sled were safely brought to rest in 1.40 s. Determine (a) the negative acceleration he experienced and (b) the distance he traveled during this negative acceleration.

38. An electron in a cathode-ray tube (CRT) accelerates uniformly from 2.00×10^4 m/s to 6.00×10^6 m/s over 1.50 cm. (a) How long does the electron take to travel this 1.50 cm? (b) What is its acceleration?

39. A ball starts from rest and accelerates at 0.500 m/s² while moving down an inclined plane 9.00 m long. When it reaches the bottom, the ball rolls up another plane, where, after moving 15.0 m, it comes to rest.

(a) What is the speed of the ball at the bottom of the first plane? (b) How long does it take to roll down the first plane? (c) What is the acceleration along the second plane? (d) What is the ball's speed 8.00 m along the second plane?

40. Speedy Sue, driving at 30.0 m/s, enters a one-lane tunnel. She then observes a slow-moving van 155 m ahead traveling at 5.00 m/s. Sue applies her brakes but can accelerate only at − 2.00 m/s² because the road is wet. Will there be a collision? If so, determine how far into the tunnel and at what time the collision occurs. If not, determine the distance of closest approach between Sue's car and the van.

Section 2.6   Freely Falling Objects

Note: In all problems in this section, ignore the effects of air resistance.

 41. A golf ball is released from rest from the top of a very tall building. Calculate (a) the position and (b) the velocity of the ball after 1.00 s, 2.00 s, and 3.00 s.

42. *Every morning at seven o'clock*
    There's twenty terriers drilling on the rock.
    The boss comes around and he says, "Keep still
    And bear down heavy on the cast-iron drill
    And drill, ye terriers, drill." And drill, ye terriers, drill.
    It's work all day for sugar in your tea . . .
    And drill, ye terriers, drill.

    One day a premature blast went off
    And a mile in the air went big Jim Goff. And drill . . .

    Then when next payday came around
    Jim Goff a dollar short was found.
    When he asked what for, came this reply:
    "You were docked for the time you were up in the sky." And
    drill . . .

    —American folksong

    What was Goff's hourly wage? State the assumptions you make in computing it.

WEB 43. A student throws a set of keys vertically upward to her sorority sister, who is in a window 4.00 m above. The keys are caught 1.50 s later by the sister's outstretched hand. (a) With what initial velocity were the keys thrown? (b) What was the velocity of the keys just before they were caught?

44. A ball is thrown directly downward with an initial speed of 8.00 m/s from a height of 30.0 m. How many seconds later does the ball strike the ground?

45. Emily challenges her friend David to catch a dollar bill as follows: She holds the bill vertically, as in Figure P2.45, with the center of the bill between David's index finger and thumb. David must catch the bill after Emily releases it without moving his hand downward. If his reaction time is 0.2 s, will he succeed? Explain your reasoning.

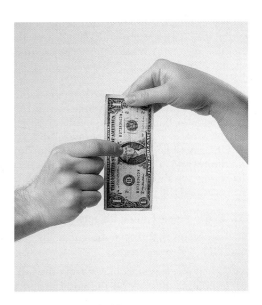

Figure P2.45   *(George Semple)*

46. A ball is dropped from rest from a height *h* above the ground. Another ball is thrown vertically upward from the ground at the instant the first ball is released. Determine the speed of the second ball if the two balls are to meet at a height *h*/2 above the ground.

47. A baseball is hit so that it travels straight upward after being struck by the bat. A fan observes that it takes 3.00 s for the ball to reach its maximum height. Find (a) its initial velocity and (b) the maximum height it reaches.

48. A woman is reported to have fallen 144 ft from the 17th floor of a building, landing on a metal ventilator box, which she crushed to a depth of 18.0 in. She suffered only minor injuries. Calculate (a) the speed of the woman just before she collided with the ventilator box, (b) her average acceleration while in contact with the box, and (c) the time it took to crush the box.

WEB 49. A daring ranch hand sitting on a tree limb wishes to drop vertically onto a horse galloping under the tree. The speed of the horse is 10.0 m/s, and the distance from the limb to the saddle is 3.00 m. (a) What must be the horizontal distance between the saddle and limb when the ranch hand makes his move? (b) How long is he in the air?

50. A ball thrown vertically upward is caught by the thrower after 20.0 s. Find (a) the initial velocity of the ball and (b) the maximum height it reaches.

51. A ball is thrown vertically upward from the ground with an initial speed of 15.0 m/s. (a) How long does it take the ball to reach its maximum altitude? (b) What is its maximum altitude? (c) Determine the velocity and acceleration of the ball at $t = 2.00$ s.

52. The height of a helicopter above the ground is given by $h = 3.00t^3$, where *h* is in meters and *t* is in seconds. After 2.00 s, the helicopter releases a small mailbag. How long after its release does the mailbag reach the ground?

(Optional)
2.7 Kinematic Equations Derived from Calculus

53. Automotive engineers refer to the time rate of change of acceleration as the "jerk." If an object moves in one dimension such that its jerk *J* is constant, (a) determine expressions for its acceleration a_x, velocity v_x, and position *x*, given that its initial acceleration, speed, and position are a_{xi}, v_{xi}, and x_i, respectively. (b) Show that $a_x^2 = a_{xi}^2 + 2J(v_x - v_{xi})$.

54. The speed of a bullet as it travels down the barrel of a rifle toward the opening is given by the expression $v = (-5.0 \times 10^7)t^2 + (3.0 \times 10^5)t$, where *v* is in meters per second and *t* is in seconds. The acceleration of the bullet just as it leaves the barrel is zero. (a) Determine the acceleration and position of the bullet as a function of time when the bullet is in the barrel. (b) Determine the length of time the bullet is accelerated. (c) Find the speed at which the bullet leaves the barrel. (d) What is the length of the barrel?

55. The acceleration of a marble in a certain fluid is proportional to the speed of the marble squared and is given (in SI units) by $a = -3.00v^2$ for $v > 0$. If the marble enters this fluid with a speed of 1.50 m/s, how long will it take before the marble's speed is reduced to half of its initial value?

ADDITIONAL PROBLEMS

56. A motorist is traveling at 18.0 m/s when he sees a deer in the road 38.0 m ahead. (a) If the maximum negative acceleration of the vehicle is -4.50 m/s², what is the maximum reaction time Δt of the motorist that will allow him to avoid hitting the deer? (b) If his reaction time is actually 0.300 s, how fast will he be traveling when he hits the deer?

57. Another scheme to catch the roadrunner has failed. A safe falls from rest from the top of a 25.0-m-high cliff toward Wile E. Coyote, who is standing at the base. Wile first notices the safe after it has fallen 15.0 m. How long does he have to get out of the way?

58. A dog's hair has been cut and is now getting longer by 1.04 mm each day. With winter coming on, this rate of hair growth is steadily increasing by 0.132 mm/day every week. By how much will the dog's hair grow during five weeks?

59. A test rocket is fired vertically upward from a well. A catapult gives it an initial velocity of 80.0 m/s at ground level. Subsequently, its engines fire and it accelerates upward at 4.00 m/s^2 until it reaches an altitude of 1000 m. At that point its engines fail, and the rocket goes into free fall, with an acceleration of -9.80 m/s^2. (a) How long is the rocket in motion above the ground? (b) What is its maximum altitude? (c) What is its velocity just before it collides with the Earth? (*Hint:* Consider the motion while the engine is operating separate from the free-fall motion.)

60. A motorist drives along a straight road at a constant speed of 15.0 m/s. Just as she passes a parked motorcycle police officer, the officer starts to accelerate at 2.00 m/s^2 to overtake her. Assuming the officer maintains this acceleration, (a) determine the time it takes the police officer to reach the motorist. Also find (b) the speed and (c) the total displacement of the officer as he overtakes the motorist.

61. In Figure 2.10a, the area under the velocity–time curve between the vertical axis and time t (vertical dashed line) represents the displacement. As shown, this area consists of a rectangle and a triangle. Compute their areas and compare the sum of the two areas with the value you calculated using Equation 2.11.

62. A commuter train travels between two downtown stations. Because the stations are only 1.00 km apart, the train never reaches its maximum possible cruising speed. The engineer minimizes the time t between the two stations by accelerating at a rate $a_1 = 0.100$ m/s^2 for a time t_1 and then by braking with acceleration $a_2 = -0.500$ m/s^2 for a time t_2. Find the minimum time of travel t and the time t_1.

63. In a 100-m race, Maggie and Judy cross the finish line in a dead heat, both taking 10.2 s. Accelerating uniformly, Maggie took 2.00 s and Judy 3.00 s to attain maximum speed, which they maintained for the rest of the race. (a) What was the acceleration of each sprinter? (b) What were their respective maximum speeds? (c) Which sprinter was ahead at the 6.00-s mark, and by how much?

64. A hard rubber ball, released at chest height, falls to the pavement and bounces back to nearly the same height. When it is in contact with the pavement, the lower side of the ball is temporarily flattened. Suppose the maximum depth of the dent is on the order of

1 cm. Compute an order-of-magnitude estimate for the maximum acceleration of the ball while it is in contact with the pavement. State your assumptions, the quantities you estimate, and the values you estimate for them.

65. A teenager has a car that speeds up at 3.00 m/s^2 and slows down at -4.50 m/s^2. On a trip to the store, he accelerates from rest to 12.0 m/s, drives at a constant speed for 5.00 s, and then comes to a momentary stop at an intersection. He then accelerates to 18.0 m/s, drives at a constant speed for 20.0 s, slows down for 2.67 s, continues for 4.00 s at this speed, and then comes to a stop. (a) How long does the trip take? (b) How far has he traveled? (c) What is his average speed for the trip? (d) How long would it take to walk to the store and back if he walks at 1.50 m/s?

66. A rock is dropped from rest into a well. (a) If the sound of the splash is heard 2.40 s later, how far below the top of the well is the surface of the water? The speed of sound in air (at the ambient temperature) is 336 m/s. (b) If the travel time for the sound is neglected, what percentage error is introduced when the depth of the well is calculated?

67. An inquisitive physics student and mountain climber climbs a 50.0-m cliff that overhangs a calm pool of water. He throws two stones vertically downward, 1.00 s apart, and observes that they cause a single splash. The first stone has an initial speed of 2.00 m/s. (a) How long after release of the first stone do the two stones hit the water? (b) What was the initial velocity of the second stone? (c) What is the velocity of each stone at the instant the two hit the water?

68. A car and train move together along parallel paths at 25.0 m/s, with the car adjacent to the rear of the train. Then, because of a red light, the car undergoes a uniform acceleration of -2.50 m/s^2 and comes to rest. It remains at rest for 45.0 s and then accelerates back to a speed of 25.0 m/s at a rate of 2.50 m/s^2. How far behind the rear of the train is the car when it reaches the speed of 25.0 m/s, assuming that the speed of the train has remained 25.0 m/s?

69. Kathy Kool buys a sports car that can accelerate at the rate of 4.90 m/s^2. She decides to test the car by racing with another speedster, Stan Speedy. Both start from rest, but experienced Stan leaves the starting line 1.00 s before Kathy. If Stan moves with a constant acceleration of 3.50 m/s^2 and Kathy maintains an acceleration of 4.90 m/s^2, find (a) the time it takes Kathy to overtake Stan, (b) the distance she travels before she catches up with him, and (c) the speeds of both cars at the instant she overtakes him.

70. To protect his food from hungry bears, a boy scout raises his food pack with a rope that is thrown over a tree limb at height h above his hands. He walks away from the vertical rope with constant velocity v_{boy}, holding the free end of the rope in his hands (Fig. P2.70).

Figure P2.70

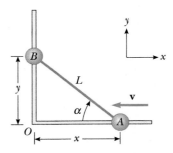

TABLE P2.72 Height of a Rock versus Time			
Time (s)	Height (m)	Time (s)	Height (m)
0.00	5.00	2.75	7.62
0.25	5.75	3.00	7.25
0.50	6.40	3.25	6.77
0.75	6.94	3.50	6.20
1.00	7.38	3.75	5.52
1.25	7.72	4.00	4.73
1.50	7.96	4.25	3.85
1.75	8.10	4.50	2.86
2.00	8.13	4.75	1.77
2.25	8.07	5.00	0.58
2.50	7.90		

(a) Show that the speed v of the food pack is $x(x^2 + h^2)^{-1/2} v_{boy}$, where x is the distance he has walked away from the vertical rope. (b) Show that the acceleration a of the food pack is $h^2(x^2 + h^2)^{-3/2} v_{boy}^2$. (c) What values do the acceleration and velocity have shortly after he leaves the point under the pack ($x = 0$)? (d) What values do the pack's velocity and acceleration approach as the distance x continues to increase?

71. In Problem 70, let the height h equal 6.00 m and the speed v_{boy} equal 2.00 m/s. Assume that the food pack starts from rest. (a) Tabulate and graph the speed–time graph. (b) Tabulate and graph the acceleration–time graph. (Let the range of time be from 0 to 5.00 s and the time intervals be 0.500 s.)

72. Astronauts on a distant planet toss a rock into the air. With the aid of a camera that takes pictures at a steady rate, they record the height of the rock as a function of time as given in Table P2.72. (a) Find the average velocity of the rock in the time interval between each measurement and the next. (b) Using these average veloci-ties to approximate instantaneous velocities at the midpoints of the time intervals, make a graph of velocity as a function of time. Does the rock move with constant acceleration? If so, plot a straight line of best fit on the graph and calculate its slope to find the acceleration.

73. Two objects, A and B, are connected by a rigid rod that has a length L. The objects slide along perpendicular guide rails, as shown in Figure P2.73. If A slides to the left with a constant speed v, find the speed of B when $\alpha = 60.0°$.

Figure P2.73

ANSWERS TO QUICK QUIZZES

2.1 Your graph should look something like the one in (a). This v_x-t graph shows that the maximum speed is about 5.0 m/s, which is 18 km/h (= 11 mi/h), and so the driver was not speeding. Can you derive the acceleration–time graph from the velocity–time graph? It should look something like the one in (b).

2.2 (a) Yes. This occurs when the car is slowing down, so that the direction of its acceleration is opposite the direction of its motion. (b) Yes. If the motion is in the direction chosen as negative, a positive acceleration causes a decrease in speed.

2.3 The left side represents the final velocity of an object. The first term on the right side is the velocity that the object had initially when we started watching it. The second term is the change in that initial velocity that is caused by the acceleration. If this second term is positive, then the initial velocity has increased ($v_{xf} > v_{xi}$). If this term is negative, then the initial velocity has decreased ($v_{xf} < v_{xi}$).

(a)

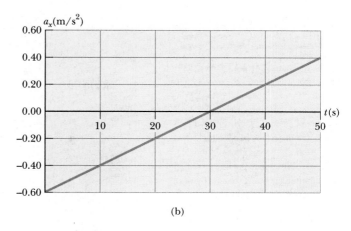

(b)

2.4 Graph (a) has a constant slope, indicating a constant acceleration; this is represented by graph (e).

Graph (b) represents a speed that is increasing constantly but not at a uniform rate. Thus, the acceleration must be increasing, and the graph that best indicates this is (d).

Graph (c) depicts a velocity that first increases at a constant rate, indicating constant acceleration. Then the velocity stops increasing and becomes constant, indicating zero acceleration. The best match to this situation is graph (f).

2.5 (c). As can be seen from Figure 2.13b, the ball is at rest for an infinitesimally short time at these three points. Nonetheless, gravity continues to act even though the ball is instantaneously not moving.

c h a p t e r

Vectors

Chapter Outline

e often need to work with physical quantities that have both numerical and directional properties. As noted in Section 2.1, quantities of this nature are represented by vectors. This chapter is primarily concerned with vector algebra and with some general properties of vector quantities. We discuss the addition and subtraction of vector quantities, together with some common applications to physical situations.

Vector quantities are used throughout this text, and it is therefore imperative that you master both their graphical and their algebraic properties.

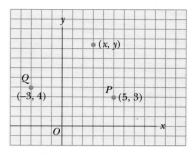

Figure 3.1 Designation of points in a cartesian coordinate system. Every point is labeled with coordinates (x, y).

3.1 COORDINATE SYSTEMS

Many aspects of physics deal in some form or other with locations in space. In Chapter 2, for example, we saw that the mathematical description of an object's motion requires a method for describing the object's position at various times. This description is accomplished with the use of coordinates, and in Chapter 2 we used the cartesian coordinate system, in which horizontal and vertical axes intersect at a point taken to be the origin (Fig. 3.1). Cartesian coordinates are also called *rectangular coordinates*.

2.2

Sometimes it is more convenient to represent a point in a plane by its *plane polar coordinates* (r, θ), as shown in Figure 3.2a. In this *polar coordinate system*, r is the distance from the origin to the point having cartesian coordinates (x, y), and θ is the angle between r and a fixed axis. This fixed axis is usually the positive x axis, and θ is usually measured counterclockwise from it. From the right triangle in Figure 3.2b, we find that $\sin \theta = y/r$ and that $\cos \theta = x/r$. (A review of trigonometric functions is given in Appendix B.4.) Therefore, starting with the plane polar coordinates of any point, we can obtain the cartesian coordinates, using the equations

$$x = r \cos \theta \qquad (3.1)$$

$$y = r \sin \theta \qquad (3.2)$$

Furthermore, the definitions of trigonometry tell us that

$$\tan \theta = \frac{y}{x} \qquad (3.3)$$

$$r = \sqrt{x^2 + y^2} \qquad (3.4)$$

These four expressions relating the coordinates (x, y) to the coordinates (r, θ) apply only when θ is defined, as shown in Figure 3.2a—in other words, when positive θ is an angle measured *counterclockwise* from the positive x axis. (Some scientific calculators perform conversions between cartesian and polar coordinates based on these standard conventions.) If the reference axis for the polar angle θ is chosen to be one other than the positive x axis or if the sense of increasing θ is chosen differently, then the expressions relating the two sets of coordinates will change.

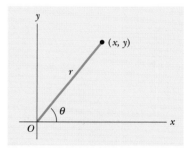

(a)

$\sin \theta = \dfrac{y}{r}$

$\cos \theta = \dfrac{x}{r}$

$\tan \theta = \dfrac{y}{x}$

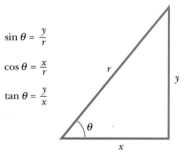

(b)

Figure 3.2 (a) The plane polar coordinates of a point are represented by the distance r and the angle θ, where θ is measured counterclockwise from the positive x axis. (b) The right triangle used to relate (x, y) to (r, θ).

 Quick Quiz 3.1

Would the honeybee at the beginning of the chapter use cartesian or polar coordinates when specifying the location of the flower? Why? What is the honeybee using as an origin of coordinates?

You may want to read *Talking Apes and Dancing Bees* (1997) by Betsy Wyckoff.

EXAMPLE 3.1 ▶ Polar Coordinates

The cartesian coordinates of a point in the *xy* plane are $(x, y) = (-3.50, -2.50)$ m, as shown in Figure 3.3. Find the polar coordinates of this point.

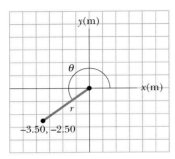

Figure 3.3   Finding polar coordinates when cartesian coordinates are given.

Solution

$$r = \sqrt{x^2 + y^2} = \sqrt{(-3.50 \text{ m})^2 + (-2.50 \text{ m})^2} = \boxed{4.30 \text{ m}}$$

$$\tan \theta = \frac{y}{x} = \frac{-2.50 \text{ m}}{-3.50 \text{ m}} = 0.714$$

$$\theta = \boxed{216°}$$

Note that you must use the signs of *x* and *y* to find that the point lies in the third quadrant of the coordinate system. That is, $\theta = 216°$ and not $35.5°$.

3.2 ▶ VECTOR AND SCALAR QUANTITIES

As noted in Chapter 2, some physical quantities are scalar quantities whereas others are vector quantities. When you want to know the temperature outside so that you will know how to dress, the only information you need is a number and the unit "degrees C" or "degrees F." Temperature is therefore an example of a **scalar quantity,** which is defined as a quantity that is completely specified by a number and appropriate units. That is,

> A scalar quantity is specified by a single value with an appropriate unit and has no direction.

Other examples of scalar quantities are volume, mass, and time intervals. The rules of ordinary arithmetic are used to manipulate scalar quantities.

If you are getting ready to pilot a small plane and need to know the wind velocity, you must know both the speed of the wind and its direction. Because direction is part of the information it gives, velocity is a **vector quantity,** which is defined as a physical quantity that is completely specified by a number and appropriate units plus a direction. That is,

> A vector quantity has both magnitude and direction.

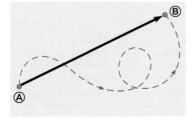

Figure 3.4   As a particle moves from Ⓐ to Ⓑ along an arbitrary path represented by the broken line, its displacement is a vector quantity shown by the arrow drawn from Ⓐ to Ⓑ.

Another example of a vector quantity is displacement, as you know from Chapter 2. Suppose a particle moves from some point Ⓐ to some point Ⓑ along a straight path, as shown in Figure 3.4. We represent this displacement by drawing an arrow from Ⓐ to Ⓑ, with the tip of the arrow pointing away from the starting point. The direction of the arrowhead represents the direction of the displacement, and the length of the arrow represents the magnitude of the displacement. If the particle travels along some other path from Ⓐ to Ⓑ, such as the broken line in Figure 3.4, its displacement is still the arrow drawn from Ⓐ to Ⓑ.

(a)   (b)   (c)

(a) The number of apples in the basket is one example of a scalar quantity. Can you think of other examples? *(Superstock)* (b) Jennifer pointing to the right. A vector quantity is one that must be specified by both magnitude and direction. *(Photo by Ray Serway)* (c) An anemometer is a device meteorologists use in weather forecasting. The cups spin around and reveal the magnitude of the wind velocity. The pointer indicates the direction. *(Courtesy of Peet Bros. Company, 1308 Doris Avenue, Ocean, NJ 07712)*

In this text, we use a boldface letter, such as **A**, to represent a vector quantity. Another common method for vector notation that you should be aware of is the use of an arrow over a letter, such as $\vec{A}$. The magnitude of the vector **A** is written either A or $|\mathbf{A}|$. The magnitude of a vector has physical units, such as meters for displacement or meters per second for velocity.

3.3 ▶ SOME PROPERTIES OF VECTORS

Equality of Two Vectors

For many purposes, two vectors **A** and **B** may be defined to be equal if they have the same magnitude and point in the same direction. That is, **A** = **B** only if $A = B$ *and* if **A** and **B** point in the same direction along parallel lines. For example, all the vectors in Figure 3.5 are equal even though they have different starting points. This property allows us to move a vector to a position parallel to itself in a diagram without affecting the vector.

Adding Vectors

2.4 The rules for adding vectors are conveniently described by geometric methods. To add vector **B** to vector **A**, first draw vector **A**, with its magnitude represented by a convenient scale, on graph paper and then draw vector **B** to the same scale with its tail starting from the tip of **A**, as shown in Figure 3.6. The **resultant vector R** = **A** + **B** is the vector drawn from the tail of **A** to the tip of **B**. This procedure is known as the **triangle method of addition.**

For example, if you walked 3.0 m toward the east and then 4.0 m toward the north, as shown in Figure 3.7, you would find yourself 5.0 m from where you

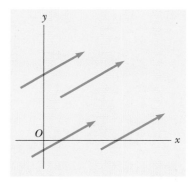

Figure 3.5 These four vectors are equal because they have equal lengths and point in the same direction.

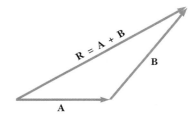

Figure 3.6 When vector **B** is added to vector **A**, the resultant **R** is the vector that runs from the tail of **A** to the tip of **B**.

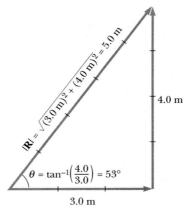

Figure 3.7 Vector addition. Walking first 3.0 m due east and then 4.0 m due north leaves you $|\mathbf{R}| = 5.0$ m from your starting point.

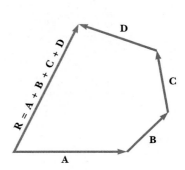

Figure 3.8 Geometric construction for summing four vectors. The resultant vector $\mathbf{R}$ is by definition the one that completes the polygon.

started, measured at an angle of 53° north of east. Your total displacement is the vector sum of the individual displacements.

A geometric construction can also be used to add more than two vectors. This is shown in Figure 3.8 for the case of four vectors. The resultant vector $\mathbf{R} = \mathbf{A} + \mathbf{B} + \mathbf{C} + \mathbf{D}$ is the vector that completes the polygon. In other words, **R is the vector drawn from the tail of the first vector to the tip of the last vector.**

An alternative graphical procedure for adding two vectors, known as the **parallelogram rule of addition,** is shown in Figure 3.9a. In this construction, the tails of the two vectors $\mathbf{A}$ and $\mathbf{B}$ are joined together and the resultant vector $\mathbf{R}$ is the diagonal of a parallelogram formed with $\mathbf{A}$ and $\mathbf{B}$ as two of its four sides.

When two vectors are added, the sum is independent of the order of the addition. (This fact may seem trivial, but as you will see in Chapter 11, the order is important when vectors are multiplied). This can be seen from the geometric construction in Figure 3.9b and is known as the **commutative law of addition:**

Commutative law

$$\mathbf{A} + \mathbf{B} = \mathbf{B} + \mathbf{A} \tag{3.5}$$

When three or more vectors are added, their sum is independent of the way in which the individual vectors are grouped together. A geometric proof of this rule

Commutative Law

Figure 3.9 (a) In this construction, the resultant $\mathbf{R}$ is the diagonal of a parallelogram having sides $\mathbf{A}$ and $\mathbf{B}$. (b) This construction shows that $\mathbf{A} + \mathbf{B} = \mathbf{B} + \mathbf{A}$—in other words, that vector addition is commutative.

(a)   (b)

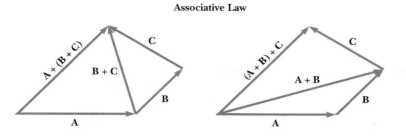

Associative Law

Figure 3.10 Geometric constructions for verifying the associative law of addition.

for three vectors is given in Figure 3.10. This is called the **associative law of addition:**

$$A + (B + C) = (A + B) + C \qquad (3.6)$$

In summary, **a vector quantity has both magnitude and direction and also obeys the laws of vector addition** as described in Figures 3.6 to 3.10. When two or more vectors are added together, *all* of them *must* have the same units. It would be meaningless to add a velocity vector (for example, 60 km/h to the east) to a displacement vector (for example, 200 km to the north) because they represent different physical quantities. The same rule also applies to scalars. For example, it would be meaningless to add time intervals to temperatures.

Negative of a Vector

The negative of the vector **A** is defined as the vector that when added to **A** gives zero for the vector sum. That is, $A + (-A) = 0$. The vectors **A** and $-A$ have the same magnitude but point in opposite directions.

Subtracting Vectors

The operation of vector subtraction makes use of the definition of the negative of a vector. We define the operation $A - B$ as vector $-B$ added to vector **A**:

$$A - B = A + (-B) \qquad (3.7)$$

The geometric construction for subtracting two vectors in this way is illustrated in Figure 3.11a.

Another way of looking at vector subtraction is to note that the difference $A - B$ between two vectors **A** and **B** is what you have to add to the second vector to obtain the first. In this case, the vector $A - B$ points from the tip of the second vector to the tip of the first, as Figure 3.11b shows.

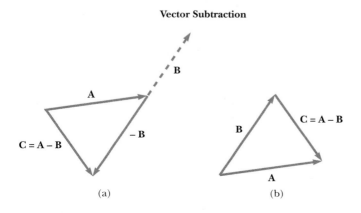

Vector Subtraction

(a)

(b)

Figure 3.11 (a) This construction shows how to subtract vector **B** from vector **A**. The vector $-B$ is equal in magnitude to vector **B** and points in the opposite direction. To subtract **B** from **A**, apply the rule of vector addition to the combination of **A** and $-B$: Draw **A** along some convenient axis, place the tail of $-B$ at the tip of **A**, and **C** is the difference $A - B$. (b) A second way of looking at vector subtraction. The difference vector $C = A - B$ is the vector that we must add to **B** to obtain **A**.

EXAMPLE 3.2   **A Vacation Trip**

A car travels 20.0 km due north and then 35.0 km in a direction 60.0° west of north, as shown in Figure 3.12. Find the magnitude and direction of the car's resultant displacement.

Solution   In this example, we show two ways to find the resultant of two vectors. We can solve the problem geometrically, using graph paper and a protractor, as shown in Figure 3.12. (In fact, even when you know you are going to be carry-

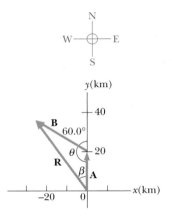

Figure 3.12   Graphical method for finding the resultant displacement vector **R** = **A** + **B**.

ing out a calculation, you should sketch the vectors to check your results.) The displacement **R** is the resultant when the two individual displacements **A** and **B** are added.

To solve the problem algebraically, we note that the magnitude of **R** can be obtained from the law of cosines as applied to the triangle (see Appendix B.4). With $\theta = 180° - 60° = 120°$ and $R^2 = A^2 + B^2 - 2AB\cos\theta$, we find that

$$R = \sqrt{A^2 + B^2 - 2AB\cos\theta}$$
$$= \sqrt{(20.0\text{ km})^2 + (35.0\text{ km})^2 - 2(20.0\text{ km})(35.0\text{ km})\cos 120°}$$
$$= \boxed{48.2\text{ km}}$$

The direction of **R** measured from the northerly direction can be obtained from the law of sines (Appendix B.4):

$$\frac{\sin\beta}{B} = \frac{\sin\theta}{R}$$

$$\sin\beta = \frac{B}{R}\sin\theta = \frac{35.0\text{ km}}{48.2\text{ km}}\sin 120° = 0.629$$

$$\beta = \boxed{38.9°}$$

The resultant displacement of the car is 48.2 km in a direction 38.9° west of north. This result matches what we found graphically.

Multiplying a Vector by a Scalar

If vector **A** is multiplied by a positive scalar quantity m, then the product $m\mathbf{A}$ is a vector that has the same direction as **A** and magnitude mA. If vector **A** is multiplied by a negative scalar quantity $-m$, then the product $-m\mathbf{A}$ is directed opposite **A**. For example, the vector 5**A** is five times as long as **A** and points in the same direction as **A**; the vector $-\frac{1}{3}\mathbf{A}$ is one-third the length of **A** and points in the direction opposite **A**.

Quick Quiz 3.2

If vector **B** is added to vector **A**, under what condition does the resultant vector **A** + **B** have magnitude $A + B$? Under what conditions is the resultant vector equal to zero?

3.4   **COMPONENTS OF A VECTOR AND UNIT VECTORS**

The geometric method of adding vectors is not recommended whenever great accuracy is required or in three-dimensional problems. In this section, we describe a method of adding vectors that makes use of the *projections* of vectors along coordinate axes. These projections are called the **components** of the vector. Any vector can be completely described by its components.

Consider a vector **A** lying in the xy plane and making an arbitrary angle θ with the positive x axis, as shown in Figure 3.13. This vector can be expressed as the

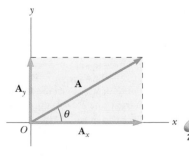

Figure 3.13   Any vector **A** lying in the xy plane can be represented by a vector $\mathbf{A}_x$ lying along the x axis and by a vector $\mathbf{A}_y$ lying along the y axis, where $\mathbf{A} = \mathbf{A}_x + \mathbf{A}_y$.

sum of two other vectors $\mathbf{A}_x$ and $\mathbf{A}_y$. From Figure 3.13, we see that the three vectors form a right triangle and that $\mathbf{A} = \mathbf{A}_x + \mathbf{A}_y$. (If you cannot see why this equality holds, go back to Figure 3.9 and review the parallelogram rule.) We shall often refer to the "components of a vector $\mathbf{A}$," written A_x and A_y (without the boldface notation). The component A_x represents the projection of $\mathbf{A}$ along the x axis, and the component A_y represents the projection of $\mathbf{A}$ along the y axis. These components can be positive or negative. The component A_x is positive if $\mathbf{A}_x$ points in the positive x direction and is negative if $\mathbf{A}_x$ points in the negative x direction. The same is true for the component A_y.

From Figure 3.13 and the definition of sine and cosine, we see that $\cos\theta = A_x/A$ and that $\sin\theta = A_y/A$. Hence, the components of $\mathbf{A}$ are

$$A_x = A \cos\theta \qquad (3.8)$$

$$A_y = A \sin\theta \qquad (3.9)$$

| Components of the vector $\mathbf{A}$ |

These components form two sides of a right triangle with a hypotenuse of length A. Thus, it follows that the magnitude and direction of $\mathbf{A}$ are related to its components through the expressions

$$A = \sqrt{A_x^2 + A_y^2} \qquad (3.10)$$

| Magnitude of $\mathbf{A}$ |

$$\theta = \tan^{-1}\left(\frac{A_y}{A_x}\right) \qquad (3.11)$$

| Direction of $\mathbf{A}$ |

Note that **the signs of the components A_x and A_y depend on the angle θ.** For example, if $\theta = 120°$, then A_x is negative and A_y is positive. If $\theta = 225°$, then both A_x and A_y are negative. Figure 3.14 summarizes the signs of the components when $\mathbf{A}$ lies in the various quadrants.

When solving problems, you can specify a vector $\mathbf{A}$ either with its components A_x and A_y or with its magnitude and direction A and θ.

	y	
A_x negative		A_x positive
A_y positive		A_y positive
A_x negative		A_x positive
A_y negative		A_y negative

(with x axis labeled to the right)

Figure 3.14 The signs of the components of a vector $\mathbf{A}$ depend on the quadrant in which the vector is located.

Quick Quiz 3.3

Can the component of a vector ever be greater than the magnitude of the vector?

Suppose you are working a physics problem that requires resolving a vector into its components. In many applications it is convenient to express the components in a coordinate system having axes that are not horizontal and vertical but are still perpendicular to each other. If you choose reference axes or an angle other than the axes and angle shown in Figure 3.13, the components must be modified accordingly. Suppose a vector $\mathbf{B}$ makes an angle θ' with the x' axis defined in Figure 3.15. The components of $\mathbf{B}$ along the x' and y' axes are $B_{x'} = B \cos\theta'$ and $B_{y'} = B \sin\theta'$, as specified by Equations 3.8 and 3.9. The magnitude and direction of $\mathbf{B}$ are obtained from expressions equivalent to Equations 3.10 and 3.11. Thus, we can express the components of a vector in *any* coordinate system that is convenient for a particular situation.

Unit Vectors

Vector quantities often are expressed in terms of unit vectors. **A unit vector is a dimensionless vector having a magnitude of exactly 1.** Unit vectors are used to specify a given direction and have no other physical significance. They are used solely as a convenience in describing a direction in space. We shall use the symbols

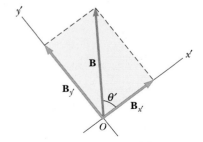

Figure 3.15 The component vectors of $\mathbf{B}$ in a coordinate system that is tilted.

i, **j**, and **k** to represent unit vectors pointing in the positive x, y, and z directions, respectively. The unit vectors **i**, **j**, and **k** form a set of mutually perpendicular vectors in a right-handed coordinate system, as shown in Figure 3.16a. The magnitude of each unit vector equals 1; that is, $|\mathbf{i}| = |\mathbf{j}| = |\mathbf{k}| = 1$.

Consider a vector **A** lying in the xy plane, as shown in Figure 3.16b. The product of the component A_x and the unit vector **i** is the vector $A_x\mathbf{i}$, which lies on the x axis and has magnitude $|A_x|$. (The vector $A_x\mathbf{i}$ is an alternative representation of vector $\mathbf{A}_x$.) Likewise, $A_y\mathbf{j}$ is a vector of magnitude $|A_y|$ lying on the y axis. (Again, vector $A_y\mathbf{j}$ is an alternative representation of vector $\mathbf{A}_y$.) Thus, the unit–vector notation for the vector **A** is

$$\mathbf{A} = A_x\mathbf{i} + A_y\mathbf{j} \tag{3.12}$$

For example, consider a point lying in the xy plane and having cartesian coordinates (x, y), as in Figure 3.17. The point can be specified by the **position vector r**, which in unit–vector form is given by

$$\mathbf{r} = x\mathbf{i} + y\mathbf{j} \tag{3.13}$$

This notation tells us that the components of **r** are the lengths x and y.

Now let us see how to use components to add vectors when the geometric method is not sufficiently accurate. Suppose we wish to add vector **B** to vector **A**, where vector **B** has components B_x and B_y. All we do is add the x and y components separately. The resultant vector $\mathbf{R} = \mathbf{A} + \mathbf{B}$ is therefore

$$\mathbf{R} = (A_x\mathbf{i} + A_y\mathbf{j}) + (B_x\mathbf{i} + B_y\mathbf{j})$$

or

$$\mathbf{R} = (A_x + B_x)\mathbf{i} + (A_y + B_y)\mathbf{j} \tag{3.14}$$

Because $\mathbf{R} = R_x\mathbf{i} + R_y\mathbf{j}$, we see that the components of the resultant vector are

$$R_x = A_x + B_x$$
$$R_y = A_y + B_y \tag{3.15}$$

Position vector

(a)

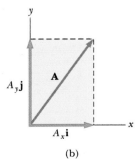

(b)

Figure 3.16 (a) The unit vectors **i**, **j**, and **k** are directed along the x, y, and z axes, respectively. (b) Vector $\mathbf{A} = A_x\mathbf{i} + A_y\mathbf{j}$ lying in the xy plane has components A_x and A_y.

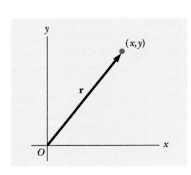

Figure 3.17 The point whose cartesian coordinates are (x, y) can be represented by the position vector $\mathbf{r} = x\mathbf{i} + y\mathbf{j}$.

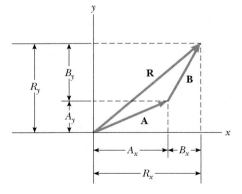

Figure 3.18 This geometric construction for the sum of two vectors shows the relationship between the components of the resultant **R** and the components of the individual vectors.

We obtain the magnitude of **R** and the angle it makes with the x axis from its components, using the relationships

$$R = \sqrt{R_x{}^2 + R_y{}^2} = \sqrt{(A_x + B_x)^2 + (A_y + B_y)^2} \qquad \textbf{(3.16)}$$

$$\tan \theta = \frac{R_y}{R_x} = \frac{A_y + B_y}{A_x + B_x} \qquad \textbf{(3.17)}$$

We can check this addition by components with a geometric construction, as shown in Figure 3.18. Remember that you must note the *signs* of the components when using either the algebraic or the geometric method.

At times, we need to consider situations involving motion in three component directions. The extension of our methods to three-dimensional vectors is straightforward. If **A** and **B** both have x, y, and z components, we express them in the form

$$\mathbf{A} = A_x\mathbf{i} + A_y\mathbf{j} + A_z\mathbf{k} \qquad \textbf{(3.18)}$$

$$\mathbf{B} = B_x\mathbf{i} + B_y\mathbf{j} + B_z\mathbf{k} \qquad \textbf{(3.19)}$$

The sum of **A** and **B** is

$$\mathbf{R} = (A_x + B_x)\mathbf{i} + (A_y + B_y)\mathbf{j} + (A_z + B_z)\mathbf{k} \qquad \textbf{(3.20)}$$

Note that Equation 3.20 differs from Equation 3.14: in Equation 3.20, the resultant vector also has a z component $R_z = A_z + B_z$.

QuickLab

Write an expression for the vector describing the displacement of a fly that moves from one corner of the floor of the room that you are in to the opposite corner of the room, near the ceiling.

Quick Quiz 3.4

If one component of a vector is not zero, can the magnitude of the vector be zero? Explain.

Quick Quiz 3.5

If **A** + **B** = 0, what can you say about the components of the two vectors?

Problem-Solving Hints

Adding Vectors

When you need to add two or more vectors, use this step-by-step procedure:

- Select a coordinate system that is convenient. (Try to reduce the number of components you need to find by choosing axes that line up with as many vectors as possible.)
- Draw a labeled sketch of the vectors described in the problem.
- Find the x and y components of all vectors and the resultant components (the algebraic sum of the components) in the x and y directions.
- If necessary, use the Pythagorean theorem to find the magnitude of the resultant vector and select a suitable trigonometric function to find the angle that the resultant vector makes with the x axis.

EXAMPLE 3.3   The Sum of Two Vectors

Find the sum of two vectors **A** and **B** lying in the *xy* plane and given by

$$\mathbf{A} = (2.0\mathbf{i} + 2.0\mathbf{j}) \text{ m} \quad \text{and} \quad \mathbf{B} = (2.0\mathbf{i} - 4.0\mathbf{j}) \text{ m}$$

Solution Comparing this expression for **A** with the general expression $\mathbf{A} = A_x\mathbf{i} + A_y\mathbf{j}$, we see that $A_x = 2.0$ m and that $A_y = 2.0$ m. Likewise, $B_x = 2.0$ m and $B_y = -4.0$ m. We obtain the resultant vector **R**, using Equation 3.14:

$$\mathbf{R} = \mathbf{A} + \mathbf{B} = (2.0 + 2.0)\mathbf{i} \text{ m} + (2.0 - 4.0)\mathbf{j} \text{ m}$$
$$= (4.0\mathbf{i} - 2.0\mathbf{j}) \text{ m}$$

or

$$R_x = 4.0 \text{ m} \qquad R_y = -2.0 \text{ m}$$

The magnitude of **R** is given by Equation 3.16:

$$R = \sqrt{R_x^2 + R_y^2} = \sqrt{(4.0 \text{ m})^2 + (-2.0 \text{ m})^2} = \sqrt{20} \text{ m}$$
$$= \boxed{4.5 \text{ m}}$$

We can find the direction of **R** from Equation 3.17:

$$\tan \theta = \frac{R_y}{R_x} = \frac{-2.0 \text{ m}}{4.0 \text{ m}} = -0.50$$

Your calculator likely gives the answer $-27°$ for $\theta = \tan^{-1}(-0.50)$. This answer is correct if we interpret it to mean 27° clockwise from the *x* axis. Our standard form has been to quote the angles measured counterclockwise from the $+ x$ axis, and that angle for this vector is $\theta = \boxed{333°}$.

EXAMPLE 3.4   The Resultant Displacement

A particle undergoes three consecutive displacements: $\mathbf{d}_1 = (15\mathbf{i} + 30\mathbf{j} + 12\mathbf{k})$ cm, $\mathbf{d}_2 = (23\mathbf{i} - 14\mathbf{j} - 5.0\mathbf{k})$ cm, and $\mathbf{d}_3 = (-13\mathbf{i} + 15\mathbf{j})$ cm. Find the components of the resultant displacement and its magnitude.

Solution Rather than looking at a sketch on flat paper, visualize the problem as follows: Start with your fingertip at the bottom left corner of your desk. Move your fingertip 15 cm to the right, then 30 cm toward the top of the desk, then 12 cm vertically toward you, then 23 cm to the right, then 14 cm toward the bottom of the desk, then 5.0 cm vertically toward the desk, then 13 cm to the left, and (finally!) 15 cm toward the top of the desk. The mathematical calculation keeps track of this motion along the three perpendicular axes:

$$\mathbf{R} = \mathbf{d}_1 + \mathbf{d}_2 + \mathbf{d}_3$$
$$= (15 + 23 - 13)\mathbf{i} \text{ cm} + (30 - 14 + 15)\mathbf{j} \text{ cm}$$
$$+ (12 - 5.0 + 0)\mathbf{k} \text{ cm}$$
$$= (25\mathbf{i} + 31\mathbf{j} + 7.0\mathbf{k}) \text{ cm}$$

The resultant displacement has components $R_x = 25$ cm, $R_y = 31$ cm, and $R_z = 7.0$ cm. Its magnitude is

$$R = \sqrt{R_x^2 + R_y^2 + R_z^2}$$
$$= \sqrt{(25 \text{ cm})^2 + (31 \text{ cm})^2 + (7.0 \text{ cm})^2} = \boxed{40 \text{ cm}}$$

EXAMPLE 3.5   Taking a Hike

A hiker begins a trip by first walking 25.0 km southeast from her car. She stops and sets up her tent for the night. On the second day, she walks 40.0 km in a direction 60.0° north of east, at which point she discovers a forest ranger's tower. (a) Determine the components of the hiker's displacement for each day.

Solution If we denote the displacement vectors on the first and second days by **A** and **B**, respectively, and use the car as the origin of coordinates, we obtain the vectors shown in Figure 3.19. Displacement **A** has a magnitude of 25.0 km and is directed 45.0° below the positive *x* axis. From Equations 3.8 and 3.9, its components are

$$A_x = A \cos(-45.0°) = (25.0 \text{ km})(0.707) = \boxed{17.7 \text{ km}}$$

$$A_y = A \sin(-45.0°) = -(25.0 \text{ km})(0.707) = \boxed{-17.7 \text{ km}}$$

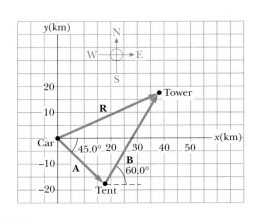

Figure 3.19   The total displacement of the hiker is the vector **R = A + B**.

The negative value of A_y indicates that the hiker walks in the negative y direction on the first day. The signs of A_x and A_y also are evident from Figure 3.19.

The second displacement **B** has a magnitude of 40.0 km and is 60.0° north of east. Its components are

$$B_x = B \cos 60.0° = (40.0 \text{ km})(0.500) = \boxed{20.0 \text{ km}}$$

$$B_y = B \sin 60.0° = (40.0 \text{ km})(0.866) = \boxed{34.6 \text{ km}}$$

(b) Determine the components of the hiker's resultant displacement **R** for the trip. Find an expression for **R** in terms of unit vectors.

Solution The resultant displacement for the trip **R** = **A** + **B** has components given by Equation 3.15:

$$R_x = A_x + B_x = 17.7 \text{ km} + 20.0 \text{ km} = \boxed{37.7 \text{ km}}$$

$$R_y = A_y + B_y = -17.7 \text{ km} + 34.6 \text{ km} = \boxed{16.9 \text{ km}}$$

In unit–vector form, we can write the total displacement as

$$\boxed{\mathbf{R} = (37.7\mathbf{i} + 16.9\mathbf{j}) \text{ km}}$$

Exercise Determine the magnitude and direction of the total displacement.

Answer 41.3 km, 24.1° north of east from the car.

EXAMPLE 3.6 ▸ **Let's Fly Away!**

A commuter airplane takes the route shown in Figure 3.20. First, it flies from the origin of the coordinate system shown to city A, located 175 km in a direction 30.0° north of east. Next, it flies 153 km 20.0° west of north to city B. Finally, it flies 195 km due west to city C. Find the location of city C relative to the origin.

Solution It is convenient to choose the coordinate system shown in Figure 3.20, where the x axis points to the east and the y axis points to the north. Let us denote the three consecutive displacements by the vectors **a**, **b**, and **c**. Displacement **a** has a magnitude of 175 km and the components

$$a_x = a \cos(30.0°) = (175 \text{ km})(0.866) = 152 \text{ km}$$

$$a_y = a \sin(30.0°) = (175 \text{ km})(0.500) = 87.5 \text{ km}$$

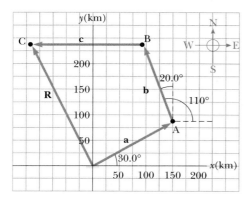

Figure 3.20 The airplane starts at the origin, flies first to city A, then to city B, and finally to city C.

Displacement **b**, whose magnitude is 153 km, has the components

$$b_x = b \cos(110°) = (153 \text{ km})(-0.342) = -52.3 \text{ km}$$

$$b_y = b \sin(110°) = (153 \text{ km})(0.940) = 144 \text{ km}$$

Finally, displacement **c**, whose magnitude is 195 km, has the components

$$c_x = c \cos(180°) = (195 \text{ km})(-1) = -195 \text{ km}$$

$$c_y = c \sin(180°) = 0$$

Therefore, the components of the position vector **R** from the starting point to city C are

$$R_x = a_x + b_x + c_x = 152 \text{ km} - 52.3 \text{ km} - 195 \text{ km}$$

$$= \boxed{-95.3 \text{ km}}$$

$$R_y = a_y + b_y + c_y = 87.5 \text{ km} + 144 \text{ km} + 0$$

$$= \boxed{232 \text{ km}}$$

In unit–vector notation, $\boxed{\mathbf{R} = (-95.3\mathbf{i} + 232\mathbf{j}) \text{ km.}}$ That is, the airplane can reach city C from the starting point by first traveling 95.3 km due west and then by traveling 232 km due north.

Exercise Find the magnitude and direction of **R**.

Answer 251 km, 22.3° west of north.

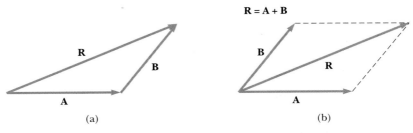

Figure 3.21   (a) Vector addition by the triangle method. (b) Vector addition by the parallelogram rule.

SUMMARY

Scalar quantities are those that have only magnitude and no associated direction. **Vector quantities** have both magnitude and direction and obey the laws of vector addition.

We can add two vectors **A** and **B** graphically, using either the triangle method or the parallelogram rule. In the triangle method (Fig. 3.21a), the resultant vector **R** = **A** + **B** runs from the tail of **A** to the tip of **B**. In the parallelogram method (Fig. 3.21b), **R** is the diagonal of a parallelogram having **A** and **B** as two of its sides. You should be able to add or subtract vectors, using these graphical methods.

The x component A_x of the vector **A** is equal to the projection of **A** along the x axis of a coordinate system, as shown in Figure 3.22, where $A_x = A \cos \theta$. The y component A_y of **A** is the projection of **A** along the y axis, where $A_y = A \sin \theta$. Be sure you can determine which trigonometric functions you should use in all situations, especially when θ is defined as something other than the counterclockwise angle from the positive x axis.

If a vector **A** has an x component A_x and a y component A_y, the vector can be expressed in unit–vector form as **A** = A_x**i** + A_y**j**. In this notation, **i** is a unit vector pointing in the positive x direction, and **j** is a unit vector pointing in the positive y direction. Because **i** and **j** are unit vectors, $|\mathbf{i}| = |\mathbf{j}| = 1$.

We can find the resultant of two or more vectors by resolving all vectors into their x and y components, adding their resultant x and y components, and then using the Pythagorean theorem to find the magnitude of the resultant vector. We can find the angle that the resultant vector makes with respect to the x axis by using a suitable trigonometric function.

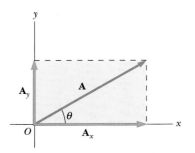

Figure 3.22   The addition of the two vectors $\mathbf{A}_x$ and $\mathbf{A}_y$ gives vector **A**. Note that $\mathbf{A}_x = A_x\mathbf{i}$ and $\mathbf{A}_y = A_y\mathbf{j}$, where A_x and A_y are the *components* of vector **A**.

QUESTIONS

1. Two vectors have unequal magnitudes. Can their sum be zero? Explain.
2. Can the magnitude of a particle's displacement be greater than the distance traveled? Explain.
3. The magnitudes of two vectors **A** and **B** are $A = 5$ units and $B = 2$ units. Find the largest and smallest values possible for the resultant vector **R** = **A** + **B**.
4. Vector **A** lies in the xy plane. For what orientations of vector **A** will both of its components be negative? For what orientations will its components have opposite signs?
5. If the component of vector **A** along the direction of vector

   B is zero, what can you conclude about these two vectors?
6. Can the magnitude of a vector have a negative value? Explain.
7. Which of the following are vectors and which are not: force, temperature, volume, ratings of a television show, height, velocity, age?
8. Under what circumstances would a nonzero vector lying in the xy plane ever have components that are equal in magnitude?
9. Is it possible to add a vector quantity to a scalar quantity? Explain.

PROBLEMS

1, 2, 3 = straightforward, intermediate, challenging ☐ = full solution available in the *Student Solutions Manual and Study Guide*
WEB = solution posted at **http://www.saunderscollege.com/physics/** 🖥 = Computer useful in solving problem 🖼 = Interactive Physics
☐ = paired numerical/symbolic problems

Section 3.1   Coordinate Systems

WEB 1. The polar coordinates of a point are $r = 5.50$ m and $\theta = 240°$. What are the cartesian coordinates of this point?

2. Two points in the xy plane have cartesian coordinates $(2.00, -4.00)$ m and $(-3.00, 3.00)$ m. Determine (a) the distance between these points and (b) their polar coordinates.

3. If the cartesian coordinates of a point are given by $(2, y)$ and its polar coordinates are $(r, 30°)$, determine y and r.

4. Two points in a plane have polar coordinates $(2.50$ m, $30.0°)$ and $(3.80$ m, $120.0°)$. Determine (a) the cartesian coordinates of these points and (b) the distance between them.

5. A fly lands on one wall of a room. The lower left-hand corner of the wall is selected as the origin of a two-dimensional cartesian coordinate system. If the fly is located at the point having coordinates $(2.00, 1.00)$ m, (a) how far is it from the corner of the room? (b) what is its location in polar coordinates?

6. If the polar coordinates of the point (x, y) are (r, θ), determine the polar coordinates for the points (a) $(-x, y)$, (b) $(-2x, -2y)$, and (c) $(3x, -3y)$.

Section 3.2   Vector and Scalar Quantities

Section 3.3   Some Properties of Vectors

7. An airplane flies 200 km due west from city A to city B and then 300 km in the direction 30.0° north of west from city B to city C. (a) In straight-line distance, how far is city C from city A? (b) Relative to city A, in what direction is city C?

8. A pedestrian moves 6.00 km east and then 13.0 km north. Using the graphical method, find the magnitude and direction of the resultant displacement vector.

9. A surveyor measures the distance across a straight river by the following method: Starting directly across from a tree on the opposite bank, she walks 100 m along the riverbank to establish a baseline. Then she sights across to the tree. The angle from her baseline to the tree is 35.0°. How wide is the river?

10. A plane flies from base camp to lake A, a distance of 280 km at a direction 20.0° north of east. After dropping off supplies, it flies to lake B, which is 190 km and 30.0° west of north from lake A. Graphically determine the distance and direction from lake B to the base camp.

11. Vector **A** has a magnitude of 8.00 units and makes an angle of 45.0° with the positive x axis. Vector **B** also has a magnitude of 8.00 units and is directed along the neg-

ative x axis. Using graphical methods, find (a) the vector sum **A** + **B** and (b) the vector difference **A** − **B**.

12. A force $\mathbf{F}_1$ of magnitude 6.00 units acts at the origin in a direction 30.0° above the positive x axis. A second force $\mathbf{F}_2$ of magnitude 5.00 units acts at the origin in the direction of the positive y axis. Find graphically the magnitude and direction of the resultant force $\mathbf{F}_1 + \mathbf{F}_2$.

WEB 13. A person walks along a circular path of radius 5.00 m. If the person walks around one half of the circle, find (a) the magnitude of the displacement vector and (b) how far the person walked. (c) What is the magnitude of the displacement if the person walks all the way around the circle?

14. A dog searching for a bone walks 3.50 m south, then 8.20 m at an angle 30.0° north of east, and finally 15.0 m west. Using graphical techniques, find the dog's resultant displacement vector.

WEB 15. Each of the displacement vectors **A** and **B** shown in Figure P3.15 has a magnitude of 3.00 m. Find graphically (a) **A** + **B**, (b) **A** − **B**, (c) **B** − **A**, (d) **A** − 2**B**. Report all angles counterclockwise from the positive x axis.

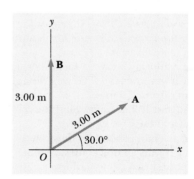

Figure P3.15   Problems 15 and 39.

16. Arbitrarily define the "instantaneous vector height" of a person as the displacement vector from the point halfway between the feet to the top of the head. Make an order-of-magnitude estimate of the total vector height of all the people in a city of population 100 000 (a) at 10 a.m. on a Tuesday and (b) at 5 a.m. on a Saturday. Explain your reasoning.

17. A roller coaster moves 200 ft horizontally and then rises 135 ft at an angle of 30.0° above the horizontal. It then travels 135 ft at an angle of 40.0° downward. What is its displacement from its starting point? Use graphical techniques.

18. The driver of a car drives 3.00 km north, 2.00 km northeast (45.0° east of north), 4.00 km west, and then

3.00 km southeast (45.0° east of south). Where does he end up relative to his starting point? Work out your answer graphically. Check by using components. (The car is not near the North Pole or the South Pole.)

19. Fox Mulder is trapped in a maze. To find his way out, he walks 10.0 m, makes a 90.0° right turn, walks 5.00 m, makes another 90.0° right turn, and walks 7.00 m. What is his displacement from his initial position?

Section 3.4   Components of a Vector and Unit Vectors

20. Find the horizontal and vertical components of the 100-m displacement of a superhero who flies from the top of a tall building following the path shown in Figure P3.20.

Figure P3.20

21. A person walks 25.0° north of east for 3.10 km. How far would she have to walk due north and due east to arrive at the same location?

22. While exploring a cave, a spelunker starts at the entrance and moves the following distances: She goes 75.0 m north, 250 m east, 125 m at an angle 30.0° north of east, and 150 m south. Find the resultant displacement from the cave entrance.

23. In the assembly operation illustrated in Figure P3.23, a robot first lifts an object upward along an arc that forms one quarter of a circle having a radius of 4.80 cm and

Figure P3.23

lying in an east–west vertical plane. The robot then moves the object upward along a second arc that forms one quarter of a circle having a radius of 3.70 cm and lying in a north–south vertical plane. Find (a) the magnitude of the total displacement of the object and (b) the angle the total displacement makes with the vertical.

24. Vector **B** has x, y, and z components of 4.00, 6.00, and 3.00 units, respectively. Calculate the magnitude of **B** and the angles that **B** makes with the coordinate axes.

WEB **25.** A vector has an x component of -25.0 units and a y component of 40.0 units. Find the magnitude and direction of this vector.

26. A map suggests that Atlanta is 730 mi in a direction 5.00° north of east from Dallas. The same map shows that Chicago is 560 mi in a direction 21.0° west of north from Atlanta. Assuming that the Earth is flat, use this information to find the displacement from Dallas to Chicago.

27. A displacement vector lying in the xy plane has a magnitude of 50.0 m and is directed at an angle of 120° to the positive x axis. Find the x and y components of this vector and express the vector in unit–vector notation.

28. If **A** = 2.00**i** + 6.00**j** and **B** = 3.00**i** $-$ 2.00**j**, (a) sketch the vector sum **C** = **A** + **B** and the vector difference **D** = **A** $-$ **B**. (b) Find solutions for **C** and **D**, first in terms of unit vectors and then in terms of polar coordinates, with angles measured with respect to the $+ x$ axis.

29. Find the magnitude and direction of the resultant of three displacements having x and y components (3.00, 2.00) m, $(-5.00, 3.00)$ m, and (6.00, 1.00) m.

30. Vector **A** has x and y components of -8.70 cm and 15.0 cm, respectively; vector **B** has x and y components of 13.2 cm and -6.60 cm, respectively. If **A** $-$ **B** + 3**C** = 0, what are the components of **C**?

31. Consider two vectors **A** = 3**i** $-$ 2**j** and **B** = $-$**i** $-$ 4**j**. Calculate (a) **A** + **B**, (b) **A** $-$ **B**, (c) $|$**A** + **B**$|$, (d) $|$**A** $-$ **B**$|$, (e) the directions of **A** + **B** and **A** $-$ **B**.

32. A boy runs 3.00 blocks north, 4.00 blocks northeast, and 5.00 blocks west. Determine the length and direction of the displacement-vector that goes from the starting point to his final position.

33. Obtain expressions in component form for the position vectors having polar coordinates (a) 12.8 m, 150°; (b) 3.30 cm, 60.0°; (c) 22.0 in., 215°.

34. Consider the displacement vectors **A** = (3**i** + 3**j**) m, **B** = (**i** $-$ 4**j**) m, and **C** = ($-$2**i** + 5**j**) m. Use the component method to determine (a) the magnitude and direction of the vector **D** = **A** + **B** + **C** and (b) the magnitude and direction of **E** = $-$**A** $-$ **B** + **C**.

35. A particle undergoes the following consecutive displacements: 3.50 m south, 8.20 m northeast, and 15.0 m west. What is the resultant displacement?

36. In a game of American football, a quarterback takes the ball from the line of scrimmage, runs backward for 10.0 yards, and then sideways parallel to the line of scrimmage for 15.0 yards. At this point, he throws a forward pass 50.0 yards straight downfield perpendicular to the line of scrimmage. What is the magnitude of the football's resultant displacement?

37. The helicopter view in Figure P3.37 shows two people pulling on a stubborn mule. Find (a) the single force that is equivalent to the two forces shown and (b) the force that a third person would have to exert on the mule to make the resultant force equal to zero. The forces are measured in units of newtons.

Figure P3.37

38. A novice golfer on the green takes three strokes to sink the ball. The successive displacements are 4.00 m to the north, 2.00 m northeast, and 1.00 m 30.0° west of south. Starting at the same initial point, an expert golfer could make the hole in what single displacement?

39. Find the x and y components of the vectors **A** and **B** shown in Figure P3.15; then derive an expression for the resultant vector **A** + **B** in unit–vector notation.

40. You are standing on the ground at the origin of a coordinate system. An airplane flies over you with constant velocity parallel to the x axis and at a constant height of 7.60×10^3 m. At $t = 0$, the airplane is directly above you, so that the vector from you to it is given by $\mathbf{P}_0 = (7.60 \times 10^3 \text{ m})\mathbf{j}$. At $t = 30.0$ s, the position vector leading from you to the airplane is $\mathbf{P}_{30} = (8.04 \times 10^3 \text{ m})\mathbf{i} + (7.60 \times 10^3 \text{ m})\mathbf{j}$. Determine the magnitude and orientation of the airplane's position vector at $t = 45.0$ s.

41. A particle undergoes two displacements. The first has a magnitude of 150 cm and makes an angle of 120° with the positive x axis. The *resultant* displacement has a magnitude of 140 cm and is directed at an angle of 35.0° to the positive x axis. Find the magnitude and direction of the second displacement.

42. Vectors **A** and **B** have equal magnitudes of 5.00. If the sum of **A** and **B** is the vector $6.00\,\mathbf{j}$, determine the angle between **A** and **B**.

43. The vector **A** has x, y, and z components of 8.00, 12.0, and -4.00 units, respectively. (a) Write a vector expression for **A** in unit–vector notation. (b) Obtain a unit–vector expression for a vector **B** one-fourth the length of **A** pointing in the same direction as **A**. (c) Obtain a unit–vector expression for a vector **C** three times the length of **A** pointing in the direction opposite the direction of **A**.

44. Instructions for finding a buried treasure include the following: Go 75.0 paces at 240°, turn to 135° and walk 125 paces, then travel 100 paces at 160°. The angles are measured counterclockwise from an axis pointing to the east, the $+x$ direction. Determine the resultant displacement from the starting point.

45. Given the displacement vectors $\mathbf{A} = (3\mathbf{i} - 4\mathbf{j} + 4\mathbf{k})$ m and $\mathbf{B} = (2\mathbf{i} + 3\mathbf{j} - 7\mathbf{k})$ m, find the magnitudes of the vectors (a) $\mathbf{C} = \mathbf{A} + \mathbf{B}$ and (b) $\mathbf{D} = 2\mathbf{A} - \mathbf{B}$, also expressing each in terms of its x, y, and z components.

46. A radar station locates a sinking ship at range 17.3 km and bearing 136° clockwise from north. From the same station a rescue plane is at horizontal range 19.6 km, 153° clockwise from north, with elevation 2.20 km. (a) Write the vector displacement from plane to ship, letting **i** represent east, **j** north, and **k** up. (b) How far apart are the plane and ship?

47. As it passes over Grand Bahama Island, the eye of a hurricane is moving in a direction 60.0° north of west with a speed of 41.0 km/h. Three hours later, the course of the hurricane suddenly shifts due north and its speed slows to 25.0 km/h. How far from Grand Bahama is the eye 4.50 h after it passes over the island?

48. (a) Vector **E** has magnitude 17.0 cm and is directed 27.0° counterclockwise from the $+x$ axis. Express it in unit–vector notation. (b) Vector **F** has magnitude 17.0 cm and is directed 27.0° counterclockwise from the $+y$ axis. Express it in unit–vector notation. (c) Vector **G** has magnitude 17.0 cm and is directed 27.0° clockwise from the $+y$ axis. Express it in unit–vector notation.

49. Vector **A** has a negative x component 3.00 units in length and a positive y component 2.00 units in length. (a) Determine an expression for **A** in unit–vector notation. (b) Determine the magnitude and direction of **A**. (c) What vector **B**, when added to vector **A**, gives a resultant vector with no x component and a negative y component 4.00 units in length?

50. An airplane starting from airport A flies 300 km east, then 350 km at 30.0° west of north, and then 150 km north to arrive finally at airport B. (a) The next day, another plane flies directly from airport A to airport B in a straight line. In what direction should the pilot travel in this direct flight? (b) How far will the pilot travel in this direct flight? Assume there is no wind during these flights.

WEB 51. Three vectors are oriented as shown in Figure P3.51, where $|\mathbf{A}| = 20.0$ units, $|\mathbf{B}| = 40.0$ units, and $|\mathbf{C}| = 30.0$ units. Find (a) the x and y components of the resultant vector (expressed in unit–vector notation) and (b) the magnitude and direction of the resultant vector.

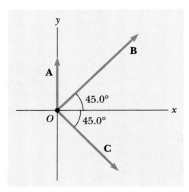

Figure P3.51

52. If $\mathbf{A} = (6.00\mathbf{i} - 8.00\mathbf{j})$ units, $\mathbf{B} = (-8.00\mathbf{i} + 3.00\mathbf{j})$ units, and $\mathbf{C} = (26.0\mathbf{i} + 19.0\mathbf{j})$ units, determine a and b such that $a\mathbf{A} + b\mathbf{B} + \mathbf{C} = 0$.

ADDITIONAL PROBLEMS

53. Two vectors **A** and **B** have precisely equal magnitudes. For the magnitude of $\mathbf{A} + \mathbf{B}$ to be 100 times greater than the magnitude of $\mathbf{A} - \mathbf{B}$, what must be the angle between them?

54. Two vectors **A** and **B** have precisely equal magnitudes. For the magnitude of $\mathbf{A} + \mathbf{B}$ to be greater than the magnitude of $\mathbf{A} - \mathbf{B}$ by the factor n, what must be the angle between them?

55. A vector is given by $\mathbf{R} = 2.00\mathbf{i} + 1.00\mathbf{j} + 3.00\mathbf{k}$. Find (a) the magnitudes of the x, y, and z components, (b) the magnitude of **R**, and (c) the angles between **R** and the x, y, and z axes.

56. Find the sum of these four vector forces: 12.0 N to the right at 35.0° above the horizontal, 31.0 N to the left at 55.0° above the horizontal, 8.40 N to the left at 35.0° below the horizontal, and 24.0 N to the right at 55.0° below the horizontal. (*Hint:* Make a drawing of this situation and select the best axes for x and y so that you have the least number of components. Then add the vectors, using the component method.)

57. A person going for a walk follows the path shown in Figure P3.57. The total trip consists of four straight-line paths. At the end of the walk, what is the person's resultant displacement measured from the starting point?

58. In general, the instantaneous position of an object is specified by its position vector **P** leading from a fixed

Figure P3.57

origin to the location of the object. Suppose that for a certain object the position vector is a function of time, given by $\mathbf{P} = 4\mathbf{i} + 3\mathbf{j} - 2t\mathbf{j}$, where P is in meters and t is in seconds. Evaluate $d\mathbf{P}/dt$. What does this derivative represent about the object?

59. A jet airliner, moving initially at 300 mi/h to the east, suddenly enters a region where the wind is blowing at 100 mi/h in a direction 30.0° north of east. What are the new speed and direction of the aircraft relative to the ground?

60. A pirate has buried his treasure on an island with five trees located at the following points: A(30.0 m, -20.0 m), B(60.0 m, 80.0 m), C(-10.0 m, -10.0 m), D(40.0 m, -30.0 m), and E(-70.0 m, 60.0 m). All points are measured relative to some origin, as in Figure P3.60. Instructions on the map tell you to start at A and move toward B, but to cover only one-half the distance between A and B. Then, move toward C, covering one-third the distance between your current location and C. Next, move toward D, covering one-fourth the distance between where you are and D. Finally, move toward E, covering one-fifth the distance between you and E, stop, and dig. (a) What are the coordinates of the point where the pirate's treasure is buried? (b) Re-

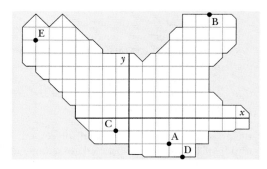

Figure P3.60

arrange the order of the trees, (for instance, B(30.0 m, − 20.0 m), A(60.0 m, 80.0 m), E(− 10.0 m, − 10.0 m), C(40.0 m, − 30.0 m), and D(− 70.0 m, 60.0 m), and repeat the calculation to show that the answer does not depend on the order of the trees.

61. A rectangular parallelepiped has dimensions a, b, and c, as in Figure P3.61. (a) Obtain a vector expression for the face diagonal vector $\mathbf{R}_1$. What is the magnitude of this vector? (b) Obtain a vector expression for the body diagonal vector $\mathbf{R}_2$. Note that $\mathbf{R}_1$, $c\mathbf{k}$, and $\mathbf{R}_2$ make a right triangle, and prove that the magnitude of $\mathbf{R}_2$ is $\sqrt{a^2 + b^2 + c^2}$.

62. A point lying in the xy plane and having coordinates (x, y) can be described by the position vector given by $\mathbf{r} = x\mathbf{i} + y\mathbf{j}$. (a) Show that the displacement vector for a particle moving from (x_1, y_1) to (x_2, y_2) is given by $\mathbf{d} = (x_2 - x_1)\mathbf{i} + (y_2 - y_1)\mathbf{j}$. (b) Plot the position vectors $\mathbf{r}_1$ and $\mathbf{r}_2$ and the displacement vector $\mathbf{d}$, and verify by the graphical method that $\mathbf{d} = \mathbf{r}_2 - \mathbf{r}_1$.

63. A point P is described by the coordinates (x, y) with respect to the normal cartesian coordinate system shown in Figure P3.63. Show that (x', y'), the coordinates of this point in the rotated coordinate system, are related to (x, y) and the rotation angle α by the expressions

$$x' = x \cos \alpha + y \sin \alpha$$

$$y' = -x \sin \alpha + y \cos \alpha$$

Figure P3.61

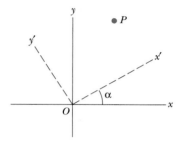

Figure P3.63

Answers to Quick Quizzes

3.1 The honeybee needs to communicate to the other honeybees how far it is to the flower and in what direction they must fly. This is exactly the kind of information that polar coordinates convey, as long as the origin of the coordinates is the beehive.

3.2 The resultant has magnitude $A + B$ when vector $\mathbf{A}$ is oriented in the same direction as vector $\mathbf{B}$. The resultant vector is $\mathbf{A} + \mathbf{B} = 0$ when vector $\mathbf{A}$ is oriented in the direction opposite vector $\mathbf{B}$ and $A = B$.

3.3 No. In two dimensions, a vector and its components form a right triangle. The vector is the hypotenuse and must be longer than either side. Problem 61 extends this concept to three dimensions.

3.4 No. The magnitude of a vector $\mathbf{A}$ is equal to $\sqrt{A_x^2 + A_y^2 + A_z^2}$. Therefore, if any component is nonzero, A cannot be zero. This generalization of the Pythagorean theorem is left for you to prove in Problem 61.

3.5 The fact that $\mathbf{A} + \mathbf{B} = 0$ tells you that $\mathbf{A} = -\mathbf{B}$. Therefore, the components of the two vectors must have opposite signs and equal magnitudes: $A_x = -B_x$, $A_y = -B_y$, and $A_z = -B_z$.

chapter

4

Motion in Two Dimensions

Chapter Outline

4.1 The Displacement, Velocity, and Acceleration Vectors

4.2 Two-Dimensional Motion with Constant Acceleration

4.3 Projectile Motion

4.4 Uniform Circular Motion

4.5 Tangential and Radial Acceleration

4.6 Relative Velocity and Relative Acceleration

*I*n this chapter we deal with the kinematics of a particle moving in two dimensions. Knowing the basics of two-dimensional motion will allow us to examine — in future chapters — a wide variety of motions, ranging from the motion of satellites in orbit to the motion of electrons in a uniform electric field. We begin by studying in greater detail the vector nature of displacement, velocity, and acceleration. As in the case of one-dimensional motion, we derive the kinematic equations for two-dimensional motion from the fundamental definitions of these three quantities. We then treat projectile motion and uniform circular motion as special cases of motion in two dimensions. We also discuss the concept of relative motion, which shows why observers in different frames of reference may measure different displacements, velocities, and accelerations for a given particle.

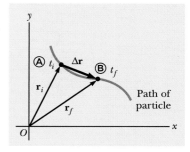

Figure 4.1   A particle moving in the *xy* plane is located with the position vector **r** drawn from the origin to the particle. The displacement of the particle as it moves from Ⓐ to Ⓑ in the time interval $\Delta t = t_f - t_i$ is equal to the vector $\Delta \mathbf{r} = \mathbf{r}_f - \mathbf{r}_i$.

 ### THE DISPLACEMENT, VELOCITY, AND ACCELERATION VECTORS

In Chapter 2 we found that the motion of a particle moving along a straight line is completely known if its position is known as a function of time. Now let us extend this idea to motion in the *xy* plane. We begin by describing the position of a particle by its position vector **r**, drawn from the origin of some coordinate system to the particle located in the *xy* plane, as in Figure 4.1. At time t_i the particle is at point Ⓐ, and at some later time t_f it is at point Ⓑ. The path from Ⓐ to Ⓑ is not necessarily a straight line. As the particle moves from Ⓐ to Ⓑ in the time interval $\Delta t = t_f - t_i$, its position vector changes from $\mathbf{r}_i$ to $\mathbf{r}_f$. As we learned in Chapter 2, displacement is a vector, and the displacement of the particle is the difference between its final position and its initial position. We now formally define the **displacement vector** $\Delta \mathbf{r}$ for the particle of Figure 4.1 as being the difference between its final position vector and its initial position vector:

$$\Delta \mathbf{r} \equiv \mathbf{r}_f - \mathbf{r}_i \tag{4.1}$$

| Displacement vector |

The direction of $\Delta \mathbf{r}$ is indicated in Figure 4.1. As we see from the figure, the magnitude of $\Delta \mathbf{r}$ is *less* than the distance traveled along the curved path followed by the particle.

As we saw in Chapter 2, it is often useful to quantify motion by looking at the ratio of a displacement divided by the time interval during which that displacement occurred. In two-dimensional (or three-dimensional) kinematics, everything is the same as in one-dimensional kinematics except that we must now use vectors rather than plus and minus signs to indicate the direction of motion.

We define the **average velocity** of a particle during the time interval Δt as the displacement of the particle divided by that time interval:

$$\bar{\mathbf{v}} \equiv \frac{\Delta \mathbf{r}}{\Delta t} \tag{4.2}$$

| Average velocity |

Multiplying or dividing a vector quantity by a scalar quantity changes only the magnitude of the vector, not its direction. Because displacement is a vector quantity and the time interval is a scalar quantity, we conclude that the average velocity is a vector quantity directed along $\Delta \mathbf{r}$.

Note that the average velocity between points is *independent of the path* taken. This is because average velocity is proportional to displacement, which depends

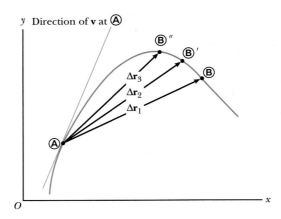

Figure 4.2   As a particle moves between two points, its average velocity is in the direction of the displacement vector $\Delta\mathbf{r}$. As the end point of the path is moved from Ⓑ to Ⓑ′ to Ⓑ″, the respective displacements and corresponding time intervals become smaller and smaller. In the limit that the end point approaches Ⓐ, Δt approaches zero, and the direction of $\Delta\mathbf{r}$ approaches that of the line tangent to the curve at Ⓐ. By definition, the instantaneous velocity at Ⓐ is in the direction of this tangent line.

only on the initial and final position vectors and not on the path taken. As we did with one-dimensional motion, we conclude that if a particle starts its motion at some point and returns to this point via any path, its average velocity is zero for this trip because its displacement is zero.

Consider again the motion of a particle between two points in the xy plane, as shown in Figure 4.2. As the time interval over which we observe the motion becomes smaller and smaller, the direction of the displacement approaches that of the line tangent to the path at Ⓐ.

> The **instantaneous velocity v** is defined as the limit of the average velocity $\Delta\mathbf{r}/\Delta t$ as Δt approaches zero:
>
> $$\mathbf{v} \equiv \lim_{\Delta t \to 0} \frac{\Delta\mathbf{r}}{\Delta t} = \frac{d\mathbf{r}}{dt} \qquad (4.3)$$

Instantaneous velocity

That is, the instantaneous velocity equals the derivative of the position vector with respect to time. The direction of the instantaneous velocity vector at any point in a particle's path is along a line tangent to the path at that point and in the direction of motion (Fig. 4.3).

The magnitude of the instantaneous velocity vector $v = |\mathbf{v}|$ is called the *speed,* which, as you should remember, is a scalar quantity.

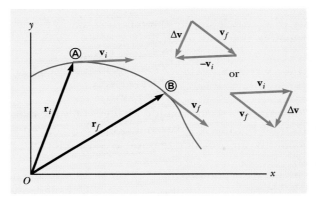

Figure 4.3   A particle moves from position Ⓐ to position Ⓑ. Its velocity vector changes from v_i to v_f. The vector diagrams at the upper right show two ways of determining the vector $\Delta\mathbf{v}$ from the initial and final velocities.

As a particle moves from one point to another along some path, its instantaneous velocity vector changes from $\mathbf{v}_i$ at time t_i to $\mathbf{v}_f$ at time t_f. Knowing the velocity at these points allows us to determine the average acceleration of the particle:

The **average acceleration** of a particle as it moves from one position to another is defined as the change in the instantaneous velocity vector $\Delta\mathbf{v}$ divided by the time Δt during which that change occurred:

$$\overline{\mathbf{a}} \equiv \frac{\mathbf{v}_f - \mathbf{v}_i}{t_f - t_i} = \frac{\Delta\mathbf{v}}{\Delta t} \tag{4.4}$$

Average acceleration

Because it is the ratio of a vector quantity $\Delta\mathbf{v}$ and a scalar quantity Δt, we conclude that average acceleration $\overline{\mathbf{a}}$ is a vector quantity directed along $\Delta\mathbf{v}$. As indicated in Figure 4.3, the direction of $\Delta\mathbf{v}$ is found by adding the vector $-\mathbf{v}_i$ (the negative of $\mathbf{v}_i$) to the vector $\mathbf{v}_f$, because by definition $\Delta\mathbf{v} = \mathbf{v}_f - \mathbf{v}_i$.

When the average acceleration of a particle changes during different time intervals, it is useful to define its instantaneous acceleration $\mathbf{a}$:

The **instantaneous acceleration a** is defined as the limiting value of the ratio $\Delta\mathbf{v}/\Delta t$ as Δt approaches zero:

$$\mathbf{a} \equiv \lim_{\Delta t \to 0} \frac{\Delta\mathbf{v}}{\Delta t} = \frac{d\mathbf{v}}{dt} \tag{4.5}$$

Instantaneous acceleration

3.5 In other words, the instantaneous acceleration equals the derivative of the velocity vector with respect to time.

It is important to recognize that various changes can occur when a particle accelerates. First, the magnitude of the velocity vector (the speed) may change with time as in straight-line (one-dimensional) motion. Second, the direction of the velocity vector may change with time even if its magnitude (speed) remains constant, as in curved-path (two-dimensional) motion. Finally, both the magnitude and the direction of the velocity vector may change simultaneously.

Quick Quiz 4.1

The gas pedal in an automobile is called the *accelerator.* (a) Are there any other controls in an automobile that can be considered accelerators? (b) When is the gas pedal not an accelerator?

4.2   TWO-DIMENSIONAL MOTION WITH CONSTANT ACCELERATION

Let us consider two-dimensional motion during which the acceleration remains constant in both magnitude and direction.

The position vector for a particle moving in the xy plane can be written

$$\mathbf{r} = x\mathbf{i} + y\mathbf{j} \tag{4.6}$$

where x, y, and $\mathbf{r}$ change with time as the particle moves while $\mathbf{i}$ and $\mathbf{j}$ remain constant. If the position vector is known, the velocity of the particle can be obtained from Equations 4.3 and 4.6, which give

$$\mathbf{v} = v_x\mathbf{i} + v_y\mathbf{j} \tag{4.7}$$

Because **a** is assumed constant, its components a_x and a_y also are constants. Therefore, we can apply the equations of kinematics to the x and y components of the velocity vector. Substituting $v_{xf} = v_{xi} + a_x t$ and $v_{yf} = v_{yi} + a_y t$ into Equation 4.7 to determine the final velocity at any time t, we obtain

$$\mathbf{v}_f = (v_{xi} + a_x t)\mathbf{i} + (v_{yi} + a_y t)\mathbf{j}$$
$$= (v_{xi}\mathbf{i} + v_{yi}\mathbf{j}) + (a_x\mathbf{i} + a_y\mathbf{j})t$$

Velocity vector as a function of time

$$\mathbf{v}_f = \mathbf{v}_i + \mathbf{a}t \tag{4.8}$$

This result states that the velocity of a particle at some time t equals the vector sum of its initial velocity $\mathbf{v}_i$ and the additional velocity $\mathbf{a}t$ acquired in the time t as a result of constant acceleration.

Similarly, from Equation 2.11 we know that the x and y coordinates of a particle moving with constant acceleration are

$$x_f = x_i + v_{xi}t + \tfrac{1}{2}a_x t^2 \qquad y_f = y_i + v_{yi}t + \tfrac{1}{2}a_y t^2$$

Substituting these expressions into Equation 4.6 (and labeling the final position vector $\mathbf{r}_f$) gives

$$\mathbf{r}_f = (x_i + v_{xi}t + \tfrac{1}{2}a_x t^2)\mathbf{i} + (y_i + v_{yi}t + \tfrac{1}{2}a_y t^2)\mathbf{j}$$
$$= (x_i\mathbf{i} + y_i\mathbf{j}) + (v_{xi}\mathbf{i} + v_{yi}\mathbf{j})t + \tfrac{1}{2}(a_x\mathbf{i} + a_y\mathbf{j})t^2$$

Position vector as a function of time

$$\mathbf{r}_f = \mathbf{r}_i + \mathbf{v}_i t + \tfrac{1}{2}\mathbf{a}t^2 \tag{4.9}$$

This equation tells us that the displacement vector $\Delta\mathbf{r} = \mathbf{r}_f - \mathbf{r}_i$ is the vector sum of a displacement $\mathbf{v}_i t$ arising from the initial velocity of the particle and a displacement $\tfrac{1}{2}\mathbf{a}t^2$ resulting from the uniform acceleration of the particle.

Graphical representations of Equations 4.8 and 4.9 are shown in Figure 4.4. For simplicity in drawing the figure, we have taken $\mathbf{r}_i = 0$ in Figure 4.4a. That is, we assume the particle is at the origin at $t = t_i = 0$. Note from Figure 4.4a that $\mathbf{r}_f$ is generally not along the direction of either $\mathbf{v}_i$ or $\mathbf{a}$ because the relationship between these quantities is a vector expression. For the same reason, from Figure 4.4b we see that $\mathbf{v}_f$ is generally not along the direction of $\mathbf{v}_i$ or $\mathbf{a}$. Finally, note that $\mathbf{v}_f$ and $\mathbf{r}_f$ are generally not in the same direction.

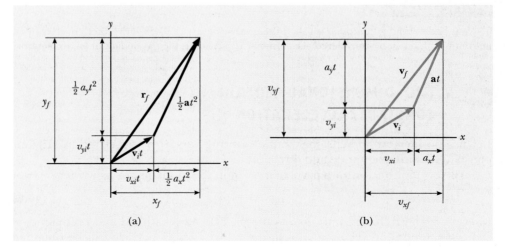

Figure 4.4 Vector representations and components of (a) the displacement and (b) the velocity of a particle moving with a uniform acceleration **a**. To simplify the drawing, we have set $\mathbf{r}_i = 0$.

Because Equations 4.8 and 4.9 are vector expressions, we may write them in component form:

$$\mathbf{v}_f = \mathbf{v}_i + \mathbf{a}t \qquad \begin{cases} v_{xf} = v_{xi} + a_x t \\ v_{yf} = v_{yi} + a_y t \end{cases} \qquad \textbf{(4.8a)}$$

$$\mathbf{r}_f = \mathbf{r}_i + \mathbf{v}_i t + \tfrac{1}{2}\mathbf{a}t^2 \qquad \begin{cases} x_f = x_i + v_{xi}t + \tfrac{1}{2}a_x t^2 \\ y_f = y_i + v_{yi}t + \tfrac{1}{2}a_y t^2 \end{cases} \qquad \textbf{(4.9a)}$$

These components are illustrated in Figure 4.4. The component form of the equations for $\mathbf{v}_f$ and $\mathbf{r}_f$ show us that two-dimensional motion at constant acceleration is equivalent to two *independent* motions—one in the x direction and one in the y direction—having constant accelerations a_x and a_y.

EXAMPLE 4.1 ▸ Motion in a Plane

A particle starts from the origin at $t = 0$ with an initial velocity having an x component of 20 m/s and a y component of -15 m/s. The particle moves in the xy plane with an x component of acceleration only, given by $a_x = 4.0$ m/s^2. (a) Determine the components of the velocity vector at any time and the total velocity vector at any time.

Solution   After carefully reading the problem, we realize we can set $v_{xi} = 20$ m/s, $v_{yi} = -15$ m/s, $a_x = 4.0$ m/s^2, and $a_y = 0$. This allows us to sketch a rough motion diagram of the situation. The x component of velocity starts at 20 m/s and increases by 4.0 m/s every second. The y component of velocity never changes from its initial value of -15 m/s. From this information we sketch some velocity vectors as shown in Figure 4.5. Note that the spacing between successive images increases as time goes on because the velocity is increasing.

The equations of kinematics give

$$v_{xf} = v_{xi} + a_x t = (20 + 4.0t)\ \text{m/s}$$

$$v_{yf} = v_{yi} + a_y t = -15\ \text{m/s} + 0 = -15\ \text{m/s}$$

Therefore,

$$\mathbf{v}_f = v_{xf}\mathbf{i} + v_{yf}\mathbf{j} = \boxed{[(20 + 4.0t)\mathbf{i} - 15\mathbf{j}]\ \text{m/s}}$$

Figure 4.5   Motion diagram for the particle.

We could also obtain this result using Equation 4.8 directly, noting that $\mathbf{a} = 4.0\mathbf{i}$ m/s^2 and $\mathbf{v}_i = (20\mathbf{i} - 15\mathbf{j})$ m/s. According to this result, the x component of velocity increases while the y component remains constant; this is consistent with what we predicted. After a long time, the x component will be so great that the y component will be negligible. If we were to extend the object's path in Figure 4.5, eventually it would become nearly parallel to the x axis. It is always helpful to make comparisons between final answers and initial stated conditions.

(b) Calculate the velocity and speed of the particle at $t = 5.0$ s.

Solution   With $t = 5.0$ s, the result from part (a) gives

$$\mathbf{v}_f = \{[20 + 4.0(5.0)]\mathbf{i} - 15\mathbf{j}\}\ \text{m/s} = \boxed{(40\mathbf{i} - 15\mathbf{j})\ \text{m/s}}$$

This result tells us that at $t = 5.0$ s, $v_{xf} = 40$ m/s and $v_{yf} = -15$ m/s. Knowing these two components for this two-dimensional motion, we can find both the direction and the magnitude of the velocity vector. To determine the angle θ that $\mathbf{v}$ makes with the x axis at $t = 5.0$ s, we use the fact that $\tan\theta = v_{yf}/v_{xf}$:

$$\theta = \tan^{-1}\left(\frac{v_{yf}}{v_{xf}}\right) = \tan^{-1}\left(\frac{-15\ \text{m/s}}{40\ \text{m/s}}\right) = \boxed{-21°}$$

where the minus sign indicates an angle of 21° below the positive x axis. The speed is the magnitude of $\mathbf{v}_f$:

$$v_f = |\mathbf{v}_f| = \sqrt{v_{xf}^2 + v_{yf}^2} = \sqrt{(40)^2 + (-15)^2}\ \text{m/s} = \boxed{43\ \text{m/s}}$$

In looking over our result, we notice that if we calculate v_i from the x and y components of $\mathbf{v}_i$, we find that $v_f > v_i$. Does this make sense?

(c) Determine the x and y coordinates of the particle at any time t and the position vector at this time.

Solution   Because $x_i = y_i = 0$ at $t = 0$, Equation 2.11 gives

$$x_f = v_{xi}t + \tfrac{1}{2}a_x t^2 = \boxed{(20t + 2.0t^2)\ \text{m}}$$

$$y_f = v_{yi}t = \boxed{(-15t)\ \text{m}}$$

Therefore, the position vector at any time t is

$$\mathbf{r}_f = x_f\mathbf{i} + y_f\mathbf{j} = \boxed{[(20t + 2.0t^2)\mathbf{i} - 15t\mathbf{j}]\ \text{m}}$$

(Alternatively, we could obtain $\mathbf{r}_f$ by applying Equation 4.9 directly, with $\mathbf{v}_i = (20\mathbf{i} - 15\mathbf{j})$ m/s and $\mathbf{a} = 4.0\mathbf{i}$ m/s^2. Try it!) Thus, for example, at $t = 5.0$ s, $x = 150$ m, $y = -75$ m, and $\mathbf{r}_f = (150\mathbf{i} - 75\mathbf{j})$ m. The magnitude of the displacement of the particle from the origin at $t = 5.0$ s is the magnitude of $\mathbf{r}_f$ at this time:

$$r_f = |\mathbf{r}_f| = \sqrt{(150)^2 + (-75)^2}\ \text{m} = 170\ \text{m}$$

Note that this is *not* the distance that the particle travels in this time! Can you determine this distance from the available data?

4.3 PROJECTILE MOTION

Anyone who has observed a baseball in motion (or, for that matter, any other object thrown into the air) has observed projectile motion. The ball moves in a curved path, and its motion is simple to analyze if we make two assumptions:

Assumptions of projectile motion

(1) the free-fall acceleration $\mathbf{g}$ is constant over the range of motion and is directed downward,[1] and (2) the effect of air resistance is negligible.[2] With these assumptions, we find that the path of a projectile, which we call its *trajectory*, is *always* a parabola. **We use these assumptions throughout this chapter.**

To show that the trajectory of a projectile is a parabola, let us choose our reference frame such that the y direction is vertical and positive is upward. Because air resistance is neglected, we know that $a_y = -g$ (as in one-dimensional free fall) and that $a_x = 0$. Furthermore, let us assume that at $t = 0$, the projectile leaves the origin ($x_i = y_i = 0$) with speed v_i, as shown in Figure 4.6. The vector $\mathbf{v}_i$ makes an angle θ_i with the horizontal, where θ_i is the angle at which the projectile leaves the origin. From the definitions of the cosine and sine functions we have

3.5

$$\cos\theta_i = v_{xi}/v_i \qquad \sin\theta_i = v_{yi}/v_i$$

Therefore, the initial x and y components of velocity are

$$v_{xi} = v_i\cos\theta_i \qquad v_{yi} = v_i\sin\theta_i$$

Substituting the x component into Equation 4.9a with $x_i = 0$ and $a_x = 0$, we find that

Horizontal position component

$$x_f = v_{xi}t = (v_i\cos\theta_i)t \tag{4.10}$$

Repeating with the y component and using $y_i = 0$ and $a_y = -g$, we obtain

Vertical position component

$$y_f = v_{yi}t + \tfrac{1}{2}a_y t^2 = (v_i\sin\theta_i)t - \tfrac{1}{2}gt^2 \tag{4.11}$$

Next, we solve Equation 4.10 for $t = x_f/(v_i\cos\theta_i)$ and substitute this expression for t into Equation 4.11; this gives

$$y = (\tan\theta_i)x - \left(\frac{g}{2v_i^2\cos^2\theta_i}\right)x^2 \tag{4.12}$$

---

[1] This assumption is reasonable as long as the range of motion is small compared with the radius of the Earth (6.4×10^6 m). In effect, this assumption is equivalent to assuming that the Earth is flat over the range of motion considered.

[2] This assumption is generally *not* justified, especially at high velocities. In addition, any spin imparted to a projectile, such as that applied when a pitcher throws a curve ball, can give rise to some very interesting effects associated with aerodynamic forces, which will be discussed in Chapter 15.

Figure 4.6 The parabolic path of a projectile that leaves the origin with a velocity $\mathbf{v}_i$. The velocity vector $\mathbf{v}$ changes with time in both magnitude and direction. This change is the result of acceleration in the negative y direction. The x component of velocity remains constant in time because there is no acceleration along the horizontal direction. The y component of velocity is zero at the peak of the path.

A welder cuts holes through a heavy metal construction beam with a hot torch. The sparks generated in the process follow parabolic paths. *(©The Telegraph Colour Library/FPG)*

This equation is valid for launch angles in the range $0 < \theta_i < \pi/2$. We have left the subscripts off the x and y because the equation is valid for any point (x, y) along the path of the projectile. The equation is of the form $y = ax - bx^2$, which is the equation of a parabola that passes through the origin. Thus, we have shown that the trajectory of a projectile is a parabola. Note that the trajectory is completely specified if both the initial speed v_i and the launch angle θ_i are known.

The vector expression for the position vector of the projectile as a function of time follows directly from Equation 4.9, with $\mathbf{r}_i = 0$ and $\mathbf{a} = \mathbf{g}$:

$$\mathbf{r} = \mathbf{v}_i t + \tfrac{1}{2}\mathbf{g}t^2$$

This expression is plotted in Figure 4.7.

QuickLab

Place two tennis balls at the edge of a tabletop. Sharply snap one ball horizontally off the table with one hand while gently tapping the second ball off with your other hand. Compare how long it takes the two to reach the floor. Explain your results.

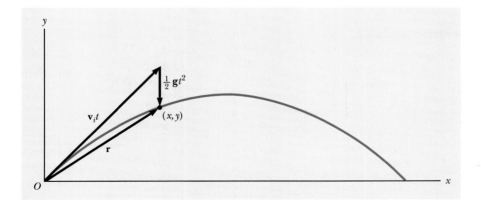

Figure 4.7 The position vector $\mathbf{r}$ of a projectile whose initial velocity at the origin is $\mathbf{v}_i$. The vector $\mathbf{v}_i t$ would be the displacement of the projectile if gravity were absent, and the vector $\frac{1}{2}\mathbf{g}t^2$ is its vertical displacement due to its downward gravitational acceleration.

Multiflash exposure of a tennis player executing a forehand swing. Note that the ball follows a parabolic path characteristic of a projectile. Such photographs can be used to study the quality of sports equipment and the performance of an athlete. *(© Zimmerman, FPG International)*

It is interesting to realize that the motion of a particle can be considered the superposition of the term $\mathbf{v}_i t$, the displacement if no acceleration were present, and the term $\frac{1}{2}\mathbf{g}t^2$, which arises from the acceleration due to gravity. In other words, if there were no gravitational acceleration, the particle would continue to move along a straight path in the direction of $\mathbf{v}_i$. Therefore, the vertical distance $\frac{1}{2}\mathbf{g}t^2$ through which the particle "falls" off the straight-line path is the same distance that a freely falling body would fall during the same time interval. We conclude that **projectile motion is the superposition of two motions: (1) constant-velocity motion in the horizontal direction** and (2) **free-fall motion in the vertical direction.** Except for t, the time of flight, the horizontal and vertical components of a projectile's motion are completely independent of each other.

EXAMPLE 4.2   Approximating Projectile Motion

A ball is thrown in such a way that its initial vertical and horizontal components of velocity are 40 m/s and 20 m/s, respectively. Estimate the total time of flight and the distance the ball is from its starting point when it lands.

Solution We start by remembering that the two velocity components are independent of each other. By considering the vertical motion first, we can determine how long the ball remains in the air. Then, we can use the time of flight to estimate the horizontal distance covered.

A motion diagram like Figure 4.8 helps us organize what we know about the problem. The acceleration vectors are all the same, pointing downward with a magnitude of nearly 10 m/s². The velocity vectors change direction. Their hori-

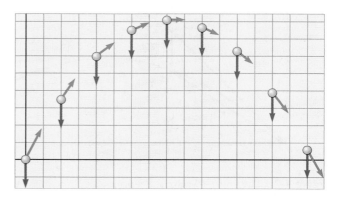

Figure 4.8   Motion diagram for a projectile.

zontal components are all the same: 20 m/s. Because the vertical motion is free fall, the vertical components of the velocity vectors change, second by second, from 40 m/s to roughly 30, 20, and 10 m/s in the upward direction, and then to 0 m/s. Subsequently, its velocity becomes 10, 20, 30, and 40 m/s in the downward direction. Thus it takes the ball about 4 s to go up and another 4 s to come back down, for a total time of flight of approximately 8 s. Because the horizontal component of velocity is 20 m/s, and because the ball travels at this speed for 8 s, it ends up approximately 160 m from its starting point.

Horizontal Range and Maximum Height of a Projectile

Let us assume that a projectile is fired from the origin at $t_i = 0$ with a positive v_{yi} component, as shown in Figure 4.9. Two points are especially interesting to analyze: the peak point Ⓐ, which has cartesian coordinates $(R/2, h)$, and the point Ⓑ, which has coordinates $(R, 0)$. The distance R is called the *horizontal range* of the projectile, and the distance h is its *maximum height*. Let us find h and R in terms of v_i, θ_i, and g.

We can determine h by noting that at the peak, $v_{yA} = 0$. Therefore, we can use Equation 4.8a to determine the time t_A it takes the projectile to reach the peak:

$$v_{yf} = v_{yi} + a_y t$$

$$0 = v_i \sin \theta_i - g t_A$$

$$t_A = \frac{v_i \sin \theta_i}{g}$$

Substituting this expression for t_A into the y part of Equation 4.9a and replacing $y_f = y_A$ with h, we obtain an expression for h in terms of the magnitude and direction of the initial velocity vector:

$$h = (v_i \sin \theta_i) \frac{v_i \sin \theta_i}{g} - \frac{1}{2} g \left(\frac{v_i \sin \theta_i}{g} \right)^2$$

$$h = \frac{v_i^2 \sin^2 \theta_i}{2g} \qquad (4.13)$$

The range R is the horizontal distance that the projectile travels in twice the time it takes to reach its peak, that is, in a time $t_B = 2t_A$. Using the x part of Equation 4.9a, noting that $v_{xi} = v_{xB} = v_i \cos \theta_i$, and setting $R \equiv x_B$ at $t = 2t_A$, we find that

$$R = v_{xi} t_B = (v_i \cos \theta_i) 2 t_A$$

$$= (v_i \cos \theta_i) \frac{2 v_i \sin \theta_i}{g} = \frac{2 v_i^2 \sin \theta_i \cos \theta_i}{g}$$

Using the identity $\sin 2\theta = 2 \sin \theta \cos \theta$ (see Appendix B.4), we write R in the more compact form

$$R = \frac{v_i^2 \sin 2\theta_i}{g} \qquad (4.14)$$

Keep in mind that Equations 4.13 and 4.14 are useful for calculating h and R only if v_i and θ_i are known (which means that only $\mathbf{v}_i$ has to be specified) and if the projectile lands at the same height from which it started, as it does in Figure 4.9.

The maximum value of R from Equation 4.14 is $R_{max} = v_i^2/g$. This result follows from the fact that the maximum value of $\sin 2\theta_i$ is 1, which occurs when $2\theta_i = 90°$. Therefore, R is a maximum when $\theta_i = 45°$.

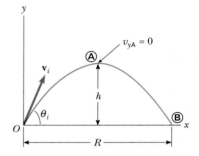

Figure 4.9 A projectile fired from the origin at $t_i = 0$ with an initial velocity $\mathbf{v}_i$. The maximum height of the projectile is h, and the horizontal range is R. At Ⓐ, the peak of the trajectory, the particle has coordinates $(R/2, h)$.

Maximum height of projectile

Range of projectile

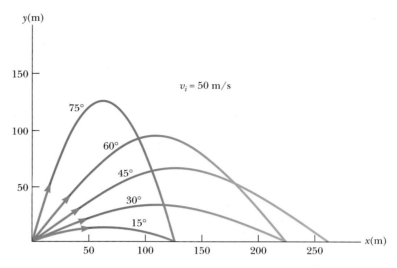

Figure 4.10   A projectile fired from the origin with an initial speed of 50 m/s at various angles of projection. Note that complementary values of θ_i result in the same value of x (range of the projectile).

Figure 4.10 illustrates various trajectories for a projectile having a given initial speed but launched at different angles. As you can see, the range is a maximum for $\theta_i = 45°$. In addition, for any θ_i other than 45°, a point having cartesian coordinates $(R, 0)$ can be reached by using either one of two complementary values of θ_i, such as 75° and 15°. Of course, the maximum height and time of flight for one of these values of θ_i are different from the maximum height and time of flight for the complementary value.

Quick Quiz 4.2

As a projectile moves in its parabolic path, is there any point along the path where the velocity and acceleration vectors are (a) perpendicular to each other? (b) parallel to each other? (c) Rank the five paths in Figure 4.10 with respect to time of flight, from the shortest to the longest.

Problem–Solving Hints
Projectile Motion

We suggest that you use the following approach to solving projectile motion problems:

- Select a coordinate system and resolve the initial velocity vector into x and y components.
- Follow the techniques for solving constant-velocity problems to analyze the horizontal motion. Follow the techniques for solving constant-acceleration problems to analyze the vertical motion. The x and y motions share the same time of flight t.

EXAMPLE 4.3 ▸ The Long-Jump

A long-jumper leaves the ground at an angle of 20.0° above the horizontal and at a speed of 11.0 m/s. (a) How far does he jump in the horizontal direction? (Assume his motion is equivalent to that of a particle.)

Solution   Because the initial speed and launch angle are given, the most direct way of solving to this problem is to use the range formula given by Equation 4.14. However, it is more instructive to take a more general approach and use Figure 4.9. As before, we set our origin of coordinates at the

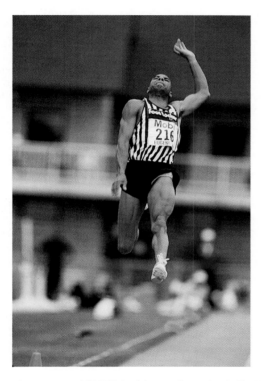

In a long-jump event, 1993 United States champion Mike Powell can leap horizontal distances of at least 8 m. *(Chuck Muhlstock/FPG International)*

takeoff point and label the peak as Ⓐ and the landing point as Ⓑ. The horizontal motion is described by Equation 4.10:

$$x_f = x_B = (v_i \cos \theta_i)\, t_B = (11.0 \text{ m/s})(\cos 20.0°)\, t_B$$

The value of x_B can be found if the total time of the jump is known. We are able to find t_B by remembering that $a_y = -g$ and by using the y part of Equation 4.8a. We also note that at the top of the jump the vertical component of velocity v_{yA} is zero:

$$v_{yf} = v_{yA} = v_i \sin \theta_i - gt_A$$
$$0 = (11.0 \text{ m/s}) \sin 20.0° - (9.80 \text{ m/s}^2)\, t_A$$
$$t_A = 0.384 \text{ s}$$

This is the time needed to reach the *top* of the jump. Because of the symmetry of the vertical motion, an identical time interval passes before the jumper returns to the ground. Therefore, the *total time* in the air is $t_B = 2t_A = 0.768$ s. Substituting this value into the above expression for x_f gives

$$x_f = x_B = (11.0 \text{ m/s})(\cos 20.0°)(0.768 \text{ s}) = \boxed{7.94 \text{ m}}$$

This is a reasonable distance for a world-class athlete.

(b) What is the maximum height reached?

Solution   We find the maximum height reached by using Equation 4.11:

$$y_{max} = y_A = (v_i \sin \theta_i)\, t_A - \tfrac{1}{2} g t_A^2$$
$$= (11.0 \text{ m/s})(\sin 20.0°)(0.384 \text{ s})$$
$$\quad - \tfrac{1}{2}(9.80 \text{ m/s}^2)(0.384 \text{ s})^2$$
$$= \boxed{0.722 \text{ m}}$$

Treating the long-jumper as a particle is an oversimplification. Nevertheless, the values obtained are reasonable.

Exercise   To check these calculations, use Equations 4.13 and 4.14 to find the maximum height and horizontal range.

EXAMPLE 4.4 ▸ A Bull's-Eye Every Time

In a popular lecture demonstration, a projectile is fired at a target in such a way that the projectile leaves the gun at the same time the target is dropped from rest, as shown in Figure 4.11. Show that if the gun is initially aimed at the stationary target, the projectile hits the target.

Solution   We can argue that a collision results under the conditions stated by noting that, as soon as they are released, the projectile and the target experience the same accelera-

tion $a_y = -g$. First, note from Figure 4.11b that the initial y coordinate of the target is $x_T \tan \theta_i$ and that it falls through a distance $\tfrac{1}{2} g t^2$ in a time t. Therefore, the y coordinate of the target at any moment after release is

$$y_T = x_T \tan \theta_i - \tfrac{1}{2} g t^2$$

Now if we use Equation 4.9a to write an expression for the y coordinate of the projectile at any moment, we obtain

$$y_P = x_P \tan \theta_i - \tfrac{1}{2} g t^2$$

(a)   (b)

Figure 4.11   (a) Multiflash photograph of projectile–target demonstration. If the gun is aimed directly at the target and is fired at the same instant the target begins to fall, the projectile will hit the target. Note that the velocity of the projectile (red arrows) changes in direction and magnitude, while the downward acceleration (violet arrows) remains constant. *(Central Scientific Company.)* (b) Schematic diagram of the projectile–target demonstration. Both projectile and target fall through the same vertical distance in a time *t* because both experience the same acceleration $a_y = -g$.

Thus, by comparing the two previous equations, we see that when the *y* coordinates of the projectile and target are the same, their *x* coordinates are the same and a collision results. That is, when $y_P = y_T$, $x_P = x_T$. You can obtain the same result, using expressions for the position vectors for the projectile and target.

Note that a collision will *not* always take place owing to a further restriction: A collision can result only when $v_i \sin \theta_i \geq \sqrt{gd/2}$, where *d* is the initial elevation of the target above the *floor*. If $v_i \sin \theta_i$ is less than this value, the projectile will strike the floor before reaching the target.

EXAMPLE 4.5   That's Quite an Arm!

A stone is thrown from the top of a building upward at an angle of 30.0° to the horizontal and with an initial speed of 20.0 m/s, as shown in Figure 4.12. If the height of the building is 45.0 m, (a) how long is it before the stone hits the ground?

Solution   We have indicated the various parameters in Figure 4.12. When working problems on your own, you should always make a sketch such as this and label it.

The initial *x* and *y* components of the stone's velocity are

$$v_{xi} = v_i \cos \theta_i = (20.0 \text{ m/s})(\cos 30.0°) = 17.3 \text{ m/s}$$

$$v_{yi} = v_i \sin \theta_i = (20.0 \text{ m/s})(\sin 30.0°) = 10.0 \text{ m/s}$$

To find *t*, we can use $y_f = v_{yi}t + \frac{1}{2}a_y t^2$ (Eq. 4.9a) with $y_f = -45.0$ m, $a_y = -g$, and $v_{yi} = 10.0$ m/s (there is a minus sign on the numerical value of y_f because we have chosen the top of the building as the origin):

$$-45.0 \text{ m} = (10.0 \text{ m/s})t - \tfrac{1}{2}(9.80 \text{ m/s}^2)t^2$$

Solving the quadratic equation for *t* gives, for the positive root, $t =$ 4.22 s.   Does the negative root have any physical

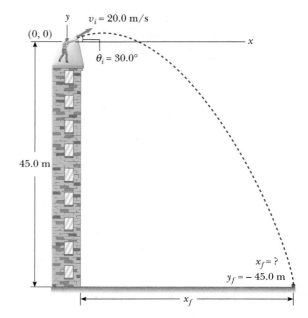

Figure 4.12

meaning? (Can you think of another way of finding t from the information given?)

(b) What is the speed of the stone just before it strikes the ground?

Solution We can use Equation 4.8a, $v_{yf} = v_{yi} + a_y t$, with $t = 4.22$ s to obtain the x and y components of the velocity just before the stone strikes the ground:

$$v_{yf} = 10.0 \text{ m/s} - (9.80 \text{ m/s}^2)(4.22 \text{ s}) = -31.4 \text{ m/s}$$

The negative sign indicates that the stone is moving downward. Because $v_{xf} = v_{xi} = 17.3$ m/s, the required speed is

$$v_f = \sqrt{v_{xf}^2 + v_{yf}^2} = \sqrt{(17.3)^2 + (-31.4)^2} \text{ m/s} = \boxed{35.9 \text{ m/s}}$$

Exercise Where does the stone strike the ground?

Answer 73.0 m from the base of the building.

EXAMPLE 4.6   The Stranded Explorers

An Alaskan rescue plane drops a package of emergency rations to a stranded party of explorers, as shown in Figure 4.13. If the plane is traveling horizontally at 40.0 m/s and is 100 m above the ground, where does the package strike the ground relative to the point at which it was released?

Solution For this problem we choose the coordinate system shown in Figure 4.13, in which the origin is at the point of release of the package. Consider first the horizontal motion of the package. The only equation available to us for finding the distance traveled along the horizontal direction is $x_f = v_{xi}t$ (Eq. 4.9a). The initial x component of the package

velocity is the same as that of the plane when the package is released: 40.0 m/s. Thus, we have

$$x_f = (40.0 \text{ m/s})t$$

If we know t, the length of time the package is in the air, then we can determine x_f, the distance the package travels in the horizontal direction. To find t, we use the equations that describe the vertical motion of the package. We know that at the instant the package hits the ground, its y coordinate is $y_f = -100$ m. We also know that the initial vertical component of the package velocity v_{yi} is zero because at the moment of release, the package had only a horizontal component of velocity.

From Equation 4.9a, we have

$$y_f = -\tfrac{1}{2}gt^2$$
$$-100 \text{ m} = -\tfrac{1}{2}(9.80 \text{ m/s}^2)t^2$$
$$t = 4.52 \text{ s}$$

Substitution of this value for the time of flight into the equation for the x coordinate gives

$$x_f = (40.0 \text{ m/s})(4.52 \text{ s}) = \boxed{181 \text{ m}}$$

The package hits the ground 181 m to the right of the drop point.

Exercise What are the horizontal and vertical components of the velocity of the package just before it hits the ground?

Answer $v_{xf} = 40.0$ m/s; $v_{yf} = -44.3$ m/s.

Exercise Where is the plane when the package hits the ground? (Assume that the plane does not change its speed or course.)

Answer Directly over the package.

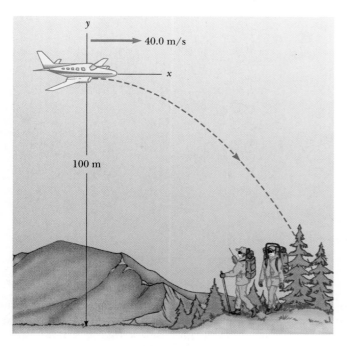

y

40.0 m/s

x

100 m

Figure 4.13

EXAMPLE 4.7   The End of the Ski Jump

A ski jumper leaves the ski track moving in the horizontal direction with a speed of 25.0 m/s, as shown in Figure 4.14. The landing incline below him falls off with a slope of 35.0°. (a) Where does he land on the incline?

Solution It is reasonable to expect the skier to be airborne for less than 10 s, and so he will not go farther than 250 m horizontally. We should expect the value of d, the distance traveled along the incline, to be of the same order of magnitude. It is convenient to select the beginning of the jump as the origin ($x_i = 0$, $y_i = 0$). Because $v_{xi} = 25.0$ m/s and $v_{yi} = 0$, the x and y component forms of Equation 4.9a are

$$(1) \qquad x_f = v_{xi}t = (25.0 \text{ m/s})t$$

$$(2) \qquad y_f = \tfrac{1}{2}a_y t^2 = -\tfrac{1}{2}(9.80 \text{ m/s}^2)t^2$$

From the right triangle in Figure 4.14, we see that the jumper's x and y coordinates at the landing point are $x_f = d \cos 35.0°$ and $y_f = -d \sin 35.0°$. Substituting these relationships into (1) and (2), we obtain

$$(3) \qquad d \cos 35.0° = (25.0 \text{ m/s})t$$

$$(4) \qquad -d \sin 35.0° = -\tfrac{1}{2}(9.80 \text{ m/s}^2)t^2$$

Solving (3) for t and substituting the result into (4), we find that $d = 109$ m. Hence, the x and y coordinates of the point at which he lands are

$$x_f = d \cos 35.0° = (109 \text{ m}) \cos 35.0° = \boxed{89.3 \text{ m}}$$

$$y_f = -d \sin 35.0° = -(109 \text{ m}) \sin 35.0° = \boxed{-62.5 \text{ m}}$$

Exercise Determine how long the jumper is airborne and his vertical component of velocity just before he lands.

Answer 3.57 s;   −35.0 m/s.

25.0 m/s

(0, 0)

$\theta = 35.0°$

y

d

x

Figure 4.14

     What would have occurred if the skier in the last example happened to be carrying a stone and let go of it while in midair? Because the stone has the same initial velocity as the skier, it will stay near him as he moves—that is, it floats alongside him. This is a technique that NASA uses to train astronauts. The plane pictured at the beginning of the chapter flies in the same type of projectile path that the skier and stone follow. The passengers and cargo in the plane fall along-

QuickLab

Armed with nothing but a ruler and the knowledge that the time between images was 1/30 s, find the horizontal speed of the yellow ball in Figure 4.15. (*Hint:* Start by analyzing the motion of the red ball. Because you know its vertical acceleration, you can calibrate the distances depicted in the photograph. Then you can find the horizontal speed of the yellow ball.)

Figure 4.15 This multiflash photograph of two balls released simultaneously illustrates both free fall (red ball) and projectile motion (yellow ball). The yellow ball was projected horizontally, while the red ball was released from rest. *(Richard Megna/Fundamental Photographs)*

side each other; that is, they have the same trajectory. An astronaut can release a piece of equipment and it will float freely alongside her hand. The same thing happens in the space shuttle. The craft and everything in it are falling as they orbit the Earth.

4.4 UNIFORM CIRCULAR MOTION

3.6   Figure 4.16a shows a car moving in a circular path with constant linear speed v. Such motion is called **uniform circular motion.** Because the car's direction of motion changes, the car has an acceleration, as we learned in Section 4.1. For any motion, the velocity vector is tangent to the path. Consequently, when an object moves in a circular path, its velocity vector is perpendicular to the radius of the circle.

We now show that the acceleration vector in uniform circular motion is always perpendicular to the path and always points toward the center of the circle. An ac-

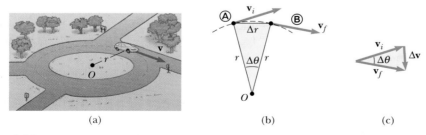

(a)   (b)   (c)

Figure 4.16 (a) A car moving along a circular path at constant speed experiences uniform circular motion. (b) As a particle moves from Ⓐ to Ⓑ, its velocity vector changes from $\mathbf{v}_i$ to $\mathbf{v}_f$. (c) The construction for determining the direction of the change in velocity $\Delta\mathbf{v}$, which is toward the center of the circle for small $\Delta\mathbf{r}$.

celeration of this nature is called a **centripetal** (center-seeking) acceleration, and its magnitude is

$$a_r = \frac{v^2}{r} \tag{4.15}$$

where r is the radius of the circle and the notation a_r is used to indicate that the centripetal acceleration is along the radial direction.

To derive Equation 4.15, consider Figure 4.16b, which shows a particle first at point Ⓐ and then at point Ⓑ. The particle is at Ⓐ at time t_i, and its velocity at that time is $\mathbf{v}_i$. It is at Ⓑ at some later time t_f, and its velocity at that time is $\mathbf{v}_f$. Let us assume here that $\mathbf{v}_i$ and $\mathbf{v}_f$ differ only in direction; their magnitudes (speeds) are the same (that is, $v_i = v_f = v$). To calculate the acceleration of the particle, let us begin with the defining equation for average acceleration (Eq. 4.4):

$$\overline{\mathbf{a}} = \frac{\mathbf{v}_f - \mathbf{v}_i}{t_f - t_i} = \frac{\Delta\mathbf{v}}{\Delta t}$$

This equation indicates that we must subtract $\mathbf{v}_i$ from $\mathbf{v}_f$, being sure to treat them as vectors, where $\Delta\mathbf{v} = \mathbf{v}_f - \mathbf{v}_i$ is the change in the velocity. Because $\mathbf{v}_i + \Delta\mathbf{v} = \mathbf{v}_f$, we can find the vector $\Delta\mathbf{v}$, using the vector triangle in Figure 4.16c.

Now consider the triangle in Figure 4.16b, which has sides Δr and r. This triangle and the one in Figure 4.16c, which has sides Δv and v, are similar. This fact enables us to write a relationship between the lengths of the sides:

$$\frac{\Delta v}{v} = \frac{\Delta r}{r}$$

This equation can be solved for Δv and the expression so obtained substituted into $\overline{a} = \Delta v/\Delta t$ (Eq. 4.4) to give

$$\overline{a} = \frac{v\,\Delta r}{r\,\Delta t}$$

Now imagine that points Ⓐ and Ⓑ in Figure 4.16b are extremely close together. In this case $\Delta\mathbf{v}$ points toward the center of the circular path, and because the acceleration is in the direction of $\Delta\mathbf{v}$, it too points toward the center. Furthermore, as Ⓐ and Ⓑ approach each other, Δt approaches zero, and the ratio $\Delta r/\Delta t$ approaches the speed v. Hence, in the limit $\Delta t \to 0$, the magnitude of the acceleration is

$$a_r = \frac{v^2}{r}$$

Thus, we conclude that in uniform circular motion, the acceleration is directed toward the center of the circle and has a magnitude given by v^2/r, where v is the speed of the particle and r is the radius of the circle. You should be able to show that the dimensions of a_r are L/T^2. We shall return to the discussion of circular motion in Section 6.1.

4.5 ▶ TANGENTIAL AND RADIAL ACCELERATION

3.6 Now let us consider a particle moving along a curved path where the velocity changes both in direction and in magnitude, as shown in Figure 4.17. As is always the case, the velocity vector is tangent to the path, but now the direction of the ac-

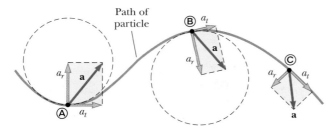

Figure 4.17 The motion of a particle along an arbitrary curved path lying in the xy plane. If the velocity vector **v** (always tangent to the path) changes in direction and magnitude, the component vectors of the acceleration **a** are a tangential component a_t and a radial component a_r.

celeration vector **a** changes from point to point. This vector can be resolved into two component vectors: a radial component vector $\mathbf{a}_r$ and a tangential component vector $\mathbf{a}_t$. Thus, **a** can be written as the vector sum of these component vectors:

$$\mathbf{a} = \mathbf{a}_r + \mathbf{a}_t \tag{4.16}$$

Total acceleration

The tangential acceleration causes the change in the speed of the particle. It is parallel to the instantaneous velocity, and its magnitude is

$$a_t = \frac{d|\mathbf{v}|}{dt} \tag{4.17}$$

Tangential acceleration

The radial acceleration arises from the change in direction of the velocity vector as described earlier and has an absolute magnitude given by

$$a_r = \frac{v^2}{r} \tag{4.18}$$

Radial acceleration

where r is the radius of curvature of the path at the point in question. Because $\mathbf{a}_r$ and $\mathbf{a}_t$ are mutually perpendicular component vectors of **a**, it follows that $a = \sqrt{a_r{}^2 + a_t{}^2}$. As in the case of uniform circular motion, $\mathbf{a}_r$ in nonuniform circular motion always points toward the center of curvature, as shown in Figure 4.17. Also, at a given speed, a_r is large when the radius of curvature is small (as at points Ⓐ and Ⓑ in Figure 4.17) and small when r is large (such as at point Ⓒ). The direction of $\mathbf{a}_t$ is either in the same direction as **v** (if v is increasing) or opposite **v** (if v is decreasing).

In uniform circular motion, where v is constant, $a_t = 0$ and the acceleration is always completely radial, as we described in Section 4.4. (*Note:* Eq. 4.18 is identical to Eq. 4.15.) In other words, uniform circular motion is a special case of motion along a curved path. Furthermore, if the direction of **v** does not change, then there is no radial acceleration and the motion is one-dimensional (in this case, $a_r = 0$, but a_t may not be zero).

Quick Quiz 4.3

(a) Draw a motion diagram showing velocity and acceleration vectors for an object moving with constant speed counterclockwise around a circle. Draw similar diagrams for an object moving counterclockwise around a circle but (b) slowing down at constant tangential acceleration and (c) speeding up at constant tangential acceleration.

It is convenient to write the acceleration of a particle moving in a circular path in terms of unit vectors. We do this by defining the unit vectors $\hat{\mathbf{r}}$ and $\hat{\boldsymbol{\theta}}$ shown in

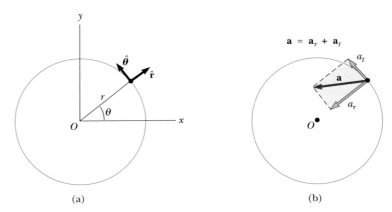

Figure 4.18 (a) Descriptions of the unit vectors $\hat{\mathbf{r}}$ and $\hat{\boldsymbol{\theta}}$. (b) The total acceleration $\mathbf{a}$ of a particle moving along a curved path (which at any instant is part of a circle of radius r) is the sum of radial and tangential components. The radial component is directed toward the center of curvature. If the tangential component of acceleration becomes zero, the particle follows uniform circular motion.

Figure 4.18a, where $\hat{\mathbf{r}}$ is a unit vector lying along the radius vector and directed radially outward from the center of the circle and $\hat{\boldsymbol{\theta}}$ is a unit vector tangent to the circle. The direction of $\hat{\boldsymbol{\theta}}$ is in the direction of increasing θ, where θ is measured counterclockwise from the positive x axis. Note that both $\hat{\mathbf{r}}$ and $\hat{\boldsymbol{\theta}}$ "move along with the particle" and so vary in time. Using this notation, we can express the total acceleration as

$$\mathbf{a} = \mathbf{a}_t + \mathbf{a}_r = \frac{d|\mathbf{v}|}{dt}\,\hat{\boldsymbol{\theta}} - \frac{v^2}{r}\,\hat{\mathbf{r}} \qquad \textbf{(4.19)}$$

These vectors are described in Figure 4.18b. The negative sign on the v^2/r term in Equation 4.19 indicates that the radial acceleration is always directed radially inward, *opposite* $\hat{\mathbf{r}}$.

Quick Quiz 4.4

Based on your experience, draw a motion diagram showing the position, velocity, and acceleration vectors for a pendulum that, from an initial position 45° to the right of a central vertical line, swings in an arc that carries it to a final position 45° to the left of the central vertical line. The arc is part of a circle, and you should use the center of this circle as the origin for the position vectors.

EXAMPLE 4.8   The Swinging Ball

A ball tied to the end of a string 0.50 m in length swings in a vertical circle under the influence of gravity, as shown in Figure 4.19. When the string makes an angle $\theta = 20°$ with the vertical, the ball has a speed of 1.5 m/s. (a) Find the magnitude of the radial component of acceleration at this instant.

Solution The diagram from the answer to Quick Quiz 4.4 (p. 109) applies to this situation, and so we have a good idea of how the acceleration vector varies during the motion. Fig-

ure 4.19 lets us take a closer look at the situation. The radial acceleration is given by Equation 4.18. With $v = 1.5$ m/s and $r = 0.50$ m, we find that

$$a_r = \frac{v^2}{r} = \frac{(1.5 \text{ m/s})^2}{0.50 \text{ m}} = \boxed{4.5 \text{ m/s}^2}$$

(b) What is the magnitude of the tangential acceleration when $\theta = 20°$?

Solution When the ball is at an angle θ to the vertical, it has a tangential acceleration of magnitude $g \sin \theta$ (the component of **g** tangent to the circle). Therefore, at $\theta = 20°$,

$$a_t = g \sin 20° = \boxed{3.4 \text{ m/s}^2}.$$

(c) Find the magnitude and direction of the total acceleration **a** at $\theta = 20°$.

Solution Because $\mathbf{a} = \mathbf{a}_r + \mathbf{a}_t$, the magnitude of **a** at $\theta = 20°$ is

$$a = \sqrt{a_r^2 + a_t^2} = \sqrt{(4.5)^2 + (3.4)^2} \text{ m/s}^2 = \boxed{5.6 \text{ m/s}^2}$$

If ϕ is the angle between **a** and the string, then

$$\phi = \tan^{-1} \frac{a_t}{a_r} = \tan^{-1} \left(\frac{3.4 \text{ m/s}^2}{4.5 \text{ m/s}^2} \right) = \boxed{37°}$$

Note that **a**, $\mathbf{a}_t$, and $\mathbf{a}_r$ all change in direction *and* magnitude as the ball swings through the circle. When the ball is at its lowest elevation ($\theta = 0$), $a_t = 0$ because there is no tangential component of **g** at this angle; also, a_r is a *maximum* because v is a maximum. If the ball has enough speed to reach its highest position ($\theta = 180°$), then a_t is again zero but a_r is a minimum because v is now a minimum. Finally, in the two

horizontal positions ($\theta = 90°$ and $270°$), $|\mathbf{a}_t| = g$ and a_r has a value between its minimum and maximum values.

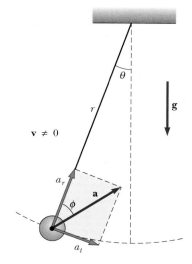

Figure 4.19 Motion of a ball suspended by a string of length r. The ball swings with nonuniform circular motion in a vertical plane, and its acceleration **a** has a radial component a_r and a tangential component a_t.

4.6 RELATIVE VELOCITY AND RELATIVE ACCELERATION

In this section, we describe how observations made by different observers in different frames of reference are related to each other. We find that observers in different frames of reference may measure different displacements, velocities, and accelerations for a given particle. That is, two observers moving relative to each other generally do not agree on the outcome of a measurement.

For example, suppose two cars are moving in the same direction with speeds of 50 mi/h and 60 mi/h. To a passenger in the slower car, the speed of the faster car is 10 mi/h. Of course, a stationary observer will measure the speed of the faster car to be 60 mi/h, not 10 mi/h. Which observer is correct? They both are! This simple example demonstrates that the velocity of an object depends on the frame of reference in which it is measured.

Suppose a person riding on a skateboard (observer A) throws a ball in such a way that it appears in this person's frame of reference to move first straight upward and then straight downward along the same vertical line, as shown in Figure 4.20a. A stationary observer B sees the path of the ball as a parabola, as illustrated in Figure 4.20b. Relative to observer B, the ball has a vertical component of velocity (resulting from the initial upward velocity and the downward acceleration of gravity) *and* a horizontal component.

Another example of this concept that of is a package dropped from an airplane flying with a constant velocity; this is the situation we studied in Example 4.6. An observer on the airplane sees the motion of the package as a straight line toward the Earth. The stranded explorer on the ground, however, sees the trajectory of the package as a parabola. If, once it drops the package, the airplane con-

Figure 4.20   (a) Observer A on a moving vehicle throws a ball upward and sees it rise and fall in a straight-line path. (b) Stationary observer B sees a parabolic path for the same ball.

tinues to move horizontally with the same velocity, then the package hits the ground directly beneath the airplane (if we assume that air resistance is neglected)!

In a more general situation, consider a particle located at point Ⓐ in Figure 4.21. Imagine that the motion of this particle is being described by two observers, one in reference frame S, fixed relative to the Earth, and another in reference frame S', moving to the right relative to S (and therefore relative to the Earth) with a constant velocity $\mathbf{v}_0$. (Relative to an observer in S', S moves to the left with a velocity $-\mathbf{v}_0$.) Where an observer stands in a reference frame is irrelevant in this discussion, but for purposes of this discussion let us place each observer at her or his respective origin.

We label the position of the particle relative to the S frame with the position vector $\mathbf{r}$ and that relative to the S' frame with the position vector $\mathbf{r}'$, both after some time t. The vectors $\mathbf{r}$ and $\mathbf{r}'$ are related to each other through the expression $\mathbf{r} = \mathbf{r}' + \mathbf{v}_0 t$, or

Galilean coordinate transformation

$$\mathbf{r}' = \mathbf{r} - \mathbf{v}_0 t \qquad (4.20)$$

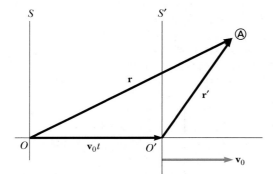

Figure 4.21   A particle located at Ⓐ is described by two observers, one in the fixed frame of reference S, and the other in the frame S', which moves to the right with a constant velocity $\mathbf{v}_0$. The vector $\mathbf{r}$ is the particle's position vector relative to S, and $\mathbf{r}'$ is its position vector relative to S'.

That is, after a time t, the S' frame is displaced to the right of the S frame by an amount $\mathbf{v}_0 t$.

If we differentiate Equation 4.20 with respect to time and note that $\mathbf{v}_0$ is constant, we obtain

$$\frac{d\mathbf{r}'}{dt} = \frac{d\mathbf{r}}{dt} - \mathbf{v}_0$$

$$\mathbf{v}' = \mathbf{v} - \mathbf{v}_0 \qquad \textbf{(4.21)}$$

Galilean velocity transformation

where $\mathbf{v}'$ is the velocity of the particle observed in the S' frame and $\mathbf{v}$ is its velocity observed in the S frame. Equations 4.20 and 4.21 are known as **Galilean transformation equations.** They relate the coordinates and velocity of a particle as measured in a frame fixed relative to the Earth to those measured in a frame moving with uniform motion relative to the Earth.

Although observers in two frames measure different velocities for the particle, they measure the *same acceleration* when $\mathbf{v}_0$ is constant. We can verify this by taking the time derivative of Equation 4.21:

$$\frac{d\mathbf{v}'}{dt} = \frac{d\mathbf{v}}{dt} - \frac{d\mathbf{v}_0}{dt}$$

Because $\mathbf{v}_0$ is constant, $d\mathbf{v}_0/dt = 0$. Therefore, we conclude that $\mathbf{a}' = \mathbf{a}$ because $\mathbf{a}' = d\mathbf{v}'/dt$ and $\mathbf{a} = d\mathbf{v}/dt$. That is, **the acceleration of the particle measured by an observer in the Earth's frame of reference is the same as that measured by any other observer moving with constant velocity relative to the Earth's frame.**

Quick Quiz 4.5

A passenger in a car traveling at 60 mi/h pours a cup of coffee for the tired driver. Describe the path of the coffee as it moves from a Thermos bottle into a cup as seen by (a) the passenger and (b) someone standing beside the road and looking in the window of the car as it drives past. (c) What happens if the car accelerates while the coffee is being poured?

The woman standing on the beltway sees the walking man pass by at a slower speed than the woman standing on the stationary floor does.

LE 4.9 ▶ A Boat Crossing a River

...ading due north crosses a wide river with a speed of 10.0 km/h relative to the water. The water in the river has a uniform speed of 5.00 km/h due east relative to the Earth. Determine the velocity of the boat relative to an observer standing on either bank.

Solution We know $\mathbf{v}_{br}$, the velocity of the *boat* relative to the *river*, and $\mathbf{v}_{rE}$, the velocity of the *river* relative to the *Earth*. What we need to find is $\mathbf{v}_{bE}$, the velocity of the *boat* relative to the *Earth*. The relationship between these three quantities is

$$\mathbf{v}_{bE} = \mathbf{v}_{br} + \mathbf{v}_{rE}$$

The terms in the equation must be manipulated as vector quantities; the vectors are shown in Figure 4.22. The quantity $\mathbf{v}_{br}$ is due north, $\mathbf{v}_{rE}$ is due east, and the vector sum of the two, $\mathbf{v}_{bE}$, is at an angle θ, as defined in Figure 4.22. Thus, we can find the speed v_{bE} of the boat relative to the Earth by using the Pythagorean theorem:

$$v_{bE} = \sqrt{v_{br}^2 + v_{rE}^2} = \sqrt{(10.0)^2 + (5.00)^2}\ \text{km/h}$$

$$= \boxed{11.2\ \text{km/h}}$$

The direction of $\mathbf{v}_{bE}$ is

$$\theta = \tan^{-1}\left(\frac{v_{rE}}{v_{br}}\right) = \tan^{-1}\left(\frac{5.00}{10.0}\right) = 26.6°$$

The boat is moving at a speed of 11.2 km/h in the direction 26.6° east of north relative to the Earth.

Exercise If the width of the river is 3.0 km, find the time it takes the boat to cross it.

Answer 18 min.

Figure 4.22

EXAMPLE 4.10 ▶ Which Way Should We Head?

If the boat of the preceding example travels with the same speed of 10.0 km/h relative to the river and is to travel due north, as shown in Figure 4.23, what should its heading be?

Solution As in the previous example, we know $\mathbf{v}_{rE}$ and the magnitude of the vector $\mathbf{v}_{br}$, and we want $\mathbf{v}_{bE}$ to be directed across the river. Figure 4.23 shows that the boat must head upstream in order to travel directly northward across the river. Note the difference between the triangle in Figure 4.22 and the one in Figure 4.23—specifically, that the hypotenuse in Figure 4.23 is no longer $\mathbf{v}_{bE}$. Therefore, when we use the Pythagorean theorem to find $\mathbf{v}_{bE}$ this time, we obtain

$$v_{bE} = \sqrt{v_{br}^2 - v_{rE}^2} = \sqrt{(10.0)^2 - (5.00)^2}\ \text{km/h} = 8.66\ \text{km/h}$$

Now that we know the magnitude of $\mathbf{v}_{bE}$, we can find the direction in which the boat is heading:

$$\theta = \tan^{-1}\left(\frac{v_{rE}}{v_{bE}}\right) = \tan^{-1}\left(\frac{5.00}{8.66}\right) = \boxed{30.0°}$$

The boat must steer a course 30.0° west of north.

Exercise If the width of the river is 3.0 km, find the time it takes the boat to cross it.

Answer 21 min.

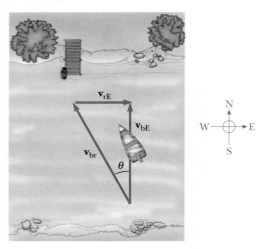

Figure 4.23

SUMMARY

If a particle moves with *constant* acceleration **a** and has velocity $\mathbf{v}_i$ and position $\mathbf{r}_i$ at $t = 0$, its velocity and position vectors at some later time t are

$$\mathbf{v}_f = \mathbf{v}_i + \mathbf{a}t \tag{4.8}$$

$$\mathbf{r}_f = \mathbf{r}_i + \mathbf{v}_i t + \tfrac{1}{2}\mathbf{a}t^2 \tag{4.9}$$

For two-dimensional motion in the xy plane under constant acceleration, each of these vector expressions is equivalent to two component expressions—one for the motion in the x direction and one for the motion in the y direction. You should be able to break the two-dimensional motion of any object into these two components.

Projectile motion is one type of two-dimensional motion under constant acceleration, where $a_x = 0$ and $a_y = -g$. It is useful to think of projectile motion as the superposition of two motions: (1) constant-velocity motion in the x direction and (2) free-fall motion in the vertical direction subject to a constant downward acceleration of magnitude $g = 9.80$ m/s². You should be able to analyze the motion in terms of separate horizontal and vertical components of velocity, as shown in Figure 4.24.

A particle moving in a circle of radius r with constant speed v is in **uniform circular motion**. It undergoes a centripetal (or radial) acceleration $\mathbf{a}_r$ because the direction of **v** changes in time. The magnitude of $\mathbf{a}_r$ is

$$a_r = \frac{v^2}{r} \tag{4.18}$$

and its direction is always toward the center of the circle.

If a particle moves along a curved path in such a way that both the magnitude and the direction of **v** change in time, then the particle has an acceleration vector that can be described by two component vectors: (1) a radial component vector $\mathbf{a}_r$ that causes the change in direction of **v** and (2) a tangential component vector $\mathbf{a}_t$ that causes the change in magnitude of **v**. The magnitude of $\mathbf{a}_r$ is v^2/r, and the magnitude of $\mathbf{a}_t$ is $d|\mathbf{v}|/dt$. You should be able to sketch motion diagrams for an object following a curved path and show how the velocity and acceleration vectors change as the object's motion varies.

The velocity **v** of a particle measured in a fixed frame of reference S can be related to the velocity $\mathbf{v}'$ of the same particle measured in a moving frame of reference S' by

$$\mathbf{v}' = \mathbf{v} - \mathbf{v}_0 \tag{4.21}$$

where $\mathbf{v}_0$ is the velocity of S' relative to S. You should be able to translate back and forth between different frames of reference.

Figure 4.24 Analyzing projectile motion in terms of horizontal and vertical components.

QUESTIONS

1. Can an object accelerate if its speed is constant? Can an object accelerate if its velocity is constant?

2. If the average velocity of a particle is zero in some time interval, what can you say about the displacement of the particle for that interval?

3. If you know the position vectors of a particle at two points along its path and also know the time it took to get from one point to the other, can you determine the particle's instantaneous velocity? Its average velocity? Explain.

4. Describe a situation in which the velocity of a particle is always perpendicular to the position vector.

5. Explain whether or not the following particles have an acceleration: (a) a particle moving in a straight line with constant speed and (b) a particle moving around a curve with constant speed.

6. Correct the following statement: "The racing car rounds the turn at a constant velocity of 90 mi/h."

7. Determine which of the following moving objects have an approximately parabolic trajectory: (a) a ball thrown in an arbitrary direction, (b) a jet airplane, (c) a rocket leaving the launching pad, (d) a rocket whose engines fail a few minutes after launch, (e) a tossed stone moving to the bottom of a pond.

8. A rock is dropped at the same instant that a ball at the same initial elevation is thrown horizontally. Which will have the greater speed when it reaches ground level?

9. A spacecraft drifts through space at a constant velocity. Suddenly, a gas leak in the side of the spacecraft causes a constant acceleration of the spacecraft in a direction perpendicular to the initial velocity. The orientation of the spacecraft does not change, and so the acceleration remains perpendicular to the original direction of the velocity. What is the shape of the path followed by the spacecraft in this situation?

10. A ball is projected horizontally from the top of a building. One second later another ball is projected horizontally from the same point with the same velocity. At what point in the motion will the balls be closest to each other? Will the first ball always be traveling faster than the second ball? How much time passes between the moment the first ball hits the ground and the moment the second one hits the ground? Can the horizontal projection velocity of the second ball be changed so that the balls arrive at the ground at the same time?

11. A student argues that as a satellite orbits the Earth in a circular path, the satellite moves with a constant velocity and therefore has no acceleration. The professor claims that the student is wrong because the satellite must have a centripetal acceleration as it moves in its circular orbit. What is wrong with the student's argument?

12. What is the fundamental difference between the unit vectors $\hat{\mathbf{r}}$ and $\hat{\boldsymbol{\theta}}$ and the unit vectors $\mathbf{i}$ and $\mathbf{j}$?

13. At the end of its arc, the velocity of a pendulum is zero. Is its acceleration also zero at this point?

14. If a rock is dropped from the top of a sailboat's mast, will it hit the deck at the same point regardless of whether the boat is at rest or in motion at constant velocity?

15. A stone is thrown upward from the top of a building. Does the stone's displacement depend on the location of the origin of the coordinate system? Does the stone's velocity depend on the location of the origin?

16. Is it possible for a vehicle to travel around a curve without accelerating? Explain.

17. A baseball is thrown with an initial velocity of $(10\mathbf{i} + 15\mathbf{j})$ m/s. When it reaches the top of its trajectory, what are (a) its velocity and (b) its acceleration? Neglect the effect of air resistance.

18. An object moves in a circular path with constant speed v. (a) Is the velocity of the object constant? (b) Is its acceleration constant? Explain.

19. A projectile is fired at some angle to the horizontal with some initial speed v_i, and air resistance is neglected. Is the projectile a freely falling body? What is its acceleration in the vertical direction? What is its acceleration in the horizontal direction?

20. A projectile is fired at an angle of 30° from the horizontal with some initial speed. Firing at what other projectile angle results in the same range if the initial speed is the same in both cases? Neglect air resistance.

21. A projectile is fired on the Earth with some initial velocity. Another projectile is fired on the Moon with the same initial velocity. If air resistance is neglected, which projectile has the greater range? Which reaches the greater altitude? (Note that the free-fall acceleration on the Moon is about 1.6 m/s².)

22. As a projectile moves through its parabolic trajectory, which of these quantities, if any, remain constant: (a) speed, (b) acceleration, (c) horizontal component of velocity, (d) vertical component of velocity?

23. A passenger on a train that is moving with constant velocity drops a spoon. What is the acceleration of the spoon relative to (a) the train and (b) the Earth?

PROBLEMS

1, **2**, **3** = straightforward, intermediate, challenging   ☐ = full solution available in the *Student Solutions Manual and Study Guide*
WEB = solution posted at **http://www.saunderscollege.com/physics/**   💻 = Computer useful in solving problem   🖥 = Interactive Physics
☐ = paired numerical/symbolic problems

Section 4.1   The Displacement, Velocity, and Acceleration Vectors

WEB **1.** A motorist drives south at 20.0 m/s for 3.00 min, then turns west and travels at 25.0 m/s for 2.00 min, and finally travels northwest at 30.0 m/s for 1.00 min. For this 6.00-min trip, find (a) the total vector displacement, (b) the average speed, and (c) the average velocity. Use a coordinate system in which east is the positive x axis.

2. Suppose that the position vector for a particle is given as $\mathbf{r} = x\mathbf{i} + y\mathbf{j}$, with $x = at + b$ and $y = ct^2 + d$, where $a = 1.00$ m/s, $b = 1.00$ m, $c = 0.125$ m/s^2, and $d = 1.00$ m. (a) Calculate the average velocity during the time interval from $t = 2.00$ s to $t = 4.00$ s. (b) Determine the velocity and the speed at $t = 2.00$ s.

3. A golf ball is hit off a tee at the edge of a cliff. Its x and y coordinates versus time are given by the following expressions:

$$x = (18.0 \text{ m/s})t$$

and

$$y = (4.00 \text{ m/s})t - (4.90 \text{ m/s}^2)t^2$$

(a) Write a vector expression for the ball's position as a function of time, using the unit vectors $\mathbf{i}$ and $\mathbf{j}$. By taking derivatives of your results, write expressions for (b) the velocity vector as a function of time and (c) the acceleration vector as a function of time. Now use unit vector notation to write expressions for (d) the position, (e) the velocity, and (f) the acceleration of the ball, all at $t = 3.00$ s.

4. The coordinates of an object moving in the xy plane vary with time according to the equations

$$x = -(5.00 \text{ m}) \sin \omega t$$

and

$$y = (4.00 \text{ m}) - (5.00 \text{ m})\cos \omega t$$

where t is in seconds and ω has units of seconds^{-1}. (a) Determine the components of velocity and components of acceleration at $t = 0$. (b) Write expressions for the position vector, the velocity vector, and the acceleration vector at any time $t > 0$. (c) Describe the path of the object on an xy graph.

Section 4.2   Two-Dimensional Motion with Constant Acceleration

5. At $t = 0$, a particle moving in the xy plane with constant acceleration has a velocity of $\mathbf{v}_i = (3.00\mathbf{i} - 2.00\mathbf{j})$ m/s when it is at the origin. At $t = 3.00$ s, the particle's velocity is $\mathbf{v} = (9.00\mathbf{i} + 7.00\mathbf{j})$ m/s. Find (a) the acceleration of the particle and (b) its coordinates at any time t.

6. The vector position of a particle varies in time according to the expression $\mathbf{r} = (3.00\mathbf{i} - 6.00t^2\mathbf{j})$ m. (a) Find expressions for the velocity and acceleration as functions of time. (b) Determine the particle's position and velocity at $t = 1.00$ s.

7. A fish swimming in a horizontal plane has velocity $\mathbf{v}_i = (4.00\mathbf{i} + 1.00\mathbf{j})$ m/s at a point in the ocean whose displacement from a certain rock is $\mathbf{r}_i = (10.0\mathbf{i} - 4.00\mathbf{j})$ m. After the fish swims with constant acceleration for 20.0 s, its velocity is $\mathbf{v} = (20.0\mathbf{i} - 5.00\mathbf{j})$ m/s. (a) What are the components of the acceleration? (b) What is the direction of the acceleration with respect to the unit vector $\mathbf{i}$? (c) Where is the fish at $t = 25.0$ s if it maintains its original acceleration and in what direction is it moving?

8. A particle initially located at the origin has an acceleration of $\mathbf{a} = 3.00\mathbf{j}$ m/s^2 and an initial velocity of $\mathbf{v}_i = 5.00\mathbf{i}$ m/s. Find (a) the vector position and velocity at any time t and (b) the coordinates and speed of the particle at $t = 2.00$ s.

Section 4.3   Projectile Motion

(Neglect air resistance in all problems and take $g = 9.80$ m/s^2.)

WEB **9.** In a local bar, a customer slides an empty beer mug down the counter for a refill. The bartender is momentarily distracted and does not see the mug, which slides off the counter and strikes the floor 1.40 m from the base of the counter. If the height of the counter is 0.860 m, (a) with what velocity did the mug leave the counter and (b) what was the direction of the mug's velocity just before it hit the floor?

10. In a local bar, a customer slides an empty beer mug down the counter for a refill. The bartender is momentarily distracted and does not see the mug, which slides off the counter and strikes the floor at distance d from the base of the counter. If the height of the counter is h, (a) with what velocity did the mug leave the counter and (b) what was the direction of the mug's velocity just before it hit the floor?

11. One strategy in a snowball fight is to throw a first snowball at a high angle over level ground. While your opponent is watching the first one, you throw a second one at a low angle and timed to arrive before or at the same time as the first one. Assume both snowballs are thrown with a speed of 25.0 m/s. The first one is thrown at an angle of 70.0° with respect to the horizontal. (a) At what angle should the second (low-angle) snowball be thrown if it is to land at the same point as the first? (b) How many seconds later should the second snowball be thrown if it is to land at the same time as the first?

12. A tennis player standing 12.6 m from the net hits the ball at 3.00° above the horizontal. To clear the net, the ball must rise at least 0.330 m. If the ball just clears the net at the apex of its trajectory, how fast was the ball moving when it left the racket?

13. An artillery shell is fired with an initial velocity of 300 m/s at 55.0° above the horizontal. It explodes on a mountainside 42.0 s after firing. What are the x and y coordinates of the shell where it explodes, relative to its firing point?

14. An astronaut on a strange planet finds that she can jump a maximum horizontal distance of 15.0 m if her initial speed is 3.00 m/s. What is the free-fall acceleration on the planet?

15. A projectile is fired in such a way that its horizontal range is equal to three times its maximum height. What is the angle of projection? Give your answer to three significant figures.

16. A ball is tossed from an upper-story window of a building. The ball is given an initial velocity of 8.00 m/s at an angle of 20.0° below the horizontal. It strikes the ground 3.00 s later. (a) How far horizontally from the base of the building does the ball strike the ground? (b) Find the height from which the ball was thrown. (c) How long does it take the ball to reach a point 10.0 m below the level of launching?

17. A cannon with a muzzle speed of 1 000 m/s is used to start an avalanche on a mountain slope. The target is 2 000 m from the cannon horizontally and 800 m above the cannon. At what angle, above the horizontal, should the cannon be fired?

18. Consider a projectile that is launched from the origin of an xy coordinate system with speed v_i at initial angle θ_i above the horizontal. Note that at the apex of its trajectory the projectile is moving horizontally, so that the slope of its path is zero. Use the expression for the trajectory given in Equation 4.12 to find the x coordinate that corresponds to the maximum height. Use this x coordinate and the symmetry of the trajectory to determine the horizontal range of the projectile.

19. A placekicker must kick a football from a point 36.0 m (about 40 yards) from the goal, and half the crowd hopes the ball will clear the crossbar, which is 3.05 m high. When kicked, the ball leaves the ground with a speed of 20.0 m/s at an angle of 53.0° to the horizontal. (a) By how much does the ball clear or fall short of clearing the crossbar? (b) Does the ball approach the crossbar while still rising or while falling?

20. A firefighter 50.0 m away from a burning building directs a stream of water from a fire hose at an angle of 30.0° above the horizontal, as in Figure P4.20. If the speed of the stream is 40.0 m/s, at what height will the water strike the building?

Figure P4.20   Problems 20 and 21. *(Photo by Frederick McKinney/FPG International)*

21. A firefighter a distance d from a burning building directs a stream of water from a fire hose at angle θ_i above the horizontal as in Figure P4.20. If the initial speed of the stream is v_i, at what height h does the water strike the building?

22. A soccer player kicks a rock horizontally off a cliff 40.0 m high into a pool of water. If the player hears the sound of the splash 3.00 s later, what was the initial speed given to the rock? Assume the speed of sound in air to be 343 m/s.

23. A basketball star covers 2.80 m horizontally in a jump to dunk the ball (Fig. P4.23). His motion through space can be modeled as that of a particle at a point called his center of mass (which we shall define in Chapter 9). His center of mass is at elevation 1.02 m when he leaves the floor. It reaches a maximum height of 1.85 m above the floor and is at elevation 0.900 m when he touches down again. Determine (a) his time of flight (his "hang time"), (b) his horizontal and (c) vertical velocity components at the instant of takeoff, and (d) his takeoff angle. (e) For comparison, determine the hang time of a whitetail deer making a jump with center-of-mass elevations $y_i = 1.20$ m, $y_{max} = 2.50$ m, $y_f = 0.700$ m.

Figure P4.23 *(Top, Ron Chapple/FPG International; bottom, Bill Lea/Dembinsky Photo Associates)*

Section 4.4 **Uniform Circular Motion**

24. The orbit of the Moon about the Earth is approximately circular, with a mean radius of 3.84×10^8 m. It takes 27.3 days for the Moon to complete one revolution about the Earth. Find (a) the mean orbital speed of the Moon and (b) its centripetal acceleration.

WEB 25. The athlete shown in Figure P4.25 rotates a 1.00-kg discus along a circular path of radius 1.06 m. The maximum speed of the discus is 20.0 m/s. Determine the magnitude of the maximum radial acceleration of the discus.

Figure P4.25 *(Sam Sargent/Liaison International)*

26. From information on the endsheets of this book, compute, for a point located on the surface of the Earth at the equator, the radial acceleration due to the rotation of the Earth about its axis.

27. A tire 0.500 m in radius rotates at a constant rate of 200 rev/min. Find the speed and acceleration of a small stone lodged in the tread of the tire (on its outer edge). (*Hint:* In one revolution, the stone travels a distance equal to the circumference of its path, $2\pi r$.)

28. During liftoff, Space Shuttle astronauts typically feel accelerations up to 1.4g, where $g = 9.80$ m/s². In their training, astronauts ride in a device where they experience such an acceleration as a centripetal acceleration. Specifically, the astronaut is fastened securely at the end of a mechanical arm that then turns at constant speed in a horizontal circle. Determine the rotation rate, in revolutions per second, required to give an astronaut a centripetal acceleration of 1.40g while the astronaut moves in a circle of radius 10.0 m.

29. Young David who slew Goliath experimented with slings before tackling the giant. He found that he could revolve a sling of length 0.600 m at the rate of 8.00 rev/s. If he increased the length to 0.900 m, he could revolve the sling only 6.00 times per second. (a) Which rate of rotation gives the greater speed for the stone at the end of the sling? (b) What is the centripetal acceleration of the stone at 8.00 rev/s? (c) What is the centripetal acceleration at 6.00 rev/s?

30. The astronaut orbiting the Earth in Figure P4.30 is preparing to dock with a Westar VI satellite. The satellite is in a circular orbit 600 km above the Earth's surface, where the free-fall acceleration is 8.21 m/s². The radius of the Earth is 6 400 km. Determine the speed of the satellite and the time required to complete one orbit around the Earth.

Figure P4.30   (Courtesy of NASA)

Section 4.5   Tangential and Radial Acceleration

31. A train slows down as it rounds a sharp horizontal curve, slowing from 90.0 km/h to 50.0 km/h in the 15.0 s that it takes to round the curve. The radius of the curve is 150 m. Compute the acceleration at the moment the train speed reaches 50.0 km/h. Assume that the train slows down at a uniform rate during the 15.0-s interval.

32. An automobile whose speed is increasing at a rate of 0.600 m/s² travels along a circular road of radius 20.0 m. When the instantaneous speed of the automobile is 4.00 m/s, find (a) the tangential acceleration component, (b) the radial acceleration component, and (c) the magnitude and direction of the total acceleration.

33. Figure P4.33 shows the total acceleration and velocity of a particle moving clockwise in a circle of radius 2.50 m

$a = 15.0 \text{ m/s}^2$

2.50 m

30.0°

v

a

Figure P4.33

at a given instant of time. At this instant, find (a) the radial acceleration, (b) the speed of the particle, and (c) its tangential acceleration.

34. A student attaches a ball to the end of a string 0.600 m in length and then swings the ball in a vertical circle. The speed of the ball is 4.30 m/s at its highest point and 6.50 m/s at its lowest point. Find the acceleration of the ball when the string is vertical and the ball is at (a) its highest point and (b) its lowest point.

35. A ball swings in a vertical circle at the end of a rope 1.50 m long. When the ball is 36.9° past the lowest point and on its way up, its total acceleration is $(-22.5\mathbf{i} + 20.2\mathbf{j})$ m/s². At that instant, (a) sketch a vector diagram showing the components of this acceleration, (b) determine the magnitude of its radial acceleration, and (c) determine the speed and velocity of the ball.

Section 4.6   Relative Velocity and Relative Acceleration

36. Heather in her Corvette accelerates at the rate of $(3.00\mathbf{i} - 2.00\mathbf{j})$ m/s², while Jill in her Jaguar accelerates at $(1.00\mathbf{i} + 3.00\mathbf{j})$ m/s². They both start from rest at the origin of an xy coordinate system. After 5.00 s, (a) what is Heather's speed with respect to Jill, (b) how far apart are they, and (c) what is Heather's acceleration relative to Jill?

37. A river has a steady speed of 0.500 m/s. A student swims upstream a distance of 1.00 km and swims back to the starting point. If the student can swim at a speed of 1.20 m/s in still water, how long does the trip take? Compare this with the time the trip would take if the water were still.

38. How long does it take an automobile traveling in the left lane at 60.0 km/h to pull alongside a car traveling in the right lane at 40.0 km/h if the cars' front bumpers are initially 100 m apart?

39. The pilot of an airplane notes that the compass indicates a heading due west. The airplane's speed relative to the air is 150 km/h. If there is a wind of 30.0 km/h toward the north, find the velocity of the airplane relative to the ground.

40. Two swimmers, Alan and Beth, start at the same point in a stream that flows with a speed v. Both move at the same speed c $(c > v)$ relative to the stream. Alan swims downstream a distance L and then upstream the same distance. Beth swims such that her motion relative to the ground is perpendicular to the banks of the stream. She swims a distance L in this direction and then back. The result of the motions of Alan and Beth is that they both return to the starting point. Which swimmer returns first? (*Note:* First guess at the answer.)

41. A child in danger of drowning in a river is being carried downstream by a current that has a speed of 2.50 km/h. The child is 0.600 km from shore and 0.800 km upstream of a boat landing when a rescue boat sets out. (a) If the boat proceeds at its maximum speed of 20.0 km/h relative to the water, what heading relative to the shore should the pilot take? (b) What angle does

the boat velocity make with the shore? (c) How long does it take the boat to reach the child?

42. A bolt drops from the ceiling of a train car that is accelerating northward at a rate of 2.50 m/s². What is the acceleration of the bolt relative to (a) the train car and (b) the Earth?

43. A science student is riding on a flatcar of a train traveling along a straight horizontal track at a constant speed of 10.0 m/s. The student throws a ball into the air along a path that he judges to make an initial angle of 60.0° with the horizontal and to be in line with the track. The student's professor, who is standing on the ground nearby, observes the ball to rise vertically. How high does she see the ball rise?

ADDITIONAL PROBLEMS

44. A ball is thrown with an initial speed v_i at an angle θ_i with the horizontal. The horizontal range of the ball is R, and the ball reaches a maximum height $R/6$. In terms of R and g, find (a) the time the ball is in motion, (b) the ball's speed at the peak of its path, (c) the initial vertical component of its velocity, (d) its initial speed, and (e) the angle θ_i. (f) Suppose the ball is thrown at the same initial speed found in part (d) but at the angle appropriate for reaching the maximum height. Find this height. (g) Suppose the ball is thrown at the same initial speed but at the angle necessary for maximum range. Find this range.

45. As some molten metal splashes, one droplet flies off to the east with initial speed v_i at angle θ_i above the horizontal, and another droplet flies off to the west with the same speed at the same angle above the horizontal, as in Figure P4.45. In terms of v_i and θ_i, find the distance between the droplets as a function of time.

Figure P4.45

46. A ball on the end of a string is whirled around in a horizontal circle of radius 0.300 m. The plane of the circle is 1.20 m above the ground. The string breaks and the ball lands 2.00 m (horizontally) away from the point on the ground directly beneath the ball's location when the string breaks. Find the radial acceleration of the ball during its circular motion.

47. A projectile is fired up an incline (incline angle ϕ) with an initial speed v_i at an angle θ_i with respect to the horizontal ($\theta_i > \phi$), as shown in Figure P4.47. (a) Show that the projectile travels a distance d up the incline, where

$$d = \frac{2v_i^2 \cos \theta_i \sin(\theta_i - \phi)}{g \cos^2 \phi}$$

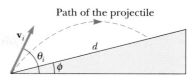

Figure P4.47

(b) For what value of θ_i is d a maximum, and what is that maximum value of d?

48. A student decides to measure the muzzle velocity of the pellets from his BB gun. He points the gun horizontally. On a vertical wall a distance x away from the gun, a target is placed. The shots hit the target a vertical distance y below the gun. (a) Show that the vertical displacement component of the pellets when traveling through the air is given by $y = Ax^2$, where A is a constant. (b) Express the constant A in terms of the initial velocity and the free-fall acceleration. (c) If $x = 3.00$ m and $y = 0.210$ m, what is the initial speed of the pellets?

49. A home run is hit in such a way that the baseball just clears a wall 21.0 m high, located 130 m from home plate. The ball is hit at an angle of 35.0° to the horizontal, and air resistance is negligible. Find (a) the initial speed of the ball, (b) the time it takes the ball to reach the wall, and (c) the velocity components and the speed of the ball when it reaches the wall. (Assume the ball is hit at a height of 1.00 m above the ground.)

50. An astronaut standing on the Moon fires a gun so that the bullet leaves the barrel initially moving in a horizontal direction. (a) What must be the muzzle speed of the bullet so that it travels completely around the Moon and returns to its original location? (b) How long does this trip around the Moon take? Assume that the free-fall acceleration on the Moon is one-sixth that on the Earth.

51. A pendulum of length 1.00 m swings in a vertical plane (Fig. 4.19). When the pendulum is in the two horizontal positions $\theta = 90°$ and $\theta = 270°$, its speed is 5.00 m/s. (a) Find the magnitude of the radial acceleration and tangential acceleration for these positions. (b) Draw a vector diagram to determine the direction of the total acceleration for these two positions. (c) Calculate the magnitude and direction of the total acceleration.

52. A basketball player who is 2.00 m tall is standing on the floor 10.0 m from the basket, as in Figure P4.52. If he shoots the ball at a 40.0° angle with the horizontal, at what initial speed must he throw so that it goes through the hoop without striking the backboard? The basket height is 3.05 m.

53. A particle has velocity components

$$v_x = +4 \text{ m/s} \qquad v_y = -(6 \text{ m/s}^2)t + 4 \text{ m/s}$$

Calculate the speed of the particle and the direction $\theta = \tan^{-1}(v_y/v_x)$ of the velocity vector at $t = 2.00$ s.

54. When baseball players throw the ball in from the outfield, they usually allow it to take one bounce before it reaches the infielder on the theory that the ball arrives

Figure P4.52

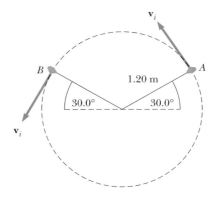

Figure P4.57

sooner that way. Suppose that the angle at which a bounced ball leaves the ground is the same as the angle at which the outfielder launched it, as in Figure P4.54, but that the ball's speed after the bounce is one half of what it was before the bounce. (a) Assuming the ball is always thrown with the same initial speed, at what angle θ should the ball be thrown in order to go the same distance D with one bounce (blue path) as a ball thrown upward at 45.0° with no bounce (green path)? (b) Determine the ratio of the times for the one-bounce and no-bounce throws.

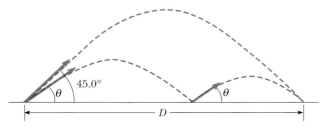

Figure P4.54

55. A boy can throw a ball a maximum horizontal distance of 40.0 m on a level field. How far can he throw the same ball vertically upward? Assume that his muscles give the ball the same speed in each case.

56. A boy can throw a ball a maximum horizontal distance of R on a level field. How far can he throw the same ball vertically upward? Assume that his muscles give the ball the same speed in each case.

57. A stone at the end of a sling is whirled in a vertical circle of radius 1.20 m at a constant speed $v_i = 1.50$ m/s as in Figure P4.57. The center of the string is 1.50 m above the ground. What is the range of the stone if it is released when the sling is inclined at 30.0° with the horizontal (a) at A? (b) at B? What is the acceleration of the stone (c) just before it is released at A? (d) just after it is released at A?

58. A quarterback throws a football straight toward a receiver with an initial speed of 20.0 m/s, at an angle of 30.0° above the horizontal. At that instant, the receiver is 20.0 m from the quarterback. In what direction and with what constant speed should the receiver run to catch the football at the level at which it was thrown?

59. A bomber is flying horizontally over level terrain, with a speed of 275 m/s relative to the ground, at an altitude of 3 000 m. Neglect the effects of air resistance. (a) How far will a bomb travel horizontally between its release from the plane and its impact on the ground? (b) If the plane maintains its original course and speed, where will it be when the bomb hits the ground? (c) At what angle from the vertical should the telescopic bombsight be set so that the bomb will hit the target seen in the sight at the time of release?

60. A person standing at the top of a hemispherical rock of radius R kicks a ball (initially at rest on the top of the rock) to give it horizontal velocity $\mathbf{v}_i$ as in Figure P4.60. (a) What must be its minimum initial speed if the ball is never to hit the rock after it is kicked? (b) With this initial speed, how far from the base of the rock does the ball hit the ground?

Figure P4.60

61. A hawk is flying horizontally at 10.0 m/s in a straight line, 200 m above the ground. A mouse it has been carrying struggles free from its grasp. The hawk continues on its path at the same speed for 2.00 s before attempting to retrieve its prey. To accomplish the retrieval, it dives in a straight line at constant speed and recaptures the mouse 3.00 m above the ground. (a) Assuming no air resistance, find the diving speed of the hawk. (b) What angle did the hawk make with the horizontal during its descent? (c) For how long did the mouse "enjoy" free fall?

62. A truck loaded with cannonball watermelons stops suddenly to avoid running over the edge of a washed-out bridge (Fig. P4.62). The quick stop causes a number of melons to fly off the truck. One melon rolls over the edge with an initial speed $v_i = 10.0$ m/s in the horizontal direction. A cross-section of the bank has the shape of the bottom half of a parabola with its vertex at the edge of the road, and with the equation $y^2 = 16x$, where x and y are measured in meters. What are the x and y coordinates of the melon when it splatters on the bank?

Figure P4.62

63. A catapult launches a rocket at an angle of 53.0° above the horizontal with an initial speed of 100 m/s. The rocket engine immediately starts a burn, and for 3.00 s the rocket moves along its initial line of motion with an acceleration of 30.0 m/s². Then its engine fails, and the rocket proceeds to move in free fall. Find (a) the maximum altitude reached by the rocket, (b) its total time of flight, and (c) its horizontal range.

64. A river flows with a uniform velocity **v**. A person in a motorboat travels 1.00 km upstream, at which time she passes a log floating by. Always with the same throttle setting, the boater continues to travel upstream for another 60.0 min and then returns downstream to her starting point, which she reaches just as the same log does. Find the velocity of the river. (*Hint:* The time of travel of the boat after it meets the log equals the time of travel of the log.)

WEB **65.** A car is parked on a steep incline overlooking the ocean, where the incline makes an angle of 37.0° below the horizontal. The negligent driver leaves the car in neutral, and the parking brakes are defective. The car rolls from rest down the incline with a constant acceleration of 4.00 m/s², traveling 50.0 m to the edge of a vertical cliff. The cliff is 30.0 m above the ocean. Find (a) the speed of the car when it reaches the edge of the cliff and the time it takes to get there, (b) the velocity of the car when it lands in the ocean, (c) the total time the car is in motion, and (d) the position of the car when it lands in the ocean, relative to the base of the cliff.

66. The determined coyote is out once more to try to capture the elusive roadrunner. The coyote wears a pair of Acme jet-powered roller skates, which provide a constant horizontal acceleration of 15.0 m/s² (Fig. P4.66). The coyote starts off at rest 70.0 m from the edge of a cliff at the instant the roadrunner zips past him in the direction of the cliff. (a) If the roadrunner moves with constant speed, determine the minimum speed he must have to reach the cliff before the coyote. At the brink of the cliff, the roadrunner escapes by making a sudden turn, while the coyote continues straight ahead. (b) If the cliff is 100 m above the floor of a canyon, determine where the coyote lands in the canyon (assume his skates remain horizontal and continue to operate when he is in "flight"). (c) Determine the components of the coyote's impact velocity.

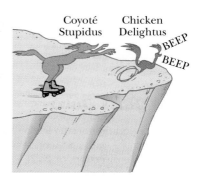

Coyoté   Chicken
Stupidus   Delightus   *BEEP*
                               BEEP

Figure P4.66

67. A skier leaves the ramp of a ski jump with a velocity of 10.0 m/s, 15.0° above the horizontal, as in Figure P4.67. The slope is inclined at 50.0°, and air resistance is negligible. Find (a) the distance from the ramp to where the jumper lands and (b) the velocity components just before the landing. (How do you think the results might be affected if air resistance were included? Note that jumpers lean forward in the shape of an airfoil, with their hands at their sides, to increase their distance. Why does this work?)

Figure P4.67

68. Two soccer players, Mary and Jane, begin running from nearly the same point at the same time. Mary runs in an easterly direction at 4.00 m/s, while Jane takes off in a direction 60.0° north of east at 5.40 m/s. (a) How long is it before they are 25.0 m apart? (b) What is the velocity of Jane relative to Mary? (c) How far apart are they after 4.00 s?

69. Do not hurt yourself; do not strike your hand against anything. Within these limitations, describe what you do to give your hand a large acceleration. Compute an order-of-magnitude estimate of this acceleration, stating the quantities you measure or estimate and their values.

70. An enemy ship is on the western side of a mountain island, as shown in Figure P4.70. The enemy ship can maneuver to within 2 500 m of the 1 800-m-high mountain peak and can shoot projectiles with an initial speed of 250 m/s. If the eastern shoreline is horizontally 300 m from the peak, what are the distances from the eastern shore at which a ship can be safe from the bombardment of the enemy ship?

Figure P4.70

ANSWERS TO QUICK QUIZZES

4.1 (a) Because acceleration occurs whenever the velocity changes in any way—with an increase or decrease in speed, a change in direction, or both—the brake pedal can also be considered an accelerator because it causes the car to slow down. The steering wheel is also an accelerator because it changes the direction of the velocity vector. (b) When the car is moving with constant speed, the gas pedal is not causing an acceleration; it is an accelerator only when it causes a change in the speedometer reading.

4.2 (a) At only one point—the peak of the trajectory—are the velocity and acceleration vectors perpendicular to each other. (b) If the object is thrown straight up or down, **v** and **a** are parallel to each other throughout the downward motion. Otherwise, the velocity and acceleration vectors are never parallel to each other. (c) The greater the maximum height, the longer it takes the projectile to reach that altitude and then fall back down from

it. So, as the angle increases from 0° to 90°, the time of flight increases. Therefore, the 15° angle gives the shortest time of flight, and the 75° angle gives the longest.

4.3 (a) Because the object is moving with a constant speed, the velocity vector is always the same length; because the motion is circular, this vector is always tangent to the circle. The only acceleration is that which changes the direction of the velocity vector; it points radially inward.

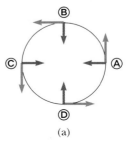

(a)

(b) Now there is a component of the acceleration vector that is tangent to the circle and points in the direction opposite the velocity. As a result, the acceleration vector does not point toward the center. The object is slowing down, and so the velocity vectors become shorter and shorter.

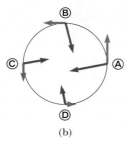

(b)

(c) Now the tangential component of the acceleration points in the same direction as the velocity. The object is speeding up, and so the velocity vectors become longer and longer. Because the speed changes rapidly here, but gradually in part (b), the acceleration vectors are longer here than in part (b).

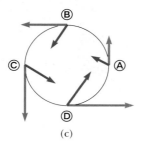

(c)

4.4 The motion diagram is as shown below. Note that each position vector points from the pivot point at the center of the circle to the position of the ball.

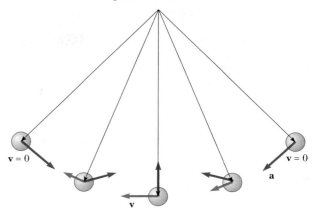

4.5 (a) The passenger sees the coffee pouring nearly vertically into the cup, just as if she were standing on the ground pouring it. (b) The stationary observer sees the coffee moving in a parabolic path with a constant horizontal velocity of 60 mi/h ($=88$ ft/s) and a downward acceleration of $-g$. If it takes the coffee 0.10 s to reach the cup, the stationary observer sees the coffee moving 8.8 ft horizontally before it hits the cup! (c) If the car slows suddenly, the coffee reaches the place where the cup *would have been* had there been no change in velocity and continues falling because the cup has not yet reached that location. If the car rapidly speeds up, the coffee falls behind the cup. If the car accelerates sideways, the coffee again ends up somewhere other than the cup.

PUZZLER

The *Spirit of Akron* is an airship that is more than 60 m long. When it is parked at an airport, one person can easily support it overhead using a single hand. Nonetheless, it is impossible for even a very strong adult to move the ship abruptly. What property of this huge airship makes it very difficult to cause any sudden changes in its motion? *(Courtesy of Edward E. Ogden)*

web

For more information about the airship, visit

http://www.goodyear.com/about/blimp

c h a p t e r

5

The Laws of Motion

In Chapters 2 and 4, we described motion in terms of displacement, velocity, and acceleration without considering what might cause that motion. What might cause one particle to remain at rest and another particle to accelerate? In this chapter, we investigate what causes changes in motion. The two main factors we need to consider are the forces acting on an object and the mass of the object. We discuss the three basic laws of motion, which deal with forces and masses and were formulated more than three centuries ago by Isaac Newton. Once we understand these laws, we can answer such questions as "What mechanism changes motion?" and "Why do some objects accelerate more than others?"

5.1 THE CONCEPT OF FORCE

Everyone has a basic understanding of the concept of force from everyday experience. When you push your empty dinner plate away, you exert a force on it. Similarly, you exert a force on a ball when you throw or kick it. In these examples, the word *force* is associated with muscular activity and some change in the velocity of an object. Forces do not always cause motion, however. For example, as you sit reading this book, the force of gravity acts on your body and yet you remain stationary. As a second example, you can push (in other words, exert a force) on a large boulder and not be able to move it.

What force (if any) causes the Moon to orbit the Earth? Newton answered this and related questions by stating that forces are what cause any change in the velocity of an object. Therefore, if an object moves with uniform motion (constant velocity), no force is required for the motion to be maintained. The Moon's velocity is not constant because it moves in a nearly circular orbit around the Earth. We now know that this change in velocity is caused by the force exerted on the Moon by the Earth. Because only a force can cause a change in velocity, we can think of force as *that which causes a body to accelerate.* In this chapter, we are concerned with the relationship between the force exerted on an object and the acceleration of that object.

A body accelerates because of an external force

What happens when several forces act simultaneously on an object? In this case, the object accelerates only if the net force acting on it is not equal to zero. The **net force** acting on an object is defined as the vector sum of all forces acting on the object. (We sometimes refer to the net force as the *total force,* the *resultant force,* or the *unbalanced force.*) **If the net force exerted on an object is zero, then the acceleration of the object is zero and its velocity remains constant.** That is, if the net force acting on the object is zero, then the object either remains at rest or continues to move with constant velocity. When the velocity of an object is constant (including the case in which the object remains at rest), the object is said to be in **equilibrium.**

Definition of equilibrium

When a coiled spring is pulled, as in Figure 5.1a, the spring stretches. When a stationary cart is pulled sufficently hard that friction is overcome, as in Figure 5.1b, the cart moves. When a football is kicked, as in Figure 5.1c, it is both deformed and set in motion. These situations are all examples of a class of forces called *contact forces.* That is, they involve physical contact between two objects. Other examples of contact forces are the force exerted by gas molecules on the walls of a container and the force exerted by your feet on the floor.

Another class of forces, known as *field forces,* do not involve physical contact between two objects but instead act through empty space. The force of gravitational attraction between two objects, illustrated in Figure 5.1d, is an example of this class of force. This gravitational force keeps objects bound to the Earth. The plan-

Figure 5.1 Some examples of applied forces. In each case a force is exerted on the object within the boxed area. Some agent in the environment external to the boxed area exerts a force on the object.

ets of our Solar System are bound to the Sun by the action of gravitational forces. Another common example of a field force is the electric force that one electric charge exerts on another, as shown in Figure 5.1e. These charges might be those of the electron and proton that form a hydrogen atom. A third example of a field force is the force a bar magnet exerts on a piece of iron, as shown in Figure 5.1f. The forces holding an atomic nucleus together also are field forces but are very short in range. They are the dominating interaction for particle separations of the order of 10^{-15} m.

Early scientists, including Newton, were uneasy with the idea that a force can act between two disconnected objects. To overcome this conceptual problem, Michael Faraday (1791–1867) introduced the concept of a *field*. According to this approach, when object 1 is placed at some point *P* near object 2, we say that object 1 interacts with object 2 by virtue of the gravitational field that exists at *P*. The gravitational field at *P* is created by object 2. Likewise, a gravitational field created by object 1 exists at the position of object 2. In fact, all objects create a gravitational field in the space around themselves.

The distinction between contact forces and field forces is not as sharp as you may have been led to believe by the previous discussion. When examined at the atomic level, all the forces we classify as contact forces turn out to be caused by electric (field) forces of the type illustrated in Figure 5.1e. Nevertheless, in developing models for macroscopic phenomena, it is convenient to use both classifications of forces. The only known *fundamental* forces in nature are all field forces: (1) gravitational forces between objects, (2) electromagnetic forces between electric charges, (3) strong nuclear forces between subatomic particles, and (4) weak nuclear forces that arise in certain radioactive decay processes. In classical physics, we are concerned only with gravitational and electromagnetic forces.

Measuring the Strength of a Force

It is convenient to use the deformation of a spring to measure force. Suppose we apply a vertical force to a spring scale that has a fixed upper end, as shown in Figure 5.2a. The spring elongates when the force is applied, and a pointer on the scale reads the value of the applied force. We can calibrate the spring by defining the unit force $\mathbf{F}_1$ as the force that produces a pointer reading of 1.00 cm. (Because force is a vector quantity, we use the bold-faced symbol $\mathbf{F}$.) If we now apply a different downward force $\mathbf{F}_2$ whose magnitude is 2 units, as seen in Figure 5.2b, the pointer moves to 2.00 cm. Figure 5.2c shows that the combined effect of the two collinear forces is the sum of the effects of the individual forces.

Now suppose the two forces are applied simultaneously with $\mathbf{F}_1$ downward and $\mathbf{F}_2$ horizontal, as illustrated in Figure 5.2d. In this case, the pointer reads $\sqrt{5} \text{ cm}^2 = 2.24$ cm. The single force $\mathbf{F}$ that would produce this same reading is the sum of the two vectors $\mathbf{F}_1$ and $\mathbf{F}_2$, as described in Figure 5.2d. That is, $|\mathbf{F}| = \sqrt{F_1{}^2 + F_2{}^2} = 2.24$ units, and its direction is $\theta = \tan^{-1}(-0.500) = -26.6°$. **Because forces are vector quantities, you must use the rules of vector addition to obtain the net force acting on an object.**

Figure 5.2 The vector nature of a force is tested with a spring scale. (a) A downward force $\mathbf{F}_1$ elongates the spring 1 cm. (b) A downward force $\mathbf{F}_2$ elongates the spring 2 cm. (c) When $\mathbf{F}_1$ and $\mathbf{F}_2$ are applied simultaneously, the spring elongates by 3 cm. (d) When $\mathbf{F}_1$ is downward and $\mathbf{F}_2$ is horizontal, the combination of the two forces elongates the spring $\sqrt{1^2 + 2^2}$ cm $= \sqrt{5}$ cm.

5.2 NEWTON'S FIRST LAW AND INERTIAL FRAMES

4.2   Before we state Newton's first law, consider the following simple experiment. Suppose a book is lying on a table. Obviously, the book remains at rest. Now imagine that you push the book with a horizontal force great enough to overcome the force of friction between book and table. (This force you exert, the force of friction, and any other forces exerted on the book by other objects are referred to as *external forces*.) You can keep the book in motion with constant velocity by applying a force that is just equal in magnitude to the force of friction and acts in the opposite direction. If you then push harder so that the magnitude of your applied force exceeds the magnitude of the force of friction, the book accelerates. If you stop pushing, the book stops after moving a short distance because the force of friction retards its motion. Suppose you now push the book across a smooth, highly waxed floor. The book again comes to rest after you stop pushing but not as quickly as before. Now imagine a floor so highly polished that friction is absent; in this case, the book, once set in motion, moves until it hits a wall.

Before about 1600, scientists felt that the natural state of matter was the state of rest. Galileo was the first to take a different approach to motion and the natural state of matter. He devised thought experiments, such as the one we just discussed for a book on a frictionless surface, and concluded that it is not the nature of an object to stop once set in motion: rather, it is its nature to *resist changes in its motion*. In his words, "Any velocity once imparted to a moving body will be rigidly maintained as long as the external causes of retardation are removed."

This new approach to motion was later formalized by Newton in a form that has come to be known as **Newton's first law of motion:**

> In the absence of external forces, an object at rest remains at rest and an object in motion continues in motion with a constant velocity (that is, with a constant speed in a straight line).

In simpler terms, we can say that **when no force acts on an object, the acceleration of the object is zero.** If nothing acts to change the object's motion, then its velocity does not change. From the first law, we conclude that any *isolated object* (one that does not interact with its environment) is either at rest or moving with constant velocity. The tendency of an object to resist any attempt to change its velocity is called the **inertia** of the object. Figure 5.3 shows one dramatic example of a consequence of Newton's first law.

Another example of uniform (constant-velocity) motion on a nearly frictionless surface is the motion of a light disk on a film of air (the lubricant), as shown in Figure 5.4. If the disk is given an initial velocity, it coasts a great distance before stopping.

Finally, consider a spaceship traveling in space and far removed from any planets or other matter. The spaceship requires some propulsion system to change its velocity. However, if the propulsion system is turned off when the spaceship reaches a velocity **v**, the ship coasts at that constant velocity and the astronauts get a free ride (that is, no propulsion system is required to keep them moving at the velocity **v**).

Inertial Frames

As we saw in Section 4.6, a moving object can be observed from any number of reference frames. Newton's first law, sometimes called the *law of inertia*, defines a special set of reference frames called *inertial frames*. **An inertial frame of reference**

QuickLab

Use a drinking straw to impart a strong, short-duration burst of air against a tennis ball as it rolls along a tabletop. Make the force perpendicular to the ball's path. What happens to the ball's motion? What is different if you apply a continuous force (constant magnitude and direction) that is directed along the direction of motion?

Newton's first law

Definition of inertia

Definition of inertial frame

Figure 5.3 Unless a net external force acts on it, an object at rest remains at rest and an object in motion continues in motion with constant velocity. In this case, the wall of the building did not exert a force on the moving train that was large enough to stop it. *(Roger Viollet, Mill Valley, CA, University Science Books, 1982)*

Isaac Newton **English physicist and mathematician (1642–1727)** Isaac Newton was one of the most brilliant scientists in history. Before the age of 30, he formulated the basic concepts and laws of mechanics, discovered the law of universal gravitation, and invented the mathematical methods of calculus. As a consequence of his theories, Newton was able to explain the motions of the planets, the ebb and flow of the tides, and many special features of the motions of the Moon and the Earth. He also interpreted many fundamental observations concerning the nature of light. His contributions to physical theories dominated scientific thought for two centuries and remain important today. *(Giraudon/Art Resource)*

is one that is not accelerating. Because Newton's first law deals only with objects that are not accelerating, it holds only in inertial frames. Any reference frame that moves with constant velocity relative to an inertial frame is itself an inertial frame. (The Galilean transformations given by Equations 4.20 and 4.21 relate positions and velocities between two inertial frames.)

A reference frame that moves with constant velocity relative to the distant stars is the best approximation of an inertial frame, and for our purposes we can consider planet Earth as being such a frame. The Earth is not really an inertial frame because of its orbital motion around the Sun and its rotational motion about its own axis. As the Earth travels in its nearly circular orbit around the Sun, it experiences an acceleration of about 4.4×10^{-3} m/s^2 directed toward the Sun. In addition, because the Earth rotates about its own axis once every 24 h, a point on the equator experiences an additional acceleration of 3.37×10^{-2} m/s^2 directed toward the center of the Earth. However, these accelerations are small compared with g and can often be neglected. For this reason, we assume that the Earth is an inertial frame, as is any other frame attached to it.

If an object is moving with constant velocity, an observer in one inertial frame (say, one at rest relative to the object) claims that the acceleration of the object and the resultant force acting on it are zero. An observer in *any other* inertial frame also finds that $\mathbf{a} = 0$ and $\Sigma\mathbf{F} = 0$ for the object. According to the first law, a body at rest and one moving with constant velocity are equivalent. A passenger in a car moving along a straight road at a constant speed of 100 km/h can easily pour coffee into a cup. But if the driver steps on the gas or brake pedal or turns the steering wheel while the coffee is being poured, the car accelerates and it is no longer an inertial frame. The laws of motion do not work as expected, and the coffee ends up in the passenger's lap!

Figure 5.4 Air hockey takes advantage of Newton's first law to make the game more exciting.

Quick Quiz 5.1

True or false: (a) It is possible to have motion in the absence of a force. (b) It is possible to have force in the absence of motion.

5.3 MASS

Imagine playing catch with either a basketball or a bowling ball. Which ball is more likely to keep moving when you try to catch it? Which ball has the greater tendency to remain motionless when you try to throw it? Because the bowling ball is more resistant to changes in its velocity, we say it has greater inertia than the basketball. As noted in the preceding section, inertia is a measure of how an object responds to an external force.

Definition of mass

Mass is that property of an object that specifies how much inertia the object has, and as we learned in Section 1.1, the SI unit of mass is the kilogram. The greater the mass of an object, the less that object accelerates under the action of an applied force. For example, if a given force acting on a 3-kg mass produces an acceleration of 4 m/s^2, then the same force applied to a 6-kg mass produces an acceleration of 2 m/s^2.

To describe mass quantitatively, we begin by comparing the accelerations a given force produces on different objects. Suppose a force acting on an object of mass m_1 produces an acceleration $\mathbf{a}_1$, and the *same force* acting on an object of mass m_2 produces an acceleration $\mathbf{a}_2$. The ratio of the two masses is defined as the *inverse* ratio of the magnitudes of the accelerations produced by the force:

$$\frac{m_1}{m_2} \equiv \frac{a_2}{a_1} \tag{5.1}$$

If one object has a known mass, the mass of the other object can be obtained from acceleration measurements.

Mass is an inherent property of an object and is independent of the object's surroundings and of the method used to measure it. Also, **mass is a scalar quantity** and thus obeys the rules of ordinary arithmetic. That is, several masses can be combined in simple numerical fashion. For example, if you combine a 3-kg mass with a 5-kg mass, their total mass is 8 kg. We can verify this result experimentally by comparing the acceleration that a known force gives to several objects separately with the acceleration that the same force gives to the same objects combined as one unit.

Mass and weight are different quantities

Mass should not be confused with weight. **Mass and weight are two different quantities.** As we see later in this chapter, the weight of an object is equal to the magnitude of the gravitational force exerted on the object and varies with location. For example, a person who weighs 180 lb on the Earth weighs only about 30 lb on the Moon. On the other hand, the mass of a body is the same everywhere: an object having a mass of 2 kg on the Earth also has a mass of 2 kg on the Moon.

5.4 NEWTON'S SECOND LAW

Newton's first law explains what happens to an object when no forces act on it. It either remains at rest or moves in a straight line with constant speed. Newton's second law answers the question of what happens to an object that has a nonzero resultant force acting on it.

Imagine pushing a block of ice across a frictionless horizontal surface. When you exert some horizontal force **F**, the block moves with some acceleration **a**. If you apply a force twice as great, the acceleration doubles. If you increase the applied force to 3**F**, the acceleration triples, and so on. From such observations, we conclude that **the acceleration of an object is directly proportional to the resultant force acting on it.**

The acceleration of an object also depends on its mass, as stated in the preceding section. We can understand this by considering the following experiment. If you apply a force **F** to a block of ice on a frictionless surface, then the block undergoes some acceleration **a**. If the mass of the block is doubled, then the same applied force produces an acceleration **a**/2. If the mass is tripled, then the same applied force produces an acceleration **a**/3, and so on. According to this observation, we conclude that **the magnitude of the acceleration of an object is inversely proportional to its mass.**

These observations are summarized in **Newton's second law:**

> The acceleration of an object is directly proportional to the net force acting on it and inversely proportional to its mass.

Newton's second law

Thus, we can relate mass and force through the following mathematical statement of Newton's second law:[1]

$$\sum \mathbf{F} = m\mathbf{a} \tag{5.2}$$

Note that this equation is a vector expression and hence is equivalent to three component equations:

$$\sum F_x = ma_x \qquad \sum F_y = ma_y \qquad \sum F_z = ma_z \tag{5.3}$$

Newton's second law— component form

Quick Quiz 5.2

Is there any relationship between the net force acting on an object and the direction in which the object moves?

Unit of Force

The SI unit of force is the **newton,** which is defined as the force that, when acting on a 1-kg mass, produces an acceleration of 1 m/s². From this definition and Newton's second law, we see that the newton can be expressed in terms of the following fundamental units of mass, length, and time:

$$1 \text{ N} \equiv 1 \text{ kg} \cdot \text{m/s}^2 \tag{5.4}$$

Definition of newton

In the British engineering system, the unit of force is the **pound,** which is defined as the force that, when acting on a 1-slug mass,[2] produces an acceleration of 1 ft/s²:

$$1 \text{ lb} \equiv 1 \text{ slug} \cdot \text{ft/s}^2 \tag{5.5}$$

A convenient approximation is that $1 \text{ N} \approx \frac{1}{4} \text{ lb}$.

[1] Equation 5.2 is valid only when the speed of the object is much less than the speed of light. We treat the relativistic situation in Chapter 39.

[2] The *slug* is the unit of mass in the British engineering system and is that system's counterpart of the SI unit the *kilogram.* Because most of the calculations in our study of classical mechanics are in SI units, the slug is seldom used in this text.

TABLE 5.1   Units of Force, Mass, and Acceleration[a]

System of Units	Mass	Acceleration	Force
SI	kg	m/s^2	$N = kg \cdot m/s^2$
British engineering	slug	ft/s^2	$lb = slug \cdot ft/s^2$

[a] 1 N = 0.225 lb.

The units of force, mass, and acceleration are summarized in Table 5.1.

We can now understand how a single person can hold up an airship but is not able to change its motion abruptly, as stated at the beginning of the chapter. The mass of the blimp is greater than 6 800 kg. In order to make this large mass accelerate appreciably, a very large force is required—certainly one much greater than a human can provide.

EXAMPLE 5.1 ▶ An Accelerating Hockey Puck

A hockey puck having a mass of 0.30 kg slides on the horizontal, frictionless surface of an ice rink. Two forces act on the puck, as shown in Figure 5.5. The force $\mathbf{F}_1$ has a magnitude of 5.0 N, and the force $\mathbf{F}_2$ has a magnitude of 8.0 N. Determine both the magnitude and the direction of the puck's acceleration.

Solution   The resultant force in the x direction is

$$\sum F_x = F_{1x} + F_{2x} = F_1 \cos(-20°) + F_2 \cos 60°$$
$$= (5.0 \text{ N})(0.940) + (8.0 \text{ N})(0.500) = 8.7 \text{ N}$$

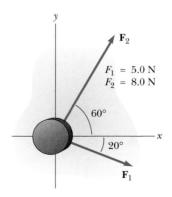

Figure 5.5   A hockey puck moving on a frictionless surface accelerates in the direction of the resultant force $\mathbf{F}_1 + \mathbf{F}_2$.

The resultant force in the y direction is

$$\sum F_y = F_{1y} + F_{2y} = F_1 \sin(-20°) + F_2 \sin 60°$$
$$= (5.0 \text{ N})(-0.342) + (8.0 \text{ N})(0.866) = 5.2 \text{ N}$$

Now we use Newton's second law in component form to find the x and y components of acceleration:

$$a_x = \frac{\sum F_x}{m} = \frac{8.7 \text{ N}}{0.30 \text{ kg}} = 29 \text{ m/s}^2$$

$$a_y = \frac{\sum F_y}{m} = \frac{5.2 \text{ N}}{0.30 \text{ kg}} = 17 \text{ m/s}^2$$

The acceleration has a magnitude of

$$a = \sqrt{(29)^2 + (17)^2} \text{ m/s}^2 = \boxed{34 \text{ m/s}^2}$$

and its direction relative to the positive x axis is

$$\theta = \tan^{-1}\left(\frac{a_y}{a_x}\right) = \tan^{-1}\left(\frac{17}{29}\right) = \boxed{30°}$$

We can graphically add the vectors in Figure 5.5 to check the reasonableness of our answer. Because the acceleration vector is along the direction of the resultant force, a drawing showing the resultant force helps us check the validity of the answer.

Exercise   Determine the components of a third force that, when applied to the puck, causes it to have zero acceleration.

Answer $F_{3x} = -8.7 \text{ N}, F_{3y} = -5.2 \text{ N}$.

5.5 ▷ THE FORCE OF GRAVITY AND WEIGHT

We are well aware that all objects are attracted to the Earth. The attractive force exerted by the Earth on an object is called the **force of gravity** $\mathbf{F}_g$. This force is directed toward the center of the Earth,[3] and its magnitude is called the **weight** of the object.

We saw in Section 2.6 that a freely falling object experiences an acceleration $\mathbf{g}$ acting toward the center of the Earth. Applying Newton's second law $\Sigma\mathbf{F} = m\mathbf{a}$ to a freely falling object of mass m, with $\mathbf{a} = \mathbf{g}$ and $\Sigma\mathbf{F} = \mathbf{F}_g$, we obtain

$$\mathbf{F}_g = m\mathbf{g} \tag{5.6}$$

Thus, the weight of an object, being defined as the magnitude of $\mathbf{F}_g$, is mg. (You should not confuse the italicized symbol g for gravitational acceleration with the nonitalicized symbol g used as the abbreviation for "gram.")

Because it depends on g, weight varies with geographic location. Hence, weight, unlike mass, is not an inherent property of an object. Because g decreases with increasing distance from the center of the Earth, bodies weigh less at higher altitudes than at sea level. For example, a 1 000-kg palette of bricks used in the construction of the Empire State Building in New York City weighed about 1 N less by the time it was lifted from sidewalk level to the top of the building. As another example, suppose an object has a mass of 70.0 kg. Its weight in a location where $g = 9.80$ m/s^2 is $F_g = mg = 686$ N (about 150 lb). At the top of a mountain, however, where $g = 9.77$ m/s^2, its weight is only 684 N. Therefore, if you want to lose weight without going on a diet, climb a mountain or weigh yourself at 30 000 ft during an airplane flight!

Because weight $= F_g = mg$, we can compare the masses of two objects by measuring their weights on a spring scale. At a given location, the ratio of the weights of two objects equals the ratio of their masses.

Definition of weight

QuickLab ▷

Drop a pen and your textbook simultaneously from the same height and watch as they fall. How can they have the same acceleration when their weights are so different?

The life-support unit strapped to the back of astronaut Edwin Aldrin weighed 300 lb on the Earth. During his training, a 50-lb mock-up was used. Although this effectively simulated the reduced weight the unit would have on the Moon, it did not correctly mimic the unchanging mass. It was just as difficult to accelerate the unit (perhaps by jumping or twisting suddenly) on the Moon as on the Earth. *(Courtesy of NASA)*

---

[3] This statement ignores the fact that the mass distribution of the Earth is not perfectly spherical.

CONCEPTUAL EXAMPLE 5.2   How Much Do You Weigh in an Elevator?

You have most likely had the experience of standing in an elevator that accelerates upward as it moves toward a higher floor. In this case, you feel heavier. In fact, if you are standing on a bathroom scale at the time, the scale measures a force magnitude that is greater than your weight. Thus, you have tactile and measured evidence that leads you to believe you are heavier in this situation. *Are* you heavier?

Solution   No, your weight is unchanged. To provide the acceleration upward, the floor or scale must exert on your feet an upward force that is greater in magnitude than your weight. It is this greater force that you feel, which you interpret as feeling heavier. The scale reads this upward force, not your weight, and so its reading increases.

Quick Quiz 5.3

A baseball of mass m is thrown upward with some initial speed. If air resistance is neglected, what forces are acting on the ball when it reaches (a) half its maximum height and (b) its maximum height?

5.6   NEWTON'S THIRD LAW

4.5   If you press against a corner of this textbook with your fingertip, the book pushes back and makes a small dent in your skin. If you push harder, the book does the same and the dent in your skin gets a little larger. This simple experiment illustrates a general principle of critical importance known as **Newton's third law:**

Newton's third law

> If two objects interact, the force $\mathbf{F}_{12}$ exerted by object 1 on object 2 is equal in magnitude to and opposite in direction to the force $\mathbf{F}_{21}$ exerted by object 2 on object 1:
>
> $$\mathbf{F}_{12} = -\mathbf{F}_{21} \tag{5.7}$$

This law, which is illustrated in Figure 5.6a, states that a force that affects the motion of an object must come from a second, *external,* object. The external object, in turn, is subject to an equal-magnitude but oppositely directed force exerted on it.

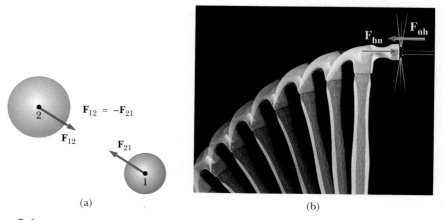

Figure 5.6   Newton's third law. (a) The force $\mathbf{F}_{12}$ exerted by object 1 on object 2 is equal in magnitude to and opposite in direction to the force $\mathbf{F}_{21}$ exerted by object 2 on object 1. (b) The force $\mathbf{F}_{hn}$ exerted by the hammer on the nail is equal to and opposite the force $\mathbf{F}_{nh}$ exerted by the nail on the hammer. *(John Gillmoure/The Stock Market)*

This is equivalent to stating that **a single isolated force cannot exist.** The force that object 1 exerts on object 2 is sometimes called the *action force,* while the force object 2 exerts on object 1 is called the *reaction force.* In reality, either force can be labeled the action or the reaction force. **The action force is equal in magnitude to the reaction force and opposite in direction. In all cases, the action and reaction forces act on different objects.** For example, the force acting on a freely falling projectile is $\mathbf{F}_g = m\mathbf{g}$, which is the force of gravity exerted by the Earth on the projectile. The reaction to this force is the force exerted by the projectile on the Earth, $\mathbf{F}_g' = -\mathbf{F}_g$. The reaction force $\mathbf{F}_g'$ accelerates the Earth toward the projectile just as the action force $\mathbf{F}_g$ accelerates the projectile toward the Earth. However, because the Earth has such a great mass, its acceleration due to this reaction force is negligibly small.

Another example of Newton's third law is shown in Figure 5.6b. The force exerted by the hammer on the nail (the action force $\mathbf{F}_{hn}$) is equal in magnitude and opposite in direction to the force exerted by the nail on the hammer (the reaction force $\mathbf{F}_{nh}$). It is this latter force that causes the hammer to stop its rapid forward motion when it strikes the nail.

You experience Newton's third law directly whenever you slam your fist against a wall or kick a football. You should be able to identify the action and reaction forces in these cases.

Compression of a football as the force exerted by a player's foot sets the ball in motion. *(Ralph Cowan/ Tony Stone Images)*

Quick Quiz 5.4

A person steps from a boat toward a dock. Unfortunately, he forgot to tie the boat to the dock, and the boat scoots away as he steps from it. Analyze this situation in terms of Newton's third law.

The force of gravity $\mathbf{F}_g$ was defined as the attractive force the Earth exerts on an object. If the object is a TV at rest on a table, as shown in Figure 5.7a, why does the TV not accelerate in the direction of $\mathbf{F}_g$? The TV does not accelerate because the table holds it up. What is happening is that the table exerts on the TV an upward force $\mathbf{n}$ called the **normal force.**[4] The normal force is a contact force that prevents the TV from falling through the table and can have any magnitude needed to balance the downward force $\mathbf{F}_g$, up to the point of breaking the table. If someone stacks books on the TV, the normal force exerted by the table on the TV increases. If someone lifts up on the TV, the normal force exerted by the table on the TV decreases. (The normal force becomes zero if the TV is raised off the table.)

The two forces in an action–reaction pair **always act on different objects.** For the hammer-and-nail situation shown in Figure 5.6b, one force of the pair acts on the hammer and the other acts on the nail. For the unfortunate person stepping out of the boat in Quick Quiz 5.4, one force of the pair acts on the person, and the other acts on the boat.

For the TV in Figure 5.7, the force of gravity $\mathbf{F}_g$ and the normal force $\mathbf{n}$ are *not* an action–reaction pair because they act on the same body—the TV. The two reaction forces in this situation—$\mathbf{F}_g'$ and $\mathbf{n}'$—are exerted on objects other than the TV. Because the reaction to $\mathbf{F}_g$ is the force $\mathbf{F}_g'$ exerted by the TV on the Earth and the reaction to $\mathbf{n}$ is the force $\mathbf{n}'$ exerted by the TV on the table, we conclude that

$$\mathbf{F}_g = -\mathbf{F}_g' \qquad \text{and} \qquad \mathbf{n} = -\mathbf{n}'$$

Definition of normal force

---

[4] *Normal* in this context means *perpendicular.*

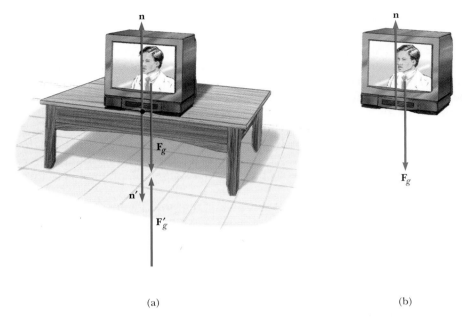

(a)   (b)

Figure 5.7   When a TV is at rest on a table, the forces acting on the TV are the normal force **n** and the force of gravity $\mathbf{F}_g$, as illustrated in part (b). The reaction to **n** is the force **n**′ exerted by the TV on the table. The reaction to $\mathbf{F}_g$ is the force $\mathbf{F}_g'$ exerted by the TV on the Earth.

The forces **n** and **n**′ have the same magnitude, which is the same as that of $\mathbf{F}_g$ until the table breaks. From the second law, we see that, because the TV is in equilibrium (**a** = 0), it follows[5] that $F_g = n = mg$.

Quick Quiz 5.5

If a fly collides with the windshield of a fast-moving bus, (a) which experiences the greater impact force: the fly or the bus, or is the same force experienced by both? (b) Which experiences the greater acceleration: the fly or the bus, or is the same acceleration experienced by both?

CONCEPTUAL EXAMPLE 5.3 ▸ You Push Me and I'll Push You

A large man and a small boy stand facing each other on frictionless ice. They put their hands together and push against each other so that they move apart. (a) Who moves away with the higher speed?

Solution   This situation is similar to what we saw in Quick Quiz 5.5. According to Newton's third law, the force exerted by the man on the boy and the force exerted by the boy on the man are an action–reaction pair, and so they must be equal in magnitude. (A bathroom scale placed between their hands would read the same, regardless of which way it faced.)

Therefore, the boy, having the lesser mass, experiences the greater acceleration. Both individuals accelerate for the same amount of time, but the greater acceleration of the boy over this time interval results in his moving away from the interaction with the higher speed.

(b) Who moves farther while their hands are in contact?

Solution   Because the boy has the greater acceleration, he moves farther during the interval in which the hands are in contact.

---

[5] Technically, we should write this equation in the component form $F_{gy} = n_y = mg_y$. This component notation is cumbersome, however, and so in situations in which a vector is parallel to a coordinate axis, we usually do not include the subscript for that axis because there is no other component.

5.7 ▷ SOME APPLICATIONS OF NEWTON'S LAWS

4.6 In this section we apply Newton's laws to objects that are either in equilibrium ($\mathbf{a} = 0$) or accelerating along a straight line under the action of constant external forces. We assume that the objects behave as particles so that we need not worry about rotational motion. We also neglect the effects of friction in those problems involving motion; this is equivalent to stating that the surfaces are *frictionless*. Finally, we usually neglect the mass of any ropes involved. In this approximation, the magnitude of the force exerted at any point along a rope is the same at all points along the rope. In problem statements, the synonymous terms *light, lightweight,* and *of negligible mass* are used to indicate that a mass is to be ignored when you work the problems.

When we apply Newton's laws to an object, we are interested only in external forces that act on the object. For example, in Figure 5.7 the only external forces acting on the TV are $\mathbf{n}$ and $\mathbf{F}_g$. The reactions to these forces, $\mathbf{n}'$ and $\mathbf{F}'_g$, act on the table and on the Earth, respectively, and therefore do not appear in Newton's second law applied to the TV.

When a rope attached to an object is pulling on the object, the rope exerts a force $\mathbf{T}$ on the object, and the magnitude of that force is called the **tension** in the rope. Because it is the magnitude of a vector quantity, tension is a scalar quantity.

Consider a crate being pulled to the right on a frictionless, horizontal surface, as shown in Figure 5.8a. Suppose you are asked to find the acceleration of the crate and the force the floor exerts on it. First, note that the horizontal force being applied to the crate acts through the rope. Use the symbol $\mathbf{T}$ to denote the force exerted by the rope on the crate. The magnitude of $\mathbf{T}$ is equal to the tension in the rope. A dotted circle is drawn around the crate in Figure 5.8a to remind you that you are interested only in the forces acting on the crate. These are illustrated in Figure 5.8b. In addition to the force $\mathbf{T}$, this force diagram for the crate includes the force of gravity $\mathbf{F}_g$ and the normal force $\mathbf{n}$ exerted by the floor on the crate. Such a force diagram, referred to as a **free-body diagram,** shows all external forces acting on the object. The construction of a correct free-body diagram is an important step in applying Newton's laws. The *reactions* to the forces we have listed—namely, the force exerted by the crate on the rope, the force exerted by the crate on the Earth, and the force exerted by the crate on the floor—are *not* included in the free-body diagram because they act on *other* bodies and not on the crate.

We can now apply Newton's second law in component form to the crate. The only force acting in the *x* direction is $\mathbf{T}$. Applying $\Sigma F_x = ma_x$ to the horizontal motion gives

$$\sum F_x = T = ma_x \qquad \text{or} \qquad a_x = \frac{T}{m}$$

No acceleration occurs in the *y* direction. Applying $\Sigma F_y = ma_y$ with $a_y = 0$ yields

$$n + (-F_g) = 0 \qquad \text{or} \qquad n = F_g$$

That is, the normal force has the same magnitude as the force of gravity but is in the opposite direction.

If $\mathbf{T}$ is a constant force, then the acceleration $a_x = T/m$ also is constant. Hence, the constant-acceleration equations of kinematics from Chapter 2 can be used to obtain the crate's displacement Δx and velocity v_x as functions of time. Be-

Tension

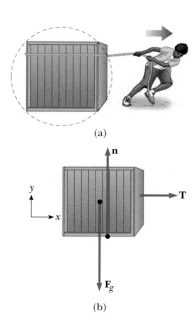

(a)

(b)

Figure 5.8 (a) A crate being pulled to the right on a frictionless surface. (b) The free-body diagram representing the external forces acting on the crate.

Figure 5.9 When one object pushes downward on another object with a force **F**, the normal force **n** is greater than the force of gravity: $n = F_g + F$.

cause $a_x = T/m$ = constant, Equations 2.8 and 2.11 can be written as

$$v_{xf} = v_{xi} + \left(\frac{T}{m}\right)t$$

$$\Delta x = v_{xi}t + \frac{1}{2}\left(\frac{T}{m}\right)t^2$$

In the situation just described, the magnitude of the normal force **n** is equal to the magnitude of $\mathbf{F}_g$, but this is not always the case. For example, suppose a book is lying on a table and you push down on the book with a force **F**, as shown in Figure 5.9. Because the book is at rest and therefore not accelerating, $\Sigma F_y = 0$, which gives $n - F_g - F = 0$, or $n = F_g + F$. Other examples in which $n \neq F_g$ are presented later.

Consider a lamp suspended from a light chain fastened to the ceiling, as in Figure 5.10a. The free-body diagram for the lamp (Figure 5.10b) shows that the forces acting on the lamp are the downward force of gravity $\mathbf{F}_g$ and the upward force **T** exerted by the chain. If we apply the second law to the lamp, noting that $\mathbf{a} = 0$, we see that because there are no forces in the x direction, $\Sigma F_x = 0$ provides no helpful information. The condition $\Sigma F_y = ma_y = 0$ gives

$$\sum F_y = T - F_g = 0 \qquad \text{or} \qquad T = F_g$$

Again, note that **T** and $\mathbf{F}_g$ are *not* an action–reaction pair because they act on the same object—the lamp. The reaction force to **T** is **T**′, the downward force exerted by the lamp on the chain, as shown in Figure 5.10c. The ceiling exerts on the chain a force **T**″ that is equal in magnitude to the magnitude of **T**′ and points in the opposite direction.

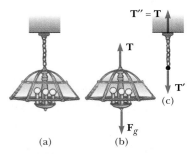

Figure 5.10 (a) A lamp suspended from a ceiling by a chain of negligible mass. (b) The forces acting on the lamp are the force of gravity $\mathbf{F}_g$ and the force exerted by the chain **T**. (c) The forces acting on the chain are the force exerted by the lamp **T**′ and the force exerted by the ceiling **T**″.

Problem-Solving Hints

Applying Newton's Laws

The following procedure is recommended when dealing with problems involving Newton's laws:

- Draw a simple, neat diagram of the system.
- Isolate the object whose motion is being analyzed. Draw a free-body diagram for this object. For systems containing more than one object, draw *separate* free-body diagrams for each object. *Do not* include in the free-body diagram forces exerted by the object on its surroundings. Establish convenient coordinate axes for each object and find the components of the forces along these axes.
- Apply Newton's second law, $\Sigma\mathbf{F} = m\mathbf{a}$, in component form. Check your dimensions to make sure that all terms have units of force.
- Solve the component equations for the unknowns. Remember that you must have as many independent equations as you have unknowns to obtain a complete solution.
- Make sure your results are consistent with the free-body diagram. Also check the predictions of your solutions for extreme values of the variables. By doing so, you can often detect errors in your results.

EXAMPLE 5.4   A Traffic Light at Rest

A traffic light weighing 125 N hangs from a cable tied to two other cables fastened to a support. The upper cables make angles of 37.0° and 53.0° with the horizontal. Find the tension in the three cables.

Solution   Figure 5.11a shows the type of drawing we might make of this situation. We then construct two free-body diagrams—one for the traffic light, shown in Figure 5.11b, and one for the knot that holds the three cables together, as seen in Figure 5.11c. This knot is a convenient object to choose because all the forces we are interested in act through it. Because the acceleration of the system is zero, we know that the net force on the light and the net force on the knot are both zero.

In Figure 5.11b the force $\mathbf{T}_3$ exerted by the vertical cable supports the light, and so $T_3 = F_g = \boxed{125 \text{ N.}}$ Next, we choose the coordinate axes shown in Figure 5.11c and resolve the forces acting on the knot into their components:

Force	*x* Component	*y* Component
$\mathbf{T}_1$	$- T_1 \cos 37.0°$	$T_1 \sin 37.0°$
$\mathbf{T}_2$	$T_2 \cos 53.0°$	$T_2 \sin 53.0°$
$\mathbf{T}_3$	0	-125 N

Knowing that the knot is in equilibrium ($\mathbf{a} = 0$) allows us to write

(1) $\sum F_x = - T_1 \cos 37.0° + T_2 \cos 53.0° = 0$

(2) $\sum F_y = T_1 \sin 37.0° + T_2 \sin 53.0°$
    $+ (- 125 \text{ N}) = 0$

From (1) we see that the horizontal components of $\mathbf{T}_1$ and $\mathbf{T}_2$ must be equal in magnitude, and from (2) we see that the sum of the vertical components of $\mathbf{T}_1$ and $\mathbf{T}_2$ must balance the weight of the light. We solve (1) for T_2 in terms of T_1 to obtain

$$T_2 = T_1 \left(\frac{\cos 37.0°}{\cos 53.0°} \right) = 1.33 T_1$$

This value for T_2 is substituted into (2) to yield

$$T_1 \sin 37.0° + (1.33 T_1)(\sin 53.0°) - 125 \text{ N} = 0$$

$$T_1 = \boxed{75.1 \text{ N}}$$

$$T_2 = 1.33 T_1 = \boxed{99.9 \text{ N}}$$

This problem is important because it combines what we have learned about vectors with the new topic of forces. The general approach taken here is very powerful, and we will repeat it many times.

Exercise   In what situation does $T_1 = T_2$?

Answer   When the two cables attached to the support make equal angles with the horizontal.

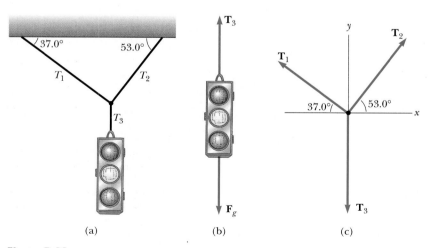

(a)   (b)   (c)

Figure 5.11   (a) A traffic light suspended by cables. (b) Free-body diagram for the traffic light. (c) Free-body diagram for the knot where the three cables are joined.

CONCEPTUAL EXAMPLE 5.5   Forces Between Cars in a Train

In a train, the cars are connected by *couplers*, which are under tension as the locomotive pulls the train. As you move down the train from locomotive to caboose, does the tension in the couplers increase, decrease, or stay the same as the train speeds up? When the engineer applies the brakes, the couplers are under compression. How does this compression force vary from locomotive to caboose? (Assume that only the brakes on the wheels of the engine are applied.)

Solution   As the train speeds up, the tension decreases from the front of the train to the back. The coupler between the locomotive and the first car must apply enough force to accelerate all of the remaining cars. As you move back along the train, each coupler is accelerating less mass behind it. The last coupler has to accelerate only the caboose, and so it is under the least tension.

When the brakes are applied, the force again decreases from front to back. The coupler connecting the locomotive to the first car must apply a large force to slow down all the remaining cars. The final coupler must apply a force large enough to slow down only the caboose.

EXAMPLE 5.6   Crate on a Frictionless Incline

A crate of mass m is placed on a frictionless inclined plane of angle θ. (a) Determine the acceleration of the crate after it is released.

Solution   Because we know the forces acting on the crate, we can use Newton's second law to determine its acceleration. (In other words, we have classified the problem; this gives us a hint as to the approach to take.) We make a sketch as in Figure 5.12a and then construct the free-body diagram for the crate, as shown in Figure 5.12b. The only forces acting on the crate are the normal force **n** exerted by the inclined plane, which acts perpendicular to the plane, and the force of gravity $\mathbf{F}_g = m\mathbf{g}$, which acts vertically downward. For problems involving inclined planes, it is convenient to choose the coordinate axes with x downward along the incline and y perpendicular to it, as shown in Figure 5.12b. (It is possible to solve the problem with "standard" horizontal and vertical axes. You may want to try this, just for practice.) Then, we re-place the force of gravity by a component of magnitude $mg \sin \theta$ along the positive x axis and by one of magnitude $mg \cos \theta$ along the negative y axis.

Now we apply Newton's second law in component form, noting that $a_y = 0$:

$$(1) \qquad \sum F_x = mg \sin \theta = ma_x$$

$$(2) \qquad \sum F_y = n - mg \cos \theta = 0$$

Solving (1) for a_x, we see that the acceleration along the incline is caused by the component of $\mathbf{F}_g$ directed down the incline:

$$(3) \qquad \boxed{a_x = g \sin \theta}$$

Note that this acceleration component is independent of the mass of the crate! It depends only on the angle of inclination and on g.

From (2) we conclude that the component of $\mathbf{F}_g$ perpendicular to the incline is balanced by the normal force; that is, $n = mg \cos \theta$. This is one example of a situation in which the normal force is *not* equal in magnitude to the weight of the object.

Special Cases   Looking over our results, we see that in the extreme case of $\theta = 90°$, $a_x = g$ and $n = 0$. This condition corresponds to the crate's being in free fall. When $\theta = 0$, $a_x = 0$ and $n = mg$ (its maximum value); in this case, the crate is sitting on a horizontal surface.

(b) Suppose the crate is released from rest at the top of the incline, and the distance from the front edge of the crate to the bottom is d. How long does it take the front edge to reach the bottom, and what is its speed just as it gets there?

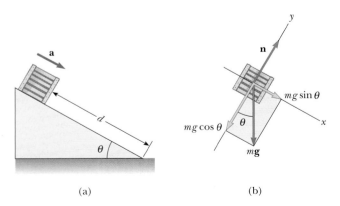

(a)   (b)

Figure 5.12   (a) A crate of mass m sliding down a frictionless incline. (b) The free-body diagram for the crate. Note that its acceleration along the incline is $g \sin \theta$.

Solution   Because $a_x =$ constant, we can apply Equation 2.11, $x_f - x_i = v_{xi}t + \frac{1}{2}a_x t^2$, to analyze the crate's motion.

With the displacement $x_f - x_i = d$ and $v_{xi} = 0$, we obtain

$$d = \tfrac{1}{2}a_x t^2$$

$$(4) \qquad t = \sqrt{\frac{2d}{a_x}} = \boxed{\sqrt{\frac{2d}{g \sin \theta}}}$$

Using Equation 2.12, $v_{xf}^2 = v_{xi}^2 + 2a_x(x_f - x_i)$, with $v_{xi} = 0$, we find that

$$v_{xf}^2 = 2a_x d$$

$$(5) \qquad v_{xf} = \sqrt{2a_x d} = \boxed{\sqrt{2gd \sin \theta}}$$

We see from equations (4) and (5) that the time t needed to reach the bottom and the speed v_{xf}, like acceleration, are independent of the crate's mass. This suggests a simple method you can use to measure g, using an inclined air track: Measure the angle of inclination, some distance traveled by a cart along the incline, and the time needed to travel that distance. The value of g can then be calculated from (4).

EXAMPLE 5.7   One Block Pushes Another

Two blocks of masses m_1 and m_2 are placed in contact with each other on a frictionless horizontal surface. A constant horizontal force **F** is applied to the block of mass m_1. (a) Determine the magnitude of the acceleration of the two-block system.

Solution Common sense tells us that both blocks must experience the same acceleration because they remain in contact with each other. Just as in the preceding example, we make a labeled sketch and free-body diagrams, which are shown in Figure 5.13. In Figure 5.13a the dashed line indicates that we treat the two blocks together as a system. Because **F** is the only external horizontal force acting on the system (the two blocks), we have

$$\sum F_x(\text{system}) = F = (m_1 + m_2)a_x$$

$$(1) \qquad \boxed{a_x = \frac{F}{m_1 + m_2}}$$

(a)

(b)   (c)

Figure 5.13

Treating the two blocks together as a system simplifies the solution but does not provide information about internal forces.

(b) Determine the magnitude of the contact force between the two blocks.

Solution To solve this part of the problem, we must treat each block separately with its own free-body diagram, as in Figures 5.13b and 5.13c. We denote the contact force by **P**. From Figure 5.13c, we see that the only horizontal force acting on block 2 is the contact force **P** (the force exerted by block 1 on block 2), which is directed to the right. Applying Newton's second law to block 2 gives

$$(2) \qquad \sum F_x = P = m_2 a_x$$

Substituting into (2) the value of a_x given by (1), we obtain

$$(3) \qquad \boxed{P = m_2 a_x = \left(\frac{m_2}{m_1 + m_2}\right)F}$$

From this result, we see that the contact force **P** exerted by block 1 on block 2 is *less* than the applied force **F**. This is consistent with the fact that the force required to accelerate block 2 alone must be less than the force required to produce the same acceleration for the two-block system.

It is instructive to check this expression for P by considering the forces acting on block 1, shown in Figure 5.13b. The horizontal forces acting on this block are the applied force **F** to the right and the contact force **P'** to the left (the force exerted by block 2 on block 1). From Newton's third law, **P'** is the reaction to **P**, so that $|\mathbf{P'}| = |\mathbf{P}|$. Applying Newton's second law to block 1 produces

$$(4) \qquad \sum F_x = F - P' = F - P = m_1 a_x$$

Substituting into (4) the value of a_x from (1), we obtain

$$P = F - m_1 a_x = F - \frac{m_1 F}{m_1 + m_2} = \left(\frac{m_2}{m_1 + m_2} \right) F$$

This agrees with (3), as it must.

Exercise   If $m_1 = 4.00$ kg, $m_2 = 3.00$ kg, and $F = 9.00$ N, find the magnitude of the acceleration of the system and the magnitude of the contact force.

Answer $a_x = 1.29$ m/s^2; $P = 3.86$ N.

EXAMPLE 5.8   Weighing a Fish in an Elevator

A person weighs a fish of mass m on a spring scale attached to the ceiling of an elevator, as illustrated in Figure 5.14. Show that if the elevator accelerates either upward or downward, the spring scale gives a reading that is different from the weight of the fish.

Solution   The external forces acting on the fish are the downward force of gravity $\mathbf{F}_g = m\mathbf{g}$ and the force $\mathbf{T}$ exerted by the scale. By Newton's third law, the tension T is also the reading of the scale. If the elevator is either at rest or moving at constant velocity, the fish is not accelerating, and so $\Sigma F_y = T - mg = 0$ or $T = mg$ (remember that the scalar mg is the weight of the fish).

If the elevator moves upward with an acceleration $\mathbf{a}$ relative to an observer standing outside the elevator in an inertial frame (see Fig. 5.14a), Newton's second law applied to the fish gives the net force on the fish:

$$(1) \qquad \Sigma F_y = T - mg = ma_y$$

where we have chosen upward as the positive direction. Thus, we conclude from (1) that the scale reading T is greater than the weight mg if $\mathbf{a}$ is upward, so that a_y is positive, and that the reading is less than mg if $\mathbf{a}$ is downward, so that a_y is negative.

For example, if the weight of the fish is 40.0 N and $\mathbf{a}$ is upward, so that $a_y = +2.00$ m/s^2, the scale reading from (1) is

(a)   (b)

Observer in inertial frame

Figure 5.14   Apparent weight versus true weight. (a) When the elevator accelerates upward, the spring scale reads a value greater than the weight of the fish. (b) When the elevator accelerates downward, the spring scale reads a value less than the weight of the fish.

(2) $T = ma_y + mg = mg\left(\dfrac{a_y}{g} + 1\right)$

$= (40.0 \text{ N})\left(\dfrac{2.00 \text{ m/s}^2}{9.80 \text{ m/s}^2} + 1\right)$

$= \boxed{48.2 \text{ N}}$

If **a** is downward so that $a_y = -2.00 \text{ m/s}^2$, then (2) gives us

$T = mg\left(\dfrac{a_y}{g} + 1\right) = (40.0 \text{ N})\left(\dfrac{-2.00 \text{ m/s}^2}{9.80 \text{ m/s}^2} + 1\right)$

$= \boxed{31.8 \text{ N}}$

Hence, if you buy a fish by weight in an elevator, make sure the fish is weighed while the elevator is either at rest or accelerating downward! Furthermore, note that from the information given here one cannot determine the direction of motion of the elevator.

Special Cases   If the elevator cable breaks, the elevator falls freely and $a_y = -g$. We see from (2) that the scale reading T is zero in this case; that is, the fish appears to be weightless. If the elevator accelerates downward with an acceleration greater than g, the fish (along with the person in the elevator) eventually hits the ceiling because the acceleration of fish and person is still that of a freely falling object relative to an outside observer.

EXAMPLE 5.9   Atwood's Machine

When two objects of unequal mass are hung vertically over a frictionless pulley of negligible mass, as shown in Figure 5.15a, the arrangement is called an *Atwood machine*. The device is sometimes used in the laboratory to measure the free-fall acceleration. Determine the magnitude of the acceleration of the two objects and the tension in the lightweight cord.

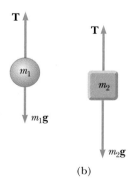

(a)

(b)

Figure 5.15   Atwood's machine. (a) Two objects ($m_2 > m_1$) connected by a cord of negligible mass strung over a frictionless pulley. (b) Free-body diagrams for the two objects.

Solution   If we were to define our system as being made up of both objects, as we did in Example 5.7, we would have to determine an *internal* force (tension in the cord). We must define two systems here—one for each object—and apply Newton's second law to each. The free-body diagrams for the two objects are shown in Figure 5.15b. Two forces act on each object: the upward force **T** exerted by the cord and the downward force of gravity.

We need to be very careful with signs in problems such as this, in which a string or rope passes over a pulley or some other structure that causes the string or rope to bend. In Figure 5.15a, notice that if object 1 accelerates upward, then object 2 accelerates downward. Thus, for consistency with signs, if we define the upward direction as positive for object 1, we must define the downward direction as positive for object 2. With this sign convention, both objects accelerate in the same direction as defined by the choice of sign. With this sign convention applied to the forces, the y component of the net force exerted on object 1 is $T - m_1g$, and the y component of the net force exerted on object 2 is $m_2g - T$. Because the objects are connected by a cord, their accelerations must be equal in magnitude. (Otherwise the cord would stretch or break as the distance between the objects increased.) If we assume $m_2 > m_1$, then object 1 must accelerate upward and object 2 downward.

When Newton's second law is applied to object 1, we obtain

(1) $\sum F_y = T - m_1g = m_1a_y$

Similarly, for object 2 we find

(2) $\sum F_y = m_2g - T = m_2a_y$

When (2) is added to (1), T drops out and we get

$$-m_1g + m_2g = m_1a_y + m_2a_y$$

(3) $$a_y = \left(\frac{m_2 - m_1}{m_1 + m_2}\right)g$$

When (3) is substituted into (1), we obtain

(4) $$T = \left(\frac{2m_1m_2}{m_1 + m_2}\right)g$$

The result for the acceleration in (3) can be interpreted as the ratio of the unbalanced force on the system $(m_2g - m_1g)$ to the total mass of the system $(m_1 + m_2)$, as expected from Newton's second law.

Special Cases   When $m_1 = m_2$, then $a_y = 0$ and $T = m_1g$, as we would expect for this balanced case. If $m_2 \gg m_1$, then $a_y \approx g$ (a freely falling body) and $T \approx 2m_1g$.

Exercise   Find the magnitude of the acceleration and the string tension for an Atwood machine in which $m_1 = 2.00$ kg and $m_2 = 4.00$ kg.

Answer $a_y = 3.27$ m/s^2, $T = 26.1$ N.

EXAMPLE 5.10   Acceleration of Two Objects Connected by a Cord

A ball of mass m_1 and a block of mass m_2 are attached by a lightweight cord that passes over a frictionless pulley of negligible mass, as shown in Figure 5.16a. The block lies on a frictionless incline of angle θ. Find the magnitude of the acceleration of the two objects and the tension in the cord.

Solution   Because the objects are connected by a cord (which we assume does not stretch), their accelerations have the same magnitude. The free-body diagrams are shown in Figures 5.16b and 5.16c. Applying Newton's second law in component form to the ball, with the choice of the upward direction as positive, yields

(1) $$\sum F_x = 0$$
(2) $$\sum F_y = T - m_1g = m_1a_y = m_1a$$

Note that in order for the ball to accelerate upward, it is necessary that $T > m_1g$. In (2) we have replaced a_y with a because the acceleration has only a y component.

For the block it is convenient to choose the positive x' axis along the incline, as shown in Figure 5.16c. Here we choose the positive direction to be down the incline, in the $+x'$ direction. Applying Newton's second law in component form to the block gives

(3) $$\sum F_{x'} = m_2g \sin\theta - T = m_2a_{x'} = m_2a$$
(4) $$\sum F_{y'} = n - m_2g \cos\theta = 0$$

In (3) we have replaced $a_{x'}$ with a because that is the acceleration's only component. In other words, the two objects have accelerations of the same magnitude a, which is what we are trying to find. Equations (1) and (4) provide no information regarding the acceleration. However, if we solve (2) for T and then substitute this value for T into (3) and solve for a, we obtain

(5) $$a = \frac{m_2g \sin\theta - m_1g}{m_1 + m_2}$$

When this value for a is substituted into (2), we find

(6) $$T = \frac{m_1m_2g(\sin\theta + 1)}{m_1 + m_2}$$

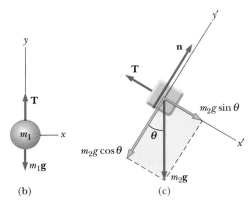

Figure 5.16   (a) Two objects connected by a lightweight cord strung over a frictionless pulley. (b) Free-body diagram for the ball. (c) Free-body diagram for the block. (The incline is frictionless.)

Note that the block accelerates down the incline only if $m_2 \sin \theta > m_1$ (that is, if **a** is in the direction we assumed). If $m_1 > m_2 \sin \theta$, then the acceleration is up the incline for the block and downward for the ball. Also note that the result for the acceleration (5) can be interpreted as the resultant force acting on the system divided by the total mass of the system; this is consistent with Newton's second law. Finally, if $\theta = 90°$, then the results for a and T are identical to those of Example 5.9.

Exercise   If $m_1 = 10.0$ kg, $m_2 = 5.00$ kg, and $\theta = 45.0°$, find the acceleration of each object.

Answer $a = -4.22$ m/s^2, where the negative sign indicates that the block accelerates up the incline and the ball accelerates downward.

5.8 FORCES OF FRICTION

When a body is in motion either on a surface or in a viscous medium such as air or water, there is resistance to the motion because the body interacts with its surroundings. We call such resistance a **force of friction**. Forces of friction are very important in our everyday lives. They allow us to walk or run and are necessary for the motion of wheeled vehicles.

Have you ever tried to move a heavy desk across a rough floor? You push harder and harder until all of a sudden the desk seems to "break free" and subsequently moves relatively easily. It takes a greater force to start the desk moving than it does to keep it going once it has started sliding. To understand why this happens, consider a book on a table, as shown in Figure 5.17a. If we apply an external horizontal force **F** to the book, acting to the right, the book remains stationary if **F** is not too great. The force that counteracts **F** and keeps the book from moving acts to the left and is called the **frictional force f.**

As long as the book is not moving, $f = F$. Because the book is stationary, we call this frictional force the **force of static friction f_s**. Experiments show that this force arises from contacting points that protrude beyond the general level of the surfaces in contact, even for surfaces that are apparently very smooth, as shown in the magnified view in Figure 5.17a. (If the surfaces are clean and smooth at the atomic level, they are likely to weld together when contact is made.) The frictional force arises in part from one peak's physically blocking the motion of a peak from the opposing surface, and in part from chemical bonding of opposing points as they come into contact. If the surfaces are rough, bouncing is likely to occur, further complicating the analysis. Although the details of friction are quite complex at the atomic level, this force ultimately involves an electrical interaction between atoms or molecules.

Force of static friction

If we increase the magnitude of **F**, as shown in Figure 5.17b, the magnitude of f_s increases along with it, keeping the book in place. The force f_s cannot increase indefinitely, however. Eventually the surfaces in contact can no longer supply sufficient frictional force to counteract **F**, and the book accelerates. When it is on the verge of moving, f_s is a maximum, as shown in Figure 5.17c. When F exceeds $f_{s,max}$, the book accelerates to the right. Once the book is in motion, the retarding frictional force becomes less than $f_{s,max}$ (see Fig. 5.17c). When the book is in motion, we call the retarding force the **force of kinetic friction f_k**. If $F = f_k$, then the book moves to the right with constant speed. If $F > f_k$, then there is an unbalanced force $F - f_k$ in the positive x direction, and this force accelerates the book to the right. If the applied force **F** is removed, then the frictional force f_k acting to the left accelerates the book in the negative x direction and eventually brings it to rest.

Force of kinetic friction

Experimentally, we find that, to a good approximation, both $f_{s,max}$ and f_k are proportional to the normal force acting on the book. The following empirical laws of friction summarize the experimental observations:

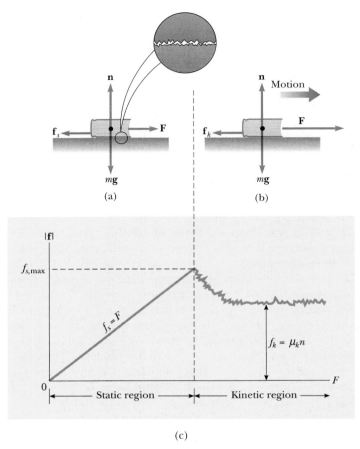

Figure 5.17 The direction of the force of friction **f** between a book and a rough surface is opposite the direction of the applied force **F**. Because the two surfaces are both rough, contact is made only at a few points, as illustrated in the "magnified" view. (a) The magnitude of the force of static friction equals the magnitude of the applied force. (b) When the magnitude of the applied force exceeds the magnitude of the force of kinetic friction, the book accelerates to the right. (c) A graph of frictional force versus applied force. Note that $f_{s,\max} > f_k$.

- The direction of the force of static friction between any two surfaces in contact with each other is opposite the direction of relative motion and can have values

$$f_s \leq \mu_s n \qquad (5.8)$$

where the dimensionless constant μ_s is called the **coefficient of static friction** and n is the magnitude of the normal force. The equality in Equation 5.8 holds when one object is on the verge of moving, that is, when $f_s = f_{s,\max} = \mu_s n$. The inequality holds when the applied force is less than $\mu_s n$.

- The direction of the force of kinetic friction acting on an object is opposite the direction of the object's sliding motion relative to the surface applying the frictional force and is given by

$$f_k = \mu_k n \qquad (5.9)$$

where μ_k is the **coefficient of kinetic friction.**

- The values of μ_k and μ_s depend on the nature of the surfaces, but μ_k is generally less than μ_s. Typical values range from around 0.03 to 1.0. Table 5.2 lists some reported values.

TABLE 5.2	Coefficients of Friction[a]	
	μ_s	μ_k
Steel on steel	0.74	0.57
Aluminum on steel	0.61	0.47
Copper on steel	0.53	0.36
Rubber on concrete	1.0	0.8
Wood on wood	0.25–0.5	0.2
Glass on glass	0.94	0.4
Waxed wood on wet snow	0.14	0.1
Waxed wood on dry snow	—	0.04
Metal on metal (lubricated)	0.15	0.06
Ice on ice	0.1	0.03
Teflon on Teflon	0.04	0.04
Synovial joints in humans	0.01	0.003

[a] All values are approximate. In some cases, the coefficient of friction can exceed 1.0.

- The coefficients of friction are nearly independent of the area of contact between the surfaces. To understand why, we must examine the difference between the *apparent contact area,* which is the area we see with our eyes, and the *real contact area,* represented by two irregular surfaces touching, as shown in the magnified view in Figure 5.17a. It seems that increasing the apparent contact area does not increase the real contact area. When we increase the apparent area (without changing anything else), there is less force per unit area driving the jagged points together. This decrease in force counteracts the effect of having more points involved.

Although the coefficient of kinetic friction can vary with speed, we shall usually neglect any such variations in this text. We can easily demonstrate the approximate nature of the equations by trying to get a block to slip down an incline at constant speed. Especially at low speeds, the motion is likely to be characterized by alternate episodes of sticking and movement.

If you would like to learn more about this subject, read the article "Friction at the Atomic Scale" by J. Krim in the October 1996 issue of *Scientific American.*

Quick Quiz 5.6

A crate is sitting in the center of a flatbed truck. The truck accelerates to the right, and the crate moves with it, not sliding at all. What is the direction of the frictional force exerted by the truck on the crate? (a) To the left. (b) To the right. (c) No frictional force because the crate is not sliding.

QuickLab

Can you apply the ideas of Example 5.12 to determine the coefficients of static and kinetic friction between the cover of your book and a quarter? What should happen to those coefficients if you make the measurements between your book and *two* quarters taped one on top of the other?

CONCEPTUAL EXAMPLE 5.11 ▸ Why Does the Sled Accelerate?

A horse pulls a sled along a level, snow-covered road, causing the sled to accelerate, as shown in Figure 5.18a. Newton's third law states that the sled exerts an equal and opposite force on the horse. In view of this, how can the sled accelerate? Under what condition does the system (horse plus sled) move with constant velocity?

Solution It is important to remember that the forces described in Newton's third law act on different objects—the horse exerts a force on the sled, and the sled exerts an equal-magnitude and oppositely directed force on the horse. Because we are interested only in the motion of the sled, we do not consider the forces it exerts on the horse. When deter-

(a)   (b)   (c)

Figure 5.18

mining the motion of an object, you must add only the forces on that object. The horizontal forces exerted on the sled are the forward force T exerted by the horse and the backward force of friction $\mathbf{f}_{\text{sled}}$ between sled and snow (see Fig. 5.18b). When the forward force exceeds the backward force, the sled accelerates to the right.

The force that accelerates the system (horse plus sled) is the frictional force $\mathbf{f}_{\text{horse}}$ exerted by the Earth on the horse's feet. The horizontal forces exerted on the horse are the forward force $\mathbf{f}_{\text{horse}}$ exerted by the Earth and the backward tension force $\mathbf{T}$ exerted by the sled (Fig. 5.18c). The resultant of

these two forces causes the horse to accelerate. When $\mathbf{f}_{\text{horse}}$ balances $\mathbf{f}_{\text{sled}}$, the system moves with constant velocity.

Exercise   Are the normal force exerted by the snow on the horse and the gravitational force exerted by the Earth on the horse a third-law pair?

Answer   No, because they act on the same object. Third-law force pairs are equal in magnitude and opposite in direction, and the forces act on *different* objects.

EXAMPLE 5.12   Experimental Determination of μ_s and μ_k

The following is a simple method of measuring coefficients of friction: Suppose a block is placed on a rough surface inclined relative to the horizontal, as shown in Figure 5.19. The incline angle is increased until the block starts to move. Let us show that by measuring the critical angle θ_c at which this slipping just occurs, we can obtain μ_s.

Solution   The only forces acting on the block are the force of gravity $m\mathbf{g}$, the normal force $\mathbf{n}$, and the force of static friction $\mathbf{f}_s$. These forces balance when the block is on the verge

of slipping but has not yet moved. When we take x to be parallel to the plane and y perpendicular to it, Newton's second law applied to the block for this balanced situation gives

Static case: (1) $\sum F_x = mg \sin \theta - f_s = ma_x = 0$

              (2) $\sum F_y = n - mg \cos \theta = ma_y = 0$

We can eliminate mg by substituting $mg = n/\cos \theta$ from (2) into (1) to get

(3) $f_s = mg \sin \theta = \left(\dfrac{n}{\cos \theta} \right) \sin \theta = n \tan \theta$

When the incline is at the critical angle θ_c, we know that $f_s = f_{s,\text{max}} = \mu_s n$, and so at this angle, (3) becomes

$$\mu_s n = n \tan \theta_c$$

Static case: $\mu_s = \tan \theta_c$

For example, if the block just slips at $\theta_c = 20°$, then we find that $\mu_s = \tan 20° = 0.364$.

Once the block starts to move at $\theta \geq \theta_c$, it accelerates down the incline and the force of friction is $f_k = \mu_k n$. However, if θ is reduced to a value less than θ_c, it may be possible to find an angle θ_c' such that the block moves down the incline with constant speed ($a_x = 0$). In this case, using (1) and (2) with f_s replaced by f_k gives

Kinetic case: $\mu_k = \tan \theta_c'$

where $\theta_c' < \theta_c$.

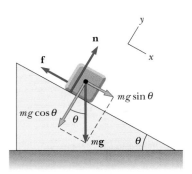

Figure 5.19   The external forces exerted on a block lying on a rough incline are the force of gravity $m\mathbf{g}$, the normal force $\mathbf{n}$, and the force of friction $\mathbf{f}$. For convenience, the force of gravity is resolved into a component along the incline $mg \sin \theta$ and a component perpendicular to the incline $mg \cos \theta$.

EXAMPLE 5.13 The Sliding Hockey Puck

A hockey puck on a frozen pond is given an initial speed of 20.0 m/s. If the puck always remains on the ice and slides 115 m before coming to rest, determine the coefficient of kinetic friction between the puck and ice.

Solution The forces acting on the puck after it is in motion are shown in Figure 5.20. If we assume that the force of kinetic friction f_k remains constant, then this force produces a uniform acceleration of the puck in the direction opposite its velocity, causing the puck to slow down. First, we find this acceleration in terms of the coefficient of kinetic friction, using Newton's second law. Knowing the acceleration of the puck and the distance it travels, we can then use the equations of kinematics to find the coefficient of kinetic friction.

Figure 5.20 *After* the puck is given an initial velocity to the right, the only external forces acting on it are the force of gravity $m\mathbf{g}$, the normal force $\mathbf{n}$, and the force of kinetic friction $\mathbf{f}_k$.

Defining rightward and upward as our positive directions, we apply Newton's second law in component form to the puck and obtain

$$(1) \qquad \sum F_x = -f_k = ma_x$$

$$(2) \qquad \sum F_y = n - mg = 0 \qquad (a_y = 0)$$

But $f_k = \mu_k n$, and from (2) we see that $n = mg$. Therefore, (1) becomes

$$-\mu_k n = -\mu_k mg = ma_x$$

$$a_x = -\mu_k g$$

The negative sign means the acceleration is to the left; this means that the puck is slowing down. The acceleration is independent of the mass of the puck and is constant because we assume that μ_k remains constant.

Because the acceleration is constant, we can use Equation 2.12, $v_{xf}^2 = v_{xi}^2 + 2a_x(x_f - x_i)$, with $x_i = 0$ and $v_{xf} = 0$:

$$v_{xi}^2 + 2ax_f = v_{xi}^2 - 2\mu_k gx_f = 0$$

$$\mu_k = \frac{v_{xi}^2}{2gx_f}$$

$$\mu_k = \frac{(20.0 \text{ m/s})^2}{2(9.80 \text{ m/s}^2)(115 \text{ m})} = \boxed{0.177}$$

Note that μ_k is dimensionless.

EXAMPLE 5.14 Acceleration of Two Connected Objects When Friction Is Present

A block of mass m_1 on a rough, horizontal surface is connected to a ball of mass m_2 by a lightweight cord over a lightweight, frictionless pulley, as shown in Figure 5.21a. A force of magnitude F at an angle θ with the horizontal is applied to the block as shown. The coefficient of kinetic friction between the block and surface is μ_k. Determine the magnitude of the acceleration of the two objects.

Solution We start by drawing free-body diagrams for the two objects, as shown in Figures 5.21b and 5.21c. (Are you beginning to see the similarities in all these examples?) Next, we apply Newton's second law in component form to each object and use Equation 5.9, $f_k = \mu_k n$. Then we can solve for the acceleration in terms of the parameters given.

The applied force $\mathbf{F}$ has x and y components $F \cos \theta$ and $F \sin \theta$, respectively. Applying Newton's second law to both objects and assuming the motion of the block is to the right, we obtain

Motion of block: $\quad (1) \quad \sum F_x = F\cos\theta - f_k - T = m_1 a_x$
$$= m_1 a$$

$$(2) \quad \sum F_y = n + F\sin\theta - m_1 g$$
$$= m_1 a_y = 0$$

Motion of ball: $\qquad\qquad \sum F_x = m_2 a_x = 0$

$$(3) \quad \sum F_y = T - m_2 g = m_2 a_y = m_2 a$$

Note that because the two objects are connected, we can equate the magnitudes of the x component of the acceleration of the block and the y component of the acceleration of the ball. From Equation 5.9 we know that $f_k = \mu_k n$, and from (2) we know that $n = m_1 g - F\sin\theta$ (note that in this case n is *not* equal to $m_1 g$); therefore,

$$(4) \qquad f_k = \mu_k(m_1 g - F\sin\theta)$$

That is, the frictional force is reduced because of the positive

y component of **F**. Substituting (4) and the value of T from (3) into (1) gives

$$F \cos \theta - \mu_k(m_1 g - F \sin \theta) - m_2(a + g) = m_1 a$$

Solving for a, we obtain

(5) $$a = \frac{F(\cos \theta + \mu_k \sin \theta) - g(m_2 + \mu_k m_1)}{m_1 + m_2}$$

Note that the acceleration of the block can be either to the right or to the left,[6] depending on the sign of the numerator in (5). If the motion is to the left, then we must reverse the sign of f_k in (1) because the force of kinetic friction must oppose the motion. In this case, the value of a is the same as in (5), with μ_k replaced by $-\mu_k$.

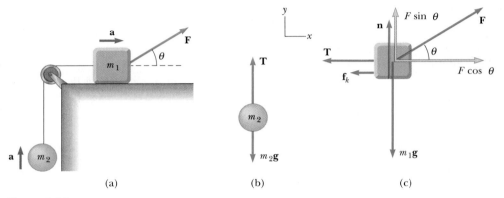

(a)   (b)   (c)

Figure 5.21   (a) The external force **F** applied as shown can cause the block to accelerate to the right. (b) and (c) The free-body diagrams, under the assumption that the block accelerates to the right and the ball accelerates upward. The magnitude of the force of kinetic friction in this case is given by $f_k = \mu_k n = \mu_k(m_1 g - F \sin \theta)$.

APPLICATION   Automobile Antilock Braking Systems (ABS)

If an automobile tire is rolling and not slipping on a road surface, then the maximum frictional force that the road can exert on the tire is the force of static friction $\mu_s n$. One must use static friction in this situation because at the point of contact between the tire and the road, no sliding of one surface over the other occurs if the tire is not skidding. However, if the tire starts to skid, the frictional force exerted on it is reduced to the force of kinetic friction $\mu_k n$. Thus, to maximize the frictional force and minimize stopping distance, the wheels must maintain pure rolling motion and not skid. An additional benefit of maintaining wheel rotation is that directional control is not lost as it is in skidding.

Unfortunately, in emergency situations drivers typically press down as hard as they can on the brake pedal, "locking the brakes." This stops the wheels from rotating, ensuring a skid and reducing the frictional force from the static to the kinetic case. To address this problem, automotive engineers have developed antilock braking systems (ABS) that very briefly release the brakes when a wheel is just about to stop turning. This maintains rolling contact between the tire and the pavement. When the brakes are released momentarily, the stopping distance is greater than it would be if the brakes were being applied continuously. However, through the use of computer control, the "brake-off" time is kept to a minimum. As a result, the stopping distance is much less than what it would be if the wheels were to skid.

Let us model the stopping of a car by examining real data. In a recent issue of *AutoWeek*,[7] the braking performance for a Toyota Corolla was measured. These data correspond to the braking force acquired by a highly trained, professional driver. We begin by assuming constant acceleration. (Why do we need to make this assumption?) The magazine provided the initial speed and stopping distance in non-SI units. After converting these values to SI we use $v_{xf}^2 = v_{xi}^2 + 2a_x x$ to deter-

---

[6] Equation 5 shows that when $\mu_k m_1 > m_2$, there is a range of values of F for which no motion occurs at a given angle θ.

[7] *AutoWeek* magazine, 48:22–23, 1998.

mine the acceleration at different speeds. These do not vary greatly, and so our assumption of constant acceleration is reasonable.

Initial Speed		Stopping Distance		Acceleration
(mi/h)	(m/s)	(ft)	(m)	(m/s²)
30	13.4	34	10.4	− 8.67
60	26.8	143	43.6	− 8.25
80	35.8	251	76.5	− 8.36

We take an average value of acceleration of -8.4 m/s², which is approximately $0.86g$. We then calculate the coefficient of friction from $\Sigma F = \mu_s mg = ma$; this gives $\mu_s = 0.86$ for the Toyota. This is lower than the rubber-to-concrete value given in Table 5.2. Can you think of any reasons for this?

Let us now estimate the stopping distance of the car if the wheels were skidding. Examining Table 5.2 again, we see that the difference between the coefficients of static and kinetic friction for rubber against concrete is about 0.2. Let us therefore assume that our coefficients differ by the same amount, so that $\mu_k \approx 0.66$. This allows us to calculate estimated stopping distances for the case in which the wheels are locked and the car skids across the pavement. The results illustrate the advantage of not allowing the wheels to skid.

Initial Speed (mi/h)	Stopping Distance no skid (m)	Stopping distance skidding (m)
30	10.4	13.9
60	43.6	55.5
80	76.5	98.9

An ABS keeps the wheels rotating, with the result that the higher coefficient of static friction is maintained between the tires and road. This approximates the technique of a professional driver who is able to maintain the wheels at the point of maximum frictional force. Let us estimate the ABS performance by assuming that the magnitude of the acceleration is not quite as good as that achieved by the professional driver but instead is reduced by 5%.

We now plot in Figure 5.22 vehicle speed versus distance from where the brakes were applied (at an initial speed of 80 mi/h = 37.5 m/s) for the three cases of amateur driver, professional driver, and estimated ABS performance (amateur driver). We find that a markedly shorter distance is necessary for stopping without locking the wheels and skidding and a satisfactory value of stopping distance when the ABS computer maintains tire rotation.

The purpose of the ABS is to help typical drivers (whose tendency is to lock the wheels in an emergency) to better maintain control of their automobiles and minimize stopping distance.

Speed (m/s)

——— Amateur driver
——— Professional driver
——— ABS, amateur driver

Distance from point of application of brakes (m)

Figure 5.22 This plot of vehicle speed versus distance from where the brakes were applied shows that an antilock braking system (ABS) approaches the performance of a trained professional driver.

SUMMARY

Newton's first law states that, in the absence of an external force, a body at rest remains at rest and a body in uniform motion in a straight line maintains that motion. An **inertial frame** is one that is not accelerating.

Newton's second law states that the acceleration of an object is directly proportional to the net force acting on it and inversely proportional to its mass. The net force acting on an object equals the product of its mass and its acceleration: $\Sigma \mathbf{F} = m\mathbf{a}$. You should be able to apply the x and y component forms of this equation to describe the acceleration of any object acting under the influence of speci-

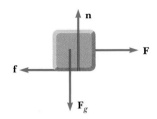

A block pulled to the right on a
rough horizontal surface

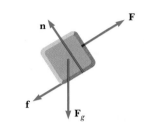

A block pulled up a rough incline

F

Two blocks in contact, pushed to the
right on a frictionless surface

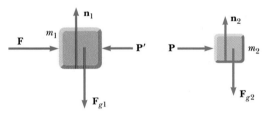

Note: **P** = −**P′** because they are an action–reaction pair

Two masses connected by a light cord. The
surface is rough, and the pulley is frictionless.

Figure 5.23   Various systems *(left)* and the corresponding free-body diagrams *(right)*.

fied forces. If the object is either stationary or moving with constant velocity, then the forces must vectorially cancel each other.

The **force of gravity** exerted on an object is equal to the product of its mass (a scalar quantity) and the free-fall acceleration: $\mathbf{F}_g = m\mathbf{g}$. The **weight** of an object is the magnitude of the force of gravity acting on the object.

Newton's third law states that if two objects interact, then the force exerted by object 1 on object 2 is equal in magnitude to and opposite in direction to the force exerted by object 2 on object 1. Thus, an isolated force cannot exist in nature. Make sure you can identify third-law pairs and the two objects upon which they act.

The **maximum force of static friction $f_{s,\max}$** between an object and a surface is proportional to the normal force acting on the object. In general, $f_s \leq \mu_s n$, where μ_s is the **coefficient of static friction** and n is the magnitude of the normal force. When an object slides over a surface, the direction of the **force of kinetic friction $\mathbf{f}_k$** is opposite the direction of sliding motion and is also proportional to the magnitude of the normal force. The magnitude of this force is given by $f_k = \mu_k n$, where μ_k is the **coefficient of kinetic friction.**

More on Free-Body Diagrams

To be successful in applying Newton's second law to a system, you must be able to recognize all the forces acting on the system. That is, you must be able to construct the correct free-body diagram. The importance of constructing the free-body diagram cannot be overemphasized. In Figure 5.23 a number of systems are presented together with their free-body diagrams. You should examine these carefully and then construct free-body diagrams for other systems described in the end-of-chapter problems. When a system contains more than one element, it is important that you construct a separate free-body diagram for *each* element.

As usual, $\mathbf{F}$ denotes some applied force, $\mathbf{F}_g = m\mathbf{g}$ is the force of gravity, $\mathbf{n}$ denotes a normal force, $\mathbf{f}$ is the force of friction, and $\mathbf{T}$ is the force whose magnitude is the tension exerted on an object.

QUESTIONS

1. A passenger sitting in the rear of a bus claims that he was injured when the driver slammed on the brakes, causing a suitcase to come flying toward the passenger from the front of the bus. If you were the judge in this case, what disposition would you make? Why?

2. A space explorer is in a spaceship moving through space far from any planet or star. She notices a large rock, taken as a specimen from an alien planet, floating around the cabin of the spaceship. Should she push it gently toward a storage compartment or kick it toward the compartment? Why?

3. A massive metal object on a rough metal surface may undergo contact welding to that surface. Discuss how this affects the frictional force between object and surface.

4. The observer in the elevator of Example 5.8 would claim that the weight of the fish is T, the scale reading. This claim is obviously wrong. Why does this observation differ from that of a person in an inertial frame outside the elevator?

5. Identify the action–reaction pairs in the following situa-

tions: a man takes a step; a snowball hits a woman in the back; a baseball player catches a ball; a gust of wind strikes a window.

6. A ball is held in a person's hand. (a) Identify all the external forces acting on the ball and the reaction to each. (b) If the ball is dropped, what force is exerted on it while it is falling? Identify the reaction force in this case. (Neglect air resistance.)

7. If a car is traveling westward with a constant speed of 20 m/s, what is the resultant force acting on it?

8. "When the locomotive in Figure 5.3 broke through the wall of the train station, the force exerted by the locomotive on the wall was greater than the force the wall could exert on the locomotive." Is this statement true or in need of correction? Explain your answer.

9. A rubber ball is dropped onto the floor. What force causes the ball to bounce?

10. What is wrong with the statement, "Because the car is at rest, no forces are acting on it"? How would you correct this statement?

11. Suppose you are driving a car along a highway at a high speed. Why should you avoid slamming on your brakes if you want to stop in the shortest distance? That is, why should you keep the wheels turning as you brake?

12. If you have ever taken a ride in an elevator of a high-rise building, you may have experienced a nauseating sensation of "heaviness" and "lightness" depending on the direction of the acceleration. Explain these sensations. Are we truly weightless in free-fall?

13. The driver of a speeding empty truck slams on the brakes and skids to a stop through a distance *d*. (a) If the truck carried a heavy load such that its mass were doubled, what would be its skidding distance? (b) If the initial speed of the truck is halved, what would be its skidding distance?

14. In an attempt to define Newton's third law, a student states that the action and reaction forces are equal in magnitude to and opposite in direction to each other. If this is the case, how can there ever be a net force on an object?

15. What force causes (a) a propeller-driven airplane to move? (b) a rocket? (c) a person walking?

16. Suppose a large and spirited Freshman team is beating the Sophomores in a tug-of-war contest. The center of the rope being tugged is gradually accelerating toward the Freshman team. State the relationship between the strengths of these two forces: First, the force the Freshmen exert on the Sophomores; and second, the force the Sophomores exert on the Freshmen.

17. If you push on a heavy box that is at rest, you must exert some force to start its motion. However, once the box is sliding, you can apply a smaller force to maintain that motion. Why?

18. A weight lifter stands on a bathroom scale. He pumps a barbell up and down. What happens to the reading on the scale as this is done? Suppose he is strong enough to actually *throw* the barbell upward. How does the reading on the scale vary now?

19. As a rocket is fired from a launching pad, its speed *and* acceleration increase with time as its engines continue to operate. Explain why this occurs even though the force of the engines exerted on the rocket remains constant.

20. In the motion picture *It Happened One Night* (Columbia Pictures, 1934), Clark Gable is standing inside a stationary bus in front of Claudette Colbert, who is seated. The bus suddenly starts moving forward, and Clark falls into Claudette's lap. Why did this happen?

PROBLEMS

1, 2, 3 = straightforward, intermediate, challenging   ☐ = full solution available in the *Student Solutions Manual and Study Guide*
WEB = solution posted at **http://www.saunderscollege.com/physics/**   🖥 = Computer useful in solving problem   🖱 = Interactive Physics
☐ = paired numerical/symbolic problems

Sections 5.1 through 5.6

1. A force **F** applied to an object of mass m_1 produces an acceleration of 3.00 m/s^2. The same force applied to a second object of mass m_2 produces an acceleration of 1.00 m/s^2. (a) What is the value of the ratio m_1/m_2? (b) If m_1 and m_2 are combined, find their acceleration under the action of the force **F**.

2. A force of 10.0 N acts on a body of mass 2.00 kg. What are (a) the body's acceleration, (b) its weight in newtons, and (c) its acceleration if the force is doubled?

3. A 3.00-kg mass undergoes an acceleration given by **a** = $(2.00\mathbf{i} + 5.00\mathbf{j})$ m/s^2. Find the resultant force Σ**F** and its magnitude.

4. A heavy freight train has a mass of 15 000 metric tons. If the locomotive can pull with a force of 750 000 N, how long does it take to increase the speed from 0 to 80.0 km/h?

5. A 5.00-g bullet leaves the muzzle of a rifle with a speed of 320 m/s. The expanding gases behind it exert what force on the bullet while it is traveling down the barrel of the rifle, 0.820 m long? Assume constant acceleration and negligible friction.

6. After uniformly accelerating his arm for 0.090 0 s, a pitcher releases a baseball of weight 1.40 N with a velocity of 32.0 m/s horizontally forward. If the ball starts from rest, (a) through what distance does the ball accelerate before its release? (b) What force does the pitcher exert on the ball?

7. After uniformly accelerating his arm for a time *t*, a pitcher releases a baseball of weight $-F_g\mathbf{j}$ with a velocity *v***i**. If the ball starts from rest, (a) through what distance does the ball accelerate before its release? (b) What force does the pitcher exert on the ball?

8. Define one pound as the weight of an object of mass 0.453 592 37 kg at a location where the acceleration due to gravity is 32.174 0 ft/s^2. Express the pound as one quantity with one SI unit.

9. A 4.00-kg object has a velocity of 3.00**i** m/s at one instant. Eight seconds later, its velocity has increased to $(8.00\mathbf{i} + 10.0\mathbf{j})$ m/s. Assuming the object was subject to a constant total force, find (a) the components of the force and (b) its magnitude.

10. The average speed of a nitrogen molecule in air is about 6.70×10^2 m/s, and its mass is 4.68×10^{-26} kg. (a) If it takes 3.00×10^{-13} s for a nitrogen molecule to hit a wall and rebound with the same speed but moving in the opposite direction, what is the average acceleration of the molecule during this time interval? (b) What average force does the molecule exert on the wall?

11. An electron of mass 9.11×10^{-31} kg has an initial speed of 3.00×10^5 m/s. It travels in a straight line, and its speed increases to 7.00×10^5 m/s in a distance of 5.00 cm. Assuming its acceleration is constant, (a) determine the force exerted on the electron and (b) compare this force with the weight of the electron, which we neglected.

12. A woman weighs 120 lb. Determine (a) her weight in newtons and (b) her mass in kilograms.

13. If a man weighs 900 N on the Earth, what would he weigh on Jupiter, where the acceleration due to gravity is 25.9 m/s²?

14. The distinction between mass and weight was discovered after Jean Richer transported pendulum clocks from Paris to French Guiana in 1671. He found that they ran slower there quite systematically. The effect was reversed when the clocks returned to Paris. How much weight would you personally lose in traveling from Paris, where $g = 9.809\ 5$ m/s², to Cayenne, where $g = 9.780\ 8$ m/s²? (We shall consider how the free-fall acceleration influences the period of a pendulum in Section 13.4.)

15. Two forces $\mathbf{F}_1$ and $\mathbf{F}_2$ act on a 5.00-kg mass. If $F_1 = 20.0$ N and $F_2 = 15.0$ N, find the accelerations in (a) and (b) of Figure P5.15.

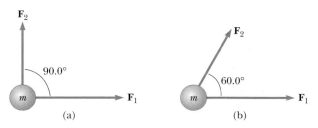

(a)

(b)

Figure P5.15

16. Besides its weight, a 2.80-kg object is subjected to one other constant force. The object starts from rest and in 1.20 s experiences a displacement of $(4.20\ \text{m})\mathbf{i} - (3.30\ \text{m})\mathbf{j}$, where the direction of $\mathbf{j}$ is the upward vertical direction. Determine the other force.

17. You stand on the seat of a chair and then hop off. (a) During the time you are in flight down to the floor, the Earth is lurching up toward you with an acceleration of what order of magnitude? In your solution explain your logic. Visualize the Earth as a perfectly solid object. (b) The Earth moves up through a distance of what order of magnitude?

18. Forces of 10.0 N north, 20.0 N east, and 15.0 N south are simultaneously applied to a 4.00-kg mass as it rests on an air table. Obtain the object's acceleration.

19. A boat moves through the water with two horizontal forces acting on it. One is a 2000-N forward push caused by the motor; the other is a constant 1800-N resistive force caused by the water. (a) What is the acceler-

ation of the 1 000-kg boat? (b) If it starts from rest, how far will it move in 10.0 s? (c) What will be its speed at the end of this time?

20. Three forces, given by $\mathbf{F}_1 = (-2.00\mathbf{i} + 2.00\mathbf{j})$ N, $\mathbf{F}_2 = (5.00\mathbf{i} - 3.00\mathbf{j})$ N, and $\mathbf{F}_3 = (-45.0\mathbf{i})$ N, act on an object to give it an acceleration of magnitude 3.75 m/s². (a) What is the direction of the acceleration? (b) What is the mass of the object? (c) If the object is initially at rest, what is its speed after 10.0 s? (d) What are the velocity components of the object after 10.0 s?

21. A 15.0-lb block rests on the floor. (a) What force does the floor exert on the block? (b) If a rope is tied to the block and run vertically over a pulley, and the other end is attached to a free-hanging 10.0-lb weight, what is the force exerted by the floor on the 15.0-lb block? (c) If we replace the 10.0-lb weight in part (b) with a 20.0-lb weight, what is the force exerted by the floor on the 15.0-lb block?

Section 5.7   Some Applications of Newton's Laws

22. A 3.00-kg mass is moving in a plane, with its x and y coordinates given by $x = 5t^2 - 1$ and $y = 3t^3 + 2$, where x and y are in meters and t is in seconds. Find the magnitude of the net force acting on this mass at $t = 2.00$ s.

23. The distance between two telephone poles is 50.0 m. When a 1.00-kg bird lands on the telephone wire midway between the poles, the wire sags 0.200 m. Draw a free-body diagram of the bird. How much tension does the bird produce in the wire? Ignore the weight of the wire.

24. A bag of cement of weight 325 N hangs from three wires as shown in Figure P5.24. Two of the wires make angles $\theta_1 = 60.0°$ and $\theta_2 = 25.0°$ with the horizontal. If the system is in equilibrium, find the tensions T_1, T_2, and T_3 in the wires.

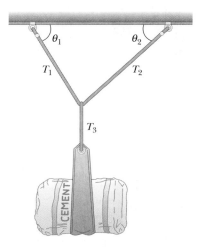

Figure P5.24 Problems 24 and 25.

25. A bag of cement of weight F_g hangs from three wires as shown in Figure P5.24. Two of the wires make angles θ_1 and θ_2 with the horizontal. If the system is in equilibrium, show that the tension in the left-hand wire is

$$T_1 = F_g \cos \theta_2 / \sin(\theta_1 + \theta_2)$$

26. You are a judge in a children's kite-flying contest, and two children will win prizes for the kites that pull most strongly and least strongly on their strings. To measure string tensions, you borrow a weight hanger, some slotted weights, and a protractor from your physics teacher and use the following protocol, illustrated in Figure P5.26: Wait for a child to get her kite well-controlled, hook the hanger onto the kite string about 30 cm from her hand, pile on weights until that section of string is horizontal, record the mass required, and record the angle between the horizontal and the string running up to the kite. (a) Explain how this method works. As you construct your explanation, imagine that the children's parents ask you about your method, that they might make false assumptions about your ability without concrete evidence, and that your explanation is an opportunity to give them confidence in your evaluation technique. (b) Find the string tension if the mass required to make the string horizontal is 132 g and the angle of the kite string is 46.3°.

Figure P5.26

27. The systems shown in Figure P5.27 are in equilibrium. If the spring scales are calibrated in newtons, what do they read? (Neglect the masses of the pulleys and strings, and assume the incline is frictionless.)

28. A fire helicopter carries a 620-kg bucket of water at the end of a cable 20.0 m long. As the aircraft flies back from a fire at a constant speed of 40.0 m/s, the cable makes an angle of 40.0° with respect to the vertical.
(a) Determine the force of air resistance on the bucket.
(b) After filling the bucket with sea water, the pilot re-

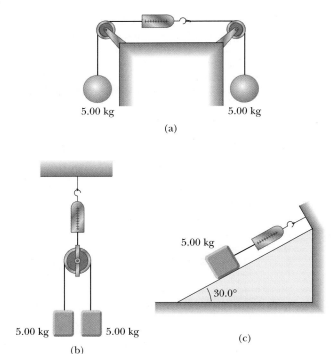

5.00 kg   5.00 kg

(a)

5.00 kg

30.0°

(c)

5.00 kg   5.00 kg

(b)

Figure P5.27

turns to the fire at the same speed with the bucket now making an angle of 7.00° with the vertical. What is the mass of the water in the bucket?

WEB **29.** A 1.00-kg mass is observed to accelerate at 10.0 m/s² in a direction 30.0° north of east (Fig. P5.29). The force $\mathbf{F}_2$ acting on the mass has a magnitude of 5.00 N and is directed north. Determine the magnitude and direction of the force $\mathbf{F}_1$ acting on the mass.

$\mathbf{F}_2$

$a = 10.0 \text{ m/s}^2$

30.0°

1.00 kg

$\mathbf{F}_1$

Figure P5.29

30. A simple accelerometer is constructed by suspending a mass m from a string of length L that is tied to the top of a cart. As the cart is accelerated the string-mass system makes a constant angle θ with the vertical.
(a) Assuming that the string mass is negligible compared with m, derive an expression for the cart's acceleration in terms of θ and show that it is independent of

the mass m and the length L. (b) Determine the acceleration of the cart when $\theta = 23.0°$.

31. Two people pull as hard as they can on ropes attached to a boat that has a mass of 200 kg. If they pull in the same direction, the boat has an acceleration of 1.52 m/s^2 to the right. If they pull in opposite directions, the boat has an acceleration of 0.518 m/s^2 to the left. What is the force exerted by each person on the boat? (Disregard any other forces on the boat.)

32. Draw a free-body diagram for a block that slides down a frictionless plane having an inclination of $\theta = 15.0°$ (Fig. P5.32). If the block starts from rest at the top and the length of the incline is 2.00 m, find (a) the acceleration of the block and (b) its speed when it reaches the bottom of the incline.

Figure P5.32

WEB 33. A block is given an initial velocity of 5.00 m/s up a frictionless 20.0° incline. How far up the incline does the block slide before coming to rest?

34. Two masses are connected by a light string that passes over a frictionless pulley, as in Figure P5.34. If the incline is frictionless and if $m_1 = 2.00$ kg, $m_2 = 6.00$ kg, and $\theta = 55.0°$, find (a) the accelerations of the masses, (b) the tension in the string, and (c) the speed of each mass 2.00 s after being released from rest.

Figure P5.34

 35. Two masses m_1 and m_2 situated on a frictionless, horizontal surface are connected by a light string. A force **F** is exerted on one of the masses to the right (Fig. P5.35). Determine the acceleration of the system and the tension T in the string.

Figure P5.35   Problems 35 and 51.

36. Two masses of 3.00 kg and 5.00 kg are connected by a light string that passes over a frictionless pulley, as was shown in Figure 5.15a. Determine (a) the tension in the string, (b) the acceleration of each mass, and (c) the distance each mass will move in the first second of motion if they start from rest.

37. In the system shown in Figure P5.37, a horizontal force F_x acts on the 8.00-kg mass. The horizontal surface is frictionless. (a) For what values of F_x does the 2.00-kg mass accelerate upward? (b) For what values of F_x is the tension in the cord zero? (c) Plot the acceleration of the 8.00-kg mass versus F_x. Include values of F_x from -100 N to $+100$ N.

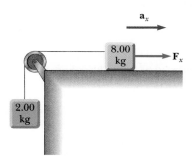

Figure P5.37

38. Mass m_1 on a frictionless horizontal table is connected to mass m_2 by means of a very light pulley P_1 and a light fixed pulley P_2 as shown in Figure P5.38. (a) If a_1 and a_2

Figure P5.38

are the accelerations of m_1 and m_2, respectively, what is the relationship between these accelerations? Express (b) the tensions in the strings and (c) the accelerations a_1 and a_2 in terms of the masses m_1 and m_2 and g.

39. A 72.0-kg man stands on a spring scale in an elevator. Starting from rest, the elevator ascends, attaining its maximum speed of 1.20 m/s in 0.800 s. It travels with this constant speed for the next 5.00 s. The elevator then undergoes a uniform acceleration in the negative y direction for 1.50 s and comes to rest. What does the spring scale register (a) before the elevator starts to move? (b) during the first 0.800 s? (c) while the elevator is traveling at constant speed? (d) during the time it is slowing down?

Section 5.8   Forces of Friction

40. The coefficient of static friction is 0.800 between the soles of a sprinter's running shoes and the level track surface on which she is running. Determine the maximum acceleration she can achieve. Do you need to know that her mass is 60.0 kg?

41. A 25.0-kg block is initially at rest on a horizontal surface. A horizontal force of 75.0 N is required to set the block in motion. After it is in motion, a horizontal force of 60.0 N is required to keep the block moving with constant speed. Find the coefficients of static and kinetic friction from this information.

42. A racing car accelerates uniformly from 0 to 80.0 mi/h in 8.00 s. The external force that accelerates the car is the frictional force between the tires and the road. If the tires do not slip, determine the minimum coefficient of friction between the tires and the road.

43. A car is traveling at 50.0 mi/h on a horizontal highway. (a) If the coefficient of friction between road and tires on a rainy day is 0.100, what is the minimum distance in which the car will stop? (b) What is the stopping distance when the surface is dry and $\mu_s = 0.600$?

44. A woman at an airport is towing her 20.0-kg suitcase at constant speed by pulling on a strap at an angle of θ above the horizontal (Fig. P5.44). She pulls on the strap with a 35.0-N force, and the frictional force on the suitcase is 20.0 N. Draw a free-body diagram for the suitcase. (a) What angle does the strap make with the horizontal? (b) What normal force does the ground exert on the suitcase?

WEB 45. A 3.00-kg block starts from rest at the top of a 30.0° incline and slides a distance of 2.00 m down the incline in 1.50 s. Find (a) the magnitude of the acceleration of the block, (b) the coefficient of kinetic friction between block and plane, (c) the frictional force acting on the block, and (d) the speed of the block after it has slid 2.00 m.

46. To determine the coefficients of friction between rubber and various surfaces, a student uses a rubber eraser and an incline. In one experiment the eraser begins to slip down the incline when the angle of inclination is

Figure P5.44

36.0° and then moves down the incline with constant speed when the angle is reduced to 30.0°. From these data, determine the coefficients of static and kinetic friction for this experiment.

47. A boy drags his 60.0-N sled at constant speed up a 15.0° hill. He does so by pulling with a 25.0-N force on a rope attached to the sled. If the rope is inclined at 35.0° to the horizontal, (a) what is the coefficient of kinetic friction between sled and snow? (b) At the top of the hill, he jumps on the sled and slides down the hill. What is the magnitude of his acceleration down the slope?

48. Determine the stopping distance for a skier moving down a slope with friction with an initial speed of 20.0 m/s (Fig. P5.48). Assume $\mu_k = 0.180$ and $\theta = 5.00°$.

Figure P5.48

49. A 9.00-kg hanging weight is connected by a string over a pulley to a 5.00-kg block that is sliding on a flat table (Fig. P5.49). If the coefficient of kinetic friction is 0.200, find the tension in the string.

50. Three blocks are connected on a table as shown in Figure P5.50. The table is rough and has a coefficient of ki-

5.00 kg

9.00 kg

Figure P5.49

1.00 kg

4.00 kg

2.00 kg

Figure P5.50

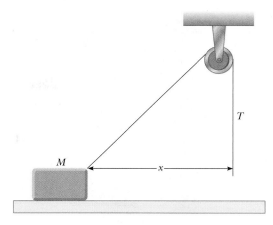

T

M

x

Figure P5.52

50.0°

P

Figure P5.53

netic friction of 0.350. The three masses are 4.00 kg, 1.00 kg, and 2.00 kg, and the pulleys are frictionless. Draw a free-body diagram for each block. (a) Determine the magnitude and direction of the acceleration of each block. (b) Determine the tensions in the two cords.

 51. Two blocks connected by a rope of negligible mass are being dragged by a horizontal force **F** (see Fig. P5.35). Suppose that $F = 68.0$ N, $m_1 = 12.0$ kg, $m_2 = 18.0$ kg, and the coefficient of kinetic friction between each block and the surface is 0.100. (a) Draw a free-body diagram for each block. (b) Determine the tension T and the magnitude of the acceleration of the system.

52. A block of mass 2.20 kg is accelerated across a rough surface by a rope passing over a pulley, as shown in Figure P5.52. The tension in the rope is 10.0 N, and the pulley is 10.0 cm above the top of the block. The coefficient of kinetic friction is 0.400. (a) Determine the acceleration of the block when $x = 0.400$ m. (b) Find the value of x at which the acceleration becomes zero.

53. A block of mass 3.00 kg is pushed up against a wall by a force **P** that makes a 50.0° angle with the horizontal as shown in Figure P5.53. The coefficient of static friction between the block and the wall is 0.250. Determine the possible values for the magnitude of **P** that allow the block to remain stationary.

ADDITIONAL PROBLEMS

54. A time-dependent force $\mathbf{F} = (8.00\mathbf{i} - 4.00t\mathbf{j})$ N (where t is in seconds) is applied to a 2.00-kg object initially at rest. (a) At what time will the object be moving with a speed of 15.0 m/s? (b) How far is the object from its initial position when its speed is 15.0 m/s? (c) What is the object's displacement at the time calculated in (a)?

55. An inventive child named Pat wants to reach an apple in a tree without climbing the tree. Sitting in a chair connected to a rope that passes over a frictionless pulley (Fig. P5.55), Pat pulls on the loose end of the rope with such a force that the spring scale reads 250 N. Pat's weight is 320 N, and the chair weighs 160 N. (a) Draw free-body diagrams for Pat and the chair considered as separate systems, and draw another diagram for Pat and the chair considered as one system. (b) Show that the acceleration of the system is *upward* and find its magnitude. (c) Find the force Pat exerts on the chair.

56. Three blocks are in contact with each other on a frictionless, horizontal surface, as in Figure P5.56. A horizontal force **F** is applied to m_1. If $m_1 = 2.00$ kg, $m_2 = 3.00$ kg, $m_3 = 4.00$ kg, and $F = 18.0$ N, draw a separate free-body diagram for each block and find (a) the acceleration of the blocks, (b) the *resultant* force on each block, and (c) the magnitudes of the contact forces between the blocks.

Figure P5.55

Figure P5.56

57. A high diver of mass 70.0 kg jumps off a board 10.0 m above the water. If his downward motion is stopped 2.00 s after he enters the water, what average upward force did the water exert on him?

58. Review Problem.   Consider the three connected objects shown in Figure P5.58. If the inclined plane is frictionless and the system is in equilibrium, find (in terms of m, g, and θ) (a) the mass M and (b) the tensions T_1 and T_2. If the value of M is double the value found in part (a), find (c) the acceleration of each object, and (d) the tensions T_1 and T_2. If the coefficient of static friction between m and $2m$ and the inclined plane is μ_s,

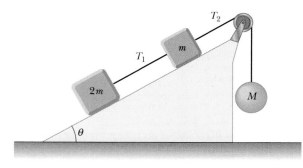

Figure P5.58

and the system is in equilibrium, find (e) the minimum value of M and (f) the maximum value of M. (g) Compare the values of T_2 when M has its minimum and maximum values.

WEB **59.** A mass M is held in place by an applied force $\mathbf{F}$ and a pulley system as shown in Figure P5.59. The pulleys are massless and frictionless. Find (a) the tension in each section of rope, T_1, T_2, T_3, T_4, and T_5 and (b) the magnitude of $\mathbf{F}$. (*Hint:* Draw a free-body diagram for each pulley.)

Figure P5.59

60. Two forces, given by $\mathbf{F}_1 = (-6.00\mathbf{i} - 4.00\mathbf{j})$ N and $\mathbf{F}_2 = (-3.00\mathbf{i} + 7.00\mathbf{j})$ N, act on a particle of mass 2.00 kg that is initially at rest at coordinates $(-2.00$ m, $+4.00$ m). (a) What are the components of the particle's velocity at $t = 10.0$ s? (b) In what direction is the particle moving at $t = 10.0$ s? (c) What displacement does the particle undergo during the first 10.0 s? (d) What are the coordinates of the particle at $t = 10.0$ s?

61. A crate of weight $\mathbf{F}_g$ is pushed by a force $\mathbf{P}$ on a horizontal floor. (a) If the coefficient of static friction is μ_s and $\mathbf{P}$ is directed at an angle θ below the horizontal, show that the minimum value of P that will move the crate is given by

$$P = \mu_s F_g \sec \theta (1 - \mu_s \tan \theta)^{-1}$$

(b) Find the minimum value of P that can produce motion when $\mu_s = 0.400$, $F_g = 100$ N, and $\theta = 0°$, 15.0°, 30.0°, 45.0°, and 60.0°.

62. **Review Problem.** A block of mass $m = 2.00$ kg is released from rest $h = 0.500$ m from the surface of a table, at the top of a $\theta = 30.0°$ incline as shown in Figure P5.62. The frictionless incline is fixed on a table of height $H = 2.00$ m. (a) Determine the acceleration of the block as it slides down the incline. (b) What is the velocity of the block as it leaves the incline? (c) How far from the table will the block hit the floor? (d) How much time has elapsed between when the block is released and when it hits the floor? (e) Does the mass of the block affect any of the above calculations?

65. A block of mass $m = 2.00$ kg rests on the left edge of a block of larger mass $M = 8.00$ kg. The coefficient of kinetic friction between the two blocks is 0.300, and the surface on which the 8.00-kg block rests is frictionless. A constant horizontal force of magnitude $F = 10.0$ N is applied to the 2.00-kg block, setting it in motion as shown in Figure P5.65a. If the length L that the leading edge of the smaller block travels on the larger block is 3.00 m, (a) how long will it take before this block makes it to the right side of the 8.00-kg block, as shown in Figure P5.65b? (*Note:* Both blocks are set in motion when **F** is applied.) (b) How far does the 8.00-kg block move in the process?

(a)

(b)

Figure P5.65

Figure P5.62

63. A 1.30-kg toaster is not plugged in. The coefficient of static friction between the toaster and a horizontal countertop is 0.350. To make the toaster start moving, you carelessly pull on its electric cord. (a) For the cord tension to be as small as possible, you should pull at what angle above the horizontal? (b) With this angle, how large must the tension be?

64. A 2.00-kg aluminum block and a 6.00-kg copper block are connected by a light string over a frictionless pulley. They sit on a steel surface, as shown in Figure P5.64, and $\theta = 30.0°$. Do they start to move once any holding mechanism is released? If so, determine (a) their acceleration and (b) the tension in the string. If not, determine the sum of the magnitudes of the forces of friction acting on the blocks.

66. A student is asked to measure the acceleration of a cart on a "frictionless" inclined plane as seen in Figure P5.32, using an air track, a stopwatch, and a meter stick. The height of the incline is measured to be 1.774 cm, and the total length of the incline is measured to be $d = 127.1$ cm. Hence, the angle of inclination θ is determined from the relation $\sin \theta = 1.774/127.1$. The cart is released from rest at the top of the incline, and its displacement x along the incline is measured versus time, where $x = 0$ refers to the initial position of the cart. For x values of 10.0 cm, 20.0 cm, 35.0 cm, 50.0 cm, 75.0 cm, and 100 cm, the measured times to undergo these displacements (averaged over five runs) are 1.02 s, 1.53 s, 2.01 s, 2.64 s, 3.30 s, and 3.75 s, respectively. Construct a graph of x versus t^2, and perform a linear least-squares fit to the data. Determine the acceleration of the cart from the slope of this graph, and compare it with the value you would get using $a' = g \sin \theta$, where $g = 9.80$ m/s².

67. A 2.00-kg block is placed on top of a 5.00-kg block as in Figure P5.67. The coefficient of kinetic friction between the 5.00-kg block and the surface is 0.200. A horizontal force **F** is applied to the 5.00-kg block. (a) Draw a free-body diagram for each block. What force accelerates the 2.00-kg block? (b) Calculate the magnitude of the force necessary to pull both blocks to the right with an

Figure P5.64

Figure P5.67

acceleration of 3.00 m/s². (c) Find the minimum coefficient of static friction between the blocks such that the 2.00-kg block does not slip under an acceleration of 3.00 m/s².

68. A 5.00-kg block is placed on top of a 10.0-kg block (Fig. P5.68). A horizontal force of 45.0 N is applied to the 10.0-kg block, and the 5.00-kg block is tied to the wall. The coefficient of kinetic friction between all surfaces is 0.200. (a) Draw a free-body diagram for each block and identify the action–reaction forces between the blocks. (b) Determine the tension in the string and the magnitude of the acceleration of the 10.0-kg block.

Figure P5.68

69. What horizontal force must be applied to the cart shown in Figure P5.69 so that the blocks remain stationary relative to the cart? Assume all surfaces, wheels, and pulley are frictionless. (*Hint:* Note that the force exerted by the string accelerates m_1.)

Figure P5.69   Problems 69 and 70.

70. Initially the system of masses shown in Figure P5.69 is held motionless. All surfaces, pulley, and wheels are frictionless. Let the force **F** be zero and assume that m_2 can move only vertically. At the instant after the system of masses is released, find (a) the tension T in the string, (b) the acceleration of m_2, (c) the acceleration of M, and (d) the acceleration of m_1. (*Note:* The pulley accelerates along with the cart.)

71. A block of mass 5.00 kg sits on top of a second block of mass 15.0 kg, which in turn sits on a horizontal table. The coefficients of friction between the two blocks are $\mu_s = 0.300$ and $\mu_k = 0.100$. The coefficients of friction between the lower block and the rough table are $\mu_s = 0.500$ and $\mu_k = 0.400$. You apply a constant horizontal force to the lower block, just large enough to make this block start sliding out from between the upper block and the table. (a) Draw a free-body diagram of each block, naming the forces acting on each. (b) Determine the magnitude of each force on each block at the instant when you have started pushing but motion has not yet started. (c) Determine the acceleration you measure for each block.

72. Two blocks of mass 3.50 kg and 8.00 kg are connected by a string of negligible mass that passes over a frictionless pulley (Fig. P5.72). The inclines are frictionless. Find (a) the magnitude of the acceleration of each block and (b) the tension in the string.

Figure P5.72   Problems 72 and 73.

73. The system shown in Figure P5.72 has an acceleration of magnitude 1.50 m/s². Assume the coefficients of kinetic friction between block and incline are the same for both inclines. Find (a) the coefficient of kinetic friction and (b) the tension in the string.

74. In Figure P5.74, a 500-kg horse pulls a sledge of mass 100 kg. The system (horse plus sledge) has a forward acceleration of 1.00 m/s² when the frictional force exerted on the sledge is 500 N. Find (a) the tension in the connecting rope and (b) the magnitude and direction of the force of friction exerted on the horse. (c) Verify that the total forces of friction the ground exerts on the system will give the system an acceleration of 1.00 m/s².

75. A van accelerates down a hill (Fig. P5.75), going from rest to 30.0 m/s in 6.00 s. During the acceleration, a toy ($m = 0.100$ kg) hangs by a string from the van's ceiling. The acceleration is such that the string remains perpendicular to the ceiling. Determine (a) the angle θ and (b) the tension in the string.

Figure P5.74

Figure P5.75

76. A mobile is formed by supporting four metal butterflies of equal mass m from a string of length L. The points of support are evenly spaced a distance ℓ apart as shown in Figure P5.76. The string forms an angle θ_1 with the ceiling at each end point. The center section of string is horizontal. (a) Find the tension in each section of string in terms of θ_1, m, and g. (b) Find the angle θ_2, in terms of θ_1, that the sections of string between the outside butterflies and the inside butterflies form with the horizontal. (c) Show that the distance D between the end points of the string is

$$D = \frac{L}{5}\left\{2\cos\theta_1 + 2\cos\left[\tan^{-1}\left(\frac{1}{2}\tan\theta_1\right)\right] + 1\right\}$$

77. Before 1960 it was believed that the maximum attainable coefficient of static friction for an automobile tire was less than 1. Then about 1962, three companies independently developed racing tires with coefficients of 1.6. Since then, tires have improved, as illustrated in this problem. According to the 1990 Guinness Book of Records, the fastest time in which a piston-engine car initially at rest has covered a distance of one-quarter mile is 4.96 s. This record was set by Shirley Muldowney in September 1989 (Fig. P5.77). (a) Assuming that the rear wheels nearly lifted the front wheels off the pavement, what minimum value of μ_s is necessary to achieve the record time? (b) Suppose Muldowney were able to double her engine power, keeping other things equal. How would this change affect the elapsed time?

Figure P5.77 *(Mike Powell/AllSport USA)*

78. An 8.40-kg mass slides down a fixed, frictionless inclined plane. Use a computer to determine and tabulate the normal force exerted on the mass and its acceleration for a series of incline angles (measured from the horizontal) ranging from 0 to 90° in 5° increments. Plot a graph of the normal force and the acceleration as functions of the incline angle. In the limiting cases of 0 and 90°, are your results consistent with the known behavior?

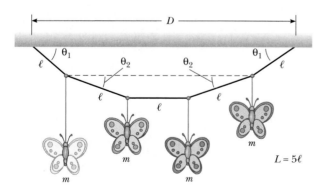

Figure P5.76

$L = 5\ell$

ANSWERS TO QUICK QUIZZES

5.1 (a) True. Newton's first law tells us that motion requires no force: An object in motion continues to move at constant velocity in the absence of external forces. (b) True. A stationary object can have several forces acting on it, but if the vector sum of all these external forces is zero, there is no net force and the object remains stationary. It also is possible to have a net force and no motion, but only for an instant. A ball tossed vertically upward stops at the peak of its path for an infinitesimally short time, but the force of gravity is still acting on it. Thus, although **v** = 0 at the peak, the net force acting on the ball is *not* zero.

5.2 No. Direction of motion is part of an object's *velocity,* and force determines the direction of acceleration, not that of velocity.

5.3 (a) Force of gravity. (b) Force of gravity. The only external force acting on the ball at *all* points in its trajectory is the downward force of gravity.

5.4 As the person steps out of the boat, he pushes against it with his foot, expecting the boat to push back on him so that he accelerates toward the dock. However, because the boat is untied, the force exerted by the foot causes the boat to scoot away from the dock. As a result, the person is not able to exert a very large force on the boat before it moves out of reach. Therefore, the boat does not exert a very large reaction force on him, and he ends up not being accelerated sufficiently to make it to the dock. Consequently, he falls into the water instead. If a small dog were to jump from the untied boat toward the dock, the force exerted by the boat on the dog would probably be enough to ensure the dog's successful landing because of the dog's small mass.

5.5 (a) The same force is experienced by both. The fly and bus experience forces that are equal in magnitude but opposite in direction. (b) The fly. Because the fly has such a small mass, it undergoes a very large acceleration. The huge mass of the bus means that it more effectively resists any change in its motion.

5.6 (b) The crate accelerates to the right. Because the only horizontal force acting on it is the force of static friction between its bottom surface and the truck bed, that force must also be directed to the right.

Calvin and Hobbes   by Bill Watterson

This sky diver is falling at more than 50 m/s (120 mi/h), but once her parachute opens, her downward velocity will be greatly reduced. Why does she slow down rapidly when her chute opens, enabling her to fall safely to the ground? If the chute does not function properly, the sky diver will almost certainly be seriously injured. What force exerted on her limits her maximum speed? *(Guy Savage/Photo Researchers, Inc.)*

c h a p t e r

Circular Motion and Other Applications of Newton's Laws

Chapter Outline

In the preceding chapter we introduced Newton's laws of motion and applied them to situations involving linear motion. Now we discuss motion that is slightly more complicated. For example, we shall apply Newton's laws to objects traveling in circular paths. Also, we shall discuss motion observed from an accelerating frame of reference and motion in a viscous medium. For the most part, this chapter is a series of examples selected to illustrate the application of Newton's laws to a wide variety of circumstances.

6.1   NEWTON'S SECOND LAW APPLIED TO UNIFORM CIRCULAR MOTION

In Section 4.4 we found that a particle moving with uniform speed v in a circular path of radius r experiences an acceleration $\mathbf{a}_r$ that has a magnitude

$$a_r = \frac{v^2}{r}$$

The acceleration is called the *centripetal acceleration* because $\mathbf{a}_r$ is directed toward the center of the circle. Furthermore, $\mathbf{a}_r$ is *always* perpendicular to $\mathbf{v}$. (If there were a component of acceleration parallel to $\mathbf{v}$, the particle's speed would be changing.)

Consider a ball of mass m that is tied to a string of length r and is being whirled at constant speed in a horizontal circular path, as illustrated in Figure 6.1. Its weight is supported by a low-friction table. Why does the ball move in a circle? Because of its inertia, the tendency of the ball is to move in a straight line; however, the string prevents motion along a straight line by exerting on the ball a force that makes it follow the circular path. This force is directed along the string toward the center of the circle, as shown in Figure 6.1. This force can be any one of our familiar forces causing an object to follow a circular path.

If we apply Newton's second law along the radial direction, we find that the value of the net force causing the centripetal acceleration can be evaluated:

Force causing centripetal acceleration

$$\sum F_r = ma_r = m\frac{v^2}{r} \tag{6.1}$$

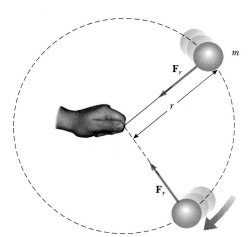

Figure 6.1   Overhead view of a ball moving in a circular path in a horizontal plane. A force $\mathbf{F}_r$ directed toward the center of the circle keeps the ball moving in its circular path.

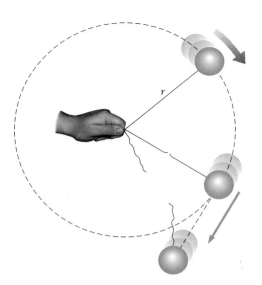

Figure 6.2 When the string breaks, the ball moves in the direction tangent to the circle.

A force causing a centripetal acceleration acts toward the center of the circular path and causes a change in the direction of the velocity vector. If that force should vanish, the object would no longer move in its circular path; instead, it would move along a straight-line path tangent to the circle. This idea is illustrated in Figure 6.2 for the ball whirling at the end of a string. If the string breaks at some instant, the ball moves along the straight-line path tangent to the circle at the point where the string broke.

An athlete in the process of throwing the hammer at the 1996 Olympic Games in Atlanta, Georgia. The force exerted by the chain is the force causing the circular motion. Only when the athlete releases the hammer will it move along a straight-line path tangent to the circle. *(Mike Powell/AllSport USA)*

Quick Quiz 6.1

Is it possible for a car to move in a circular path in such a way that it has a tangential acceleration but no centripetal acceleration?

CONCEPTUAL EXAMPLE 6.1   Forces That Cause Centripetal Acceleration

The force causing centripetal acceleration is sometimes called a *centripetal force*. We are familiar with a variety of forces in nature—friction, gravity, normal forces, tension, and so forth. Should we add *centripetal* force to this list?

Solution   No; centripetal force *should not* be added to this list. This is a pitfall for many students. Giving the force causing circular motion a name—centripetal force—leads many students to consider it a new kind of force rather than a new *role* for force. A common mistake in force diagrams is to draw all the usual forces and then to add another vector for the centripetal force. But it is not a separate force—it is simply one of our familiar forces *acting in the role of a force that causes a circular motion*.

Consider some examples. For the motion of the Earth around the Sun, the centripetal force is *gravity*. For an object sitting on a rotating turntable, the centripetal force is *friction*. For a rock whirled on the end of a string, the centripetal force is the force of *tension* in the string. For an amusement-park patron pressed against the inner wall of a rapidly rotating circular room, the centripetal force is the *normal force* exerted by the wall. What's more, the centripetal force could be a combination of two or more forces. For example, as a Ferris-wheel rider passes through the lowest point, the centripetal force on her is the difference between the normal force exerted by the seat and her weight.

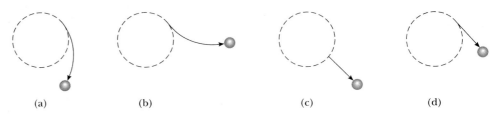

Figure 6.3   A ball that had been moving in a circular path is acted on by various external forces that change its path.

Quick Quiz 6.2

QuickLab

Tie a string to a tennis ball, swing it in a circle, and then, while it is swinging, let go of the string to verify your answer to the last part of Quick Quiz 6.2.

A ball is following the dotted circular path shown in Figure 6.3 under the influence of a force. At a certain instant of time, the force on the ball changes abruptly to a new force, and the ball follows the paths indicated by the solid line with an arrowhead in each of the four parts of the figure. For each part of the figure, describe the magnitude and direction of the force required to make the ball move in the solid path. If the dotted line represents the path of a ball being whirled on the end of a string, which path does the ball follow if the string breaks?

Let us consider some examples of uniform circular motion. In each case, be sure to recognize the external force (or forces) that causes the body to move in its circular path.

EXAMPLE 6.2   **How Fast Can It Spin?**

A ball of mass 0.500 kg is attached to the end of a cord 1.50 m long. The ball is whirled in a horizontal circle as was shown in Figure 6.1. If the cord can withstand a maximum tension of 50.0 N, what is the maximum speed the ball can attain before the cord breaks? Assume that the string remains horizontal during the motion.

Solution   It is difficult to know what might be a reasonable value for the answer. Nonetheless, we know that it cannot be too large, say 100 m/s, because a person cannot make a ball move so quickly. It makes sense that the stronger the cord, the faster the ball can twirl before the cord breaks. Also, we expect a more massive ball to break the cord at a lower speed. (Imagine whirling a bowling ball!)

Because the force causing the centripetal acceleration in this case is the force **T** exerted by the cord on the ball, Equation 6.1 yields for $\Sigma F_r = ma_r$

$$T = m\frac{v^2}{r}$$

Solving for v, we have

$$v = \sqrt{\frac{Tr}{m}}$$

This shows that v increases with T and decreases with larger m, as we expect to see—for a given v, a large mass requires a large tension and a small mass needs only a small tension. The maximum speed the ball can have corresponds to the maximum tension. Hence, we find

$$v_{\text{max}} = \sqrt{\frac{T_{\text{max}}r}{m}} = \sqrt{\frac{(50.0\ \text{N})(1.50\ \text{m})}{0.500\ \text{kg}}}$$

$$= \boxed{12.2\ \text{m/s}}$$

Exercise   Calculate the tension in the cord if the speed of the ball is 5.00 m/s.

Answer   8.33 N.

EXAMPLE 6.3   **The Conical Pendulum**

A small object of mass m is suspended from a string of length L. The object revolves with constant speed v in a horizontal circle of radius r, as shown in Figure 6.4. (Because the string sweeps out the surface of a cone, the system is known as a *conical pendulum*.) Find an expression for v.

Solution   Let us choose θ to represent the angle between string and vertical. In the free-body diagram shown in Figure 6.4, the force **T** exerted by the string has been resolved into a vertical component $T \cos\theta$ and a horizontal component $T \sin\theta$ acting toward the center of revolution. Because the object does not accelerate in the vertical direction, $\Sigma F_y = ma_y = 0$, and the upward vertical component of **T** must balance the downward force of gravity. Therefore,

$$(1) \qquad T\cos\theta = mg$$

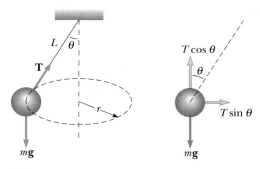

T

mg

$T\cos\theta$

θ

$T\sin\theta$

mg

Figure 6.4 The conical pendulum and its free-body diagram.

Because the force providing the centripetal acceleration in this example is the component $T\sin\theta$, we can use Newton's second law and Equation 6.1 to obtain

$$(2) \qquad \Sigma F_r = T\sin\theta = ma_r = \frac{mv^2}{r}$$

Dividing (2) by (1) and remembering that $\sin\theta/\cos\theta = \tan\theta$, we eliminate T and find that

$$\tan\theta = \frac{v^2}{rg}$$

$$v = \sqrt{rg\tan\theta}$$

From the geometry in Figure 6.4, we note that $r = L\sin\theta$; therefore,

$$v = \boxed{\sqrt{Lg\sin\theta\tan\theta}}$$

Note that the speed is independent of the mass of the object.

 EXAMPLE 6.4 ▶ **What Is the Maximum Speed of the Car?**

A 1 500-kg car moving on a flat, horizontal road negotiates a curve, as illustrated in Figure 6.5. If the radius of the curve is 35.0 m and the coefficient of static friction between the tires and dry pavement is 0.500, find the maximum speed the car can have and still make the turn successfully.

f_s

(a)

n

f_s

mg

(b)

Figure 6.5 (a) The force of static friction directed toward the center of the curve keeps the car moving in a circular path. (b) The free-body diagram for the car.

Solution From experience, we should expect a maximum speed less than 50 m/s. (A convenient mental conversion is that 1 m/s is roughly 2 mi/h.) In this case, the force that enables the car to remain in its circular path is the force of static friction. (Because no slipping occurs at the point of contact between road and tires, the acting force is a force of static friction directed toward the center of the curve. If this force of static friction were zero—for example, if the car were on an icy road—the car would continue in a straight line and slide off the road.) Hence, from Equation 6.1 we have

$$(1) \qquad f_s = m\frac{v^2}{r}$$

The maximum speed the car can have around the curve is the speed at which it is on the verge of skidding outward. At this point, the friction force has its maximum value $f_{s,\text{max}} = \mu_s n$. Because the car is on a horizontal road, the magnitude of the normal force equals the weight ($n = mg$) and thus $f_{s,\text{max}} = \mu_s mg$. Substituting this value for f_s into (1), we find that the maximum speed is

$$v_{\text{max}} = \sqrt{\frac{f_{s,\text{max}}r}{m}} = \sqrt{\frac{\mu_s mgr}{m}} = \sqrt{\mu_s gr}$$

$$= \sqrt{(0.500)(9.80\text{ m/s}^2)(35.0\text{ m})} = \boxed{13.1\text{ m/s}}$$

Note that the maximum speed does not depend on the mass of the car. That is why curved highways do not need multiple speed limit signs to cover the various masses of vehicles using the road.

Exercise On a wet day, the car begins to skid on the curve when its speed reaches 8.00 m/s. What is the coefficient of static friction in this case?

Answer 0.187.

EXAMPLE 6.5   The Banked Exit Ramp

A civil engineer wishes to design a curved exit ramp for a highway in such a way that a car will not have to rely on friction to round the curve without skidding. In other words, a car moving at the designated speed can negotiate the curve even when the road is covered with ice. Such a ramp is usually *banked;* this means the roadway is tilted toward the inside of the curve. Suppose the designated speed for the ramp is to be 13.4 m/s (30.0 mi/h) and the radius of the curve is 50.0 m. At what angle should the curve be banked?

Solution On a level (unbanked) road, the force that causes the centripetal acceleration is the force of static friction between car and road, as we saw in the previous example. However, if the road is banked at an angle θ, as shown in Figure 6.6, the normal force **n** has a horizontal component n

sin θ pointing toward the center of the curve. Because the ramp is to be designed so that the force of static friction is zero, only the component $n \sin \theta$ causes the centripetal acceleration. Hence, Newton's second law written for the radial direction gives

$$(1) \qquad \sum F_r = n \sin \theta = \frac{mv^2}{r}$$

The car is in equilibrium in the vertical direction. Thus, from $\sum F_y = 0$, we have

$$(2) \qquad n \cos \theta = mg$$

Dividing (1) by (2) gives

$$\tan \theta = \frac{v^2}{rg}$$

$$\theta = \tan^{-1}\left[\frac{(13.4 \text{ m/s})^2}{(50.0 \text{ m})(9.80 \text{ m/s}^2)}\right] = \boxed{20.1°}$$

If a car rounds the curve at a speed less than 13.4 m/s, friction is needed to keep it from sliding down the bank (to the left in Fig. 6.6). A driver who attempts to negotiate the curve at a speed greater than 13.4 m/s has to depend on friction to keep from sliding up the bank (to the right in Fig. 6.6). The banking angle is independent of the mass of the vehicle negotiating the curve.

Exercise Write Newton's second law applied to the radial direction when a frictional force $\mathbf{f}_s$ is directed down the bank, toward the center of the curve.

Answer $n \sin \theta + f_s \cos \theta = \dfrac{mv^2}{r}$

Figure 6.6 Car rounding a curve on a road banked at an angle θ to the horizontal. When friction is neglected, the force that causes the centripetal acceleration and keeps the car moving in its circular path is the horizontal component of the normal force. Note that **n** is the *sum* of the forces exerted by the road on the wheels.

EXAMPLE 6.6   Satellite Motion

This example treats a satellite moving in a circular orbit around the Earth. To understand this situation, you must know that the gravitational force between spherical objects and small objects that can be modeled as particles having

masses m_1 and m_2 and separated by a distance r is attractive and has a magnitude

$$F_g = G\frac{m_1 m_2}{r^2}$$

where $G = 6.673 \times 10^{-11}$ N·m²/kg². This is Newton's law of gravitation, which we study in Chapter 14.

Consider a satellite of mass m moving in a circular orbit around the Earth at a constant speed v and at an altitude h above the Earth's surface, as illustrated in Figure 6.7. Determine the speed of the satellite in terms of G, h, R_E (the radius of the Earth), and M_E (the mass of the Earth).

Solution The only external force acting on the satellite is the force of gravity, which acts toward the center of the Earth

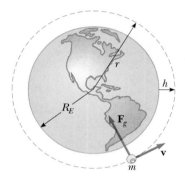

Figure 6.7 A satellite of mass m moving around the Earth at a constant speed v in a circular orbit of radius $r = R_E + h$. The force $\mathbf{F}_g$ acting on the satellite that causes the centripetal acceleration is the gravitational force exerted by the Earth on the satellite.

and keeps the satellite in its circular orbit. Therefore,

$$F_r = F_g = G\frac{M_E m}{r^2}$$

From Newton's second law and Equation 6.1 we obtain

$$G\frac{M_E m}{r^2} = m\frac{v^2}{r}$$

Solving for v and remembering that the distance r from the center of the Earth to the satellite is $r = R_E + h$, we obtain

$$(1) \qquad v = \sqrt{\frac{GM_E}{r}} = \boxed{\sqrt{\frac{GM_E}{R_E + h}}}$$

If the satellite were orbiting a different planet, its velocity would increase with the mass of the planet and decrease as the satellite's distance from the center of the planet increased.

Exercise A satellite is in a circular orbit around the Earth at an altitude of 1 000 km. The radius of the Earth is equal to 6.37×10^6 m, and its mass is 5.98×10^{24} kg. Find the speed of the satellite, and then find the *period*, which is the time it needs to make one complete revolution.

Answer 7.36×10^3 m/s; 6.29×10^3 s = 105 min.

EXAMPLE 6.7   Let's Go Loop-the-Loop!

A pilot of mass m in a jet aircraft executes a loop-the-loop, as shown in Figure 6.8a. In this maneuver, the aircraft moves in a vertical circle of radius 2.70 km at a constant speed of 225 m/s. Determine the force exerted by the seat on the pilot (a) at the bottom of the loop and (b) at the top of the loop. Express your answers in terms of the weight of the pilot mg.

Solution We expect the answer for (a) to be greater than that for (b) because at the bottom of the loop the normal and gravitational forces act in opposite directions, whereas at the top of the loop these two forces act in the same direction. It is the vector sum of these two forces that gives the force of constant magnitude that keeps the pilot moving in a circular path. To yield net force vectors with the same magnitude, the normal force at the bottom (where the normal and gravitational forces are in opposite directions) must be greater than that at the top (where the normal and gravitational forces are in the same direction). (a) The free-body diagram for the pilot at the bottom of the loop is shown in Figure 6.8b. The only forces acting on him are the downward force of gravity $\mathbf{F}_g = m\mathbf{g}$ and the upward force $\mathbf{n}_{bot}$ exerted by the seat. Because the net upward force that provides the centripetal ac-

celeration has a magnitude $n_{bot} - mg$, Newton's second law for the radial direction combined with Equation 6.1 gives

$$\sum F_r = n_{bot} - mg = m\frac{v^2}{r}$$

$$n_{bot} = mg + m\frac{v^2}{r} = mg\left(1 + \frac{v^2}{rg}\right)$$

Substituting the values given for the speed and radius gives

$$n_{bot} = mg\left[1 + \frac{(225 \text{ m/s})^2}{(2.70 \times 10^3 \text{ m})(9.80 \text{ m/s}^2)}\right] = \boxed{2.91\,mg}$$

Hence, the magnitude of the force $\mathbf{n}_{bot}$ exerted by the seat on the pilot is *greater* than the weight of the pilot by a factor of 2.91. This means that the pilot experiences an apparent weight that is greater than his true weight by a factor of 2.91.

(b) The free-body diagram for the pilot at the top of the loop is shown in Figure 6.8c. As we noted earlier, both the gravitational force exerted by the Earth and the force $\mathbf{n}_{top}$ exerted by the seat on the pilot act downward, and so the net downward force that provides the centripetal acceleration has

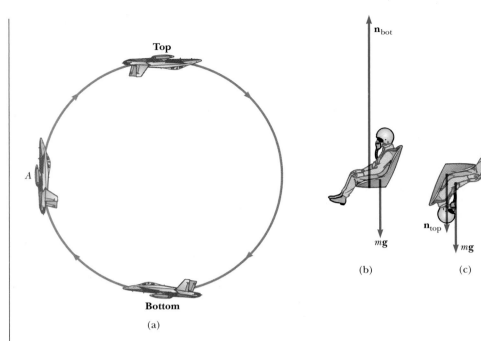

Top

A

Bottom

(a)

$\mathbf{n}_{bot}$

$m\mathbf{g}$

(b)

$\mathbf{n}_{top}$

$m\mathbf{g}$

(c)

Figure 6.8 (a) An aircraft executes a loop-the-loop maneuver as it moves in a vertical circle at constant speed. (b) Free-body diagram for the pilot at the bottom on the loop. In this position the pilot experiences an apparent weight greater than his true weight. (c) Free-body diagram for the pilot at the top of the loop.

a magnitude $n_{top} + mg$. Applying Newton's second law yields

$$\sum F_r = n_{top} + mg = m\frac{v^2}{r}$$

$$n_{top} = m\frac{v^2}{r} - mg = mg\left(\frac{v^2}{rg} - 1\right)$$

$$n_{top} = mg\left[\frac{(225 \text{ m/s})^2}{(2.70 \times 10^3 \text{ m})(9.80 \text{ m/s}^2)} - 1\right] = \boxed{0.913\,mg}$$

In this case, the magnitude of the force exerted by the seat on the pilot is *less* than his true weight by a factor of 0.913, and the pilot feels lighter.

Exercise Determine the magnitude of the radially directed force exerted on the pilot by the seat when the aircraft is at point *A* in Figure 6.8a, midway up the loop.

Answer $n_A = 1.913\,mg$ directed to the right.

Quick Quiz 6.3

A bead slides freely along a curved wire at constant speed, as shown in the overhead view of Figure 6.9. At each of the points Ⓐ, Ⓑ, and Ⓒ, draw the vector representing the force that the wire exerts on the bead in order to cause it to follow the path of the wire at that point.

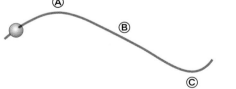

Ⓐ

Ⓑ

Ⓒ   ***Figure 6.9***

6.2   **NONUNIFORM CIRCULAR MOTION**

In Chapter 4 we found that if a particle moves with varying speed in a circular path, there is, in addition to the centripetal (radial) component of acceleration, a tangential component having magnitude dv/dt. Therefore, the force acting on the

Some examples of forces acting during circular motion. *(Left)* As these speed skaters round a curve, the force exerted by the ice on their skates provides the centripetal acceleration. *(Right)* Passengers on a "corkscrew" roller coaster. What are the origins of the forces in this example? *(Left: Bill Bachman/Photo Researchers, Inc.; Right: Robin Smith/Tony Stone Images)*

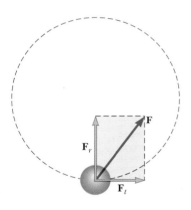

Figure 6.10 When the force acting on a particle moving in a circular path has a tangential component F_t, the particle's speed changes. The total force exerted on the particle in this case is the vector sum of the radial force and the tangential force. That is, $\mathbf{F} = \mathbf{F}_r + \mathbf{F}_t$.

particle must also have a tangential and a radial component. Because the total acceleration is $\mathbf{a} = \mathbf{a}_r + \mathbf{a}_t$, the total force exerted on the particle is $\mathbf{F} = \mathbf{F}_r + \mathbf{F}_t$, as shown in Figure 6.10. The vector $\mathbf{F}_r$ is directed toward the center of the circle and is responsible for the centripetal acceleration. The vector $\mathbf{F}_t$ tangent to the circle is responsible for the tangential acceleration, which represents a change in the speed of the particle with time. The following example demonstrates this type of motion.

EXAMPLE 6.8 ▶ Keep Your Eye on the Ball

A small sphere of mass m is attached to the end of a cord of length R and whirls in a *vertical* circle about a fixed point O, as illustrated in Figure 6.11a. Determine the tension in the cord at any instant when the speed of the sphere is v and the cord makes an angle θ with the vertical.

Solution Unlike the situation in Example 6.7, the speed is *not* uniform in this example because, at most points along the path, a tangential component of acceleration arises from the gravitational force exerted on the sphere. From the free-body diagram in Figure 6.11b, we see that the only forces acting on

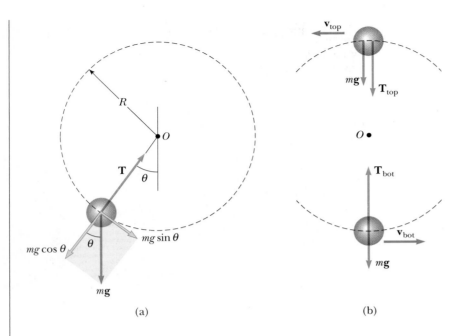

Figure 6.11 (a) Forces acting on a sphere of mass m connected to a cord of length R and rotating in a vertical circle centered at O. (b) Forces acting on the sphere at the top and bottom of the circle. The tension is a maximum at the bottom and a minimum at the top.

the sphere are the gravitational force $\mathbf{F}_g = m\mathbf{g}$ exerted by the Earth and the force $\mathbf{T}$ exerted by the cord. Now we resolve $\mathbf{F}_g$ into a tangential component $mg \sin \theta$ and a radial component $mg \cos \theta$. Applying Newton's second law to the forces acting on the sphere in the tangential direction yields

$$\sum F_t = mg \sin \theta = ma_t$$

$$a_t = g \sin \theta$$

This tangential component of the acceleration causes v to change in time because $a_t = dv/dt$.

Applying Newton's second law to the forces acting on the sphere in the radial direction and noting that both $\mathbf{T}$ and $\mathbf{a}_r$ are directed toward O, we obtain

$$\sum F_r = T - mg \cos \theta = \frac{mv^2}{R}$$

$$T = \boxed{m\left(\frac{v^2}{R} + g \cos \theta\right)}$$

Special Cases   At the top of the path, where $\theta = 180°$, we have $\cos 180° = -1$, and the tension equation becomes

$$T_{\text{top}} = m\left(\frac{v_{\text{top}}^2}{R} - g\right)$$

This is the minimum value of T. Note that at this point $a_t = 0$ and therefore the acceleration is purely radial and directed downward.

At the bottom of the path, where $\theta = 0$, we see that, because $\cos 0 = 1$,

$$T_{\text{bot}} = m\left(\frac{v_{\text{bot}}^2}{R} + g\right)$$

This is the maximum value of T. At this point, a_t is again 0 and the acceleration is now purely radial and directed upward.

Exercise   At what position of the sphere would the cord most likely break if the average speed were to increase?

Answer   At the bottom, where T has its maximum value.

Optional Section

6.3 MOTION IN ACCELERATED FRAMES

When Newton's laws of motion were introduced in Chapter 5, we emphasized that they are valid only when observations are made in an inertial frame of reference. In this section, we analyze how an observer in a noninertial frame of reference (one that is accelerating) applies Newton's second law.

To understand the motion of a system that is noninertial because an object is moving along a curved path, consider a car traveling along a highway at a high speed and approaching a curved exit ramp, as shown in Figure 6.12a. As the car takes the sharp left turn onto the ramp, a person sitting in the passenger seat slides to the right and hits the door. At that point, the force exerted on her by the door keeps her from being ejected from the car. What causes her to move toward the door? A popular, but improper, explanation is that some mysterious force acting from left to right pushes her outward. (This is often called the "centrifugal" force, but we shall not use this term because it often creates confusion.) The passenger invents this **fictitious force** to explain what is going on in her accelerated frame of reference, as shown in Figure 6.12b. (The driver also experiences this effect but holds on to the steering wheel to keep from sliding to the right.)

The phenomenon is correctly explained as follows. Before the car enters the ramp, the passenger is moving in a straight-line path. As the car enters the ramp and travels a curved path, the passenger tends to move along the original straight-line path. This is in accordance with Newton's first law: The natural tendency of a body is to continue moving in a straight line. However, if a sufficiently large force (toward the center of curvature) acts on the passenger, as in Figure 6.12c, she will move in a curved path along with the car. The origin of this force is the force of friction between her and the car seat. If this frictional force is not large enough, she will slide to the right as the car turns to the left under her. Eventually, she encounters the door, which provides a force large enough to enable her to follow the same curved path as the car. She slides toward the door not because of some mysterious outward force but because **the force of friction is not sufficiently great to allow her to travel along the circular path followed by the car.**

4.8

In general, if a particle moves with an acceleration **a** relative to an observer in an inertial frame, that observer may use Newton's second law and correctly claim that $\Sigma \mathbf{F} = m\mathbf{a}$. If another observer in an accelerated frame tries to apply Newton's second law to the motion of the particle, the person must introduce fictitious forces to make Newton's second law work. These forces "invented" by the observer in the accelerating frame appear to be real. However, we emphasize that **these fictitious forces do not exist when the motion is observed in an inertial frame.** Fictitious forces are used only in an accelerating frame and do not represent "real" forces acting on the particle. (By real forces, we mean the interaction of the particle with its environment.) If the fictitious forces are properly defined in the accelerating frame, the description of motion in this frame is equivalent to the description given by an inertial observer who considers only real forces. Usually, we analyze motions using inertial reference frames, but there are cases in which it is more convenient to use an accelerating frame.

Figure 6.12 (a) A car approaching a curved exit ramp. What causes a front-seat passenger to move toward the right-hand door? (b) From the frame of reference of the passenger, a (fictitious) force pushes her toward the right door. (c) Relative to the reference frame of the Earth, the car seat applies a leftward force to the passenger, causing her to change direction along with the rest of the car.

QuickLab

Use a string, a small weight, and a protractor to measure your acceleration as you start sprinting from a standing position.

Fictitious forces

(a)

(b)

(c)

EXAMPLE 6.9 ▶ Fictitious Forces in Linear Motion

A small sphere of mass m is hung by a cord from the ceiling of a boxcar that is accelerating to the right, as shown in Figure 6.13. According to the inertial observer at rest (Fig. 6.13a), the forces on the sphere are the force $\mathbf{T}$ exerted by the cord and the force of gravity. The inertial observer concludes that the acceleration of the sphere is the same as that of the boxcar and that this acceleration is provided by the horizontal component of $\mathbf{T}$. Also, the vertical component of $\mathbf{T}$ balances the force of gravity. Therefore, she writes Newton's second law as $\Sigma\mathbf{F} = \mathbf{T} + m\mathbf{g} = m\mathbf{a}$, which in component form becomes

Inertial observer
$$\begin{cases} (1) & \sum F_x = T \sin\theta = ma \\ (2) & \sum F_y = T \cos\theta - mg = 0 \end{cases}$$

Thus, by solving (1) and (2) simultaneously for a, the inertial observer can determine the magnitude of the car's acceleration through the relationship

$$a = g \tan\theta$$

Because the deflection of the cord from the vertical serves as a measure of acceleration, *a simple pendulum can be used as an accelerometer.*

According to the noninertial observer riding in the car (Fig. 6.13b), the cord still makes an angle θ with the vertical; however, to her the sphere is at rest and so its acceleration is zero. Therefore, she introduces a fictitious force to balance the horizontal component of $\mathbf{T}$ and claims that the net force on the sphere is *zero*! In this noninertial frame of reference, Newton's second law in component form yields

Noninertial observer
$$\begin{cases} \sum F_x' = T \sin\theta - F_{\text{fictitious}} = 0 \\ \sum F_y' = T \cos\theta - mg = 0 \end{cases}$$

If we recognize that $F_{\text{fictitious}} = ma_{\text{inertial}} = ma$, then these expressions are equivalent to (1) and (2); therefore, the noninertial observer obtains the same mathematical results as the inertial observer does. However, the physical interpretation of the deflection of the cord differs in the two frames of reference.

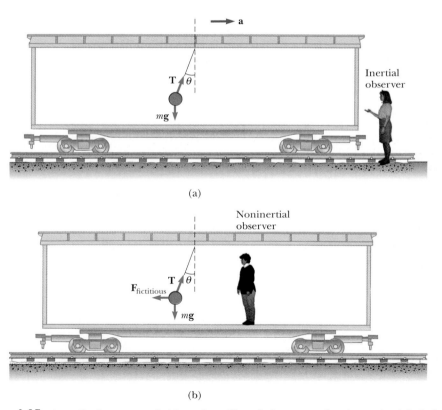

(a)

(b)

Figure 6.13 A small sphere suspended from the ceiling of a boxcar accelerating to the right is deflected as shown. (a) An inertial observer at rest outside the car claims that the acceleration of the sphere is provided by the horizontal component of $\mathbf{T}$. (b) A noninertial observer riding in the car says that the net force on the sphere is zero and that the deflection of the cord from the vertical is due to a fictitious force $\mathbf{F}_{\text{fictitious}}$ that balances the horizontal component of $\mathbf{T}$.

EXAMPLE 6.10 Fictitious Force in a Rotating System

Suppose a block of mass m lying on a horizontal, frictionless turntable is connected to a string attached to the center of the turntable, as shown in Figure 6.14. According to an inertial observer, if the block rotates uniformly, it undergoes an acceleration of magnitude v^2/r, where v is its linear speed. The inertial observer concludes that this centripetal acceleration is provided by the force **T** exerted by the string and writes Newton's second law as $T = mv^2/r$.

According to a noninertial observer attached to the turntable, the block is at rest and its acceleration is zero. Therefore, she must introduce a fictitious outward force of magnitude mv^2/r to balance the inward force exerted by the string. According to her, the net force on the block is zero, and she writes Newton's second law as $T - mv^2/r = 0$.

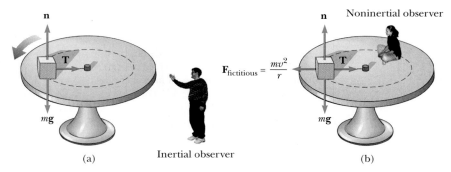

Figure 6.14 A block of mass m connected to a string tied to the center of a rotating turntable. (a) The inertial observer claims that the force causing the circular motion is provided by the force **T** exerted by the string on the block. (b) The noninertial observer claims that the block is not accelerating, and therefore she introduces a fictitious force of magnitude mv^2/r that acts outward and balances the force **T**.

Optional Section

6.4 MOTION IN THE PRESENCE OF RESISTIVE FORCES

In the preceding chapter we described the force of kinetic friction exerted on an object moving on some surface. We completely ignored any interaction between the object and the medium through which it moves. Now let us consider the effect of that medium, which can be either a liquid or a gas. The medium exerts a **resistive force R** on the object moving through it. Some examples are the air resistance associated with moving vehicles (sometimes called *air drag*) and the viscous forces that act on objects moving through a liquid. The magnitude of **R** depends on such factors as the speed of the object, and the direction of **R** is always opposite the direction of motion of the object relative to the medium. The magnitude of **R** nearly always increases with increasing speed.

The magnitude of the resistive force can depend on speed in a complex way, and here we consider only two situations. In the first situation, we assume the resistive force is proportional to the speed of the moving object; this assumption is valid for objects falling slowly through a liquid and for very small objects, such as dust particles, moving through air. In the second situation, we assume a resistive force that is proportional to the square of the speed of the moving object; large objects, such as a skydiver moving through air in free fall, experience such a force.

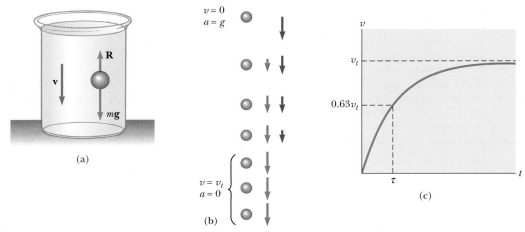

Figure 6.15 (a) A small sphere falling through a liquid. (b) Motion diagram of the sphere as it falls. (c) Speed–time graph for the sphere. The sphere reaches a maximum, or terminal, speed v_t, and the time constant τ is the time it takes to reach $0.63v_t$.

Resistive Force Proportional to Object Speed

If we assume that the resistive force acting on an object moving through a liquid or gas is proportional to the object's speed, then the magnitude of the resistive force can be expressed as

$$R = bv \tag{6.2}$$

where v is the speed of the object and b is a constant whose value depends on the properties of the medium and on the shape and dimensions of the object. If the object is a sphere of radius r, then b is proportional to r.

Consider a small sphere of mass m released from rest in a liquid, as in Figure 6.15a. Assuming that the only forces acting on the sphere are the resistive force bv and the force of gravity F_g, let us describe its motion.[1] Applying Newton's second law to the vertical motion, choosing the downward direction to be positive, and noting that $\Sigma F_y = mg - bv$, we obtain

$$mg - bv = ma = m\frac{dv}{dt} \tag{6.3}$$

where the acceleration dv/dt is downward. Solving this expression for the acceleration gives

$$\frac{dv}{dt} = g - \frac{b}{m}v \tag{6.4}$$

This equation is called a *differential equation*, and the methods of solving it may not be familiar to you as yet. However, note that initially, when $v = 0$, the resistive force $-bv$ is also zero and the acceleration dv/dt is simply g. As t increases, the resistive force increases and the acceleration decreases. Eventually, the acceleration becomes zero when the magnitude of the resistive force equals the sphere's weight. At this point, the sphere reaches its **terminal speed** v_t, and from then on

Terminal speed

[1] There is also a *buoyant force* acting on the submerged object. This force is constant, and its magnitude is equal to the weight of the displaced liquid. This force changes the apparent weight of the sphere by a constant factor, so we will ignore the force here. We discuss buoyant forces in Chapter 15.

it continues to move at this speed with zero acceleration, as shown in Figure 6.15b. We can obtain the terminal speed from Equation 6.3 by setting $a = dv/dt = 0$. This gives

$$mg - bv_t = 0 \qquad \text{or} \qquad v_t = mg/b$$

The expression for v that satisfies Equation 6.4 with $v = 0$ at $t = 0$ is

$$v = \frac{mg}{b}(1 - e^{-bt/m}) = v_t(1 - e^{-t/\tau}) \tag{6.5}$$

This function is plotted in Figure 6.15c. The **time constant** $\tau = m/b$ (Greek letter tau) is the time it takes the sphere to reach 63.2% ($= 1 - 1/e$) of its terminal speed. This can be seen by noting that when $t = \tau$, Equation 6.5 yields $v = 0.632v_t$.

We can check that Equation 6.5 is a solution to Equation 6.4 by direct differentiation:

$$\frac{dv}{dt} = \frac{d}{dt}\left(\frac{mg}{b} - \frac{mg}{b}e^{-bt/m}\right) = -\frac{mg}{b}\frac{d}{dt}e^{-bt/m} = ge^{-bt/m}$$

(See Appendix Table B.4 for the derivative of e raised to some power.) Substituting into Equation 6.4 both this expression for dv/dt and the expression for v given by Equation 6.5 shows that our solution satisfies the differential equation.

Aerodynamic car. A streamlined body reduces air drag and increases fuel efficiency. (*Courtesy of Ford Motor Company*)

EXAMPLE 6.11 ▸ Sphere Falling in Oil

A small sphere of mass 2.00 g is released from rest in a large vessel filled with oil, where it experiences a resistive force proportional to its speed. The sphere reaches a terminal speed of 5.00 cm/s. Determine the time constant τ and the time it takes the sphere to reach 90% of its terminal speed.

Solution Because the terminal speed is given by $v_t = mg/b$, the coefficient b is

$$b = \frac{mg}{v_t} = \frac{(2.00 \text{ g})(980 \text{ cm/s}^2)}{5.00 \text{ cm/s}} = 392 \text{ g/s}$$

Therefore, the time constant τ is

$$\tau = \frac{m}{b} = \frac{2.00 \text{ g}}{392 \text{ g/s}} = \boxed{5.10 \times 10^{-3} \text{ s}}$$

The speed of the sphere as a function of time is given by Equation 6.5. To find the time t it takes the sphere to reach a speed of $0.900v_t$, we set $v = 0.900v_t$ in Equation 6.5 and solve for t:

$$0.900v_t = v_t(1 - e^{-t/\tau})$$

$$1 - e^{-t/\tau} = 0.900$$

$$e^{-t/\tau} = 0.100$$

$$-\frac{t}{\tau} = \ln(0.100) = -2.30$$

$$t = 2.30\tau = 2.30(5.10 \times 10^{-3} \text{ s}) = 11.7 \times 10^{-3} \text{ s}$$

$$= \boxed{11.7 \text{ ms}}$$

Thus, the sphere reaches 90% of its terminal (maximum) speed in a very short time.

Exercise What is the sphere's speed through the oil at $t = 11.7$ ms? Compare this value with the speed the sphere would have if it were falling in a vacuum and so were influenced only by gravity.

Answer 4.50 cm/s in oil versus 11.5 cm/s in free fall.

Air Drag at High Speeds

For objects moving at high speeds through air, such as airplanes, sky divers, cars, and baseballs, the resistive force is approximately proportional to the square of the speed. In these situations, the magnitude of the resistive force can be expressed as

$$R = \tfrac{1}{2}D\rho Av^2 \tag{6.6}$$

where ρ is the density of air, A is the cross-sectional area of the falling object measured in a plane perpendicular to its motion, and D is a dimensionless empirical quantity called the *drag coefficient*. The drag coefficient has a value of about 0.5 for spherical objects but can have a value as great as 2 for irregularly shaped objects.

Let us analyze the motion of an object in free fall subject to an upward air resistive force of magnitude $R = \frac{1}{2}D\rho A v^2$. Suppose an object of mass m is released from rest. As Figure 6.16 shows, the object experiences two external forces: the downward force of gravity $\mathbf{F}_g = m\mathbf{g}$ and the upward resistive force $\mathbf{R}$. (There is also an upward buoyant force that we neglect.) Hence, the magnitude of the net force is

$$\sum F = mg - \tfrac{1}{2}D\rho A v^2 \qquad \textbf{(6.7)}$$

where we have taken downward to be the positive vertical direction. Substituting $\Sigma F = ma$ into Equation 6.7, we find that the object has a downward acceleration of magnitude

$$a = g - \left(\frac{D\rho A}{2m}\right)v^2 \qquad \textbf{(6.8)}$$

We can calculate the terminal speed v_t by using the fact that when the force of gravity is balanced by the resistive force, the net force on the object is zero and therefore its acceleration is zero. Setting $a = 0$ in Equation 6.8 gives

$$g - \left(\frac{D\rho A}{2m}\right)v_t^2 = 0$$

$$v_t = \sqrt{\frac{2mg}{D\rho A}} \qquad \textbf{(6.9)}$$

Using this expression, we can determine how the terminal speed depends on the dimensions of the object. Suppose the object is a sphere of radius r. In this case, $A \propto r^2$ (from $A = \pi r^2$) and $m \propto r^3$ (because the mass is proportional to the volume of the sphere, which is $V = \frac{4}{3}\pi r^3$). Therefore, $v_t \propto \sqrt{r}$.

Table 6.1 lists the terminal speeds for several objects falling through air.

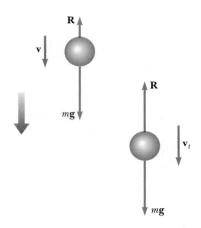

Figure 6.16 An object falling through air experiences a resistive force $\mathbf{R}$ and a gravitational force $\mathbf{F}_g = m\mathbf{g}$. The object reaches terminal speed (on the right) when the net force acting on it is zero, that is, when $\mathbf{R} = -\mathbf{F}_g$ or $R = mg$. Before this occurs, the acceleration varies with speed according to Equation 6.8.

The high cost of fuel has prompted many truck owners to install wind deflectors on their cabs to reduce drag.

TABLE 6.1   Terminal Speed for Various Objects Falling Through Air

Object	Mass (kg)	Cross-Sectional Area (m^2)	v_t (m/s)
Sky diver	75	0.70	60
Baseball (radius 3.7 cm)	0.145	4.2×10^{-3}	43
Golf ball (radius 2.1 cm)	0.046	1.4×10^{-3}	44
Hailstone (radius 0.50 cm)	4.8×10^{-4}	7.9×10^{-5}	14
Raindrop (radius 0.20 cm)	3.4×10^{-5}	1.3×10^{-5}	9.0

web
For more information about the sport of sky surfing, visit **http://espn.sportszone.com /editors/xgames/surf/index.html**

CONCEPTUAL EXAMPLE 6.12

Consider a sky surfer who jumps from a plane with her feet attached firmly to her surfboard, does some tricks, and then opens her parachute. Describe the forces acting on her during these maneuvers.

Solution   When the surfer first steps out of the plane, she has no vertical velocity. The downward force of gravity causes her to accelerate toward the ground. As her downward speed increases, so does the upward resistive force exerted by the air on her body and the board. This upward force reduces their acceleration, and so their speed increases more slowly. Eventually, they are going so fast that the upward resistive force matches the downward force of gravity. Now the net force is zero and they no longer accelerate, but reach their terminal speed. At some point after reaching terminal speed, she opens her parachute, resulting in a drastic increase in the upward resistive force. The net force (and thus the acceleration) is now upward, in the direction opposite the direction of the velocity. This causes the downward velocity to decrease rapidly; this means the resistive force on the chute also decreases. Eventually the upward resistive force and the downward force of gravity balance each other and a much smaller terminal speed is reached, permitting a safe landing.

(Contrary to popular belief, the velocity vector of a sky diver never points upward. You may have seen a videotape in which a sky diver appeared to "rocket" upward once the chute opened. In fact, what happened is that the diver slowed down while the person holding the camera continued falling at high speed.)

A sky surfer takes advantage of the upward force of the air on her board. (*Jump Run Productions/Image Bank*)

EXAMPLE 6.13   Falling Coffee Filters

The dependence of resistive force on speed is an empirical relationship. In other words, it is based on observation rather than on a theoretical model. A series of stacked filters is dropped, and the terminal speeds are measured. Table 6.2 presents data for these coffee filters as they fall through the air. The time constant τ is small, so that a dropped filter quickly reaches terminal speed. Each filter has a mass of 1.64 g. When the filters are nested together, they stack in such a way that the front-facing surface area does not increase. Determine the relationship between the resistive force exerted by the air and the speed of the falling filters.

Solution   At terminal speed, the upward resistive force balances the downward force of gravity. So, a single filter falling at its terminal speed experiences a resistive force of

$$R = mg = \left(\frac{1.64 \text{ g}}{1000 \text{ g/kg}}\right)(9.80 \text{ m/s}^2) = 0.016 \text{ 1 N}$$

Two filters nested together experience 0.032 2 N of resistive force, and so forth. A graph of the resistive force on the filters as a function of terminal speed is shown in Figure 6.17a. A straight line would not be a good fit, indicating that the resistive force is not proportional to the speed. The curved line is for a second-order polynomial, indicating a proportionality of the resistive force to the square of the speed. This proportionality is more clearly seen in Figure 6.17b, in which the resistive force is plotted as a function of the square of the terminal speed.

TABLE 6.2 Terminal Speed for Stacked Coffee Filters	
Number of Filters	v_t **(m/s)**[a]
1	1.01
2	1.40
3	1.63
4	2.00
5	2.25
6	2.40
7	2.57
8	2.80
9	3.05
10	3.22

[a] All values of v_t are approximate.

Pleated coffee filters can be nested together so that the force of air resistance can be studied. (*Charles D. Winters*)

(a)

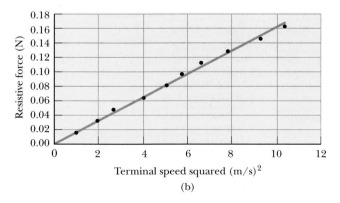

(b)

Figure 6.17   (a) Relationship between the resistive force acting on falling coffee filters and their terminal speed. The curved line is a second-order polynomial fit. (b) Graph relating the resistive force to the square of the terminal speed. The fit of the straight line to the data points indicates that the resistive force is proportional to the terminal speed squared. Can you find the proportionality constant?

EXAMPLE 6.14   Resistive Force Exerted on a Baseball

A pitcher hurls a 0.145-kg baseball past a batter at 40.2 m/s (= 90 mi/h). Find the resistive force acting on the ball at this speed.

Solution   We do not expect the air to exert a huge force on the ball, and so the resistive force we calculate from Equation 6.6 should not be more than a few newtons. First, we must determine the drag coefficient D. We do this by imagining that we drop the baseball and allow it to reach terminal speed. We solve Equation 6.9 for D and substitute the appropriate values for m, v_t, and A from Table 6.1. Taking the density of air as 1.29 kg/m^3, we obtain

$$D = \frac{2\,mg}{v_t^2\,\rho A} = \frac{2(0.145 \text{ kg})(9.80 \text{ m/s}^2)}{(43 \text{ m/s})^2\,(1.29 \text{ kg/m}^3)(4.2 \times 10^{-3} \text{ m}^2)}$$

$$= 0.284$$

This number has no dimensions. We have kept an extra digit beyond the two that are significant and will drop it at the end of our calculation.

We can now use this value for D in Equation 6.6 to find the magnitude of the resistive force:

$$R = \tfrac{1}{2} D\rho A v^2$$

$$= \tfrac{1}{2}(0.284)(1.29 \text{ kg/m}^3)(4.2 \times 10^{-3} \text{ m}^2)(40.2 \text{ m/s})^2$$

$$= \boxed{1.2 \text{ N}}$$

Optional Section

NUMERICAL MODELING IN PARTICLE DYNAMICS[2]

As we have seen in this and the preceding chapter, the study of the dynamics of a particle focuses on describing the position, velocity, and acceleration as functions of time. Cause-and-effect relationships exist among these quantities: Velocity causes position to change, and acceleration causes velocity to change. Because acceleration is the direct result of applied forces, any analysis of the dynamics of a particle usually begins with an evaluation of the net force being exerted on the particle.

Up till now, we have used what is called the *analytical method* to investigate the position, velocity, and acceleration of a moving particle. Let us review this method briefly before learning about a second way of approaching problems in dynamics. (Because we confine our discussion to one-dimensional motion in this section, boldface notation will not be used for vector quantities.)

If a particle of mass m moves under the influence of a net force ΣF, Newton's second law tells us that the acceleration of the particle is $a = \Sigma F/m$. In general, we apply the analytical method to a dynamics problem using the following procedure:

1. Sum all the forces acting on the particle to get the net force ΣF.
2. Use this net force to determine the acceleration from the relationship $a = \Sigma F/m$.
3. Use this acceleration to determine the velocity from the relationship $dv/dt = a$.
4. Use this velocity to determine the position from the relationship $dx/dt = v$.

The following straightforward example illustrates this method.

EXAMPLE 6.15   An Object Falling in a Vacuum—Analytical Method

Consider a particle falling in a vacuum under the influence of the force of gravity, as shown in Figure 6.18. Use the analytical method to find the acceleration, velocity, and position of the particle.

Solution   The only force acting on the particle is the downward force of gravity of magnitude F_g, which is also the net force. Applying Newton's second law, we set the net force acting on the particle equal to the mass of the particle times

[2] The authors are most grateful to Colonel James Head of the U.S. Air Force Academy for preparing this section. See the *Student Tools CD-ROM* for some assistance with numerical modeling.

its acceleration (taking upward to be the positive y direction):

$$F_g = ma_y = -mg$$

Thus, $a_y = -g$, which means the acceleration is constant. Because $dv_y/dt = a_y$, we see that $dv_y/dt = -g$, which may be integrated to yield

$$v_y(t) = v_{yi} - gt$$

Then, because $v_y = dy/dt$, the position of the particle is obtained from another integration, which yields the well-known result

$$y(t) = y_i + v_{yi}t - \tfrac{1}{2}gt^2$$

In these expressions, y_i and v_{yi} represent the position and speed of the particle at $t_i = 0$.

Figure 6.18   An object falling in vacuum under the influence of gravity.

The analytical method is straightforward for many physical situations. In the "real world," however, complications often arise that make analytical solutions difficult and perhaps beyond the mathematical abilities of most students taking introductory physics. For example, the net force acting on a particle may depend on the particle's position, as in cases where the gravitational acceleration varies with height. Or the force may vary with velocity, as in cases of resistive forces caused by motion through a liquid or gas.

Another complication arises because the expressions relating acceleration, velocity, position, and time are differential equations rather than algebraic ones. Differential equations are usually solved using integral calculus and other special techniques that introductory students may not have mastered.

When such situations arise, scientists often use a procedure called *numerical modeling* to study motion. The simplest numerical model is called the Euler method, after the Swiss mathematician Leonhard Euler (1707–1783).

The Euler Method

In the **Euler method** for solving differential equations, derivatives are approximated as ratios of finite differences. Considering a small increment of time Δt, we can approximate the relationship between a particle's speed and the magnitude of its acceleration as

$$a(t) \approx \frac{\Delta v}{\Delta t} = \frac{v(t + \Delta t) - v(t)}{\Delta t}$$

Then the speed $v(t + \Delta t)$ of the particle at the end of the time interval Δt is approximately equal to the speed $v(t)$ at the beginning of the time interval plus the magnitude of the acceleration during the interval multiplied by Δt:

$$v(t + \Delta t) \approx v(t) + a(t)\Delta t \qquad \textbf{(6.10)}$$

Because the acceleration is a function of time, this estimate of $v(t + \Delta t)$ is accurate only if the time interval Δt is short enough that the change in acceleration during it is very small (as is discussed later). Of course, Equation 6.10 is exact if the acceleration is constant.

The position $x(t + \Delta t)$ of the particle at the end of the interval Δt can be found in the same manner:

$$v(t) \approx \frac{\Delta x}{\Delta t} = \frac{x(t + \Delta t) - x(t)}{\Delta t}$$

$$x(t + \Delta t) \approx x(t) + v(t)\Delta t \tag{6.11}$$

You may be tempted to add the term $\frac{1}{2} a(\Delta t)^2$ to this result to make it look like the familiar kinematics equation, but this term is not included in the Euler method because Δt is assumed to be so small that Δt^2 is nearly zero.

If the acceleration at any instant t is known, the particle's velocity and position at a time $t + \Delta t$ can be calculated from Equations 6.10 and 6.11. The calculation then proceeds in a series of finite steps to determine the velocity and position at any later time. The acceleration is determined from the net force acting on the particle, and this force may depend on position, velocity, or time:

$$a(x, v, t) = \frac{\sum F(x, v, t)}{m} \tag{6.12}$$

It is convenient to set up the numerical solution to this kind of problem by numbering the steps and entering the calculations in a table, a procedure that is illustrated in Table 6.3.

The equations in the table can be entered into a spreadsheet and the calculations performed row by row to determine the velocity, position, and acceleration as functions of time. The calculations can also be carried out by using a program written in either BASIC, C++, or FORTRAN or by using commercially available mathematics packages for personal computers. Many small increments can be taken, and accurate results can usually be obtained with the help of a computer. Graphs of velocity versus time or position versus time can be displayed to help you visualize the motion.

One advantage of the Euler method is that the dynamics is not obscured—the fundamental relationships between acceleration and force, velocity and acceleration, and position and velocity are clearly evident. Indeed, these relationships form the heart of the calculations. There is no need to use advanced mathematics, and the basic physics governs the dynamics.

The Euler method is completely reliable for infinitesimally small time increments, but for practical reasons a finite increment size must be chosen. For the finite difference approximation of Equation 6.10 to be valid, the time increment must be small enough that the acceleration can be approximated as being constant during the increment. We can determine an appropriate size for the time in-

See the spreadsheet file "Baseball with Drag" in the *Student Tools CD-ROM* for an example of how this technique can be applied to find the initial speed of the baseball described in Example 6.14. We cannot use our regular approach because our kinematics equations assume constant acceleration. Euler's method provides a way to circumvent this difficulty.

A detailed solution to Problem 41 involving interative integration appears in the *Student Solutions Manual and Study Guide* and is posted on the Web at **http:/ www.saunderscollege.com/physics**

TABLE 6.3 **The Euler Method for Solving Dynamics Problems**				
Step	**Time**	**Position**	**Velocity**	**Acceleration**
0	t_0	x_0	v_0	$a_0 = F(x_0, v_0, t_0)/m$
1	$t_1 = t_0 + \Delta t$	$x_1 = x_0 + v_0 \Delta t$	$v_1 = v_0 + a_0 \Delta t$	$a_1 = F(x_1, v_1, t_1)/m$
2	$t_2 = t_1 + \Delta t$	$x_2 = x_1 + v_1 \Delta t$	$v_2 = v_1 + a_1 \Delta t$	$a_2 = F(x_2, v_2, t_2)/m$
3	$t_3 = t_2 + \Delta t$	$x_3 = x_2 + v_2 \Delta t$	$v_3 = v_2 + a_2 \Delta t$	$a_3 = F(x_3, v_3, t_3)/m$
$\vdots$	$\vdots$	$\vdots$	$\vdots$	$\vdots$
n	t_n	x_n	v_n	a_n

crement by examining the particular problem being investigated. The criterion for the size of the time increment may need to be changed during the course of the motion. In practice, however, we usually choose a time increment appropriate to the initial conditions and use the same value throughout the calculations.

The size of the time increment influences the accuracy of the result, but unfortunately it is not easy to determine the accuracy of a Euler-method solution without a knowledge of the correct analytical solution. One method of determining the accuracy of the numerical solution is to repeat the calculations with a smaller time increment and compare results. If the two calculations agree to a certain number of significant figures, you can assume that the results are correct to that precision.

SUMMARY

Newton's second law applied to a particle moving in uniform circular motion states that the net force causing the particle to undergo a centripetal acceleration is

$$\sum F_r = ma_r = \frac{mv^2}{r} \tag{6.1}$$

You should be able to use this formula in situations where the force providing the centripetal acceleration could be the force of gravity, a force of friction, a force of string tension, or a normal force.

A particle moving in nonuniform circular motion has both a centripetal component of acceleration and a nonzero tangential component of acceleration. In the case of a particle rotating in a vertical circle, the force of gravity provides the tangential component of acceleration and part or all of the centripetal component of acceleration. Be sure you understand the directions and magnitudes of the velocity and acceleration vectors for nonuniform circular motion.

An observer in a noninertial (accelerating) frame of reference must introduce **fictitious forces** when applying Newton's second law in that frame. If these fictitious forces are properly defined, the description of motion in the noninertial frame is equivalent to that made by an observer in an inertial frame. However, the observers in the two frames do not agree on the causes of the motion. You should be able to distinguish between inertial and noninertial frames and identify the fictitious forces acting in a noninertial frame.

A body moving through a liquid or gas experiences a **resistive force** that is speed-dependent. This resistive force, which opposes the motion, generally increases with speed. The magnitude of the resistive force depends on the shape of the body and on the properties of the medium through which the body is moving. In the limiting case for a falling body, when the magnitude of the resistive force equals the body's weight, the body reaches its **terminal speed.** You should be able to apply Newton's laws to analyze the motion of objects moving under the influence of resistive forces. You may need to apply **Euler's method** if the force depends on velocity, as it does for air drag.

QUESTIONS

1. Because the Earth rotates about its axis and revolves around the Sun, it is a noninertial frame of reference. Assuming the Earth is a uniform sphere, why would the ap-

parent weight of an object be greater at the poles than at the equator?

2. Explain why the Earth bulges at the equator.

3. Why is it that an astronaut in a space capsule orbiting the Earth experiences a feeling of weightlessness?

4. Why does mud fly off a rapidly turning automobile tire?

5. Imagine that you attach a heavy object to one end of a spring and then whirl the spring and object in a horizontal circle (by holding the free end of the spring). Does the spring stretch? If so, why? Discuss this in terms of the force causing the circular motion.

6. It has been suggested that rotating cylinders about 10 mi in length and 5 mi in diameter be placed in space and used as colonies. The purpose of the rotation is to simulate gravity for the inhabitants. Explain this concept for producing an effective gravity.

7. Why does a pilot tend to black out when pulling out of a steep dive?

8. Describe a situation in which a car driver can have a centripetal acceleration but no tangential acceleration.

9. Describe the path of a moving object if its acceleration is constant in magnitude at all times and (a) perpendicular to the velocity; (b) parrallel to the velocity.

10. Analyze the motion of a rock falling through water in terms of its speed and acceleration as it falls. Assume that the resistive force acting on the rock increases as the speed increases.

11. Consider a small raindrop and a large raindrop falling through the atmosphere. Compare their terminal speeds. What are their accelerations when they reach terminal speed?

PROBLEMS

1, 2, 3 = straightforward, intermediate, challenging ☐ = full solution available in the *Student Solutions Manual and Study Guide*

WEB = solution posted at **http://www.saunderscollege.com/physics/** 🖥 = Computer useful in solving problem 🖥 = Interactive Physics

☐ = paired numerical/symbolic problems

Section 6.1   Newton's Second Law
Applied to Uniform Circular Motion

1. A toy car moving at constant speed completes one lap around a circular track (a distance of 200 m) in 25.0 s. (a) What is its average speed? (b) If the mass of the car is 1.50 kg, what is the magnitude of the force that keeps it in a circle?

2. A 55.0-kg ice skater is moving at 4.00 m/s when she grabs the loose end of a rope, the opposite end of which is tied to a pole. She then moves in a circle of radius 0.800 m around the pole. (a) Determine the force exerted by the rope on her arms. (b) Compare this force with her weight.

3. A light string can support a stationary hanging load of 25.0 kg before breaking. A 3.00-kg mass attached to the string rotates on a horizontal, frictionless table in a circle of radius 0.800 m. What range of speeds can the mass have before the string breaks?

4. In the Bohr model of the hydrogen atom, the speed of the electron is approximately 2.20×10^6 m/s. Find (a) the force acting on the electron as it revolves in a circular orbit of radius 0.530×10^{-10} m and (b) the centripetal acceleration of the electron.

5. In a cyclotron (one type of particle accelerator), a deuteron (of atomic mass 2.00 u) reaches a final speed of 10.0% of the speed of light while moving in a circular path of radius 0.480 m. The deuteron is maintained in the circular path by a magnetic force. What magnitude of force is required?

6. A satellite of mass 300 kg is in a circular orbit around the Earth at an altitude equal to the Earth's mean radius (see Example 6.6). Find (a) the satellite's orbital speed, (b) the period of its revolution, and (c) the gravitational force acting on it.

7. Whenever two Apollo astronauts were on the surface of the Moon, a third astronaut orbited the Moon. Assume the orbit to be circular and 100 km above the surface of the Moon. If the mass of the Moon is 7.40×10^{22} kg and its radius is 1.70×10^6 m, determine (a) the orbiting astronaut's acceleration, (b) his orbital speed, and (c) the period of the orbit.

8. The speed of the tip of the minute hand on a town clock is 1.75×10^{-3} m/s. (a) What is the speed of the tip of the second hand of the same length? (b) What is the centripetal acceleration of the tip of the second hand?

9. A coin placed 30.0 cm from the center of a rotating, horizontal turntable slips when its speed is 50.0 cm/s. (a) What provides the force in the radial direction when the coin is stationary relative to the turntable? (b) What is the coefficient of static friction between coin and turntable?

10. The cornering performance of an automobile is evaluated on a skid pad, where the maximum speed that a car can maintain around a circular path on a dry, flat surface is measured. The centripetal acceleration, also called the lateral acceleration, is then calculated as a multiple of the free-fall acceleration g. The main factors affecting the performance are the tire characteristics and the suspension system of the car. A Dodge Viper GTS can negotiate a skid pad of radius 61.0 m at 86.5 km/h. Calculate its maximum lateral acceleration.

11. A crate of eggs is located in the middle of the flatbed of a pickup truck as the truck negotiates an unbanked

curve in the road. The curve may be regarded as an arc of a circle of radius 35.0 m. If the coefficient of static friction between crate and truck is 0.600, how fast can the truck be moving without the crate sliding?

12. A car initially traveling eastward turns north by traveling in a circular path at uniform speed as in Figure P6.12. The length of the arc *ABC* is 235 m, and the car completes the turn in 36.0 s. (a) What is the acceleration when the car is at *B* located at an angle of 35.0°? Express your answer in terms of the unit vectors **i** and **j**. Determine (b) the car's average speed and (c) its average acceleration during the 36.0-s interval.

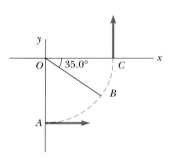

Figure P6.12

13. Consider a conical pendulum with an 80.0-kg bob on a 10.0-m wire making an angle of $\theta = 5.00°$ with the vertical (Fig. P6.13). Determine (a) the horizontal and vertical components of the force exerted by the wire on the pendulum and (b) the radial acceleration of the bob.

Figure P6.13

Section 6.2   Nonuniform Circular Motion

14. A car traveling on a straight road at 9.00 m/s goes over a hump in the road. The hump may be regarded as an arc of a circle of radius 11.0 m. (a) What is the apparent weight of a 600-N woman in the car as she rides over the

hump? (b) What must be the speed of the car over the hump if she is to experience weightlessness? (That is, if her apparent weight is zero.)

WEB 15. Tarzan ($m = 85.0$ kg) tries to cross a river by swinging from a vine. The vine is 10.0 m long, and his speed at the bottom of the swing (as he just clears the water) is 8.00 m/s. Tarzan doesn't know that the vine has a breaking strength of 1 000 N. Does he make it safely across the river?

16. A hawk flies in a horizontal arc of radius 12.0 m at a constant speed of 4.00 m/s. (a) Find its centripetal acceleration. (b) It continues to fly along the same horizontal arc but steadily increases its speed at the rate of 1.20 m/s². Find the acceleration (magnitude and direction) under these conditions.

17. A 40.0-kg child sits in a swing supported by two chains, each 3.00 m long. If the tension in each chain at the lowest point is 350 N, find (a) the child's speed at the lowest point and (b) the force exerted by the seat on the child at the lowest point. (Neglect the mass of the seat.)

18. A child of mass *m* sits in a swing supported by two chains, each of length *R*. If the tension in each chain at the lowest point is *T*, find (a) the child's speed at the lowest point and (b) the force exerted by the seat on the child at the lowest point. (Neglect the mass of the seat.)

WEB 19. A pail of water is rotated in a vertical circle of radius 1.00 m. What must be the minimum speed of the pail at the top of the circle if no water is to spill out?

20. A 0.400-kg object is swung in a vertical circular path on a string 0.500 m long. If its speed is 4.00 m/s at the top of the circle, what is the tension in the string there?

21. A roller-coaster car has a mass of 500 kg when fully loaded with passengers (Fig. P6.21). (a) If the car has a speed of 20.0 m/s at point *A*, what is the force exerted by the track on the car at this point? (b) What is the maximum speed the car can have at *B* and still remain on the track?

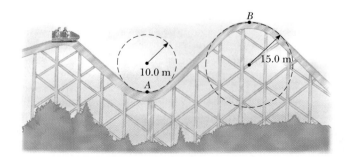

Figure P6.21

22. A roller coaster at the Six Flags Great America amusement park in Gurnee, Illinois, incorporates some of the latest design technology and some basic physics. Each vertical loop, instead of being circular, is shaped like a teardrop (Fig. P6.22). The cars ride on the inside of the loop at the top, and the speeds are high enough to ensure that the cars remain on the track. The biggest loop is 40.0 m high, with a maximum speed of 31.0 m/s (nearly 70 mi/h) at the bottom. Suppose the speed at the top is 13.0 m/s and the corresponding centripetal acceleration is 2*g*. (a) What is the radius of the arc of the teardrop at the top? (b) If the total mass of the cars plus people is *M*, what force does the rail exert on this total mass at the top? (c) Suppose the roller coaster had a loop of radius 20.0 m. If the cars have the same speed, 13.0 m/s at the top, what is the centripetal acceleration at the top? Comment on the normal force at the top in this situation.

Figure P6.22   *(Frank Cezus/FPG International)*

(Optional)
Section 6.3   Motion in Accelerated Frames

23. A merry-go-round makes one complete revolution in 12.0 s. If a 45.0-kg child sits on the horizontal floor of the merry-go-round 3.00 m from the center, find (a) the child's acceleration and (b) the horizontal force of friction that acts on the child. (c) What minimum coefficient of static friction is necessary to keep the child from slipping?

24. A 5.00-kg mass attached to a spring scale rests on a frictionless, horizontal surface as in Figure P6.24. The spring scale, attached to the front end of a boxcar, reads 18.0 N when the car is in motion. (a) If the spring scale reads zero when the car is at rest, determine the acceleration of the car. (b) What will the spring scale read if the car moves with constant velocity? (c) Describe the forces acting on the mass as observed by someone in the car and by someone at rest outside the car.

5.00 kg

Figure P6.24

25. A 0.500-kg object is suspended from the ceiling of an accelerating boxcar as was seen in Figure 6.13. If *a* = 3.00 m/s², find (a) the angle that the string makes with the vertical and (b) the tension in the string.

26. The Earth rotates about its axis with a period of 24.0 h. Imagine that the rotational speed can be increased. If an object at the equator is to have zero apparent weight, (a) what must the new period be? (b) By what factor would the speed of the object be increased when the planet is rotating at the higher speed? (*Hint:* See Problem 53 and note that the apparent weight of the object becomes zero when the normal force exerted on it is zero. Also, the distance traveled during one period is $2\pi R$, where *R* is the Earth's radius.)

27. A person stands on a scale in an elevator. As the elevator starts, the scale has a constant reading of 591 N. As the elevator later stops, the scale reading is 391 N. Assume the magnitude of the acceleration is the same during starting and stopping, and determine (a) the weight of the person, (b) the person's mass, and (c) the acceleration of the elevator.

28. A child on vacation wakes up. She is lying on her back. The tension in the muscles on both sides of her neck is 55.0 N as she raises her head to look past her toes and out the motel window. Finally, it is not raining! Ten minutes later she is screaming and sliding feet first down a water slide at a constant speed of 5.70 m/s, riding high on the outside wall of a horizontal curve of radius 2.40 m (Fig. P6.28). She raises her head to look forward past her toes; find the tension in the muscles on both sides of her neck.

Figure P6.28

Figure P6.34

29. A plumb bob does not hang exactly along a line directed to the center of the Earth, because of the Earth's rotation. How much does the plumb bob deviate from a radial line at 35.0° north latitude? Assume that the Earth is spherical.

(Optional)

Section 6.4   Motion in the Presence of Resistive Forces

30. A sky diver of mass 80.0 kg jumps from a slow-moving aircraft and reaches a terminal speed of 50.0 m/s. (a) What is the acceleration of the sky diver when her speed is 30.0 m/s? What is the drag force exerted on the diver when her speed is (b) 50.0 m/s? (c) 30.0 m/s?

31. A small piece of Styrofoam packing material is dropped from a height of 2.00 m above the ground. Until it reaches terminal speed, the magnitude of its acceleration is given by $a = g - bv$. After falling 0.500 m, the Styrofoam effectively reaches its terminal speed, and then takes 5.00 s more to reach the ground. (a) What is the value of the constant b? (b) What is the acceleration at $t = 0$? (c) What is the acceleration when the speed is 0.150 m/s?

32. (a) Estimate the terminal speed of a wooden sphere (density 0.830 g/cm³) falling through the air if its radius is 8.00 cm. (b) From what height would a freely falling object reach this speed in the absence of air resistance?

33. Calculate the force required to pull a copper ball of radius 2.00 cm upward through a fluid at the constant speed 9.00 cm/s. Take the drag force to be proportional to the speed, with proportionality constant 0.950 kg/s. Ignore the buoyant force.

34. A fire helicopter carries a 620-kg bucket at the end of a cable 20.0 m long as in Figure P6.34. As the helicopter flies to a fire at a constant speed of 40.0 m/s, the cable makes an angle of 40.0° with respect to the vertical. The bucket presents a cross-sectional area of 3.80 m² in a plane perpendicular to the air moving past it. Determine the drag coefficient assuming that the resistive force is proportional to the square of the bucket's speed.

35. A small, spherical bead of mass 3.00 g is released from rest at $t = 0$ in a bottle of liquid shampoo. The terminal speed is observed to be $v_t = 2.00$ cm/s. Find (a) the value of the constant b in Equation 6.4, (b) the time τ the bead takes to reach $0.632v_t$, and (c) the value of the resistive force when the bead reaches terminal speed.

36. The mass of a sports car is 1 200 kg. The shape of the car is such that the aerodynamic drag coefficient is 0.250 and the frontal area is 2.20 m². Neglecting all other sources of friction, calculate the initial acceleration of the car if, after traveling at 100 km/h, it is shifted into neutral and is allowed to coast.

37. A motorboat cuts its engine when its speed is 10.0 m/s and coasts to rest. The equation governing the motion of the motorboat during this period is $v = v_i e^{-ct}$, where v is the speed at time t, v_i is the initial speed, and c is a constant. At $t = 20.0$ s, the speed is 5.00 m/s. (a) Find the constant c. (b) What is the speed at $t = 40.0$ s? (c) Differentiate the expression for $v(t)$ and thus show that the acceleration of the boat is proportional to the speed at any time.

38. Assume that the resistive force acting on a speed skater is $f = -kmv^2$, where k is a constant and m is the skater's mass. The skater crosses the finish line of a straight-line race with speed v_f and then slows down by coasting on his skates. Show that the skater's speed at any time t after crossing the finish line is $v(t) = v_f / (1 + ktv_f)$.

39. You can feel a force of air drag on your hand if you stretch your arm out of the open window of a speeding car. (*Note:* Do not get hurt.) What is the order of magnitude of this force? In your solution, state the quantities you measure or estimate and their values.

(Optional)

6.5   Numerical Modeling in Particle Dynamics

40. A 3.00-g leaf is dropped from a height of 2.00 m above the ground. Assume the net downward force exerted on the leaf is $F = mg - bv$, where the drag factor is $b = 0.030\ 0$ kg/s. (a) Calculate the terminal speed of the leaf. (b) Use Euler's method of numerical analysis to find the speed and position of the leaf as functions of

time, from the instant it is released until 99% of terminal speed is reached. (*Hint:* Try $\Delta t = 0.005$ s.)

WEB 🖳 41. A hailstone of mass 4.80×10^{-4} kg falls through the air and experiences a net force given by

$$F = -mg + Cv^2$$

where $C = 2.50 \times 10^{-5}$ kg/m. (a) Calculate the terminal speed of the hailstone. (b) Use Euler's method of numerical analysis to find the speed and position of the hailstone at 0.2-s intervals, taking the initial speed to be zero. Continue the calculation until the hailstone reaches 99% of terminal speed.

🖳 42. A 0.142-kg baseball has a terminal speed of 42.5 m/s (95 mi/h). (a) If a baseball experiences a drag force of magnitude $R = Cv^2$, what is the value of the constant C? (b) What is the magnitude of the drag force when the speed of the baseball is 36.0 m/s? (c) Use a computer to determine the motion of a baseball thrown vertically upward at an initial speed of 36.0 m/s. What maximum height does the ball reach? How long is it in the air? What is its speed just before it hits the ground?

🖳 43. A 50.0-kg parachutist jumps from an airplane and falls with a drag force proportional to the square of the speed $R = Cv^2$. Take $C = 0.200$ kg/m with the parachute closed and $C = 20.0$ kg/m with the chute open. (a) Determine the terminal speed of the parachutist in both configurations, before and after the chute is opened. (b) Set up a numerical analysis of the motion and compute the speed and position as functions of time, assuming the jumper begins the descent at 1 000 m above the ground and is in free fall for 10.0 s before opening the parachute. (*Hint:* When the parachute opens, a sudden large acceleration takes place; a smaller time step may be necessary in this region.)

🖳 44. Consider a 10.0-kg projectile launched with an initial speed of 100 m/s, at an angle of 35.0° elevation. The resistive force is $\mathbf{R} = -b\mathbf{v}$, where $b = 10.0$ kg/s. (a) Use a numerical method to determine the horizontal and vertical positions of the projectile as functions of time. (b) What is the range of this projectile? (c) Determine the elevation angle that gives the maximum range for the projectile. (*Hint:* Adjust the elevation angle by trial and error to find the greatest range.)

🖳 45. A professional golfer hits a golf ball of mass 46.0 g with her 5-iron, and the ball first strikes the ground 155 m (170 yards) away. The ball experiences a drag force of magnitude $R = Cv^2$ and has a terminal speed of 44.0 m/s. (a) Calculate the drag constant C for the golf ball. (b) Use a numerical method to analyze the trajectory of this shot. If the initial velocity of the ball makes an angle of 31.0° (the loft angle) with the horizontal, what initial speed must the ball have to reach the 155-m distance? (c) If the same golfer hits the ball with her 9-iron (47.0° loft) and it first strikes the ground 119 m away, what is the initial speed of the ball? Discuss the differences in trajectories between the two shots.

ADDITIONAL PROBLEMS

46. An 1 800-kg car passes over a bump in a road that follows the arc of a circle of radius 42.0 m as in Figure P6.46. (a) What force does the road exert on the car as the car passes the highest point of the bump if the car travels at 16.0 m/s? (b) What is the maximum speed the car can have as it passes this highest point before losing contact with the road?

47. A car of mass m passes over a bump in a road that follows the arc of a circle of radius R as in Figure P6.46. (a) What force does the road exert on the car as the car passes the highest point of the bump if the car travels at a speed v? (b) What is the maximum speed the car can have as it passes this highest point before losing contact with the road?

Figure P6.46 Problems 46 and 47.

48. In one model of a hydrogen atom, the electron in orbit around the proton experiences an attractive force of about 8.20×10^{-8} N. If the radius of the orbit is 5.30×10^{-11} m, how many revolutions does the electron make each second? (This number of revolutions per unit time is called the *frequency* of the motion.) See the inside front cover for additional data.

49. A student builds and calibrates an accelerometer, which she uses to determine the speed of her car around a certain unbanked highway curve. The accelerometer is a plumb bob with a protractor that she attaches to the roof of her car. A friend riding in the car with her observes that the plumb bob hangs at an angle of 15.0° from the vertical when the car has a speed of 23.0 m/s. (a) What is the centripetal acceleration of the car rounding the curve? (b) What is the radius of the curve? (c) What is the speed of the car if the plumb bob deflection is 9.00° while the car is rounding the same curve?

50. Suppose the boxcar shown in Figure 6.13 is moving with constant acceleration a up a hill that makes an angle ϕ with the horizontal. If the hanging pendulum makes a constant angle θ with the perpendicular to the ceiling, what is a?

51. An air puck of mass 0.250 kg is tied to a string and allowed to revolve in a circle of radius 1.00 m on a fric-

tionless horizontal table. The other end of the string passes through a hole in the center of the table, and a mass of 1.00 kg is tied to it (Fig. P6.51). The suspended mass remains in equilibrium while the puck on the tabletop revolves. What are (a) the tension in the string, (b) the force exerted by the string on the puck, and (c) the speed of the puck?

52. An air puck of mass m_1 is tied to a string and allowed to revolve in a circle of radius R on a frictionless horizontal table. The other end of the string passes through a hole in the center of the table, and a mass m_2 is tied to it (Fig. P6.51). The suspended mass remains in equilibrium while the puck on the tabletop revolves. What are (a) the tension in the string? (b) the central force exerted on the puck? (c) the speed of the puck?

Figure P6.51   Problems 51 and 52.

WEB 53. Because the Earth rotates about its axis, a point on the equator experiences a centripetal acceleration of 0.033 7 m/s², while a point at one of the poles experiences no centripetal acceleration. (a) Show that at the equator the gravitational force acting on an object (the true weight) must exceed the object's apparent weight. (b) What is the apparent weight at the equator and at the poles of a person having a mass of 75.0 kg? (Assume the Earth is a uniform sphere and take $g = 9.800$ m/s².)

54. A string under a tension of 50.0 N is used to whirl a rock in a horizontal circle of radius 2.50 m at a speed of 20.4 m/s. The string is pulled in and the speed of the rock increases. When the string is 1.00 m long and the speed of the rock is 51.0 m/s, the string breaks. What is the breaking strength (in newtons) of the string?

55. A child's toy consists of a small wedge that has an acute angle θ (Fig. P6.55). The sloping side of the wedge is frictionless, and a mass m on it remains at constant height if the wedge is spun at a certain constant speed. The wedge is spun by rotating a vertical rod that is firmly attached to the wedge at the bottom end. Show

that, when the mass sits a distance L up along the sloping side, the speed of the mass must be

$$v = (g L \sin \theta)^{1/2}$$

Figure P6.55

56. The pilot of an airplane executes a constant-speed loop-the-loop maneuver. His path is a vertical circle. The speed of the airplane is 300 mi/h, and the radius of the circle is 1 200 ft. (a) What is the pilot's apparent weight at the lowest point if his true weight is 160 lb? (b) What is his apparent weight at the highest point? (c) Describe how the pilot could experience apparent weightlessness if both the radius and the speed can be varied. (*Note:* His apparent weight is equal to the force that the seat exerts on his body.)

57. For a satellite to move in a stable circular orbit at a constant speed, its centripetal acceleration must be inversely proportional to the square of the radius r of the orbit. (a) Show that the tangential speed of a satellite is proportional to $r^{-1/2}$. (b) Show that the time required to complete one orbit is proportional to $r^{3/2}$.

58. A penny of mass 3.10 g rests on a small 20.0-g block supported by a spinning disk (Fig. P6.58). If the coeffi-

Figure P6.58

cients of friction between block and disk are 0.750 (static) and 0.640 (kinetic) while those for the penny and block are 0.450 (kinetic) and 0.520 (static), what is the maximum rate of rotation (in revolutions per minute) that the disk can have before either the block or the penny starts to slip?

59. Figure P6.59 shows a Ferris wheel that rotates four times each minute and has a diameter of 18.0 m. (a) What is the centripetal acceleration of a rider? What force does the seat exert on a 40.0-kg rider (b) at the lowest point of the ride and (c) at the highest point of the ride? (d) What force (magnitude and direction) does the seat exert on a rider when the rider is halfway between top and bottom?

Figure P6.59   *(Color Box/FPG)*

60. A space station, in the form of a large wheel 120 m in diameter, rotates to provide an "artificial gravity" of 3.00 m/s^2 for persons situated at the outer rim. Find the rotational frequency of the wheel (in revolutions per minute) that will produce this effect.

61. An amusement park ride consists of a rotating circular platform 8.00 m in diameter from which 10.0-kg seats are suspended at the end of 2.50-m massless chains (Fig. P6.61). When the system rotates, the chains make an angle $\theta = 28.0°$ with the vertical. (a) What is the speed of each seat? (b) Draw a free-body diagram of a 40.0-kg child riding in a seat and find the tension in the chain.

62. A piece of putty is initially located at point A on the rim of a grinding wheel rotating about a horizontal axis. The putty is dislodged from point A when the diameter through A is horizontal. The putty then rises vertically and returns to A the instant the wheel completes one revolution. (a) Find the speed of a point on the rim of the wheel in terms of the acceleration due to gravity and the radius R of the wheel. (b) If the mass of the putty is m, what is the magnitude of the force that held it to the wheel?

Figure P6.61

63. An amusement park ride consists of a large vertical cylinder that spins about its axis fast enough that any person inside is held up against the wall when the floor drops away (Fig. P6.63). The coefficient of static friction between person and wall is μ_s, and the radius of the cylinder is R. (a) Show that the maximum period of revolution necessary to keep the person from falling is $T = (4\pi^2 R \mu_s / g)^{1/2}$. (b) Obtain a numerical value for T

Figure P6.63

if $R = 4.00$ m and $\mu_s = 0.400$. How many revolutions per minute does the cylinder make?

64. *An example of the Coriolis effect.* Suppose air resistance is negligible for a golf ball. A golfer tees off from a location precisely at $\phi_i = 35.0°$ north latitude. He hits the ball due south, with range 285 m. The ball's initial velocity is at $48.0°$ above the horizontal. (a) For what length of time is the ball in flight? The cup is due south of the golfer's location, and he would have a hole-in-one if the Earth were not rotating. As shown in Figure P6.64, the Earth's rotation makes the tee move in a circle of radius $R_E \cos \phi_i = (6.37 \times 10^6 \text{ m}) \cos 35.0°$, completing one revolution each day. (b) Find the eastward speed of the tee, relative to the stars. The hole is also moving eastward, but it is 285 m farther south and thus at a slightly lower latitude ϕ_f. Because the hole moves eastward in a slightly larger circle, its speed must be greater than that of the tee. (c) By how much does the hole's speed exceed that of the tee? During the time the ball is in flight, it moves both upward and downward, as well as southward with the projectile motion you studied in Chapter 4, but it also moves eastward with the speed you found in part (b). The hole moves to the east at a faster speed, however, pulling ahead of the ball with the relative speed you found in part (c). (d) How far to the west of the hole does the ball land?

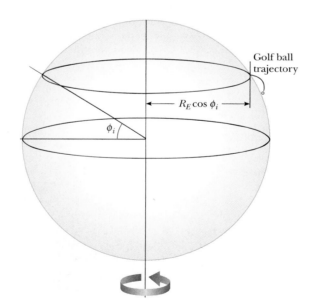

Figure P6.64

65. A curve in a road forms part of a horizontal circle. As a car goes around it at constant speed 14.0 m/s, the total force exerted on the driver has magnitude 130 N. What are the magnitude and direction of the total force exerted on the driver if the speed is 18.0 m/s instead?

66. A car rounds a banked curve as shown in Figure 6.6. The radius of curvature of the road is R, the banking angle is θ, and the coefficient of static friction is μ_s. (a) Determine the range of speeds the car can have without slipping up or down the banked surface. (b) Find the minimum value for μ_s such that the minimum speed is zero. (c) What is the range of speeds possible if $R = 100$ m, $\theta = 10.0°$, and $\mu_s = 0.100$ (slippery conditions)?

67. A single bead can slide with negligible friction on a wire that is bent into a circle of radius 15.0 cm, as in Figure P6.67. The circle is always in a vertical plane and rotates steadily about its vertical diameter with a period of 0.450 s. The position of the bead is described by the angle θ that the radial line from the center of the loop to the bead makes with the vertical. (a) At what angle up from the lowest point can the bead stay motionless relative to the turning circle? (b) Repeat the problem if the period of the circle's rotation is 0.850 s.

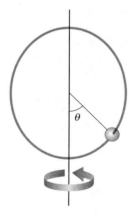

Figure P6.67

68. The expression $F = arv + br^2v^2$ gives the magnitude of the resistive force (in newtons) exerted on a sphere of radius r (in meters) by a stream of air moving at speed v (in meters per second), where a and b are constants with appropriate SI units. Their numerical values are $a = 3.10 \times 10^{-4}$ and $b = 0.870$. Using this formula, find the terminal speed for water droplets falling under their own weight in air, taking the following values for the drop radii: (a) 10.0 μm, (b) 100 μm, (c) 1.00 mm. Note that for (a) and (c) you can obtain accurate answers without solving a quadratic equation, by considering which of the two contributions to the air resistance is dominant and ignoring the lesser contribution.

69. A model airplane of mass 0.750 kg flies in a horizontal circle at the end of a 60.0-m control wire, with a speed of 35.0 m/s. Compute the tension in the wire if it makes a constant angle of 20.0° with the horizontal. The forces exerted on the airplane are the pull of the control wire,

its own weight, and aerodynamic lift, which acts at 20.0° inward from the vertical as shown in Figure P6.69.

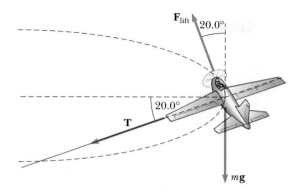

Figure P6.69

70. A 9.00-kg object starting from rest falls through a viscous medium and experiences a resistive force $\mathbf{R} = -b\mathbf{v}$, where $\mathbf{v}$ is the velocity of the object. If the object's speed reaches one-half its terminal speed in 5.54 s, (a) determine the terminal speed. (b) At what time is the speed of the object three-fourths the terminal speed? (c) How far has the object traveled in the first 5.54 s of motion?

71. Members of a skydiving club were given the following data to use in planning their jumps. In the table, d is the distance fallen from rest by a sky diver in a "free-fall stable spread position" versus the time of fall t. (a) Convert the distances in feet into meters. (b) Graph d (in meters) versus t. (c) Determine the value of the terminal speed v_t by finding the slope of the straight portion of the curve. Use a least-squares fit to determine this slope.

t (s)	d (ft)
1	16
2	62
3	138
4	242
5	366
6	504
7	652
8	808
9	971
10	1 138
11	1 309
12	1 483
13	1 657
14	1 831
15	2 005
16	2 179
17	2 353
18	2 527
19	2 701
20	2 875

ANSWERS TO QUICK QUIZZES

6.1 No. The tangential acceleration changes just the speed part of the velocity vector. For the car to move in a circle, the *direction* of its velocity vector must change, and the only way this can happen is for there to be a centripetal acceleration.

6.2 (a) The ball travels in a circular path that has a larger radius than the original circular path, and so there must be some external force causing the change in the velocity vector's direction. The external force must not be as strong as the original tension in the string because if it were, the ball would follow the original path. (b) The ball again travels in an arc, implying some kind of external force. As in part (a), the external force is directed toward the center of the new arc and not toward the center of the original circular path. (c) The ball undergoes an abrupt change in velocity—from tangent to the circle to perpendicular to it—and so must have experienced a large force that had one component opposite the ball's velocity (tangent to the circle) and another component radially outward. (d) The ball travels in a straight line tangent to the original path. If there is an external force, it cannot have a component perpendicular to this line because if it did, the path would curve. In

fact, if the string breaks and there is no other force acting on the ball, Newton's first law says the ball will travel along such a tangent line at constant speed.

6.3 At Ⓐ the path is along the circumference of the larger circle. Therefore, the wire must be exerting a force on the bead directed toward the center of the circle. Because the speed is constant, there is no tangential force component. At Ⓑ the path is not curved, and so the wire exerts no force on the bead. At Ⓒ the path is again curved, and so the wire is again exerting a force on the bead. This time the force is directed toward the center of the smaller circle. Because the radius of this circle is smaller, the magnitude of the force exerted on the bead is larger here than at Ⓐ.

c h a p t e r

Work and Kinetic Energy

The concept of energy is one of the most important topics in science and engineering. In everyday life, we think of energy in terms of fuel for transportation and heating, electricity for lights and appliances, and foods for consumption. However, these ideas do not really define energy. They merely tell us that fuels are needed to do a job and that those fuels provide us with something we call *energy*.

In this chapter, we first introduce the concept of work. Work is done by a force acting on an object when the point of application of that force moves through some distance and the force has a component along the line of motion. Next, we define kinetic energy, which is energy an object possesses because of its motion. In general, we can think of *energy* as the capacity that an object has for performing work. We shall see that the concepts of work and kinetic energy can be applied to the dynamics of a mechanical system without resorting to Newton's laws. In a complex situation, in fact, the "energy approach" can often allow a much simpler analysis than the direct application of Newton's second law. However, it is important to note that the work–energy concepts are based on Newton's laws and therefore allow us to make predictions that are always in agreement with these laws.

This alternative method of describing motion is especially useful when the force acting on a particle varies with the position of the particle. In this case, the acceleration is not constant, and we cannot apply the kinematic equations developed in Chapter 2. Often, a particle in nature is subject to a force that varies with the position of the particle. Such forces include the gravitational force and the force exerted on an object attached to a spring. Although we could analyze situations like these by applying numerical methods such as those discussed in Section 6.5, utilizing the ideas of work and energy is often much simpler. We describe techniques for treating complicated systems with the help of an extremely important theorem called the *work–kinetic energy theorem*, which is the central topic of this chapter.

7.1 ▷ WORK DONE BY A CONSTANT FORCE

5.1 Almost all the terms we have used thus far—velocity, acceleration, force, and so on—convey nearly the same meaning in physics as they do in everyday life. Now, however, we encounter a term whose meaning in physics is distinctly different from its everyday meaning. That new term is *work*.

To understand what *work* means to the physicist, consider the situation illustrated in Figure 7.1. A force is applied to a chalkboard eraser, and the eraser slides along the tray. If we are interested in how effective the force is in moving the

(a)

(b)

(c)

Figure 7.1 An eraser being pushed along a chalkboard tray. *(Charles D. Winters)*

Figure 7.2   If an object undergoes a displacement **d** under the action of a constant force **F**, the work done by the force is $(F \cos \theta)d$.

eraser, we need to consider not only the magnitude of the force but also its direction. If we assume that the magnitude of the applied force is the same in all three photographs, it is clear that the push applied in Figure 7.1b does more to move the eraser than the push in Figure 7.1a. On the other hand, Figure 7.1c shows a situation in which the applied force does not move the eraser at all, regardless of how hard it is pushed. (Unless, of course, we apply a force so great that we break something.) So, in analyzing forces to determine the work they do, we must consider the vector nature of forces. We also need to know how far the eraser moves along the tray if we want to determine the work required to cause that motion. Moving the eraser 3 m requires more work than moving it 2 cm.

Let us examine the situation in Figure 7.2, where an object undergoes a displacement **d** along a straight line while acted on by a constant force **F** that makes an angle θ with **d**.

> The **work** W done on an object by an agent exerting a constant force on the object is the product of the component of the force in the direction of the displacement and the magnitude of the displacement:
>
> $$W = Fd \cos \theta \qquad \textbf{(7.1)}$$

Work done by a constant force

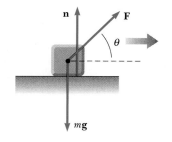

Figure 7.3   When an object is displaced on a frictionless, horizontal, surface, the normal force **n** and the force of gravity *m***g** do no work on the object. In the situation shown here, **F** is the only force doing work on the object.

As an example of the distinction between this definition of work and our everyday understanding of the word, consider holding a heavy chair at arm's length for 3 min. At the end of this time interval, your tired arms may lead you to think that you have done a considerable amount of work on the chair. According to our definition, however, you have done no work on it whatsoever.[1] You exert a force to support the chair, but you do not move it. A force does no work on an object if the object does not move. This can be seen by noting that if $d = 0$, Equation 7.1 gives $W = 0$—the situation depicted in Figure 7.1c.

Also note from Equation 7.1 that the work done by a force on a moving object is zero when the force applied is perpendicular to the object's displacement. That is, if $\theta = 90°$, then $W = 0$ because $\cos 90° = 0$. For example, in Figure 7.3, the work done by the normal force on the object and the work done by the force of gravity on the object are both zero because both forces are perpendicular to the displacement and have zero components in the direction of **d**.

The sign of the work also depends on the direction of **F** relative to **d**. The work done by the applied force is positive when the vector associated with the component $F \cos \theta$ is in the same direction as the displacement. For example, when an object is lifted, the work done by the applied force is positive because the direction of that force is upward, that is, in the same direction as the displacement. When the vector associated with the component $F \cos \theta$ is in the direction opposite the displacement, W is negative. For example, as an object is lifted, the work done by the gravitational force on the object is negative. The factor $\cos \theta$ in the definition of W (Eq. 7.1) automatically takes care of the sign. It is important to note that **work is an energy transfer;** if energy is transferred *to* the system (object), W is positive; if energy is transferred *from* the system, W is negative.

5.3

---

[1] Actually, you do work while holding the chair at arm's length because your muscles are continuously contracting and relaxing; this means that they are exerting internal forces on your arm. Thus, work is being done by your body—but internally on itself rather than on the chair.

If an applied force **F** acts along the direction of the displacement, then $\theta = 0$ and $\cos 0 = 1$. In this case, Equation 7.1 gives

$$W = Fd$$

Work is a scalar quantity, and its units are force multiplied by length. Therefore, the SI unit of work is the **newton·meter** (N·m). This combination of units is used so frequently that it has been given a name of its own: the **joule** (J).

Quick Quiz 7.1

Can the component of a force that gives an object a centripetal acceleration do any work on the object? (One such force is that exerted by the Sun on the Earth that holds the Earth in a circular orbit around the Sun.)

In general, a particle may be moving with either a constant or a varying velocity under the influence of several forces. In these cases, because work is a scalar quantity, the total work done as the particle undergoes some displacement is the algebraic sum of the amounts of work done by all the forces.

EXAMPLE 7.1   Mr. Clean

A man cleaning a floor pulls a vacuum cleaner with a force of magnitude $F = 50.0$ N at an angle of $30.0°$ with the horizontal (Fig. 7.4a). Calculate the work done by the force on the vacuum cleaner as the vacuum cleaner is displaced 3.00 m to the right.

Solution   Because they aid us in clarifying which forces are acting on the object being considered, drawings like Figure 7.4b are helpful when we are gathering information and organizing a solution. For our analysis, we use the definition of work (Eq. 7.1):

$$W = (F \cos \theta)\, d$$
$$= (50.0 \text{ N})(\cos 30.0°)(3.00 \text{ m}) = 130 \text{ N·m}$$
$$= \boxed{130 \text{ J}}$$

One thing we should learn from this problem is that the normal force **n**, the force of gravity $\mathbf{F}_g = m\mathbf{g}$, and the upward component of the applied force $(50.0 \text{ N})(\sin 30.0°)$ do no work on the vacuum cleaner because these forces are perpendicular to its displacement.

Exercise   Find the work done by the man on the vacuum cleaner if he pulls it 3.0 m with a horizontal force of 32 N.

Answer   96 J.

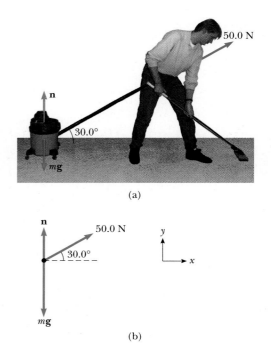

Figure 7.4   (a) A vacuum cleaner being pulled at an angle of $30.0°$ with the horizontal. (b) Free-body diagram of the forces acting on the vacuum cleaner.

The weightlifter does no work on the weights as he holds them on his shoulders. (If he could rest the bar on his shoulders and lock his knees, he would be able to support the weights for quite some time.) Did he do any work when he raised the weights to this height? *(Gerard Vandystadt/Photo Researchers, Inc.)*

Figure 7.5 A person lifts a box of mass *m* a vertical distance *h* and then walks horizontally a distance *d*.

Quick Quiz 7.2 ▶

A person lifts a heavy box of mass *m* a vertical distance *h* and then walks horizontally a distance *d* while holding the box, as shown in Figure 7.5. Determine (a) the work he does on the box and (b) the work done on the box by the force of gravity.

7.2 ▷ THE SCALAR PRODUCT OF TWO VECTORS

2.6 Because of the way the force and displacement vectors are combined in Equation 7.1, it is helpful to use a convenient mathematical tool called the **scalar product.** This tool allows us to indicate how **F** and **d** interact in a way that depends on how close to parallel they happen to be. We write this scalar product **F·d.** (Because of the dot symbol, the scalar product is often called the **dot product.**) Thus, we can express Equation 7.1 as a scalar product:

Work expressed as a dot product

$$W = \mathbf{F \cdot d} = Fd \cos \theta \qquad (7.2)$$

In other words, **F·d** (read "F dot d") is a shorthand notation for $Fd \cos \theta$.

Scalar product of any two vectors **A** *and* **B**

    In general, the scalar product of any two vectors **A** and **B** is a scalar quantity equal to the product of the magnitudes of the two vectors and the cosine of the angle θ between them:

$$\mathbf{A \cdot B} \equiv AB \cos \theta \qquad (7.3)$$

This relationship is shown in Figure 7.6. Note that **A** and **B** need not have the same units.

In Figure 7.6, $B \cos \theta$ is the projection of **B** onto **A**. Therefore, Equation 7.3 says that $\mathbf{A} \cdot \mathbf{B}$ is the product of the magnitude of **A** and the projection of **B** onto **A**.[2]

From the right-hand side of Equation 7.3 we also see that the scalar product is **commutative**.[3] That is,

$$\mathbf{A} \cdot \mathbf{B} = \mathbf{B} \cdot \mathbf{A}$$

Finally, the scalar product obeys the **distributive law of multiplication,** so that

$$\mathbf{A} \cdot (\mathbf{B} + \mathbf{C}) = \mathbf{A} \cdot \mathbf{B} + \mathbf{A} \cdot \mathbf{C}$$

The dot product is simple to evaluate from Equation 7.3 when **A** is either perpendicular or parallel to **B**. If **A** is perpendicular to **B** ($\theta = 90°$), then $\mathbf{A} \cdot \mathbf{B} = 0$. (The equality $\mathbf{A} \cdot \mathbf{B} = 0$ also holds in the more trivial case when either **A** or **B** is zero.) If vector **A** is parallel to vector **B** and the two point in the same direction ($\theta = 0$), then $\mathbf{A} \cdot \mathbf{B} = AB$. If vector **A** is parallel to vector **B** but the two point in opposite directions ($\theta = 180°$), then $\mathbf{A} \cdot \mathbf{B} = -AB$. The scalar product is negative when $90° < \theta < 180°$.

The unit vectors **i**, **j**, and **k**, which were defined in Chapter 3, lie in the positive x, y, and z directions, respectively, of a right-handed coordinate system. Therefore, it follows from the definition of $\mathbf{A} \cdot \mathbf{B}$ that the scalar products of these unit vectors are

$$\mathbf{i} \cdot \mathbf{i} = \mathbf{j} \cdot \mathbf{j} = \mathbf{k} \cdot \mathbf{k} = 1 \qquad \textbf{(7.4)}$$

$$\mathbf{i} \cdot \mathbf{j} = \mathbf{i} \cdot \mathbf{k} = \mathbf{j} \cdot \mathbf{k} = 0 \qquad \textbf{(7.5)}$$

Equations 3.18 and 3.19 state that two vectors **A** and **B** can be expressed in component vector form as

$$\mathbf{A} = A_x \mathbf{i} + A_y \mathbf{j} + A_z \mathbf{k}$$

$$\mathbf{B} = B_x \mathbf{i} + B_y \mathbf{j} + B_z \mathbf{k}$$

Using the information given in Equations 7.4 and 7.5 shows that the scalar product of **A** and **B** reduces to

$$\mathbf{A} \cdot \mathbf{B} = A_x B_x + A_y B_y + A_z B_z \qquad \textbf{(7.6)}$$

(Details of the derivation are left for you in Problem 7.10.) In the special case in which $\mathbf{A} = \mathbf{B}$, we see that

$$\mathbf{A} \cdot \mathbf{A} = A_x{}^2 + A_y{}^2 + A_z{}^2 = A^2$$

The order of the dot product can be reversed

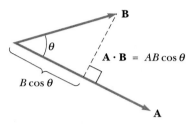

Figure 7.6 The scalar product $\mathbf{A} \cdot \mathbf{B}$ equals the magnitude of **A** multiplied by $B \cos \theta$, which is the projection of **B** onto **A**.

Dot products of unit vectors

Quick Quiz 7.3

If the dot product of two vectors is positive, must the vectors have positive rectangular components?

---

[2] This is equivalent to stating that $\mathbf{A} \cdot \mathbf{B}$ equals the product of the magnitude of **B** and the projection of **A** onto **B**.

[3] This may seem obvious, but in Chapter 11 you will see another way of combining vectors that proves useful in physics and is not commutative.

EXAMPLE 7.2   The Scalar Product

The vectors **A** and **B** are given by $\mathbf{A} = 2\mathbf{i} + 3\mathbf{j}$ and $\mathbf{B} = -\mathbf{i} + 2\mathbf{j}$. (a) Determine the scalar product $\mathbf{A} \cdot \mathbf{B}$.

Solution

$$
\begin{aligned}
\mathbf{A}\cdot\mathbf{B} &= (2\mathbf{i} + 3\mathbf{j}) \cdot (-\mathbf{i} + 2\mathbf{j}) \\
&= -2\mathbf{i}\cdot\mathbf{i} + 2\mathbf{i}\cdot 2\mathbf{j} - 3\mathbf{j}\cdot\mathbf{i} + 3\mathbf{j}\cdot 2\mathbf{j} \\
&= -2(1) + 4(0) - 3(0) + 6(1) \\
&= -2 + 6 = \boxed{4}
\end{aligned}
$$

where we have used the facts that $\mathbf{i}\cdot\mathbf{i} = \mathbf{j}\cdot\mathbf{j} = 1$ and $\mathbf{i}\cdot\mathbf{j} = \mathbf{j}\cdot\mathbf{i} = 0$. The same result is obtained when we use Equation 7.6 directly, where $A_x = 2$, $A_y = 3$, $B_x = -1$, and $B_y = 2$.

(b) Find the angle θ between **A** and **B**.

Solution   The magnitudes of **A** and **B** are

$$
A = \sqrt{A_x^2 + A_y^2} = \sqrt{(2)^2 + (3)^2} = \sqrt{13}
$$
$$
B = \sqrt{B_x^2 + B_y^2} = \sqrt{(-1)^2 + (2)^2} = \sqrt{5}
$$

Using Equation 7.3 and the result from part (a) we find that

$$
\cos\theta = \frac{\mathbf{A}\cdot\mathbf{B}}{AB} = \frac{4}{\sqrt{13}\sqrt{5}} = \frac{4}{\sqrt{65}}
$$

$$
\theta = \cos^{-1}\frac{4}{8.06} = \boxed{60.2°}
$$

EXAMPLE 7.3   Work Done by a Constant Force

A particle moving in the xy plane undergoes a displacement $\mathbf{d} = (2.0\mathbf{i} + 3.0\mathbf{j})$ m as a constant force $\mathbf{F} = (5.0\mathbf{i} + 2.0\mathbf{j})$ N acts on the particle. (a) Calculate the magnitude of the displacement and that of the force.

Solution

$$
d = \sqrt{x^2 + y^2} = \sqrt{(2.0)^2 + (3.0)^2} = \boxed{3.6 \text{ m}}
$$

$$
F = \sqrt{F_x^2 + F_y^2} = \sqrt{(5.0)^2 + (2.0)^2} = \boxed{5.4 \text{ N}}
$$

(b) Calculate the work done by **F**.

Solution   Substituting the expressions for **F** and **d** into Equations 7.4 and 7.5, we obtain

$$
\begin{aligned}
W = \mathbf{F}\cdot\mathbf{d} &= (5.0\mathbf{i} + 2.0\mathbf{j})\cdot(2.0\mathbf{i} + 3.0\mathbf{j}) \text{ N}\cdot\text{m} \\
&= 5.0\mathbf{i}\cdot 2.0\mathbf{i} + 5.0\mathbf{i}\cdot 3.0\mathbf{j} + 2.0\mathbf{j}\cdot 2.0\mathbf{i} + 2.0\mathbf{j}\cdot 3.0\mathbf{j} \\
&= 10 + 0 + 0 + 6 = 16 \text{ N}\cdot\text{m} = \boxed{16 \text{ J}}
\end{aligned}
$$

Exercise   Calculate the angle between **F** and **d**.

Answer   35°.

7.3   WORK DONE BY A VARYING FORCE

Consider a particle being displaced along the x axis under the action of a varying force. The particle is displaced in the direction of increasing x from $x = x_i$ to $x = x_f$. In such a situation, we cannot use $W = (F\cos\theta)d$ to calculate the work done by the force because this relationship applies only when **F** is constant in magnitude and direction. However, if we imagine that the particle undergoes a very small displacement Δx, shown in Figure 7.7a, then the x component of the force F_x is approximately constant over this interval; for this small displacement, we can express the work done by the force as

$$
\Delta W = F_x \Delta x
$$

This is just the area of the shaded rectangle in Figure 7.7a. If we imagine that the F_x versus x curve is divided into a large number of such intervals, then the total work done for the displacement from x_i to x_f is approximately equal to the sum of a large number of such terms:

$$
W \approx \sum_{x_i}^{x_f} F_x \Delta x
$$

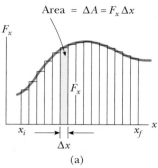

Area = $\Delta A = F_x \Delta x$

(a)

Work

(b)

Figure 7.7   (a) The work done by the force component F_x for the small displacement Δx is $F_x \Delta x$, which equals the area of the shaded rectangle. The total work done for the displacement from x_i to x_f is approximately equal to the sum of the areas of all the rectangles. (b) The work done by the component F_x of the varying force as the particle moves from x_i to x_f is exactly equal to the area under this curve.

If the displacements are allowed to approach zero, then the number of terms in the sum increases without limit but the value of the sum approaches a definite value equal to the area bounded by the F_x curve and the x axis:

$$\lim_{\Delta x \to 0} \sum_{x_i}^{x_f} F_x \,\Delta x = \int_{x_i}^{x_f} F_x \, dx$$

This definite integral is numerically equal to the area under the F_x-versus-x curve between x_i and x_f. Therefore, we can express the work done by F_x as the particle moves from x_i to x_f as

$$W = \int_{x_i}^{x_f} F_x \, dx \qquad (7.7)$$

Work done by a varying force

This equation reduces to Equation 7.1 when the component $F_x = F \cos \theta$ is constant.

If more than one force acts on a particle, the total work done is just the work done by the resultant force. If we express the resultant force in the x direction as ΣF_x, then the total work, or *net work*, done as the particle moves from x_i to x_f is

$$\sum W = W_{\text{net}} = \int_{x_i}^{x_f} \left(\sum F_x \right) dx \qquad (7.8)$$

EXAMPLE 7.4   **Calculating Total Work Done from a Graph**

A force acting on a particle varies with x, as shown in Figure 7.8. Calculate the work done by the force as the particle moves from $x = 0$ to $x = 6.0$ m.

Solution   The work done by the force is equal to the area under the curve from $x_A = 0$ to $x_C = 6.0$ m. This area is equal to the area of the rectangular section from Ⓐ to Ⓑ plus

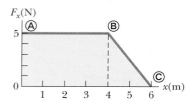

Figure 7.8 The force acting on a particle is constant for the first 4.0 m of motion and then decreases linearly with x from $x_B = 4.0$ m to $x_C = 6.0$ m. The net work done by this force is the area under the curve.

the area of the triangular section from Ⓑ to Ⓒ. The area of the rectangle is $(4.0)(5.0)$ N·m = 20 J, and the area of the triangle is $\frac{1}{2}(2.0)(5.0)$ N·m = 5.0 J. Therefore, the total work done is   25 J.

EXAMPLE 7.5 ▶ Work Done by the Sun on a Probe

The interplanetary probe shown in Figure 7.9a is attracted to the Sun by a force of magnitude

$$F = -1.3 \times 10^{22}/x^2$$

where x is the distance measured outward from the Sun to the probe. Graphically and analytically determine how much work is done by the Sun on the probe as the probe–Sun separation changes from 1.5×10^{11} m to 2.3×10^{11} m.

Graphical Solution The minus sign in the formula for the force indicates that the probe is attracted to the Sun. Because the probe is moving away from the Sun, we expect to calculate a negative value for the work done on it.

A spreadsheet or other numerical means can be used to generate a graph like that in Figure 7.9b. Each small square of the grid corresponds to an area $(0.05 \text{ N})(0.1 \times 10^{11} \text{ m}) = 5 \times 10^8$ N·m. The work done is equal to the shaded area in Figure 7.9b. Because there are approximately 60 squares shaded, the total area (which is negative because it is below the x axis) is about -3×10^{10} N·m. This is the work done by the Sun on the probe.

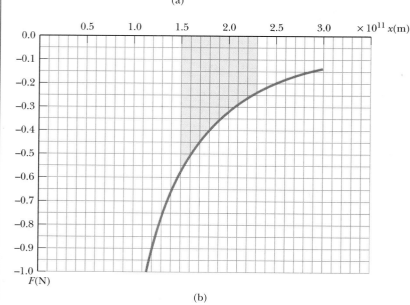

Figure 7.9 (a) An interplanetary probe moves from a position near the Earth's orbit radially outward from the Sun, ending up near the orbit of Mars. (b) Attractive force versus distance for the interplanetary probe.

Analytical Solution   We can use Equation 7.7 to calculate a more precise value for the work done on the probe by the Sun. To solve this integral, we use the first formula of Table B.5 in Appendix B with $n = -2$:

$$W = \int_{1.5\times10^{11}}^{2.3\times10^{11}} \left(\frac{-1.3 \times 10^{22}}{x^2} \right) dx$$

$$= (-1.3 \times 10^{22}) \int_{1.5\times10^{11}}^{2.3\times10^{11}} x^{-2}\, dx$$

$$= (-1.3 \times 10^{22})(-x^{-1}) \Big|_{1.5\times10^{11}}^{2.3\times10^{11}}$$

$$= (-1.3 \times 10^{22})\left(\frac{-1}{2.3 \times 10^{11}} - \frac{-1}{1.5 \times 10^{11}} \right)$$

$$= \boxed{-3.0 \times 10^{10}\,\text{J}}$$

Exercise   Does it matter whether the path of the probe is not directed along a radial line away from the Sun?

Answer   No; the value of W depends only on the initial and final positions, not on the path taken between these points.

Work Done by a Spring

A common physical system for which the force varies with position is shown in Figure 7.10. A block on a horizontal, frictionless surface is connected to a spring. If the spring is either stretched or compressed a small distance from its unstretched (equilibrium) configuration, it exerts on the block a force of magnitude

$$F_s = -kx \qquad\qquad \textbf{(7.9)}$$

Spring force

where x is the displacement of the block from its unstretched ($x = 0$) position and k is a positive constant called the **force constant** of the spring. In other words, the force required to stretch or compress a spring is proportional to the amount of stretch or compression x. This force law for springs, known as **Hooke's law,** is valid only in the limiting case of small displacements. The value of k is a measure of the *stiffness* of the spring. Stiff springs have large k values, and soft springs have small k values.

FORCE EXERTED
BY SPRING ON BLOCK

Quick Quiz 7.4

What are the units for k, the force constant in Hooke's law?

The negative sign in Equation 7.9 signifies that the force exerted by the spring is always directed *opposite* the displacement. When $x > 0$ as in Figure 7.10a, the spring force is directed to the left, in the negative x direction. When $x < 0$ as in Figure 7.10c, the spring force is directed to the right, in the positive x direction. When $x = 0$ as in Figure 7.10b, the spring is unstretched and $F_s = 0$. Because the spring force always acts toward the equilibrium position ($x = 0$), it sometimes is called a *restoring force*. If the spring is compressed until the block is at the point $-x_{\text{max}}$ and is then released, the block moves from $-x_{\text{max}}$ through zero to $+x_{\text{max}}$. If the spring is instead stretched until the block is at the point x_{max} and is then released, the block moves from $+x_{\text{max}}$ through zero to $-x_{\text{max}}$. It then reverses direction, returns to $+x_{\text{max}}$, and continues oscillating back and forth.

Suppose the block has been pushed to the left a distance x_{max} from equilibrium and is then released. Let us calculate the work W_s done by the spring force as the block moves from $x_i = -x_{\text{max}}$ to $x_f = 0$. Applying Equation 7.7 and assuming the block may be treated as a particle, we obtain

$$W_s = \int_{x_i}^{x_f} F_s\, dx = \int_{-x_{\text{max}}}^{0} (-kx)\, dx = \tfrac{1}{2} k x_{\text{max}}^2 \qquad\qquad \textbf{(7.10)}$$

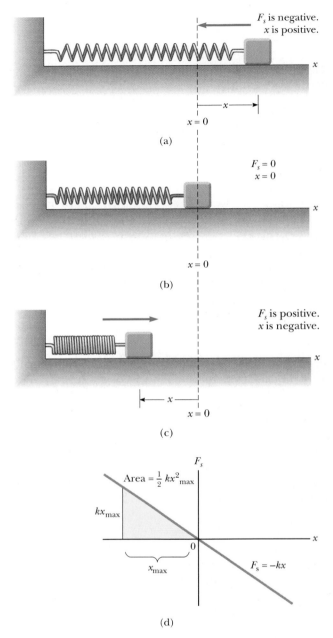

Figure 7.10   The force exerted by a spring on a block varies with the block's displacement x from the equilibrium position $x = 0$. (a) When x is positive (stretched spring), the spring force is directed to the left. (b) When x is zero (natural length of the spring), the spring force is zero. (c) When x is negative (compressed spring), the spring force is directed to the right. (d) Graph of F_s versus x for the block–spring system. The work done by the spring force as the block moves from $-x_{max}$ to 0 is the area of the shaded triangle, $\frac{1}{2}kx_{max}^2$.

where we have used the indefinite integral $\int x^n dx = x^{n+1}/(n + 1)$ with $n = 1$. The work done by the spring force is positive because the force is in the same direction as the displacement (both are to the right). When we consider the work done by the spring force as the block moves from $x_i = 0$ to $x_f = x_{max}$, we find that

$W_s = -\frac{1}{2}kx_{max}^2$ because for this part of the motion the displacement is to the right and the spring force is to the left. Therefore, the *net* work done by the spring force as the block moves from $x_i = -x_{max}$ to $x_f = x_{max}$ is *zero*.

Figure 7.10d is a plot of F_s versus x. The work calculated in Equation 7.10 is the area of the shaded triangle, corresponding to the displacement from $-x_{max}$ to 0. Because the triangle has base x_{max} and height kx_{max}, its area is $\frac{1}{2}kx_{max}^2$, the work done by the spring as given by Equation 7.10.

If the block undergoes an arbitrary displacement from $x = x_i$ to $x = x_f$, the work done by the spring force is

$$W_s = \int_{x_i}^{x_f}(-kx)\,dx = \frac{1}{2}kx_i^2 - \frac{1}{2}kx_f^2 \qquad \textbf{(7.11)}$$

Work done by a spring

For example, if the spring has a force constant of 80 N/m and is compressed 3.0 cm from equilibrium, the work done by the spring force as the block moves from $x_i = -3.0$ cm to its unstretched position $x_f = 0$ is 3.6×10^{-2} J. From Equation 7.11 we also see that the work done by the spring force is zero for any motion that ends where it began ($x_i = x_f$). We shall make use of this important result in Chapter 8, in which we describe the motion of this system in greater detail.

Equations 7.10 and 7.11 describe the work done by the spring on the block. Now let us consider the work done on the spring by an *external agent* that stretches the spring very slowly from $x_i = 0$ to $x_f = x_{max}$, as in Figure 7.11. We can calculate this work by noting that at any value of the displacement, the *applied force* $\mathbf{F}_{app}$ is equal to and opposite the spring force $\mathbf{F}_s$, so that $F_{app} = -(-kx) = kx$. Therefore, the work done by this applied force (the external agent) is

$$W_{F_{app}} = \int_0^{x_{max}} F_{app}\,dx = \int_0^{x_{max}} kx\,dx = \frac{1}{2}kx_{max}^2$$

This work is equal to the negative of the work done by the spring force for this displacement.

Figure 7.11 A block being pulled from $x_i = 0$ to $x_f = x_{max}$ on a frictionless surface by a force $\mathbf{F}_{app}$. If the process is carried out very slowly, the applied force is equal to and opposite the spring force at all times.

EXAMPLE 7.6 ▸ Measuring *k* for a Spring

A common technique used to measure the force constant of a spring is described in Figure 7.12. The spring is hung vertically, and an object of mass m is attached to its lower end. Under the action of the "load" mg, the spring stretches a distance d from its equilibrium position. Because the spring force is upward (opposite the displacement), it must balance the downward force of gravity $m\mathbf{g}$ when the system is at rest. In this case, we can apply Hooke's law to give $|\mathbf{F}_s| = kd = mg$, or

$$k = \frac{mg}{d}$$

For example, if a spring is stretched 2.0 cm by a suspended object having a mass of 0.55 kg, then the force constant is

$$k = \frac{mg}{d} = \frac{(0.55 \text{ kg})(9.80 \text{ m/s}^2)}{2.0 \times 10^{-2} \text{ m}} = \boxed{2.7 \times 10^2 \text{ N/m}}$$

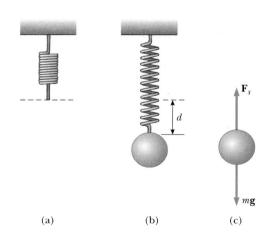

(a)   (b)   (c)

Figure 7.12 Determining the force constant k of a spring. The elongation d is caused by the attached object, which has a weight mg. Because the spring force balances the force of gravity, it follows that $k = mg/d$.

7.4 ▸ KINETIC ENERGY AND THE WORK–KINETIC ENERGY THEOREM

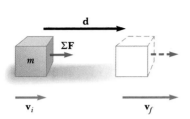

Figure 7.13 A particle undergoing a displacement **d** and a change in velocity under the action of a constant net force Σ**F**.

It can be difficult to use Newton's second law to solve motion problems involving complex forces. An alternative approach is to relate the speed of a moving particle to its displacement under the influence of some net force. If the work done by the net force on a particle can be calculated for a given displacement, then the change in the particle's speed can be easily evaluated.

Figure 7.13 shows a particle of mass m moving to the right under the action of a constant net force Σ**F**. Because the force is constant, we know from Newton's second law that the particle moves with a constant acceleration **a**. If the particle is displaced a distance d, the net work done by the total force Σ**F** is

$$\sum W = \left(\sum F\right)d = (ma)d \tag{7.12}$$

In Chapter 2 we found that the following relationships are valid when a particle undergoes constant acceleration:

$$d = \tfrac{1}{2}(v_i + v_f)t \qquad a = \frac{v_f - v_i}{t}$$

where v_i is the speed at $t = 0$ and v_f is the speed at time t. Substituting these expressions into Equation 7.12 gives

$$\sum W = m\left(\frac{v_f - v_i}{t}\right)\tfrac{1}{2}(v_i + v_f)t$$

$$\sum W = \tfrac{1}{2}mv_f^2 - \tfrac{1}{2}mv_i^2 \tag{7.13}$$

The quantity $\tfrac{1}{2}mv^2$ represents the energy associated with the motion of the particle. This quantity is so important that it has been given a special name—**kinetic energy**. The net work done on a particle by a constant net force Σ**F** acting on it equals the change in kinetic energy of the particle.

In general, the kinetic energy K of a particle of mass m moving with a speed v is defined as

Kinetic energy is energy associated with the motion of a body

$$K \equiv \tfrac{1}{2}mv^2 \tag{7.14}$$

TABLE 7.1   Kinetic Energies for Various Objects

Object	Mass (kg)	Speed (m/s)	Kinetic Energy (J)
Earth orbiting the Sun	5.98×10^{24}	2.98×10^4	2.65×10^{33}
Moon orbiting the Earth	7.35×10^{22}	1.02×10^3	3.82×10^{28}
Rocket moving at escape speed[a]	500	1.12×10^4	3.14×10^{10}
Automobile at 55 mi/h	2 000	25	6.3×10^5
Running athlete	70	10	3.5×10^3
Stone dropped from 10 m	1.0	14	9.8×10^1
Golf ball at terminal speed	0.046	44	4.5×10^1
Raindrop at terminal speed	3.5×10^{-5}	9.0	1.4×10^{-3}
Oxygen molecule in air	5.3×10^{-26}	500	6.6×10^{-21}

[a] *Escape speed* is the minimum speed an object must attain near the Earth's surface if it is to escape the Earth's gravitational force.

Kinetic energy is a scalar quantity and has the same units as work. For example, a 2.0-kg object moving with a speed of 4.0 m/s has a kinetic energy of 16 J. Table 7.1 lists the kinetic energies for various objects.

It is often convenient to write Equation 7.13 in the form

$$\sum W = K_f - K_i = \Delta K \tag{7.15}$$

Work–kinetic energy theorem

That is, $K_i + \sum W = K_f$.

Equation 7.15 is an important result known as the **work–kinetic energy theorem.** It is important to note that when we use this theorem, we must include *all* of the forces that do work on the particle in the calculation of the net work done. From this theorem, we see that the speed of a particle increases if the net work done on it is positive because the final kinetic energy is greater than the initial kinetic energy. The particle's speed decreases if the net work done is negative because the final kinetic energy is less than the initial kinetic energy.

The work–kinetic energy theorem as expressed by Equation 7.15 allows us to think of kinetic energy as the work a particle can do in coming to rest, or the amount of energy stored in the particle. For example, suppose a hammer (our particle) is on the verge of striking a nail, as shown in Figure 7.14. The moving hammer has kinetic energy and so can do work on the nail. The work done on the nail is equal to Fd, where F is the average force exerted on the nail by the hammer and d is the distance the nail is driven into the wall.[4]

We derived the work–kinetic energy theorem under the assumption of a constant net force, but it also is valid when the force varies. To see this, suppose the net force acting on a particle in the x direction is $\sum F_x$. We can apply Newton's second law, $\sum F_x = ma_x$, and use Equation 7.8 to express the net work done as

$$\sum W = \int_{x_i}^{x_f} \left(\sum F_x \right) dx = \int_{x_i}^{x_f} ma_x \, dx$$

If the resultant force varies with x, the acceleration and speed also depend on x. Because we normally consider acceleration as a function of t, we now use the following chain rule to express a in a slightly different way:

$$a = \frac{dv}{dt} = \frac{dv}{dx}\frac{dx}{dt} = v\frac{dv}{dx}$$

Substituting this expression for a into the above equation for $\sum W$ gives

$$\sum W = \int_{x_i}^{x_f} mv\frac{dv}{dx} \, dx = \int_{v_i}^{v_f} mv \, dv$$

$$\sum W = \tfrac{1}{2}mv_f^2 - \tfrac{1}{2}mv_i^2 \tag{7.16}$$

Figure 7.14 The moving hammer has kinetic energy and thus can do work on the nail, driving it into the wall.

The net work done on a particle equals the change in its kinetic energy

The limits of the integration were changed from x values to v values because the variable was changed from x to v. Thus, we conclude that the net work done on a particle by the net force acting on it is equal to the change in the kinetic energy of the particle. This is true whether or not the net force is constant.

---

[4] Note that because the nail and the hammer are *systems* of particles rather than single particles, part of the hammer's kinetic energy goes into warming the hammer and the nail upon impact. Also, as the nail moves into the wall in response to the impact, the large frictional force between the nail and the wood results in the continuous transformation of the kinetic energy of the nail into further temperature increases in the nail and the wood, as well as in deformation of the wall. Energy associated with temperature changes is called *internal energy* and will be studied in detail in Chapter 20.

Situations Involving Kinetic Friction

One way to include frictional forces in analyzing the motion of an object sliding on a *horizontal* surface is to describe the kinetic energy lost because of friction. Suppose a book moving on a horizontal surface is given an initial horizontal velocity $\mathbf{v}_i$ and slides a distance d before reaching a final velocity $\mathbf{v}_f$ as shown in Figure 7.15. The external force that causes the book to undergo an acceleration in the negative x direction is the force of kinetic friction $\mathbf{f}_k$ acting to the left, opposite the motion. The initial kinetic energy of the book is $\frac{1}{2}mv_i^2$, and its final kinetic energy is $\frac{1}{2}mv_f^2$. Applying Newton's second law to the book can show this. Because the only force acting on the book in the x direction is the friction force, Newton's second law gives $-f_k = ma_x$. Multiplying both sides of this expression by d and using Equation 2.12 in the form $v_{xf}^2 - v_{xi}^2 = 2a_x d$ for motion under constant acceleration give $-f_k d = (ma_x)d = \frac{1}{2}mv_{xf}^2 - \frac{1}{2}mv_{xi}^2$ or

$$\Delta K_{\text{friction}} = -f_k d \tag{7.17a}$$

This result specifies that the amount by which the force of kinetic friction changes the kinetic energy of the book is equal to $-f_k d$. Part of this lost kinetic energy goes into warming up the book, and the rest goes into warming up the surface over which the book slides. In effect, the quantity $-f_k d$ is equal to the work done by kinetic friction on the book *plus* the work done by kinetic friction on the surface. (We shall study the relationship between temperature and energy in Part III of this text.) When friction—as well as other forces—acts on an object, the work–kinetic energy theorem reads

$$K_i + \sum W_{\text{other}} - f_k d = K_f \tag{7.17b}$$

Here, $\sum W_{\text{other}}$ represents the sum of the amounts of work done on the object by forces other than kinetic friction.

Loss in kinetic energy due to friction

Figure 7.15 A book sliding to the right on a horizontal surface slows down in the presence of a force of kinetic friction acting to the left. The initial velocity of the book is $\mathbf{v}_i$, and its final velocity is $\mathbf{v}_f$. The normal force and the force of gravity are not included in the diagram because they are perpendicular to the direction of motion and therefore do not influence the book's velocity.

Quick Quiz 7.5

Can frictional forces ever *increase* an object's kinetic energy? p 213

YES — STATIC FRICTION!

EXAMPLE 7.7 A Block Pulled on a Frictionless Surface

A 6.0-kg block initially at rest is pulled to the right along a horizontal, frictionless surface by a constant horizontal force of 12 N. Find the speed of the block after it has moved 3.0 m.

Solution We have made a drawing of this situation in Figure 7.16a. We could apply the equations of kinematics to determine the answer, but let us use the energy approach for

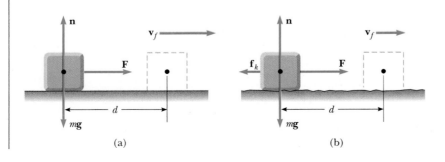

Figure 7.16 A block pulled to the right by a constant horizontal force. (a) Frictionless surface. (b) Rough surface.

practice. The normal force balances the force of gravity on the block, and neither of these vertically acting forces does work on the block because the displacement is horizontal. Because there is no friction, the net external force acting on the block is the 12-N force. The work done by this force is

$$W = Fd = (12 \text{ N})(3.0 \text{ m}) = 36 \text{ N·m} = 36 \text{ J}$$

Using the work–kinetic energy theorem and noting that the initial kinetic energy is zero, we obtain

$$W = K_f - K_i = \tfrac{1}{2} m v_f^2 - 0$$

$$v_f^2 = \frac{2W}{m} = \frac{2(36 \text{ J})}{6.0 \text{ kg}} = 12 \text{ m}^2/\text{s}^2$$

$$v_f = \boxed{3.5 \text{ m/s}}$$

Exercise Find the acceleration of the block and determine its final speed, using the kinematics equation $v_{xf}^2 = v_{xi}^2 + 2a_x d$.

Answer $a_x = 2.0 \text{ m/s}^2$; $v_f = 3.5 \text{ m/s}$.

EXAMPLE 7.8 ▸ A Block Pulled on a Rough Surface

Find the final speed of the block described in Example 7.7 if the surface is not frictionless but instead has a coefficient of kinetic friction of 0.15.

Solution The applied force does work just as in Example 7.7:

$$W = Fd = (12 \text{ N})(3.0 \text{ m}) = 36 \text{ J}$$

In this case we must use Equation 7.17a to calculate the kinetic energy lost to friction $\Delta K_{\text{friction}}$. The magnitude of the frictional force is

$$f_k = \mu_k n = \mu_k mg = (0.15)(6.0 \text{ kg})(9.80 \text{ m/s}^2) = 8.82 \text{ N}$$

The change in kinetic energy due to friction is

$$\Delta K_{\text{friction}} = -f_k d = -(8.82 \text{ N})(3.0 \text{ m}) = -26.5 \text{ J}$$

The final speed of the block follows from Equation 7.17b:

$$\tfrac{1}{2} m v_i^2 + \sum W_{\text{other}} - f_k d = \tfrac{1}{2} m v_f^2$$

$$0 + 36 \text{ J} - 26.5 \text{ J} = \tfrac{1}{2}(6.0 \text{ kg}) v_f^2$$

$$v_f^2 = 2(9.5 \text{ J})/(6.0 \text{ kg}) = 3.18 \text{ m}^2/\text{s}^2$$

$$v_f = \boxed{1.8 \text{ m/s}}$$

After sliding the 3-m distance on the rough surface, the block is moving at a speed of 1.8 m/s; in contrast, after covering the same distance on a frictionless surface (see Example 7.7), its speed was 3.5 m/s.

Exercise Find the acceleration of the block from Newton's second law and determine its final speed, using equations of kinematics.

Answer $a_x = 0.53 \text{ m/s}^2$; $v_f = 1.8 \text{ m/s}$.

CONCEPTUAL EXAMPLE 7.9 ▸ Does the Ramp Lessen the Work Required?

A man wishes to load a refrigerator onto a truck using a ramp, as shown in Figure 7.17. He claims that less work would be required to load the truck if the length L of the ramp were increased. Is his statement valid?

Solution No. Although less force is required with a longer ramp, that force must act over a greater distance if the same amount of work is to be done. Suppose the refrigerator is wheeled on a dolly up the ramp at constant speed. The

Figure 7.17 A refrigerator attached to a frictionless wheeled dolly is moved up a ramp at constant speed.

normal force exerted by the ramp on the refrigerator is directed 90° to the motion and so does no work on the refrigerator. Because $\Delta K = 0$, the work–kinetic energy theorem gives

$$\sum W = W_{\text{by man}} + W_{\text{by gravity}} = 0$$

The work done by the force of gravity equals the weight of the refrigerator mg times the vertical height h through which it is displaced times cos 180°, or $W_{\text{by gravity}} = -mgh$. (The minus sign arises because the downward force of gravity is opposite the displacement.) Thus, the man must do work mgh on the refrigerator, regardless of the length of the ramp.

QuickLab

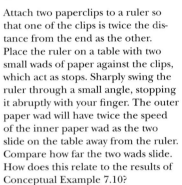

Attach two paperclips to a ruler so that one of the clips is twice the distance from the end as the other. Place the ruler on a table with two small wads of paper against the clips, which act as stops. Sharply swing the ruler through a small angle, stopping it abruptly with your finger. The outer paper wad will have twice the speed of the inner paper wad as the two slide on the table away from the ruler. Compare how far the two wads slide. How does this relate to the results of Conceptual Example 7.10?

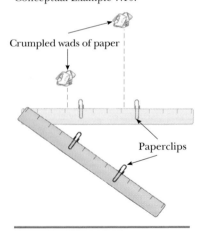

Crumpled wads of paper

Paperclips

Consider the chum salmon attempting to swim upstream in the photograph at the beginning of this chapter. The "steps" of a fish ladder built around a dam do not change the total amount of work that must be done by the salmon as they leap through some vertical distance. However, the ladder allows the fish to perform that work in a series of smaller jumps, and the net effect is to raise the vertical position of the fish by the height of the dam.

These cyclists are working hard and expending energy as they pedal uphill in Marin County, CA.
(David Madison/Tony Stone Images)

CONCEPTUAL EXAMPLE 7.10   Useful Physics for Safer Driving

A certain car traveling at an initial speed v slides a distance d to a halt after its brakes lock. Assuming that the car's initial speed is instead $2v$ at the moment the brakes lock, estimate the distance it slides.

Solution Let us assume that the force of kinetic friction between the car and the road surface is constant and the same for both speeds. The net force multiplied by the displacement of the car is equal to the initial kinetic energy of the car (because $K_f = 0$). If the speed is doubled, as it is in this example, the kinetic energy is quadrupled. For a given constant applied force (in this case, the frictional force), the distance traveled is four times as great when the initial speed is doubled, and so the estimated distance that the car slides is $4d$.

EXAMPLE 7.11 A Block–Spring System

A block of mass 1.6 kg is attached to a horizontal spring that has a force constant of 1.0×10^3 N/m, as shown in Figure 7.10. The spring is compressed 2.0 cm and is then released from rest. (a) Calculate the speed of the block as it passes through the equilibrium position $x = 0$ if the surface is frictionless.

Solution In this situation, the block starts with $v_i = 0$ at $x_i = -2.0$ cm, and we want to find v_f at $x_f = 0$. We use Equation 7.10 to find the work done by the spring with $x_{max} = x_i = -2.0$ cm $= -2.0 \times 10^{-2}$ m:

$$W_s = \tfrac{1}{2}kx_{max}^2 = \tfrac{1}{2}(1.0 \times 10^3 \text{ N/m})(-2.0 \times 10^{-2} \text{ m})^2 = 0.20 \text{ J}$$

Using the work–kinetic energy theorem with $v_i = 0$, we obtain the change in kinetic energy of the block due to the work done on it by the spring:

$$W_s = \tfrac{1}{2}mv_f^2 - \tfrac{1}{2}mv_i^2$$

$$0.20 \text{ J} = \tfrac{1}{2}(1.6 \text{ kg})v_f^2 - 0$$

$$v_f^2 = \frac{0.40 \text{ J}}{1.6 \text{ kg}} = 0.25 \text{ m}^2/\text{s}^2$$

$$v_f = \boxed{0.50 \text{ m/s}}$$

(b) Calculate the speed of the block as it passes through the equilibrium position if a constant frictional force of 4.0 N retards its motion from the moment it is released.

Solution Certainly, the answer has to be less than what we found in part (a) because the frictional force retards the motion. We use Equation 7.17 to calculate the kinetic energy lost because of friction and add this negative value to the kinetic energy found in the absence of friction. The kinetic energy lost due to friction is

$$\Delta K = -f_k d = -(4.0 \text{ N})(2.0 \times 10^{-2} \text{ m}) = -0.080 \text{ J}$$

In part (a), the final kinetic energy without this loss was found to be 0.20 J. Therefore, the final kinetic energy in the presence of friction is

$$K_f = 0.20 \text{ J} - 0.080 \text{ J} = 0.12 \text{ J} = \tfrac{1}{2}mv_f^2$$

$$\tfrac{1}{2}(1.6 \text{ kg})v_f^2 = 0.12 \text{ J}$$

$$v_f^2 = \frac{0.24 \text{ J}}{1.6 \text{ kg}} = 0.15 \text{ m}^2/\text{s}^2$$

$$v_f = \boxed{0.39 \text{ m/s}}$$

As expected, this value is somewhat less than the 0.50 m/s we found in part (a). If the frictional force were greater, then the value we obtained as our answer would have been even smaller.

7.5 POWER

Imagine two identical models of an automobile: one with a base-priced four-cylinder engine; and the other with the highest-priced optional engine, a mighty eight-cylinder powerplant. Despite the differences in engines, the two cars have the same mass. Both cars climb a roadway up a hill, but the car with the optional engine takes much less time to reach the top. Both cars have done the same amount of work against gravity, but in different time periods. From a practical viewpoint, it is interesting to know not only the work done by the vehicles but also the *rate* at which it is done. In taking the ratio of the amount of work done to the time taken to do it, we have a way of quantifying this concept. The time rate of doing work is called **power.**

If an external force is applied to an object (which we assume acts as a particle), and if the work done by this force in the time interval Δt is W, then the **average power** expended during this interval is defined as

$$\overline{\mathcal{P}} \equiv \frac{W}{\Delta t}$$

Average power

The work done on the object contributes to the increase in the energy of the object. Therefore, a more general definition of power is the *time rate of energy transfer.* In a manner similar to how we approached the definition of velocity and acceleration, we can define the **instantaneous power** $\mathcal{P}$ as the limiting value of the average power as Δt approaches zero:

$$\mathcal{P} \equiv \lim_{\Delta t \to 0} \frac{W}{\Delta t} = \frac{dW}{dt}$$

where we have represented the increment of work done by dW. We find from Equation 7.2, letting the displacement be expressed as $d\mathbf{s}$, that $dW = \mathbf{F} \cdot d\mathbf{s}$. Therefore, the instantaneous power can be written

Instantaneous power

$$\mathcal{P} = \frac{dW}{dt} = \mathbf{F} \cdot \frac{d\mathbf{s}}{dt} = \mathbf{F} \cdot \mathbf{v} \tag{7.18}$$

where we use the fact that $\mathbf{v} = d\mathbf{s}/dt$.

The SI unit of power is joules per second (J/s), also called the *watt* (W) (after James Watt, the inventor of the steam engine):

The watt

$$1\ \text{W} = 1\ \text{J/s} = 1\ \text{kg·m}^2/\text{s}^3$$

The symbol W (not italic) for watt should not be confused with the symbol W (italic) for work.

A unit of power in the British engineering system is the horsepower (hp):

$$1\ \text{hp} = 746\ \text{W}$$

A unit of energy (or work) can now be defined in terms of the unit of power. One **kilowatt hour** (kWh) is the energy converted or consumed in 1 h at the constant rate of 1 kW = 1 000 J/s. The numerical value of 1 kWh is

The kilowatt hour is a unit of energy

$$1\ \text{kWh} = (10^3\ \text{W})(3\ 600\ \text{s}) = 3.60 \times 10^6\ \text{J}$$

It is important to realize that a kilowatt hour is a unit of energy, not power. When you pay your electric bill, you pay the power company for the total electrical energy you used during the billing period. This energy is the power used multiplied by the time during which it was used. For example, a 300-W lightbulb run for 12 h would convert $(0.300\ \text{kW})(12\ \text{h}) = 3.6\ \text{kWh}$ of electrical energy.

Quick Quiz 7.6

Suppose that an old truck and a sports car do the same amount of work as they climb a hill but that the truck takes much longer to accomplish this work. How would graphs of $\mathcal{P}$ versus t compare for the two vehicles?

EXAMPLE 7.12   Power Delivered by an Elevator Motor

An elevator car has a mass of 1 000 kg and is carrying passengers having a combined mass of 800 kg. A constant frictional force of 4 000 N retards its motion upward, as shown in Figure 7.18a. (a) What must be the minimum power delivered by the motor to lift the elevator car at a constant speed of 3.00 m/s?

Solution   The motor must supply the force of magnitude T that pulls the elevator car upward. Reading that the speed is constant provides the hint that $a = 0$, and therefore we know from Newton's second law that $\Sigma F_y = 0$. We have drawn

a free-body diagram in Figure 7.18b and have arbitrarily specified that the upward direction is positive. From Newton's second law we obtain

$$\sum F_y = T - f - Mg = 0$$

where M is the *total* mass of the system (car plus passengers), equal to 1 800 kg. Therefore,

$$T = f + Mg$$
$$= 4.00 \times 10^3\ \text{N} + (1.80 \times 10^3\ \text{kg})(9.80\ \text{m/s}^2)$$
$$= 2.16 \times 10^4\ \text{N}$$

Using Equation 7.18 and the fact that **T** is in the same direction as **v**, we find that

$$\mathcal{P} = \mathbf{T}\cdot\mathbf{v} = Tv$$

$$= (2.16 \times 10^4 \text{ N})(3.00 \text{ m/s}) = \boxed{6.48 \times 10^4 \text{ W}}$$

(b) What power must the motor deliver at the instant its speed is v if it is designed to provide an upward acceleration of 1.00 m/s^2?

Solution   Now we expect to obtain a value greater than we did in part (a), where the speed was constant, because the motor must now perform the additional task of accelerating the car. The only change in the setup of the problem is that now $a > 0$. Applying Newton's second law to the car gives

$$\sum F_y = T - f - Mg = Ma$$
$$T = M(a + g) + f$$
$$= (1.80 \times 10^3 \text{ kg})(1.00 + 9.80)\text{m/s}^2 + 4.00 \times 10^3 \text{ N}$$
$$= 2.34 \times 10^4 \text{ N}$$

Therefore, using Equation 7.18, we obtain for the required power

$$\mathcal{P} = Tv = \boxed{(2.34 \times 10^4 v) \text{ W}}$$

where v is the instantaneous speed of the car in meters per second. The power is less than that obtained in part (a) as long as the speed is less than $\mathcal{P}/T = 2.77 \text{ m/s}$, but it is greater when the elevator's speed exceeds this value.

(a)   (b)

Figure 7.18   (a) The motor exerts an upward force **T** on the elevator car. The magnitude of this force is the tension T in the cable connecting the car and motor. The downward forces acting on the car are a frictional force **f** and the force of gravity $\mathbf{F}_g = M\mathbf{g}$. (b) The free-body diagram for the elevator car.

CONCEPTUAL EXAMPLE 7.13

In part (a) of the preceding example, the motor delivers power to lift the car, and yet the car moves at constant speed. A student analyzing this situation notes that the kinetic energy of the car does not change because its speed does not change. This student then reasons that, according to the work–kinetic energy theorem, $W = \Delta K = 0$. Knowing that $\mathcal{P} = W/t$, the student concludes that the power delivered by the motor also must be zero. How would you explain this apparent paradox?

Solution   The work–kinetic energy theorem tells us that the *net* force acting on the system multiplied by the displacement is equal to the change in the kinetic energy of the system. In our elevator case, the net force is indeed zero (that is, $T - Mg - f = 0$), and so $W = (\Sigma F_y)d = 0$. However, the power from the motor is calculated not from the *net* force but rather from the force exerted by the motor acting in the direction of motion, which in this case is T and not zero.

Optional Section

7.6   ENERGY AND THE AUTOMOBILE

Automobiles powered by gasoline engines are very inefficient machines. Even under ideal conditions, less than 15% of the chemical energy in the fuel is used to power the vehicle. The situation is much worse under stop-and-go driving conditions in a city. In this section, we use the concepts of energy, power, and friction to analyze automobile fuel consumption.

Many mechanisms contribute to energy loss in an automobile. About 67% of the energy available from the fuel is lost in the engine. This energy ends up in the atmosphere, partly via the exhaust system and partly via the cooling system. (As we shall see in Chapter 22, the great energy loss from the exhaust and cooling systems is required by a fundamental law of thermodynamics.) Approximately 10% of the available energy is lost to friction in the transmission, drive shaft, wheel and axle bearings, and differential. Friction in other moving parts dissipates approximately 6% of the energy, and 4% of the energy is used to operate fuel and oil pumps and such accessories as power steering and air conditioning. This leaves a mere 13% of the available energy to propel the automobile! This energy is used mainly to balance the energy loss due to flexing of the tires and the friction caused by the air, which is more commonly referred to as *air resistance.*

Let us examine the power required to provide a force in the forward direction that balances the combination of the two frictional forces. The coefficient of rolling friction μ between the tires and the road is about 0.016. For a 1 450-kg car, the weight is 14 200 N and the force of rolling friction has a magnitude of $\mu n = \mu mg = 227$ N. As the speed of the car increases, a small reduction in the normal force occurs as a result of a decrease in atmospheric pressure as air flows over the top of the car. (This phenomenon is discussed in Chapter 15.) This reduction in the normal force causes a slight reduction in the force of rolling friction f_r with increasing speed, as the data in Table 7.2 indicate.

Now let us consider the effect of the resistive force that results from the movement of air past the car. For large objects, the resistive force f_a associated with air friction is proportional to the square of the speed (in meters per second; see Section 6.4) and is given by Equation 6.6:

$$f_a = \tfrac{1}{2}D\rho Av^2$$

where D is the drag coefficient, ρ is the density of air, and A is the cross-sectional area of the moving object. We can use this expression to calculate the f_a values in Table 7.2, using $D = 0.50$, $\rho = 1.293$ kg/m^3, and $A \approx 2$ m^2.

The magnitude of the total frictional force f_t is the sum of the rolling frictional force and the air resistive force:

$$f_t = f_r + f_a$$

At low speeds, road friction is the predominant resistive force, but at high speeds air drag predominates, as shown in Table 7.2. Road friction can be decreased by a reduction in tire flexing (for example, by an increase in the air pres-

TABLE 7.2   **Frictional Forces and Power Requirements for a Typical Car**[a]

v (m/s)	n (N)	f_r (N)	f_a (N)	f_t (N)	$\mathcal{P} = f_t v$ (kW)
0	14 200	227	0	227	0
8.9	14 100	226	51	277	2.5
17.8	13 900	222	204	426	7.6
26.8	13 600	218	465	683	18.3
35.9	13 200	211	830	1 041	37.3
44.8	12 600	202	1 293	1 495	67.0

[a] In this table, n is the normal force, f_r is road friction, f_a is air friction, f_t is total friction, and $\mathcal{P}$ is the power delivered to the wheels.

sure slightly above recommended values) and by the use of radial tires. Air drag can be reduced through the use of a smaller cross-sectional area and by streamlining the car. Although driving a car with the windows open increases air drag and thus results in a 3% decrease in mileage, driving with the windows closed and the air conditioner running results in a 12% decrease in mileage.

The total power needed to maintain a constant speed v is $f_t v$, and it is this power that must be delivered to the wheels. For example, from Table 7.2 we see that at $v = 26.8$ m/s (60 mi/h) the required power is

$$\mathcal{P} = f_t v = (683 \text{ N})\left(26.8\frac{\text{m}}{\text{s}}\right) = 18.3 \text{ kW}$$

This power can be broken down into two parts: (1) the power $f_r v$ needed to compensate for road friction, and (2) the power $f_a v$ needed to compensate for air drag. At $v = 26.8$ m/s, we obtain the values

$$\mathcal{P}_r = f_r v = (218 \text{ N})\left(26.8\frac{\text{m}}{\text{s}}\right) = 5.84 \text{ kW}$$

$$\mathcal{P}_a = f_a v = (465 \text{ N})\left(26.8\frac{\text{m}}{\text{s}}\right) = 12.5 \text{ kW}$$

Note that $\mathcal{P} = \mathcal{P}_r + \mathcal{P}_a$.

On the other hand, at $v = 44.8$ m/s (100 mi/h), $\mathcal{P}_r = 9.05$ kW, $\mathcal{P}_a = 57.9$ kW, and $\mathcal{P} = 67.0$ kW. This shows the importance of air drag at high speeds.

EXAMPLE 7.14 ▶ Gas Consumed by a Compact Car

A compact car has a mass of 800 kg, and its efficiency is rated at 18%. (That is, 18% of the available fuel energy is delivered to the wheels.) Find the amount of gasoline used to accelerate the car from rest to 27 m/s (60 mi/h). Use the fact that the energy equivalent of 1 gal of gasoline is 1.3×10^8 J.

Solution The energy required to accelerate the car from rest to a speed v is its final kinetic energy $\frac{1}{2}mv^2$:

$$K = \tfrac{1}{2}mv^2 = \tfrac{1}{2}(800 \text{ kg})(27 \text{ m/s})^2 = 2.9 \times 10^5 \text{ J}$$

If the engine were 100% efficient, each gallon of gasoline would supply 1.3×10^8 J of energy. Because the engine is only 18% efficient, each gallon delivers only $(0.18)(1.3 \times 10^8 \text{ J}) = 2.3 \times 10^7$ J. Hence, the number of gallons used to accelerate the car is

$$\text{Number of gallons} = \frac{2.9 \times 10^5 \text{ J}}{2.3 \times 10^7 \text{ J/gal}} = \boxed{0.013 \text{ gal}}$$

At cruising speed, this much gasoline is sufficient to propel the car nearly 0.5 mi. This demonstrates the extreme energy requirements of stop-and-start driving.

EXAMPLE 7.15 ▶ Power Delivered to Wheels

Suppose the compact car in Example 7.14 gets 35 mi/gal at 60 mi/h. How much power is delivered to the wheels?

Solution By simply canceling units, we determine that the car consumes 60 mi/h ÷ 35 mi/gal = 1.7 gal/h. Using the fact that each gallon is equivalent to 1.3×10^8 J, we find that the total power used is

$$\mathcal{P} = \frac{(1.7 \text{ gal/h})(1.3 \times 10^8 \text{ J/gal})}{3.6 \times 10^3 \text{ s/h}}$$

$$= \frac{2.2 \times 10^8 \text{ J}}{3.6 \times 10^3 \text{ s}} = 62 \text{ kW}$$

Because 18% of the available power is used to propel the car, the power delivered to the wheels is $(0.18)(62 \text{ kW}) = $ 11 kW. This is 40% less than the 18.3-kW value obtained for the 1 450-kg car discussed in the text. Vehicle mass is clearly an important factor in power-loss mechanisms.

EXAMPLE 7.16   Car Accelerating Up a Hill

Consider a car of mass m that is accelerating up a hill, as shown in Figure 7.19. An automotive engineer has measured the magnitude of the total resistive force to be

$$f_t = (218 + 0.70v^2) \text{ N}$$

where v is the speed in meters per second. Determine the power the engine must deliver to the wheels as a function of speed.

Solution   The forces on the car are shown in Figure 7.19, in which $\mathbf{F}$ is the force of friction from the road that propels the car; the remaining forces have their usual meaning. Applying Newton's second law to the motion along the road surface, we find that

$$\sum F_x = F - f_t - mg \sin \theta = ma$$

$$F = ma + mg \sin \theta + f_t$$

$$= ma + mg \sin \theta + (218 + 0.70v^2)$$

Therefore, the power required to move the car forward is

$$\mathcal{P} = Fv = mva + mvg \sin \theta + 218v + 0.70v^3$$

The term mva represents the power that the engine must deliver to accelerate the car. If the car moves at constant speed, this term is zero and the total power requirement is reduced. The term $mvg \sin \theta$ is the power required to provide a force to balance a component of the force of gravity as the car moves up the incline. This term would be zero for motion on a horizontal surface. The term $218v$ is the power required to provide a force to balance road friction, and the term $0.70v^3$ is the power needed to do work on the air.

If we take $m = 1\,450$ kg, $v = 27$ m/s ($= 60$ mi/h), $a =$

Figure 7.19

1.0 m/s², and $\theta = 10°$, then the various terms in $\mathcal{P}$ are calculated to be

$$mva = (1\,450 \text{ kg})(27 \text{ m/s})(1.0 \text{ m/s}^2)$$
$$= 39 \text{ kW} = 52 \text{ hp}$$

$$mvg \sin \theta = (1\,450 \text{ kg})(27 \text{ m/s})(9.80 \text{ m/s}^2)(\sin 10°)$$
$$= 67 \text{ kW} = 89 \text{ hp}$$

$$218v = 218(27 \text{ m/s}) = 5.9 \text{ kW} = 7.9 \text{ hp}$$

$$0.70v^3 = 0.70(27 \text{ m/s})^3 = 14 \text{ kW} = 19 \text{ hp}$$

Hence, the total power required is 126 kW, or   168 hp.

Note that the power requirements for traveling at constant speed on a horizontal surface are only 20 kW, or 27 hp (the sum of the last two terms). Furthermore, if the mass were halved (as in the case of a compact car), then the power required also is reduced by almost the same factor.

Optional Section

7.7   KINETIC ENERGY AT HIGH SPEEDS

The laws of Newtonian mechanics are valid only for describing the motion of particles moving at speeds that are small compared with the speed of light in a vacuum c ($= 3.00 \times 10^8$ m/s). When speeds are comparable to c, the equations of Newtonian mechanics must be replaced by the more general equations predicted by the theory of relativity. One consequence of the theory of relativity is that the kinetic energy of a particle of mass m moving with a speed v is no longer given by $K = mv^2/2$. Instead, one must use the relativistic form of the kinetic energy:

Relativistic kinetic energy

$$K = mc^2 \left(\frac{1}{\sqrt{1 - (v/c)^2}} - 1 \right) \qquad \textbf{(7.19)}$$

According to this expression, speeds greater than c are not allowed because, as v approaches c, K approaches ∞. This limitation is consistent with experimental observations on subatomic particles, which have shown that no particles travel at speeds greater than c. (In other words, c is the ultimate speed.) From this relativistic point of view, the work–kinetic energy theorem says that v can only approach c because it would take an infinite amount of work to attain the speed $v = c$.

All formulas in the theory of relativity must reduce to those in Newtonian mechanics at low particle speeds. It is instructive to show that this is the case for the kinetic energy relationship by analyzing Equation 7.19 when v is small compared with c. In this case, we expect K to reduce to the Newtonian expression. We can check this by using the binomial expansion (Appendix B.5) applied to the quantity $[1 - (v/c)^2]^{-1/2}$, with $v/c \ll 1$. If we let $x = (v/c)^2$, the expansion gives

$$\frac{1}{(1 - x)^{1/2}} = 1 + \frac{x}{2} + \frac{3}{8}x^2 + \cdots$$

Making use of this expansion in Equation 7.19 gives

$$K = mc^2 \left(1 + \frac{v^2}{2c^2} + \frac{3}{8}\frac{v^4}{c^4} + \cdots - 1 \right)$$

$$= \frac{1}{2}mv^2 + \frac{3}{8}m\frac{v^4}{c^2} + \cdots$$

$$= \frac{1}{2}mv^2 \qquad \text{for} \qquad \frac{v}{c} \ll 1$$

Thus, we see that the relativistic kinetic energy expression does indeed reduce to the Newtonian expression for speeds that are small compared with c. We shall return to the subject of relativity in Chapter 39.

SUMMARY

The work done by a constant force **F** acting on a particle is defined as the product of the component of the force in the direction of the particle's displacement and the magnitude of the displacement. Given a force **F** that makes an angle θ with the displacement vector **d** of a particle acted on by the force, you should be able to determine the work done by **F** using the equation

$$W \equiv Fd \cos \theta \tag{7.1}$$

The **scalar product** (dot product) of two vectors **A** and **B** is defined by the relationship

$$\mathbf{A} \cdot \mathbf{B} \equiv AB \cos \theta \tag{7.3}$$

where the result is a scalar quantity and θ is the angle between the two vectors. The scalar product obeys the commutative and distributive laws.

If a varying force does work on a particle as the particle moves along the x axis from x_i to x_f, you must use the expression

$$W \equiv \int_{x_i}^{x_f} F_x \, dx \tag{7.7}$$

where F_x is the component of force in the x direction. If several forces are acting on the particle, the net work done by all of the forces is the sum of the amounts of work done by all of the forces.

The **kinetic energy** of a particle of mass m moving with a speed v (where v is small compared with the speed of light) is

$$K \equiv \tfrac{1}{2} mv^2 \qquad \textbf{(7.14)}$$

The **work–kinetic energy theorem** states that the net work done on a particle by external forces equals the change in kinetic energy of the particle:

$$\sum W = K_f - K_i = \tfrac{1}{2} mv_f^2 - \tfrac{1}{2} mv_i^2 \qquad \textbf{(7.16)}$$

If a frictional force acts, then the work–kinetic energy theorem can be modified to give

$$K_i + \sum W_{\text{other}} - f_k d = K_f \qquad \textbf{(7.17b)}$$

The **instantaneous power** $\mathcal{P}$ is defined as the time rate of energy transfer. If an agent applies a force $\mathbf{F}$ to an object moving with a velocity $\mathbf{v}$, the power delivered by that agent is

$$\mathcal{P} \equiv \frac{dW}{dt} = \mathbf{F} \cdot \mathbf{v} \qquad \textbf{(7.18)}$$

QUESTIONS

1. Consider a tug-of-war in which two teams pulling on a rope are evenly matched so that no motion takes place. Assume that the rope does not stretch. Is work done on the rope? On the pullers? On the ground? Is work done on anything?

2. For what values of θ is the scalar product (a) positive and (b) negative?

3. As the load on a spring hung vertically is increased, one would not expect the F_s-versus-x curve to always remain linear, as shown in Figure 7.10d. Explain qualitatively what you would expect for this curve as m is increased.

4. Can the kinetic energy of an object be negative? Explain.

5. (a) If the speed of a particle is doubled, what happens to its kinetic energy? (b) If the net work done on a particle is zero, what can be said about the speed?

6. In Example 7.16, does the required power increase or decrease as the force of friction is reduced?

7. An automobile sales representative claims that a "souped-up" 300-hp engine is a necessary option in a compact car (instead of a conventional 130-hp engine). Suppose you intend to drive the car within speed limits (≤ 55 mi/h) and on flat terrain. How would you counter this sales pitch?

8. One bullet has twice the mass of another bullet. If both bullets are fired so that they have the same speed, which has the greater kinetic energy? What is the ratio of the kinetic energies of the two bullets?

9. When a punter kicks a football, is he doing any work on the ball while his toe is in contact with it? Is he doing any work on the ball after it loses contact with his toe? Are any forces doing work on the ball while it is in flight?

10. Discuss the work done by a pitcher throwing a baseball. What is the approximate distance through which the force acts as the ball is thrown?

11. Two sharpshooters fire 0.30-caliber rifles using identical shells. The barrel of rifle A is 2.00 cm longer than that of rifle B. Which rifle will have the higher muzzle speed? (*Hint:* The force of the expanding gases in the barrel accelerates the bullets.)

12. As a simple pendulum swings back and forth, the forces acting on the suspended mass are the force of gravity, the tension in the supporting cord, and air resistance. (a) Which of these forces, if any, does no work on the pendulum? (b) Which of these forces does negative work at all times during its motion? (c) Describe the work done by the force of gravity while the pendulum is swinging.

13. The kinetic energy of an object depends on the frame of reference in which its motion is measured. Give an example to illustrate this point.

14. An older model car accelerates from 0 to a speed v in 10 s. A newer, more powerful sports car accelerates from 0 to $2v$ in the same time period. What is the ratio of powers expended by the two cars? Consider the energy coming from the engines to appear only as kinetic energy of the cars.

PROBLEMS

1, 2, 3 = straightforward, intermediate, challenging   ☐ = full solution available in the *Student Solutions Manual and Study Guide*
WEB = solution posted at **http://www.saunderscollege.com/physics/**   🖥 = Computer useful in solving problem   🎮 = Interactive Physics
☐ = paired numerical/symbolic problems

Section 7.1   Work Done by a Constant Force

1. A tugboat exerts a constant force of 5 000 N on a ship moving at constant speed through a harbor. How much work does the tugboat do on the ship in a distance of 3.00 km?

2. A shopper in a supermarket pushes a cart with a force of 35.0 N directed at an angle of 25.0° downward from the horizontal. Find the work done by the shopper as she moves down an aisle 50.0 m in length.

3. A raindrop ($m = 3.35 \times 10^{-5}$ kg) falls vertically at constant speed under the influence of gravity and air resistance. After the drop has fallen 100 m, what is the work done (a) by gravity and (b) by air resistance?

4. A sledge loaded with bricks has a total mass of 18.0 kg and is pulled at constant speed by a rope. The rope is inclined at 20.0° above the horizontal, and the sledge moves a distance of 20.0 m on a horizontal surface. The coefficient of kinetic friction between the sledge and the surface is 0.500. (a) What is the tension of the rope? (b) How much work is done on the sledge by the rope? (c) What is the energy lost due to friction?

5. A block of mass 2.50 kg is pushed 2.20 m along a frictionless horizontal table by a constant 16.0-N force directed 25.0° below the horizontal. Determine the work done by (a) the applied force, (b) the normal force exerted by the table, and (c) the force of gravity. (d) Determine the total work done on the block.

6. A 15.0-kg block is dragged over a rough, horizontal surface by a 70.0-N force acting at 20.0° above the horizontal. The block is displaced 5.00 m, and the coefficient of kinetic friction is 0.300. Find the work done by (a) the 70-N force, (b) the normal force, and (c) the force of gravity. (d) What is the energy loss due to friction? (e) Find the total change in the block's kinetic energy.

WEB 7. Batman, whose mass is 80.0 kg, is holding onto the free end of a 12.0-m rope, the other end of which is fixed to a tree limb above. He is able to get the rope in motion as only Batman knows how, eventually getting it to swing enough so that he can reach a ledge when the rope makes a 60.0° angle with the vertical. How much work was done against the force of gravity in this maneuver?

Section 7.2   The Scalar Product of Two Vectors

In Problems 8 to 14, calculate all numerical answers to three significant figures.

8. Vector **A** has a magnitude of 5.00 units, and vector **B** has a magnitude of 9.00 units. The two vectors make an angle of 50.0° with each other. Find **A · B**.

9. Vector **A** extends from the origin to a point having polar coordinates (7, 70°), and vector **B** extends from the origin to a point having polar coordinates (4, 130°). Find **A · B**.

10. Given two arbitrary vectors **A** and **B**, show that **A·B** = $A_x B_x + A_y B_y + A_z B_z$. (*Hint:* Write **A** and **B** in unit vector form and use Equations 7.4 and 7.5.)

WEB 11. A force **F** = $(6\mathbf{i} - 2\mathbf{j})$ N acts on a particle that undergoes a displacement **d** = $(3\mathbf{i} + \mathbf{j})$ m. Find (a) the work done by the force on the particle and (b) the angle between **F** and **d**.

12. For **A** = $3\mathbf{i} + \mathbf{j} - \mathbf{k}$, **B** = $-\mathbf{i} + 2\mathbf{j} + 5\mathbf{k}$, and **C** = $2\mathbf{j} - 3\mathbf{k}$, find **C·(A − B)**.

13. Using the definition of the scalar product, find the angles between (a) **A** = $3\mathbf{i} - 2\mathbf{j}$ and **B** = $4\mathbf{i} - 4\mathbf{j}$; (b) **A** = $-2\mathbf{i} + 4\mathbf{j}$ and **B** = $3\mathbf{i} - 4\mathbf{j} + 2\mathbf{k}$; (c) **A** = $\mathbf{i} - 2\mathbf{j} + 2\mathbf{k}$ and **B** = $3\mathbf{j} + 4\mathbf{k}$.

14. Find the scalar product of the vectors in Figure P7.14.

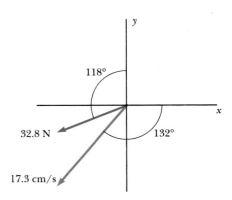

Figure P7.14

Section 7.3   Work Done by a Varying Force

15. The force acting on a particle varies as shown in Figure P7.15. Find the work done by the force as the particle moves (a) from $x = 0$ to $x = 8.00$ m, (b) from $x = 8.00$ m to $x = 10.0$ m, and (c) from $x = 0$ to $x = 10.0$ m.

16. The force acting on a particle is $F_x = (8x - 16)$ N, where x is in meters. (a) Make a plot of this force versus x from $x = 0$ to $x = 3.00$ m. (b) From your graph, find the net work done by this force as the particle moves from $x = 0$ to $x = 3.00$ m.

WEB 17. A particle is subject to a force F_x that varies with position as in Figure P7.17. Find the work done by the force on the body as it moves (a) from $x = 0$ to $x = 5.00$ m,

Figure P7.15

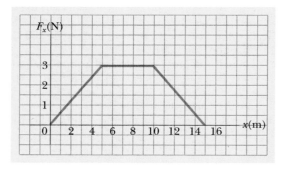

Figure P7.17   Problems 17 and 32.

Figure P7.21

(b) from $x = 5.00$ m to $x = 10.0$ m, and (c) from $x = 10.0$ m to $x = 15.0$ m. (d) What is the total work done by the force over the distance $x = 0$ to $x = 15.0$ m?

18. A force $\mathbf{F} = (4x\mathbf{i} + 3y\mathbf{j})$ N acts on an object as it moves in the x direction from the origin to $x = 5.00$ m. Find the work done on the object by the force.

19. When a 4.00-kg mass is hung vertically on a certain light spring that obeys Hooke's law, the spring stretches 2.50 cm. If the 4.00-kg mass is removed, (a) how far will the spring stretch if a 1.50-kg mass is hung on it and (b) how much work must an external agent do to stretch the same spring 4.00 cm from its unstretched position?

20. An archer pulls her bow string back 0.400 m by exerting a force that increases uniformly from zero to 230 N. (a) What is the equivalent spring constant of the bow? (b) How much work is done by the archer in pulling the bow?

21. A 6 000-kg freight car rolls along rails with negligible friction. The car is brought to rest by a combination of two coiled springs, as illustrated in Figure P7.21. Both springs obey Hooke's law with $k_1 = 1\ 600$ N/m and $k_2 = 3\ 400$ N/m. After the first spring compresses a distance of 30.0 cm, the second spring (acting with the first) increases the force so that additional compression occurs, as shown in the graph. If the car is brought to

rest 50.0 cm after first contacting the two-spring system, find the car's initial speed.

22. A 100-g bullet is fired from a rifle having a barrel 0.600 m long. Assuming the origin is placed where the bullet begins to move, the force (in newtons) exerted on the bullet by the expanding gas is 15 000 + 10 000x − 25 000x^2, where x is in meters. (a) Determine the work done by the gas on the bullet as the bullet travels the length of the barrel. (b) If the barrel is 1.00 m long, how much work is done and how does this value compare with the work calculated in part (a)?

23. If it takes 4.00 J of work to stretch a Hooke's-law spring 10.0 cm from its unstressed length, determine the extra work required to stretch it an additional 10.0 cm.

24. If it takes work W to stretch a Hooke's-law spring a distance d from its unstressed length, determine the extra work required to stretch it an additional distance d.

25. A small mass m is pulled to the top of a frictionless half-cylinder (of radius R) by a cord that passes over the top of the cylinder, as illustrated in Figure P7.25. (a) If the mass moves at a constant speed, show that $F = mg \cos \theta$. (*Hint:* If the mass moves at a constant speed, the component of its acceleration tangent to the cylinder must be zero at all times.) (b) By directly integrating $W = \int \mathbf{F} \cdot d\mathbf{s}$, find the work done in moving the mass at constant speed from the bottom to the top of the half-

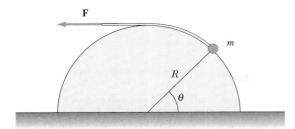

Figure P7.25

cylinder. Here $d\mathbf{s}$ represents an incremental displacement of the small mass.

26. Express the unit of the force constant of a spring in terms of the basic units meter, kilogram, and second.

Section 7.4 **Kinetic Energy and the Work–Kinetic Energy Theorem**

27. A 0.600-kg particle has a speed of 2.00 m/s at point A and kinetic energy of 7.50 J at point B. What is (a) its kinetic energy at A? (b) its speed at B? (c) the total work done on the particle as it moves from A to B?

28. A 0.300-kg ball has a speed of 15.0 m/s. (a) What is its kinetic energy? (b) If its speed were doubled, what would be its kinetic energy?

29. A 3.00-kg mass has an initial velocity $\mathbf{v}_i = (6.00\mathbf{i} - 2.00\mathbf{j})$ m/s. (a) What is its kinetic energy at this time? (b) Find the total work done on the object if its velocity changes to $(8.00\mathbf{i} + 4.00\mathbf{j})$ m/s. (*Hint:* Remember that $v^2 = \mathbf{v} \cdot \mathbf{v}$.)

30. A mechanic pushes a 2 500-kg car, moving it from rest and making it accelerate from rest to a speed v. He does 5 000 J of work in the process. During this time, the car moves 25.0 m. If friction between the car and the road is negligible, (a) what is the final speed v of the car? (b) What constant horizontal force did he exert on the car?

31. A mechanic pushes a car of mass m, doing work W in making it accelerate from rest. If friction between the car and the road is negligible, (a) what is the final speed of the car? During the time the mechanic pushes the car, the car moves a distance d. (b) What constant horizontal force did the mechanic exert on the car?

32. A 4.00-kg particle is subject to a total force that varies with position, as shown in Figure P7.17. The particle starts from rest at $x = 0$. What is its speed at (a) $x = 5.00$ m, (b) $x = 10.0$ m, (c) $x = 15.0$ m?

33. A 40.0-kg box initially at rest is pushed 5.00 m along a rough, horizontal floor with a constant applied horizontal force of 130 N. If the coefficient of friction between the box and the floor is 0.300, find (a) the work done by the applied force, (b) the energy loss due to friction, (c) the work done by the normal force, (d) the work done by gravity, (e) the change in kinetic energy of the box, and (f) the final speed of the box.

34. You can think of the work–kinetic energy theorem as a second theory of motion, parallel to Newton's laws in describing how outside influences affect the motion of an object. In this problem, work out parts (a) and (b) separately from parts (c) and (d) to compare the predictions of the two theories. In a rifle barrel, a 15.0-g bullet is accelerated from rest to a speed of 780 m/s. (a) Find the work that is done on the bullet. (b) If the rifle barrel is 72.0 cm long, find the magnitude of the average total force that acted on it, as $F = W/(d \cos \theta)$. (c) Find the constant acceleration of a bullet that starts from rest and gains a speed of 780 m/s over a distance of 72.0 cm. (d) Find the total force that acted on it as $\Sigma F = ma$.

35. A crate of mass 10.0 kg is pulled up a rough incline with an initial speed of 1.50 m/s. The pulling force is 100 N parallel to the incline, which makes an angle of 20.0° with the horizontal. The coefficient of kinetic friction is 0.400, and the crate is pulled 5.00 m. (a) How much work is done by gravity? (b) How much energy is lost because of friction? (c) How much work is done by the 100-N force? (d) What is the change in kinetic energy of the crate? (e) What is the speed of the crate after it has been pulled 5.00 m?

36. A block of mass 12.0 kg slides from rest down a frictionless 35.0° incline and is stopped by a strong spring with $k = 3.00 \times 10^4$ N/m. The block slides 3.00 m from the point of release to the point where it comes to rest against the spring. When the block comes to rest, how far has the spring been compressed?

WEB 37. A sled of mass m is given a kick on a frozen pond. The kick imparts to it an initial speed $v_i = 2.00$ m/s. The coefficient of kinetic friction between the sled and the ice is $\mu_k = 0.100$. Utilizing energy considerations, find the distance the sled moves before it stops.

38. A picture tube in a certain television set is 36.0 cm long. The electrical force accelerates an electron in the tube from rest to 1.00% of the speed of light over this distance. Determine (a) the kinetic energy of the electron as it strikes the screen at the end of the tube, (b) the magnitude of the average electrical force acting on the electron over this distance, (c) the magnitude of the average acceleration of the electron over this distance, and (d) the time of flight.

39. A bullet with a mass of 5.00 g and a speed of 600 m/s penetrates a tree to a depth of 4.00 cm. (a) Use work and energy considerations to find the average frictional force that stops the bullet. (b) Assuming that the frictional force is constant, determine how much time elapsed between the moment the bullet entered the tree and the moment it stopped.

40. An Atwood's machine (see Fig. 5.15) supports masses of 0.200 kg and 0.300 kg. The masses are held at rest beside each other and then released. Neglecting friction, what is the speed of each mass the instant it has moved 0.400 m?

41. A 2.00-kg block is attached to a spring of force constant 500 N/m, as shown in Figure 7.10. The block is pulled 5.00 cm to the right of equilibrium and is then released from rest. Find the speed of the block as it passes through equilibrium if (a) the horizontal surface is frictionless and (b) the coefficient of friction between the block and the surface is 0.350.

Section 7.5   Power

42. Make an order-of-magnitude estimate of the power a car engine contributes to speeding up the car to highway speed. For concreteness, consider your own car (if you use one). In your solution, state the physical quantities you take as data and the values you measure or estimate for them. The mass of the vehicle is given in the owner's manual. If you do not wish to consider a car, think about a bus or truck for which you specify the necessary physical quantities.

WEB 43. A 700-N Marine in basic training climbs a 10.0-m vertical rope at a constant speed in 8.00 s. What is his power output?

44. If a certain horse can maintain 1.00 hp of output for 2.00 h, how many 70.0-kg bundles of shingles can the horse hoist (using some pulley arrangement) to the roof of a house 8.00 m tall, assuming 70.0% efficiency?

45. A certain automobile engine delivers 2.24×10^4 W (30.0 hp) to its wheels when moving at a constant speed of 27.0 m/s ($\approx$ 60 mi/h). What is the resistive force acting on the automobile at that speed?

46. A skier of mass 70.0 kg is pulled up a slope by a motor-driven cable. (a) How much work is required for him to be pulled a distance of 60.0 m up a 30.0° slope (assumed frictionless) at a constant speed of 2.00 m/s? (b) A motor of what power is required to perform this task?

47. A 650-kg elevator starts from rest. It moves upward for 3.00 s with constant acceleration until it reaches its cruising speed of 1.75 m/s. (a) What is the average power of the elevator motor during this period? (b) How does this power compare with its power when it moves at its cruising speed?

48. An energy-efficient lightbulb, taking in 28.0 W of power, can produce the same level of brightness as a conventional bulb operating at 100-W power. The lifetime of the energy-efficient bulb is 10 000 h and its purchase price is $17.00, whereas the conventional bulb has a lifetime of 750 h and costs $0.420 per bulb. Determine the total savings obtained through the use of one energy-efficient bulb over its lifetime as opposed to the use of conventional bulbs over the same time period. Assume an energy cost of $0.080 0 per kilowatt hour.

(Optional)
Section 7.6   Energy and the Automobile

49. A compact car of mass 900 kg has an overall motor efficiency of 15.0%. (That is, 15.0% of the energy supplied by the fuel is delivered to the wheels of the car.) (a) If burning 1 gal of gasoline supplies 1.34×10^8 J of energy, find the amount of gasoline used by the car in accelerating from rest to 55.0 mi/h. Here you may ignore the effects of air resistance and rolling resistance. (b) How many such accelerations will 1 gal provide? (c) The mileage claimed for the car is 38.0 mi/gal at 55 mi/h. What power is delivered to the wheels (to overcome frictional effects) when the car is driven at this speed?

50. Suppose the empty car described in Table 7.2 has a fuel economy of 6.40 km/L (15 mi/gal) when traveling at 26.8 m/s (60 mi/h). Assuming constant efficiency, determine the fuel economy of the car if the total mass of the passengers and the driver is 350 kg.

51. When an air conditioner is added to the car described in Problem 50, the additional output power required to operate the air conditioner is 1.54 kW. If the fuel economy of the car is 6.40 km/L without the air conditioner, what is it when the air conditioner is operating?

(Optional)
Section 7.7   Kinetic Energy at High Speeds

52. An electron moves with a speed of $0.995c$. (a) What is its kinetic energy? (b) If you use the classical expression to calculate its kinetic energy, what percentage error results?

53. A proton in a high-energy accelerator moves with a speed of $c/2$. Using the work–kinetic energy theorem, find the work required to increase its speed to (a) $0.750c$ and (b) $0.995c$.

54. Find the kinetic energy of a 78.0-kg spacecraft launched out of the Solar System with a speed of 106 km/s using (a) the classical equation $K = \frac{1}{2}mv^2$ and (b) the relativistic equation.

ADDITIONAL PROBLEMS

55. A baseball outfielder throws a 0.150-kg baseball at a speed of 40.0 m/s and an initial angle of 30.0°. What is the kinetic energy of the baseball at the highest point of the trajectory?

56. While running, a person dissipates about 0.600 J of mechanical energy per step per kilogram of body mass. If a 60.0-kg runner dissipates a power of 70.0 W during a race, how fast is the person running? Assume a running step is 1.50 m in length.

57. A particle of mass m moves with a constant acceleration **a**. If the initial position vector and velocity of the particle are $\mathbf{r}_i$ and $\mathbf{v}_i$, respectively, use energy arguments to show that its speed v at any time satisfies the equation

$$v_f{}^2 = v_i{}^2 + 2\mathbf{a} \cdot (\mathbf{r}_f - \mathbf{r}_i)$$

where $\mathbf{r}_f$ is the position vector of the particle at that same time.

58. The direction of an arbitrary vector **A** can be completely specified with the angles α, β, and γ that the vec-

tor makes with the x, y, and z axes, respectively. If $\mathbf{A} = A_x\mathbf{i} + A_y\mathbf{j} + A_z\mathbf{k}$, (a) find expressions for $\cos \alpha$, $\cos \beta$, and $\cos \gamma$ (known as *direction cosines*) and (b) show that these angles satisfy the relation $\cos^2 \alpha + \cos^2 \beta + \cos^2 \gamma = 1$. (*Hint:* Take the scalar product of $\mathbf{A}$ with $\mathbf{i}$, $\mathbf{j}$, and $\mathbf{k}$ separately.)

59. A 4.00-kg particle moves along the x axis. Its position varies with time according to $x = t + 2.0t^3$, where x is in meters and t is in seconds. Find (a) the kinetic energy at any time t, (b) the acceleration of the particle and the force acting on it at time t, (c) the power being delivered to the particle at time t, and (d) the work done on the particle in the interval $t = 0$ to $t = 2.00$ s.

60. A traveler at an airport takes an escalator up one floor (Fig. P7.60). The moving staircase would itself carry him upward with vertical velocity component v between entry and exit points separated by height h. However, while the escalator is moving, the hurried traveler climbs the steps of the escalator at a rate of n steps/s. Assume that the height of each step is h_s. (a) Determine the amount of work done by the traveler during his escalator ride, given that his mass is m. (b) Determine the work the escalator motor does on this person.

Figure P7.60   (*©Ron Chapple/FPG*)

61. When a certain spring is stretched beyond its proportional limit, the restoring force satisfies the equation $F = -kx + \beta x^3$. If $k = 10.0$ N/m and $\beta = 100$ N/m^3,

calculate the work done by this force when the spring is stretched 0.100 m.

62. In a control system, an accelerometer consists of a 4.70-g mass sliding on a low-friction horizontal rail. A low-mass spring attaches the mass to a flange at one end of the rail. When subject to a steady acceleration of $0.800g$, the mass is to assume a location 0.500 cm away from its equilibrium position. Find the stiffness constant required for the spring.

63. A 2 100-kg pile driver is used to drive a steel I-beam into the ground. The pile driver falls 5.00 m before coming into contact with the beam, and it drives the beam 12.0 cm into the ground before coming to rest. Using energy considerations, calculate the average force the beam exerts on the pile driver while the pile driver is brought to rest.

64. A cyclist and her bicycle have a combined mass of 75.0 kg. She coasts down a road inclined at 2.00° with the horizontal at 4.00 m/s and down a road inclined at 4.00° at 8.00 m/s. She then holds on to a moving vehicle and coasts on a level road. What power must the vehicle expend to maintain her speed at 3.00 m/s? Assume that the force of air resistance is proportional to her speed and that other frictional forces remain constant. (*Warning:* You must *not* attempt this dangerous maneuver.)

65. A single constant force $\mathbf{F}$ acts on a particle of mass m. The particle starts at rest at $t = 0$. (a) Show that the instantaneous power delivered by the force at any time t is $(F^2/m)t$. (b) If $F = 20.0$ N and $m = 5.00$ kg, what is the power delivered at $t = 3.00$ s?

66. A particle is attached between two identical springs on a horizontal frictionless table. Both springs have spring constant k and are initially unstressed. (a) If the particle is pulled a distance x along a direction perpendicular to the initial configuration of the springs, as in Figure P7.66, show that the force exerted on the particle by the springs is

$$\mathbf{F} = -2kx\left(1 - \frac{L}{\sqrt{x^2 + L^2}}\right)\mathbf{i}$$

(b) Determine the amount of work done by this force in moving the particle from $x = A$ to $x = 0$.

Top view

Figure P7.66

67. **Review Problem.**   Two constant forces act on a 5.00-kg object moving in the *xy* plane, as shown in Figure P7.67. Force $\mathbf{F}_1$ is 25.0 N at 35.0°, while $\mathbf{F}_2 = 42.0$ N at 150°. At time $t = 0$, the object is at the origin and has velocity $(4.0\mathbf{i} + 2.5\mathbf{j})$ m/s. (a) Express the two forces in unit–vector notation. Use unit–vector notation for your other answers. (b) Find the total force on the object. (c) Find the object's acceleration. Now, considering the instant $t = 3.00$ s, (d) find the object's velocity, (e) its location, (f) its kinetic energy from $\frac{1}{2}mv_f^2$, and (g) its kinetic energy from $\frac{1}{2}mv_i^2 + \Sigma\mathbf{F}\cdot\mathbf{d}$.

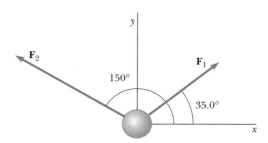

Figure P7.67

68. When different weights are hung on a spring, the spring stretches to different lengths as shown in the following table. (a) Make a graph of the applied force versus the extension of the spring. By least-squares fitting, determine the straight line that best fits the data. (You may not want to use all the data points.) (b) From the slope of the best-fit line, find the spring constant k. (c) If the spring is extended to 105 mm, what force does it exert on the suspended weight?

F (N)	2.0	4.0	6.0	8.0	10	12	14	16	18
L (mm)	15	32	49	64	79	98	112	126	149

69. A 200-g block is pressed against a spring of force constant 1.40 kN/m until the block compresses the spring 10.0 cm. The spring rests at the bottom of a ramp inclined at 60.0° to the horizontal. Using energy considerations, determine how far up the incline the block moves before it stops (a) if there is no friction between the block and the ramp and (b) if the coefficient of kinetic friction is 0.400.

70. A 0.400-kg particle slides around a horizontal track. The track has a smooth, vertical outer wall forming a circle with a radius of 1.50 m. The particle is given an initial speed of 8.00 m/s. After one revolution, its speed has dropped to 6.00 m/s because of friction with the rough floor of the track. (a) Find the energy loss due to friction in one revolution. (b) Calculate the coefficient of kinetic friction. (c) What is the total number of revolutions the particle makes before stopping?

Figure P7.71

71. The ball launcher in a pinball machine has a spring that has a force constant of 1.20 N/cm (Fig. P7.71). The surface on which the ball moves is inclined 10.0° with respect to the horizontal. If the spring is initially compressed 5.00 cm, find the launching speed of a 100-g ball when the plunger is released. Friction and the mass of the plunger are negligible.

72. In diatomic molecules, the constituent atoms exert attractive forces on each other at great distances and repulsive forces at short distances. For many molecules, the Lennard–Jones law is a good approximation to the magnitude of these forces:

$$F = F_0\left[2\left(\frac{\sigma}{r}\right)^{13} - \left(\frac{\sigma}{r}\right)^7\right]$$

where r is the center-to-center distance between the atoms in the molecule, σ is a length parameter, and F_0 is the force when $r = \sigma$. For an oxygen molecule, $F_0 = 9.60 \times 10^{-11}$ N and $\sigma = 3.50 \times 10^{-10}$ m. Determine the work done by this force if the atoms are pulled apart from $r = 4.00 \times 10^{-10}$ m to $r = 9.00 \times 10^{-10}$ m.

73. A horizontal string is attached to a 0.250-kg mass lying on a rough, horizontal table. The string passes over a light, frictionless pulley, and a 0.400-kg mass is then attached to its free end. The coefficient of sliding friction between the 0.250-kg mass and the table is 0.200. Using the work–kinetic energy theorem, determine (a) the speed of the masses after each has moved 20.0 m from rest and (b) the mass that must be added to the 0.250-kg mass so that, given an initial velocity, the masses continue to move at a constant speed. (c) What mass must be removed from the 0.400-kg mass so that the same outcome as in part (b) is achieved?

74. Suppose a car is modeled as a cylinder moving with a speed v, as in Figure P7.74. In a time Δt, a column of air

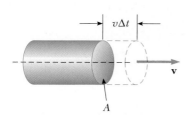

Figure P7.74

of mass Δm must be moved a distance $v\,\Delta t$ and, hence, must be given a kinetic energy $\frac{1}{2}(\Delta m)v^2$. Using this model, show that the power loss due to air resistance is $\frac{1}{2}\rho Av^3$ and that the resistive force is $\frac{1}{2}\rho Av^2$, where ρ is the density of air.

 75. A particle moves along the x axis from $x = 12.8$ m to $x = 23.7$ m under the influence of a force

$$F = \frac{375}{x^3 + 3.75\,x}$$

where F is in newtons and x is in meters. Using numerical integration, determine the total work done by this force during this displacement. Your result should be accurate to within 2%.

76. More than 2 300 years ago the Greek teacher Aristotle wrote the first book called *Physics*. The following passage, rephrased with more precise terminology, is from the end of the book's Section Eta:

Let $\mathcal{P}$ be the power of an agent causing motion; w, the thing moved; d, the distance covered; and t, the time taken. Then (1) a power equal to $\mathcal{P}$ will in a period of time equal to t move $w/2$ a distance $2d$; or (2) it will move $w/2$ the given distance d in time $t/2$. Also, if (3) the given power $\mathcal{P}$ moves the given object w a distance $d/2$ in time $t/2$, then (4) $\mathcal{P}/2$ will move $w/2$ the given distance d in the given time t.

(a) Show that Aristotle's proportions are included in the equation $\mathcal{P}t = bwd$, where b is a proportionality constant. (b) Show that our theory of motion includes this part of Aristotle's theory as one special case. In particular, describe a situation in which it is true, derive the equation representing Aristotle's proportions, and determine the proportionality constant.

ANSWERS TO QUICK QUIZZES

7.1 No. The force does no work on the object because the force is pointed toward the center of the circle and is therefore perpendicular to the motion.

7.2 (a) Assuming the person lifts with a force of magnitude mg, the weight of the box, the work he does during the vertical displacement is mgh because the force is in the direction of the displacement. The work he does during the horizontal displacement is zero because now the force he exerts on the box is perpendicular to the displacement. The net work he does is $mgh + 0 = mgh$. (b) The work done by the gravitational force on the box as the box is displaced vertically is $-mgh$ because the direction of this force is opposite the direction of the displacement. The work done by the gravitational force is zero during the horizontal displacement because now the direction of this force is perpendicular to the direction of the displacement. The net work done by the gravitational force $-mgh + 0 = -mgh$. The total work done on the box is $+mgh - mgh = 0$.

7.3 No. For example, consider the two vectors $\mathbf{A} = 3\mathbf{i} - 2\mathbf{j}$ and $\mathbf{B} = 2\mathbf{i} - \mathbf{j}$. Their dot product is $\mathbf{A} \cdot \mathbf{B} = 8$, yet both vectors have negative y components.

7.4 Force divided by displacement, which in SI units is newtons per meter (N/m).

7.5 Yes, whenever the frictional force has a component along the direction of motion. Consider a crate sitting on the bed of a truck as the truck accelerates to the east. The static friction force exerted on the crate by the truck acts to the east to give the crate the same acceleration as the truck (assuming that the crate does not slip). Because the crate accelerates, its kinetic energy must increase.

7.6 Because the two vehicles perform the same amount of work, the areas under the two graphs are equal. However, the graph for the low-power truck extends over a longer time interval and does not extend as high on the $\mathcal{P}$ axis as the graph for the sports car does.

c h a p t e r

Potential Energy and Conservation of Energy

*I*n Chapter 7 we introduced the concept of kinetic energy, which is the energy associated with the motion of an object. In this chapter we introduce another form of energy—*potential energy,* which is the energy associated with the arrangement of a system of objects that exert forces on each other. Potential energy can be thought of as stored energy that can either do work or be converted to kinetic energy.

The potential energy concept can be used only when dealing with a special class of forces called *conservative forces.* When only conservative forces act within an isolated system, the kinetic energy gained (or lost) by the system as its members change their relative positions is balanced by an equal loss (or gain) in potential energy. This balancing of the two forms of energy is known as the *principle of conservation of mechanical energy.*

Energy is present in the Universe in various forms, including mechanical, electromagnetic, chemical, and nuclear. Furthermore, one form of energy can be converted to another. For example, when an electric motor is connected to a battery, the chemical energy in the battery is converted to electrical energy in the motor, which in turn is converted to mechanical energy as the motor turns some device. The transformation of energy from one form to another is an essential part of the study of physics, engineering, chemistry, biology, geology, and astronomy.

When energy is changed from one form to another, the total amount present does not change. Conservation of energy means that although the form of energy may change, if an object (or system) loses energy, that same amount of energy appears in another object or in the object's surroundings.

8.1 ▶ POTENTIAL ENERGY

5.3 An object that possesses kinetic energy can do work on another object—for example, a moving hammer driving a nail into a wall. Now we shall introduce another form of energy. This energy, called **potential energy U,** is the energy associated with a system of objects.

Before we describe specific forms of potential energy, we must first define a *system,* which consists of two or more objects that exert forces on one another. **If the arrangement of the system changes, then the potential energy of the system changes.** If the system consists of only two particle-like objects that exert forces on each other, then the work done by the force acting on one of the objects causes a transformation of energy between the object's kinetic energy and other forms of the system's energy.

Gravitational Potential Energy

As an object falls toward the Earth, the Earth exerts a gravitational force $m\mathbf{g}$ on the object, with the direction of the force being the same as the direction of the object's motion. The gravitational force does work on the object and thereby increases the object's kinetic energy. Imagine that a brick is dropped from rest directly above a nail in a board lying on the ground. When the brick is released, it falls toward the ground, gaining speed and therefore gaining kinetic energy. The brick–Earth system has potential energy when the brick is at any distance above the ground (that is, it has the *potential* to do work), and this potential energy is converted to kinetic energy as the brick falls. The conversion from potential energy to kinetic energy occurs continuously over the entire fall. When the brick reaches the nail and the board lying on the ground, it does work on the nail, driving it into the board. What determines how much work the brick is able to do on the nail? It is easy to see that the heavier the brick, the farther in it drives the nail; also the higher the brick is before it is released, the more work it does when it strikes the nail.

The product of the magnitude of the gravitational force mg acting on an object and the height y of the object is so important in physics that we give it a name: the **gravitational potential energy.** The symbol for gravitational potential energy is U_g, and so the defining equation for gravitational potential energy is

Gravitational potential energy

$$U_g \equiv mgy \qquad\qquad \textbf{(8.1)}$$

Gravitational potential energy is the potential energy of the object–Earth system. This potential energy is transformed into kinetic energy of the system by the gravitational force. In this type of system, in which one of the members (the Earth) is much more massive than the other (the object), the massive object can be modeled as stationary, and the kinetic energy of the system can be represented entirely by the kinetic energy of the lighter object. Thus, the kinetic energy of the system is represented by that of the object falling toward the Earth. Also note that Equation 8.1 is valid only for objects near the surface of the Earth, where **g** is approximately constant.[1]

Let us now directly relate the work done on an object by the gravitational force to the gravitational potential energy of the object–Earth system. To do this, let us consider a brick of mass m at an initial height y_i above the ground, as shown in Figure 8.1. If we neglect air resistance, then the only force that does work on the brick as it falls is the gravitational force exerted on the brick $m\mathbf{g}$. The work W_g done by the gravitational force as the brick undergoes a downward displacement **d** is

$$W_g = (m\mathbf{g}) \cdot \mathbf{d} = (-mg\mathbf{j}) \cdot (y_f - y_i)\,\mathbf{j} = mgy_i - mgy_f$$

where we have used the fact that $\mathbf{j} \cdot \mathbf{j} = 1$ (Eq. 7.4). If an object undergoes both a horizontal and a vertical displacement, so that $\mathbf{d} = (x_f - x_i)\mathbf{i} + (y_f - y_i)\mathbf{j}$, then the work done by the gravitational force is still $mgy_i - mgy_f$ because $-mg\mathbf{j} \cdot (x_f - x_i)\mathbf{i} = 0$. Thus, the work done by the gravitational force depends only on the change in y and not on any change in the horizontal position x.

We just learned that the quantity mgy is the gravitational potential energy of the system U_g, and so we have

$$W_g = U_i - U_f = -(U_f - U_i) = -\Delta U_g \qquad\qquad \textbf{(8.2)}$$

From this result, we see that the work done on any object by the gravitational force is equal to the negative of the change in the system's gravitational potential energy. Also, this result demonstrates that it is only the *difference* in the gravitational potential energy at the initial and final locations that matters. This means that we are free to place the origin of coordinates in any convenient location. Finally, the work done by the gravitational force on an object as the object falls to the Earth is the same as the work done were the object to start at the same point and slide down an incline to the Earth. Horizontal motion does not affect the value of W_g.

The unit of gravitational potential energy is the same as that of work—the joule. Potential energy, like work and kinetic energy, is a scalar quantity.

Figure 8.1 The work done on the brick by the gravitational force as the brick falls from a height y_i to a height y_f is equal to $mgy_i - mgy_f$.

---

[1] The assumption that the force of gravity is constant is a good one as long as the vertical displacement is small compared with the Earth's radius.

Quick Quiz 8.1

Can the gravitational potential energy of a system ever be negative?

EXAMPLE 8.1   The Bowler and the Sore Toe

A bowling ball held by a careless bowler slips from the bowler's hands and drops on the bowler's toe. Choosing floor level as the $y = 0$ point of your coordinate system, estimate the total work done on the ball by the force of gravity as the ball falls. Repeat the calculation, using the top of the bowler's head as the origin of coordinates.

Solution   First, we need to estimate a few values. A bowling ball has a mass of approximately 7 kg, and the top of a person's toe is about 0.03 m above the floor. Also, we shall assume the ball falls from a height of 0.5 m. Holding nonsignificant digits until we finish the problem, we calculate the gravitational potential energy of the ball–Earth system just before the ball is released to be $U_i = mgy_i = (7$ kg$)$ $(9.80$ m/s$^2)(0.5$ m$) = 34.3$ J. A similar calculation for when

the ball reaches his toe gives $U_f = mgy_f = (7$ kg$)$ $(9.80$ m/s$^2)(0.03$ m$) = 2.06$ J. So, the work done by the gravitational force is $W_g = U_i - U_f = 32.24$ J. We should probably keep only one digit because of the roughness of our estimates; thus, we estimate that the gravitational force does 30 J of work on the bowling ball as it falls. The system had 30 J of gravitational potential energy relative to the top of the toe before the ball began its fall.

When we use the bowler's head (which we estimate to be 1.50 m above the floor) as our origin of coordinates, we find that $U_i = mgy_i = (7$ kg$)$ $(9.80$ m/s$^2)(-1$ m$) = -68.6$ J and that $U_f = mgy_f = (7$ kg$)$ $(9.80$ m/s$^2)(-1.47$ m$) = -100.8$ J. The work being done by the gravitational force is still

$$W_g = U_i - U_f = 32.24 \text{ J} \approx \boxed{30 \text{ J.}}$$

Elastic Potential Energy

Now consider a system consisting of a block plus a spring, as shown in Figure 8.2. The force that the spring exerts on the block is given by $F_s = -kx$. In the previous chapter, we learned that the work done by the spring force on a block connected to the spring is given by Equation 7.11:

$$W_s = \tfrac{1}{2}kx_i^2 - \tfrac{1}{2}kx_f^2 \tag{8.3}$$

In this situation, the initial and final x coordinates of the block are measured from its equilibrium position, $x = 0$. Again we see that W_s depends only on the initial and final x coordinates of the object and is zero for any closed path. The **elastic potential energy** function associated with the system is defined by

$$U_s \equiv \tfrac{1}{2}kx^2 \tag{8.4}$$

Elastic potential energy stored in a spring

The elastic potential energy of the system can be thought of as the energy stored in the deformed spring (one that is either compressed or stretched from its equilibrium position). To visualize this, consider Figure 8.2, which shows a spring on a frictionless, horizontal surface. When a block is pushed against the spring (Fig. 8.2b) and the spring is compressed a distance x, the elastic potential energy stored in the spring is $kx^2/2$. When the block is released from rest, the spring snaps back to its original length and the stored elastic potential energy is transformed into kinetic energy of the block (Fig. 8.2c). The elastic potential energy stored in the spring is zero whenever the spring is undeformed ($x = 0$). Energy is stored in the spring only when the spring is either stretched or compressed. Furthermore, the elastic potential energy is a maximum when the spring has reached its maximum compression or extension (that is, when $|x|$ is a maximum). Finally, because the elastic potential energy is proportional to x^2, we see that U_s is always positive in a deformed spring.

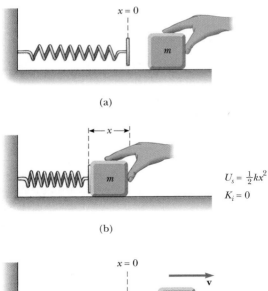

(a)

(b)

$U_s = \frac{1}{2}kx^2$
$K_i = 0$

$U_s = 0$
$K_f = \frac{1}{2}mv^2$

(c)

Figure 8.2   (a) An undeformed spring on a frictionless horizontal surface. (b) A block of mass m is pushed against the spring, compressing it a distance x. (c) When the block is released from rest, the elastic potential energy stored in the spring is transferred to the block in the form of kinetic energy.

8.2   CONSERVATIVE AND NONCONSERVATIVE FORCES

The work done by the gravitational force does not depend on whether an object falls vertically or slides down a sloping incline. All that matters is the change in the object's elevation. On the other hand, the energy loss due to friction on that incline depends on the distance the object slides. In other words, the path makes no difference when we consider the work done by the gravitational force, but it does make a difference when we consider the energy loss due to frictional forces. We can use this varying dependence on path to classify forces as either conservative or nonconservative.

Of the two forces just mentioned, the gravitational force is conservative and the frictional force is nonconservative.

Conservative Forces

Properties of a conservative force

Conservative forces have two important properties:

1. A force is conservative if the work it does on a particle moving between any two points is independent of the path taken by the particle.
2. The work done by a conservative force on a particle moving through any closed path is zero. (A closed path is one in which the beginning and end points are identical.)

The gravitational force is one example of a conservative force, and the force that a spring exerts on any object attached to the spring is another. As we learned in the preceding section, the work done by the gravitational force on an object moving between any two points near the Earth's surface is $W_g = mgy_i - mgy_f$. From this equation we see that W_g depends only on the initial and final y coordinates of the object and hence is independent of the path. Furthermore, W_g is zero when the object moves over any closed path (where $y_i = y_f$).

For the case of the object–spring system, the work W_s done by the spring force is given by $W_s = \frac{1}{2}kx_i^2 - \frac{1}{2}kx_f^2$ (Eq. 8.3). Again, we see that the spring force is conservative because W_s depends only on the initial and final x coordinates of the object and is zero for any closed path.

We can associate a potential energy with any conservative force and can do this *only* for conservative forces. In the previous section, the potential energy associated with the gravitational force was defined as $U_g \equiv mgy$. In general, the work W_c done on an object by a conservative force is equal to the initial value of the potential energy associated with the object minus the final value:

$$W_c = U_i - U_f = -\Delta U \quad (8.5)$$

| Work done by a conservative force |

This equation should look familiar to you. It is the general form of the equation for work done by the gravitational force (Eq. 8.2) and that for the work done by the spring force (Eq. 8.3).

Nonconservative Forces

A force is nonconservative if it causes a change in mechanical energy E, which we define as the sum of kinetic and potential energies. For example, if a book is sent sliding on a horizontal surface that is not frictionless, the force of kinetic friction reduces the book's kinetic energy. As the book slows down, its kinetic energy decreases. As a result of the frictional force, the temperatures of the book and surface increase. The type of energy associated with temperature is *internal energy*, which we will study in detail in Chapter 20. Experience tells us that this internal energy cannot be transferred back to the kinetic energy of the book. In other words, the energy transformation is not reversible. Because the force of kinetic friction changes the mechanical energy of a system, it is a nonconservative force.

| Properties of a nonconservative force |

From the work–kinetic energy theorem, we see that the work done by a conservative force on an object causes a change in the kinetic energy of the object. The change in kinetic energy depends only on the initial and final positions of the object, and not on the path connecting these points. Let us compare this to the sliding book example, in which the nonconservative force of friction is acting between the book and the surface. According to Equation 7.17a, the change in kinetic energy of the book due to friction is $\Delta K_{friction} = -f_k d$, where d is the length of the path over which the friction force acts. Imagine that the book slides from A to B over the straight-line path of length d in Figure 8.3. The change in kinetic energy is $-f_k d$. Now, suppose the book slides over the semicircular path from A to B. In this case, the path is longer and, as a result, the change in kinetic energy is greater in magnitude than that in the straight-line case. For this particular path, the change in kinetic energy is $-f_k \pi d/2$, since d is the diameter of the semicircle. Thus, we see that for a nonconservative force, the change in kinetic energy depends on the path followed between the initial and final points. If a potential energy is involved, then the change in the total mechanical energy depends on the path followed. We shall return to this point in Section 8.5.

Figure 8.3 The loss in mechanical energy due to the force of kinetic friction depends on the path taken as the book is moved from A to B. The loss in mechanical energy is greater along the red path than along the blue path.

8.3 ▶ CONSERVATIVE FORCES AND POTENTIAL ENERGY

In the preceding section we found that the work done on a particle by a conservative force does not depend on the path taken by the particle. The work depends only on the particle's initial and final coordinates. As a consequence, we can define a **potential energy function** U such that the work done by a conservative force equals the decrease in the potential energy of the system. The work done by a conservative force **F** as a particle moves along the x axis is[2]

$$W_c = \int_{x_i}^{x_f} F_x \, dx = -\Delta U \tag{8.6}$$

where F_x is the component of **F** in the direction of the displacement. That is, **the work done by a conservative force equals the negative of the change in the potential energy associated with that force,** where the change in the potential energy is defined as $\Delta U = U_f - U_i$.

We can also express Equation 8.6 as

$$\Delta U = U_f - U_i = -\int_{x_i}^{x_f} F_x \, dx \tag{8.7}$$

Therefore, ΔU is negative when F_x and dx are in the same direction, as when an object is lowered in a gravitational field or when a spring pushes an object toward equilibrium.

The term *potential energy* implies that the object has the potential, or capability, of either gaining kinetic energy or doing work when it is released from some point under the influence of a conservative force exerted on the object by some other member of the system. It is often convenient to establish some particular location x_i as a reference point and measure all potential energy differences with respect to it. We can then define the potential energy function as

$$U_f(x) = -\int_{x_i}^{x_f} F_x \, dx + U_i \tag{8.8}$$

The value of U_i is often taken to be zero at the reference point. It really does not matter what value we assign to U_i, because any nonzero value merely shifts $U_f(x)$ by a constant amount, and only the *change* in potential energy is physically meaningful.

If the conservative force is known as a function of position, we can use Equation 8.8 to calculate the change in potential energy of a system as an object within the system moves from x_i to x_f. It is interesting to note that in the case of one-dimensional displacement, a force is always conservative if it is a function of position only. This is not necessarily the case for motion involving two- or three-dimensional displacements.

8.4   CONSERVATION OF MECHANICAL ENERGY

An object held at some height h above the floor has no kinetic energy. However, as we learned earlier, the gravitational potential energy of the object–Earth system is equal to mgh. If the object is dropped, it falls to the floor; as it falls, its speed and thus its kinetic energy increase, while the potential energy of the system decreases. If factors such as air resistance are ignored, whatever potential energy the system loses as the object moves downward appears as kinetic energy of the object. In other words, the sum of the kinetic and potential energies—the total mechanical energy E—remains constant. This is an example of the principle of **conservation**

---

[2] For a general displacement, the work done in two or three dimensions also equals $U_i - U_f$, where $U = U(x, y, z)$. We write this formally as $W = \int_i^f \mathbf{F} \cdot d\mathbf{s} = U_i - U_f$.

of mechanical energy. For the case of an object in free fall, this principle tells us that any increase (or decrease) in potential energy is accompanied by an equal decrease (or increase) in kinetic energy. Note that **the total mechanical energy of a system remains constant in any isolated system of objects that interact only through conservative forces.**

Because the total mechanical energy E of a system is defined as the sum of the kinetic and potential energies, we can write

$$E \equiv K + U \tag{8.9}$$

We can state the principle of conservation of energy as $E_i = E_f$, and so we have

$$K_i + U_i = K_f + U_f \tag{8.10}$$

It is important to note that Equation 8.10 is valid only when no energy is added to or removed from the system. Furthermore, there must be no nonconservative forces doing work within the system.

Consider the carnival Ring-the-Bell event illustrated at the beginning of the chapter. The participant is trying to convert the initial kinetic energy of the hammer into gravitational potential energy associated with a weight that slides on a vertical track. If the hammer has sufficient kinetic energy, the weight is lifted high enough to reach the bell at the top of the track. To maximize the hammer's kinetic energy, the player must swing the heavy hammer as rapidly as possible. The fast-moving hammer does work on the pivoted target, which in turn does work on the weight. Of course, greasing the track (so as to minimize energy loss due to friction) would also help but is probably not allowed!

If more than one conservative force acts on an object within a system, a potential energy function is associated with each force. In such a case, we can apply the principle of conservation of mechanical energy for the system as

$$K_i + \sum U_i = K_f + \sum U_f \tag{8.11}$$

where the number of terms in the sums equals the number of conservative forces present. For example, if an object connected to a spring oscillates vertically, two conservative forces act on the object: the spring force and the gravitational force.

Total mechanical energy

The mechanical energy of an isolated system remains constant

$$K_i = K_f + U_f - U_i$$
$$= K_f - \Delta U$$
$$= K_f + W_c$$

QuickLab

Dangle a shoe from its lace and use it as a pendulum. Hold it to the side, release it, and note how high it swings at the end of its arc. How does this height compare with its initial height? You may want to check Question 8.3 as part of your investigation.

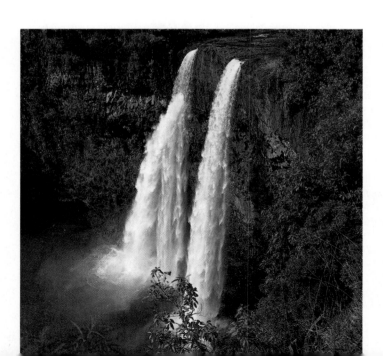

Twin Falls on the Island of Kauai, Hawaii. The gravitational potential energy of the water–Earth system when the water is at the top of the falls is converted to kinetic energy once that water begins falling. How did the water get to the top of the cliff? In other words, what was the original source of the gravitational potential energy when the water was at the top? (*Hint:* This same source powers nearly everything on the planet.) (*Bruce Byers/FPG*)

Figure 8.4 A ball connected to a massless spring suspended vertically. What forms of potential energy are associated with the ball–spring–Earth system when the ball is displaced downward?

Quick Quiz 8.2

A ball is connected to a light spring suspended vertically, as shown in Figure 8.4. When displaced downward from its equilibrium position and released, the ball oscillates up and down. If air resistance is neglected, is the total mechanical energy of the system (ball plus spring plus Earth) conserved? How many forms of potential energy are there for this situation?

Quick Quiz 8.3

Three identical balls are thrown from the top of a building, all with the same initial speed. The first is thrown horizontally, the second at some angle above the horizontal, and the third at some angle below the horizontal, as shown in Figure 8.5. Neglecting air resistance, rank the speeds of the balls at the instant each hits the ground.

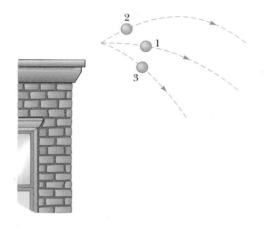

Figure 8.5 Three identical balls are thrown with the same initial speed from the top of a building.

EXAMPLE 8.2   Ball in Free Fall

A ball of mass m is dropped from a height h above the ground, as shown in Figure 8.6. (a) Neglecting air resistance, determine the speed of the ball when it is at a height y above the ground.

Solution Because the ball is in free fall, the only force acting on it is the gravitational force. Therefore, we apply the principle of conservation of mechanical energy to the ball–Earth system. Initially, the system has potential energy but no kinetic energy. As the ball falls, the total mechanical energy remains constant and equal to the initial potential energy of the system.

At the instant the ball is released, its kinetic energy is $K_i = 0$ and the potential energy of the system is $U_i = mgh$. When the ball is at a distance y above the ground, its kinetic energy is $K_f = \frac{1}{2}mv_f^2$ and the potential energy relative to the ground is $U_f = mgy$. Applying Equation 8.10, we obtain

$$K_i + U_i = K_f + U_f$$

$$0 + mgh = \tfrac{1}{2}mv_f^2 + mgy$$

$$v_f^2 = 2g(h - y)$$

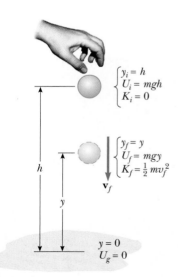

Figure 8.6 A ball is dropped from a height h above the ground. Initially, the total energy of the ball–Earth system is potential energy, equal to mgh relative to the ground. At the elevation y, the total energy is the sum of the kinetic and potential energies.

$$v_f = \boxed{\sqrt{2g(h - y)}}$$

The speed is always positive. If we had been asked to find the ball's velocity, we would use the negative value of the square root as the *y* component to indicate the downward motion.

(b) Determine the speed of the ball at *y* if at the instant of release it already has an initial speed v_i at the initial altitude *h*.

Solution   In this case, the initial energy includes kinetic energy equal to $\frac{1}{2}mv_i^2$, and Equation 8.10 gives

$$\tfrac{1}{2}mv_i^2 + mgh = \tfrac{1}{2}mv_f^2 + mgy$$

$$v_f^2 = v_i^2 + 2g(h - y)$$

$$v_f = \boxed{\sqrt{v_i^2 + 2g(h - y)}}$$

This result is consistent with the expression $v_{yf}^2 = v_{yi}^2 - 2g(y_f - y_i)$ from kinematics, where $y_i = h$. Furthermore, this result is valid even if the initial velocity is at an angle to the horizontal (the projectile situation) for two reasons: (1) energy is a scalar, and the kinetic energy depends only on the magnitude of the velocity; and (2) the change in the gravitational potential energy depends only on the change in position in the vertical direction.

EXAMPLE 8.3 ▶ The Pendulum

A pendulum consists of a sphere of mass *m* attached to a light cord of length *L*, as shown in Figure 8.7. The sphere is released from rest when the cord makes an angle θ_A with the vertical, and the pivot at *P* is frictionless. (a) Find the speed of the sphere when it is at the lowest point ⑧.

Solution   The only force that does work on the sphere is the gravitational force. (The force of tension is always perpendicular to each element of the displacement and hence does no work.) Because the gravitational force is conservative, the total mechanical energy of the pendulum–Earth system is constant. (In other words, we can classify this as an "energy conservation" problem.) As the pendulum swings, continuous transformation between potential and kinetic energy occurs. At the instant the pendulum is released, the energy of the system is entirely potential energy. At point ⑧ the pendulum has kinetic energy, but the system has lost some potential energy. At ⓒ the system has regained its initial potential energy, and the kinetic energy of the pendulum is again zero.

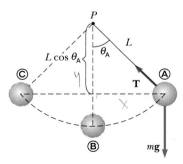

Figure 8.7   If the sphere is released from rest at the angle θ_A it will never swing above this position during its motion. At the start of the motion, position ⓐ, the energy is entirely potential. This initial potential energy is all transformed into kinetic energy at the lowest elevation ⑧. As the sphere continues to move along the arc, the energy again becomes entirely potential at ⓒ.

If we measure the *y* coordinates of the sphere from the center of rotation, then $y_A = -L \cos \theta_A$ and $y_B = -L$. Therefore, $U_A = -mgL \cos \theta_A$ and $U_B = -mgL$. Applying the principle of conservation of mechanical energy to the system gives

$$K_A + U_A = K_B + U_B$$

$$0 - mgL \cos \theta_A = \tfrac{1}{2}mv_B^2 - mgL$$

(1) $$v_B = \boxed{\sqrt{2\,gL(1 - \cos \theta_A)}}$$

(b) What is the tension T_B in the cord at ⑧?

Solution   Because the force of tension does no work, we cannot determine the tension using the energy method. To find T_B, we can apply Newton's second law to the radial direction. First, recall that the centripetal acceleration of a particle moving in a circle is equal to v^2/r directed toward the center of rotation. Because $r = L$ in this example, we obtain

(2) $$\sum F_r = T_B - mg = ma_r = m\frac{v_B^2}{L}$$

Substituting (1) into (2) gives the tension at point ⑧:

(3) $$T_B = mg + 2\,mg(1 - \cos \theta_A)$$

$$= \boxed{mg(3 - 2 \cos \theta_A)}$$

From (2) we see that the tension at ⑧ is greater than the weight of the sphere. Furthermore, (3) gives the expected result that $T_B = mg$ when the initial angle $\theta_A = 0$.

Exercise   A pendulum of length 2.00 m and mass 0.500 kg is released from rest when the cord makes an angle of 30.0° with the vertical. Find the speed of the sphere and the tension in the cord when the sphere is at its lowest point.

Answer   2.29 m/s; 6.21 N.

8.5 WORK DONE BY NONCONSERVATIVE FORCES

As we have seen, if the forces acting on objects within a system are conservative, then the mechanical energy of the system remains constant. However, if some of the forces acting on objects within the system are not conservative, then the mechanical energy of the system does not remain constant. Let us examine two types of nonconservative forces: an applied force and the force of kinetic friction.

Work Done by an Applied Force

When you lift a book through some distance by applying a force to it, the force you apply does work W_{app} on the book, while the gravitational force does work W_g on the book. If we treat the book as a particle, then the net work done on the book is related to the change in its kinetic energy as described by the work–kinetic energy theorem given by Equation 7.15:

$$W_{app} + W_g = \Delta K \tag{8.12}$$

Because the gravitational force is conservative, we can use Equation 8.2 to express the work done by the gravitational force in terms of the change in potential energy, or $W_g = -\Delta U$. Substituting this into Equation 8.12 gives

$$W_{app} = \Delta K + \Delta U \tag{8.13}$$

Note that the right side of this equation represents the change in the mechanical energy of the book–Earth system. This result indicates that your applied force transfers energy to the system in the form of kinetic energy of the book and gravitational potential energy of the book–Earth system. Thus, we conclude that if an object is part of a system, then **an applied force can transfer energy into or out of the system.**

Situations Involving Kinetic Friction

Kinetic friction is an example of a nonconservative force. If a book is given some initial velocity on a horizontal surface that is not frictionless, then the force of kinetic friction acting on the book opposes its motion and the book slows down and eventually stops. The force of kinetic friction reduces the kinetic energy of the book by transforming kinetic energy to internal energy of the book and part of the horizontal surface. Only part of the book's kinetic energy is transformed to internal energy in the book. The rest appears as internal energy in the surface. (When you trip and fall while running across a gymnasium floor, not only does the skin on your knees warm up but so does the floor!)

As the book moves through a distance d, the only force that does work is the force of kinetic friction. This force causes a decrease in the kinetic energy of the book. This decrease was calculated in Chapter 7, leading to Equation 7.17a, which we repeat here:

$$\Delta K_{friction} = -f_k d \tag{8.14}$$

If the book moves on an incline that is not frictionless, a change in the gravitational potential energy of the book–Earth system also occurs, and $-f_k d$ is the amount by which the mechanical energy of the system changes because of the force of kinetic friction. In such cases,

$$\Delta E = \Delta K + \Delta U = -f_k d \tag{8.15}$$

where $E_i + \Delta E = E_f$.

QuickLab

Find a friend and play a game of raquetball. After a long volley, feel the ball and note that it is warm. Why is that?

Quick Quiz 8.4

Write down the work–kinetic energy theorem for the general case of two objects that are connected by a spring and acted upon by gravity and some other external applied force. Include the effects of friction as $\Delta E_{\text{friction}}$.

Problem-Solving Hints

Conservation of Energy

We can solve many problems in physics using the principle of conservation of energy. You should incorporate the following procedure when you apply this principle:

- Define your system, which may include two or more interacting particles, as well as springs or other systems in which elastic potential energy can be stored. Choose the initial and final points.
- Identify zero points for potential energy (both gravitational and spring). If there is more than one conservative force, write an expression for the potential energy associated with each force.
- Determine whether any nonconservative forces are present. Remember that if friction or air resistance is present, mechanical energy *is not conserved*.
- If mechanical energy is *conserved*, you can write the total initial energy $E_i = K_i + U_i$ at some point. Then, write an expression for the total final energy $E_f = K_f + U_f$ at the final point that is of interest. Because mechanical energy is *conserved*, you can equate the two total energies and solve for the quantity that is unknown.
- If frictional forces are present (and thus mechanical energy is *not conserved*), first write expressions for the total initial and total final energies. In this case, the difference between the total final mechanical energy and the total initial mechanical energy equals the change in mechanical energy in the system due to friction.

EXAMPLE 8.4   Crate Sliding Down a Ramp

A 3.00-kg crate slides down a ramp. The ramp is 1.00 m in length and inclined at an angle of 30.0°, as shown in Figure 8.8. The crate starts from rest at the top, experiences a constant frictional force of magnitude 5.00 N, and continues to move a short distance on the flat floor after it leaves the ramp. Use energy methods to determine the speed of the crate at the bottom of the ramp.

Solution   Because $v_i = 0$, the initial kinetic energy at the top of the ramp is zero. If the y coordinate is measured from the bottom of the ramp (the final position where the potential energy is zero) with the upward direction being positive, then $y_i = 0.500$ m. Therefore, the total mechanical energy of the crate–Earth system at the top is all potential energy:

$$E_i = K_i + U_i = 0 + U_i = mgy_i$$
$$= (3.00 \text{ kg})(9.80 \text{ m/s}^2)(0.500 \text{ m}) = 14.7 \text{ J}$$

Figure 8.8   A crate slides down a ramp under the influence of gravity. The potential energy decreases while the kinetic energy increases.

When the crate reaches the bottom of the ramp, the potential energy of the system is *zero* because the elevation of the crate is $y_f = 0$. Therefore, the total mechanical energy of the system when the crate reaches the bottom is all kinetic energy:

$$E_f = K_f + U_f = \tfrac{1}{2}mv_f^2 + 0$$

We cannot say that $E_i = E_f$ because a nonconservative force reduces the mechanical energy of the system: the force of kinetic friction acting on the crate. In this case, Equation 8.15 gives $\Delta E = -f_k d$, where d is the displacement along the ramp. (Remember that the forces normal to the ramp do no work on the crate because they are perpendicular to the displacement.) With $f_k = 5.00$ N and $d = 1.00$ m, we have

$$\Delta E = -f_k d = -(5.00 \text{ N})(1.00 \text{ m}) = -5.00 \text{ J}$$

This result indicates that the system loses some mechanical energy because of the presence of the nonconservative frictional force. Applying Equation 8.15 gives

$$E_f - E_i = \tfrac{1}{2}mv_f^2 - mgy_i = -f_k d$$
$$\tfrac{1}{2}mv_f^2 = 14.7 \text{ J} - 5.00 \text{ J} = 9.70 \text{ J}$$
$$v_f^2 = \frac{19.4 \text{ J}}{3.00 \text{ kg}} = 6.47 \text{ m}^2/\text{s}^2$$
$$v_f = \boxed{2.54 \text{ m/s}}$$

Exercise Use Newton's second law to find the acceleration of the crate along the ramp, and use the equations of kinematics to determine the final speed of the crate.

Answer 3.23 m/s²; 2.54 m/s.

Exercise Assuming the ramp to be frictionless, find the final speed of the crate and its acceleration along the ramp.

Answer 3.13 m/s; 4.90 m/s².

EXAMPLE 8.5 Motion on a Curved Track

A child of mass m rides on an irregularly curved slide of height $h = 2.00$ m, as shown in Figure 8.9. The child starts from rest at the top. (a) Determine his speed at the bottom, assuming no friction is present.

Solution The normal force $\mathbf{n}$ does no work on the child because this force is always perpendicular to each element of the displacement. Because there is no friction, the mechanical energy of the child–Earth system is conserved. If we measure the y coordinate in the upward direction from the bottom of the slide, then $y_i = h$, $y_f = 0$, and we obtain

Figure 8.9 If the slide is frictionless, the speed of the child at the bottom depends only on the height of the slide.

$$K_i + U_i = K_f + U_f$$
$$0 + mgh = \tfrac{1}{2}mv_f^2 + 0$$
$$v_f = \sqrt{2gh}$$

Note that the result is the same as it would be had the child fallen vertically through a distance h! In this example, $h = 2.00$ m, giving

$$v_f = \sqrt{2gh} = \sqrt{2(9.80 \text{ m/s}^2)(2.00 \text{ m})} = \boxed{6.26 \text{ m/s}}$$

(b) If a force of kinetic friction acts on the child, how much mechanical energy does the system lose? Assume that $v_f = 3.00$ m/s and $m = 20.0$ kg.

Solution In this case, mechanical energy is *not* conserved, and so we must use Equation 8.15 to find the loss of mechanical energy due to friction:

$$\Delta E = E_f - E_i = (K_f + U_f) - (K_i + U_i)$$
$$= (\tfrac{1}{2}mv_f^2 + 0) - (0 + mgh) = \tfrac{1}{2}mv_f^2 - mgh$$
$$= \tfrac{1}{2}(20.0 \text{ kg})(3.00 \text{ m/s})^2 - (20.0 \text{ kg})(9.80 \text{ m/s}^2)(2.00 \text{ m})$$
$$= \boxed{-302 \text{ J}}$$

Again, ΔE is negative because friction is reducing mechanical energy of the system (the final mechanical energy is less than the initial mechanical energy). Because the slide is curved, the normal force changes in magnitude and direction during the motion. Therefore, the frictional force, which is proportional to n, also changes during the motion. Given this changing frictional force, do you think it is possible to determine μ_k from these data?

EXAMPLE 8.6 ▸ Let's Go Skiing!

A skier starts from rest at the top of a frictionless incline of height 20.0 m, as shown in Figure 8.10. At the bottom of the incline, she encounters a horizontal surface where the coefficient of kinetic friction between the skis and the snow is 0.210. How far does she travel on the horizontal surface before coming to rest?

Solution First, let us calculate her speed at the bottom of the incline, which we choose as our zero point of potential energy. Because the incline is frictionless, the mechanical energy of the skier–Earth system remains constant, and we find, as we did in the previous example, that

$$v_{\text{B}} = \sqrt{2gh} = \sqrt{2(9.80 \text{ m/s}^2)(20.0 \text{ m})} = 19.8 \text{ m/s}$$

Now we apply Equation 8.15 as the skier moves along the rough horizontal surface from Ⓑ to Ⓒ. The change in mechanical energy along the horizontal is $\Delta E = -f_k d$, where d is the horizontal displacement.

To find the distance the skier travels before coming to rest, we take $K_{\text{C}} = 0$. With $v_{\text{B}} = 19.8$ m/s and the frictional force given by $f_k = \mu_k n = \mu_k mg$, we obtain

$$\Delta E = E_{\text{C}} - E_{\text{B}} = -\mu_k mgd$$

$$(K_{\text{C}} + U_{\text{C}}) - (K_{\text{B}} + U_{\text{B}}) = (0 + 0) - (\tfrac{1}{2}mv_{\text{B}}{}^2 + 0)$$

$$-\tfrac{1}{2}mv_{\text{B}}{}^2 = -\mu_k mgd$$

$$d = \frac{v_{\text{B}}{}^2}{2\mu_k g} = \frac{(19.8 \text{ m/s})^2}{2(0.210)(9.80 \text{ m/s}^2)}$$

$$= \boxed{95.2 \text{ m}}$$

Exercise Find the horizontal distance the skier travels before coming to rest if the incline also has a coefficient of kinetic friction equal to 0.210.

Answer 40.3 m.

Figure 8.10 The skier slides down the slope and onto a level surface, stopping after a distance d from the bottom of the hill.

EXAMPLE 8.7 ▸ The Spring-Loaded Popgun

The launching mechanism of a toy gun consists of a spring of unknown spring constant (Fig. 8.11a). When the spring is compressed 0.120 m, the gun, when fired vertically, is able to launch a 35.0-g projectile to a maximum height of 20.0 m above the position of the projectile before firing. (a) Neglecting all resistive forces, determine the spring constant.

Solution Because the projectile starts from rest, the initial kinetic energy is zero. If we take the zero point for the gravita-

tional potential energy of the projectile–Earth system to be at the lowest position of the projectile x_{A}, then the initial gravitational potential energy also is zero. The mechanical energy of this system is constant because no nonconservative forces are present.

Initially, the only mechanical energy in the system is the elastic potential energy stored in the spring of the gun, $U_{s\text{A}} = kx^2/2$, where the compression of the spring is $x = 0.120$ m. The projectile rises to a maximum height

(a)   (b)

Figure 8.11   A spring-loaded popgun.

$x_C = h = 20.0$ m, and so the final gravitational potential energy when the projectile reaches its peak is mgh. The final kinetic energy of the projectile is zero, and the final elastic potential energy stored in the spring is zero. Because the mechanical energy of the system is constant, we find that

$$E_A = E_C$$

$$K_A + U_{gA} + U_{sA} = K_C + U_{gC} + U_{sC}$$

$$0 + 0 + \tfrac{1}{2}kx^2 = 0 + mgh + 0$$

$$\tfrac{1}{2}k(0.120 \text{ m})^2 = (0.0350 \text{ kg})(9.80 \text{ m/s}^2)(20.0 \text{ m})$$

$$k = \boxed{953 \text{ N/m}}$$

(b) Find the speed of the projectile as it moves through the equilibrium position of the spring (where $x_B = 0.120$ m) as shown in Figure 8.11b.

Solution   As already noted, the only mechanical energy in the system at Ⓐ is the elastic potential energy $kx^2/2$. The total energy of the system as the projectile moves through the equilibrium position of the spring comprises the kinetic energy of the projectile $mv_B^2/2$, and the gravitational potential energy mgx_B. Hence, the principle of the conservation of mechanical energy in this case gives

$$E_A = E_B$$

$$K_A + U_{gA} + U_{sA} = K_B + U_{gB} + U_{sB}$$

$$0 + 0 + \tfrac{1}{2}kx^2 = \tfrac{1}{2}mv_B^2 + mgx_B + 0$$

Solving for v_B gives

$$v_B = \sqrt{\frac{kx^2}{m} - 2gx_B}$$

$$= \sqrt{\frac{(953 \text{ N/m})(0.120 \text{ m})^2}{0.0350 \text{ kg}} - 2(9.80 \text{ m/s}^2)(0.120 \text{ m})}$$

$$= \boxed{19.7 \text{ m/s}}$$

You should compare the different examples we have presented so far in this chapter. Note how breaking the problem into a sequence of labeled events helps in the analysis.

Exercise   What is the speed of the projectile when it is at a height of 10.0 m?

Answer   14.0 m/s.

EXAMPLE 8.8   **Block–Spring Collision**

A block having a mass of 0.80 kg is given an initial velocity $v_A = 1.2$ m/s to the right and collides with a spring of negligible mass and force constant $k = 50$ N/m, as shown in Figure 8.12. (a) Assuming the surface to be frictionless, calculate the maximum compression of the spring after the collision.

Solution   Our system in this example consists of the block and spring. Before the collision, at Ⓐ, the block has kinetic

energy and the spring is uncompressed, so that the elastic potential energy stored in the spring is zero. Thus, the total mechanical energy of the system before the collision is just $\tfrac{1}{2}mv_A^2$. After the collision, at Ⓒ, the spring is fully compressed; now the block is at rest and so has zero kinetic energy, while the energy stored in the spring has its maximum value $\tfrac{1}{2}kx^2 = \tfrac{1}{2}kx_m^2$, where the origin of coordinates $x = 0$ is chosen to be the equilibrium position of the spring and x_m is

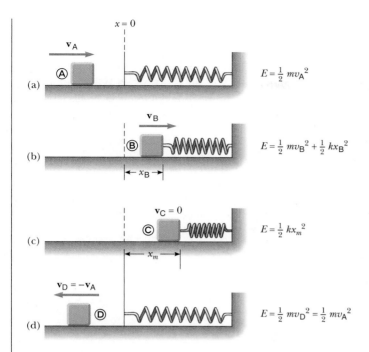

(a) $E = \frac{1}{2} m{v_A}^2$

(b) $E = \frac{1}{2} m{v_B}^2 + \frac{1}{2} k{x_B}^2$

(c) $E = \frac{1}{2} k{x_m}^2$

(d) $E = \frac{1}{2} m{v_D}^2 = \frac{1}{2} m{v_A}^2$

Figure 8.12   A block sliding on a smooth, horizontal surface collides with a light spring. (a) Initially the mechanical energy is all kinetic energy. (b) The mechanical energy is the sum of the kinetic energy of the block and the elastic potential energy in the spring. (c) The energy is entirely potential energy. (d) The energy is transformed back to the kinetic energy of the block. The total energy remains constant throughout the motion.

Multiflash photograph of a pole vault event. How many forms of energy can you identify in this picture?
(© *Harold E. Edgerton/Courtesy of Palm Press, Inc.*)

the maximum compression of the spring, which in this case happens to be x_C. The total mechanical energy of the system is conserved because no nonconservative forces act on objects within the system.

Because mechanical energy is conserved, the kinetic energy of the block before the collision must equal the maximum potential energy stored in the fully compressed spring:

$$E_A = E_C$$

$$K_A + U_{sA} = K_C + U_{sC}$$

$$\tfrac{1}{2}m{v_A}^2 + 0 = 0 + \tfrac{1}{2}k{x_m}^2$$

$$x_m = \sqrt{\frac{m}{k}}\, v_A = \sqrt{\frac{0.80\ \text{kg}}{50\ \text{N/m}}}\ (1.2\ \text{m/s})$$

$$= \boxed{0.15\ \text{m}}$$

Note that we have not included U_g terms because no change in vertical position occurred.

(b) Suppose a constant force of kinetic friction acts between the block and the surface, with $\mu_k = 0.50$. If the speed

of the block at the moment it collides with the spring is $v_A = 1.2$ m/s, what is the maximum compression in the spring?

Solution   In this case, mechanical energy is *not* conserved because a frictional force acts on the block. The magnitude of the frictional force is

$$f_k = \mu_k n = \mu_k mg = 0.50(0.80\ \text{kg})(9.80\ \text{m/s}^2)\ = 3.92\ \text{N}$$

Therefore, the change in the block's mechanical energy due to friction as the block is displaced from the equilibrium position of the spring (where we have set our origin) to x_B is

$$\Delta E = -f_k x_B = -3.92 x_B$$

Substituting this into Equation 8.15 gives

$$\Delta E = E_f - E_i = (0 + \tfrac{1}{2}k{x_B}^2) - (\tfrac{1}{2}m{v_A}^2 + 0) = -f_k x_B$$

$$\tfrac{1}{2}(50){x_B}^2 - \tfrac{1}{2}(0.80)(1.2)^2 = -3.92 x_B$$

$$25{x_B}^2 + 3.92 x_B - 0.576 = 0$$

Solving the quadratic equation for x_B gives $x_B = 0.092$ m and $x_B = -0.25$ m. The physically meaningful root is $x_B = \boxed{0.092\ \text{m}}$. The negative root does not apply to this situation because the block must be to the right of the origin (positive value of x) when it comes to rest. Note that 0.092 m is less than the distance obtained in the frictionless case of part (a). This result is what we expect because friction retards the motion of the system.

EXAMPLE 8.9 ▶ Connected Blocks in Motion

Two blocks are connected by a light string that passes over a frictionless pulley, as shown in Figure 8.13. The block of mass m_1 lies on a horizontal surface and is connected to a spring of force constant k. The system is released from rest when the spring is unstretched. If the hanging block of mass m_2 falls a distance h before coming to rest, calculate the coefficient of kinetic friction between the block of mass m_1 and the surface.

Solution The key word *rest* appears twice in the problem statement, telling us that the initial and final velocities and kinetic energies are zero. (Also note that because we are concerned only with the beginning and ending points of the motion, we do not need to label events with circled letters as we did in the previous two examples. Simply using i and f is sufficient to keep track of the situation.) In this situation, the system consists of the two blocks, the spring, and the Earth. We need to consider two forms of potential energy: gravitational and elastic. Because the initial and final kinetic energies of the system are zero, $\Delta K = 0$, and we can write

(1) $$\Delta E = \Delta U_g + \Delta U_s = -f_k h$$

Figure 8.13 As the hanging block moves from its highest elevation to its lowest, the system loses gravitational potential energy but gains elastic potential energy in the spring. Some mechanical energy is lost because of friction between the sliding block and the surface.

where $\Delta U_g = U_{gf} - U_{gi}$ is the change in the system's gravitational potential energy and $\Delta U_s = U_{sf} - U_{si}$ is the change in the system's elastic potential energy. As the hanging block falls a distance h, the horizontally moving block moves the same distance h to the right. Therefore, using Equation 8.15, we find that the loss in energy due to friction between the horizontally sliding block and the surface is

(2) $$\Delta E = -f_k h = -\mu_k m_1 g h$$

The change in the gravitational potential energy of the system is associated with only the falling block because the vertical coordinate of the horizontally sliding block does not change. Therefore, we obtain

(3) $$\Delta U_g = U_{gf} - U_{gi} = 0 - m_2 g h$$

where the coordinates have been measured from the lowest position of the falling block.

The change in the elastic potential energy stored in the spring is

(4) $$\Delta U_s = U_{sf} - U_{si} = \tfrac{1}{2}kh^2 - 0$$

Substituting Equations (2), (3), and (4) into Equation (1) gives

$$-\mu_k m_1 g h = -m_2 g h + \tfrac{1}{2}kh^2$$

$$\mu_k = \frac{m_2 g - \tfrac{1}{2}kh}{m_1 g}$$

This setup represents a way of measuring the coefficient of kinetic friction between an object and some surface. As you can see from the problem, sometimes it is easier to work with the changes in the various types of energy rather than the actual values. For example, if we wanted to calculate the numerical value of the gravitational potential energy associated with the horizontally sliding block, we would need to specify the height of the horizontal surface relative to the lowest position of the falling block. Fortunately, this is not necessary because the gravitational potential energy associated with the first block does not change.

EXAMPLE 8.10 ▶ A Grand Entrance

You are designing apparatus to support an actor of mass 65 kg who is to "fly" down to the stage during the performance of a play. You decide to attach the actor's harness to a 130-kg sandbag by means of a lightweight steel cable running smoothly over two frictionless pulleys, as shown in Figure 8.14a. You need 3.0 m of cable between the harness and the nearest pulley so that the pulley can be hidden behind a curtain. For the apparatus to work successfully, the sandbag must never lift above the floor as the actor swings from above the

stage to the floor. Let us call the angle that the actor's cable makes with the vertical θ. What is the maximum value θ can have before the sandbag lifts off the floor?

Solution We need to draw on several concepts to solve this problem. First, we use the principle of the conservation of mechanical energy to find the actor's speed as he hits the floor as a function of θ and the radius R of the circular path through which he swings. Next, we apply Newton's second law to the actor at the bottom of his path to find the cable tension as a function of the given parameters. Finally, we note that the sandbag lifts off the floor when the upward force exerted on it by the cable exceeds the gravitational force acting on it; the normal force is zero when this happens.

Applying conservation of energy to the actor–Earth system gives

$$K_i + U_i = K_f + U_f$$

$$(1) \qquad 0 + m_{actor}\,gy_i = \tfrac{1}{2}m_{actor}v_f^2 + 0$$

Actor

Sandbag

(a)

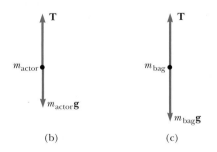

m_{actor}

$m_{actor}\mathbf{g}$

(b)

m_{bag}

$m_{bag}\mathbf{g}$

(c)

Figure 8.14 (a) An actor uses some clever staging to make his entrance. (b) Free-body diagram for actor at the bottom of the circular path. (c) Free-body diagram for sandbag.

where y_i is the initial height of the actor above the floor and v_f is the speed of the actor at the instant before he lands. (Note that $K_i = 0$ because he starts from rest and that $U_f = 0$ because we set the level of the actor's harness when he is standing on the floor as the zero level of potential energy.) From the geometry in Figure 8.14a, we see that $y_i = R - R\cos\theta = R(1 - \cos\theta)$. Using this relationship in Equation (1), we obtain

$$(2) \qquad v_f^2 = 2gR(1 - \cos\theta)$$

Now we apply Newton's second law to the actor when he is at the bottom of the circular path, using the free-body diagram in Figure 8.14b as a guide:

$$\sum F_y = T - m_{actor}g = m_{actor}\frac{v_f^2}{R}$$

$$(3) \qquad T = m_{actor}g + m_{actor}\frac{v_f^2}{R}$$

A force of the same magnitude as T is transmitted to the sandbag. If it is to be just lifted off the floor, the normal force on it becomes zero, and we require that $T = m_{bag}g$, as shown in Figure 8.14c. Using this condition together with Equations (2) and (3), we find that

$$m_{bag}g = m_{actor}g + m_{actor}\frac{2gR(1 - \cos\theta)}{R}$$

Solving for θ and substituting in the given parameters, we obtain

$$\cos\theta = \frac{3m_{actor} - m_{bag}}{2m_{actor}} = \frac{3(65\ \text{kg}) - 130\ \text{kg}}{2(65\ \text{kg})} = \frac{1}{2}$$

$$\theta = \boxed{60°}$$

Notice that we did not need to be concerned with the length R of the cable from the actor's harness to the leftmost pulley. The important point to be made from this problem is that it is sometimes necessary to combine energy considerations with Newton's laws of motion.

Exercise If the initial angle $\theta = 40°$, find the speed of the actor and the tension in the cable just before he reaches the floor. (*Hint:* You cannot ignore the length $R = 3.0$ m in this calculation.)

Answer 3.7 m/s; 940 N.

8.6 RELATIONSHIP BETWEEN CONSERVATIVE FORCES AND POTENTIAL ENERGY

Once again let us consider a particle that is part of a system. Suppose that the particle moves along the x axis, and assume that a conservative force with an x component F_x acts on the particle. Earlier in this chapter, we showed how to determine the change in potential energy of a system when we are given the conservative force. We now show how to find F_x if the potential energy of the system is known.

In Section 8.2 we learned that the work done by the conservative force as its point of application undergoes a displacement Δx equals the negative of the change in the potential energy associated with that force; that is, $W = F_x \Delta x = -\Delta U$. If the point of application of the force undergoes an infinitesimal displacement dx, we can express the infinitesimal change in the potential energy of the system dU as

$$dU = -F_x \, dx$$

Therefore, the conservative force is related to the potential energy function through the relationship[3]

Relationship between force and potential energy

$$F_x = -\frac{dU}{dx} \tag{8.16}$$

That is, **any conservative force acting on an object within a system equals the negative derivative of the potential energy of the system with respect to x.**

We can easily check this relationship for the two examples already discussed. In the case of the deformed spring, $U_s = \frac{1}{2}kx^2$, and therefore

$$F_s = -\frac{dU_s}{dx} = -\frac{d}{dx}(\tfrac{1}{2}kx^2) = -kx$$

which corresponds to the restoring force in the spring. Because the gravitational potential energy function is $U_g = mgy$, it follows from Equation 8.16 that $F_g = -mg$ when we differentiate U_g with respect to y instead of x.

We now see that U is an important function because a conservative force can be derived from it. Furthermore, Equation 8.16 should clarify the fact that adding a constant to the potential energy is unimportant because the derivative of a constant is zero.

Quick Quiz 8.5

What does the slope of a graph of $U(x)$ versus x represent?

Optional Section

8.7 ENERGY DIAGRAMS AND THE EQUILIBRIUM OF A SYSTEM

The motion of a system can often be understood qualitatively through a graph of its potential energy versus the separation distance between the objects in the system. Consider the potential energy function for a block–spring system, given by $U_s = \frac{1}{2}kx^2$. This function is plotted versus x in Figure 8.15a. (A common mistake is to think that potential energy on the graph represents height. This is clearly not

---

[3] In three dimensions, the expression is $\mathbf{F} = -\mathbf{i}\dfrac{\partial U}{\partial x} - \mathbf{j}\dfrac{\partial U}{\partial y} - \mathbf{k}\dfrac{\partial U}{\partial z}$, where $\dfrac{\partial U}{\partial x}$, and so forth, are partial derivatives. In the language of vector calculus, $\mathbf{F}$ equals the negative of the gradient of the scalar quantity $U(x, y, z)$.

(a)

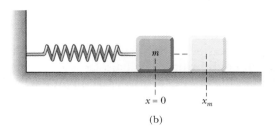

(b)

Figure 8.15 (a) Potential energy as a function of x for the block–spring system shown in (b). The block oscillates between the turning points, which have the coordinates $x = \pm x_m$. Note that the restoring force exerted by the spring always acts toward $x = 0$, the position of stable equilibrium.

the case here, where the block is only moving horizontally.) The force F_s exerted by the spring on the block is related to U_s through Equation 8.16:

$$F_s = -\frac{dU_s}{dx} = -kx$$

As we saw in Quick Quiz 8.5, the force is equal to the negative of the slope of the U versus x curve. When the block is placed at rest at the equilibrium position of the spring ($x = 0$), where $F_s = 0$, it will remain there unless some external force F_{ext} acts on it. If this external force stretches the spring from equilibrium, x is positive and the slope dU/dx is positive; therefore, the force F_s exerted by the spring is negative, and the block accelerates back toward $x = 0$ when released. If the external force compresses the spring, then x is negative and the slope is negative; therefore, F_s is positive, and again the mass accelerates toward $x = 0$ upon release.

From this analysis, we conclude that the $x = 0$ position for a block–spring system is one of **stable equilibrium.** That is, any movement away from this position results in a force directed back toward $x = 0$. In general, **positions of stable equilibrium correspond to points for which $U(x)$ is a minimum.**

From Figure 8.15 we see that if the block is given an initial displacement x_m and is released from rest, its total energy initially is the potential energy stored in the spring $\frac{1}{2}kx_m^2$. As the block starts to move, the system acquires kinetic energy and loses an equal amount of potential energy. Because the total energy must remain constant, the block oscillates (moves back and forth) between the two points $x = -x_m$ and $x = +x_m$, called the *turning points.* In fact, because no energy is lost (no friction), the block will oscillate between $-x_m$ and $+x_m$ forever. (We discuss these oscillations further in Chapter 13.) From an energy viewpoint, the energy of the system cannot exceed $\frac{1}{2}kx_m^2$; therefore, the block must stop at these points and, because of the spring force, must accelerate toward $x = 0$.

Another simple mechanical system that has a position of stable equilibrium is a ball rolling about in the bottom of a bowl. Anytime the ball is displaced from its lowest position, it tends to return to that position when released.

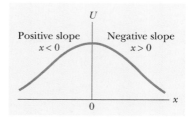

Figure 8.16 A plot of U versus x for a particle that has a position of unstable equilibrium located at $x = 0$. For any finite displacement of the particle, the force on the particle is directed away from $x = 0$.

Now consider a particle moving along the x axis under the influence of a conservative force F_x, where the U versus x curve is as shown in Figure 8.16. Once again, $F_x = 0$ at $x = 0$, and so the particle is in equilibrium at this point. However, this is a position of **unstable equilibrium** for the following reason: Suppose that the particle is displaced to the right ($x > 0$). Because the slope is negative for $x > 0$, $F_x = -dU/dx$ is positive and the particle accelerates *away from* $x = 0$. If instead the particle is at $x = 0$ and is displaced to the left ($x < 0$), the force is negative because the slope is positive for $x < 0$, and the particle again accelerates *away from* the equilibrium position. The position $x = 0$ in this situation is one of unstable equilibrium because for any displacement from this point, the force pushes the particle *farther away from* equilibrium. The force pushes the particle toward a position of lower potential energy. A pencil balanced on its point is in a position of unstable equilibrium. If the pencil is displaced slightly from its absolutely vertical position and is then released, it will surely fall over. In general, **positions of unstable equilibrium correspond to points for which $U(x)$ is a maximum.**

Finally, a situation may arise where U is constant over some region and hence $F_x = 0$. This is called a position of **neutral equilibrium.** Small displacements from this position produce neither restoring nor disrupting forces. A ball lying on a flat horizontal surface is an example of an object in neutral equilibrium.

EXAMPLE 8.11   Force and Energy on an Atomic Scale

The potential energy associated with the force between two neutral atoms in a molecule can be modeled by the Lennard–Jones potential energy function:

$$U(x) = 4\epsilon\left[\left(\frac{\sigma}{x}\right)^{12} - \left(\frac{\sigma}{x}\right)^{6}\right]$$

where x is the separation of the atoms. The function $U(x)$ contains two parameters σ and ϵ that are determined from experiments. Sample values for the interaction between two atoms in a molecule are $\sigma = 0.263$ nm and $\epsilon = 1.51 \times 10^{-22}$ J. (a) Using a spreadsheet or similar tool, graph this function and find the most likely distance between the two atoms.

Solution We expect to find stable equilibrium when the two atoms are separated by some equilibrium distance and the potential energy of the system of two atoms (the molecule) is a minimum. One can minimize the function $U(x)$ by taking its derivative and setting it equal to zero:

$$\frac{dU(x)}{dx} = 4\epsilon\frac{d}{dx}\left[\left(\frac{\sigma}{x}\right)^{12} - \left(\frac{\sigma}{x}\right)^{6}\right] = 0$$

$$= 4\epsilon\left[\frac{-12\sigma^{12}}{x^{13}} - \frac{-6\sigma^{6}}{x^{7}}\right] = 0$$

Solving for x—the equilibrium separation of the two atoms in the molecule—and inserting the given information yield

$$x = 2.95 \times 10^{-10} \text{ m.}$$

We graph the Lennard–Jones function on both sides of this critical value to create our energy diagram, as shown in Figure 8.17a. Notice how $U(x)$ is extremely large when the atoms are very close together, is a minimum when the atoms

are at their critical separation, and then increases again as the atoms move apart. When $U(x)$ is a minimum, the atoms are in stable equilibrium; this indicates that this is the most likely separation between them.

(b) Determine $F_x(x)$—the force that one atom exerts on the other in the molecule as a function of separation—and argue that the way this force behaves is physically plausible when the atoms are close together and far apart.

Solution Because the atoms combine to form a molecule, we reason that the force must be attractive when the atoms are far apart. On the other hand, the force must be repulsive when the two atoms get very close together. Otherwise, the molecule would collapse in on itself. Thus, the force must change sign at the critical separation, similar to the way spring forces switch sign in the change from extension to compression. Applying Equation 8.16 to the Lennard–Jones potential energy function gives

$$F_x = -\frac{dU(x)}{dx} = -4\epsilon\frac{d}{dx}\left[\left(\frac{\sigma}{x}\right)^{12} - \left(\frac{\sigma}{x}\right)^{6}\right]$$

$$= 4\epsilon\left[\frac{12\sigma^{12}}{x^{13}} - \frac{6\sigma^{6}}{x^{7}}\right]$$

This result is graphed in Figure 8.17b. As expected, the force is positive (repulsive) at small atomic separations, zero when the atoms are at the position of stable equilibrium [recall how we found the minimum of $U(x)$], and negative (attractive) at greater separations. Note that the force approaches zero as the separation between the atoms becomes very great.

Figure 8.17   (a) Potential energy curve associated with a molecule. The distance *x* is the separation between the two atoms making up the molecule. (b) Force exerted on one atom by the other.

8.8   CONSERVATION OF ENERGY IN GENERAL

We have seen that the total mechanical energy of a system is constant when only conservative forces act within the system. Furthermore, we can associate a potential energy function with each conservative force. On the other hand, as we saw in Section 8.5, mechanical energy is lost when nonconservative forces such as friction are present.

In our study of thermodynamics later in this course, we shall find that mechanical energy can be transformed into energy stored *inside* the various objects that make up the system. This form of energy is called *internal energy*. For example, when a block slides over a rough surface, the mechanical energy lost because of friction is transformed into internal energy that is stored temporarily inside the block and inside the surface, as evidenced by a measurable increase in the temperature of both block and surface. We shall see that on a submicroscopic scale, this internal energy is associated with the vibration of atoms about their equilibrium positions. Such internal atomic motion involves both kinetic and potential energy. Therefore, if we include in our energy expression this increase in the internal energy of the objects that make up the system, then energy is conserved.

This is just one example of how you can analyze an isolated system and always find that the total amount of energy it contains does not change, as long as you account for all forms of energy. That is, **energy can never be created or destroyed. Energy may be transformed from one form to another, but the**

Total energy is always conserved

total energy of an isolated system is always constant. From a universal point of view, we can say that the **total energy of the Universe is constant.** If one part of the Universe gains energy in some form, then another part must lose an equal amount of energy. No violation of this principle has ever been found.

Optional Section

8.9 ▸ MASS–ENERGY EQUIVALENCE

This chapter has been concerned with the important principle of energy conservation and its application to various physical phenomena. Another important principle, **conservation of mass,** states that **in any physical or chemical process, mass is neither created nor destroyed.** That is, the mass before the process equals the mass after the process.

For centuries, scientists believed that energy and mass were two quantities that were separately conserved. However, in 1905 Einstein made the brilliant discovery that the mass of any system is a measure of the energy of that system. Hence, energy and mass are related concepts. The relationship between the two is given by Einstein's most famous formula:

$$E_R = mc^2 \qquad\qquad \textbf{(8.17)}$$

where c is the speed of light and E_R is the energy equivalent of a mass m. The subscript R on the energy refers to the **rest energy** of an object of mass m—that is, the energy of the object when its speed is $v = 0$.

The rest energy associated with even a small amount of matter is enormous. For example, the rest energy of 1 kg of any substance is

$$E_R = mc^2 = (1 \text{ kg})(3 \times 10^8 \text{ m/s})^2 = 9 \times 10^{16} \text{ J}$$

This is equivalent to the energy content of about 15 million barrels of crude oil—about one day's consumption in the United States! If this energy could easily be released as useful work, our energy resources would be unlimited.

In reality, only a small fraction of the energy contained in a material sample can be released through chemical or nuclear processes. The effects are greatest in nuclear reactions, in which fractional changes in energy, and hence mass, of approximately 10^{-3} are routinely observed. A good example is the enormous amount of energy released when the uranium-235 nucleus splits into two smaller nuclei. This happens because the sum of the masses of the product nuclei is slightly less than the mass of the original ^{235}U nucleus. The awesome nature of the energy released in such reactions is vividly demonstrated in the explosion of a nuclear weapon.

Equation 8.17 indicates that *energy has mass.* Whenever the energy of an object changes in any way, its mass changes as well. If ΔE is the change in energy of an object, then its change in mass is

$$\Delta m = \frac{\Delta E}{c^2} \qquad\qquad \textbf{(8.18)}$$

Anytime energy ΔE in any form is supplied to an object, the change in the mass of the object is $\Delta m = \Delta E / c^2$. However, because c^2 is so large, the changes in mass in any ordinary mechanical experiment or chemical reaction are too small to be detected.

EXAMPLE 8.12 ▶ **Here Comes the Sun**

The Sun converts an enormous amount of matter to energy. Each second, 4.19×10^9 kg—approximately the capacity of 400 average-sized cargo ships—is changed to energy. What is the power output of the Sun?

Solution We find the energy liberated per second by means of a straightforward conversion:

$$E_R = (4.19 \times 10^9 \text{ kg})(3.00 \times 10^8 \text{ m/s})^2 = 3.77 \times 10^{26} \text{ J}$$

We then apply the definition of power:

$$\mathscr{P} = \frac{3.77 \times 10^{26} \text{ J}}{1.00 \text{ s}} = \boxed{3.77 \times 10^{26} \text{ W}}$$

The Sun radiates uniformly in all directions, and so only a very tiny fraction of its total output is collected by the Earth. Nonetheless this amount is sufficient to supply energy to nearly everything on the Earth. (Nuclear and geothermal energy are the only alternatives.) Plants absorb solar energy and convert it to chemical potential energy (energy stored in the plant's molecules). When an animal eats the plant, this chemical potential energy can be turned into kinetic and other forms of energy. You are reading this book with solar-powered eyes!

Optional Section

8.10 ▶ QUANTIZATION OF ENERGY

Certain physical quantities such as electric charge are *quantized;* that is, the quantities have discrete values rather than continuous values. The quantized nature of energy is especially important in the atomic and subatomic world. As an example, let us consider the energy levels of the hydrogen atom (which consists of an electron orbiting around a proton). The atom can occupy only certain energy levels, called *quantum states,* as shown in Figure 8.18a. The atom cannot have any energy values lying between these quantum states. The lowest energy level E_1 is called the

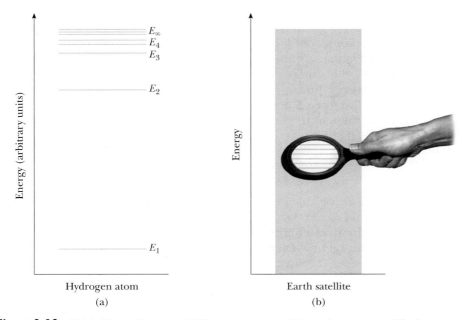

Figure 8.18   Energy-level diagrams: (a) Quantum states of the hydrogen atom. The lowest state E_1 is the ground state. (b) The energy levels of an Earth satellite are also quantized but are so close together that they cannot be distinguished from one another.

ground state of the atom. The ground state corresponds to the state that an isolated atom usually occupies. The atom can move to higher energy states by absorbing energy from some external source or by colliding with other atoms. The highest energy on the scale shown in Figure 8.18a, E_∞, corresponds to the energy of the atom when the electron is completely removed from the proton. The energy difference $E_\infty - E_1$ is called the **ionization energy.** Note that the energy levels get closer together at the high end of the scale.

Next, consider a satellite in orbit about the Earth. If you were asked to describe the possible energies that the satellite could have, it would be reasonable (but incorrect) to say that it could have any arbitrary energy value. Just like that of the hydrogen atom, however, **the energy of the satellite is quantized.** If you were to construct an energy level diagram for the satellite showing its allowed energies, the levels would be so close to one another, as shown in Figure 8.18b, that it would be difficult to discern that they were not continuous. In other words, we have no way of experiencing quantization of energy in the macroscopic world; hence, we can ignore it in describing everyday experiences.

SUMMARY

If a particle of mass m is at a distance y above the Earth's surface, the **gravitational potential energy** of the particle–Earth system is

$$U_g = mgy \tag{8.1}$$

The **elastic potential energy** stored in a spring of force constant k is

$$U_s \equiv \tfrac{1}{2}kx^2 \tag{8.4}$$

You should be able to apply these two equations in a variety of situations to determine the potential an object has to perform work.

A force is **conservative** if the work it does on a particle moving between two points is independent of the path the particle takes between the two points. Furthermore, a force is conservative if the work it does on a particle is zero when the particle moves through an arbitrary closed path and returns to its initial position. A force that does not meet these criteria is said to be **nonconservative.**

A **potential energy** function U can be associated only with a conservative force. If a conservative force $\mathbf{F}$ acts on a particle that moves along the x axis from x_i to x_f, then the change in the potential energy of the system equals the negative of the work done by that force:

$$U_f - U_i = -\int_{x_i}^{x_f} F_x \, dx \tag{8.7}$$

You should be able to use calculus to find the potential energy associated with a conservative force and vice versa.

The **total mechanical energy of a system** is defined as the sum of the kinetic energy and the potential energy:

$$E \equiv K + U \tag{8.9}$$

If no external forces do work on a system and if no nonconservative forces are acting on objects inside the system, then the total mechanical energy of the system is constant:

$$K_i + U_i = K_f + U_f \tag{8.10}$$

If nonconservative forces (such as friction) act on objects inside a system, then mechanical energy is not conserved. In these situations, the difference between the total final mechanical energy and the total initial mechanical energy of the system equals the energy transferred to or from the system by the nonconservative forces.

QUESTIONS

1. Many mountain roads are constructed so that they spiral around a mountain rather than go straight up the slope. Discuss this design from the viewpoint of energy and power.

2. A ball is thrown straight up into the air. At what position is its kinetic energy a maximum? At what position is the gravitational potential energy a maximum?

3. A bowling ball is suspended from the ceiling of a lecture hall by a strong cord. The bowling ball is drawn away from its equilibrium position and released from rest at the tip

Figure Q8.3

of the student's nose as in Figure Q8.3. If the student remains stationary, explain why she will not be struck by the ball on its return swing. Would the student be safe if she pushed the ball as she released it?

4. One person drops a ball from the top of a building, while another person at the bottom observes its motion. Will these two people agree on the value of the potential energy of the ball–Earth system? on its change in potential energy? on the kinetic energy of the ball?

5. When a person runs in a track event at constant velocity, is any work done? (*Note:* Although the runner moves with constant velocity, the legs and arms accelerate.) How does air resistance enter into the picture? Does the center of mass of the runner move horizontally?

6. Our body muscles exert forces when we lift, push, run, jump, and so forth. Are these forces conservative?

7. If three conservative forces and one nonconservative force act on a system, how many potential energy terms appear in the equation that describes this system?

8. Consider a ball fixed to one end of a rigid rod whose other end pivots on a horizontal axis so that the rod can rotate in a vertical plane. What are the positions of stable and unstable equilibrium?

9. Is it physically possible to have a situation where $E - U < 0$?

10. What would the curve of U versus x look like if a particle were in a region of neutral equilibrium?

11. Explain the energy transformations that occur during (a) the pole vault, (b) the shot put, (c) the high jump. What is the source of energy in each case?

12. Discuss some of the energy transformations that occur during the operation of an automobile.

13. If only one external force acts on a particle, does it necessarily change the particle's (a) kinetic energy? (b) velocity?

PROBLEMS

1, 2, 3 = straightforward, intermediate, challenging ☐ = full solution available in the *Student Solutions Manual and Study Guide*
WEB = solution posted at **http://www.saunderscollege.com/physics/** 🖥 = Computer useful in solving problem 🕹 = Interactive Physics
☐ = paired numerical/symbolic problems

Section 8.1 Potential Energy
Section 8.2 Conservative and Nonconservative Forces

1. A 1 000-kg roller coaster is initially at the top of a rise, at point A. It then moves 135 ft, at an angle of 40.0° below the horizontal, to a lower point B. (a) Choose point B to be the zero level for gravitational potential energy. Find the potential energy of the roller coaster–Earth system at points A and B and the change in its potential energy as the coaster moves. (b) Repeat part (a), setting the zero reference level at point A.

2. A 40.0-N child is in a swing that is attached to ropes 2.00 m long. Find the gravitational potential energy of the child–Earth system relative to the child's lowest position when (a) the ropes are horizontal, (b) the ropes make a 30.0° angle with the vertical, and (c) the child is at the bottom of the circular arc.

3. A 4.00-kg particle moves from the origin to position C, which has coordinates $x = 5.00$ m and $y = 5.00$ m (Fig. P8.3). One force on it is the force of gravity acting in the negative y direction. Using Equation 7.2, calculate the work done by gravity as the particle moves from O to C along (a) OAC, (b) OBC, and (c) OC. Your results should all be identical. Why?

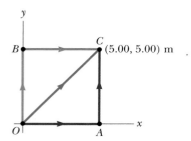

Figure P8.3   Problems 3, 4, and 5.

4. (a) Suppose that a constant force acts on an object. The force does not vary with time, nor with the position or velocity of the object. Start with the general definition for work done by a force

$$W = \int_i^f \mathbf{F} \cdot d\mathbf{s}$$

and show that the force is conservative. (b) As a special case, suppose that the force $\mathbf{F} = (3\mathbf{i} + 4\mathbf{j})$ N acts on a particle that moves from O to C in Figure P8.3. Calculate the work done by $\mathbf{F}$ if the particle moves along each one of the three paths OAC, OBC, and OC. (Your three answers should be identical.)

5. A force acting on a particle moving in the xy plane is given by $\mathbf{F} = (2 y\mathbf{i} + x^2 \mathbf{j})$ N, where x and y are in meters. The particle moves from the origin to a final position having coordinates $x = 5.00$ m and $y = 5.00$ m, as in Figure P8.3. Calculate the work done by $\mathbf{F}$ along (a) OAC, (b) OBC, (c) OC. (d) Is $\mathbf{F}$ conservative or non-conservative? Explain.

Section 8.3   Conservative Forces and Potential Energy
Section 8.4   Conservation of Mechanical Energy

6. At time t_i, the kinetic energy of a particle in a system is 30.0 J and the potential energy of the system is 10.0 J. At some later time t_f, the kinetic energy of the particle is 18.0 J. (a) If only conservative forces act on the particle, what are the potential energy and the total energy at

time t_f? (b) If the potential energy of the system at time t_f is 5.00 J, are any nonconservative forces acting on the particle? Explain.

WEB 7. A single conservative force acts on a 5.00-kg particle. The equation $F_x = (2x + 4)$ N, where x is in meters, describes this force. As the particle moves along the x axis from $x = 1.00$ m to $x = 5.00$ m, calculate (a) the work done by this force, (b) the change in the potential energy of the system, and (c) the kinetic energy of the particle at $x = 5.00$ m if its speed at $x = 1.00$ m is 3.00 m/s.

8. A single constant force $\mathbf{F} = (3\mathbf{i} + 5\mathbf{j})$ N acts on a 4.00-kg particle. (a) Calculate the work done by this force if the particle moves from the origin to the point having the vector position $\mathbf{r} = (2\mathbf{i} - 3\mathbf{j})$ m. Does this result depend on the path? Explain. (b) What is the speed of the particle at $\mathbf{r}$ if its speed at the origin is 4.00 m/s? (c) What is the change in the potential energy of the system?

9. A single conservative force acting on a particle varies as $\mathbf{F} = (- Ax + Bx^2)\mathbf{i}$ N, where A and B are constants and x is in meters. (a) Calculate the potential energy function $U(x)$ associated with this force, taking $U = 0$ at $x = 0$. (b) Find the change in potential energy and change in kinetic energy as the particle moves from $x = 2.00$ m to $x = 3.00$ m.

10. A particle of mass 0.500 kg is shot from P as shown in Figure P8.10. The particle has an initial velocity $\mathbf{v}_i$ with a horizontal component of 30.0 m/s. The particle rises to a maximum height of 20.0 m above P. Using the law of conservation of energy, determine (a) the vertical component of $\mathbf{v}_i$, (b) the work done by the gravitational force on the particle during its motion from P to B, and (c) the horizontal and the vertical components of the velocity vector when the particle reaches B.

Figure P8.10

11. A 3.00-kg mass starts from rest and slides a distance d down a frictionless 30.0° incline. While sliding, it comes into contact with an unstressed spring of negligible mass, as shown in Figure P8.11. The mass slides an additional 0.200 m as it is brought momentarily to rest by compression of the spring ($k = 400$ N/m). Find the initial separation d between the mass and the spring.

12. A mass m starts from rest and slides a distance d down a frictionless incline of angle θ. While sliding, it contacts an unstressed spring of negligible mass, as shown in Figure P8.11. The mass slides an additional distance x as it is brought momentarily to rest by compression of the spring (of force constant k). Find the initial separation d between the mass and the spring.

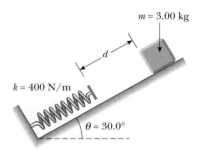

Figure P8.11   Problems 11 and 12.

13. A particle of mass $m = 5.00$ kg is released from point Ⓐ and slides on the frictionless track shown in Figure P8.13. Determine (a) the particle's speed at points Ⓑ and Ⓒ and (b) the net work done by the force of gravity in moving the particle from Ⓐ to Ⓒ.

Figure P8.13

14. A simple, 2.00-m-long pendulum is released from rest when the support string is at an angle of $25.0°$ from the vertical. What is the speed of the suspended mass at the bottom of the swing?

15. A bead slides without friction around a loop-the-loop (Fig. P8.15). If the bead is released from a height $h = 3.50R$, what is its speed at point A? How great is the normal force on it if its mass is 5.00 g?

16. A 120-g mass is attached to the bottom end of an unstressed spring. The spring is hanging vertically and has a spring constant of 40.0 N/m. The mass is dropped. (a) What is its maximum speed? (b) How far does it drop before coming to rest momentarily?

17. A block of mass 0.250 kg is placed on top of a light verti-

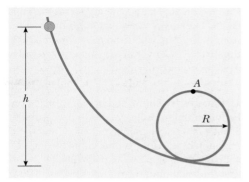

Figure P8.15

cal spring of constant $k = 5\ 000$ N/m and is pushed downward so that the spring is compressed 0.100 m. After the block is released, it travels upward and then leaves the spring. To what maximum height above the point of release does it rise?

18. Dave Johnson, the bronze medalist at the 1992 Olympic decathlon in Barcelona, leaves the ground for his high jump with a vertical velocity component of 6.00 m/s. How far up does his center of gravity move as he makes the jump?

19. A 0.400-kg ball is thrown straight up into the air and reaches a maximum altitude of 20.0 m. Taking its initial position as the point of zero potential energy and using energy methods, find (a) its initial speed, (b) its total mechanical energy, and (c) the ratio of its kinetic energy to the potential energy of the ball–Earth system when the ball is at an altitude of 10.0 m.

20. In the dangerous "sport" of bungee-jumping, a daring student jumps from a balloon with a specially designed

Figure P8.20   Bungee-jumping. *(Gamma)*

elastic cord attached to his ankles, as shown in Figure P8.20. The unstretched length of the cord is 25.0 m, the student weighs 700 N, and the balloon is 36.0 m above the surface of a river below. Assuming that Hooke's law describes the cord, calculate the required force constant if the student is to stop safely 4.00 m above the river.

21. Two masses are connected by a light string passing over a light frictionless pulley, as shown in Figure P8.21. The 5.00-kg mass is released from rest. Using the law of conservation of energy, (a) determine the speed of the 3.00-kg mass just as the 5.00-kg mass hits the ground and (b) find the maximum height to which the 3.00-kg mass rises.

22. Two masses are connected by a light string passing over a light frictionless pulley, as shown in Figure P8.21. The mass m_1 (which is greater than m_2) is released from rest. Using the law of conservation of energy, (a) determine the speed of m_2 just as m_1 hits the ground in terms of m_1, m_2, and h, and (b) find the maximum height to which m_2 rises.

$m_1 = 5.00$ kg

$m_2 = 3.00$ kg $h = 4.00$ m

Figure P8.21   Problems 21 and 22.

23. A 20.0-kg cannon ball is fired from a cannon with a muzzle speed of 1 000 m/s at an angle of 37.0° with the horizontal. A second ball is fired at an angle of 90.0°. Use the law of conservation of mechanical energy to find (a) the maximum height reached by each ball and (b) the total mechanical energy at the maximum height for each ball. Let $y = 0$ at the cannon.

24. A 2.00-kg ball is attached to the bottom end of a length of 10-lb (44.5-N) fishing line. The top end of the fishing line is held stationary. The ball is released from rest while the line is taut and horizontal ($\theta = 90.0°$). At what angle θ (measured from the vertical) will the fishing line break?

25. The circus apparatus known as the *trapeze* consists of a bar suspended by two parallel ropes, each of length ℓ. The trapeze allows circus performers to swing in a verti-

cal circular arc (Fig. P8.25). Suppose a performer with mass m and holding the bar steps off an elevated platform, starting from rest with the ropes at an angle of θ_i with respect to the vertical. Suppose the size of the performer's body is small compared with the length ℓ, that she does not pump the trapeze to swing higher, and that air resistance is negligible. (a) Show that when the ropes make an angle of θ with respect to the vertical, the performer must exert a force

$$F = mg \, (3 \cos \theta - 2 \cos \theta_i)$$

in order to hang on. (b) Determine the angle θ_i at which the force required to hang on at the bottom of the swing is twice the performer's weight.

θ

ℓ

Figure P8.25

26. After its release at the top of the first rise, a roller-coaster car moves freely with negligible friction. The roller coaster shown in Figure P8.26 has a circular loop of radius 20.0 m. The car barely makes it around the loop: At the top of the loop, the riders are upside down and feel weightless. (a) Find the speed of the roller coaster car at the top of the loop (position 3). Find the speed of the roller coaster car (b) at position 1 and (c) at position 2. (d) Find the difference in height between positions 1 and 4 if the speed at position 4 is 10.0 m/s.

27. A light rigid rod is 77.0 cm long. Its top end is pivoted on a low-friction horizontal axle. The rod hangs straight down at rest, with a small massive ball attached to its bottom end. You strike the ball, suddenly giving it a horizontal velocity so that it swings around in a full circle. What minimum speed at the bottom is required to make the ball go over the top of the circle?

Figure P8.26

Section 8.5 Work Done by Nonconservative Forces

28. A 70.0-kg diver steps off a 10.0-m tower and drops straight down into the water. If he comes to rest 5.00 m beneath the surface of the water, determine the average resistance force that the water exerts on the diver.

29. A force F_x, shown as a function of distance in Figure P8.29, acts on a 5.00-kg mass. If the particle starts from rest at $x = 0$ m, determine the speed of the particle at $x = 2.00$, 4.00, and 6.00 m.

Figure P8.29

30. A softball pitcher swings a ball of mass 0.250 kg around a vertical circular path of radius 60.0 cm before releasing it from her hand. The pitcher maintains a component of force on the ball of constant magnitude 30.0 N in the direction of motion around the complete path. The speed of the ball at the top of the circle is 15.0 m/s. If the ball is released at the bottom of the circle, what is its speed upon release?

WEB 31. The coefficient of friction between the 3.00-kg block and the surface in Figure P8.31 is 0.400. The system starts from rest. What is the speed of the 5.00-kg ball when it has fallen 1.50 m?

Figure P8.31

32. A 2 000-kg car starts from rest and coasts down from the top of a 5.00-m-long driveway that is sloped at an angle of 20.0° with the horizontal. If an average friction force of 4 000 N impedes the motion of the car, find the speed of the car at the bottom of the driveway.

33. A 5.00-kg block is set into motion up an inclined plane with an initial speed of 8.00 m/s (Fig. P8.33). The block comes to rest after traveling 3.00 m along the plane, which is inclined at an angle of 30.0° to the horizontal. For this motion determine (a) the change in the block's kinetic energy, (b) the change in the potential energy, and (c) the frictional force exerted on it (assumed to be constant). (d) What is the coefficient of kinetic friction?

Figure P8.33

34. A boy in a wheelchair (total mass, 47.0 kg) wins a race with a skateboarder. He has a speed of 1.40 m/s at the crest of a slope 2.60 m high and 12.4 m long. At the bottom of the slope, his speed is 6.20 m/s. If air resistance and rolling resistance can be modeled as a constant frictional force of 41.0 N, find the work he did in pushing forward on his wheels during the downhill ride.

35. A parachutist of mass 50.0 kg jumps out of a balloon at a height of 1 000 m and lands on the ground with a speed of 5.00 m/s. How much energy was lost to air friction during this jump?

36. An 80.0-kg sky diver jumps out of a balloon at an altitude of 1 000 m and opens the parachute at an altitude of 200.0 m. (a) Assuming that the total retarding force

on the diver is constant at 50.0 N with the parachute closed and constant at 3 600 N with the parachute open, what is the speed of the diver when he lands on the ground? (b) Do you think the sky diver will get hurt? Explain. (c) At what height should the parachute be opened so that the final speed of the sky diver when he hits the ground is 5.00 m/s? (d) How realistic is the assumption that the total retarding force is constant? Explain.

37. A toy cannon uses a spring to project a 5.30-g soft rubber ball. The spring is originally compressed by 5.00 cm and has a stiffness constant of 8.00 N/m. When it is fired, the ball moves 15.0 cm through the barrel of the cannon, and there is a constant frictional force of 0.032 0 N between the barrel and the ball. (a) With what speed does the projectile leave the barrel of the cannon? (b) At what point does the ball have maximum speed? (c) What is this maximum speed?

38. A 1.50-kg mass is held 1.20 m above a relaxed, massless vertical spring with a spring constant of 320 N/m. The mass is dropped onto the spring. (a) How far does it compress the spring? (b) How far would it compress the spring if the same experiment were performed on the Moon, where $g = 1.63$ m/s^2? (c) Repeat part (a), but this time assume that a constant air-resistance force of 0.700 N acts on the mass during its motion.

39. A 3.00-kg block starts at a height $h = 60.0$ cm on a plane that has an inclination angle of 30.0°, as shown in Figure P8.39. Upon reaching the bottom, the block slides along a horizontal surface. If the coefficient of friction on both surfaces is $\mu_k = 0.200$, how far does the block slide on the horizontal surface before coming to rest? (*Hint:* Divide the path into two straight-line parts.)

$m = 3.00$ kg

$h = 60.0$ cm

$\theta = 30.0°$

Figure P8.39

40. A 75.0-kg sky diver is falling with a terminal speed of 60.0 m/s. Determine the rate at which he is losing mechanical energy.

Section 8.6 Relationship Between Conservative Forces and Potential Energy

WEB 41. The potential energy of a two-particle system separated by a distance r is given by $U(r) = A/r$, where A is a constant. Find the radial force $\mathbf{F}_r$ that each particle exerts on the other.

42. A potential energy function for a two-dimensional force is of the form $U = 3x^3y - 7x$. Find the force that acts at the point (x, y).

(Optional)
Section 8.7 Energy Diagrams and the Equilibrium of a System

43. A particle moves along a line where the potential energy depends on its position r, as graphed in Figure P8.43. In the limit as r increases without bound, $U(r)$ approaches $+1$ J. (a) Identify each equilibrium position for this particle. Indicate whether each is a point of stable, unstable, or neutral equilibrium. (b) The particle will be bound if its total energy is in what range? Now suppose the particle has energy -3 J. Determine (c) the range of positions where it can be found, (d) its maximum kinetic energy, (e) the location at which it has maximum kinetic energy, and (f) its *binding energy*—that is, the additional energy that it would have to be given in order for it to move out to $r \rightarrow \infty$.

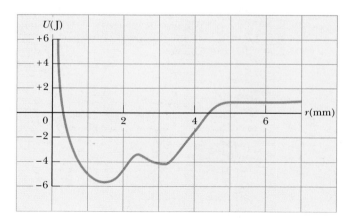

Figure P8.43

44. A right circular cone can be balanced on a horizontal surface in three different ways. Sketch these three equilibrium configurations and identify them as positions of stable, unstable, or neutral equilibrium.

45. For the potential energy curve shown in Figure P8.45, (a) determine whether the force F_x is positive, negative, or zero at the five points indicated. (b) Indicate points of stable, unstable, and neutral equilibrium. (c) Sketch the curve for F_x versus x from $x = 0$ to $x = 9.5$ m.

46. A hollow pipe has one or two weights attached to its inner surface, as shown in Figure P8.46. Characterize each configuration as being stable, unstable, or neutral equilibrium and explain each of your choices ("CM" indicates center of mass).

47. A particle of mass m is attached between two identical springs on a horizontal frictionless tabletop. The

Figure P8.45

(a) (b) (c)

Figure P8.46

springs have spring constant k, and each is initially un-stressed. (a) If the mass is pulled a distance x along a direction perpendicular to the initial configuration of the springs, as in Figure P8.47, show that the potential energy of the system is

$$U(x) = kx^2 + 2kL(L - \sqrt{x^2 + L^2})$$

(*Hint:* See Problem 66 in Chapter 7.) (b) Make a plot of $U(x)$ versus x and identify all equilibrium points. Assume that $L = 1.20$ m and $k = 40.0$ N/m. (c) If the mass is pulled 0.500 m to the right and then released, what is its speed when it reaches the equilibrium point $x = 0$?

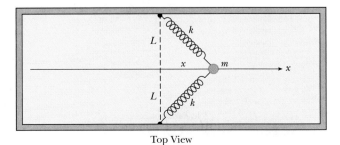

Top View

Figure P8.47

(Optional)

Section 8.9 Mass–Energy Equivalence

48. Find the energy equivalents of (a) an electron of mass 9.11×10^{-31} kg, (b) a uranium atom with a mass of 4.00×10^{-25} kg, (c) a paper clip of mass 2.00 g, and (d) the Earth (of mass 5.99×10^{24} kg).

49. The expression for the kinetic energy of a particle moving with speed v is given by Equation 7.19, which can be written as $K = \gamma mc^2 - mc^2$, where $\gamma = [1 - (v/c)^2]^{-1/2}$. The term γmc^2 is the total energy of the particle, and the term mc^2 is its rest energy. A proton moves with a speed of $0.990c$, where c is the speed of light. Find (a) its rest energy, (b) its total energy, and (c) its kinetic energy.

ADDITIONAL PROBLEMS

50. A block slides down a curved frictionless track and then up an inclined plane as in Figure P8.50. The coefficient of kinetic friction between the block and the incline is μ_k. Use energy methods to show that the maximum height reached by the block is

$$y_{max} = \frac{h}{1 + \mu_k \cot \theta}$$

Figure P8.50

51. Close to the center of a campus is a tall silo topped with a hemispherical cap. The cap is frictionless when wet. Someone has somehow balanced a pumpkin at the highest point. The line from the center of curvature of the cap to the pumpkin makes an angle $\theta_i = 0°$ with the vertical. On a rainy night, a breath of wind makes the pumpkin start sliding downward from rest. It loses contact with the cap when the line from the center of the hemisphere to the pumpkin makes a certain angle with the vertical; what is this angle?

52. A 200-g particle is released from rest at point Ⓐ along the horizontal diameter on the inside of a frictionless, hemispherical bowl of radius $R = 30.0$ cm (Fig. P8.52). Calculate (a) the gravitational potential energy when the particle is at point Ⓐ relative to point Ⓑ, (b) the kinetic energy of the particle at point Ⓑ, (c) its speed at point Ⓑ, and (d) its kinetic energy and the potential energy at point Ⓒ.

Figure P8.52   Problems 52 and 53.

Figure P8.56

WEB **53.** The particle described in Problem 52 (Fig. P8.52) is released from rest at Ⓐ, and the surface of the bowl is
rough. The speed of the particle at Ⓑ is 1.50 m/s.
(a) What is its kinetic energy at Ⓑ? (b) How much energy is lost owing to friction as the particle moves from
Ⓐ to Ⓑ? (c) Is it possible to determine μ from these results in any simple manner? Explain.

54. **Review Problem.**   The mass of a car is 1 500 kg. The
shape of the body is such that its aerodynamic drag coefficient is $D = 0.330$ and the frontal area is 2.50 m². Assuming that the drag force is proportional to v^2 and neglecting other sources of friction, calculate the power
the car requires to maintain a speed of 100 km/h as it
climbs a long hill sloping at 3.20°.

55. Make an order-of-magnitude estimate of your power
output as you climb stairs. In your solution, state the
physical quantities you take as data and the values you
measure or estimate for them. Do you consider your
peak power or your sustainable power?

56. A child's pogo stick (Fig. P8.56) stores energy in a
spring ($k = 2.50 \times 10^4$ N/m). At position Ⓐ ($x_A =
-0.100$ m), the spring compression is a maximum and
the child is momentarily at rest. At position Ⓑ ($x_B = 0$),
the spring is relaxed and the child is moving upward. At
position Ⓒ, the child is again momentarily at rest at the
top of the jump. Assuming that the combined mass of
the child and the pogo stick is 25.0 kg, (a) calculate the
total energy of the system if both potential energies are
zero at $x = 0$, (b) determine x_C, (c) calculate the speed
of the child at $x = 0$, (d) determine the value of x for

which the kinetic energy of the system is a maximum,
and (e) calculate the child's maximum upward speed.

57. A 10.0-kg block is released from point Ⓐ in Figure
P8.57. The track is frictionless except for the portion
between Ⓑ and Ⓒ, which has a length of 6.00 m. The
block travels down the track, hits a spring of force constant $k = 2\,250$ N/m, and compresses the spring
0.300 m from its equilibrium position before coming to
rest momentarily. Determine the coefficient of kinetic
friction between the block and the rough surface between Ⓑ and Ⓒ.

58. A 2.00-kg block situated on a rough incline is connected
to a spring of negligible mass having a spring constant
of 100 N/m (Fig. P8.58). The pulley is frictionless. The
block is released from rest when the spring is unstretched. The block moves 20.0 cm down the incline
before coming to rest. Find the coefficient of kinetic
friction between block and incline.

Figure P8.57

Figure P8.58 Problems 58 and 59.

59. **Review Problem.** Suppose the incline is frictionless for the system described in Problem 58 (see Fig. P8.58). The block is released from rest with the spring initially unstretched. (a) How far does it move down the incline before coming to rest? (b) What is its acceleration at its lowest point? Is the acceleration constant? (c) Describe the energy transformations that occur during the descent.

60. The potential energy function for a system is given by $U(x) = -x^3 + 2x^2 + 3x$. (a) Determine the force F_x as a function of x. (b) For what values of x is the force equal to zero? (c) Plot $U(x)$ versus x and F_x versus x, and indicate points of stable and unstable equilibrium.

61. A 20.0-kg block is connected to a 30.0-kg block by a string that passes over a frictionless pulley. The 30.0-kg block is connected to a spring that has negligible mass and a force constant of 250 N/m, as shown in Figure P8.61. The spring is unstretched when the system is as shown in the figure, and the incline is frictionless. The 20.0-kg block is pulled 20.0 cm down the incline (so that the 30.0-kg block is 40.0 cm above the floor) and is released from rest. Find the speed of each block when the 30.0-kg block is 20.0 cm above the floor (that is, when the spring is unstretched).

Figure P8.61

62. A 1.00-kg mass slides to the right on a surface having a coefficient of friction $\mu = 0.250$ (Fig. P8.62). The mass has a speed of $v_i = 3.00$ m/s when it makes contact with a light spring that has a spring constant $k = 50.0$ N/m. The mass comes to rest after the spring has been compressed a distance d. The mass is then forced toward the

left by the spring and continues to move in that direction beyond the spring's unstretched position. Finally, the mass comes to rest at a distance D to the left of the unstretched spring. Find (a) the distance of compression d, (b) the speed v of the mass at the unstretched position when the mass is moving to the left, and (c) the distance D between the unstretched spring and the point at which the mass comes to rest.

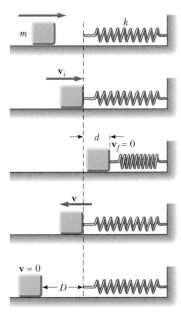

Figure P8.62

WEB 63. A block of mass 0.500 kg is pushed against a horizontal spring of negligible mass until the spring is compressed a distance Δx (Fig. P8.63). The spring constant is 450 N/m. When it is released, the block travels along a frictionless, horizontal surface to point B, at the bottom of a vertical circular track of radius $R = 1.00$ m, and continues to move up the track. The speed of the block at the bottom of the track is $v_B = 12.0$ m/s, and the block experiences an average frictional force of 7.00 N while sliding up the track. (a) What is Δx? (b) What speed do you predict for the block at the top of the track? (c) Does the block actually reach the top of the track, or does it fall off before reaching the top?

64. A uniform chain of length 8.00 m initially lies stretched out on a horizontal table. (a) If the coefficient of static friction between the chain and the table is 0.600, show that the chain will begin to slide off the table if at least 3.00 m of it hangs over the edge of the table. (b) Determine the speed of the chain as all of it leaves the table, given that the coefficient of kinetic friction between the chain and the table is 0.400.

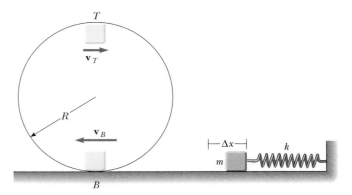

Figure P8.63

65. An object of mass m is suspended from a post on top of a cart by a string of length L as in Figure P8.65a. The cart and object are initially moving to the right at constant speed v_i. The cart comes to rest after colliding and sticking to a bumper as in Figure P8.65b, and the suspended object swings through an angle θ. (a) Show that the speed is $v_i = \sqrt{2gL(1 - \cos\theta)}$. (b) If $L = 1.20$ m and $\theta = 35.0°$, find the initial speed of the cart. (*Hint:* The force exerted by the string on the object does no work on the object.)

Figure P8.65

66. A child slides without friction from a height h along a curved water slide (Fig. P8.66). She is launched from a height $h/5$ into the pool. Determine her maximum airborne height y in terms of h and θ.

Figure P8.66

67. A ball having mass m is connected by a strong string of length L to a pivot point and held in place in a vertical position. A wind exerting constant force of magnitude F is blowing from left to right as in Figure P8.67a. (a) If the ball is released from rest, show that the maximum height H it reaches, as measured from its initial height, is

$$H = \frac{2L}{1 + (mg/F)^2}$$

Check that the above formula is valid both when $0 \le H \le L$ and when $L \le H \le 2L$. (*Hint:* First determine the potential energy associated with the *constant* wind force.) (b) Compute the value of H using the values $m = 2.00$ kg, $L = 2.00$ m, and $F = 14.7$ N. (c) Using these same values, determine the *equilibrium* height of the ball. (d) Could the equilibrium height ever be greater than L? Explain.

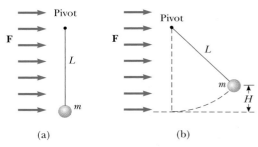

Figure P8.67

68. A ball is tied to one end of a string. The other end of the string is fixed. The ball is set in motion around a vertical circle without friction. At the top of the circle, the ball has a speed of $v_i = \sqrt{Rg}$, as shown in Figure P8.68. At what angle θ should the string be cut so that the ball will travel through the center of the circle?

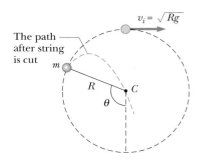

Figure P8.68

69. A ball at the end of a string whirls around in a vertical circle. If the ball's total energy remains constant, show that the tension in the string at the bottom is greater than the tension at the top by a value six times the weight of the ball.

70. A pendulum comprising a string of length L and a sphere swings in the vertical plane. The string hits a peg located a distance d below the point of suspension (Fig. P8.70). (a) Show that if the sphere is released from a height below that of the peg, it will return to this height after striking the peg. (b) Show that if the pendulum is released from the horizontal position ($\theta = 90°$) and is to swing in a complete circle centered on the peg, then the minimum value of d must be $3L/5$.

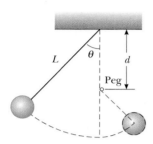

Figure P8.70

71. Jane, whose mass is 50.0 kg, needs to swing across a river (having width D) filled with man-eating crocodiles to save Tarzan from danger. However, she must swing into a wind exerting constant horizontal force $\mathbf{F}$ on a vine having length L and initially making an angle θ with the vertical (Fig. P8.71). Taking $D = 50.0$ m, $F = 110$ N, $L = 40.0$ m, and $\theta = 50.0°$, (a) with what minimum speed must Jane begin her swing to just make it to

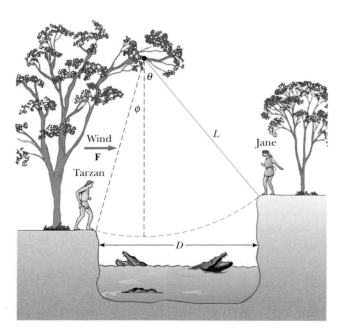

Figure P8.71

the other side? (*Hint:* First determine the potential energy associated with the wind force.) (b) Once the rescue is complete, Tarzan and Jane must swing back across the river. With what minimum speed must they begin their swing? Assume that Tarzan has a mass of 80.0 kg.

72. A child starts from rest and slides down the frictionless slide shown in Figure P8.72. In terms of R and H, at what height h will he lose contact with the section of radius R?

Figure P8.72

73. A 5.00-kg block free to move on a horizontal, frictionless surface is attached to one end of a light horizontal spring. The other end of the spring is fixed. The spring is compressed 0.100 m from equilibrium and is then released. The speed of the block is 1.20 m/s when it passes the equilibrium position of the spring. The same experiment is now repeated with the frictionless surface replaced by a surface for which $\mu_k = 0.300$. Determine the speed of the block at the equilibrium position of the spring.

74. A 50.0-kg block and a 100-kg block are connected by a string as in Figure P8.74. The pulley is frictionless and of negligible mass. The coefficient of kinetic friction between the 50.0-kg block and the incline is $\mu_k = 0.250$. Determine the change in the kinetic energy of the 50.0-kg block as it moves from Ⓐ to Ⓑ, a distance of 20.0 m.

Figure P8.74

ANSWERS TO QUICK QUIZZES

8.1 Yes, because we are free to choose any point whatsoever as our origin of coordinates, which is the $U_g = 0$ point. If the object is below the origin of coordinates that we choose, then $U_g < 0$ for the object–Earth system.

8.2 Yes, the total mechanical energy of the system is conserved because the only forces acting are conservative: the force of gravity and the spring force. There are two forms of potential energy: (1) gravitational potential energy and (2) elastic potential energy stored in the spring.

8.3 The first and third balls speed up after they are thrown, while the second ball initially slows down but then speeds up after reaching its peak. The paths of all three balls are parabolas, and the balls take different times to reach the ground because they have different initial velocities. However, all three balls have the same speed at the moment they hit the ground because all start with the same kinetic energy and undergo the same change in gravitational potential energy. In other words, $E_{total} = \frac{1}{2}mv^2 + mgh$ is the same for all three balls at the start of the motion.

8.4 Designate one object as No. 1 and the other as No. 2. The external force does work W_{app} on the system. If $W_{app} > 0$, then the system energy increases. If $W_{app} < 0$, then the system energy decreases. The effect of friction is to decrease the total system energy. Equation 8.15 then becomes

$$\Delta E = W_{app} - \Delta E_{friction}$$
$$= \Delta K + \Delta U$$
$$= [K_{1f} + K_{2f}) - (K_{1i} + K_{2i})]$$
$$+ [(U_{g1f} + U_{g2f} + U_{sf}) - (U_{g1i} + U_{g2i} + U_{si})]$$

You may find it easier to think of this equation with its terms in a different order, saying

total initial energy + net change = total final energy

$$K_{1i} + K_{2i} + U_{g1i} + U_{g2i} + U_{si} + W_{app} - f_k d =$$
$$K_{1f} + K_{2f} + U_{g1f} + U_{g2f} + U_{sf}$$

8.5 The slope of a $U(x)$-versus-x graph is by definition $dU(x)/dx$. From Equation 8.16, we see that this expression is equal to the negative of the x component of the conservative force acting on an object that is part of the system.

PUZZLER

Airbags have saved countless lives by reducing the forces exerted on vehicle occupants during collisions. How can airbags change the force needed to bring a person from a high speed to a complete stop? Why are they usually safer than seat belts alone? *(Courtesy of Saab)*

c h a p t e r

Linear Momentum and Collisions

Consider what happens when a golf ball is struck by a club. The ball is given a very large initial velocity as a result of the collision; consequently, it is able to travel more than 100 m through the air. The ball experiences a large acceleration. Furthermore, because the ball experiences this acceleration over a very short time interval, the average force exerted on it during the collision is very great. According to Newton's third law, the ball exerts on the club a reaction force that is equal in magnitude to and opposite in direction to the force exerted by the club on the ball. This reaction force causes the club to accelerate. Because the club is much more massive than the ball, however, the acceleration of the club is much less than the acceleration of the ball.

One of the main objectives of this chapter is to enable you to understand and analyze such events. As a first step, we introduce the concept of *momentum,* which is useful for describing objects in motion and as an alternate and more general means of applying Newton's laws. For example, a very massive football player is often said to have a great deal of momentum as he runs down the field. A much less massive player, such as a halfback, can have equal or greater momentum if his speed is greater than that of the more massive player. This follows from the fact that momentum is defined as the product of mass and velocity. The concept of momentum leads us to a second conservation law, that of conservation of momentum. This law is especially useful for treating problems that involve collisions between objects and for analyzing rocket propulsion. The concept of the center of mass of a system of particles also is introduced, and we shall see that the motion of a system of particles can be described by the motion of one representative particle located at the center of mass.

9.1 LINEAR MOMENTUM AND ITS CONSERVATION

In the preceding two chapters we studied situations too complex to analyze easily with Newton's laws. In fact, Newton himself used a form of his second law slightly different from $\Sigma \mathbf{F} = m\mathbf{a}$ (Eq. 5.2) — a form that is considerably easier to apply in complicated circumstances. Physicists use this form to study everything from subatomic particles to rocket propulsion. In studying situations such as these, it is often useful to know both something about the object and something about its motion. We start by defining a new term that incorporates this information:

Definition of linear momentum of a particle

> The **linear momentum** of a particle of mass m moving with a velocity $\mathbf{v}$ is defined to be the product of the mass and velocity:
>
> $$\mathbf{p} \equiv m\mathbf{v} \tag{9.1}$$

6.2 Linear momentum is a vector quantity because it equals the product of a scalar quantity m and a vector quantity $\mathbf{v}$. Its direction is along $\mathbf{v}$, it has dimensions ML/T, and its SI unit is kg·m/s.

If a particle is moving in an arbitrary direction, $\mathbf{p}$ must have three components, and Equation 9.1 is equivalent to the component equations

$$p_x = mv_x \qquad p_y = mv_y \qquad p_z = mv_z \tag{9.2}$$

As you can see from its definition, the concept of momentum provides a quantitative distinction between heavy and light particles moving at the same velocity. For example, the momentum of a bowling ball moving at 10 m/s is much greater than that of a tennis ball moving at the same speed. Newton called the product $m\mathbf{v}$

quantity of motion; this is perhaps a more graphic description than our present-day word *momentum,* which comes from the Latin word for movement.

Two objects have equal kinetic energies. How do the magnitudes of their momenta compare? (a) $p_1 < p_2$, (b) $p_1 = p_2$, (c) $p_1 > p_2$, (d) not enough information to tell.

Using Newton's second law of motion, we can relate the linear momentum of a particle to the resultant force acting on the particle: **The time rate of change of the linear momentum of a particle is equal to the net force acting on the particle:**

$$\sum \mathbf{F} = \frac{d\mathbf{p}}{dt} = \frac{d(m\mathbf{v})}{dt} \qquad \textbf{(9.3)}$$

| Newton's second law for a particle |

In addition to situations in which the velocity vector varies with time, we can use Equation 9.3 to study phenomena in which the mass changes. The real value of Equation 9.3 as a tool for analysis, however, stems from the fact that when the net force acting on a particle is zero, the time derivative of the momentum of the particle is zero, and therefore its linear momentum[1] is constant. Of course, if the particle is *isolated,* then by necessity $\sum \mathbf{F} = 0$ and $\mathbf{p}$ remains unchanged. This means that $\mathbf{p}$ is conserved. Just as the law of conservation of energy is useful in solving complex motion problems, the law of conservation of momentum can greatly simplify the analysis of other types of complicated motion.

Conservation of Momentum for a Two-Particle System

Consider two particles 1 and 2 that can interact with each other but are isolated from their surroundings (Fig. 9.1). That is, the particles may exert a force on each other, but no external forces are present. It is important to note the impact of Newton's third law on this analysis. If an internal force from particle 1 (for example, a gravitational force) acts on particle 2, then there must be a second internal force—equal in magnitude but opposite in direction—that particle 2 exerts on particle 1.

Suppose that at some instant, the momentum of particle 1 is $\mathbf{p}_1$ and that of particle 2 is $\mathbf{p}_2$. Applying Newton's second law to each particle, we can write

$$\mathbf{F}_{21} = \frac{d\mathbf{p}_1}{dt} \qquad \text{and} \qquad \mathbf{F}_{12} = \frac{d\mathbf{p}_2}{dt}$$

where $\mathbf{F}_{21}$ is the force exerted by particle 2 on particle 1 and $\mathbf{F}_{12}$ is the force exerted by particle 1 on particle 2. Newton's third law tells us that $\mathbf{F}_{12}$ and $\mathbf{F}_{21}$ are equal in magnitude and opposite in direction. That is, they form an action–reaction pair $\mathbf{F}_{12} = -\mathbf{F}_{21}$. We can express this condition as

$$\mathbf{F}_{21} + \mathbf{F}_{12} = 0$$

or as

$$\frac{d\mathbf{p}_1}{dt} + \frac{d\mathbf{p}_2}{dt} = \frac{d}{dt}(\mathbf{p}_1 + \mathbf{p}_2) = 0$$

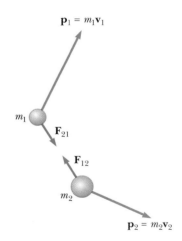

Figure 9.1 At some instant, the momentum of particle 1 is $\mathbf{p}_1 = m_1\mathbf{v}_1$ and the momentum of particle 2 is $\mathbf{p}_2 = m_2\mathbf{v}_2$. Note that $\mathbf{F}_{12} = -\mathbf{F}_{21}$. The total momentum of the system $\mathbf{p}_{\text{tot}}$ is equal to the vector sum $\mathbf{p}_1 + \mathbf{p}_2$.

[1] In this chapter, the terms *momentum* and *linear momentum* have the same meaning. Later, in Chapter 11, we shall use the term *angular momentum* when dealing with rotational motion.

Because the time derivative of the total momentum $\mathbf{p}_{tot} = \mathbf{p}_1 + \mathbf{p}_2$ is *zero*, we conclude that the *total* momentum of the system must remain constant:

$$\mathbf{p}_{tot} = \sum_{\text{system}} \mathbf{p} = \mathbf{p}_1 + \mathbf{p}_2 = \text{constant} \qquad (9.4)$$

or, equivalently,

$$\mathbf{p}_{1i} + \mathbf{p}_{2i} = \mathbf{p}_{1f} + \mathbf{p}_{2f} \qquad (9.5)$$

where $\mathbf{p}_{1i}$ and $\mathbf{p}_{2i}$ are the initial values and $\mathbf{p}_{1f}$ and $\mathbf{p}_{2f}$ the final values of the momentum during the time interval dt over which the reaction pair interacts. Equation 9.5 in component form demonstrates that the total momenta in the x, y, and z directions are all independently conserved:

$$\sum_{\text{system}} p_{ix} = \sum_{\text{system}} p_{fx} \qquad \sum_{\text{system}} p_{iy} = \sum_{\text{system}} p_{fy} \qquad \sum_{\text{system}} p_{iz} = \sum_{\text{system}} p_{fz} \qquad (9.6)$$

This result, known as the **law of conservation of linear momentum,** can be extended to any number of particles in an isolated system. It is considered one of the most important laws of mechanics. We can state it as follows:

Conservation of momentum

> Whenever two or more particles in an isolated system interact, the total momentum of the system remains constant.

This law tells us that **the total momentum of an isolated system at all times equals its initial momentum.**

Notice that we have made no statement concerning the nature of the forces acting on the particles of the system. The only requirement is that the forces must be *internal* to the system.

Quick Quiz 9.2

Your physical education teacher throws a baseball to you at a certain speed, and you catch it. The teacher is next going to throw you a medicine ball whose mass is ten times the mass of the baseball. You are given the following choices: You can have the medicine ball thrown with (a) the same speed as the baseball, (b) the same momentum, or (c) the same kinetic energy. Rank these choices from easiest to hardest to catch.

EXAMPLE 9.1   The Floating Astronaut

A SkyLab astronaut discovered that while concentrating on writing some notes, he had gradually floated to the middle of an open area in the spacecraft. Not wanting to wait until he floated to the opposite side, he asked his colleagues for a push. Laughing at his predicament, they decided not to help, and so he had to take off his uniform and throw it in one direction so that he would be propelled in the opposite direction. Estimate his resulting velocity.

Figure 9.2 A hapless astronaut has discarded his uniform to get somewhere.

Solution We begin by making some reasonable guesses of relevant data. Let us assume we have a 70-kg astronaut who threw his 1-kg uniform at a speed of 20 m/s. For convenience, we set the positive direction of the x axis to be the direction of the throw (Fig. 9.2). Let us also assume that the x axis is tangent to the circular path of the spacecraft.

We take the system to consist of the astronaut and the uniform. Because of the gravitational force (which keeps the astronaut, his uniform, and the entire spacecraft in orbit), the system is not really isolated. However, this force is directed perpendicular to the motion of the system. Therefore, momentum is constant in the x direction because there are no external forces in this direction.

The total momentum of the system before the throw is zero ($m_1\mathbf{v}_{1i} + m_2\mathbf{v}_{2i} = 0$). Therefore, the total momentum after the throw must be zero; that is,

$$m_1\mathbf{v}_{1f} + m_2\mathbf{v}_{2f} = 0$$

With $m_1 = 70$ kg, $\mathbf{v}_{2f} = 20\mathbf{i}$ m/s, and $m_2 = 1$ kg, solving for $\mathbf{v}_{1f}$, we find the recoil velocity of the astronaut to be

$$\mathbf{v}_{1f} = -\frac{m_2}{m_1}\mathbf{v}_{2f} = -\left(\frac{1\text{ kg}}{70\text{ kg}}\right)(20\mathbf{i}\text{ m/s}) = \boxed{-0.3\mathbf{i}\text{ m/s}}$$

The negative sign for $\mathbf{v}_{1f}$ indicates that the astronaut is moving to the left after the throw, in the direction opposite the direction of motion of the uniform, in accordance with Newton's third law. Because the astronaut is much more massive than his uniform, his acceleration and consequent velocity are much smaller than the acceleration and velocity of the uniform.

EXAMPLE 9.2 ▶ Breakup of a Kaon at Rest

One type of nuclear particle, called the *neutral kaon* (K^0), breaks up into a pair of other particles called *pions* (π^+ and π^-) that are oppositely charged but equal in mass, as illustrated in Figure 9.3. Assuming the kaon is initially at rest, prove that the two pions must have momenta that are equal in magnitude and opposite in direction.

Solution   The breakup of the kaon can be written

$$\text{K}^0 \longrightarrow \pi^+ + \pi^-$$

If we let $\mathbf{p}^+$ be the momentum of the positive pion and $\mathbf{p}^-$ the momentum of the negative pion, the final momentum of the system consisting of the two pions can be written

$$\mathbf{p}_f = \mathbf{p}^+ + \mathbf{p}^-$$

Because the kaon is at rest before the breakup, we know that $\mathbf{p}_i = 0$. Because momentum is conserved, $\mathbf{p}_i = \mathbf{p}_f = 0$, so that $\mathbf{p}^+ + \mathbf{p}^- = 0$, or

$$\mathbf{p}^+ = -\mathbf{p}^-$$

The important point behind this problem is that even though it deals with objects that are very different from those in the preceding example, the physics is identical: Linear momentum is conserved in an isolated system.

Figure 9.3   A kaon at rest breaks up spontaneously into a pair of oppositely charged pions. The pions move apart with momenta that are equal in magnitude but opposite in direction.

9.2 ▶ IMPULSE AND MOMENTUM

As we have seen, the momentum of a particle changes if a net force acts on the particle. Knowing the change in momentum caused by a force is useful in solving some types of problems. To begin building a better understanding of this important concept, let us assume that a single force $\mathbf{F}$ acts on a particle and that this force may vary with time. According to Newton's second law, $\mathbf{F} = d\mathbf{p}/dt$, or

$$d\mathbf{p} = \mathbf{F}\,dt \tag{9.7}$$

We can integrate[2] this expression to find the change in the momentum of a particle when the force acts over some time interval. If the momentum of the particle

---

[2]Note that here we are integrating force with respect to time. Compare this with our efforts in Chapter 7, where we integrated force with respect to position to express the work done by the force.

changes from $\mathbf{p}_i$ at time t_i to $\mathbf{p}_f$ at time t_f, integrating Equation 9.7 gives

$$\Delta \mathbf{p} = \mathbf{p}_f - \mathbf{p}_i = \int_{t_i}^{t_f} \mathbf{F} \, dt \tag{9.8}$$

To evaluate the integral, we need to know how the force varies with time. The quantity on the right side of this equation is called the **impulse** of the force $\mathbf{F}$ acting on a particle over the time interval $\Delta t = t_f - t_i$. Impulse is a vector defined by

Impulse of a force

$$\mathbf{I} \equiv \int_{t_i}^{t_f} \mathbf{F} \, dt = \Delta \mathbf{p} \tag{9.9}$$

Impulse–momentum theorem

> The impulse of the force $\mathbf{F}$ acting on a particle equals the change in the momentum of the particle caused by that force.

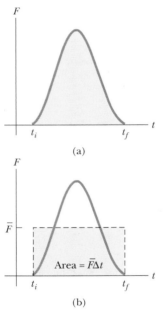

Figure 9.4 (a) A force acting on a particle may vary in time. The impulse imparted to the particle by the force is the area under the force versus time curve. (b) In the time interval Δt, the time-averaged force (horizontal dashed line) gives the same impulse to a particle as does the time-varying force described in part (a).

This statement, known as the **impulse–momentum theorem,**[3] is equivalent to Newton's second law. From this definition, we see that impulse is a vector quantity having a magnitude equal to the area under the force–time curve, as described in Figure 9.4a. In this figure, it is assumed that the force varies in time in the general manner shown and is nonzero in the time interval $\Delta t = t_f - t_i$. The direction of the impulse vector is the same as the direction of the change in momentum. Impulse has the dimensions of momentum—that is, ML/T. Note that impulse is *not* a property of a particle; rather, it is a measure of the degree to which an external force changes the momentum of the particle. Therefore, when we say that an impulse is given to a particle, we mean that momentum is transferred from an external agent to that particle.

Because the force imparting an impulse can generally vary in time, it is convenient to define a time-averaged force

$$\overline{\mathbf{F}} \equiv \frac{1}{\Delta t} \int_{t_i}^{t_f} \mathbf{F} \, dt \tag{9.10}$$

where $\Delta t = t_f - t_i$. (This is an application of the mean value theorem of calculus.) Therefore, we can express Equation 9.9 as

$$\mathbf{I} \equiv \overline{\mathbf{F}} \, \Delta t \tag{9.11}$$

This time-averaged force, described in Figure 9.4b, can be thought of as the constant force that would give to the particle in the time interval Δt the same impulse that the time-varying force gives over this same interval.

In principle, if $\mathbf{F}$ is known as a function of time, the impulse can be calculated from Equation 9.9. The calculation becomes especially simple if the force acting on the particle is constant. In this case, $\overline{\mathbf{F}} = \mathbf{F}$ and Equation 9.11 becomes

$$\mathbf{I} = \mathbf{F} \, \Delta t \tag{9.12}$$

In many physical situations, we shall use what is called the **impulse approximation, in which we assume that one of the forces exerted on a particle acts for a short time but is much greater than any other force present.** This approximation is especially useful in treating collisions in which the duration of the

---

[3]Although we assumed that only a single force acts on the particle, the impulse–momentum theorem is valid when several forces act; in this case, we replace $\mathbf{F}$ in Equation 9.9 with $\Sigma\mathbf{F}$.

During the brief time the club is in contact with the ball, the ball gains momentum as a result of the collision, and the club loses the same amount of momentum. *(Courtesy of Michel Hans/Photo Researchers, Inc.)*

collision is very short. When this approximation is made, we refer to the force as an *impulsive force.* For example, when a baseball is struck with a bat, the time of the collision is about 0.01 s and the average force that the bat exerts on the ball in this time is typically several thousand newtons. Because this is much greater than the magnitude of the gravitational force, the impulse approximation justifies our ignoring the weight of the ball and bat. When we use this approximation, it is important to remember that $\mathbf{p}_i$ and $\mathbf{p}_f$ represent the momenta *immediately* before and after the collision, respectively. Therefore, in any situation in which it is proper to use the impulse approximation, the particle moves very little during the collision.

QuickLab

If you can find someone willing, play catch with an egg. What is the best way to move your hands so that the egg does not break when you change its momentum to zero?

Quick Quiz 9.3

Two objects are at rest on a frictionless surface. Object 1 has a greater mass than object 2. When a force is applied to object 1, it accelerates through a distance d. The force is removed from object 1 and is applied to object 2. At the moment when object 2 has accelerated through the same distance d, which statements are true? (a) $p_1 < p_2$, (b) $p_1 = p_2$, (c) $p_1 > p_2$, (d) $K_1 < K_2$, (e) $K_1 = K_2$, (f) $K_1 > K_2$.

EXAMPLE 9.3   Teeing Off

A golf ball of mass 50 g is struck with a club (Fig. 9.5). The force exerted on the ball by the club varies from zero, at the instant before contact, up to some maximum value (at which the ball is deformed) and then back to zero when the ball leaves the club. Thus, the force–time curve is qualitatively described by Figure 9.4. Assuming that the ball travels 200 m, estimate the magnitude of the impulse caused by the collision.

Solution   Let us use Ⓐ to denote the moment when the club first contacts the ball, Ⓑ to denote the moment when

the club loses contact with the ball as the ball starts on its trajectory, and Ⓒ to denote its landing. Neglecting air resistance, we can use Equation 4.14 for the range of a projectile:

$$R = x_C = \frac{v_B^2}{g} \sin 2\theta_B$$

Let us assume that the launch angle θ_B is 45°, the angle that provides the maximum range for any given launch velocity. This assumption gives $\sin 2\theta_B = 1$, and the launch velocity of

the ball is

$$v_B = \sqrt{x_{CG}g} = \sqrt{(200 \text{ m})(9.80 \text{ m/s}^2)} = 44 \text{ m/s}$$

Considering the time interval for the collision, $v_i = v_A = 0$ and $v_f = v_B$ for the ball. Hence, the magnitude of the impulse imparted to the ball is

$$I = \Delta p = mv_B - mv_A = (50 \times 10^{-3} \text{ kg})(44 \text{ m/s}) - 0$$

$$= \boxed{2.2 \text{ kg} \cdot \text{m/s}}$$

Exercise   If the club is in contact with the ball for a time of 4.5×10^{-4} s, estimate the magnitude of the average force exerted by the club on the ball.

Answer 4.9×10^3 N, a value that is extremely large when compared with the weight of the ball, 0.49 N.

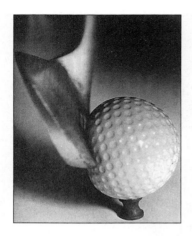

Figure 9.5   A golf ball being struck by a club. *(© Harold E. Edgerton/ Courtesy of Palm Press, Inc.)*

EXAMPLE 9.4 ▸ How Good Are the Bumpers?

In a particular crash test, an automobile of mass 1 500 kg collides with a wall, as shown in Figure 9.6. The initial and final velocities of the automobile are $\mathbf{v}_i = -15.0\mathbf{i}$ m/s and $\mathbf{v}_f = 2.60\mathbf{i}$ m/s, respectively. If the collision lasts for 0.150 s, find the impulse caused by the collision and the average force exerted on the automobile.

Solution   Let us assume that the force exerted on the car by the wall is large compared with other forces on the car so that we can apply the impulse approximation. Furthermore, we note that the force of gravity and the normal force exerted by the road on the car are perpendicular to the motion and therefore do not affect the horizontal momentum.

The initial and final momenta of the automobile are

$$\mathbf{p}_i = m\mathbf{v}_i = (1\,500 \text{ kg})(-15.0\mathbf{i} \text{ m/s}) = -2.25 \times 10^4\mathbf{i} \text{ kg} \cdot \text{m/s}$$

$$\mathbf{p}_f = m\mathbf{v}_f = (1\,500 \text{ kg})(2.60\mathbf{i} \text{ m/s}) = 0.39 \times 10^4\mathbf{i} \text{ kg} \cdot \text{m/s}$$

Hence, the impulse is

$$\mathbf{I} = \Delta\mathbf{p} = \mathbf{p}_f - \mathbf{p}_i = 0.39 \times 10^4\mathbf{i} \text{ kg} \cdot \text{m/s} - (-2.25 \times 10^4\mathbf{i} \text{ kg} \cdot \text{m/s})$$

$$\mathbf{I} = \boxed{2.64 \times 10^4\mathbf{i} \text{ kg} \cdot \text{m/s}}$$

The average force exerted on the automobile is

$$\overline{\mathbf{F}} = \frac{\Delta\mathbf{p}}{\Delta t} = \frac{2.64 \times 10^4\mathbf{i} \text{ kg} \cdot \text{m/s}}{0.150 \text{ s}} = \boxed{1.76 \times 10^5\mathbf{i} \text{ N}}$$

Before

−15.0 m/s

After

2.60 m/s

Figure 9.6   (a) This car's momentum changes as a result of its collision with the wall. (b) In a crash test, much of the car's initial kinetic energy is transformed into energy used to damage the car. *(b, courtesy of General Motors)*

(a)

(b)

Note that the magnitude of this force is large compared with the weight of the car ($mg = 1.47 \times 10^4$ N), which justifies our initial assumption. Of note in this problem is how the signs of the velocities indicated the reversal of directions. What would the mathematics be describing if both the initial and final velocities had the same sign?

Quick Quiz 9.4

Rank an automobile dashboard, seatbelt, and airbag in terms of (a) the impulse and (b) the average force they deliver to a front-seat passenger during a collision.

9.3 COLLISIONS

6.5 & 6.6 In this section we use the law of conservation of linear momentum to describe what happens when two particles collide. We use the term **collision** to represent the event of two particles' coming together for a short time and thereby producing impulsive forces on each other. **These forces are assumed to be much greater than any external forces present.**

(a)

A collision may entail physical contact between two macroscopic objects, as described in Figure 9.7a, but the notion of what we mean by collision must be generalized because "physical contact" on a submicroscopic scale is ill-defined and hence meaningless. To understand this, consider a collision on an atomic scale (Fig. 9.7b), such as the collision of a proton with an alpha particle (the nucleus of a helium atom). Because the particles are both positively charged, they never come into physical contact with each other; instead, they repel each other because of the strong electrostatic force between them at close separations. When two particles 1 and 2 of masses m_1 and m_2 collide as shown in Figure 9.7, the impulsive forces may vary in time in complicated ways, one of which is described in Figure 9.8. If $\mathbf{F}_{21}$ is the force exerted by particle 2 on particle 1, and if we assume that no external forces act on the particles, then the change in momentum of particle 1 due to the collision is given by Equation 9.8:

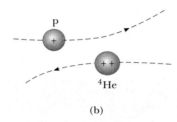

(b)

Figure 9.7 (a) The collision between two objects as the result of direct contact. (b) The "collision" between two charged particles.

$$\Delta\mathbf{p}_1 = \int_{t_i}^{t_f} \mathbf{F}_{21}\, dt$$

Likewise, if $\mathbf{F}_{12}$ is the force exerted by particle 1 on particle 2, then the change in momentum of particle 2 is

$$\Delta\mathbf{p}_2 = \int_{t_i}^{t_f} \mathbf{F}_{12}\, dt$$

From Newton's third law, we conclude that

$$\Delta\mathbf{p}_1 = -\Delta\mathbf{p}_2$$

$$\Delta\mathbf{p}_1 + \Delta\mathbf{p}_2 = 0$$

Because the total momentum of the system is $\mathbf{p}_{\text{system}} = \mathbf{p}_1 + \mathbf{p}_2$, we conclude that the *change* in the momentum of the system due to the collision is zero:

$$\mathbf{p}_{\text{system}} = \mathbf{p}_1 + \mathbf{p}_2 = \text{constant}$$

This is precisely what we expect because no external forces are acting on the system (see Section 9.2). Because the impulsive forces are internal, they do not change the total momentum of the system (only external forces can do that).

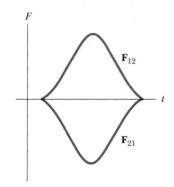

Figure 9.8 The impulse force as a function of time for the two colliding particles described in Figure 9.7a. Note that $\mathbf{F}_{12} = -\mathbf{F}_{21}$.

Momentum is conserved for any collision

Therefore, we conclude that **the total momentum of an isolated system just before a collision equals the total momentum of the system just after the collision.**

EXAMPLE 9.5   Carry Collision Insurance!

A car of mass 1800 kg stopped at a traffic light is struck from the rear by a 900-kg car, and the two become entangled. If the smaller car was moving at 20.0 m/s before the collision, what is the velocity of the entangled cars after the collision?

Solution   We can guess that the final speed is less than 20.0 m/s, the initial speed of the smaller car. The total momentum of the system (the two cars) before the collision must equal the total momentum immediately after the collision because momentum is conserved in any type of collision. The magnitude of the total momentum before the collision is equal to that of the smaller car because the larger car is initially at rest:

$$p_i = m_1 v_{1i} = (900 \text{ kg})(20.0 \text{ m/s}) = 1.80 \times 10^4 \text{ kg} \cdot \text{m/s}$$

After the collision, the magnitude of the momentum of the entangled cars is

$$p_f = (m_1 + m_2)v_f = (2\,700 \text{ kg})v_f$$

Equating the momentum before to the momentum after and solving for v_f, the final velocity of the entangled cars, we have

$$v_f = \frac{p_i}{m_1 + m_2} = \frac{1.80 \times 10^4 \text{ kg} \cdot \text{m/s}}{2\,700 \text{ kg}} = \boxed{6.67 \text{ m/s}}$$

The direction of the final velocity is the same as the velocity of the initially moving car.

Exercise   What would be the final speed if the two cars each had a mass of 900 kg?

Answer   10.0 m/s.

When the bowling ball and pin collide, part of the ball's momentum is transferred to the pin. Consequently, the pin acquires momentum and kinetic energy, and the ball loses momentum and kinetic energy. However, the total momentum of the system (ball and pin) remains constant. *(Ben Rose/The Image Bank)*

Elastic collision

Quick Quiz 9.5

As a ball falls toward the Earth, the ball's momentum increases because its speed increases. Does this mean that momentum is not conserved in this situation?

Quick Quiz 9.6

A skater is using very low-friction rollerblades. A friend throws a Frisbee straight at her. In which case does the Frisbee impart the greatest impulse to the skater: (a) she catches the Frisbee and holds it, (b) she catches it momentarily but drops it, (c) she catches it and at once throws it back to her friend?

9.4   ELASTIC AND INELASTIC COLLISIONS IN ONE DIMENSION

As we have seen, momentum is conserved in any collision in which external forces are negligible. In contrast, kinetic energy may or may not be constant, depending on the type of collision. In fact, whether or not kinetic energy is the same before and after the collision is used to classify collisions as being either elastic or inelastic.

An **elastic collision** between two objects is one in which *total kinetic energy (as well as total momentum) is the same before and after the collision.* Billiard-ball collisions and the collisions of air molecules with the walls of a container at ordinary temperatures are approximately elastic. Truly elastic collisions do occur, however, between atomic and subatomic particles. Collisions between certain objects in the macroscopic world, such as billiard-ball collisions, are only approximately elastic because some deformation and loss of kinetic energy take place.

An **inelastic collision** is one in which *total kinetic energy is not the same before and after the collision (even though momentum is constant).* Inelastic collisions are of two types. When the colliding objects stick together after the collision, as happens when a meteorite collides with the Earth, the collision is called **perfectly inelastic.** When the colliding objects do not stick together, but some kinetic energy is lost, as in the case of a rubber ball colliding with a hard surface, the collision is called **inelastic** (with no modifying adverb). For example, when a rubber ball collides with a hard surface, the collision is inelastic because some of the kinetic energy of the ball is lost when the ball is deformed while it is in contact with the surface.

In most collisions, kinetic energy is *not* the same before and after the collision because some of it is converted to internal energy, to elastic potential energy when the objects are deformed, and to rotational energy. Elastic and perfectly inelastic collisions are limiting cases; most collisions fall somewhere between them.

In the remainder of this section, we treat collisions in one dimension and consider the two extreme cases—perfectly inelastic and elastic collisions. The important distinction between these two types of collisions is that **momentum is constant in all collisions, but kinetic energy is constant only in elastic collisions.**

Perfectly Inelastic Collisions

Consider two particles of masses m_1 and m_2 moving with initial velocities $\mathbf{v}_{1i}$ and $\mathbf{v}_{2i}$ along a straight line, as shown in Figure 9.9. The two particles collide head-on, stick together, and then move with some common velocity $\mathbf{v}_f$ after the collision. Because momentum is conserved in any collision, we can say that the total momentum before the collision equals the total momentum of the composite system after the collision:

$$m_1\mathbf{v}_{1i} + m_2\mathbf{v}_{2i} = (m_1 + m_2)\mathbf{v}_f \tag{9.13}$$

$$\mathbf{v}_f = \frac{m_1\mathbf{v}_{1i} + m_2\mathbf{v}_{2i}}{m_1 + m_2} \tag{9.14}$$

Quick Quiz 9.7

Which is worse, crashing into a brick wall at 40 mi/h or crashing head-on into an oncoming car that is identical to yours and also moving at 40 mi/h?

Elastic Collisions

Now consider two particles that undergo an elastic head-on collision (Fig. 9.10). In this case, both momentum and kinetic energy are conserved; therefore, we have

6.6

$$m_1 v_{1i} + m_2 v_{2i} = m_1 v_{1f} + m_2 v_{2f} \tag{9.15}$$

$$\tfrac{1}{2}m_1 v_{1i}{}^2 + \tfrac{1}{2}m_2 v_{2i}{}^2 = \tfrac{1}{2}m_1 v_{1f}{}^2 + \tfrac{1}{2}m_2 v_{2f}{}^2 \tag{9.16}$$

Because all velocities in Figure 9.10 are either to the left or the right, they can be represented by the corresponding speeds along with algebraic signs indicating direction. We shall indicate v as positive if a particle moves to the right and negative

Inelastic collision

QuickLab

Hold a Ping-Pong ball or tennis ball on top of a basketball. Drop them both at the same time so that the basketball hits the floor, bounces up, and hits the smaller falling ball. What happens and why?

Before collision

(a)

After collision

(b)

Figure 9.9 Schematic representation of a perfectly inelastic head-on collision between two particles: (a) before collision and (b) after collision.

Before collision

After collision

Figure 9.10   Schematic representation of an elastic head-on collision between two particles: (a) before collision and (b) after collision.

if it moves to the left. As has been seen in earlier chapters, it is common practice to call these values "speed" even though this term technically refers to the magnitude of the velocity vector, which does not have an algebraic sign.

In a typical problem involving elastic collisions, there are two unknown quantities, and Equations 9.15 and 9.16 can be solved simultaneously to find these. An alternative approach, however—one that involves a little mathematical manipulation of Equation 9.16—often simplifies this process. To see how, let us cancel the factor $\frac{1}{2}$ in Equation 9.16 and rewrite it as

$$m_1(v_{1i}^2 - v_{1f}^2) = m_2(v_{2f}^2 - v_{2i}^2)$$

and then factor both sides:

$$m_1(v_{1i} - v_{1f})(v_{1i} + v_{1f}) = m_2(v_{2f} - v_{2i})(v_{2f} + v_{2i}) \tag{9.17}$$

Next, let us separate the terms containing m_1 and m_2 in Equation 9.15 to get

$$m_1(v_{1i} - v_{1f}) = m_2(v_{2f} - v_{2i}) \tag{9.18}$$

To obtain our final result, we divide Equation 9.17 by Equation 9.18 and get

$$v_{1i} + v_{1f} = v_{2f} + v_{2i}$$

$$v_{1i} - v_{2i} = -(v_{1f} - v_{2f}) \tag{9.19}$$

This equation, in combination with Equation 9.15, can be used to solve problems dealing with elastic collisions. According to Equation 9.19, the relative speed of the two particles before the collision $v_{1i} - v_{2i}$ equals the negative of their relative speed after the collision, $-(v_{1f} - v_{2f})$.

Suppose that the masses and initial velocities of both particles are known. Equations 9.15 and 9.19 can be solved for the final speeds in terms of the initial speeds because there are two equations and two unknowns:

Elastic collision: relationships between final and initial velocities

$$v_{1f} = \left(\frac{m_1 - m_2}{m_1 + m_2}\right)v_{1i} + \left(\frac{2m_2}{m_1 + m_2}\right)v_{2i} \tag{9.20}$$

$$v_{2f} = \left(\frac{2m_1}{m_1 + m_2}\right)v_{1i} + \left(\frac{m_2 - m_1}{m_1 + m_2}\right)v_{2i} \tag{9.21}$$

It is important to remember that the appropriate signs for v_{1i} and v_{2i} must be included in Equations 9.20 and 9.21. For example, if particle 2 is moving to the left initially, then v_{2i} is negative.

Let us consider some special cases: If $m_1 = m_2$, then $v_{1f} = v_{2i}$ and $v_{2f} = v_{1i}$. That is, the particles exchange speeds if they have equal masses. This is approximately what one observes in head-on billiard ball collisions—the cue ball stops, and the struck ball moves away from the collision with the same speed that the cue ball had.

If particle 2 is initially at rest, then $v_{2i} = 0$ and Equations 9.20 and 9.21 become

Elastic collision: particle 2 initially at rest

$$v_{1f} = \left(\frac{m_1 - m_2}{m_1 + m_2}\right)v_{1i} \tag{9.22}$$

$$v_{2f} = \left(\frac{2m_1}{m_1 + m_2}\right)v_{1i} \tag{9.23}$$

If m_1 is much greater than m_2 and $v_{2i} = 0$, we see from Equations 9.22 and 9.23 that $v_{1f} \approx v_{1i}$ and $v_{2f} \approx 2v_{1i}$. That is, when a very heavy particle collides head-on with a very light one that is initially at rest, the heavy particle continues its motion unaltered after the collision, and the light particle rebounds with a speed equal to about twice the initial speed of the heavy particle. An example of such a collision would be that of a moving heavy atom, such as uranium, with a light atom, such as hydrogen.

If m_2 is much greater than m_1 and particle 2 is initially at rest, then $v_{1f} \approx -v_{1i}$ and $v_{2f} \approx v_{2i} = 0$. That is, when a very light particle collides head-on with a very heavy particle that is initially at rest, the light particle has its velocity reversed and the heavy one remains approximately at rest.

EXAMPLE 9.6 ▶ The Ballistic Pendulum

The ballistic pendulum (Fig. 9.11) is a system used to measure the speed of a fast-moving projectile, such as a bullet. The bullet is fired into a large block of wood suspended from some light wires. The bullet embeds in the block, and the entire system swings through a height h. The collision is perfectly inelastic, and because momentum is conserved, Equation 9.14 gives the speed of the system right after the collision, when we assume the impulse approximation. If we call the bullet particle 1 and the block particle 2, the total kinetic energy right after the collision is

$$(1) \qquad K_f = \tfrac{1}{2}(m_1 + m_2)v_f^{\,2}$$

With $v_{2i} = 0$, Equation 9.14 becomes

$$(2) \qquad v_f = \frac{m_1 v_{1i}}{m_1 + m_2}$$

Substituting this value of v_f into (1) gives

$$K_f = \frac{m_1^{\,2}v_{1i}^{\,2}}{2(m_1 + m_2)}$$

Note that this kinetic energy immediately after the collision is less than the initial kinetic energy of the bullet. In all the energy changes that take place *after* the collision, however, the total amount of mechanical energy remains constant; thus, we can say that after the collision, the kinetic energy of the block and bullet at the bottom is transformed to potential energy at the height h:

$$\frac{m_1^{\,2}v_{1i}^{\,2}}{2(m_1 + m_2)} = (m_1 + m_2)gh$$

Solving for v_{1i}, we obtain

$$v_{1i} = \left(\frac{m_1 + m_2}{m_1}\right)\sqrt{2gh}$$

This expression tells us that it is possible to obtain the initial speed of the bullet by measuring h and the two masses.

Because the collision is perfectly inelastic, some mechanical energy is converted to internal energy and it would be *incorrect* to equate the initial kinetic energy of the incoming bullet to the final gravitational potential energy of the bullet–block combination.

Exercise In a ballistic pendulum experiment, suppose that $h = 5.00$ cm, $m_1 = 5.00$ g, and $m_2 = 1.00$ kg. Find (a) the initial speed of the bullet and (b) the loss in mechanical energy due to the collision.

Answer 199 m/s; 98.5 J.

(a)

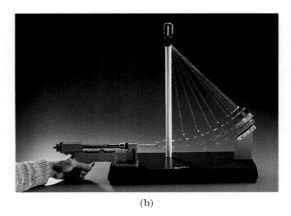

(b)

Figure 9.11 (a) Diagram of a ballistic pendulum. Note that $\mathbf{v}_{1i}$ is the velocity of the bullet just before the collision and $\mathbf{v}_f = \mathbf{v}_{1f} = \mathbf{v}_{2f}$ is the velocity of the bullet + block system just after the perfectly inelastic collision. (b) Multiflash photograph of a ballistic pendulum used in the laboratory. *(Courtesy of Central Scientific Company.)*

EXAMPLE 9.7   A Two-Body Collision with a Spring

A block of mass $m_1 = 1.60$ kg initially moving to the right with a speed of 4.00 m/s on a frictionless horizontal track collides with a spring attached to a second block of mass $m_2 = 2.10$ kg initially moving to the left with a speed of 2.50 m/s, as shown in Figure 9.12a. The spring constant is 600 N/m. (a) At the instant block 1 is moving to the right with a speed of 3.00 m/s, as in Figure 9.12b, determine the velocity of block 2.

Solution   First, note that the initial velocity of block 2 is -2.50 m/s because its direction is to the left. Because momentum is conserved for the system of two blocks, we have

$$m_1 v_{1i} + m_2 v_{2i} = m_1 v_{1f} + m_2 v_{2f}$$

$$(1.60 \text{ kg})(4.00 \text{ m/s}) + (2.10 \text{ kg})(-2.50 \text{ m/s})$$

$$= (1.60 \text{ kg})(3.00 \text{ m/s}) + (2.10 \text{ kg})v_{2f}$$

$$v_{2f} = \boxed{-1.74 \text{ m/s}}$$

The negative value for v_{2f} means that block 2 is still moving to the left at the instant we are considering just after the collision.

(b) Determine the distance the spring is compressed at that instant.

Solution   To determine the distance that the spring is compressed, shown as x in Figure 9.12b, we can use the concept of conservation of mechanical energy because no friction or other nonconservative forces are acting on the system. Thus, we have

$$\tfrac{1}{2}m_1 v_{1i}^2 + \tfrac{1}{2}m_2 v_{2i}^2 = \tfrac{1}{2}m_1 v_{1f}^2 + \tfrac{1}{2}m_2 v_{2f}^2 + \tfrac{1}{2}kx^2$$

Substituting the given values and the result to part (a) into this expression gives

$$x = \boxed{0.173 \text{ m}}$$

It is important to note that we needed to use the principles of both conservation of momentum and conservation of mechanical energy to solve the two parts of this problem.

Exercise   Find the velocity of block 1 and the compression in the spring at the instant that block 2 is at rest.

Answer   0.719 m/s to the right; 0.251 m

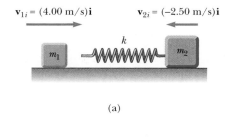

$\mathbf{v}_{1i} = (4.00 \text{ m/s})\mathbf{i}$ $\mathbf{v}_{2i} = (-2.50 \text{ m/s})\mathbf{i}$

k

m_1 m_2

(a)

$\mathbf{v}_{1f} = (3.00 \text{ m/s})\mathbf{i}$ $\mathbf{v}_{2f}$

k

m_1 m_2

$|\leftarrow x \rightarrow|$

(b)

Figure 9.12

EXAMPLE 9.8   Slowing Down Neutrons by Collisions

In a nuclear reactor, neutrons are produced when a $^{235}_{92}\text{U}$ atom splits in a process called *fission*. These neutrons are moving at about 10^7 m/s and must be slowed down to about 10^3 m/s before they take part in another fission event. They are slowed down by being passed through a solid or liquid material called a *moderator*. The slowing-down process involves elastic collisions. Let us show that a neutron can lose most of its kinetic energy if it collides elastically with a moderator containing light nuclei, such as deuterium (in "heavy water," D_2O) or carbon (in graphite).

Solution   Let us assume that the moderator nucleus of mass m_m is at rest initially and that a neutron of mass m_n and initial speed v_{ni} collides with it head-on.

Because these are elastic collisions, the first thing we do is recognize that both momentum and kinetic energy are constant. Therefore, Equations 9.22 and 9.23 can be applied to the head-on collision of a neutron with a moderator nucleus. We can represent this process by a drawing such as Figure 9.10.

The initial kinetic energy of the neutron is

$$K_{ni} = \tfrac{1}{2} m_n v_{ni}^2$$

After the collision, the neutron has kinetic energy $\tfrac{1}{2} m_n v_{nf}^2$, and we can substitute into this the value for v_{nf} given by Equation 9.22:

$$K_{nf} = \tfrac{1}{2} m_n v_{nf}^2 = \frac{m_n}{2} \left(\frac{m_n - m_m}{m_n + m_m} \right)^2 v_{ni}^2$$

Therefore, the fraction f_n of the initial kinetic energy possessed by the neutron after the collision is

$$(1) \qquad f_n = \frac{K_{nf}}{K_{ni}} = \left(\frac{m_n - m_m}{m_n + m_m} \right)^2$$

From this result, we see that the final kinetic energy of the neutron is small when m_m is close to m_n and zero when $m_n = m_m$.

We can use Equation 9.23, which gives the final speed of the particle that was initially at rest, to calculate the kinetic energy of the moderator nucleus after the collision:

$$K_{mf} = \tfrac{1}{2} m_m v_{mf}^2 = \frac{2 m_n^2 m_m}{(m_n + m_m)^2} v_{ni}^2$$

Hence, the fraction f_m of the initial kinetic energy transferred to the moderator nucleus is

$$(2) \qquad f_m = \frac{K_{mf}}{K_{ni}} = \frac{4 m_n m_m}{(m_n + m_m)^2}$$

Because the total kinetic energy of the system is conserved, (2) can also be obtained from (1) with the condition that $f_n + f_m = 1$, so that $f_m = 1 - f_n$.

Suppose that heavy water is used for the moderator. For collisions of the neutrons with deuterium nuclei in D_2O ($m_m = 2 m_n$), $f_n = 1/9$ and $f_m = 8/9$. That is, 89% of the neutron's kinetic energy is transferred to the deuterium nucleus. In practice, the moderator efficiency is reduced because head-on collisions are very unlikely.

How do the results differ when graphite (^{12}C, as found in pencil lead) is used as the moderator?

Quick Quiz 9.8

An ingenious device that illustrates conservation of momentum and kinetic energy is shown in Figure 9.13a. It consists of five identical hard balls supported by strings of equal lengths. When ball 1 is pulled out and released, after the almost-elastic collision between it and ball 2, ball 5 moves out, as shown in Figure 9.13b. If balls 1 and 2 are pulled out and released, balls 4 and 5 swing out, and so forth. Is it ever possible that, when ball 1 is released, balls 4 and 5 will swing out on the opposite side and travel with half the speed of ball 1, as in Figure 9.13c?

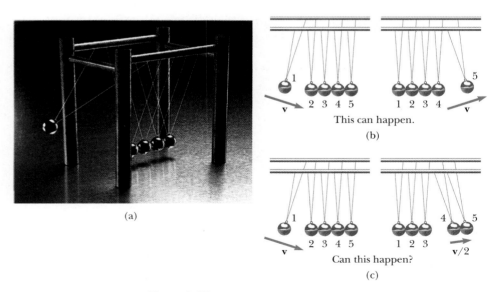

Figure 9.13   An executive stress reliever.

9.5 ▷ TWO-DIMENSIONAL COLLISIONS

In Sections 9.1 and 9.3, we showed that the momentum of a system of two particles is constant when the system is isolated. For any collision of two particles, this result implies that the momentum in each of the directions x, y, and z is constant. However, an important subset of collisions takes place in a plane. The game of billiards is a familiar example involving multiple collisions of objects moving on a two-dimensional surface. For such two-dimensional collisions, we obtain two component equations for conservation of momentum:

$$m_1 v_{1ix} + m_2 v_{2ix} = m_1 v_{1fx} + m_2 v_{2fx}$$

$$m_1 v_{1iy} + m_2 v_{2iy} = m_1 v_{1fy} + m_2 v_{2fy}$$

Let us consider a two-dimensional problem in which particle 1 of mass m_1 collides with particle 2 of mass m_2, where particle 2 is initially at rest, as shown in Figure 9.14. After the collision, particle 1 moves at an angle θ with respect to the horizontal and particle 2 moves at an angle ϕ with respect to the horizontal. This is called a *glancing* collision. Applying the law of conservation of momentum in component form, and noting that the initial y component of the momentum of the two-particle system is zero, we obtain

$$m_1 v_{1i} = m_1 v_{1f} \cos \theta + m_2 v_{2f} \cos \phi \qquad \textbf{(9.24)}$$

$$0 = m_1 v_{1f} \sin \theta - m_2 v_{2f} \sin \phi \qquad \textbf{(9.25)}$$

where the minus sign in Equation 9.25 comes from the fact that after the collision, particle 2 has a y component of velocity that is downward. We now have two independent equations. As long as no more than two of the seven quantities in Equations 9.24 and 9.25 are unknown, we can solve the problem.

If the collision is elastic, we can also use Equation 9.16 (conservation of kinetic energy), with $v_{2i} = 0$, to give

$$\tfrac{1}{2} m_1 v_{1i}{}^2 = \tfrac{1}{2} m_1 v_{1f}{}^2 + \tfrac{1}{2} m_2 v_{2f}{}^2 \qquad \textbf{(9.26)}$$

Knowing the initial speed of particle 1 and both masses, we are left with four unknowns (v_{1f}, v_{2f}, θ, ϕ). Because we have only three equations, one of the four remaining quantities must be given if we are to determine the motion after the collision from conservation principles alone.

If the collision is inelastic, kinetic energy is *not* conserved and Equation 9.26 does *not* apply.

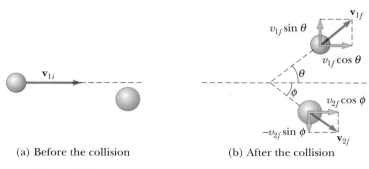

(a) Before the collision   (b) After the collision

Figure 9.14   An elastic glancing collision between two particles.

Problem-Solving Hints

Collisions

The following procedure is recommended when dealing with problems involving collisions between two objects:

- Set up a coordinate system and define your velocities with respect to that system. It is usually convenient to have the x axis coincide with one of the initial velocities.
- In your sketch of the coordinate system, draw and label all velocity vectors and include all the given information.
- Write expressions for the x and y components of the momentum of each object before and after the collision. Remember to include the appropriate signs for the components of the velocity vectors.
- Write expressions for the total momentum in the x direction before and after the collision and equate the two. Repeat this procedure for the total momentum in the y direction. These steps follow from the fact that, because the momentum of the *system* is conserved in any collision, the total momentum along any direction must also be constant. Remember, it is the momentum of the *system* that is constant, not the momenta of the individual objects.
- If the collision is inelastic, kinetic energy is not conserved, and additional information is probably required. If the collision is perfectly inelastic, the final velocities of the two objects are equal. Solve the momentum equations for the unknown quantities.
- If the collision is elastic, kinetic energy is conserved, and you can equate the total kinetic energy before the collision to the total kinetic energy after the collision to get an additional relationship between the velocities.

EXAMPLE 9.9   Collision at an Intersection

A 1 500-kg car traveling east with a speed of 25.0 m/s collides at an intersection with a 2 500-kg van traveling north at a speed of 20.0 m/s, as shown in Figure 9.15. Find the direction and magnitude of the velocity of the wreckage after the collision, assuming that the vehicles undergo a perfectly inelastic collision (that is, they stick together).

Solution Let us choose east to be along the positive x direction and north to be along the positive y direction. Before the collision, the only object having momentum in the x direction is the car. Thus, the magnitude of the total initial momentum of the system (car plus van) in the x direction is

$$\sum p_{xi} = (1\ 500\ \text{kg})(25.0\ \text{m/s}) = 3.75 \times 10^4\ \text{kg}\cdot\text{m/s}$$

Let us assume that the wreckage moves at an angle θ and speed v_f after the collision. The magnitude of the total momentum in the x direction after the collision is

$$\sum p_{xf} = (4\ 000\ \text{kg})v_f \cos\theta$$

Because the total momentum in the x direction is constant, we can equate these two equations to obtain

$$(1)\qquad 3.75 \times 10^4\ \text{kg}\cdot\text{m/s} = (4\ 000\ \text{kg})v_f \cos\theta$$

Figure 9.15 An eastbound car colliding with a northbound van.

Similarly, the total initial momentum of the system in the y direction is that of the van, and the magnitude of this momentum is $(2\ 500\ \text{kg})(20.0\ \text{m/s})$. Applying conservation of

momentum to the y direction, we have

$$\sum p_{yi} = \sum p_{yf}$$

$$(2\ 500 \text{ kg})(20.0 \text{ m/s}) = (4\ 000 \text{ kg})v_f \sin \theta$$

(2) $$5.00 \times 10^4 \text{ kg} \cdot \text{m/s} = (4\ 000 \text{ kg})v_f \sin \theta$$

If we divide (2) by (1), we get

$$\frac{\sin \theta}{\cos \theta} = \tan \theta = \frac{5.00 \times 10^4}{3.75 \times 10^4} = 1.33$$

$$\theta = \boxed{53.1°}$$

When this angle is substituted into (2), the value of v_f is

$$v_f = \frac{5.00 \times 10^4 \text{ kg} \cdot \text{m/s}}{(4\ 000 \text{ kg})\sin 53.1°} = \boxed{15.6 \text{ m/s}}$$

It might be instructive for you to draw the momentum vectors of each vehicle before the collision and the two vehicles together after the collision.

EXAMPLE 9.10   Proton–Proton Collision

Proton 1 collides elastically with proton 2 that is initially at rest. Proton 1 has an initial speed of 3.50×10^5 m/s and makes a glancing collision with proton 2, as was shown in Figure 9.14. After the collision, proton 1 moves at an angle of $37.0°$ to the horizontal axis, and proton 2 deflects at an angle ϕ to the same axis. Find the final speeds of the two protons and the angle ϕ.

Solution   Because both particles are protons, we know that $m_1 = m_2$. We also know that $\theta = 37.0°$ and $v_{1i} = 3.50 \times 10^5$ m/s. Equations 9.24, 9.25, and 9.26 become

$$v_{1f} \cos 37.0° + v_{2f} \cos \phi = 3.50 \times 10^5 \text{ m/s}$$

$$v_{1f} \sin 37.0° - v_{2f} \sin \phi = 0$$

$$v_{1f}^2 + v_{2f}^2 = (3.50 \times 10^5 \text{ m/s})^2$$

Solving these three equations with three unknowns simultaneously gives

$$v_{1f} = \boxed{2.80 \times 10^5 \text{ m/s}}$$

$$v_{2f} = \boxed{2.11 \times 10^5 \text{ m/s}}$$

$$\phi = \boxed{53.0°}$$

Note that $\theta + \phi = 90°$. This result is not accidental. **Whenever two equal masses collide elastically in a glancing collision and one of them is initially at rest, their final velocities are always at right angles to each other.** The next example illustrates this point in more detail.

EXAMPLE 9.11   Billiard Ball Collision

In a game of billiards, a player wishes to sink a target ball 2 in the corner pocket, as shown in Figure 9.16. If the angle to the corner pocket is $35°$, at what angle θ is the cue ball 1 deflected? Assume that friction and rotational motion are unimportant and that the collision is elastic.

Solution   Because the target ball is initially at rest, conservation of energy (Eq. 9.16) gives

$$\tfrac{1}{2} m_1 v_{1i}^2 = \tfrac{1}{2} m_1 v_{1f}^2 + \tfrac{1}{2} m_2 v_{2f}^2$$

But $m_1 = m_2$, so that

(1) $$v_{1i}^2 = v_{1f}^2 + v_{2f}^2$$

Applying conservation of momentum to the two-dimensional collision gives

(2) $$\mathbf{v}_{1i} = \mathbf{v}_{1f} + \mathbf{v}_{2f}$$

Note that because $m_1 = m_2$, the masses also cancel in (2). If we square both sides of (2) and use the definition of the dot

Figure 9.16

product of two vectors from Section 7.2, we get

$$v_{1i}{}^2 = (\mathbf{v}_{1f} + \mathbf{v}_{2f}) \cdot (\mathbf{v}_{1f} + \mathbf{v}_{2f}) = v_{1f}{}^2 + v_{2f}{}^2 + 2\mathbf{v}_{1f}\cdot\mathbf{v}_{2f}$$

Because the angle between $\mathbf{v}_{1f}$ and $\mathbf{v}_{2f}$ is $\theta + 35°$, $\mathbf{v}_{1f}\cdot\mathbf{v}_{2f} = v_{1f}v_{2f}\cos(\theta + 35°)$, and so

$$(3) \qquad v_{1i}{}^2 = v_{1f}{}^2 + v_{2f}{}^2 + 2v_{1f}v_{2f}\cos(\theta + 35°)$$

Subtracting (1) from (3) gives

$$0 = 2v_{1f}v_{2f}\cos(\theta + 35°)$$

$$0 = \cos(\theta + 35°)$$

$$\theta + 35° = 90° \qquad \text{or} \qquad \theta = \boxed{55°}$$

This result shows that whenever two equal masses undergo a glancing elastic collision and one of them is initially at rest, they move at right angles to each other after the collision. The same physics describes two very different situations, protons in Example 9.10 and billiard balls in this example.

9.6   THE CENTER OF MASS

In this section we describe the overall motion of a mechanical system in terms of a special point called the **center of mass** of the system. The mechanical system can be either a system of particles, such as a collection of atoms in a container, or an extended object, such as a gymnast leaping through the air. We shall see that the center of mass of the system moves as if all the mass of the system were concentrated at that point. Furthermore, if the resultant external force on the system is $\Sigma\mathbf{F}_{\text{ext}}$ and the total mass of the system is M, the center of mass moves with an acceleration given by $\mathbf{a} = \Sigma\mathbf{F}_{\text{ext}}/M$. That is, the system moves as if the resultant external force were applied to a single particle of mass M located at the center of mass. This behavior is independent of other motion, such as rotation or vibration of the system. This result was implicitly assumed in earlier chapters because many examples referred to the motion of extended objects that were treated as particles.

Consider a mechanical system consisting of a pair of particles that have different masses and are connected by a light, rigid rod (Fig. 9.17). One can describe the position of the center of mass of a system as being the *average position* of the system's mass. The center of mass of the system is located somewhere on the line joining the

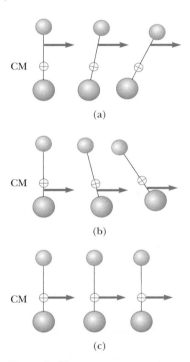

Figure 9.17 Two particles of unequal mass are connected by a light, rigid rod. (a) The system rotates clockwise when a force is applied between the less massive particle and the center of mass. (b) The system rotates counterclockwise when a force is applied between the more massive particle and the center of mass. (c) The system moves in the direction of the force without rotating when a force is applied at the center of mass.

This multiflash photograph shows that as the acrobat executes a somersault, his center of mass follows a parabolic path, the same path that a particle would follow. *(Globus Bros. Studios/The Stock Market)*

Figure 9.18 The center of mass of two particles of unequal mass on the *x* axis is located at x_{CM}, a point between the particles, closer to the one having the larger mass.

particles and is closer to the particle having the larger mass. If a single force is applied at some point on the rod somewhere between the center of mass and the less massive particle, the system rotates clockwise (see Fig. 9.17a). If the force is applied at a point on the rod somewhere between the center of mass and the more massive particle, the system rotates counterclockwise (see Fig. 9.17b). If the force is applied at the center of mass, the system moves in the direction of **F** without rotating (see Fig. 9.17c). Thus, the center of mass can be easily located.

The center of mass of the pair of particles described in Figure 9.18 is located on the *x* axis and lies somewhere between the particles. Its *x* coordinate is

$$x_{CM} \equiv \frac{m_1 x_1 + m_2 x_2}{m_1 + m_2} \tag{9.27}$$

For example, if $x_1 = 0$, $x_2 = d$, and $m_2 = 2m_1$, we find that $x_{CM} = \frac{2}{3}d$. That is, the center of mass lies closer to the more massive particle. If the two masses are equal, the center of mass lies midway between the particles.

We can extend this concept to a system of many particles in three dimensions. The *x* coordinate of the center of mass of *n* particles is defined to be

$$x_{CM} \equiv \frac{m_1 x_1 + m_2 x_2 + m_3 x_3 + \cdots + m_n x_n}{m_1 + m_2 + m_3 + \cdots + m_n} = \frac{\sum_i m_i x_i}{\sum_i m_i} \tag{9.28}$$

where x_i is the *x* coordinate of the *i*th particle. For convenience, we express the total mass as $M \equiv \sum_i m_i$, where the sum runs over all *n* particles. The *y* and *z* coordinates of the center of mass are similarly defined by the equations

$$y_{CM} \equiv \frac{\sum_i m_i y_i}{M} \quad \text{and} \quad z_{CM} \equiv \frac{\sum_i m_i z_i}{M} \tag{9.29}$$

The center of mass can also be located by its position vector, $\mathbf{r}_{CM}$. The cartesian coordinates of this vector are x_{CM}, y_{CM}, and z_{CM}, defined in Equations 9.28 and 9.29. Therefore,

$$\mathbf{r}_{CM} = x_{CM}\mathbf{i} + y_{CM}\mathbf{j} + z_{CM}\mathbf{k}$$
$$= \frac{\sum_i m_i x_i \mathbf{i} + \sum_i m_i y_i \mathbf{j} + \sum_i m_i z_i \mathbf{k}}{M}$$

Vector position of the center of mass for a system of particles

$$\mathbf{r}_{CM} \equiv \frac{\sum_i m_i \mathbf{r}_i}{M} \tag{9.30}$$

where $\mathbf{r}_i$ is the position vector of the *i*th particle, defined by

$$\mathbf{r}_i \equiv x_i\mathbf{i} + y_i\mathbf{j} + z_i\mathbf{k}$$

Although locating the center of mass for an extended object is somewhat more cumbersome than locating the center of mass of a system of particles, the basic ideas we have discussed still apply. We can think of an extended object as a system containing a large number of particles (Fig. 9.19). The particle separation is very small, and so the object can be considered to have a continuous mass distribution. By dividing the object into elements of mass Δm_i, with coordinates x_i, y_i, z_i, we see that the *x* coordinate of the center of mass is approximately

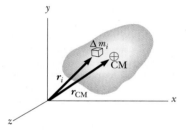

Figure 9.19 An extended object can be considered a distribution of small elements of mass Δm_i. The center of mass is located at the vector position $\mathbf{r}_{CM}$, which has coordinates x_{CM}, y_{CM}, and z_{CM}.

$$x_{CM} \approx \frac{\sum_i x_i \, \Delta m_i}{M}$$

with similar expressions for y_{CM} and z_{CM}. If we let the number of elements *n* approach infinity, then x_{CM} is given precisely. In this limit, we replace the sum by an integral and Δm_i by the differential element dm:

$$x_{CM} = \lim_{\Delta m_i \to 0} \frac{\sum_i x_i \, \Delta m_i}{M} = \frac{1}{M} \int x \, dm \qquad \textbf{(9.31)}$$

Likewise, for y_{CM} and z_{CM} we obtain

$$y_{CM} = \frac{1}{M} \int y \, dm \qquad \text{and} \qquad z_{CM} = \frac{1}{M} \int z \, dm \qquad \textbf{(9.32)}$$

We can express the vector position of the center of mass of an extended object in the form

$$\mathbf{r}_{CM} = \frac{1}{M} \int \mathbf{r} \, dm \qquad \textbf{(9.33)}$$

which is equivalent to the three expressions given by Equations 9.31 and 9.32.

The center of mass of any symmetric object lies on an axis of symmetry and on any plane of symmetry.[4] For example, the center of mass of a rod lies in the rod, midway between its ends. The center of mass of a sphere or a cube lies at its geometric center.

One can determine the center of mass of an irregularly shaped object by suspending the object first from one point and then from another. In Figure 9.20, a wrench is hung from point A, and a vertical line AB (which can be established with a plumb bob) is drawn when the wrench has stopped swinging. The wrench is then hung from point C, and a second vertical line CD is drawn. The center of mass is halfway through the thickness of the wrench, under the intersection of these two lines. In general, if the wrench is hung freely from any point, the vertical line through this point must pass through the center of mass.

Because an extended object is a continuous distribution of mass, each small mass element is acted upon by the force of gravity. The net effect of all these forces is equivalent to the effect of a single force, $M\mathbf{g}$, acting through a special point, called the **center of gravity.** If $\mathbf{g}$ is constant over the mass distribution, then the center of gravity coincides with the center of mass. If an extended object is pivoted at its center of gravity, it balances in any orientation.

Figure 9.20 An experimental technique for determining the center of mass of a wrench. The wrench is hung freely first from point A and then from point C. The intersection of the two lines AB and CD locates the center of mass.

Quick Quiz 9.9

If a baseball bat is cut at the location of its center of mass as shown in Figure 9.21, do the two pieces have the same mass?

Figure 9.21 A baseball bat cut at the location of its center of mass.

QuickLab

Cut a triangle from a piece of cardboard and draw a set of adjacent strips inside it, parallel to one of the sides. Put a dot at the approximate location of the center of mass of each strip and then draw a straight line through the dots and into the angle opposite your starting side. The center of mass for the triangle must lie on this bisector of the angle. Repeat these steps for the other two sides. The three angle bisectors you have drawn will intersect at the center of mass of the triangle. If you poke a hole anywhere in the triangle and hang the cardboard from a string attached at that hole, the center of mass will be vertically aligned with the hole.

[4]This statement is valid only for objects that have a uniform mass per unit volume.

EXAMPLE 9.12 The Center of Mass of Three Particles

A system consists of three particles located as shown in Figure 9.22a. Find the center of mass of the system.

Solution We set up the problem by labeling the masses of the particles as shown in the figure, with $m_1 = m_2 = 1.0$ kg and $m_3 = 2.0$ kg. Using the basic defining equations for the coordinates of the center of mass and noting that $z_{CM} = 0$, we obtain

$$x_{CM} = \frac{\sum_i m_i x_i}{M} = \frac{m_1 x_1 + m_2 x_2 + m_3 x_3}{m_1 + m_2 + m_3}$$

$$= \frac{(1.0 \text{ kg})(1.0 \text{ m}) + (1.0 \text{ kg})(2.0 \text{ m}) + (2.0 \text{ kg})(0 \text{ m})}{1.0 \text{ kg} + 1.0 \text{ kg} + 2.0 \text{ kg}}$$

$$= \frac{3.0 \text{ kg} \cdot \text{m}}{4.0 \text{ kg}} = 0.75 \text{ m}$$

$$y_{CM} = \frac{\sum_i m_i y_i}{M} = \frac{m_1 y_1 + m_2 y_2 + m_3 y_3}{m_1 + m_2 + m_3}$$

$$= \frac{(1.0 \text{ kg})(0) + (1.0 \text{ kg})(0) + (2.0 \text{ kg})(2.0 \text{ m})}{4.0 \text{ kg}}$$

$$= \frac{4.0 \text{ kg} \cdot \text{m}}{4.0 \text{ kg}} = 1.0 \text{ m}$$

The position vector to the center of mass measured from the origin is therefore

$$\mathbf{r}_{CM} = x_{CM}\mathbf{i} + y_{CM}\mathbf{j} = \boxed{0.75\mathbf{i} \text{ m} + 1.0\mathbf{j} \text{ m}}$$

We can verify this result graphically by adding together $m_1\mathbf{r}_1 + m_2\mathbf{r}_2 + m_3\mathbf{r}_3$ and dividing the vector sum by M, the total mass. This is shown in Figure 9.22b.

Figure 9.22 (a) Two 1-kg masses and a single 2-kg mass are located as shown. The vector indicates the location of the system's center of mass. (b) The vector sum of $m_i\mathbf{r}_i$.

(a)

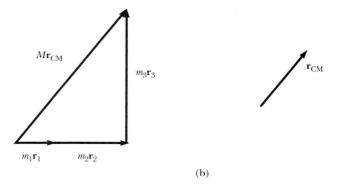

(b)

EXAMPLE 9.13 The Center of Mass of a Rod

(a) Show that the center of mass of a rod of mass M and length L lies midway between its ends, assuming the rod has a uniform mass per unit length.

Solution The rod is shown aligned along the x axis in Figure 9.23, so that $y_{CM} = z_{CM} = 0$. Furthermore, if we call the mass per unit length λ (this quantity is called the *linear mass density*), then $\lambda = M/L$ for the uniform rod we assume here. If we divide the rod into elements of length dx, then the mass of each element is $dm = \lambda \, dx$. For an arbitrary element located a distance x from the origin, Equation 9.31 gives

$$x_{CM} = \frac{1}{M} \int x \, dm = \frac{1}{M} \int_0^L x\lambda \, dx = \frac{\lambda}{M} \frac{x^2}{2} \bigg|_0^L = \frac{\lambda L^2}{2M}$$

Because $\lambda = M/L$, this reduces to

$$x_{CM} = \frac{L^2}{2M}\left(\frac{M}{L}\right) = \boxed{\frac{L}{2}}$$

One can also use symmetry arguments to obtain the same result.

(b) Suppose a rod is *nonuniform* such that its mass per unit length varies linearly with x according to the expression $\lambda = \alpha x$, where α is a constant. Find the x coordinate of the center of mass as a fraction of L.

Solution In this case, we replace dm by λdx where λ is not constant. Therefore, x_{CM} is

$$x_{CM} = \frac{1}{M} \int x \, dm = \frac{1}{M} \int_0^L x\lambda \, dx = \frac{1}{M} \int_0^L x\alpha x \, dx$$

$$= \frac{\alpha}{M} \int_0^L x^2 \, dx = \frac{\alpha L^3}{3M}$$

We can eliminate α by noting that the total mass of the rod is related to α through the relationship

$$M = \int dm = \int_0^L \lambda \, dx = \int_0^L \alpha x \, dx = \frac{\alpha L^2}{2}$$

Substituting this into the expression for x_{CM} gives

$$x_{CM} = \frac{\alpha L^3}{3\alpha L^2/2} = \boxed{\frac{2}{3} L}$$

Figure 9.23 The center of mass of a uniform rod of length L is located at $x_{CM} = L/2$.

EXAMPLE 9.14 ▶ The Center of Mass of a Right Triangle

An object of mass M is in the shape of a right triangle whose dimensions are shown in Figure 9.24. Locate the coordinates of the center of mass, assuming the object has a uniform mass per unit area.

Solution By inspection we can estimate that the x coordinate of the center of mass must be past the center of the base, that is, greater than $a/2$, because the largest part of the triangle lies beyond that point. A similar argument indicates that its y coordinate must be less than $b/2$. To evaluate the x coordinate, we divide the triangle into narrow strips of width dx and height y as in Figure 9.24. The mass dm of each strip is

$$dm = \frac{\text{total mass of object}}{\text{total area of object}} \times \text{area of strip}$$

$$= \frac{M}{1/2\,ab}(y \, dx) = \left(\frac{2M}{ab}\right) y \, dx$$

Therefore, the x coordinate of the center of mass is

$$x_{CM} = \frac{1}{M} \int x \, dm = \frac{1}{M} \int_0^a x\left(\frac{2M}{ab}\right) y \, dx = \frac{2}{ab} \int_0^a xy \, dx$$

To evaluate this integral, we must express y in terms of x. From similar triangles in Figure 9.24, we see that

$$\frac{y}{x} = \frac{b}{a} \qquad \text{or} \qquad y = \frac{b}{a} x$$

With this substitution, x_{CM} becomes

$$x_{CM} = \frac{2}{ab} \int_0^a x\left(\frac{b}{a} x\right) dx = \frac{2}{a^2} \int_0^a x^2 \, dx = \frac{2}{a^2}\left[\frac{x^3}{3}\right]_0^a$$

$$= \boxed{\frac{2}{3} a}$$

By a similar calculation, we get for the y coordinate of the center of mass

$$y_{CM} = \boxed{\frac{1}{3} b}$$

These values fit our original estimates.

Figure 9.24

9.7 ▶ MOTION OF A SYSTEM OF PARTICLES

6.8

We can begin to understand the physical significance and utility of the center of mass concept by taking the time derivative of the position vector given by Equation 9.30. From Section 4.1 we know that the time derivative of a position vector is by definition a velocity. Assuming M remains constant for a system of particles, that is, no particles enter or leave the system, we get the following expression for the **velocity of the center of mass** of the system:

Velocity of the center of mass

$$\mathbf{v}_{CM} = \frac{d\mathbf{r}_{CM}}{dt} = \frac{1}{M} \sum_i m_i \frac{d\mathbf{r}_i}{dt} = \frac{\sum_i m_i \mathbf{v}_i}{M} \tag{9.34}$$

where $\mathbf{v}_i$ is the velocity of the ith particle. Rearranging Equation 9.34 gives

Total momentum of a system of particles

$$M\mathbf{v}_{CM} = \sum_i m_i \mathbf{v}_i = \sum_i \mathbf{p}_i = \mathbf{p}_{tot} \tag{9.35}$$

Therefore, we conclude that the **total linear momentum of the system** equals the total mass multiplied by the velocity of the center of mass. In other words, the total linear momentum of the system is equal to that of a single particle of mass M moving with a velocity $\mathbf{v}_{CM}$.

If we now differentiate Equation 9.34 with respect to time, we get the **acceleration of the center of mass** of the system:

Acceleration of the center of mass

$$\mathbf{a}_{CM} = \frac{d\mathbf{v}_{CM}}{dt} = \frac{1}{M} \sum_i m_i \frac{d\mathbf{v}_i}{dt} = \frac{1}{M} \sum_i m_i \mathbf{a}_i \tag{9.36}$$

Rearranging this expression and using Newton's second law, we obtain

$$M\mathbf{a}_{CM} = \sum_i m_i \mathbf{a}_i = \sum_i \mathbf{F}_i \tag{9.37}$$

where $\mathbf{F}_i$ is the net force on particle i.

The forces on any particle in the system may include both external forces (from outside the system) and internal forces (from within the system). However, by Newton's third law, the internal force exerted by particle 1 on particle 2, for example, is equal in magnitude and opposite in direction to the internal force exerted by particle 2 on particle 1. Thus, when we sum over all internal forces in Equation 9.37, they cancel in pairs and the net force on the system is caused *only* by external forces. Thus, we can write Equation 9.37 in the form

Newton's second law for a system of particles

$$\sum \mathbf{F}_{ext} = M\mathbf{a}_{CM} = \frac{d\mathbf{p}_{tot}}{dt} \tag{9.38}$$

That is, the resultant external force on a system of particles equals the total mass of the system multiplied by the acceleration of the center of mass. If we compare this with Newton's second law for a single particle, we see that

The center of mass of a system of particles of combined mass M moves like an equivalent particle of mass M would move under the influence of the resultant external force on the system.

Finally, we see that if the resultant external force is zero, then from Equation 9.38 it follows that

$$\frac{d\mathbf{p}_{tot}}{dt} = M\mathbf{a}_{CM} = 0$$

Figure 9.25 Multiflash photograph showing an overhead view of a wrench moving on a horizontal surface. The center of mass of the wrench moves in a straight line as the wrench rotates about this point, shown by the white dots. *(Richard Megna/Fundamental Photographs)*

so that

$$\mathbf{p}_{tot} = M\mathbf{v}_{CM} = \text{constant} \qquad (\text{when } \Sigma \mathbf{F}_{ext} = 0) \qquad \textbf{(9.39)}$$

That is, the total linear momentum of a system of particles is conserved if no net external force is acting on the system. It follows that for an isolated system of particles, both the total momentum and the velocity of the center of mass are constant in time, as shown in Figure 9.25. This is a generalization to a many-particle system of the law of conservation of momentum discussed in Section 9.1 for a two-particle system.

Suppose an isolated system consisting of two or more members is at rest. The center of mass of such a system remains at rest unless acted upon by an external force. For example, consider a system made up of a swimmer standing on a raft, with the system initially at rest. When the swimmer dives horizontally off the raft, the center of mass of the system remains at rest (if we neglect friction between raft and water). Furthermore, the linear momentum of the diver is equal in magnitude to that of the raft but opposite in direction.

As another example, suppose an unstable atom initially at rest suddenly breaks up into two fragments of masses M_A and M_B, with velocities $\mathbf{v}_A$ and $\mathbf{v}_B$, respectively. Because the total momentum of the system before the breakup is zero, the total momentum of the system after the breakup must also be zero. Therefore, $M_A\mathbf{v}_A + M_B\mathbf{v}_B = 0$. If the velocity of one of the fragments is known, the recoil velocity of the other fragment can be calculated.

EXAMPLE 9.15 ▸ The Sliding Bear

Suppose you tranquilize a polar bear on a smooth glacier as part of a research effort. How might you estimate the bear's mass using a measuring tape, a rope, and knowledge of your own mass?

Solution Tie one end of the rope around the bear, and then lay out the tape measure on the ice with one end at the bear's original position, as shown in Figure 9.26. Grab hold of the free end of the rope and position yourself as shown, noting your location. Take off your spiked shoes and pull on the rope hand over hand. Both you and the bear will slide over the ice until you meet. From the tape, observe how far you have slid, x_p, and how far the bear has slid, x_b. The point where you meet the bear is the constant location of the center of mass of the system (bear plus you), and so you can determine the mass of the bear from $m_b x_b = m_p x_p$. (Unfortunately, you cannot get back to your spiked shoes and so are in big trouble if the bear wakes up!)

Figure 9.26   The center of mass of an isolated system remains at rest unless acted on by an external force. How can you determine the mass of the polar bear?

CONCEPTUAL EXAMPLE 9.16 ▶ Exploding Projectile

A projectile fired into the air suddenly explodes into several fragments (Fig. 9.27). What can be said about the motion of the center of mass of the system made up of all the fragments after the explosion?

Solution   Neglecting air resistance, the only external force on the projectile is the gravitational force. Thus, if the projectile did not explode, it would continue to move along the parabolic path indicated by the broken line in Figure 9.27. Because the forces caused by the explosion are internal, they do not affect the motion of the center of mass. Thus, after the explosion the center of mass of the system (the fragments) follows the same parabolic path the projectile would have followed if there had been no explosion.

Motion of center of mass

Figure 9.27   When a projectile explodes into several fragments, the center of mass of the system made up of all the fragments follows the same parabolic path the projectile would have taken had there been no explosion.

EXAMPLE 9.17 ▶ The Exploding Rocket

A rocket is fired vertically upward. At the instant it reaches an altitude of 1 000 m and a speed of 300 m/s, it explodes into three equal fragments. One fragment continues to move upward with a speed of 450 m/s following the explosion. The second fragment has a speed of 240 m/s and is moving east right after the explosion. What is the velocity of the third fragment right after the explosion?

Solution   Let us call the total mass of the rocket M; hence, the mass of each fragment is $M/3$. Because the forces of the explosion are internal to the system and cannot affect its total momentum, the total momentum $\mathbf{p}_i$ of the rocket just before the explosion must equal the total momentum $\mathbf{p}_f$ of the fragments right after the explosion.

Before the explosion:

$$\mathbf{p}_i = M\mathbf{v}_i = M(300 \text{ m/s})\mathbf{j}$$

After the explosion:

$$\mathbf{p}_f = \frac{M}{3}(240 \text{ m/s})\mathbf{i} + \frac{M}{3}(450 \text{ m/s})\mathbf{j} + \frac{M}{3}\mathbf{v}_f$$

where $\mathbf{v}_f$ is the unknown velocity of the third fragment. Equating these two expressions (because $\mathbf{p}_i = \mathbf{p}_f$) gives

$$\frac{M}{3}\mathbf{v}_f + M(80 \text{ m/s})\mathbf{i} + M(150 \text{ m/s})\mathbf{j} = M(300 \text{ m/s})\mathbf{j}$$

$$\mathbf{v}_f = \boxed{(-240\mathbf{i} + 450\mathbf{j}) \text{ m/s}}$$

What does the sum of the momentum vectors for all the fragments look like?

Exercise Find the position of the center of mass of the system of fragments relative to the ground 3.00 s after the explosion. Assume the rocket engine is nonoperative after the explosion.

Answer The *x* coordinate does not change; $y_{CM} = 1.86$ km.

Optional Section

ROCKET PROPULSION

When ordinary vehicles, such as automobiles and locomotives, are propelled, the driving force for the motion is friction. In the case of the automobile, the driving force is the force exerted by the road on the car. A locomotive "pushes" against the tracks; hence, the driving force is the force exerted by the tracks on the locomotive. However, a rocket moving in space has no road or tracks to push against. Therefore, the source of the propulsion of a rocket must be something other than friction. Figure 9.28 is a dramatic photograph of a spacecraft at liftoff. **The operation of a rocket depends upon the law of conservation of linear momentum as applied to a system of particles, where the system is the rocket plus its ejected fuel.**

Rocket propulsion can be understood by first considering the mechanical system consisting of a machine gun mounted on a cart on wheels. As the gun is fired,

Figure 9.28 Liftoff of the space shuttle *Columbia*. Enormous thrust is generated by the shuttle's liquid-fuel engines, aided by the two solid-fuel boosters. Many physical principles from mechanics, thermodynamics, and electricity and magnetism are involved in such a launch. *(Courtesy of NASA)*

The force from a nitrogen-propelled, hand-controlled device allows an astronaut to move about freely in space without restrictive tethers. *(Courtesy of NASA)*

$\mathbf{p}_i = (M + \Delta m)\mathbf{v}$

(a)

(b)

Figure 9.29   Rocket propulsion. (a) The initial mass of the rocket plus all its fuel is $M + \Delta m$ at a time t, and its speed is v. (b) At a time $t + \Delta t$, the rocket's mass has been reduced to M and an amount of fuel Δm has been ejected. The rocket's speed increases by an amount Δv.

Expression for rocket propulsion

each bullet receives a momentum $m\mathbf{v}$ in some direction, where $\mathbf{v}$ is measured with respect to a stationary Earth frame. The momentum of the system made up of cart, gun, and bullets must be conserved. Hence, for each bullet fired, the gun and cart must receive a compensating momentum in the opposite direction. That is, the reaction force exerted by the bullet on the gun accelerates the cart and gun, and the cart moves in the direction opposite that of the bullets. If n is the number of bullets fired each second, then the average force exerted on the gun is $\mathbf{F}_{av} = nm\mathbf{v}$.

In a similar manner, as a rocket moves in free space, its linear momentum changes when some of its mass is released in the form of ejected gases. **Because the gases are given momentum when they are ejected out of the engine, the rocket receives a compensating momentum in the opposite direction.** Therefore, the rocket is accelerated as a result of the "push," or thrust, from the exhaust gases. In free space, the center of mass of the system (rocket plus expelled gases) moves uniformly, independent of the propulsion process.[5]

Suppose that at some time t, the magnitude of the momentum of a rocket plus its fuel is $(M + \Delta m)v$, where v is the speed of the rocket relative to the Earth (Fig. 9.29a). Over a short time interval Δt, the rocket ejects fuel of mass Δm, and so at the end of this interval the rocket's speed is $v + \Delta v$, where Δv is the change in speed of the rocket (Fig. 9.29b). If the fuel is ejected with a speed v_e relative to the rocket (the subscript "e" stands for *exhaust*, and v_e is usually called the *exhaust speed*), the velocity of the fuel relative to a stationary frame of reference is $v - v_e$. Thus, if we equate the total initial momentum of the system to the total final momentum, we obtain

$$(M + \Delta m)v = M(v + \Delta v) + \Delta m(v - v_e)$$

where M represents the mass of the rocket and its remaining fuel after an amount of fuel having mass Δm has been ejected. Simplifying this expression gives

$$M\,\Delta v = v_e\,\Delta m$$

We also could have arrived at this result by considering the system in the center-of-mass frame of reference, which is a frame having the same velocity as the center of mass of the system. In this frame, the total momentum of the system is zero; therefore, if the rocket gains a momentum $M\,\Delta v$ by ejecting some fuel, the exhausted fuel obtains a momentum $v_e\,\Delta m$ in the *opposite* direction, so that $M\,\Delta v - v_e\,\Delta m = 0$. If we now take the limit as Δt goes to zero, we get $\Delta v \rightarrow dv$ and $\Delta m \rightarrow dm$. Futhermore, the increase in the exhaust mass dm corresponds to an equal decrease in the rocket mass, so that $dm = -dM$. Note that dM is given a negative sign because it represents a decrease in mass. Using this fact, we obtain

$$M\,dv = v_e\,dm = -v_e\,dM \qquad \textbf{(9.40)}$$

Integrating this equation and taking the initial mass of the rocket plus fuel to be M_i and the final mass of the rocket plus its remaining fuel to be M_f, we obtain

$$\int_{v_i}^{v_f} dv = -v_e \int_{M_i}^{M_f} \frac{dM}{M}$$

$$v_f - v_i = v_e \ln\!\left(\frac{M_i}{M_f}\right) \qquad \textbf{(9.41)}$$

---

[5]It is interesting to note that the rocket and machine gun represent cases of the reverse of a perfectly inelastic collision: Momentum is conserved, but the kinetic energy of the system increases (at the expense of chemical potential energy in the fuel).

This is the basic expression of rocket propulsion. First, it tells us that the increase in rocket speed is proportional to the exhaust speed of the ejected gases, v_e. Therefore, the exhaust speed should be very high. Second, the increase in rocket speed is proportional to the natural logarithm of the ratio M_i/M_f. Therefore, this ratio should be as large as possible, which means that the mass of the rocket without its fuel should be as small as possible and the rocket should carry as much fuel as possible.

The **thrust** on the rocket is the force exerted on it by the ejected exhaust gases. We can obtain an expression for the thrust from Equation 9.40:

$$\text{Thrust} = M\frac{dv}{dt} = \left| v_e\frac{dM}{dt} \right| \qquad \textbf{(9.42)}$$

This expression shows us that the thrust increases as the exhaust speed increases and as the rate of change of mass (called the *burn rate*) increases.

EXAMPLE 9.18   A Rocket in Space

A rocket moving in free space has a speed of 3.0×10^3 m/s relative to the Earth. Its engines are turned on, and fuel is ejected in a direction opposite the rocket's motion at a speed of 5.0×10^3 m/s relative to the rocket. (a) What is the speed of the rocket relative to the Earth once the rocket's mass is reduced to one-half its mass before ignition?

Solution   We can guess that the speed we are looking for must be greater than the original speed because the rocket is accelerating. Applying Equation 9.41, we obtain

$$v_f = v_i + v_e\ln\left(\frac{M_i}{M_f}\right)$$

$$= 3.0 \times 10^3 \text{ m/s} + (5.0 \times 10^3 \text{ m/s})\ln\left(\frac{M_i}{0.5\,M_i}\right)$$

$$= \boxed{6.5 \times 10^3 \text{ m/s}}$$

(b) What is the thrust on the rocket if it burns fuel at the rate of 50 kg/s?

Solution

$$\text{Thrust} = \left| v_e\frac{dM}{dt} \right| = (5.0 \times 10^3 \text{ m/s})(50 \text{ kg/s})$$

$$= \boxed{2.5 \times 10^5 \text{ N}}$$

EXAMPLE 9.19   Fighting a Fire

Two firefighters must apply a total force of 600 N to steady a hose that is discharging water at 3 600 L/min. Estimate the speed of the water as it exits the nozzle.

Solution   The water is exiting at 3 600 L/min, which is 60 L/s. Knowing that 1 L of water has a mass of 1 kg, we can say that about 60 kg of water leaves the nozzle every second. As the water leaves the hose, it exerts on the hose a thrust that must be counteracted by the 600-N force exerted on the hose by the firefighters. So, applying Equation 9.42 gives

$$\text{Thrust} = \left| v_e\frac{dM}{dt} \right|$$

$$600 \text{ N} = \left| v_e(60 \text{ kg/s}) \right|$$

$$v_e = \boxed{10 \text{ m/s}}$$

Firefighting is dangerous work. If the nozzle should slip from

their hands, the movement of the hose due to the thrust it receives from the rapidly exiting water could injure the firefighters.

Firefighters attack a burning house with a hose line. *(© Bill Stormont/The Stock Market)*

SUMMARY

The **linear momentum p** of a particle of mass m moving with a velocity $\mathbf{v}$ is

$$\mathbf{p} \equiv m\mathbf{v} \tag{9.1}$$

The law of **conservation of linear momentum** indicates that the total momentum of an isolated system is conserved. If two particles form an isolated system, their total momentum is conserved regardless of the nature of the force between them. Therefore, the total momentum of the system at all times equals its initial total momentum, or

$$\mathbf{p}_{1i} + \mathbf{p}_{2i} = \mathbf{p}_{1f} + \mathbf{p}_{2f} \tag{9.5}$$

The **impulse** imparted to a particle by a force $\mathbf{F}$ is equal to the change in the momentum of the particle:

$$\mathbf{I} \equiv \int_{t_i}^{t_f} \mathbf{F}\, dt = \Delta\mathbf{p} \tag{9.9}$$

This is known as the **impulse–momentum theorem.**

Impulsive forces are often very strong compared with other forces on the system and usually act for a very short time, as in the case of collisions.

When two particles collide, the total momentum of the system before the collision always equals the total momentum after the collision, regardless of the nature of the collision. An **inelastic collision** is one for which the total kinetic energy is not conserved. A **perfectly inelastic collision** is one in which the colliding bodies stick together after the collision. An **elastic collision** is one in which kinetic energy is constant.

In a two- or three-dimensional collision, the components of momentum in each of the three directions (x, y, and z) are conserved independently.

The position vector of the center of mass of a system of particles is defined as

$$\mathbf{r}_{CM} \equiv \frac{\sum_i m_i \mathbf{r}_i}{M} \tag{9.30}$$

where $M = \sum_i m_i$ is the total mass of the system and $\mathbf{r}_i$ is the position vector of the ith particle.

The position vector of the center of mass of a rigid body can be obtained from the integral expression

$$\mathbf{r}_{CM} = \frac{1}{M} \int \mathbf{r}\, dm \tag{9.33}$$

The velocity of the center of mass for a system of particles is

$$\mathbf{v}_{CM} = \frac{\sum_i m_i \mathbf{v}_i}{M} \tag{9.34}$$

The total momentum of a system of particles equals the total mass multiplied by the velocity of the center of mass.

Newton's second law applied to a system of particles is

$$\sum \mathbf{F}_{ext} = M\mathbf{a}_{CM} = \frac{d\mathbf{p}_{tot}}{dt} \tag{9.38}$$

where $\mathbf{a}_{CM}$ is the acceleration of the center of mass and the sum is over all external forces. The center of mass moves like an imaginary particle of mass M under the influence of the resultant external force on the system. It follows from Equation 9.38 that the total momentum of the system is conserved if there are no external forces acting on it.

QUESTIONS

1. If the kinetic energy of a particle is zero, what is its linear momentum?

2. If the speed of a particle is doubled, by what factor is its momentum changed? By what factor is its kinetic energy changed?

3. If two particles have equal kinetic energies, are their momenta necessarily equal? Explain.

4. If two particles have equal momenta, are their kinetic energies necessarily equal? Explain.

5. An isolated system is initially at rest. Is it possible for parts of the system to be in motion at some later time? If so, explain how this might occur.

6. If two objects collide and one is initially at rest, is it possible for both to be at rest after the collision? Is it possible for one to be at rest after the collision? Explain.

7. Explain how linear momentum is conserved when a ball bounces from a floor.

8. Is it possible to have a collision in which all of the kinetic energy is lost? If so, cite an example.

9. In a perfectly elastic collision between two particles, does the kinetic energy of each particle change as a result of the collision?

10. When a ball rolls down an incline, its linear momentum increases. Does this imply that momentum is not conserved? Explain.

11. Consider a perfectly inelastic collision between a car and a large truck. Which vehicle loses more kinetic energy as a result of the collision?

12. Can the center of mass of a body lie outside the body? If so, give examples.

13. Three balls are thrown into the air simultaneously. What is the acceleration of their center of mass while they are in motion?

14. A meter stick is balanced in a horizontal position with the index fingers of the right and left hands. If the two fingers are slowly brought together, the stick remains balanced and the two fingers always meet at the 50-cm mark regardless of their original positions (try it!). Explain.

15. A sharpshooter fires a rifle while standing with the butt of the gun against his shoulder. If the forward momentum of a bullet is the same as the backward momentum of the gun, why is it not as dangerous to be hit by the gun as by the bullet?

16. A piece of mud is thrown against a brick wall and sticks to the wall. What happens to the momentum of the mud? Is momentum conserved? Explain.

17. Early in this century, Robert Goddard proposed sending a rocket to the Moon. Critics took the position that in a vacuum, such as exists between the Earth and the Moon, the gases emitted by the rocket would have nothing to push against to propel the rocket. According to *Scientific American* (January 1975), Goddard placed a gun in a vacuum and fired a blank cartridge from it. (A blank cartridge fires only the wadding and hot gases of the burning gunpowder.) What happened when the gun was fired?

18. A pole-vaulter falls from a height of 6.0 m onto a foam rubber pad. Can you calculate his speed just before he reaches the pad? Can you estimate the force exerted on him due to the collision? Explain.

19. Explain how you would use a balloon to demonstrate the mechanism responsible for rocket propulsion.

20. Does the center of mass of a rocket in free space accelerate? Explain. Can the speed of a rocket exceed the exhaust speed of the fuel? Explain.

21. A ball is dropped from a tall building. Identify the system for which linear momentum is conserved.

22. A bomb, initially at rest, explodes into several pieces. (a) Is linear momentum conserved? (b) Is kinetic energy conserved? Explain.

23. NASA often uses the gravity of a planet to "slingshot" a probe on its way to a more distant planet. This is actually a collision where the two objects do not touch. How can the probe have its speed increased in this manner?

24. The Moon revolves around the Earth. Is the Moon's linear momentum conserved? Is its kinetic energy conserved? Assume that the Moon's orbit is circular.

25. A raw egg dropped to the floor breaks apart upon impact. However, a raw egg dropped onto a thick foam rubber cushion from a height of about 1 m rebounds without breaking. Why is this possible? (If you try this experiment, be sure to catch the egg after the first bounce.)

26. From among the following positions, identify the one that reflects your own view and argue to support it: (a) The best theory of motion is that force causes acceleration. (b) The true measure of a force's effectiveness is the work it does, and the best theory of motion is that work on an object changes its energy. (c) The true measure of a force's effect is impulse, and the best theory of motion is that impulse injected into an object changes its momentum.

PROBLEMS

1, 2, 3 = straightforward, intermediate, challenging   ☐ = full solution available in the *Student Solutions Manual and Study Guide*
WEB = solution posted at **http://www.saunderscollege.com/physics/**   🖥 = Computer useful in solving problem   📼 = Interactive Physics
☐ = paired numerical/symbolic problems

Section 9.1   Linear Momentum and Its Conservation

1. A 3.00-kg particle has a velocity of $(3.00\mathbf{i} - 4.00\mathbf{j})$ m/s. (a) Find its x and y components of momentum. (b) Find the magnitude and direction of its momentum.

2. A 0.100-kg ball is thrown straight up into the air with an initial speed of 15.0 m/s. Find the momentum of the ball (a) at its maximum height and (b) halfway up to its maximum height.

3. A 40.0-kg child standing on a frozen pond throws a 0.500-kg stone to the east with a speed of 5.00 m/s. Neglecting friction between child and ice, find the recoil velocity of the child.

4. A pitcher claims he can throw a baseball with as much momentum as a 3.00-g bullet moving with a speed of 1 500 m/s. A baseball has a mass of 0.145 kg. What must be its speed if the pitcher's claim is valid?

5. How fast can you set the Earth moving? In particular, when you jump straight up as high as you can, you give the Earth a maximum recoil speed of what order of magnitude? Model the Earth as a perfectly solid object. In your solution, state the physical quantities you take as data and the values you measure or estimate for them.

6. Two blocks of masses M and $3M$ are placed on a horizontal, frictionless surface. A light spring is attached to one of them, and the blocks are pushed together with the spring between them (Fig. P9.6). A cord initially holding the blocks together is burned; after this, the block of mass $3M$ moves to the right with a speed of 2.00 m/s. (a) What is the speed of the block of mass M? (b) Find the original elastic energy in the spring if $M = 0.350$ kg.

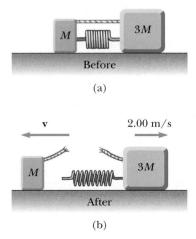

Figure P9.6

7. (a) A particle of mass m moves with momentum p. Show that the kinetic energy of the particle is given by $K = p^2/2m$. (b) Express the magnitude of the particle's momentum in terms of its kinetic energy and mass.

Section 9.2   Impulse and Momentum

8. A car is stopped for a traffic signal. When the light turns green, the car accelerates, increasing its speed from zero to 5.20 m/s in 0.832 s. What linear impulse and average force does a 70.0-kg passenger in the car experience?

9. An estimated force–time curve for a baseball struck by a bat is shown in Figure P9.9. From this curve, determine (a) the impulse delivered to the ball, (b) the average force exerted on the ball, and (c) the peak force exerted on the ball.

Figure P9.9

10. A tennis player receives a shot with the ball (0.060 0 kg) traveling horizontally at 50.0 m/s and returns the shot with the ball traveling horizontally at 40.0 m/s in the opposite direction. (a) What is the impulse delivered to the ball by the racket? (b) What work does the racket do on the ball?

WEB 11. A 3.00-kg steel ball strikes a wall with a speed of 10.0 m/s at an angle of 60.0° with the surface. It bounces off with the same speed and angle (Fig. P9.11). If the ball is in contact with the wall for 0.200 s, what is the average force exerted on the ball by the wall?

12. In a slow-pitch softball game, a 0.200-kg softball crossed the plate at 15.0 m/s at an angle of 45.0° below the horizontal. The ball was hit at 40.0 m/s, 30.0° above the horizontal. (a) Determine the impulse delivered to the ball. (b) If the force on the ball increased linearly for 4.00 ms, held constant for 20.0 ms, and then decreased to zero linearly in another 4.00 ms, what was the maximum force on the ball?

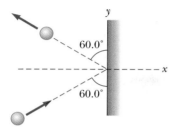

Figure P9.11

13. A garden hose is held in the manner shown in Figure P9.13. What force is necessary to hold the nozzle stationary if the discharge rate is 0.600 kg/s with a speed of 25.0 m/s?

Figure P9.13

14. A professional diver performs a dive from a platform 10 m above the water surface. Estimate the order of magnitude of the average impact force she experiences in her collision with the water. State the quantities you take as data and their values.

Section 9.3 Collisions

Section 9.4 Elastic and Inelastic Collisions in One Dimension

15. High-speed stroboscopic photographs show that the head of a golf club of mass 200 g is traveling at 55.0 m/s just before it strikes a 46.0-g golf ball at rest on a tee. After the collision, the club head travels (in the same direction) at 40.0 m/s. Find the speed of the golf ball just after impact.

16. A 75.0-kg ice skater, moving at 10.0 m/s, crashes into a stationary skater of equal mass. After the collision, the two skaters move as a unit at 5.00 m/s. Suppose the average force a skater can experience without breaking a bone is 4 500 N. If the impact time is 0.100 s, does a bone break?

17. A 10.0-g bullet is fired into a stationary block of wood ($m = 5.00$ kg). The relative motion of the bullet stops

inside the block. The speed of the bullet-plus-wood combination immediately after the collision is measured as 0.600 m/s. What was the original speed of the bullet?

18. As shown in Figure P9.18, a bullet of mass m and speed v passes completely through a pendulum bob of mass M. The bullet emerges with a speed of $v/2$. The pendulum bob is suspended by a stiff rod of length ℓ and negligible mass. What is the minimum value of v such that the pendulum bob will barely swing through a complete vertical circle?

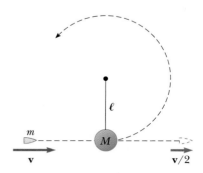

Figure P9.18

19. A 45.0-kg girl is standing on a plank that has a mass of 150 kg. The plank, originally at rest, is free to slide on a frozen lake, which is a flat, frictionless supporting surface. The girl begins to walk along the plank at a constant speed of 1.50 m/s relative to the plank. (a) What is her speed relative to the ice surface? (b) What is the speed of the plank relative to the ice surface?

20. Gayle runs at a speed of 4.00 m/s and dives on a sled, which is initially at rest on the top of a frictionless snow-covered hill. After she has descended a vertical distance of 5.00 m, her brother, who is initially at rest, hops on her back and together they continue down the hill. What is their speed at the bottom of the hill if the total vertical drop is 15.0 m? Gayle's mass is 50.0 kg, the sled has a mass of 5.00 kg and her brother has a mass of 30.0 kg.

21. A 1 200-kg car traveling initially with a speed of 25.0 m/s in an easterly direction crashes into the rear end of a 9 000-kg truck moving in the same direction at 20.0 m/s (Fig. P9.21). The velocity of the car right after the collision is 18.0 m/s to the east. (a) What is the velocity of the truck right after the collision? (b) How much mechanical energy is lost in the collision? Account for this loss in energy.

22. A railroad car of mass 2.50×10^4 kg is moving with a speed of 4.00 m/s. It collides and couples with three other coupled railroad cars, each of the same mass as the single car and moving in the same direction with an initial speed of 2.00 m/s. (a) What is the speed of the four cars after the collision? (b) How much energy is lost in the collision?

Figure P9.21

23. Four railroad cars, each of mass 2.50×10^4 kg, are coupled together and coasting along horizontal tracks at a speed of v_i toward the south. A very strong but foolish movie actor, riding on the second car, uncouples the front car and gives it a big push, increasing its speed to 4.00 m/s southward. The remaining three cars continue moving toward the south, now at 2.00 m/s. (a) Find the initial speed of the cars. (b) How much work did the actor do? (c) State the relationship between the process described here and the process in Problem 22.

24. A 7.00-kg bowling ball collides head-on with a 2.00-kg bowling pin. The pin flies forward with a speed of 3.00 m/s. If the ball continues forward with a speed of 1.80 m/s, what was the initial speed of the ball? Ignore rotation of the ball.

WEB **25.** A neutron in a reactor makes an elastic head-on collision with the nucleus of a carbon atom initially at rest. (a) What fraction of the neutron's kinetic energy is transferred to the carbon nucleus? (b) If the initial kinetic energy of the neutron is 1.60×10^{-13} J, find its final kinetic energy and the kinetic energy of the carbon nucleus after the collision. (The mass of the carbon nucleus is about 12.0 times greater than the mass of the neutron.)

26. Consider a frictionless track *ABC* as shown in Figure P9.26. A block of mass $m_1 = 5.00$ kg is released from *A*. It makes a head-on elastic collision at *B* with a block of mass $m_2 = 10.0$ kg that is initially at rest. Calculate the maximum height to which m_1 rises after the collision.

Figure P9.26

WEB **27.** A 12.0-g bullet is fired into a 100-g wooden block initially at rest on a horizontal surface. After impact, the block slides 7.50 m before coming to rest. If the coefficient of friction between the block and the surface is

0.650, what was the speed of the bullet immediately before impact?

28. A 7.00-g bullet, when fired from a gun into a 1.00-kg block of wood held in a vise, would penetrate the block to a depth of 8.00 cm. This block of wood is placed on a frictionless horizontal surface, and a 7.00-g bullet is fired from the gun into the block. To what depth will the bullet penetrate the block in this case?

Section 9.5   Two-Dimensional Collisions

29. A 90.0-kg fullback running east with a speed of 5.00 m/s is tackled by a 95.0-kg opponent running north with a speed of 3.00 m/s. If the collision is perfectly inelastic, (a) calculate the speed and direction of the players just after the tackle and (b) determine the energy lost as a result of the collision. Account for the missing energy.

30. The mass of the blue puck in Figure P9.30 is 20.0% greater than the mass of the green one. Before colliding, the pucks approach each other with equal and opposite momenta, and the green puck has an initial speed of 10.0 m/s. Find the speeds of the pucks after the collision if half the kinetic energy is lost during the collision.

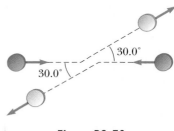

Figure P9.30

31. Two automobiles of equal mass approach an intersection. One vehicle is traveling with velocity 13.0 m/s toward the east and the other is traveling north with a speed of v_{2i}. Neither driver sees the other. The vehicles collide in the intersection and stick together, leaving parallel skid marks at an angle of 55.0° north of east. The speed limit for both roads is 35 mi/h, and the driver of the northward-moving vehicle claims he was within the speed limit when the collision occurred. Is he telling the truth?

32. A proton, moving with a velocity of $v_i\mathbf{i}$, collides elastically with another proton that is initially at rest. If the two protons have equal speeds after the collision, find (a) the speed of each proton after the collision in terms of v_i and (b) the direction of the velocity vectors after the collision.

33. A billiard ball moving at 5.00 m/s strikes a stationary ball of the same mass. After the collision, the first ball moves at 4.33 m/s and at an angle of 30.0° with respect to the original line of motion. Assuming an elastic collision (and ignoring friction and rotational motion), find the struck ball's velocity.

34. A 0.300-kg puck, initially at rest on a horizontal, frictionless surface, is struck by a 0.200-kg puck moving initially along the x axis with a speed of 2.00 m/s. After the collision, the 0.200-kg puck has a speed of 1.00 m/s at an angle of $\theta = 53.0°$ to the positive x axis (see Fig. 9.14). (a) Determine the velocity of the 0.300-kg puck after the collision. (b) Find the fraction of kinetic energy lost in the collision.

35. A 3.00-kg mass with an initial velocity of $5.00\mathbf{i}$ m/s collides with and sticks to a 2.00-kg mass with an initial velocity of $-3.00\mathbf{j}$ m/s. Find the final velocity of the composite mass.

36. Two shuffleboard disks of equal mass, one orange and the other yellow, are involved in a perfectly elastic, glancing collision. The yellow disk is initially at rest and is struck by the orange disk moving with a speed of 5.00 m/s. After the collision, the orange disk moves along a direction that makes an angle of 37.0° with its initial direction of motion, and the velocity of the yellow disk is perpendicular to that of the orange disk (after the collision). Determine the final speed of each disk.

37. Two shuffleboard disks of equal mass, one orange and the other yellow, are involved in a perfectly elastic, glancing collision. The yellow disk is initially at rest and is struck by the orange disk moving with a speed v_i. After the collision, the orange disk moves along a direction that makes an angle θ with its initial direction of motion, and the velocity of the yellow disk is perpendicular to that of the orange disk (after the collision). Determine the final speed of each disk.

38. During the battle of Gettysburg, the gunfire was so intense that several bullets collided in midair and fused together. Assume a 5.00-g Union musket ball was moving to the right at a speed of 250 m/s, 20.0° above the horizontal, and that a 3.00-g Confederate ball was moving to the left at a speed of 280 m/s, 15.0° above the horizontal. Immediately after they fuse together, what is their velocity?

WEB 39. An unstable nucleus of mass 17.0×10^{-27} kg initially at rest disintegrates into three particles. One of the particles, of mass 5.00×10^{-27} kg, moves along the y axis with a velocity of 6.00×10^6 m/s. Another particle, of mass 8.40×10^{-27} kg, moves along the x axis with a speed of 4.00×10^6 m/s. Find (a) the velocity of the third particle and (b) the total kinetic energy increase in the process.

Section 9.6 The Center of Mass

40. Four objects are situated along the y axis as follows: A 2.00-kg object is at $+3.00$ m, a 3.00-kg object is at $+2.50$ m, a 2.50-kg object is at the origin, and a 4.00-kg object is at -0.500 m. Where is the center of mass of these objects?

41. A uniform piece of sheet steel is shaped as shown in Figure P9.41. Compute the x and y coordinates of the center of mass of the piece.

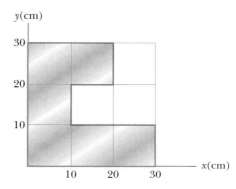

Figure P9.41

42. The mass of the Earth is 5.98×10^{24} kg, and the mass of the Moon is 7.36×10^{22} kg. The distance of separation, measured between their centers, is 3.84×10^8 m. Locate the center of mass of the Earth–Moon system as measured from the center of the Earth.

43. A water molecule consists of an oxygen atom with two hydrogen atoms bound to it (Fig. P9.43). The angle between the two bonds is 106°. If the bonds are 0.100 nm long, where is the center of mass of the molecule?

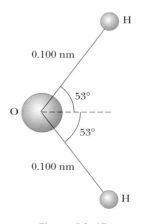

Figure P9.43

44. A 0.400-kg mass m_1 has position $\mathbf{r}_1 = 12.0\mathbf{j}$ cm. A 0.800-kg mass m_2 has position $\mathbf{r}_2 = -12.0\mathbf{i}$ cm. Another 0.800-kg mass m_3 has position $\mathbf{r}_3 = (12.0\mathbf{i} - 12.0\mathbf{j})$ cm. Make a drawing of the masses. Start from the origin and, to the scale 1 cm = 1 kg·cm, construct the vector $m_1\mathbf{r}_1$, then the vector $m_1\mathbf{r}_1 + m_2\mathbf{r}_2$, then the vector $m_1\mathbf{r}_1 + m_2\mathbf{r}_2 + m_3\mathbf{r}_3$, and at last $\mathbf{r}_{CM} = (m_1\mathbf{r}_1 + m_2\mathbf{r}_2 + m_3\mathbf{r}_3)/(m_1 + m_2 + m_3)$. Observe that the head of the vector $\mathbf{r}_{CM}$ indicates the position of the center of mass.

45. A rod of length 30.0 cm has linear density (mass-per-length) given by

$$\lambda = 50.0 \text{ g/m} + 20.0x \text{ g/m}^2$$

where x is the distance from one end, measured in meters. (a) What is the mass of the rod? (b) How far from the $x = 0$ end is its center of mass?

Section 9.7   Motion of a System of Particles

46. Consider a system of two particles in the xy plane: $m_1 = 2.00$ kg is at $\mathbf{r}_1 = (1.00\mathbf{i} + 2.00\mathbf{j})$ m and has velocity $(3.00\mathbf{i} + 0.500\mathbf{j})$ m/s; $m_2 = 3.00$ kg is at $\mathbf{r}_2 = (-4.00\mathbf{i} - 3.00\mathbf{j})$ m and has velocity $(3.00\mathbf{i} - 2.00\mathbf{j})$ m/s. (a) Plot these particles on a grid or graph paper. Draw their position vectors and show their velocities. (b) Find the position of the center of mass of the system and mark it on the grid. (c) Determine the velocity of the center of mass and also show it on the diagram. (d) What is the total linear momentum of the system?

47. Romeo (77.0 kg) entertains Juliet (55.0 kg) by playing his guitar from the rear of their boat at rest in still water, 2.70 m away from Juliet who is in the front of the boat. After the serenade, Juliet carefully moves to the rear of the boat (away from shore) to plant a kiss on Romeo's cheek. How far does the 80.0-kg boat move toward the shore it is facing?

48. Two masses, 0.600 kg and 0.300 kg, begin uniform motion at the same speed, 0.800 m/s, from the origin at $t = 0$ and travel in the directions shown in Figure P9.48. (a) Find the velocity of the center of mass in unit–vector notation. (b) Find the magnitude and direction

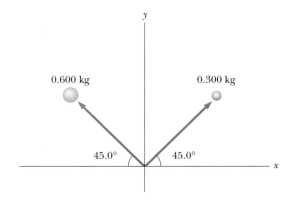

Figure P9.48

of the velocity of the center of mass. (c) Write the position vector of the center of mass as a function of time.

49. A 2.00-kg particle has a velocity of $(2.00\mathbf{i} - 3.00\mathbf{j})$ m/s, and a 3.00-kg particle has a velocity of $(1.00\mathbf{i} + 6.00\mathbf{j})$ m/s. Find (a) the velocity of the center of mass and (b) the total momentum of the system.

50. A ball of mass 0.200 kg has a velocity of $1.50\mathbf{i}$ m/s; a ball of mass 0.300 kg has a velocity of $-0.400\mathbf{i}$ m/s. They meet in a head-on elastic collision. (a) Find their velocities after the collision. (b) Find the velocity of their center of mass before and after the collision.

(Optional)
Section 9.8   Rocket Propulsion

WEB 51. The first stage of a Saturn V space vehicle consumes fuel and oxidizer at the rate of 1.50×10^4 kg/s, with an exhaust speed of 2.60×10^3 m/s. (a) Calculate the thrust produced by these engines. (b) Find the initial acceleration of the vehicle on the launch pad if its initial mass is 3.00×10^6 kg. [*Hint:* You must include the force of gravity to solve part (b).]

52. A large rocket with an exhaust speed of $v_e = 3\,000$ m/s develops a thrust of 24.0 million newtons. (a) How much mass is being blasted out of the rocket exhaust per second? (b) What is the maximum speed the rocket can attain if it starts from rest in a force-free environment with $v_e = 3.00$ km/s and if 90.0% of its initial mass is fuel and oxidizer?

53. A rocket for use in deep space is to have the capability of boosting a total load (payload plus rocket frame and engine) of 3.00 metric tons to a speed of 10 000 m/s. (a) It has an engine and fuel designed to produce an exhaust speed of 2 000 m/s. How much fuel plus oxidizer is required? (b) If a different fuel and engine design could give an exhaust speed of 5 000 m/s, what amount of fuel and oxidizer would be required for the same task?

54. A rocket car has a mass of 2 000 kg unfueled and a mass of 5 000 kg when completely fueled. The exhaust velocity is 2 500 m/s. (a) Calculate the amount of fuel used to accelerate the completely fueled car from rest to 225 m/s (about 500 mi/h). (b) If the burn rate is constant at 30.0 kg/s, calculate the time it takes the car to reach this speed. Neglect friction and air resistance.

ADDITIONAL PROBLEMS

55. **Review Problem.**   A 60.0-kg person running at an initial speed of 4.00 m/s jumps onto a 120-kg cart initially at rest. (Fig. P9.55) The person slides on the cart's top surface and finally comes to rest relative to the cart. The coefficient of kinetic friction between the person and the cart is 0.400. Friction between the cart and ground can be neglected. (a) Find the final velocity of the person and cart relative to the ground. (b) Find the frictional force acting on the person while he is sliding

across the top surface of the cart. (c) How long does the frictional force act on the person? (d) Find the change in momentum of the person and the change in momentum of the cart. (e) Determine the displacement of the person relative to the ground while he is sliding on the cart. (f) Determine the displacement of the cart relative to the ground while the person is sliding. (g) Find the change in kinetic energy of the person. (h) Find the change in kinetic energy of the cart. (i) Explain why the answers to parts (g) and (h) differ. (What kind of collision is this, and what accounts for the loss of mechanical energy?)

60.0 kg   4.00 m/s

120 kg

Figure P9.55

56. A golf ball ($m = 46.0$ g) is struck a blow that makes an angle of 45.0° with the horizontal. The ball lands 200 m away on a flat fairway. If the golf club and ball are in contact for 7.00 ms, what is the average force of impact? (Neglect air resistance.)

57. An 8.00-g bullet is fired into a 2.50-kg block that is initially at rest at the edge of a frictionless table of height 1.00 m (Fig. P9.57). The bullet remains in the block, and after impact the block lands 2.00 m from the bottom of the table. Determine the initial speed of the bullet.

58. A bullet of mass m is fired into a block of mass M that is initially at rest at the edge of a frictionless table of height h (see Fig. P9.57). The bullet remains in the block, and after impact the block lands a distance d from the bottom of the table. Determine the initial speed of the bullet.

59. An 80.0-kg astronaut is working on the engines of his ship, which is drifting through space with a constant velocity. The astronaut, wishing to get a better view of the Universe, pushes against the ship and much later finds himself 30.0 m behind the ship and at rest with respect to it. Without a thruster, the only way to return to the ship is to throw his 0.500-kg wrench directly away from the ship. If he throws the wrench with a speed of 20.0 m/s relative to the ship, how long does it take the astronaut to reach the ship?

60. A small block of mass $m_1 = 0.500$ kg is released from rest at the top of a curve-shaped frictionless wedge of mass $m_2 = 3.00$ kg, which sits on a frictionless horizontal surface, as shown in Figure P9.60a. When the block leaves the wedge, its velocity is measured to be 4.00 m/s to the right, as in Figure P9.60b. (a) What is the velocity of the wedge after the block reaches the horizontal surface? (b) What is the height h of the wedge?

m_1

h

m_2 v_2 m_2

4.00 m/s

(a)   (b)

Figure P9.60

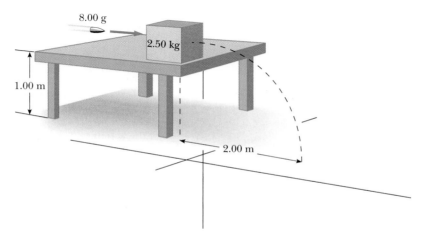

8.00 g

2.50 kg

1.00 m

2.00 m

Figure P9.57   Problems 57 and 58.

61. Tarzan, whose mass is 80.0 kg, swings from a 3.00-m vine that is horizontal when he starts. At the bottom of his arc, he picks up 60.0-kg Jane in a completely inelastic collision. What is the height of the highest tree limb they can reach on their upward swing?

62. A jet aircraft is traveling at 500 mi/h (223 m/s) in horizontal flight. The engine takes in air at a rate of 80.0 kg/s and burns fuel at a rate of 3.00 kg/s. If the exhaust gases are ejected at 600 m/s relative to the aircraft, find the thrust of the jet engine and the delivered horsepower.

63. A 75.0-kg firefighter slides down a pole while a constant frictional force of 300 N retards her motion. A horizontal 20.0-kg platform is supported by a spring at the bottom of the pole to cushion the fall. The firefighter starts from rest 4.00 m above the platform, and the spring constant is 4 000 N/m. Find (a) the firefighter's speed just before she collides with the platform and (b) the maximum distance the spring is compressed. (Assume the frictional force acts during the entire motion.)

64. A cannon is rigidly attached to a carriage, which can move along horizontal rails but is connected to a post by a large spring, initially unstretched and with force constant $k = 2.00 \times 10^4$ N/m, as shown in Figure P9.64. The cannon fires a 200-kg projectile at a velocity of 125 m/s directed 45.0° above the horizontal. (a) If the mass of the cannon and its carriage is 5 000 kg, find the recoil speed of the cannon. (b) Determine the maximum extension of the spring. (c) Find the maximum force the spring exerts on the carriage. (d) Consider the system consisting of the cannon, carriage, and shell. Is the momentum of this system conserved during the firing? Why or why not?

45.0°

Figure P9.64

65. A chain of length L and total mass M is released from rest with its lower end just touching the top of a table, as shown in Figure P9.65a. Find the force exerted by the table on the chain after the chain has fallen through a distance x, as shown in Figure P9.65b. (Assume each link comes to rest the instant it reaches the table.)

L

x

$L - x$

(a)   (b)

Figure P9.65

66. Two gliders are set in motion on an air track. A spring of force constant k is attached to the near side of one glider. The first glider of mass m_1 has a velocity of $\mathbf{v}_1$, and the second glider of mass m_2 has a velocity of $\mathbf{v}_2$, as shown in Figure P9.66 ($v_1 > v_2$). When m_1 collides with the spring attached to m_2 and compresses the spring to its maximum compression x_m, the velocity of the gliders is $\mathbf{v}$. In terms of $\mathbf{v}_1$, $\mathbf{v}_2$, m_1, m_2, and k, find (a) the velocity $\mathbf{v}$ at maximum compression, (b) the maximum compression x_m, and (c) the velocities of each glider after m_1 has lost contact with the spring.

$\mathbf{v}_2$

k

m_2

$\mathbf{v}_1$

m_1

Figure P9.66

67. Sand from a stationary hopper falls onto a moving conveyor belt at the rate of 5.00 kg/s, as shown in Figure P9.67. The conveyor belt is supported by frictionless rollers and moves at a constant speed of 0.750 m/s under the action of a constant horizontal external force $\mathbf{F}_{ext}$ supplied by the motor that drives the belt. Find (a) the sand's rate of change of momentum in the horizontal direction, (b) the force of friction exerted by the belt on the sand, (c) the external force $\mathbf{F}_{ext}$, (d) the work done by $\mathbf{F}_{ext}$ in 1 s, and (e) the kinetic energy acquired by the falling sand each second due to the change in its horizontal motion. (f) Why are the answers to parts (d) and (e) different?

Figure P9.67

68. A rocket has total mass $M_i = 360$ kg, including 330 kg of fuel and oxidizer. In interstellar space it starts from rest, turns on its engine at time $t = 0$, and puts out exhaust with a relative speed of $v_e = 1\,500$ m/s at the constant rate $k = 2.50$ kg/s. Although the fuel will last for an actual burn time of 330 kg/$(2.5$ kg/s$) = 132$ s, define a "projected depletion time" as $T_p = M_i/k = 360$ kg/$(2.5$ kg/s$) = 144$ s. (This would be the burn time if the rocket could use its payload, fuel tanks, and even the walls of the combustion chamber as fuel.)
(a) Show that during the burn the velocity of the rocket is given as a function of time by

$$v(t) = -v_e \ln(1 - t/T_p)$$

(b) Make a graph of the velocity of the rocket as a function of time for times running from 0 to 132 s. (c) Show that the acceleration of the rocket is

$$a(t) = v_e/(T_p - t)$$

(d) Graph the acceleration as a function of time.
(e) Show that the displacement of the rocket from its initial position at $t = 0$ is

$$x(t) = v_e(T_p - t)\ln(1 - t/T_p) + v_e t$$

(f) Graph the displacement during the burn.
69. A 40.0-kg child stands at one end of a 70.0-kg boat that is 4.00 m in length (Fig. P9.69). The boat is initially 3.00 m from the pier. The child notices a turtle on a rock near the far end of the boat and proceeds to walk to that end to catch the turtle. Neglecting friction be-

tween the boat and the water, (a) describe the subsequent motion of the system (child plus boat). (b) Where is the child *relative to the pier* when he reaches the far end of the boat? (c) Will he catch the turtle? (Assume he can reach out 1.00 m from the end of the boat.)

70. A student performs a ballistic pendulum experiment, using an apparatus similar to that shown in Figure 9.11b. She obtains the following average data: $h = 8.68$ cm, $m_1 = 68.8$ g, and $m_2 = 263$ g. The symbols refer to the quantities in Figure 9.11a. (a) Determine the initial speed v_{1i} of the projectile. (b) In the second part of her experiment she is to obtain v_{1i} by firing the same projectile horizontally (with the pendulum removed from the path) and measuring its horizontal displacement x and vertical displacement y (Fig. P9.70). Show that the initial speed of the projectile is related to x and y through the relationship

$$v_{1i} = \frac{x}{\sqrt{2y/g}}$$

What numerical value does she obtain for v_{1i} on the basis of her measured values of $x = 257$ cm and $y = 85.3$ cm? What factors might account for the difference in this value compared with that obtained in part (a)?

(b)

Figure P9.70

Figure P9.69

71. A 5.00-g bullet moving with an initial speed of 400 m/s is fired into and passes through a 1.00-kg block, as shown in Figure P9.71. The block, initially at rest on a

frictionless, horizontal surface, is connected to a spring of force constant 900 N/m. If the block moves 5.00 cm to the right after impact, find (a) the speed at which the bullet emerges from the block and (b) the energy lost in the collision.

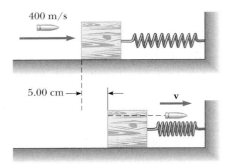

Figure P9.71

72. Two masses m and $3m$ are moving toward each other along the x axis with the same initial speeds v_i. Mass m is traveling to the left, while mass $3m$ is traveling to the right. They undergo a head-on elastic collision and each rebounds along the same line as it approached. Find the final speeds of the masses.

73. Two masses m and $3m$ are moving toward each other along the x axis with the same initial speeds v_i. Mass m is traveling to the left, while mass $3m$ is traveling to the right. They undergo an elastic glancing collision such that mass m is moving downward after the collision at right angles from its initial direction. (a) Find the final speeds of the two masses. (b) What is the angle θ at which the mass $3m$ is scattered?

74. **Review Problem.** There are (one can say) three co-equal theories of motion: Newton's second law, stating that the total force on an object causes its acceleration; the work–kinetic energy theorem, stating that the total work on an object causes its change in kinetic energy; and the impulse–momentum theorem, stating that the total impulse on an object causes its change in momentum. In this problem, you compare predictions of the three theories in one particular case. A 3.00-kg object has a velocity of $7.00\mathbf{j}$ m/s. Then, a total force $12.0\mathbf{i}$ N acts on the object for 5.00 s. (a) Calculate the object's final velocity, using the impulse–momentum theorem. (b) Calculate its acceleration from $\mathbf{a} = (\mathbf{v}_f - \mathbf{v}_i)/t$. (c) Calculate its acceleration from $\mathbf{a} = \Sigma\mathbf{F}/m$. (d) Find the object's vector displacement from $\mathbf{r} = \mathbf{v}_i t + \frac{1}{2}\mathbf{a}\, t^2$. (e) Find the work done on the object from $W = \mathbf{F}\cdot\mathbf{r}$. (f) Find the final kinetic energy from $\frac{1}{2}mv_f^2 = \frac{1}{2}m\mathbf{v}_f\cdot\mathbf{v}_f$. (g) Find the final kinetic energy from $\frac{1}{2}mv_i^2 + W$.

75. A rocket has a total mass of $M_i = 360$ kg, including 330 kg of fuel and oxidizer. In interstellar space it starts from rest. Its engine is turned on at time $t = 0$, and it puts out exhaust with a relative speed of $v_e = 1\,500$ m/s at the constant rate 2.50 kg/s. The burn lasts until the fuel runs out at time 330 kg/$(2.5$ kg/s$) = 132$ s. Set up and carry out a computer analysis of the motion according to Euler's method. Find (a) the final velocity of the rocket and (b) the distance it travels during the burn.

ANSWERS TO QUICK QUIZZES

9.1 (d). Two identical objects ($m_1 = m_2$) traveling in the same direction at the same speed ($v_1 = v_2$) have the same kinetic energies and the same momenta. However, this is not true if the two objects are moving at the same speed but in different directions. In the latter case, $K_1 = K_2$, but the differing velocity directions indicate that $\mathbf{p}_1 \neq \mathbf{p}_2$ because momentum is a vector quantity.

It also is possible for particular combinations of masses and velocities to satisfy $K_1 = K_2$ but not $p_1 = p_2$. For example, a 1-kg object moving at 2 m/s has the same kinetic energy as a 4-kg object moving at 1 m/s, but the two clearly do not have the same momenta.

9.2 (b), (c), (a). The slower the ball, the easier it is to catch. If the momentum of the medicine ball is the same as the momentum of the baseball, the speed of the medicine ball must be 1/10 the speed of the baseball because the medicine ball has 10 times the mass. If the kinetic energies are the same, the speed of the medicine ball must be $1/\sqrt{10}$ the speed of the baseball because of the squared speed term in the formula for K. The medicine ball is hardest to catch when it has the same speed as the baseball.

9.3 (c) and (e). Object 2 has a greater acceleration because of its smaller mass. Therefore, it takes less time to travel the distance d. Thus, even though the force applied to objects 1 and 2 is the same, the change in momentum is less for object 2 because Δt is smaller. Therefore, because the initial momenta were the same (both zero), $p_1 > p_2$. The work $W = Fd$ done on both objects is the same because both F and d are the same in the two cases. Therefore, $K_1 = K_2$.

9.4 Because the passenger is brought from the car's initial speed to a full stop, the change in momentum (the impulse) is the same regardless of whether the passenger is stopped by dashboard, seatbelt, or airbag. However, the dashboard stops the passenger very quickly, the seatbelt takes a little more time, and the airbag takes the most time. Therefore, the dashboard applies the greatest force, the seatbelt an intermediate force, and the airbag the least force. Airbags are designed to work in conjunction with seatbelts, so make sure you wear your seatbelt at all times while in a moving vehicle.

9.5 If we define the ball as our system, momentum is not conserved. The ball's speed—and hence its momentum—continually increase. This is consistent with the fact that the gravitational force is external to this chosen system. However, if we define our system as the ball and the Earth, momentum is conserved, for the Earth also has momentum because the ball exerts a gravitational force on it. As the ball falls, the Earth moves up to meet it (although the Earth's speed is on the order of 10^{25} times less than that of the ball!). This upward movement changes the Earth's momentum. The change in the Earth's momentum is numerically equal to the change in the ball's momentum but is in the opposite direction. Therefore, the total momentum of the Earth–ball system is conserved. Because the Earth's mass is so great, its upward motion is negligibly small.

9.6 (c). The greatest impulse (greatest change in momentum) is imparted to the Frisbee when the skater reverses its momentum vector by catching it and throwing it back. Since this is when the skater imparts the greatest impulse to the Frisbee, then this also is when the Frisbee imparts the greatest impulse to her.

9.7 Both are equally bad. Imagine watching the collision from a safer location alongside the road. As the "crush zones" of the two cars are compressed, you will see that the actual point of contact is stationary. You would see the same thing if your car were to collide with a solid wall.

9.8 No, such movement can never occur if we assume the collisions are elastic. The momentum of the system before the collision is mv, where m is the mass of ball 1 and v is its speed just before the collision. After the collision, we would have two balls, each of mass m and moving with a speed of $v/2$. Thus, the total momentum of the system after the collision would be $m(v/2) + m(v/2) = mv$. Thus, momentum is conserved. However, the kinetic energy just before the collision is $K_i = \frac{1}{2}mv^2$, and that after the collision is $K_f = \frac{1}{2}m(v/2)^2 + \frac{1}{2}m(v/2)^2 = \frac{1}{4}mv^2$. Thus, kinetic energy is *not* conserved. Both momentum and kinetic energy are conserved only when one ball moves out when one ball is released, two balls move out when two are released, and so on.

9.9 No they will not! The piece with the handle will have less mass than the piece made up of the end of the bat. To see why this is so, take the origin of coordinates as the center of mass before the bat was cut. Replace each cut piece by a small sphere located at the center of mass for each piece. The sphere representing the handle piece is farther from the origin, but the product of lesser mass and greater distance balances the product of greater mass and lesser distance for the end piece:

c h a p t e r

Rotation of a Rigid Object About a Fixed Axis

When an extended object, such as a wheel, rotates about its axis, the motion cannot be analyzed by treating the object as a particle because at any given time different parts of the object have different linear velocities and linear accelerations. For this reason, it is convenient to consider an extended object as a large number of particles, each of which has its own linear velocity and linear acceleration.

In dealing with a rotating object, analysis is greatly simplified by assuming that the object is rigid. A **rigid object** is one that is nondeformable—that is, it is an object in which the separations between all pairs of particles remain constant. All real bodies are deformable to some extent; however, our rigid-object model is useful in many situations in which deformation is negligible.

Rigid object

In this chapter, we treat the rotation of a rigid object about a fixed axis, which is commonly referred to as *pure rotational motion.*

10.1 ANGULAR DISPLACEMENT, VELOCITY, AND ACCELERATION

Figure 10.1 illustrates a planar (flat), rigid object of arbitrary shape confined to the *xy* plane and rotating about a fixed axis through *O*. The axis is perpendicular to the plane of the figure, and *O* is the origin of an *xy* coordinate system. Let us look at the motion of only one of the millions of "particles" making up this object. A particle at *P* is at a fixed distance *r* from the origin and rotates about it in a circle of radius *r*. (In fact, *every* particle on the object undergoes circular motion about *O*.) It is convenient to represent the position of *P* with its polar coordinates (r, θ), where *r* is the distance from the origin to *P* and θ is measured *counterclockwise* from some preferred direction—in this case, the positive *x* axis. In this representation, the only coordinate that changes in time is the angle θ; *r* remains constant. (In cartesian coordinates, both *x* and *y* vary in time.) As the particle moves along the circle from the positive *x* axis ($\theta = 0$) to *P*, it moves through an arc of length *s*, which is related to the angular position θ through the relationship

$$s = r\theta \tag{10.1a}$$

$$\theta = \frac{s}{r} \tag{10.1b}$$

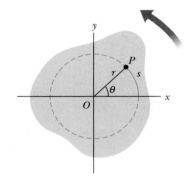

Figure 10.1 A rigid object rotating about a fixed axis through *O* perpendicular to the plane of the figure. (In other words, the axis of rotation is the *z* axis.) A particle at *P* rotates in a circle of radius *r* centered at *O*.

It is important to note the units of θ in Equation 10.1b. Because θ is the ratio of an arc length and the radius of the circle, it is a pure number. However, we commonly give θ the artificial unit **radian** (rad), where

one radian is the angle subtended by an arc length equal to the radius of the arc.

Radian

Because the circumference of a circle is $2\pi r$, it follows from Equation 10.1b that 360° corresponds to an angle of $2\pi r/r$ rad $= 2\pi$ rad (one revolution). Hence, 1 rad $= 360°/2\pi \approx 57.3°$. To convert an angle in degrees to an angle in radians, we use the fact that 2π rad $= 360°$:

$$\theta \text{ (rad)} = \frac{\pi}{180°} \, \theta \text{ (deg)}$$

For example, 60° equals $\pi/3$ rad, and 45° equals $\pi/4$ rad.

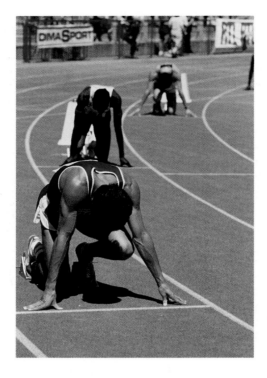

In a short track event, such as a 200-m or 400-m sprint, the runners begin from staggered positions on the track. Why don't they all begin from the same line? *(© Gerard Vandystadt/Photo Researchers, Inc.)*

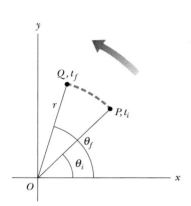

Figure 10.2 A particle on a rotating rigid object moves from P to Q along the arc of a circle. In the time interval $\Delta t = t_f - t_i$, the radius vector sweeps out an angle $\Delta\theta = \theta_f - \theta_i$.

As the particle in question on our rigid object travels from position P to position Q in a time Δt as shown in Figure 10.2, the radius vector sweeps out an angle $\Delta\theta = \theta_f - \theta_i$. This quantity $\Delta\theta$ is defined as the **angular displacement** of the particle:

$$\Delta\theta = \theta_f - \theta_i \qquad \textbf{(10.2)}$$

We define the **average angular speed** $\overline{\omega}$ (omega) as the ratio of this angular displacement to the time interval Δt:

$$\overline{\omega} \equiv \frac{\theta_f - \theta_i}{t_f - t_i} = \frac{\Delta\theta}{\Delta t} \qquad \textbf{(10.3)}$$

In analogy to linear speed, the **instantaneous angular speed** ω is defined as the limit of the ratio $\Delta\theta/\Delta t$ as Δt approaches zero:

$$\omega \equiv \lim_{\Delta t \to 0} \frac{\Delta\theta}{\Delta t} = \frac{d\theta}{dt} \qquad \textbf{(10.4)}$$

Angular speed has units of radians per second (rad/s), or rather second^{-1} (s^{-1}) because radians are not dimensional. We take ω to be positive when θ is increasing (counterclockwise motion) and negative when θ is decreasing (clockwise motion).

If the instantaneous angular speed of an object changes from ω_i to ω_f in the time interval Δt, the object has an angular acceleration. The **average angular acceleration** $\overline{\alpha}$ (alpha) of a rotating object is defined as the ratio of the change in the angular speed to the time interval Δt:

$$\overline{\alpha} \equiv \frac{\omega_f - \omega_i}{t_f - t_i} = \frac{\Delta\omega}{\Delta t} \qquad \textbf{(10.5)}$$

Figure 10.3 The right-hand rule for determining the direction of the angular velocity vector.

In analogy to linear acceleration, the **instantaneous angular acceleration** is defined as the limit of the ratio $\Delta\omega/\Delta t$ as Δt approaches zero:

$$\alpha \equiv \lim_{\Delta t \to 0} \frac{\Delta\omega}{\Delta t} = \frac{d\omega}{dt} \qquad (10.6)$$

Instantaneous angular acceleration

Angular acceleration has units of radians per second squared (rad/s^2), or just second^{-2} (s^{-2}). Note that α is positive when the rate of counterclockwise rotation is increasing or when the rate of clockwise rotation is decreasing.

When rotating about a fixed axis, every particle on a rigid object rotates through the same angle and has the same angular speed and the same angular acceleration. That is, the quantities θ, ω, and α characterize the rotational motion of the entire rigid object. Using these quantities, we can greatly simplify the analysis of rigid-body rotation.

Angular position (θ), angular speed (ω), and angular acceleration (α) are analogous to linear position (x), linear speed (v), and linear acceleration (a). The variables θ, ω, and α differ dimensionally from the variables x, v, and a only by a factor having the unit of length.

We have not specified any direction for ω and α. Strictly speaking, these variables are the magnitudes of the angular velocity and the angular acceleration vectors $\boldsymbol{\omega}$ and $\boldsymbol{\alpha}$, respectively, and they should always be positive. Because we are considering rotation about a fixed axis, however, we can indicate the directions of the vectors by assigning a positive or negative sign to ω and α, as discussed earlier with regard to Equations 10.4 and 10.6. For rotation about a fixed axis, the only direction that uniquely specifies the rotational motion is the direction along the axis of rotation. Therefore, the directions of $\boldsymbol{\omega}$ and $\boldsymbol{\alpha}$ are along this axis. If an object rotates in the xy plane as in Figure 10.1, the direction of $\boldsymbol{\omega}$ is out of the plane of the diagram when the rotation is counterclockwise and into the plane of the diagram when the rotation is clockwise. To illustrate this convention, it is convenient to use the *right-hand rule* demonstrated in Figure 10.3. When the four fingers of the right hand are wrapped in the direction of rotation, the extended right thumb points in the direction of $\boldsymbol{\omega}$. The direction of $\boldsymbol{\alpha}$ follows from its definition $d\boldsymbol{\omega}/dt$. It is the same as the direction of $\boldsymbol{\omega}$ if the angular speed is increasing in time, and it is antiparallel to $\boldsymbol{\omega}$ if the angular speed is decreasing in time.

Quick Quiz 10.1

Describe a situation in which $\omega < 0$ and $\boldsymbol{\omega}$ and $\boldsymbol{\alpha}$ are antiparallel.

10.2 ROTATIONAL KINEMATICS: ROTATIONAL MOTION WITH CONSTANT ANGULAR ACCELERATION

In our study of linear motion, we found that the simplest form of accelerated motion to analyze is motion under constant linear acceleration. Likewise, for rotational motion about a fixed axis, the simplest accelerated motion to analyze is motion under constant angular acceleration. Therefore, we next develop kinematic relationships for this type of motion. If we write Equation 10.6 in the form $d\omega = \alpha\,dt$, and let $t_i = 0$ and $t_f = t$, we can integrate this expression directly:

$$\omega_f = \omega_i + \alpha t \qquad \text{(for constant } \alpha\text{)} \qquad \textbf{(10.7)}$$

Substituting Equation 10.7 into Equation 10.4 and integrating once more we obtain

Rotational kinematic equations

$$\theta_f = \theta_i + \omega_i t + \tfrac{1}{2}\alpha t^2 \qquad \text{(for constant } \alpha\text{)} \qquad \textbf{(10.8)}$$

If we eliminate t from Equations 10.7 and 10.8, we obtain

$$\omega_f^2 = \omega_i^2 + 2\alpha(\theta_f - \theta_i) \qquad \text{(for constant } \alpha\text{)} \qquad \textbf{(10.9)}$$

Notice that these kinematic expressions for rotational motion under constant angular acceleration are of the same form as those for linear motion under constant linear acceleration with the substitutions $x \rightarrow \theta$, $v \rightarrow \omega$, and $a \rightarrow \alpha$. Table 10.1 compares the kinematic equations for rotational and linear motion.

EXAMPLE 10.1   Rotating Wheel

A wheel rotates with a constant angular acceleration of 3.50 rad/s². If the angular speed of the wheel is 2.00 rad/s at $t_i = 0$, (a) through what angle does the wheel rotate in 2.00 s?

Solution   We can use Figure 10.2 to represent the wheel, and so we do not need a new drawing. This is a straightforward application of an equation from Table 10.1:

$$\theta_f - \theta_i = \omega_i t + \tfrac{1}{2}\alpha t^2 = (2.00 \text{ rad/s})(2.00 \text{ s})$$
$$+ \tfrac{1}{2}(3.50 \text{ rad/s}^2)(2.00 \text{ s})^2$$

$$= 11.0 \text{ rad} = (11.0 \text{ rad})(57.3°/\text{rad}) = \boxed{630°}$$

$$= \frac{630°}{360°/\text{rev}} = \boxed{1.75 \text{ rev}}$$

(b) What is the angular speed at $t = 2.00$ s?

Solution   Because the angular acceleration and the angular speed are both positive, we can be sure our answer must be greater than 2.00 rad/s.

$$\omega_f = \omega_i + \alpha t = 2.00 \text{ rad/s} + (3.50 \text{ rad/s}^2)(2.00 \text{ s})$$

$$= \boxed{9.00 \text{ rad/s}}$$

We could also obtain this result using Equation 10.9 and the results of part (a). Try it! You also may want to see if you can formulate the linear motion analog to this problem.

Exercise   Find the angle through which the wheel rotates between $t = 2.00$ s and $t = 3.00$ s.

Answer   10.8 rad.

TABLE 10.1 **Kinematic Equations for Rotational and Linear Motion Under Constant Acceleration**

Rotational Motion About a Fixed Axis	Linear Motion
$\omega_f = \omega_i + \alpha t$	$v_f = v_i + at$
$\theta_f = \theta_i + \omega_i t + \frac{1}{2}\alpha t^2$	$x_f = x_i + v_i t + \frac{1}{2}at^2$
$\omega_f{}^2 = \omega_i{}^2 + 2\alpha(\theta_f - \theta_i)$	$v_f{}^2 = v_i{}^2 + 2a(x_f - x_i)$

10.3 ANGULAR AND LINEAR QUANTITIES

In this section we derive some useful relationships between the angular speed and acceleration of a rotating rigid object and the linear speed and acceleration of an arbitrary point in the object. To do so, we must keep in mind that when a rigid object rotates about a fixed axis, as in Figure 10.4, every particle of the object moves in a circle whose center is the axis of rotation.

We can relate the angular speed of the rotating object to the tangential speed of a point P on the object. Because point P moves in a circle, the linear velocity vector $\mathbf{v}$ is always tangent to the circular path and hence is called *tangential velocity*. The magnitude of the tangential velocity of the point P is by definition the tangential speed $v = ds/dt$, where s is the distance traveled by this point measured along the circular path. Recalling that $s = r\theta$ (Eq. 10.1a) and noting that r is constant, we obtain

$$v = \frac{ds}{dt} = r\frac{d\theta}{dt}$$

Because $d\theta/dt = \omega$ (see Eq. 10.4), we can say

$$v = r\omega \qquad \text{(10.10)}$$

That is, the tangential speed of a point on a rotating rigid object equals the perpendicular distance of that point from the axis of rotation multiplied by the angular speed. Therefore, although every point on the rigid object has the same *angular* speed, not every point has the same *linear* speed because r is not the same for all points on the object. Equation 10.10 shows that the linear speed of a point on the rotating object increases as one moves outward from the center of rotation, as we would intuitively expect. The outer end of a swinging baseball bat moves much faster than the handle.

We can relate the angular acceleration of the rotating rigid object to the tangential acceleration of the point P by taking the time derivative of v:

$$a_t = \frac{dv}{dt} = r\frac{d\omega}{dt}$$

$$a_t = r\alpha \qquad \text{(10.11)}$$

That is, the tangential component of the linear acceleration of a point on a rotating rigid object equals the point's distance from the axis of rotation multiplied by the angular acceleration.

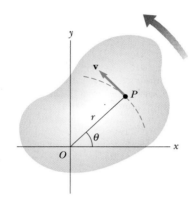

Figure 10.4 As a rigid object rotates about the fixed axis through O, the point P has a linear velocity $\mathbf{v}$ that is always tangent to the circular path of radius r.

Relationship between linear and angular speed

QuickLab

Spin a tennis ball or basketball and watch it gradually slow down and stop. Estimate α and a_t as accurately as you can.

Relationship between linear and angular acceleration

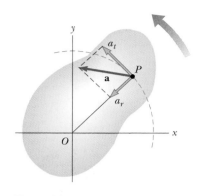

Figure 10.5 As a rigid object rotates about a fixed axis through O, the point P experiences a tangential component of linear acceleration a_t and a radial component of linear acceleration a_r. The total linear acceleration of this point is $\mathbf{a} = \mathbf{a}_t + \mathbf{a}_r$.

In Section 4.4 we found that a point rotating in a circular path undergoes a centripetal, or radial, acceleration $\mathbf{a}_r$ of magnitude v^2/r directed toward the center of rotation (Fig. 10.5). Because $v = r\omega$ for a point P on a rotating object, we can express the radial acceleration of that point as

$$a_r = \frac{v^2}{r} = r\omega^2 \tag{10.12}$$

The total linear acceleration vector of the point is $\mathbf{a} = \mathbf{a}_t + \mathbf{a}_r$. ($\mathbf{a}_t$ describes the change in how fast the point is moving, and $\mathbf{a}_r$ represents the change in its direction of travel.) Because $\mathbf{a}$ is a vector having a radial and a tangential component, the magnitude of $\mathbf{a}$ for the point P on the rotating rigid object is

$$a = \sqrt{a_t^2 + a_r^2} = \sqrt{r^2\alpha^2 + r^2\omega^4} = r\sqrt{\alpha^2 + \omega^4} \tag{10.13}$$

Quick Quiz 10.2

When a wheel of radius R rotates about a fixed axis, do all points on the wheel have (a) the same angular speed and (b) the same linear speed? If the angular speed is constant and equal to ω, describe the linear speeds and linear accelerations of the points located at (c) $r = 0$, (d) $r = R/2$, and (e) $r = R$, all measured from the center of the wheel.

EXAMPLE 10.2   CD Player

On a compact disc, audio information is stored in a series of pits and flat areas on the surface of the disc. The information is stored digitally, and the alternations between pits and flat areas on the surface represent binary ones and zeroes to be read by the compact disc player and converted back to sound waves. The pits and flat areas are detected by a system consisting of a laser and lenses. The length of a certain number of ones and zeroes is the same everywhere on the disc, whether the information is near the center of the disc or near its outer edge. In order that this length of ones and zeroes always passes by the laser–lens system in the same time period, the linear speed of the disc surface at the location of the lens must be constant. This requires, according to Equation 10.10, that the angular speed vary as the laser–lens system moves radially along the disc. In a typical compact disc player, the disc spins counterclockwise (Fig. 10.6), and the constant speed of the surface at the point of the laser–lens system is 1.3 m/s. (a) Find the angular speed of the disc in revolutions per minute when information is being read from the innermost first track ($r = 23$ mm) and the outermost final track ($r = 58$ mm).

Solution Using Equation 10.10, we can find the angular speed; this will give us the required linear speed at the position of the inner track,

$$\omega_i = \frac{v}{r_i} = \frac{1.3\ \text{m/s}}{2.3 \times 10^{-2}\ \text{m}} = 56.5\ \text{rad/s}$$

$$= (56.5\ \text{rad/s})\left(\frac{1}{2\pi}\ \text{rev/rad}\right)(60\ \text{s/min})$$

$$= 5.4 \times 10^2\ \text{rev/min}$$

Figure 10.6 A compact disc. *(George Semple)*

For the outer track,

$$\omega_f = \frac{v}{r_f} = \frac{1.3 \text{ m/s}}{5.8 \times 10^{-2} \text{ m}} = 22.4 \text{ rad/s}$$

$$= 2.1 \times 10^2 \text{ rev/min}$$

The player adjusts the angular speed ω of the disc within this range so that information moves past the objective lens at a constant rate. These angular velocity values are positive because the direction of rotation is counterclockwise.

(b) The maximum playing time of a standard music CD is 74 minutes and 33 seconds. How many revolutions does the disc make during that time?

Solution We know that the angular speed is always decreasing, and we assume that it is decreasing steadily, with α constant. The time interval t is $(74 \text{ min})(60 \text{ s/min}) + 33 \text{ s} = 4\,473 \text{ s}$. We are looking for the angular position θ_f, where we set the initial angular position $\theta_i = 0$. We can use Equation 10.3, replacing the average angular speed $\overline{\omega}$ with its mathematical equivalent $(\omega_i + \omega_f)/2$:

$$\theta_f = \theta_i + \tfrac{1}{2}(\omega_i + \omega_f)t$$
$$= 0 + \tfrac{1}{2}(540 \text{ rev/min} + 210 \text{ rev/min})$$
$$(1 \text{ min}/60 \text{ s})(4\,473 \text{ s})$$

$$= 2.8 \times 10^4 \text{ rev}$$

(c) What total length of track moves past the objective lens during this time?

Solution Because we know the (constant) linear velocity and the time interval, this is a straightforward calculation:

$$x_f = v_i t = (1.3 \text{ m/s})(4\,473 \text{ s}) = 5.8 \times 10^3 \text{ m}$$

More than 3.6 miles of track spins past the objective lens!

(d) What is the angular acceleration of the CD over the 4 473-s time interval? Assume that α is constant.

Solution We have several choices for approaching this problem. Let us use the most direct approach by utilizing Equation 10.5, which is based on the definition of the term we are seeking. We should obtain a negative number for the angular acceleration because the disc spins more and more slowly in the positive direction as time goes on. Our answer should also be fairly small because it takes such a long time— more than an hour—for the change in angular speed to be accomplished:

$$\alpha = \frac{\omega_f - \omega_i}{t} = \frac{22.4 \text{ rad/s} - 56.5 \text{ rad/s}}{4\,473 \text{ s}}$$

$$= -7.6 \times 10^{-3} \text{ rad/s}^2$$

The disc experiences a very gradual decrease in its rotation rate, as expected.

10.4 ROTATIONAL ENERGY

Let us now look at the kinetic energy of a rotating rigid object, considering the object as a collection of particles and assuming it rotates about a fixed z axis with an angular speed ω (Fig. 10.7). Each particle has kinetic energy determined by its mass and linear speed. If the mass of the ith particle is m_i and its linear speed is v_i, its kinetic energy is

$$K_i = \tfrac{1}{2}m_i v_i^2$$

To proceed further, we must recall that although every particle in the rigid object has the same angular speed ω, the individual linear speeds depend on the distance r_i from the axis of rotation according to the expression $v_i = r_i \omega$ (see Eq. 10.10). The *total* kinetic energy of the rotating rigid object is the sum of the kinetic energies of the individual particles:

$$K_{\text{R}} = \sum_i K_i = \sum_i \tfrac{1}{2}m_i v_i^2 = \tfrac{1}{2}\sum_i m_i r_i^2 \omega^2$$

We can write this expression in the form

$$K_{\text{R}} = \tfrac{1}{2}\left(\sum_i m_i r_i^2\right)\omega^2 \qquad \textbf{(10.14)}$$

where we have factored ω^2 from the sum because it is common to every particle.

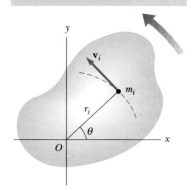

web

If you want to learn more about the physics of CD players, visit the Special Interest Group on CD Applications and Technology at **www.sigcat.org**

Figure 10.7 A rigid object rotating about a z axis with angular speed ω. The kinetic energy of the particle of mass m_i is $\tfrac{1}{2}m_i v_i^2$. The total kinetic energy of the object is called its rotational kinetic energy

We simplify this expression by defining the quantity in parentheses as the **moment of inertia** I:

Moment of inertia

$$I \equiv \sum_i m_i r_i^2 \qquad\qquad \textbf{(10.15)}$$

From the definition of moment of inertia, we see that it has dimensions of ML^2 ($kg \cdot m^2$ in SI units).[1] With this notation, Equation 10.14 becomes

Rotational kinetic energy

$$K_R = \tfrac{1}{2} I \omega^2 \qquad\qquad \textbf{(10.16)}$$

Although we commonly refer to the quantity $\frac{1}{2} I \omega^2$ as **rotational kinetic energy,** it is not a new form of energy. It is ordinary kinetic energy because it is derived from a sum over individual kinetic energies of the particles contained in the rigid object. However, the mathematical form of the kinetic energy given by Equation 10.16 is a convenient one when we are dealing with rotational motion, provided we know how to calculate I.

It is important that you recognize the analogy between kinetic energy associated with linear motion $\frac{1}{2} mv^2$ and rotational kinetic energy $\frac{1}{2} I \omega^2$. The quantities I and ω in rotational motion are analogous to m and v in linear motion, respectively. (In fact, I takes the place of m every time we compare a linear-motion equation with its rotational counterpart.) The moment of inertia is a measure of the resistance of an object to changes in its rotational motion, just as mass is a measure of the tendency of an object to resist changes in its linear motion. Note, however, that mass is an intrinsic property of an object, whereas I depends on the physical arrangement of that mass. Can you think of a situation in which an object's moment of inertia changes even though its mass does not?

EXAMPLE 10.3   The Oxygen Molecule

Consider an oxygen molecule (O_2) rotating in the xy plane about the z axis. The axis passes through the center of the molecule, perpendicular to its length. The mass of each oxygen atom is 2.66×10^{-26} kg, and at room temperature the average separation between the two atoms is $d = 1.21 \times 10^{-10}$ m (the atoms are treated as point masses). (a) Calculate the moment of inertia of the molecule about the z axis.

Solution   This is a straightforward application of the definition of I. Because each atom is a distance $d/2$ from the z axis, the moment of inertia about the axis is

$$I = \sum_i m_i r_i^2 = m \left(\frac{d}{2} \right)^2 + m \left(\frac{d}{2} \right)^2 = \tfrac{1}{2} md^2$$

$$= \tfrac{1}{2} (2.66 \times 10^{-26} \text{ kg})(1.21 \times 10^{-10} \text{ m})^2$$

$$= 1.95 \times 10^{-46} \text{ kg} \cdot \text{m}^2$$

This is a very small number, consistent with the minuscule masses and distances involved.

(b) If the angular speed of the molecule about the z axis is 4.60×10^{12} rad/s, what is its rotational kinetic energy?

Solution   We apply the result we just calculated for the moment of inertia in the formula for K_R:

$$K_R = \tfrac{1}{2} I \omega^2$$

$$= \tfrac{1}{2} (1.95 \times 10^{-46} \text{ kg} \cdot \text{m}^2)(4.60 \times 10^{12} \text{ rad/s})^2$$

$$= 2.06 \times 10^{-21} \text{ J}$$

---

[1] Civil engineers use moment of inertia to characterize the elastic properties (rigidity) of such structures as loaded beams. Hence, it is often useful even in a nonrotational context.

EXAMPLE 10.4   **Four Rotating Masses**

Four tiny spheres are fastened to the corners of a frame of negligible mass lying in the xy plane (Fig. 10.8). We shall assume that the spheres' radii are small compared with the dimensions of the frame. (a) If the system rotates about the y axis with an angular speed ω, find the moment of inertia and the rotational kinetic energy about this axis.

Solution   First, note that the two spheres of mass m, which lie on the y axis, do not contribute to I_y (that is, $r_i = 0$ for these spheres about this axis). Applying Equation 10.15, we obtain

$$I_y = \sum_i m_i r_i^2 = Ma^2 + Ma^2 = \boxed{2Ma^2}$$

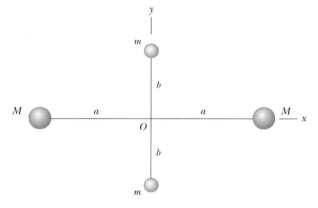

Figure 10.8   The four spheres are at a fixed separation as shown. The moment of inertia of the system depends on the axis about which it is evaluated.

Therefore, the rotational kinetic energy about the y axis is

$$K_R = \tfrac{1}{2}I_y\omega^2 = \tfrac{1}{2}(2Ma^2)\omega^2 = \boxed{Ma^2\omega^2}$$

The fact that the two spheres of mass m do not enter into this result makes sense because they have no motion about the axis of rotation; hence, they have no rotational kinetic energy. By similar logic, we expect the moment of inertia about the x axis to be $I_x = 2mb^2$ with a rotational kinetic energy about that axis of $K_R = mb^2\omega^2$.

(b) Suppose the system rotates in the xy plane about an axis through O (the z axis). Calculate the moment of inertia and rotational kinetic energy about this axis.

Solution   Because r_i in Equation 10.15 is the *perpendicular* distance to the axis of rotation, we obtain

$$I_z = \sum_i m_i r_i^2 = Ma^2 + Ma^2 + mb^2 + mb^2 = \boxed{2Ma^2 + 2mb^2}$$

$$K_R = \tfrac{1}{2}I_z\omega^2 = \tfrac{1}{2}(2Ma^2 + 2mb^2)\omega^2 = \boxed{(Ma^2 + mb^2)\omega^2}$$

Comparing the results for parts (a) and (b), we conclude that the moment of inertia and therefore the rotational kinetic energy associated with a given angular speed depend on the axis of rotation. In part (b), we expect the result to include all four spheres and distances because all four spheres are rotating in the xy plane. Furthermore, the fact that the rotational kinetic energy in part (a) is smaller than that in part (b) indicates that it would take less effort (work) to set the system into rotation about the y axis than about the z axis.

10.5   CALCULATION OF MOMENTS OF INERTIA

We can evaluate the moment of inertia of an extended rigid object by imagining the object divided into many small volume elements, each of which has mass Δm. We use the definition $I = \sum_i r_i^2 \Delta m_i$ and take the limit of this sum as $\Delta m \to 0$. In this limit, the sum becomes an integral over the whole object:

$$I = \lim_{\Delta m_i \to 0} \sum_i r_i^2 \Delta m_i = \int r^2 \, dm \qquad \textbf{(10.17)}$$

It is usually easier to calculate moments of inertia in terms of the volume of the elements rather than their mass, and we can easily make that change by using Equation 1.1, $\rho = m/V$, where ρ is the density of the object and V is its volume. We want this expression in its differential form $\rho = dm/dV$ because the volumes we are dealing with are very small. Solving for $dm = \rho \, dV$ and substituting the result into Equation 10.17 gives

$$I = \int \rho r^2 \, dV$$

If the object is homogeneous, then ρ is constant and the integral can be evaluated for a known geometry. If ρ is not constant, then its variation with position must be known to complete the integration.

The density given by $\rho = m/V$ sometimes is referred to as *volume density* for the obvious reason that it relates to volume. Often we use other ways of expressing density. For instance, when dealing with a sheet of uniform thickness t, we can define a *surface density* $\sigma = \rho t$, which signifies *mass per unit area*. Finally, when mass is distributed along a uniform rod of cross-sectional area A, we sometimes use *linear density* $\lambda = M/L = \rho A$, which is the *mass per unit length*.

EXAMPLE 10.5   Uniform Hoop

Find the moment of inertia of a uniform hoop of mass M and radius R about an axis perpendicular to the plane of the hoop and passing through its center (Fig. 10.9).

Solution   All mass elements dm are the same distance $r = R$ from the axis, and so, applying Equation 10.17, we obtain for the moment of inertia about the z axis through O:

$$I_z = \int r^2 \, dm = R^2 \int dm = \boxed{MR^2}$$

Note that this moment of inertia is the same as that of a single particle of mass M located a distance R from the axis of rotation.

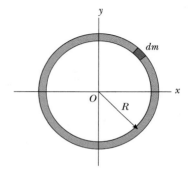

Figure 10.9   The mass elements dm of a uniform hoop are all the same distance from O.

Quick Quiz 10.3

(a) Based on what you have learned from Example 10.5, what do you expect to find for the moment of inertia of two particles, each of mass $M/2$, located anywhere on a circle of radius R around the axis of rotation? (b) How about the moment of inertia of four particles, each of mass $M/4$, again located a distance R from the rotation axis?

EXAMPLE 10.6   Uniform Rigid Rod

Calculate the moment of inertia of a uniform rigid rod of length L and mass M (Fig. 10.10) about an axis perpendicular to the rod (the y axis) and passing through its center of mass.

Solution   The shaded length element dx has a mass dm equal to the mass per unit length λ multiplied by dx:

$$dm = \lambda \, dx = \frac{M}{L} \, dx$$

Substituting this expression for dm into Equation 10.17, with $r = x$, we obtain

$$I_y = \int r^2 \, dm = \int_{-L/2}^{L/2} x^2 \, \frac{M}{L} \, dx = \frac{M}{L} \int_{-L/2}^{L/2} x^2 \, dx$$

$$= \frac{M}{L} \left[\frac{x^3}{3} \right]_{-L/2}^{L/2} = \boxed{\tfrac{1}{12} ML^2}$$

Figure 10.10 A uniform rigid rod of length L. The moment of inertia about the y axis is less than that about the y' axis. The latter axis is examined in Example 10.8.

EXAMPLE 10.7 ▶ Uniform Solid Cylinder

A uniform solid cylinder has a radius R, mass M, and length L. Calculate its moment of inertia about its central axis (the z axis in Fig. 10.11).

Solution It is convenient to divide the cylinder into many

Figure 10.11 Calculating I about the z axis for a uniform solid cylinder.

cylindrical shells, each of which has radius r, thickness dr, and length L, as shown in Figure 10.11. The volume dV of each shell is its cross-sectional area multiplied by its length: $dV = dA \cdot L = (2\pi r\, dr)L$. If the mass per unit volume is ρ, then the mass of this differential volume element is $dm = \rho dV = \rho 2\pi rL\, dr$. Substituting this expression for dm into Equation 10.17, we obtain

$$I_z = \int r^2\, dm = 2\pi\rho L \int_0^R r^3\, dr = \tfrac{1}{2}\pi\rho LR^4$$

Because the total volume of the cylinder is $\pi R^2 L$, we see that $\rho = M/V = M/\pi R^2 L$. Substituting this value for ρ into the above result gives

$$(1) \qquad I_z = \boxed{\tfrac{1}{2}MR^2}$$

Note that this result does not depend on L, the length of the cylinder. In other words, it applies equally well to a long cylinder and a flat disc. Also note that this is exactly half the value we would expect were all the mass concentrated at the outer edge of the cylinder or disc. (See Example 10.5.)

Table 10.2 gives the moments of inertia for a number of bodies about specific axes. The moments of inertia of rigid bodies with simple geometry (high symmetry) are relatively easy to calculate provided the rotation axis coincides with an axis of symmetry. The calculation of moments of inertia about an arbitrary axis can be cumbersome, however, even for a highly symmetric object. Fortunately, use of an important theorem, called the **parallel-axis theorem,** often simplifies the calculation. Suppose the moment of inertia about an axis through the center of mass of an object is I_{CM}. The parallel-axis theorem states that the moment of inertia about any axis parallel to and a distance D away from this axis is

$$I = I_{\mathrm{CM}} + MD^2 \qquad\qquad \textbf{(10.18)}$$

Parallel-axis theorem

TABLE 10.2   Moments of Inertia of Homogeneous Rigid Bodies with Different Geometries

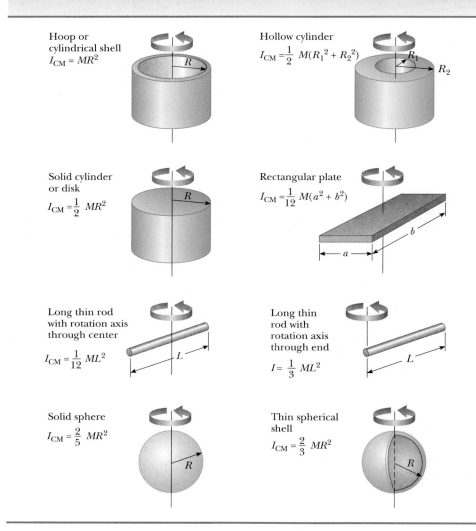

Hoop or cylindrical shell
$I_{CM} = MR^2$

Hollow cylinder
$I_{CM} = \frac{1}{2} M(R_1^2 + R_2^2)$

Solid cylinder or disk
$I_{CM} = \frac{1}{2} MR^2$

Rectangular plate
$I_{CM} = \frac{1}{12} M(a^2 + b^2)$

Long thin rod with rotation axis through center
$I_{CM} = \frac{1}{12} ML^2$

Long thin rod with rotation axis through end
$I = \frac{1}{3} ML^2$

Solid sphere
$I_{CM} = \frac{2}{5} MR^2$

Thin spherical shell
$I_{CM} = \frac{2}{3} MR^2$

Proof of the Parallel-Axis Theorem *(Optional).* Suppose that an object rotates in the xy plane about the z axis, as shown in Figure 10.12, and that the coordinates of the center of mass are x_{CM}, y_{CM}. Let the mass element dm have coordinates x, y. Because this element is a distance $r = \sqrt{x^2 + y^2}$ from the z axis, the moment of inertia about the z axis is

$$I = \int r^2 \, dm = \int (x^2 + y^2) \, dm$$

However, we can relate the coordinates x, y of the mass element dm to the coordinates of this same element located in a coordinate system having the object's center of mass as its origin. If the coordinates of the center of mass are x_{CM}, y_{CM} in the original coordinate system centered on O, then from Figure 10.12a we see that the relationships between the unprimed and primed coordinates are $x = x' + x_{CM}$

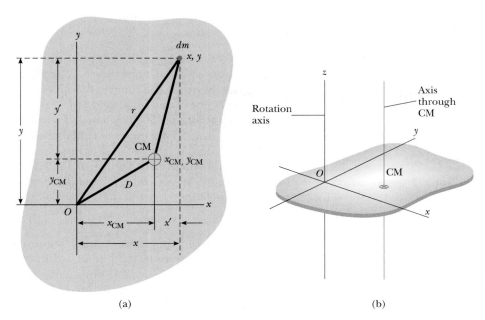

(a)   (b)

Figure 10.12 (a) The parallel-axis theorem: If the moment of inertia about an axis perpendicular to the figure through the center of mass is I_{CM}, then the moment of inertia about the z axis is $I_z = I_{CM} + MD^2$. (b) Perspective drawing showing the z axis (the axis of rotation) and the parallel axis through the CM.

and $y = y' + y_{CM}$. Therefore,

$$I = \int [(x' + x_{CM})^2 + (y' + y_{CM})^2]\, dm$$

$$= \int [(x')^2 + (y')^2]\, dm + 2x_{CM} \int x'\, dm + 2y_{CM} \int y'\, dm + (x_{CM}{}^2 + y_{CM}{}^2) \int dm$$

The first integral is, by definition, the moment of inertia about an axis that is parallel to the z axis and passes through the center of mass. The second two integrals are zero because, by definition of the center of mass, $\int x'\, dm = \int y'\, dm = 0$. The last integral is simply MD^2 because $\int dm = M$ and $D^2 = x_{CM}{}^2 + y_{CM}{}^2$. Therefore, we conclude that

$$I = I_{CM} + MD^2$$

EXAMPLE 10.8 ▶ **Applying the Parallel-Axis Theorem**

Consider once again the uniform rigid rod of mass M and length L shown in Figure 10.10. Find the moment of inertia of the rod about an axis perpendicular to the rod through one end (the y′ axis in Fig. 10.10).

Solution Intuitively, we expect the moment of inertia to be greater than $I_{CM} = \frac{1}{12}ML^2$ because it should be more difficult to change the rotational motion of a rod spinning about an axis at one end than one that is spinning about its center. Because the distance between the center-of-mass axis and the y′ axis is $D = L/2$, the parallel-axis theorem gives

$$I = I_{CM} + MD^2 = \tfrac{1}{12} ML^2 + M\left(\frac{L}{2}\right)^2 = \boxed{\tfrac{1}{3} ML^2}$$

So, it is four times more difficult to change the rotation of a rod spinning about its end than it is to change the motion of one spinning about its center.

Exercise Calculate the moment of inertia of the rod about a perpendicular axis through the point $x = L/4$.

Answer $I = \tfrac{7}{48} ML^2$.

10.6 TORQUE

Why are a door's doorknob and hinges placed near opposite edges of the door?
This question actually has an answer based on common sense ideas. The harder
we push against the door and the farther we are from the hinges, the more likely
we are to open or close the door. When a force is exerted on a rigid object pivoted
about an axis, the object tends to rotate about that axis. The tendency of a force to
rotate an object about some axis is measured by a vector quantity called **torque τ**
(tau).

Consider the wrench pivoted on the axis through O in Figure 10.13. The applied force **F** acts at an angle ϕ to the horizontal. We define the magnitude of the
torque associated with the force **F** by the expression

Definition of torque

$$\tau \equiv rF \sin \phi = Fd \tag{10.19}$$

where r is the distance between the pivot point and the point of application of **F**
and d is the perpendicular distance from the pivot point to the line of action of **F**.
(The *line of action* of a force is an imaginary line extending out both ends of the
vector representing the force. The dashed line extending from the tail of **F** in Figure 10.13 is part of the line of action of **F**.) From the right triangle in Figure 10.13
that has the wrench as its hypotenuse, we see that $d = r \sin \phi$. This quantity d is
called the **moment arm** (or *lever arm*) of **F**.

Moment arm

It is very important that you recognize that *torque is defined only when a reference
axis is specified.* Torque is the product of a force and the moment arm of that force,
and moment arm is defined only in terms of an axis of rotation.

In Figure 10.13, the only component of **F** that tends to cause rotation is $F \sin$
ϕ, the component perpendicular to r. The horizontal component $F \cos \phi$, because
it passes through O, has no tendency to produce rotation. From the definition of
torque, we see that the rotating tendency increases as **F** increases and as d increases. This explains the observation that it is easier to close a door if we push at
the doorknob rather than at a point close to the hinge. We also want to apply our
push as close to perpendicular to the door as we can. Pushing sideways on the
doorknob will not cause the door to rotate.

If two or more forces are acting on a rigid object, as shown in Figure 10.14,
each tends to produce rotation about the pivot at O. In this example, $\mathbf{F}_2$ tends to

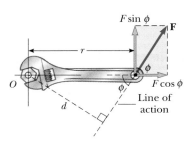

Figure 10.13 The force **F** has a
greater rotating tendency about O
as F increases and as the moment
arm d increases. It is the component $F \sin \phi$ that tends to rotate the
wrench about O.

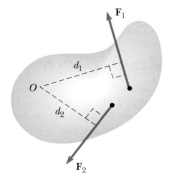

Figure 10.14 The force $\mathbf{F}_1$ tends
to rotate the object counterclock-
wise about O, and $\mathbf{F}_2$ tends to rotate it clockwise.

rotate the object clockwise, and $\mathbf{F}_1$ tends to rotate it counterclockwise. We use the convention that the sign of the torque resulting from a force is positive if the turning tendency of the force is counterclockwise and is negative if the turning tendency is clockwise. For example, in Figure 10.14, the torque resulting from $\mathbf{F}_1$, which has a moment arm d_1, is positive and equal to $+F_1 d_1$; the torque from $\mathbf{F}_2$ is negative and equal to $-F_2 d_2$. Hence, the net torque about O is

$$\sum \tau = \tau_1 + \tau_2 = F_1 d_1 - F_2 d_2$$

Torque should not be confused with force. Forces can cause a change in linear motion, as described by Newton's second law. Forces can also cause a change in rotational motion, but the effectiveness of the forces in causing this change depends on both the forces and the moment arms of the forces, in the combination that we call *torque*. Torque has units of force times length—newton · meters in SI units—and should be reported in these units. Do not confuse torque and work, which have the same units but are very different concepts.

EXAMPLE 10.9   **The Net Torque on a Cylinder**

A one-piece cylinder is shaped as shown in Figure 10.15, with a core section protruding from the larger drum. The cylinder is free to rotate around the central axis shown in the drawing. A rope wrapped around the drum, which has radius R_1, exerts a force $\mathbf{F}_1$ to the right on the cylinder. A rope wrapped around the core, which has radius R_2, exerts a force $\mathbf{F}_2$ downward on the cylinder. (a) What is the net torque acting on the cylinder about the rotation axis (which is the z axis in Figure 10.15)?

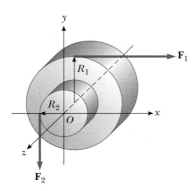

Figure 10.15 A solid cylinder pivoted about the z axis through O. The moment arm of $\mathbf{F}_1$ is R_1, and the moment arm of $\mathbf{F}_2$ is R_2.

Solution The torque due to $\mathbf{F}_1$ is $-R_1 F_1$ (the sign is negative because the torque tends to produce clockwise rotation). The torque due to $\mathbf{F}_2$ is $+R_2 F_2$ (the sign is positive because the torque tends to produce counterclockwise rotation). Therefore, the net torque about the rotation axis is

$$\sum \tau = \tau_1 + \tau_2 = \boxed{-R_1 F_1 + R_2 F_2}$$

We can make a quick check by noting that if the two forces are of equal magnitude, the net torque is negative because $R_1 > R_2$. Starting from rest with both forces acting on it, the cylinder would rotate clockwise because $\mathbf{F}_1$ would be more effective at turning it than would $\mathbf{F}_2$.

(b) Suppose $F_1 = 5.0$ N, $R_1 = 1.0$ m, $F_2 = 15.0$ N, and $R_2 = 0.50$ m. What is the net torque about the rotation axis, and which way does the cylinder rotate from rest?

$$\sum \tau = -(5.0 \text{ N})(1.0 \text{ m}) + (15.0 \text{ N})(0.50 \text{ m}) = \boxed{2.5 \text{ N} \cdot \text{m}}$$

Because the net torque is positive, if the cylinder starts from rest, it will commence rotating counterclockwise with increasing angular velocity. (If the cylinder's initial rotation is clockwise, it will slow to a stop and then rotate counterclockwise with increasing angular speed.)

10.7 RELATIONSHIP BETWEEN TORQUE AND ANGULAR ACCELERATION

In this section we show that the angular acceleration of a rigid object rotating about a fixed axis is proportional to the net torque acting about that axis. Before discussing the more complex case of rigid-body rotation, however, it is instructive

Figure 10.16 A particle rotating in a circle under the influence of a tangential force $\mathbf{F}_t$. A force $\mathbf{F}_r$ in the radial direction also must be present to maintain the circular motion.

Relationship between torque and angular acceleration

first to discuss the case of a particle rotating about some fixed point under the influence of an external force.

Consider a particle of mass m rotating in a circle of radius r under the influence of a tangential force $\mathbf{F}_t$ and a radial force $\mathbf{F}_r$, as shown in Figure 10.16. (As we learned in Chapter 6, the radial force must be present to keep the particle moving in its circular path.) The tangential force provides a tangential acceleration $\mathbf{a}_t$, and

$$F_t = ma_t$$

The torque about the center of the circle due to $\mathbf{F}_t$ is

$$\tau = F_t r = (ma_t)r$$

Because the tangential acceleration is related to the angular acceleration through the relationship $a_t = r\alpha$ (see Eq. 10.11), the torque can be expressed as

$$\tau = (mr\alpha)r = (mr^2)\alpha$$

Recall from Equation 10.15 that mr^2 is the moment of inertia of the rotating particle about the z axis passing through the origin, so that

$$\tau = I\alpha \qquad \text{(10.20)}$$

That is, **the torque acting on the particle is proportional to its angular acceleration,** and the proportionality constant is the moment of inertia. It is important to note that $\tau = I\alpha$ is the rotational analog of Newton's second law of motion, $F = ma$.

Now let us extend this discussion to a rigid object of arbitrary shape rotating about a fixed axis, as shown in Figure 10.17. The object can be regarded as an infinite number of mass elements dm of infinitesimal size. If we impose a cartesian coordinate system on the object, then each mass element rotates in a circle about the origin, and each has a tangential acceleration $\mathbf{a}_t$ produced by an external tangential force $d\mathbf{F}_t$. For any given element, we know from Newton's second law that

$$dF_t = (dm)a_t$$

The torque $d\tau$ associated with the force $d\mathbf{F}_t$ acts about the origin and is given by

$$d\tau = r\, dF_t = (r\, dm)a_t$$

Because $a_t = r\alpha$, the expression for $d\tau$ becomes

$$d\tau = (r\, dm)r\alpha = (r^2\, dm)\alpha$$

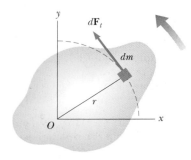

Figure 10.17 A rigid object rotating about an axis through O. Each mass element dm rotates about O with the same angular acceleration α, and the net torque on the object is proportional to α.

It is important to recognize that although each mass element of the rigid object may have a different linear acceleration $\mathbf{a}_t$, they all have the *same* angular acceleration α. With this in mind, we can integrate the above expression to obtain the net torque about O due to the external forces:

$$\sum \tau = \int (r^2\, dm)\alpha = \alpha \int r^2\, dm$$

where α can be taken outside the integral because it is common to all mass elements. From Equation 10.17, we know that $\int r^2\, dm$ is the moment of inertia of the object about the rotation axis through O, and so the expression for $\sum \tau$ becomes

$$\sum \tau = I\alpha \qquad \text{(10.21)}$$

Torque is proportional to angular acceleration

Note that this is the same relationship we found for a particle rotating in a circle (see Eq. 10.20). So, again we see that the net torque about the rotation axis is proportional to the angular acceleration of the object, with the proportionality factor being I, a quantity that depends upon the axis of rotation and upon the size and shape of the object. In view of the complex nature of the system, it is interesting to note that the relationship $\Sigma\tau = I\alpha$ is strikingly simple and in complete agreement with experimental observations. The simplicity of the result lies in the manner in which the motion is described.

Although each point on a rigid object rotating about a fixed axis may not experience the same force, linear acceleration, or linear speed, each point experiences the same angular acceleration and angular speed at any instant. Therefore, at any instant the rotating rigid object as a whole is characterized by specific values for angular acceleration, net torque, and angular speed.

Finally, note that the result $\Sigma\tau = I\alpha$ also applies when the forces acting on the mass elements have radial components as well as tangential components. This is because the line of action of all radial components must pass through the axis of rotation, and hence all radial components produce zero torque about that axis.

Every point has the same ω and α

QuickLab

Tip over a child's tall tower of blocks. Try this several times. Does the tower "break" at the same place each time? What affects where the tower comes apart as it tips? If the tower were made of toy bricks that snap together, what would happen? (Refer to Conceptual Example 10.11.)

EXAMPLE 10.10 ▶ Rotating Rod

A uniform rod of length L and mass M is attached at one end to a frictionless pivot and is free to rotate about the pivot in the vertical plane, as shown in Figure 10.18. The rod is released from rest in the horizontal position. What is the initial angular acceleration of the rod and the initial linear acceleration of its right end?

Solution We cannot use our kinematic equations to find α or a because the torque exerted on the rod varies with its position, and so neither acceleration is constant. We have enough information to find the torque, however, which we can then use in the torque–angular acceleration relationship (Eq. 10.20) to find α and then a.

The only force contributing to torque about an axis through the pivot is the gravitational force $M\mathbf{g}$ exerted on the rod. (The force exerted by the pivot on the rod has zero torque about the pivot because its moment arm is zero.) To

compute the torque on the rod, we can assume that the gravitational force acts at the center of mass of the rod, as shown in Figure 10.18. The torque due to this force about an axis through the pivot is

$$\tau = Mg\left(\frac{L}{2}\right)$$

With $\Sigma\tau = I\alpha$, and $I = \frac{1}{3}ML^2$ for this axis of rotation (see Table 10.2), we obtain

$$\alpha = \frac{\tau}{I} = \frac{Mg(L/2)}{1/3\ ML^2} = \boxed{\frac{3g}{2L}}$$

All points on the rod have this angular acceleration.

To find the linear acceleration of the right end of the rod, we use the relationship $a_t = r\alpha$ (Eq. 10.11), with $r = L$:

$$a_t = L\alpha = \boxed{\frac{3}{2}g}$$

This result—that $a_t > g$ for the free end of the rod—is rather interesting. It means that if we place a coin at the tip of the rod, hold the rod in the horizontal position, and then release the rod, the tip of the rod falls faster than the coin does!

Other points on the rod have a linear acceleration that is less than $\frac{3}{2}g$. For example, the middle of the rod has an acceleration of $\frac{3}{4}g$.

Figure 10.18 The uniform rod is pivoted at the left end.

CONCEPTUAL EXAMPLE 10.11 ▶ Falling Smokestacks and Tumbling Blocks

When a tall smokestack falls over, it often breaks somewhere along its length before it hits the ground, as shown in Figure 10.19. The same thing happens with a tall tower of children's toy blocks. Why does this happen?

Solution   As the smokestack rotates around its base, each higher portion of the smokestack falls with an increasing tangential acceleration. (The tangential acceleration of a given point on the smokestack is proportional to the distance of that portion from the base.) As the acceleration increases, higher portions of the smokestack experience an acceleration greater than that which could result from gravity alone; this is similar to the situation described in Example 10.10. This can happen only if these portions are being pulled downward by a force in addition to the gravitational force. The force that causes this to occur is the shear force from lower portions of the smokestack. Eventually the shear force that provides this acceleration is greater than the smokestack can withstand, and the smokestack breaks.

Figure 10.19   A falling smokestack.

EXAMPLE 10.12 ▶ Angular Acceleration of a Wheel

A wheel of radius R, mass M, and moment of inertia I is mounted on a frictionless, horizontal axle, as shown in Figure 10.20. A light cord wrapped around the wheel supports an object of mass m. Calculate the angular acceleration of the wheel, the linear acceleration of the object, and the tension in the cord.

Solution   The torque acting on the wheel about its axis of rotation is $\tau = TR$, where T is the force exerted by the cord on the rim of the wheel. (The gravitational force exerted by the Earth on the wheel and the normal force exerted by the axle on the wheel both pass through the axis of rotation and thus produce no torque.) Because $\Sigma\tau = I\alpha$, we obtain

$$\Sigma\tau = I\alpha = TR$$

$$(1) \qquad \alpha = \frac{TR}{I}$$

Now let us apply Newton's second law to the motion of the object, taking the downward direction to be positive:

$$\Sigma F_y = mg - T = ma$$

$$(2) \qquad a = \frac{mg - T}{m}$$

Equations (1) and (2) have three unknowns, α, a, and T. Because the object and wheel are connected by a string that does not slip, the linear acceleration of the suspended object is equal to the linear acceleration of a point on the rim of the

Figure 10.20   The tension in the cord produces a torque about the axle passing through O.

wheel. Therefore, the angular acceleration of the wheel and this linear acceleration are related by $a = R\alpha$. Using this fact together with Equations (1) and (2), we obtain

$$(3) \qquad a = R\alpha = \frac{TR^2}{I} = \frac{mg - T}{m}$$

$$(4) \qquad T = \frac{mg}{1 + \dfrac{mR^2}{I}}$$

Substituting Equation (4) into Equation (2), and solving for a and α, we find that

$$a = \frac{g}{1 + I/mR^2}$$

$$\alpha = \frac{a}{R} = \frac{g}{R + I/mR}$$

Exercise   The wheel in Figure 10.20 is a solid disc of $M = 2.00$ kg, $R = 30.0$ cm, and $I = 0.090\,0$ kg·m². The suspended object has a mass of $m = 0.500$ kg. Find the tension in the cord and the angular acceleration of the wheel.

Answer   3.27 N; 10.9 rad/s².

EXAMPLE 10.13   Atwood's Machine Revisited

Two blocks having masses m_1 and m_2 are connected to each other by a light cord that passes over two identical, friction-less pulleys, each having a moment of inertia I and radius R, as shown in Figure 10.21a. Find the acceleration of each block and the tensions T_1, T_2, and T_3 in the cord. (Assume no slipping between cord and pulleys.)

Solution   We shall define the downward direction as posi-tive for m_1 and upward as the positive direction for m_2. This allows us to represent the acceleration of both masses by a single variable a and also enables us to relate a positive a to a positive (counterclockwise) angular acceleration α. Let us write Newton's second law of motion for each block, using the free-body diagrams for the two blocks as shown in Figure 10.21b:

$$(1) \qquad m_1 g - T_1 = m_1 a$$

$$(2) \qquad T_3 - m_2 g = m_2 a$$

Next, we must include the effect of the pulleys on the mo-tion. Free-body diagrams for the pulleys are shown in Figure 10.21c. The net torque about the axle for the pulley on the left is $(T_1 - T_2)R$, while the net torque for the pulley on the right is $(T_2 - T_3)R$. Using the relation $\Sigma\tau = I\alpha$ for each pul-ley and noting that each pulley has the same angular acceler-ation α, we obtain

$$(3) \qquad (T_1 - T_2)R = I\alpha$$

$$(4) \qquad (T_2 - T_3)R = I\alpha$$

We now have four equations with four unknowns: a, T_1, T_2, and T_3. These can be solved simultaneously. Adding Equations (3) and (4) gives

$$(5) \qquad (T_1 - T_3)R = 2I\alpha$$

Adding Equations (1) and (2) gives

$$T_3 - T_1 + m_1 g - m_2 g = (m_1 + m_2)a$$

$$(6) \qquad T_1 - T_3 = (m_1 - m_2)g - (m_1 + m_2)a$$

Substituting Equation (6) into Equation (5), we have

$$[(m_1 - m_2)g - (m_1 + m_2)a]R = 2I\alpha$$

Because $\alpha = a/R$, this expression can be simplified to

$$(m_1 - m_2)g - (m_1 + m_2)a = 2I\frac{a}{R^2}$$

$$(7) \qquad a = \frac{(m_1 - m_2)g}{m_1 + m_2 + 2\dfrac{I}{R^2}}$$

This value of a can then be substituted into Equations (1)

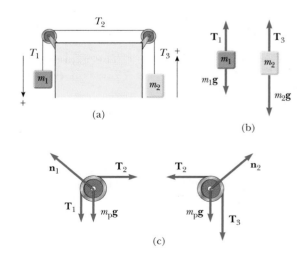

Figure 10.21   (a) Another look at Atwood's machine. (b) Free-body diagrams for the blocks. (c) Free-body diagrams for the pulleys, where $m_p\mathbf{g}$ represents the force of gravity acting on each pulley.

and (2) to give T_1 and T_3. Finally, T_2 can be found from Equation (3) or Equation (4). Note that if $m_1 > m_2$, the acceleration is positive; this means that the left block accelerates downward, the right block accelerates upward, and both pulleys accelerate counterclockwise. If $m_1 < m_2$, then all the values are negative and the motions are reversed. If $m_1 = m_2$, then no acceleration occurs at all. You should compare these results with those found in Example 5.9 on page 129.

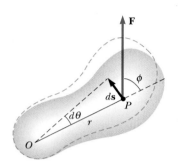

Figure 10.22 A rigid object rotates about an axis through O under the action of an external force $\mathbf{F}$ applied at P.

Power delivered to a rigid object

10.8 ▸ WORK, POWER, AND ENERGY IN ROTATIONAL MOTION

In this section, we consider the relationship between the torque acting on a rigid object and its resulting rotational motion in order to generate expressions for the power and a rotational analog to the work–kinetic energy theorem. Consider the rigid object pivoted at O in Figure 10.22. Suppose a single external force $\mathbf{F}$ is applied at P, where $\mathbf{F}$ lies in the plane of the page. The work done by $\mathbf{F}$ as the object rotates through an infinitesimal distance $ds = r \, d\theta$ in a time dt is

$$dW = \mathbf{F} \cdot d\mathbf{s} = (F \sin \phi) r \, d\theta$$

where $F \sin \phi$ is the tangential component of $\mathbf{F}$, or, in other words, the component of the force along the displacement. Note that *the radial component of $\mathbf{F}$ does no work because it is perpendicular to the displacement.*

Because the magnitude of the torque due to $\mathbf{F}$ about O is defined as $rF \sin \phi$ by Equation 10.19, we can write the work done for the infinitesimal rotation as

$$dW = \tau \, d\theta \tag{10.22}$$

The rate at which work is being done by $\mathbf{F}$ as the object rotates about the fixed axis is

$$\frac{dW}{dt} = \tau \frac{d\theta}{dt}$$

Because dW/dt is the instantaneous power $\mathcal{P}$ (see Section 7.5) delivered by the force, and because $d\theta/dt = \omega$, this expression reduces to

$$\mathcal{P} = \frac{dW}{dt} = \tau \omega \tag{10.23}$$

This expression is analogous to $\mathcal{P} = Fv$ in the case of linear motion, and the expression $dW = \tau \, d\theta$ is analogous to $dW = F_x \, dx$.

Work and Energy in Rotational Motion

In studying linear motion, we found the energy concept—and, in particular, the work–kinetic energy theorem—extremely useful in describing the motion of a system. The energy concept can be equally useful in describing rotational motion. From what we learned of linear motion, we expect that when a symmetric object rotates about a fixed axis, the work done by external forces equals the change in the rotational energy.

To show that this is in fact the case, let us begin with $\Sigma \tau = I\alpha$. Using the chain rule from the calculus, we can express the resultant torque as

$$\sum \tau = I\alpha = I\frac{d\omega}{dt} = I\frac{d\omega}{d\theta}\frac{d\theta}{dt} = I\frac{d\omega}{d\theta}\omega$$

TABLE 10.3   Useful Equations in Rotational and Linear Motion

Rotational Motion About a Fixed Axis	Linear Motion
Angular speed $\omega = d\theta/dt$	Linear speed $v = dx/dt$
Angular acceleration $\alpha = d\omega/dt$	Linear acceleration $a = dv/dt$
Resultant torque $\Sigma\tau = I\alpha$	Resultant force $\Sigma F = ma$
If $\alpha =$ constant $\begin{cases} \omega_f = \omega_i + \alpha t \\ \theta_f - \theta_i = \omega_i t + \frac{1}{2}\alpha t^2 \\ \omega_f^2 = \omega_i^2 + 2\alpha(\theta_f - \theta_i) \end{cases}$	If $a =$ constant $\begin{cases} v_f = v_i + at \\ x_f - x_i = v_i t + \frac{1}{2}at^2 \\ v_f^2 = v_i^2 + 2a(x_f - x_i) \end{cases}$
Work $W = \int_{\theta_i}^{\theta_f} \tau \, d\theta$	Work $W = \int_{x_i}^{x_f} F_x \, dx$
Rotational kinetic energy $K_R = \frac{1}{2}I\omega^2$	Kinetic energy $K = \frac{1}{2}mv^2$
Power $\mathcal{P} = \tau\omega$	Power $\mathcal{P} = Fv$
Angular momentum $L = I\omega$	Linear momentum $p = mv$
Resultant torque $\Sigma\tau = dL/dt$	Resultant force $\Sigma F = dp/dt$

Rearranging this expression and noting that $\Sigma\tau \, d\theta = dW$, we obtain

$$\sum \tau \, d\theta = dW = I\omega \, d\omega$$

Integrating this expression, we get for the total work done by the net external force acting on a rotating system

$$\sum W = \int_{\theta_i}^{\theta_f} \sum \tau \, d\theta = \int_{\omega_i}^{\omega_f} I\omega \, d\omega = \tfrac{1}{2}I\omega_f^2 - \tfrac{1}{2}I\omega_i^2 \qquad \textbf{(10.24)}$$

> Work–kinetic energy theorem for rotational motion

where the angular speed changes from ω_i to ω_f as the angular position changes from θ_i to θ_f. That is,

the net work done by external forces in rotating a symmetric rigid object about a fixed axis equals the change in the object's rotational energy.

Table 10.3 lists the various equations we have discussed pertaining to rotational motion, together with the analogous expressions for linear motion. The last two equations in Table 10.3, involving angular momentum L, are discussed in Chapter 11 and are included here only for the sake of completeness.

Quick Quiz 10.4

For a hoop lying in the xy plane, which of the following requires that more work be done by an external agent to accelerate the hoop from rest to an angular speed ω: (a) rotation about the z axis through the center of the hoop, or (b) rotation about an axis parallel to z passing through a point P on the hoop rim?

EXAMPLE 10.14   Rotating Rod Revisited

A uniform rod of length L and mass M is free to rotate on a frictionless pin passing through one end (Fig 10.23). The rod is released from rest in the horizontal position. (a) What is its angular speed when it reaches its lowest position?

Solution   The question can be answered by considering the mechanical energy of the system. When the rod is horizontal, it has no rotational energy. The potential energy relative to the lowest position of the center of mass of the rod (O') is $MgL/2$. When the rod reaches its lowest position, the

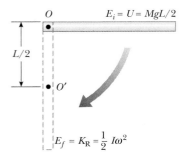

Figure 10.23   A uniform rigid rod pivoted at O rotates in a vertical plane under the action of gravity.

energy is entirely rotational energy, $\frac{1}{2}I\omega^2$, where I is the moment of inertia about the pivot. Because $I = \frac{1}{3}ML^2$ (see Table 10.2) and because mechanical energy is constant, we have $E_i = E_f$ or

$$\tfrac{1}{2}MgL = \tfrac{1}{2}I\omega^2 = \tfrac{1}{2}(\tfrac{1}{3}ML^2)\omega^2$$

$$\omega = \sqrt{\frac{3g}{L}}$$

(b) Determine the linear speed of the center of mass and the linear speed of the lowest point on the rod when it is in the vertical position.

Solution   These two values can be determined from the relationship between linear and angular speeds. We know ω from part (a), and so the linear speed of the center of mass is

$$v_{CM} = r\omega = \frac{L}{2}\omega = \tfrac{1}{2}\sqrt{3gL}$$

Because r for the lowest point on the rod is twice what it is for the center of mass, the lowest point has a linear speed equal to

$$2v_{CM} = \sqrt{3gL}$$

EXAMPLE 10.15   Connected Cylinders

Consider two cylinders having masses m_1 and m_2, where $m_1 \neq m_2$, connected by a string passing over a pulley, as shown in Figure 10.24. The pulley has a radius R and moment of

Figure 10.24

inertia I about its axis of rotation. The string does not slip on the pulley, and the system is released from rest. Find the linear speeds of the cylinders after cylinder 2 descends through a distance h, and the angular speed of the pulley at this time.

Solution   We are now able to account for the effect of a massive pulley. Because the string does not slip, the pulley rotates. We neglect friction in the axle about which the pulley rotates for the following reason: Because the axle's radius is small relative to that of the pulley, the frictional torque is much smaller than the torque applied by the two cylinders, provided that their masses are quite different. Mechanical energy is constant; hence, the increase in the system's kinetic energy (the system being the two cylinders, the pulley, and the Earth) equals the decrease in its potential energy. Because $K_i = 0$ (the system is initially at rest), we have

$$\Delta K = K_f - K_i = (\tfrac{1}{2}m_1v_f^2 + \tfrac{1}{2}m_2v_f^2 + \tfrac{1}{2}I\omega_f^2) - 0$$

where v_f is the same for both blocks. Because $v_f = R\omega_f$, this expression becomes

$$\Delta K = \tfrac{1}{2}\left(m_1 + m_2 + \frac{I}{R^2}\right)v_f^2$$

From Figure 10.24, we see that the system loses potential energy as cylinder 2 descends and gains potential energy as cylinder 1 rises. That is, $\Delta U_2 = -m_2 gh$ and $\Delta U_1 = m_1 gh$. Applying the principle of conservation of energy in the form $\Delta K + \Delta U_1 + \Delta U_2 = 0$ gives

$$\tfrac{1}{2}\left(m_1 + m_2 + \frac{I}{R^2}\right)v_f^2 + m_1 gh - m_2 gh = 0$$

$$v_f = \left[\frac{2(m_2 - m_1)gh}{\left(m_1 + m_2 + \dfrac{I}{R^2}\right)}\right]^{1/2}$$

Because $v_f = R\omega_f$, the angular speed of the pulley at this instant is

$$\omega_f = \frac{v_f}{R} = \frac{1}{R}\left[\frac{2(m_2 - m_1)gh}{\left(m_1 + m_2 + \dfrac{I}{R^2}\right)}\right]^{1/2}$$

Exercise Repeat the calculation of v_f, using $\Sigma\tau = I\alpha$ applied to the pulley and Newton's second law applied to the two cylinders. Use the procedures presented in Examples 10.12 and 10.13.

SUMMARY

If a particle rotates in a circle of radius r through an angle θ (measured in radians), the arc length it moves through is $s = r\theta$.

The **angular displacement** of a particle rotating in a circle or of a rigid object rotating about a fixed axis is

$$\Delta\theta = \theta_f - \theta_i \tag{10.2}$$

The **instantaneous angular speed** of a particle rotating in a circle or of a rigid object rotating about a fixed axis is

$$\omega = \frac{d\theta}{dt} \tag{10.4}$$

The **instantaneous angular acceleration** of a rotating object is

$$\alpha = \frac{d\omega}{dt} \tag{10.6}$$

When a rigid object rotates about a fixed axis, every part of the object has the same angular speed and the same angular acceleration.

If a particle or object rotates about a fixed axis under constant angular acceleration, one can apply equations of kinematics that are analogous to those for linear motion under constant linear acceleration:

$$\omega_f = \omega_i + \alpha t \tag{10.7}$$

$$\theta_f = \theta_i + \omega_i t + \tfrac{1}{2}\alpha t^2 \tag{10.8}$$

$$\omega_f^2 = \omega_i^2 + 2\alpha(\theta_f - \theta_i) \tag{10.9}$$

A useful technique in solving problems dealing with rotation is to visualize a linear version of the same problem.

When a rigid object rotates about a fixed axis, the angular position, angular speed, and angular acceleration are related to the linear position, linear speed, and linear acceleration through the relationships

$$s = r\upsilon \tag{10.1a}$$

$$v = r\omega \tag{10.10}$$

$$a_t = r\alpha \tag{10.11}$$

You must be able to easily alternate between the linear and rotational variables that describe a given situation.

The **moment of inertia of a system of particles** is

$$I \equiv \sum_i m_i r_i{}^2 \tag{10.15}$$

If a rigid object rotates about a fixed axis with angular speed ω, its **rotational energy** can be written

$$K_R = \tfrac{1}{2} I \omega^2 \tag{10.16}$$

where I is the moment of inertia about the axis of rotation.

The **moment of inertia of a rigid object** is

$$I = \int r^2 \, dm \tag{10.17}$$

where r is the distance from the mass element dm to the axis of rotation.

The magnitude of the **torque** associated with a force $\mathbf{F}$ acting on an object is

$$\tau = Fd \tag{10.19}$$

where d is the moment arm of the force, which is the perpendicular distance from some origin to the line of action of the force. Torque is a measure of the tendency of the force to change the rotation of the object about some axis.

If a rigid object free to rotate about a fixed axis has a **net external torque** acting on it, the object undergoes an angular acceleration α, where

$$\sum \tau = I\alpha \tag{10.21}$$

The rate at which work is done by an external force in rotating a rigid object about a fixed axis, or the **power** delivered, is

$$\mathcal{P} = \tau \omega \tag{10.23}$$

The net work done by external forces in rotating a rigid object about a fixed axis equals the change in the rotational kinetic energy of the object:

$$\sum W = \tfrac{1}{2} I \omega_f{}^2 - \tfrac{1}{2} I \omega_i{}^2 \tag{10.24}$$

QUESTIONS

1. What is the angular speed of the second hand of a clock? What is the direction of $\boldsymbol{\omega}$ as you view a clock hanging vertically? What is the magnitude of the angular acceleration vector $\boldsymbol{\alpha}$ of the second hand?

2. A wheel rotates counterclockwise in the xy plane. What is the direction of $\boldsymbol{\omega}$? What is the direction of $\boldsymbol{\alpha}$ if the angular velocity is decreasing in time?

3. Are the kinematic expressions for θ, ω, and α valid when the angular displacement is measured in degrees instead of in radians?

4. A turntable rotates at a constant rate of 45 rev/min. What is its angular speed in radians per second? What is the magnitude of its angular acceleration?

5. Suppose $a = b$ and $M > m$ for the system of particles described in Figure 10.8. About what axis (x, y, or z) does the moment of inertia have the smallest value? the largest value?

6. Suppose the rod in Figure 10.10 has a nonuniform mass distribution. In general, would the moment of inertia about the y axis still equal $ML^2/12$? If not, could the moment of inertia be calculated without knowledge of the manner in which the mass is distributed?

7. Suppose that only two external forces act on a rigid body, and the two forces are equal in magnitude but opposite in direction. Under what condition does the body rotate?

8. Explain how you might use the apparatus described in Example 10.12 to determine the moment of inertia of the wheel. (If the wheel does not have a uniform mass density, the moment of inertia is not necessarily equal to $\tfrac{1}{2} MR^2$.)

9. Using the results from Example 10.12, how would you calculate the angular speed of the wheel and the linear speed of the suspended mass at $t = 2$ s, if the system is released from rest at $t = 0$? Is the expression $v = R\omega$ valid in this situation?

10. If a small sphere of mass M were placed at the end of the rod in Figure 10.23, would the result for ω be greater than, less than, or equal to the value obtained in Example 10.14?

11. Explain why changing the axis of rotation of an object changes its moment of inertia.

12. Is it possible to change the translational kinetic energy of an object without changing its rotational energy?

13. Two cylinders having the same dimensions are set into rotation about their long axes with the same angular speed. One is hollow, and the other is filled with water. Which cylinder will be easier to stop rotating? Explain your answer.

14. Must an object be rotating to have a nonzero moment of inertia?

15. If you see an object rotating, is there necessarily a net torque acting on it?

16. Can a (momentarily) stationary object have a nonzero angular acceleration?

17. The polar diameter of the Earth is slightly less than the equatorial diameter. How would the moment of inertia of the Earth change if some mass from near the equator were removed and transferred to the polar regions to make the Earth a perfect sphere?

PROBLEMS

1, 2, 3 = straightforward, intermediate, challenging ☐ = full solution available in the *Student Solutions Manual and Study Guide*
WEB = solution posted at **http://www.saunderscollege.com/physics/** 🖥 = Computer useful in solving problem 📱 = Interactive Physics
☐ = paired numerical/symbolic problems

Section 10.2 Rotational Kinematics: Rotational Motion with Constant Angular Acceleration

1. A wheel starts from rest and rotates with constant angular acceleration and reaches an angular speed of 12.0 rad/s in 3.00 s. Find (a) the magnitude of the angular acceleration of the wheel and (b) the angle (in radians) through which it rotates in this time.

2. What is the angular speed in radians per second of (a) the Earth in its orbit about the Sun and (b) the Moon in its orbit about the Earth?

3. An airliner arrives at the terminal, and its engines are shut off. The rotor of one of the engines has an initial clockwise angular speed of 2 000 rad/s. The engine's rotation slows with an angular acceleration of magnitude 80.0 rad/s². (a) Determine the angular speed after 10.0 s. (b) How long does it take for the rotor to come to rest?

4. (a) The positions of the hour and minute hand on a clock face coincide at 12 o'clock. Determine all other times (up to the second) at which the positions of the hands coincide. (b) If the clock also has a second hand, determine all times at which the positions of all three hands coincide, given that they all coincide at 12 o'clock.

WEB 5. An electric motor rotating a grinding wheel at 100 rev/min is switched off. Assuming constant negative acceleration of magnitude 2.00 rad/s², (a) how long does it take the wheel to stop? (b) Through how many radians does it turn during the time found in part (a)?

6. A centrifuge in a medical laboratory rotates at a rotational speed of 3 600 rev/min. When switched off, it rotates 50.0 times before coming to rest. Find the constant angular acceleration of the centrifuge.

7. The angular position of a swinging door is described by $\theta = 5.00 + 10.0t + 2.00t^2$ rad. Determine the angular position, angular speed, and angular acceleration of the door (a) at $t = 0$ and (b) at $t = 3.00$ s.

8. The tub of a washer goes into its spin cycle, starting from rest and gaining angular speed steadily for 8.00 s, when it is turning at 5.00 rev/s. At this point the person doing the laundry opens the lid, and a safety switch turns off the washer. The tub smoothly slows to rest in 12.0 s. Through how many revolutions does the tub turn while it is in motion?

9. A rotating wheel requires 3.00 s to complete 37.0 revolutions. Its angular speed at the end of the 3.00-s interval is 98.0 rad/s. What is the constant angular acceleration of the wheel?

10. (a) Find the angular speed of the Earth's rotation on its axis. As the Earth turns toward the east, we see the sky turning toward the west at this same rate.
(b)   *The rainy Pleiads wester*
      And seek beyond the sea
   The head that I shall dream of
      That shall not dream of me.

   A. E. Housman   (© *Robert E. Symons*)

   Cambridge, England, is at longitude 0°, and Saskatoon, Saskatchewan, is at longitude 107° west. How much time elapses after the Pleiades set in Cambridge until these stars fall below the western horizon in Saskatoon?

Section 10.3 Angular and Linear Quantities

11. Make an order-of-magnitude estimate of the number of revolutions through which a typical automobile tire

turns in 1 yr. State the quantities you measure or estimate and their values.

12. The diameters of the main rotor and tail rotor of a single-engine helicopter are 7.60 m and 1.02 m, respectively. The respective rotational speeds are 450 rev/min and 4 138 rev/min. Calculate the speeds of the tips of both rotors. Compare these speeds with the speed of sound, 343 m/s.

Figure P10.12   *(Ross Harrrison Koty/Tony Stone Images)*

13. A racing car travels on a circular track with a radius of 250 m. If the car moves with a constant linear speed of 45.0 m/s, find (a) its angular speed and (b) the magnitude and direction of its acceleration.

14. A car is traveling at 36.0 km/h on a straight road. The radius of its tires is 25.0 cm. Find the angular speed of one of the tires, with its axle taken as the axis of rotation.

15. A wheel 2.00 m in diameter lies in a vertical plane and rotates with a constant angular acceleration of 4.00 rad/s². The wheel starts at rest at $t = 0$, and the radius vector of point P on the rim makes an angle of 57.3° with the horizontal at this time. At $t = 2.00$ s, find (a) the angular speed of the wheel, (b) the linear speed and acceleration of the point P, and (c) the angular position of the point P.

16. A discus thrower accelerates a discus from rest to a speed of 25.0 m/s by whirling it through 1.25 rev. As-

Figure P10.16   *(Bruce Ayers/Tony Stone Images)*

sume the discus moves on the arc of a circle 1.00 m in radius. (a) Calculate the final angular speed of the discus. (b) Determine the magnitude of the angular acceleration of the discus, assuming it to be constant. (c) Calculate the acceleration time.

17. A car accelerates uniformly from rest and reaches a speed of 22.0 m/s in 9.00 s. If the diameter of a tire is 58.0 cm, find (a) the number of revolutions the tire makes during this motion, assuming that no slipping occurs. (b) What is the final rotational speed of a tire in revolutions per second?

18. A 6.00-kg block is released from *A* on the frictionless track shown in Figure P10.18. Determine the radial and tangential components of acceleration for the block at *P*.

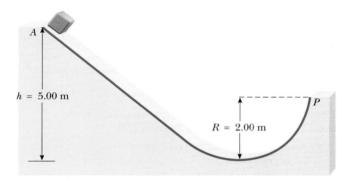

Figure P10.18

19. A disc 8.00 cm in radius rotates at a constant rate of 1 200 rev/min about its central axis. Determine (a) its angular speed, (b) the linear speed at a point 3.00 cm from its center, (c) the radial acceleration of a point on the rim, and (d) the total distance a point on the rim moves in 2.00 s.

20. A car traveling on a flat (unbanked) circular track accelerates uniformly from rest with a tangential acceleration of 1.70 m/s². The car makes it one quarter of the way around the circle before it skids off the track. Determine the coefficient of static friction between the car and track from these data.

21. A small object with mass 4.00 kg moves counterclockwise with constant speed 4.50 m/s in a circle of radius 3.00 m centered at the origin. (a) It started at the point with cartesian coordinates (3 m, 0). When its angular displacement is 9.00 rad, what is its position vector, in cartesian unit-vector notation? (b) In what quadrant is the particle located, and what angle does its position vector make with the positive *x* axis? (c) What is its velocity vector, in unit–vector notation? (d) In what direction is it moving? Make a sketch of the position and velocity vectors. (e) What is its acceleration, expressed in unit–vector notation? (f) What total force acts on the object? (Express your answer in unit vector notation.)

22. A standard cassette tape is placed in a standard cassette player. Each side plays for 30 min. The two tape wheels of the cassette fit onto two spindles in the player. Suppose that a motor drives one spindle at a constant angular speed of ~ 1 rad/s and that the other spindle is free to rotate at any angular speed. Estimate the order of magnitude of the thickness of the tape.

Section 10.4   Rotational Energy

23. Three small particles are connected by rigid rods of negligible mass lying along the y axis (Fig. P10.23). If the system rotates about the x axis with an angular speed of 2.00 rad/s, find (a) the moment of inertia about the x axis and the total rotational kinetic energy evaluated from $\frac{1}{2}I\omega^2$ and (b) the linear speed of each particle and the total kinetic energy evaluated from $\Sigma \frac{1}{2}m_iv_i^2$.

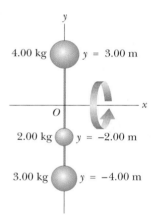

Figure P10.23

24. The center of mass of a pitched baseball (3.80-cm radius) moves at 38.0 m/s. The ball spins about an axis through its center of mass with an angular speed of 125 rad/s. Calculate the ratio of the rotational energy to the translational kinetic energy. Treat the ball as a uniform sphere.

WEB **25.** The four particles in Figure P10.25 are connected by rigid rods of negligible mass. The origin is at the center of the rectangle. If the system rotates in the xy plane about the z axis with an angular speed of 6.00 rad/s, calculate (a) the moment of inertia of the system about the z axis and (b) the rotational energy of the system.

26. The hour hand and the minute hand of Big Ben, the famous Parliament tower clock in London, are 2.70 m long and 4.50 m long and have masses of 60.0 kg and 100 kg, respectively. Calculate the total rotational kinetic energy of the two hands about the axis of rotation. (You may model the hands as long thin rods.)

Figure P10.25

Figure P10.26   Problems 26 and 74. (*John Lawrence/Tony Stone Images*)

27. Two masses M and m are connected by a rigid rod of length L and of negligible mass, as shown in Figure P10.27. For an axis perpendicular to the rod, show that the system has the minimum moment of inertia when the axis passes through the center of mass. Show that this moment of inertia is $I = \mu L^2$, where $\mu = mM/(m + M)$.

Figure P10.27

Section 10.5   Calculation of Moments of Inertia

28. Three identical thin rods, each of length L and mass m, are welded perpendicular to each other, as shown in Figure P10.28. The entire setup is rotated about an axis

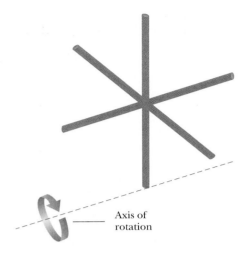

Axis of
rotation

Figure P10.28

that passes through the end of one rod and is parallel to
another. Determine the moment of inertia of this
arrangement.

29. Figure P10.29 shows a side view of a car tire and its radial dimensions. The rubber tire has two sidewalls of
uniform thickness 0.635 cm and a tread wall of uniform
thickness 2.50 cm and width 20.0 cm. Suppose its density is uniform, with the value 1.10×10^3 kg/m³. Find
its moment of inertia about an axis through its center
perpendicular to the plane of the sidewalls.

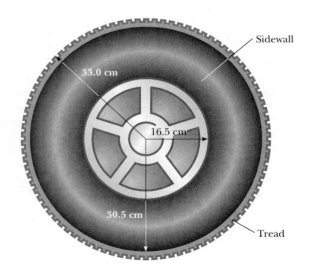

Sidewall

33.0 cm

16.5 cm

30.5 cm

Tread

Figure P10.29

30. Use the parallel-axis theorem and Table 10.2 to find the
moments of inertia of (a) a solid cylinder about an axis
parallel to the center-of-mass axis and passing through
the edge of the cylinder and (b) a solid sphere about an
axis tangent to its surface.

31. *Attention! About face!* Compute an order-of-magnitude estimate for the moment of inertia of your body as you
stand tall and turn around a vertical axis passing
through the top of your head and the point halfway between your ankles. In your solution state the quantities
you measure or estimate and their values.

Section 10.6 Torque

32. Find the mass m needed to balance the 1 500-kg truck
on the incline shown in Figure P10.32. Assume all pulleys are frictionless and massless.

3r

r

1500 kg

$\theta = 45°$

m

Figure P10.32

WEB **33.** Find the net torque on the wheel in Figure P10.33
about the axle through O if $a = 10.0$ cm and $b = 25.0$ cm.

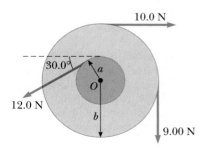

10.0 N

30.0°

a

O

12.0 N

b

9.00 N

Figure P10.33

34. The fishing pole in Figure P10.34 makes an angle of
20.0° with the horizontal. What is the torque exerted by

Figure P10.34

Figure P10.39

the fish about an axis perpendicular to the page and passing through the fisher's hand?

35. The tires of a 1 500-kg car are 0.600 m in diameter, and the coefficients of friction with the road surface are $\mu_s = 0.800$ and $\mu_k = 0.600$. Assuming that the weight is evenly distributed on the four wheels, calculate the maximum torque that can be exerted by the engine on a driving wheel such that the wheel does not spin. If you wish, you may suppose that the car is at rest.

36. Suppose that the car in Problem 35 has a disk brake system. Each wheel is slowed by the frictional force between a single brake pad and the disk-shaped rotor. On this particular car, the brake pad comes into contact with the rotor at an average distance of 22.0 cm from the axis. The coefficients of friction between the brake pad and the disk are $\mu_s = 0.600$ and $\mu_k = 0.500$. Calculate the normal force that must be applied to the rotor such that the car slows as quickly as possible.

Section 10.7   Relationship Between Torque and Angular Acceleration

WEB 37. A model airplane having a mass of 0.750 kg is tethered by a wire so that it flies in a circle 30.0 m in radius. The airplane engine provides a net thrust of 0.800 N perpendicular to the tethering wire. (a) Find the torque the net thrust produces about the center of the circle. (b) Find the angular acceleration of the airplane when it is in level flight. (c) Find the linear acceleration of the airplane tangent to its flight path.

38. The combination of an applied force and a frictional force produces a constant total torque of 36.0 N·m on a wheel rotating about a fixed axis. The applied force acts for 6.00 s; during this time the angular speed of the wheel increases from 0 to 10.0 rad/s. The applied force is then removed, and the wheel comes to rest in 60.0 s. Find (a) the moment of inertia of the wheel, (b) the magnitude of the frictional torque, and (c) the total number of revolutions of the wheel.

39. A block of mass $m_1 = 2.00$ kg and a block of mass $m_2 = 6.00$ kg are connected by a massless string over a pulley

in the shape of a disk having radius $R = 0.250$ m and mass $M = 10.0$ kg. These blocks are allowed to move on a fixed block–wedge of angle $\theta = 30.0°$, as shown in Figure P10.39. The coefficient of kinetic friction for both blocks is 0.360. Draw free-body diagrams of both blocks and of the pulley. Determine (a) the acceleration of the two blocks and (b) the tensions in the string on both sides of the pulley.

40. A potter's wheel—a thick stone disk with a radius of 0.500 m and a mass of 100 kg—is freely rotating at 50.0 rev/min. The potter can stop the wheel in 6.00 s by pressing a wet rag against the rim and exerting a radially inward force of 70.0 N. Find the effective coefficient of kinetic friction between the wheel and the rag.

41. A bicycle wheel has a diameter of 64.0 cm and a mass of 1.80 kg. Assume that the wheel is a hoop with all of its mass concentrated on the outside radius. The bicycle is placed on a stationary stand on rollers, and a resistive force of 120 N is applied tangent to the rim of the tire. (a) What force must be applied by a chain passing over a 9.00-cm-diameter sprocket if the wheel is to attain an acceleration of 4.50 rad/s²? (b) What force is required if the chain shifts to a 5.60-cm-diameter sprocket?

Section 10.8   Work , Power, and Energy in Rotational Motion

42. A cylindrical rod 24.0 cm long with a mass of 1.20 kg and a radius of 1.50 cm has a ball with a diameter of 8.00 cm and a mass of 2.00 kg attached to one end. The arrangement is originally vertical and stationary, with the ball at the top. The apparatus is free to pivot about the bottom end of the rod. (a) After it falls through 90°, what is its rotational kinetic energy? (b) What is the angular speed of the rod and ball? (c) What is the linear speed of the ball? (d) How does this compare with the speed if the ball had fallen freely through the same distance of 28 cm?

43. A 15.0-kg mass and a 10.0-kg mass are suspended by a pulley that has a radius of 10.0 cm and a mass of 3.00 kg (Fig. P10.43). The cord has a negligible mass and causes the pulley to rotate without slipping. The pulley

rotates without friction. The masses start from rest 3.00 m apart. Treating the pulley as a uniform disk, determine the speeds of the two masses as they pass each other.

44. A mass m_1 and a mass m_2 are suspended by a pulley that has a radius R and a mass M (see Fig. P10.43). The cord has a negligible mass and causes the pulley to rotate without slipping. The pulley rotates without friction. The masses start from rest a distance d apart. Treating the pulley as a uniform disk, determine the speeds of the two masses as they pass each other.

$M = 3.00$ kg
$R = 10.0$ cm
$m_1 = 15.0$ kg
$m_2 = 10.0$ kg

3.00 m $h = 1.5m$

Figure P10.43   Problems 43 and 44.

45. A weight of 50.0 N is attached to the free end of a light string wrapped around a reel with a radius of 0.250 m and a mass of 3.00 kg. The reel is a solid disk, free to rotate in a vertical plane about the horizontal axis passing through its center. The weight is released 6.00 m above the floor. (a) Determine the tension in the string, the acceleration of the mass, and the speed with which the weight hits the floor. (b) Find the speed calculated in part (a), using the principle of conservation of energy.

46. A constant torque of 25.0 N·m is applied to a grindstone whose moment of inertia is 0.130 kg·m². Using energy principles, find the angular speed after the grindstone has made 15.0 revolutions. (Neglect friction.)

47. This problem describes one experimental method of determining the moment of inertia of an irregularly shaped object such as the payload for a satellite. Figure P10.47 shows a mass m suspended by a cord wound around a spool of radius r, forming part of a turntable supporting the object. When the mass is released from rest, it descends through a distance h, acquiring a speed

Figure P10.47

v. Show that the moment of inertia I of the equipment (including the turntable) is $mr^2(2gh/v^2 - 1)$.

48. A bus is designed to draw its power from a rotating flywheel that is brought up to its maximum rate of rotation (3 000 rev/min) by an electric motor. The flywheel is a solid cylinder with a mass of 1 000 kg and a diameter of 1.00 m. If the bus requires an average power of 10.0 kW, how long does the flywheel rotate?

49. (a) A uniform, solid disk of radius R and mass M is free to rotate on a frictionless pivot through a point on its rim (Fig. P10.49). If the disk is released from rest in the position shown by the blue circle, what is the speed of its center of mass when the disk reaches the position indicated by the dashed circle? (b) What is the speed of the lowest point on the disk in the dashed position? (c) Repeat part (a), using a uniform hoop.

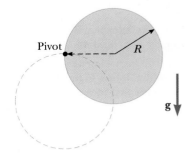

Figure P10.49

50. A horizontal 800-N merry-go-round is a solid disk of radius 1.50 m and is started from rest by a constant horizontal force of 50.0 N applied tangentially to the cylinder. Find the kinetic energy of the solid cylinder after 3.00 s.

ADDITIONAL PROBLEMS

51. Toppling chimneys often break apart in mid-fall (Fig. P10.51) because the mortar between the bricks cannot

withstand much shear stress. As the chimney begins to fall, shear forces must act on the topmost sections to accelerate them tangentially so that they can keep up with the rotation of the lower part of the stack. For simplicity, let us model the chimney as a uniform rod of length ℓ pivoted at the lower end. The rod starts at rest in a vertical position (with the frictionless pivot at the bottom) and falls over under the influence of gravity. What fraction of the length of the rod has a tangential acceleration greater than $g \sin \theta$, where θ is the angle the chimney makes with the vertical?

Figure P10.51 A building demolition site in Baltimore, MD. At the left is a chimney, mostly concealed by the building, that has broken apart on its way down. Compare with Figure 10.19. (*Jerry Wachter/Photo Researchers, Inc.*)

52. **Review Problem.**   A mixing beater consists of three thin rods: Each is 10.0 cm long, diverges from a central hub, and is separated from the others by 120°. All turn in the same plane. A ball is attached to the end of each rod. Each ball has a cross-sectional area of 4.00 cm² and is shaped so that it has a drag coefficient of 0.600. Calculate the power input required to spin the beater at 1 000 rev/min (a) in air and (b) in water.

53. A grinding wheel is in the form of a uniform solid disk having a radius of 7.00 cm and a mass of 2.00 kg. It starts from rest and accelerates uniformly under the action of the constant torque of 0.600 N·m that the motor

exerts on the wheel. (a) How long does the wheel take to reach its final rotational speed of 1 200 rev/min? (b) Through how many revolutions does it turn while accelerating?

54. The density of the Earth, at any distance r from its center, is approximately

$$\rho = [14.2 - 11.6\, r/R] \times 10^3 \text{ kg/m}^3$$

where R is the radius of the Earth. Show that this density leads to a moment of inertia $I = 0.330MR^2$ about an axis through the center, where M is the mass of the Earth.

55. A 4.00-m length of light nylon cord is wound around a uniform cylindrical spool of radius 0.500 m and mass 1.00 kg. The spool is mounted on a frictionless axle and is initially at rest. The cord is pulled from the spool with a constant acceleration of magnitude 2.50 m/s². (a) How much work has been done on the spool when it reaches an angular speed of 8.00 rad/s? (b) Assuming that there is enough cord on the spool, how long does it take the spool to reach this angular speed? (c) Is there enough cord on the spool?

56. A flywheel in the form of a heavy circular disk of diameter 0.600 m and mass 200 kg is mounted on a frictionless bearing. A motor connected to the flywheel accelerates it from rest to 1 000 rev/min. (a) What is the moment of inertia of the flywheel? (b) How much work is done on it during this acceleration? (c) When the angular speed reaches 1 000 rev/min, the motor is disengaged. A friction brake is used to slow the rotational rate to 500 rev/min. How much energy is dissipated as internal energy in the friction brake?

57. A shaft is turning at 65.0 rad/s at time zero. Thereafter, its angular acceleration is given by

$$\alpha = -10 \text{ rad/s}^2 - 5t\, \text{rad/s}^3$$

where t is the elapsed time. (a) Find its angular speed at $t = 3.00$ s. (b) How far does it turn in these 3 s?

58. For any given rotational axis, the *radius of gyration K* of a rigid body is defined by the expression $K^2 = I/M$, where M is the total mass of the body and I is its moment of inertia. Thus, the radius of gyration is equal to the distance between an imaginary point mass M and the axis of rotation such that I for the point mass about that axis is the same as that for the rigid body. Find the radius of gyration of (a) a solid disk of radius R, (b) a uniform rod of length L, and (c) a solid sphere of radius R, all three of which are rotating about a central axis.

59. A long, uniform rod of length L and mass M is pivoted about a horizontal, frictionless pin passing through one end. The rod is released from rest in a vertical position, as shown in Figure P10.59. At the instant the rod is horizontal, find (a) its angular speed, (b) the magnitude of its angular acceleration, (c) the x and y components of the acceleration of its center of mass, and (d) the components of the reaction force at the pivot.

Figure P10.59

Figure P10.60   Problems 60 and 61.

60. A bicycle is turned upside down while its owner repairs a flat tire. A friend spins the other wheel, of radius 0.381 m, and observes that drops of water fly off tangentially. She measures the height reached by drops moving vertically (Fig. P10.60). A drop that breaks loose from the tire on one turn rises $h = 54.0$ cm above the tangent point. A drop that breaks loose on the next turn rises 51.0 cm above the tangent point. The height to which the drops rise decreases because the angular speed of the wheel decreases. From this information, determine the magnitude of the average angular acceleration of the wheel.

61. A bicycle is turned upside down while its owner repairs a flat tire. A friend spins the other wheel of radius R and observes that drops of water fly off tangentially. She measures the height reached by drops moving vertically (see Fig. P10.60). A drop that breaks loose from the tire on one turn rises a distance h_1 above the tangent point.

A drop that breaks loose on the next turn rises a distance $h_2 < h_1$ above the tangent point. The height to which the drops rise decreases because the angular speed of the wheel decreases. From this information, determine the magnitude of the average angular acceleration of the wheel.

62. The top shown in Figure P10.62 has a moment of inertia of 4.00×10^{-4} kg·m² and is initially at rest. It is free to rotate about the stationary axis AA'. A string, wrapped around a peg along the axis of the top, is pulled in such a manner that a constant tension of 5.57 N is maintained. If the string does not slip while it is unwound from the peg, what is the angular speed of the top after 80.0 cm of string has been pulled off the peg?

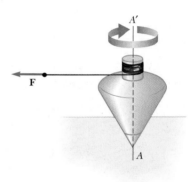

Figure P10.62

63. A cord is wrapped around a pulley of mass m and of radius r. The free end of the cord is connected to a block of mass M. The block starts from rest and then slides down an incline that makes an angle θ with the horizontal. The coefficient of kinetic friction between block and incline is μ. (a) Use energy methods to show that the block's speed as a function of displacement d down the incline is

$$v = [4gdM(m + 2M)^{-1}(\sin \theta - \mu \cos \theta)]^{1/2}$$

(b) Find the magnitude of the acceleration of the block in terms of μ, m, M, g, and θ.

64. (a) What is the rotational energy of the Earth about its spin axis? The radius of the Earth is 6 370 km, and its mass is 5.98×10^{24} kg. Treat the Earth as a sphere of moment of inertia $\frac{2}{5}MR^2$. (b) The rotational energy of the Earth is decreasing steadily because of tidal friction. Estimate the change in one day, given that the rotational period increases by about 10 μs each year.

65. The speed of a moving bullet can be determined by allowing the bullet to pass through two rotating paper disks mounted a distance d apart on the same axle (Fig. P10.65). From the angular displacement $\Delta \theta$ of the two

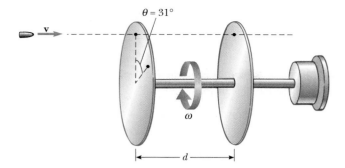

Figure P10.65

bullet holes in the disks and the rotational speed of the disks, we can determine the speed v of the bullet. Find the bullet speed for the following data: $d = 80$ cm, $\omega = 900$ rev/min, and $\Delta\theta = 31.0°$.

66. A wheel is formed from a hoop and n equally spaced spokes extending from the center of the hoop to its rim. The mass of the hoop is M, and the radius of the hoop (and hence the length of each spoke) is R. The mass of each spoke is m. Determine (a) the moment of inertia of the wheel about an axis through its center and perpendicular to the plane of the wheel and (b) the moment of inertia of the wheel about an axis through its rim and perpendicular to the plane of the wheel.

67. A uniform, thin, solid door has a height of 2.20 m, a width of 0.870 m, and a mass of 23.0 kg. Find its moment of inertia for rotation on its hinges. Are any of the data unnecessary?

68. A uniform, hollow, cylindrical spool has inside radius $R/2$, outside radius R, and mass M (Fig. P10.68). It is mounted so that it rotates on a massless horizontal axle. A mass m is connected to the end of a string wound around the spool. The mass m falls from rest through a distance y in time t. Show that the torque due to the frictional forces between spool and axle is
$$\tau_f = R[m(g - 2y/t^2) - M(5y/4t^2)]$$

69. An electric motor can accelerate a Ferris wheel of moment of inertia $I = 20\ 000$ kg·m² from rest to 10.0 rev/min in 12.0 s. When the motor is turned off, friction causes the wheel to slow down from 10.0 to 8.00 rev/min in 10.0 s. Determine (a) the torque generated by the motor to bring the wheel to 10.0 rev/min and (b) the power that would be needed to maintain this rotational speed.

70. The pulley shown in Figure P10.70 has radius R and moment of inertia I. One end of the mass m is connected to a spring of force constant k, and the other end is fastened to a cord wrapped around the pulley. The pulley axle and the incline are frictionless. If the pulley is wound counterclockwise so that the spring is stretched a distance d from its unstretched position and is then released from rest, find (a) the angular speed of the pulley when the spring is again unstretched and (b) a numerical value for the angular speed at this point if $I = 1.00$ kg·m², $R = 0.300$ m, $k = 50.0$ N/m, $m = 0.500$ kg, $d = 0.200$ m, and $\theta = 37.0°$.

Figure P10.70

71. Two blocks, as shown in Figure P10.71, are connected by a string of negligible mass passing over a pulley of radius 0.250 m and moment of inertia I. The block on the frictionless incline is moving upward with a constant acceleration of 2.00 m/s². (a) Determine T_1 and T_2, the tensions in the two parts of the string. (b) Find the moment of inertia of the pulley.

72. A common demonstration, illustrated in Figure P10.72, consists of a ball resting at one end of a uniform board

Figure P10.68

Figure P10.71

of length ℓ, hinged at the other end, and elevated at an angle θ. A light cup is attached to the board at r_c so that it will catch the ball when the support stick is suddenly

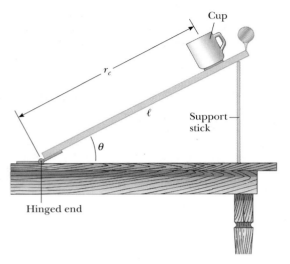

Figure P10.72

removed. (a) Show that the ball will lag behind the falling board when θ is less than 35.3° ; and that (b) the ball will fall into the cup when the board is supported at

this limiting angle and the cup is placed at

$$r_c = \frac{2\,\ell}{3\cos\theta}$$

(c) If a ball is at the end of a 1.00-m stick at this critical angle, show that the cup must be 18.4 cm from the moving end.

73. As a result of friction, the angular speed of a wheel changes with time according to the relationship

$$d\theta/dt = \omega_0 e^{-\sigma t}$$

where ω_0 and σ are constants. The angular speed changes from 3.50 rad/s at $t = 0$ to 2.00 rad/s at $t = 9.30$ s. Use this information to determine σ and ω_0. Then, determine (a) the magnitude of the angular acceleration at $t = 3.00$ s, (b) the number of revolutions the wheel makes in the first 2.50 s, and (c) the number of revolutions it makes before coming to rest.

74. The hour hand and the minute hand of Big Ben, the famous Parliament tower clock in London, are 2.70 m long and 4.50 m long and have masses of 60.0 kg and 100 kg, respectively (see Fig. P10.26). (a) Determine the total torque due to the weight of these hands about the axis of rotation when the time reads (i) 3:00, (ii) 5:15, (iii) 6:00, (iv) 8:20, and (v) 9:45. (You may model the hands as long thin rods.) (b) Determine all times at which the total torque about the axis of rotation is zero. Determine the times to the nearest second, solving a transcendental equation numerically.

ANSWERS TO QUICK QUIZZES

10.1 The fact that ω is negative indicates that we are dealing with an object that is rotating in the clockwise direction. We also know that when $\boldsymbol{\omega}$ and $\boldsymbol{\alpha}$ are antiparallel, ω must be decreasing—the object is slowing down. Therefore, the object is spinning more and more slowly (with less and less angular speed) in the clockwise, or negative, direction. This has a linear analogy to a sky diver opening her parachute. The velocity is negative—downward. When the sky diver opens the parachute, a large upward force causes an upward acceleration. As a result, the acceleration and velocity vectors are in opposite directions. Consequently, the parachutist slows down.

10.2 (a) Yes, all points on the wheel have the same angular speed. This is why we use angular quantities to describe

rotational motion. (b) No, not all points on the wheel have the same linear speed. (c) $v = 0$, $a = 0$. (d) $v = R\omega/2$, $a = a_r = v^2/(R/2) = R\omega^2/2$ (a_t is zero at all points because ω is constant).(e) $v = R\omega$, $a = R\omega^2$.

10.3 (a) $I = MR^2$. (b) $I = MR^2$. The moment of inertia of a system of masses equidistant from an axis of rotation is always the sum of the masses multiplied by the square of the distance from the axis.

10.4 (b) Rotation about the axis through point P requires more work. The moment of inertia of the hoop about the center axis is $I_{CM} = MR^2$, whereas, by the parallel-axis theorem, the moment of inertia about the axis through point P is $I_P = I_{CM} + MR^2 = MR^2 + MR^2 = 2MR^2$.

c h a p t e r

11

Rolling Motion and Angular Momentum

Chapter Outline

n the preceding chapter we learned how to treat a rigid body rotating about a fixed axis; in the present chapter, we move on to the more general case in which the axis of rotation is not fixed in space. We begin by describing such motion, which is called *rolling motion*. The central topic of this chapter is, however, angular momentum, a quantity that plays a key role in rotational dynamics. In analogy to the conservation of linear momentum, we find that the angular momentum of a rigid object is always conserved if no external torques act on the object. Like the law of conservation of linear momentum, the law of conservation of angular momentum is a fundamental law of physics, equally valid for relativistic and quantum systems.

11.1 ROLLING MOTION OF A RIGID OBJECT

In this section we treat the motion of a rigid object rotating about a moving axis. In general, such motion is very complex. However, we can simplify matters by restricting our discussion to a homogeneous rigid object having a high degree of symmetry, such as a cylinder, sphere, or hoop. Furthermore, we assume that the object undergoes rolling motion along a flat surface. We shall see that if an object such as a cylinder rolls without slipping on the surface (we call this *pure rolling motion*), a simple relationship exists between its rotational and translational motions.

Suppose a cylinder is rolling on a straight path. As Figure 11.1 shows, the center of mass moves in a straight line, but a point on the rim moves in a more complex path called a *cycloid*. This means that the axis of rotation remains parallel to its initial orientation in space. Consider a uniform cylinder of radius R rolling without slipping on a horizontal surface (Fig. 11.2). As the cylinder rotates through an angle θ, its center of mass moves a linear distance $s = R\theta$ (see Eq. 10.1a). Therefore, the linear speed of the center of mass for pure rolling motion is given by

$$v_{\text{CM}} = \frac{ds}{dt} = R\frac{d\theta}{dt} = R\omega \tag{11.1}$$

where ω is the angular velocity of the cylinder. Equation 11.1 holds whenever a cylinder or sphere rolls without slipping and is the **condition for pure rolling**

Figure 11.1   One light source at the center of a rolling cylinder and another at one point on the rim illustrate the different paths these two points take. The center moves in a straight line (green line), whereas the point on the rim moves in the path called a *cycloid* (red curve). *(Henry Leap and Jim Lehman)*

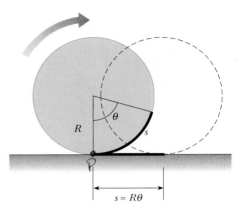

Figure 11.2 In pure rolling motion, as the cylinder rotates through an angle θ, its center of mass moves a linear distance $s = R\theta$.

motion. The magnitude of the linear acceleration of the center of mass for pure rolling motion is

$$a_{CM} = \frac{dv_{CM}}{dt} = R\frac{d\omega}{dt} = R\alpha \qquad \textbf{(11.2)}$$

where α is the angular acceleration of the cylinder.

The linear velocities of the center of mass and of various points on and within the cylinder are illustrated in Figure 11.3. A short time after the moment shown in the drawing, the rim point labeled P will have rotated from the six o'clock position to, say, the seven o'clock position, the point Q will have rotated from the ten o'clock position to the eleven o'clock position, and so on. Note that the linear velocity of any point is in a direction perpendicular to the line from that point to the contact point P. At any instant, the part of the rim that is at point P is at rest relative to the surface because slipping does not occur.

All points on the cylinder have the same angular speed. Therefore, because the distance from P' to P is twice the distance from P to the center of mass, P' has a speed $2v_{CM} = 2R\omega$. To see why this is so, let us model the rolling motion of the cylinder in Figure 11.4 as a combination of translational (linear) motion and rotational motion. For the pure translational motion shown in Figure 11.4a, imagine that the cylinder does not rotate, so that each point on it moves to the right with speed v_{CM}. For the pure rotational motion shown in Figure 11.4b, imagine that a rotation axis through the center of mass is stationary, so that each point on the cylinder has the same rotational speed ω. The combination of these two motions represents the rolling motion shown in Figure 11.4c. Note in Figure 11.4c that the top of the cylinder has linear speed $v_{CM} + R\omega = v_{CM} + v_{CM} = 2v_{CM}$, which is greater than the linear speed of any other point on the cylinder. As noted earlier, the center of mass moves with linear speed v_{CM} while the contact point between the surface and cylinder has a linear speed of zero.

We can express the total kinetic energy of the rolling cylinder as

$$K = \tfrac{1}{2}I_P\omega^2 \qquad \textbf{(11.3)}$$

where I_P is the moment of inertia about a rotation axis through P. Applying the parallel-axis theorem, we can substitute $I_P = I_{CM} + MR^2$ into Equation 11.3 to obtain

$$K = \tfrac{1}{2}I_{CM}\omega^2 + \tfrac{1}{2}MR^2\omega^2$$

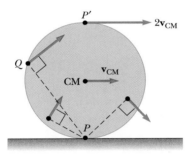

Figure 11.3 All points on a rolling object move in a direction perpendicular to an axis through the instantaneous point of contact P. In other words, all points rotate about P. The center of mass of the object moves with a velocity $\mathbf{v}_{CM}$, and the point P' moves with a velocity $2\mathbf{v}_{CM}$.

7.2

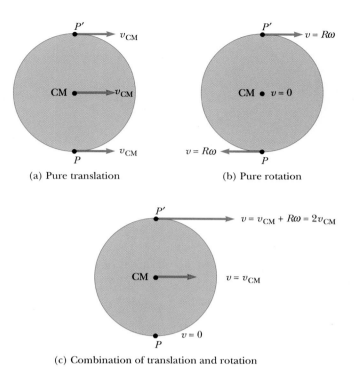

(a) Pure translation

(b) Pure rotation

(c) Combination of translation and rotation

Figure 11.4   The motion of a rolling object can be modeled as a combination of pure translation and pure rotation.

or, because $v_{CM} = R\omega$,

| Total kinetic energy of a rolling body |

$$K = \tfrac{1}{2}I_{CM}\omega^2 + \tfrac{1}{2}Mv_{CM}{}^2 \tag{11.4}$$

The term $\tfrac{1}{2}I_{CM}\omega^2$ represents the rotational kinetic energy of the cylinder about its center of mass, and the term $\tfrac{1}{2}Mv_{CM}{}^2$ represents the kinetic energy the cylinder would have if it were just translating through space without rotating. Thus, we can say that the **total kinetic energy of a rolling object is the sum of the rotational kinetic energy about the center of mass and the translational kinetic energy of the center of mass.**

We can use energy methods to treat a class of problems concerning the rolling motion of a sphere down a rough incline. (The analysis that follows also applies to the rolling motion of a cylinder or hoop.) We assume that the sphere in Figure 11.5 rolls without slipping and is released from rest at the top of the incline. Note that accelerated rolling motion is possible only if a frictional force is present between the sphere and the incline to produce a net torque about the center of mass. Despite the presence of friction, no loss of mechanical energy occurs because the contact point is at rest relative to the surface at any instant. On the other hand, if the sphere were to slip, mechanical energy would be lost as motion progressed.

Using the fact that $v_{CM} = R\omega$ for pure rolling motion, we can express Equation 11.4 as

$$K = \tfrac{1}{2}I_{CM}\left(\frac{v_{CM}}{R}\right)^2 + \tfrac{1}{2}Mv_{CM}{}^2$$

$$K = \tfrac{1}{2}\left(\frac{I_{CM}}{R^2} + M\right)v_{CM}{}^2 \tag{11.5}$$

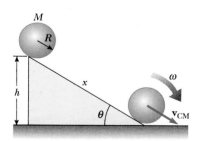

Figure 11.5 A sphere rolling down an incline. Mechanical energy is conserved if no slipping occurs.

By the time the sphere reaches the bottom of the incline, work equal to Mgh has been done on it by the gravitational field, where h is the height of the incline. Because the sphere starts from rest at the top, its kinetic energy at the bottom, given by Equation 11.5, must equal this work done. Therefore, the speed of the center of mass at the bottom can be obtained by equating these two quantities:

$$\tfrac{1}{2}\left(\frac{I_{CM}}{R^2} + M\right)v_{CM}{}^2 = Mgh$$

$$v_{CM} = \left(\frac{2gh}{1 + I_{CM}/MR^2}\right)^{1/2} \qquad \textbf{(11.6)}$$

Quick Quiz 11.1

Imagine that you slide your textbook across a gymnasium floor with a certain initial speed. It quickly stops moving because of friction between it and the floor. Yet, if you were to start a basketball rolling with the same initial speed, it would probably keep rolling from one end of the gym to the other. Why does a basketball roll so far? Doesn't friction affect its motion?

EXAMPLE 11.1 Sphere Rolling Down an Incline

For the solid sphere shown in Figure 11.5, calculate the linear speed of the center of mass at the bottom of the incline and the magnitude of the linear acceleration of the center of mass.

Solution The sphere starts from the top of the incline with potential energy $U_g = Mgh$ and kinetic energy $K = 0$. As we have seen before, if it fell vertically from that height, it would have a linear speed of $\sqrt{2gh}$ at the moment before it hit the floor. After rolling down the incline, the linear speed of the center of mass must be less than this value because some of the initial potential energy is diverted into rotational kinetic energy rather than all being converted into translational kinetic energy. For a uniform solid sphere, $I_{CM} = \tfrac{2}{5}MR^2$ (see Table 10.2), and therefore Equation 11.6 gives

$$v_{CM} = \left(\frac{2gh}{1 + \dfrac{2/5MR^2}{MR^2}}\right)^{1/2} = \left(\frac{10}{7}gh\right)^{1/2}$$

which is less than $\sqrt{2gh}$.

To calculate the linear acceleration of the center of mass, we note that the vertical displacement is related to the displacement x along the incline through the relationship $h =$ $x\sin\theta$. Hence, after squaring both sides, we can express the equation above as

$$v_{CM}{}^2 = \tfrac{10}{7}gx\sin\theta$$

Comparing this with the expression from kinematics, $v_{CM}{}^2 = 2a_{CM}x$ (see Eq. 2.12), we see that the acceleration of the center of mass is

$$a_{CM} = \tfrac{5}{7}g\sin\theta$$

These results are quite interesting in that both the speed and the acceleration of the center of mass are *independent* of the mass and the radius of the sphere! That is, **all homogeneous solid spheres experience the same speed and acceleration on a given incline.**

If we repeated the calculations for a hollow sphere, a solid cylinder, or a hoop, we would obtain similar results in which only the factor in front of $g\sin\theta$ would differ. The constant factors that appear in the expressions for v_{CM} and a_{CM} depend only on the moment of inertia about the center of mass for the specific body. In all cases, the acceleration of the center of mass is *less* than $g\sin\theta$, the value the acceleration would have if the incline were frictionless and no rolling occurred.

EXAMPLE 11.2 Another Look at the Rolling Sphere

In this example, let us use dynamic methods to verify the results of Example 11.1. The free-body diagram for the sphere is illustrated in Figure 11.6.

Solution Newton's second law applied to the center of mass gives

$$(1) \qquad \Sigma F_x = Mg\sin\theta - f = Ma_{CM}$$

$$\Sigma F_y = n - Mg\cos\theta = 0$$

where x is measured along the slanted surface of the incline.

Now let us write an expression for the torque acting on the sphere. A convenient axis to choose is the one that passes

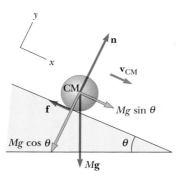

Figure 11.6   Free-body diagram for a solid sphere rolling down an incline.

clockwise direction,

$$\tau_{CM} = fR = I_{CM}\alpha$$

Because $I_{CM} = \frac{2}{5}MR^2$ and $\alpha = a_{CM}/R$, we obtain

$$(2) \qquad f = \frac{I_{CM}\alpha}{R} = \left(\frac{\frac{2}{5}MR^2}{R}\right)\frac{a_{CM}}{R} = \frac{2}{5}Ma_{CM}$$

Substituting Equation (2) into Equation (1) gives

$$a_{CM} = \boxed{\frac{5}{7}g\sin\theta}$$

which agrees with the result of Example 11.1.

Note that $\Sigma\mathbf{F} = m\mathbf{a}$ applies only if $\Sigma\mathbf{F}$ is the net external force exerted on the sphere and $\mathbf{a}$ is the acceleration of its center of mass. In the case of our sphere rolling down an incline, even though the frictional force does not change the total kinetic energy of the sphere, it does contribute to $\Sigma\mathbf{F}$ and thus decreases the acceleration of the center of mass. As a result, the final translational kinetic energy is less than it would be in the absence of friction. As mentioned in Example 11.1, some of the initial potential energy is converted to rotational kinetic energy.

through the center of the sphere and is perpendicular to the plane of the figure.[1] Because $\mathbf{n}$ and $M\mathbf{g}$ go through the center of mass, they have zero moment arm about this axis and thus do not contribute to the torque. However, the force of static friction produces a torque about this axis equal to fR in the clockwise direction; therefore, because $\boldsymbol{\tau}$ is also in the

QuickLab

Hold a basketball and a tennis ball side by side at the top of a ramp and release them at the same time. Which reaches the bottom first? Does the outcome depend on the angle of the ramp? What if the angle were 90° (that is, if the balls were in free fall)?

Quick Quiz 11.2

Which gets to the bottom first: a ball rolling without sliding down incline A or a box sliding down a frictionless incline B having the same dimensions as incline A?

11.2   THE VECTOR PRODUCT AND TORQUE

Consider a force $\mathbf{F}$ acting on a rigid body at the vector position $\mathbf{r}$ (Fig. 11.7). **The origin O is assumed to be in an inertial frame, so Newton's first law is valid in this case.** As we saw in Section 10.6, the *magnitude* of the torque due to this force relative to the origin is, by definition, $rF\sin\phi$, where ϕ is the angle between $\mathbf{r}$ and $\mathbf{F}$. The axis about which $\mathbf{F}$ tends to produce rotation is perpendicular to the plane formed by $\mathbf{r}$ and $\mathbf{F}$. If the force lies in the xy plane, as it does in Figure 11.7, the torque $\boldsymbol{\tau}$ is represented by a vector parallel to the z axis. The force in Figure 11.7 creates a torque that tends to rotate the body counterclockwise about the z axis; thus the direction of $\boldsymbol{\tau}$ is toward increasing z, and $\boldsymbol{\tau}$ is therefore in the positive z direction. If we reversed the direction of $\mathbf{F}$ in Figure 11.7, then $\boldsymbol{\tau}$ would be in the negative z direction.

The torque $\boldsymbol{\tau}$ involves the two vectors $\mathbf{r}$ and $\mathbf{F}$, and its direction is perpendicular to the plane of $\mathbf{r}$ and $\mathbf{F}$. We can establish a mathematical relationship between $\boldsymbol{\tau}$, $\mathbf{r}$, and $\mathbf{F}$, using a new mathematical operation called the **vector product,** or **cross product:**

Torque

$$\boldsymbol{\tau} \equiv \mathbf{r} \times \mathbf{F} \tag{11.7}$$

[1] Although a coordinate system whose origin is at the center of mass of a rolling object is not an inertial frame, the expression $\tau_{CM} = I\alpha$ still applies in the center-of-mass frame.

We now give a formal definition of the vector product. Given any two vectors **A** and **B**, the **vector product A × B** is defined as a third vector **C**, the magnitude of which is $AB \sin \theta$, where θ is the angle between **A** and **B**. That is, if **C** is given by

$$\mathbf{C} = \mathbf{A} \times \mathbf{B} \tag{11.8}$$

then its magnitude is

$$C \equiv AB \sin \theta \tag{11.9}$$

The quantity $AB \sin \theta$ is equal to the area of the parallelogram formed by **A** and **B**, as shown in Figure 11.8. The *direction* of **C** is perpendicular to the plane formed by **A** and **B**, and the best way to determine this direction is to use the right-hand rule illustrated in Figure 11.8. The four fingers of the right hand are pointed along **A** and then "wrapped" into **B** through the angle θ. The direction of the erect right thumb is the direction of **A × B = C**. Because of the notation, **A × B** is often read "**A** cross **B**"; hence, the term *cross product*.

Some properties of the vector product that follow from its definition are as follows:

1. Unlike the scalar product, the vector product is *not* commutative. Instead, the order in which the two vectors are multiplied in a cross product is important:

$$\mathbf{A} \times \mathbf{B} = -\mathbf{B} \times \mathbf{A} \tag{11.10}$$

Therefore, if you change the order of the vectors in a cross product, you must change the sign. You could easily verify this relationship with the right-hand rule.

2. If **A** is parallel to **B** ($\theta = 0°$ or $180°$), then **A × B** = 0; therefore, it follows that **A × A** = 0.

3. If **A** is perpendicular to **B**, then $|\mathbf{A} \times \mathbf{B}| = AB$.

4. The vector product obeys the distributive law:

$$\mathbf{A} \times (\mathbf{B} + \mathbf{C}) = \mathbf{A} \times \mathbf{B} + \mathbf{A} \times \mathbf{C} \tag{11.11}$$

5. The derivative of the cross product with respect to some variable such as t is

$$\frac{d}{dt}(\mathbf{A} \times \mathbf{B}) = \mathbf{A} \times \frac{d\mathbf{B}}{dt} + \frac{d\mathbf{A}}{dt} \times \mathbf{B} \tag{11.12}$$

where it is important to preserve the multiplicative order of **A** and **B**, in view of Equation 11.10.

It is left as an exercise to show from Equations 11.9 and 11.10 and from the definition of unit vectors that the cross products of the rectangular unit vectors **i**,

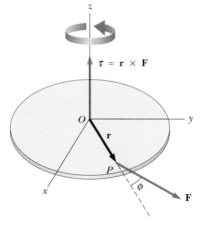

Figure 11.7 The torque vector τ lies in a direction perpendicular to the plane formed by the position vector **r** and the applied force vector **F**.

| Properties of the vector product |

Right-hand rule

Figure 11.8 The vector product **A × B** is a third vector **C** having a magnitude $AB \sin \theta$ equal to the area of the parallelogram shown. The direction of **C** is perpendicular to the plane formed by **A** and **B**, and this direction is determined by the right-hand rule.

j, and **k** obey the following rules:

$$\mathbf{i} \times \mathbf{i} = \mathbf{j} \times \mathbf{j} = \mathbf{k} \times \mathbf{k} = 0 \qquad \text{(11.13a)}$$

$$\mathbf{i} \times \mathbf{j} = -\mathbf{j} \times \mathbf{i} = \mathbf{k} \qquad \text{(11.13b)}$$

$$\mathbf{j} \times \mathbf{k} = -\mathbf{k} \times \mathbf{j} = \mathbf{i} \qquad \text{(11.13c)}$$

$$\mathbf{k} \times \mathbf{i} = -\mathbf{i} \times \mathbf{k} = \mathbf{j} \qquad \text{(11.13d)}$$

Signs are interchangeable in cross products. For example, $\mathbf{A} \times (-\mathbf{B}) = -\mathbf{A} \times \mathbf{B}$ and $\mathbf{i} \times (-\mathbf{j}) = -\mathbf{i} \times \mathbf{j}.$

The cross product of any two vectors **A** and **B** can be expressed in the following determinant form:

$$\mathbf{A} \times \mathbf{B} = \begin{vmatrix} \mathbf{i} & \mathbf{j} & \mathbf{k} \\ A_x & A_y & A_z \\ B_x & B_y & B_z \end{vmatrix} = \mathbf{i} \begin{vmatrix} A_y & A_z \\ B_y & B_z \end{vmatrix} - \mathbf{j} \begin{vmatrix} A_x & A_z \\ B_x & B_z \end{vmatrix} + \mathbf{k} \begin{vmatrix} A_x & A_y \\ B_x & B_y \end{vmatrix}$$

Expanding these determinants gives the result

$$\mathbf{A} \times \mathbf{B} = (A_y B_z - A_z B_y)\mathbf{i} - (A_x B_z - A_z B_x)\mathbf{j} + (A_x B_y - A_y B_x)\mathbf{k} \qquad \text{(11.14)}$$

EXAMPLE 11.3 ▶ The Cross Product

Two vectors lying in the xy plane are given by the equations $\mathbf{A} = 2\mathbf{i} + 3\mathbf{j}$ and $\mathbf{B} = -\mathbf{i} + 2\mathbf{j}$. Find $\mathbf{A} \times \mathbf{B}$ and verify that $\mathbf{A} \times \mathbf{B} = -\mathbf{B} \times \mathbf{A}.$

Solution Using Equations 11.13a through 11.13d, we obtain

$$\mathbf{A} \times \mathbf{B} = (2\mathbf{i} + 3\mathbf{j}) \times (-\mathbf{i} + 2\mathbf{j})$$

$$= 2\mathbf{i} \times 2\mathbf{j} + 3\mathbf{j} \times (-\mathbf{i}) = 4\mathbf{k} + 3\mathbf{k} = \boxed{7\mathbf{k}}$$

(We have omitted the terms containing $\mathbf{i} \times \mathbf{i}$ and $\mathbf{j} \times \mathbf{j}$ because, as Equation 11.13a shows, they are equal to zero.)

We can show that $\mathbf{A} \times \mathbf{B} = -\mathbf{B} \times \mathbf{A}$, since

$$\mathbf{B} \times \mathbf{A} = (-\mathbf{i} + 2\mathbf{j}) \times (2\mathbf{i} + 3\mathbf{j})$$

$$= -\mathbf{i} \times 3\mathbf{j} + 2\mathbf{j} \times 2\mathbf{i} = -3\mathbf{k} - 4\mathbf{k} = \boxed{-7\mathbf{k}}$$

Therefore, $\mathbf{A} \times \mathbf{B} = -\mathbf{B} \times \mathbf{A}.$

As an alternative method for finding $\mathbf{A} \times \mathbf{B}$, we could use Equation 11.14, with $A_x = 2$, $A_y = 3$, $A_z = 0$ and $B_x = -1$, $B_y = 2$, $B_z = 0$:

$$\mathbf{A} \times \mathbf{B} = (0)\mathbf{i} - (0)\mathbf{j} + [(2)(2) - (3)(-1)]\mathbf{k} = 7\mathbf{k}$$

Exercise Use the results to this example and Equation 11.9 to find the angle between **A** and **B**.

Answer 60.3°

11.3 ▶ ANGULAR MOMENTUM OF A PARTICLE

Imagine a rigid pole sticking up through the ice on a frozen pond (Fig. 11.9). A skater glides rapidly toward the pole, aiming a little to the side so that she does not hit it. As she approaches a point beside the pole, she reaches out and grabs the pole, an action that whips her rapidly into a circular path around the pole. Just as the idea of linear momentum helps us analyze translational motion, a rotational analog—*angular momentum*—helps us describe this skater and other objects undergoing rotational motion.

To analyze the motion of the skater, we need to know her mass and her velocity, as well as her position relative to the pole. In more general terms, consider a particle of mass m located at the vector position $\mathbf{r}$ and moving with linear velocity $\mathbf{v}$ (Fig. 11.10).

> The instantaneous angular momentum $\mathbf{L}$ of the particle relative to the origin O is defined as the cross product of the particle's instantaneous position vector $\mathbf{r}$ and its instantaneous linear momentum $\mathbf{p}$:
>
> $$\mathbf{L} \equiv \mathbf{r} \times \mathbf{p} \tag{11.15}$$

Angular momentum of a particle

The SI unit of angular momentum is kg·m²/s. It is important to note that both the magnitude and the direction of $\mathbf{L}$ depend on the choice of origin. Following the right-hand rule, note that the direction of $\mathbf{L}$ is perpendicular to the plane formed by $\mathbf{r}$ and $\mathbf{p}$. In Figure 11.10, $\mathbf{r}$ and $\mathbf{p}$ are in the xy plane, and so $\mathbf{L}$ points in the z direction. Because $\mathbf{p} = m\mathbf{v}$, the magnitude of $\mathbf{L}$ is

$$L = mvr\sin\phi \tag{11.16}$$

where ϕ is the angle between $\mathbf{r}$ and $\mathbf{p}$. It follows that L is zero when $\mathbf{r}$ is parallel to $\mathbf{p}$ ($\phi = 0$ or $180°$). In other words, when the linear velocity of the particle is along a line that passes through the origin, the particle has zero angular momentum with respect to the origin. On the other hand, if $\mathbf{r}$ is perpendicular to $\mathbf{p}$ ($\phi = 90°$), then $L = mvr$. At that instant, the particle moves exactly as if it were on the rim of a wheel rotating about the origin in a plane defined by $\mathbf{r}$ and $\mathbf{p}$.

Quick Quiz 11.3

Recall the skater described at the beginning of this section. What would be her angular momentum relative to the pole if she were skating directly toward it?

Figure 11.9 As the skater passes the pole, she grabs hold of it. This causes her to swing around the pole rapidly in a circular path.

In describing linear motion, we found that the net force on a particle equals the time rate of change of its linear momentum, $\Sigma\mathbf{F} = d\mathbf{p}/dt$ (see Eq. 9.3). We now show that the net torque acting on a particle equals the time rate of change of its angular momentum. Let us start by writing the net torque on the particle in the form

$$\sum\boldsymbol{\tau} = \mathbf{r} \times \sum\mathbf{F} = \mathbf{r} \times \frac{d\mathbf{p}}{dt} \tag{11.17}$$

Now let us differentiate Equation 11.15 with respect to time, using the rule given by Equation 11.12:

$$\frac{d\mathbf{L}}{dt} = \frac{d}{dt}(\mathbf{r} \times \mathbf{p}) = \mathbf{r} \times \frac{d\mathbf{p}}{dt} + \frac{d\mathbf{r}}{dt} \times \mathbf{p}$$

Remember, it is important to adhere to the order of terms because $\mathbf{A} \times \mathbf{B} = -\mathbf{B} \times \mathbf{A}$. The last term on the right in the above equation is zero because $\mathbf{v} = d\mathbf{r}/dt$ is parallel to $\mathbf{p} = m\mathbf{v}$ (property 2 of the vector product). Therefore,

$$\frac{d\mathbf{L}}{dt} = \mathbf{r} \times \frac{d\mathbf{p}}{dt} \tag{11.18}$$

Comparing Equations 11.17 and 11.18, we see that

$$\sum\boldsymbol{\tau} = \frac{d\mathbf{L}}{dt} \tag{11.19}$$

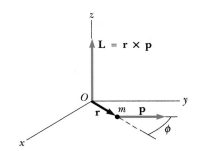

Figure 11.10 The angular momentum $\mathbf{L}$ of a particle of mass m and linear momentum $\mathbf{p}$ located at the vector position $\mathbf{r}$ is a vector given by $\mathbf{L} = \mathbf{r} \times \mathbf{p}$. The value of $\mathbf{L}$ depends on the origin about which it is measured and is a vector perpendicular to both $\mathbf{r}$ and $\mathbf{p}$.

The net torque equals time rate of change of angular momentum which is the rotational analog of Newton's second law, $\Sigma\mathbf{F} = d\mathbf{p}/dt$. Note that torque causes the angular momentum $\mathbf{L}$ to change just as force causes linear momentum $\mathbf{p}$ to change. This rotational result, Equation 11.19, states that

the net torque acting on a particle is equal to the time rate of change of the particle's angular momentum.

It is important to note that Equation 11.19 is valid only if $\Sigma\boldsymbol{\tau}$ and $\mathbf{L}$ are measured about the same origin. (Of course, the same origin must be used in calculating all of the torques.) Furthermore, **the expression is valid for any origin fixed in an inertial frame.**

Angular Momentum of a System of Particles

The total angular momentum of a system of particles about some point is defined as the vector sum of the angular momenta of the individual particles:

$$\mathbf{L} = \mathbf{L}_1 + \mathbf{L}_2 + \cdots + \mathbf{L}_n = \sum_i \mathbf{L}_i$$

where the vector sum is over all n particles in the system.

Because individual angular momenta can change with time, so can the total angular momentum. In fact, from Equations 11.18 and 11.19, we find that the time rate of change of the total angular momentum equals the vector sum of all torques acting on the system, both those associated with internal forces between particles and those associated with external forces. However, the net torque associated with all internal forces is zero. To understand this, recall that Newton's third law tells us that internal forces between particles of the system are equal in magnitude and opposite in direction. If we assume that these forces lie along the line of separation of each pair of particles, then the torque due to each action–reaction force pair is zero. That is, the moment arm d from O to the line of action of the forces is equal for both particles. In the summation, therefore, we see that the net internal torque vanishes. We conclude that the total angular momentum of a system can vary with time only if a net external torque is acting on the system, so that we have

$$\sum \boldsymbol{\tau}_{\text{ext}} = \sum_i \frac{d\mathbf{L}_i}{dt} = \frac{d}{dt}\sum_i \mathbf{L}_i = \frac{d\mathbf{L}}{dt} \tag{11.20}$$

That is,

the time rate of change of the total angular momentum of a system about some origin in an inertial frame equals the net external torque acting on the system about that origin.

Note that Equation 11.20 is the rotational analog of Equation 9.38, $\Sigma\mathbf{F}_{\text{ext}} = d\mathbf{p}/dt$, for a system of particles.

EXAMPLE 11.4 Circular Motion

A particle moves in the xy plane in a circular path of radius r, as shown in Figure 11.11. (a) Find the magnitude and direction of its angular momentum relative to O when its linear velocity is **v**.

Solution You might guess that because the linear momentum of the particle is always changing (in direction, not magnitude), the direction of the angular momentum must also change. In this example, however, this is not the case. The magnitude of **L** is given by

$$L = mvr \sin 90° = \boxed{mvr} \qquad \text{(for } \mathbf{r} \text{ perpendicular to } \mathbf{v}\text{)}$$

This value of L is constant because all three factors on the right are constant. The direction of **L** also is constant, even

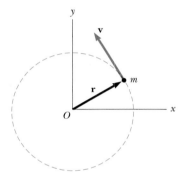

Figure 11.11 A particle moving in a circle of radius r has an angular momentum about O that has magnitude mvr. The vector $\mathbf{L} = \mathbf{r} \times \mathbf{p}$ points *out* of the diagram.

though the direction of $\mathbf{p} = m\mathbf{v}$ keeps changing. You can visualize this by sliding the vector **v** in Figure 11.11 parallel to itself until its tail meets the tail of **r** and by then applying the right-hand rule. (You can use **v** to determine the direction of $\mathbf{L} = \mathbf{r} \times \mathbf{p}$ because the direction of **p** is the same as the direction of **v**.) Line up your fingers so that they point along **r** and wrap your fingers into the vector **v**. Your thumb points upward and away from the page; this is the direction of **L**. Hence, we can write the vector expression $\mathbf{L} = (mvr)\mathbf{k}$. If the particle were to move clockwise, **L** would point downward and into the page.

(b) Find the magnitude and direction of **L** in terms of the particle's angular speed ω.

Solution Because $v = r\omega$ for a particle rotating in a circle, we can express L as

$$L = mvr = mr^2\omega = \boxed{I\omega}$$

where I is the moment of inertia of the particle about the z axis through O. Because the rotation is counterclockwise, the direction of $\boldsymbol{\omega}$ is along the z axis (see Section 10.1). The direction of **L** is the same as that of $\boldsymbol{\omega}$, and so we can write the angular momentum as $\mathbf{L} = I\boldsymbol{\omega} = I\omega\mathbf{k}$.

Exercise A car of mass 1 500 kg moves with a linear speed of 40 m/s on a circular race track of radius 50 m. What is the magnitude of its angular momentum relative to the center of the track?

Answer 3.0×10^6 kg·m²/s

11.4 ANGULAR MOMENTUM OF A ROTATING RIGID OBJECT

Consider a rigid object rotating about a fixed axis that coincides with the z axis of a coordinate system, as shown in Figure 11.12. Let us determine the angular momentum of this object. Each particle of the object rotates in the xy plane about the z axis with an angular speed ω. The magnitude of the angular momentum of a particle of mass m_i about the origin O is $m_i v_i r_i$. Because $v_i = r_i\omega$, we can express the magnitude of the angular momentum of this particle as

$$L_i = m_i r_i^2 \omega$$

The vector $\mathbf{L}_i$ is directed along the z axis, as is the vector $\boldsymbol{\omega}$.

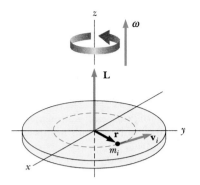

Figure 11.12 When a rigid body rotates about an axis, the angular momentum **L** is in the same direction as the angular velocity $\boldsymbol{\omega}$, according to the expression $\mathbf{L} = I\boldsymbol{\omega}$.

We can now find the angular momentum (which in this situation has only a z component) of the whole object by taking the sum of L_i over all particles:

$$L_z = \sum_i m_i r_i^2 \omega = \left(\sum_i m_i r_i^2 \right) \omega$$

$$L_z = I\omega \quad \text{see note below} \qquad (11.21)$$

where I is the moment of inertia of the object about the z axis.

Now let us differentiate Equation 11.21 with respect to time, noting that I is constant for a rigid body:

$$\frac{dL_z}{dt} = I\frac{d\omega}{dt} = I\alpha \qquad (11.22)$$

where α is the angular acceleration relative to the axis of rotation. Because dL_z/dt is equal to the net external torque (see Eq. 11.20), we can express Equation 11.22 as

$$\sum \tau_{\text{ext}} = \frac{dL_z}{dt} = I\alpha \qquad (11.23)$$

That is, the net external torque acting on a rigid object rotating about a fixed axis equals the moment of inertia about the rotation axis multiplied by the object's angular acceleration relative to that axis.

Equation 11.23 also is valid for a rigid object rotating about a moving axis provided the moving axis (1) passes through the center of mass and (2) is a symmetry axis.

You should note that if a symmetrical object rotates about a fixed axis passing through its center of mass, you can write Equation 11.21 in vector form as $\mathbf{L} = I\boldsymbol{\omega}$, where **L** is the total angular momentum of the object measured with respect to the axis of rotation. Furthermore, the expression is valid for any object, regardless of its symmetry, if **L** stands for the component of angular momentum along the axis of rotation.[2]

EXAMPLE 11.5   Bowling Ball

Estimate the magnitude of the angular momentum of a bowling ball spinning at 10 rev/s, as shown in Figure 11.13.

Solution We start by making some estimates of the relevant physical parameters and model the ball as a uniform solid sphere. A typical bowling ball might have a mass of 6 kg and a radius of 12 cm. The moment of inertia of a solid sphere about an axis through its center is, from Table 10.2,

$$I = \tfrac{2}{5}MR^2 = \tfrac{2}{5}(6 \text{ kg})(0.12 \text{ m})^2 = 0.035 \text{ kg} \cdot \text{m}^2$$

Therefore, the magnitude of the angular momentum is

[2] In general, the expression $\mathbf{L} = I\boldsymbol{\omega}$ is not always valid. If a rigid object rotates about an arbitrary axis, **L** and $\boldsymbol{\omega}$ may point in different directions. In this case, the moment of inertia cannot be treated as a scalar. Strictly speaking, $\mathbf{L} = I\boldsymbol{\omega}$ applies only to rigid objects of any shape that rotate about one of three mutually perpendicular axes (called *principal axes*) through the center of mass. This is discussed in more advanced texts on mechanics.

$$L = I\omega = (0.035 \text{ kg} \cdot \text{m}^2)(10 \text{ rev/s})(2\pi \text{ rad/rev})$$
$$= 2.2 \text{ kg} \cdot \text{m}^2/\text{s}$$

Because of the roughness of our estimates, we probably want to keep only one significant figure, and so $L \approx \boxed{2 \text{ kg} \cdot \text{m}^2/\text{s}}$.

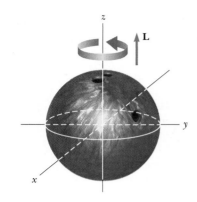

Figure 11.13 A bowling ball that rotates about the z axis in the direction shown has an angular momentum $\mathbf{L}$ in the positive z direction. If the direction of rotation is reversed, $\mathbf{L}$ points in the negative z direction.

 ### EXAMPLE 11.6 ▶ Rotating Rod

A rigid rod of mass M and length ℓ is pivoted without friction at its center (Fig. 11.14). Two particles of masses m_1 and m_2 are connected to its ends. The combination rotates in a vertical plane with an angular speed ω. (a) Find an expression for the magnitude of the angular momentum of the system.

Solution This is different from the last example in that we now must account for the motion of more than one object. The moment of inertia of the system equals the sum of the moments of inertia of the three components: the rod and the two particles. Referring to Table 10.2 to obtain the expression for the moment of inertia of the rod, and using the expression $I = mr^2$ for each particle, we find that the total moment of inertia about the z axis through O is

$$I = \frac{1}{12}M\ell^2 + m_1\left(\frac{\ell}{2}\right)^2 + m_2\left(\frac{\ell}{2}\right)^2$$
$$= \frac{\ell^2}{4}\left(\frac{M}{3} + m_1 + m_2\right)$$

Therefore, the magnitude of the angular momentum is

$$L = I\omega = \boxed{\frac{\ell^2}{4}\left(\frac{M}{3} + m_1 + m_2\right)\omega}$$

(b) Find an expression for the magnitude of the angular acceleration of the system when the rod makes an angle θ with the horizontal.

Solution If the masses of the two particles are equal, then the system has no angular acceleration because the net torque on the system is zero when $m_1 = m_2$. If the initial angle θ is exactly $\pi/2$ or $-\pi/2$ (vertical position), then the rod will be in equilibrium. To find the angular acceleration of the system at any angle θ, we first calculate the net torque on the system and then use $\Sigma\tau_{\text{ext}} = I\alpha$ to obtain an expression for α.

The torque due to the force $m_1 g$ about the pivot is

$$\tau_1 = m_1 g \frac{\ell}{2}\cos\theta \qquad (\tau_1 \text{ out of page})$$

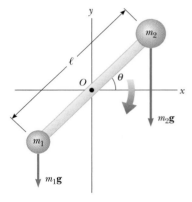

Figure 11.14 Because gravitational forces act on the rotating rod, there is in general a net nonzero torque about O when $m_1 \neq m_2$. This net torque produces an angular acceleration given by $\alpha = \Sigma\tau_{\text{ext}}/I$.

The torque due to the force $m_2 g$ about the pivot is

$$\tau_2 = -m_2 g \frac{\ell}{2}\cos\theta \qquad (\tau_2 \text{ into page})$$

Hence, the net torque exerted on the system about O is

$$\Sigma\tau_{\text{ext}} = \tau_1 + \tau_2 = \tfrac{1}{2}(m_1 - m_2)g\ell\cos\theta$$

The direction of $\Sigma\tau_{\text{ext}}$ is out of the page if $m_1 > m_2$ and is into the page if $m_2 > m_1$.

To find α, we use $\Sigma\tau_{\text{ext}} = I\alpha$, where I was obtained in part (a):

$$\alpha = \frac{\Sigma\tau_{\text{ext}}}{I} = \frac{2(m_1 - m_2)g\cos\theta}{\ell(M/3 + m_1 + m_2)}$$

Note that α is zero when θ is $\pi/2$ or $-\pi/2$ (vertical position) and is a maximum when θ is 0 or π (horizontal position).

Exercise If $m_2 > m_1$, at what value of θ is ω a maximum?

Answer $\theta = -\pi/2$.

EXAMPLE 11.7 Two Connected Masses

A sphere of mass m_1 and a block of mass m_2 are connected by a light cord that passes over a pulley, as shown in Figure 11.15. The radius of the pulley is R, and the moment of inertia about its axle is I. The block slides on a frictionless, horizontal surface. Find an expression for the linear acceleration of the two objects, using the concepts of angular momentum and torque.

Solution We need to determine the angular momentum of the system, which consists of the two objects and the pulley. Let us calculate the angular momentum about an axis that coincides with the axle of the pulley.

At the instant the sphere and block have a common speed v, the angular momentum of the sphere is $m_1 vR$, and that of the block is $m_2 vR$. At the same instant, the angular momentum of the pulley is $I\omega = Iv/R$. Hence, the total angular momentum of the system is

$$(1) \qquad L = m_1 vR + m_2 vR + I\frac{v}{R}$$

Figure 11.15

Now let us evaluate the total external torque acting on the system about the pulley axle. Because it has a moment arm of zero, the force exerted by the axle on the pulley does not contribute to the torque. Furthermore, the normal force acting on the block is balanced by the force of gravity $m_2\mathbf{g}$, and so these forces do not contribute to the torque. The force of gravity $m_1\mathbf{g}$ acting on the sphere produces a torque about the axle equal in magnitude to $m_1 gR$, where R is the moment arm of the force about the axle. (Note that in this situation, the tension is *not* equal to $m_1 g$.) This is the total external torque about the pulley axle; that is, $\Sigma\tau_{\text{ext}} = m_1 gR$. Using this result, together with Equation (1) and Equation 11.23, we find

$$\Sigma\tau_{\text{ext}} = \frac{dL}{dt}$$

$$m_1 gR = \frac{d}{dt}\left[(m_1 + m_2)Rv + I\frac{v}{R}\right]$$

$$(2) \qquad m_1 gR = (m_1 + m_2)R\frac{dv}{dt} + \frac{I}{R}\frac{dv}{dt}$$

Because $dv/dt = a$, we can solve this for a to obtain

$$a = \frac{m_1 g}{(m_1 + m_2) + I/R^2}$$

You may wonder why we did not include the forces that the cord exerts on the objects in evaluating the net torque about the axle. The reason is that these forces are internal to the system under consideration, and we analyzed the system as a whole. Only external torques contribute to the change in the

---

11.5 CONSERVATION OF ANGULAR MOMENTUM

In Chapter 9 we found that the total linear momentum of a system of particles remains constant when the resultant external force acting on the system is zero. We have an analogous conservation law in rotational motion:

Conservation of angular momentum

> The total angular momentum of a system is constant in both magnitude and direction if the resultant external torque acting on the system is zero.

This follows directly from Equation 11.20, which indicates that if

$$\Sigma\tau_{\text{ext}} = \frac{d\mathbf{L}}{dt} = 0 \tag{11.24}$$

then

$$\mathbf{L} = \text{constant} \tag{11.25}$$

For a system of particles, we write this conservation law as $\Sigma\mathbf{L}_n = \text{constant}$, where the index n denotes the nth particle in the system.

If the mass of an object undergoes redistribution in some way, then the object's moment of inertia changes; hence, its angular speed must change because $L = I\omega$. In this case we express the conservation of angular momentum in the form

$$\mathbf{L}_i = \mathbf{L}_f = \text{constant} \tag{11.26}$$

If the system is an object rotating about a *fixed* axis, such as the z axis, we can write $L_z = I\omega$, where L_z is the component of $\mathbf{L}$ along the axis of rotation and I is the moment of inertia about this axis. In this case, we can express the conservation of angular momentum as

$$I_i\omega_i = I_f\omega_f = \text{constant} \tag{11.27}$$

This expression is valid both for rotation about a fixed axis and for rotation about an axis through the center of mass of a moving system as long as that axis remains parallel to itself. We require only that the net external torque be zero.

Although we do not prove it here, there is an important theorem concerning the angular momentum of an object relative to the object's center of mass:

> The resultant torque acting on an object about an axis through the center of mass equals the time rate of change of angular momentum regardless of the motion of the center of mass.

This theorem applies even if the center of mass is accelerating, provided τ and $\mathbf{L}$ are evaluated relative to the center of mass.

In Equation 11.26 we have a third conservation law to add to our list. We can now state that the energy, linear momentum, and angular momentum of an isolated system all remain constant:

$$\left.\begin{array}{c} K_i + U_i = K_f + U_f \\ \mathbf{p}_i = \mathbf{p}_f \\ \mathbf{L}_i = \mathbf{L}_f \end{array}\right\} \quad \text{For an isolated system}$$

There are many examples that demonstrate conservation of angular momentum. You may have observed a figure skater spinning in the finale of a program. The angular speed of the skater increases when the skater pulls his hands and feet close to his body, thereby decreasing I. Neglecting friction between skates and ice, no external torques act on the skater. The change in angular speed is due to the fact that, because angular momentum is conserved, the product $I\omega$ remains constant, and a decrease in the moment of inertia of the skater causes an increase in the angular speed. Similarly, when divers or acrobats wish to make several somersaults, they pull their hands and feet close to their bodies to rotate at a higher rate. In these cases, the external force due to gravity acts through the center of mass and hence exerts no torque about this point. Therefore, the angular momentum about the center of mass must be conserved—that is, $I_i\omega_i = I_f\omega_f$. For example, when divers wish to double their angular speed, they must reduce their moment of inertia to one-half its initial value.

Angular momentum is conserved as figure skater Todd Eldredge pulls his arms toward his body. (© *1998 David Madison*)

Quick Quiz 11.4

A particle moves in a straight line, and you are told that the net torque acting on it is zero about some unspecified point. Decide whether the following statements are true or false: (a) The net force on the particle must be zero. (b) The particle's velocity must be constant.

A color-enhanced, infrared image of Hurricane Mitch, which devastated large areas of Honduras and Nicaragua in October 1998. The spiral, nonrigid mass of air undergoes rotation and has angular momentum. *(Courtesy of NOAA)*

EXAMPLE 11.8   Formation of a Neutron Star

A star rotates with a period of 30 days about an axis through its center. After the star undergoes a supernova explosion, the stellar core, which had a radius of 1.0×10^4 km, collapses into a neutron star of radius 3.0 km. Determine the period of rotation of the neutron star.

Solution   The same physics that makes a skater spin faster with his arms pulled in describes the motion of the stellar core. Let us assume that during the collapse of the star, (1) no torque acts on it, (2) it remains spherical, and (3) its mass remains constant. Also, let us use the symbol T for the period, with T_i being the initial period of the star and T_f being the period of the neutron star. The period is the length of

time a point on the star's equator takes to make one complete circle around the axis of rotation. The angular speed of a star is given by $\omega = 2\pi/T$. Therefore, because I is proportional to r^2, Equation 11.27 gives

$$T_f = T_i \left(\frac{r_f}{r_i} \right)^2 = (30 \text{ days}) \left(\frac{3.0 \text{ km}}{1.0 \times 10^4 \text{ km}} \right)^2$$

$$= 2.7 \times 10^{-6} \text{ days} = \boxed{0.23 \text{ s}}$$

Thus, the neutron star rotates about four times each second; this result is approximately the same as that for a spinning figure skater.

EXAMPLE 11.9   The Merry-Go-Round

A horizontal platform in the shape of a circular disk rotates in a horizontal plane about a frictionless vertical axle (Fig. 11.16). The platform has a mass $M = 100$ kg and a radius $R = 2.0$ m. A student whose mass is $m = 60$ kg walks slowly from the rim of the disk toward its center. If the angular speed of the system is 2.0 rad/s when the student is at the rim, what is the angular speed when he has reached a point $r = 0.50$ m from the center?

Solution   The speed change here is similar to the increase in angular speed of the spinning skater when he pulls his arms inward. Let us denote the moment of inertia of the platform as I_p and that of the student as I_s. Treating the student as a point mass, we can write the initial moment of inertia I_i of the system (student plus platform) about the axis of rotation:

$$I_i = I_{pi} + I_{si} = \tfrac{1}{2}MR^2 + mR^2$$

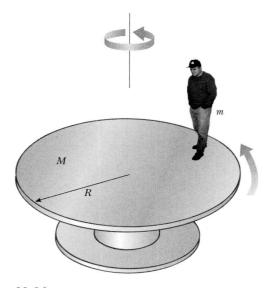

Figure 11.16 As the student walks toward the center of the rotating platform, the angular speed of the system increases because the angular momentum must remain constant.

When the student has walked to the position $r < R$, the moment of inertia of the system reduces to

$$I_f = I_{pf} + I_{sf} = \frac{1}{2}MR^2 + mr^2$$

Note that we still use the greater radius R when calculating I_{pf} because the radius of the platform has not changed. Because no external torques act on the system about the axis of rotation, we can apply the law of conservation of angular momentum:

$$I_i\omega_i = I_f\omega_f$$

$$\left(\tfrac{1}{2}MR^2 + mR^2\right)\omega_i = \left(\tfrac{1}{2}MR^2 + mr^2\right)\omega_f$$

$$\omega_f = \left(\frac{\tfrac{1}{2}MR^2 + mR^2}{\tfrac{1}{2}MR^2 + mr^2}\right)\omega_i$$

$$\omega_f = \left(\frac{200 + 240}{200 + 15}\right)(2.0\,\text{rad/s}) = \boxed{4.1\,\text{rad/s}}$$

As expected, the angular speed has increased.

Exercise Calculate the initial and final rotational energies of the system.

Answer $K_i = 880\,\text{J}$; $K_f = 1.8 \times 10^3\,\text{J}$.

Quick Quiz 11.5

Note that the rotational energy of the system described in Example 11.9 increases. What accounts for this increase in energy?

EXAMPLE 11.10 ▶ The Spinning Bicycle Wheel

In a favorite classroom demonstration, a student holds the axle of a spinning bicycle wheel while seated on a stool that is free to rotate (Fig. 11.17). The student and stool are initially at rest while the wheel is spinning in a horizontal plane with an initial angular momentum $\mathbf{L}_i$ that points upward. When the wheel is inverted about its center by 180°, the student and stool start rotating. In terms of $\mathbf{L}_i$, what are the magnitude and the direction of $\mathbf{L}$ for the student plus stool?

Solution The system consists of the student, the wheel, and the stool. Initially, the total angular momentum of the system $\mathbf{L}_i$ comes entirely from the spinning wheel. As the wheel is inverted, the student applies a torque to the wheel, but this torque is internal to the system. No external torque is acting on the system about the vertical axis. Therefore, the angular momentum of the system is conserved. Initially, we have

$$\mathbf{L}_{\text{system}} = \mathbf{L}_i = \mathbf{L}_{\text{wheel}} \quad \text{(upward)}$$

After the wheel is inverted, we have $\mathbf{L}_{\text{inverted wheel}} = -\mathbf{L}_i$. For angular momentum to be conserved, some other part of the system has to start rotating so that the total angular momentum remains the initial angular momentum $\mathbf{L}_i$. That other part of the system is the student plus the stool she is sitting on. So, we can now state that

$$\mathbf{L}_f = \mathbf{L}_i = \mathbf{L}_{\text{student+stool}} - \mathbf{L}_i$$

$$\mathbf{L}_{\text{student+stool}} = \boxed{2\mathbf{L}_i}$$

Figure 11.17 The wheel is initially spinning when the student is at rest. What happens when the wheel is inverted?

EXAMPLE 11.11   **Disk and Stick**

A 2.0-kg disk traveling at 3.0 m/s strikes a 1.0-kg stick that is lying flat on nearly frictionless ice, as shown in Figure 11.18. Assume that the collision is elastic. Find the translational speed of the disk, the translational speed of the stick, and the rotational speed of the stick after the collision. The moment of inertia of the stick about its center of mass is 1.33 kg·m².

Solution   Because the disk and stick form an isolated system, we can assume that total energy, linear momentum, and angular momentum are all conserved. We have three unknowns, and so we need three equations to solve simultaneously. The first comes from the law of the conservation of linear momentum:

$$p_i = p_f$$

$$m_d v_{di} = m_d v_{df} + m_s v_s$$

$$(2.0 \text{ kg})(3.0 \text{ m/s}) = (2.0 \text{ kg}) v_{df} + (1.0 \text{ kg}) v_s$$

$$(1) \qquad 6.0 \text{ kg} \cdot \text{m/s} - (2.0 \text{ kg}) v_{df} = (1.0 \text{ kg}) v_s$$

Now we apply the law of conservation of angular momentum, using the initial position of the center of the stick as our reference point. We know that the component of angular momentum of the disk along the axis perpendicular to the plane of the ice is negative (the right-hand rule shows that $\mathbf{L}_d$ points into the ice).

$$L_i = L_f$$

$$-r m_d v_{di} = -r m_d v_{df} + I\omega$$

$$-(2.0 \text{ m})(2.0 \text{ kg})(3.0 \text{ m/s}) = -(2.0 \text{ m})(2.0 \text{ kg}) v_{df}$$
$$+ (1.33 \text{ kg} \cdot \text{m}^2)\omega$$

$$-12 \text{ kg} \cdot \text{m}^2/\text{s} = -(4.0 \text{ kg} \cdot \text{m}) v_{df}$$
$$+ (1.33 \text{ kg} \cdot \text{m}^2)\omega$$

$$(2) \qquad -9.0 \text{ rad/s} + (3.0 \text{ rad/m}) v_{df} = \omega$$

We used the fact that radians are dimensionless to ensure consistent units for each term.

Finally, the elastic nature of the collision reminds us that kinetic energy is conserved; in this case, the kinetic energy consists of translational and rotational forms:

$$K_i = K_f$$

$$\tfrac{1}{2} m_d v_{di}^2 = \tfrac{1}{2} m_d v_{df}^2 + \tfrac{1}{2} m_s v_s^2 + \tfrac{1}{2} I \omega^2$$

$$\tfrac{1}{2}(2.0 \text{ kg})(3.0 \text{ m/s})^2 = \tfrac{1}{2}(2.0 \text{ kg}) v_{df}^2 + \tfrac{1}{2}(1.0 \text{ kg}) v_s^2$$
$$+ \tfrac{1}{2}(1.33 \text{ kg} \cdot \text{m}^2/\text{s})\omega^2$$

$$(3) \qquad 54 \text{ m}^2/\text{s}^2 = 6.0 v_{df}^2 + 3.0 v_s^2 + (4.0 \text{ m}^2)\omega^2$$

In solving Equations (1), (2), and (3) simultaneously, we find that $v_{df} = 2.3$ m/s, $v_s = 1.3$ m/s, and $\omega = -2.0$ rad/s. These values seem reasonable. The disk is moving more slowly than it was before the collision, and the stick has a small translational speed. Table 11.1 summarizes the initial and final values of variables for the disk and the stick and verifies the conservation of linear momentum, angular momentum, and kinetic energy.

Exercise   Verify the values in Table 11.1.

Figure 11.18   Overhead view of a disk striking a stick in an elastic collision, which causes the stick to rotate.

TABLE 11.1   Comparison of Values in Example 11.11 Before and After the Collision[a]

	v (m/s)	ω (rad/s)	p (kg·m/s)	L (kg·m²/s)	K_{trans} (J)	K_{rot} (J)
Before						
Disk	3.0	—	6.0	−12	9.0	—
Stick	0	0	0	0	0	0
Total	—	—	6.0	−12	9.0	0
After						
Disk	2.3	—	4.7	−9.3	5.4	—
Stick	1.3	−2.0	1.3	−2.7	0.9	2.7
Total	—	—	6.0	−12	6.3	2.7

[a]Notice that linear momentum, angular momentum, and total kinetic energy are conserved.

Optional Section

11.6 ▶ THE MOTION OF GYROSCOPES AND TOPS

A very unusual and fascinating type of motion you probably have observed is that of a top spinning about its axis of symmetry, as shown in Figure 11.19a. If the top spins very rapidly, the axis rotates about the z axis, sweeping out a cone (see Fig. 11.19b). The motion of the axis about the vertical—known as **precessional motion**—is usually slow relative to the spin motion of the top.

It is quite natural to wonder why the top does not fall over. Because the center of mass is not directly above the pivot point O, a net torque is clearly acting on the top about O—a torque resulting from the force of gravity $M\mathbf{g}$. The top would certainly fall over if it were not spinning. Because it is spinning, however, it has an angular momentum $\mathbf{L}$ directed along its symmetry axis. As we shall show, the motion of this symmetry axis about the z axis (the precessional motion) occurs because the torque produces a change in the *direction* of the symmetry axis. This is an excellent example of the importance of the directional nature of angular momentum.

The two forces acting on the top are the downward force of gravity $M\mathbf{g}$ and the normal force $\mathbf{n}$ acting upward at the pivot point O. The normal force produces no torque about the pivot because its moment arm through that point is zero. However, the force of gravity produces a torque $\boldsymbol{\tau} = \mathbf{r} \times M\mathbf{g}$ about O, where the direction of $\boldsymbol{\tau}$ is perpendicular to the plane formed by $\mathbf{r}$ and $M\mathbf{g}$. By necessity, the vector $\boldsymbol{\tau}$ lies in a horizontal xy plane perpendicular to the angular momentum vector. The net torque and angular momentum of the top are related through Equation 11.19:

$$\boldsymbol{\tau} = \frac{d\mathbf{L}}{dt}$$

From this expression, we see that the nonzero torque produces a change in angular momentum $d\mathbf{L}$—a change that is in the same direction as $\boldsymbol{\tau}$. Therefore, like the torque vector, $d\mathbf{L}$ must also be at right angles to $\mathbf{L}$. Figure 11.19b illustrates the resulting precessional motion of the symmetry axis of the top. In a time Δt, the change in angular momentum is $\Delta\mathbf{L} = \mathbf{L}_f - \mathbf{L}_i = \boldsymbol{\tau}\,\Delta t$. Because $\Delta\mathbf{L}$ is perpendicular to $\mathbf{L}$, the magnitude of $\mathbf{L}$ does not change ($|\mathbf{L}_i| = |\mathbf{L}_f|$). Rather, what is changing is the *direction* of $\mathbf{L}$. Because the change in angular momentum $\Delta\mathbf{L}$ is in the direction of $\boldsymbol{\tau}$, which lies in the xy plane, the top undergoes precessional motion.

The essential features of precessional motion can be illustrated by considering the simple gyroscope shown in Figure 11.20a. This device consists of a wheel free to spin about an axle that is pivoted at a distance h from the center of mass of the wheel. When given an angular velocity $\boldsymbol{\omega}$ about the axle, the wheel has an angular momentum $\mathbf{L} = I\boldsymbol{\omega}$ directed along the axle as shown. Let us consider the torque acting on the wheel about the pivot O. Again, the force $\mathbf{n}$ exerted by the support on the axle produces no torque about O, and the force of gravity $M\mathbf{g}$ produces a torque of magnitude Mgh about O, where the axle is perpendicular to the support. The direction of this torque is perpendicular to the axle (and perpendicular to $\mathbf{L}$), as shown in Figure 11.20a. This torque causes the angular momentum to change in the direction perpendicular to the axle. Hence, the axle moves in the direction of the torque—that is, in the horizontal plane.

To simplify the description of the system, we must make an assumption: The total angular momentum of the precessing wheel is the sum of the angular momentum $I\boldsymbol{\omega}$ due to the spinning and the angular momentum due to the motion of

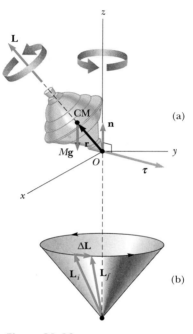

Figure 11.19 Precessional motion of a top spinning about its symmetry axis. (a) The only external forces acting on the top are the normal force $\mathbf{n}$ and the force of gravity $M\mathbf{g}$. The direction of the angular momentum $\mathbf{L}$ is along the axis of symmetry. The right-hand rule indicates that $\boldsymbol{\tau} = \mathbf{r} \times \mathbf{F} = \mathbf{r} \times M\mathbf{g}$ is in the xy plane. (b). The direction of $\Delta\mathbf{L}$ is parallel to that of $\boldsymbol{\tau}$ in part (a). The fact that $\mathbf{L}_f = \Delta\mathbf{L} + \mathbf{L}_i$ indicates that the top precesses about the z axis.

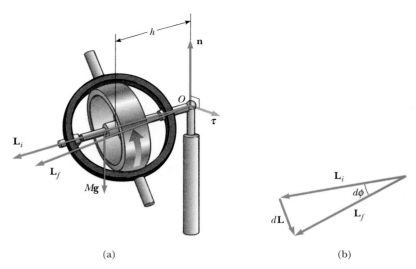

(a)   (b)

Figure 11.20 (a) The motion of a simple gyroscope pivoted a distance h from its center of mass. The force of gravity $M\mathbf{g}$ produces a torque about the pivot, and this torque is perpendicular to the axle. (b) This torque results in a change in angular momentum $d\mathbf{L}$ in a direction perpendicular to the axle. The axle sweeps out an angle $d\phi$ in a time dt.

This toy gyroscope undergoes precessional motion about the vertical axis as it spins about its axis of symmetry. The only forces acting on it are the force of gravity $M\mathbf{g}$ and the upward force of the pivot **n**. The direction of its angular momentum **L** is along the axis of symmetry. The torque and $\Delta\mathbf{L}$ are directed into the page. *(Courtesy of Central Scientific Company)*

the center of mass about the pivot. In our treatment, we shall neglect the contribution from the center-of-mass motion and take the total angular momentum to be just $I\boldsymbol{\omega}$. In practice, this is a good approximation if $\boldsymbol{\omega}$ is made very large.

In a time dt, the torque due to the gravitational force changes the angular momentum of the system by $d\mathbf{L} = \boldsymbol{\tau} \, dt$. When added vectorially to the original total angular momentum $I\omega$, this additional angular momentum causes a shift in the direction of the total angular momentum.

The vector diagram in Figure 11.20b shows that in the time dt, the angular momentum vector rotates through an angle $d\phi$, which is also the angle through which the axle rotates. From the vector triangle formed by the vectors $\mathbf{L}_i$, $\mathbf{L}_f$, and $d\mathbf{L}$, we see that

$$\sin(d\phi) \approx d\phi = \frac{dL}{L} = \frac{(Mgh)\,dt}{L}$$

where we have used the fact that, for small values of any angle θ, $\sin \theta \approx \theta$. Dividing through by dt and using the relationship $L = I\omega$, we find that the rate at which the axle rotates about the vertical axis is

$$\omega_p = \frac{d\phi}{dt} = \frac{Mgh}{I\omega} \qquad \textbf{(11.28)}$$

The angular speed ω_p is called the **precessional frequency.** This result is valid only when $\omega_p \ll \omega$. Otherwise, a much more complicated motion is involved. As you can see from Equation 11.28, the condition $\omega_p \ll \omega$ is met when $I\omega$ is great compared with Mgh. Furthermore, note that the precessional frequency decreases as ω increases—that is, as the wheel spins faster about its axis of symmetry.

> Precessional frequency

Quick Quiz 11.6

How much work is done by the force of gravity when a top precesses through one complete circle?

Optional Section

11.7 ANGULAR MOMENTUM AS A FUNDAMENTAL QUANTITY

We have seen that the concept of angular momentum is very useful for describing the motion of macroscopic systems. However, the concept also is valid on a submicroscopic scale and has been used extensively in the development of modern theories of atomic, molecular, and nuclear physics. In these developments, it was found that the angular momentum of a system is a fundamental quantity. The word *fundamental* in this context implies that angular momentum is an intrinsic property of atoms, molecules, and their constituents, a property that is a part of their very nature.

To explain the results of a variety of experiments on atomic and molecular systems, we rely on the fact that the angular momentum has discrete values. These discrete values are multiples of the fundamental unit of angular momentum $\hbar = h/2\pi$, where h is called Planck's constant:

$$\text{Fundamental unit of angular momentum} = \hbar = 1.054 \times 10^{-34} \text{ kg} \cdot \text{m}^2/\text{s}$$

Let us accept this postulate without proof for the time being and show how it can be used to estimate the angular speed of a diatomic molecule. Consider the O_2 molecule as a rigid rotor, that is, two atoms separated by a fixed distance d and rotating about the center of mass (Fig. 11.21). Equating the angular momentum to the fundamental unit $\hbar$, we can estimate the lowest angular speed:

$$I_{CM}\omega \approx \hbar \qquad \text{or} \qquad \omega \approx \frac{\hbar}{I_{CM}}$$

Figure 11.21 The rigid-rotor model of a diatomic molecule. The rotation occurs about the center of mass in the plane of the page.

In Example 10.3, we found that the moment of inertia of the O_2 molecule about this axis of rotation is 1.95×10^{-46} kg·m². Therefore,

$$\omega \approx \frac{\hbar}{I_{CM}} = \frac{1.054 \times 10^{-34} \text{ kg·m}^2/\text{s}}{1.95 \times 10^{-46} \text{ kg·m}^2} = 5.41 \times 10^{11} \text{ rad/s}$$

Actual angular speeds are multiples of this smallest possible value.

This simple example shows that certain classical concepts and models, when properly modified, might be useful in describing some features of atomic and molecular systems. A wide variety of phenomena on the submicroscopic scale can be explained only if we assume discrete values of the angular momentum associated with a particular type of motion.

The Danish physicist Niels Bohr (1885–1962) accepted and adopted this radical idea of discrete angular momentum values in developing his theory of the hydrogen atom. Strictly classical models were unsuccessful in describing many properties of the hydrogen atom. Bohr postulated that the electron could occupy only those circular orbits about the proton for which the orbital angular momentum was equal to $n\hbar$, where n is an integer. That is, he made the bold assumption that orbital angular momentum is quantized. From this simple model, the rotational frequencies of the electron in the various orbits can be estimated (see Problem 43).

SUMMARY

The **total kinetic energy** of a rigid object rolling on a rough surface without slipping equals the rotational kinetic energy about its center of mass, $\frac{1}{2}I_{CM}\omega^2$, plus the translational kinetic energy of the center of mass, $\frac{1}{2}Mv_{CM}^2$:

$$K = \tfrac{1}{2}I_{CM}\omega^2 + \tfrac{1}{2}Mv_{CM}^2 \tag{11.4}$$

The **torque** τ due to a force **F** about an origin in an inertial frame is defined to be

$$\tau \equiv \mathbf{r} \times \mathbf{F} \tag{11.7}$$

Given two vectors **A** and **B**, the **cross product** $\mathbf{A} \times \mathbf{B}$ is a vector **C** having a magnitude

$$C \equiv AB \sin \theta \tag{11.9}$$

where θ is the angle between **A** and **B**. The direction of the vector $\mathbf{C} = \mathbf{A} \times \mathbf{B}$ is perpendicular to the plane formed by **A** and **B**, and this direction is determined by the right-hand rule.

The **angular momentum L** of a particle having linear momentum $\mathbf{p} = m\mathbf{v}$ is

$$\mathbf{L} \equiv \mathbf{r} \times \mathbf{p} \tag{11.15}$$

where **r** is the vector position of the particle relative to an origin in an inertial frame.

The **net external torque** acting on a particle or rigid object is equal to the time rate of change of its angular momentum:

$$\sum \tau_{ext} = \frac{d\mathbf{L}}{dt} \tag{11.20}$$

The z *component* of **angular momentum** of a rigid object rotating about a fixed z axis is

$$L_z = I\omega \tag{11.21}$$

where I is the moment of inertia of the object about the axis of rotation and ω is its angular speed.

The **net external torque** acting on a rigid object equals the product of its moment of inertia about the axis of rotation and its angular acceleration:

$$\sum \tau_{\text{ext}} = I\alpha \qquad\qquad \textbf{(11.23)}$$

If the net external torque acting on a system is zero, then the total angular momentum of the system is constant. Applying this **law of conservation of angular momentum** to a system whose moment of inertia changes gives

$$I_i\omega_i = I_f\omega_f = \text{constant} \qquad\qquad \textbf{(11.27)}$$

QUESTIONS

1. Is it possible to calculate the torque acting on a rigid body without specifying a center of rotation? Is the torque independent of the location of the center of rotation?
2. Is the triple product defined by $\mathbf{A}\cdot(\mathbf{B}\times\mathbf{C})$ a scalar or a vector quantity? Explain why the operation $(\mathbf{A}\cdot\mathbf{B})\times\mathbf{C}$ has no meaning.
3. In some motorcycle races, the riders drive over small hills, and the motorcycles become airborne for a short time. If a motorcycle racer keeps the throttle open while leaving the hill and going into the air, the motorcycle tends to nose upward. Why does this happen?
4. If the torque acting on a particle about a certain origin is zero, what can you say about its angular momentum about that origin?
5. Suppose that the velocity vector of a particle is completely specified. What can you conclude about the direction of its angular momentum vector with respect to the direction of motion?
6. If a single force acts on an object, and the torque caused by that force is nonzero about some point, is there any other point about which the torque is zero?
7. If a system of particles is in motion, is it possible for the total angular momentum to be zero about some origin? Explain.
8. A ball is thrown in such a way that it does not spin about its own axis. Does this mean that the angular momentum is zero about an arbitrary origin? Explain.
9. In a tape recorder, the tape is pulled past the read-and-write heads at a constant speed by the drive mechanism. Consider the reel from which the tape is pulled—as the tape is pulled off it, the radius of the roll of remaining tape decreases. How does the torque on the reel change with time? How does the angular speed of the reel change with time? If the tape mechanism is suddenly turned on so that the tape is quickly pulled with a great force, is the tape more likely to break when pulled from a nearly full reel or a nearly empty reel?
10. A scientist at a hotel sought assistance from a bellhop to carry a mysterious suitcase. When the unaware bellhop rounded a corner carrying the suitcase, it suddenly moved away from him for some unknown reason. At this point, the alarmed bellhop dropped the suitcase and ran off. What do you suppose might have been in the suitcase?
11. When a cylinder rolls on a horizontal surface as in Figure 11.3, do any points on the cylinder have only a vertical component of velocity at some instant? If so, where are they?
12. Three objects of uniform density—a solid sphere, a solid cylinder, and a hollow cylinder—are placed at the top of an incline (Fig. Q11.12). If they all are released from rest at the same elevation and roll without slipping, which object reaches the bottom first? Which reaches it last? You should try this at home and note that the result is independent of the masses and the radii of the objects.

Figure Q11.12   Which object wins the race?

13. A mouse is initially at rest on a horizontal turntable mounted on a frictionless vertical axle. If the mouse begins to walk around the perimeter, what happens to the turntable? Explain.
14. Stars originate as large bodies of slowly rotating gas. Because of gravity, these regions of gas slowly decrease in size. What happens to the angular speed of a star as it shrinks? Explain.
15. Often, when a high diver wants to execute a flip in midair, she draws her legs up against her chest. Why does this make her rotate faster? What should she do when she wants to come out of her flip?
16. As a tether ball winds around a thin pole, what happens to its angular speed? Explain.

17. Two solid spheres—a large, massive sphere and a small sphere with low mass—are rolled down a hill. Which sphere reaches the bottom of the hill first? Next, a large, low-density sphere and a small, high-density sphere having the same mass are rolled down the hill. Which one reaches the bottom first in this case?

18. Suppose you are designing a car for a coasting race—the cars in this race have no engines; they simply coast down a hill. Do you want to use large wheels or small wheels? Do you want to use solid, disk-like wheels or hoop-like wheels? Should the wheels be heavy or light?

19. Why do tightrope walkers carry a long pole to help themselves keep their balance?

20. Two balls have the same size and mass. One is hollow, whereas the other is solid. How would you determine which is which without breaking them apart?

21. A particle is moving in a circle with constant speed. Locate one point about which the particle's angular momentum is constant and another about which it changes with time.

22. If global warming occurs over the next century, it is likely that some polar ice will melt and the water will be distributed closer to the equator. How would this change the moment of inertia of the Earth? Would the length of the day (one revolution) increase or decrease?

PROBLEMS

1, 2, 3 = straightforward, intermediate, challenging ☐ = full solution available in the *Student Solutions Manual and Study Guide*
WEB = solution posted at **http://www.saunderscollege.com/physics/** 🖥 = Computer useful in solving problem 📱 = Interactive Physics
☐ = paired numerical/symbolic problems

Section 11.1 Rolling Motion of a Rigid Object

WEB 1. A cylinder of mass 10.0 kg rolls without slipping on a horizontal surface. At the instant its center of mass has a speed of 10.0 m/s, determine (a) the translational kinetic energy of its center of mass, (b) the rotational energy about its center of mass, and (c) its total energy.

2. A bowling ball has a mass of 4.00 kg, a moment of inertia of 1.60×10^{-2} kg·m², and a radius of 0.100 m. If it rolls down the lane without slipping at a linear speed of 4.00 m/s, what is its total energy?

3. A bowling ball has a mass M, a radius R, and a moment of inertia $\frac{2}{5}MR^2$. If it starts from rest, how much work must be done on it to set it rolling without slipping at a linear speed v? Express the work in terms of M and v.

4. A uniform solid disk and a uniform hoop are placed side by side at the top of an incline of height h. If they are released from rest and roll without slipping, determine their speeds when they reach the bottom. Which object reaches the bottom first?

5. (a) Determine the acceleration of the center of mass of a uniform solid disk rolling down an incline making an angle θ with the horizontal. Compare this acceleration with that of a uniform hoop. (b) What is the minimum coefficient of friction required to maintain pure rolling motion for the disk?

6. A ring of mass 2.40 kg, inner radius 6.00 cm, and outer radius 8.00 cm rolls (without slipping) up an inclined plane that makes an angle of $\theta = 36.9°$ (Fig. P11.6). At the moment the ring is at position $x = 2.00$ m up the plane, its speed is 2.80 m/s. The ring continues up the plane for some additional distance and then rolls back down. It does not roll off the top end. How far up the plane does it go?

Figure P11.6

7. A metal can containing condensed mushroom soup has a mass of 215 g, a height of 10.8 cm, and a diameter of 6.38 cm. It is placed at rest on its side at the top of a 3.00-m-long incline that is at an angle of 25.0° to the horizontal and is then released to roll straight down. Assuming energy conservation, calculate the moment of inertia of the can if it takes 1.50 s to reach the bottom of the incline. Which pieces of data, if any, are unnecessary for calculating the solution?

8. A tennis ball is a hollow sphere with a thin wall. It is set rolling without slipping at 4.03 m/s on the horizontal section of a track, as shown in Figure P11.8. It rolls around the inside of a vertical circular loop 90.0 cm in diameter and finally leaves the track at a point 20.0 cm below the horizontal section. (a) Find the speed of the ball at the top of the loop. Demonstrate that it will not fall from the track. (b) Find its speed as it leaves the track. (c) Suppose that static friction between the ball and the track was negligible, so that the ball slid instead of rolling. Would its speed

Figure P11.8

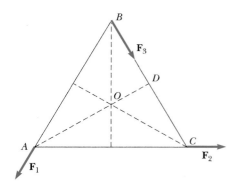

Figure P11.18

then be higher, lower, or the same at the top of the loop? Explain.

Section 11.2   The Vector Product and Torque

9. Given $\mathbf{M} = 6\mathbf{i} + 2\mathbf{j} - \mathbf{k}$ and $\mathbf{N} = 2\mathbf{i} - \mathbf{j} - 3\mathbf{k}$, calculate the vector product $\mathbf{M} \times \mathbf{N}$.

10. The vectors 42.0 cm at 15.0° and 23.0 cm at 65.0° both start from the origin. Both angles are measured counterclockwise from the x axis. The vectors form two sides of a parallelogram. (a) Find the area of the parallelogram. (b) Find the length of its longer diagonal.

WEB 11. Two vectors are given by $\mathbf{A} = -3\mathbf{i} + 4\mathbf{j}$ and $\mathbf{B} = 2\mathbf{i} + 3\mathbf{j}$. Find (a) $\mathbf{A} \times \mathbf{B}$ and (b) the angle between $\mathbf{A}$ and $\mathbf{B}$.

12. For the vectors $\mathbf{A} = -3\mathbf{i} + 7\mathbf{j} - 4\mathbf{k}$ and $\mathbf{B} = 6\mathbf{i} - 10\mathbf{j} + 9\mathbf{k}$, evaluate the expressions (a) $\cos^{-1}(\mathbf{A} \cdot \mathbf{B}/AB)$ and (b) $\sin^{-1}(|\mathbf{A} \times \mathbf{B}|/AB)$. (c) Which give(s) the angle between the vectors?

13. A force of $\mathbf{F} = 2.00\mathbf{i} + 3.00\mathbf{j}$ N is applied to an object that is pivoted about a fixed axis aligned along the z coordinate axis. If the force is applied at the point $\mathbf{r} = (4.00\mathbf{i} + 5.00\mathbf{j} + 0\mathbf{k})$ m, find (a) the magnitude of the net torque about the z axis and (b) the direction of the torque vector $\boldsymbol{\tau}$.

14. A student claims that she has found a vector $\mathbf{A}$ such that $(2\mathbf{i} - 3\mathbf{j} + 4\mathbf{k}) \times \mathbf{A} = (4\mathbf{i} + 3\mathbf{j} - \mathbf{k})$. Do you believe this claim? Explain.

15. Vector $\mathbf{A}$ is in the negative y direction, and vector $\mathbf{B}$ is in the negative x direction. What are the directions of (a) $\mathbf{A} \times \mathbf{B}$ and (b) $\mathbf{B} \times \mathbf{A}$?

16. A particle is located at the vector position $\mathbf{r} = (\mathbf{i} + 3\mathbf{j})$ m, and the force acting on it is $\mathbf{F} = (3\mathbf{i} + 2\mathbf{j})$ N. What is the torque about (a) the origin and (b) the point having coordinates (0, 6) m?

17. If $|\mathbf{A} \times \mathbf{B}| = \mathbf{A} \cdot \mathbf{B}$, what is the angle between $\mathbf{A}$ and $\mathbf{B}$?

18. Two forces $\mathbf{F}_1$ and $\mathbf{F}_2$ act along the two sides of an equilateral triangle, as shown in Figure P11.18. Point O is the intersection of the altitudes of the triangle. Find a third force $\mathbf{F}_3$ to be applied at B and along BC that will make the total torque about the point O be zero. Will the total torque change if $\mathbf{F}_3$ is applied not at B, but rather at any other point along BC?

Section 11.3   Angular Momentum of a Particle

19. A light, rigid rod 1.00 m in length joins two particles—with masses 4.00 kg and 3.00 kg—at its ends. The combination rotates in the xy plane about a pivot through the center of the rod (Fig. P11.19). Determine the angular momentum of the system about the origin when the speed of each particle is 5.00 m/s.

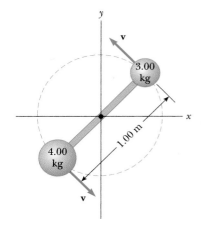

Figure P11.19

20. A 1.50-kg particle moves in the xy plane with a velocity of $\mathbf{v} = (4.20\mathbf{i} - 3.60\mathbf{j})$ m/s. Determine the particle's angular momentum when its position vector is $\mathbf{r} = (1.50\mathbf{i} + 2.20\mathbf{j})$ m.

WEB 21. The position vector of a particle of mass 2.00 kg is given as a function of time by $\mathbf{r} = (6.00\mathbf{i} + 5.00t\mathbf{j})$ m. Determine the angular momentum of the particle about the origin as a function of time.

22. A conical pendulum consists of a bob of mass m in motion in a circular path in a horizontal plane, as shown in Figure P11.22. During the motion, the supporting wire of length ℓ maintains the constant angle θ with the vertical. Show that the magnitude of the angular momen-

tum of the mass about the center of the circle is

$$L = (m^2 g \ell^3 \sin^4 \theta / \cos \theta)^{1/2}$$

Figure P11.22

23. A particle of mass m moves in a circle of radius R at a constant speed v, as shown in Figure P11.23. If the motion begins at point Q, determine the angular momentum of the particle about point P as a function of time.

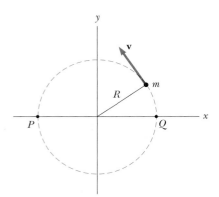

Figure P11.23

24. A 4.00-kg mass is attached to a light cord that is wound around a pulley (see Fig. 10.20). The pulley is a uniform solid cylinder with a radius of 8.00 cm and a mass of 2.00 kg. (a) What is the net torque on the system about the point O? (b) When the mass has a speed v, the pulley has an angular speed $\omega = v/R$. Determine the total angular momentum of the system about O. (c) Using the fact that $\tau = d\mathbf{L}/dt$ and your result from part (b), calculate the acceleration of the mass.

25. A particle of mass m is shot with an initial velocity $\mathbf{v}_i$ and makes an angle θ with the horizontal, as shown in Figure P11.25. The particle moves in the gravitational field of the Earth. Find the angular momentum of the parti-

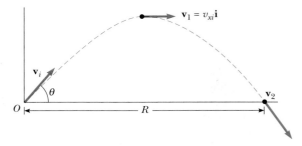

Figure P11.25

cle about the origin when the particle is (a) at the origin, (b) at the highest point of its trajectory, and (c) just before it hits the ground. (d) What torque causes its angular momentum to change?

26. Heading straight toward the summit of Pike's Peak, an airplane of mass 12 000 kg flies over the plains of Kansas at a nearly constant altitude of 4.30 km and with a constant velocity of 175 m/s westward. (a) What is the airplane's vector angular momentum relative to a wheat farmer on the ground directly below the airplane? (b) Does this value change as the airplane continues its motion along a straight line? (c) What is its angular momentum relative to the summit of Pike's Peak?

27. A ball of mass m is fastened at the end of a flagpole connected to the side of a tall building at point P, as shown in Figure P11.27. The length of the flagpole is ℓ, and θ is the angle the flagpole makes with the horizontal. Suppose that the ball becomes loose and starts to fall. Determine the angular momentum (as a function of time) of the ball about point P. Neglect air resistance.

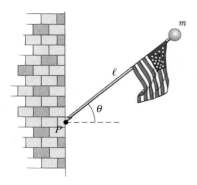

Figure P11.27

28. A fireman clings to a vertical ladder and directs the nozzle of a hose horizontally toward a burning building. The rate of water flow is 6.31 kg/s, and the nozzle speed is 12.5 m/s. The hose passes between the fireman's feet, which are 1.30 m vertically below the nozzle. Choose the origin to be inside the hose between the fireman's

feet. What torque must the fireman exert on the hose? That is, what is the rate of change of angular momentum of the water?

Section 11.4 Angular Momentum of a Rotating Rigid Object

29. A uniform solid sphere with a radius of 0.500 m and a mass of 15.0 kg turns counterclockwise about a vertical axis through its center. Find its vector angular momentum when its angular speed is 3.00 rad/s.

30. A uniform solid disk with a mass of 3.00 kg and a radius of 0.200 m rotates about a fixed axis perpendicular to its face. If the angular speed is 6.00 rad/s, calculate the angular momentum of the disk when the axis of rotation (a) passes through its center of mass and (b) passes through a point midway between the center and the rim.

31. A particle with a mass of 0.400 kg is attached to the 100-cm mark of a meter stick with a mass of 0.100 kg. The meter stick rotates on a horizontal, frictionless table with an angular speed of 4.00 rad/s. Calculate the angular momentum of the system when the stick is pivoted about an axis (a) perpendicular to the table through the 50.0-cm mark and (b) perpendicular to the table through the 0-cm mark.

32. The hour and minute hands of Big Ben, the famous Parliament Building tower clock in London, are 2.70 m and 4.50 m long and have masses of 60.0 kg and 100 kg, respectively. Calculate the total angular momentum of these hands about the center point. Treat the hands as long thin rods.

Section 11.5 Conservation of Angular Momentum

33. A cylinder with a moment of inertia of I_1 rotates about a vertical, frictionless axle with angular velocity ω_i. A second cylinder that has a moment of inertia of I_2 and initially is not rotating drops onto the first cylinder (Fig. P11.33). Because of friction between the surfaces, the two eventually reach the same angular speed ω_f. (a) Calculate ω_f. (b) Show that the kinetic energy of the system decreases in this interaction and calculate

the ratio of the final rotational energy to the initial rotational energy.

34. A playground merry-go-round of radius $R = 2.00$ m has a moment of inertia of $I = 250$ kg·m^2 and is rotating at 10.0 rev/min about a frictionless vertical axle. Facing the axle, a 25.0-kg child hops onto the merry-go-round and manages to sit down on its edge. What is the new angular speed of the merry-go-round?

35. A student sits on a freely rotating stool holding two weights, each of which has a mass of 3.00 kg. When his arms are extended horizontally, the weights are 1.00 m from the axis of rotation and he rotates with an angular speed of 0.750 rad/s. The moment of inertia of the student plus stool is 3.00 kg·m^2 and is assumed to be constant. The student pulls the weights inward horizontally to a position 0.300 m from the rotation axis. (a) Find the new angular speed of the student. (b) Find the kinetic energy of the student before and after he pulls the weights inward.

36. A uniform rod with a mass of 100 g and a length of 50.0 cm rotates in a horizontal plane about a fixed, vertical, frictionless pin passing through its center. Two small beads, each having a mass 30.0 g, are mounted on the rod so that they are able to slide without friction along its length. Initially, the beads are held by catches at positions 10.0 cm on each side of center; at this time, the system rotates at an angular speed of 20.0 rad/s. Suddenly, the catches are released, and the small beads slide outward along the rod. Find (a) the angular speed of the system at the instant the beads reach the ends of the rod and (b) the angular speed of the rod after the beads fly off the rod's ends.

WEB 37. A 60.0-kg woman stands at the rim of a horizontal turntable having a moment of inertia of 500 kg·m^2 and a radius of 2.00 m. The turntable is initially at rest and is free to rotate about a frictionless, vertical axle through its center. The woman then starts walking around the rim clockwise (as viewed from above the system) at a constant speed of 1.50 m/s relative to the Earth. (a) In what direction and with what angular speed does the turntable rotate? (b) How much work does the woman do to set herself and the turntable into motion?

38. A puck with a mass of 80.0 g and a radius of 4.00 cm slides along an air table at a speed of 1.50 m/s, as shown in Figure P11.38a. It makes a glancing collision

Before   After

Figure P11.33

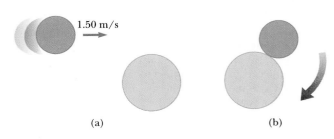

1.50 m/s

(a)   (b)

Figure P11.38

with a second puck having a radius of 6.00 cm and a mass of 120 g (initially at rest) such that their rims just touch. Because their rims are coated with instant-acting glue, the pucks stick together and spin after the collision (Fig. P11.38b). (a) What is the angular momentum of the system relative to the center of mass? (b) What is the angular speed about the center of mass?

39. A wooden block of mass M resting on a frictionless horizontal surface is attached to a rigid rod of length ℓ and of negligible mass (Fig. P11.39). The rod is pivoted at the other end. A bullet of mass m traveling parallel to the horizontal surface and normal to the rod with speed v hits the block and becomes embedded in it. (a) What is the angular momentum of the bullet–block system? (b) What fraction of the original kinetic energy is lost in the collision?

Figure P11.40

Figure P11.39

40. A space station shaped like a giant wheel has a radius of 100 m and a moment of inertia of 5.00×10^8 kg· m². A crew of 150 are living on the rim, and the station's rotation causes the crew to experience an acceleration of $1g$ (Fig. P11.40). When 100 people move to the center of the station for a union meeting, the angular speed changes. What acceleration is experienced by the managers remaining at the rim? Assume that the average mass of each inhabitant is 65.0 kg.

41. A wad of sticky clay of mass m and velocity $\mathbf{v}_i$ is fired at a solid cylinder of mass M and radius R (Fig. P11.41). The cylinder is initially at rest and is mounted on a fixed horizontal axle that runs through the center of mass. The line of motion of the projectile is perpendicular to the axle and at a distance d, less than R, from the center. (a) Find the angular speed of the system just after the clay strikes and sticks to the surface of the cylinder. (b) Is mechanical energy conserved in this process? Explain your answer.

42. Suppose a meteor with a mass of 3.00×10^{13} kg is moving at 30.0 km/s relative to the center of the Earth and strikes the Earth. What is the order of magnitude of the

Figure P11.41

maximum possible decrease in the angular speed of the Earth due to this collision? Explain your answer.

(Optional)
Section 11.7   Angular Momentum as a Fundamental Quantity

43. In the Bohr model of the hydrogen atom, the electron moves in a circular orbit of radius 0.529×10^{-10} m around the proton. Assuming that the orbital angular momentum of the electron is equal to $h/2\pi$, calculate (a) the orbital speed of the electron, (b) the kinetic energy of the electron, and (c) the angular speed of the electron's motion.

ADDITIONAL PROBLEMS

44. **Review Problem.**   A rigid, massless rod has three equal masses attached to it, as shown in Figure P11.44. The rod is free to rotate in a vertical plane about a frictionless axle perpendicular to the rod through the point P, and it is released from rest in the horizontal position at $t = 0$. Assuming m and d are known, find (a) the moment of inertia of the system about the pivot, (b) the torque acting on the system at $t = 0$, (c) the angular acceleration of the system at $t = 0$, (d) the linear acceleration of the mass labeled "3" at $t = 0$, (e) the maximum

kinetic energy of the system, (f) the maximum angular speed attained by the rod, (g) the maximum angular momentum of the system, and (h) the maximum speed attained by the mass labeled "2."

Figure P11.44

45. A uniform solid sphere of radius r is placed on the inside surface of a hemispherical bowl having a much greater radius R. The sphere is released from rest at an angle θ to the vertical and rolls without slipping (Fig. P11.45). Determine the angular speed of the sphere when it reaches the bottom of the bowl.

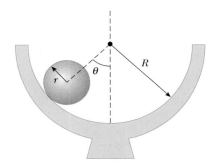

Figure P11.45

46. A 100-kg uniform horizontal disk of radius 5.50 m turns without friction at 2.50 rev/s on a vertical axis through its center, as shown in Figure P11.46. A feedback mechanism senses the angular speed of the disk, and a drive motor at A ensures that the angular speed remains constant. While the disk turns, a 1.20-kg mass on top of the disk slides outward in a radial slot. The 1.20-kg mass starts at the center of the disk at time $t = 0$ and moves outward with a constant speed of 1.25 cm/s relative to the disk until it reaches the edge at $t = 440$ s. The sliding mass experiences no friction. Its motion is constrained by a brake at B so that its radial speed remains constant. The constraint produces tension in a light string tied to the mass. (a) Find the torque as a function of time that the drive motor must provide while the mass is sliding. (b) Find the value of this torque at $t = 440$ s, just before the sliding mass finishes its motion. (c) Find the power that the drive motor must deliver as a function of time. (d) Find the value of the power when the sliding mass is just reaching the end of the slot. (e) Find the string tension as a function of

time. (f) Find the work done by the drive motor during the 440-s motion. (g) Find the work done by the string brake on the sliding mass. (h) Find the total work done on the system consisting of the disk and the sliding mass.

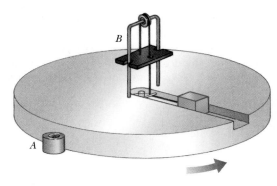

Figure P11.46

47. A string is wound around a uniform disk of radius R and mass M. The disk is released from rest when the string is vertical and its top end is tied to a fixed bar (Fig. P11.47). Show that (a) the tension in the string is one-third the weight of the disk, (b) the magnitude of the acceleration of the center of mass is $2g/3$, and (c) the speed of the center of mass is $(4gh/3)^{1/2}$ as the disk descends. Verify your answer to part (c) using the energy approach.

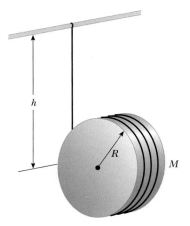

Figure P11.47

48. Comet Halley moves about the Sun in an elliptical orbit, with its closest approach to the Sun being about 0.590 AU and its greatest distance from the Sun being 35.0 AU (1 AU = the average Earth–Sun distance). If the comet's speed at its closest approach is 54.0 km/s,

what is its speed when it is farthest from the Sun? The angular momentum of the comet about the Sun is conserved because no torque acts on the comet. The gravitational force exerted by the Sun on the comet has a moment arm of zero.

49. A constant horizontal force **F** is applied to a lawn roller having the form of a uniform solid cylinder of radius R and mass M (Fig. P11.49). If the roller rolls without slipping on the horizontal surface, show that (a) the acceleration of the center of mass is $2\mathbf{F}/3M$ and that (b) the minimum coefficient of friction necessary to prevent slipping is $F/3Mg$. (*Hint:* Consider the torque with respect to the center of mass.)

Figure P11.49

50. A light rope passes over a light, frictionless pulley. A bunch of bananas of mass M is fastened at one end, and a monkey of mass M clings to the other (Fig. P11.50).

The monkey climbs the rope in an attempt to reach the bananas. (a) Treating the system as consisting of the monkey, bananas, rope, and pulley, evaluate the net torque about the pulley axis. (b) Using the results to part (a), determine the total angular momentum about the pulley axis and describe the motion of the system. Will the monkey reach the bananas?

51. A solid sphere of mass m and radius r rolls without slipping along the track shown in Figure P11.51. The sphere starts from rest with its lowest point at height h above the bottom of a loop of radius R, which is much larger than r. (a) What is the minimum value that h can have (in terms of R) if the sphere is to complete the loop? (b) What are the force components on the sphere at point P if $h = 3R$?

Figure P11.51

52. A thin rod with a mass of 0.630 kg and a length of 1.24 m is at rest, hanging vertically from a strong fixed hinge at its top end. Suddenly, a horizontal impulsive force $(14.7 \text{ N})\mathbf{i}$ is applied to it. (a) Suppose that the force acts at the bottom end of the rod. Find the acceleration of the rod's center of mass and the horizontal force that the hinge exerts. (b) Suppose that the force acts at the midpoint of the rod. Find the acceleration of this point and the horizontal hinge reaction. (c) Where can the impulse be applied so that the hinge exerts no horizontal force? (This point is called the *center of percussion.*)

53. At one moment, a bowling ball is both sliding and spinning on a horizontal surface such that its rotational kinetic energy equals its translational kinetic energy. Let v_{CM} represent the ball's center-of-mass speed relative to the surface. Let v_r represent the speed of the topmost point on the ball's surface relative to the center of mass. Find the ratio v_{CM}/v_r.

54. A projectile of mass m moves to the right with speed v_i (Fig. P11.54a). The projectile strikes and sticks to the end of a stationary rod of mass M and length d that is pivoted about a frictionless axle through its center (Fig. P11.54b). (a) Find the angular speed of the system right after the collision. (b) Determine the fractional loss in mechanical energy due to the collision.

Figure P11.50

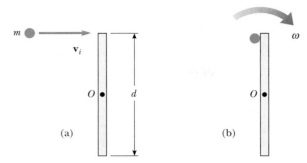

Figure P11.54

55. A mass m is attached to a cord passing through a small hole in a frictionless, horizontal surface (Fig. P11.55). The mass is initially orbiting with speed v_i in a circle of radius r_i. The cord is then slowly pulled from below, and the radius of the circle decreases to r. (a) What is the speed of the mass when the radius is r? (b) Find the tension in the cord as a function of r. (c) How much work W is done in moving m from r_i to r? (*Note:* The tension depends on r.) (d) Obtain numerical values for v, T, and W when $r = 0.100$ m, $m = 50.0$ g, $r_i = 0.300$ m, and $v_i = 1.50$ m/s.

Figure P11.55

56. A bowler releases a bowling ball with no spin, sending it sliding straight down the alley toward the pins. The ball continues to slide for some distance before its motion becomes rolling without slipping; of what order of magnitude is this distance? State the quantities you take as data, the values you measure or estimate for them, and your reasoning.

57. Following Thanksgiving dinner, your uncle falls into a deep sleep while sitting straight up and facing the television set. A naughty grandchild balances a small spheri-

cal grape at the top of his bald head, which itself has the shape of a sphere. After all of the children have had time to giggle, the grape starts from rest and rolls down your uncle's head without slipping. It loses contact with your uncle's scalp when the radial line joining it to the center of curvature makes an angle θ with the vertical. What is the measure of angle θ?

58. A thin rod of length h and mass M is held vertically with its lower end resting on a frictionless horizontal surface. The rod is then released to fall freely. (a) Determine the speed of its center of mass just before it hits the horizontal surface. (b) Now suppose that the rod has a fixed pivot at its lower end. Determine the speed of the rod's center of mass just before it hits the surface.

WEB 59. Two astronauts (Fig. P11.59), each having a mass of 75.0 kg, are connected by a 10.0-m rope of negligible mass. They are isolated in space, orbiting their center of mass at speeds of 5.00 m/s. (a) Treating the astronauts as particles, calculate the magnitude of the angular momentum and (b) the rotational energy of the system. By pulling on the rope, one of the astronauts shortens the distance between them to 5.00 m. (c) What is the new angular momentum of the system? (d) What are the astronauts' new speeds? (e) What is the new rotational energy of the system? (f) How much work is done by the astronaut in shortening the rope?

60. Two astronauts (see Fig. P11.59), each having a mass M, are connected by a rope of length d having negligible mass. They are isolated in space, orbiting their center of mass at speeds v. Treating the astronauts as particles, calculate (a) the magnitude of the angular momentum and (b) the rotational energy of the system. By pulling on the rope, one of the astronauts shortens the distance between them to $d/2$. (c) What is the new angular momentum of the system? (d) What are the astronauts' new speeds? (e) What is the new rotational energy of the system? (f) How much work is done by the astronaut in shortening the rope?

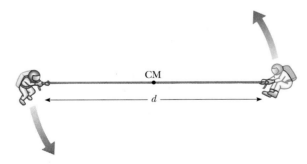

Figure P11.59   Problems 59 and 60.

61. A solid cube of wood of side $2a$ and mass M is resting on a horizontal surface. The cube is constrained to ro-

tate about an axis *AB* (Fig. P11.61). A bullet of mass *m* and speed *v* is shot at the face opposite *ABCD* at a height of $4a/3$. The bullet becomes embedded in the cube. Find the minimum value of *v* required to tip the cube so that it falls on face *ABCD*. Assume $m \ll M$.

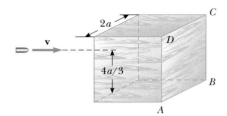

Figure P11.61

62. A large, cylindrical roll of paper of initial radius *R* lies on a long, horizontal surface with the open end of the paper nailed to the surface. The roll is given a slight shove ($v_i \approx 0$) and begins to unroll. (a) Determine the speed of the center of mass of the roll when its radius has diminished to *r*. (b) Calculate a numerical value for this speed at $r = 1.00$ mm, assuming $R = 6.00$ m. (c) What happens to the energy of the system when the paper is completely unrolled? (*Hint:* Assume that the roll has a uniform density and apply energy methods.)

63. A spool of wire of mass *M* and radius *R* is unwound under a constant force **F** (Fig. P11.63). Assuming that the spool is a uniform solid cylinder that does not slip, show that (a) the acceleration of the center of mass is $4F/3M$ and that (b) the force of friction is to the *right* and is equal in magnitude to $F/3$. (c) If the cylinder starts from rest and rolls without slipping, what is the speed of its center of mass after it has rolled through a distance *d*?

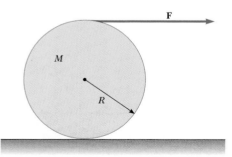

Figure P11.63

64. A uniform solid disk is set into rotation with an angular speed ω_i about an axis through its center. While still rotating at this speed, the disk is placed into contact with

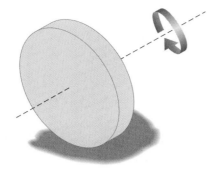

Figure P11.64   Problems 64 and 65.

a horizontal surface and released, as shown in Figure P11.64. (a) What is the angular speed of the disk once pure rolling takes place? (b) Find the fractional loss in kinetic energy from the time the disk is released until the time pure rolling occurs. (*Hint:* Consider torques about the center of mass.)

65. Suppose a solid disk of radius *R* is given an angular speed ω_i about an axis through its center and is then lowered to a horizontal surface and released, as shown in Problem 64 (see Fig. P11.64). Furthermore, assume that the coefficient of friction between the disk and the surface is μ. (a) Show that the time it takes for pure rolling motion to occur is $R\omega_i/3\mu g$. (b) Show that the distance the disk travels before pure rolling occurs is $R^2\omega_i^2/18\mu g$.

66. A solid cube of side $2a$ and mass *M* is sliding on a frictionless surface with uniform velocity **v**, as shown in Figure P11.66a. It hits a small obstacle at the end of the table; this causes the cube to tilt, as shown in Figure

(a)

Figure P11.66

P11.66b. Find the minimum value of **v** such that the cube falls off the table. Note that the moment of inertia of the cube about an axis along one of its edges is $8Ma^2/3$. (*Hint:* The cube undergoes an inelastic collision at the edge.)

67. A plank with a mass $M = 6.00$ kg rides on top of two identical solid cylindrical rollers that have $R = 5.00$ cm and $m = 2.00$ kg (Fig. P11.67). The plank is pulled by a constant horizontal force of magnitude $F = 6.00$ N applied to the end of the plank and perpendicular to the axes of the cylinders (which are parallel). The cylinders roll without slipping on a flat surface. Also, no slipping occurs between the cylinders and the plank. (a) Find the acceleration of the plank and that of the rollers. (b) What frictional forces are acting?

Figure P11.67

68. A spool of wire rests on a horizontal surface as in Figure P11.68. As the wire is pulled, the spool does not slip at the contact point P. On separate trials, each one of the forces $\mathbf{F}_1$, $\mathbf{F}_2$, $\mathbf{F}_3$, and $\mathbf{F}_4$ is applied to the spool. For each one of these forces, determine the direction in which the spool will roll. Note that the line of action of $\mathbf{F}_2$ passes through P.

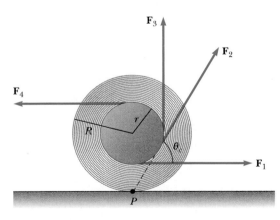

Figure P11.68 Problems 68 and 69.

69. The spool of wire shown in Figure P11.68 has an inner radius r and an outer radius R. The angle θ between the applied force and the horizontal can be varied. Show

that the critical angle for which the spool does not slip and remains stationary is

$$\cos \theta_c = \frac{r}{R}$$

(*Hint:* At the critical angle, the line of action of the applied force passes through the contact point.)

 70. In a demonstration that employs a ballistics cart, a ball is projected vertically upward from a cart moving with constant velocity along the horizontal direction. The ball lands in the catching cup of the cart because both the cart and the ball have the same horizontal component of velocity. Now consider a ballistics cart on an incline making an angle θ with the horizontal, as shown in Figure P11.70. The cart (including its wheels) has a mass M, and the moment of inertia of each of the two wheels is $mR^2/2$. (a) Using conservation of energy considerations (assuming that there is no friction between the cart and the axles) and assuming pure rolling motion (that is, no slipping), show that the acceleration of the cart along the incline is

$$a_x = \left(\frac{M}{M + 2m}\right) g \sin \theta$$

(b) Note that the x component of acceleration of the ball released by the cart is $g \sin \theta$. Thus, the x component of the cart's acceleration is *smaller* than that of the ball by the factor $M/(M + 2m)$. Use this fact and kinematic equations to show that the ball overshoots the cart by an amount Δx, where

$$\Delta x = \left(\frac{4m}{M + 2m}\right)\left(\frac{\sin \theta}{\cos^2 \theta}\right)\frac{v_{yi}^2}{g}$$

and v_{yi} is the initial speed of the ball imparted to it by the spring in the cart. (c) Show that the distance d that the ball travels measured along the incline is

$$d = \frac{2v_{yi}^2}{g}\frac{\sin \theta}{\cos^2 \theta}$$

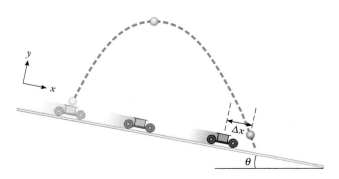

Figure P11.70

ANSWERS TO QUICK QUIZZES

11.1 There is very little resistance to motion that can reduce the kinetic energy of the rolling ball. Even though there is friction between the ball and the floor (if there were not, then no rotation would occur, and the ball would slide), there is no relative motion of the two surfaces (by the definition of "rolling"), and so kinetic friction cannot reduce K. (Air drag and friction associated with deformation of the ball eventually stop the ball.)

11.2 The box. Because none of the box's initial potential energy is converted to rotational kinetic energy, at any time $t > 0$ its translational kinetic energy is greater than that of the rolling ball.

11.3 Zero. If she were moving directly toward the pole, **r** and **p** would be antiparallel to each other, and the sine of the angle between them is zero; therefore, $L = 0$.

11.4 Both (a) and (b) are false. The net force is not necessarily zero. If the line of action of the net force passes through the point, then the net torque about an axis passing through that point is zero even though the net force is not zero. Because the net force is not necessarily zero, you cannot conclude that the particle's velocity is constant.

11.5 The student does work as he walks from the rim of the platform toward its center.

11.6 Because it is perpendicular to the precessional motion of the top, the force of gravity does no work. This is a situation in which a force causes motion but does no work.

P U Z Z L E R

This one-bottle wine holder is an interesting example of a mechanical system that seems to defy gravity. The system (holder plus bottle) is balanced when its center of gravity is directly over the lowest support point. What two conditions are necessary for an object to exhibit this kind of stability? *(Charles D. Winters)*

c h a p t e r

Static Equilibrium and Elasticity

12

*I*n Chapters 10 and 11 we studied the dynamics of rigid objects — that is, objects whose parts remain at a fixed separation with respect to each other when subjected to external forces. Part of this chapter addresses the conditions under which a rigid object is in equilibrium. The term *equilibrium* implies either that the object is at rest or that its center of mass moves with constant velocity. We deal here only with the former case, in which the object is described as being in *static equilibrium*. Static equilibrium represents a common situation in engineering practice, and the principles it involves are of special interest to civil engineers, architects, and mechanical engineers. If you are an engineering student you will undoubtedly take an advanced course in statics in the future.

The last section of this chapter deals with how objects deform under load conditions. Such deformations are usually elastic and do not affect the conditions for equilibrium. An *elastic* object returns to its original shape when the deforming forces are removed. Several elastic constants are defined, each corresponding to a different type of deformation.

12.1 THE CONDITIONS FOR EQUILIBRIUM

In Chapter 5 we stated that one necessary condition for equilibrium is that the net force acting on an object be zero. If the object is treated as a particle, then this is the only condition that must be satisfied for equilibrium. The situation with real (extended) objects is more complex, however, because these objects cannot be treated as particles. For an extended object to be in static equilibrium, a second condition must be satisfied. This second condition involves the net torque acting on the extended object. Note that equilibrium does not require the absence of motion. For example, a rotating object can have constant angular velocity and still be in equilibrium.

Consider a single force **F** acting on a rigid object, as shown in Figure 12.1. The effect of the force depends on its point of application *P*. If **r** is the position vector of this point relative to *O*, the torque associated with the force **F** about *O* is given by Equation 11.7:

$$\boldsymbol{\tau} = \mathbf{r} \times \mathbf{F}$$

Recall from the discussion of the vector product in Section 11.2 that the vector $\boldsymbol{\tau}$ is perpendicular to the plane formed by **r** and **F**. You can use the right-hand rule to determine the direction of $\boldsymbol{\tau}$: Curl the fingers of your right hand in the direction of rotation that **F** tends to cause about an axis through *O*: your thumb then points in the direction of $\boldsymbol{\tau}$. Hence, in Figure 12.1 $\boldsymbol{\tau}$ is directed toward you out of the page.

As you can see from Figure 12.1, the tendency of **F** to rotate the object about an axis through *O* depends on the moment arm *d*, as well as on the magnitude of **F**. Recall that the magnitude of $\boldsymbol{\tau}$ is *Fd* (see Eq. 10.19). Now suppose a rigid object is acted on first by force $\mathbf{F}_1$ and later by force $\mathbf{F}_2$. If the two forces have the same magnitude, they will produce the same effect on the object only if they have the same direction and the same line of action. In other words,

Figure 12.1 A single force **F** acts on a rigid object at the point *P*.

Equivalent forces

two forces $\mathbf{F}_1$ and $\mathbf{F}_2$ are **equivalent** if and only if $F_1 = F_2$ and if and only if the two produce the same torque about any axis.

The two forces shown in Figure 12.2 are equal in magnitude and opposite in direction. They are *not* equivalent. The force directed to the right tends to rotate the object clockwise about an axis perpendicular to the diagram through O, whereas the force directed to the left tends to rotate it counterclockwise about that axis.

Suppose an object is pivoted about an axis through its center of mass, as shown in Figure 12.3. Two forces of equal magnitude act in opposite directions along parallel lines of action. A pair of forces acting in this manner form what is called a **couple.** (The two forces shown in Figure 12.2 also form a couple.) Do not make the mistake of thinking that the forces in a couple are a result of Newton's third law. They cannot be third-law forces because they act on the same object. Third-law force pairs act on different objects. Because each force produces the same torque Fd, the net torque has a magnitude of $2Fd$. Clearly, the object rotates clockwise and undergoes an angular acceleration about the axis. With respect to rotational motion, this is a nonequilibrium situation. The net torque on the object gives rise to an angular acceleration α according to the relationship $\Sigma \tau = 2Fd = I\alpha$ (see Eq. 10.21).

In general, an object is in rotational equilibrium only if its angular acceleration $\alpha = 0$. Because $\Sigma \tau = I\alpha$ for rotation about a fixed axis, our second necessary condition for equilibrium is that **the net torque about any axis must be zero.** We now have two necessary conditions for equilibrium of an object:

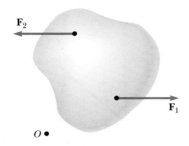

Figure 12.2 The forces $\mathbf{F}_1$ and $\mathbf{F}_2$ are not equivalent because they do not produce the same torque about some axis, even though they are equal in magnitude and opposite in direction.

1. The resultant external force must equal zero. $\quad \sum \mathbf{F} = 0 \qquad$ **(12.1)**

2. The resultant external torque about *any* axis must be zero. $\quad \sum \tau = 0 \qquad$ **(12.2)**

Conditions for equilibrium

The first condition is a statement of translational equilibrium; it tells us that the linear acceleration of the center of mass of the object must be zero when viewed from an inertial reference frame. The second condition is a statement of rotational equilibrium and tells us that the angular acceleration about any axis must be zero. In the special case of **static equilibrium,** which is the main subject of this chapter, the object is at rest and so has no linear or angular speed (that is, $v_{CM} = 0$ and $\omega = 0$).

Quick Quiz 12.1

(a) Is it possible for a situation to exist in which Equation 12.1 is satisfied while Equation 12.2 is not? (b) Can Equation 12.2 be satisfied while Equation 12.1 is not?

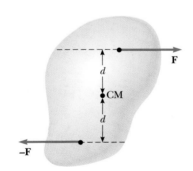

Figure 12.3 Two forces of equal magnitude form a couple if their lines of action are different parallel lines. In this case, the object rotates clockwise. The net torque about any axis is $2Fd$.

The two vector expressions given by Equations 12.1 and 12.2 are equivalent, in general, to six scalar equations: three from the first condition for equilibrium, and three from the second (corresponding to x, y, and z components). Hence, in a complex system involving several forces acting in various directions, you could be faced with solving a set of equations with many unknowns. Here, we restrict our discussion to situations in which all the forces lie in the xy plane. (Forces whose vector representations are in the same plane are said to be *coplanar.*) With this restriction, we must deal with only three scalar equations. Two of these come from balancing the forces in the x and y directions. The third comes from the torque equation—namely, that the net torque about *any* point in the xy plane must be zero. Hence, the two conditions of equilibrium provide the equations

$$\sum F_x = 0 \qquad \sum F_y = 0 \qquad \sum \tau_z = 0 \qquad \textbf{(12.3)}$$

where the axis of the torque equation is arbitrary, as we now show.

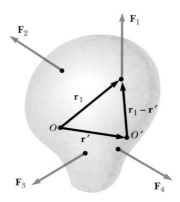

Figure 12.4 Construction showing that if the net torque is zero about origin O, it is also zero about any other origin, such as O'.

Regardless of the number of forces that are acting, if an object is in translational equilibrium and if the net torque is zero about one axis, then the net torque must also be zero about any other axis. The point can be inside or outside the boundaries of the object. Consider an object being acted on by several forces such that the resultant force $\Sigma \mathbf{F} = \mathbf{F}_1 + \mathbf{F}_2 + \mathbf{F}_3 + \cdots = 0$. Figure 12.4 describes this situation (for clarity, only four forces are shown). The point of application of $\mathbf{F}_1$ relative to O is specified by the position vector $\mathbf{r}_1$. Similarly, the points of application of $\mathbf{F}_2$, $\mathbf{F}_3$, . . . are specified by $\mathbf{r}_2$, $\mathbf{r}_3$, . . . (not shown). The net torque about an axis through O is

$$\sum \tau_O = \mathbf{r}_1 \times \mathbf{F}_1 + \mathbf{r}_2 \times \mathbf{F}_2 + \mathbf{r}_3 \times \mathbf{F}_3 + \cdots$$

Now consider another arbitrary point O' having a position vector $\mathbf{r}'$ relative to O. The point of application of $\mathbf{F}_1$ relative to O' is identified by the vector $\mathbf{r}_1 - \mathbf{r}'$. Likewise, the point of application of $\mathbf{F}_2$ relative to O' is $\mathbf{r}_2 - \mathbf{r}'$, and so forth. Therefore, the torque about an axis through O' is

$$\sum \tau_{O'} = (\mathbf{r}_1 - \mathbf{r}') \times \mathbf{F}_1 + (\mathbf{r}_2 - \mathbf{r}') \times \mathbf{F}_2 + (\mathbf{r}_3 - \mathbf{r}') \times \mathbf{F}_3 + \cdots$$
$$= \mathbf{r}_1 \times \mathbf{F}_1 + \mathbf{r}_2 \times \mathbf{F}_2 + \mathbf{r}_3 \times \mathbf{F}_3 + \cdots - \mathbf{r}' \times (\mathbf{F}_1 + \mathbf{F}_2 + \mathbf{F}_3 + \cdots)$$

Because the net force is assumed to be zero (given that the object is in translational equilibrium), the last term vanishes, and we see that the torque about O' is equal to the torque about O. Hence, **if an object is in translational equilibrium and the net torque is zero about one point, then the net torque must be zero about any other point.**

12.2 MORE ON THE CENTER OF GRAVITY

We have seen that the point at which a force is applied can be critical in determining how an object responds to that force. For example, two equal-magnitude but oppositely directed forces result in equilibrium if they are applied at the same point on an object. However, if the point of application of one of the forces is moved, so that the two forces no longer act along the same line of action, then a force couple results and the object undergoes an angular acceleration. (This is the situation shown in Figure 12.3.)

Whenever we deal with a rigid object, one of the forces we must consider is the force of gravity acting on it, and we must know the point of application of this force. As we learned in Section 9.6, on every object is a special point called its center of gravity. All the various gravitational forces acting on all the various mass elements of the object are equivalent to a single gravitational force acting through this point. Thus, to compute the torque due to the gravitational force on an object of mass M, we need only consider the force $M\mathbf{g}$ acting at the center of gravity of the object.

How do we find this special point? As we mentioned in Section 9.6, if we assume that $\mathbf{g}$ is uniform over the object, then the center of gravity of the object coincides with its center of mass. To see that this is so, consider an object of arbitrary shape lying in the xy plane, as illustrated in Figure 12.5. Suppose the object is divided into a large number of particles of masses m_1, m_2, m_3, . . . having coordinates (x_1, y_1), (x_2, y_2), (x_3, y_3), ... In

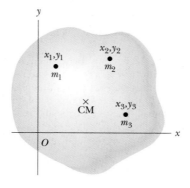

Figure 12.5 An object can be divided into many small particles each having a specific mass and specific coordinates. These particles can be used to locate the center of mass.

Equation 9.28 we defined the x coordinate of the center of mass of such an object to be

$$x_{CM} = \frac{m_1 x_1 + m_2 x_2 + m_3 x_3 + \cdots}{m_1 + m_2 + m_3 + \cdots} = \frac{\sum_i m_i x_i}{\sum_i m_i}$$

We use a similar equation to define the y coordinate of the center of mass, replacing each x with its y counterpart.

Let us now examine the situation from another point of view by considering the force of gravity exerted on each particle, as shown in Figure 12.6. Each particle contributes a torque about the origin equal in magnitude to the particle's weight mg multiplied by its moment arm. For example, the torque due to the force $m_1\mathbf{g}_1$ is $m_1 g_1 x_1$, where g_1 is the magnitude of the gravitational field at the position of the particle of mass m_1. We wish to locate the center of gravity, the point at which application of the single gravitational force $M\mathbf{g}$ (where $M = m_1 + m_2 + m_3 + \cdots$ is the total mass of the object) has the same effect on rotation as does the combined effect of all the individual gravitational forces $m_i\mathbf{g}_i$. Equating the torque resulting from $M\mathbf{g}$ acting at the center of gravity to the sum of the torques acting on the individual particles gives

$$(m_1 g_1 + m_2 g_2 + m_3 g_3 + \cdots)x_{CG} = m_1 g_1 x_1 + m_2 g_2 x_2 + m_3 g_3 x_3 + \cdots$$

This expression accounts for the fact that the gravitational field strength g can in general vary over the object. If we assume uniform g over the object (as is usually the case), then the g terms cancel and we obtain

$$x_{CG} = \frac{m_1 x_1 + m_2 x_2 + m_3 x_3 + \cdots}{m_1 + m_2 + m_3 + \cdots} \tag{12.4}$$

Comparing this result with Equation 9.28, we see that **the center of gravity is located at the center of mass as long as the object is in a uniform gravitational field.**

In several examples presented in the next section, we are concerned with homogeneous, symmetric objects. The center of gravity for any such object coincides with its geometric center.

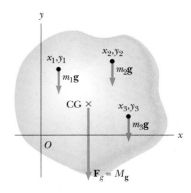

Figure 12.6 The center of gravity of an object is located at the center of mass if $\mathbf{g}$ is constant over the object.

12.3 EXAMPLES OF RIGID OBJECTS IN STATIC EQUILIBRIUM

The photograph of the one-bottle wine holder on the first page of this chapter shows one example of a balanced mechanical system that seems to defy gravity. For the system (wine holder plus bottle) to be in equilibrium, the net external force must be zero (see Eq. 12.1) and the net external torque must be zero (see Eq. 12.2). The second condition can be satisfied only when the center of gravity of the system is directly over the support point.

In working static equilibrium problems, it is important to recognize all the external forces acting on the object. Failure to do so results in an incorrect analysis. When analyzing an object in equilibrium under the action of several external forces, use the following procedure.

A large balanced rock at the Garden of the Gods in Colorado Springs, Colorado—an example of stable equilibrium. *(David Serway)*

Problem-Solving Hints
Objects in Static Equilibrium

- Draw a simple, neat diagram of the system.
- Isolate the object being analyzed. Draw a free-body diagram and then show and label all external forces acting on the object, indicating where those forces are applied. Do not include forces exerted by the object on its surroundings. (For systems that contain more than one object, draw a *separate* free-body diagram for each one.) Try to guess the correct direction for each force. If the direction you select leads to a negative force, do not be alarmed; this merely means that the direction of the force is the opposite of what you guessed.
- Establish a convenient coordinate system for the object and find the components of the forces along the two axes. Then apply the first condition for equilibrium. Remember to keep track of the signs of all force components.
- Choose a convenient axis for calculating the net torque on the object. Remember that the choice of origin for the torque equation is arbitrary; therefore, choose an origin that simplifies your calculation as much as possible. Note that a force that acts along a line passing through the point chosen as the origin gives zero contribution to the torque and thus can be ignored.

The first and second conditions for equilibrium give a set of linear equations containing several unknowns, and these equations can be solved simultaneously.

EXAMPLE 12.1   The Seesaw

A uniform 40.0-N board supports a father and daughter weighing 800 N and 350 N, respectively, as shown in Figure 12.7. If the support (called the *fulcrum*) is under the center of gravity of the board and if the father is 1.00 m from the center, (a) determine the magnitude of the upward force **n** exerted on the board by the support.

Solution   First note that, in addition to **n**, the external forces acting on the board are the downward forces exerted by each person and the force of gravity acting on the board. We know that the board's center of gravity is at its geometric center because we were told the board is uniform. Because the system is in static equilibrium, the upward force **n** must balance all the downward forces. From $\Sigma F_y = 0$, we have, once we define upward as the positive y direction,

$$n - 800 \text{ N} - 350 \text{ N} - 40.0 \text{ N} = 0$$

$$n = \boxed{1\ 190 \text{ N}}$$

(The equation $\Sigma F_x = 0$ also applies, but we do not need to consider it because no forces act horizontally on the board.)

(b) Determine where the child should sit to balance the system.

Solution   To find this position, we must invoke the second condition for equilibrium. Taking an axis perpendicular to the page through the center of gravity of the board as the axis for our torque equation (this means that the torques

Figure 12.7   A balanced system.

produced by **n** and the force of gravity acting on the board about this axis are zero), we see from $\Sigma \tau = 0$ that

$$(800\ \text{N})(1.00\ \text{m}) - (350\ \text{N})x = 0$$

$$x = \boxed{2.29\ \text{m}}$$

(c) Repeat part (b) for another axis.

Solution To illustrate that the choice of axis is arbitrary, let us choose an axis perpendicular to the page and passing

through the location of the father. Recall that the sign of the torque associated with a force is positive if that force tends to rotate the system counterclockwise, while the sign of the torque is negative if the force tends to rotate the system clockwise. In this case, $\Sigma \tau = 0$ yields

$$n(1.00\ \text{m}) - (40.0\ \text{N})(1.00\ \text{m}) - (350\ \text{N})(1.00\ \text{m} + x) = 0$$

From part (a) we know that $n = 1\ 190$ N. Thus, we can solve for x to find $x = 2.29$ m. This result is in agreement with the one we obtained in part (b).

Quick Quiz 12.2

In Example 12.1, if the fulcrum did not lie under the board's center of gravity, what other information would you need to solve the problem?

EXAMPLE 12.2   A Weighted Hand

A person holds a 50.0-N sphere in his hand. The forearm is horizontal, as shown in Figure 12.8a. The biceps muscle is attached 3.00 cm from the joint, and the sphere is 35.0 cm from the joint. Find the upward force exerted by the biceps on the forearm and the downward force exerted by the upper arm on the forearm and acting at the joint. Neglect the weight of the forearm.

Figure 12.8   (a) The biceps muscle pulls upward with a force **F** that is essentially at right angles to the forearm. (b) The mechanical model for the system described in part (a).

Solution We simplify the situation by modeling the forearm as a bar as shown in Figure 12.8b, where **F** is the upward force exerted by the biceps and **R** is the downward force exerted by the upper arm at the joint. From the first condition for equilibrium, we have, with upward as the positive y direction,

$$(1) \qquad \Sigma F_y = F - R - 50.0\ \text{N} = 0$$

From the second condition for equilibrium, we know that the sum of the torques about any point must be zero. With the joint O as the axis, we have

$$Fd - mg\ell = 0$$

$$F(3.00\ \text{cm}) - (50.0\ \text{N})(35.0\ \text{cm}) = 0$$

$$F = \boxed{583\ \text{N}}$$

This value for F can be substituted into Equation (1) to give $R = 533$ N. As this example shows, the forces at joints and in muscles can be extremely large.

Exercise In reality, the biceps makes an angle of 15.0° with the vertical; thus, **F** has both a vertical and a horizontal component. Find the magnitude of **F** and the components of **R** when you include this fact in your analysis.

Answer $F = 604$ N, $R_x = 156$ N, $R_y = 533$ N.

EXAMPLE 12.3   Standing on a Horizontal Beam

A uniform horizontal beam with a length of 8.00 m and a weight of 200 N is attached to a wall by a pin connection. Its far end is supported by a cable that makes an angle of 53.0° with

the horizontal (Fig. 12.9a). If a 600-N person stands 2.00 m from the wall, find the tension in the cable, as well as the magnitude and direction of the force exerted by the wall on the beam.

Figure 12.9 (a) A uniform beam supported by a cable. (b) The free-body diagram for the beam. (c) The free-body diagram for the beam showing the components of **R** and **T**.

Solution First we must identify all the external forces acting on the beam: They are the 200-N force of gravity, the force **T** exerted by the cable, the force **R** exerted by the wall at the pivot, and the 600-N force that the person exerts on the beam. These forces are all indicated in the free-body diagram for the beam shown in Figure 12.9b. When we consider directions for forces, it sometimes is helpful if we imagine what would happen if a force were suddenly removed. For example, if the wall were to vanish suddenly,

the left end of the beam would probably move to the left as it begins to fall. This tells us that the wall is not only holding the beam up but is also pressing outward against it. Thus, we draw the vector **R** as shown in Figure 12.9b. If we resolve **T** and **R** into horizontal and vertical components, as shown in Figure 12.9c, and apply the first condition for equilibrium, we obtain

$$(1) \qquad \sum F_x = R \cos \theta - T \cos 53.0° = 0$$

$$(2) \qquad \sum F_y = R \sin \theta + T \sin 53.0°$$
$$- 600 \text{ N} - 200 \text{ N} = 0$$

where we have chosen rightward and upward as our positive directions. Because R, T, and θ are all unknown, we cannot obtain a solution from these expressions alone. (The number of simultaneous equations must equal the number of unknowns for us to be able to solve for the unknowns.)

Now let us invoke the condition for rotational equilibrium. A convenient axis to choose for our torque equation is the one that passes through the pin connection. The feature that makes this point so convenient is that the force **R** and the horizontal component of **T** both have a moment arm of zero; hence, these forces provide no torque about this point. Recalling our counterclockwise-equals-positive convention for the sign of the torque about an axis and noting that the moment arms of the 600-N, 200-N, and $T \sin 53.0°$ forces are 2.00 m, 4.00 m, and 8.00 m, respectively, we obtain

$$\sum \tau = (T \sin 53.0°)(8.00 \text{ m})$$
$$- (600 \text{ N})(2.00 \text{ m}) - (200 \text{ N})(4.00 \text{ m}) = 0$$

$$T = \boxed{313 \text{ N}}$$

Thus, the torque equation with this axis gives us one of the unknowns directly! We now substitute this value into Equations (1) and (2) and find that

$$R \cos \theta = 188 \text{ N}$$
$$R \sin \theta = 550 \text{ N}$$

We divide the second equation by the first and, recalling the trigonometric identity $\sin \theta / \cos \theta = \tan \theta$, we obtain

$$\tan \theta = \frac{550 \text{ N}}{188 \text{ N}} = 2.93$$

$$\theta = \boxed{71.1°}$$

This positive value indicates that our estimate of the direction of **R** was accurate.

Finally,

$$R = \frac{188 \text{ N}}{\cos \theta} = \frac{188 \text{ N}}{\cos 71.1°} = \boxed{580 \text{ N}}$$

If we had selected some other axis for the torque equation, the solution would have been the same. For example, if we had chosen an axis through the center of gravity of the beam, the torque equation would involve both T and R. However, this equation, coupled with Equations (1) and (2), could still be solved for the unknowns. Try it!

When many forces are involved in a problem of this nature, it is convenient to set up a table. For instance, for the example just given, we could construct the following table. Setting the sum of the terms in the last column equal to zero represents the condition of rotational equilibrium.

Force Component	Moment Arm Relative to O (m)	Torque About O (N·m)
$T \sin 53.0°$	8.00	$(8.00)\,T \sin 53.0°$
$T \cos 53.0°$	0	0
200 N	4.00	$-(4.00)(200)$
600 N	2.00	$-(2.00)(600)$
$R \sin \theta$	0	0
$R \cos \theta$	0	0

EXAMPLE 12.4   The Leaning Ladder

A uniform ladder of length ℓ and weight $mg = 50$ N rests against a smooth, vertical wall (Fig. 12.10a). If the coefficient of static friction between the ladder and the ground is $\mu_s = 0.40$, find the minimum angle θ_{min} at which the ladder does not slip.

Solution   The free-body diagram showing all the external forces acting on the ladder is illustrated in Figure 12.10b. The reaction force $\mathbf{R}$ exerted by the ground on the ladder is the vector sum of a normal force $\mathbf{n}$ and the force of static friction $\mathbf{f}_s$. The reaction force $\mathbf{P}$ exerted by the wall on the ladder is horizontal because the wall is frictionless. Notice how we have included only forces that act on the ladder. For example, the forces exerted by the ladder on the ground and on the wall are not part of the problem and thus do not appear in the free-body diagram. Applying the first condition

for equilibrium to the ladder, we have

$$\sum F_x = f - P = 0$$
$$\sum F_y = n - mg = 0$$

From the second equation we see that $n = mg = 50$ N. Furthermore, when the ladder is on the verge of slipping, the force of friction must be a maximum, which is given by $f_{s,max} = \mu_s n = 0.40(50 \text{ N}) = 20$ N. (Recall Eq. 5.8: $f_s \le \mu_s n$.) Thus, at this angle, $P = 20$ N.

To find θ_{min}, we must use the second condition for equilibrium. When we take the torques about an axis through the origin O at the bottom of the ladder, we have

$$\sum \tau_O = P\ell \sin \theta - mg \frac{\ell}{2} \cos \theta = 0$$

Because $P = 20$ N when the ladder is about to slip, and because $mg = 50$ N, this expression gives

$$\tan \theta_{min} = \frac{mg}{2P} = \frac{50 \text{ N}}{40 \text{ N}} = 1.25$$

$$\theta_{min} = \boxed{51°}$$

An alternative approach is to consider the intersection O' of the lines of action of forces $m\mathbf{g}$ and $\mathbf{P}$. Because the torque about any origin must be zero, the torque about O' must be zero. This requires that the line of action of $\mathbf{R}$ (the resultant of $\mathbf{n}$ and $\mathbf{f}$) pass through O'. In other words, because the ladder is stationary, the three forces acting on it must all pass through some common point. (We say that such forces are *concurrent*.) With this condition, you could then obtain the angle ϕ that $\mathbf{R}$ makes with the horizontal (where ϕ is greater than θ). Because this approach depends on the length of the ladder, you would have to know the value of ℓ to obtain a value for θ_{min}.

Exercise   For the angles labeled in Figure 12.10, show that $\tan \phi = 2 \tan \theta$.

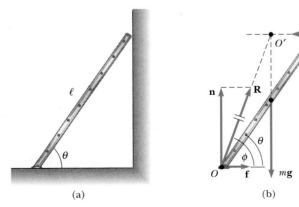

Figure 12.10   (a) A uniform ladder at rest, leaning against a smooth wall. The ground is rough. (b) The free-body diagram for the ladder. Note that the forces $\mathbf{R}$, $m\mathbf{g}$, and $\mathbf{P}$ pass through a common point O'.

EXAMPLE 12.5 ▶ Negotiating a Curb

(a) Estimate the magnitude of the force **F** a person must apply to a wheelchair's main wheel to roll up over a sidewalk curb (Fig. 12.11a). This main wheel, which is the one that comes in contact with the curb, has a radius R, and the height of the curb is h.

Solution   Normally, the person's hands supply the required force to a slightly smaller wheel that is concentric with the main wheel. We assume that the radius of the smaller wheel is the same as the radius of the main wheel, and so we can use R for our radius. Let us estimate a combined weight of $mg = 1\,400$ N for the person and the wheelchair and choose a wheel radius of $R = 30$ cm, as shown in Figure 12.11b. We also pick a curb height of $h = 10$ cm. We assume that the wheelchair and occupant are symmetric, and that each wheel supports a weight of 700 N. We then proceed to analyze only one of the wheels.

When the wheel is just about to be raised from the street, the reaction force exerted by the ground on the wheel at point Q goes to zero. Hence, at this time only three forces act on the wheel, as shown in Figure 12.11c. However, the force **n**, which is the force exerted on the wheel by the curb, acts at point P, and so if we choose to have our axis of rotation pass through point P, we do not need to include **n** in our torque equation. From the triangle OPQ shown in Figure 12.11b, we see that the moment arm d of the gravitational force $m\mathbf{g}$ acting on the wheel relative to point P is

$$d = \sqrt{R^2 - (R - h)^2} = \sqrt{2Rh - h^2}$$

The moment arm of **F** relative to point P is $2R - h$. Therefore, the net torque acting on the wheel about point P is

$$mgd - F(2R - h) = 0$$

$$mg\sqrt{2Rh - h^2} - F(2R - h) = 0$$

$$F = \frac{mg\sqrt{2Rh - h^2}}{2R - h}$$

$$F = \frac{(700\text{ N})\sqrt{2(0.3\text{ m})(0.1\text{ m}) - (0.1\text{ m})^2}}{2(0.3\text{ m}) - 0.1\text{ m}} = \boxed{300\text{ N}}$$

(Notice that we have kept only one digit as significant.) This result indicates that the force that must be applied to each wheel is substantial. You may want to estimate the force required to roll a wheelchair up a typical sidewalk accessibility ramp for comparison.

(b) Determine the magnitude and direction of **n**.

Solution   We use the first condition for equilibrium to determine the direction:

$$\sum F_x = F - n\cos\theta = 0$$

$$\sum F_y = n\sin\theta - mg = 0$$

Dividing the second equation by the first gives

$$\tan\theta = \frac{mg}{F} = \frac{700\text{ N}}{300\text{ N}}; \quad \theta = \boxed{70°}$$

(a)

(b)

(c)

Figure 12.11   (a) A wheelchair and person of total weight mg being raised over a curb by a force **F**. (b) Details of the wheel and curb. (c) The free-body diagram for the wheel when it is just about to be raised. Three forces act on the wheel at this instant: **F**, which is exerted by the hand; **n**, which is exerted by the curb; and the gravitational force $m\mathbf{g}$. (d) The vector sum of the three external forces acting on the wheel is zero.

We can use the right triangle shown in Figure 12.11d to obtain n:

$$n = \sqrt{(mg)^2 + F^2} = \sqrt{(700 \text{ N})^2 + (300 \text{ N})^2} = \boxed{800 \text{ N}}$$

Exercise Solve this problem by noting that the three forces acting on the wheel are concurrent (that is, that all three pass through the point C). The three forces form the sides of the triangle shown in Figure 12.11d.

APPLICATION ▶ Analysis of a Truss

Roofs, bridges, and other structures that must be both strong and lightweight often are made of trusses similar to the one shown in Figure 12.12a. Imagine that this truss structure represents part of a bridge. To approach this problem, we assume that the structural components are connected by pin joints. We also assume that the entire structure is free to slide horizontally because it sits on "rockers" on each end, which allow it to move back and forth as it undergoes thermal expansion and contraction. Assuming the mass of the bridge structure is negligible compared with the load, let us calculate the forces of tension or compression in all the structural components when it is supporting a 7 200-N load at the center (see Problem 58).

The force notation that we use here is not of our usual format. Until now, we have used the notation F_{AB} to mean "the force exerted by A on B." For this application, however, all double-letter subscripts on F indicate only the body exerting the force. The body on which a given force acts is not named in the subscript. For example, in Figure 12.12, F_{AB} is the force exerted by strut AB on the pin at A.

First, we apply Newton's second law to the truss as a whole in the vertical direction. Internal forces do not enter into this accounting. We balance the weight of the load with the normal forces exerted at the two ends by the supports on which the bridge rests:

$$\sum F_y = n_A + n_E - F_g = 0$$
$$n_A + n_E = 7\ 200 \text{ N}$$

Next, we calculate the torque about A, noting that the overall length of the bridge structure is $L = 50$ m:

$$\sum \tau = Ln_E - (L/2)F_g = 0$$
$$n_E = F_g/2 = 3\ 600 \text{ N}$$

Although we could repeat the torque calculation for the right end (point E), it should be clear from symmetry arguments that $n_A = 3\ 600$ N.

Now let us balance the vertical forces acting on the pin at point A. If we assume that strut AB is in compression, then the force F_{AB} that the strut exerts on the pin at point A has a negative y component. (If the strut is actually in tension, our calculations will result in a negative value for the magnitude of the force, still of the correct size):

$$\sum F_y = n_A - F_{AB} \sin 30° = 0$$
$$F_{AB} = 7\ 200 \text{ N}$$

The positive result shows that our assumption of compression was correct.

We can now find the forces acting on the beam between A and C by considering the horizontal forces acting on the pin at point A. Because point A is not accelerating, we can safely assume that F_{AC} must point toward the right (Fig. 12.12b); this indicates that the beam between points A and C is under tension:

$$\sum F_x = F_{AC} - F_{AB} \cos 30° = 0$$
$$F_{AC} = (7\ 200 \text{ N}) \cos 30° = 6\ 200 \text{ N}$$

Now let us consider the vertical forces acting on the pin at point C. We shall assume that strut BC is in tension. (Imagine the subsequent motion of the pin at point C if strut BC were to break suddenly.) On the basis of symmetry, we assert that $F_{BC} = F_{DC}$ and that $F_{AC} = F_{EC}$:

$$\sum F_y = 2 F_{BC} \sin 30° - 7\ 200 \text{ N} = 0$$
$$F_{BC} = 7\ 200 \text{ N}$$

Finally, we balance the horizontal forces on B, assuming that strut BD is in compression:

$$\sum F_x = F_{AB} \cos 30° + F_{BC} \cos 30° - F_{BD} = 0$$
$$(7\ 200 \text{ N}) \cos 30° + (7\ 200 \text{ N}) \cos 30° - F_{BD} = 0$$
$$F_{BD} = 12\ 000 \text{ N}$$

Thus, the top beam in a bridge of this design must be very strong.

Figure 12.12 (a) Truss structure from a bridge. (b) The forces acting on the pins at points A, C, and E. As an exercise, you should diagram the forces acting on the pin at point B.

12.4 ELASTIC PROPERTIES OF SOLIDS

In our study of mechanics thus far, we have assumed that objects remain undeformed when external forces act on them. In reality, all objects are deformable. That is, it is possible to change the shape or the size of an object (or both) by applying external forces. As these changes take place, however, internal forces in the object resist the deformation.

We shall discuss the deformation of solids in terms of the concepts of stress and strain. **Stress** is a quantity that is proportional to the force causing a deformation; more specifically, stress is the external force acting on an object per unit cross-sectional area. **Strain** is a measure of the degree of deformation. It is found that, for sufficiently small stresses, **strain is proportional to stress;** the constant of proportionality depends on the material being deformed and on the nature of the deformation. We call this proportionality constant the **elastic modulus.** The elastic modulus is therefore the ratio of the stress to the resulting strain:

$$\text{Elastic modulus} \equiv \frac{\text{stress}}{\text{strain}} \qquad (12.5)$$

In a very real sense it is a comparison of what is done to a solid object (a force is applied) and how that object responds (it deforms to some extent).

A plastic model of an arch structure under load conditions. The wavy lines indicate regions where the stresses are greatest. Such models are useful in designing architectural components. *(Peter Aprahamian/Sharples Stress Engineers Ltd./Science Photo Library)*

We consider three types of deformation and define an elastic modulus for each:

1. **Young's modulus,** which measures the resistance of a solid to a change in its length
2. **Shear modulus,** which measures the resistance to motion of the planes of a solid sliding past each other
3. **Bulk modulus,** which measures the resistance of solids or liquids to changes in their volume

Young's Modulus: Elasticity in Length

Consider a long bar of cross-sectional area A and initial length L_i that is clamped at one end, as in Figure 12.13. When an external force is applied perpendicular to the cross section, internal forces in the bar resist distortion ("stretching"), but the bar attains an equilibrium in which its length L_f is greater than L_i and in which the external force is exactly balanced by internal forces. In such a situation, the bar is said to be stressed. We define the **tensile stress** as the ratio of the magnitude of the external force F to the cross-sectional area A. The **tensile strain** in this case is defined as the ratio of the change in length ΔL to the original length L_i. We define **Young's modulus** by a combination of these two ratios:

$$Y = \frac{\text{tensile stress}}{\text{tensile strain}} = \frac{F/A}{\Delta L/L_i} \qquad \textbf{(12.6)}$$

Young's modulus is typically used to characterize a rod or wire stressed under either tension or compression. Note that because strain is a dimensionless quantity, Y has units of force per unit area. Typical values are given in Table 12.1. Experiments show (a) that for a fixed applied force, the change in length is proportional to the original length and (b) that the force necessary to produce a given strain is proportional to the cross-sectional area. Both of these observations are in accord with Equation 12.6.

The **elastic limit** of a substance is defined as the maximum stress that can be applied to the substance before it becomes permanently deformed. It is possible to exceed the elastic limit of a substance by applying a sufficiently large stress, as seen in Figure 12.14. Initially, a stress–strain curve is a straight line. As the stress increases, however, the curve is no longer straight. When the stress exceeds the elas-

Figure 12.13 A long bar clamped at one end is stretched by an amount ΔL under the action of a force **F**.

Young's modulus

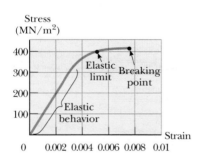

Figure 12.14 Stress-versus-strain curve for an elastic solid.

TABLE 12.1	Typical Values for Elastic Modulus		
Substance	Young's Modulus (N/m^2)	Shear Modulus (N/m^2)	Bulk Modulus (N/m^2)
Tungsten	35×10^{10}	14×10^{10}	20×10^{10}
Steel	20×10^{10}	8.4×10^{10}	6×10^{10}
Copper	11×10^{10}	4.2×10^{10}	14×10^{10}
Brass	9.1×10^{10}	3.5×10^{10}	6.1×10^{10}
Aluminum	7.0×10^{10}	2.5×10^{10}	7.0×10^{10}
Glass	$6.5–7.8 \times 10^{10}$	$2.6–3.2 \times 10^{10}$	$5.0–5.5 \times 10^{10}$
Quartz	5.6×10^{10}	2.6×10^{10}	2.7×10^{10}
Water	—	—	0.21×10^{10}
Mercury	—	—	2.8×10^{10}

Figure 12.15 (a) A shear deformation in which a rectangular block is distorted by two forces of equal magnitude but opposite directions applied to two parallel faces. (b) A book under shear stress.

Shear modulus

QuickLab

Estimate the shear modulus for the pages of your textbook. Does the thickness of the book have any effect on the modulus value?

Bulk modulus

tic limit, the object is permanently distorted and does not return to its original shape after the stress is removed. Hence, the shape of the object is permanently changed. As the stress is increased even further, the material ultimately breaks.

Quick Quiz 12.3

What is Young's modulus for the elastic solid whose stress–strain curve is depicted in Figure 12.14?

Quick Quiz 12.4

A material is said to be *ductile* if it can be stressed well beyond its elastic limit without breaking. A *brittle* material is one that breaks soon after the elastic limit is reached. How would you classify the material in Figure 12.14?

Shear Modulus: Elasticity of Shape

Another type of deformation occurs when an object is subjected to a force tangential to one of its faces while the opposite face is held fixed by another force (Fig. 12.15a). The stress in this case is called a shear stress. If the object is originally a rectangular block, a shear stress results in a shape whose cross-section is a parallelogram. A book pushed sideways, as shown in Figure 12.15b, is an example of an object subjected to a shear stress. To a first approximation (for small distortions), no change in volume occurs with this deformation.

We define the **shear stress** as F/A, the ratio of the tangential force to the area A of the face being sheared. The **shear strain** is defined as the ratio $\Delta x/h$, where Δx is the horizontal distance that the sheared face moves and h is the height of the object. In terms of these quantities, the **shear modulus** is

$$S = \frac{\text{shear stress}}{\text{shear strain}} = \frac{F/A}{\Delta x/h} \qquad \textbf{(12.7)}$$

Values of the shear modulus for some representative materials are given in Table 12.1. The unit of shear modulus is force per unit area.

Bulk Modulus: Volume Elasticity

Bulk modulus characterizes the response of a substance to uniform squeezing or to a reduction in pressure when the object is placed in a partial vacuum. Suppose that the external forces acting on an object are at right angles to all its faces, as shown in Figure 12.16, and that they are distributed uniformly over all the faces. As we shall see in Chapter 15, such a uniform distribution of forces occurs when an object is immersed in a fluid. An object subject to this type of deformation undergoes a change in volume but no change in shape. The **volume stress** is defined as the ratio of the magnitude of the normal force F to the area A. The quantity $P = F/A$ is called the **pressure.** If the pressure on an object changes by an amount $\Delta P = \Delta F/A$, then the object will experience a volume change ΔV. The **volume strain** is equal to the change in volume ΔV divided by the initial volume V_i. Thus, from Equation 12.5, we can characterize a volume ("bulk") compression in terms of the **bulk modulus,** which is defined as

$$B \equiv \frac{\text{volume stress}}{\text{volume strain}} = -\frac{\Delta F/A}{\Delta V/V_i} = -\frac{\Delta P}{\Delta V/V_i} \qquad \textbf{(12.8)}$$

Figure 12.16 When a solid is under uniform pressure, it undergoes a change in volume but no change in shape. This cube is compressed on all sides by forces normal to its six faces.

A negative sign is inserted in this defining equation so that B is a positive number. This maneuver is necessary because an increase in pressure (positive ΔP) causes a decrease in volume (negative ΔV) and vice versa.

Table 12.1 lists bulk moduli for some materials. If you look up such values in a different source, you often find that the reciprocal of the bulk modulus is listed. The reciprocal of the bulk modulus is called the **compressibility** of the material.

Note from Table 12.1 that both solids and liquids have a bulk modulus. However, no shear modulus and no Young's modulus are given for liquids because a liquid does not sustain a shearing stress or a tensile stress (it flows instead).

Prestressed Concrete

If the stress on a solid object exceeds a certain value, the object fractures. The maximum stress that can be applied before fracture occurs depends on the nature of the material and on the type of applied stress. For example, concrete has a tensile strength of about 2×10^6 N/m^2, a compressive strength of 20×10^6 N/m^2, and a shear strength of 2×10^6 N/m^2. If the applied stress exceeds these values, the concrete fractures. It is common practice to use large safety factors to prevent failure in concrete structures.

Concrete is normally very brittle when it is cast in thin sections. Thus, concrete slabs tend to sag and crack at unsupported areas, as shown in Figure 12.17a. The slab can be strengthened by the use of steel rods to reinforce the concrete, as illustrated in Figure 12.17b. Because concrete is much stronger under compression (squeezing) than under tension (stretching) or shear, vertical columns of concrete can support very heavy loads, whereas horizontal beams of concrete tend to sag and crack. However, a significant increase in shear strength is achieved if the reinforced concrete is prestressed, as shown in Figure 12.17c. As the concrete is being poured, the steel rods are held under tension by external forces. The external

QuickLab

Support a new flat eraser (art gum or Pink Pearl will do) on two parallel pencils at least 3 cm apart. Press down on the middle of the top surface just enough to make the top face of the eraser curve a bit. Is the top face under tension or compression? How about the bottom? Why does a flat slab of concrete supported at the ends tend to crack on the bottom face and not the top?

Figure 12.17 (a) A concrete slab with no reinforcement tends to crack under a heavy load. (b) The strength of the concrete is increased by using steel reinforcement rods. (c) The concrete is further strengthened by prestressing it with steel rods under tension.

forces are released after the concrete cures; this results in a permanent tension in the steel and hence a compressive stress on the concrete. This enables the concrete slab to support a much heavier load.

EXAMPLE 12.6   Stage Design

Recall Example 8.10, in which we analyzed a cable used to support an actor as he swung onto the stage. The tension in the cable was 940 N. What diameter should a 10-m-long steel wire have if we do not want it to stretch more than 0.5 cm under these conditions?

Solution   From the definition of Young's modulus, we can solve for the required cross-sectional area. Assuming that the cross section is circular, we can determine the diameter of the wire. From Equation 12.6, we have

$$Y = \frac{F/A}{\Delta L/L_i}$$

$$A = \frac{FL_i}{Y\Delta L} = \frac{(940 \text{ N})(10 \text{ m})}{(20 \times 10^{10} \text{ N/m}^2)(0.005 \text{ m})} = 9.4 \times 10^{-6} \text{ m}^2$$

The radius of the wire can be found from $A = \pi r^2$:

$$r = \sqrt{\frac{A}{\pi}} = \sqrt{\frac{9.4 \times 10^{-6} \text{ m}^2}{\pi}} = 1.7 \times 10^{-3} \text{ m} = 1.7 \text{ mm}$$

$$d = 2r = 2(1.7 \text{ mm}) = \boxed{3.4 \text{ mm}}$$

To provide a large margin of safety, we would probably use a flexible cable made up of many smaller wires having a total cross-sectional area substantially greater than our calculated value.

EXAMPLE 12.7   Squeezing a Brass Sphere

A solid brass sphere is initially surrounded by air, and the air pressure exerted on it is 1.0×10^5 N/m^2 (normal atmospheric pressure). The sphere is lowered into the ocean to a depth at which the pressure is 2.0×10^7 N/m^2. The volume of the sphere in air is 0.50 m^3. By how much does this volume change once the sphere is submerged?

Solution   From the definition of bulk modulus, we have

$$B = -\frac{\Delta P}{\Delta V/V_i}$$

$$\Delta V = -\frac{V_i \Delta P}{B}$$

Because the final pressure is so much greater than the initial pressure, we can neglect the initial pressure and state that $\Delta P = P_f - P_i \approx P_f = 2.0 \times 10^7$ N/m^2. Therefore,

$$\Delta V = -\frac{(0.50 \text{ m}^3)(2.0 \times 10^7 \text{ N/m}^2)}{6.1 \times 10^{10} \text{ N/m}^2} = \boxed{-1.6 \times 10^{-4} \text{ m}^3}$$

The negative sign indicates a decrease in volume.

SUMMARY

A rigid object is in **equilibrium** if and only if **the resultant external force acting on it is zero and the resultant external torque on it is zero about any axis:**

$$\sum \mathbf{F} = 0 \tag{12.1}$$

$$\sum \boldsymbol{\tau} = 0 \tag{12.2}$$

The first condition is the condition for translational equilibrium, and the second is the condition for rotational equilibrium. These two equations allow you to analyze a great variety of problems. Make sure you can identify forces unambiguously, create a free-body diagram, and then apply Equations 12.1 and 12.2 and solve for the unknowns.

The force of gravity exerted on an object can be considered as acting at a single point called the **center of gravity.** The center of gravity of an object coincides with its center of mass if the object is in a uniform gravitational field.

We can describe the elastic properties of a substance using the concepts of stress and strain. **Stress** is a quantity proportional to the force producing a deformation; **strain** is a measure of the degree of deformation. Strain is proportional to stress, and the constant of proportionality is the **elastic modulus:**

$$\text{Elastic modulus} \equiv \frac{\text{stress}}{\text{strain}} \qquad \textbf{(12.5)}$$

Three common types of deformation are (1) the resistance of a solid to elongation under a load, characterized by **Young's modulus** Y; (2) the resistance of a solid to the motion of internal planes sliding past each other, characterized by the **shear modulus** S; and (3) the resistance of a solid or fluid to a volume change, characterized by the **bulk modulus** B.

QUESTIONS

1. Can a body be in equilibrium if only one external force acts on it? Explain.
2. Can a body be in equilibrium if it is in motion? Explain.
3. Locate the center of gravity for the following uniform objects: (a) sphere, (b) cube, (c) right circular cylinder.
4. The center of gravity of an object may be located outside the object. Give a few examples for which this is the case.
5. You are given an arbitrarily shaped piece of plywood, together with a hammer, nail, and plumb bob. How could you use these items to locate the center of gravity of the plywood? (*Hint:* Use the nail to suspend the plywood.)
6. For a chair to be balanced on one leg, where must the center of gravity of the chair be located?
7. Can an object be in equilibrium if the only torques acting on it produce clockwise rotation?
8. A tall crate and a short crate of equal mass are placed side by side on an incline (without touching each other). As the incline angle is increased, which crate will topple first? Explain.
9. When lifting a heavy object, why is it recommended to keep the back as vertical as possible, lifting from the knees, rather than bending over and lifting from the waist?
10. Give a few examples in which several forces are acting on a system in such a way that their sum is zero but the system is not in equilibrium.
11. If you measure the net torque and the net force on a system to be zero, (a) could the system still be rotating with respect to you? (b) Could it be translating with respect to you?
12. A ladder is resting inclined against a wall. Would you feel safer climbing up the ladder if you were told that the ground is frictionless but the wall is rough or that the wall is frictionless but the ground is rough? Justify your answer.
13. What kind of deformation does a cube of Jell-O exhibit when it "jiggles"?
14. Ruins of ancient Greek temples often have intact vertical columns, but few horizontal slabs of stone are still in place. Can you think of a reason why this is so?

PROBLEMS

1, 2, 3 = straightforward, intermediate, challenging ☐ = full solution available in the *Student Solutions Manual and Study Guide*
WEB = solution posted at **http://www.saunderscollege.com/physics/** ☐ = Computer useful in solving problem ☐ = Interactive Physics
☐ = paired numerical/symbolic problems

Section 12.1 The Conditions for Equilibrium

1. A baseball player holds a 36-oz bat (weight = 10.0 N) with one hand at the point O (Fig. P12.1). The bat is in equilibrium. The weight of the bat acts along a line 60.0 cm to the right of O. Determine the force and the torque exerted on the bat by the player.

O

←—60.0 cm—→

$m\mathbf{g}$

Figure P12.1

2. Write the necessary conditions of equilibrium for the body shown in Figure P12.2. Take the origin of the torque equation at the point O.

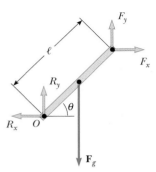

Figure P12.2

WEB **3.** A uniform beam of mass m_b and length ℓ supports blocks of masses m_1 and m_2 at two positions, as shown in Figure P12.3. The beam rests on two points. For what value of x will the beam be balanced at P such that the normal force at O is zero?

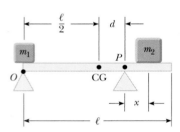

Figure P12.3

4. A student gets his car stuck in a snow drift. Not at a loss, having studied physics, he attaches one end of a stout rope to the vehicle and the other end to the trunk of a nearby tree, allowing for a very small amount of slack. The student then exerts a force **F** on the center of the rope in the direction perpendicular to the car–tree line, as shown in Figure P12.4. If the rope is inextensible and if the magnitude of the applied force is 500 N, what is the force on the car? (Assume equilibrium conditions.)

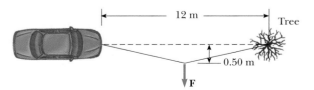

Figure P12.4

Section 12.2   More on the Center of Gravity

5. A 3.00-kg particle is located on the x axis at $x = -5.00$ m, and a 4.00-kg particle is located on the x axis at $x = 3.00$ m. Find the center of gravity of this two-particle system.

6. A circular pizza of radius R has a circular piece of radius $R/2$ removed from one side, as shown in Figure P12.6. Clearly, the center of gravity has moved from C to C' along the x axis. Show that the distance from C to C' is $R/6$. (Assume that the thickness and density of the pizza are uniform throughout.)

Figure P12.6

7. A carpenter's square has the shape of an L, as shown in Figure P12.7. Locate its center of gravity.

Figure P12.7

8. Pat builds a track for his model car out of wood, as illustrated in Figure P12.8. The track is 5.00 cm wide, 1.00 m high, and 3.00 m long, and it is solid. The runway is cut so that it forms a parabola described by the equation $y = (x - 3)^2/9$. Locate the horizontal position of the center of gravity of this track.

WEB **9.** Consider the following mass distribution: 5.00 kg at $(0, 0)$ m, 3.00 kg at $(0, 4.00)$ m, and 4.00 kg at $(3.00, 0)$ m. Where should a fourth mass of 8.00 kg be placed so that the center of gravity of the four-mass arrangement will be at $(0, 0)$?

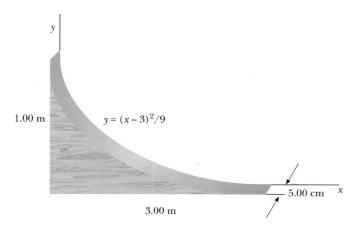

$y = (x - 3)^2/9$

1.00 m

5.00 cm x

3.00 m

Figure P12.8

10. Figure P12.10 shows three uniform objects: a rod, a right triangle, and a square. Their masses in kilograms and their coordinates in meters are given. Determine the center of gravity for the three-object system.

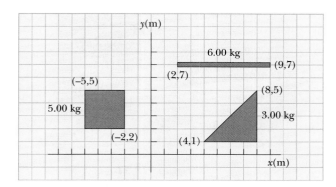

6.00 kg

(9,7)

(2,7)

(−5,5)

(8,5)

5.00 kg

3.00 kg

(−2,2)

(4,1)

y(m)

x(m)

Figure P12.10

Section 12.3   Examples of Rigid Objects in Static Equilibrium

11. Stephen is pushing his sister Joyce in a wheelbarrow when it is stopped by a brick 8.00 cm high (Fig. P12.11). The handles make an angle of 15.0° from the horizontal. A downward force of 400 N is exerted on the wheel, which has a radius of 20.0 cm. (a) What force must Stephen apply along the handles to just start the wheel over the brick? (b) What is the force (magnitude and direction) that the brick exerts on the wheel just as the wheel begins to lift over the brick? Assume in both parts (a) and (b) that the brick remains fixed and does not slide along the ground.

12. Two pans of a balance are 50.0 cm apart. The fulcrum of the balance has been shifted 1.00 cm away from the center by a dishonest shopkeeper. By what percentage is the true weight of the goods being marked up by the shopkeeper? (Assume that the balance has negligible mass.)

15.0°

Figure P12.11

13. A 15.0-m uniform ladder weighing 500 N rests against a frictionless wall. The ladder makes a 60.0° angle with the horizontal. (a) Find the horizontal and vertical forces that the ground exerts on the base of the ladder when an 800-N firefighter is 4.00 m from the bottom. (b) If the ladder is just on the verge of slipping when the firefighter is 9.00 m up, what is the coefficient of static friction between the ladder and the ground?

14. A uniform ladder of length L and mass m_1 rests against a frictionless wall. The ladder makes an angle θ with the horizontal. (a) Find the horizontal and vertical forces that the ground exerts on the base of the ladder when a firefighter of mass m_2 is a distance x from the bottom. (b) If the ladder is just on the verge of slipping when the firefighter is a distance d from the bottom, what is the coefficient of static friction between the ladder and the ground?

15. Figure P12.15 shows a claw hammer as it is being used to pull a nail out of a horizontal surface. If a force of magnitude 150 N is exerted horizontally as shown, find

F

30.0 cm

Single point of contact

30.0°

5.00 cm

Figure P12.15

(a) the force exerted by the hammer claws on the nail and (b) the force exerted by the surface on the point of contact with the hammer head. Assume that the force the hammer exerts on the nail is parallel to the nail.

16. A uniform plank with a length of 6.00 m and a mass of 30.0 kg rests horizontally across two horizontal bars of a scaffold. The bars are 4.50 m apart, and 1.50 m of the plank hangs over one side of the scaffold. Draw a free-body diagram for the plank. How far can a painter with a mass of 70.0 kg walk on the overhanging part of the plank before it tips?

17. A 1 500-kg automobile has a wheel base (the distance between the axles) of 3.00 m. The center of mass of the automobile is on the center line at a point 1.20 m behind the front axle. Find the force exerted by the ground on each wheel.

18. A vertical post with a square cross section is 10.0 m tall. Its bottom end is encased in a base 1.50 m tall that is precisely square but slightly loose. A force of 5.50 N to the right acts on the top of the post. The base maintains the post in equilibrium. Find the force that the top of the right sidewall of the base exerts on the post. Find the force that the bottom of the left sidewall of the base exerts on the post.

19. A flexible chain weighing 40.0 N hangs between two hooks located at the same height (Fig. P12.19). At each hook, the tangent to the chain makes an angle $\theta = 42.0°$ with the horizontal. Find (a) the magnitude of the force each hook exerts on the chain and (b) the tension in the chain at its midpoint. (*Hint:* For part (b), make a free-body diagram for half the chain.)

Figure P12.19

20. A hemispherical sign 1.00 m in diameter and of uniform mass density is supported by two strings, as shown in Figure P12.20. What fraction of the sign's weight is supported by each string?

21. Sir Lost-a-Lot dons his armor and sets out from the castle on his trusty steed in his quest to improve communication between damsels and dragons (Fig. P12.21). Unfortunately, his squire lowered the draw bridge too far and finally stopped lowering it when it was 20.0° below the horizontal. Lost-a-Lot and his horse stop when their combined center of mass is 1.00 m from the end of the bridge. The bridge is 8.00 m long and has a mass of 2 000 kg. The lift cable is attached to the bridge 5.00 m from the hinge at the castle end and to a point on the castle wall 12.0 m above the bridge. Lost-a-Lot's mass

Figure P12.20

Figure P12.21

combined with that of his armor and steed is 1 000 kg. Determine (a) the tension in the cable, as well as (b) the horizontal and (c) the vertical force components acting on the bridge at the hinge.

22. Two identical, uniform bricks of length L are placed in a stack over the edge of a horizontal surface such that the maximum possible overhang without falling is achieved, as shown in Figure P12.22. Find the distance x.

Figure P12.22

23. A vaulter holds a 29.4-N pole in equilibrium by exerting an upward force **U** with her leading hand and a downward force **D** with her trailing hand, as shown in Figure P12.23. Point C is the center of gravity of the pole. What are the magnitudes of **U** and **D**?

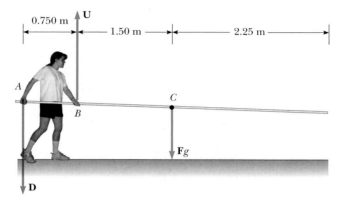

Figure P12.23

Section 12.4   Elastic Properties of Solids

24. Assume that Young's modulus for bone is 1.50×10^{10} N/m² and that a bone will fracture if more than 1.50×10^8 N/m² is exerted. (a) What is the maximum force that can be exerted on the femur bone in the leg if it has a minimum effective diameter of 2.50 cm? (b) If a force of this magnitude is applied compressively, by how much does the 25.0-cm-long bone shorten?

25. A 200-kg load is hung on a wire with a length of 4.00 m, a cross-sectional area of 0.200×10^{-4} m², and a Young's modulus of 8.00×10^{10} N/m². What is its increase in length?

26. A steel wire 1 mm in diameter can support a tension of 0.2 kN. Suppose you need a cable made of these wires to support a tension of 20 kN. The cable's diameter should be of what order of magnitude?

27. A child slides across a floor in a pair of rubber-soled shoes. The frictional force acting on each foot is 20.0 N. The footprint area of each shoe's sole is 14.0 cm², and the thickness of each sole is 5.00 mm. Find the horizontal distance by which the upper and lower surfaces of each sole are offset. The shear modulus of the rubber is 3.00×10^6 N/m².

28. Review Problem. A 30.0-kg hammer strikes a steel spike 2.30 cm in diameter while moving with a speed of 20.0 m/s. The hammer rebounds with a speed of 10.0 m/s after 0.110 s. What is the average strain in the spike during the impact?

29. If the elastic limit of copper is 1.50×10^8 N/m², determine the minimum diameter a copper wire can have under a load of 10.0 kg if its elastic limit is not to be exceeded.

30. Review Problem. A 2.00-m-long cylindrical steel wire with a cross-sectional diameter of 4.00 mm is placed over a light frictionless pulley, with one end of the wire connected to a 5.00-kg mass and the other end connected to a 3.00-kg mass. By how much does the wire stretch while the masses are in motion?

31. Review Problem. A cylindrical steel wire of length L_i with a cross-sectional diameter d is placed over a light frictionless pulley, with one end of the wire connected to a mass m_1 and the other end connected to a mass m_2. By how much does the wire stretch while the masses are in motion?

32. Calculate the density of sea water at a depth of 1 000 m, where the water pressure is about 1.00×10^7 N/m². (The density of sea water is 1.030×10^3 kg/m³ at the surface.)

WEB **33.** If the shear stress exceeds about 4.00×10^8 N/m², steel ruptures. Determine the shearing force necessary (a) to shear a steel bolt 1.00 cm in diameter and (b) to punch a 1.00-cm-diameter hole in a steel plate 0.500 cm thick.

34. (a) Find the minimum diameter of a steel wire 18.0 m long that elongates no more than 9.00 mm when a load of 380 kg is hung on its lower end. (b) If the elastic limit for this steel is 3.00×10^8 N/m², does permanent deformation occur with this load?

35. When water freezes, it expands by about 9.00%. What would be the pressure increase inside your automobile's engine block if the water in it froze? (The bulk modulus of ice is 2.00×10^9 N/m².)

36. For safety in climbing, a mountaineer uses a 50.0-m nylon rope that is 10.0 mm in diameter. When supporting the 90.0-kg climber on one end, the rope elongates by 1.60 m. Find Young's modulus for the rope material.

ADDITIONAL PROBLEMS

37. A bridge with a length of 50.0 m and a mass of 8.00×10^4 kg is supported on a smooth pier at each end, as illustrated in Figure P12.37. A truck of mass 3.00×10^4 kg

Figure P12.37

is located 15.0 m from one end. What are the forces on the bridge at the points of support?

38. A frame in the shape of the letter A is formed from two uniform pieces of metal, each of weight 26.0 N and length 1.00 m. They are hinged at the top and held together by a horizontal wire 1.20 m in length (Fig. P12.38). The structure rests on a frictionless surface. If the wire is connected at points a distance of 0.650 m from the top of the frame, determine the tension in the wire.

Figure P12.38

39. Refer to Figure 12.17c. A lintel of prestressed reinforced concrete is 1.50 m long. The cross-sectional area of the concrete is 50.0 cm². The concrete encloses one steel reinforcing rod with a cross-sectional area of 1.50 cm². The rod joins two strong end plates. Young's modulus for the concrete is $30.0 \times 10^9 \text{ N/m}^2$. After the concrete cures and the original tension T_1 in the rod is released, the concrete will be under a compressive stress of $8.00 \times 10^6 \text{ N/m}^2$. (a) By what distance will the rod compress the concrete when the original tension in the rod is released? (b) Under what tension T_2 will the rod still be? (c) How much longer than its unstressed length will the rod then be? (d) When the concrete was poured, the rod should have been stretched by what extension distance from its unstressed length? (e) Find the required original tension T_1 in the rod.

40. A solid sphere of radius R and mass M is placed in a trough, as shown in Figure P12.40. The inner surfaces of the trough are frictionless. Determine the forces exerted by the trough on the sphere at the two contact points.

41. A 10.0-kg monkey climbs up a 120-N uniform ladder of length L, as shown in Figure P12.41. The upper and

Figure P12.41

lower ends of the ladder rest on frictionless surfaces. The lower end is fastened to the wall by a horizontal rope that can support a maximum tension of 110 N. (a) Draw a free-body diagram for the ladder. (b) Find the tension in the rope when the monkey is one third the way up the ladder. (c) Find the maximum distance d that the monkey can climb up the ladder before the rope breaks. Express your answer as a fraction of L.

42. A hungry bear weighing 700 N walks out on a beam in an attempt to retrieve a basket of food hanging at the end of the beam (Fig. P12.42). The beam is uniform, weighs 200 N, and is 6.00 m long; the basket weighs 80.0 N. (a) Draw a free-body diagram for the beam. (b) When the bear is at $x = 1.00$ m, find the tension in the wire and the components of the force exerted by the wall on the left end of the beam. (c) If the wire can withstand a maximum tension of 900 N, what is the maximum distance that the bear can walk before the wire breaks?

Goodies

Figure P12.42

Figure P12.40

43. Old MacDonald had a farm, and on that farm he had a gate (Fig. P12.43). The gate is 3.00 m wide and 1.80 m

1.80 m

30.0°

3.00 m

Figure P12.43

Ye Olde Greasy Spoon

d

2L

Figure P12.45

high, with hinges attached to the top and bottom. The guy wire makes an angle of 30.0° with the top of the gate and is tightened by a turn buckle to a tension of 200 N. The mass of the gate is 40.0 kg. (a) Determine the horizontal force exerted on the gate by the bottom hinge. (b) Find the horizontal force exerted by the upper hinge. (c) Determine the combined vertical force exerted by both hinges. (d) What must the tension in the guy wire be so that the horizontal force exerted by the upper hinge is zero?

44. A 1 200-N uniform boom is supported by a cable, as illustrated in Figure P12.44. The boom is pivoted at the bottom, and a 2 000-N object hangs from its top. Find the tension in the cable and the components of the reaction force exerted on the boom by the floor.

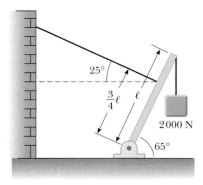

25°

$\frac{3}{4}\ell$ ℓ

2000 N

65°

Figure P12.44

45. A uniform sign of weight F_g and width $2L$ hangs from a light, horizontal beam hinged at the wall and supported by a cable (Fig. P12.45). Determine (a) the tension in the cable and (b) the components of the reaction force exerted by the wall on the beam in terms of F_g, d, L, and θ.

 46. A crane of mass 3 000 kg supports a load of 10 000 kg as illustrated in Figure P12.46. The crane is pivoted with a frictionless pin at A and rests against a smooth support at B. Find the reaction forces at A and B.

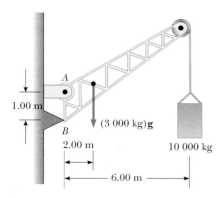

A

1.00 m

$(3\,000\text{ kg})\mathbf{g}$

B

2.00 m

10 000 kg

6.00 m

Figure P12.46

47. A ladder having a uniform density and a mass m rests against a frictionless vertical wall, making an angle 60.0° with the horizontal. The lower end rests on a flat surface, where the coefficient of static friction is $\mu_s = 0.400$. A window cleaner having a mass $M = 2m$ attempts to climb the ladder. What fraction of the length L of the ladder will the worker have reached when the ladder begins to slip?

48. A uniform ladder weighing 200 N is leaning against a wall (see Fig. 12.10). The ladder slips when $\theta = 60.0°$. Assuming that the coefficients of static friction at the wall and the ground are the same, obtain a value for μ_s.

49. A 10 000-N shark is supported by a cable attached to a 4.00-m rod that can pivot at its base. Calculate the tension that the cable must have if it is to hold the system in the position shown in Figure P12.49. Find the horizontal and vertical forces exerted on the base of the rod. (Neglect the weight of the rod.)

Figure P12.49

50. When a person stands on tiptoe (a strenuous position), the position of the foot is as shown in Figure P12.50a. The total weight of the body $\mathbf{F}_g$ is supported by the force $\mathbf{n}$ exerted by the floor on the toe. A mechanical model for the situation is shown in Figure P12.50b,

(a)

(b)

Figure P12.50

where $\mathbf{T}$ is the force exerted by the Achilles tendon on the foot and $\mathbf{R}$ is the force exerted by the tibia on the foot. Find the values of T, R, and θ when $F_g = 700$ N.

51. A person bends over and lifts a 200-N object as shown in Figure P12.51a, with his back in a horizontal position (a terrible way to lift an object). The back muscle attached at a point two thirds the way up the spine maintains the position of the back, and the angle between the spine and this muscle is 12.0°. Using the mechanical model shown in Figure P12.51b and taking the weight of the upper body to be 350 N, find the tension in the back muscle and the compressional force in the spine.

(a)   (b)

Figure P12.51

52. Two 200-N traffic lights are suspended from a single cable, as shown in Figure 12.52. Neglecting the cable's weight, (a) prove that if $\theta_1 = \theta_2$, then $T_1 = T_2$. (b) Determine the three tensions T_1, T_2, and T_3 if $\theta_1 = \theta_2 = 8.00°$.

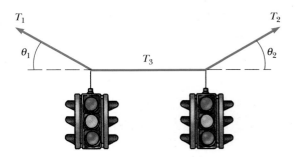

Figure P12.52

53. A force acts on a rectangular cabinet weighing 400 N, as illustrated in Figure P12.53. (a) If the cabinet slides with constant speed when $F = 200$ N and $h = 0.400$ m,

find the coefficient of kinetic friction and the position of the resultant normal force. (b) If $F = 300$ N, find the value of h for which the cabinet just begins to tip.

Figure P12.53 Problems 53 and 54.

54. Consider the rectangular cabinet of Problem 53, but with a force **F** applied horizontally at its upper edge. (a) What is the minimum force that must be applied for the cabinet to start tipping? (b) What is the minimum coefficient of static friction required to prevent the cabinet from sliding with the application of a force of this magnitude? (c) Find the magnitude and direction of the minimum force required to tip the cabinet if the point of application can be chosen anywhere on it.

55. A uniform rod of weight F_g and length L is supported at its ends by a frictionless trough, as shown in Figure P12.55. (a) Show that the center of gravity of the rod is directly over point O when the rod is in equilibrium. (b) Determine the equilibrium value of the angle θ.

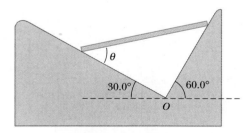

Figure P12.55

56. **Review Problem.** A cue stick strikes a cue ball and delivers a horizontal impulse in such a way that the ball rolls without slipping as it starts to move. At what height above the ball's center (in terms of the radius of the ball) was the blow struck?

57. A uniform beam of mass m is inclined at an angle θ to the horizontal. Its upper end produces a 90° bend in a very rough rope tied to a wall, and its lower end rests on a rough floor (Fig. P12.57). (a) If the coefficient of static friction between the beam and the floor is μ_s, determine an expression for the maximum mass M that can

be suspended from the top before the beam slips. (b) Determine the magnitude of the reaction force at the floor and the magnitude of the force exerted by the beam on the rope at P in terms of m, M, and μ_s.

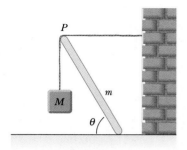

Figure P12.57

58. Figure P12.58 shows a truss that supports a downward force of 1 000 N applied at the point B. The truss has negligible weight. The piers at A and C are smooth. (a) Apply the conditions of equilibrium to prove that $n_A = 366$ N and that $n_C = 634$ N. (b) Show that, because forces act on the light truss only at the hinge joints, each bar of the truss must exert on each hinge pin only a force along the length of that bar—a force of tension or compression. (c) Find the force of tension or compression in each of the three bars.

Figure P12.58

59. A stepladder of negligible weight is constructed as shown in Figure P12.59. A painter with a mass of 70.0 kg stands on the ladder 3.00 m from the bottom. Assuming that the floor is frictionless, find (a) the tension in the horizontal bar connecting the two halves of the ladder, (b) the normal forces at A and B, and (c) the components of the reaction force at the single hinge C that the left half of the ladder exerts on the right half. (*Hint:* Treat each half of the ladder separately.)

Figure P12.59

60. A flat dance floor of dimensions 20.0 m by 20.0 m has a mass of 1 000 kg. Three dance couples, each of mass 125 kg, start in the top left, top right, and bottom left corners. (a) Where is the initial center of gravity? (b) The couple in the bottom left corner moves 10.0 m to the right. Where is the new center of gravity? (c) What was the average velocity of the center of gravity if it took that couple 8.00 s to change position?

61. A shelf bracket is mounted on a vertical wall by a single screw, as shown in Figure P12.61. Neglecting the weight of the bracket, find the horizontal component of the force that the screw exerts on the bracket when an 80.0-N vertical force is applied as shown. (*Hint:* Imagine that the bracket is slightly loose.)

Figure P12.61

62. Figure P12.62 shows a vertical force applied tangentially to a uniform cylinder of weight F_g. The coefficient of

Figure P12.62

static friction between the cylinder and all surfaces is 0.500. In terms of F_g, find the maximum force **P** that can be applied that does not cause the cylinder to rotate. (*Hint:* When the cylinder is on the verge of slipping, both friction forces are at their maximum values. Why?)

WEB 63. **Review Problem.** A wire of length L_i, Young's modulus Y, and cross-sectional area A is stretched elastically by an amount ΔL. According to Hooke's law, the restoring force is $-k\,\Delta L$. (a) Show that $k = YA/L_i$. (b) Show that the work done in stretching the wire by an amount ΔL is $W = YA(\Delta L)^2/2L_i$.

64. Two racquetballs are placed in a glass jar, as shown in Figure P12.64. Their centers and the point A lie on a straight line. (a) Assuming that the walls are frictionless, determine P_1, P_2, and P_3. (b) Determine the magnitude of the force exerted on the right ball by the left ball. Assume each ball has a mass of 170 g.

Figure P12.64

65. In Figure P12.65, the scales read $F_{g1} = 380$ N and $F_{g2} = 320$ N. Neglecting the weight of the supporting plank,

Figure P12.65

how far from the woman's feet is her center of mass, given that her height is 2.00 m?

66. A steel cable 3.00 cm² in cross-sectional area has a mass of 2.40 kg per meter of length. If 500 m of the cable is hung over a vertical cliff, how much does the cable stretch under its own weight? (For Young's modulus for steel, refer to Table 12.1.)

67. (a) Estimate the force with which a karate master strikes a board if the hand's speed at time of impact is 10.0 m/s and decreases to 1.00 m/s during a 0.002 00-s time-of-contact with the board. The mass of coordinated hand-and-arm is 1.00 kg. (b) Estimate the shear stress if this force,is exerted on a 1.00-cm-thick pine board that is 10.0 cm wide. (c) If the maximum shear stress a pine board can receive before breaking is $3.60 \times 10^6 \text{ N/m}^2$, will the board break?

68. A bucket is made from thin sheet metal. The bottom and top of the bucket have radii of 25.0 cm and 35.0 cm, respectively. The bucket is 30.0 cm high and filled with water. Where is the center of gravity? (Ignore the weight of the bucket itself.)

69. **Review Problem.** A trailer with a loaded weight of F_g is being pulled by a vehicle with a force **P**, as illustrated in Figure P12.69. The trailer is loaded such that its center of mass is located as shown. Neglect the force of rolling friction and let a represent the x component of the acceleration of the trailer. (a) Find the vertical component of **P** in terms of the given parameters. (b) If $a = 2.00 \text{ m/s}^2$ and $h = 1.50$ m, what must be the value of d

so that $P_y = 0$ (that is, no vertical load on the vehicle)? (c) Find the values of P_x and P_y given that $F_g = 1\ 500$ N, $d = 0.800$ m, $L = 3.00$ m, $h = 1.50$ m, and $a = -2.00 \text{ m/s}^2$.

70. **Review Problem.** An aluminum wire is 0.850 m long and has a circular cross section of diameter 0.780 mm. Fixed at the top end, the wire supports a 1.20-kg mass that swings in a horizontal circle. Determine the angular velocity required to produce strain 1.00×10^{-3}.

71. A 200-m-long bridge truss extends across a river (Fig. P12.71). Calculate the force of tension or compression in each structural component when a 1 360-kg car is at the center of the bridge. Assume that the structure is free to slide horizontally to permit thermal expansion and contraction, that the structural components are connected by pin joints, and that the masses of the structural components are small compared with the mass of the car.

Figure P12.71

72. A 100-m-long bridge truss is supported at its ends so that it can slide freely (Fig. P12.72). A 1 500-kg car is halfway between points A and C. Show that the weight of the car is evenly distributed between points A and C, and calculate the force in each structural component. Specify whether each strut is under tension or compression. Assume that the structural components are connected by pin joints and that the masses of the components are small compared with the mass of the car.

Figure P12.72

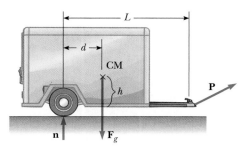

Figure P12.69

Answers to Quick Quizzes

12.1 (a) Yes, as Figure 12.3 shows. The unbalanced torques cause an angular acceleration even though the linear acceleration is zero. (b) Yes, again. This happens when the lines of action of all the forces intersect at a common point. If a net force acts on the object, then the object has a translational acceleration. However, because there is no net torque on the object, the object has no angular acceleration. There are other instances in which torques cancel but the forces do not. You should be able to draw at least two.

12.2 The location of the board's center of gravity relative to the fulcrum.

12.3 Young's modulus is given by the ratio of stress to strain, which is the slope of the elastic behavior section of the graph in Figure 12.14. Reading from the graph, we note that a stress of approximately $3 \times 10^8 \, \text{N/m}^2$ results in a strain of 0.003. The slope, and hence Young's modulus, are therefore $10 \times 10^{10} \, \text{N/m}^2$.

12.4 A substantial part of the graph extends beyond the elastic limit, indicating permanent deformation. Thus, the material is ductile.

PUZZLER✦

Inside the pocket watch is a small disk (called a torsional pendulum) that oscillates back and forth at a very precise rate and controls the watch gears. A grandfather clock keeps accurate time because of its pendulum. The tall wooden case provides the space needed by the long pendulum as it advances the clock gears with each swing. In both of these timepieces, the vibration of a carefully shaped component is critical to accurate operation. What properties of oscillating objects make them so useful in timing devices? *(Photograph of pocket watch, George Semple; photograph of grandfather clock, Charles D. Winters)*

c h a p t e r

13

Oscillatory Motion

A very special kind of motion occurs when the force acting on a body is proportional to the displacement of the body from some equilibrium position. If this force is always directed toward the equilibrium position, repetitive back-and-forth motion occurs about this position. Such motion is called *periodic motion, harmonic motion, oscillation,* or *vibration* (the four terms are completely equivalent).

You are most likely familiar with several examples of periodic motion, such as the oscillations of a block attached to a spring, the swinging of a child on a playground swing, the motion of a pendulum, and the vibrations of a stringed musical instrument. In addition to these everyday examples, numerous other systems exhibit periodic motion. For example, the molecules in a solid oscillate about their equilibrium positions; electromagnetic waves, such as light waves, radar, and radio waves, are characterized by oscillating electric and magnetic field vectors; and in alternating-current electrical circuits, voltage, current, and electrical charge vary periodically with time.

Most of the material in this chapter deals with *simple harmonic motion,* in which an object oscillates such that its position is specified by a sinusoidal function of time with no loss in mechanical energy. In real mechanical systems, damping (frictional) forces are often present. These forces are considered in optional Section 13.6 at the end of this chapter.

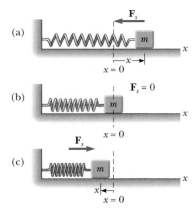

Figure 13.1 A block attached to a spring moving on a frictionless surface. (a) When the block is displaced to the right of equilibrium ($x > 0$), the force exerted by the spring acts to the left. (b) When the block is at its equilibrium position ($x = 0$), the force exerted by the spring is zero. (c) When the block is displaced to the left of equilibrium ($x < 0$), the force exerted by the spring acts to the right.

13.1 SIMPLE HARMONIC MOTION

Consider a physical system that consists of a block of mass m attached to the end of a spring, with the block free to move on a horizontal, frictionless surface (Fig. 13.1). When the spring is neither stretched nor compressed, the block is at the position $x = 0$, called the *equilibrium position* of the system. We know from experience that such a system oscillates back and forth if disturbed from its equilibrium position.

We can understand the motion in Figure 13.1 qualitatively by first recalling that when the block is displaced a small distance x from equilibrium, the spring exerts on the block a force that is proportional to the displacement and given by Hooke's law (see Section 7.3):

$$F_s = -kx \qquad \text{(13.1)}$$

We call this a **restoring force** because it is is always directed toward the equilibrium position and therefore *opposite* the displacement. That is, when the block is displaced to the right of $x = 0$ in Figure 13.1, then the displacement is positive and the restoring force is directed to the left. When the block is displaced to the left of $x = 0$, then the displacement is negative and the restoring force is directed to the right.

Applying Newton's second law to the motion of the block, together with Equation 13.1, we obtain

$$F_s = -kx = ma$$

$$a = -\frac{k}{m}x \qquad \text{(13.2)}$$

That is, the acceleration is proportional to the displacement of the block, and its direction is opposite the direction of the displacement. Systems that behave in this way are said to exhibit **simple harmonic motion. An object moves with simple harmonic motion whenever its acceleration is proportional to its displacement from some equilibrium position and is oppositely directed.**

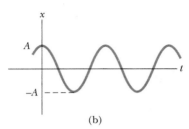

Figure 13.2 An experimental apparatus for demonstrating simple harmonic motion. A pen attached to the oscillating mass traces out a wavelike pattern on the moving chart paper.

An experimental arrangement that exhibits simple harmonic motion is illustrated in Figure 13.2. A mass oscillating vertically on a spring has a pen attached to it. While the mass is oscillating, a sheet of paper is moved perpendicular to the direction of motion of the spring, and the pen traces out a wavelike pattern.

8.2 & 8.3

In general, a particle moving along the x axis exhibits simple harmonic motion when x, the particle's displacement from equilibrium, varies in time according to the relationship

$$x = A \cos(\omega t + \phi) \qquad (13.3)$$

where A, ω, and ϕ are constants. To give physical significance to these constants, we have labeled a plot of x as a function of t in Figure 13.3a. This is just the pattern that is observed with the experimental apparatus shown in Figure 13.2. The **amplitude** A of the motion is the maximum displacement of the particle in either the positive or negative x direction. The constant ω is called the **angular frequency** of the motion and has units of radians per second. (We shall discuss the geometric significance of ω in Section 13.2.) The constant angle ϕ, called the **phase constant** (or phase angle), is determined by the initial displacement and velocity of the particle. If the particle is at its maximum position $x = A$ at $t = 0$, then $\phi = 0$ and the curve of x versus t is as shown in Figure 13.3b. If the particle is at some other position at $t = 0$, the constants ϕ and A tell us what the position was at time $t = 0$. The quantity $(\omega t + \phi)$ is called the **phase** of the motion and is useful in comparing the motions of two oscillators.

Note from Equation 13.3 that the trigonometric function x is *periodic* and repeats itself every time ωt increases by 2π rad. **The period T of the motion is the time it takes for the particle to go through one full cycle.** We say that the particle has made *one oscillation*. This definition of T tells us that the value of x at time t equals the value of x at time $t + T$. We can show that $T = 2\pi/\omega$ by using the preceding observation that the phase $(\omega t + \phi)$ increases by 2π rad in a time T:

$$\omega t + \phi + 2\pi = \omega(t + T) + \phi$$

Hence, $\omega T = 2\pi$, or

$$T = \frac{2\pi}{\omega} \qquad (13.4)$$

Displacement versus time for simple harmonic motion

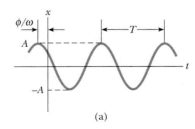

(a)

(b)

Figure 13.3 (a) An x–t curve for a particle undergoing simple harmonic motion. The amplitude of the motion is A, the period is T, and the phase constant is ϕ. (b) The x–t curve in the special case in which $x = A$ at $t = 0$ and hence $\phi = 0$.

The inverse of the period is called the **frequency** f of the motion. **The frequency represents the number of oscillations that the particle makes per unit time:**

Frequency

$$f = \frac{1}{T} = \frac{\omega}{2\pi} \qquad (13.5)$$

The units of f are cycles per second $= s^{-1}$, or **hertz** (Hz).

Rearranging Equation 13.5, we obtain the angular frequency:

Angular frequency

$$\omega = 2\pi f = \frac{2\pi}{T} \qquad (13.6)$$

Quick Quiz 13.1

What would the phase constant ϕ have to be in Equation 13.3 if we were describing an oscillating object that happened to be at the origin at $t = 0$?

Quick Quiz 13.2

An object undergoes simple harmonic motion of amplitude A. Through what total distance does the object move during one complete cycle of its motion? (a) $A/2$. (b) A. (c) $2A$. (d) $4A$.

$$x = A\cos(\omega t + \phi)$$

We can obtain the linear velocity of a particle undergoing simple harmonic motion by differentiating Equation 13.3 with respect to time:

Velocity in simple harmonic motion

$$v = \frac{dx}{dt} = -\omega A \sin(\omega t + \phi) \qquad (13.7)$$

The acceleration of the particle is

Acceleration in simple harmonic motion

$$a = \frac{dv}{dt} = -\omega^2 A \cos(\omega t + \phi) \qquad (13.8)$$

Because $x = A\cos(\omega t + \phi)$, we can express Equation 13.8 in the form

$$a = -\omega^2 x \qquad (13.9)$$

From Equation 13.7 we see that, because the sine function oscillates between ± 1, the extreme values of v are $\pm \omega A$. Because the cosine function also oscillates between ± 1, Equation 13.8 tells us that the extreme values of a are $\pm \omega^2 A$. Therefore, the maximum speed and the magnitude of the maximum acceleration of a particle moving in simple harmonic motion are

Maximum values of speed and acceleration in simple harmonic motion

$$v_{max} = \omega A \qquad (13.10)$$

$$a_{max} = \omega^2 A \qquad (13.11)$$

Figure 13.4a represents the displacement versus time for an arbitrary value of the phase constant. The velocity and acceleration curves are illustrated in Figure 13.4b and c. These curves show that the phase of the velocity differs from the phase of the displacement by $\pi/2$ rad, or 90°. That is, when x is a maximum or a minimum, the velocity is zero. Likewise, when x is zero, the speed is a maximum.

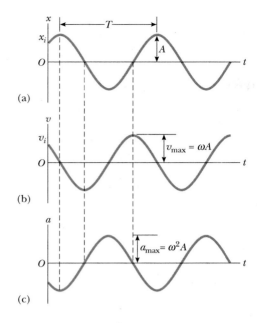

Figure 13.4 Graphical representation of simple harmonic motion. (a) Displacement versus time. (b) Velocity versus time. (c) Acceleration versus time. Note that at any specified time the velocity is 90° out of phase with the displacement and the acceleration is 180° out of phase with the displacement.

Furthermore, note that the phase of the acceleration differs from the phase of the displacement by π rad, or 180°. That is, when x is a maximum, a is a maximum in the opposite direction.

The phase constant ϕ is important when we compare the motion of two or more oscillating objects. Imagine two identical pendulum bobs swinging side by side in simple harmonic motion, with one having been released later than the other. The pendulum bobs have different phase constants. Let us show how the phase constant and the amplitude of any particle moving in simple harmonic motion can be determined if we know the particle's initial speed and position and the angular frequency of its motion.

Suppose that at $t = 0$ the initial position of a single oscillator is $x = x_i$ and its initial speed is $v = v_i$. Under these conditions, Equations 13.3 and 13.7 give

$$x_i = A \cos \phi \qquad\qquad \text{(13.12)}$$

$$v_i = -\omega A \sin \phi \qquad\qquad \text{(13.13)}$$

Dividing Equation 13.13 by Equation 13.12 eliminates A, giving $v_i/x_i = -\omega \tan \phi$, or

$$\tan \phi = -\frac{v_i}{\omega x_i} \qquad\qquad \text{(13.14)}$$

Furthermore, if we square Equations 13.12 and 13.13, divide the velocity equation by ω^2, and then add terms, we obtain

$$x_i^2 + \left(\frac{v_i}{\omega}\right)^2 = A^2 \cos^2 \phi + A^2 \sin^2 \phi$$

Using the identity $\sin^2 \phi + \cos^2 \phi = 1$, we can solve for A:

$$A = \sqrt{x_i^2 + \left(\frac{v_i}{\omega}\right)^2} \qquad\qquad \text{(13.15)}$$

The following properties of a particle moving in simple harmonic motion are important:

- The acceleration of the particle is proportional to the displacement but is in the opposite direction. This is the *necessary and sufficient condition for simple harmonic motion,* as opposed to all other kinds of vibration.
- The displacement from the equilibrium position, velocity, and acceleration all vary sinusoidally with time but are not in phase, as shown in Figure 13.4.
- The frequency and the period of the motion are independent of the amplitude. (We show this explicitly in the next section.)

Quick Quiz 13.3

Can we use Equations 2.8, 2.10, 2.11, and 2.12 (see pages 35 and 36) to describe the motion of a simple harmonic oscillator?

EXAMPLE 13.1   An Oscillating Object

An object oscillates with simple harmonic motion along the x axis. Its displacement from the origin varies with time according to the equation

$$x = (4.00 \text{ m}) \cos\left(\pi t + \frac{\pi}{4}\right)$$

where t is in seconds and the angles in the parentheses are in radians. (a) Determine the amplitude, frequency, and period of the motion.

Solution By comparing this equation with Equation 13.3, the general equation for simple harmonic motion — $x = A \cos(\omega t + \phi)$ — we see that $A = 4.00$ m and $\omega = \pi$ rad/s. Therefore, $f = \omega/2\pi = \pi/2\pi = 0.500$ Hz and $T = 1/f = 2.00$ s.

(b) Calculate the velocity and acceleration of the object at any time t.

Solution

$$v = \frac{dx}{dt} = -(4.00 \text{ m/s}) \sin\left(\pi t + \frac{\pi}{4}\right) \frac{d}{dt}(\pi t)$$

$$= -(4.00\pi \text{ m/s}) \sin\left(\pi t + \frac{\pi}{4}\right)$$

$$a = \frac{dv}{dt} = -(4.00\pi \text{ m/s}^2) \cos\left(\pi t + \frac{\pi}{4}\right) \frac{d}{dt}(\pi t)$$

$$= -(4.00\pi^2 \text{ m/s}^2) \cos\left(\pi t + \frac{\pi}{4}\right)$$

(c) Using the results of part (b), determine the position, velocity, and acceleration of the object at $t = 1.00$ s.

Solution Noting that the angles in the trigonometric functions are in radians, we obtain, at $t = 1.00$ s,

$$x = (4.00 \text{ m}) \cos\left(\pi + \frac{\pi}{4}\right) = (4.00 \text{ m}) \cos\left(\frac{5\pi}{4}\right)$$

$$= (4.00 \text{ m})(-0.707) = \boxed{-2.83 \text{ m}}$$

$$v = -(4.00\pi \text{ m/s}) \sin\left(\frac{5\pi}{4}\right) = -(4.00\pi \text{ m/s})(-0.707)$$

$$= \boxed{8.89 \text{ m/s}}$$

$$a = -(4.00\pi^2 \text{ m/s}^2) \cos\left(\frac{5\pi}{4}\right)$$

$$= -(4.00\pi^2 \text{ m/s}^2)(-0.707) = \boxed{27.9 \text{ m/s}^2}$$

(d) Determine the maximum speed and maximum acceleration of the object.

Solution In the general expressions for v and a found in part (b), we use the fact that the maximum values of the sine and cosine functions are unity. Therefore, v varies between $\pm 4.00\pi$ m/s, and a varies between $\pm 4.00\pi^2$ m/s^2. Thus,

$$v_{\text{max}} = 4.00\pi \text{ m/s} = \boxed{12.6 \text{ m/s}}$$

$$a_{\text{max}} = 4.00\pi^2 \text{ m/s} = \boxed{39.5 \text{ m/s}^2}$$

We obtain the same results using $v_{\text{max}} = \omega A$ and $a_{\text{max}} = \omega^2 A$, where $A = 4.00$ m and $\omega = \pi$ rad/s.

(e) Find the displacement of the object between $t = 0$ and $t = 1.00$ s.

Solution   The x coordinate at $t = 0$ is

$$x_i = (4.00 \text{ m}) \cos\left(0 + \frac{\pi}{4}\right) = (4.00 \text{ m})(0.707) = 2.83 \text{ m}$$

In part (c), we found that the x coordinate at $t = 1.00$ s is -2.83 m; therefore, the displacement between $t = 0$ and $t = 1.00$ s is

$$\Delta x = x_f - x_i = -2.83 \text{ m} - 2.83 \text{ m} = \boxed{-5.66 \text{ m}}$$

Because the object's velocity changes sign during the first second, the magnitude of Δx is not the same as the distance traveled in the first second. (By the time the first second is over, the object has been through the point $x = -2.83$ m once, traveled to $x = -4.00$ m, and come back to $x = -2.83$ m.)

Exercise   What is the phase of the motion at $t = 2.00$ s?

Answer $9\pi/4$ rad.

13.2 THE BLOCK–SPRING SYSTEM REVISITED

Let us return to the block–spring system (Fig. 13.5). Again we assume that the surface is frictionless; hence, when the block is displaced from equilibrium, the only force acting on it is the restoring force of the spring. As we saw in Equation 13.2, when the block is displaced a distance x from equilibrium, it experiences an acceleration $a = -(k/m)x$. If the block is displaced a maximum distance $x = A$ at some initial time and then released from rest, its initial acceleration at that instant is $-kA/m$ (its extreme negative value). When the block passes through the equilibrium position $x = 0$, its acceleration is zero. At this instant, its speed is a maximum. The block then continues to travel to the left of equilibrium and finally reaches $x = -A$, at which time its acceleration is kA/m (maximum positive) and its speed is again zero. Thus, we see that the block oscillates between the turning points $x = \pm A$.

Let us now describe the oscillating motion in a quantitative fashion. Recall that $a = dv/dt = d^2x/dt^2$, and so we can express Equation 13.2 as

$$\frac{d^2x}{dt^2} = -\frac{k}{m}x \qquad \textbf{(13.16)}$$

If we denote the ratio k/m with the symbol ω^2, this equation becomes

$$\frac{d^2x}{dt^2} = -\omega^2 x \qquad \textbf{(13.17)}$$

Now we require a solution to Equation 13.17—that is, a function $x(t)$ that satisfies this second-order differential equation. Because Equations 13.17 and 13.9 are equivalent, each solution must be that of simple harmonic motion:

$$x = A\cos(\omega t + \phi)$$

To see this explicitly, assume that $x = A\cos(\omega t + \phi)$. Then

$$\frac{dx}{dt} = A\frac{d}{dt}\cos(\omega t + \phi) = -\omega A\sin(\omega t + \phi)$$

$$\frac{d^2x}{dt^2} = -\omega A\frac{d}{dt}\sin(\omega t + \phi) = -\omega^2 A\cos(\omega t + \phi)$$

Comparing the expressions for x and d^2x/dt^2, we see that $d^2x/dt^2 = -\omega^2 x$, and Equation 13.17 is satisfied. We conclude that **whenever the force acting on a particle is linearly proportional to the displacement from some equilibrium**

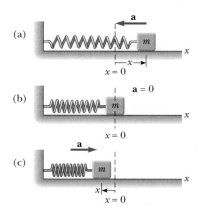

Figure 13.5   A block of mass m attached to a spring on a frictionless surface undergoes simple harmonic motion. (a) When the block is displaced to the right of equilibrium, the displacement is positive and the acceleration is negative. (b) At the equilibrium position, $x = 0$, the acceleration is zero and the speed is a maximum. (c) When the block is displaced to the left of equilibrium, the displacement is negative and the acceleration is positive.

position and in the opposite direction ($F = -kx$), the particle moves in simple harmonic motion.

Recall that the period of any simple harmonic oscillator is $T = 2\pi/\omega$ (Eq. 13.4) and that the frequency is the inverse of the period. We know from Equations 13.16 and 13.17 that $\omega = \sqrt{k/m}$, so we can express the period and frequency of the block–spring system as

Period and frequency for a block–spring system

$$T = \frac{2\pi}{\omega} = 2\pi \sqrt{\frac{m}{k}}$$ **(13.18)**

$$f = \frac{1}{T} = \frac{1}{2\pi}\sqrt{\frac{k}{m}} = \frac{\omega}{2\pi}$$ **(13.19)**

That is, **the frequency and period depend only on the mass of the block and on the force constant of the spring.** Furthermore, the frequency and period are independent of the amplitude of the motion. As we might expect, the frequency is greater for a stiffer spring (the stiffer the spring, the greater the value of k) and decreases with increasing mass.

Special Case 1.   Let us consider a special case to better understand the physical significance of Equation 13.3, the defining expression for simple harmonic motion. We shall use this equation to describe the motion of an oscillating block–spring system. Suppose we pull the block a distance A from equilibrium and then release it from rest at this stretched position, as shown in Figure 13.6. Our solution for x must obey the initial conditions that $x_i = A$ and $v_i = 0$ at $t = 0$. It does if we choose $\phi = 0$, which gives $x = A \cos \omega t$ as the solution. To check this solution, we note that it satisfies the condition that $x_i = A$ at $t = 0$ because $\cos 0 = 1$. Thus, we see that A and ϕ contain the information on initial conditions.

Now let us investigate the behavior of the velocity and acceleration for this special case. Because $x = A \cos \omega t$,

$$v = \frac{dx}{dt} = -\omega A \sin \omega t$$

$$a = \frac{dv}{dt} = -\omega^2 A \cos \omega t$$

From the velocity expression we see that, because $\sin 0 = 0$, $v_i = 0$ at $t = 0$, as we require. The expression for the acceleration tells us that $a = -\omega^2 A$ at $t = 0$. Physically, this negative acceleration makes sense because the force acting on the block is directed to the left when the displacement is positive. In fact, at the extreme po-

QuickLab

Hang an object from a rubber band and start it oscillating. Measure T. Now tie four identical rubber bands together, end to end. How should k for this longer band compare with k for the single band? Again, time the oscillations with the same object. Can you verify Equation 13.19?

Figure 13.6   A block–spring system that starts from rest at $x_i = A$. In this case, $\phi = 0$ and thus $x = A \cos \omega t$.

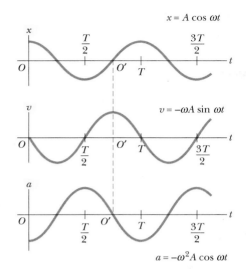

$x = A \cos \omega t$

$v = -\omega A \sin \omega t$

$a = -\omega^2 A \cos \omega t$

Figure 13.7 Displacement, velocity, and acceleration versus time for a block–spring system like the one shown in Figure 13.6, undergoing simple harmonic motion under the initial conditions that at $t = 0$, $x_i = A$ and $v_i = 0$ (Special Case 1). The origins at O' correspond to Special Case 2, the block–spring system under the initial conditions shown in Figure 13.8.

sition shown in Figure 13.6, $F_s = -kA$ (to the left) and the initial acceleration is $-\omega^2 A = -kA/m$.

Another approach to showing that $x = A \cos \omega t$ is the correct solution involves using the relationship $\tan \phi = -v_i/\omega x_i$ (Eq. 13.14). Because $v_i = 0$ at $t = 0$, $\tan \phi = 0$ and thus $\phi = 0$. (The tangent of π also equals zero, but $\phi = \pi$ gives the wrong value for x_i.)

Figure 13.7 is a plot of displacement, velocity, and acceleration versus time for this special case. Note that the acceleration reaches extreme values of $\pm \omega^2 A$ while the displacement has extreme values of $\pm A$ because the force is maximal at those positions. Furthermore, the velocity has extreme values of $\pm \omega A$, which both occur at $x = 0$. Hence, the quantitative solution agrees with our qualitative description of this system.

Special Case 2. Now suppose that the block is given an initial velocity $\mathbf{v}_i$ to the right at the instant it is at the equilibrium position, so that $x_i = 0$ and $v = v_i$ at $t = 0$ (Fig. 13.8). The expression for x must now satisfy these initial conditions. Because the block is moving in the positive x direction at $t = 0$ and because $x_i = 0$ at $t = 0$, the expression for x must have the form $x = A \sin \omega t$.

Applying Equation 13.14 and the initial condition that $x_i = 0$ at $t = 0$, we find that $\tan \phi = -\infty$ and $\phi = -\pi/2$. Hence, Equation 13.3 becomes $x = A \cos (\omega t - \pi/2)$, which can be written $x = A \sin \omega t$. Furthermore, from Equation 13.15 we see that $A = v_i/\omega$; therefore, we can express x as

$$x = \frac{v_i}{\omega} \sin \omega t$$

The velocity and acceleration in this case are

$$v = \frac{dx}{dt} = v_i \cos \omega t$$

$$a = \frac{dv}{dt} = -\omega v_i \sin \omega t$$

These results are consistent with the facts that (1) the block always has a maximum

$x_i = 0$
$t = 0$
$v = v_i$

$x = 0$

$\mathbf{v}_i$

m

$x = A \sin \omega t$

Figure 13.8 The block–spring system starts its motion at the equilibrium position at $t = 0$. If its initial velocity is v_i to the right, the block's x coordinate varies as $x = (v_i/\omega) \sin \omega t$.

speed at $x = 0$ and (2) the force and acceleration are zero at this position. The graphs of these functions versus time in Figure 13.7 correspond to the origin at O'.

Quick Quiz 13.4

What is the solution for x if the block is initially moving to the left in Figure 13.8?

EXAMPLE 13.2   Watch Out for Potholes!

A car with a mass of 1 300 kg is constructed so that its frame is supported by four springs. Each spring has a force constant of 20 000 N/m. If two people riding in the car have a combined mass of 160 kg, find the frequency of vibration of the car after it is driven over a pothole in the road.

Solution We assume that the mass is evenly distributed. Thus, each spring supports one fourth of the load. The total mass is 1 460 kg, and therefore each spring supports 365 kg.

Hence, the frequency of vibration is, from Equation 13.19,

$$f = \frac{1}{2\pi}\sqrt{\frac{k}{m}} = \frac{1}{2\pi}\sqrt{\frac{20\ 000\ \text{N/m}}{365\ \text{kg}}} = \boxed{1.18\ \text{Hz}}$$

Exercise How long does it take the car to execute two complete vibrations?

Answer 1.70 s.

EXAMPLE 13.3   A Block–Spring System

A block with a mass of 200 g is connected to a light spring for which the force constant is 5.00 N/m and is free to oscillate on a horizontal, frictionless surface. The block is displaced 5.00 cm from equilibrium and released from rest, as shown in Figure 13.6. (a) Find the period of its motion.

Solution From Equations 13.16 and 13.17, we know that the angular frequency of any block–spring system is

$$\omega = \sqrt{\frac{k}{m}} = \sqrt{\frac{5.00\ \text{N/m}}{200 \times 10^{-3}\ \text{kg}}} = 5.00\ \text{rad/s}$$

and the period is

$$T = \frac{2\pi}{\omega} = \frac{2\pi}{5.00\ \text{rad/s}} = \boxed{1.26\ \text{s}}$$

(b) Determine the maximum speed of the block.

Solution We use Equation 13.10:

$$v_{max} = \omega A = (5.00\ \text{rad/s})(5.00 \times 10^{-2}\ \text{m}) = \boxed{0.250\ \text{m/s}}$$

(c) What is the maximum acceleration of the block?

Solution We use Equation 13.11:

$$a_{max} = \omega^2 A = (5.00\ \text{rad/s})^2(5.00 \times 10^{-2}\ \text{m}) = \boxed{1.25\ \text{m/s}^2}$$

(d) Express the displacement, speed, and acceleration as functions of time.

Solution This situation corresponds to Special Case 1, where our solution is $x = A \cos \omega t$. Using this expression and the results from (a), (b), and (c), we find that

$$x = A \cos \omega t = \boxed{(0.050\ \text{m}) \cos 5.00t}$$

$$v = \omega A \sin \omega t = \boxed{-(0.250\ \text{m/s}) \sin 5.00t}$$

$$a = \omega^2 A \cos \omega t = \boxed{-(1.25\ \text{m/s}^2) \cos 5.00t}$$

13.3   ENERGY OF THE SIMPLE HARMONIC OSCILLATOR

Let us examine the mechanical energy of the block–spring system illustrated in Figure 13.6. Because the surface is frictionless, we expect the total mechanical energy to be constant, as was shown in Chapter 8. We can use Equation 13.7 to express the kinetic energy as

$$K = \tfrac{1}{2} mv^2 = \tfrac{1}{2} m\omega^2 A^2 \sin^2(\omega t + \phi) \qquad \textbf{(13.20)}$$

Kinetic energy of a simple harmonic oscillator

The elastic potential energy stored in the spring for any elongation x is given by $\tfrac{1}{2} kx^2$ (see Eq. 8.4). Using Equation 13.3, we obtain

$$U = \tfrac{1}{2} kx^2 = \tfrac{1}{2} kA^2 \cos^2(\omega t + \phi) \qquad \textbf{(13.21)}$$

Potential energy of a simple harmonic oscillator

We see that K and U are *always* positive quantities. Because $\omega^2 = k/m$, we can express the total mechanical energy of the simple harmonic oscillator as

$$E = K + U = \tfrac{1}{2} kA^2[\sin^2(\omega t + \phi) + \cos^2(\omega t + \phi)]$$

From the identity $\sin^2 \theta + \cos^2 \theta = 1$, we see that the quantity in square brackets is unity. Therefore, this equation reduces to

$$E = \tfrac{1}{2} kA^2 \qquad \textbf{(13.22)}$$

Total energy of a simple harmonic oscillator

That is, **the total mechanical energy of a simple harmonic oscillator is a constant of the motion and is proportional to the square of the amplitude.** Note that U is small when K is large, and vice versa, because the sum must be constant. In fact, the total mechanical energy is equal to the maximum potential energy stored in the spring when $x = \pm A$ because $v = 0$ at these points and thus there is no kinetic energy. At the equilibrium position, where $U = 0$ because $x = 0$, the total energy, all in the form of kinetic energy, is again $\tfrac{1}{2} kA^2$. That is,

$$E = \tfrac{1}{2} mv^2_{\text{max}} = \tfrac{1}{2} m\omega^2 A^2 = \tfrac{1}{2} m \frac{k}{m} A^2 = \tfrac{1}{2} kA^2 \qquad (\text{at } x = 0)$$

Plots of the kinetic and potential energies versus time appear in Figure 13.9a, where we have taken $\phi = 0$. As already mentioned, both K and U are always positive, and at all times their sum is a constant equal to $\tfrac{1}{2} kA^2$, the total energy of the system. The variations of K and U with the displacement x of the block are plotted

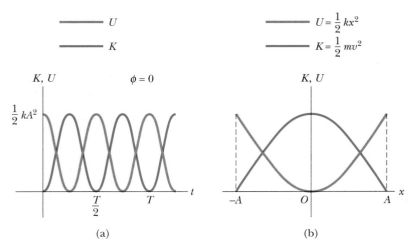

Figure 13.9 (a) Kinetic energy and potential energy versus time for a simple harmonic oscillator with $\phi = 0$. (b) Kinetic energy and potential energy versus displacement for a simple harmonic oscillator. In either plot, note that $K + U =$ constant.

in Figure 13.9b. Energy is continuously being transformed between potential energy stored in the spring and kinetic energy of the block.

Figure 13.10 illustrates the position, velocity, acceleration, kinetic energy, and potential energy of the block–spring system for one full period of the motion. Most of the ideas discussed so far are incorporated in this important figure. Study it carefully.

Finally, we can use the principle of conservation of energy to obtain the velocity for an arbitrary displacement by expressing the total energy at some arbitrary position x as

$$E = K + U = \tfrac{1}{2} mv^2 + \tfrac{1}{2} kx^2 = \tfrac{1}{2} kA^2$$

Velocity as a function of position for a simple harmonic oscillator

$$v = \pm \sqrt{\frac{k}{m} (A^2 - x^2)} = \pm \omega \sqrt{A^2 - x^2} \qquad \textbf{(13.23)}$$

When we check Equation 13.23 to see whether it agrees with known cases, we find that it substantiates the fact that the speed is a maximum at $x = 0$ and is zero at the turning points $x = \pm A$.

t	x	v	a	K	U
0	A	0	$-\omega^2 A$	0	$\tfrac{1}{2} kA^2$
$T/4$	0	$-\omega A$	0	$\tfrac{1}{2} kA^2$	0
$T/2$	$-A$	0	$\omega^2 A$	0	$\tfrac{1}{2} kA^2$
$3T/4$	0	ωA	0	$\tfrac{1}{2} kA^2$	0
T	A	0	$-\omega^2 A$	0	$\tfrac{1}{2} kA^2$

Figure 13.10   Simple harmonic motion for a block–spring system and its relationship to the motion of a simple pendulum. The parameters in the table refer to the block–spring system, assuming that $x = A$ at $t = 0$; thus, $x = A \cos \omega t$ (see Special Case 1).

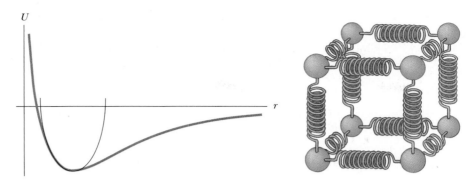

Figure 13.11 (a) If the atoms in a molecule do not move too far from their equilibrium positions, a graph of potential energy versus separation distance between atoms is similar to the graph of potential energy versus position for a simple harmonic oscillator. (b) Tiny springs approximate the forces holding atoms together.

You may wonder why we are spending so much time studying simple harmonic oscillators. We do so because they are good models of a wide variety of physical phenomena. For example, recall the Lennard–Jones potential discussed in Example 8.11. This complicated function describes the forces holding atoms together. Figure 13.11a shows that, for small displacements from the equilibrium position, the potential energy curve for this function approximates a parabola, which represents the potential energy function for a simple harmonic oscillator. Thus, we can approximate the complex atomic binding forces as tiny springs, as depicted in Figure 13.11b.

The ideas presented in this chapter apply not only to block–spring systems and atoms, but also to a wide range of situations that include bungee jumping, tuning in a television station, and viewing the light emitted by a laser. You will see more examples of simple harmonic oscillators as you work through this book.

 EXAMPLE 13.4 **Oscillations on a Horizontal Surface**

A 0.500-kg cube connected to a light spring for which the force constant is 20.0 N/m oscillates on a horizontal, frictionless track. (a) Calculate the total energy of the system and the maximum speed of the cube if the amplitude of the motion is 3.00 cm.

Solution Using Equation 13.22, we obtain

$$E = K + U = \tfrac{1}{2} kA^2 = \tfrac{1}{2} (20.0 \text{ N/m}) (3.00 \times 10^{-2} \text{ m})^2$$

$$= \boxed{9.00 \times 10^{-3} \text{ J}}$$

When the cube is at $x = 0$, we know that $U = 0$ and $E = \tfrac{1}{2} m v_{max}^2$; therefore,

$$\tfrac{1}{2} m v_{max}^2 = 9.00 \times 10^{-3} \text{ J}$$

$$v_{max} = \sqrt{\frac{18.0 \times 10^{-3} \text{ J}}{0.500 \text{ kg}}} = \boxed{0.190 \text{ m/s}}$$

(b) What is the velocity of the cube when the displacement is 2.00 cm?

Solution We can apply Equation 13.23 directly:

$$v = \pm \sqrt{\frac{k}{m} (A^2 - x^2)}$$

$$= \pm \sqrt{\frac{20.0 \text{ N/m}}{0.500 \text{ kg}} [(0.030 \, 0 \text{ m})^2 - (0.020 \, 0 \text{ m})^2]}$$

$$= \boxed{\pm 0.141 \text{ m/s}}$$

The positive and negative signs indicate that the cube could be moving to either the right or the left at this instant.

(c) Compute the kinetic and potential energies of the system when the displacement is 2.00 cm.

Solution   Using the result of (b), we find that

$$K = \tfrac{1}{2} mv^2 = \tfrac{1}{2} (0.500 \text{ kg})(0.141 \text{ m/s})^2 = \boxed{5.00 \times 10^{-3} \text{ J}}$$

$$U = \tfrac{1}{2} kx^2 = \tfrac{1}{2} (20.0 \text{ N/m})(0.020\ 0 \text{ m})^2 = \boxed{4.00 \times 10^{-3} \text{ J}}$$

Note that $K + U = E$.

Exercise   For what values of x is the speed of the cube 0.100 m/s?

Answer ± 2.55 cm.

13.4 THE PENDULUM

8.11
&
8.12

The **simple pendulum** is another mechanical system that exhibits periodic motion. It consists of a particle-like bob of mass m suspended by a light string of length L that is fixed at the upper end, as shown in Figure 13.12. The motion occurs in the vertical plane and is driven by the force of gravity. We shall show that, provided the angle θ is small (less than about 10°), the motion is that of a simple harmonic oscillator.

The forces acting on the bob are the force **T** exerted by the string and the gravitational force $m\mathbf{g}$. The tangential component of the gravitational force, $mg \sin \theta$, always acts toward $\theta = 0$, opposite the displacement. Therefore, the tangential force is a restoring force, and we can apply Newton's second law for motion in the tangential direction:

$$\sum F_t = - mg \sin \theta = m \frac{d^2s}{dt^2}$$

where s is the bob's displacement measured along the arc and the minus sign indicates that the tangential force acts toward the equilibrium (vertical) position. Because $s = L\theta$ (Eq. 10.1a) and L is constant, this equation reduces to

$$\frac{d^2\theta}{dt^2} = - \frac{g}{L} \sin \theta$$

The right side is proportional to $\sin \theta$ rather than to θ; hence, with $\sin \theta$ present, we would not expect simple harmonic motion because this expression is not of the form of Equation 13.17. However, if we assume that θ is small, we can use the approximation $\sin \theta \approx \theta$; thus the equation of motion for the simple pen-

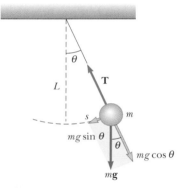

Figure 13.12   When θ is small, a simple pendulum oscillates in simple harmonic motion about the equilibrium position $\theta = 0$. The restoring force is $mg \sin \theta$, the component of the gravitational force tangent to the arc.

The motion of a simple pendulum, captured with multiflash photography. Is the oscillating motion simple harmonic in this case?   *(Richard Megna/Fundamental Photographs)*

dulum becomes

$$\frac{d^2\theta}{dt^2} = -\frac{g}{L}\theta \tag{13.24}$$

Equation of motion for a simple pendulum (small θ)

Now we have an expression of the same form as Equation 13.17, and we conclude that the motion for small amplitudes of oscillation is simple harmonic motion. Therefore, θ can be written as $\theta = \theta_{max}\cos(\omega t + \phi)$, where θ_{max} is the *maximum angular displacement* and the angular frequency ω is

$$\omega = \sqrt{\frac{g}{L}} \tag{13.25}$$

Angular frequency of motion for a simple pendulum

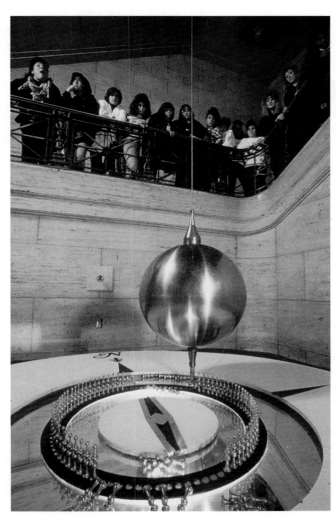

The Foucault pendulum at the Franklin Institute in Philadelphia. This type of pendulum was first used by the French physicist Jean Foucault to verify the Earth's rotation experimentally. As the pendulum swings, the vertical plane in which it oscillates appears to rotate as the bob successively knocks over the indicators arranged in a circle on the floor. In reality, the plane of oscillation is fixed in space, and the Earth rotating beneath the swinging pendulum moves the indicators into position to be knocked down, one after the other. *(© Bob Emott, Photographer)*

The period of the motion is

Period of motion for a simple pendulum

$$T = \frac{2\pi}{\omega} = 2\pi \sqrt{\frac{L}{g}} \qquad (13.26)$$

In other words, **the period and frequency of a simple pendulum depend only on the length of the string and the acceleration due to gravity.** Because the period is independent of the mass, we conclude that all simple pendulums that are of equal length and are at the same location (so that *g* is constant) oscillate with the same period. The analogy between the motion of a simple pendulum and that of a block–spring system is illustrated in Figure 13.10.

 The simple pendulum can be used as a timekeeper because its period depends only on its length and the local value of *g*. It is also a convenient device for making precise measurements of the free-fall acceleration. Such measurements are important because variations in local values of *g* can provide information on the location of oil and of other valuable underground resources.

Quick Quiz 13.5

A block of mass *m* is first allowed to hang from a spring in static equilibrium. It stretches the spring a distance *L* beyond the spring's unstressed length. The block and spring are then set into oscillation. Is the period of this system less than, equal to, or greater than the period of a simple pendulum having a length *L* and a bob mass *m*?

EXAMPLE 13.5   **A Connection Between Length and Time**

Christian Huygens (1629–1695), the greatest clockmaker in history, suggested that an international unit of length could be defined as the length of a simple pendulum having a period of exactly 1 s. How much shorter would our length unit be had his suggestion been followed?

Thus, the meter's length would be slightly less than one-fourth its current length. Note that the number of significant digits depends only on how precisely we know *g* because the time has been defined to be exactly 1 s.

Solution   Solving Equation 13.26 for the length gives

$$L = \frac{T^2 g}{4\pi^2} = \frac{(1\ \text{s})^2 (9.80\ \text{m/s}^2)}{4\pi^2} = \boxed{0.248\ \text{m}}$$

Physical Pendulum

Suppose you balance a wire coat hanger so that the hook is supported by your extended index finger. When you give the hanger a small displacement (with your other hand) and then release it, it oscillates. If a hanging object oscillates about a fixed axis that does not pass through its center of mass and the object cannot be approximated as a point mass, we cannot treat the system as a simple pendulum. In this case the system is called a **physical pendulum.**

Consider a rigid body pivoted at a point *O* that is a distance *d* from the center of mass (Fig. 13.13). The force of gravity provides a torque about an axis through *O*, and the magnitude of that torque is *mgd* sin *θ*, where *θ* is as shown in Figure 13.13. Using the law of motion $\Sigma \tau = I\alpha$, where *I* is the moment of inertia about

QuickLab

Firmly hold a ruler so that about half of it is over the edge of your desk. With your other hand, pull down and then release the free end, watching how it vibrates. Now slide the ruler so that only about a quarter of it is free to vibrate. This time when you release it, how does the vibrational period compare with its earlier value? Why?

the axis through O, we obtain

$$- mgd \sin \theta = I \frac{d^2\theta}{dt^2}$$

The minus sign indicates that the torque about O tends to decrease θ. That is, the force of gravity produces a restoring torque. Because this equation gives us the angular acceleration $d^2\theta/dt^2$ of the pivoted body, we can consider it the equation of motion for the system. If we again assume that θ is small, the approximation $\sin \theta \approx \theta$ is valid, and the equation of motion reduces to

$$\frac{d^2\theta}{dt^2} = -\left(\frac{mgd}{I}\right)\theta = -\omega^2\theta \qquad \textbf{(13.27)}$$

Because this equation is of the same form as Equation 13.17, the motion is simple harmonic motion. That is, the solution of Equation 13.27 is $\theta = \theta_{max} \cos(\omega t + \phi)$, where θ_{max} is the maximum angular displacement and

$$\omega = \sqrt{\frac{mgd}{I}}$$

The period is

$$T = \frac{2\pi}{\omega} = 2\pi\sqrt{\frac{I}{mgd}} \qquad \textbf{(13.28)}$$

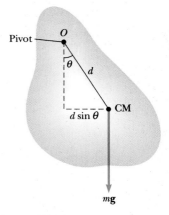

Pivot O

CM

$d \sin \theta$

mg

Figure 13.13 A physical pendulum.

> Period of motion for a physical pendulum

One can use this result to measure the moment of inertia of a flat rigid body. If the location of the center of mass—and hence the value of d—are known, the moment of inertia can be obtained by measuring the period. Finally, note that Equation 13.28 reduces to the period of a simple pendulum (Eq. 13.26) when $I = md^2$—that is, when all the mass is concentrated at the center of mass.

 EXAMPLE 13.6 ▶ **A Swinging Rod**

A uniform rod of mass M and length L is pivoted about one end and oscillates in a vertical plane (Fig. 13.14). Find the period of oscillation if the amplitude of the motion is small.

Solution In Chapter 10 we found that the moment of inertia of a uniform rod about an axis through one end is $\frac{1}{3}ML^2$. The distance d from the pivot to the center of mass is $L/2$. Substituting these quantities into Equation 13.28 gives

$$T = 2\pi\sqrt{\frac{\frac{1}{3}ML^2}{Mg\frac{L}{2}}} = 2\pi\sqrt{\frac{2L}{3g}}$$

Comment In one of the Moon landings, an astronaut walking on the Moon's surface had a belt hanging from his space suit, and the belt oscillated as a physical pendulum. A scientist on the Earth observed this motion on television and used it to estimate the free-fall acceleration on the Moon. How did the scientist make this calculation?

Exercise Calculate the period of a meter stick that is pivoted about one end and is oscillating in a vertical plane.

Answer 1.64 s.

O
Pivot

L

CM

Mg

Figure 13.14 A rigid rod oscillating about a pivot through one end is a physical pendulum with $d = L/2$ and, from Table 10.2, $I = \frac{1}{3}ML^2$.

Figure 13.15 A torsional pendulum consists of a rigid body suspended by a wire attached to a rigid support. The body oscillates about the line *OP* with an amplitude θ_{max}.

Figure 13.16 The balance wheel of this antique pocket watch is a torsional pendulum and regulates the time-keeping mechanism. *(George Semple)*

Torsional Pendulum

Figure 13.15 shows a rigid body suspended by a wire attached at the top to a fixed support. When the body is twisted through some small angle θ, the twisted wire exerts on the body a restoring torque that is proportional to the angular displacement. That is,

$$\tau = -\kappa\theta$$

where κ (kappa) is called the *torsion constant* of the support wire. The value of κ can be obtained by applying a known torque to twist the wire through a measurable angle θ. Applying Newton's second law for rotational motion, we find

$$\tau = -\kappa\theta = I\frac{d^2\theta}{dt^2}$$

$$\frac{d^2\theta}{dt^2} = -\frac{\kappa}{I}\theta \qquad \textbf{(13.29)}$$

Again, this is the equation of motion for a simple harmonic oscillator, with $\omega = \sqrt{\kappa/I}$ and a period

> Period of motion for a torsional pendulum

$$T = 2\pi\sqrt{\frac{I}{\kappa}} \qquad \textbf{(13.30)}$$

This system is called a *torsional pendulum*. There is no small-angle restriction in this situation as long as the elastic limit of the wire is not exceeded. Figure 13.16 shows the balance wheel of a watch oscillating as a torsional pendulum, energized by the mainspring.

13.5 COMPARING SIMPLE HARMONIC MOTION WITH UNIFORM CIRCULAR MOTION

We can better understand and visualize many aspects of simple harmonic motion by studying its relationship to uniform circular motion. Figure 13.17 is an overhead view of an experimental arrangement that shows this relationship. A ball is attached to the rim of a turntable of radius *A*, which is illuminated from the side by a lamp. The ball casts a shadow on a screen. We find that as the turntable rotates with constant angular speed, the shadow of the ball moves back and forth in simple harmonic motion.

Consider a particle located at point P on the circumference of a circle of radius A, as shown in Figure 13.18a, with the line OP making an angle ϕ with the x axis at $t = 0$. We call this circle a *reference circle* for comparing simple harmonic motion and uniform circular motion, and we take the position of P at $t = 0$ as our reference position. If the particle moves along the circle with constant angular speed ω until OP makes an angle θ with the x axis, as illustrated in Figure 13.18b, then at some time $t > 0$, the angle between OP and the x axis is $\theta = \omega t + \phi$. As the particle moves along the circle, the projection of P on the x axis, labeled point Q, moves back and forth along the x axis, between the limits $x = \pm A$.

Note that points P and Q always have the same x coordinate. From the right triangle OPQ, we see that this x coordinate is

$$x = A \cos(\omega t + \phi) \qquad \textbf{(13.31)}$$

This expression shows that the point Q moves with simple harmonic motion along the x axis. Therefore, we conclude that

> simple harmonic motion along a straight line can be represented by the projection of uniform circular motion along a diameter of a reference circle.

We can make a similar argument by noting from Figure 13.18b that the projection of P along the y axis also exhibits simple harmonic motion. Therefore, **uniform circular motion can be considered a combination of two simple harmonic motions,** one along the x axis and one along the y axis, with the two differing in phase by $90°$.

This geometric interpretation shows that the time for one complete revolution of the point P on the reference circle is equal to the period of motion T for simple harmonic motion between $x = \pm A$. That is, the angular speed ω of P is the same as the angular frequency ω of simple harmonic motion along the x axis (this is why we use the same symbol). The phase constant ϕ for simple harmonic motion corresponds to the initial angle that OP makes with the x axis. The radius A of the reference circle equals the amplitude of the simple harmonic motion.

Figure 13.17 An experimental setup for demonstrating the connection between simple harmonic motion and uniform circular motion. As the ball rotates on the turntable with constant angular speed, its shadow on the screen moves back and forth in simple harmonic motion.

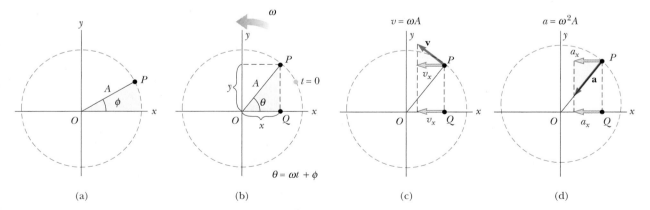

Figure 13.18 Relationship between the uniform circular motion of a point P and the simple harmonic motion of a point Q. A particle at P moves in a circle of radius A with constant angular speed ω. (a) A reference circle showing the position of P at $t = 0$. (b) The x coordinates of points P and Q are equal and vary in time as $x = A \cos(\omega t + \phi)$. (c) The x component of the velocity of P equals the velocity of Q. (d) The x component of the acceleration of P equals the acceleration of Q.

Because the relationship between linear and angular speed for circular motion is $v = r\omega$ (see Eq. 10.10), the particle moving on the reference circle of radius A has a velocity of magnitude ωA. From the geometry in Figure 13.18c, we see that the x component of this velocity is $-\omega A \sin(\omega t + \phi)$. By definition, the point Q has a velocity given by dx/dt. Differentiating Equation 13.31 with respect to time, we find that the velocity of Q is the same as the x component of the velocity of P.

The acceleration of P on the reference circle is directed radially inward toward O and has a magnitude $v^2/A = \omega^2 A$. From the geometry in Figure 13.18d, we see that the x component of this acceleration is $-\omega^2 A \cos(\omega t + \phi)$. This value is also the acceleration of the projected point Q along the x axis, as you can verify by taking the second derivative of Equation 13.31.

EXAMPLE 13.7 ▸ Circular Motion with Constant Angular Speed

A particle rotates counterclockwise in a circle of radius 3.00 m with a constant angular speed of 8.00 rad/s. At $t = 0$, the particle has an x coordinate of 2.00 m and is moving to the right. (a) Determine the x coordinate as a function of time.

Solution Because the amplitude of the particle's motion equals the radius of the circle and $\omega = 8.00$ rad/s, we have

$$x = A\cos(\omega t + \phi) = (3.00 \text{ m})\cos(8.00t + \phi)$$

We can evaluate ϕ by using the initial condition that $x = 2.00$ m at $t = 0$:

$$2.00 \text{ m} = (3.00 \text{ m})\cos(0 + \phi)$$

$$\phi = \cos^{-1}\left(\frac{2.00 \text{ m}}{3.00 \text{ m}}\right)$$

If we were to take our answer as $\phi = 48.2°$, then the coordinate $x = (3.00 \text{ m})\cos(8.00t + 48.2°)$ would be decreasing at time $t = 0$ (that is, moving to the left). Because our particle is first moving to the right, we must choose $\phi = -48.2° = -0.841$ rad. The x coordinate as a function of time is then

$$x = (3.00 \text{ m})\cos(8.00t - 0.841)$$

Note that ϕ in the cosine function must be in radians.

(b) Find the x components of the particle's velocity and acceleration at any time t.

Solution

$$v_x = \frac{dx}{dt} = (-3.00 \text{ m})(8.00 \text{ rad/s})\sin(8.00t - 0.841)$$

$$= -(24.0 \text{ m/s})\sin(8.00t - 0.841)$$

$$a_x = \frac{dv_x}{dt} = (-24.0 \text{ m/s})(8.00 \text{ rad/s})\cos(8.00t - 0.841)$$

$$= -(192 \text{ m/s}^2)\cos(8.00t - 0.841)$$

From these results, we conclude that $v_{max} = 24.0$ m/s and that $a_{max} = 192$ m/s^2. Note that these values also equal the tangential speed ωA and the centripetal acceleration $\omega^2 A$.

Optional Section

13.6 ▸ DAMPED OSCILLATIONS

The oscillatory motions we have considered so far have been for ideal systems—that is, systems that oscillate indefinitely under the action of a linear restoring force. In many real systems, dissipative forces, such as friction, retard the motion. Consequently, the mechanical energy of the system diminishes in time, and the motion is said to be *damped*.

One common type of retarding force is the one discussed in Section 6.4, where the force is proportional to the speed of the moving object and acts in the direction opposite the motion. This retarding force is often observed when an object moves through air, for instance. Because the retarding force can be expressed as $\mathbf{R} = -b\mathbf{v}$ (where b is a constant called the *damping coefficient*) and the restoring force of the system is $-kx$, we can write Newton's second law as

$$\sum F_x = -kx - bv = ma_x$$

$$-kx - b\frac{dx}{dt} = m\frac{d^2x}{dt^2} \qquad \textbf{(13.32)}$$

The solution of this equation requires mathematics that may not be familiar to you yet; we simply state it here without proof. When the retarding force is small compared with the maximum restoring force—that is, when b is small—the solution to Equation 13.32 is

$$x = Ae^{-\frac{b}{2m}t}\cos(\omega t + \phi) \qquad \textbf{(13.33)}$$

where the angular frequency of oscillation is

$$\omega = \sqrt{\frac{k}{m} - \left(\frac{b}{2m}\right)^2} \qquad \textbf{(13.34)}$$

This result can be verified by substituting Equation 13.33 into Equation 13.32.

Figure 13.19a shows the displacement as a function of time for an object oscillating in the presence of a retarding force, and Figure 13.19b depicts one such system: a block attached to a spring and submersed in a viscous liquid. We see that **when the retarding force is much smaller than the restoring force, the oscillatory character of the motion is preserved but the amplitude decreases in time, with the result that the motion ultimately ceases.** Any system that behaves in this way is known as a **damped oscillator.** The dashed blue lines in Figure 13.19a, which define the *envelope* of the oscillatory curve, represent the exponential factor in Equation 13.33. This envelope shows that **the amplitude decays exponentially with time.** For motion with a given spring constant and block mass, the oscillations dampen more rapidly as the maximum value of the retarding force approaches the maximum value of the restoring force.

It is convenient to express the angular frequency of a damped oscillator in the form

$$\omega = \sqrt{\omega_0^2 - \left(\frac{b}{2m}\right)^2}$$

where $\omega_0 = \sqrt{k/m}$ represents the angular frequency in the absence of a retarding force (the undamped oscillator) and is called the **natural frequency** of the system. When the magnitude of the maximum retarding force $R_{max} = bv_{max} < kA$, the system is said to be **underdamped.** As the value of R approaches kA, the amplitudes of the oscillations decrease more and more rapidly. This motion is represented by the blue curve in Figure 13.20. When b reaches a critical value b_c such that $b_c/2m = \omega_0$, the system does not oscillate and is said to be **critically damped.** In this case the system, once released from rest at some nonequilibrium position, returns to equilibrium and then stays there. The graph of displacement versus time for this case is the red curve in Figure 13.20.

If the medium is so viscous that the retarding force is greater than the restoring force—that is, if $R_{max} = bv_{max} > kA$ and $b/2m > \omega_0$—the system is **overdamped.** Again, the displaced system, when free to move, does not oscillate but simply returns to its equilibrium position. As the damping increases, the time it takes the system to approach equilibrium also increases, as indicated by the black curve in Figure 13.20.

In any case in which friction is present, whether the system is overdamped or underdamped, the energy of the oscillator eventually falls to zero. The lost mechanical energy dissipates into internal energy in the retarding medium.

(a)

(b)

Figure 13.19 (a) Graph of displacement versus time for a damped oscillator. Note the decrease in amplitude with time. (b) One example of a damped oscillator is a mass attached to a spring and submersed in a viscous liquid.

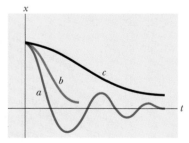

Figure 13.20 Graphs of displacement versus time for (a) an underdamped oscillator, (b) a critically damped oscillator, and (c) an overdamped oscillator.

Figure 13.21   (a) A shock absorber consists of a piston oscillating in a chamber filled with oil. As the piston oscillates, the oil is squeezed through holes between the piston and the chamber, causing a damping of the piston's oscillations. (b) One type of automotive suspension system, in which a shock absorber is placed inside a coil spring at each wheel.

web

To learn more about shock absorbers, visit
http://www.monroe.com

Quick Quiz 13.6

An automotive suspension system consists of a combination of springs and shock absorbers, as shown in Figure 13.21. If you were an automotive engineer, would you design a suspension system that was underdamped, critically damped, or overdamped? Discuss each case.

Optional Section

13.7 ▸ FORCED OSCILLATIONS

It is possible to compensate for energy loss in a damped system by applying an external force that does positive work on the system. At any instant, energy can be put into the system by an applied force that acts in the direction of motion of the oscillator. For example, a child on a swing can be kept in motion by appropriately timed pushes. The amplitude of motion remains constant if the energy input per cycle exactly equals the energy lost as a result of damping. Any motion of this type is called **forced oscillation.**

A common example of a forced oscillator is a damped oscillator driven by an external force that varies periodically, such as $F = F_{ext} \cos \omega t$, where ω is the angular frequency of the periodic force and F_{ext} is a constant. Adding this driving force to the left side of Equation 13.32 gives

$$F_{ext} \cos \omega t - kx - b\frac{dx}{dt} = m\frac{d^2x}{dt^2} \qquad \textbf{(13.35)}$$

(As earlier, we present the solution of this equation without proof.) After a sufficiently long period of time, when the energy input per cycle equals the energy lost per cycle, a steady-state condition is reached in which the oscillations proceed with constant amplitude. At this time, when the system is in a steady state, the solution of Equation 13.35 is

$$x = A \cos(\omega t + \phi) \qquad \textbf{(13.36)}$$

where

$$A = \frac{F_{\text{ext}}/m}{\sqrt{(\omega^2 - \omega_0{}^2)^2 + \left(\dfrac{b\omega}{m}\right)^2}}$$ **(13.37)**

and where $\omega_0 = \sqrt{k/m}$ is the angular frequency of the undamped oscillator ($b = 0$). One could argue that in steady state the oscillator must physically have the same frequency as the driving force, and thus the solution given by Equation 13.36 is expected. In fact, when this solution is substituted into Equation 13.35, one finds that it is indeed a solution, provided the amplitude is given by Equation 13.37.

Equation 13.37 shows that, because an external force is driving it, the motion of the forced oscillator is not damped. The external agent provides the necessary energy to overcome the losses due to the retarding force. Note that the system oscillates at the angular frequency ω of the driving force. For small damping, the amplitude becomes very large when the frequency of the driving force is near the natural frequency of oscillation. The dramatic increase in amplitude near the natural frequency ω_0 is called **resonance,** and for this reason ω_0 is sometimes called the **resonance frequency** of the system.

The reason for large-amplitude oscillations at the resonance frequency is that energy is being transferred to the system under the most favorable conditions. We can better understand this by taking the first time derivative of x in Equation 13.36, which gives an expression for the velocity of the oscillator. We find that v is proportional to $\sin(\omega t + \phi)$. When the applied force **F** is in phase with the velocity, the rate at which work is done on the oscillator by **F** equals the dot product **F · v**. Remember that "rate at which work is done" is the definition of power. Because the product **F · v** is a maximum when **F** and **v** are in phase, we conclude that **at resonance the applied force is in phase with the velocity and that the power transferred to the oscillator is a maximum.**

Figure 13.22 is a graph of amplitude as a function of frequency for a forced oscillator with and without damping. Note that the amplitude increases with decreasing damping ($b \rightarrow 0$) and that the resonance curve broadens as the damping increases. Under steady-state conditions and at any driving frequency, the energy transferred into the system equals the energy lost because of the damping force; hence, the average total energy of the oscillator remains constant. In the absence of a damping force ($b = 0$), we see from Equation 13.37 that the steady-state amplitude approaches infinity as $\omega \rightarrow \omega_0$. In other words, if there are no losses in the system and if we continue to drive an initially motionless oscillator with a periodic force that is in phase with the velocity, the amplitude of motion builds without limit (see the red curve in Fig. 13.22). This limitless building does not occur in practice because some damping is always present.

The behavior of a driven oscillating system after the driving force is removed depends on b and on how close ω was to ω_0. This behavior is sometimes quantified by a parameter called the *quality factor Q*. The closer a system is to being undamped, the greater its Q. The amplitude of oscillation drops by a factor of $e\ (= 2.718 \ldots)$ in Q/π cycles.

Later in this book we shall see that resonance appears in other areas of physics. For example, certain electrical circuits have natural frequencies. A bridge has natural frequencies that can be set into resonance by an appropriate driving force. A dramatic example of such resonance occurred in 1940, when the Tacoma Narrows Bridge in the state of Washington was destroyed by resonant vibrations. Although the winds were not particularly strong on that occasion, the bridge ultimately collapsed (Fig. 13.23) because the bridge design had no built-in safety features.

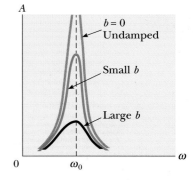

Figure 13.22 Graph of amplitude versus frequency for a damped oscillator when a periodic driving force is present. When the frequency of the driving force equals the natural frequency ω_0, resonance occurs. Note that the shape of the resonance curve depends on the size of the damping coefficient b.

QuickLab

Tie several objects to strings and suspend them from a horizontal string, as illustrated in the figure. Make two of the hanging strings approximately the same length. If one of this pair, such as P, is set into sideways motion, all the others begin to oscillate. But Q, whose length is the same as that of P, oscillates with the greatest amplitude. Must all the masses have the same value?

(a)

(b)

Figure 13.23   (a) In 1940 turbulent winds set up torsional vibrations in the Tacoma Narrows Bridge, causing it to oscillate at a frequency near one of the natural frequencies of the bridge structure. (b) Once established, this resonance condition led to the bridge's collapse. *(UPI/Bettmann Newsphotos)*

Many other examples of resonant vibrations can be cited. A resonant vibration that you may have experienced is the "singing" of telephone wires in the wind. Machines often break if one vibrating part is at resonance with some other moving part. Soldiers marching in cadence across a bridge have been known to set up resonant vibrations in the structure and thereby cause it to collapse. Whenever any real physical system is driven near its resonance frequency, you can expect oscillations of very large amplitudes.

SUMMARY

When the acceleration of an object is proportional to its displacement from some equilibrium position and is in the direction opposite the displacement, the object moves with simple harmonic motion. The position x of a simple harmonic oscillator varies periodically in time according to the expression

$$x = A \cos(\omega t + \phi) \tag{13.3}$$

where A is the **amplitude** of the motion, ω is the **angular frequency,** and ϕ is the **phase constant.** The value of ϕ depends on the initial position and initial velocity of the oscillator. You should be able to use this formula to describe the motion of an object undergoing simple harmonic motion.

The time T needed for one complete oscillation is defined as the **period** of the motion:

$$T = \frac{2\pi}{\omega} \tag{13.4}$$

The inverse of the period is the **frequency** of the motion, which equals the number of oscillations per second.

The velocity and acceleration of a simple harmonic oscillator are

$$v = \frac{dx}{dt} = -\omega A \sin(\omega t + \phi) \tag{13.7}$$

$$a = \frac{dv}{dt} = -\omega^2 A \cos(\omega t + \phi) \tag{13.8}$$

$$v = \pm \omega \sqrt{A^2 - x^2} \tag{13.23}$$

$$\omega = \sqrt{\frac{k}{m}} = \sqrt{\frac{g}{L}}$$

Thus, the maximum speed is ωA, and the maximum acceleration is $\omega^2 A$. The speed is zero when the oscillator is at its turning points, $x = \pm A$, and is a maximum when the oscillator is at the equilibrium position $x = 0$. The magnitude of the acceleration is a maximum at the turning points and zero at the equilibrium position. You should be able to find the velocity and acceleration of an oscillating object at any time if you know the amplitude, angular frequency, and phase constant.

A block–spring system moves in simple harmonic motion on a frictionless surface, with a period

$$T = \frac{2\pi}{\omega} = 2\pi\sqrt{\frac{m}{k}} \tag{13.18}$$

The kinetic energy and potential energy for a simple harmonic oscillator vary with time and are given by

$$K = \tfrac{1}{2}mv^2 = \tfrac{1}{2}m\omega^2 A^2 \sin^2(\omega t + \phi) \tag{13.20}$$

$$U = \tfrac{1}{2}kx^2 = \tfrac{1}{2}kA^2 \cos^2(\omega t + \phi) \tag{13.21}$$

These three formulas allow you to analyze a wide variety of situations involving oscillations. Be sure you recognize how the mass of the block and the spring constant of the spring enter into the calculations.

The total energy of a simple harmonic oscillator is a constant of the motion and is given by

$$E = \tfrac{1}{2}kA^2 \tag{13.22}$$

The potential energy of the oscillator is a maximum when the oscillator is at its turning points and is zero when the oscillator is at the equilibrium position. The kinetic energy is zero at the turning points and a maximum at the equilibrium position. You should be able to determine the division of energy between potential and kinetic forms at any time t.

A **simple pendulum** of length L moves in simple harmonic motion. For small angular displacements from the vertical, its period is

$$T = 2\pi\sqrt{\frac{L}{g}} \tag{13.26}$$

For small angular displacements from the vertical, a **physical pendulum** moves in simple harmonic motion about a pivot that does not go through the center of mass. The period of this motion is

$$T = 2\pi\sqrt{\frac{I}{mgd}} \tag{13.28}$$

where I is the moment of inertia about an axis through the pivot and d is the distance from the pivot to the center of mass. You should be able to distinguish when to use the simple-pendulum formula and when the system must be considered a physical pendulum.

Uniform circular motion can be considered a combination of two simple harmonic motions, one along the x axis and the other along the y axis, with the two differing in phase by $90°$.

QUESTIONS

1. Is a bouncing ball an example of simple harmonic motion? Is the daily movement of a student from home to school and back simple harmonic motion? Why or why not?

2. If the coordinate of a particle varies as $x = -A\cos \omega t$, what is the phase constant in Equation 13.3? At what position does the particle begin its motion?

3. Does the displacement of an oscillating particle between $t = 0$ and a later time t necessarily equal the position of the particle at time t? Explain.

4. Determine whether the following quantities can be in the same direction for a simple harmonic oscillator: (a) displacement and velocity, (b) velocity and acceleration, (c) displacement and acceleration.

5. Can the amplitude A and the phase constant ϕ be determined for an oscillator if only the position is specified at $t = 0$? Explain.

6. Describe qualitatively the motion of a mass–spring system when the mass of the spring is not neglected.

7. Make a graph showing the potential energy of a stationary block hanging from a spring, $U = \frac{1}{2}ky^2 + mgy$. Why is the lowest part of the graph offset from the origin?

8. A block–spring system undergoes simple harmonic motion with an amplitude A. Does the total energy change if the mass is doubled but the amplitude is not changed? Do the kinetic and potential energies depend on the mass? Explain.

9. What happens to the period of a simple pendulum if the pendulum's length is doubled? What happens to the period if the mass of the suspended bob is doubled?

10. A simple pendulum is suspended from the ceiling of a stationary elevator, and the period is determined. Describe the changes, if any, in the period when the elevator

(a) accelerates upward, (b) accelerates downward, and (c) moves with constant velocity.

11. A simple pendulum undergoes simple harmonic motion when θ is small. Is the motion periodic when θ is large? How does the period of motion change as θ increases?

12. Will damped oscillations occur for any values of b and k? Explain.

13. As it possible to have damped oscillations when a system is at resonance? Explain.

14. At resonance, what does the phase constant ϕ equal in Equation 13.36? (*Hint:* Compare this equation with the expression for the driving force, which must be in phase with the velocity at resonance.)

15. Some parachutes have holes in them to allow air to move smoothly through them. Without such holes, sometimes the air that has gathered beneath the chute as a parachutist falls is released from under its edges alternately and periodically, at one side and then at the other. Why might this periodic release of air cause a problem?

16. If a grandfather clock were running slowly, how could we adjust the length of the pendulum to correct the time?

17. A pendulum bob is made from a sphere filled with water. What would happen to the frequency of vibration of this pendulum if the sphere had a hole in it that allowed the water to leak out slowly?

PROBLEMS

1, 2, 3 = straightforward, intermediate, challenging   ☐ = full solution available in the *Student Solutions Manual and Study Guide*
WEB = solution posted at **http://www.saunderscollege.com/physics/**   🖥 = Computer useful in solving problem   📲 = Interactive Physics
☐ = paired numerical/symbolic problems

Section 13.1   Simple Harmonic Motion

1. The displacement of a particle at $t = 0.250$ s is given by the expression $x = (4.00 \text{ m}) \cos(3.00\pi t + \pi)$, where x is in meters and t is in seconds. Determine (a) the frequency and period of the motion, (b) the amplitude of the motion, (c) the phase constant, and (d) the displacement of the particle at $t = 0.250$ s.

2. A ball dropped from a height of 4.00 m makes a perfectly elastic collision with the ground. Assuming that no energy is lost due to air resistance, (a) show that the motion is periodic and (b) determine the period of the motion. (c) Is the motion simple harmonic? Explain.

3. A particle moves in simple harmonic motion with a frequency of 3.00 oscillations/s and an amplitude of 5.00 cm. (a) Through what total distance does the particle move during one cycle of its motion? (b) What is its maximum speed? Where does this occur? (c) Find the maximum acceleration of the particle. Where in the motion does the maximum acceleration occur?

4. In an engine, a piston oscillates with simple harmonic motion so that its displacement varies according to the expression

$$x = (5.00 \text{ cm}) \cos(2t + \pi/6)$$

where x is in centimeters and t is in seconds. At $t = 0$,

find (a) the displacement of the particle, (b) its velocity, and (c) its acceleration. (d) Find the period and amplitude of the motion.

WEB 5. A particle moving along the x axis in simple harmonic motion starts from its equilibrium position, the origin, at $t = 0$ and moves to the right. The amplitude of its motion is 2.00 cm, and the frequency is 1.50 Hz. (a) Show that the displacement of the particle is given by $x = (2.00 \text{ cm}) \sin(3.00\pi t)$. Determine (b) the maximum speed and the earliest time $(t > 0)$ at which the particle has this speed, (c) the maximum acceleration and the earliest time $(t > 0)$ at which the particle has this acceleration, and (d) the total distance traveled between $t = 0$ and $t = 1.00$ s.

6. The initial position and initial velocity of an object moving in simple harmonic motion are x_i and v_i; the angular frequency of oscillation is ω. (a) Show that the position and velocity of the object for all time can be written as

$$x(t) = x_i \cos \omega t + \left(\frac{v_i}{\omega}\right) \sin \omega t$$

$$v(t) = -x_i\omega \sin \omega t + v_i \cos \omega t$$

(b) If the amplitude of the motion is A, show that

$$v^2 - ax = v_i{}^2 - a_i x_i = \omega^2 A^2$$

Section 13.2 The Block–Spring System Revisited

Note: Neglect the mass of the spring in all problems in this section.

7. A spring stretches by 3.90 cm when a 10.0-g mass is hung from it. If a 25.0-g mass attached to this spring oscillates in simple harmonic motion, calculate the period of the motion.

8. A simple harmonic oscillator takes 12.0 s to undergo five complete vibrations. Find (a) the period of its motion, (b) the frequency in hertz, and (c) the angular frequency in radians per second.

9. A 0.500-kg mass attached to a spring with a force constant of 8.00 N/m vibrates in simple harmonic motion with an amplitude of 10.0 cm. Calculate (a) the maximum value of its speed and acceleration, (b) the speed and acceleration when the mass is 6.00 cm from the equilibrium position, and (c) the time it takes the mass to move from $x = 0$ to $x = 8.00$ cm.

10. A 1.00-kg mass attached to a spring with a force constant of 25.0 N/m oscillates on a horizontal, frictionless track. At $t = 0$, the mass is released from rest at $x = -3.00$ cm. (That is, the spring is compressed by 3.00 cm.) Find (a) the period of its motion; (b) the maximum values of its speed and acceleration; and (c) the displacement, velocity, and acceleration as functions of time.

11. A 7.00-kg mass is hung from the bottom end of a vertical spring fastened to an overhead beam. The mass is set into vertical oscillations with a period of 2.60 s. Find the force constant of the spring.

12. A block of unknown mass is attached to a spring with a spring constant of 6.50 N/m and undergoes simple harmonic motion with an amplitude of 10.0 cm. When the mass is halfway between its equilibrium position and the end point, its speed is measured to be $+30.0$ cm/s. Calculate (a) the mass of the block, (b) the period of the motion, and (c) the maximum acceleration of the block.

13. A particle that hangs from a spring oscillates with an angular frequency of 2.00 rad/s. The spring–particle system is suspended from the ceiling of an elevator car and hangs motionless (relative to the elevator car) as the car descends at a constant speed of 1.50 m/s. The car then stops suddenly. (a) With what amplitude does the particle oscillate? (b) What is the equation of motion for the particle? (Choose upward as the positive direction.)

14. A particle that hangs from a spring oscillates with an angular frequency ω. The spring–particle system is suspended from the ceiling of an elevator car and hangs motionless (relative to the elevator car) as the car descends at a constant speed v. The car then stops suddenly. (a) With what amplitude does the particle oscillate? (b) What is the equation of motion for the particle? (Choose upward as the positive direction.)

15. A 1.00-kg mass is attached to a horizontal spring. The spring is initially stretched by 0.100 m, and the mass is released from rest there. It proceeds to move without friction. After 0.500 s, the speed of the mass is zero. What is the maximum speed of the mass?

Section 13.3 Energy of the Simple Harmonic Oscillator

Note: Neglect the mass of the spring in all problems in this section.

16. A 200-g mass is attached to a spring and undergoes simple harmonic motion with a period of 0.250 s. If the total energy of the system is 2.00 J, find (a) the force constant of the spring and (b) the amplitude of the motion.

WEB **17.** An automobile having a mass of 1 000 kg is driven into a brick wall in a safety test. The bumper behaves as a spring of constant 5.00×10^6 N/m and compresses 3.16 cm as the car is brought to rest. What was the speed of the car before impact, assuming that no energy is lost during impact with the wall?

18. A mass–spring system oscillates with an amplitude of 3.50 cm. If the spring constant is 250 N/m and the mass is 0.500 kg, determine (a) the mechanical energy of the system, (b) the maximum speed of the mass, and (c) the maximum acceleration.

19. A 50.0-g mass connected to a spring with a force constant of 35.0 N/m oscillates on a horizontal, frictionless surface with an amplitude of 4.00 cm. Find (a) the total energy of the system and (b) the speed of the mass when the displacement is 1.00 cm. Find (c) the kinetic energy and (d) the potential energy when the displacement is 3.00 cm.

20. A 2.00-kg mass is attached to a spring and placed on a horizontal, smooth surface. A horizontal force of 20.0 N is required to hold the mass at rest when it is pulled 0.200 m from its equilibrium position (the origin of the x axis). The mass is now released from rest with an initial displacement of $x_i = 0.200$ m, and it subsequently undergoes simple harmonic oscillations. Find (a) the force constant of the spring, (b) the frequency of the oscillations, and (c) the maximum speed of the mass. Where does this maximum speed occur? (d) Find the maximum acceleration of the mass. Where does it occur? (e) Find the total energy of the oscillating system. Find (f) the speed and (g) the acceleration when the displacement equals one third of the maximum value.

21. A 1.50-kg block at rest on a tabletop is attached to a horizontal spring having force constant of 19.6 N/m. The spring is initially unstretched. A constant 20.0-N horizontal force is applied to the object, causing the spring to stretch. (a) Determine the speed of the block after it has moved 0.300 m from equilibrium, assuming that the surface between the block and the tabletop is frictionless. (b) Answer part (a) for a coefficient of kinetic friction of 0.200 between the block and the tabletop.

22. The amplitude of a system moving in simple harmonic motion is doubled. Determine the change in (a) the total energy, (b) the maximum speed, (c) the maximum acceleration, and (d) the period.

23. A particle executes simple harmonic motion with an amplitude of 3.00 cm. At what displacement from the midpoint of its motion does its speed equal one half of its maximum speed?

24. A mass on a spring with a constant of 3.24 N/m vibrates, with its position given by the equation $x = (5.00 \text{ cm}) \cos(3.60t \text{ rad/s})$. (a) During the first cycle, for $0 < t < 1.75$ s, when is the potential energy of the system changing most rapidly into kinetic energy? (b) What is the maximum rate of energy transformation?

Section 13.4 The Pendulum

25. A man enters a tall tower, needing to know its height. He notes that a long pendulum extends from the ceiling almost to the floor and that its period is 12.0 s. (a) How tall is the tower? (b) If this pendulum is taken to the Moon, where the free-fall acceleration is 1.67 m/s^2, what is its period there?

26. A "seconds" pendulum is one that moves through its equilibrium position once each second. (The period of the pendulum is 2.000 s.) The length of a seconds pendulum is 0.992 7 m at Tokyo and 0.994 2 m at Cambridge, England. What is the ratio of the free-fall accelerations at these two locations?

27. A rigid steel frame above a street intersection supports standard traffic lights, each of which is hinged to hang immediately below the frame. A gust of wind sets a light swinging in a vertical plane. Find the order of magnitude of its period. State the quantities you take as data and their values.

28. The angular displacement of a pendulum is represented by the equation $\theta = (0.320 \text{ rad}) \cos \omega t$, where θ is in radians and $\omega = 4.43 \text{ rad/s}$. Determine the period and length of the pendulum.

WEB **29.** A simple pendulum has a mass of 0.250 kg and a length of 1.00 m. It is displaced through an angle of 15.0° and then released. What are (a) the maximum speed, (b) the maximum angular acceleration, and (c) the maximum restoring force?

30. A simple pendulum is 5.00 m long. (a) What is the period of simple harmonic motion for this pendulum if it is hanging in an elevator that is accelerating upward at 5.00 m/s^2? (b) What is its period if the elevator is accelerating downward at 5.00 m/s^2? (c) What is the period of simple harmonic motion for this pendulum if it is placed in a truck that is accelerating horizontally at 5.00 m/s^2?

31. A particle of mass m slides without friction inside a hemispherical bowl of radius R. Show that, if it starts from rest with a small displacement from equilibrium, the particle moves in simple harmonic motion with an angular frequency equal to that of a simple pendulum of length R. That is, $\omega = \sqrt{g/R}$.

32. A mass is attached to the end of a string to form a simple pendulum. The period of its harmonic motion is

measured for small angular displacements and three lengths; in each case, the motion is clocked with a stopwatch for 50 oscillations. For lengths of 1.000 m, 0.750 m, and 0.500 m, total times of 99.8 s, 86.6 s, and 71.1 s, respectively, are measured for the 50 oscillations. (a) Determine the period of motion for each length. (b) Determine the mean value of g obtained from these three independent measurements, and compare it with the accepted value. (c) Plot T^2 versus L, and obtain a value for g from the slope of your best-fit straight-line graph. Compare this value with that obtained in part (b).

33. A physical pendulum in the form of a planar body moves in simple harmonic motion with a frequency of 0.450 Hz. If the pendulum has a mass of 2.20 kg and the pivot is located 0.350 m from the center of mass, determine the moment of inertia of the pendulum.

34. A very light, rigid rod with a length of 0.500 m extends straight out from one end of a meter stick. The stick is suspended from a pivot at the far end of the rod and is set into oscillation. (a) Determine the period of oscillation. (b) By what percentage does this differ from a 1.00-m-long simple pendulum?

35. Consider the physical pendulum of Figure 13.13. (a) If I_{CM} is its moment of inertia about an axis that passes through its center of mass and is parallel to the axis that passes through its pivot point, show that its period is

$$T = 2\pi \sqrt{\frac{I_{CM} + md^2}{mgd}}$$

where d is the distance between the pivot point and the center of mass. (b) Show that the period has a minimum value when d satisfies $md^2 = I_{CM}$.

36. A torsional pendulum is formed by attaching a wire to the center of a meter stick with a mass of 2.00 kg. If the resulting period is 3.00 min, what is the torsion constant for the wire?

37. A clock balance wheel has a period of oscillation of 0.250 s. The wheel is constructed so that 20.0 g of mass is concentrated around a rim of radius 0.500 cm. What are (a) the wheel's moment of inertia and (b) the torsion constant of the attached spring?

Section 13.5 Comparing Simple Harmonic Motion with Uniform Circular Motion

38. While riding behind a car that is traveling at 3.00 m/s, you notice that one of the car's tires has a small hemispherical boss on its rim, as shown in Figure P13.38. (a) Explain why the boss, from your viewpoint behind the car, executes simple harmonic motion. (b) If the radius of the car's tires is 0.300 m, what is the boss's period of oscillation?

39. Consider the simplified single-piston engine shown in Figure P13.39. If the wheel rotates with constant angular speed, explain why the piston rod oscillates in simple harmonic motion.

Figure P13.38

Figure P13.39

(*Optional*)
Section 13.6 Damped Oscillations

40. Show that the time rate of change of mechanical energy for a damped, undriven oscillator is given by $dE/dt = -bv^2$ and hence is always negative. (*Hint:* Differentiate the expression for the mechanical energy of an oscillator, $E = \frac{1}{2} mv^2 + \frac{1}{2} kx^2$, and use Eq. 13.32.)

41. A pendulum with a length of 1.00 m is released from an initial angle of 15.0°. After 1 000 s, its amplitude is reduced by friction to 5.50°. What is the value of $b/2m$?

42. Show that Equation 13.33 is a solution of Equation 13.32 provided that $b^2 < 4mk$.

(*Optional*)
Section 13.7 Forced Oscillations

43. A baby rejoices in the day by crowing and jumping up and down in her crib. Her mass is 12.5 kg, and the crib mattress can be modeled as a light spring with a force constant of 4.30 kN/m. (a) The baby soon learns to bounce with maximum amplitude and minimum effort by bending her knees at what frequency? (b) She learns to use the mattress as a trampoline—losing contact with it for part of each cycle—when her amplitude exceeds what value?

44. A 2.00-kg mass attached to a spring is driven by an external force $F = (3.00 \text{ N}) \cos(2\pi t)$. If the force constant of the spring is 20.0 N/m, determine (a) the pe-

riod and (b) the amplitude of the motion. (*Hint:* Assume that there is no damping—that is, that $b = 0$—and use Eq. 13.37.)

45. Considering an *undamped*, forced oscillator ($b = 0$), show that Equation 13.36 is a solution of Equation 13.35, with an amplitude given by Equation 13.37.

46. A weight of 40.0 N is suspended from a spring that has a force constant of 200 N/m. The system is undamped and is subjected to a harmonic force with a frequency of 10.0 Hz, which results in a forced-motion amplitude of 2.00 cm. Determine the maximum value of the force.

47. Damping is negligible for a 0.150-kg mass hanging from a light 6.30-N/m spring. The system is driven by a force oscillating with an amplitude of 1.70 N. At what frequency will the force make the mass vibrate with an amplitude of 0.440 m?

48. You are a research biologist. Before dining at a fine restaurant, you set your pager to vibrate instead of beep, and you place it in the side pocket of your suit coat. The arm of your chair presses the light cloth against your body at one spot. Fabric with a length of 8.21 cm hangs freely below that spot, with the pager at the bottom. A co-worker telephones you. The motion of the vibrating pager makes the hanging part of your coat swing back and forth with remarkably large amplitude. The waiter, maître d', wine steward, and nearby diners notice immediately and fall silent. Your daughter pipes up and says, "Daddy, look! Your cockroaches must have gotten out again!" Find the frequency at which your pager vibrates.

ADDITIONAL PROBLEMS

49. A car with bad shock absorbers bounces up and down with a period of 1.50 s after hitting a bump. The car has a mass of 1 500 kg and is supported by four springs of equal force constant k. Determine the value of k.

50. A large passenger with a mass of 150 kg sits in the middle of the car described in Problem 49. What is the new period of oscillation?

51. A compact mass M is attached to the end of a uniform rod, of equal mass M and length L, that is pivoted at the top (Fig. P13.51). (a) Determine the tensions in the rod

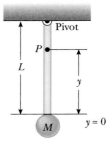

Figure P13.51

at the pivot and at the point *P* when the system is stationary. (b) Calculate the period of oscillation for small displacements from equilibrium, and determine this period for *L* = 2.00 m. (*Hint:* Assume that the mass at the end of the rod is a point mass, and use Eq. 13.28.)

52. A mass, m_1 = 9.00 kg, is in equilibrium while connected to a light spring of constant *k* = 100 N/m that is fastened to a wall, as shown in Figure P13.52a. A second mass, m_2 = 7.00 kg, is slowly pushed up against mass m_1, compressing the spring by the amount *A* = 0.200 m (see Fig. P13.52b). The system is then released, and both masses start moving to the right on the frictionless surface. (a) When m_1 reaches the equilibrium point, m_2 loses contact with m_1 (see Fig. P13.52c) and moves to the right with speed *v*. Determine the value of *v*. (b) How far apart are the masses when the spring is fully stretched for the first time (*D* in Fig. P13.52d)? (*Hint:* First determine the period of oscillation and the amplitude of the m_1–spring system after m_2 loses contact with m_1.)

(a)

(b)

(c)

(d)

Figure P13.52

WEB 53. A large block *P* executes horizontal simple harmonic motion as it slides across a frictionless surface with a frequency of *f* = 1.50 Hz. Block *B* rests on it, as shown

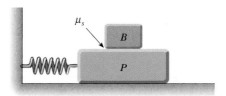

Figure P13.53   Problems 53 and 54.

in Figure P13.53, and the coefficient of static friction between the two is μ_s = 0.600. What maximum amplitude of oscillation can the system have if block *B* is not to slip?

54. A large block *P* executes horizontal simple harmonic motion as it slides across a frictionless surface with a frequency *f*. Block *B* rests on it, as shown in Figure P13.53, and the coefficient of static friction between the two is μ_s. What maximum amplitude of oscillation can the system have if the upper block is not to slip?

55. The mass of the deuterium molecule (D_2) is twice that of the hydrogen molecule (H_2). If the vibrational frequency of H_2 is 1.30×10^{14} Hz, what is the vibrational frequency of D_2? Assume that the "spring constant" of attracting forces is the same for the two molecules.

56. A solid sphere (radius = *R*) rolls without slipping in a cylindrical trough (radius = 5*R*), as shown in Figure P13.56. Show that, for small displacements from equilibrium perpendicular to the length of the trough, the sphere executes simple harmonic motion with a period $T = 2\pi\sqrt{28R/5g}$.

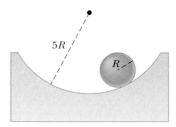

Figure P13.56

57. A light cubical container of volume a^3 is initially filled with a liquid of mass density ρ. The container is initially supported by a light string to form a pendulum of length L_i, measured from the center of mass of the filled container. The liquid is allowed to flow from the bottom of the container at a constant rate (dM/dt). At any time *t*, the level of the liquid in the container is *h*

and the length of the pendulum is L (measured relative to the instantaneous center of mass). (a) Sketch the apparatus and label the dimensions a, h, L_i, and L. (b) Find the time rate of change of the period as a function of time t. (c) Find the period as a function of time.

58. After a thrilling plunge, bungee-jumpers bounce freely on the bungee cords through many cycles. Your little brother can make a pest of himself by figuring out the mass of each person, using a proportion he set up by solving this problem: A mass m is oscillating freely on a vertical spring with a period T (Fig. P13.58a). An unknown mass m' on the same spring oscillates with a period T'. Determine (a) the spring constant k and (b) the unknown mass m'.

Figure P13.59

Figure P13.60

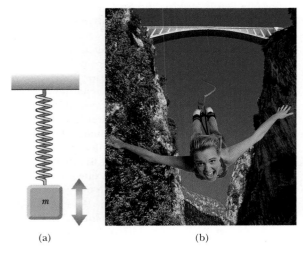

(a)   (b)

Figure P13.58   (a) Mass–spring system for Problems 58 and 68. (b) Bungee-jumping from a bridge. *(Telegraph Colour Library/ FPG International)*

59. A pendulum of length L and mass M has a spring of force constant k connected to it at a distance h below its point of suspension (Fig. P13.59). Find the frequency of vibration of the system for small values of the amplitude (small θ). (Assume that the vertical suspension of length L is rigid, but neglect its mass.)

60. A horizontal plank of mass m and length L is pivoted at one end. The plank's other end is supported by a spring of force constant k (Fig. P13.60). The moment of inertia of the plank about the pivot is $\frac{1}{3}mL^2$. (a) Show that the plank, after being displaced a small angle θ from its horizontal equilibrium position and released, moves with simple harmonic motion of angular frequency $\omega = \sqrt{3k/m}$. (b) Evaluate the frequency if the mass is 5.00 kg and the spring has a force constant of 100 N/m.

61. One end of a light spring with a force constant of 100 N/m is attached to a vertical wall. A light string is tied to the other end of the horizontal spring. The string changes from horizontal to vertical as it passes over a 4.00-cm-diameter solid pulley that is free to turn on a fixed smooth axle. The vertical section of the string supports a 200-g mass. The string does not slip at its contact with the pulley. Find the frequency of oscillation of the mass if the mass of the pulley is (a) negligible, (b) 250 g, and (c) 750 g.

62. A 2.00-kg block hangs without vibrating at the end of a spring ($k = 500$ N/m) that is attached to the ceiling of an elevator car. The car is rising with an upward acceleration of $g/3$ when the acceleration suddenly ceases (at $t = 0$). (a) What is the angular frequency of oscillation of the block after the acceleration ceases? (b) By what amount is the spring stretched during the acceleration of the elevator car? (c) What are the amplitude of the oscillation and the initial phase angle observed by a rider in the car? Take the upward direction to be positive.

63. A simple pendulum with a length of 2.23 m and a mass of 6.74 kg is given an initial speed of 2.06 m/s at its equilibrium position. Assume that it undergoes simple harmonic motion, and determine its (a) period, (b) total energy, and (c) maximum angular displacement.

64. People who ride motorcycles and bicycles learn to look out for bumps in the road and especially for *washboarding*, which is a condition of many equally spaced ridges worn into the road. What is so bad about washboarding? A motorcycle has several springs and shock absorbers in its suspension, but you can model it as a single spring supporting a mass. You can estimate the spring constant by thinking about how far the spring compresses when a big biker sits down on the seat. A motorcyclist traveling at highway speed must be particularly careful of washboard bumps that are a certain distance apart. What is the order of magnitude of their separation distance? State the quantities you take as data and the values you estimate or measure for them.

65. A wire is bent into the shape of one cycle of a cosine curve. It is held in a vertical plane so that the height y of the wire at any horizontal distance x from the center is given by $y = 20.0$ cm$[1 - \cos(0.160x \text{ rad/m})]$. A bead can slide without friction on the stationary wire. Show that if its excursion away from $x = 0$ is never large, the bead moves with simple harmonic motion. Determine its angular frequency. (*Hint:* $\cos \theta \cong 1 - \theta^2/2$ for small θ.)

66. A block of mass M is connected to a spring of mass m and oscillates in simple harmonic motion on a horizontal, frictionless track (Fig. P13.66). The force constant of the spring is k, and the equilibrium length is ℓ. Find (a) the kinetic energy of the system when the block has a speed v, and (b) the period of oscillation. (*Hint:* Assume that all portions of the spring oscillate in phase and that the velocity of a segment dx is proportional to the distance x from the fixed end; that is, $v_x = [x/\ell]v$. Also, note that the mass of a segment of the spring is $dm = [m/\ell]dx$.)

Figure P13.66

WEB 67. A ball of mass m is connected to two rubber bands of length L, each under tension T, as in Figure P13.67. The ball is displaced by a small distance y perpendicular to the length of the rubber bands. Assuming that the tension does not change, show that (a) the restoring force is $-(2T/L)y$ and (b) the system exhibits simple harmonic motion with an angular frequency $\omega = \sqrt{2T/mL}$.

Figure P13.67

68. When a mass M, connected to the end of a spring of mass $m_s = 7.40$ g and force constant k, is set into simple harmonic motion, the period of its motion is

$$T = 2\pi\sqrt{\frac{M + (m_s/3)}{k}}$$

A two-part experiment is conducted with the use of various masses suspended vertically from the spring, as shown in Figure P13.58a. (a) Static extensions of 17.0, 29.3, 35.3, 41.3, 47.1, and 49.3 cm are measured for M values of 20.0, 40.0, 50.0, 60.0, 70.0, and 80.0 g, respectively. Construct a graph of Mg versus x, and perform a linear least-squares fit to the data. From the slope of your graph, determine a value for k for this spring. (b) The system is now set into simple harmonic motion, and periods are measured with a stopwatch. With $M = 80.0$ g, the total time for 10 oscillations is measured to be 13.41 s. The experiment is repeated with M values of 70.0, 60.0, 50.0, 40.0, and 20.0 g, with corresponding times for 10 oscillations of 12.52, 11.67, 10.67, 9.62, and 7.03 s. Compute the experimental value for T for each of these measurements. Plot a graph of T^2 versus M, and determine a value for k from the slope of the linear least-squares fit through the data points. Compare this value of k with that obtained in part (a). (c) Obtain a value for m_s from your graph, and compare it with the given value of 7.40 g.

69. A small, thin disk of radius r and mass m is attached rigidly to the face of a second thin disk of radius R and mass M, as shown in Figure P13.69. The center of the small disk is located at the edge of the large disk. The large disk is mounted at its center on a frictionless axle. The assembly is rotated through a small angle θ from its equilibrium position and released. (a) Show that the

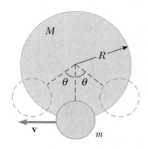

Figure P13.69

speed of the center of the small disk as it passes through the equilibrium position is

$$v = 2\left[\frac{Rg(1 - \cos\theta)}{(M/m) + (r/R)^2 + 2}\right]^{1/2}$$

(b) Show that the period of the motion is

$$T = 2\pi\left[\frac{(M + 2m)R^2 + mr^2}{2mgR}\right]^{1/2}$$

70. Consider the damped oscillator illustrated in Figure 13.19. Assume that the mass is 375 g, the spring constant is 100 N/m, and $b = 0.100$ kg/s. (a) How long does it takes for the amplitude to drop to half its initial value? (b) How long does it take for the mechanical energy to drop to half its initial value? (c) Show that, in general, the fractional rate at which the amplitude decreases in a damped harmonic oscillator is one-half the fractional rate at which the mechanical energy decreases.

71. A mass m is connected to two springs of force constants k_1 and k_2, as shown in Figure P13.71a and b. In each case, the mass moves on a frictionless table and is displaced from equilibrium and then released. Show that in the two cases the mass exhibits simple harmonic motion with periods

(a) $$T = 2\pi\sqrt{\frac{m(k_1 + k_2)}{k_1 k_2}}$$

(b) $$T = 2\pi\sqrt{\frac{m}{k_1 + k_2}}$$

72. Consider a simple pendulum of length $L = 1.20$ m that is displaced from the vertical by an angle θ_{max} and then released. You are to predict the subsequent angular displacements when θ_{max} is small and also when it is large. Set up and carry out a numerical method to integrate

(a)

(b)

Figure P13.71

the equation of motion for the simple pendulum:

$$\frac{d^2\theta}{dt^2} = -\frac{g}{L}\sin\theta$$

Take the initial conditions to be $\theta = \theta_{max}$ and $d\theta/dt = 0$ at $t = 0$. On one trial choose $\theta_{max} = 5.00°$, and on another trial take $\theta_{max} = 100°$. In each case, find the displacement θ as a function of time. Using the same values for θ_{max}, compare your results for θ with those obtained from $\theta_{max}\cos\omega t$. How does the period for the large value of θ_{max} compare with that for the small value of θ_{max}? *Note:* Using the Euler method to solve this differential equation, you may find that the amplitude tends to increase with time. The fourth-order Runge–Kutta method would be a better choice to solve the differential equation. However, if you choose Δt small enough, the solution that you obtain using Euler's method can still be good.

ANSWERS TO QUICK QUIZZES

13.1 Because A can never be zero, ϕ must be any value that results in the cosine function's being zero at $t = 0$. In other words, $\phi = \cos^{-1}(0)$. This is true at $\phi = \pi/2$, $3\pi/2$ or, more generally, $\phi = \pm n\pi/2$, where n is any nonzero odd integer. If we want to restrict our choices of ϕ to values between 0 and 2π, we need to know whether the object was moving to the right or to the left at $t = 0$. If it was moving with a positive velocity, then $\phi = 3\pi/2$. If $v_i < 0$, then $\phi = \pi/2$.

13.2 (d) $4A$. From its maximum positive position to the equilibrium position, it travels a distance A, by definition of *amplitude*. It then goes an equal distance past the equilibrium position to its maximum negative position. It then repeats these two motions in the reverse direction to return to its original position and complete one cycle.

13.3 No, because in simple harmonic motion, the acceleration is not constant.

13.4 $x = -A\sin\omega t$, where $A = v_i/\omega$.

13.5 From Hooke's law, the spring constant must be $k = mg/L$. If we substitute this value for k into Equation 13.18, we find that

$$T = 2\pi\sqrt{\frac{m}{k}} = 2\pi\sqrt{\frac{m}{mg/L}} = 2\pi\sqrt{\frac{L}{g}}$$

This is the same as Equation 13.26, which gives the period of a simple pendulum. Thus, when an object stretches a vertically hung spring, the period of the system is the same as that of a simple pendulum having a length equal to the amount of static extension of the spring.

13.6 If your goal is simply to stop the bounce from an absorbed shock as rapidly as possible, you should critically damp the suspension. Unfortunately, the stiffness of this design makes for an uncomfortable ride. If you underdamp the suspension, the ride is more comfortable but the car bounces. If you overdamp the suspension, the wheel is displaced from its equilibrium position longer than it should be. (For example, after hitting a bump, the spring stays compressed for a short time and the wheel does not quickly drop back down into contact with the road after the wheel is past the bump—a dangerous situation.) Because of all these considerations, automotive engineers usually design suspensions to be slightly underdamped. This allows the suspension to absorb a shock rapidly (minimizing the roughness of the ride) and then return to equilibrium after only one or two noticeable oscillations.

PUZZLER

More than 300 years ago, Isaac Newton realized that the same gravitational force that causes apples to fall to the Earth also holds the Moon in its orbit. In recent years, scientists have used the Hubble Space Telescope to collect evidence of the gravitational force acting even farther away, such as at this protoplanetary disk in the constellation Taurus. What properties of an object such as a protoplanet or the Moon determine the strength of its gravitational attraction to another object? *(Left, Larry West/FPG International; right, Courtesy of NASA)*

web

For more information about the Hubble, visit the Space Telescope Science Institute at **http://www.stsci.edu/**

c h a p t e r

14

The Law of Gravity

Chapter Outline

423

Before 1687, a large amount of data had been collected on the motions of the Moon and the planets, but a clear understanding of the forces causing these motions was not available. In that year, Isaac Newton provided the key that unlocked the secrets of the heavens. He knew, from his first law, that a net force had to be acting on the Moon because without such a force the Moon would move in a straight-line path rather than in its almost circular orbit. Newton reasoned that this force was the gravitational attraction exerted by the Earth on the Moon. He realized that the forces involved in the Earth–Moon attraction and in the Sun–planet attraction were not something special to those systems, but rather were particular cases of a general and universal attraction between objects. In other words, Newton saw that the same force of attraction that causes the Moon to follow its path around the Earth also causes an apple to fall from a tree. As he put it, "I deduced that the forces which keep the planets in their orbs must be reciprocally as the squares of their distances from the centers about which they revolve; and thereby compared the force requisite to keep the Moon in her orb with the force of gravity at the surface of the Earth; and found them answer pretty nearly."

In this chapter we study the law of gravity. We place emphasis on describing the motion of the planets because astronomical data provide an important test of the validity of the law of gravity. We show that the laws of planetary motion developed by Johannes Kepler follow from the law of gravity and the concept of conservation of angular momentum. We then derive a general expression for gravitational potential energy and examine the energetics of planetary and satellite motion. We close by showing how the law of gravity is also used to determine the force between a particle and an extended object.

14.1   NEWTON'S LAW OF UNIVERSAL GRAVITATION

You may have heard the legend that Newton was struck on the head by a falling apple while napping under a tree. This alleged accident supposedly prompted him to imagine that perhaps all bodies in the Universe were attracted to each other in the same way the apple was attracted to the Earth. Newton analyzed astronomical data on the motion of the Moon around the Earth. From that analysis, he made the bold assertion that the force law governing the motion of planets was the *same* as the force law that attracted a falling apple to the Earth. This was the first time that "earthly" and "heavenly" motions were unified. We shall look at the mathematical details of Newton's analysis in Section 14.5.

In 1687 Newton published his work on the law of gravity in his treatise *Mathematical Principles of Natural Philosophy*. **Newton's law of universal gravitation** states that

every particle in the Universe attracts every other particle with a force that is directly proportional to the product of their masses and inversely proportional to the square of the distance between them.

If the particles have masses m_1 and m_2 and are separated by a distance r, the magnitude of this gravitational force is

The law of gravity

$$F_g = G\frac{m_1 m_2}{r^2}$$

(14.1)

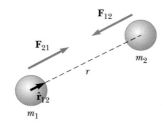

Figure 14.1 The gravitational force between two particles is attractive. The unit vector $\hat{\mathbf{r}}_{12}$ is directed from particle 1 to particle 2. Note that $\mathbf{F}_{21} = -\mathbf{F}_{12}$.

where G is a constant, called the *universal gravitational constant,* that has been measured experimentally. As noted in Example 6.6, its value in SI units is

$$G = 6.673 \times 10^{-11} \ \text{N} \cdot \text{m}^2/\text{kg}^2 \tag{14.2}$$

The form of the force law given by Equation 14.1 is often referred to as an **inverse-square law** because the magnitude of the force varies as the inverse square of the separation of the particles.[1] We shall see other examples of this type of force law in subsequent chapters. We can express this force in vector form by defining a unit vector $\hat{\mathbf{r}}_{12}$ (Fig. 14.1). Because this unit vector is directed from particle 1 to particle 2, the force exerted by particle 1 on particle 2 is

$$\mathbf{F}_{12} = -G \frac{m_1 m_2}{r^2} \hat{\mathbf{r}}_{12} \tag{14.3}$$

where the minus sign indicates that particle 2 is attracted to particle 1, and hence the force must be directed toward particle 1. By Newton's third law, the force exerted by particle 2 on particle 1, designated $\mathbf{F}_{21}$, is equal in magnitude to $\mathbf{F}_{12}$ and in the opposite direction. That is, these forces form an action–reaction pair, and $\mathbf{F}_{21} = -\mathbf{F}_{12}$.

Several features of Equation 14.3 deserve mention. The gravitational force is a field force that always exists between two particles, regardless of the medium that separates them. Because the force varies as the inverse square of the distance between the particles, it decreases rapidly with increasing separation. We can relate this fact to the geometry of the situation by noting that the intensity of light emanating from a point source drops off in the same $1/r^2$ manner, as shown in Figure 14.2.

Another important point about Equation 14.3 is that **the gravitational force exerted by a finite-size, spherically symmetric mass distribution on a particle outside the distribution is the same as if the entire mass of the distribution were concentrated at the center.** For example, the force exerted by the

Properties of the gravitational force

QuickLab

Inflate a balloon just enough to form a small sphere. Measure its diameter. Use a marker to color in a 1-cm square on its surface. Now continue inflating the balloon until it reaches twice the original diameter. Measure the size of the square you have drawn. Also note how the color of the marked area has changed. Have you verified what is shown in Figure 14.2?

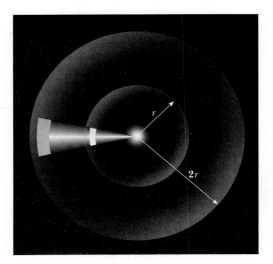

Figure 14.2 Light radiating from a point source drops off as $1/r^2$, a relationship that matches the way the gravitational force depends on distance. When the distance from the light source is doubled, the light has to cover four times the area and thus is one fourth as bright.

[1] An inverse relationship between two quantities x and y is one in which $y = k/x$, where k is a constant. A direct proportion between x and y exists when $y = kx$.

Earth on a particle of mass m near the Earth's surface has the magnitude

$$F_g = G\frac{M_E m}{R_E{}^2} \qquad\qquad (14.4)$$

where M_E is the Earth's mass and R_E its radius. This force is directed toward the center of the Earth.

We have evidence of the fact that the gravitational force acting on an object is directly proportional to its mass from our observations of falling objects, discussed in Chapter 2. All objects, regardless of mass, fall in the absence of air resistance at the same acceleration g near the surface of the Earth. According to Newton's second law, this acceleration is given by $g = F_g/m$, where m is the mass of the falling object. If this ratio is to be the same for all falling objects, then F_g must be directly proportional to m, so that the mass cancels in the ratio. If we consider the more general situation of a gravitational force between any two objects with mass, such as two planets, this same argument can be applied to show that the gravitational force is proportional to one of the masses. We can choose *either* of the masses in the argument, however; thus, the gravitational force must be directly proportional to *both* masses, as can be seen in Equation 14.3.

14.2 MEASURING THE GRAVITATIONAL CONSTANT

The universal gravitational constant G was measured in an important experiment by Henry Cavendish (1731–1810) in 1798. The Cavendish apparatus consists of two small spheres, each of mass m, fixed to the ends of a light horizontal rod suspended by a fine fiber or thin metal wire, as illustrated in Figure 14.3. When two large spheres, each of mass M, are placed near the smaller ones, the attractive force between smaller and larger spheres causes the rod to rotate and twist the wire suspension to a new equilibrium orientation. The angle of rotation is measured by the deflection of a light beam reflected from a mirror attached to the vertical suspension. The deflection of the light is an effective technique for amplifying the motion. The experiment is carefully repeated with different masses at various separations. In addition to providing a value for G, the results show experimentally that the force is attractive, proportional to the product mM, and inversely proportional to the square of the distance r.

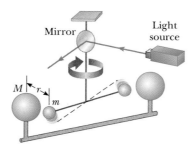

Figure 14.3 Schematic diagram of the Cavendish apparatus for measuring G. As the small spheres of mass M are attracted to the large spheres of mass M, the rod between the two small spheres rotates through a small angle. A light beam reflected from a mirror on the rotating apparatus measures the angle of rotation. The dashed line represents the original position of the rod.

EXAMPLE 14.1   Billiards, Anyone?

Three 0.300-kg billiard balls are placed on a table at the corners of a right triangle, as shown in Figure 14.4. Calculate the gravitational force on the cue ball (designated m_1) resulting from the other two balls.

Solution First we calculate separately the individual forces on the cue ball due to the other two balls, and then we find the vector sum to get the resultant force. We can see graphically that this force should point upward and toward the right. We locate our coordinate axes as shown in Figure 14.4, placing our origin at the position of the cue ball.

The force exerted by m_2 on the cue ball is directed upward and is given by

$$\mathbf{F}_{21} = G\frac{m_2 m_1}{r_{21}{}^2}\mathbf{j}$$

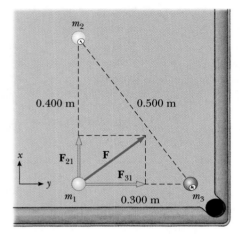

Figure 14.4 The resultant gravitational force acting on the cue ball is the vector sum $\mathbf{F}_{21} + \mathbf{F}_{31}$.

$$= \left(6.67 \times 10^{-11}\,\frac{\mathrm{N \cdot m^2}}{\mathrm{kg^2}}\right)\frac{(0.300\;\mathrm{kg})(0.300\;\mathrm{kg})}{(0.400\;\mathrm{m})^2}\mathbf{j}$$

$$= 3.75 \times 10^{-11}\mathbf{j}\;\mathrm{N}$$

This result shows that the gravitational forces between everyday objects have extremely small magnitudes. The force exerted by m_3 on the cue ball is directed to the right:

$$\mathbf{F}_{31} = G\frac{m_3 m_1}{r_{31}{}^2}\mathbf{i}$$

$$= \left(6.67 \times 10^{-11}\,\frac{\mathrm{N \cdot m^2}}{\mathrm{kg^2}}\right)\frac{(0.300\;\mathrm{kg})(0.300\;\mathrm{kg})}{(0.300\;\mathrm{m})^2}\mathbf{i}$$

$$= 6.67 \times 10^{-11}\,\mathbf{i}\;\mathrm{N}$$

Therefore, the resultant force on the cue ball is

$$\mathbf{F} = \mathbf{F}_{21} + \mathbf{F}_{31} = \boxed{(3.75\mathbf{j} + 6.67\mathbf{i}) \times 10^{-11}\;\mathrm{N}}$$

and the magnitude of this force is

$$F = \sqrt{F_{21}{}^2 + F_{31}{}^2} = \sqrt{(3.75)^2 + (6.67)^2} \times 10^{-11}$$

$$= 7.65 \times 10^{-11}\;\mathrm{N}$$

Exercise Find the direction of $\mathbf{F}$.

Answer 29.3° counterclockwise from the positive x axis.

14.3 FREE-FALL ACCELERATION AND THE GRAVITATIONAL FORCE

In Chapter 5, when defining mg as the weight of an object of mass m, we referred to g as the magnitude of the free-fall acceleration. Now we are in a position to obtain a more fundamental description of g. Because the force acting on a freely falling object of mass m near the Earth's surface is given by Equation 14.4, we can equate mg to this force to obtain

$$mg = G\frac{M_E m}{R_E{}^2}$$

$$g = G\frac{M_E}{R_E{}^2} \qquad\qquad \textbf{(14.5)}$$

Free-fall acceleration near the Earth's surface

Now consider an object of mass m located a distance h above the Earth's surface or a distance r from the Earth's center, where $r = R_E + h$. The magnitude of the gravitational force acting on this object is

$$F_g = G\frac{M_E m}{r^2} = G\frac{M_E m}{(R_E + h)^2}$$

The gravitational force acting on the object at this position is also $F_g = mg'$, where g' is the value of the free-fall acceleration at the altitude h. Substituting this expression for F_g into the last equation shows that g' is

Variation of g with altitude

$$g' = \frac{GM_E}{r^2} = \frac{GM_E}{(R_E + h)^2}$$ **(14.6)**

Thus, it follows that g' *decreases* with *increasing altitude*. Because the weight of a body is mg', we see that as $r \rightarrow \infty$, its weight approaches zero.

EXAMPLE 14.2 ▸ Variation of g with Altitude h

The International Space Station is designed to operate at an altitude of 350 km. When completed, it will have a weight (measured at the Earth's surface) of 4.22×10^6 N. What is its weight when in orbit?

Solution Because the station is above the surface of the Earth, we expect its weight in orbit to be less than its weight on Earth, 4.22×10^6 N. Using Equation 14.6 with $h = 350$ km, we obtain

$$g' = \frac{GM_E}{(R_E + h)^2}$$

$$= \frac{(6.67 \times 10^{-11}\,\text{N}\cdot\text{m}^2/\text{kg}^2)(5.98 \times 10^{24}\,\text{kg})}{(6.37 \times 10^6\,\text{m} + 0.350 \times 10^6\,\text{m})^2}$$

$$= 8.83\ \text{m/s}^2$$

Because $g'/g = 8.83/9.80 = 0.901$, we conclude that the weight of the station at an altitude of 350 km is 90.1% of the value at the Earth's surface. So the station's weight in orbit is

$(0.901)(4.22 \times 10^6\,\text{N}) = $ $3.80 \times 10^6\,\text{N}$

Values of g' at other altitudes are listed in Table 14.1.

TABLE 14.1   Free-Fall Acceleration g' at Various Altitudes Above the Earth's Surface

Altitude h (km)	g' (m/s^2)
1 000	7.33
2 000	5.68
3 000	4.53
4 000	3.70
5 000	3.08
6 000	2.60
7 000	2.23
8 000	1.93
9 000	1.69
10 000	1.49
50 000	0.13
∞	0

web

The official web site for the International Space Station is **www.station.nasa.gov**

EXAMPLE 14.3 ▸ The Density of the Earth

Using the fact that $g = 9.80$ m/s^2 at the Earth's surface, find the average density of the Earth.

Solution Using $g = 9.80$ m/s^2 and $R_E = 6.37 \times 10^6$ m, we find from Equation 14.5 that $M_E = 5.96 \times 10^{24}$ kg. From this result, and using the definition of density from Chapter 1, we obtain

$$\rho_E = \frac{M_E}{V_E} = \frac{M_E}{\frac{4}{3}\pi R_E^3} = \frac{5.96 \times 10^{24}\,\text{kg}}{\frac{4}{3}\pi(6.37 \times 10^6\,\text{m})^3}$$

$$= 5.50 \times 10^3\ \text{kg/m}^3$$

Because this value is about twice the density of most rocks at the Earth's surface, we conclude that the inner core of the Earth has a density much higher than the average value. It is most amazing that the Cavendish experiment, which determines G (and can be done on a tabletop), combined with simple free-fall measurements of g, provides information about the core of the Earth.

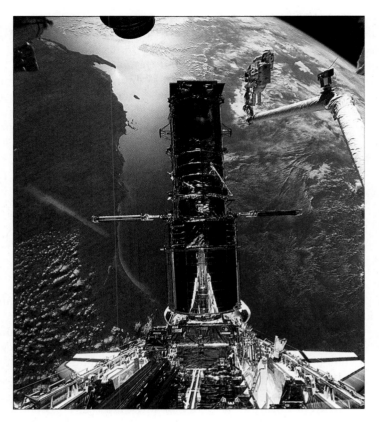

Astronauts F. Story Musgrave and Jeffrey A. Hoffman, along with the Hubble Space Telescope and the space shuttle *Endeavor,* are all falling around the Earth. *(Courtesy of NASA)*

14.4 ▶ KEPLER'S LAWS

People have observed the movements of the planets, stars, and other celestial bodies for thousands of years. In early history, scientists regarded the Earth as the center of the Universe. This so-called geocentric model was elaborated and formalized by the Greek astronomer Claudius Ptolemy (c. 100–c. 170) in the second century A.D. and was accepted for the next 1 400 years. In 1543 the Polish astronomer Nicolaus Copernicus (1473–1543) suggested that the Earth and the other planets revolved in circular orbits around the Sun (the heliocentric model).

The Danish astronomer Tycho Brahe (1546–1601) wanted to determine how the heavens were constructed, and thus he developed a program to determine the positions of both stars and planets. It is interesting to note that those observations of the planets and 777 stars visible to the naked eye were carried out with only a large sextant and a compass. (The telescope had not yet been invented.)

The German astronomer Johannes Kepler was Brahe's assistant for a short while before Brahe's death, whereupon he acquired his mentor's astronomical data and spent 16 years trying to deduce a mathematical model for the motion of the planets. Such data are difficult to sort out because the Earth is also in motion around the Sun. After many laborious calculations, Kepler found that Brahe's data on the revolution of Mars around the Sun provided the answer.

Johannes Kepler **German astronomer (1571–1630)** The German astronomer Johannes Kepler is best known for developing the laws of planetary motion based on the careful observations of Tycho Brahe. *(Art Resource)*

For more information about Johannes Kepler, visit our Web site at
www.saunderscollege.com/physics/

Kepler's analysis first showed that the concept of circular orbits around the Sun had to be abandoned. He eventually discovered that the orbit of Mars could be accurately described by an **ellipse.** Figure 14.5 shows the geometric description of an ellipse. The longest dimension is called the major axis and is of length $2a$, where a is the **semimajor axis.** The shortest dimension is the minor axis, of length $2b$, where b is the **semiminor axis.** On either side of the center is a **focal point,** a distance c from the center, where $a^2 = b^2 + c^2$. The Sun is located at one of the focal points of Mars's orbit. Kepler generalized his analysis to include the motions of all planets. The complete analysis is summarized in three statements known as **Kepler's laws:**

Kepler's laws

1. All planets move in elliptical orbits with the Sun at one focal point.
2. The radius vector drawn from the Sun to a planet sweeps out equal areas in equal time intervals.
3. The square of the orbital period of any planet is proportional to the cube of the semimajor axis of the elliptical orbit.

Most of the planetary orbits are close to circular in shape; for example, the semimajor and semiminor axes of the orbit of Mars differ by only 0.4%. Mercury and Pluto have the most elliptical orbits of the nine planets. In addition to the planets, there are many asteroids and comets orbiting the Sun that obey Kepler's laws. Comet Halley is such an object; it becomes visible when it is close to the Sun every 76 years. Its orbit is very elliptical, with a semiminor axis 76% smaller than its semimajor axis.

Although we do not prove it here, Kepler's first law is a direct consequence of the fact that the gravitational force varies as $1/r^2$. That is, under an inverse-square gravitational-force law, the orbit of a planet can be shown mathematically to be an ellipse with the Sun at one focal point. Indeed, half a century after Kepler developed his laws, Newton demonstrated that these laws are a consequence of the gravitational force that exists between any two masses. Newton's law of universal gravitation, together with his development of the laws of motion, provides the basis for a full mathematical solution to the motion of planets and satellites.

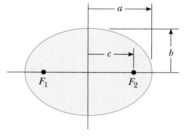

Figure 14.5 Plot of an ellipse. The semimajor axis has a length a, and the semiminor axis has a length b. The focal points are located at a distance c from the center, where $a^2 = b^2 + c^2$.

14.5 > THE LAW OF GRAVITY AND THE MOTION OF PLANETS

In formulating his law of gravity, Newton used the following reasoning, which supports the assumption that the gravitational force is proportional to the inverse square of the separation between the two interacting bodies. He compared the acceleration of the Moon in its orbit with the acceleration of an object falling near the Earth's surface, such as the legendary apple (Fig. 14.6). Assuming that both accelerations had the same cause—namely, the gravitational attraction of the Earth—Newton used the inverse-square law to reason that the acceleration of the Moon toward the Earth (centripetal acceleration) should be proportional to $1/r_M^2$, where r_M is the distance between the centers of the Earth and the Moon. Furthermore, the acceleration of the apple toward the Earth should be proportional to $1/R_E^2$, where R_E is the radius of the Earth, or the distance between the centers of the Earth and the apple. Using the values $r_M = 3.84 \times 10^8$ m and

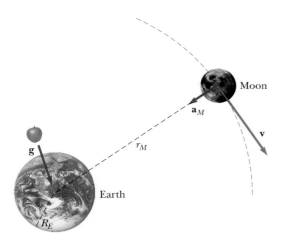

Figure 14.6 As it revolves around the Earth, the Moon experiences a centripetal acceleration $\mathbf{a}_M$ directed toward the Earth. An object near the Earth's surface, such as the apple shown here, experiences an acceleration $\mathbf{g}$. (Dimensions are not to scale.)

$R_E = 6.37 \times 10^6$ m, Newton predicted that the ratio of the Moon's acceleration a_M to the apple's acceleration g would be

$$\frac{a_M}{g} = \frac{(1/r_M)^2}{(1/R_E)^2} = \left(\frac{R_E}{r_M}\right)^2 = \left(\frac{6.37 \times 10^6 \text{ m}}{3.84 \times 10^8 \text{ m}}\right)^2 = 2.75 \times 10^{-4}$$

Therefore, the centripetal acceleration of the Moon is

$$a_M = (2.75 \times 10^{-4})(9.80 \text{ m/s}^2) = 2.70 \times 10^{-3} \text{ m/s}^2$$

Acceleration of the Moon

Newton also calculated the centripetal acceleration of the Moon from a knowledge of its mean distance from the Earth and its orbital period, $T = 27.32$ days $= 2.36 \times 10^6$ s. In a time T, the Moon travels a distance $2\pi r_M$, which equals the circumference of its orbit. Therefore, its orbital speed is $2\pi r_M/T$ and its centripetal acceleration is

$$a_M = \frac{v^2}{r_M} = \frac{(2\pi r_M/T)^2}{r_M} = \frac{4\pi^2 r_M}{T^2} = \frac{4\pi^2(3.84 \times 10^8 \text{ m})}{(2.36 \times 10^6 \text{ s})^2}$$

$$= 2.72 \times 10^{-3} \text{ m/s}^2 \approx \frac{9.80 \text{ m/s}^2}{60^2}$$

In other words, because the Moon is roughly 60 Earth radii away, the gravitational acceleration at that distance should be about $1/60^2$ of its value at the Earth's surface. This is just the acceleration needed to account for the circular motion of the Moon around the Earth. The nearly perfect agreement between this value and the value Newton obtained using g provides strong evidence of the inverse-square nature of the gravitational force law.

Although these results must have been very encouraging to Newton, he was deeply troubled by an assumption he made in the analysis. To evaluate the acceleration of an object at the Earth's surface, Newton treated the Earth as if its mass were all concentrated at its center. That is, he assumed that the Earth acted as a particle as far as its influence on an exterior object was concerned. Several years later, in 1687, on the basis of his pioneering work in the development of calculus, Newton proved that this assumption was valid and was a natural consequence of the law of universal gravitation.

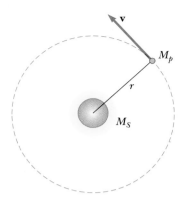

Figure 14.7 A planet of mass M_p moving in a circular orbit around the Sun. The orbits of all planets except Mercury and Pluto are nearly circular.

Kepler's third law

Kepler's Third Law

It is informative to show that Kepler's third law can be predicted from the inverse-square law for circular orbits.[2] Consider a planet of mass M_p moving around the Sun of mass M_S in a circular orbit, as shown in Figure 14.7. Because the gravitational force exerted by the Sun on the planet is a radially directed force that keeps the planet moving in a circle, we can apply Newton's second law ($\Sigma F = ma$) to the planet:

$$\frac{GM_SM_p}{r^2} = \frac{M_pv^2}{r}$$

Because the orbital speed v of the planet is simply $2\pi r/T$, where T is its period of revolution, the preceding expression becomes

$$\frac{GM_S}{r^2} = \frac{(2\pi r/T)^2}{r}$$

$$T^2 = \left(\frac{4\pi^2}{GM_S}\right)r^3 = K_Sr^3 \qquad (14.7)$$

where K_S is a constant given by

$$K_S = \frac{4\pi^2}{GM_S} = 2.97 \times 10^{-19}\ \text{s}^2/\text{m}^3$$

Equation 14.7 is Kepler's third law. It can be shown that the law is also valid for elliptical orbits if we replace r with the length of the semimajor axis a. Note that the constant of proportionality K_S is independent of the mass of the planet. Therefore, Equation 14.7 is valid for *any* planet.[3] Table 14.2 contains a collection of useful planetary data. The last column verifies that T^2/r^3 is a constant. The small variations in the values in this column reflect uncertainties in the measured values of the periods and semimajor axes of the planets.

If we were to consider the orbit around the Earth of a satellite such as the Moon, then the proportionality constant would have a different value, with the Sun's mass replaced by the Earth's mass.

EXAMPLE 14.4   **The Mass of the Sun**

Calculate the mass of the Sun using the fact that the period of the Earth's orbit around the Sun is 3.156×10^7 s and its distance from the Sun is 1.496×10^{11} m.

Solution   Using Equation 14.7, we find that

$$M_S = \frac{4\pi^2r^3}{GT^2} = \frac{4\pi^2(1.496 \times 10^{11}\ \text{m})^3}{(6.67 \times 10^{-11}\ \text{N}\cdot\text{m}^2/\text{kg}^2)(3.156 \times 10^7\ \text{s})^2}$$

$$= \boxed{1.99 \times 10^{30}\ \text{kg}}$$

In Example 14.3, an understanding of gravitational forces enabled us to find out something about the density of the Earth's core, and now we have used this understanding to determine the mass of the Sun.

---

[2] The orbits of all planets except Mercury and Pluto are very close to being circular; hence, we do not introduce much error with this assumption. For example, the ratio of the semiminor axis to the semimajor axis for the Earth's orbit is $b/a = 0.999\ 86$.

[3] Equation 14.7 is indeed a proportion because the ratio of the two quantities T^2 and r^3 is a constant. The variables in a proportion are not required to be limited to the first power only.

TABLE 14.2 Useful Planetary Data

Body	Mass (kg)	Mean Radius (m)	Period of Revolution (s)	Mean Distance from Sun (m)	$\dfrac{T^2}{r^3}$ (s²/m³)
Mercury	3.18×10^{23}	2.43×10^6	7.60×10^6	5.79×10^{10}	2.97×10^{-19}
Venus	4.88×10^{24}	6.06×10^6	1.94×10^7	1.08×10^{11}	2.99×10^{-19}
Earth	5.98×10^{24}	6.37×10^6	3.156×10^7	1.496×10^{11}	2.97×10^{-19}
Mars	6.42×10^{23}	3.37×10^6	5.94×10^7	2.28×10^{11}	2.98×10^{-19}
Jupiter	1.90×10^{27}	6.99×10^7	3.74×10^8	7.78×10^{11}	2.97×10^{-19}
Saturn	5.68×10^{26}	5.85×10^7	9.35×10^8	1.43×10^{12}	2.99×10^{-19}
Uranus	8.68×10^{25}	2.33×10^7	2.64×10^9	2.87×10^{12}	2.95×10^{-19}
Neptune	1.03×10^{26}	2.21×10^7	5.22×10^9	4.50×10^{12}	2.99×10^{-19}
Pluto	$\approx 1.4 \times 10^{22}$	$\approx 1.5 \times 10^6$	7.82×10^9	5.91×10^{12}	2.96×10^{-19}
Moon	7.36×10^{22}	1.74×10^6	—	—	—
Sun	1.991×10^{30}	6.96×10^8	—	—	—

Kepler's Second Law and Conservation of Angular Momentum

Consider a planet of mass M_p moving around the Sun in an elliptical orbit (Fig. 14.8). The gravitational force acting on the planet is always along the radius vector, directed toward the Sun, as shown in Figure 14.9a. When a force is directed toward or away from a fixed point and is a function of r only, it is called a **central force.** The torque acting on the planet due to this force is clearly zero; that is, because **F** is parallel to **r**,

$$\boldsymbol{\tau} = \mathbf{r} \times \mathbf{F} = \mathbf{r} \times F\hat{\mathbf{r}} = 0$$

(You may want to revisit Section 11.2 to refresh your memory on the vector product.) Recall from Equation 11.19, however, that torque equals the time rate of change of angular momentum: $\boldsymbol{\tau} = d\mathbf{L}/dt$. Therefore, **because the gravitational**

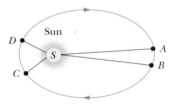

Figure 14.8 Kepler's second law is called the law of equal areas. When the time interval required for a planet to travel from A to B is equal to the time interval required for it to go from C to D, the two areas swept out by the planet's radius vector are equal. Note that in order for this to be true, the planet must be moving faster between C and D than between A and B.

Separate views of Jupiter and of Periodic Comet Shoemaker–Levy 9—both taken with the Hubble Space Telescope about two months before Jupiter and the comet collided in July 1994—were put together with the use of a computer. Their relative sizes and distances were altered. The black spot on Jupiter is the shadow of its moon Io. *(Courtesy of NASA)*

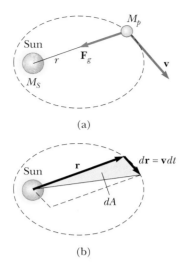

(a)

(b)

Figure 14.9 (a) The gravitational force acting on a planet is directed toward the Sun, along the radius vector. (b) As a planet orbits the Sun, the area swept out by the radius vector in a time dt is equal to one-half the area of the parallelogram formed by the vectors $\mathbf{r}$ and $d\mathbf{r} = \mathbf{v}dt$.

force exerted by the Sun on a planet results in no torque on the planet, the angular momentum L of the planet is constant:

$$\mathbf{L} = \mathbf{r} \times \mathbf{p} = \mathbf{r} \times M_p\mathbf{v} = M_p\mathbf{r} \times \mathbf{v} = \text{constant} \qquad (14.8)$$

Because **L** remains constant, the planet's motion at any instant is restricted to the plane formed by **r** and **v**.

We can relate this result to the following geometric consideration. The radius vector **r** in Figure 14.9b sweeps out an area dA in a time dt. This area equals one-half the area $|\mathbf{r} \times d\mathbf{r}|$ of the parallelogram formed by the vectors **r** and $d\mathbf{r}$ (see Section 11.2). Because the displacement of the planet in a time dt is $d\mathbf{r} = \mathbf{v}dt$, we can say that

$$dA = \tfrac{1}{2}|\mathbf{r} \times d\mathbf{r}| = \tfrac{1}{2}|\mathbf{r} \times \mathbf{v}\,dt| = \frac{L}{2M_p}\,dt$$

$$\frac{dA}{dt} = \frac{L}{2M_p} = \text{constant} \qquad (14.9)$$

where L and M_p are both constants. Thus, we conclude that

the radius vector from the Sun to a planet sweeps out equal areas in equal time intervals.

It is important to recognize that this result, which is Kepler's second law, is a consequence of the fact that the force of gravity is a central force, which in turn implies that angular momentum is constant. Therefore, Kepler's second law applies to *any* situation involving a central force, whether inverse-square or not.

EXAMPLE 14.5 **Motion in an Elliptical Orbit**

A satellite of mass m moves in an elliptical orbit around the Earth (Fig. 14.10). The minimum distance of the satellite from the Earth is called the *perigee* (indicated by p in Fig. 14.10), and the maximum distance is called the *apogee* (indicated by a). If the speed of the satellite at p is v_p, what is its speed at a?

Figure 14.10 As a satellite moves around the Earth in an elliptical orbit, its angular momentum is constant. Therefore, $mv_a r_a = mv_p r_p$, where the subscripts a and p represent apogee and perigee, respectively.

Solution As the satellite moves from perigee toward apogee, it is moving farther from the Earth. Thus, a component of the gravitational force exerted by the Earth on the satellite is opposite the velocity vector. Negative work is done on the satellite, which causes it to slow down, according to the work–kinetic energy theorem. As a result, we expect the speed at apogee to be lower than the speed at perigee.

The angular momentum of the satellite relative to the Earth is $\mathbf{r} \times m\mathbf{v} = m\mathbf{r} \times \mathbf{v}$. At the points a and p, **v** is perpendicular to **r**. Therefore, the magnitude of the angular momentum at these positions is $L_a = mv_a r_a$ and $L_p = mv_p r_p$. Because angular momentum is constant, we see that

$$mv_a r_a = mv_p r_p$$

$$v_a = \frac{r_p}{r_a}v_p$$

How would you explain the fact that Saturn and Jupiter have periods much greater than one year?

14.6   THE GRAVITATIONAL FIELD

When Newton published his theory of universal gravitation, it was considered a success because it satisfactorily explained the motion of the planets. Since 1687 the same theory has been used to account for the motions of comets, the deflection of a Cavendish balance, the orbits of binary stars, and the rotation of galaxies. Nevertheless, both Newton's contemporaries and his successors found it difficult to accept the concept of a force that acts through a distance, as mentioned in Section 5.1. They asked how it was possible for two objects to interact when they were not in contact with each other. Newton himself could not answer that question.

An approach to describing interactions between objects that are not in contact came well after Newton's death, and it enables us to look at the gravitational interaction in a different way. As described in Section 5.1, this alternative approach uses the concept of a **gravitational field** that exists at every point in space. When a particle of mass m is placed at a point where the gravitational field is $\mathbf{g}$, the particle experiences a force $\mathbf{F}_g = m\mathbf{g}$. In other words, the field exerts a force on the particle. Hence, the gravitational field $\mathbf{g}$ is defined as

$$\mathbf{g} \equiv \frac{\mathbf{F}_g}{m} \qquad (14.10)$$

Gravitational field

That is, the gravitational field at a point in space equals the gravitational force experienced by a *test particle* placed at that point divided by the mass of the test particle. Notice that the presence of the test particle is not necessary for the field to exist—the Earth creates the gravitational field. We call the object creating the field the *source particle* (although the Earth is clearly not a particle; we shall discuss shortly the fact that we can approximate the Earth as a particle for the purpose of finding the gravitational field that it creates). We can detect the presence of the field and measure its strength by placing a test particle in the field and noting the force exerted on it.

Although the gravitational force is inherently an interaction between two objects, the concept of a gravitational field allows us to "factor out" the mass of one of the objects. In essence, we are describing the "effect" that any object (in this case, the Earth) has on the empty space around itself in terms of the force that *would* be present *if* a second object were somewhere in that space.[4]

As an example of how the field concept works, consider an object of mass m near the Earth's surface. Because the gravitational force acting on the object has a magnitude $GM_E m/r^2$ (see Eq. 14.4), the field $\mathbf{g}$ at a distance r from the center of the Earth is

$$\mathbf{g} = \frac{\mathbf{F}_g}{m} = -\frac{GM_E}{r^2}\hat{\mathbf{r}} \qquad (14.11)$$

where $\hat{\mathbf{r}}$ is a unit vector pointing radially outward from the Earth and the minus

---

[4] We shall return to this idea of mass affecting the space around it when we discuss Einstein's theory of gravitation in Chapter 39.

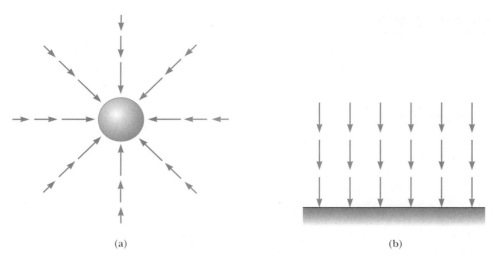

Figure 14.11   (a) The gravitational field vectors in the vicinity of a uniform spherical mass such as the Earth vary in both direction and magnitude. The vectors point in the direction of the acceleration a particle would experience if it were placed in the field. The magnitude of the field vector at any location is the magnitude of the free-fall acceleration at that location. (b) The gravitational field vectors in a small region near the Earth's surface are uniform in both direction and magnitude.

sign indicates that the field points toward the center of the Earth, as illustrated in Figure 14.11a. Note that the field vectors at different points surrounding the Earth vary in both direction and magnitude. In a small region near the Earth's surface, the downward field **g** is approximately constant and uniform, as indicated in Figure 14.11b. Equation 14.11 is valid at all points *outside* the Earth's surface, assuming that the Earth is spherical. At the Earth's surface, where $r = R_E$, **g** has a magnitude of 9.80 N/kg.

14.7 GRAVITATIONAL POTENTIAL ENERGY

In Chapter 8 we introduced the concept of gravitational potential energy, which is the energy associated with the position of a particle. We emphasized that the gravitational potential energy function $U = mgy$ is valid only when the particle is near the Earth's surface, where the gravitational force is constant. Because the gravitational force between two particles varies as $1/r^2$, we expect that a more general potential energy function—one that is valid without the restriction of having to be near the Earth's surface—will be significantly different from $U = mgy$.

Before we calculate this general form for the gravitational potential energy function, let us first verify that *the gravitational force is conservative*. (Recall from Section 8.2 that a force is conservative if the work it does on an object moving between any two points is independent of the path taken by the object.) To do this, we first note that the gravitational force is a central force. By definition, a central force is any force that is directed along a radial line to a fixed center and has a magnitude that depends only on the radial coordinate r. Hence, a central force can be represented by $F(r)\hat{\mathbf{r}}$, where $\hat{\mathbf{r}}$ is a unit vector directed from the origin to the particle, as shown in Figure 14.12.

Consider a central force acting on a particle moving along the general path P to Q in Figure 14.12. The path from P to Q can be approximated by a series of steps according to the following procedure. In Figure 14.12, we draw several thin wedges, which are shown as dashed lines. The outer boundary of our set of wedges is a path consisting of short radial line segments and arcs (gray in the figure). We select the length of the radial dimension of each wedge such that the short arc at the wedge's wide end intersects the actual path of the particle. Then we can approximate the actual path with a series of zigzag movements that alternate between moving along an arc and moving along a radial line.

By definition, a central force is always directed along one of the radial segments; therefore, the work done by **F** along any radial segment is

$$dW = \mathbf{F} \cdot d\mathbf{r} = F(r)\ dr$$

You should recall that, by definition, the work done by a force that is perpendicular to the displacement is zero. Hence, the work done in moving along any arc is zero because **F** is perpendicular to the displacement along these segments. Therefore, the total work done by **F** is the sum of the contributions along the radial segments:

$$W = \int_{r_i}^{r_f} F(r)\ dr$$

where the subscripts i and f refer to the initial and final positions. Because the integrand is a function only of the radial position, this integral depends only on the initial and final values of r. Thus, the work done is the same over *any* path from P to Q. Because the work done is independent of the path and depends only on the end points, we conclude that *any central force is conservative.* We are now assured that a potential energy function can be obtained once the form of the central force is specified.

Recall from Equation 8.2 that the change in the gravitational potential energy associated with a given displacement is defined as the negative of the work done by the gravitational force during that displacement:

$$\Delta U = U_f - U_i = -\int_{r_i}^{r_f} F(r)\ dr \qquad \textbf{(14.12)}$$

We can use this result to evaluate the gravitational potential energy function. Consider a particle of mass m moving between two points P and Q above the Earth's surface (Fig. 14.13). The particle is subject to the gravitational force given by Equation 14.1. We can express this force as

$$F(r) = -\frac{GM_E m}{r^2}$$

where the negative sign indicates that the force is attractive. Substituting this expression for $F(r)$ into Equation 14.12, we can compute the change in the gravita-

> Work done by a central force

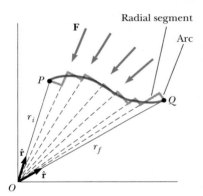

Figure 14.12 A particle moves from P to Q while acted on by a central force **F**, which is directed radially. The path is broken into a series of radial segments and arcs. Because the work done along the arcs is zero, the work done is independent of the path and depends only on r_f and r_i.

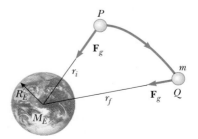

Figure 14.13 As a particle of mass m moves from P to Q above the Earth's surface, the gravitational potential energy changes according to Equation 14.12.

tional potential energy function:

$$U_f - U_i = GM_E m \int_{r_i}^{r_f} \frac{dr}{r^2} = GM_E m \left[-\frac{1}{r} \right]_{r_i}^{r_f}$$

$$U_f - U_i = -GM_E m \left(\frac{1}{r_f} - \frac{1}{r_i} \right) \tag{14.13}$$

As always, the choice of a reference point for the potential energy is completely arbitrary. It is customary to choose the reference point where the force is zero. Taking $U_i = 0$ at $r_i = \infty$, we obtain the important result

$$U = -\frac{GM_E m}{r} \tag{14.14}$$

This expression applies to the Earth–particle system where the two masses are separated by a distance r, provided that $r \geq R_E$. The result is not valid for particles inside the Earth, where $r < R_E$. (The situation in which $r < R_E$ is treated in Section 14.10.) Because of our choice of U_i, the function U is always negative (Fig. 14.14).

Although Equation 14.14 was derived for the particle–Earth system, it can be applied to any two particles. That is, the gravitational potential energy associated with any pair of particles of masses m_1 and m_2 separated by a distance r is

$$U = -\frac{Gm_1 m_2}{r} \tag{14.15}$$

This expression shows that the gravitational potential energy for any pair of particles varies as $1/r$, whereas the force between them varies as $1/r^2$. Furthermore, the potential energy is negative because the force is attractive and we have taken the potential energy as zero when the particle separation is infinite. Because the force between the particles is attractive, we know that an external agent must do positive work to increase the separation between them. The work done by the external agent produces an increase in the potential energy as the two particles are separated. That is, U becomes less negative as r increases.

When two particles are at rest and separated by a distance r, an external agent has to supply an energy at least equal to $+Gm_1 m_2/r$ in order to separate the particles to an infinite distance. It is therefore convenient to think of the absolute value of the potential energy as the *binding energy* of the system. If the external agent supplies an energy greater than the binding energy, the excess energy of the system will be in the form of kinetic energy when the particles are at an infinite separation.

We can extend this concept to three or more particles. In this case, the total potential energy of the system is the sum over all pairs of particles.[5] Each pair contributes a term of the form given by Equation 14.15. For example, if the system contains three particles, as in Figure 14.15, we find that

$$U_{\text{total}} = U_{12} + U_{13} + U_{23} = -G \left(\frac{m_1 m_2}{r_{12}} + \frac{m_1 m_3}{r_{13}} + \frac{m_2 m_3}{r_{23}} \right) \tag{14.16}$$

The absolute value of U_{total} represents the work needed to separate the particles by an infinite distance.

Change in gravitational potential energy

Gravitational potential energy of the Earth–particle system for $r \geq R_E$

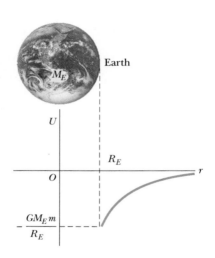

Figure 14.14 Graph of the gravitational potential energy U versus r for a particle above the Earth's surface. The potential energy goes to zero as r approaches infinity.

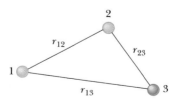

Figure 14.15 Three interacting particles.

---

[5] The fact that potential energy terms can be added for all pairs of particles stems from the experimental fact that gravitational forces obey the superposition principle.

EXAMPLE 14.6   **The Change in Potential Energy**

A particle of mass m is displaced through a small vertical distance Δy near the Earth's surface. Show that in this situation the general expression for the change in gravitational potential energy given by Equation 14.13 reduces to the familiar relationship $\Delta U = mg\,\Delta y$.

Solution   We can express Equation 14.13 in the form

$$\Delta U = -GM_E m \left(\frac{1}{r_f} - \frac{1}{r_i} \right) = GM_E m \left(\frac{r_f - r_i}{r_i r_f} \right)$$

If both the initial and final positions of the particle are close to the Earth's surface, then $r_f - r_i = \Delta y$ and $r_i r_f \approx R_E^2$. (Recall that r is measured from the center of the Earth.) Therefore, the change in potential energy becomes

$$\Delta U \approx \frac{GM_E m}{R_E^2}\,\Delta y = mg\,\Delta y$$

where we have used the fact that $g = GM_E/R_E^2$ (Eq. 14.5). Keep in mind that the reference point is arbitrary because it is the *change* in potential energy that is meaningful.

14.8   ENERGY CONSIDERATIONS IN PLANETARY AND SATELLITE MOTION

Consider a body of mass m moving with a speed v in the vicinity of a massive body of mass M, where $M \gg m$. The system might be a planet moving around the Sun, a satellite in orbit around the Earth, or a comet making a one-time flyby of the Sun. If we assume that the body of mass M is at rest in an inertial reference frame, then the total mechanical energy E of the two-body system when the bodies are separated by a distance r is the sum of the kinetic energy of the body of mass m and the potential energy of the system, given by Equation 14.15:[6]

$$E = K + U$$

$$E = \tfrac{1}{2}mv^2 - \frac{GMm}{r} \qquad \textbf{(14.17)}$$

This equation shows that E may be positive, negative, or zero, depending on the value of v. However, for a bound system,[7] such as the Earth–Sun system, E is necessarily *less than zero* because we have chosen the convention that $U \to 0$ as $r \to \infty$.

We can easily establish that $E < 0$ for the system consisting of a body of mass m moving in a circular orbit about a body of mass $M \gg m$ (Fig. 14.16). Newton's second law applied to the body of mass m gives

$$\frac{GMm}{r^2} = ma = \frac{mv^2}{r}$$

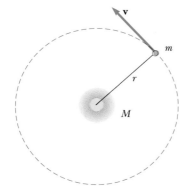

Figure 14.16   A body of mass m moving in a circular orbit about a much larger body of mass M.

---

[6] You might recognize that we have ignored the acceleration and kinetic energy of the larger body. To see that this simplification is reasonable, consider an object of mass m falling toward the Earth. Because the center of mass of the object–Earth system is effectively stationary, it follows that $mv = M_E v_E$. Thus, the Earth acquires a kinetic energy equal to

$$\tfrac{1}{2}M_E v_E^2 = \tfrac{1}{2}\frac{m^2}{M_E}v^2 = \frac{m}{M_E}K$$

where K is the kinetic energy of the object. Because $M_E \gg m$, this result shows that the kinetic energy of the Earth is negligible.

[7] Of the three examples provided at the beginning of this section, the planet moving around the Sun and a satellite in orbit around the Earth are bound systems—the Earth will always stay near the Sun, and the satellite will always stay near the Earth. The one-time comet flyby represents an unbound system—the comet interacts once with the Sun but is not bound to it. Thus, in theory the comet can move infinitely far away from the Sun.

Multiplying both sides by r and dividing by 2 gives

$$\tfrac{1}{2}mv^2 = \frac{GMm}{2r} \tag{14.18}$$

Substituting this into Equation 14.17, we obtain

$$E = \frac{GMm}{2r} - \frac{GMm}{r}$$

Total energy for circular orbits

$$E = -\frac{GMm}{2r} \tag{14.19}$$

This result clearly shows that **the total mechanical energy is negative in the case of circular orbits.** Note that **the kinetic energy is positive and equal to one-half the absolute value of the potential energy.** The absolute value of E is also equal to the binding energy of the system, because this amount of energy must be provided to the system to move the two masses infinitely far apart.

The total mechanical energy is also negative in the case of elliptical orbits. The expression for E for elliptical orbits is the same as Equation 14.19 with r replaced by the semimajor axis length a. Furthermore, the total energy is constant if we assume that the system is isolated. Therefore, as the body of mass m moves from P to Q in Figure 14.13, the total energy remains constant and Equation 14.17 gives

$$E = \tfrac{1}{2}mv_i{}^2 - \frac{GMm}{r_i} = \tfrac{1}{2}mv_f{}^2 - \frac{GMm}{r_f} \tag{14.20}$$

Combining this statement of energy conservation with our earlier discussion of conservation of angular momentum, we see that **both the total energy and the total angular momentum of a gravitationally bound, two-body system are constants of the motion.**

EXAMPLE 14.7   **Changing the Orbit of a Satellite**

The space shuttle releases a 470-kg communications satellite while in an orbit that is 280 km above the surface of the Earth. A rocket engine on the satellite boosts it into a geosynchronous orbit, which is an orbit in which the satellite stays directly over a single location on the Earth. How much energy did the engine have to provide?

Solution   First we must determine the radius of a geosynchronous orbit. Then we can calculate the change in energy needed to boost the satellite into orbit.

The period of the orbit T must be one day (86 400 s), so that the satellite travels once around the Earth in the same time that the Earth spins once on its axis. Knowing the period, we can then apply Kepler's third law (Eq. 14.7) to find the radius, once we replace K_S with $K_E = 4\pi^2/GM_E = 9.89 \times 10^{-14}\ \text{s}^2/\text{m}^3$:

$$T^2 = K_E r^3$$

$$r = \sqrt[3]{\frac{T^2}{K_E}} = \sqrt[3]{\frac{(86\ 400\ \text{s})^2}{9.89 \times 10^{-14}\ \text{s}^2/\text{m}^3}} = 4.23 \times 10^7\ \text{m} = R_f$$

This is a little more than 26 000 mi above the Earth's surface.

We must also determine the initial radius (not the altitude above the Earth's surface) of the satellite's orbit when it was still in the shuttle's cargo bay. This is simply

$$R_E + 280\ \text{km} = 6.65 \times 10^6\ \text{m} = R_i$$

Now, applying Equation 14.19, we obtain, for the total initial and final energies,

$$E_i = -\frac{GM_E m}{2R_i} \qquad E_f = -\frac{GM_E m}{2R_f}$$

The energy required from the engine to boost the satellite is

$$E_\text{engine} = E_f - E_i = -\frac{GM_E m}{2}\left(\frac{1}{R_f} - \frac{1}{R_i}\right)$$

$$= -\frac{(6.67 \times 10^{-11}\ \text{N} \cdot \text{m}^2/\text{kg}^2)(5.98 \times 10^{24}\ \text{kg})(470\ \text{kg})}{2}$$

$$\times \left(\frac{1}{4.23 \times 10^7\ \text{m}} - \frac{1}{6.65 \times 10^6\ \text{m}}\right)$$

$$= \boxed{1.19 \times 10^{10}\ \text{J}}$$

This is the energy equivalent of 89 gal of gasoline. NASA engineers must account for the changing mass of the spacecraft as it ejects burned fuel, something we have not done here. Would you expect the calculation that includes the effect of this changing mass to yield a greater or lesser amount of energy required from the engine?

If we wish to determine how the energy is distributed after the engine is fired, we find from Equation 14.18 that the change in kinetic energy is $\Delta K = (GM_E m/2)(1/R_f - 1/R_i) = -1.19 \times 10^{10}$ J (a decrease),

and the corresponding change in potential energy is $\Delta U = -GM_E m(1/R_f - 1/R_i) = 2.38 \times 10^{10}$ J (an increase). Thus, the change in mechanical energy of the system is $\Delta E = \Delta K + \Delta U = 1.19 \times 10^{10}$ J, as we already calculated. The firing of the engine results in an increase in the total mechanical energy of the system. Because an increase in potential energy is accompanied by a decrease in kinetic energy, we conclude that the speed of an orbiting satellite decreases as its altitude increases.

Escape Speed

Suppose an object of mass m is projected vertically upward from the Earth's surface with an initial speed v_i, as illustrated in Figure 14.17. We can use energy considerations to find the minimum value of the initial speed needed to allow the object to escape the Earth's gravitational field. Equation 14.17 gives the total energy of the object at any point. At the surface of the Earth, $v = v_i$ and $r = r_i = R_E$. When the object reaches its maximum altitude, $v = v_f = 0$ and $r = r_f = r_{\max}$. Because the total energy of the system is constant, substituting these conditions into Equation 14.20 gives

$$\tfrac{1}{2}mv_i^2 - \frac{GM_E m}{R_E} = -\frac{GM_E m}{r_{\max}}$$

Solving for v_i^2 gives

$$v_i^2 = 2GM_E\left(\frac{1}{R_E} - \frac{1}{r_{\max}}\right) \tag{14.21}$$

Therefore, if the initial speed is known, this expression can be used to calculate the maximum altitude h because we know that

$$h = r_{\max} - R_E$$

We are now in a position to calculate **escape speed,** which is the minimum speed the object must have at the Earth's surface in order to escape from the influence of the Earth's gravitational field. Traveling at this minimum speed, the object continues to move farther and farther away from the Earth as its speed asymptotically approaches zero. Letting $r_{\max} \to \infty$ in Equation 14.21 and taking $v_i = v_{\mathrm{esc}}$, we obtain

$$v_{\mathrm{esc}} = \sqrt{\frac{2GM_E}{R_E}} \tag{14.22}$$

Escape speed

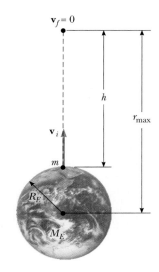

Figure 14.17 An object of mass m projected upward from the Earth's surface with an initial speed v_i reaches a maximum altitude h.

Note that this expression for v_{esc} is independent of the mass of the object. In other words, a spacecraft has the same escape speed as a molecule. Furthermore, the result is independent of the direction of the velocity and ignores air resistance.

If the object is given an initial speed equal to v_{esc}, its total energy is equal to zero. This can be seen by noting that when $r \to \infty$, the object's kinetic energy and its potential energy are both zero. If v_i is greater than v_{esc}, the total energy is greater than zero and the object has some residual kinetic energy as $r \to \infty$.

EXAMPLE 14.8 **Escape Speed of a Rocket**

Calculate the escape speed from the Earth for a 5 000-kg spacecraft, and determine the kinetic energy it must have at the Earth's surface in order to escape the Earth's gravitational field.

Solution Using Equation 14.22 gives

$$v_{esc} = \sqrt{\frac{2GM_E}{R_E}}$$

$$= \sqrt{\frac{2(6.67 \times 10^{-11}\,\text{N}\cdot\text{m}^2/\text{kg}^2)(5.98 \times 10^{24}\,\text{kg})}{6.37 \times 10^6\,\text{m}}}$$

$$= \boxed{1.12 \times 10^4\,\text{m/s}}$$

This corresponds to about 25 000 mi/h.

The kinetic energy of the spacecraft is

$$K = \tfrac{1}{2}mv_{esc}^2 = \tfrac{1}{2}(5.00 \times 10^3\,\text{kg})(1.12 \times 10^4\,\text{m/s})^2$$

$$= \boxed{3.14 \times 10^{11}\,\text{J}}$$

This is equivalent to about 2 300 gal of gasoline.

Equations 14.21 and 14.22 can be applied to objects projected from any planet. That is, in general, the escape speed from the surface of any planet of mass M and radius R is

$$v_{esc} = \sqrt{\frac{2GM}{R}}$$

Escape speeds for the planets, the Moon, and the Sun are provided in Table 14.3. Note that the values vary from 1.1 km/s for Pluto to about 618 km/s for the Sun. These results, together with some ideas from the kinetic theory of gases (see Chapter 21), explain why some planets have atmospheres and others do not. As we shall see later, a gas molecule has an average kinetic energy that depends on the temperature of the gas. Hence, lighter molecules, such as hydrogen and helium, have a higher average speed than heavier species at the same temperature. When the average speed of the lighter molecules is not much less than the escape speed of a planet, a significant fraction of them have a chance to escape from the planet.

This mechanism also explains why the Earth does not retain hydrogen molecules and helium atoms in its atmosphere but does retain heavier molecules, such as oxygen and nitrogen. On the other hand, the very large escape speed for Jupiter enables that planet to retain hydrogen, the primary constituent of its atmosphere.

TABLE 14.3
Escape Speeds from the Surfaces of the Planets, Moon, and Sun

Body	v_{esc} (km/s)
Mercury	4.3
Venus	10.3
Earth	11.2
Moon	2.3
Mars	5.0
Jupiter	60
Saturn	36
Uranus	22
Neptune	24
Pluto	1.1
Sun	618

Quick Quiz 14.2

If you were a space prospector and discovered gold on an asteroid, it probably would not be a good idea to jump up and down in excitement over your find. Why?

Quick Quiz 14.3

Figure 14.18 is a drawing by Newton showing the path of a stone thrown from a mountaintop. He shows the stone landing farther and farther away when thrown at higher and higher speeds (at points *D, E, F,* and *G*), until finally it is thrown all the way around the Earth. Why didn't Newton show the stone landing at *B* and *A* before it was going fast enough to complete an orbit?

Figure 14.18   "The greater the velocity . . . with which [a stone] is projected, the farther it goes before it falls to the Earth. We may therefore suppose the velocity to be so increased, that it would describe an arc of 1, 2, 5, 10, 100, 1000 miles before it arrived at the Earth, till at last, exceeding the limits of the Earth, it should pass into space without touching." Sir Isaac Newton, *System of the World*.

Optional Section

14.9   THE GRAVITATIONAL FORCE BETWEEN AN EXTENDED OBJECT AND A PARTICLE

We have emphasized that the law of universal gravitation given by Equation 14.3 is valid only if the interacting objects are treated as particles. In view of this, how can we calculate the force between a particle and an object having finite dimensions? This is accomplished by treating the extended object as a collection of particles and making use of integral calculus. We first evaluate the potential energy function, and then calculate the gravitational force from that function.

We obtain the potential energy associated with a system consisting of a particle of mass m and an extended object of mass M by dividing the object into many elements, each having a mass ΔM_i (Fig. 14.19). The potential energy associated with the system consisting of any one element and the particle is $U = -Gm\,\Delta M_i/r_i$, where r_i is the distance from the particle to the element ΔM_i. The total potential energy of the overall system is obtained by taking the sum over all elements as $\Delta M_i \rightarrow 0$. In this limit, we can express U in integral form as

$$U = -Gm \int \frac{dM}{r} \qquad (14.23)$$

Once U has been evaluated, we obtain the force exerted by the extended object on the particle by taking the negative derivative of this scalar function (see Section 8.6). If the extended object has spherical symmetry, the function U depends only on r, and the force is given by $-dU/dr$. We treat this situation in Section 14.10. In principle, one can evaluate U for any geometry; however, the integration can be cumbersome.

An alternative approach to evaluating the gravitational force between a particle and an extended object is to perform a vector sum over all mass elements of the object. Using the procedure outlined in evaluating U and the law of universal gravitation in the form shown in Equation 14.3, we obtain, for the total force exerted on the particle

$$\mathbf{F}_g = -Gm \int \frac{dM}{r^2}\,\hat{\mathbf{r}} \qquad (14.24)$$

where $\hat{\mathbf{r}}$ is a unit vector directed from the element dM toward the particle (see Fig. 14.19) and the minus sign indicates that the direction of the force is opposite that of $\hat{\mathbf{r}}$. This procedure is not always recommended because working with a vector function is more difficult than working with the scalar potential energy function. However, if the geometry is simple, as in the following example, the evaluation of $\mathbf{F}$ can be straightforward.

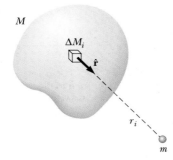

Figure 14.19   A particle of mass m interacting with an extended object of mass M. The total gravitational force exerted by the object on the particle can be obtained by dividing the object into numerous elements, each having a mass ΔM_i, and then taking a vector sum over the forces exerted by all elements.

Total force exerted on a particle by an extended object

EXAMPLE 14.9   Gravitational Force Between a Particle and a Bar

The left end of a homogeneous bar of length L and mass M is at a distance h from a particle of mass m (Fig. 14.20). Calculate the total gravitational force exerted by the bar on the particle.

Solution   The arbitrary segment of the bar of length dx has a mass dM. Because the mass per unit length is constant, it follows that the ratio of masses dM/M is equal to the ratio

y

m
O
x

Figure 14.20   The gravitational force exerted by the bar on the particle is directed to the right. Note that the bar is *not* equivalent to a particle of mass M located at the center of mass of the bar.

of lengths dx/L, and so $dM = (M/L)\, dx$. In this problem, the variable r in Equation 14.24 is the distance x shown in Figure 14.20, the unit vector $\hat{\mathbf{r}}$ is $\hat{\mathbf{r}} = -\mathbf{i}$, and the force acting on the particle is to the right; therefore, Equation 14.24 gives us

$$\mathbf{F}_g = -Gm \int_h^{h+L} \frac{M\,dx}{L}\frac{1}{x^2}(-\mathbf{i}) = Gm\frac{M}{L}\int_h^{h+L}\frac{dx}{x^2}\mathbf{i}$$

$$\mathbf{F}_g = \frac{GmM}{L}\left[-\frac{1}{x}\right]_h^{h+L}\mathbf{i} = \frac{GmM}{h(h+L)}\mathbf{i}$$

We see that the force exerted on the particle is in the positive x direction, which is what we expect because the gravitational force is attractive.

Note that in the limit $L \to 0$, the force varies as $1/h^2$, which is what we expect for the force between two point masses. Furthermore, if $h \gg L$, the force also varies as $1/h^2$. This can be seen by noting that the denominator of the expression for $\mathbf{F}_g$ can be expressed in the form $h^2(1 + L/h)$, which is approximately equal to h^2 when $h \gg L$. Thus, when bodies are separated by distances that are great relative to their characteristic dimensions, they behave like particles.

Optional Section

14.10   THE GRAVITATIONAL FORCE BETWEEN A PARTICLE AND A SPHERICAL MASS

We have already stated that a large sphere attracts a particle outside it as if the total mass of the sphere were concentrated at its center. We now describe the force acting on a particle when the extended object is either a spherical shell or a solid sphere, and then apply these facts to some interesting systems.

Spherical Shell

Case 1.   If a particle of mass m is located outside a spherical shell of mass M at, for instance, point P in Figure 14.21a, the shell attracts the particle as though the mass of the shell were concentrated at its center. We can show this, as Newton did, with integral calculus. Thus, as far as the gravitational force acting on a particle outside the shell is concerned, a spherical shell acts no differently from the solid spherical distributions of mass we have seen.

Case 2.   If the particle is located inside the shell (at point P in Fig. 14.21b), the gravitational force acting on it can be shown to be zero.

We can express these two important results in the following way:

Force on a particle due to a spherical shell

$$\mathbf{F}_g = -\frac{GMm}{r^2}\hat{\mathbf{r}} \qquad \text{for } r \geq R \tag{14.25a}$$

$$\mathbf{F}_g = 0 \qquad \text{for } r < R \tag{14.25b}$$

The gravitational force as a function of the distance r is plotted in Figure 14.21c.

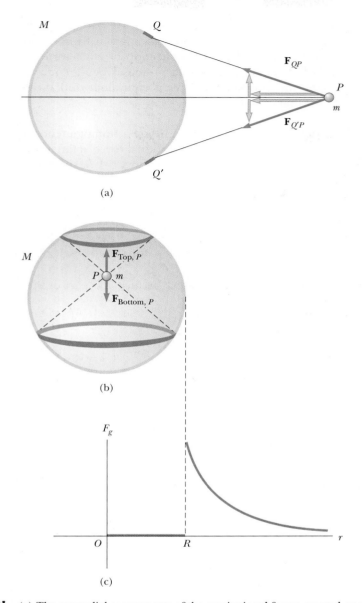

Figure 14.21 (a) The nonradial components of the gravitational forces exerted on a particle of mass m located at point P outside a spherical shell of mass M cancel out. (b) The spherical shell can be broken into rings. Even though point P is closer to the top ring than to the bottom ring, the bottom ring is larger, and the gravitational forces exerted on the particle at P by the matter in the two rings cancel each other. Thus, for a particle located at any point P inside the shell, there is no gravitational force exerted on the particle by the mass M of the shell. (c) The magnitude of the gravitational force versus the radial distance r from the center of the shell.

The shell does not act as a gravitational shield, which means that a particle inside a shell may experience forces exerted by bodies outside the shell.

Solid Sphere

Case 1. If a particle of mass m is located outside a homogeneous solid sphere of mass M (at point P in Fig. 14.22), the sphere attracts the particle as though the mass of the sphere were concentrated at its center. We have used this notion at several places in this chapter already, and we can argue it from Equation 14.25a. A solid sphere can be considered to be a collection of concentric spherical shells. The masses of all of the shells can be interpreted as being concentrated at their common center, and the gravitational force is equivalent to that due to a particle of mass M located at that center.

Case 2.   If a particle of mass m is located inside a homogeneous solid sphere of mass M (at point Q in Fig. 14.22), the gravitational force acting on it is due *only* to the mass M' contained within the sphere of radius $r < R$, shown in Figure 14.22. In other words,

Force on a particle due to a solid sphere

$$\mathbf{F}_g = -\frac{GmM}{r^2}\,\hat{\mathbf{r}} \qquad \text{for } r \geq R \tag{14.26a}$$

$$\mathbf{F}_g = -\frac{GmM'}{r^2}\,\hat{\mathbf{r}} \qquad \text{for } r < R \tag{14.26b}$$

This also follows from spherical-shell Case 1 because the part of the sphere that is

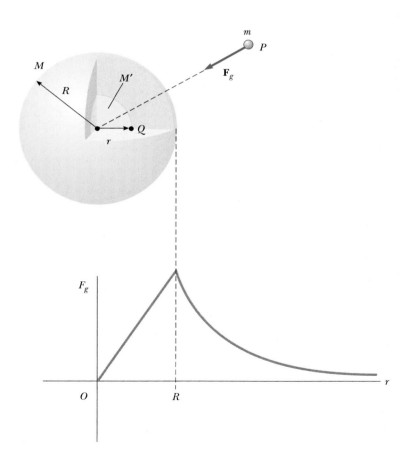

Figure 14.22   The gravitational force acting on a particle when it is outside a uniform solid sphere is GMm/r^2 and is directed toward the center of the sphere. The gravitational force acting on the particle when it is inside such a sphere is proportional to r and goes to zero at the center.

farther from the center than Q can be treated as a series of concentric spherical shells that do not exert a net force on the particle because the particle is inside them. Because the sphere is assumed to have a uniform density, it follows that the ratio of masses M'/M is equal to the ratio of volumes V'/V, where V is the total volume of the sphere and V' is the volume within the sphere of radius r only:

$$\frac{M'}{M} = \frac{V'}{V} = \frac{\frac{4}{3}\pi r^3}{\frac{4}{3}\pi R^3} = \frac{r^3}{R^3}$$

Solving this equation for M' and substituting the value obtained into Equation 14.26b, we have

$$\mathbf{F}_g = -\frac{GmM}{R^3}\, r\hat{\mathbf{r}} \qquad \text{for } r < R \qquad\qquad \textbf{(14.27)}$$

This equation tells us that at the center of the solid sphere, where $r = 0$, the gravitational force goes to zero, as we intuitively expect. The force as a function of r is plotted in Figure 14.22.

Case 3. If a particle is located inside a solid sphere having a density ρ that is spherically symmetric but not uniform, then M' in Equation 14.26b is given by an integral of the form $M' = \int \rho\, dV$, where the integration is taken over the volume contained within the sphere of radius r in Figure 14.22. We can evaluate this integral if the radial variation of ρ is given. In this case, we take the volume element dV as the volume of a spherical shell of radius r and thickness dr, and thus $dV = 4\pi r^2\, dr$. For example, if $\rho = Ar$, where A is a constant, it is left to a problem (Problem 63) to show that $M' = \pi A r^4$. Hence, we see from Equation 14.26b that F is proportional to r^2 in this case and is zero at the center.

Quick Quiz 14.4

A particle is projected through a small hole into the interior of a spherical shell. Describe the motion of the particle inside the shell.

EXAMPLE 14.10   A Free Ride, Thanks to Gravity

An object of mass m moves in a smooth, straight tunnel dug between two points on the Earth's surface (Fig. 14.23). Show that the object moves with simple harmonic motion, and find the period of its motion. Assume that the Earth's density is uniform.

Solution The gravitational force exerted on the object acts toward the Earth's center and is given by Equation 14.27:

$$\mathbf{F}_g = -\frac{GmM}{R^3}\, r\hat{\mathbf{r}}$$

We receive our first indication that this force should result in simple harmonic motion by comparing it to Hooke's law, first seen in Section 7.3. Because the gravitational force on the object is linearly proportional to the displacement, the object experiences a Hooke's law force.

The y component of the gravitational force on the object is balanced by the normal force exerted by the tunnel wall, and the x component is

$$F_x = -\frac{GmM_E}{R_E^3}\, r\cos\theta$$

Because the x coordinate of the object is $x = r\cos\theta$, we can write

$$F_x = -\frac{GmM_E}{R_E^3}\, x$$

Applying Newton's second law to the motion along the x direction gives

$$F_x = -\frac{GmM_E}{R_E^3}\, x = ma_x$$

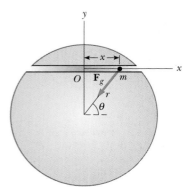

Figure 14.23 An object moves along a tunnel dug through the Earth. The component of the gravitational force $\mathbf{F}_g$ along the x axis is the driving force for the motion. Note that this component always acts toward O.

Solving for a_x, we obtain

$$a_x = -\frac{GM_E}{R_E^3}\,x$$

If we use the symbol ω^2 for the coefficient of x — $GM_E/R_E^3 = \omega^2$ — we see that

$$(1) \qquad a_x = -\omega^2 x$$

an expression that matches the mathematical form of Equation 13.9, which gives the acceleration of a particle in simple harmonic motion: $a_x = -\omega^2 x$. Therefore, Equation (1),

which we have derived for the acceleration of our object in the tunnel, is the acceleration equation for simple harmonic motion at angular speed ω with

$$\omega = \sqrt{\frac{GM_E}{R_E^3}}$$

Thus, the object in the tunnel moves in the same way as a block hanging from a spring! The period of oscillation is

$$T = \frac{2\pi}{\omega} = 2\pi\sqrt{\frac{R_E^3}{GM_E}}$$

$$= 2\pi\sqrt{\frac{(6.37\times10^6\,\text{m})^3}{(6.67\times10^{-11}\,\text{N}\cdot\text{m}^2/\text{kg}^2)(5.98\times10^{24}\,\text{kg})}}$$

$$= 5.06\times10^3\,\text{s} = \boxed{84.3\ \text{min}}$$

This period is the same as that of a satellite traveling in a circular orbit just above the Earth's surface (ignoring any trees, buildings, or other objects in the way). Note that the result is independent of the length of the tunnel.

A proposal has been made to operate a mass-transit system between any two cities, using the principle described in this example. A one-way trip would take about 42 min. A more precise calculation of the motion must account for the fact that the Earth's density is not uniform. More important, there are many practical problems to consider. For instance, it would be impossible to achieve a frictionless tunnel, and so some auxiliary power source would be required. Can you think of other problems?

SUMMARY

Newton's law of universal gravitation states that the gravitational force of attraction between any two particles of masses m_1 and m_2 separated by a distance r has the magnitude

$$F_g = G\frac{m_1 m_2}{r^2} \tag{14.1}$$

where $G = 6.673\times10^{-11}\,\text{N}\cdot\text{m}^2/\text{kg}^2$ is the universal gravitational constant. This equation enables us to calculate the force of attraction between masses under a wide variety of circumstances.

An object at a distance h above the Earth's surface experiences a gravitational force of magnitude mg', where g' is the free-fall acceleration at that elevation:

$$g' = \frac{GM_E}{r^2} = \frac{GM_E}{(R_E + h)^2} \tag{14.6}$$

In this expression, M_E is the mass of the Earth and R_E is its radius. Thus, the weight of an object decreases as the object moves away from the Earth's surface.

Kepler's laws of planetary motion state that

1. All planets move in elliptical orbits with the Sun at one focal point.
2. The radius vector drawn from the Sun to a planet sweeps out equal areas in equal time intervals.
3. The square of the orbital period of any planet is proportional to the cube of the semimajor axis of the elliptical orbit.

Kepler's third law can be expressed as

$$T^2 = \left(\frac{4\pi^2}{GM_S}\right)r^3 \tag{14.7}$$

where M_S is the mass of the Sun and r is the orbital radius. For elliptical orbits, Equation 14.7 is valid if r is replaced by the semimajor axis a. Most planets have nearly circular orbits around the Sun.

The **gravitational field** at a point in space equals the gravitational force experienced by any test particle located at that point divided by the mass of the test particle:

$$\mathbf{g} = \frac{\mathbf{F}_g}{m} \tag{14.10}$$

The gravitational force is conservative, and therefore a potential energy function can be defined. The **gravitational potential energy** associated with two particles separated by a distance r is

$$U = -\frac{Gm_1m_2}{r} \tag{14.15}$$

where U is taken to be zero as $r \rightarrow \infty$. The total potential energy for a system of particles is the sum of energies for all pairs of particles, with each pair represented by a term of the form given by Equation 14.15.

If an isolated system consists of a particle of mass m moving with a speed v in the vicinity of a massive body of mass M, the total energy E of the system is the sum of the kinetic and potential energies:

$$E = \tfrac{1}{2}mv^2 - \frac{GMm}{r} \tag{14.17}$$

The total energy is a constant of the motion. If the particle moves in a circular orbit of radius r around the massive body and if $M \gg m$, the total energy of the system is

$$E = -\frac{GMm}{2r} \tag{14.19}$$

The total energy is negative for any bound system.

The **escape speed** for an object projected from the surface of the Earth is

$$v_{\text{esc}} = \sqrt{\frac{2GM_E}{R_E}} \tag{14.22}$$

QUESTIONS

1. Use Kepler's second law to convince yourself that the Earth must move faster in its orbit during December, when it is closest to the Sun, than during June, when it is farthest from the Sun.

2. The gravitational force that the Sun exerts on the Moon is about twice as great as the gravitational force that the Earth exerts on the Moon. Why doesn't the Sun pull the Moon away from the Earth during a total eclipse of the Sun?

3. If a system consists of five particles, how many terms appear in the expression for the total potential energy? How many terms appear if the system consists of *N* particles?

4. Is it possible to calculate the potential energy function associated with a particle and an extended body without knowing the geometry or mass distribution of the extended body?

5. Does the escape speed of a rocket depend on its mass? Explain.

6. Compare the energies required to reach the Moon for a 10^5-kg spacecraft and a 10^3-kg satellite.

7. Explain why it takes more fuel for a spacecraft to travel from the Earth to the Moon than for the return trip. Estimate the difference.

8. Why don't we put a geosynchronous weather satellite in orbit around the 45th parallel? Wouldn't this be more useful for the United States than such a satellite in orbit around the equator?

9. Is the potential energy associated with the Earth–Moon system greater than, less than, or equal to the kinetic energy of the Moon relative to the Earth?

10. Explain why no work is done on a planet as it moves in a circular orbit around the Sun, even though a gravitational force is acting on the planet. What is the net work done on a planet during each revolution as it moves around the Sun in an elliptical orbit?

11. Explain why the force exerted on a particle by a uniform sphere must be directed toward the center of the sphere. Would this be the case if the mass distribution of the sphere were not spherically symmetric?

12. Neglecting the density variation of the Earth, what would be the period of a particle moving in a smooth hole dug between opposite points on the Earth's surface, passing through its center?

13. At what position in its elliptical orbit is the speed of a planet a maximum? At what position is the speed a minimum?

14. If you were given the mass and radius of planet X, how would you calculate the free-fall acceleration on the surface of this planet?

15. If a hole could be dug to the center of the Earth, do you think that the force on a mass *m* would still obey Equation 14.1 there? What do you think the force on *m* would be at the center of the Earth?

16. In his 1798 experiment, Cavendish was said to have "weighed the Earth." Explain this statement.

17. The gravitational force exerted on the *Voyager* spacecraft by Jupiter accelerated it toward escape speed from the Sun. How is this possible?

18. How would you find the mass of the Moon?

19. The *Apollo 13* spaceship developed trouble in the oxygen system about halfway to the Moon. Why did the spaceship continue on around the Moon and then return home, rather than immediately turn back to Earth?

PROBLEMS

1, 2, 3 = straightforward, intermediate, challenging   ☐ = full solution available in the *Student Solutions Manual and Study Guide*
WEB = solution posted at **http://www.saunderscollege.com/physics/**   🖥 = Computer useful in solving problem   🕹 = Interactive Physics
☐ = paired numerical/symbolic problems

Section 14.1   Newton's Law of Universal Gravitation
Section 14.2   Measuring the Gravitational Constant
Section 14.3   Free-Fall Acceleration and the Gravitational Force

1. Determine the order of magnitude of the gravitational force that you exert on another person 2 m away. In your solution, state the quantities that you measure or estimate and their values.

2. A 200-kg mass and a 500-kg mass are separated by 0.400 m. (a) Find the net gravitational force exerted by these masses on a 50.0-kg mass placed midway between them. (b) At what position (other than infinitely remote ones) can the 50.0-kg mass be placed so as to experience a net force of zero?

3. Three equal masses are located at three corners of a square of edge length ℓ, as shown in Figure P14.3. Find the gravitational field **g** at the fourth corner due to these masses.

4. Two objects attract each other with a gravitational force of magnitude 1.00×10^{-8} N when separated by 20.0 cm. If the total mass of the two objects is 5.00 kg, what is the mass of each?

5. Three uniform spheres of masses 2.00 kg, 4.00 kg, and 6.00 kg are placed at the corners of a right triangle, as illustrated in Figure P14.5. Calculate the resultant gravi-

Figure P14.3

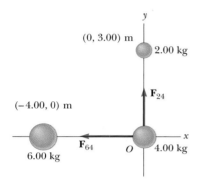

Figure P14.5

tational force on the 4.00-kg mass, assuming that the spheres are isolated from the rest of the Universe.

6. The free-fall acceleration on the surface of the Moon is about one-sixth that on the surface of the Earth. If the radius of the Moon is about $0.250 R_E$, find the ratio of their average densities, $\rho_{\text{Moon}}/\rho_{\text{Earth}}$.

7. During a solar eclipse, the Moon, Earth, and Sun all lie on the same line, with the Moon between the Earth and the Sun. (a) What force is exerted by the Sun on the Moon? (b) What force is exerted by the Earth on the Moon? (c) What force is exerted by the Sun on the Earth?

8. The center-to-center distance between the Earth and the Moon is 384 400 km. The Moon completes an orbit in 27.3 days. (a) Determine the Moon's orbital speed. (b) If gravity were switched off, the Moon would move along a straight line tangent to its orbit, as described by Newton's first law. In its actual orbit in 1.00 s, how far does the Moon fall below the tangent line and toward the Earth?

WEB 9. When a falling meteoroid is at a distance above the Earth's surface of 3.00 times the Earth's radius, what is its acceleration due to the Earth's gravity?

10. Two ocean liners, each with a mass of 40 000 metric tons, are moving on parallel courses, 100 m apart. What is the magnitude of the acceleration of one of the liners toward the other due to their mutual gravitational attraction? (Treat the ships as point masses.)

11. A student proposes to measure the gravitational constant G by suspending two spherical masses from the ceiling of a tall cathedral and measuring the deflection of the cables from the vertical. Draw a free-body diagram of one of the masses. If two 100.0-kg masses are suspended at the end of 45.00-m-long cables, and the cables are attached to the ceiling 1.000 m apart, what is the separation of the masses?

12. On the way to the Moon, the Apollo astronauts reached a point where the Moon's gravitational pull became stronger than the Earth's. (a) Determine the distance of this point from the center of the Earth. (b) What is the acceleration due to the Earth's gravity at this point?

Section 14.4 Kepler's Laws
Section 14.5 The Law of Gravity and the Motion of Planets

13. A particle of mass m moves along a straight line with constant speed in the x direction, a distance b from the x axis (Fig. P14.13). Show that Kepler's second law is satisfied by demonstrating that the two shaded triangles in the figure have the same area when $t_4 - t_3 = t_2 - t_1$.

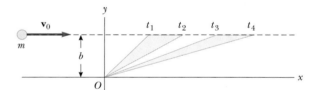

Figure P14.13

14. A communications satellite in geosynchronous orbit remains above a single point on the Earth's equator as the planet rotates on its axis. (a) Calculate the radius of its orbit. (b) The satellite relays a radio signal from a transmitter near the north pole to a receiver, also near the north pole. Traveling at the speed of light, how long is the radio wave in transit?

15. Plaskett's binary system consists of two stars that revolve in a circular orbit about a center of mass midway between them. This means that the masses of the two stars are equal (Fig. P14.15). If the orbital velocity of each star is 220 km/s and the orbital period of each is 14.4 days, find the mass M of each star. (For comparison, the mass of our Sun is 1.99×10^{30} kg.)

16. Plaskett's binary system consists of two stars that revolve in a circular orbit about a center of gravity midway between them. This means that the masses of the two stars are equal (see Fig. P14.15). If the orbital speed of each star is v and the orbital period of each is T, find the mass M of each star.

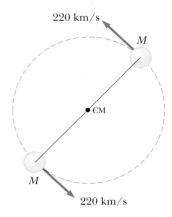

220 km/s

M

CM

M

220 km/s

Figure P14.15   Problems 15 and 16.

17. The *Explorer VIII* satellite, placed into orbit November 3, 1960, to investigate the ionosphere, had the following orbit parameters: perigee, 459 km; apogee, 2 289 km (both distances above the Earth's surface); and period, 112.7 min. Find the ratio v_p/v_a of the speed at perigee to that at apogee.

18. Comet Halley (Fig. P14.18) approaches the Sun to within 0.570 AU, and its orbital period is 75.6 years (AU is the symbol for astronomical unit, where 1 AU = 1.50×10^{11} m is the mean Earth–Sun distance). How far from the Sun will Halley's comet travel before it starts its return journey?

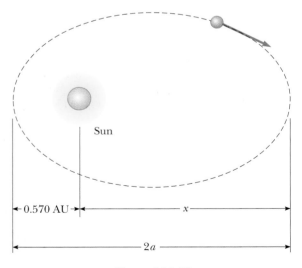

Sun

←0.570 AU→|←————— x —————→

|←————————— 2a —————————→|

Figure P14.18

WEB **19.** Io, a satellite of Jupiter, has an orbital period of 1.77 days and an orbital radius of 4.22×10^5 km. From these data, determine the mass of Jupiter.

20. Two planets, X and Y, travel counterclockwise in circular orbits about a star, as shown in Figure P14.20. The radii of their orbits are in the ratio 3:1. At some time, they are aligned as in Figure P14.20a, making a straight line with the star. During the next five years, the angular displacement of planet X is 90.0°, as shown in Figure P14.20b. Where is planet Y at this time?

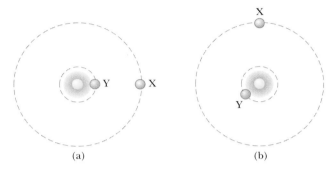

X

Y X

Y

(a)   (b)

Figure P14.20

21. A synchronous satellite, which always remains above the same point on a planet's equator, is put in orbit around Jupiter so that scientists can study the famous red spot. Jupiter rotates once every 9.84 h. Use the data in Table 14.2 to find the altitude of the satellite.

22. Neutron stars are extremely dense objects that are formed from the remnants of supernova explosions. Many rotate very rapidly. Suppose that the mass of a certain spherical neutron star is twice the mass of the Sun and that its radius is 10.0 km. Determine the greatest possible angular speed it can have for the matter at the surface of the star on its equator to be just held in orbit by the gravitational force.

23. The Solar and Heliospheric Observatory (SOHO) spacecraft has a special orbit, chosen so that its view of the Sun is never eclipsed and it is always close enough to the Earth to transmit data easily. It moves in a near-circle around the Sun that is smaller than the Earth's circular orbit. Its period, however, is not less than 1 yr but is just equal to 1 yr. It is always located between the Earth and the Sun along the line joining them. Both objects exert gravitational forces on the observatory. Show that the spacecraft's distance from the Earth must be between 1.47×10^9 m and 1.48×10^9 m. In 1772 Joseph Louis Lagrange determined theoretically the special location that allows this orbit. The SOHO spacecraft took this position on February 14, 1996. (*Hint:* Use data that are precise to four digits. The mass of the Earth is 5.983×10^{24} kg.)

Section 14.6   The Gravitational Field

24. A spacecraft in the shape of a long cylinder has a length of 100 m, and its mass with occupants is 1 000 kg. It has

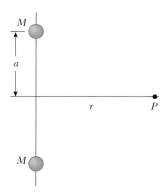

Figure P14.24

strayed too close to a 1.0-m-radius black hole having a mass 100 times that of the Sun (Fig. P14.24). The nose of the spacecraft is pointing toward the center of the black hole, and the distance between the nose and the black hole is 10.0 km. (a) Determine the total force on the spacecraft. (b) What is the difference in the gravitational fields acting on the occupants in the nose of the ship and on those in the rear of the ship, farthest from the black hole?

25. Compute the magnitude and direction of the gravitational field at a point *P* on the perpendicular bisector of two equal masses separated by a distance 2*a*, as shown in Figure P14.25.

Figure P14.25

26. Find the gravitational field at a distance *r* along the axis of a thin ring of mass *M* and radius *a*.

Section 14.7 Gravitational Potential Energy

Note: Assume that $U = 0$ as $r \rightarrow \infty$.

27. A satellite of the Earth has a mass of 100 kg and is at an altitude of 2.00×10^6 m. (a) What is the potential energy of the satellite–Earth system? (b) What is the magnitude of the gravitational force exerted by the Earth on the satellite? (c) What force does the satellite exert on the Earth?

28. How much energy is required to move a 1 000-kg mass from the Earth's surface to an altitude twice the Earth's radius?

29. After our Sun exhausts its nuclear fuel, its ultimate fate may be to collapse to a *white-dwarf* state, in which it has approximately the same mass it has now but a radius equal to the radius of the Earth. Calculate (a) the average density of the white dwarf, (b) the acceleration due to gravity at its surface, and (c) the gravitational potential energy associated with a 1.00-kg object at its surface.

30. At the Earth's surface a projectile is launched straight up at a speed of 10.0 km/s. To what height will it rise? Ignore air resistance.

31. A system consists of three particles, each of mass 5.00 g, located at the corners of an equilateral triangle with sides of 30.0 cm. (a) Calculate the potential energy of the system. (b) If the particles are released simultaneously, where will they collide?

32. How much work is done by the Moon's gravitational field as a 1 000-kg meteor comes in from outer space and impacts the Moon's surface?

Section 14.8 Energy Considerations in Planetary and Satellite Motion

33. A 500-kg satellite is in a circular orbit at an altitude of 500 km above the Earth's surface. Because of air friction, the satellite is eventually brought to the Earth's surface, and it hits the Earth with a speed of 2.00 km/s. How much energy was transformed to internal energy by means of friction?

34. (a) What is the minimum speed, relative to the Sun, that is necessary for a spacecraft to escape the Solar System if it starts at the Earth's orbit? (b) *Voyager 1* achieved a maximum speed of 125 000 km/h on its way to photograph Jupiter. Beyond what distance from the Sun is this speed sufficient for a spacecraft to escape the Solar System?

35. A satellite with a mass of 200 kg is placed in Earth orbit at a height of 200 km above the surface. (a) Assuming a circular orbit, how long does the satellite take to complete one orbit? (b) What is the satellite's speed? (c) What is the minimum energy necessary to place this satellite in orbit (assuming no air friction)?

36. A satellite of mass *m* is placed in Earth orbit at an altitude *h*. (a) Assuming a circular orbit, how long does the satellite take to complete one orbit? (b) What is the satellite's speed? (c) What is the minimum energy necessary to place this satellite in orbit (assuming no air friction)?

WEB 37. A spaceship is fired from the Earth's surface with an initial speed of 2.00×10^4 m/s. What will its speed be when it is very far from the Earth? (Neglect friction.)

38. A 1 000-kg satellite orbits the Earth at a constant altitude of 100 km. How much energy must be added to the system to move the satellite into a circular orbit at an altitude of 200 km?

39. A "treetop satellite" moves in a circular orbit just above the surface of a planet, which is assumed to offer no air resistance. Show that its orbital speed *v* and the escape speed from the planet are related by the expression $v_{\text{esc}} = \sqrt{2}v$.

40. The planet Uranus has a mass about 14 times the Earth's mass, and its radius is equal to about 3.7 Earth

radii. (a) By setting up ratios with the corresponding Earth values, find the acceleration due to gravity at the cloud tops of Uranus. (b) Ignoring the rotation of the planet, find the minimum escape speed from Uranus.

41. Determine the escape velocity for a rocket on the far side of Ganymede, the largest of Jupiter's moons. The radius of Ganymede is 2.64×10^6 m, and its mass is 1.495×10^{23} kg. The mass of Jupiter is 1.90×10^{27} kg, and the distance between Jupiter and Ganymede is 1.071×10^9 m. Be sure to include the gravitational effect due to Jupiter, but you may ignore the motions of Jupiter and Ganymede as they revolve about their center of mass (Fig. P14.41).

Jupiter

Ganymede

Figure P14.41

42. In Robert Heinlein's *The Moon is a Harsh Mistress*, the colonial inhabitants of the Moon threaten to launch rocks down onto the Earth if they are not given independence (or at least representation). Assuming that a rail gun could launch a rock of mass m at twice the lunar escape speed, calculate the speed of the rock as it enters the Earth's atmosphere. (By *lunar escape speed* we mean the speed required to escape entirely from a stationary Moon alone in the Universe.)

43. Derive an expression for the work required to move an Earth satellite of mass m from a circular orbit of radius $2R_E$ to one of radius $3R_E$.

(Optional)
Section 14.9   The Gravitational Force Between an Extended Object and a Particle

44. Consider two identical uniform rods of length L and mass m lying along the same line and having their closest points separated by a distance d (Fig. P14.44). Show that the mutual gravitational force between these rods has a magnitude

$$F = \frac{Gm^2}{L^2} \ln \left(\frac{(L+d)^2}{d(2L+d)} \right)$$

45. A uniform rod of mass M is in the shape of a semicircle of radius R (Fig. P14.45). Calculate the force on a point mass m placed at the center of the semicircle.

Figure P14.44

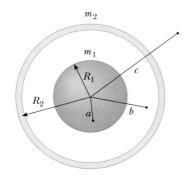

Figure P14.45

(Optional)
Section 14.10   The Gravitational Force Between a Particle and a Spherical Mass

46. (a) Show that the period calculated in Example 14.10 can be written as

$$T = 2\pi \sqrt{\frac{R_E}{g}}$$

where g is the free-fall acceleration on the surface of the Earth. (b) What would this period be if tunnels were made through the Moon? (c) What practical problem regarding these tunnels on Earth would be removed if they were built on the Moon?

47. A 500-kg uniform solid sphere has a radius of 0.400 m. Find the magnitude of the gravitational force exerted by the sphere on a 50.0-g particle located (a) 1.50 m from the center of the sphere, (b) at the surface of the sphere, and (c) 0.200 m from the center of the sphere.

48. A uniform solid sphere of mass m_1 and radius R_1 is inside and concentric with a spherical shell of mass m_2 and radius R_2 (Fig. P14.48). Find the gravitational force exerted by the spheres on a particle of mass m located at (a) $r = a$, (b) $r = b$, and (c) $r = c$, where r is measured from the center of the spheres.

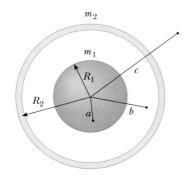

Figure P14.48

ADDITIONAL PROBLEMS

49. Let Δg_M represent the difference in the gravitational fields produced by the Moon at the points on the Earth's surface nearest to and farthest from the Moon. Find the fraction $\Delta g_M/g$, where g is the Earth's gravitational field. (This difference is responsible for the occurrence of the *lunar tides* on the Earth.)

50. Two spheres having masses M and $2M$ and radii R and $3R$, respectively, are released from rest when the distance between their centers is $12R$. How fast will each sphere be moving when they collide? Assume that the two spheres interact only with each other.

51. In Larry Niven's science-fiction novel *Ringworld,* a rigid ring of material rotates about a star (Fig. P14.51). The rotational speed of the ring is 1.25×10^6 m/s, and its radius is 1.53×10^{11} m. (a) Show that the centripetal acceleration of the inhabitants is 10.2 m/s². (b) The inhabitants of this ring world experience a normal contact force $\mathbf{n}$. Acting alone, this normal force would produce an inward acceleration of 9.90 m/s². Additionally, the star at the center of the ring exerts a gravitational force on the ring and its inhabitants. The difference between the total acceleration and the acceleration provided by the normal force is due to the gravitational attraction of the central star. Show that the mass of the star is approximately 10^{32} kg.

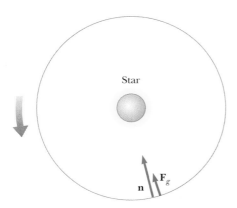

Figure P14.51

52. (a) Show that the rate of change of the free-fall acceleration with distance above the Earth's surface is

$$\frac{dg}{dr} = -\frac{2GM_E}{R_E^3}$$

This rate of change over distance is called a *gradient.* (b) If h is small compared to the radius of the Earth, show that the difference in free-fall acceleration between two points separated by vertical distance h is

$$|\Delta g| = \frac{2GM_E h}{R_E^3}$$

(c) Evaluate this difference for $h = 6.00$ m, a typical height for a two-story building.

53. A particle of mass m is located inside a uniform solid sphere of radius R and mass M, at a distance r from its center. (a) Show that the gravitational potential energy of the system is $U = (GmM/2R^3)r^2 - 3GmM/2R$. (b) Write an expression for the amount of work done by the gravitational force in bringing the particle from the surface of the sphere to its center.

54. *Voyagers 1* and *2* surveyed the surface of Jupiter's moon Io and photographed active volcanoes spewing liquid sulfur to heights of 70 km above the surface of this moon. Find the speed with which the liquid sulfur left the volcano. Io's mass is 8.9×10^{22} kg, and its radius is 1 820 km.

55. As an astronaut, you observe a small planet to be spherical. After landing on the planet, you set off, walking always straight ahead, and find yourself returning to your spacecraft from the opposite side after completing a lap of 25.0 km. You hold a hammer and a falcon feather at a height of 1.40 m, release them, and observe that they fall together to the surface in 29.2 s. Determine the mass of the planet.

56. A cylindrical habitat in space, 6.00 km in diameter and 30 km long, was proposed by G. K. O'Neill in 1974. Such a habitat would have cities, land, and lakes on the inside surface and air and clouds in the center. All of these would be held in place by the rotation of the cylinder about its long axis. How fast would the cylinder have to rotate to imitate the Earth's gravitational field at the walls of the cylinder?

WEB 57. In introductory physics laboratories, a typical Cavendish balance for measuring the gravitational constant G uses lead spheres with masses of 1.50 kg and 15.0 g whose centers are separated by about 4.50 cm. Calculate the gravitational force between these spheres, treating each as a point mass located at the center of the sphere.

58. Newton's law of universal gravitation is valid for distances covering an enormous range, but it is thought to fail for very small distances, where the structure of space itself is uncertain. The crossover distance, far less than the diameter of an atomic nucleus, is called the *Planck length.* It is determined by a combination of the constants G, c, and h, where c is the speed of light in vacuum and h is Planck's constant (introduced briefly in Chapter 11 and discussed in greater detail in Chapter 40) with units of angular momentum. (a) Use dimensional analysis to find a combination of these three universal constants that has units of length. (b) Determine the order of magnitude of the Planck length. (*Hint:* You will need to consider noninteger powers of the constants.)

59. Show that the escape speed from the surface of a planet of uniform density is directly proportional to the radius of the planet.

60. (a) Suppose that the Earth (or another object) has density $\rho(r)$, which can vary with radius but is spherically symmetric. Show that at any particular radius r inside the Earth, the gravitational field strength $g(r)$ will increase as r increases, if and only if the density there exceeds 2/3 the average density of the portion of the Earth inside the radius r. (b) The Earth as a whole has an average density of 5.5 g/cm³, while the density at the surface is 1.0 g/cm³ on the oceans and about 3 g/cm³ on land. What can you infer from this?

WEB 61. Two hypothetical planets of masses m_1 and m_2 and radii r_1 and r_2, respectively, are nearly at rest when they are an infinite distance apart. Because of their gravitational attraction, they head toward each other on a collision course. (a) When their center-to-center separation is d, find expressions for the speed of each planet and their *relative* velocity. (b) Find the kinetic energy of each planet *just* before they collide, if $m_1 = 2.00 \times 10^{24}$ kg, $m_2 = 8.00 \times 10^{24}$ kg, $r_1 = 3.00 \times 10^6$ m, and $r_2 = 5.00 \times 10^6$ m. (*Hint:* Both energy and momentum are conserved.)

62. The maximum distance from the Earth to the Sun (at our aphelion) is 1.521×10^{11} m, and the distance of closest approach (at perihelion) is 1.471×10^{11} m. If the Earth's orbital speed at perihelion is 30.27 km/s, determine (a) the Earth's orbital speed at aphelion, (b) the kinetic and potential energies at perihelion, and (c) the kinetic and potential energies at aphelion. Is the total energy constant? (Neglect the effect of the Moon and other planets.)

63. A sphere of mass M and radius R has a nonuniform density that varies with r, the distance from its center, according to the expression $\rho = Ar$, for $0 \le r \le R$. (a) What is the constant A in terms of M and R? (b) Determine an expression for the force exerted on a particle of mass m placed outside the sphere. (c) Determine an expression for the force exerted on the particle if it is inside the sphere. (*Hint:* See Section 14.10 and note that the distribution is spherically symmetric.)

64. (a) Determine the amount of work (in joules) that must be done on a 100-kg payload to elevate it to a height of 1 000 km above the Earth's surface. (b) Determine the amount of additional work that is required to put the payload into circular orbit at this elevation.

65. X-ray pulses from Cygnus X-1, a celestial x-ray source, have been recorded during high-altitude rocket flights. The signals can be interpreted as originating when a blob of ionized matter orbits a black hole with a period of 5.0 ms. If the blob is in a circular orbit about a black hole whose mass is $20 M_{Sun}$, what is the orbital radius?

66. Studies of the relationship of the Sun to its galaxy—the Milky Way—have revealed that the Sun is located near the outer edge of the galactic disk, about 30 000 lightyears from the center. Furthermore, it has been found that the Sun has an orbital speed of approximately 250 km/s around the galactic center. (a) What is the period of the Sun's galactic motion? (b) What is the order of magnitude of the mass of the Milky Way galaxy? Suppose that the galaxy is made mostly of stars, of which the Sun is typical. What is the order of magnitude of the number of stars in the Milky Way?

67. The oldest artificial satellite in orbit is *Vanguard I*, launched March 3, 1958. Its mass is 1.60 kg. In its initial orbit, its minimum distance from the center of the Earth was 7.02 Mm, and its speed at this perigee point was 8.23 km/s. (a) Find its total energy. (b) Find the magnitude of its angular momentum. (c) Find its speed at apogee and its maximum (apogee) distance from the center of the Earth. (d) Find the semimajor axis of its orbit. (e) Determine its period.

68. A rocket is given an initial speed vertically upward of $v_i = 2\sqrt{Rg}$ at the surface of the Earth, which has radius R and surface free-fall acceleration g. The rocket motors are quickly cut off, and thereafter the rocket coasts under the action of gravitational forces only. (Ignore atmospheric friction and the Earth's rotation.) Derive an expression for the subsequent speed v as a function of the distance r from the center of the Earth in terms of g, R, and r.

69. Two stars of masses M and m, separated by a distance d, revolve in circular orbits about their center of mass (Fig. P14.69). Show that each star has a period given by

$$T^2 = \frac{4\pi^2 d^3}{G(M + m)}$$

(*Hint:* Apply Newton's second law to each star, and note that the center-of-mass condition requires that $Mr_2 = mr_1$, where $r_1 + r_2 = d$.)

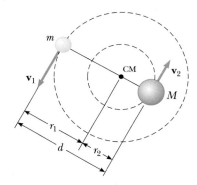

Figure P14.69

70. (a) A 5.00-kg mass is released 1.20×10^7 m from the center of the Earth. It moves with what acceleration relative to the Earth? (b) A 2.00×10^{24} kg mass is released 1.20×10^7 m from the center of the Earth. It moves with what acceleration relative to the Earth? Assume that the objects behave as pairs of particles, isolated from the rest of the Universe.

71. The acceleration of an object moving in the gravitational field of the Earth is

$$\mathbf{a} = -\frac{GM_E}{r^3}\mathbf{r}$$

where **r** is the position vector directed from the center of the Earth to the object. Choosing the origin at the center of the Earth and assuming that the small object is moving in the xy plane, we find that the rectangular (cartesian) components of its acceleration are

$$a_x = -\frac{GM_E x}{(x^2 + y^2)^{3/2}} \qquad a_y = -\frac{GM_E y}{(x^2 + y^2)^{3/2}}$$

Use a computer to set up and carry out a numerical pre-

diction of the motion of the object, according to Euler's method. Assume that the initial position of the object is $x = 0$ and $y = 2R_E$, where R_E is the radius of the Earth. Give the object an initial velocity of 5 000 m/s in the x direction. The time increment should be made as small as practical. Try 5 s. Plot the x and y coordinates of the object as time goes on. Does the object hit the Earth? Vary the initial velocity until you find a circular orbit.

ANSWERS TO QUICK QUIZZES

14.1 Kepler's third law (Eq. 14.7), which applies to all the planets, tells us that the period of a planet is proportional to $r^{3/2}$. Because Saturn and Jupiter are farther from the Sun than the Earth is, they have longer periods. The Sun's gravitational field is much weaker at Saturn and Jupiter than it is at the Earth. Thus, these planets experience much less centripetal acceleration than the Earth does, and they have correspondingly longer periods.

14.2 The mass of the asteroid might be so small that you would be able to exceed escape velocity by leg power alone. You would jump up, but you would never come back down!

14.3 Kepler's first law applies not only to planets orbiting the Sun but also to any relatively small object orbiting another under the influence of gravity. Any elliptical path that does not touch the Earth before reaching point A will continue around the other side to point B in a complete orbit (see figure in next column).

14.4 The gravitational force is zero inside the shell (Eq. 14.25b). Because the force on it is zero, the particle moves with constant velocity in the direction of its original motion outside the shell until it hits the wall opposite the entry hole. Its path thereafter depends on the nature of the collision and on the particle's original direction.

c h a p t e r

Fluid Mechanics

Chapter Outline

atter is normally classified as being in one of three states: solid, liquid, or gas. From everyday experience, we know that a solid has a definite volume and shape. A brick maintains its familiar shape and size day in and day out. We also know that a liquid has a definite volume but no definite shape. Finally, we know that an unconfined gas has neither a definite volume nor a definite shape. These definitions help us picture the states of matter, but they are somewhat artificial. For example, asphalt and plastics are normally considered solids, but over long periods of time they tend to flow like liquids. Likewise, most substances can be a solid, a liquid, or a gas (or a combination of any of these), depending on the temperature and pressure. In general, the time it takes a particular substance to change its shape in response to an external force determines whether we treat the substance as a solid, as a liquid, or as a gas.

A **fluid** is a collection of molecules that are randomly arranged and held together by weak cohesive forces and by forces exerted by the walls of a container. Both liquids and gases are fluids.

In our treatment of the mechanics of fluids, we shall see that we do not need to learn any new physical principles to explain such effects as the buoyant force acting on a submerged object and the dynamic lift acting on an airplane wing. First, we consider the mechanics of a fluid at rest—that is, *fluid statics*—and derive an expression for the pressure exerted by a fluid as a function of its density and depth. We then treat the mechanics of fluids in motion—that is, *fluid dynamics*. We can describe a fluid in motion by using a model in which we make certain simplifying assumptions. We use this model to analyze some situations of practical importance. An analysis leading to *Bernoulli's equation* enables us to determine relationships between the pressure, density, and velocity at every point in a fluid.

15.1 PRESSURE

Fluids do not sustain shearing stresses or tensile stresses; thus, the only stress that can be exerted on an object submerged in a fluid is one that tends to compress the object. In other words, the force exerted by a fluid on an object is always perpendicular to the surfaces of the object, as shown in Figure 15.1.

The pressure in a fluid can be measured with the device pictured in Figure 15.2. The device consists of an evacuated cylinder that encloses a light piston connected to a spring. As the device is submerged in a fluid, the fluid presses on the top of the piston and compresses the spring until the inward force exerted by the fluid is balanced by the outward force exerted by the spring. The fluid pressure can be measured directly if the spring is calibrated in advance. If F is the magnitude of the force exerted on the piston and A is the surface area of the piston,

Figure 15.1 At any point on the surface of a submerged object, the force exerted by the fluid is perpendicular to the surface of the object. The force exerted by the fluid on the walls of the container is perpendicular to the walls at all points.

Figure 15.2 A simple device for measuring the pressure exerted by a fluid.

then the **pressure** P of the fluid at the level to which the device has been submerged is defined as the ratio F/A:

$$P \equiv \frac{F}{A} \qquad \textbf{(15.1)}$$

Note that pressure is a scalar quantity because it is proportional to the magnitude of the force on the piston.

To define the pressure at a specific point, we consider a fluid acting on the device shown in Figure 15.2. If the force exerted by the fluid over an infinitesimal surface element of area dA containing the point in question is dF, then the pressure at that point is

$$P = \frac{dF}{dA} \qquad \textbf{(15.2)}$$

As we shall see in the next section, the pressure exerted by a fluid varies with depth. Therefore, to calculate the total force exerted on a flat wall of a container, we must integrate Equation 15.2 over the surface area of the wall.

Because pressure is force per unit area, it has units of newtons per square meter (N/m^2) in the SI system. Another name for the SI unit of pressure is **pascal** (Pa):

$$1 \text{ Pa} \equiv 1 \text{ N/m}^2 \qquad \textbf{(15.3)}$$

Snowshoes keep you from sinking into soft snow because they spread the downward force you exert on the snow over a large area, reducing the pressure on the snow's surface. *(Earl Young/FPG)*

Quick Quiz 15.1

Suppose you are standing directly behind someone who steps back and accidentally stomps on your foot with the heel of one shoe. Would you be better off if that person were a professional basketball player wearing sneakers or a petite woman wearing spike-heeled shoes? Explain.

Quick Quiz 15.2

After a long lecture, the daring physics professor stretches out for a nap on a bed of nails, as shown in Figure 15.3. How is this possible?

QuickLab

Place a tack between your thumb and index finger, as shown in the figure. Now very gently squeeze the tack and note the sensation. The pointed end of the tack causes pain, and the blunt end does not. According to Newton's third law, the force exerted by the tack on the thumb is equal in magnitude and opposite in direction to the force exerted by the tack on the index finger. However, the pressure at the pointed end of the tack is much greater than the pressure at the blunt end. (Remember that pressure is force per unit area.)

Tack

Figure 15.3 *(Jim Lehman)*

EXAMPLE **15.1**   The Water Bed

The mattress of a water bed is 2.00 m long by 2.00 m wide and 30.0 cm deep. (a) Find the weight of the water in the mattress.

Solution   The density of water is 1 000 kg/m^3 (Table 15.1), and so the mass of the water is

$$M = \rho V = (1\ 000\ \text{kg/m}^3)(1.20\ \text{m}^3) = 1.20 \times 10^3\ \text{kg}$$

and its weight is

$$Mg = (1.20 \times 10^3\ \text{kg})(9.80\ \text{m/s}^2) = \boxed{1.18 \times 10^4\ \text{N}}$$

This is approximately 2 650 lb. (A regular bed weighs approx-

imately 300 lb.) Because this load is so great, such a water bed is best placed in the basement or on a sturdy, well-supported floor.

(b) Find the pressure exerted by the water on the floor when the bed rests in its normal position. Assume that the entire lower surface of the bed makes contact with the floor.

Solution   When the bed is in its normal position, the cross-sectional area is 4.00 m^2; thus, from Equation 15.1, we find that

$$P = \frac{1.18 \times 10^4\ \text{N}}{4.00\ \text{m}^2} = \boxed{2.95 \times 10^3\ \text{Pa}}$$

TABLE 15.1   Densities of Some Common Substances at Standard Temperature (0°C) and Pressure (Atmospheric)

Substance	ρ (kg/m^3)	Substance	ρ (kg/m^3)
Air	1.29	Ice	0.917×10^3
Aluminum	2.70×10^3	Iron	7.86×10^3
Benzene	0.879×10^3	Lead	11.3×10^3
Copper	8.92×10^3	Mercury	13.6×10^3
Ethyl alcohol	0.806×10^3	Oak	0.710×10^3
Fresh water	1.00×10^3	Oxygen gas	1.43
Glycerine	1.26×10^3	Pine	0.373×10^3
Gold	19.3×10^3	Platinum	21.4×10^3
Helium gas	1.79×10^{-1}	Seawater	1.03×10^3
Hydrogen gas	8.99×10^{-2}	Silver	10.5×10^3

15.2   VARIATION OF PRESSURE WITH DEPTH

As divers well know, water pressure increases with depth. Likewise, atmospheric pressure decreases with increasing altitude; it is for this reason that aircraft flying at high altitudes must have pressurized cabins.

We now show how the pressure in a liquid increases linearly with depth. As Equation 1.1 describes, the *density* of a substance is defined as its mass per unit volume: $\rho \equiv m/V$. Table 15.1 lists the densities of various substances. These values vary slightly with temperature because the volume of a substance is temperature dependent (as we shall see in Chapter 19). Note that under standard conditions (at 0°C and at atmospheric pressure) the densities of gases are about 1/1 000 the densities of solids and liquids. This difference implies that the average molecular spacing in a gas under these conditions is about ten times greater than that in a solid or liquid.

Now let us consider a fluid of density ρ at rest and open to the atmosphere, as shown in Figure 15.4. We assume that ρ is constant; this means that the fluid is incompressible. Let us select a sample of the liquid contained within an imaginary cylinder of cross-sectional area A extending from the surface to a depth h. The

Figure 15.4   How pressure varies with depth in a fluid. The net force exerted on the volume of water within the darker region must be zero.

QuickLab

Poke two holes in the side of a paper or polystyrene cup—one near the top and the other near the bottom. Fill the cup with water and watch the water flow out of the holes. Why does water exit from the bottom hole at a higher speed than it does from the top hole?

Variation of pressure with depth

pressure exerted by the outside liquid on the bottom face of the cylinder is P, and the pressure exerted on the top face of the cylinder is the atmospheric pressure P_0. Therefore, the upward force exerted by the outside fluid on the bottom of the cylinder is PA, and the downward force exerted by the atmosphere on the top is P_0A. The mass of liquid in the cylinder is $M = \rho V = \rho Ah$; therefore, the weight of the liquid in the cylinder is $Mg = \rho Ahg$. Because the cylinder is in equilibrium, the net force acting on it must be zero. Choosing upward to be the positive y direction, we see that

$$\sum F_y = PA - P_0A - Mg = 0$$

or

$$PA - P_0A - \rho Ahg = 0$$

$$PA - P_0A = \rho Ahg$$

$$P = P_0 + \rho gh \qquad (15.4)$$

That is, **the pressure P at a depth h below the surface of a liquid open to the atmosphere is *greater* than atmospheric pressure by an amount ρgh.** In our calculations and working of end-of-chapter problems, we usually take atmospheric pressure to be

$$P_0 = 1.00 \text{ atm} = 1.013 \times 10^5 \text{ Pa}$$

Equation 15.4 implies that the pressure is the same at all points having the same depth, independent of the shape of the container.

Quick Quiz 15.3

In the derivation of Equation 15.4, why were we able to ignore the pressure that the liquid exerts on the sides of the cylinder?

This arrangement of interconnected tubes demonstrates that the pressure in a liquid is the same at all points having the same elevation. For example, the pressure is the same at points A, B, C, and D. *(Courtesy of Central Scientific Company)*

In view of the fact that the pressure in a fluid depends on depth and on the value of P_0, any increase in pressure at the surface must be transmitted to every other point in the fluid. This concept was first recognized by the French scientist Blaise Pascal (1623–1662) and is called **Pascal's law: A change in the pressure applied to a fluid is transmitted undiminished to every point of the fluid and to the walls of the container.**

An important application of Pascal's law is the hydraulic press illustrated in Figure 15.5a. A force of magnitude F_1 is applied to a small piston of surface area A_1. The pressure is transmitted through a liquid to a larger piston of surface area A_2. Because the pressure must be the same on both sides, $P = F_1/A_1 = F_2/A_2$. Therefore, the force F_2 is greater than the force F_1 by a factor A_2/A_1, which is called the *force-multiplying factor*. Because liquid is neither added nor removed, the volume pushed down on the left as the piston moves down a distance d_1 equals the volume pushed up on the right as the right piston moves up a distance d_2. That is, $A_1d_1 = A_2d_2$; thus, the force-multiplying factor can also be written as d_1/d_2. Note that $F_1d_1 = F_2d_2$. Hydraulic brakes, car lifts, hydraulic jacks, and forklifts all make use of this principle (Fig. 15.5b).

Quick Quiz 15.4

A grain silo has many bands wrapped around its perimeter (Fig. 15.6). Why is the spacing between successive bands smaller at the lower portions of the silo, as shown in the photograph?

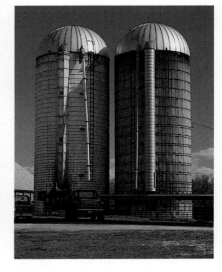

Figure 15.5 (a) Diagram of a hydraulic press. Because the increase in pressure is the same on the two sides, a small force **F**₁ at the left produces a much greater force **F**₂ at the right. (b) A vehicle undergoing repair is supported by a hydraulic lift in a garage. *(David Frazier)*

Figure 15.6 *(Henry Leap)*

EXAMPLE 15.2   The Car Lift

In a car lift used in a service station, compressed air exerts a force on a small piston that has a circular cross section and a radius of 5.00 cm. This pressure is transmitted by a liquid to a piston that has a radius of 15.0 cm. What force must the compressed air exert to lift a car weighing 13 300 N? What air pressure produces this force?

Solution Because the pressure exerted by the compressed air is transmitted undiminished throughout the liquid, we have

$$F_1 = \left(\frac{A_1}{A_2}\right) F_2 = \frac{\pi (5.00 \times 10^{-2}\,\text{m})^2}{\pi (15.0 \times 10^{-2}\,\text{m})^2}\,(1.33 \times 10^4\,\text{N})$$

$$= 1.48 \times 10^3\,\text{N}$$

The air pressure that produces this force is

$$P = \frac{F_1}{A_1} = \frac{1.48 \times 10^3\,\text{N}}{\pi (5.00 \times 10^{-2}\,\text{m})^2} = \boxed{1.88 \times 10^5\,\text{Pa}}$$

This pressure is approximately twice atmospheric pressure.

The input work (the work done by **F**₁) is equal to the output work (the work done by **F**₂), in accordance with the principle of conservation of energy.

EXAMPLE 15.3   A Pain in the Ear

Estimate the force exerted on your eardrum due to the water above when you are swimming at the bottom of a pool that is 5.0 m deep.

Solution First, we must find the unbalanced pressure on the eardrum; then, after estimating the eardrum's surface area, we can determine the force that the water exerts on it.

The air inside the middle ear is normally at atmospheric pressure P_0. Therefore, to find the net force on the eardrum, we must consider the difference between the total pressure at the bottom of the pool and atmospheric pressure:

$$P_{bot} - P_0 = \rho gh$$
$$= (1.00 \times 10^3 \text{ kg/m}^3)(9.80 \text{ m/s}^2)(5.0 \text{ m})$$
$$= 4.9 \times 10^4 \text{ Pa}$$

We estimate the surface area of the eardrum to be approximately 1 cm² = 1 × 10⁻⁴ m². This means that the force on it

is $F = (P_{bot} - P_0)A \approx 5$ N. Because a force on the eardrum of this magnitude is extremely uncomfortable, swimmers often "pop their ears" while under water, an action that pushes air from the lungs into the middle ear. Using this technique equalizes the pressure on the two sides of the eardrum and relieves the discomfort.

EXAMPLE 15.4   The Force on a Dam

Water is filled to a height H behind a dam of width w (Fig. 15.7). Determine the resultant force exerted by the water on the dam.

Solution Because pressure varies with depth, we cannot calculate the force simply by multiplying the area by the pressure. We can solve the problem by finding the force dF ex-

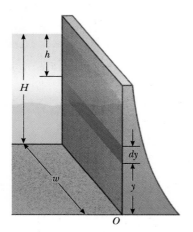

Figure 15.7 Because pressure varies with depth, the total force exerted on a dam must be obtained from the expression $F = \int P \, dA$, where dA is the area of the dark strip.

erted on a narrow horizontal strip at depth h and then integrating the expression to find the total force. Let us imagine a vertical y axis, with $y = 0$ at the bottom of the dam and our strip a distance y above the bottom.

We can use Equation 15.4 to calculate the pressure at the depth h; we omit atmospheric pressure because it acts on both sides of the dam:

$$P = \rho gh = \rho g(H - y)$$

Using Equation 15.2, we find that the force exerted on the shaded strip of area $dA = w \, dy$ is

$$dF = P \, dA = \rho g(H - y)w \, dy$$

Therefore, the total force on the dam is

$$F = \int P \, dA = \int_0^H \rho g(H - y)w \, dy = \boxed{\tfrac{1}{2}\rho gwH^2}$$

Note that the thickness of the dam shown in Figure 15.7 increases with depth. This design accounts for the greater and greater pressure that the water exerts on the dam at greater depths.

Exercise Find an expression for the average pressure on the dam from the total force exerted by the water on the dam.

Answer $\tfrac{1}{2}\rho gH$.

15.3   PRESSURE MEASUREMENTS

One simple device for measuring pressure is the open-tube manometer illustrated in Figure 15.8a. One end of a U-shaped tube containing a liquid is open to the atmosphere, and the other end is connected to a system of unknown pressure P. The difference in pressure $P - P_0$ is equal to ρgh; hence, $P = P_0 + \rho gh$. The pressure P is called the **absolute pressure,** and the difference $P - P_0$ is called the **gauge pressure.** The latter is the value that normally appears on a pressure gauge. For example, the pressure you measure in your bicycle tire is the gauge pressure.

Another instrument used to measure pressure is the common *barometer,* which was invented by Evangelista Torricelli (1608–1647). The barometer consists of a

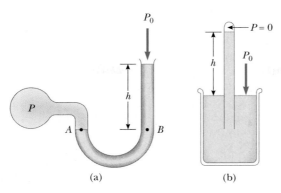

Figure 15.8 Two devices for measuring pressure: (a) an open-tube manometer and (b) a mercury barometer.

long, mercury-filled tube closed at one end and inverted into an open container of mercury (Fig. 15.8b). The closed end of the tube is nearly a vacuum, and so its pressure can be taken as zero. Therefore, it follows that $P_0 = \rho g h$, where h is the height of the mercury column.

One atmosphere ($P_0 = 1$ atm) of pressure is defined as the pressure that causes the column of mercury in a barometer tube to be exactly 0.760 0 m in height at 0°C, with $g = 9.806\ 65$ m/s². At this temperature, mercury has a density of 13.595×10^3 kg/m³; therefore,

$$P_0 = \rho g h = (13.595 \times 10^3\ \text{kg/m}^3)(9.806\ 65\ \text{m/s}^2)(0.760\ 0\ \text{m})$$
$$= 1.013 \times 10^5\ \text{Pa} = 1\ \text{atm}$$

Quick Quiz 15.5

Other than the obvious problem that occurs with freezing, why don't we use water in a barometer in the place of mercury?

15.4 BUOYANT FORCES AND ARCHIMEDES'S PRINCIPLE

Have you ever tried to push a beach ball under water? This is extremely difficult to do because of the large upward force exerted by the water on the ball. The upward force exerted by water on any immersed object is called a **buoyant force.** We can determine the magnitude of a buoyant force by applying some logic and Newton's second law. Imagine that, instead of air, the beach ball is filled with water. If you were standing on land, it would be difficult to hold the water-filled ball in your arms. If you held the ball while standing neck deep in a pool, however, the force you would need to hold it would almost disappear. In fact, the required force would be zero if we were to ignore the thin layer of plastic of which the beach ball is made. Because the water-filled ball is in equilibrium while it is submerged, the magnitude of the upward buoyant force must equal its weight.

If the submerged ball were filled with air rather than water, then the upward buoyant force exerted by the surrounding water would still be present. However, because the weight of the water is now replaced by the much smaller weight of that volume of air, the net force is upward and quite great; as a result, the ball is pushed to the surface.

Archimedes (c. 287–212 B.C.)
Archimedes, a Greek mathematician, physicist, and engineer, was perhaps the greatest scientist of antiquity. He was the first to compute accurately the ratio of a circle's circumference to its diameter, and he showed how to calculate the volume and surface area of spheres, cylinders, and other geometric shapes. He is well known for discovering the nature of the buoyant force.

Archimedes was also a gifted inventor. One of his practical inventions, still in use today, is Archimedes's screw – an inclined, rotating, coiled tube originally used to lift water from the holds of ships. He also invented the catapult and devised systems of levers, pulleys, and weights for raising heavy loads. Such inventions were successfully used to defend his native city Syracuse during a two-year siege by the Romans.

Figure 15.9 The external forces acting on the cube of liquid are the force of gravity $\mathbf{F}_g$ and the buoyant force $\mathbf{B}$. Under equilibrium conditions, $B = F_g$.

The manner in which buoyant forces act is summarized by **Archimedes's principle,** which states that **the magnitude of the buoyant force always equals the weight of the fluid displaced by the object.** The buoyant force acts vertically upward through the point that was the center of gravity of the displaced fluid.

Note that Archimedes's principle does not refer to the makeup of the object experiencing the buoyant force. The object's composition is not a factor in the buoyant force. We can verify this in the following manner: Suppose we focus our attention on the indicated cube of liquid in the container illustrated in Figure 15.9. This cube is in equilibrium as it is acted on by two forces. One of these forces is the gravitational force $\mathbf{F}_g$. What cancels this downward force? Apparently, the rest of the liquid in the container is holding the cube in equilibrium. Thus, the magnitude of the buoyant force $\mathbf{B}$ exerted on the cube is exactly equal to the magnitude of $\mathbf{F}_g$, which is the weight of the liquid inside the cube:

$$B = F_g$$

Now imagine that the cube of liquid is replaced by a cube of steel of the same dimensions. What is the buoyant force acting on the steel? The liquid surrounding a cube behaves in the same way no matter what the cube is made of. Therefore, **the buoyant force acting on the steel cube is the same as the buoyant force acting on a cube of liquid of the same dimensions.** In other words, the magnitude of the buoyant force is the same as the weight of the *liquid* cube, not the steel cube. Although mathematically more complicated, this same principle applies to submerged objects of any shape, size, or density.

Although we have described the magnitude and direction of the buoyant force, we still do not know its origin. Why would a fluid exert such a strange force, almost as if the fluid were trying to expel a foreign body? To understand why, look again at Figure 15.9. The pressure at the bottom of the cube is greater than the pressure at the top by an amount $\rho g h$, where h is the length of any side of the cube. The pressure difference ΔP between the bottom and top faces of the cube is equal to the buoyant force per unit area of those faces—that is, $\Delta P = B/A$. Therefore, $B = (\Delta P)A = (\rho g h)A = \rho g V$, where V is the volume of the cube. Because the mass of the fluid in the cube is $M = \rho V$, we see that

$$B = F_g = \rho V g = Mg \qquad \textbf{(15.5)}$$

where Mg is the weight of the fluid in the cube. Thus, the buoyant force is a result of the pressure differential on a submerged or partly submerged object.

Before we proceed with a few examples, it is instructive for us to compare the forces acting on a totally submerged object with those acting on a floating (partly submerged) object.

Case 1: Totally Submerged Object   When an object is totally submerged in a fluid of density ρ_f, the magnitude of the upward buoyant force is $B = \rho_f V_o g$, where V_o is the volume of the object. If the object has a mass M and density ρ_o, its weight is equal to $F_g = Mg = \rho_o V_o g$, and the net force on it is $B - F_g = (\rho_f - \rho_o) V_o g$. Hence, if the density of the object is less than the density of the fluid, then the downward force of gravity is less than the buoyant force, and the unconstrained object accelerates upward (Fig. 15.10a). If the density of the object is greater than the density of the fluid, then the upward buoyant force is less than the downward force of gravity, and the unsupported object sinks (Fig. 15.10b).

Case 2: Floating Object   Now consider an object of volume V_o in static equilibrium floating on a fluid—that is, an object that is only partially submerged. In this

Figure 15.10 (a) A totally submerged object that is less dense than the fluid in which it is submerged experiences a net upward force. (b) A totally submerged object that is denser than the fluid sinks.

case, the upward buoyant force is balanced by the downward gravitational force acting on the object. If V_f is the volume of the fluid displaced by the object (this volume is the same as the volume of that part of the object that is beneath the fluid level), the buoyant force has a magnitude $B = \rho_f V_f g$. Because the weight of the object is $F_g = Mg = \rho_o V_o g$, and because $F_g = B$, we see that $\rho_f V_f g = \rho_o V_o g$, or

$$\frac{\rho_o}{\rho_f} = \frac{V_f}{V_o} \tag{15.6}$$

Under normal conditions, the average density of a fish is slightly greater than the density of water. It follows that the fish would sink if it did not have some mechanism for adjusting its density. The fish accomplishes this by internally regulating the size of its air-filled swim bladder to balance the change in the magnitude of the buoyant force acting on it. In this manner, fish are able to swim to various depths. Unlike a fish, a scuba diver cannot achieve neutral buoyancy (at which the buoyant force just balances the weight) by adjusting the magnitude of the buoyant force B. Instead, the diver adjusts F_g by manipulating lead weights.

Hot-air balloons. Because hot air is less dense than cold air, a net upward buoyant force acts on the balloons. *(Richard Megna/Fundamental Photographs)*

Quick Quiz 15.6

Steel is much denser than water. In view of this fact, how do steel ships float?

Quick Quiz 15.7

A glass of water contains a single floating ice cube (Fig. 15.11). When the ice melts, does the water level go up, go down, or remain the same?

Quick Quiz 15.8

When a person in a rowboat in a small pond throws an anchor overboard, does the water level of the pond go up, go down, or remain the same?

Figure 15.11

EXAMPLE 15.5 Eureka!

Archimedes supposedly was asked to determine whether a crown made for the king consisted of pure gold. Legend has it that he solved this problem by weighing the crown first in air and then in water, as shown in Figure 15.12. Suppose the

scale read 7.84 N in air and 6.86 N in water. What should Archimedes have told the king?

Solution When the crown is suspended in air, the scale

reads the true weight $T_1 = F_g$ (neglecting the buoyancy of air). When it is immersed in water, the buoyant force **B** reduces the scale reading to an apparent weight of $T_2 = F_g - B$. Hence, the buoyant force exerted on the crown is the difference between its weight in air and its weight in water:

$$B = F_g - T_2 = 7.84 \text{ N} - 6.86 \text{ N} = 0.98 \text{ N}$$

Because this buoyant force is equal in magnitude to the weight of the displaced water, we have $\rho_w g V_w = 0.98$ N, where V_w is the volume of the displaced water and ρ_w is its density. Also, the volume of the crown V_c is equal to the volume of the displaced water because the crown is completely submerged. Therefore,

$$V_c = V_w = \frac{0.98 \text{ N}}{g\rho_w} = \frac{0.98 \text{ N}}{(9.8 \text{ m/s}^2)(1\,000 \text{ kg/m}^3)}$$

$$= 1.0 \times 10^{-4} \text{ m}^3$$

Finally, the density of the crown is

$$\rho_c = \frac{m_c}{V_c} = \frac{m_c g}{V_c g} = \frac{7.84 \text{ N}}{(1.0 \times 10^{-4} \text{ m}^3)(9.8 \text{ m/s}^2)}$$

$$= 8.0 \times 10^3 \text{ kg/m}^3$$

From Table 15.1 we see that the density of gold is 19.3×10^3 kg/m³. Thus, Archimedes should have told the king that

he had been cheated. Either the crown was hollow, or it was not made of pure gold.

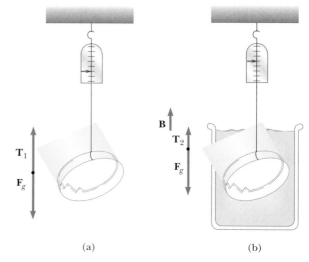

Figure 15.12   (a) When the crown is suspended in air, the scale reads its true weight $T_1 = F_g$ (the buoyancy of air is negligible). (b) When the crown is immersed in water, the buoyant force **B** reduces the scale reading to the apparent weight $T_2 = F_g - B$.

EXAMPLE 15.6   A Titanic Surprise

An iceberg floating in seawater, as shown in Figure 15.13a, is extremely dangerous because much of the ice is below the surface. This hidden ice can damage a ship that is still a considerable distance from the visible ice. What fraction of the iceberg lies below the water level?

Solution   This problem corresponds to Case 2. The weight of the iceberg is $F_{gi} = \rho_i V_i g$, where $\rho_i = 917$ kg/m³ and V_i is the volume of the whole iceberg. The magnitude of the up-

ward buoyant force equals the weight of the displaced water: $B = \rho_w V_w g$, where V_w, the volume of the displaced water, is equal to the volume of the ice beneath the water (the shaded region in Fig. 15.13b) and ρ_w is the density of seawater, $\rho_w = 1\,030$ kg/m³. Because $\rho_i V_i g = \rho_w V_w g$, the fraction of ice beneath the water's surface is

$$f = \frac{V_w}{V_i} = \frac{\rho_i}{\rho_w} = \frac{917 \text{ kg/m}^3}{1\,030 \text{ kg/m}^3} = 0.890 \quad \text{or} \quad \boxed{89.0\%}$$

(a)

(b)

Figure 15.13   (a) Much of the volume of this iceberg is beneath the water. *(Geraldine Prentice/Tony Stone Images)* (b) A ship can be damaged even when it is not near the exposed ice.

15.5 ▶ FLUID DYNAMICS

Thus far, our study of fluids has been restricted to fluids at rest. We now turn our attention to fluids in motion. Instead of trying to study the motion of each particle of the fluid as a function of time, we describe the properties of a moving fluid at each point as a function of time.

Flow Characteristics

When fluid is in motion, its flow can be characterized as being one of two main types. The flow is said to be **steady,** or **laminar,** if each particle of the fluid follows a smooth path, such that the paths of different particles never cross each other, as shown in Figure 15.14. In steady flow, the velocity of the fluid at any point remains constant in time.

Above a certain critical speed, fluid flow becomes **turbulent;** turbulent flow is irregular flow characterized by small whirlpool-like regions, as shown in Figure 15.15.

The term **viscosity** is commonly used in the description of fluid flow to characterize the degree of internal friction in the fluid. This internal friction, or *viscous force,* is associated with the resistance that two adjacent layers of fluid have to moving relative to each other. Viscosity causes part of the kinetic energy of a fluid to be converted to internal energy. This mechanism is similar to the one by which an object sliding on a rough horizontal surface loses kinetic energy.

Because the motion of real fluids is very complex and not fully understood, we make some simplifying assumptions in our approach. In our model of an **ideal fluid,** we make the following four assumptions:

1. **The fluid is nonviscous.** In a nonviscous fluid, internal friction is neglected. An object moving through the fluid experiences no viscous force.
2. **The flow is steady.** In steady (laminar) flow, the velocity of the fluid at each point remains constant.

Properties of an ideal fluid

Figure 15.14   Laminar flow around an automobile in a test wind tunnel. *(Andy Sacks/Tony Stone Images)*

Figure 15.15   Hot gases from a cigarette made visible by smoke particles. The smoke first moves in laminar flow at the bottom and then in turbulent flow above. *(Werner Wolff/Black Star)*

3. **The fluid is incompressible.** The density of an incompressible fluid is constant.
4. **The flow is irrotational.** In irrotational flow, the fluid has no angular momentum about any point. If a small paddle wheel placed anywhere in the fluid does not rotate about the wheel's center of mass, then the flow is irrotational.

15.6   STREAMLINES AND THE EQUATION OF CONTINUITY

The path taken by a fluid particle under steady flow is called a **streamline.** The velocity of the particle is always tangent to the streamline, as shown in Figure 15.16. A set of streamlines like the ones shown in Figure 15.16 form a *tube of flow.* Note that fluid particles cannot flow into or out of the sides of this tube; if they could, then the streamlines would cross each other.

Consider an ideal fluid flowing through a pipe of nonuniform size, as illustrated in Figure 15.17. The particles in the fluid move along streamlines in steady flow. In a time t, the fluid at the bottom end of the pipe moves a distance $\Delta x_1 = v_1 t$. If A_1 is the cross-sectional area in this region, then the mass of fluid contained in the left shaded region in Figure 15.17 is $m_1 = \rho A_1 \Delta x_1 = \rho A_1 v_1 t$, where ρ is the (nonchanging) density of the ideal fluid. Similarly, the fluid that moves through the upper end of the pipe in the time t has a mass $m_2 = \rho A_2 v_2 t$. However, because *mass is conserved* and because the flow is steady, the mass that crosses A_1 in a time t must equal the mass that crosses A_2 in the time t. That is, $m_1 = m_2$, or $\rho A_1 v_1 t = \rho A_2 v_2 t$; this means that

$$A_1 v_1 = A_2 v_2 = \text{constant} \qquad (15.7)$$

Figure 15.16 A particle in laminar flow follows a streamline, and at each point along its path the particle's velocity is tangent to the streamline.

Equation of continuity

This expression is called the **equation of continuity.** It states that

> the product of the area and the fluid speed at all points along the pipe is a constant for an incompressible fluid.

This equation tells us that the speed is high where the tube is constricted (small A) and low where the tube is wide (large A). The product Av, which has the dimensions of volume per unit time, is called either the *volume flux* or the *flow rate.* The condition $Av = $ constant is equivalent to the statement that the volume of fluid that enters one end of a tube in a given time interval equals the volume leaving the other end of the tube in the same time interval if no leaks are present.

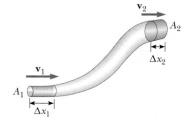

Figure 15.17 A fluid moving with steady flow through a pipe of varying cross-sectional area. The volume of fluid flowing through area A_1 in a time interval t must equal the volume flowing through area A_2 in the same time interval. Therefore, $A_1 v_1 = A_2 v_2$.

Figure 15.18 *(George Semple)*

Quick Quiz 15.9

As water flows from a faucet, as shown in Figure 15.18, why does the stream of water become narrower as it descends?

EXAMPLE 15.7 Niagara Falls

Each second, 5 525 m³ of water flows over the 670-m-wide cliff of the Horseshoe Falls portion of Niagara Falls. The water is approximately 2 m deep as it reaches the cliff. What is its speed at that instant?

Solution   The cross-sectional area of the water as it reaches the edge of the cliff is $A = (670\text{ m})(2\text{ m}) = 1\,340\text{ m}^2$. The flow rate of 5 525 m³/s is equal to Av. This gives

$$v = \frac{5\,525\text{ m}^3/\text{s}}{A} = \frac{5\,525\text{ m}^3/\text{s}}{1\,340\text{ m}^2} = \boxed{4\text{ m/s}}$$

Note that we have kept only one significant figure because our value for the depth has only one significant figure.

Exercise   A barrel floating along in the river plunges over the Falls. How far from the base of the cliff is the barrel when it reaches the water 49 m below?

Answer $13\text{ m} \approx 10\text{ m}$.

15.7 BERNOULLI'S EQUATION

When you press your thumb over the end of a garden hose so that the opening becomes a small slit, the water comes out at high speed, as shown in Figure 15.19. Is the water under greater pressure when it is inside the hose or when it is out in the air? You can answer this question by noting how hard you have to push your thumb against the water inside the end of the hose. The pressure inside the hose is definitely greater than atmospheric pressure.

The relationship between fluid speed, pressure, and elevation was first derived in 1738 by the Swiss physicist Daniel Bernoulli. Consider the flow of an ideal fluid through a nonuniform pipe in a time t, as illustrated in Figure 15.20. Let us call the lower shaded part section 1 and the upper shaded part section 2. The force exerted by the fluid in section 1 has a magnitude P_1A_1. The work done by this force in a time t is $W_1 = F_1\Delta x_1 = P_1A_1\Delta x_1 = P_1V$, where V is the volume of section 1. In a similar manner, the work done by the fluid in section 2 in the same time t is $W_2 = -P_2A_2\Delta x_2 = -P_2V$. (The volume that passes through section 1 in a time t equals the volume that passes through section 2 in the same time.) This work is negative because the fluid force opposes the displacement. Thus, the net work done by these forces in the time t is

$$W = (P_1 - P_2)V$$

Daniel Bernoulli (1700–1782)
Daniel Bernoulli, a Swiss physicist and mathematician, made important discoveries in fluid dynamics. Born into a family of mathematicians, he was the only member of the family to make a mark in physics.

Bernoulli's most famous work, *Hydrodynamica*, was published in 1738; it is both a theoretical and a practical study of equilibrium, pressure, and speed in fluids. He showed that as the speed of a fluid increases, its pressure decreases.

In *Hydrodynamica* Bernoulli also attempted the first explanation of the behavior of gases with changing pressure and temperature; this was the beginning of the kinetic theory of gases, a topic we study in Chapter 21. *(Corbis–Bettmann)*

Figure 15.19   The speed of water spraying from the end of a hose increases as the size of the opening is decreased with the thumb. *(George Semple)*

Figure 15.20   A fluid in laminar flow through a constricted pipe. The volume of the shaded section on the left is equal to the volume of the shaded section on the right.

Part of this work goes into changing the kinetic energy of the fluid, and part goes into changing the gravitational potential energy. If m is the mass that enters one end and leaves the other in a time t, then the change in the kinetic energy of this mass is

$$\Delta K = \tfrac{1}{2}mv_2{}^2 - \tfrac{1}{2}mv_1{}^2$$

The change in gravitational potential energy is

$$\Delta U = mgy_2 - mgy_1$$

We can apply Equation 8.13, $W = \Delta K + \Delta U$, to this volume of fluid to obtain

$$(P_1 - P_2)V = \tfrac{1}{2}mv_2{}^2 - \tfrac{1}{2}mv_1{}^2 + mgy_2 - mgy_1$$

If we divide each term by V and recall that $\rho = m/V$, this expression reduces to

$$P_1 - P_2 = \tfrac{1}{2}\rho v_2{}^2 - \tfrac{1}{2}\rho v_1{}^2 + \rho gy_2 - \rho gy_1$$

Rearranging terms, we obtain

$$P_1 + \tfrac{1}{2}\rho v_1{}^2 + \rho gy_1 = P_2 + \tfrac{1}{2}\rho v_2{}^2 + \rho gy_2 \qquad \textbf{(15.8)}$$

This is **Bernoulli's equation** as applied to an ideal fluid. It is often expressed as

Bernoulli's equation

$$P + \tfrac{1}{2}\rho v^2 + \rho gy = \text{constant} \qquad \textbf{(15.9)}$$

This expression specifies that, in laminar flow, the sum of the pressure (P), kinetic energy per unit volume ($\tfrac{1}{2}\rho v^2$), and gravitational potential energy per unit volume (ρgy) has the same value at all points along a streamline.

When the fluid is at rest, $v_1 = v_2 = 0$ and Equation 15.8 becomes

$$P_1 - P_2 = \rho g(y_2 - y_1) = \rho gh$$

This is in agreement with Equation 15.4.

EXAMPLE 15.8   The Venturi Tube

The horizontal constricted pipe illustrated in Figure 15.21, known as a *Venturi tube*, can be used to measure the flow speed of an incompressible fluid. Let us determine the flow speed at point 2 if the pressure difference $P_1 - P_2$ is known.

Solution   Because the pipe is horizontal, $y_1 = y_2$, and applying Equation 15.8 to points 1 and 2 gives

$$(1) \qquad P_1 + \tfrac{1}{2}\rho v_1{}^2 = P_2 + \tfrac{1}{2}\rho v_2{}^2$$

Figure 15.21   (a) Pressure P_1 is greater than pressure P_2 because $v_1 < v_2$. This device can be used to measure the speed of fluid flow. (b) A Venturi tube. *(Courtesy of Central Scientific Company)*

(a)

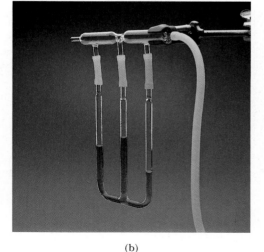

(b)

From the equation of continuity, $A_1v_1 = A_2v_2$, we find that

$$(2) \quad v_1 = \frac{A_2}{A_1} v_2$$

Substituting this expression into equation (1) gives

$$P_1 + \frac{1}{2}\rho \left(\frac{A_2}{A_1}\right)^2 v_2{}^2 = P_2 + \frac{1}{2}\rho v_2{}^2$$

$$v_2 = A_1 \sqrt{\frac{2(P_1 - P_2)}{\rho(A_1{}^2 - A_2{}^2)}}$$

We can use this result and the continuity equation to obtain an expression for v_1. Because $A_2 < A_1$, Equation (2) shows us that $v_2 > v_1$. This result, together with equation (1), indicates that $P_1 > P_2$. In other words, the pressure is reduced in the constricted part of the pipe. This result is somewhat analogous to the following situation: Consider a very crowded room in which people are squeezed together. As soon as a door is opened and people begin to exit, the squeezing (pressure) is least near the door, where the motion (flow) is greatest.

EXAMPLE 15.9 ▸ A Good Trick

It is possible to blow a dime off a table and into a tumbler. Place the dime about 2 cm from the edge of the table. Place the tumbler on the table horizontally with its open edge about 2 cm from the dime, as shown in Figure 15.22a. If you blow forcefully across the top of the dime, it will rise, be caught in the airstream, and end up in the tumbler. The

mass of a dime is $m = 2.24\,\text{g}$, and its surface area is $A = 2.50 \times 10^{-4}\,\text{m}^2$. How hard are you blowing when the dime rises and travels into the tumbler?

Solution Figure 15.22b indicates we must calculate the upward force acting on the dime. First, note that a thin stationary layer of air is present between the dime and the table. When you blow across the dime, it deflects most of the moving air from your breath across its top, so that the air above the dime has a greater speed than the air beneath it. This fact, together with Bernoulli's equation, demonstrates that the air moving across the top of the dime is at a lower pressure than the air beneath the dime. If we neglect the small thickness of the dime, we can apply Equation 15.8 to obtain

$$P_{\text{above}} + \frac{1}{2}\rho v_{\text{above}}^2 = P_{\text{beneath}} + \frac{1}{2}\rho v_{\text{beneath}}^2$$

Because the air beneath the dime is almost stationary, we can neglect the last term in this expression and write the difference as $P_{\text{beneath}} - P_{\text{above}} = \frac{1}{2}\rho v_{\text{above}}^2$. If we multiply this pressure difference by the surface area of the dime, we obtain the upward force acting on the dime. At the very least, this upward force must balance the gravitational force acting on the dime, and so, taking the density of air from Table 15.1, we can state that

$$F_g = mg = (P_{\text{beneath}} - P_{\text{above}})A = (\tfrac{1}{2}\rho v_{\text{above}}^2)A$$

$$v_{\text{above}} = \sqrt{\frac{2mg}{\rho A}} = \sqrt{\frac{2(2.24 \times 10^{-3}\,\text{kg})(9.80\,\text{m/s}^2)}{(1.29\,\text{kg/m}^3)(2.50 \times 10^{-4}\,\text{m}^2)}}$$

$$v_{\text{above}} = 11.7\,\text{m/s}$$

The air you blow must be moving faster than this if the upward force is to exceed the weight of the dime. Practice this trick a few times and then impress all your friends!

2 cm   2 cm

(a)

$\mathbf{F}_{\text{Bernoulli}}$

$\mathbf{F}_g$

(b)

Figure 15.22

EXAMPLE 15.10 ▸ Torricelli's Law

An enclosed tank containing a liquid of density ρ has a hole in its side at a distance y_1 from the tank's bottom (Fig. 15.23). The hole is open to the atmosphere, and its diameter is much smaller than the diameter of the tank. The air above the liquid is maintained at a pressure P. Determine the speed at which the liquid leaves the hole when the liquid's level is a distance h above the hole.

Solution Because $A_2 \gg A_1$, the liquid is approximately at rest at the top of the tank, where the pressure is P. Applying Bernoulli's equation to points 1 and 2 and noting that at the hole P_1 is equal to atmospheric pressure P_0, we find that

$$P_0 + \tfrac{1}{2}\rho v_1{}^2 + \rho g y_1 = P + \rho g y_2$$

But $y_2 - y_1 = h$; thus, this expression reduces to

$$v_1 = \sqrt{\frac{2(P - P_0)}{\rho} + 2gh}$$

When P is much greater than P_0 (so that the term $2gh$ can be neglected), the exit speed of the water is mainly a function of P. If the tank is open to the atmosphere, then $P = P_0$ and $v_1 = \sqrt{2gh}$. In other words, for an open tank, the speed of liquid coming out through a hole a distance h below the surface is equal to that acquired by an object falling freely through a vertical distance h. This phenomenon is known as **Torricelli's law.**

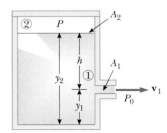

Figure 15.23 When P is much larger than atmospheric pressure P_0, the liquid speed as the liquid passes through the hole in the side of the container is given approximately by $v_1 = \sqrt{2(P - P_0)/\rho}$.

Optional Section

15.8 ▸ OTHER APPLICATIONS OF BERNOULLI'S EQUATION

The lift on an aircraft wing can be explained, in part, by the Bernoulli effect. Airplane wings are designed so that the air speed above the wing is greater than that below the wing. As a result, the air pressure above the wing is less than the pressure below, and a net upward force on the wing, called *lift*, results.

Another factor influencing the lift on a wing is shown in Figure 15.24. The wing has a slight upward tilt that causes air molecules striking its bottom to be deflected downward. This deflection means that the wing is exerting a downward force on the air. According to Newton's third law, the air must exert an equal and opposite force on the wing.

Finally, turbulence also has an effect. If the wing is tilted too much, the flow of air across the upper surface becomes turbulent, and the pressure difference across the wing is not as great as that predicted by Bernoulli's equation. In an extreme case, this turbulence may cause the aircraft to stall.

In general, an object moving through a fluid experiences lift as the result of any effect that causes the fluid to change its direction as it flows past the object. Some factors that influence lift are the shape of the object, its orientation with respect to the fluid flow, any spinning motion it might have, and the texture of its surface. For example, a golf ball struck with a club is given a rapid backspin, as shown in Figure 15.25a. The dimples on the ball help "entrain" the air to follow the curvature of the ball's surface. This effect is most pronounced on the top half of the ball, where the ball's surface is moving in the same direction as the air flow. Figure 15.25b shows a thin layer of air wrapping part way around the ball and being deflected downward as a result. Because the ball pushes the air down, the air must push up on the ball. Without the dimples, the air is not as well entrained,

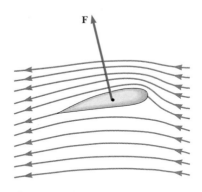

Figure 15.24 Streamline flow around an airplane wing. The pressure above the wing is less than the pressure below, and a dynamic lift upward results.

(a)

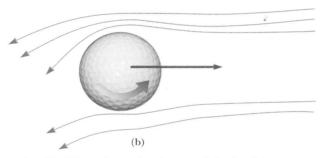

(b)

Figure 15.25   (a) A golf ball is made to spin when struck by the club. *(© Harold E. Edgerton/Courtesy of Palm Press, Inc.)*   (b) The spinning ball experiences a lifting force that allows it to travel much farther than it would if it were not spinning.

and the golf ball does not travel as far. For the same reason, a tennis ball's fuzz helps the spinning ball "grab" the air rushing by and helps deflect it.

A number of devices operate by means of the pressure differentials that result from differences in a fluid's speed. For example, a stream of air passing over one end of an open tube, the other end of which is immersed in a liquid, reduces the pressure above the tube, as illustrated in Figure 15.26. This reduction in pressure causes the liquid to rise into the air stream. The liquid is then dispersed into a fine spray of droplets. You might recognize that this so-called atomizer is used in perfume bottles and paint sprayers. The same principle is used in the carburetor of a gasoline engine. In this case, the low-pressure region in the carburetor is produced by air drawn in by the piston through the air filter. The gasoline vaporizes in that region, mixes with the air, and enters the cylinder of the engine, where combustion occurs.

Figure 15.26   A stream of air passing over a tube dipped into a liquid causes the liquid to rise in the tube.

Quick Quiz 15.10

People in buildings threatened by a tornado are often told to open the windows to minimize damage. Why?

SUMMARY

The **pressure** P in a fluid is the force per unit area exerted by the fluid on a surface:

$$P \equiv \frac{F}{A} \tag{15.1}$$

In the SI system, pressure has units of newtons per square meter (N/m^2), and $1 \, N/m^2 = 1$ pascal (Pa).

The pressure in a fluid at rest varies with depth h in the fluid according to the expression

$$P = P_0 + \rho g h \tag{15.4}$$

where P_0 is atmospheric pressure ($= 1.013 \times 10^5 \, N/m^2$) and ρ is the density of the fluid, assumed uniform.

Pascal's law states that when pressure is applied to an enclosed fluid, the pressure is transmitted undiminished to every point in the fluid and to every point on the walls of the container.

When an object is partially or fully submerged in a fluid, the fluid exerts on the object an upward force called the **buoyant force.** According to **Archimedes's principle,** the magnitude of the buoyant force is equal to the weight of the fluid displaced by the object. Be sure you can apply this principle to a wide variety of situations, including sinking objects, floating ones, and neutrally buoyant ones.

You can understand various aspects of a fluid's dynamics by assuming that the fluid is nonviscous and incompressible and that the fluid's motion is a steady flow with no rotation.

Two important concepts regarding ideal fluid flow through a pipe of nonuniform size are as follows:

1. The flow rate (volume flux) through the pipe is constant; this is equivalent to stating that the product of the cross-sectional area A and the speed v at any point is a constant. This result is expressed in the **equation of continuity:**

$$A_1 v_1 = A_2 v_2 = \text{constant} \tag{15.7}$$

You can use this expression to calculate how the velocity of a fluid changes as the fluid is constricted or as it flows into a more open area.

2. The sum of the pressure, kinetic energy per unit volume, and gravitational potential energy per unit volume has the same value at all points along a streamline. This result is summarized in **Bernoulli's equation:**

$$P + \tfrac{1}{2}\rho v^2 + \rho g y = \text{constant} \tag{15.9}$$

QUESTIONS

1. Two drinking glasses of the same weight but of different shape and different cross-sectional area are filled to the same level with water. According to the expression $P = P_0 + \rho g h$, the pressure at the bottom of both glasses is the same. In view of this, why does one glass weigh more than the other?

2. If the top of your head has a surface area of 100 cm^2, what is the weight of the air above your head?

3. When you drink a liquid through a straw, you reduce the pressure in your mouth and let the atmosphere move the liquid. Explain why this is so. Can you use a straw to sip a drink on the Moon?

4. A helium-filled balloon rises until its density becomes the same as that of the surrounding air. If a sealed submarine begins to sink, will it go all the way to the bottom of the ocean or will it stop when its density becomes the same as that of the surrounding water?

5. A fish rests on the bottom of a bucket of water while the

bucket is being weighed. When the fish begins to swim around, does the weight change?

6. Does a ship ride higher in the water of an inland lake or in the ocean? Why?

7. Lead has a greater density than iron, and both metals are denser than water. Is the buoyant force on a lead object greater than, less than, or equal to the buoyant force on an iron object of the same volume?

8. The water supply for a city is often provided by reservoirs built on high ground. Water flows from the reservoir, through pipes, and into your home when you turn the tap on your faucet. Why is the flow of water more rapid out of a faucet on the first floor of a building than it is in an apartment on a higher floor?

9. Smoke rises in a chimney faster when a breeze is blowing than when there is no breeze at all. Use Bernoulli's equation to explain this phenomenon.

10. If a Ping–Pong ball is above a hair dryer, the ball can be suspended in the air column emitted by the dryer. Explain.

11. When ski jumpers are airborne (Fig. Q15.11), why do they bend their bodies forward and keep their hands at their sides?

Figure Q15.11   (*Galen Powell/Peter Arnold, Inc.*)

12. Explain why a sealed bottle partially filled with a liquid can float.

13. When is the buoyant force on a swimmer greater—after exhaling or after inhaling?

14. A piece of unpainted wood barely floats in a container partly filled with water. If the container is sealed and then pressurized above atmospheric pressure, does the wood rise, sink, or remain at the same level? (*Hint:* Wood is porous.)

15. A flat plate is immersed in a liquid at rest. For what orientation of the plate is the pressure on its flat surface uniform?

16. Because atmospheric pressure is about 10^5 N/m^2 and the area of a person's chest is about 0.13 m^2, the force of the atmosphere on one's chest is around 13 000 N. In view of this enormous force, why don't our bodies collapse?

17. How would you determine the density of an irregularly shaped rock?

18. Why do airplane pilots prefer to take off into the wind?

19. If you release a ball while inside a freely falling elevator, the ball remains in front of you rather than falling to the floor because the ball, the elevator, and you all experience the same downward acceleration **g**. What happens if you repeat this experiment with a helium-filled balloon? (This one is tricky.)

20. Two identical ships set out to sea. One is loaded with a cargo of Styrofoam, and the other is empty. Which ship is more submerged?

21. A small piece of steel is tied to a block of wood. When the wood is placed in a tub of water with the steel on top, half of the block is submerged. If the block is inverted so that the steel is underwater, does the amount of the block submerged increase, decrease, or remain the same? What happens to the water level in the tub when the block is inverted?

22. Prairie dogs (Fig. Q15.22) ventilate their burrows by building a mound over one entrance, which is open to a stream of air. A second entrance at ground level is open to almost stagnant air. How does this construction create an air flow through the burrow?

Figure Q15.22   (*Pamela Zilly/The Image Bank*)

23. An unopened can of diet cola floats when placed in a tank of water, whereas a can of regular cola of the same brand sinks in the tank. What do you suppose could explain this phenomenon?

24. Figure Q15.24 shows a glass cylinder containing four liquids of different densities. From top to bottom, the liquids are oil (orange), water (yellow), salt water (green), and mercury (silver). The cylinder also contains, from top to bottom, a Ping–Pong ball, a piece of wood, an egg, and a steel ball. (a) Which of these liquids has the lowest density, and which has the greatest? (b) What can you conclude about the density of each object?

Figure Q15.24   *(Henry Leap and Jim Lehman)*

Figure Q15.25   *(Henry Leap and Jim Lehman)*

25. In Figure Q15.25, an air stream moves from right to left through a tube that is constricted at the middle. Three Ping–Pong balls are levitated in equilibrium above the vertical columns through which the air escapes. (a) Why is the ball at the right higher than the one in the middle?

(b) Why is the ball at the left lower than the ball at the right even though the horizontal tube has the same dimensions at these two points?

26. You are a passenger on a spacecraft. For your comfort, the interior contains air just like that at the surface of the Earth. The craft is coasting through a very empty region of space. That is, a nearly perfect vacuum exists just outside the wall. Suddenly a meteoroid pokes a hole, smaller than the palm of your hand, right through the wall next to your seat. What will happen? Is there anything you can or should do about it?

PROBLEMS

1, 2, 3 = straightforward, intermediate, challenging   ☐ = full solution available in the *Student Solutions Manual and Study Guide*
WEB = solution posted at **http://www.saunderscollege.com/physics/**   🖥 = Computer useful in solving problem   📱 = Interactive Physics
☐ = paired numerical/symbolic problems

Section 15.1   Pressure

1. Calculate the mass of a solid iron sphere that has a diameter of 3.00 cm.

2. Find the order of magnitude of the density of the *nucleus* of an atom. What does this result suggest concerning the structure of matter? (Visualize a nucleus as protons and neutrons closely packed together. Each has mass 1.67×10^{-27} kg and radius on the order of 10^{-15} m.)

3. A 50.0-kg woman balances on one heel of a pair of high-heeled shoes. If the heel is circular and has a radius of 0.500 cm, what pressure does she exert on the floor?

4. The four tires of an automobile are inflated to a gauge pressure of 200 kPa. Each tire has an area of 0.024 0 m^2 in contact with the ground. Determine the weight of the automobile.

5. What is the total mass of the Earth's atmosphere? (The radius of the Earth is 6.37×10^6 m, and atmospheric pressure at the Earth's surface is 1.013×10^5 N/m^2.)

Section 15.2   Variation of Pressure with Depth

6. (a) Calculate the absolute pressure at an ocean depth of 1 000 m. Assume the density of seawater is 1 024 kg/m^3 and that the air above exerts a pressure of 101.3 kPa. (b) At this depth, what force must the frame around a circular submarine porthole having a diameter of 30.0 cm exert to counterbalance the force exerted by the water?

7. The spring of the pressure gauge shown in Figure 15.2 has a force constant of 1 000 N/m, and the piston has a diameter of 2.00 cm. When the gauge is lowered into water, at what depth does the piston move in by 0.500 cm?

8. The small piston of a hydraulic lift has a cross-sectional area of 3.00 cm^2, and its large piston has a cross-sectional area of 200 cm^2 (see Fig. 15.5a). What force must be applied to the small piston for it to raise a load of 15.0 kN? (In service stations, this force is usually generated with the use of compressed air.)

(a)   (b)

Figure P15.10

WEB **9.** What must be the contact area between a suction cup (completely exhausted) and a ceiling if the cup is to support the weight of an 80.0-kg student?

10. (a) A very powerful vacuum cleaner has a hose 2.86 cm in diameter. With no nozzle on the hose, what is the weight of the heaviest brick that the cleaner can lift (Fig. P15.10)? (b) A very powerful octopus uses one sucker of diameter 2.86 cm on each of the two shells of a clam in an attempt to pull the shells apart. Find the greatest force that the octopus can exert in salt water 32.3 m in depth. (*Caution:* Experimental verification can be interesting, but do not drop a brick on your foot. Do not overheat the motor of a vacuum cleaner. Do not get an octopus mad at you.)

11. For the cellar of a new house, a hole with vertical sides descending 2.40 m is dug in the ground. A concrete foundation wall is built all the way across the 9.60-m width of the excavation. This foundation wall is 0.183 m away from the front of the cellar hole. During a rainstorm, drainage from the street fills up the space in front of the concrete wall but not the cellar behind the wall. The water does not soak into the clay soil. Find the force that the water causes on the foundation wall. For comparison, the weight of the water is given by

$$2.40 \text{ m} \times 9.60 \text{ m} \times 0.183 \text{ m} \times 1\ 000 \text{ kg/m}^3$$
$$\times 9.80 \text{ m/s}^2 = 41.3 \text{ kN}$$

12. A swimming pool has dimensions 30.0 m × 10.0 m and a flat bottom. When the pool is filled to a depth of 2.00 m with fresh water, what is the force caused by the water on the bottom? On each end? On each side?

13. A sealed spherical shell of diameter d is rigidly attached to a cart that is moving horizontally with an acceleration a, as shown in Figure P15.13. The sphere is nearly filled with a fluid having density ρ and also contains one small bubble of air at atmospheric pressure. Find an expression for the pressure P at the center of the sphere.

Figure P15.13

14. The tank shown in Figure P15.14 is filled with water to a depth of 2.00 m. At the bottom of one of the side walls is a rectangular hatch 1.00 m high and 2.00 m wide. The hatch is hinged at its top. (a) Determine the force that the water exerts on the hatch. (b) Find the torque exerted about the hinges.

Figure P15.14

15. Review Problem. A solid copper ball with a diameter of 3.00 m at sea level is placed at the bottom of the ocean (at a depth of 10.0 km). If the density of seawater is 1 030 kg/m³, by how much (approximately) does the diameter of the ball decrease when it reaches bottom? Take the bulk modulus of copper as $14.0 \times 10^{10} \text{ N/m}^2$.

Section 15.3   Pressure Measurements

16. Normal atmospheric pressure is 1.013×10^5 Pa. The approach of a storm causes the height of a mercury barometer to drop by 20.0 mm from the normal height. What is the atmospheric pressure? (The density of mercury is 13.59 g/cm^3.)

WEB 17. Blaise Pascal duplicated Torricelli's barometer, using a red Bordeaux wine, of density 984 kg/m^3, as the working liquid (Fig. P15.17). What was the height h of the wine column for normal atmospheric pressure? Would you expect the vacuum above the column to be as good as that for mercury?

Figure P15.17

18. Mercury is poured into a U-tube, as shown in Figure P15.18a. The left arm of the tube has a cross-sectional area A_1 of 10.0 cm^2, and the right arm has a cross-sectional area A_2 of 5.00 cm^2. One-hundred grams of water are then poured into the right arm, as shown in Figure P15.18b. (a) Determine the length of the water column

in the right arm of the U-tube. (b) Given that the density of mercury is 13.6 g/cm^3, what distance h does the mercury rise in the left arm?

19. A U-tube of uniform cross-sectional area and open to the atmosphere is partially filled with mercury. Water is then poured into both arms. If the equilibrium configuration of the tube is as shown in Figure P15.19, with $h_2 = 1.00$ cm, determine the value of h_1.

Figure P15.19

Section 15.4   Buoyant Forces and Archimedes's Principle

20. (a) A light balloon is filled with 400 m^3 of helium. At 0°C, what is the mass of the payload that the balloon can lift? (b) In Table 15.1, note that the density of hydrogen is nearly one-half the density of helium. What load can the balloon lift if it is filled with hydrogen?

21. A Styrofoam slab has a thickness of 10.0 cm and a density of 300 kg/m^3. When a 75.0-kg swimmer is resting on it, the slab floats in fresh water with its top at the same level as the water's surface. Find the area of the slab.

22. A Styrofoam slab has thickness h and density ρ_S. What is the area of the slab if it floats with its upper surface just awash in fresh water, when a swimmer of mass m is on top?

23. A piece of aluminum with mass 1.00 kg and density 2 700 kg/m^3 is suspended from a string and then completely immersed in a container of water (Fig. P15.23). Calculate the tension in the string (a) before and (b) after the metal is immersed.

24. A 10.0-kg block of metal measuring 12.0 cm × 10.0 cm × 10.0 cm is suspended from a scale and immersed in water, as shown in Figure P15.23b. The 12.0-cm dimension is vertical, and the top of the block is 5.00 cm from the surface of the water. (a) What are the forces acting on the top and on the bottom of the block? (Take $P_0 = 1.013\ 0 \times 10^5$ N/m^2.) (b) What is the reading of the spring scale? (c) Show that the buoyant force equals the difference between the forces at the top and bottom of the block.

(a)   (b)

Figure P15.18

Figure P15.23 Problems 23 and 24.

WEB **25.** A cube of wood having a side dimension of 20.0 cm and a density of 650 kg/m³ floats on water. (a) What is the distance from the horizontal top surface of the cube to the water level? (b) How much lead weight must be placed on top of the cube so that its top is just level with the water?

26. To an order of magnitude, how many helium-filled toy balloons would be required to lift you? Because helium is an irreplaceable resource, develop a theoretical answer rather than an experimental answer. In your solution, state what physical quantities you take as data and the values you measure or estimate for them.

27. A plastic sphere floats in water with 50.0% of its volume submerged. This same sphere floats in oil with 40.0% of its volume submerged. Determine the densities of the oil and the sphere.

28. A frog in a hemispherical pod finds that he just floats without sinking into a sea of blue-green ooze having a density of 1.35 g/cm³ (Fig. P15.28). If the pod has a radius of 6.00 cm and a negligible mass, what is the mass of the frog?

Figure P15.28

29. How many cubic meters of helium are required to lift a balloon with a 400-kg payload to a height of 8 000 m? (Take $\rho_{He} = 0.180$ kg/m³.) Assume that the balloon

maintains a constant volume and that the density of air decreases with the altitude z according to the expression $\rho_{air} = \rho_0 e^{-z/8\,000}$, where z is in meters and $\rho_0 = 1.25$ kg/m³ is the density of air at sea level.

30. **Review Problem.** A long cylindrical tube of radius r is weighted on one end so that it floats upright in a fluid having a density ρ. It is pushed downward a distance x from its equilibrium position and then released. Show that the tube will execute simple harmonic motion if the resistive effects of the water are neglected, and determine the period of the oscillations.

31. A bathysphere used for deep-sea exploration has a radius of 1.50 m and a mass of 1.20×10^4 kg. To dive, this submarine takes on mass in the form of seawater. Determine the amount of mass that the submarine must take on if it is to descend at a constant speed of 1.20 m/s, when the resistive force on it is 1 100 N in the upward direction. Take 1.03×10^3 kg/m³ as the density of seawater.

32. The United States possesses the eight largest warships in the world—aircraft carriers of the *Nimitz* class—and it is building one more. Suppose that one of the ships bobs up to float 11.0 cm higher in the water when 50 fighters take off from it at a location where $g = 9.78$ m/s². The planes have an average mass of 29 000 kg. Find the horizontal area enclosed by the waterline of the ship. (By comparison, its flight deck has an area of 18 000 m².)

Section 15.5 Fluid Dynamics
Section 15.6 Streamlines and the Equation of Continuity
Section 15.7 Bernoulli's Equation

33. (a) A water hose 2.00 cm in diameter is used to fill a 20.0-L bucket. If it takes 1.00 min to fill the bucket, what is the speed v at which water moves through the hose? (*Note:* 1 L = 1 000 cm³.) (b) If the hose has a nozzle 1.00 cm in diameter, find the speed of the water at the nozzle.

34. A horizontal pipe 10.0 cm in diameter has a smooth reduction to a pipe 5.00 cm in diameter. If the pressure of the water in the larger pipe is 8.00×10^4 Pa and the pressure in the smaller pipe is 6.00×10^4 Pa, at what rate does water flow through the pipes?

35. A large storage tank, open at the top and filled with water, develops a small hole in its side at a point 16.0 m below the water level. If the rate of flow from the leak is 2.50×10^{-3} m³/min, determine (a) the speed at which the water leaves the hole and (b) the diameter of the hole.

36. Through a pipe of diameter 15.0 cm, water is pumped from the Colorado River up to Grand Canyon Village, located on the rim of the canyon. The river is at an elevation of 564 m, and the village is at an elevation of 2 096 m. (a) What is the minimum pressure at which the water must be pumped if it is to arrive at the village?

(b) If 4 500 m³ are pumped per day, what is the speed of the water in the pipe? (c) What additional pressure is necessary to deliver this flow? (*Note:* You may assume that the acceleration due to gravity and the density of air are constant over this range of elevations.)

37. Water flows through a fire hose of diameter 6.35 cm at a rate of 0.012 0 m³/s. The fire hose ends in a nozzle with an inner diameter of 2.20 cm. What is the speed at which the water exits the nozzle?

38. Old Faithful Geyser in Yellowstone National Park erupts at approximately 1-h intervals, and the height of the water column reaches 40.0 m (Fig. P15.38). (a) Consider the rising stream as a series of separate drops. Analyze the free-fall motion of one of these drops to determine the speed at which the water leaves the ground. (b) Treating the rising stream as an ideal fluid in streamline flow, use Bernoulli's equation to determine the speed of the water as it leaves ground level. (c) What is the pressure (above atmospheric) in the heated underground chamber if its depth is 175 m? You may assume that the chamber is large compared with the geyser's vent.

Figure P15.38 *(Stan Osolinski/Dembinsky Photo Associates)*

(Optional)
Section 15.8 Other Applications of Bernoulli's Equation

39. An airplane has a mass of 1.60×10^4 kg, and each wing has an area of 40.0 m². During level flight, the pressure on the lower wing surface is 7.00×10^4 Pa. Determine the pressure on the upper wing surface.

40. A Venturi tube may be used as a fluid flow meter (see Fig. 15.21). If the difference in pressure is $P_1 - P_2 = 21.0$ kPa, find the fluid flow rate in cubic meters per second, given that the radius of the outlet tube is 1.00 cm, the radius of the inlet tube is 2.00 cm, and the fluid is gasoline ($\rho = 700$ kg/m³).

41. A Pitot tube can be used to determine the velocity of air flow by measuring the difference between the total pressure and the static pressure (Fig. P15.41). If the fluid in the tube is mercury, whose density is $\rho_{Hg} = 13\,600$ kg/m³, and if $\Delta h = 5.00$ cm, find the speed of

Figure P15.41

air flow. (Assume that the air is stagnant at point A, and take $\rho_{air} = 1.25$ kg/m³.)

42. An airplane is cruising at an altitude of 10 km. The pressure outside the craft is 0.287 atm; within the passenger compartment, the pressure is 1.00 atm and the temperature is 20°C. A small leak occurs in one of the window seals in the passenger compartment. Model the air as an ideal fluid to find the speed of the stream of air flowing through the leak.

43. A siphon is used to drain water from a tank, as illustrated in Figure P15.43. The siphon has a uniform diameter. Assume steady flow without friction. (a) If the distance $h = 1.00$ m, find the speed of outflow at the end of the siphon. (b) What is the limitation on the height of the top of the siphon above the water surface? (For the flow of liquid to be continuous, the pressure must not drop below the vapor pressure of the liquid.)

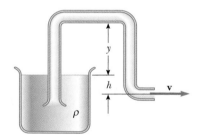

Figure P15.43

44. A hypodermic syringe contains a medicine with the density of water (Fig. P15.44). The barrel of the syringe has a cross-sectional area $A = 2.50 \times 10^{-5}$ m², and the needle has a cross-sectional area $a = 1.00 \times 10^{-8}$ m². In the absence of a force on the plunger, the pressure everywhere is 1 atm. A force **F** of magnitude 2.00 N acts on the plunger, making the medicine squirt horizontally from the needle. Determine the speed of the medicine as it leaves the needle's tip.

WEB 45. A large storage tank is filled to a height h_0. The tank is punctured at a height h above the bottom of the tank (Fig. P15.45). Find an expression for how far from the tank the exiting stream lands.

Figure P15.44

Figure P15.45   Problems 45 and 46.

46. A hole is punched at a height h in the side of a container of height h_0. The container is full of water, as shown in Figure P15.45. If the water is to shoot as far as possible horizontally, (a) how far from the bottom of the container should the hole be punched? (b) Neglecting frictional losses, how far (initially) from the side of the container will the water land?

ADDITIONAL PROBLEMS

47. A Ping–Pong ball has a diameter of 3.80 cm and an average density of $0.084\ 0\ \text{g/cm}^3$. What force would be required to hold it completely submerged under water?

48. Figure P15.48 shows a tank of water with a valve at the bottom. If this valve is opened, what is the maximum height attained by the water stream exiting the right side of the tank? Assume that $h = 10.0$ m, $L = 2.00$ m, and $\theta = 30.0°$, and that the cross-sectional area at point A is very large compared with that at point B.

Figure P15.48

49. A helium-filled balloon is tied to a 2.00-m-long, 0.050 0-kg uniform string. The balloon is spherical with a radius of 0.400 m. When released, the balloon lifts a length h of string and then remains in equilibrium, as shown in Figure P15.49. Determine the value of h. The envelope of the balloon has a mass of 0.250 kg.

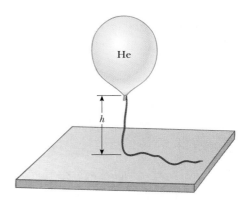

Figure P15.49

50. Water is forced out of a fire extinguisher by air pressure, as shown in Figure P15.50. How much gauge air pressure in the tank (above atmospheric) is required for the water jet to have a speed of 30.0 m/s when the water level is 0.500 m below the nozzle?

Figure P15.50

51. The true weight of an object is measured in a vacuum, where buoyant forces are absent. An object of volume V is weighed in air on a balance with the use of weights of density ρ. If the density of air is ρ_{air} and the balance reads F_g', show that the true weight F_g is

$$F_g = F_g' + \left(V - \frac{F_g'}{\rho g}\right)\rho_{air}g$$

52. Evangelista Torricelli was the first to realize that we live at the bottom of an ocean of air. He correctly surmised that the pressure of our atmosphere is attributable to the weight of the air. The density of air at 0°C at the Earth's surface is $1.29\ \text{kg/m}^3$. The density decreases with increasing altitude (as the atmosphere thins). On the other hand, if we assume that the density is constant

(1.29 kg/m^3) up to some altitude h, and zero above that altitude, then h would represent the thickness of our atmosphere. Use this model to determine the value of h that gives a pressure of 1.00 atm at the surface of the Earth. Would the peak of Mt. Everest rise above the surface of such an atmosphere?

53. A wooden dowel has a diameter of 1.20 cm. It floats in water with 0.400 cm of its diameter above water level (Fig. P15.53). Determine the density of the dowel.

0.400 cm

0.80 cm

Figure P15.53

54. A light spring of constant $k = 90.0$ N/m rests vertically on a table (Fig. P15.54a). A 2.00-g balloon is filled with helium (density = 0.180 kg/m^3) to a volume of 5.00 m^3 and is then connected to the spring, causing it to stretch as shown in Figure P15.54b. Determine the extension distance L when the balloon is in equilibrium.

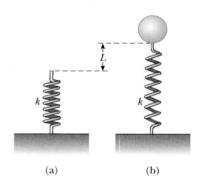

(a)   (b)

Figure P15.54

55. A 1.00-kg beaker containing 2.00 kg of oil (density = 916.0 kg/m^3) rests on a scale. A 2.00-kg block of iron is suspended from a spring scale and completely submerged in the oil, as shown in Figure P15.55. Determine the equilibrium readings of both scales.

56. A beaker of mass m_b containing oil of mass m_0 (density = ρ_0) rests on a scale. A block of iron of mass m_{Fe} is suspended from a spring scale and completely submerged in the oil, as shown in Figure P15.55. Determine the equilibrium readings of both scales.

WEB 57. **Review Problem.** With reference to Figure 15.7, show that the total torque exerted by the water behind the

Figure P15.55   Problems 55 and 56.

dam about an axis through O is $\frac{1}{6} \rho g w H^3$. Show that the effective line of action of the total force exerted by the water is at a distance $\frac{1}{3}H$ above O.

58. In about 1657 Otto von Guericke, inventor of the air pump, evacuated a sphere made of two brass hemispheres. Two teams of eight horses each could pull the hemispheres apart only on some trials, and then "with greatest difficulty," with the resulting sound likened to a cannon firing (Fig. P15.58). (a) Show that the force F

Figure P15.58   The colored engraving, dated 1672, illustrates Otto von Guericke's demonstration of the force due to air pressure as performed before Emperor Ferdinand III in 1657. (*The Granger Collection*)

required to pull the evacuated hemispheres apart is $\pi R^2(P_0 - P)$, where R is the radius of the hemispheres and P is the pressure inside the hemispheres, which is much less than P_0. (b) Determine the force if $P = 0.100P_0$ and $R = 0.300$ m.

59. In 1983 the United States began coining the cent piece out of copper-clad zinc rather than pure copper. The mass of the old copper cent is 3.083 g, whereas that of the new cent is 2.517 g. Calculate the percent of zinc (by volume) in the new cent. The density of copper is 8.960 g/cm^3, and that of zinc is 7.133 g/cm^3. The new and old coins have the same volume.

60. A thin spherical shell with a mass of 4.00 kg and a diameter of 0.200 m is filled with helium (density = 0.180 kg/m^3). It is then released from rest on the bottom of a pool of water that is 4.00 m deep. (a) Neglecting frictional effects, show that the shell rises with constant acceleration and determine the value of that acceleration. (b) How long does it take for the top of the shell to reach the water's surface?

61. An incompressible, nonviscous fluid initially rests in the vertical portion of the pipe shown in Figure P15.61a, where $L = 2.00$ m. When the valve is opened, the fluid flows into the horizontal section of the pipe. What is the speed of the fluid when all of it is in the horizontal section, as in Figure P15.61b? Assume that the cross-sectional area of the entire pipe is constant.

Figure P15.63

Figure P15.61

62. **Review Problem.** A uniform disk with a mass of 10.0 kg and a radius of 0.250 m spins at 300 rev/min on a low-friction axle. It must be brought to a stop in 1.00 min by a brake pad that makes contact with the disk at an average distance of 0.220 m from the axis. The coefficient of friction between the pad and the disk is 0.500. A piston in a cylinder with a diameter of 5.00 cm presses the brake pad against the disk. Find the pressure that the brake fluid in the cylinder must have.

63. Figure P15.63 shows Superman attempting to drink water through a very long straw. With his great strength,

he achieves maximum possible suction. The walls of the tubular straw do not collapse. (a) Find the maximum height through which he can lift the water. (b) Still thirsty, the Man of Steel repeats his attempt on the Moon, which has no atmosphere. Find the difference between the water levels inside and outside the straw.

64. Show that the variation of atmospheric pressure with altitude is given by $P = P_0 e^{-\alpha h}$, where $\alpha = \rho_0 g/P_0$, P_0 is atmospheric pressure at some reference level $y = 0$, and ρ_0 is the atmospheric density at this level. Assume that the decrease in atmospheric pressure with increasing altitude is given by Equation 15.4, so that $dP/dy = -\rho g$, and assume that the density of air is proportional to the pressure.

65. A cube of ice whose edge measures 20.0 mm is floating in a glass of ice-cold water with one of its faces parallel to the water's surface. (a) How far below the water surface is the bottom face of the block? (b) Ice-cold ethyl alcohol is gently poured onto the water's surface to form a layer 5.00 mm thick above the water. The alcohol does not mix with the water. When the ice cube again attains hydrostatic equilibrium, what is the distance from the top of the water to the bottom face of the block? (c) Additional cold ethyl alcohol is poured onto the water's surface until the top surface of the alcohol coincides with the top surface of the ice cube (in

hydrostatic equilibrium). How thick is the required layer of ethyl alcohol?

66. **Review Problem.**   A light balloon filled with helium with a density of 0.180 kg/m^3 is tied to a light string of length $L = 3.00$ m. The string is tied to the ground, forming an "inverted" simple pendulum, as shown in Figure P15.66a. If the balloon is displaced slightly from its equilibrium position as shown in Figure P15.66b, (a) show that the ensuing motion is simple harmonic and (b) determine the period of the motion. Take the density of air to be 1.29 kg/m^3 and ignore any energy loss due to air friction.

(a)   (b)

Figure P15.66

67. The water supply of a building is fed through a main 6.00-cm-diameter pipe. A 2.00-cm-diameter faucet tap located 2.00 m above the main pipe is observed to fill a 25.0-L container in 30.0 s. (a) What is the speed at which the water leaves the faucet? (b) What is the gauge pressure in the 6-cm main pipe? (Assume that the faucet is the only "leak" in the building.)

68. The *spirit-in-glass thermometer*, invented in Florence, Italy, around 1654, consists of a tube of liquid (the spirit) containing a number of submerged glass spheres with slightly different masses (Fig. P15.68). At sufficiently low temperatures, all the spheres float, but as the temperature rises, the spheres sink one after the other. The device is a crude but interesting tool for measuring temperature. Suppose that the tube is filled with ethyl alcohol, whose density is 0.789 45 g/cm^3 at 20.0°C and decreases to 0.780 97 g/cm^3 at 30.0°C. (a) If one of the spheres has a radius of 1.000 cm and is in equilibrium halfway up the tube at 20.0°C, determine its mass. (b) When the temperature increases to 30.0°C, what mass must a second sphere of the same radius have to be in equilibrium at the halfway point? (c) At 30.0°C the first sphere has fallen to the bottom of the tube. What upward force does the bottom of the tube exert on this sphere?

69. A U-tube open at both ends is partially filled with water (Fig. P15.69a). Oil having a density of 750 kg/m^3 is then poured into the right arm and forms a column $L = 5.00$ cm in height (Fig. P15.69b). (a) Determine

Figure P15.68   *(Courtesy of Jeanne Maier)*

the difference h in the heights of the two liquid surfaces. (b) The right arm is shielded from any air motion while air is blown across the top of the left arm until the surfaces of the two liquids are at the same height (Fig. P15.69c). Determine the speed of the air being blown across the left arm. (Take the density of air as 1.29 kg/m^3.)

Figure P15.69

ANSWERS TO QUICK QUIZZES

15.1 You would be better off with the basketball player. Although weight is distributed over the larger surface area, equal to about half of the total surface area of the sneaker sole, the pressure (F/A) that he applies is relatively small. The woman's lesser weight is distributed over the very small cross-sectional area of the spiked heel. Some museums make women in high-heeled shoes wear slippers or special heel attachments so that they do not damage the wood floors.

15.2 If the professor were to try to support his entire weight on a single nail, the pressure exerted on his skin would be his entire weight divided by the very small surface area of the nail point. This extremely great pressure would cause the nail to puncture his skin. However, if the professor distributes his weight over several hundred nails, as shown in the photograph, the pressure exerted on his skin is considerably reduced because the surface area that supports his weight is now the total surface area of all the nail points. (Lying on the bed of nails is much more comfortable than sitting on the bed, and standing on the bed without shoes is definitely not recommended. Do not lie on a bed of nails unless you have been shown how to do so safely.)

15.3 Because the horizontal force exerted by the outside fluid on an element of the cylinder is equal and opposite the horizontal force exerted by the fluid on another element diametrically opposite the first, the net force on the cylinder in the horizontal direction is zero.

15.4 If you think of the grain stored in the silo as a fluid, then the pressure it exerts on the walls increases with increasing depth. The spacing between bands is smaller at the lower portions so that the greater outward forces acting on the walls can be overcome. The silo on the right shows another way of accomplishing the same thing: double banding at the bottom.

15.5 Because water is so much less dense than mercury, the column for a water barometer would have to be $h = P_0/\rho g = 10.3$ m high, and such a column is inconveniently tall.

15.6 The entire hull of a ship is full of air, and the density of air is about one-thousandth the density of water. Hence, the total weight of the ship equals the weight of the volume of water that is displaced by the portion of the ship that is below sea level.

15.7 Remains the same. In effect, the ice creates a "hole" in the water, and the weight of the water displaced from the hole is the same as all the weight of the cube. When the cube changes from ice to water, the water just fills the hole.

15.8 Goes down because the anchor displaces more water while in the boat than it does in the pond. While it is in the boat, the anchor can be thought of as a floating object that displaces a volume of water weighing as much as it does. When the anchor is thrown overboard, it sinks and displaces a volume of water equal to its own volume. Because the density of the anchor is greater than that of water, the volume of water that weighs the same as the anchor is greater than the volume of the anchor.

15.9 As the water falls, its speed increases. Because the flow rate Av must remain constant at all cross sections (see Eq. 15.7), the stream must become narrower as the speed increases.

15.10 The rapidly moving air characteristic of a tornado is at a pressure below atmospheric pressure. The stationary air inside the building remains at atmospheric pressure. The pressure difference results in an outward force on the roof and walls, and this force can be great enough to lift the roof off the building. Opening the windows helps to equalize the inside and outside pressures.

(CALVIN AND HOBBES ©1992 Watterson. Reprinted with permission of UNIVERSAL PRESS SYNDICATE. All rights reserved.)

part 2

Mechanical Waves

As we learned in Chapter 13, most elastic objects oscillate when a force is applied to them and then removed. That is, once such an object is distorted, its shape tends to be restored to some equilibrium configuration. Even the atoms in a solid oscillate about some equilibrium position, as if they were connected to their neighbors by imaginary springs.

Wave motion—the subject we study next—is closely related to the phenomenon of oscillation. Sound waves, earthquake waves, waves on stretched strings, and water waves are all produced by some source of oscillation. As a sound wave travels through the air, the air molecules oscillate back and forth; as a water wave travels across a pond, the water molecules oscillate up and down and backward and forward. In general, as waves travel through any medium, the particles of the medium move in repetitive cycles. Therefore, the motion of the particles bears a strong resemblance to the periodic motion of an oscillating pendulum or a mass attached to a spring.

To explain many other phenomena in nature, we must understand the concepts of oscillations and waves. For instance, although sky-scrapers and bridges appear to be rigid, they actually oscillate, a fact that the architects and engineers who design and build them must take into account. To understand how radio and television work, we must understand the origin and nature of electromagnetic waves and how they propagate through space. Finally, much of what scientists have learned about atomic structure has come from information carried by waves. Therefore, we must first study oscillations and waves if we are to understand the concepts and theories of atomic physics.

◀ Copyright Warren Bolster/Tony Stone Images.

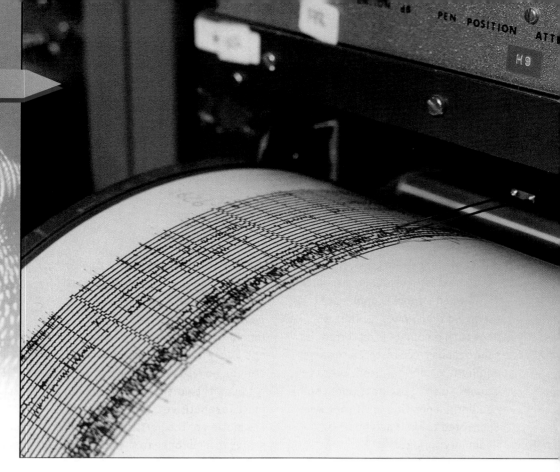

c h a p t e r

16

Wave Motion

Chapter Outline

Most of us experienced waves as children when we dropped a pebble into a pond. At the point where the pebble hits the water's surface, waves are created. These waves move outward from the creation point in expanding circles until they reach the shore. If you were to examine carefully the motion of a leaf floating on the disturbed water, you would see that the leaf moves up, down, and sideways about its original position but does not undergo any net displacement away from or toward the point where the pebble hit the water. The water molecules just beneath the leaf, as well as all the other water molecules on the pond's surface, behave in the same way. That is, the water *wave* moves from the point of origin to the shore, but the water is not carried with it.

An excerpt from a book by Einstein and Infeld gives the following remarks concerning wave phenomena:[1]

> A bit of gossip starting in Washington reaches New York [by word of mouth] very quickly, even though not a single individual who takes part in spreading it travels between these two cities. There are two quite different motions involved, that of the rumor, Washington to New York, and that of the persons who spread the rumor. The wind, passing over a field of grain, sets up a wave which spreads out across the whole field. Here again we must distinguish between the motion of the wave and the motion of the separate plants, which undergo only small oscillations... The particles constituting the medium perform only small vibrations, but the whole motion is that of a progressive wave. The essentially new thing here is that for the first time we consider the motion of something which is not matter, but energy propagated through matter.

The world is full of waves, the two main types being *mechanical* waves and *electromagnetic* waves. We have already mentioned examples of mechanical waves: sound waves, water waves, and "grain waves." In each case, some physical medium is being disturbed—in our three particular examples, air molecules, water molecules, and stalks of grain. Electromagnetic waves do not require a medium to propagate; some examples of electromagnetic waves are visible light, radio waves, television signals, and x-rays. Here, in Part 2 of this book, we study only mechanical waves.

The wave concept is abstract. When we observe what we call a water wave, what we see is a rearrangement of the water's surface. Without the water, there would be no wave. A wave traveling on a string would not exist without the string. Sound waves could not travel through air if there were no air molecules. With mechanical waves, what we interpret as a wave corresponds to the propagation of a disturbance through a medium.

Interference patterns produced by outward-spreading waves from many drops of liquid falling into a body of water. *(Martin Dohrn/ Science Photo Library/Photo Researchers, Inc.)*

[1] A. Einstein and L. Infeld, *The Evolution of Physics*, New York, Simon & Schuster, 1961. Excerpt from "What Is a Wave?"

The mechanical waves discussed in this chapter require (1) some source of disturbance, (2) a medium that can be disturbed, and (3) some physical connection through which adjacent portions of the medium can influence each other. We shall find that all waves carry energy. The amount of energy transmitted through a medium and the mechanism responsible for that transport of energy differ from case to case. For instance, the power of ocean waves during a storm is much greater than the power of sound waves generated by a single human voice.

16.1 ▶ BASIC VARIABLES OF WAVE MOTION

Imagine you are floating on a raft in a large lake. You slowly bob up and down as waves move past you. As you look out over the lake, you may be able to see the individual waves approaching. The point at which the displacement of the water from its normal level is highest is called the **crest** of the wave. The distance from one crest to the next is called the **wavelength** λ (Greek letter lambda). More generally, the wavelength is **the minimum distance between any two identical points (such as the crests) on adjacent waves,** as shown in Figure 16.1.

If you count the number of seconds between the arrivals of two adjacent waves, you are measuring the **period** T of the waves. In general, the period is **the time required for two identical points (such as the crests) of adjacent waves to pass by a point.**

The same information is more often given by the inverse of the period, which is called the **frequency** f. In general, the frequency of a periodic wave is **the number of crests (or troughs, or any other point on the wave) that pass a given point in a unit time interval.** The maximum displacement of a particle of the medium is called the **amplitude** A of the wave. For our water wave, this represents the highest distance of a water molecule above the undisturbed surface of the water as the wave passes by.

Waves travel with a specific speed, and this speed depends on the properties of the medium being disturbed. For instance, sound waves travel through room-temperature air with a speed of about 343 m/s (781 mi/h), whereas they travel through most solids with a speed greater than 343 m/s.

16.2 ▶ DIRECTION OF PARTICLE DISPLACEMENT

One way to demonstrate wave motion is to flick one end of a long rope that is under tension and has its opposite end fixed, as shown in Figure 16.2. In this manner, a single wave bump (called a *wave pulse*) is formed and travels along the rope with a definite speed. This type of disturbance is called a **traveling wave,** and Figure 16.2 represents four consecutive "snapshots" of the creation and propagation of the traveling wave. The rope is the medium through which the wave travels. Such a single pulse, in contrast to a train of pulses, has no frequency, no period, and no wavelength. However, the pulse does have definite amplitude and definite speed. As we shall see later, the properties of this particular medium that determine the speed of the wave are the tension in the rope and its mass per unit length. The shape of the wave pulse changes very little as it travels along the rope.[2]

As the wave pulse travels, each small segment of the rope, as it is disturbed, moves in a direction perpendicular to the wave motion. Figure 16.3 illustrates this

Figure 16.1   The wavelength λ of a wave is the distance between adjacent crests, adjacent troughs, or any other comparable adjacent identical points.

---

[2] Strictly speaking, the pulse changes shape and gradually spreads out during the motion. This effect is called *dispersion* and is common to many mechanical waves, as well as to electromagnetic waves. We do not consider dispersion in this chapter.

Figure 16.2 A wave pulse traveling down a stretched rope. The shape of the pulse is approximately unchanged as it travels along the rope.

Figure 16.3 A pulse traveling on a stretched rope is a transverse wave. The direction of motion of any element P of the rope (blue arrows) is perpendicular to the direction of wave motion (red arrows).

point for one particular segment, labeled P. Note that no part of the rope ever moves in the direction of the wave.

> A traveling wave that causes the particles of the disturbed medium to move perpendicular to the wave motion is called a **transverse wave.**

Transverse wave

Compare this with another type of wave—one moving down a long, stretched spring, as shown in Figure 16.4. The left end of the spring is pushed briefly to the right and then pulled briefly to the left. This movement creates a sudden compression of a region of the coils. The compressed region travels along the spring (to the right in Figure 16.4). The compressed region is followed by a region where the coils are extended. Notice that the direction of the displacement of the coils is *parallel* to the direction of propagation of the compressed region.

> A traveling wave that causes the particles of the medium to move parallel to the direction of wave motion is called a **longitudinal wave.**

Longitudinal wave

Sound waves, which we shall discuss in Chapter 17, are another example of longitudinal waves. The disturbance in a sound wave is a series of high-pressure and low-pressure regions that travel through air or any other material medium.

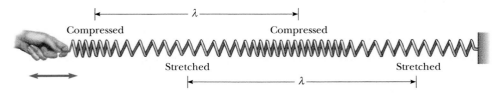

Figure 16.4 A longitudinal wave along a stretched spring. The displacement of the coils is in the direction of the wave motion. Each compressed region is followed by a stretched region.

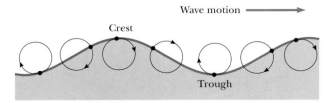

Figure 16.5 The motion of water molecules on the surface of deep water in which a wave is propagating is a combination of transverse and longitudinal displacements, with the result that molecules at the surface move in nearly circular paths. Each molecule is displaced both horizontally and vertically from its equilibrium position.

QuickLab

Make a "telephone" by poking a small hole in the bottom of two paper cups, threading a string through the holes, and tying knots in the ends of the string. If you speak into one cup while pulling the string taut, a friend can hear your voice in the other cup. What kind of wave is present in the string?

Some waves in nature exhibit a combination of transverse and longitudinal displacements. Surface water waves are a good example. When a water wave travels on the surface of deep water, water molecules at the surface move in nearly circular paths, as shown in Figure 16.5. Note that the disturbance has both transverse and longitudinal components. The transverse displacement is seen in Figure 16.5 as the variations in vertical position of the water molecules. The longitudinal displacement can be explained as follows: As the wave passes over the water's surface, water molecules at the crests move in the direction of propagation of the wave, whereas molecules at the troughs move in the direction opposite the propagation. Because the molecule at the labeled crest in Figure 16.5 will be at a trough after half a period, its movement in the direction of the propagation of the wave will be canceled by its movement in the opposite direction. This holds for every other water molecule disturbed by the wave. Thus, there is no net displacement of any water molecule during one complete cycle. Although the *molecules* experience no net displacement, the *wave* propagates along the surface of the water.

 The three-dimensional waves that travel out from the point under the Earth's surface at which an earthquake occurs are of both types—transverse and longitudinal. The longitudinal waves are the faster of the two, traveling at speeds in the range of 7 to 8 km/s near the surface. These are called **P waves,** with "P" standing for *primary* because they travel faster than the transverse waves and arrive at a seismograph first. The slower transverse waves, called **S waves** (with "S" standing for *secondary*), travel through the Earth at 4 to 5 km/s near the surface. By recording the time interval between the arrival of these two sets of waves at a seismograph, the distance from the seismograph to the point of origin of the waves can be determined. A single such measurement establishes an imaginary sphere centered on the seismograph, with the radius of the sphere determined by the difference in arrival times of the P and S waves. The origin of the waves is located somewhere on that sphere. The imaginary spheres from three or more monitoring stations located far apart from each other intersect at one region of the Earth, and this region is where the earthquake occurred.

Quick Quiz 16.1

(a) In a long line of people waiting to buy tickets, the first person leaves and a pulse of motion occurs as people step forward to fill the gap. As each person steps forward, the gap moves through the line. Is the propagation of this gap transverse or longitudinal? (b) Consider the "wave" at a baseball game: people stand up and shout as the wave arrives at their location, and the resultant pulse moves around the stadium. Is this wave transverse or longitudinal?

16.3 ONE-DIMENSIONAL TRAVELING WAVES

Consider a wave pulse traveling to the right with constant speed v on a long, taut string, as shown in Figure 16.6. The pulse moves along the x axis (the axis of the string), and the transverse (vertical) displacement of the string (the medium) is measured along the y axis. Figure 16.6a represents the shape and position of the pulse at time $t = 0$. At this time, the shape of the pulse, whatever it may be, can be represented as $y = f(x)$. That is, y, which is the vertical position of any point on the string, is some definite function of x. The displacement y, sometimes called the *wave function*, depends on both x and t. For this reason, it is often written $y(x, t)$, which is read "y as a function of x and t." Consider a particular point P on the string, identified by a specific value of its x coordinate. Before the pulse arrives at P, the y coordinate of this point is zero. As the wave passes P, the y coordinate of this point increases, reaches a maximum, and then decreases to zero. Therefore, **the wave function y represents the y coordinate of any point P of the medium at any time t.**

Because its speed is v, the wave pulse travels to the right a distance vt in a time t (see Fig. 16.6b). If the shape of the pulse does not change with time, we can represent the wave function y for all times after $t = 0$. Measured in a stationary reference frame having its origin at O, the wave function is

$$y = f(x - vt) \tag{16.1}$$

Wave traveling to the right

If the wave pulse travels to the left, the string displacement is

$$y = f(x + vt) \tag{16.2}$$

Wave traveling to the left

For any given time t, the wave function y as a function of x defines a curve representing the shape of the pulse at this time. This curve is equivalent to a "snapshot" of the wave at this time. For a pulse that moves without changing shape, the speed of the pulse is the same as that of any feature along the pulse, such as the crest shown in Figure 16.6. To find the speed of the pulse, we can calculate how far the crest moves in a short time and then divide this distance by the time interval. To follow the motion of the crest, we must substitute some particular value, say x_0, in Equation 16.1 for $x - vt$. Regardless of how x and t change individually, we must require that $x - vt = x_0$ in order to stay with the crest. This expression therefore represents the equation of motion of the crest. At $t = 0$, the crest is at $x = x_0$; at a

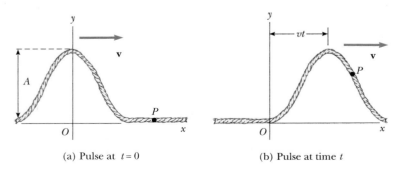

(a) Pulse at $t = 0$ (b) Pulse at time t

Figure 16.6   A one-dimensional wave pulse traveling to the right with a speed v. (a) At $t = 0$, the shape of the pulse is given by $y = f(x)$. (b) At some later time t, the shape remains unchanged and the vertical displacement of any point P of the medium is given by $y = f(x - vt)$.

time dt later, the crest is at $x = x_0 + v\,dt$. Therefore, in a time dt, the crest has moved a distance $dx = (x_0 + v\,dt) - x_0 = v\,dt$. Hence, the wave speed is

$$v = \frac{dx}{dt}$$

(16.3)

EXAMPLE 16.1   A Pulse Moving to the Right

A wave pulse moving to the right along the x axis is represented by the wave function

$$y(x, t) = \frac{2}{(x - 3.0t)^2 + 1}$$

where x and y are measured in centimeters and t is measured in seconds. Plot the wave function at $t = 0$, $t = 1.0$ s, and $t = 2.0$ s.

Solution   First, note that this function is of the form $y = f(x - vt)$. By inspection, we see that the wave speed is $v = 3.0$ cm/s. Furthermore, the wave amplitude (the maximum value of y) is given by $A = 2.0$ cm. (We find the maximum value of the function representing y by letting $x - 3.0t = 0$.) The wave function expressions are

$$y(x, 0) = \frac{2}{x^2 + 1} \qquad \text{at } t = 0$$

$$y(x, 1.0) = \frac{2}{(x - 3.0)^2 + 1} \qquad \text{at } t = 1.0 \text{ s}$$

$$y(x, 2.0) = \frac{2}{(x - 6.0)^2 + 1} \qquad \text{at } t = 2.0 \text{ s}$$

We now use these expressions to plot the wave function versus x at these times. For example, let us evaluate $y(x, 0)$ at $x = 0.50$ cm:

$$y(0.50, 0) = \frac{2}{(0.50)^2 + 1} = 1.6 \text{ cm}$$

Likewise, at $x = 1.0$ cm, $y(1.0, 0) = 1.0$ cm, and at $x = 2.0$ cm, $y(2.0, 0) = 0.40$ cm. Continuing this procedure for other values of x yields the wave function shown in Figure 16.7a. In a similar manner, we obtain the graphs of $y(x, 1.0)$ and $y(x, 2.0)$, shown in Figure 16.7b and c, respectively. These snapshots show that the wave pulse moves to the right without changing its shape and that it has a constant speed of 3.0 cm/s.

(a)

(b)

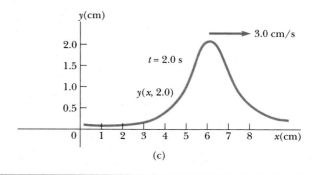

(c)

Figure 16.7   Graphs of the function $y(x, t) = 2/[(x - 3.0t)^2 + 1]$ at (a) $t = 0$, (b) $t = 1.0$ s, and (c) $t = 2.0$ s.

16.4 SUPERPOSITION AND INTERFERENCE

Many interesting wave phenomena in nature cannot be described by a single moving pulse. Instead, one must analyze complex waves in terms of a combination of many traveling waves. To analyze such wave combinations, one can make use of the **superposition principle:**

> If two or more traveling waves are moving through a medium, the resultant wave function at any point is the algebraic sum of the wave functions of the individual waves.

Linear waves obey the superposition principle

Waves that obey this principle are called *linear waves* and are generally characterized by small amplitudes. Waves that violate the superposition principle are called *nonlinear waves* and are often characterized by large amplitudes. In this book, we deal only with linear waves.

One consequence of the superposition principle is that **two traveling waves can pass through each other without being destroyed or even altered.** For instance, when two pebbles are thrown into a pond and hit the surface at different places, the expanding circular surface waves do not destroy each other but rather pass through each other. The complex pattern that is observed can be viewed as two independent sets of expanding circles. Likewise, when sound waves from two sources move through air, they pass through each other. The resulting sound that one hears at a given point is the resultant of the two disturbances.

Figure 16.8 is a pictorial representation of superposition. The wave function for the pulse moving to the right is y_1, and the wave function for the pulse moving

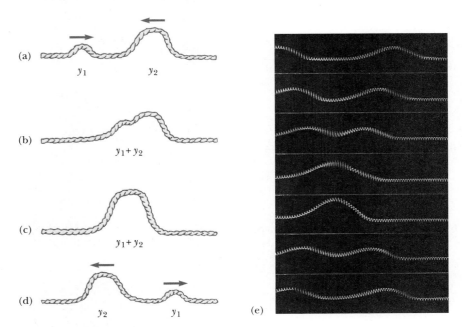

Figure 16.8 (a–d) Two wave pulses traveling on a stretched string in opposite directions pass through each other. When the pulses overlap, as shown in (b) and (c), the net displacement of the string equals the sum of the displacements produced by each pulse. Because each pulse displaces the string in the positive direction, we refer to the superposition of the two pulses as *constructive interference*. (e) Photograph of superposition of two equal, symmetric pulses traveling in opposite directions on a stretched spring. *(e, Education Development Center, Newton, MA)*

Interference of water waves produced in a ripple tank. The sources of the waves are two objects that oscillate perpendicular to the surface of the tank. *(Courtesy of Central Scientific Company)*

to the left is y_2. The pulses have the same speed but different shapes. Each pulse is assumed to be symmetric, and the displacement of the medium is in the positive y direction for both pulses. (Note, however, that the superposition principle applies even when the two pulses are not symmetric.) When the waves begin to overlap (Fig. 16.8b), the wave function for the resulting complex wave is given by $y_1 + y_2$.

Figure 16.9 (a–e) Two wave pulses traveling in opposite directions and having displacements that are inverted relative to each other. When the two overlap in (c), their displacements partially cancel each other. (f) Photograph of superposition of two symmetric pulses traveling in opposite directions, where one pulse is inverted relative to the other. *(f, Education Development Center, Newton, MA)*

When the crests of the pulses coincide (Fig. 16.8c), the resulting wave given by $y_1 + y_2$ is symmetric. The two pulses finally separate and continue moving in their original directions (Fig. 16.8d). Note that the pulse shapes remain unchanged, as if the two pulses had never met!

The combination of separate waves in the same region of space to produce a resultant wave is called **interference.** For the two pulses shown in Figure 16.8, the displacement of the medium is in the positive y direction for both pulses, and the resultant wave (created when the pulses overlap) exhibits a displacement greater than that of either individual pulse. Because the displacements caused by the two pulses are in the same direction, we refer to their superposition as **constructive interference.**

Now consider two pulses traveling in opposite directions on a taut string where one pulse is inverted relative to the other, as illustrated in Figure 16.9. In this case, when the pulses begin to overlap, the resultant wave is given by $y_1 + y_2$, but the values of the function y_2 are negative. Again, the two pulses pass through each other; however, because the displacements caused by the two pulses are in opposite directions, we refer to their superposition as **destructive interference.**

Quick Quiz 16.2

Two pulses are traveling toward each other at 10 cm/s on a long string, as shown in Figure 16.10. Sketch the shape of the string at $t = 0.6$ s.

Figure 16.10   The pulses on this string are traveling at 10 cm/s.

16.5 THE SPEED OF WAVES ON STRINGS

In this section, we focus on determining the speed of a transverse pulse traveling on a taut string. Let us first conceptually argue the parameters that determine the speed. If a string under tension is pulled sideways and then released, the tension is responsible for accelerating a particular segment of the string back toward its equilibrium position. According to Newton's second law, the acceleration of the segment increases with increasing tension. If the segment returns to equilibrium more rapidly due to this increased acceleration, we would intuitively argue that the wave speed is greater. Thus, we expect the wave speed to increase with increasing tension.

Likewise, we can argue that the wave speed decreases if the mass per unit length of the string increases. This is because it is more difficult to accelerate a massive segment of the string than a light segment. If the tension in the string is T (not to be confused with the same symbol used for the period) and its mass per

The strings of this piano vary in both tension and mass per unit length. These differences in tension and density, in combination with the different lengths of the strings, allow the instrument to produce a wide range of sounds. *(Charles D. Winters)*

unit length is μ (Greek letter mu), then, as we shall show, the wave speed is

$$v = \sqrt{\frac{T}{\mu}} \qquad \text{TENSION} \atop \text{LINEAR DENSITY}$$ **(16.4)**

First, let us verify that this expression is dimensionally correct. The dimensions of T are ML/T^2, and the dimensions of μ are M/L. Therefore, the dimensions of T/μ are L^2/T^2; hence, the dimensions of $\sqrt{T/\mu}$ are L/T—indeed, the dimensions of speed. No other combination of T and μ is dimensionally correct if we assume that they are the only variables relevant to the situation.

Now let us use a mechanical analysis to derive Equation 16.4. On our string under tension, consider a pulse moving to the right with a uniform speed v measured relative to a stationary frame of reference. Instead of staying in this reference frame, it is more convenient to choose as our reference frame one that moves along with the pulse with the same speed as the pulse, so that the pulse is at rest within the frame. This change of reference frame is permitted because Newton's laws are valid in either a stationary frame or one that moves with constant velocity. In our new reference frame, a given segment of the string initially to the right of the pulse moves to the left, rises up and follows the shape of the pulse, and then continues to move to the left. Figure 16.11a shows such a segment at the instant it is located at the top of the pulse.

The small segment of the string of length Δs shown in Figure 16.11a, and magnified in Figure 16.11b, forms an approximate arc of a circle of radius R. In our moving frame of reference (which is moving to the right at a speed v along with the pulse), the shaded segment is moving to the left with a speed v. This segment has a centripetal acceleration equal to v^2/R, which is supplied by components of the tension **T** in the string. The force **T** acts on either side of the segment and tangent to the arc, as shown in Figure 16.11b. The horizontal components of **T** cancel, and each vertical component $T \sin \theta$ acts radially toward the center of the arc. Hence, the total radial force is $2T \sin \theta$. Because the segment is small, θ is small, and we can use the small-angle approximation $\sin \theta \approx \theta$. Therefore, the total radial force is

$$\sum F_r = 2T \sin \theta \approx 2T\theta$$

The segment has a mass $m = \mu \Delta s$. Because the segment forms part of a circle and subtends an angle 2θ at the center, $\Delta s = R(2\theta)$, and hence

$$m = \mu \Delta s = 2\mu R \theta$$

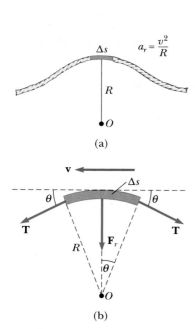

Figure 16.11 (a) To obtain the speed v of a wave on a stretched string, it is convenient to describe the motion of a small segment of the string in a moving frame of reference. (b) In the moving frame of reference, the small segment of length Δs moves to the left with speed v. The net force on the segment is in the radial direction because the horizontal components of the tension force cancel.

If we apply Newton's second law to this segment, the radial component of motion gives

$$\sum F_r = ma = \frac{mv^2}{R}$$

$$2T\theta = \frac{2\mu R\theta v^2}{R}$$

Solving for v gives Equation 16.4.

Notice that this derivation is based on the assumption that the pulse height is small relative to the length of the string. Using this assumption, we were able to use the approximation $\sin \theta \approx \theta$. Furthermore, the model assumes that the tension T is not affected by the presence of the pulse; thus, T is the same at all points on the string. Finally, this proof does *not* assume any particular shape for the pulse. Therefore, we conclude that a pulse of *any shape* travels along the string with speed $v = \sqrt{T/\mu}$ without any change in pulse shape.

EXAMPLE 16.2   The Speed of a Pulse on a Cord

A uniform cord has a mass of 0.300 kg and a length of 6.00 m (Fig. 16.12). The cord passes over a pulley and supports a 2.00-kg object. Find the speed of a pulse traveling along this cord.

Solution   The tension T in the cord is equal to the weight of the suspended 2.00-kg mass:

Figure 16.12   The tension T in the cord is maintained by the suspended object. The speed of any wave traveling along the cord is given by $v = \sqrt{T/\mu}$.

$$T = mg = (2.00 \text{ kg})(9.80 \text{ m/s}^2) = 19.6 \text{ N}$$

(This calculation of the tension neglects the small mass of the cord. Strictly speaking, the cord can never be exactly horizontal, and therefore the tension is not uniform.) The mass per unit length μ of the cord is

$$\mu = \frac{m}{\ell} = \frac{0.300 \text{ kg}}{6.00 \text{ m}} = 0.0500 \text{ kg/m}$$

Therefore, the wave speed is

$$v = \sqrt{\frac{T}{\mu}} = \sqrt{\frac{19.6 \text{ N}}{0.050 \, 0 \text{ kg/m}}} = \boxed{19.8 \text{ m/s}}$$

Exercise   Find the time it takes the pulse to travel from the wall to the pulley.

Answer   0.253 s.

Quick Quiz 16.3

Suppose you create a pulse by moving the free end of a taut string up and down once with your hand. The string is attached at its other end to a distant wall. The pulse reaches the wall in a time t. Which of the following actions, taken by itself, decreases the time it takes the pulse to reach the wall? More than one choice may be correct.
(a) Moving your hand more quickly, but still only up and down once by the same amount.
(b) Moving your hand more slowly, but still only up and down once by the same amount.
(c) Moving your hand a greater distance up and down in the same amount of time.
(d) Moving your hand a lesser distance up and down in the same amount of time.
(e) Using a heavier string of the same length and under the same tension.
(f) Using a lighter string of the same length and under the same tension.
(g) Using a string of the same linear mass density but under decreased tension.
(h) Using a string of the same linear mass density but under increased tension.

16.6 REFLECTION AND TRANSMISSION

We have discussed traveling waves moving through a uniform medium. We now consider how a traveling wave is affected when it encounters a change in the medium. For example, consider a pulse traveling on a string that is rigidly attached to a support at one end (Fig. 16.13). When the pulse reaches the support, a severe change in the medium occurs—the string ends. The result of this change is that the wave undergoes **reflection**—that is, the pulse moves back along the string in the opposite direction.

Note that the reflected pulse is inverted. This inversion can be explained as follows: When the pulse reaches the fixed end of the string, the string produces an upward force on the support. By Newton's third law, the support must exert an equal and opposite (downward) reaction force on the string. This downward force causes the pulse to invert upon reflection.

Now consider another case: this time, the pulse arrives at the end of a string that is free to move vertically, as shown in Figure 16.14. The tension at the free end is maintained because the string is tied to a ring of negligible mass that is free to slide vertically on a smooth post. Again, the pulse is reflected, but this time it is not inverted. When it reaches the post, the pulse exerts a force on the free end of the string, causing the ring to accelerate upward. The ring overshoots the height of the incoming pulse, and then the downward component of the tension force pulls the ring back down. This movement of the ring produces a reflected pulse that is not inverted and that has the same amplitude as the incoming pulse.

Finally, we may have a situation in which the boundary is intermediate between these two extremes. In this case, part of the incident pulse is reflected and part undergoes **transmission**—that is, some of the pulse passes through the boundary. For instance, suppose a light string is attached to a heavier string, as shown in Figure 16.15. When a pulse traveling on the light string reaches the boundary between the two, part of the pulse is reflected and inverted and part is transmitted to the heavier string. The reflected pulse is inverted for the same reasons described earlier in the case of the string rigidly attached to a support.

Note that the reflected pulse has a smaller amplitude than the incident pulse. In Section 16.8, we shall learn that the energy carried by a wave is related to its amplitude. Thus, according to the principle of the conservation of energy, when the pulse breaks up into a reflected pulse and a transmitted pulse at the boundary, the sum of the energies of these two pulses must equal the energy of the incident pulse. Because the reflected pulse contains only part of the energy of the incident pulse, its amplitude must be smaller.

Figure 16.13 The reflection of a traveling wave pulse at the fixed end of a stretched string. The reflected pulse is inverted, but its shape is unchanged.

Figure 16.14 The reflection of a traveling wave pulse at the free end of a stretched string. The reflected pulse is not inverted.

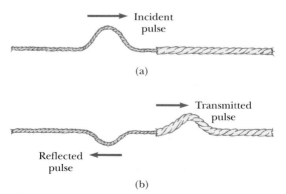

Figure 16.15 (a) A pulse traveling to the right on a light string attached to a heavier string. (b) Part of the incident pulse is reflected (and inverted), and part is transmitted to the heavier string.

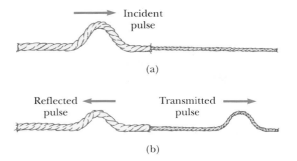

Incident pulse

(a)

Reflected pulse

Transmitted pulse

(b)

Figure 16.16 (a) A pulse traveling to the right on a heavy string attached to a lighter string. (b) The incident pulse is partially reflected and partially transmitted, and the reflected pulse is not inverted.

When a pulse traveling on a heavy string strikes the boundary between the heavy string and a lighter one, as shown in Figure 16.16, again part is reflected and part is transmitted. In this case, the reflected pulse is not inverted.

In either case, the relative heights of the reflected and transmitted pulses depend on the relative densities of the two strings. If the strings are identical, there is no discontinuity at the boundary and no reflection takes place.

According to Equation 16.4, the speed of a wave on a string increases as the mass per unit length of the string decreases. In other words, a pulse travels more slowly on a heavy string than on a light string if both are under the same tension. The following general rules apply to reflected waves: **When a wave pulse travels from medium A to medium B and $v_A > v_B$ (that is, when B is denser than A), the pulse is inverted upon reflection. When a wave pulse travels from medium A to medium B and $v_A < v_B$ (that is, when A is denser than B), the pulse is not inverted upon reflection.**

16.7 SINUSOIDAL WAVES

In this section, we introduce an important wave function whose shape is shown in Figure 16.17. The wave represented by this curve is called a **sinusoidal wave** because the curve is the same as that of the function $\sin \theta$ plotted against θ. The sinusoidal wave is the simplest example of a periodic continuous wave and can be used to build more complex waves, as we shall see in Section 18.8. The red curve represents a snapshot of a traveling sinusoidal wave at $t = 0$, and the blue curve represents a snapshot of the wave at some later time t. At $t = 0$, the function describing the positions of the particles of the medium through which the sinusoidal wave is traveling can be written

$$y = A \sin\left(\frac{2\pi}{\lambda} x\right) \tag{16.5}$$

where the constant A represents the wave amplitude and the constant λ is the wavelength. Thus, we see that the position of a particle of the medium is the same whenever x is increased by an integral multiple of λ. If the wave moves to the right with a speed v, then the wave function at some later time t is

$$y = A \sin\left[\frac{2\pi}{\lambda} (x - vt)\right] \tag{16.6}$$

That is, the traveling sinusoidal wave moves to the right a distance vt in the time t, as shown in Figure 16.17. Note that the wave function has the form $f(x - vt)$ and

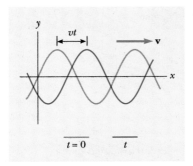

Figure 16.17 A one-dimensional sinusoidal wave traveling to the right with a speed v. The red curve represents a snapshot of the wave at $t = 0$, and the blue curve represents a snapshot at some later time t.

so represents a wave traveling to the right. If the wave were traveling to the left, the quantity $x - vt$ would be replaced by $x + vt$, as we learned when we developed Equations 16.1 and 16.2.

By definition, the wave travels a distance of one wavelength in one period T. Therefore, the wave speed, wavelength, and period are related by the expression

$$v = \frac{\lambda}{T} = \lambda f \qquad \textbf{(16.7)}$$

Substituting this expression for v into Equation 16.6, we find that

$$y = A \sin\left[2\pi\left(\frac{x}{\lambda} - \frac{t}{T}\right)\right] \qquad \textbf{(16.8)}$$

This form of the wave function clearly shows the *periodic* nature of y. At any given time t (a snapshot of the wave), y has the *same* value at the positions x, $x + \lambda$, $x + 2\lambda$, and so on. Furthermore, at any given position x, the value of y is the same at times t, $t + T$, $t + 2T$, and so on.

We can express the wave function in a convenient form by defining two other quantities, the **angular wave number** k and the **angular frequency** ω:

Angular wave number

$$k \equiv \frac{2\pi}{\lambda} \qquad \frac{rad}{m} \qquad \textbf{(16.9)}$$

Angular frequency

$$\omega \equiv \frac{2\pi}{T} \qquad \frac{rad}{s} \qquad \textbf{(16.10)}$$

Using these definitions, we see that Equation 16.8 can be written in the more compact form

Wave function for a sinusoidal wave

$$y = A \sin(kx - \omega t) \qquad \textbf{(16.11)}$$

The frequency of a sinusoidal wave is related to the period by the expression

Frequency

$$f = \frac{1}{T} \qquad \textbf{(16.12)}$$

The most common unit for frequency, as we learned in Chapter 13, is second^{-1}, or **hertz** (Hz). The corresponding unit for T is seconds.

Using Equations 16.9, 16.10, and 16.12, we can express the wave speed v originally given in Equation 16.7 in the alternative forms

Speed of a sinusoidal wave

$$v = \frac{\omega}{k} \qquad \frac{rad}{s}\,\frac{m}{rad} \qquad \textbf{(16.13)}$$

$$v = \lambda f = \frac{\lambda}{T} \qquad \textbf{(16.14)}$$

The wave function given by Equation 16.11 assumes that the vertical displacement y is zero at $x = 0$ and $t = 0$. This need not be the case. If it is not, we generally express the wave function in the form

General expression for a sinusoidal wave

$$y = A \sin(kx - \omega t + \phi) \qquad \textbf{(16.15)}$$

where ϕ is the **phase constant,** just as we learned in our study of periodic motion in Chapter 13. This constant can be determined from the initial conditions.

EXAMPLE 16.3 ▶ A Traveling Sinusoidal Wave

A sinusoidal wave traveling in the positive x direction has an amplitude of 15.0 cm, a wavelength of 40.0 cm, and a frequency of 8.00 Hz. The vertical displacement of the medium at $t = 0$ and $x = 0$ is also 15.0 cm, as shown in Figure 16.18. (a) Find the angular wave number k, period T, angular frequency ω, and speed v of the wave.

Solution   (a) Using Equations 16.9, 16.10, 16.12, and 16.14, we find the following:

$$k = \frac{2\pi}{\lambda} = \frac{2\pi \text{ rad}}{40.0 \text{ cm}} = \boxed{0.157 \text{ rad/cm}}$$

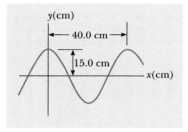

Figure 16.18   A sinusoidal wave of wavelength $\lambda = 40.0$ cm and amplitude $A = 15.0$ cm. The wave function can be written in the form $y = A \cos(kx - \omega t)$.

$$\omega = 2\pi f = 2\pi(8.00 \text{ s}^{-1}) = \boxed{50.3 \text{ rad/s}}$$

$$T = \frac{1}{f} = \frac{1}{8.00 \text{ s}^{-1}} = \boxed{0.125 \text{ s}}$$

$$v = \lambda f = (40.0 \text{ cm})(8.00 \text{ s}^{-1}) = \boxed{320 \text{ cm/s}}$$

(b) Determine the phase constant ϕ, and write a general expression for the wave function.

Solution   Because $A = 15.0$ cm and because $y = 15.0$ cm at $x = 0$ and $t = 0$, substitution into Equation 16.15 gives

$$15.0 = (15.0) \sin \phi \qquad \text{or} \qquad \sin \phi = 1$$

We may take the principal value $\phi = \pi/2$ rad (or 90°). Hence, the wave function is of the form

$$y = A \sin\left(kx - \omega t + \frac{\pi}{2}\right) = A \cos(kx - \omega t)$$

By inspection, we can see that the wave function must have this form, noting that the cosine function has the same shape as the sine function displaced by 90°. Substituting the values for A, k, and ω into this expression, we obtain

$$\boxed{y = (15.0 \text{ cm}) \cos(0.157x - 50.3t)}$$

Sinusoidal Waves on Strings

In Figure 16.2, we demonstrated how to create a pulse by jerking a taut string up and down once. To create a train of such pulses, normally referred to as a *wave train,* or just plain *wave,* we can replace the hand with an oscillating blade. If the wave consists of a train of identical cycles, whatever their shape, the relationships $f = 1/T$ and $v = f\lambda$ among speed, frequency, period, and wavelength hold true. We can make more definite statements about the wave function if the source of the waves vibrates in simple harmonic motion. Figure 16.19 represents snapshots of the wave created in this way at intervals of $T/4$. Note that because the end of the blade oscillates in simple harmonic motion, **each particle of the string, such as that at P, also oscillates vertically with simple harmonic motion.** This must be the case because each particle follows the simple harmonic motion of the blade. Therefore, every segment of the string can be treated as a simple harmonic oscillator vibrating with a frequency equal to the frequency of oscillation of the blade.[3] Note that although each segment oscillates in the y direction, the wave travels in the x direction with a speed v. Of course, this is the definition of a transverse wave.

---

[3] In this arrangement, we are assuming that a string segment always oscillates in a vertical line. The tension in the string would vary if a segment were allowed to move sideways. Such motion would make the analysis very complex.

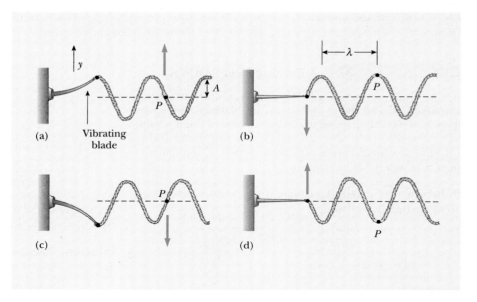

Figure 16.19   One method for producing a train of sinusoidal wave pulses on a string. The left end of the string is connected to a blade that is set into oscillation. Every segment of the string, such as the point *P*, oscillates with simple harmonic motion in the vertical direction.

If the wave at $t = 0$ is as described in Figure 16.19b, then the wave function can be written as

$$y = A \sin(kx - \omega t)$$

We can use this expression to describe the motion of any point on the string. The point *P* (or any other point on the string) moves only vertically, and so its *x* coordinate remains constant. Therefore, the **transverse speed v_y** (not to be confused with the wave speed *v*) and the **transverse acceleration a_y** are

$$v_y = \frac{dy}{dt}\bigg]_{x = \text{constant}} = \frac{\partial y}{\partial t} = -\omega A \cos(kx - \omega t) \qquad \textbf{(16.16)}$$

$$a_y = \frac{dv_y}{dt}\bigg]_{x = \text{constant}} = \frac{\partial v_y}{\partial t} = -\omega^2 A \sin(kx - \omega t) \qquad \textbf{(16.17)}$$

In these expressions, we must use partial derivatives (see Section 8.6) because *y* depends on both *x* and *t*. In the operation $\partial y/\partial t$, for example, we take a derivative with respect to *t* while holding *x* constant. The maximum values of the transverse speed and transverse acceleration are simply the absolute values of the coefficients of the cosine and sine functions:

$$v_{y,\,\text{max}} = \omega A \qquad \textbf{(16.18)}$$

$$a_{y,\,\text{max}} = \omega^2 A \qquad \textbf{(16.19)}$$

The transverse speed and transverse acceleration do not reach their maximum values simultaneously. The transverse speed reaches its maximum value (ωA) when $y = 0$, whereas the transverse acceleration reaches its maximum value ($\omega^2 A$) when $y = \pm A$. Finally, Equations 16.18 and 16.19 are identical in mathematical form to the corresponding equations for simple harmonic motion, Equations 13.10 and 13.11.

Quick Quiz 16.4

A sinusoidal wave is moving on a string. If you increase the frequency *f* of the wave, how do the transverse speed, wave speed, and wavelength change?

EXAMPLE 16.4 ▸ A Sinusoidally Driven String

The string shown in Figure 16.19 is driven at a frequency of 5.00 Hz. The amplitude of the motion is 12.0 cm, and the wave speed is 20.0 m/s. Determine the angular frequency ω and angular wave number *k* for this wave, and write an expression for the wave function.

Solution   Using Equations 16.10, 16.12, and 16.13, we find that

$$\omega = \frac{2\pi}{T} = 2\pi f = 2\pi(5.00 \text{ Hz}) = \boxed{31.4 \text{ rad/s}}$$

$$k = \frac{\omega}{v} = \frac{31.4 \text{ rad/s}}{20.0 \text{ m/s}} = \boxed{1.57 \text{ rad/m}}$$

Because $A = 12.0$ cm $= 0.120$ m, we have

$$y = A \sin(kx - \omega t) = (0.120 \text{ m}) \sin(1.57x - 31.4t)$$

Exercise   Calculate the maximum values for the transverse speed and transverse acceleration of any point on the string.

Answer   3.77 m/s; 118 m/s².

16.8 ▸ RATE OF ENERGY TRANSFER BY SINUSOIDAL WAVES ON STRINGS

As waves propagate through a medium, they transport energy. We can easily demonstrate this by hanging an object on a stretched string and then sending a pulse down the string, as shown in Figure 16.20. When the pulse meets the suspended object, the object is momentarily displaced, as illustrated in Figure 16.20b. In the process, energy is transferred to the object because work must be done for it to move upward. This section examines the rate at which energy is transported along a string. We shall assume a one-dimensional sinusoidal wave in the calculation of the energy transferred.

Consider a sinusoidal wave traveling on a string (Fig. 16.21). The source of the energy being transported by the wave is some external agent at the left end of the string; this agent does work in producing the oscillations. As the external agent performs work on the string, moving it up and down, energy enters the system of the string and propagates along its length. Let us focus our attention on a segment of the string of length Δx and mass Δm. Each such segment moves vertically with simple harmonic motion. Furthermore, all segments have the same angular frequency ω and the same amplitude *A*. As we found in Chapter 13, the elastic potential energy *U* associated with a particle in simple harmonic motion is $U = \frac{1}{2}ky^2$, where the simple harmonic motion is in the *y* direction. Using the relationship $\omega^2 = k/m$ developed in Equations 13.16 and 13.17, we can write this as

Figure 16.20   (a) A pulse traveling to the right on a stretched string on which an object has been suspended. (b) Energy is transmitted to the suspended object when the pulse arrives.

Figure 16.21   A sinusoidal wave traveling along the *x* axis on a stretched string. Every segment moves vertically, and every segment has the same total energy.

$U = \frac{1}{2}m\omega^2 y^2$. If we apply this equation to the segment of mass Δm, we see that the potential energy of this segment is

$$\Delta U = \frac{1}{2}(\Delta m)\,\omega^2 y^2$$

Because the mass per unit length of the string is $\mu = \Delta m/\Delta x$, we can express the potential energy of the segment as

$$\Delta U = \frac{1}{2}(\mu\Delta x)\,\omega^2 y^2$$

As the length of the segment shrinks to zero, $\Delta x \to dx$, and this expression becomes a differential relationship:

$$dU = \frac{1}{2}(\mu\,dx)\,\omega^2 y^2$$

We replace the general displacement y of the segment with the wave function for a sinusoidal wave:

$$dU = \frac{1}{2}\mu\omega^2 [A\sin(kx - \omega t)]^2\,dx = \frac{1}{2}\mu\omega^2 A^2 \sin^2(kx - \omega t)\,dx$$

If we take a snapshot of the wave at time $t = 0$, then the potential energy in a given segment is

$$dU = \frac{1}{2}\mu\omega^2 A^2 \sin^2 kx\,dx$$

To obtain the total potential energy in one wavelength, we integrate this expression over all the string segments in one wavelength:

$$U_\lambda = \int dU = \int_0^\lambda \frac{1}{2}\mu\omega^2 A^2 \sin^2 kx\,dx = \frac{1}{2}\mu\omega^2 A^2 \int_0^\lambda \sin^2 kx\,dx$$

$$= \frac{1}{2}\mu\omega^2 A^2 \left[\frac{1}{2}x - \frac{1}{4k}\sin 2\,kx\right]_0^\lambda = \frac{1}{2}\mu\omega^2 A^2 (\tfrac{1}{2}\lambda) = \frac{1}{4}\mu\omega^2 A^2 \lambda$$

Because it is in motion, each segment of the string also has kinetic energy. When we use this procedure to analyze the total kinetic energy in one wavelength of the string, we obtain the same result:

$$K_\lambda = \int dK = \frac{1}{4}\mu\omega^2 A^2 \lambda$$

The total energy in one wavelength of the wave is the sum of the potential and kinetic energies:

$$E_\lambda = U_\lambda + K_\lambda = \frac{1}{2}\mu\omega^2 A^2 \lambda \qquad \textbf{(16.20)}$$

As the wave moves along the string, this amount of energy passes by a given point on the string during one period of the oscillation. Thus, the power, or rate of energy transfer, associated with the wave is

$$\mathcal{P} = \frac{E_\lambda}{\Delta t} = \frac{\frac{1}{2}\mu\omega^2 A^2 \lambda}{T} = \frac{1}{2}\mu\omega^2 A^2 \left(\frac{\lambda}{T}\right)$$

Power of a wave

$$\mathcal{P} = \frac{1}{2}\mu\omega^2 A^2 v \qquad \textbf{(16.21)}$$

This shows that the rate of energy transfer by a sinusoidal wave on a string is proportional to (a) the wave speed, (b) the square of the frequency, and (c) the square of the amplitude. In fact: **the rate of energy transfer in any sinusoidal wave is proportional to the square of the angular frequency and to the square of the amplitude.**

EXAMPLE 16.5 Power Supplied to a Vibrating String

A taut string for which $\mu = 5.00 \times 10^{-2}$ kg/m is under a tension of 80.0 N. How much power must be supplied to the string to generate sinusoidal waves at a frequency of 60.0 Hz and an amplitude of 6.00 cm?

Solution The wave speed on the string is, from Equation 16.4,

$$v = \sqrt{\frac{T}{\mu}} = \sqrt{\frac{80.0 \text{ N}}{5.00 \times 10^{-2} \text{ kg/m}}} = 40.0 \text{ m/s}$$

Because $f = 60.0$ Hz, the angular frequency ω of the sinus-

oidal waves on the string has the value

$$\omega = 2\pi f = 2\pi(60.0 \text{ Hz}) = 377 \text{ s}^{-1}$$

Using these values in Equation 16.21 for the power, with $A = 6.00 \times 10^{-2}$ m, we obtain

$$\begin{aligned}
\mathcal{P} &= \tfrac{1}{2}\mu\omega^2 A^2 v \\
&= \tfrac{1}{2}(5.00 \times 10^{-2} \text{ kg/m})(377 \text{ s}^{-1})^2 \\
&= \times (6.00 \times 10^{-2} \text{ m})^2 (40.0 \text{ m/s}) \\
&= \boxed{512 \text{ W}}
\end{aligned}$$

Optional Section

16.9 THE LINEAR WAVE EQUATION

In Section 16.3 we introduced the concept of the wave function to represent waves traveling on a string. All wave functions $y(x, t)$ represent solutions of an equation called the *linear wave equation*. This equation gives a complete description of the wave motion, and from it one can derive an expression for the wave speed. Furthermore, the linear wave equation is basic to many forms of wave motion. In this section, we derive this equation as applied to waves on strings.

Suppose a traveling wave is propagating along a string that is under a tension T. Let us consider one small string segment of length Δx (Fig. 16.22). The ends of the segment make small angles θ_A and θ_B with the x axis. The net force acting on the segment in the vertical direction is

$$\sum F_y = T \sin \theta_B - T \sin \theta_A = T(\sin \theta_B - \sin \theta_A)$$

Because the angles are small, we can use the small-angle approximation $\sin \theta \approx \tan \theta$ to express the net force as

$$\sum F_y \approx T(\tan \theta_B - \tan \theta_A)$$

However, the tangents of the angles at A and B are defined as the slopes of the string segment at these points. Because the slope of a curve is given by $\partial y/\partial x$, we have

$$\sum F_y \approx T\left[\left(\frac{\partial y}{\partial x}\right)_B - \left(\frac{\partial y}{\partial x}\right)_A\right] \tag{16.22}$$

We now apply Newton's second law to the segment, with the mass of the segment given by $m = \mu\Delta x$:

$$\sum F_y = ma_y = \mu\Delta x\left(\frac{\partial^2 y}{\partial t^2}\right) \tag{16.23}$$

Combining Equation 16.22 with Equation 16.23, we obtain

$$\mu\Delta x\left(\frac{\partial^2 y}{\partial t^2}\right) = T\left[\left(\frac{\partial y}{\partial x}\right)_B - \left(\frac{\partial y}{\partial x}\right)_A\right]$$

$$\frac{\mu}{T}\frac{\partial^2 y}{\partial t^2} = \frac{(\partial y/\partial x)_B - (\partial y/\partial x)_A}{\Delta x} \tag{16.24}$$

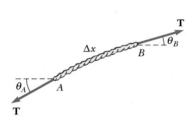

Figure 16.22 A segment of a string under tension T. The slopes at points A and B are given by tan θ_A and tan θ_B, respectively.

The right side of this equation can be expressed in a different form if we note that the partial derivative of any function is defined as

$$\frac{\partial f}{\partial x} \equiv \lim_{\Delta x \to 0} \frac{f(x + \Delta x) - f(x)}{\Delta x}$$

If we associate $f(x + \Delta x)$ with $(\partial y/\partial x)_B$ and $f(x)$ with $(\partial y/\partial x)_A$, we see that, in the limit $\Delta x \to 0$, Equation 16.24 becomes

Linear wave equation

$$\frac{\mu}{T}\frac{\partial^2 y}{\partial t^2} = \frac{\partial^2 y}{\partial x^2} \qquad \text{(16.25)}$$

This is the linear wave equation as it applies to waves on a string.

We now show that the sinusoidal wave function (Eq. 16.11) represents a solution of the linear wave equation. If we take the sinusoidal wave function to be of the form $y(x, t) = A \sin(kx - \omega t)$, then the appropriate derivatives are

$$\frac{\partial^2 y}{\partial t^2} = -\omega^2 A \sin(kx - \omega t)$$

$$\frac{\partial^2 y}{\partial x^2} = -k^2 A \sin(kx - \omega t)$$

Substituting these expressions into Equation 16.25, we obtain

$$-\frac{\mu\omega^2}{T}\sin(kx - \omega t) = -k^2 \sin(kx - \omega t)$$

This equation must be true for all values of the variables x and t in order for the sinusoidal wave function to be a solution of the wave equation. Both sides of the equation depend on x and t through the same function $\sin(kx - \omega t)$. Because this function divides out, we do indeed have an identity, provided that

$$k^2 = \frac{\mu\omega^2}{T}$$

Using the relationship $v = \omega/k$ (Eq. 16.13) in this expression, we see that

$$v^2 = \frac{\omega^2}{k^2} = \frac{T}{\mu}$$

$$v = \sqrt{\frac{T}{\mu}}$$

which is Equation 16.4. This derivation represents another proof of the expression for the wave speed on a taut string.

The linear wave equation (Eq. 16.25) is often written in the form

Linear wave equation in general

$$\frac{\partial^2 y}{\partial x^2} = \frac{1}{v^2}\frac{\partial^2 y}{\partial t^2} \qquad \text{(16.26)}$$

This expression applies in general to various types of traveling waves. For waves on strings, y represents the vertical displacement of the string. For sound waves, y corresponds to displacement of air molecules from equilibrium or variations in either the pressure or the density of the gas through which the sound waves are propagating. In the case of electromagnetic waves, y corresponds to electric or magnetic field components.

We have shown that the sinusoidal wave function (Eq. 16.11) is one solution of the linear wave equation (Eq. 16.26). Although we do not prove it here, the linear wave equation is satisfied by *any* wave function having the form $y = f(x \pm vt)$. Furthermore, we have seen that the linear wave equation is a direct consequence of Newton's second law applied to any segment of the string.

SUMMARY

A **transverse wave** is one in which the particles of the medium move in a direction *perpendicular* to the direction of the wave velocity. An example is a wave on a taut string. A **longitudinal wave** is one in which the particles of the medium move in a direction *parallel* to the direction of the wave velocity. Sound waves in fluids are longitudinal. You should be able to identify examples of both types of waves.

Any one-dimensional wave traveling with a speed v in the x direction can be represented by a wave function of the form

$$y = f(x \pm vt) \qquad \textbf{(16.1, 16.2)}$$

where the positive sign applies to a wave traveling in the negative x direction and the negative sign applies to a wave traveling in the positive x direction. The shape of the wave at any instant in time (a snapshot of the wave) is obtained by holding t constant.

The **superposition principle** specifies that when two or more waves move through a medium, the resultant wave function equals the algebraic sum of the individual wave functions. When two waves combine in space, they interfere to produce a resultant wave. The **interference** may be **constructive** (when the individual displacements are in the same direction) or **destructive** (when the displacements are in opposite directions).

The **speed** of a wave traveling on a taut string of mass per unit length μ and tension T is

$$v = \sqrt{\frac{T}{\mu}} \qquad \textbf{(16.4)}$$

A wave is totally or partially reflected when it reaches the end of the medium in which it propagates or when it reaches a boundary where its speed changes discontinuously. If a wave pulse traveling on a string meets a fixed end, the pulse is reflected and inverted. If the pulse reaches a free end, it is reflected but not inverted.

The **wave function** for a one-dimensional sinusoidal wave traveling to the right can be expressed as

$$y = A \sin\left[\frac{2\pi}{\lambda}(x - vt)\right] = A \sin(kx - \omega t) \qquad \textbf{(16.6, 16.11)}$$

where A is the **amplitude,** λ is the **wavelength,** k is the **angular wave number,** and ω is the **angular frequency.** If T is the **period** and f the **frequency,** v, k and ω can be written

$$v = \frac{\lambda}{T} = \lambda f = \frac{\omega}{k} \qquad \textbf{(16.7, 16.14)}$$

$$k \equiv \frac{2\pi}{\lambda} \qquad \textbf{(16.9)}$$

$$\omega \equiv \frac{2\pi}{T} = 2\pi f \qquad \textbf{(16.10, 16.12)}$$

You should know how to find the equation describing the motion of particles in a wave from a given set of physical parameters.

The **power** transmitted by a sinusoidal wave on a stretched string is

$$\mathcal{P} = \tfrac{1}{2}\mu\omega^2 A^2 v \qquad \textbf{(16.21)}$$

QUESTIONS

1. Why is a wave pulse traveling on a string considered a transverse wave?

2. How would you set up a longitudinal wave in a stretched spring? Would it be possible to set up a transverse wave in a spring?

3. By what factor would you have to increase the tension in a taut string to double the wave speed?

4. When traveling on a taut string, does a wave pulse always invert upon reflection? Explain.

5. Can two pulses traveling in opposite directions on the same string reflect from each other? Explain.

6. Does the vertical speed of a segment of a horizontal, taut string, through which a wave is traveling, depend on the wave speed?

7. If you were to shake one end of a taut rope periodically three times each second, what would be the period of the sinusoidal waves set up in the rope?

8. A vibrating source generates a sinusoidal wave on a string under constant tension. If the power delivered to the string is doubled, by what factor does the amplitude change? Does the wave speed change under these circumstances?

9. Consider a wave traveling on a taut rope. What is the difference, if any, between the speed of the wave and the speed of a small segment of the rope?

10. If a long rope is hung from a ceiling and waves are sent up the rope from its lower end, they do not ascend with constant speed. Explain.

11. What happens to the wavelength of a wave on a string when the frequency is doubled? Assume that the tension in the string remains the same.

12. What happens to the speed of a wave on a taut string when the frequency is doubled? Assume that the tension in the string remains the same.

13. How do transverse waves differ from longitudinal waves?

14. When all the strings on a guitar are stretched to the same tension, will the speed of a wave along the more massive bass strings be faster or slower than the speed of a wave on the lighter strings?

15. If you stretch a rubber hose and pluck it, you can observe a pulse traveling up and down the hose. What happens to the speed of the pulse if you stretch the hose more tightly? What happens to the speed if you fill the hose with water?

16. In a longitudinal wave in a spring, the coils move back and forth in the direction of wave motion. Does the speed of the wave depend on the maximum speed of each coil?

17. When two waves interfere, can the amplitude of the resultant wave be greater than either of the two original waves? Under what conditions?

18. A solid can transport both longitudinal waves and transverse waves, but a fluid can transport only longitudinal waves. Why?

PROBLEMS

1, 2, 3 = straightforward, intermediate, challenging   ☐ = full solution available in the *Student Solutions Manual and Study Guide*
WEB = solution posted at **http://www.saunderscollege.com/physics/**   🖥 = Computer useful in solving problem   🔲 = Interactive Physics
☐ = paired numerical/symbolic problems

Section 16.1   Basic Variables of Wave Motion

Section 16.2   Direction of Particle Displacement

Section 16.3   One-Dimensional Traveling Waves

1. At $t = 0$, a transverse wave pulse in a wire is described by the function

$$y = \frac{6}{x^2 + 3}$$

where x and y are in meters. Write the function $y(x, t)$ that describes this wave if it is traveling in the positive x direction with a speed of 4.50 m/s.

2. Two wave pulses A and B are moving in opposite directions along a taut string with a speed of 2.00 cm/s. The amplitude of A is twice the amplitude of B. The pulses are shown in Figure P16.2 at $t = 0$. Sketch the shape of the string at $t = 1$, 1.5, 2, 2.5, and 3 s.

Figure P16.2

3. A wave moving along the x axis is described by

$$y(x, t) = 5.00e^{-(x+5.00t)^2}$$

where x is in meters and t is in seconds. Determine (a) the direction of the wave motion and (b) the speed of the wave.

4. Ocean waves with a crest-to-crest distance of 10.0 m can be described by the equation

$$y(x, t) = (0.800 \text{ m}) \sin[0.628(x - vt)]$$

where $v = 1.20$ m/s. (a) Sketch $y(x, t)$ at $t = 0$. (b) Sketch $y(x, t)$ at $t = 2.00$ s. Note how the entire wave form has shifted 2.40 m in the positive x direction in this time interval.

5. Two points, A and B, on the surface of the Earth are at the same longitude and 60.0° apart in latitude. Suppose that an earthquake at point A sends two waves toward point B. A transverse wave travels along the surface of the Earth at 4.50 km/s, and a longitudinal wave travels straight through the body of the Earth at 7.80 km/s. (a) Which wave arrives at point B first? (b) What is the time difference between the arrivals of the two waves at point B? Take the radius of the Earth to be 6 370 km.

6. A seismographic station receives S and P waves from an earthquake, 17.3 s apart. Suppose that the waves have traveled over the same path at speeds of 4.50 km/s and 7.80 km/s, respectively. Find the distance from the seismometer to the epicenter of the quake.

Section 16.4 Superposition and Interference

WEB **7.** Two sinusoidal waves in a string are defined by the functions

$$y_1 = (2.00 \text{ cm}) \sin(20.0x - 32.0t)$$

and

$$y_2 = (2.00 \text{ cm}) \sin(25.0x - 40.0t)$$

where y and x are in centimeters and t is in seconds. (a) What is the phase difference between these two waves at the point $x = 5.00$ cm at $t = 2.00$ s? (b) What is the positive x value closest to the origin for which the two phases differ by $\pm \pi$ at $t = 2.00$ s? (This is where the sum of the two waves is zero.)

8. Two waves in one string are described by the wave functions

$$y_1 = 3.0 \cos(4.0x - 1.6t)$$

and

$$y_2 = 4.0 \sin(5.0x - 2.0t)$$

where y and x are in centimeters and t is in seconds. Find the superposition of the waves $y_1 + y_2$ at the points (a) $x = 1.00$, $t = 1.00$; (b) $x = 1.00$, $t = 0.500$; (c) $x = 0.500$, $t = 0$. (Remember that the arguments of the trigonometric functions are in radians.)

9. Two pulses traveling on the same string are described by the functions

$$y_1 = \frac{5}{(3x - 4t)^2 + 2}$$

and

$$y_2 = \frac{-5}{(3x + 4t - 6)^2 + 2}$$

(a) In which direction does each pulse travel? (b) At what time do the two cancel? (c) At what point do the two waves always cancel?

Section 16.5 The Speed of Waves on Strings

10. A phone cord is 4.00 m long. The cord has a mass of 0.200 kg. A transverse wave pulse is produced by plucking one end of the taut cord. The pulse makes four trips down and back along the cord in 0.800 s. What is the tension in the cord?

11. Transverse waves with a speed of 50.0 m/s are to be produced in a taut string. A 5.00-m length of string with a total mass of 0.060 0 kg is used. What is the required tension?

12. A piano string having a mass per unit length 5.00 × 10^{-3} kg/m is under a tension of 1 350 N. Find the speed with which a wave travels on this string.

13. An astronaut on the Moon wishes to measure the local value of g by timing pulses traveling down a wire that has a large mass suspended from it. Assume that the wire has a mass of 4.00 g and a length of 1.60 m, and that a 3.00-kg mass is suspended from it. A pulse requires 36.1 ms to traverse the length of the wire. Calculate g_{Moon} from these data. (You may neglect the mass of the wire when calculating the tension in it.)

14. Transverse pulses travel with a speed of 200 m/s along a taut copper wire whose diameter is 1.50 mm. What is the tension in the wire? (The density of copper is 8.92 g/cm³.)

15. Transverse waves travel with a speed of 20.0 m/s in a string under a tension of 6.00 N. What tension is required to produce a wave speed of 30.0 m/s in the same string?

16. A simple pendulum consists of a ball of mass M hanging from a uniform string of mass m and length L, with $m \ll M$. If the period of oscillation for the pendulum is T, determine the speed of a transverse wave in the string when the pendulum hangs at rest.

17. The elastic limit of a piece of steel wire is 2.70 × 10^9 Pa. What is the maximum speed at which transverse wave pulses can propagate along this wire before this stress is exceeded? (The density of steel is 7.86 × 10^3 kg/m³.)

18. **Review Problem.** A light string with a mass per unit length of 8.00 g/m has its ends tied to two walls separated by a distance equal to three-fourths the length of the string (Fig. P16.18). An object of mass m is sus-

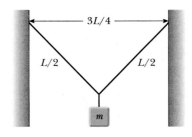

Figure P16.18

pended from the center of the string, putting a tension in the string. (a) Find an expression for the transverse wave speed in the string as a function of the hanging mass. (b) How much mass should be suspended from the string to produce a wave speed of 60.0 m/s?

19. **Review Problem.**   A light string with a mass of 10.0 g and a length L = 3.00 m has its ends tied to two walls that are separated by the distance D = 2.00 m. Two objects, each with a mass M = 2.00 kg, are suspended from the string, as shown in Figure P16.19. If a wave pulse is sent from point A, how long does it take for it to travel to point B?

20. **Review Problem.**   A light string of mass m and length L has its ends tied to two walls that are separated by the distance D. Two objects, each of mass M, are suspended from the string, as shown in Figure P16.19. If a wave pulse is sent from point A, how long does it take to travel to point B?

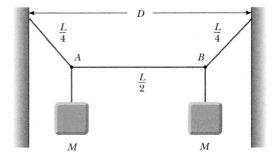

Figure P16.19   Problems 19 and 20.

WEB 21. A 30.0-m steel wire and a 20.0-m copper wire, both with 1.00-mm diameters, are connected end to end and are stretched to a tension of 150 N. How long does it take a transverse wave to travel the entire length of the two wires?

Section 16.6   Reflection and Transmission

22. A series of pulses, each of amplitude 0.150 m, are sent down a string that is attached to a post at one end. The pulses are reflected at the post and travel back along the string without loss of amplitude. What is the displacement at a point on the string where two pulses are crossing (a) if the string is rigidly attached to the post? (b) if the end at which reflection occurs is free to slide up and down?

Section 16.7   Sinusoidal Waves

23. (a) Plot y versus t at x = 0 for a sinusoidal wave of the form y = (15.0 cm) cos(0.157x − 50.3t) , where x and y are in centimeters and t is in seconds. (b) Determine

the period of vibration from this plot and compare your result with the value found in Example 16.3.

24. For a certain transverse wave, the distance between two successive crests is 1.20 m, and eight crests pass a given point along the direction of travel every 12.0 s. Calculate the wave speed.

25. A sinusoidal wave is traveling along a rope. The oscillator that generates the wave completes 40.0 vibrations in 30.0 s. Also, a given maximum travels 425 cm along the rope in 10.0 s. What is the wavelength?

26. Consider the sinusoidal wave of Example 16.3, with the wave function

$$y = (15.0 \text{ cm}) \cos(0.157x - 50.3t)$$

At a certain instant, let point A be at the origin and point B be the first point along the x axis where the wave is 60.0° out of phase with point A. What is the coordinate of point B?

27. When a particular wire is vibrating with a frequency of 4.00 Hz, a transverse wave of wavelength 60.0 cm is produced. Determine the speed of wave pulses along the wire.

28. A sinusoidal wave traveling in the − x direction (to the left) has an amplitude of 20.0 cm, a wavelength of 35.0 cm, and a frequency of 12.0 Hz. The displacement of the wave at t = 0, x = 0 is y = − 3.00 cm; at this same point, a particle of the medium has a positive velocity. (a) Sketch the wave at t = 0. (b) Find the angular wave number, period, angular frequency, and wave speed of the wave. (c) Write an expression for the wave function $y(x, t)$.

29. A sinusoidal wave train is described by the equation

$$y = (0.25 \text{ m}) \sin(0.30x - 40t)$$

where x and y are in meters and t is in seconds. Determine for this wave the (a) amplitude, (b) angular frequency, (c) angular wave number, (d) wavelength, (e) wave speed, and (f) direction of motion.

30. A transverse wave on a string is described by the expression

$$y = (0.120 \text{ m}) \sin(\pi x/8 + 4\pi t)$$

(a) Determine the transverse speed and acceleration of the string at t = 0.200 s for the point on the string located at x = 1.60 m. (b) What are the wavelength, period, and speed of propagation of this wave?

WEB 31. (a) Write the expression for y as a function of x and t for a sinusoidal wave traveling along a rope in the *negative x direction* with the following characteristics: A = 8.00 cm, λ = 80.0 cm, f = 3.00 Hz, and $y(0, t)$ = 0 at t = 0. (b) Write the expression for y as a function of x and t for the wave in part (a), assuming that $y(x, 0)$ = 0 at the point x = 10.0 cm.

32. A transverse sinusoidal wave on a string has a period T = 25.0 ms and travels in the negative x direction with a speed of 30.0 m/s. At t = 0, a particle on the string at

$x = 0$ has a displacement of 2.00 cm and travels downward with a speed of 2.00 m/s. (a) What is the amplitude of the wave? (b) What is the initial phase angle? (c) What is the maximum transverse speed of the string? (d) Write the wave function for the wave.

33. A sinusoidal wave of wavelength 2.00 m and amplitude 0.100 m travels on a string with a speed of 1.00 m/s to the right. Initially, the left end of the string is at the origin. Find (a) the frequency and angular frequency, (b) the angular wave number, and (c) the wave function for this wave. Determine the equation of motion for (d) the left end of the string and (e) the point on the string at $x = 1.50$ m to the right of the left end. (f) What is the maximum speed of any point on the string?

34. A sinusoidal wave on a string is described by the equation

$$y = (0.51 \text{ cm}) \sin(kx - \omega t)$$

where $k = 3.10$ rad/cm and $\omega = 9.30$ rad/s. How far does a wave crest move in 10.0 s? Does it move in the positive or negative x direction?

35. A wave is described by $y = (2.00 \text{ cm}) \sin(kx - \omega t)$, where $k = 2.11$ rad/m, $\omega = 3.62$ rad/s, x is in meters, and t is in seconds. Determine the amplitude, wavelength, frequency, and speed of the wave.

36. A transverse traveling wave on a taut wire has an amplitude of 0.200 mm and a frequency of 500 Hz. It travels with a speed of 196 m/s. (a) Write an equation in SI units of the form $y = A \sin(kx - \omega t)$ for this wave. (b) The mass per unit length of this wire is 4.10 g/m. Find the tension in the wire.

37. A wave on a string is described by the wave function

$$y = (0.100 \text{ m}) \sin(0.50x - 20t)$$

(a) Show that a particle in the string at $x = 2.00$ m executes simple harmonic motion. (b) Determine the frequency of oscillation of this particular point.

Section 16.8 Rate of Energy Transfer by Sinusoidal Waves on Strings

38. A taut rope has a mass of 0.180 kg and a length of 3.60 m. What power must be supplied to the rope to generate sinusoidal waves having an amplitude of 0.100 m and a wavelength of 0.500 m and traveling with a speed of 30.0 m/s?

39. A two-dimensional water wave spreads in circular wave fronts. Show that the amplitude A at a distance r from the initial disturbance is proportional to $1/\sqrt{r}$. (*Hint:* Consider the energy carried by one outward-moving ripple.)

40. Transverse waves are being generated on a rope under constant tension. By what factor is the required power increased or decreased if (a) the length of the rope is doubled and the angular frequency remains constant, (b) the amplitude is doubled and the angular fre-

quency is halved, (c) both the wavelength and the amplitude are doubled, and (d) both the length of the rope and the wavelength are halved?

WEB 41. Sinusoidal waves 5.00 cm in amplitude are to be transmitted along a string that has a linear mass density of 4.00×10^{-2} kg/m. If the source can deliver a maximum power of 300 W and the string is under a tension of 100 N, what is the highest vibrational frequency at which the source can operate?

42. It is found that a 6.00-m segment of a long string contains four complete waves and has a mass of 180 g. The string is vibrating sinusoidally with a frequency of 50.0 Hz and a peak-to-valley displacement of 15.0 cm. (The "peak-to-valley" distance is the vertical distance from the farthest positive displacement to the farthest negative displacement.) (a) Write the function that describes this wave traveling in the positive x direction. (b) Determine the power being supplied to the string.

43. A sinusoidal wave on a string is described by the equation

$$y = (0.15 \text{ m}) \sin(0.80x - 50t)$$

where x and y are in meters and t is in seconds. If the mass per unit length of this string is 12.0 g/m, determine (a) the speed of the wave, (b) the wavelength, (c) the frequency, and (d) the power transmitted to the wave.

44. A horizontal string can transmit a maximum power of $\mathcal{P}$ (without breaking) if a wave with amplitude A and angular frequency ω is traveling along it. To increase this maximum power, a student folds the string and uses the "double string" as a transmitter. Determine the maximum power that can be transmitted along the "double string," supposing that the tension is constant.

(Optional)
Section 16.9 The Linear Wave Equation

45. (a) Evaluate A in the scalar equality $(7 + 3)4 = A$. (b) Evaluate A, B, and C in the vector equality $7.00\mathbf{i} + 3.00\mathbf{k} = A\mathbf{i} + B\mathbf{j} + C\mathbf{k}$. Explain how you arrive at your answers. (c) The functional equality or identity

$$A + B\cos(Cx + Dt + E) = (7.00 \text{ mm}) \cos(3x + 4t + 2)$$

is true for all values of the variables x and t, which are measured in meters and in seconds, respectively. Evaluate the constants A, B, C, D, and E. Explain how you arrive at your answers.

46. Show that the wave function $y = e^{b(x-vt)}$ is a solution of the wave equation (Eq. 16.26), where b is a constant.

47. Show that the wave function $y = \ln[b(x - vt)]$ is a solution to Equation 16.26, where b is a constant.

48. (a) Show that the function $y(x, t) = x^2 + v^2t^2$ is a solution to the wave equation. (b) Show that the function above can be written as $f(x + vt) + g(x - vt)$, and determine the functional forms for f and g. (c) Repeat parts (a) and (b) for the function $y(x, t) = \sin(x)\cos(vt)$.

ADDITIONAL PROBLEMS

49. The "wave" is a particular type of wave pulse that can sometimes be seen propagating through a large crowd gathered at a sporting arena to watch a soccer or American football match (Fig. P16.49). The particles of the medium are the spectators, with zero displacement corresponding to their being in the seated position and maximum displacement corresponding to their being in the standing position and raising their arms. When a large fraction of the spectators participate in the wave motion, a somewhat stable pulse shape can develop. The wave speed depends on people's reaction time, which is typically on the order of 0.1 s. Estimate the order of magnitude, in minutes, of the time required for such a wave pulse to make one circuit around a large sports stadium. State the quantities you measure or estimate and their values.

Figure P16.49   *(Gregg Adams/Tony Stone Images)*

50. A traveling wave propagates according to the expression $y = (4.0 \text{ cm}) \sin(2.0x - 3.0t)$, where x is in centimeters and t is in seconds. Determine (a) the amplitude, (b) the wavelength, (c) the frequency, (d) the period, and (e) the direction of travel of the wave.

WEB 51. The wave function for a traveling wave on a taut string is (in SI units)

$$y(x, t) = (0.350 \text{ m}) \sin(10\pi t - 3\pi x + \pi/4)$$

(a) What are the speed and direction of travel of the wave? (b) What is the vertical displacement of the string at $t = 0$, $x = 0.100$ m? (c) What are the wavelength and frequency of the wave? (d) What is the maximum magnitude of the transverse speed of the string?

52. Motion picture film is projected at 24.0 frames per second. Each frame is a photograph 19.0 mm in height. At what constant speed does the film pass into the projector?

53. **Review Problem.**   A block of mass M, supported by a string, rests on an incline making an angle θ with the horizontal (Fig. P16.53). The string's length is L, and its mass is $m \ll M$. Derive an expression for the time it takes a transverse wave to travel from one end of the string to the other.

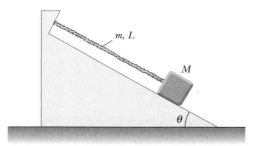

Figure P16.53

54. (a) Determine the speed of transverse waves on a string under a tension of 80.0 N if the string has a length of 2.00 m and a mass of 5.00 g. (b) Calculate the power required to generate these waves if they have a wavelength of 16.0 cm and an amplitude of 4.00 cm.

55. **Review Problem.**   A 2.00-kg block hangs from a rubber cord. The block is supported so that the cord is not stretched. The unstretched length of the cord is 0.500 m, and its mass is 5.00 g. The "spring constant" for the cord is 100 N/m. The block is released and stops at the lowest point. (a) Determine the tension in the cord when the block is at this lowest point. (b) What is the length of the cord in this "stretched" position? (c) Find the speed of a transverse wave in the cord if the block is held in this lowest position.

56. **Review Problem.**   A block of mass M hangs from a rubber cord. The block is supported so that the cord is not stretched. The unstretched length of the cord is L_0, and its mass is m, much less than M. The "spring constant" for the cord is k. The block is released and stops at the lowest point. (a) Determine the tension in the cord when the block is at this lowest point. (b) What is the length of the cord in this "stretched" position? (c) Find the speed of a transverse wave in the cord if the block is held in this lowest position.

57. A sinusoidal wave in a rope is described by the wave function

$$y = (0.20 \text{ m}) \sin(0.75\pi x + 18\pi t)$$

where x and y are in meters and t is in seconds. The rope has a linear mass density of 0.250 kg/m. If the tension in the rope is provided by an arrangement like the one illustrated in Figure 16.12, what is the value of the suspended mass?

58. A wire of density ρ is tapered so that its cross-sectional area varies with x, according to the equation

$$A = (1.0 \times 10^{-3}x + 0.010) \text{ cm}^2$$

(a) If the wire is subject to a tension T, derive a relationship for the speed of a wave as a function of position. (b) If the wire is aluminum and is subject to a tension of 24.0 N, determine the speed at the origin and at $x = 10.0$ m.

59. A rope of total mass m and length L is suspended vertically. Show that a transverse wave pulse travels the length of the rope in a time $t = 2\sqrt{L/g}$. (*Hint:* First find an expression for the wave speed at any point a distance x from the lower end by considering the tension in the rope as resulting from the weight of the segment below that point.)

60. If mass M is suspended from the bottom of the rope in Problem 59, (a) show that the time for a transverse wave to travel the length of the rope is

$$t = 2\sqrt{\frac{L}{mg}}\left[\sqrt{(M + m)} - \sqrt{M}\right]$$

(b) Show that this reduces to the result of Problem 59 when $M = 0$. (c) Show that for $m \ll M$, the expression in part (a) reduces to

$$t = \sqrt{\frac{mL}{Mg}}$$

61. It is stated in Problem 59 that a wave pulse travels from the bottom to the top of a rope of length L in a time $t = 2\sqrt{L/g}$. Use this result to answer the following questions. (It is *not* necessary to set up any new integrations.) (a) How long does it take for a wave pulse to travel halfway up the rope? (Give your answer as a fraction of the quantity $2\sqrt{L/g}$.) (b) A pulse starts traveling up the rope. How far has it traveled after a time $\sqrt{L/g}$?

62. Determine the speed and direction of propagation of each of the following sinusoidal waves, assuming that x is measured in meters and t in seconds:
(a) $y = 0.60 \cos(3.0x - 15t + 2)$
(b) $y = 0.40 \cos(3.0x + 15t - 2)$
(c) $y = 1.2 \sin(15t + 2.0x)$
(d) $y = 0.20 \sin(12t - x/2 + \pi)$

63. **Review Problem.** An aluminum wire under zero tension at room temperature is clamped at each end. The tension in the wire is increased by reducing the temperature, which results in a decrease in the wire's equilibrium length. What strain ($\Delta L/L$) results in a transverse wave speed of 100 m/s? Take the cross-sectional area of the wire to be 5.00×10^{-6} m², the density of the material to be 2.70×10^3 kg/m³, and Young's modulus to be 7.00×10^{10} N/m² .

64. (a) Show that the speed of longitudinal waves along a spring of force constant k is $v = \sqrt{kL/\mu}$, where L is the unstretched length of the spring and μ is the mass per unit length. (b) A spring with a mass of 0.400 kg has an unstretched length of 2.00 m and a force constant of 100 N/m. Using the result you obtained in (a), determine the speed of longitudinal waves along this spring.

65. A string of length L consists of two sections: The left half has mass per unit length $\mu = \mu_0/2$, whereas the right half has a mass per unit length $\mu' = 3\mu = 3\mu_0/2$. Tension in the string is T_0. Notice from the data given that this string has the same total mass as a uniform string of length L and of mass per unit length μ_0. (a) Find the speeds v and v' at which transverse wave pulses travel in the two sections. Express the speeds in terms of T_0 and μ_0, and also as multiples of the speed $v_0 = (T_0/\mu_0)^{1/2}$. (b) Find the time required for a pulse to travel from one end of the string to the other. Give your result as a multiple of $t_0 = L/v_0$.

66. A wave pulse traveling along a string of linear mass density μ is described by the relationship

$$y = [A_0 e^{-bx}] \sin(kx - \omega t)$$

where the factor in brackets before the sine function is said to be the amplitude. (a) What is the power $\mathcal{P}(x)$ carried by this wave at a point x? (b) What is the power carried by this wave at the origin? (c) Compute the ratio $\mathcal{P}(x)/\mathcal{P}(0)$.

67. An earthquake on the ocean floor in the Gulf of Alaska produces a *tsunami* (sometimes called a "tidal wave") that reaches Hilo, Hawaii, 4 450 km away, in a time of 9 h 30 min. Tsunamis have enormous wavelengths (100–200 km), and the propagation speed of these waves is $v \approx \sqrt{gd}$, where d is the average depth of the water. From the information given, find the average wave speed and the average ocean depth between Alaska and Hawaii. (This method was used in 1856 to estimate the average depth of the Pacific Ocean long before soundings were made to obtain direct measurements.)

ANSWERS TO QUICK QUIZZES

16.1 (a) It is longitudinal because the disturbance (the shift of position) is parallel to the direction in which the wave travels. (b) It is transverse because the people stand up and sit down (vertical motion), whereas the wave moves either to the left or to the right (motion perpendicular to the disturbance).

16.2

1 cm

16.3 Only answers (f) and (h) are correct. (a) and (b) affect the transverse speed of a particle of the string, but not the wave speed along the string. (c) and (d) change the amplitude. (e) and (g) increase the time by decreasing the wave speed.

16.4 The transverse speed increases because $v_{y,\,max} = \omega A = 2\pi f A$. The wave speed does not change because it depends only on the tension and mass per length of the string, neither of which has been modified. The wavelength must decrease because the wave speed $v = \lambda f$ remains constant.

PUZZLER

chapter

17

Sound Waves

Chapter Outline

Sound waves are the most important example of longitudinal waves. They can travel through any material medium with a speed that depends on the properties of the medium. As the waves travel, the particles in the medium vibrate to produce changes in density and pressure along the direction of motion of the wave. These changes result in a series of high-pressure and low-pressure regions. If the source of the sound waves vibrates sinusoidally, the pressure variations are also sinusoidal. We shall find that the mathematical description of sinusoidal sound waves is identical to that of sinusoidal string waves, which was discussed in the previous chapter.

Sound waves are divided into three categories that cover different frequency ranges. (1) *Audible waves* are waves that lie within the range of sensitivity of the human ear. They can be generated in a variety of ways, such as by musical instruments, human vocal cords, and loudspeakers. (2) *Infrasonic waves* are waves having frequencies below the audible range. Elephants can use infrasonic waves to communicate with each other, even when separated by many kilometers. (3) *Ultrasonic waves* are waves having frequencies above the audible range. You may have used a "silent" whistle to retrieve your dog. The ultrasonic sound it emits is easily heard by dogs, although humans cannot detect it at all. Ultrasonic waves are also used in medical imaging.

We begin this chapter by discussing the speed of sound waves and then wave intensity, which is a function of wave amplitude. We then provide an alternative description of the intensity of sound waves that compresses the wide range of intensities to which the ear is sensitive to a smaller range. Finally, we treat effects of the motion of sources and/or listeners.

17.1 SPEED OF SOUND WAVES

Let us describe pictorially the motion of a one-dimensional longitudinal pulse moving through a long tube containing a compressible gas (Fig. 17.1). A piston at the left end can be moved to the right to compress the gas and create the pulse. Before the piston is moved, the gas is undisturbed and of uniform density, as represented by the uniformly shaded region in Figure 17.1a. When the piston is suddenly pushed to the right (Fig. 17.1b), the gas just in front of it is compressed (as represented by the more heavily shaded region); the pressure and density in this region are now higher than they were before the piston moved. When the piston comes to rest (Fig. 17.1c), the compressed region of the gas continues to move to the right, corresponding to a longitudinal pulse traveling through the tube with

Undisturbed gas

(a)

Compressed region

(b)

v

(c)

v

(d)

Figure 17.1 Motion of a longitudinal pulse through a compressible gas. The compression (darker region) is produced by the moving piston.

An ultrasound image of a human fetus in the womb after 20 weeks of development, showing the head, body, arms, and legs in profile. (*U.H.B. Trust/Tony Stone Images*)

speed v. Note that the piston speed does *not* equal v. Furthermore, the compressed region does not "stay with" the piston as the piston moves, because the speed of the wave may be greater than the speed of the piston.

The speed of sound waves depends on the compressibility and inertia of the medium. If the medium has a bulk modulus B (see Section 12.4) and density ρ, the speed of sound waves in that medium is

$$v = \sqrt{\frac{B}{\rho}}$$ **(17.1)**

Speed of sound

It is interesting to compare this expression with Equation 16.4 for the speed of transverse waves on a string, $v = \sqrt{T/\mu}$. In both cases, the wave speed depends on an elastic property of the medium—bulk modulus B or string tension T—and on an inertial property of the medium—ρ or μ. In fact, the speed of *all mechanical waves* follows an expression of the general form

$$v = \sqrt{\frac{\text{elastic property}}{\text{inertial property}}}$$

The speed of sound also depends on the temperature of the medium. For sound traveling through air, the relationship between wave speed and medium temperature is

$$v = (331 \text{ m/s}) \sqrt{1 + \frac{T_{C}}{273°C}}$$

where 331 m/s is the speed of sound in air at 0°C, and T_{C} is the temperature in degrees Celsius. Using this equation, one finds that at 20°C the speed of sound in air is approximately 343 m/s.

This information provides a convenient way to estimate the distance to a thunderstorm, as demonstrated in the QuickLab. During a lightning flash, the temperature of a long channel of air rises rapidly as the bolt passes through it. This temperature increase causes the air in the channel to expand rapidly, and this expansion creates a sound wave. The channel produces sound throughout its entire length at essentially the same instant. If the orientation of the channel is such that all of its parts are approximately the same distance from you, sounds from the different parts reach you at the same time, and you hear a short, intense thunderclap. However, if the distances between your ear and different portions of the channel vary, sounds from different portions arrive at your ears at different times. If the channel were a straight line, the resulting sound would be a steady roar, but the zigzag shape of the path produces variations in loudness.

QuickLab

The next time a thunderstorm approaches, count the seconds between a flash of lightning (which reaches you almost instantaneously) and the following thunderclap. Divide this time by 3 to determine the approximate number of kilometers (or by 5 to estimate the miles) to the storm.

To learn more about lightning, read E. Williams, "The Electrification of Thunderstorms" *Sci. Am.* 259(5):88–89, 1988.

Quick Quiz 17.1

The speed of sound in air is a function of (a) wavelength, (b) frequency, (c) temperature, (d) amplitude.

Quick Quiz 17.2

As a result of a distant explosion, an observer first senses a ground tremor and then hears the explosion later. Explain.

EXAMPLE 17.1 ▸ **Speed of Sound in a Solid**

If a solid bar is struck at one end with a hammer, a longitudinal pulse propagates down the bar with a speed $v = \sqrt{Y/\rho}$, where Y is the Young's modulus for the material (see Section 12.4). Find the speed of sound in an aluminum bar.

Solution   From Table 12.1 we obtain $Y = 7.0 \times 10^{10}$ N/m² for aluminum, and from Table 1.5 we obtain $\rho = 2.70 \times 10^3$ kg/m³. Therefore,

$$v_{\text{Al}} = \sqrt{\frac{Y}{\rho}} = \sqrt{\frac{7.0 \times 10^{10} \text{ N/m}^2}{2.70 \times 10^3 \text{ kg/m}^3}} \approx \boxed{5.1 \text{ km/s}}$$

This typical value for the speed of sound in solids is much greater than the speed of sound in gases, as Table 17.1 shows. This difference in speeds makes sense because the molecules of a solid are bound together into a much more rigid structure than those in a gas and hence respond more rapidly to a disturbance.

EXAMPLE 17.2 ▸ **Speed of Sound in a Liquid**

(a) Find the speed of sound in water, which has a bulk modulus of 2.1×10^9 N/m² and a density of 1.00×10^3 kg/m³.

Solution   Using Equation 17.1, we find that

$$v_{\text{water}} = \sqrt{\frac{B}{\rho}} = \sqrt{\frac{2.1 \times 10^9 \text{ N/m}^2}{1.00 \times 10^3 \text{ kg/m}^3}} = \boxed{1.4 \text{ km/s}}$$

In general, sound waves travel more slowly in liquids than in solids because liquids are more compressible than solids.

(b) Dolphins use sound waves to locate food. Experiments have shown that a dolphin can detect a 7.5-cm target 110 m away, even in murky water. For a bit of "dinner" at that distance, how much time passes between the moment the dolphin emits a sound pulse and the moment the dolphin hears its reflection and thereby detects the distant target?

Solution   The total distance covered by the sound wave as it travels from dolphin to target and back is 2×110 m $=$ 220 m. From Equation 2.2, we have

$$\Delta t = \frac{\Delta x}{v_x} = \frac{220 \text{ m}}{1\,400 \text{ m/s}} = \boxed{0.16 \text{ s}}$$

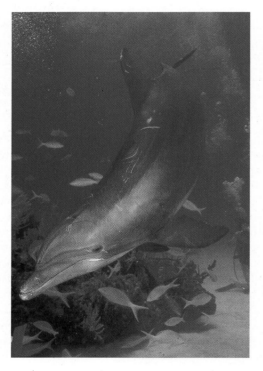

Bottle-nosed dolphin.   *(Stuart Westmoreland/Tony Stone Images)*

17.2 ▸ **PERIODIC SOUND WAVES**

This section will help you better comprehend the nature of sound waves. You will learn that pressure variations control what we hear—an important fact for understanding how our ears work.

One can produce a one-dimensional periodic sound wave in a long, narrow tube containing a gas by means of an oscillating piston at one end, as shown in Figure 17.2. The darker parts of the colored areas in this figure represent regions where the gas is compressed and thus the density and pressure are above their equilibrium values. A compressed region is formed whenever the piston is pushed into the tube. This compressed region, called a **condensation,** moves through the tube as a pulse, continuously compressing the region just in front of itself. When the piston is pulled back, the gas in front of it expands, and the pressure and density in this region fall below their equilibrium values (represented by the lighter parts of the colored areas in Fig. 17.2). These low-pressure regions, called **rarefactions,** also propagate along the tube, following the condensations. Both regions move with a speed equal to the speed of sound in the medium.

As the piston oscillates sinusoidally, regions of condensation and rarefaction are continuously set up. The distance between two successive condensations (or two successive rarefactions) equals the wavelength λ. As these regions travel through the tube, any small volume of the medium moves with simple harmonic motion parallel to the direction of the wave. If $s(x, t)$ is the displacement of a small volume element from its equilibrium position, we can express this harmonic displacement function as

$$s(x, t) = s_{max} \cos(kx - \omega t) \tag{17.2}$$

where s_{max} **is the maximum displacement of the medium from equilibrium** (in other words, the **displacement amplitude** of the wave), k is the angular wavenumber, and ω is the angular frequency of the piston. Note that the displacement of the medium is along x, in the direction of motion of the sound wave, which means we are describing a longitudinal wave.

As we shall demonstrate shortly, the variation in the gas pressure ΔP, measured from the equilibrium value, is also periodic and for the displacement function in Equation 17.2 is given by

$$\Delta P = \Delta P_{max} \sin(kx - \omega t) \tag{17.3}$$

where **the pressure amplitude** ΔP_{max}—which is the **maximum change in pres-**

TABLE 17.1
Speeds of Sound in Various Media

Medium	v (m/s)
Gases	
Hydrogen (0°C)	1 286
Helium (0°C)	972
Air (20°C)	343
Air (0°C)	331
Oxygen (0°C)	317
Liquids at 25°C	
Glycerol	1 904
Sea water	1 533
Water	1 493
Mercury	1 450
Kerosene	1 324
Methyl alcohol	1 143
Carbon tetrachloride	926
Solids	
Diamond	12 000
Pyrex glass	5 640
Iron	5 130
Aluminum	5 100
Brass	4 700
Copper	3 560
Gold	3 240
Lucite	2 680
Lead	1 322
Rubber	1 600

Figure 17.2 A sinusoidal longitudinal wave propagating through a gas-filled tube. The source of the wave is a sinusoidally oscillating piston at the left. The high-pressure and low-pressure regions are colored darkly and lightly, respectively.

sure from the equilibrium value—is given by

$$\Delta P_{max} = \rho v \omega s_{max}$$

(17.4)

Pressure amplitude

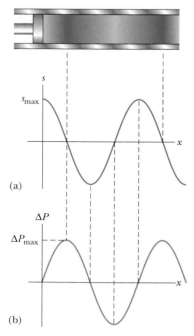

(a)

(b)

Figure 17.3 (a) Displacement amplitude versus position and (b) pressure amplitude versus position for a sinusoidal longitudinal wave. The displacement wave is 90° out of phase with the pressure wave.

Thus, we see that a sound wave may be considered as either a displacement wave or a pressure wave. A comparison of Equations 17.2 and 17.3 shows that **the pressure wave is 90° out of phase with the displacement wave.** Graphs of these functions are shown in Figure 17.3. Note that the pressure variation is a maximum when the displacement is zero, and the displacement is a maximum when the pressure variation is zero.

> ### Quick Quiz 17.3
>
> If you blow across the top of an empty soft-drink bottle, a pulse of air travels down the bottle. At the moment the pulse reaches the bottom of the bottle, compare the displacement of air molecules with the pressure variation.

Derivation of Equation 17.3

From the definition of bulk modulus (see Eq. 12.8), the pressure variation in the gas is

$$\Delta P = -B \frac{\Delta V}{V_i}$$

The volume of gas that has a thickness Δx in the horizontal direction and a cross-sectional area A is $V_i = A \Delta x$. The change in volume ΔV accompanying the pressure change is equal to $A \Delta s$, where Δs is the difference between the value of s at $x + \Delta x$ and the value of s at x. Hence, we can express ΔP as

$$\Delta P = -B \frac{\Delta V}{V_i} = -B \frac{A \Delta s}{A \Delta x} = -B \frac{\Delta s}{\Delta x}$$

As Δx approaches zero, the ratio $\Delta s/\Delta x$ becomes $\partial s/\partial x$. (The partial derivative indicates that we are interested in the variation of s with position at a *fixed* time.) Therefore,

$$\Delta P = -B \frac{\partial s}{\partial x}$$

If the displacement is the simple sinusoidal function given by Equation 17.2, we find that

$$\Delta P = -B \frac{\partial}{\partial x} [s_{max} \cos(kx - \omega t)] = Bks_{max} \sin(kx - \omega t)$$

Because the bulk modulus is given by $B = \rho v^2$ (see Eq. 17.1), the pressure variation reduces to

$$\Delta P = \rho v^2 k s_{max} \sin(kx - \omega t)$$

From Equation 16.13, we can write $k = \omega/v$; hence, ΔP can be expressed as

$$\Delta P = \rho v \omega s_{max} \sin(kx - \omega t)$$

Because the sine function has a maximum value of 1, we see that the maximum value of the pressure variation is $\Delta P_{max} = \rho v \omega s_{max}$ (see Eq. 17.4), and we arrive at Equation 17.3:

$$\Delta P = \Delta P_{max} \sin(kx - \omega t)$$

17.3 ▶ INTENSITY OF PERIODIC SOUND WAVES

In the previous chapter, we showed that a wave traveling on a taut string transports energy. The same concept applies to sound waves. Consider a volume of air of mass Δm and width Δx in front of a piston oscillating with a frequency ω, as shown in Figure 17.4. The piston transmits energy to this volume of air in the tube, and the energy is propagated away from the piston by the sound wave.[1] To evaluate the rate of energy transfer for the sound wave, we shall evaluate the kinetic energy of this volume of air, which is undergoing simple harmonic motion. We shall follow a procedure similar to that in Section 16.8, in which we evaluated the rate of energy transfer for a wave on a string.

As the sound wave propagates away from the piston, the displacement of any volume of air in front of the piston is given by Equation 17.2. To evaluate the kinetic energy of this volume of air, we need to know its speed. We find the speed by taking the time derivative of Equation 17.2:

$$v(x, t) = \frac{\partial}{\partial t} s(x, t) = \frac{\partial}{\partial t} [s_{max} \cos(kx - \omega t)] = \omega s_{max} \sin(kx - \omega t)$$

Imagine that we take a "snapshot" of the wave at $t = 0$. The kinetic energy of a given volume of air at this time is

$$\Delta K = \tfrac{1}{2} \Delta m v^2 = \tfrac{1}{2} \Delta m (\omega s_{max} \sin kx)^2 = \tfrac{1}{2} \rho A \, \Delta x (\omega s_{max} \sin kx)^2$$

$$= \tfrac{1}{2} \rho A \, \Delta x (\omega s_{max})^2 \sin^2 kx$$

where A is the cross-sectional area of the moving air and $A \, \Delta x$ is its volume. Now, as in Section 16.8, we integrate this expression over a full wavelength to find the total kinetic energy in one wavelength. Letting the volume of air shrink to infinitesimal thickness, so that $\Delta x \rightarrow dx$, we have

$$K_\lambda = \int dK = \int_0^\lambda \tfrac{1}{2} \rho A (\omega s_{max})^2 \sin^2 kx \, dx = \tfrac{1}{2} \rho A (\omega s_{max})^2 \int_0^\lambda \sin^2 kx \, dx$$

$$= \tfrac{1}{2} \rho A (\omega s_{max})^2 \left(\tfrac{1}{2}\lambda \right) = \tfrac{1}{4} \rho A (\omega s_{max})^2 \lambda$$

As in the case of the string wave in Section 16.8, the total potential energy for one wavelength has the same value as the total kinetic energy; thus, the total mechani-

Figure 17.4 An oscillating piston transfers energy to the air in the tube, initially causing the volume of air of width Δx and mass Δm to oscillate with an amplitude s_{max}.

[1] Although it is not proved here, the work done by the piston equals the energy carried away by the wave. For a detailed mathematical treatment of this concept, see Chapter 4 in Frank S. Crawford, Jr., *Waves,* Berkeley Physics Course, vol. 3, New York, McGraw-Hill Book Company, 1968.

cal energy is

$$E_\lambda = K_\lambda + U_\lambda = \tfrac{1}{2}\rho A(\omega s_{max})^2 \lambda$$

As the sound wave moves through the air, this amount of energy passes by a given point during one period of oscillation. Hence, the rate of energy transfer is

$$\mathcal{P} = \frac{E_\lambda}{\Delta t} = \frac{\tfrac{1}{2}\rho A(\omega s_{max})^2 \lambda}{T} = \tfrac{1}{2}\rho A(\omega s_{max})^2 \left(\frac{\lambda}{T}\right) = \tfrac{1}{2}\rho A v(\omega s_{max})^2$$

where v is the speed of sound in air.

> We define the **intensity I** of a wave, or the power per unit area, to be the rate at which the energy being transported by the wave flows through a unit area A perpendicular to the direction of travel of the wave.

In the present case, therefore, the intensity is

Intensity of a sound wave

$$I = \frac{\mathcal{P}}{A} = \tfrac{1}{2}\rho v(\omega s_{max})^2 \qquad (17.5)$$

Thus, we see that the intensity of a periodic sound wave is proportional to the square of the displacement amplitude and to the square of the angular frequency (as in the case of a periodic string wave). This can also be written in terms of the pressure amplitude ΔP_{max}; in this case, we use Equation 17.4 to obtain

$$I = \frac{\Delta P_{max}^2}{2\rho v} \qquad (17.6)$$

EXAMPLE 17.3   Hearing Limits

The faintest sounds the human ear can detect at a frequency of 1 000 Hz correspond to an intensity of about 1.00×10^{-12} W/m^2—the so-called *threshold of hearing*. The loudest sounds the ear can tolerate at this frequency correspond to an intensity of about 1.00 W/m^2—the *threshold of pain*. Determine the pressure amplitude and displacement amplitude associated with these two limits.

Solution   First, consider the faintest sounds. Using Equation 17.6 and taking $v = 343$ m/s as the speed of sound waves in air and $\rho = 1.20$ kg/m^3 as the density of air, we obtain

$$\Delta P_{max} = \sqrt{2\rho v I}$$
$$= \sqrt{2(1.20 \text{ kg/m}^3)(343 \text{ m/s})(1.00 \times 10^{-12} \text{ W/m}^2)}$$
$$= \boxed{2.87 \times 10^{-5} \text{ N/m}^2}$$

Because atmospheric pressure is about 10^5 N/m^2, this result

tells us that the ear can discern pressure fluctuations as small as 3 parts in 10^{10}!

We can calculate the corresponding displacement amplitude by using Equation 17.4, recalling that $\omega = 2\pi f$ (see Eqs. 16.10 and 16.12):

$$s_{max} = \frac{\Delta P_{max}}{\rho v \omega} = \frac{2.87 \times 10^{-5} \text{ N/m}^2}{(1.20 \text{ kg/m}^3)(343 \text{ m/s})(2\pi \times 1\,000 \text{ Hz})}$$
$$= \boxed{1.11 \times 10^{-11} \text{ m}}$$

This is a remarkably small number! If we compare this result for s_{max} with the diameter of a molecule (about 10^{-10} m), we see that the ear is an extremely sensitive detector of sound waves.

In a similar manner, one finds that the loudest sounds the human ear can tolerate correspond to a pressure amplitude of 28.7 N/m^2 and a displacement amplitude equal to 1.11×10^{-5} m.

Sound Level in Decibels

The example we just worked illustrates the wide range of intensities the human ear can detect. Because this range is so wide, it is convenient to use a logarithmic scale, where the **sound level** β (Greek letter beta) is defined by the equation

$$\beta = 10 \log\left(\frac{I}{I_0}\right) \tag{17.7}$$

The constant I_0 is the *reference intensity*, taken to be at the threshold of hearing ($I_0 = 1.00 \times 10^{-12}$ W/m^2), and I is the intensity, in watts per square meter, at the sound level β, where β is measured in **decibels** (dB).[2] On this scale, the threshold of pain ($I = 1.00$ W/m^2) corresponds to a sound level of $\beta = 10 \log[(1 \text{ W/m}^2)/(10^{-12} \text{ W/m}^2)] = 10 \log(10^{12}) = 120$ dB, and the threshold of hearing corresponds to $\beta = 10 \log[(10^{-12} \text{ W/m}^2)/(10^{-12} \text{ W/m}^2)] = 0$ dB.

Prolonged exposure to high sound levels may seriously damage the ear. Ear plugs are recommended whenever sound levels exceed 90 dB. Recent evidence suggests that "noise pollution" may be a contributing factor to high blood pressure, anxiety, and nervousness. Table 17.2 gives some typical sound-level values.

TABLE 17.2
Sound Levels

Source of Sound	β (dB)
Nearby jet airplane	150
Jackhammer; machine gun	130
Siren; rock concert	120
Subway; power mower	100
Busy traffic	80
Vacuum cleaner	70
Normal conversation	50
Mosquito buzzing	40
Whisper	30
Rustling leaves	10
Threshold of hearing	0

EXAMPLE 17.4 Sound Levels

Two identical machines are positioned the same distance from a worker. The intensity of sound delivered by each machine at the location of the worker is 2.0×10^{-7} W/m^2. Find the sound level heard by the worker (a) when one machine is operating and (b) when both machines are operating.

Solution (a) The sound level at the location of the worker with one machine operating is calculated from Equation 17.7:

$$\beta_1 = 10 \log\left(\frac{2.0 \times 10^{-7} \text{ W/m}^2}{1.00 \times 10^{-12} \text{ W/m}^2}\right) = 10 \log(2.0 \times 10^5)$$

$$= 53 \text{ dB}$$

(b) When both machines are operating, the intensity is doubled to 4.0×10^{-7} W/m^2; therefore, the sound level now is

$$\beta_2 = 10 \log\left(\frac{4.0 \times 10^{-7} \text{ W/m}^2}{1.00 \times 10^{-12} \text{ W/m}^2}\right) = 10 \log(4.0 \times 10^5)$$

$$= 56 \text{ dB}$$

From these results, we see that when the intensity is doubled, the sound level increases by only 3 dB.

Quick Quiz 17.4

A violin plays a melody line and is then joined by nine other violins, all playing at the same intensity as the first violin, in a repeat of the same melody. (a) When all of the violins are playing together, by how many decibels does the sound level increase? (b) If ten more violins join in, how much has the sound level increased over that for the single violin?

---

[2] The unit *bel* is named after the inventor of the telephone, Alexander Graham Bell (1847–1922). The prefix *deci-* is the SI prefix that stands for 10^{-1}.

17.4 ▷ SPHERICAL AND PLANE WAVES

If a spherical body oscillates so that its radius varies sinusoidally with time, a spherical sound wave is produced (Fig. 17.5). The wave moves outward from the source at a constant speed if the medium is uniform.

Because of this uniformity, we conclude that the energy in a spherical wave propagates equally in all directions. That is, no one direction is preferred over any other. If $\mathscr{P}_{av}$ is the average power emitted by the source, then this power at any distance r from the source must be distributed over a spherical surface of area $4\pi r^2$. Hence, the wave intensity at a distance r from the source is

$$I = \frac{\mathscr{P}_{av}}{A} = \frac{\mathscr{P}_{av}}{4\pi r^2} \tag{17.8}$$

Because $\mathscr{P}_{av}$ is the same for any spherical surface centered at the source, we see that the intensities at distances r_1 and r_2 are

$$I_1 = \frac{\mathscr{P}_{av}}{4\pi r_1^2} \quad \text{and} \quad I_2 = \frac{\mathscr{P}_{av}}{4\pi r_2^2}$$

Therefore, the ratio of intensities on these two spherical surfaces is

$$\frac{I_1}{I_2} = \frac{r_2^2}{r_1^2}$$

This inverse-square law states that the intensity decreases in proportion to the square of the distance from the source. Equation 17.5 tells us that the intensity is proportional to s_{max}^2. Setting the right side of Equation 17.5 equal to the right side

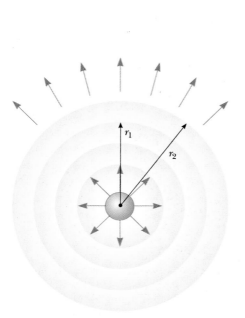

Figure 17.5   A spherical sound wave propagating radially outward from an oscillating spherical body. The intensity of the spherical wave varies as $1/r^2$.

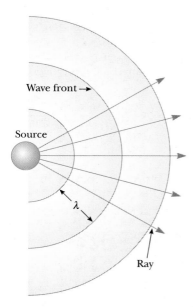

Figure 17.6   Spherical waves emitted by a point source. The circular arcs represent the spherical wave fronts that are concentric with the source. The rays are radial lines pointing outward from the source, perpendicular to the wave fronts.

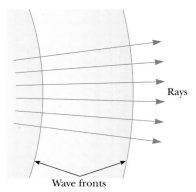

Figure 17.7   Far away from a point source, the wave fronts are nearly parallel planes, and the rays are nearly parallel lines perpendicular to the planes. Hence, a small segment of a spherical wave front is approximately a plane wave.

of Equation 17.8, we conclude that the displacement amplitude s_{max} of a spherical wave must vary as $1/r$. Therefore, we can write the wave function ψ (Greek letter psi) for an outgoing spherical wave in the form

$$\psi(r, t) = \frac{s_0}{r} \sin(kr - \omega t) \qquad \textbf{(17.9)}$$

where s_0, the displacement amplitude at unit distance from the source, is a constant parameter characterizing the whole wave.

It is useful to represent spherical waves with a series of circular arcs concentric with the source, as shown in Figure 17.6. Each arc represents a surface over which the phase of the wave is constant. We call such a surface of constant phase a **wave front.** The distance between adjacent wave fronts equals the wavelength λ. The radial lines pointing outward from the source are called **rays.**

Now consider a small portion of a wave front far from the source, as shown in Figure 17.7. In this case, the rays passing through the wave front are nearly parallel to one another, and the wave front is very close to being planar. Therefore, at distances from the source that are great compared with the wavelength, we can approximate a wave front with a plane. Any small portion of a spherical wave far from its source can be considered a **plane wave.**

Figure 17.8 illustrates a plane wave propagating along the x axis, which means that the wave fronts are parallel to the yz plane. In this case, the wave function depends only on x and t and has the form

$$\psi(x, t) = A \sin(kx - \omega t) \qquad \textbf{(17.10)}$$

That is, the wave function for a plane wave is identical in form to that for a one-dimensional traveling wave.

The intensity is the same at all points on a given wave front of a plane wave.

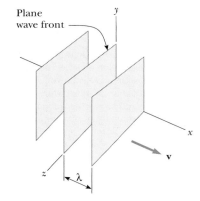

Figure 17.8   A representation of a plane wave moving in the positive x direction with a speed v. The wave fronts are planes parallel to the yz plane.

Representation of a plane wave

EXAMPLE 17.5   ▶ Intensity Variations of a Point Source

A point source emits sound waves with an average power output of 80.0 W. (a) Find the intensity 3.00 m from the source.

Solution   A point source emits energy in the form of spherical waves (see Fig. 17.5). At a distance r from the source, the power is distributed over the surface area of a sphere, $4\pi r^2$. Therefore, the intensity at the distance r is given by Equation 17.8:

$$I = \frac{\mathcal{P}_{av}}{4\pi r^2} = \frac{80.0 \text{ W}}{4\pi(3.00 \text{ m})^2} = \boxed{0.707 \text{ W/m}^2}$$

an intensity that is close to the threshold of pain.

(b) Find the distance at which the sound level is 40 dB.

Solution   We can find the intensity at the 40-dB sound level by using Equation 17.7 with $I_0 = 1.00 \times 10^{-12}$ W/m^2:

$$10 \log\left(\frac{I}{I_0}\right) = 40 \text{ dB}$$

$$\log I - \log I_0 = \frac{40}{10} = 4$$

$$\log I = 4 + \log 10^{-12}$$

$$\log I = -8$$

$$I = 1.00 \times 10^{-8} \text{ W/m}^2$$

Using this value for I in Equation 17.8 and solving for r, we obtain

$$r = \sqrt{\frac{\mathscr{P}_{av}}{4\pi I}} = \sqrt{\frac{80.0 \text{ W}}{4\pi \times 1.00 \times 10^{-8} \text{ W/m}^2}}$$

$$= 2.52 \times 10^4 \text{ m}$$

which equals about 16 miles!

17.5   THE DOPPLER EFFECT

QuickLab

(Before attempting to do this Quick-Lab, you should check to see whether it is legal to sound a horn in your area.) Sound your car horn while driving toward and away from a friend in a campus parking lot or on a country road. Try this at different speeds while driving toward and past the friend (not *at* the friend). Do the frequencies of the sounds you hear agree with what is described in the text?

Perhaps you have noticed how the sound of a vehicle's horn changes as the vehicle moves past you. The frequency of the sound you hear as the vehicle approaches you is higher than the frequency you hear as it moves away from you (see Quick-Lab). This is one example of the **Doppler effect.**[3]

To see what causes this apparent frequency change, imagine you are in a boat that is lying at anchor on a gentle sea where the waves have a period of $T = 3.0$ s. This means that every 3.0 s a crest hits your boat. Figure 17.9a shows this situation, with the water waves moving toward the left. If you set your watch to $t = 0$ just as one crest hits, the watch reads 3.0 s when the next crest hits, 6.0 s when the third crest hits, and so on. From these observations you conclude that the wave frequency is $f = 1/T = (1/3.0)$ Hz. Now suppose you start your motor and head directly into the oncoming waves, as shown in Figure 17.9b. Again you set your watch to $t = 0$ as a crest hits the front of your boat. Now, however, because you are moving toward the next wave crest as it moves toward you, it hits you less than 3.0 s after the first hit. In other words, the period you observe is shorter than the 3.0-s period you observed when you were stationary. Because $f = 1/T$, you observe a higher wave frequency than when you were at rest.

If you turn around and move in the same direction as the waves (see Fig. 17.9c), you observe the opposite effect. You set your watch to $t = 0$ as a crest hits the back of the boat. Because you are now moving away from the next crest, more than 3.0 s has elapsed on your watch by the time that crest catches you. Thus, you observe a lower frequency than when you were at rest.

These effects occur because the relative speed between your boat and the water depends on the direction of travel and on the speed of your boat. When you are moving toward the right in Figure 17.9b, this relative speed is higher than that of the wave speed, which leads to the observation of an increased frequency. When you turn around and move to the left, the relative speed is lower, as is the observed frequency of the water waves.

Let us now examine an analogous situation with sound waves, in which the water waves become sound waves, the water becomes the air, and the person on the boat becomes an observer listening to the sound. In this case, an observer O is moving and a sound source S is stationary. For simplicity, we assume that the air is also stationary and that the observer moves directly toward the source. The observer moves with a speed v_O toward a stationary point source ($v_S = 0$) (Fig. 17.10). In general, *at rest* means at rest with respect to the medium, air.

[3] Named after the Austrian physicist Christian Johann Doppler (1803–1853), who discovered the effect for light waves.

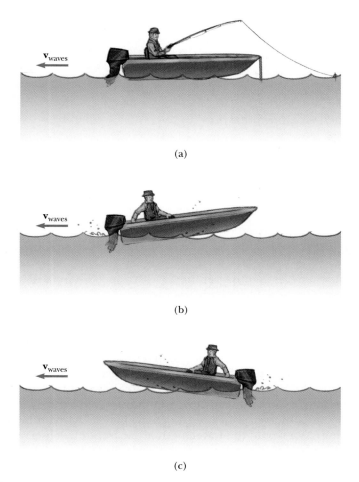

(a)

(b)

(c)

Figure 17.9 (a) Waves moving toward a stationary boat. The waves travel to the left, and their source is far to the right of the boat, out of the frame of the drawing. (b) The boat moving toward the wave source. (c) The boat moving away from the wave source.

We take the frequency of the source to be f, the wavelength to be λ, and the speed of sound to be v. If the observer were also stationary, he or she would detect f wave fronts per second. (That is, when $v_O = 0$ and $v_S = 0$, the observed frequency equals the source frequency.) When the observer moves toward the source,

Figure 17.10 An observer O (the cyclist) moves with a speed v_O toward a stationary point source S, the horn of a parked car. The observer hears a frequency f' that is greater than the source frequency.

the speed of the waves relative to the observer is $v' = v + v_O$, as in the case of the boat, but the wavelength λ is unchanged. Hence, using Equation 16.14, $v = \lambda f$, we can say that the frequency heard by the observer is *increased* and is given by

$$f' = \frac{v'}{\lambda} = \frac{v + v_O}{\lambda}$$

Because $\lambda = v/f$, we can express f' as

$$f' = \left(1 + \frac{v_O}{v}\right)f \qquad \text{(observer moving toward source)} \qquad \textbf{(17.11)}$$

If the observer is moving away from the source, the speed of the wave relative to the observer is $v' = v - v_O$. The frequency heard by the observer in this case is *decreased* and is given by

$$f' = \left(1 - \frac{v_O}{v}\right)f \qquad \text{(observer moving away from source)} \qquad \textbf{(17.12)}$$

In general, whenever an observer moves with a speed v_O relative to a stationary source, the frequency heard by the observer is

Frequency heard with an observer in motion

$$f' = \left(1 \pm \frac{v_O}{v}\right)f \qquad \textbf{(17.13)}$$

where the positive sign is used when the observer moves toward the source and the negative sign is used when the observer moves away from the source.

Now consider the situation in which the source is in motion and the observer is at rest. If the source moves directly toward observer A in Figure 17.11a, the wave fronts heard by the observer are closer together than they would be if the source were not moving. As a result, the wavelength λ' measured by observer A is shorter than the wavelength λ of the source. During each vibration, which lasts for a time T (the period), the source moves a distance $v_S T = v_S / f$ and the wavelength is

(a)

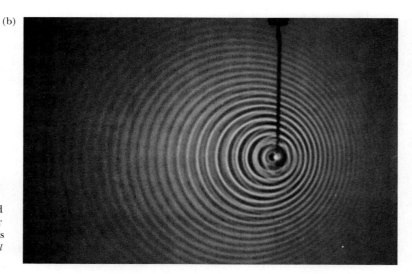

(b)

Figure 17.11   (a) A source S moving with a speed v_S toward a stationary observer A and away from a stationary observer B. Observer A hears an increased frequency, and observer B hears a decreased frequency. (b) The Doppler effect in water, observed in a ripple tank. A point source is moving to the right with speed v_S. *(Courtesy of the Educational Development Center, Newton, MA)* *shortened* by this amount. Therefore, the observed wavelength λ' is

$$\lambda' = \lambda - \Delta\lambda = \lambda - \frac{v_S}{f}$$

Because $\lambda = v/f$, the frequency heard by observer A is

$$f' = \frac{v}{\lambda'} = \frac{v}{\lambda - \dfrac{v_S}{f}} = \frac{v}{\dfrac{v}{f} - \dfrac{v_S}{f}}$$

$$f' = \left(\frac{1}{1 - \dfrac{v_S}{v}}\right)f \qquad\qquad (17.14)$$

That is, the observed frequency is *increased* whenever the source is moving toward the observer.

When the source moves away from a stationary observer, as is the case for observer B in Figure 17.11a, the observer measures a wavelength λ' that is *greater* than λ and hears a *decreased* frequency:

$$f' = \left(\frac{1}{1 + \dfrac{v_S}{v}}\right)f \qquad\qquad (17.15)$$

Combining Equations 17.14 and 17.15, we can express the general relationship for the observed frequency when a source is moving and an observer is at rest as

$$f' = \left(\frac{1}{1 \mp \dfrac{v_S}{v}}\right)f \qquad\qquad (17.16)$$

Frequency heard with source in motion

Finally, if both source and observer are in motion, we find the following general relationship for the observed frequency:

$$f' = \left(\frac{v \pm v_O}{v \mp v_S}\right)f \qquad\qquad (17.17)$$

[handwritten: V = speed of sound ≈ 343 m/s]

Frequency heard with observer and source in motion

In this expression, the upper signs ($+ v_O$ and $- v_S$) refer to motion of one toward the other, and the lower signs ($- v_O$ and $+ v_S$) refer to motion of one away from the other.

A convenient rule concerning signs for you to remember when working with all Doppler-effect problems is as follows:

The word *toward* is associated with an *increase* in observed frequency. The words *away from* are associated with a *decrease* in observed frequency.

Although the Doppler effect is most typically experienced with sound waves, it is a phenomenon that is common to all waves. For example, the relative motion of source and observer produces a frequency shift in light waves. The Doppler effect is used in police radar systems to measure the speeds of motor vehicles. Likewise, astronomers use the effect to determine the speeds of stars, galaxies, and other celestial objects relative to the Earth.

[handwritten margin notes:]

GOTTA DISTINGUISH BETWEEN ABSOLUTE & RELATIVE MOTION

LET x < y & 1d VECTORS

VELOCITIES	SIGNS $\left(\frac{v \pm v_O}{v \pm v_S}\right)$
BOTH APPROACHING	+ / −
BOTH RETREATING	− / +
x OVERTAKES y	− / −
y OUTRUNS x	+ / +

EXAMPLE 17.6 The Noisy Siren

As an ambulance travels east down a highway at a speed of $V_S = 33.5$ m/s (75 mi/h), its siren emits sound at a frequency of 400 Hz. What frequency is heard by a person in a car traveling $V_O = $ west at 24.6 m/s (55 mi/h) (a) as the car approaches the ambulance and (b) as the car moves away from the ambulance?

Solution (a) We can use Equation 17.17 in both cases, taking the speed of sound in air to be $v = 343$ m/s. As the ambulance and car approach each other, the person in the car hears the frequency

$$f' = \left(\frac{v + v_O}{v - v_S}\right)f = \left(\frac{343 \text{ m/s} + 24.6 \text{ m/s}}{343 \text{ m/s} - 33.5 \text{ m/s}}\right)(400 \text{ Hz})$$

$$= \boxed{475 \text{ Hz}}$$

(b) As the vehicles recede from each other, the person hears the frequency

$$f' = \left(\frac{v - v_O}{v + v_S}\right)f = \left(\frac{343 \text{ m/s} - 24.6 \text{ m/s}}{343 \text{ m/s} + 33.5 \text{ m/s}}\right)(400 \text{ Hz})$$

$$= \boxed{338 \text{ Hz}}$$

The *change* in frequency detected by the person in the car is $475 - 338 = 137$ Hz, which is more than 30% of the true frequency.

Exercise Suppose the car is parked on the side of the highway as the ambulance speeds by. What frequency does the person in the car hear as the ambulance (a) approaches and (b) recedes?

Answer (a) 443 Hz. (b) 364 Hz.

Shock Waves

Now let us consider what happens when the speed v_S of a source *exceeds* the wave speed v. This situation is depicted graphically in Figure 17.12a. The circles represent spherical wave fronts emitted by the source at various times during its motion. At $t = 0$, the source is at S_0, and at a later time t, the source is at S_n. In the time t,

(a)

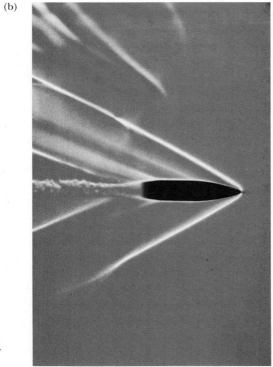

(b)

Figure 17.12 (a) A representation of a shock wave produced when a source moves from S_0 to S_n with a speed v_S, which is greater than the wave speed v in the medium. The envelope of the wave fronts forms a cone whose apex half-angle is given by $\sin \theta = v/v_S$. (b) A stroboscopic photograph of a bullet moving at supersonic speed through the hot air above a candle. Note the shock wave in the vicinity of the bullet. (©1973 Kim Vandiver & Harold E. Edgerton/Courtesy of Palm Press, Inc.)

Figure 17.13 The V-shaped bow wave of a boat is formed because the boat speed is greater than the speed of the water waves. A bow wave is analogous to a shock wave formed by an airplane traveling faster than sound. *(©1994 Comstock)*

the wave front centered at S_0 reaches a radius of vt. In this same amount of time, the source travels a distance $v_S t$ to S_n. At the instant the source is at S_n, waves are just beginning to be generated at this location, and hence the wave front has zero radius at this point. The tangent line drawn from S_n to the wave front centered on S_0 is tangent to all other wave fronts generated at intermediate times. Thus, we see that the envelope of these wave fronts is a cone whose apex half-angle θ is given by

$$\sin \theta = \frac{vt}{v_S t} = \frac{v}{v_S}$$

The ratio v_S/v is referred to as the *Mach number,* and the conical wave front produced when $v_S > v$ (supersonic speeds) is known as a *shock wave.* An interesting analogy to shock waves is the V-shaped wave fronts produced by a boat (the *bow wave*) when the boat's speed exceeds the speed of the surface-water waves (Fig. 17.13).

Jet airplanes traveling at supersonic speeds produce shock waves, which are responsible for the loud "sonic boom" one hears. The shock wave carries a great deal of energy concentrated on the surface of the cone, with correspondingly great pressure variations. Such shock waves are unpleasant to hear and can cause damage to buildings when aircraft fly supersonically at low altitudes. In fact, an airplane flying at supersonic speeds produces a double boom because two shock fronts are formed, one from the nose of the plane and one from the tail (Fig. 17.14). People near the path of the space shuttle as it glides toward its landing point often report hearing what sounds like two very closely spaced cracks of thunder.

Pressure

Atmospheric pressure

Figure 17.14 The two shock waves produced by the nose and tail of a jet airplane traveling at supersonic speeds.

Quick Quiz 17.5

An airplane flying with a constant velocity moves from a cold air mass into a warm air mass. Does the Mach number increase, decrease, or stay the same?

Quick Quiz 17.6

Suppose that an observer and a source of sound are both at rest and that a strong wind blows from the source toward the observer. Describe the effect of the wind (if any) on

(a) the observed frequency of the sound waves, (b) the observed wave speed, and (c) the observed wavelength.

SUMMARY

Sound waves are longitudinal and travel through a compressible medium with a speed that depends on the compressibility and inertia of that medium. The speed of sound in a medium having a bulk modulus B and density ρ is

$$v = \sqrt{\frac{B}{\rho}} \tag{17.1}$$

With this formula you can determine the speed of a sound wave in many different materials.

For sinusoidal sound waves, the variation in the displacement is given by

$$s(x, t) = s_{max} \cos(kx - \omega t) \tag{17.2}$$

and the variation in pressure from the equilibrium value is

$$\Delta P = \Delta P_{max} \sin(kx - \omega t) \tag{17.3}$$

where ΔP_{max} is the **pressure amplitude.** The pressure wave is 90° out of phase with the displacement wave. The relationship between s_{max} and ΔP_{max} is given by

$$\Delta P_{max} = \rho v \omega s_{max} \tag{17.4}$$

The intensity of a periodic sound wave, which is the power per unit area, is

$$I = \tfrac{1}{2}\rho v (\omega s_{max})^2 = \frac{\Delta P^2_{max}}{2\rho v} \tag{17.5, 17.6}$$

The sound level of a sound wave, in decibels, is given by

$$\beta = 10 \log\left(\frac{I}{I_0}\right) \tag{17.7}$$

The constant I_0 is a reference intensity, usually taken to be at the threshold of hearing (1.00×10^{-12} W/m^2), and I is the intensity of the sound wave in watts per square meter.

The intensity of a spherical wave produced by a point source is proportional to the average power emitted and inversely proportional to the square of the distance from the source:

$$I = \frac{\mathscr{P}_{av}}{4\pi r^2} \tag{17.8}$$

The change in frequency heard by an observer whenever there is relative motion between a source of sound waves and the observer is called the **Doppler effect.** The observed frequency is

$$f' = \left(\frac{v \pm v_O}{v \mp v_S}\right)f \tag{17.17}$$

The upper signs ($+ v_O$ and $- v_S$) are used with motion of one toward the other, and the lower signs ($- v_O$ and $+ v_S$) are used with motion of one away from the other. You can also use this formula when v_O or v_S is zero.

QUESTIONS

1. Why are sound waves characterized as longitudinal?
2. If an alarm clock is placed in a good vacuum and then activated, no sound is heard. Explain.
3. A sonic ranger is a device that determines the position of an object by sending out an ultrasonic sound pulse and measuring how long it takes for the sound wave to return after it reflects from the object. Typically, these devices cannot reliably detect an object that is less than half a meter from the sensor. Why is that?
4. In Example 17.5, we found that a point source with a power output of 80 W reduces to a sound level of 40 dB at a distance of about 16 miles. Why do you suppose you cannot normally hear a rock concert that is going on 16 miles away? (See Table 17.2.)
5. If the distance from a point source is tripled, by what factor does the intensity decrease?
6. Explain how the Doppler effect is used with microwaves to determine the speed of an automobile.
7. Explain what happens to the frequency of your echo as you move in a vehicle *toward* a canyon wall. What happens to the frequency as you move *away* from the wall?
8. Of the following sounds, which is most likely to have a sound level of 60 dB—a rock concert, the turning of a page in this text, normal conversation, or a cheering crowd at a football game?
9. Estimate the decibel level of each of the sounds in the previous question.
10. A binary star system consists of two stars revolving about their common center of mass. If we observe the light reaching us from one of these stars as it makes one complete revolution, what does the Doppler effect predict will happen to this light?
11. How can an object move with respect to an observer so that the sound from it is not shifted in frequency?
12. Why is it not possible to use sonar (sound waves) to determine the speed of an object traveling faster than the speed of sound in a given medium?
13. Why is it so quiet after a snowfall?
14. Why is the intensity of an echo less than that of the original sound?
15. If the wavelength of a sound source is reduced by a factor of 2, what happens to its frequency? Its speed?
16. In a recent discovery, a nearby star was found to have a large planet orbiting about it, although the planet could not be seen. In terms of the concept of a system rotating about its center of mass and the Doppler shift for light (which is in many ways similar to that for sound), explain how an astronomer could determine the presence of the invisible planet.
17. A friend sitting in her car far down the road waves to you and beeps her horn at the same time. How far away must her car be for you to measure the speed of sound to two significant figures by measuring the time it takes for the sound to reach you?

PROBLEMS

1, 2, 3 = straightforward, intermediate, challenging   ☐ = full solution available in the *Student Solutions Manual and Study Guide*
WEB = solution posted at **http://www.saunderscollege.com/physics/**   🖥 = Computer useful in solving problem   📷 = Interactive Physics
☐ = paired numerical/symbolic problems

Section 17.1   Speed of Sound Waves

1. Suppose that you hear a clap of thunder 16.2 s after seeing the associated lightning stroke. The speed of sound waves in air is 343 m/s, and the speed of light in air is 3.00×10^8 m/s. How far are you from the lightning stroke?
2. Find the speed of sound in mercury, which has a bulk modulus of approximately 2.80×10^{10} N/m^2 and a density of 13 600 kg/m^3.
3. A flower pot is knocked off a balcony 20.0 m above the sidewalk and falls toward an unsuspecting 1.75-m-tall man who is standing below. How close to the sidewalk can the flower pot fall before it is too late for a shouted warning from the balcony to reach the man in time? Assume that the man below requires 0.300 s to respond to the warning.
4. You are watching a pier being constructed on the far shore of a saltwater inlet when some blasting occurs.

You hear the sound in the water 4.50 s before it reaches you through the air. How wide is the inlet? (*Hint:* See Table 17.1. Assume that the air temperature is 20°C.)
5. Another approximation of the temperature dependence of the speed of sound in air (in meters per second) is given by the expression

$$v = 331.5 + 0.607 T_C$$

where T_C is the Celsius temperature. In dry air the temperature decreases about 1°C for every 150-m rise in altitude. (a) Assuming that this change is constant up to an altitude of 9 000 m, how long will it take the sound from an airplane flying at 9 000 m to reach the ground on a day when the ground temperature is 30°C? (b) Compare this to the time it would take if the air were at 30°C at all altitudes. Which interval is longer?
6. A bat can detect very small objects, such as an insect whose length is approximately equal to one wavelength

of the sound the bat makes. If bats emit a chirp at a frequency of 60.0 kHz, and if the speed of sound in air is 340 m/s, what is the smallest insect a bat can detect?

7. An airplane flies horizontally at a constant speed, piloted by rescuers who are searching for a disabled boat. When the plane is directly above the boat, the boat's crew blows a loud horn. By the time the plane's sound detector receives the horn's sound, the plane has traveled a distance equal to one-half its altitude above the ocean. If it takes the sound 2.00 s to reach the plane, determine (a) the speed of the plane and (b) its altitude. Take the speed of sound to be 343 m/s.

Section 17.2   Periodic Sound Waves

(*Note:* In this section, use the following values as needed, unless otherwise specified. The equilibrium density of air is $\rho = 1.20$ kg/m^3; the speed of sound in air is $v = 343$ m/s. Pressure variations ΔP are measured relative to atmospheric pressure, 1.013×10^5 Pa.

8. A sound wave in air has a pressure amplitude equal to 4.00×10^{-3} Pa. Calculate the displacement amplitude of the wave at a frequency of 10.0 kHz.

9. A sinusoidal sound wave is described by the displacement

$$s(x, t) = (2.00 \ \mu\text{m}) \cos[(15.7 \ \text{m}^{-1})x - (858 \ \text{s}^{-1})t]$$

(a) Find the amplitude, wavelength, and speed of this wave. (b) Determine the instantaneous displacement of the molecules at the position $x = 0.050 \ 0$ m at $t = 3.00$ ms. (c) Determine the maximum speed of a molecule's oscillatory motion.

10. As a sound wave travels through the air, it produces pressure variations (above and below atmospheric pressure) that are given by $\Delta P = 1.27 \sin(\pi x - 340\pi t)$ in SI units. Find (a) the amplitude of the pressure variations, (b) the frequency of the sound wave, (c) its wavelength in air, and (d) its speed.

11. Write an expression that describes the pressure variation as a function of position and time for a sinusoidal sound wave in air, if $\lambda = 0.100$ m and $\Delta P_{max} = 0.200$ Pa.

12. Write the function that describes the displacement wave corresponding to the pressure wave in Problem 11.

13. The tensile stress in a thick copper bar is 99.5% of its elastic breaking point of 13.0×10^{10} N/m^2. A 500-Hz sound wave is transmitted through the material. (a) What displacement amplitude will cause the bar to break? (b) What is the maximum speed of the particles at this moment?

14. Calculate the pressure amplitude of a 2.00-kHz sound wave in air if the displacement amplitude is equal to 2.00×10^{-8} m.

WEB 15. An experimenter wishes to generate in air a sound wave that has a displacement amplitude of 5.50×10^{-6} m. The pressure amplitude is to be limited to 8.40×10^{-1} Pa. What is the minimum wavelength the sound wave can have?

16. A sound wave in air has a pressure amplitude of 4.00 Pa and a frequency of 5.00 kHz. Take $\Delta P = 0$ at the point $x = 0$ when $t = 0$. (a) What is ΔP at $x = 0$ when $t = 2.00 \times 10^{-4}$ s? (b) What is ΔP at $x = 0.020 \ 0$ m when $t = 0$?

Section 17.3   Intensity of Periodic Sound Waves

17. Calculate the sound level, in decibels, of a sound wave that has an intensity of 4.00 μW/m^2.

18. A vacuum cleaner has a measured sound level of 70.0 dB. (a) What is the intensity of this sound in watts per square meter? (b) What is the pressure amplitude of the sound?

19. The intensity of a sound wave at a fixed distance from a speaker vibrating at 1.00 kHz is 0.600 W/m^2. (a) Determine the intensity if the frequency is increased to 2.50 kHz while a constant displacement amplitude is maintained. (b) Calculate the intensity if the frequency is reduced to 0.500 kHz and the displacement amplitude is doubled.

20. The intensity of a sound wave at a fixed distance from a speaker vibrating at a frequency f is I. (a) Determine the intensity if the frequency is increased to f' while a constant displacement amplitude is maintained. (b) Calculate the intensity if the frequency is reduced to $f/2$ and the displacement amplitude is doubled.

WEB 21. A family ice show is held in an enclosed arena. The skaters perform to music with a sound level of 80.0 dB. This is too loud for your baby, who consequently yells at a level of 75.0 dB. (a) What total sound intensity engulfs you? (b) What is the combined sound level?

Section 17.4   Spherical and Plane Waves

22. For sound radiating from a point source, show that the difference in sound levels, β_1 and β_2, at two receivers is related to the ratio of the distances r_1 and r_2 from the source to the receivers by the expression

$$\beta_2 - \beta_1 = 20 \log\left(\frac{r_1}{r_2}\right)$$

23. A fireworks charge is detonated many meters above the ground. At a distance of 400 m from the explosion, the acoustic pressure reaches a maximum of 10.0 N/m^2. Assume that the speed of sound is constant at 343 m/s throughout the atmosphere over the region considered, that the ground absorbs all the sound falling on it, and that the air absorbs sound energy as described by the rate 7.00 dB/km. What is the sound level (in decibels) at 4.00 km from the explosion?

24. A loudspeaker is placed between two observers who are 110 m apart, along the line connecting them. If one observer records a sound level of 60.0 dB and the other records a sound level of 80.0 dB, how far is the speaker from each observer?

25. Two small speakers emit spherical sound waves of different frequencies. Speaker A has an output of 1.00 mW, and speaker B has an output of 1.50 mW. Determine the sound level (in decibels) at point C (Fig. P17.25) if (a) only speaker A emits sound, (b) only speaker B emits sound, (c) both speakers emit sound.

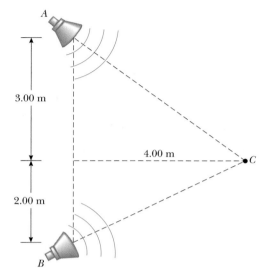

3.00 m

4.00 m

2.00 m

Figure P17.25

26. An experiment requires a sound intensity of 1.20 W/m² at a distance of 4.00 m from a speaker. What power output is required? Assume that the speakers radiate sound equally in all directions.

27. A source of sound (1 000 Hz) emits uniformly in all directions. An observer 3.00 m from the source measures a sound level of 40.0 dB. Calculate the average power output of the source.

28. A jackhammer, operated continuously at a construction site, behaves as a point source of spherical sound waves. A construction supervisor stands 50.0 m due north of this sound source and begins to walk due west. How far does she have to walk in order for the amplitude of the wave function to drop by a factor of 2.00?

29. The sound level at a distance of 3.00 m from a source is 120 dB. At what distances is the sound level (a) 100 dB and (b) 10.0 dB?

30. A fireworks rocket explodes 100 m above the ground. An observer directly under the explosion experiences an average sound intensity of 7.00×10^{-2} W/m² for 0.200 s. (a) What is the total sound energy of the explosion? (b) What sound level, in decibels, is heard by the observer?

31. As the people in a church sing on a summer morning, the sound level everywhere inside the church is 101 dB. The massive walls are opaque to sound, but all the windows and doors are open. Their total area is 22.0 m². (a) How much sound energy is radiated in 20.0 min? (b) Suppose the ground is a good reflector and sound

radiates uniformly in all horizontal and upward directions. Find the sound level 1.00 km away.

32. A spherical wave is radiating from a point source and is described by the wave function

$$\Delta P(r, t) = \left[\frac{25.0}{r}\right]\sin(1.25r - 1\,870t)$$

where ΔP is in pascals, r in meters, and t in seconds. (a) What is the pressure amplitude 4.00 m from the source? (b) Determine the speed of the wave and hence the material the wave might be traveling through. (c) Find the sound level of the wave, in decibels, 4.00 m from the source. (d) Find the instantaneous pressure 5.00 m from the source at 0.080 0 s.

Section 17.5 The Doppler Effect

33. A commuter train passes a passenger platform at a constant speed of 40.0 m/s. The train horn is sounded at its characteristic frequency of 320 Hz. (a) What change in frequency is detected by a person on the platform as the train passes? (b) What wavelength is detected by a person on the platform as the train approaches?

34. A driver travels northbound on a highway at a speed of 25.0 m/s. A police car, traveling southbound at a speed of 40.0 m/s, approaches with its siren sounding at a frequency of 2 500 Hz. (a) What frequency does the driver observe as the police car approaches? (b) What frequency does the driver detect after the police car passes him? (c) Repeat parts (a) and (b) for the case in which the police car is northbound.

WEB 35. Standing at a crosswalk, you hear a frequency of 560 Hz from the siren of an approaching police car. After the police car passes, the observed frequency of the siren is 480 Hz. Determine the car's speed from these observations.

36. Expectant parents are thrilled to hear their unborn baby's heartbeat, revealed by an ultrasonic motion detector. Suppose the fetus's ventricular wall moves in simple harmonic motion with an amplitude of 1.80 mm and a frequency of 115 per minute. (a) Find the maximum linear speed of the heart wall. Suppose the motion detector in contact with the mother's abdomen produces sound at 2 000 000.0 Hz, which travels through tissue at 1.50 km/s. (b) Find the maximum frequency at which sound arrives at the wall of the baby's heart. (c) Find the maximum frequency at which reflected sound is received by the motion detector. (By electronically "listening" for echoes at a frequency different from the broadcast frequency, the motion detector can produce beeps of audible sound in synchronization with the fetal heartbeat.)

37. A tuning fork vibrating at 512 Hz falls from rest and accelerates at 9.80 m/s². How far below the point of release is the tuning fork when waves with a frequency of 485 Hz reach the release point? Take the speed of sound in air to be 340 m/s.

38. A block with a speaker bolted to it is connected to a spring having spring constant $k = 20.0$ N/m, as shown in Figure P17.38. The total mass of the block and speaker is 5.00 kg, and the amplitude of this unit's motion is 0.500 m. (a) If the speaker emits sound waves of frequency 440 Hz, determine the highest and lowest frequencies heard by the person to the right of the speaker. (b) If the maximum sound level heard by the person is 60.0 dB when he is closest to the speaker, 1.00 m away, what is the minimum sound level heard by the observer? Assume that the speed of sound is 343 m/s.

Figure P17.38

39. A train is moving parallel to a highway with a constant speed of 20.0 m/s. A car is traveling in the same direction as the train with a speed of 40.0 m/s. The car horn sounds at a frequency of 510 Hz, and the train whistle sounds at a frequency of 320 Hz. (a) When the car is behind the train, what frequency does an occupant of the car observe for the train whistle? (b) When the car is in front of the train, what frequency does a train passenger observe for the car horn just after the car passes?

40. At the Winter Olympics, an athlete rides her luge down the track while a bell just above the wall of the chute rings continuously. When her sled passes the bell, she hears the pitch of the bell fall by the musical interval called a minor third. That is, the frequency she hears drops to five sixths of its original value. (a) Find the speed of sound in air at the ambient temperature $-10.0°$C. (b) Find the speed of the athlete.

41. A jet fighter plane travels in horizontal flight at Mach 1.20 (that is, 1.20 times the speed of sound in air). At the instant an observer on the ground hears the shock wave, what is the angle her line of sight makes with the horizontal as she looks at the plane?

42. When high-energy charged particles move through a transparent medium with a speed greater than the speed of light in that medium, a shock wave, or bow wave, of light is produced. This phenomenon is called the *Cerenkov effect* and can be observed in the vicinity of the core of a swimming-pool nuclear reactor due to

high-speed electrons moving through the water. In a particular case, the Cerenkov radiation produces a wave front with an apex half-angle of 53.0°. Calculate the speed of the electrons in the water. (The speed of light in water is 2.25×10^8 m/s.)

WEB 43. A supersonic jet traveling at Mach 3.00 at an altitude of 20 000 m is directly over a person at time $t = 0$, as in Figure P17.43. (a) How long will it be before the person encounters the shock wave? (b) Where will the plane be when it is finally heard? (Assume that the speed of sound in air is 335 m/s.)

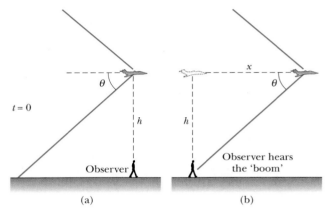

(a)   (b)

Figure P17.43

44. The tip of a circus ringmaster's whip travels at Mach 1.38 (that is, $v_S/v = 1.38$). What angle does the shock front make with the direction of the whip's motion?

ADDITIONAL PROBLEMS

45. A stone is dropped into a deep canyon and is heard to strike the bottom 10.2 s after release. The speed of sound waves in air is 343 m/s. How deep is the canyon? What would be the percentage error in the calculated depth if the time required for the sound to reach the canyon rim were ignored?

46. Unoccupied by spectators, a large set of football bleachers has solid seats and risers. You stand on the field in front of it and fire a starter's pistol or sharply clap two wooden boards together once. The sound pulse you produce has no definite pitch, no frequency, and no wavelength. You hear back from the bleachers a sound with definite pitch, which may remind you of a short toot on a trumpet, or of a buzzer or a kazoo. Account for this sound. Compute order-of-magnitude estimates for its frequency, wavelength, and duration on the basis of data that you specify.

47. Many artists sing very high notes in ornaments and cadenzas. The highest note written for a singer in a published score was F-sharp above high C, 1.480 kHz, sung

by Zerbinetta in the original version of Richard Strauss's opera *Ariadne auf Naxos*. (a) Find the wavelength of this sound in air. (b) Suppose that the people in the fourth row of seats hear this note with a level of 81.0 dB. Find the displacement amplitude of the sound. (c) In response to complaints, Strauss later transposed the note down to F above high C, 1.397 kHz. By what increment did the wavelength change?

48. A sound wave in a cylinder is described by Equations 17.2 through 17.4. Show that $\Delta P = \pm \rho v \omega \sqrt{s_{\max}^2 - s^2}$.

49. On a Saturday morning, pickup trucks carrying garbage to the town dump form a nearly steady procession on a country road, all traveling at 19.7 m/s. From this direction, two trucks arrive at the dump every three minutes. A bicyclist also is traveling toward the dump at 4.47 m/s. (a) With what frequency do the trucks pass him? (b) A hill does not slow the trucks but makes the out-of-shape cyclist's speed drop to 1.56 m/s. How often do the noisy trucks whiz past him now?

50. The ocean floor is underlain by a layer of basalt that constitutes the crust, or uppermost layer, of the Earth in that region. Below the crust is found denser peridotite rock, which forms the Earth's mantle. The boundary between these two layers is called the Mohorovicic discontinuity ("Moho" for short). If an explosive charge is set off at the surface of the basalt, it generates a seismic wave that is reflected back out at the Moho. If the speed of the wave in basalt is 6.50 km/s and the two-way travel time is 1.85 s, what is the thickness of this oceanic crust?

51. A worker strikes a steel pipeline with a hammer, generating both longitudinal and transverse waves. Reflected waves return 2.40 s apart. How far away is the reflection point? (For steel, $v_{\text{long}} = 6.20$ km/s and $v_{\text{trans}} = 3.20$ km/s.)

52. For a certain type of steel, stress is proportional to strain with Young's modulus as given in Table 12.1. The steel has the density listed for iron in Table 15.1. It bends permanently if subjected to compressive stress greater than its elastic limit, $\sigma = 400$ MPa, also called its *yield strength*. A rod 80.0 cm long, made of this steel, is projected at 12.0 m/s straight at a hard wall. (a) Find the speed of compressional waves moving along the rod. (b) After the front end of the rod hits the wall and stops, the back end of the rod keeps moving, as described by Newton's first law, until it is stopped by the excess pressure in a sound wave moving back through the rod. How much time elapses before the back end of the rod gets the message? (c) How far has the back end of the rod moved in this time? (d) Find the strain in the rod and (e) the stress. (f) If it is not to fail, show that the maximum impact speed a rod can have is given by the expression $\sigma/\sqrt{\rho Y}$.

53. To determine her own speed, a sky diver carries a buzzer that emits a steady tone at 1 800 Hz. A friend at the landing site on the ground directly below the sky diver listens to the amplified sound he receives from the buzzer. Assume that the air is calm and that the speed of sound is 343 m/s, independent of altitude. While the sky diver is falling at terminal speed, her friend on the ground receives waves with a frequency of 2 150 Hz. (a) What is the sky diver's speed of descent? (b) Suppose the sky diver is also carrying sound-receiving equipment that is sensitive enough to detect waves reflected from the ground. What frequency does she receive?

54. A train whistle ($f = 400$ Hz) sounds higher or lower in pitch depending on whether it is approaching or receding. (a) Prove that the difference in frequency between the approaching and receding train whistle is

$$\Delta f = \frac{2(u/v)}{1 - (u^2/v^2)} f$$

where u is the speed of the train and v is the speed of sound. (b) Calculate this difference for a train moving at a speed of 130 km/h. Take the speed of sound in air to be 340 m/s.

55. A bat, moving at 5.00 m/s, is chasing a flying insect. If the bat emits a 40.0-kHz chirp and receives back an echo at 40.4 kHz, at what relative speed is the bat moving toward or away from the insect? (Take the speed of sound in air to be $v = 340$ m/s.)

Figure P17.55   *(Joe McDonald/Visuals Unlimited)*

56. A supersonic aircraft is flying parallel to the ground. When the aircraft is directly overhead, an observer on the ground sees a rocket fired from the aircraft. Ten seconds later the observer hears the sonic boom, which is followed 2.80 s later by the sound of the rocket engine. What is the Mach number of the aircraft?

57. A police car is traveling east at 40.0 m/s along a straight road, overtaking a car that is moving east at 30.0 m/s. The police car has a malfunctioning siren that is stuck at 1 000 Hz. (a) Sketch the appearance of the wave fronts of the sound produced by the siren. Show the

wave fronts both to the east and to the west of the police car. (b) What would be the wavelength in air of the siren sound if the police car were at rest? (c) What is the wavelength in front of the car? (d) What is the wavelength behind the police car? (e) What frequency is heard by the driver being chased?

58. A copper bar is given a sharp compressional blow at one end. The sound of the blow, traveling through air at 0°C, reaches the opposite end of the bar 6.40 ms later than the sound transmitted through the metal of the bar. What is the length of the bar? (Refer to Table 17.1.)

59. The power output of a certain public address speaker is 6.00 W. Suppose it broadcasts equally in all directions. (a) Within what distance from the speaker would the sound be painful to the ear? (b) At what distance from the speaker would the sound be barely audible?

60. A jet flies toward higher altitude at a constant speed of 1 963 m/s in a direction that makes an angle θ with the horizontal (Fig. P17.60). An observer on the ground hears the jet for the first time when it is directly overhead. Determine the value of θ if the speed of sound in air is 340 m/s.

Figure P17.60

61. Two ships are moving along a line due east. The trailing vessel has a speed of 64.0 km/h relative to a land-based observation point, and the leading ship has a speed of 45.0 km/h relative to that point. The two ships are in a region of the ocean where the current is moving uniformly due west at 10.0 km/h. The trailing ship transmits a sonar signal at a frequency of 1 200.0 Hz. What frequency is monitored by the leading ship? (Use 1 520 m/s as the speed of sound in ocean water.)

62. A microwave oven generates a sound with intensity level 40.0 dB everywhere just outside it, when consuming 1.00 kW of power. Find the fraction of this power that is converted into the energy of sound waves. Assume the dimensions of the oven are 40.0 cm × 40.0 cm × 50.0 cm.

63. A meteoroid the size of a truck enters the Earth's atmosphere at a speed of 20.0 km/s and is not significantly slowed before entering the ocean. (a) What is the Mach angle of the shock wave from the meteoroid in the atmosphere? (Use 331 m/s as the sound speed.) (b) Assuming that the meteoroid survives the impact with the ocean surface, what is the (initial) Mach angle of the shock wave that the meteoroid produces in the water? (Use the wave speed for sea water given in Table 17.1.)

64. Consider a longitudinal (compressional) wave of wavelength λ traveling with speed v along the x direction through a medium of density ρ. The *displacement* of the molecules of the medium from their equilibrium position is

$$s = s_{max} \sin(kx - \omega t)$$

Show that the pressure variation in the medium is given by

$$\Delta P = -\left(\frac{2\pi\rho v^2}{\lambda} s_{max}\right)\cos(kx - \omega t)$$

WEB 65. By proper excitation, it is possible to produce both longitudinal and transverse waves in a long metal rod. A particular metal rod is 150 cm long and has a radius of 0.200 cm and a mass of 50.9 g. Young's modulus for the material is 6.80×10^{10} N/m². What must the tension in the rod be if the ratio of the speed of longitudinal waves to the speed of transverse waves is 8.00?

66. An interstate highway has been built though a neighborhood in a city. In the afternoon, the sound level in a rented room is 80.0 dB as 100 cars per minute pass outside the window. Late at night, the traffic flow on the freeway is only five cars per minute. What is the average late-night sound level in the room?

67. A siren creates a sound level of 60.0 dB at a location 500 m from the speaker. The siren is powered by a battery that delivers a total energy of 1.00 kJ. Assuming that the efficiency of the siren is 30.0% (that is, 30.0% of the supplied energy is transformed into sound energy), determine the total time the siren can sound.

68. A siren creates a sound level β at a distance d from the speaker. The siren is powered by a battery that delivers a total energy E. Assuming that the efficiency of the siren is e (that is, e is equal to the output sound energy divided by the supplied energy), determine the total time the siren can sound.

69. The Doppler equation presented in the text is valid when the motion between the observer and the source occurs on a straight line, so that the source and observer are moving either directly toward or directly away from each other. If this restriction is relaxed, one must use the more general Doppler equation

$$f' = \left(\frac{v + v_O \cos\theta_O}{v - v_S \cos\theta_S}\right)f$$

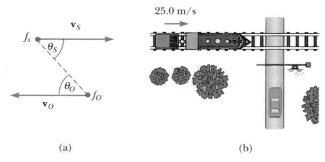

(a)   (b)

Figure P17.69

where θ_O and θ_S are defined in Figure P17.69a.
(a) Show that if the observer and source are moving away from each other, the preceding equation reduces to Equation 17.17 with lower signs. (b) Use the preceding equation to solve the following problem. A train moves at a constant speed of 25.0 m/s toward the intersection shown in Figure P17.69b. A car is stopped near the intersection, 30.0 m from the tracks. If the train's horn emits a frequency of 500 Hz, what frequency is heard by the passengers in the car when the train is 40.0 m from the intersection? Take the speed of sound to be 343 m/s.

70. Figure 17.5 illustrates that at distance r from a point source with power $\mathcal{P}_{av}$, the wave intensity is $I = \mathcal{P}_{av}/4\pi r^2$. Study Figure 17.11a and prove that at distance r straight in front of a point source with power $\mathcal{P}_{av}$, moving with constant speed v_S, the wave intensity is

$$I = \frac{\mathcal{P}_{av}}{4\pi r^2}\left(\frac{v - v_S}{v}\right)$$

71. Three metal rods are located relative to each other as shown in Figure P17.71, where $L_1 + L_2 = L_3$. The density values and Young's moduli for the three materials are $\rho_1 = 2.70 \times 10^3$ kg/m^3, $Y_1 = 7.00 \times 10^{10}$ N/m^2; $\rho_2 = 11.3 \times 10^3$ kg/m^3, $Y_2 = 1.60 \times 10^{10}$ N/m^2; $\rho_3 = 8.80 \times 10^3$ kg/m^3, $Y_3 = 11.0 \times 10^{10}$ N/m^2. (a) If $L_3 = 1.50$ m, what must the ratio L_1/L_2 be if a sound wave is to travel the combined length of rods 1 and 2 in the same time it takes to travel the length of rod 3? (b) If the frequency of the source is 4.00 kHz, determine the phase difference between the wave traveling along rods 1 and 2 and the one traveling along rod 3.

Figure P17.71

72. The volume knob on a radio has what is known as a "logarithmic taper." The electrical device connected to the knob (called a potentiometer) has a resistance R whose logarithm is proportional to the angular position of the knob: that is, $\log R \propto \theta$. If the intensity of the sound I (in watts per square meter) produced by the speaker is proportional to the resistance R, show that the sound level β (in decibels) is a linear function of θ.

73. The smallest wavelength possible for a sound wave in air is on the order of the separation distance between air molecules. Find the order of magnitude of the highest-frequency sound wave possible in air, assuming a wave speed of 343 m/s, a density of 1.20 kg/m^3, and an average molecular mass of 4.82×10^{-26} kg.

ANSWERS TO QUICK QUIZZES

17.1 The only correct answer is (c). Although the speed of a wave is given by the product of its wavelength and frequency, it is not affected by changes in either one. For example, if the sound from a musical instrument increases in frequency, the wavelength decreases, and thus $v = \lambda f$ remains constant. The amplitude of a sound wave determines the size of the oscillations of air molecules but does not affect the speed of the wave through the air.

17.2 The ground tremor represents a sound wave moving through the Earth. Sound waves move faster through the Earth than through air because rock and other ground materials are much stiffer against compression. Therefore — the vibration through the ground and the sound in the air having started together — the vibration through the ground reaches the observer first.

17.3 Because the bottom of the bottle does not allow molecular motion, the displacement in this region is at its minimum value. Because the pressure variation is a maximum when the displacement is a minimum, the pressure variation at the bottom is a maximum.

17.4 (a) 10 dB. If we call the intensity of each violin I, the total intensity when all the violins are playing is $I + 9I = 10I$. Therefore, the addition of the nine violins increases the intensity of the sound over that of one violin by a factor of 10. From Equation 17.7 we see that an increase in intensity by a factor of 10 increases the sound level by 10 dB. (b) 13 dB. The intensity is now increased by a factor of 20 over that of a single violin.

17.5 The Mach number is the ratio of the plane's speed (which does not change) to the speed of sound, which is greater in the warm air than in the cold, as we learned in Section 17.1 (see Quick Quiz 17.1). The denominator of this fraction increases while the numerator stays constant. Therefore, the fraction as a whole—the Mach number—decreases.

17.6 (a) In the reference frame of the air, the observer is moving toward the source at the wind speed through stationary air, and the source is moving away from the observer with the same speed. In Equation 17.17, therefore, a plus sign is needed in both the numerator and the denominator:

$$f' = \left(\frac{v_{sound} + v_{wind}}{v_{sound} + v_{wind}} \right) f$$

meaning the observed frequency is the same as if no wind were blowing. (b) The observer "sees" the sound waves coming toward him at a higher speed $(v_{sound} + v_{wind})$. (c) At this higher speed, he attributes a greater wavelength $\lambda' = (v_{sound} + v_{wind})/f$ to the wave.

PUZZLER

A speaker for a stereo system operates even if the wires connecting it to the amplifier are reversed, that is, + for − and − for + (or red for black and black for red). Nonetheless, the owner's manual says that for best performance you should be careful to connect the two speakers properly, so that they are "in phase." Why is this such an important consideration for the quality of the sound you hear? *(George Semple)*

c h a p t e r

18

Superposition and Standing Waves

*I*mportant in the study of waves is the combined effect of two or more waves traveling in the same medium. For instance, what happens to a string when a wave traveling along it hits a fixed end and is reflected back on itself? What is the air pressure variation at a particular seat in a theater when the instruments of an orchestra sound together?

When analyzing a linear medium—that is, one in which the restoring force acting on the particles of the medium is proportional to the displacement of the particles—we can apply the principle of superposition to determine the resultant disturbance. In Chapter 16 we discussed this principle as it applies to wave pulses. In this chapter we study the superposition principle as it applies to sinusoidal waves. If the sinusoidal waves that combine in a linear medium have the same frequency and wavelength, a stationary pattern—called a *standing wave*—can be produced at certain frequencies under certain circumstances. For example, a taut string fixed at both ends has a discrete set of oscillation patterns, called *modes of vibration*, that are related to the tension and linear mass density of the string. These modes of vibration are found in stringed musical instruments. Other musical instruments, such as the organ and the flute, make use of the natural frequencies of sound waves in hollow pipes. Such frequencies are related to the length and shape of the pipe and depend on whether the pipe is open at both ends or open at one end and closed at the other.

We also consider the superposition and interference of waves having different frequencies and wavelengths. When two sound waves having nearly the same frequency interfere, we hear variations in the loudness called *beats*. The beat frequency corresponds to the rate of alternation between constructive and destructive interference. Finally, we discuss how any non-sinusoidal periodic wave can be described as a sum of sine and cosine functions.

18.1 SUPERPOSITION AND INTERFERENCE OF SINUSOIDAL WAVES

Imagine that you are standing in a swimming pool and that a beach ball is floating a couple of meters away. You use your right hand to send a series of waves toward the beach ball, causing it to repeatedly move upward by 5 cm, return to its original position, and then move downward by 5 cm. After the water becomes still, you use your left hand to send an identical set of waves toward the beach ball and observe the same behavior. What happens if you use both hands at the same time to send two waves toward the beach ball? How the beach ball responds to the waves depends on whether the waves work together (that is, both waves make the beach ball go up at the same time and then down at the same time) or work against each other (that is, one wave tries to make the beach ball go up, while the other wave tries to make it go down). Because it is possible to have two or more waves in the same location at the same time, we have to consider how waves interact with each other and with their surroundings.

The superposition principle states that when two or more waves move in the same linear medium, the net displacement of the medium (that is, the resultant wave) at any point equals the algebraic sum of all the displacements caused by the individual waves. Let us apply this principle to two sinusoidal waves traveling in the same direction in a linear medium. If the two waves are traveling to the right and have the same frequency, wavelength, and amplitude but differ in phase, we can express their individual wave functions as

$$y_1 = A \sin(kx - \omega t) \qquad y_2 = A \sin(kx - \omega t + \phi)$$

where, as usual, $k = 2\pi/\lambda$, $\omega = 2\pi f$, and ϕ is the phase constant, which we introduced in the context of simple harmonic motion in Chapter 13. Hence, the resultant wave function y is

$$y = y_1 + y_2 = A[\sin(kx - \omega t) + \sin(kx - \omega t + \phi)]$$

To simplify this expression, we use the trigonometric identity

$$\sin a + \sin b = 2 \cos\left(\frac{a - b}{2}\right) \sin\left(\frac{a + b}{2}\right)$$

If we let $a = kx - \omega t$ and $b = kx - \omega t + \phi$, we find that the resultant wave function y reduces to

$$y = 2A \cos\left(\frac{\phi}{2}\right) \sin\left(kx - \omega t + \frac{\phi}{2}\right)$$

> Resultant of two traveling sinusoidal waves

This result has several important features. The resultant wave function y also is sinusoidal and has the same frequency and wavelength as the individual waves, since the sine function incorporates the same values of k and ω that appear in the original wave functions. The amplitude of the resultant wave is $2A \cos(\phi/2)$, and its phase is $\phi/2$. If the phase constant ϕ equals 0, then $\cos(\phi/2) = \cos 0 = 1$, and the amplitude of the resultant wave is $2A$—twice the amplitude of either individual wave. In this case, in which $\phi = 0$, the waves are said to be everywhere *in phase* and thus **interfere constructively.** That is, the crests and troughs of the individual waves y_1 and y_2 occur at the same positions and combine to form the red curve y of amplitude $2A$ shown in Figure 18.1a. Because the individual waves are in phase, they are indistinguishable in Figure 18.1a, in which they appear as a single blue curve. In general, constructive interference occurs when $\cos(\phi/2) = \pm 1$. This is true, for example, when $\phi = 0, 2\pi, 4\pi, \ldots$ rad—that is, when ϕ is an *even* multiple of π.

When ϕ is equal to π rad or to any *odd* multiple of π, then $\cos(\phi/2) = \cos(\pi/2) = 0$, and the crests of one wave occur at the same positions as the troughs of the second wave (Fig. 18.1b). Thus, the resultant wave has *zero* amplitude everywhere, as a consequence of **destructive interference.** Finally, when the phase constant has an arbitrary value other than 0 or other than an integer multiple of π rad (Fig. 18.1c), the resultant wave has an amplitude whose value is somewhere between 0 and $2A$.

> *[handwritten annotation]* ϕ = phase diff   r = path diff

> Constructive interference

> *[handwritten annotation]* ϕ = even π
> r = even $\frac{\lambda}{2}$

> Destructive interference

> *[handwritten annotation]* ϕ = odd π
> r = odd $\frac{\lambda}{2}$
> $\frac{r}{\lambda} = \frac{\phi}{2\pi}$ $\longrightarrow$ $\frac{2r}{\lambda} = \frac{\phi}{\pi}$

Interference of Sound Waves

One simple device for demonstrating interference of sound waves is illustrated in Figure 18.2. Sound from a loudspeaker S is sent into a tube at point P, where there is a T-shaped junction. Half of the sound power travels in one direction, and half travels in the opposite direction. Thus, the sound waves that reach the receiver R can travel along either of the two paths. The distance along any path from speaker to receiver is called the **path length r.** The lower path length r_1 is fixed, but the upper path length r_2 can be varied by sliding the U-shaped tube, which is similar to that on a slide trombone. When the difference in the path lengths $\Delta r = |r_2 - r_1|$ is either zero or some integer multiple of the wavelength λ (that is, $r = n\lambda$, where $n = 0, 1, 2, 3, \ldots$), the two waves reaching the receiver at any instant are in phase and interfere constructively, as shown in Figure 18.1a. For this case, a maximum in the sound intensity is detected at the receiver. If the path length r_2 is ad-

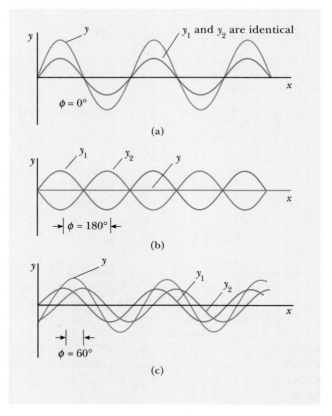

Figure 18.1   The superposition of two identical waves y_1 and y_2 (blue) to yield a resultant wave (red). (a) When y_1 and y_2 are in phase, the result is constructive interference. (b) When y_1 and y_2 are π rad out of phase, the result is destructive interference. (c) When the phase angle has a value other than 0 or π rad, the resultant wave y falls somewhere between the extremes shown in (a) and (b).

justed such that the path difference $\Delta r = \lambda/2, 3\lambda/2, \ldots, n\lambda/2$ (for n odd), the two waves are exactly π rad, or 180°, out of phase at the receiver and hence cancel each other. In this case of destructive interference, no sound is detected at the receiver. This simple experiment demonstrates that a phase difference may arise between two waves generated by the same source when they travel along paths of unequal lengths. This important phenomenon will be indispensable in our investigation of the interference of light waves in Chapter 37.

Figure 18.2   An acoustical system for demonstrating interference of sound waves. A sound wave from the speaker (S) propagates into the tube and splits into two parts at point P. The two waves, which superimpose at the opposite side, are detected at the receiver (R). The upper path length r_2 can be varied by sliding the upper section.

It is often useful to express the path difference in terms of the phase angle ϕ between the two waves. Because a path difference of one wavelength corresponds to a phase angle of 2π rad, we obtain the ratio $\phi/2\pi = \Delta r/\lambda$, or

$$\Delta r = \frac{\phi}{2\pi}\lambda \qquad \textbf{(18.1)}$$

Relationship between path difference and phase angle

Using the notion of path difference, we can express our conditions for constructive and destructive interference in a different way. If the path difference is any even multiple of $\lambda/2$, then the phase angle $\phi = 2n\pi$, where $n = 0, 1, 2, 3, \ldots$, and the interference is constructive. For path differences of odd multiples of $\lambda/2$, $\phi = (2n + 1)\pi$, where $n = 0, 1, 2, 3 \ldots$, and the interference is destructive. Thus, we have the conditions

$$\Delta r = (2n)\frac{\lambda}{2} \qquad \text{for constructive interference}$$

and $\qquad\qquad\qquad\qquad\qquad\qquad\qquad\qquad$ **(18.2)**

$$\Delta r = (2n + 1)\frac{\lambda}{2} \qquad \text{for destructive interference}$$

$\phi = \dfrac{2\pi\Delta r}{\lambda} = \dfrac{\Delta r}{\lambda}\cdot 360°$
$\quad\text{mod} \quad\quad \text{mod}$
$\quad 2\pi \quad\quad\quad 360°$

EXAMPLE 18.1 ▶ Two Speakers Driven by the Same Source

A pair of speakers placed 3.00 m apart are driven by the same oscillator (Fig. 18.3). A listener is originally at point O, which is located 8.00 m from the center of the line connecting the two speakers. The listener then walks to point P, which is a perpendicular distance 0.350 m from O, before reaching the *first minimum* in sound intensity. What is the frequency of the oscillator?

Solution To find the frequency, we need to know the wavelength of the sound coming from the speakers. With this information, combined with our knowledge of the speed of sound, we can calculate the frequency. We can determine the wavelength from the interference information given. The first minimum occurs when the two waves reaching the listener at point P are 180° out of phase—in other words, when their path difference Δr equals $\lambda/2$. To calculate the path difference, we must first find the path lengths r_1 and r_2.

Figure 18.3 shows the physical arrangement of the speakers, along with two shaded right triangles that can be drawn on the basis of the lengths described in the problem. From

these triangles, we find that the path lengths are

$$r_1 = \sqrt{(8.00\ \text{m})^2 + (1.15\ \text{m})^2} = 8.08\ \text{m}$$

and

$$r_2 = \sqrt{(8.00\ \text{m})^2 + (1.85\ \text{m})^2} = 8.21\ \text{m}$$

Hence, the path difference is $r_2 - r_1 = 0.13$ m. Because we require that this path difference be equal to $\lambda/2$ for the first minimum, we find that $\lambda = 0.26$ m.

To obtain the oscillator frequency, we use Equation 16.14, $v = \lambda f$, where v is the speed of sound in air, 343 m/s:

$$f = \frac{v}{\lambda} = \frac{343\ \text{m/s}}{0.26\ \text{m}} = \boxed{1.3\ \text{kHz}}$$

Exercise If the oscillator frequency is adjusted such that the first location at which a listener hears no sound is at a distance of 0.75 m from O, what is the new frequency?

Answer 0.63 kHz.

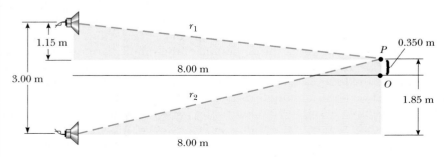

r_1

1.15 m

8.00 m

3.00 m

r_2

8.00 m

0.350 m

P

O

1.85 m

Figure 18.3

You can now understand why the speaker wires in a stereo system should be connected properly. When connected the wrong way—that is, when the positive (or red) wire is connected to the negative (or black) terminal—the speakers are said to be "out of phase" because the sound wave coming from one speaker destructively interferes with the wave coming from the other. In this situation, one speaker cone moves outward while the other moves inward. Along a line midway between the two, a rarefaction region from one speaker is superposed on a condensation region from the other speaker. Although the two sounds probably do not completely cancel each other (because the left and right stereo signals are usually not identical), a substantial loss of sound quality still occurs at points along this line.

18.2 STANDING WAVES

The sound waves from the speakers in Example 18.1 left the speakers in the forward direction, and we considered interference at a point in space in front of the speakers. Suppose that we turn the speakers so that they face each other and then have them emit sound of the same frequency and amplitude. We now have a situation in which two identical waves travel in opposite directions in the same medium. These waves combine in accordance with the superposition principle.

We can analyze such a situation by considering wave functions for two transverse sinusoidal waves having the same amplitude, frequency, and wavelength but traveling in opposite directions in the same medium:

$$y_1 = A \sin(kx - \omega t) \qquad y_2 = A \sin(kx + \omega t)$$

where y_1 represents a wave traveling to the right and y_2 represents one traveling to the left. Adding these two functions gives the resultant wave function y:

$$y = y_1 + y_2 = A \sin(kx - \omega t) + A \sin(kx + \omega t)$$

When we use the trigonometric identity $\sin(a \pm b) = \sin a \cos b \pm \cos a \sin b$, this expression reduces to

ENVELOPE WAVE

| Wave function for a standing wave |
$$y = (2A \sin kx) \cos \omega t \tag{18.3}$$

which is the wave function of a standing wave. A **standing wave,** such as the one shown in Figure 18.4, is an oscillation pattern with a stationary outline that results from the superposition of two identical waves traveling in opposite directions.

Notice that Equation 18.3 does not contain a function of $kx \pm \omega t$. Thus, it is not an expression for a traveling wave. If we observe a standing wave, we have no sense of motion in the direction of propagation of either of the original waves. If we compare this equation with Equation 13.3, we see that Equation 18.3 describes a special kind of simple harmonic motion. Every particle of the medium oscillates in simple harmonic motion with the same frequency ω (according to the cos ωt factor in the equation). However, the amplitude of the simple harmonic motion of a given particle (given by the factor $2A \sin kx$, the coefficient of the cosine function) depends on the location x of the particle in the medium. We need to distinguish carefully between the amplitude A of the individual waves and the amplitude $2A \sin kx$ of the simple harmonic motion of the particles of the medium. A given particle in a standing wave vibrates within the constraints of the *envelope* function $2A \sin kx$, where x is the particle's position in the medium. This is in contrast to the situation in a traveling sinusoidal wave, in which all particles oscillate with the

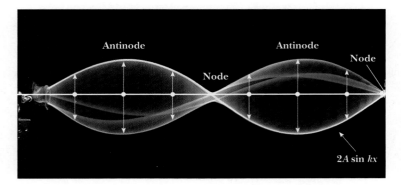

Figure 18.4 Multiflash photograph of a standing wave on a string. The vertical displacement from equilibrium of an individual particle of the string is given by $\cos \omega t$. That is, each particle vibrates at an angular frequency ω. The amplitude of the vertical oscillation of any particle on the string depends on the horizontal position of the particle. Each particle vibrates within the confines of the envelope function $2A \sin kx$. (©1991 Richard Megna/Fundamental Photographs)

same amplitude and the same frequency and in which the amplitude of the wave is the same as the amplitude of the simple harmonic motion of the particles.

The maximum displacement of a particle of the medium has a minimum value of zero when x satisfies the condition $\sin kx = 0$, that is, when

$$kx = \pi, 2\pi, 3\pi, \ldots$$

Because $k = 2\pi/\lambda$, these values for kx give

$$x = \frac{\lambda}{2}, \lambda, \frac{3\lambda}{2}, \ldots = \frac{n\lambda}{2} \qquad n = 0, 1, 2, 3, \ldots \qquad \textbf{(18.4)}$$

Position of nodes

These points of zero displacement are called **nodes.**

The particle with the greatest possible displacement from equilibrium has an amplitude of $2A$, and we define this as the amplitude of the standing wave. The positions in the medium at which this maximum displacement occurs are called **antinodes.** The antinodes are located at positions for which the coordinate x satisfies the condition $\sin kx = \pm 1$, that is, when

$$kx = \frac{\pi}{2}, \frac{3\pi}{2}, \frac{5\pi}{2}, \ldots$$

Thus, the positions of the antinodes are given by

$$x = \frac{\lambda}{4}, \frac{3\lambda}{4}, \frac{5\lambda}{4}, \ldots = \frac{n\lambda}{4} \qquad n = 1, 3, 5, \ldots \qquad \textbf{(18.5)}$$

Position of antinodes

In examining Equations 18.4 and 18.5, we note the following important features of the locations of nodes and antinodes:

The distance between adjacent antinodes is equal to $\lambda/2$.
The distance between adjacent nodes is equal to $\lambda/2$.
The distance between a node and an adjacent antinode is $\lambda/4$.

Displacement patterns of the particles of the medium produced at various times by two waves traveling in opposite directions are shown in Figure 18.5. The blue and green curves are the individual traveling waves, and the red curves are

Figure 18.5   Standing-wave patterns produced at various times by two waves of equal amplitude traveling in opposite directions. For the resultant wave *y*, the nodes (N) are points of zero displacement, and the antinodes (A) are points of maximum displacement.

Figure 18.6   A standing-wave pattern in a taut string. The five "snapshots" were taken at half-cycle intervals. (a) At $t = 0$, the string is momentarily at rest; thus, $K = 0$, and all the energy is potential energy *U* associated with the vertical displacements of the string particles. (b) At $t = T/8$, the string is in motion, as indicated by the brown arrows, and the energy is half kinetic and half potential. (c) At $t = T/4$, the string is moving but horizontal (undeformed); thus, $U = 0$, and all the energy is kinetic. (d) The motion continues as indicated. (e) At $t = T/2$, the string is again momentarily at rest, but the crests and troughs of (a) are reversed. The cycle continues until ultimately, when a time interval equal to *T* has passed, the configuration shown in (a) is repeated.

the displacement patterns. At $t = 0$ (Fig. 18.5a), the two traveling waves are in phase, giving a displacement pattern in which each particle of the medium is experiencing its maximum displacement from equilibrium. One quarter of a period later, at $t = T/4$ (Fig. 18.5b), the traveling waves have moved one quarter of a wavelength (one to the right and the other to the left). At this time, the traveling waves are out of phase, and each particle of the medium is passing through the equilibrium position in its simple harmonic motion. The result is zero displacement for particles at all values of *x*—that is, the displacement pattern is a straight line. At $t = T/2$ (Fig. 18.5c), the traveling waves are again in phase, producing a displacement pattern that is inverted relative to the $t = 0$ pattern. In the standing wave, the particles of the medium alternate in time between the extremes shown in Figure 18.5a and c.

Energy in a Standing Wave

It is instructive to describe the energy associated with the particles of a medium in which a standing wave exists. Consider a standing wave formed on a taut string fixed at both ends, as shown in Figure 18.6. Except for the nodes, which are always stationary, all points on the string oscillate vertically with the same frequency but with different amplitudes of simple harmonic motion. Figure 18.6 represents snapshots of the standing wave at various times over one half of a period.

In a traveling wave, energy is transferred along with the wave, as we discussed in Chapter 16. We can imagine this transfer to be due to work done by one segment of the string on the next segment. As one segment moves upward, it exerts a force on the next segment, moving it through a displacement—that is, work is done. A particle of the string at a node, however, experiences no displacement. Thus, it cannot do work on the neighboring segment. As a result, no energy is transmitted along the string across a node, and energy does not propagate in a standing wave. For this reason, standing waves are often called **stationary waves.**

The energy of the oscillating string continuously alternates between elastic potential energy, when the string is momentarily stationary (see Fig. 18.6a), and kinetic energy, when the string is horizontal and the particles have their maximum speed (see Fig. 18.6c). At intermediate times (see Fig. 18.6b and d), the string particles have both potential energy and kinetic energy.

Quick Quiz 18.1

A standing wave described by Equation 18.3 is set up on a string. At what points on the string do the particles move the fastest?

EXAMPLE 18.2 Formation of a Standing Wave

Two waves traveling in opposite directions produce a standing wave. The individual wave functions $y = A \sin(kx - \omega t)$ are

$$y_1 = (4.0 \text{ cm}) \sin(3.0x - 2.0t)$$

and

$$y_2 = (4.0 \text{ cm}) \sin(3.0x + 2.0t)$$

where x and y are measured in centimeters. (a) Find the amplitude of the simple harmonic motion of the particle of the medium located at $x = 2.3$ cm.

Solution The standing wave is described by Equation 18.3; in this problem, we have $A = 4.0$ cm, $k = 3.0$ rad/cm, and $\omega = 2.0$ rad/s. Thus,

$$y = (2A \sin kx) \cos \omega t = [(8.0 \text{ cm}) \sin 3.0x] \cos 2.0t$$

Thus, we obtain the amplitude of the simple harmonic motion of the particle at the position $x = 2.3$ cm by evaluating the coefficient of the cosine function at this position:

$$y_{max} = (8.0 \text{ cm}) \sin 3.0x|_{x=2.3}$$

$$= (8.0 \text{ cm}) \sin(6.9 \text{ rad}) = \boxed{4.6 \text{ cm}}$$

(b) Find the positions of the nodes and antinodes.

Solution With $k = 2\pi/\lambda = 3.0$ rad/cm, we see that $\lambda = 2\pi/3$ cm. Therefore, from Equation 18.4 we find that the nodes are located at

$$x = n\frac{\lambda}{2} = \boxed{n\left(\frac{\pi}{3}\right) \text{cm}} \qquad n = 0, 1, 2, 3 \ldots$$

and from Equation 18.5 we find that the antinodes are located at

$$x = n\frac{\lambda}{4} = \boxed{n\left(\frac{\pi}{6}\right) \text{cm}} \qquad n = 1, 3, 5, \ldots$$

(c) What is the amplitude of the simple harmonic motion of a particle located at an antinode?

Solution According to Equation 18.3, the maximum displacement of a particle at an antinode is the amplitude of the standing wave, which is twice the amplitude of the individual traveling waves:

$$y_{max} = 2A = 2(4.0 \text{ cm}) = \boxed{8.0 \text{ cm}}$$

Let us check this result by evaluating the coefficient of our standing-wave function at the positions we found for the antinodes:

$$y_{max} = (8.0 \text{ cm}) \sin 3.0x|_{x=n(\pi/6)}$$

$$= (8.0 \text{ cm}) \sin\left[3.0n\left(\frac{\pi}{6}\right) \text{rad}\right]$$

$$= (8.0 \text{ cm}) \sin\left[n\left(\frac{\pi}{2}\right) \text{rad}\right] = 8.0 \text{ cm}$$

In evaluating this expression, we have used the fact that n is an odd integer; thus, the sine function is equal to unity.

18.3 ### STANDING WAVES IN A STRING FIXED AT BOTH ENDS

Consider a string of length L fixed at both ends, as shown in Figure 18.7. Standing waves are set up in the string by a continuous superposition of waves incident on and reflected from the ends. Note that the ends of the string, because they are fixed and must necessarily have zero displacement, are nodes by definition. The string has a number of natural patterns of oscillation, called **normal modes,** each of which has a characteristic frequency that is easily calculated.

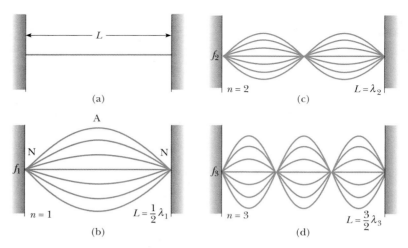

Figure 18.7   (a) A string of length L fixed at both ends. The normal modes of vibration form a harmonic series: (b) the fundamental, or first harmonic; (c) the second harmonic; (d) the third harmonic.

In general, the motion of an oscillating string fixed at both ends is described by the superposition of several normal modes. Exactly which normal modes are present depends on how the oscillation is started. For example, when a guitar string is plucked near its middle, the modes shown in Figure 18.7b and d, as well as other modes not shown, are excited.

In general, we can describe the normal modes of oscillation for the string by imposing the requirements that the ends be nodes and that the nodes and antinodes be separated by one fourth of a wavelength. The first normal mode, shown in Figure 18.7b, has nodes at its ends and one antinode in the middle. This is the longest-wavelength mode, and this is consistent with our requirements. This first normal mode occurs when the wavelength λ_1 is twice the length of the string, that is, $\lambda_1 = 2L$. The next normal mode, of wavelength λ_2 (see Fig. 18.7c), occurs when the wavelength equals the length of the string, that is, $\lambda_2 = L$. The third normal mode (see Fig. 18.7d) corresponds to the case in which $\lambda_3 = 2L/3$. In general, the wavelengths of the various normal modes for a string of length L fixed at both ends are

Wavelengths of normal modes

$$\lambda_n = \frac{2L}{n} \qquad n = 1, 2, 3, \ldots \tag{18.6}$$

where the index n refers to the nth normal mode of oscillation. These are the *possible* modes of oscillation for the string. The *actual* modes that are excited by a given pluck of the string are discussed below.

The natural frequencies associated with these modes are obtained from the relationship $f = v/\lambda$, where the wave speed v is the same for all frequencies. Using Equation 18.6, we find that the natural frequencies f_n of the normal modes are

Frequencies of normal modes as functions of wave speed and length of string

$$f_n = \frac{v}{\lambda_n} = n\frac{v}{2L} \qquad n = 1, 2, 3, \ldots \tag{18.7}$$

Because $v = \sqrt{T/\mu}$ (see Eq. 16.4), where T is the tension in the string and μ is its linear mass density, we can also express the natural frequencies of a taut string as

Frequencies of normal modes as functions of string tension and linear mass density

$$f_n = \frac{n}{2L}\sqrt{\frac{T}{\mu}} \qquad n = 1, 2, 3, \ldots \tag{18.8}$$

Multiflash photographs of standing-wave patterns in a cord driven by a vibrator at its left end. The single-loop pattern represents the first normal mode ($n = 1$). The double-loop pattern represents the second normal mode ($n = 2$), and the triple-loop pattern represents the third normal mode ($n = 3$). *(©1991 Richard Megna/Fundamental Photographs)*

The lowest frequency f_1, which corresponds to $n = 1$, is called either the **fundamental** or the **fundamental frequency** and is given by

$$f_1 = \frac{1}{2L}\sqrt{\frac{T}{\mu}}$$ **(18.9)**

The frequencies of the remaining normal modes are integer multiples of the fundamental frequency. Frequencies of normal modes that exhibit an integer-multiple relationship such as this form a **harmonic series,** and the normal modes are called **harmonics.** The fundamental frequency f_1 is the frequency of the first harmonic; the frequency $f_2 = 2f_1$ is the frequency of the second harmonic; and the frequency $f_n = nf_1$ is the frequency of the nth harmonic. Other oscillating systems, such as a drumhead, exhibit normal modes, but the frequencies are not related as integer multiples of a fundamental. Thus, we do not use the term *harmonic* in association with these types of systems.

In obtaining Equation 18.6, we used a technique based on the separation distance between nodes and antinodes. We can obtain this equation in an alternative manner. Because we require that the string be fixed at $x = 0$ and $x = L$, the wave function $y(x, t)$ given by Equation 18.3 must be zero at these points for all times. That is, the *boundary conditions* require that $y(0, t) = 0$ and that $y(L, t) = 0$ for all values of t. Because the standing wave is described by $y = (2A \sin kx) \cos \omega t$, the first boundary condition, $y(0, t) = 0$, is automatically satisfied because $\sin kx = 0$ at $x = 0$. To meet the second boundary condition, $y(L, t) = 0$, we require that $\sin kL = 0$. This condition is satisfied when the angle kL equals an integer multiple of π rad. Therefore, the allowed values of k are given by[1]

$$k_n L = n\pi \qquad n = 1, 2, 3, \ldots$$ **(18.10)**

Because $k_n = 2\pi/\lambda_n$, we find that

$$\left(\frac{2\pi}{\lambda_n}\right)L = n\pi \qquad \text{or} \qquad \lambda_n = \frac{2L}{n}$$

which is identical to Equation 18.6.

Let us now examine how these various harmonics are created in a string. If we wish to excite just a single harmonic, we need to distort the string in such a way that its distorted shape corresponded to that of the desired harmonic. After being released, the string vibrates at the frequency of that harmonic. This maneuver is difficult to perform, however, and it is not how we excite a string of a musical in-

[1] We exclude $n = 0$ because this value corresponds to the trivial case in which no wave exists ($k = 0$).

strument. If the string is distorted such that its distorted shape is not that of just one harmonic, the resulting vibration includes various harmonics. Such a distortion occurs in musical instruments when the string is plucked (as in a guitar), bowed (as in a cello), or struck (as in a piano). When the string is distorted into a non-sinusoidal shape, only waves that satisfy the boundary conditions can persist on the string. These are the harmonics.

The frequency of a stringed instrument can be varied by changing either the tension or the string's length. For example, the tension in guitar and violin strings is varied by a screw adjustment mechanism or by tuning pegs located on the neck of the instrument. As the tension is increased, the frequency of the normal modes increases in accordance with Equation 18.8. Once the instrument is "tuned," players vary the frequency by moving their fingers along the neck, thereby changing the length of the oscillating portion of the string. As the length is shortened, the frequency increases because, as Equation 18.8 specifies, the normal-mode frequencies are inversely proportional to string length.

EXAMPLE 18.3   Give Me a C Note!

Middle C on a piano has a fundamental frequency of 262 Hz, and the first A above middle C has a fundamental frequency of 440 Hz. (a) Calculate the frequencies of the next two harmonics of the C string.

Solution Knowing that the frequencies of higher harmonics are integer multiples of the fundamental frequency $f_1 = 262$ Hz, we find that

$$f_2 = 2f_1 = \boxed{524 \text{ Hz}}$$

$$f_3 = 3f_1 = \boxed{786 \text{ Hz}}$$

(b) If the A and C strings have the same linear mass density μ and length L, determine the ratio of tensions in the two strings.

Solution Using Equation 18.8 for the two strings vibrating at their fundamental frequencies gives

$$f_{1A} = \frac{1}{2L}\sqrt{\frac{T_A}{\mu}} \quad \text{and} \quad f_{1C} = \frac{1}{2L}\sqrt{\frac{T_C}{\mu}}$$

Setting up the ratio of these frequencies, we find that

$$\frac{f_{1A}}{f_{1C}} = \sqrt{\frac{T_A}{T_C}}$$

$$\frac{T_A}{T_C} = \left(\frac{f_{1A}}{f_{1C}}\right)^2 = \left(\frac{440}{262}\right)^2 = \boxed{2.82}$$

(c) With respect to a real piano, the assumption we made in (b) is only partially true. The string densities are equal, but the length of the A string is only 64 percent of the length of the C string. What is the ratio of their tensions?

Solution Using Equation 18.8 again, we set up the ratio of frequencies:

$$\frac{f_{1A}}{f_{1C}} = \frac{L_C}{L_A}\sqrt{\frac{T_A}{T_C}} = \left(\frac{100}{64}\right)\sqrt{\frac{T_A}{T_C}}$$

$$\frac{T_A}{T_C} = (0.64)^2\left(\frac{440}{262}\right)^2 = \boxed{1.16}$$

EXAMPLE 18.4   Guitar Basics

The high E string on a guitar measures 64.0 cm in length and has a fundamental frequency of 330 Hz. By pressing down on it at the first fret (Fig. 18.8), the string is shortened so that it plays an F note that has a frequency of 350 Hz. How far is the fret from the neck end of the string?

Solution Equation 18.7 relates the string's length to the fundamental frequency. With $n = 1$, we can solve for the speed of the wave on the string,

$$v = \frac{2L}{n}f_n = \frac{2(0.640 \text{ m})}{1}(330 \text{ Hz}) = 422 \text{ m/s}$$

Because we have not adjusted the tuning peg, the tension in the string, and hence the wave speed, remain constant. We can again use Equation 18.7, this time solving for L and substituting the new frequency to find the shortened string length:

$$L = n\frac{v}{2f_n} = (1)\frac{422 \text{ m/s}}{2(350 \text{ Hz})} = 0.603 \text{ m}$$

The difference between this length and the measured length of 64.0 cm is the distance from the fret to the neck end of the string, or **3.70 cm.**

Figure 18.8 Playing an F note on a guitar. *(Charles D. Winters)*

18.4 RESONANCE

We have seen that a system such as a taut string is capable of oscillating in one or more normal modes of oscillation. **If a periodic force is applied to such a system, the amplitude of the resulting motion is greater than normal when the frequency of the applied force is equal to or nearly equal to one of the natural frequencies of the system.** We discussed this phenomenon, known as *resonance,* briefly in Section 13.7. Although a block–spring system or a simple pendulum has only one natural frequency, standing-wave systems can have a whole set of natural frequencies. Because an oscillating system exhibits a large amplitude when driven at any of its natural frequencies, these frequencies are often referred to as **resonance frequencies.**

Figure 18.9 shows the response of an oscillating system to various driving frequencies, where one of the resonance frequencies of the system is denoted by f_0. Note that the amplitude of oscillation of the system is greatest when the frequency of the driving force equals the resonance frequency. The maximum amplitude is limited by friction in the system. If a driving force begins to work on an oscillating system initially at rest, the input energy is used both to increase the amplitude of the oscillation and to overcome the frictional force. Once maximum amplitude is reached, the work done by the driving force is used only to overcome friction.

A system is said to be *weakly* damped when the amount of friction to be overcome is small. Such a system has a large amplitude of motion when driven at one of its resonance frequencies, and the oscillations persist for a long time after the driving force is removed. A system in which considerable friction must be overcome is said to be *strongly damped.* For a given driving force applied at a resonance frequency, the maximum amplitude attained by a strongly damped oscillator is smaller than that attained by a comparable weakly damped oscillator. Once the driving force in a strongly damped oscillator is removed, the amplitude decreases rapidly with time.

Figure 18.9 Graph of the amplitude (response) versus driving frequency for an oscillating system. The amplitude is a maximum at the resonance frequency f_0. Note that the curve is not symmetric.

Examples of Resonance

A playground swing is a pendulum having a natural frequency that depends on its length. Whenever we use a series of regular impulses to push a child in a swing, the swing goes higher if the frequency of the periodic force equals the natural fre-

Figure 18.10 An example of resonance. If pendulum A is set into oscillation, only pendulum C, whose length matches that of A, eventually oscillates with large amplitude, or resonates. The arrows indicate motion perpendicular to the page.

Vibrating
blade

Figure 18.11 Standing waves are set up in a string when one end is connected to a vibrating blade. When the blade vibrates at one of the natural frequencies of the string, large-amplitude standing waves are created.

quency of the swing. We can demonstrate a similar effect by suspending pendulums of different lengths from a horizontal support, as shown in Figure 18.10. If pendulum A is set into oscillation, the other pendulums begin to oscillate as a result of the longitudinal waves transmitted along the beam. However, pendulum C, the length of which is close to the length of A, oscillates with a much greater amplitude than pendulums B and D, the lengths of which are much different from that of pendulum A. Pendulum C moves the way it does because its natural frequency is nearly the same as the driving frequency associated with pendulum A.

Next, consider a taut string fixed at one end and connected at the opposite end to an oscillating blade, as illustrated in Figure 18.11. The fixed end is a node, and the end connected to the blade is very nearly a node because the amplitude of the blade's motion is small compared with that of the string. As the blade oscillates, transverse waves sent down the string are reflected from the fixed end. As we learned in Section 18.3, the string has natural frequencies that are determined by its length, tension, and linear mass density (see Eq. 18.8). When the frequency of the blade equals one of the natural frequencies of the string, standing waves are produced and the string oscillates with a large amplitude. In this resonance case, the wave generated by the oscillating blade is in phase with the reflected wave, and the string absorbs energy from the blade. If the string is driven at a frequency that is not one of its natural frequencies, then the oscillations are of low amplitude and exhibit no stable pattern.

Once the amplitude of the standing-wave oscillations is a maximum, the mechanical energy delivered by the blade and absorbed by the system is lost because of the damping forces caused by friction in the system. If the applied frequency differs from one of the natural frequencies, energy is transferred to the string at first, but later the phase of the wave becomes such that it forces the blade to receive energy from the string, thereby reducing the energy in the string.

Quick Quiz 18.2

Some singers can shatter a wine glass by maintaining a certain frequency of their voice for several seconds. Figure 18.12a shows a side view of a wine glass vibrating because of a sound wave. Sketch the standing-wave pattern in the rim of the glass as seen from above. If an inte-

(a)

(b)

Figure 18.12 (a) Standing-wave pattern in a vibrating wine glass. The glass shatters if the amplitude of vibration becomes too great. *(Courtesy of Professor Thomas D. Rossing, Northern Illinois University)* (b) A wine glass shattered by the amplified sound of a human voice. *(©1992 Ben Rose/The Image Bank)* gral number of waves "fit" around the circumference of the vibrating rim, how many wavelengths fit around the rim in Figure 18.12a?

Quick Quiz 18.3

"Rumble strips" (Fig. 18.13) are sometimes placed across a road to warn drivers that they are approaching a stop sign, or laid along the sides of the road to alert drivers when they are drifting out of their lane. Why are these sets of small bumps so effective at getting a driver's attention?

Figure 18.13   Rumble strips along the side of a highway.   *(Charles D. Winters)*

18.5 STANDING WAVES IN AIR COLUMNS

Standing waves can be set up in a tube of air, such as that in an organ pipe, as the result of interference between longitudinal sound waves traveling in opposite directions. The phase relationship between the incident wave and the wave reflected from one end of the pipe depends on whether that end is open or closed. This relationship is analogous to the phase relationships between incident and reflected transverse waves at the end of a string when the end is either fixed or free to move (see Figs. 16.13 and 16.14).

In a pipe closed at one end, **the closed end is a displacement node because the wall at this end does not allow longitudinal motion of the air molecules.** As a result, at a closed end of a pipe, the reflected sound wave is 180° out of phase with the incident wave. Furthermore, because the pressure wave is 90° out of phase with the displacement wave (see Section 17.2), **the closed end of an air column corresponds to a pressure antinode** (that is, a point of maximum pressure variation).

The open end of an air column is approximately a displacement antinode[2] and a pressure node. We can understand why no pressure variation occurs at an open end by noting that here the end of the air column is open to the atmosphere; thus, the pressure at this end must remain constant at atmospheric pressure.

QuickLab

Snip off pieces at one end of a drinking straw so that the end tapers to a point. Chew on this end to flatten it, and you'll have created a double-reed instrument! Put your lips around the tapered end, press them tightly together, and blow through the straw. When you hear a steady tone, slowly snip off pieces of the straw from the other end. Be careful to maintain a constant pressure with your lips. How does the frequency change as the straw is shortened?

---

[2] Strictly speaking, the open end of an air column is not exactly a displacement antinode. A condensation reaching an open end does not reflect until it passes beyond the end. For a thin-walled tube of circular cross section, this end correction is approximately $0.6R$, where R is the tube's radius. Hence, the effective length of the tube is longer than the true length L. We ignore this end correction in this discussion.

You may wonder how a sound wave can reflect from an open end, since there may not appear to be a change in the medium at this point. It is indeed true that the medium through which the sound wave moves is air both inside and outside the pipe. Remember that sound is a pressure wave, however, and a compression region of the sound wave is constrained by the sides of the pipe as long as the region is inside the pipe. As the compression region exits at the open end of the pipe, the constraint is removed and the compressed air is free to expand into the atmosphere. Thus, there is a change in the *character* of the medium between the inside of the pipe and the outside even though there is no change in the *material* of the medium. This change in character is sufficient to allow some reflection.

The first three normal modes of oscillation of a pipe open at both ends are shown in Figure 18.14a. When air is directed against an edge at the left, longitudinal standing waves are formed, and the pipe resonates at its natural frequencies. All normal modes are excited simultaneously (although not with the same amplitude). Note that both ends are displacement antinodes (approximately). In the first normal mode, the standing wave extends between two adjacent antinodes,

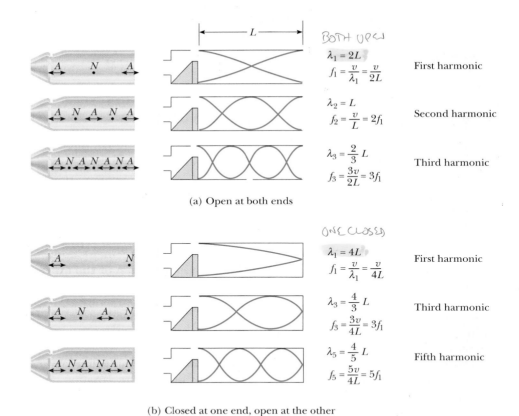

(a) Open at both ends

BOTH OPEN
$$\lambda_1 = 2L$$
$$f_1 = \frac{v}{\lambda_1} = \frac{v}{2L}$$
First harmonic

$$\lambda_2 = L$$
$$f_2 = \frac{v}{L} = 2f_1$$
Second harmonic

$$\lambda_3 = \frac{2}{3}L$$
$$f_3 = \frac{3v}{2L} = 3f_1$$
Third harmonic

ONE CLOSED
$$\lambda_1 = 4L$$
$$f_1 = \frac{v}{\lambda_1} = \frac{v}{4L}$$
First harmonic

$$\lambda_3 = \frac{4}{3}L$$
$$f_3 = \frac{3v}{4L} = 3f_1$$
Third harmonic

$$\lambda_5 = \frac{4}{5}L$$
$$f_5 = \frac{5v}{4L} = 5f_1$$
Fifth harmonic

(b) Closed at one end, open at the other

Figure 18.14 Motion of air molecules in standing longitudinal waves in a pipe, along with schematic representations of the waves. The graphs represent the displacement amplitudes, not the pressure amplitudes. (a) In a pipe open at both ends, the harmonic series created consists of all integer multiples of the fundamental frequency: $f_1, 2f_1, 3f_1, \ldots$. (b) In a pipe closed at one end and open at the other, the harmonic series created consists of only odd-integer multiples of the fundamental frequency: $f_1, 3f_1, 5f_1, \ldots$.

which is a distance of half a wavelength. Thus, the wavelength is twice the length of the pipe, and the fundamental frequency is $f_1 = v/2L$. As Figure 18.14a shows, the frequencies of the higher harmonics are $2f_1$, $3f_1$, ... Thus, we can say that

> in a pipe open at both ends, the natural frequencies of oscillation form a harmonic series that includes all integral multiples of the fundamental frequency.

Because all harmonics are present, and because the fundamental frequency is given by the same expression as that for a string (see Eq. 18.7), we can express the natural frequencies of oscillation as

$$f_n = n\frac{v}{2L} \qquad n = 1, 2, 3 \ldots \qquad \textbf{(18.11)}$$

Natural frequencies of a pipe open at both ends

Despite the similarity between Equations 18.7 and 18.11, we must remember that v in Equation 18.7 is the speed of waves on the string, whereas v in Equation 18.11 is the speed of sound in air.

If a pipe is closed at one end and open at the other, the closed end is a displacement node (see Fig. 18.14b). In this case, the standing wave for the fundamental mode extends from an antinode to the adjacent node, which is one fourth of a wavelength. Hence, the wavelength for the first normal mode is $4L$, and the fundamental frequency is $f_1 = v/4L$. As Figure 18.14b shows, the higher-frequency waves that satisfy our conditions are those that have a node at the closed end and an antinode at the open end; this means that the higher harmonics have frequencies $3f_1$, $5f_1$, . . . :

> In a pipe closed at one end and open at the other, the natural frequencies of oscillation form a harmonic series that includes only odd integer multiples of the fundamental frequency.

We express this result mathematically as

$$f_n = n\frac{v}{4L} \qquad n = 1, 3, 5, \ldots \qquad \textbf{(18.12)}$$

Natural frequencies of a pipe closed at one end and open at the other

It is interesting to investigate what happens to the frequencies of instruments based on air columns and strings during a concert as the temperature rises. The sound emitted by a flute, for example, becomes sharp (increases in frequency) as it warms up because the speed of sound increases in the increasingly warmer air inside the flute (consider Eq. 18.11). The sound produced by a violin becomes flat (decreases in frequency) as the strings expand thermally because the expansion causes their tension to decrease (see Eq. 18.8).

QuickLab

Blow across the top of an empty soda-pop bottle. From a measurement of the height of the bottle, estimate the frequency of the sound you hear. Note that the cross-sectional area of the bottle is not constant; thus, this is not a perfect model of a cylindrical air column.

Quick Quiz 18.4

A pipe open at both ends resonates at a fundamental frequency f_{open}. When one end is covered and the pipe is again made to resonate, the fundamental frequency is f_{closed}. Which of the following expressions describes how these two resonant frequencies compare?
(a) $f_{closed} = f_{open}$ (b) $f_{closed} = \frac{1}{2}f_{open}$ (c) $f_{closed} = 2f_{open}$ (d) $f_{closed} = \frac{3}{2}f_{open}$

EXAMPLE 18.5   Wind in a Culvert

A section of drainage culvert 1.23 m in length makes a howling noise when the wind blows. (a) Determine the frequencies of the first three harmonics of the culvert if it is open at both ends. Take $v = 343$ m/s as the speed of sound in air.

Solution   The frequency of the first harmonic of a pipe open at both ends is

$$f_1 = \frac{v}{2L} = \frac{343 \text{ m/s}}{2(1.23 \text{ m})} = \boxed{139 \text{ Hz}}$$

Because both ends are open, all harmonics are present; thus,

$$f_2 = 2f_1 = \boxed{278 \text{ Hz}} \quad \text{and} \quad f_3 = 3f_1 = \boxed{417 \text{ Hz}.}$$

(b) What are the three lowest natural frequencies of the culvert if it is blocked at one end?

Solution   The fundamental frequency of a pipe closed at one end is

$$f_1 = \frac{v}{4L} = \frac{343 \text{ m/s}}{4(1.23 \text{ m})} = \boxed{69.7 \text{ Hz}}$$

In this case, only odd harmonics are present; hence, the next two harmonics have frequencies $f_3 = 3f_1 = \boxed{209 \text{ Hz}}$ and $f_5 = 5f_1 = \boxed{349 \text{ Hz}.}$

(c) For the culvert open at both ends, how many of the harmonics present fall within the normal human hearing range (20 to 17 000 Hz)?

Solution   Because all harmonics are present, we can express the frequency of the highest harmonic heard as $f_n = nf_1$, where n is the number of harmonics that we can hear. For $f_n = 17\ 000$ Hz, we find that the number of harmonics present in the audible range is

$$n = \frac{17\ 000 \text{ Hz}}{139 \text{ Hz}} = \boxed{122}$$

Only the first few harmonics are of sufficient amplitude to be heard.

EXAMPLE 18.6   Measuring the Frequency of a Tuning Fork

A simple apparatus for demonstrating resonance in an air column is depicted in Figure 18.15. A vertical pipe open at both ends is partially submerged in water, and a tuning fork vibrating at an unknown frequency is placed near the top of the pipe. The length L of the air column can be adjusted by moving the pipe vertically. The sound waves generated by the fork are reinforced when L corresponds to one of the resonance frequencies of the pipe.

For a certain tube, the smallest value of L for which a peak occurs in the sound intensity is 9.00 cm. What are (a) the frequency of the tuning fork and (b) the value of L for the next two resonance frequencies?

Solution   (a) Although the pipe is open at its lower end to allow the water to enter, the water's surface acts like a wall at one end. Therefore, this setup represents a pipe closed at one end, and so the fundamental frequency is $f_1 = v/4L$. Taking $v = 343$ m/s for the speed of sound in air and $L = 0.090\ 0$ m, we obtain

$$f_1 = \frac{v}{4L} = \frac{343 \text{ m/s}}{4(0.090\ 0 \text{ m})} = \boxed{953 \text{ Hz}}$$

Because the tuning fork causes the air column to resonate at this frequency, this must be the frequency of the tuning fork.

(b) Because the pipe is closed at one end, we know from Figure 18.14b that the wavelength of the fundamental mode is $\lambda = 4L = 4(0.090\ 0 \text{ m}) = 0.360$ m. Because the frequency

of the tuning fork is constant, the next two normal modes (see Fig. 18.15b) correspond to lengths of $L = 3\lambda/4 = \boxed{0.270 \text{ m}}$ and $L = 5\lambda/4 = \boxed{0.450 \text{ m}.}$

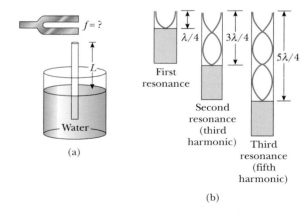

Figure 18.15   (a) Apparatus for demonstrating the resonance of sound waves in a tube closed at one end. The length L of the air column is varied by moving the tube vertically while it is partially submerged in water. (b) The first three normal modes of the system shown in part (a).

Optional Section

18.6 ► STANDING WAVES IN RODS AND PLATES

Standing waves can also be set up in rods and plates. A rod clamped in the middle and stroked at one end oscillates, as depicted in Figure 18.16a. The oscillations of the particles of the rod are longitudinal, and so the broken lines in Figure 18.16 represent *longitudinal* displacements of various parts of the rod. For clarity, we have drawn them in the transverse direction, just as we did for air columns. The midpoint is a displacement node because it is fixed by the clamp, whereas the ends are displacement antinodes because they are free to oscillate. The oscillations in this setup are analogous to those in a pipe open at both ends. The broken lines in Figure 18.16a represent the first normal mode, for which the wavelength is $2L$ and the frequency is $f = v/2L$, where v is the speed of longitudinal waves in the rod. Other normal modes may be excited by clamping the rod at different points. For example, the second normal mode (Fig. 18.16b) is excited by clamping the rod a distance $L/4$ away from one end.

Two-dimensional oscillations can be set up in a flexible membrane stretched over a circular hoop, such as that in a drumhead. As the membrane is struck at some point, wave pulses that arrive at the fixed boundary are reflected many times. The resulting sound is not harmonic because the oscillating drumhead and the drum's hollow interior together produce a set of standing waves having frequencies that are *not* related by integer multiples. Without this relationship, the sound may be more correctly described as *noise* than as music. This is in contrast to the situation in wind and stringed instruments, which produce sounds that we describe as musical.

Some possible normal modes of oscillation for a two-dimensional circular membrane are shown in Figure 18.17. The lowest normal mode, which has a frequency f_1, contains only one nodal curve; this curve runs around the outer edge of the membrane. The other possible normal modes show additional nodal curves that are circles and straight lines across the diameter of the membrane.

The sound from a tuning fork is produced by the vibrations of each of its prongs. *(Sam Dudgeon/Holt, Rinehart and Winston)*

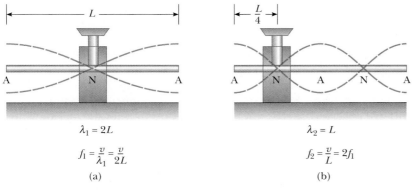

$$\lambda_1 = 2L$$

$$f_1 = \frac{v}{\lambda_1} = \frac{v}{2L}$$

(a)

$$\lambda_2 = L$$

$$f_2 = \frac{v}{L} = 2f_1$$

(b)

Figure 18.16 Normal-mode longitudinal vibrations of a rod of length L (a) clamped at the middle to produce the first normal mode and (b) clamped at a distance $L/4$ from one end to produce the second normal mode. Note that the dashed lines represent amplitudes parallel to the rod (longitudinal waves).

Wind chimes are usually designed so that the waves emanating from the vibrating rods blend into a harmonious sound. *(Joseph L. Fontenot/ Visuals Unlimited)*

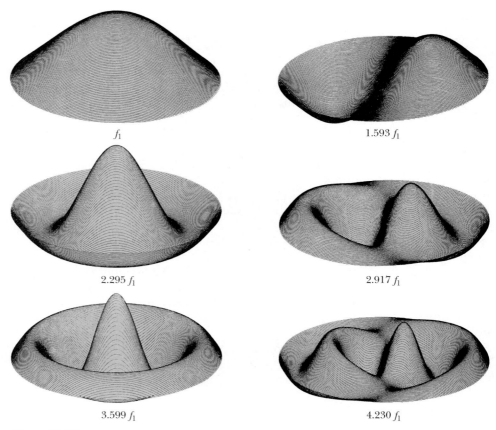

Figure 18.17   Representation of some of the normal modes possible in a circular membrane fixed at its perimeter. The frequencies of oscillation do not form a harmonic series.   *(From M. L. Warren,* Introductory Physics, *New York, W. H. Freeman & Co., Publishers, 1979, with permission.)*

18.7 ▶ BEATS: INTERFERENCE IN TIME

The interference phenomena with which we have been dealing so far involve the superposition of two or more waves having the same frequency. Because the resultant wave depends on the coordinates of the disturbed medium, we refer to the phenomenon as *spatial interference*. Standing waves in strings and pipes are common examples of spatial interference.

We now consider another type of interference, one that results from the superposition of two waves having slightly *different* frequencies. In this case, when the two waves are observed at the point of superposition, they are periodically in and out of phase. That is, there is a *temporal* (time) alternation between constructive and destructive interference. Thus, we refer to this phenomenon as *interference in time* or *temporal interference*. For example, if two tuning forks of slightly different frequencies are struck, one hears a sound of periodically varying intensity. This phenomenon is called **beating:**

Definition of beating

> Beating is the periodic variation in intensity at a given point due to the superposition of two waves having slightly different frequencies.

The number of intensity maxima one hears per second, or the *beat frequency*, equals the difference in frequency between the two sources, as we shall show below. The maximum beat frequency that the human ear can detect is about 20 beats/s. When the beat frequency exceeds this value, the beats blend indistinguishably with the compound sounds producing them.

A piano tuner can use beats to tune a stringed instrument by "beating" a note against a reference tone of known frequency. The tuner can then adjust the string tension until the frequency of the sound it emits equals the frequency of the reference tone. The tuner does this by tightening or loosening the string until the beats produced by it and the reference source become too infrequent to notice.

Consider two sound waves of equal amplitude traveling through a medium with slightly different frequencies f_1 and f_2. We use equations similar to Equation 16.11 to represent the wave functions for these two waves at a point that we choose as $x = 0$:

$$y_1 = A \cos \omega_1 t = A \cos 2\pi f_1 t$$

$$y_2 = A \cos \omega_2 t = A \cos 2\pi f_2 t$$

Using the superposition principle, we find that the resultant wave function at this point is

$$y = y_1 + y_2 = A(\cos 2\pi f_1 t + \cos 2\pi f_2 t)$$

The trigonometric identity

$$\cos a + \cos b = 2 \cos\left(\frac{a - b}{2}\right) \cos\left(\frac{a + b}{2}\right)$$

allows us to write this expression in the form

$$y = \left[2 A \cos 2\pi \left(\frac{f_1 - f_2}{2}\right) t \right] \cos 2\pi \left(\frac{f_1 + f_2}{2}\right) t \qquad \textbf{(18.13)}$$

Resultant of two waves of different frequencies but equal amplitude

Graphs of the individual waves and the resultant wave are shown in Figure 18.18. From the factors in Equation 18.13, we see that the resultant sound for a listener standing at any given point has an effective frequency equal to the average frequency $(f_1 + f_2)/2$ and an amplitude given by the expression in the square

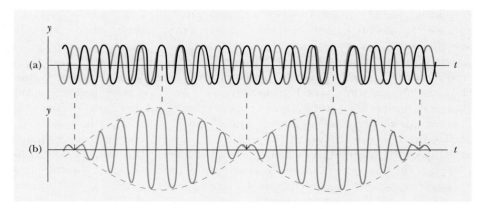

Figure 18.18 Beats are formed by the combination of two waves of slightly different frequencies. (a) The individual waves. (b) The combined wave has an amplitude (broken line) that oscillates in time.

brackets:

$$A_{\text{resultant}} = 2A \cos 2\pi \left(\frac{f_1 - f_2}{2} \right) t \tag{18.14}$$

That is, the **amplitude and therefore the intensity of the resultant sound vary in time.** The broken blue line in Figure 18.18b is a graphical representation of Equation 18.14 and is a sine wave varying with frequency $(f_1 - f_2)/2$.

Note that a maximum in the amplitude of the resultant sound wave is detected whenever

$$\cos 2\pi \left(\frac{f_1 - f_2}{2} \right) t = \pm 1$$

This means there are *two* maxima in each period of the resultant wave. Because the amplitude varies with frequency as $(f_1 - f_2)/2$, the number of beats per second, or the beat frequency f_b, is twice this value. That is,

Beat frequency

$$f_b = |f_1 - f_2| \tag{18.15}$$

For instance, if one tuning fork vibrates at 438 Hz and a second one vibrates at 442 Hz, the resultant sound wave of the combination has a frequency of 440 Hz (the musical note A) and a beat frequency of 4 Hz. A listener would hear a 440-Hz sound wave go through an intensity maximum four times every second.

Optional Section

18.8   NON-SINUSOIDAL WAVE PATTERNS

The sound-wave patterns produced by the majority of musical instruments are non-sinusoidal. Characteristic patterns produced by a tuning fork, a flute, and a clarinet, each playing the same note, are shown in Figure 18.19. Each instrument has its own characteristic pattern. Note, however, that despite the differences in the patterns, each pattern is periodic. This point is important for our analysis of these waves, which we now discuss.

We can distinguish the sounds coming from a trumpet and a saxophone even when they are both playing the same note. On the other hand, we may have difficulty distinguishing a note played on a clarinet from the same note played on an oboe. We can use the pattern of the sound waves from various sources to explain these effects.

The wave patterns produced by a musical instrument are the result of the superposition of various harmonics. This superposition results in the corresponding richness of musical tones. The human perceptive response associated with various mixtures of harmonics is the *quality* or *timbre* of the sound. For instance, the sound of the trumpet is perceived to have a "brassy" quality (that is, we have learned to associate the adjective *brassy* with that sound); this quality enables us to distinguish the sound of the trumpet from that of the saxophone, whose quality is perceived as "reedy." The clarinet and oboe, however, are both straight air columns excited by reeds; because of this similarity, it is more difficult for the ear to distinguish them on the basis of their sound quality.

The problem of analyzing non-sinusoidal wave patterns appears at first sight to be a formidable task. However, if the wave pattern is periodic, it can be represented as closely as desired by the combination of a sufficiently large number of si-

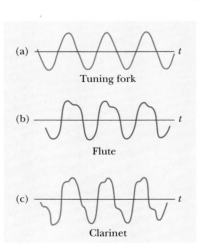

Figure 18.19 Sound wave patterns produced by (a) a tuning fork, (b) a flute, and (c) a clarinet, each at approximately the same frequency. *(Adapted from C. A. Culver, Musical Acoustics, 4th ed., New York, McGraw-Hill Book Company, 1956, p. 128.)*

Figure 18.20 Harmonics of the wave patterns shown in Figure 18.19. Note the variations in intensity of the various harmonics. *(Adapted from C. A. Culver, Musical Acoustics, 4th ed., New York, McGraw-Hill Book Company, 1956.)*

nusoidal waves that form a harmonic series. In fact, we can represent any periodic function as a series of sine and cosine terms by using a mathematical technique based on **Fourier's theorem**.[3] The corresponding sum of terms that represents the periodic wave pattern is called a **Fourier series.**

Let $y(t)$ be any function that is periodic in time with period T, such that $y(t + T) = y(t)$. Fourier's theorem states that this function can be written as

$$y(t) = \sum_n (A_n \sin 2\pi f_n t + B_n \cos 2\pi f_n t) \tag{18.16}$$

Fourier's theorem

where the lowest frequency is $f_1 = 1/T$. The higher frequencies are integer multiples of the fundamental, $f_n = nf_1$, and the coefficients A_n and B_n represent the amplitudes of the various waves. Figure 18.20 represents a harmonic analysis of the wave patterns shown in Figure 18.19. Note that a struck tuning fork produces only one harmonic (the first), whereas the flute and clarinet produce the first and many higher ones.

Note the variation in relative intensity of the various harmonics for the flute and the clarinet. In general, any musical sound consists of a fundamental frequency f plus other frequencies that are integer multiples of f, all having different intensities.

We have discussed the *analysis* of a wave pattern using Fourier's theorem. The analysis involves determining the coefficients of the harmonics in Equation 18.16 from a knowledge of the wave pattern. The reverse process, called *Fourier synthesis,* can also be performed. In this process, the various harmonics are added together to form a resultant wave pattern. As an example of Fourier synthesis, consider the building of a square wave, as shown in Figure 18.21. The symmetry of the square wave results in only odd multiples of the fundamental frequency combining in its synthesis. In Figure 18.21a, the orange curve shows the combination of f and $3f$. In Figure 18.21b, we have added $5f$ to the combination and obtained the green curve. Notice how the general shape of the square wave is approximated, even though the upper and lower portions are not flat as they should be.

[3] Developed by Jean Baptiste Joseph Fourier (1786–1830).

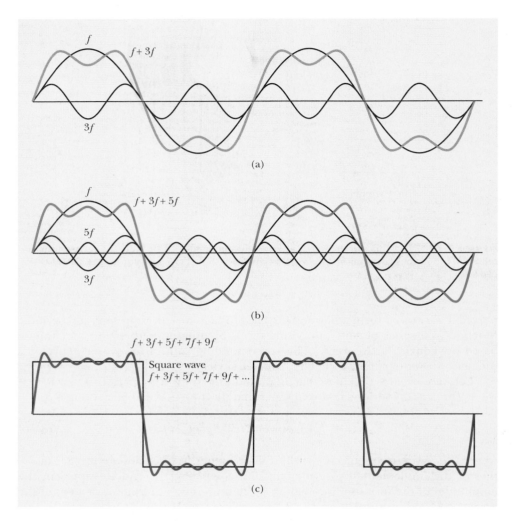

(a)

(b)

(c)

Figure 18.21   Fourier synthesis of a square wave, which is represented by the sum of odd multiples of the first harmonic, which has frequency f. (a) Waves of frequency f and $3f$ are added. (b) One more odd harmonic of frequency $5f$ is added. (c) The synthesis curve approaches the square wave when odd frequencies up to $9f$ are added.

This synthesizer can produce the characteristic sounds of different instruments by properly combining frequencies from electronic oscillators. *(Courtesy of Casio, Inc.)*

Figure 18.21c shows the result of adding odd frequencies up to $9f$. This approximation to the square wave (purple curve) is better than that in parts a and b. To approximate the square wave as closely as possible, we would need to add all odd multiples of the fundamental frequency, up to infinite frequency.

Using modern technology, we can generate musical sounds electronically by mixing different amplitudes of any number of harmonics. These widely used electronic music synthesizers are capable of producing an infinite variety of musical tones.

SUMMARY

When two traveling waves having equal amplitudes and frequencies superimpose, the resultant wave has an amplitude that depends on the phase angle ϕ between the two waves. **Constructive interference** occurs when the two waves are in phase, corresponding to $\phi = 0, 2\pi, 4\pi, \ldots$ rad. **Destructive interference** occurs when the two waves are 180° out of phase, corresponding to $\phi = \pi, 3\pi, 5\pi, \ldots$ rad. Given two wave functions, you should be able to determine which if either of these two situations applies.

Standing waves are formed from the superposition of two sinusoidal waves having the same frequency, amplitude, and wavelength but traveling in opposite directions. The resultant standing wave is described by the wave function

$$y = (2A \sin kx) \cos \omega t \qquad \textbf{(18.3)}$$

Hence, the amplitude of the standing wave is $2A$, and the amplitude of the simple harmonic motion of any particle of the medium varies according to its position as $2A \sin kx$. The points of zero amplitude (called **nodes**) occur at $x = n\lambda/2$ ($n = 0, 1, 2, 3, \ldots$). The maximum amplitude points (called **antinodes**) occur at $x = n\lambda/4$ ($n = 1, 3, 5, \ldots$). Adjacent antinodes are separated by a distance $\lambda/2$. Adjacent nodes also are separated by a distance $\lambda/2$. You should be able to sketch the standing-wave pattern resulting from the superposition of two traveling waves.

The natural frequencies of vibration of a taut string of length L and fixed at both ends are

$$f_n = \frac{n}{2L}\sqrt{\frac{T}{\mu}} \qquad n = 1, 2, 3, \ldots \qquad \textbf{(18.8)}$$

where T is the tension in the string and μ is its linear mass density. The natural frequencies of vibration $f_1, 2f_1, 3f_1, \ldots$ form a **harmonic series.**

An oscillating system is in **resonance** with some driving force whenever the frequency of the driving force matches one of the natural frequencies of the system. When the system is resonating, it responds by oscillating with a relatively large amplitude.

Standing waves can be produced in a column of air inside a pipe. If the pipe is open at both ends, all harmonics are present and the natural frequencies of oscillation are

$$f_n = n\frac{v}{2L} \qquad n = 1, 2, 3, \ldots \qquad \textbf{(18.11)}$$

If the pipe is open at one end and closed at the other, only the odd harmonics are present, and the natural frequencies of oscillation are

$$f_n = n\frac{v}{4L} \qquad n = 1, 3, 5, \ldots \qquad \textbf{(18.12)}$$

The phenomenon of **beating** is the periodic variation in intensity at a given point due to the superposition of two waves having slightly different frequencies.

QUESTIONS

1. For certain positions of the movable section shown in Figure 18.2, no sound is detected at the receiver—a situation corresponding to destructive interference. This suggests that perhaps energy is somehow lost! What happens to the energy transmitted by the speaker?

2. Does the phenomenon of wave interference apply only to sinusoidal waves?

3. When two waves interfere constructively or destructively, is there any gain or loss in energy? Explain.

4. A standing wave is set up on a string, as shown in Figure 18.6. Explain why no energy is transmitted along the string.

5. What is common to *all* points (other than the nodes) on a string supporting a standing wave?

6. What limits the amplitude of motion of a real vibrating system that is driven at one of its resonant frequencies?

7. In Balboa Park in San Diego, CA, there is a huge outdoor organ. Does the fundamental frequency of a particular

pipe of this organ change on hot and cold days? How about on days with high and low atmospheric pressure?

8. Explain why your voice seems to sound better than usual when you sing in the shower.

9. What is the purpose of the slide on a trombone or of the valves on a trumpet?

10. Explain why all harmonics are present in an organ pipe open at both ends, but only the odd harmonics are present in a pipe closed at one end.

11. Explain how a musical instrument such as a piano may be tuned by using the phenomenon of beats.

12. An airplane mechanic notices that the sound from a twin-engine aircraft rapidly varies in loudness when both engines are running. What could be causing this variation from loudness to softness?

13. Why does a vibrating guitar string sound louder when placed on the instrument than it would if it were allowed to vibrate in the air while off the instrument?

14. When the base of a vibrating tuning fork is placed against a chalkboard, the sound that it emits becomes louder. This is due to the fact that the vibrations of the tuning fork are transmitted to the chalkboard. Because it has a larger area than that of the tuning fork, the vibrating

chalkboard sets a larger number of air molecules into vibration. Thus, the chalkboard is a better radiator of sound than the tuning fork. How does this affect the length of time during which the fork vibrates? Does this agree with the principle of conservation of energy?

15. To keep animals away from their cars, some people mount short thin pipes on the front bumpers. The pipes produce a high-frequency wail when the cars are moving. How do they create this sound?

16. Guitarists sometimes play a "harmonic" by lightly touching a string at the exact center and plucking the string. The result is a clear note one octave higher than the fundamental frequency of the string, even though the string is not pressed to the fingerboard. Why does this happen?

17. If you wet your fingers and lightly run them around the rim of a fine wine glass, a high-frequency sound is heard. Why? How could you produce various musical notes with a set of wine glasses, each of which contains a different amount of water?

18. Despite a reasonably steady hand, one often spills coffee when carrying a cup of it from one place to another. Discuss resonance as a possible cause of this difficulty, and devise a means for solving the problem.

PROBLEMS

1, 2, 3 = straightforward, intermediate, challenging   ☐ = full solution available in the *Student Solutions Manual and Study Guide*
WEB = solution posted at **http://www.saunderscollege.com/physics/**   💻 = Computer useful in solving problem   📓 = Interactive Physics
☐ = paired numerical/symbolic problems

Section 18.1   Superposition and Interference of Sinusoidal Waves

WEB **1.** Two sinusoidal waves are described by the equations

$$y_1 = (5.00 \text{ m}) \sin[\pi(4.00x - 1\ 200t)]$$

and

$$y_2 = (5.00 \text{ m}) \sin[\pi(4.00x - 1\ 200t - 0.250)]$$

where x, y_1, and y_2 are in meters and t is in seconds.
(a) What is the amplitude of the resultant wave?
(b) What is the frequency of the resultant wave?

2. A sinusoidal wave is described by the equation

$$y_1 = (0.080\ 0 \text{ m}) \sin[2\pi(0.100x - 80.0t)]$$

where y_1 and x are in meters and t is in seconds. Write an expression for a wave that has the same frequency, amplitude, and wavelength as y_1 but which, when added to y_1, gives a resultant with an amplitude of $8\sqrt{3}$ cm.

3. Two waves are traveling in the same direction along a stretched string. The waves are 90.0° out of phase. Each wave has an amplitude of 4.00 cm. Find the amplitude of the resultant wave.

4. Two identical sinusoidal waves with wavelengths of 3.00 m travel in the same direction at a speed of 2.00 m/s. The second wave originates from the same

point as the first, but at a later time. Determine the minimum possible time interval between the starting moments of the two waves if the amplitude of the resultant wave is the same as that of each of the two initial waves.

5. A tuning fork generates sound waves with a frequency of 246 Hz. The waves travel in opposite directions along a hallway, are reflected by walls, and return. The hallway is 47.0 m in length, and the tuning fork is located 14.0 m from one end. What is the phase difference between the reflected waves when they meet? The speed of sound in air is 343 m/s.

6. Two identical speakers 10.0 m apart are driven by the same oscillator with a frequency of $f = 21.5$ Hz (Fig. P18.6). (a) Explain why a receiver at point A records a minimum in sound intensity from the two speakers. (b) If the receiver is moved in the plane of the speakers, what path should it take so that the intensity remains at a minimum? That is, determine the relationship between x and y (the coordinates of the receiver) that causes the receiver to record a minimum in sound intensity. Take the speed of sound to be 343 m/s.

7. Two speakers are driven by the same oscillator with frequency of 200 Hz. They are located 4.00 m apart on a

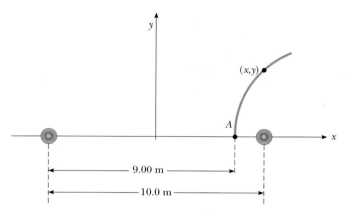

Figure P18.6

vertical pole. A man walks straight toward the lower speaker in a direction perpendicular to the pole, as shown in Figure P18.7. (a) How many times will he hear a minimum in sound intensity, and (b) how far is he from the pole at these moments? Take the speed of sound to be 330 m/s, and ignore any sound reflections coming off the ground.

8. Two speakers are driven by the same oscillator of frequency f. They are located a distance d from each other on a vertical pole. A man walks straight toward the lower speaker in a direction perpendicular to the pole, as shown in Figure P18.7. (a) How many times will he hear a minimum in sound intensity, and (b) how far is he from the pole at these moments? Take the speed of sound to be v, and ignore any sound reflections coming off the ground.

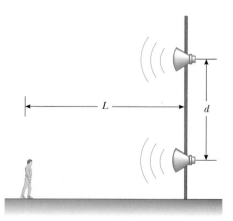

Figure P18.7   Problems 7 and 8.

Section 18.2   Standing Waves

9. Two sinusoidal waves traveling in opposite directions interfere to produce a standing wave described by the equation

$$y = (1.50 \text{ m}) \sin(0.400x) \cos(200t)$$

where x is in meters and t is in seconds. Determine the wavelength, frequency, and speed of the interfering waves.

10. Two waves in a long string are described by the equations

$$y_1 = (0.015\ 0 \text{ m}) \cos\left(\frac{x}{2} - 40t\right)$$

and

$$y_2 = (0.015\ 0 \text{ m}) \cos\left(\frac{x}{2} + 40t\right)$$

where y_1, y_2, and x are in meters and t is in seconds. (a) Determine the positions of the nodes of the resulting standing wave. (b) What is the maximum displacement at the position $x = 0.400$ m?

WEB **11.** Two speakers are driven by a common oscillator at 800 Hz and face each other at a distance of 1.25 m. Locate the points along a line joining the two speakers where relative minima would be expected. (Use $v = 343$ m/s.)

12. Two waves that set up a standing wave in a long string are given by the expressions

$$y_1 = A \sin(kx - \omega t + \phi)$$

and

$$y_2 = A \sin(kx + \omega t)$$

Show (a) that the addition of the arbitrary phase angle changes only the position of the nodes, and (b) that the distance between the nodes remains constant in time.

13. Two sinusoidal waves combining in a medium are described by the equations

$$y_1 = (3.0 \text{ cm}) \sin \pi(x + 0.60t)$$

and

$$y_2 = (3.0 \text{ cm}) \sin \pi(x - 0.60t)$$

where x is in centimeters and t is in seconds. Determine the *maximum* displacement of the medium at (a) $x = 0.250$ cm, (b) $x = 0.500$ cm, and (c) $x = 1.50$ cm. (d) Find the three smallest values of x corresponding to antinodes.

14. A standing wave is formed by the interference of two traveling waves, each of which has an amplitude $A = \pi$ cm, angular wave number $k = (\pi/2)$ cm^{-1}, and angular frequency $\omega = 10\pi$ rad/s. (a) Calculate the distance between the first two antinodes. (b) What is the amplitude of the standing wave at $x = 0.250$ cm?

15. Verify by direct substitution that the wave function for a standing wave given in Equation 18.3, $y = 2A \sin kx \cos \omega t$, is a solution of the general linear

wave equation, Equation 16.26:

$$\frac{\partial^2 y}{\partial x^2} = \frac{1}{v^2}\frac{\partial^2 y}{\partial t^2}$$

Section 18.3   Standing Waves in a String Fixed at Both Ends

16. A 2.00-m-long wire having a mass of 0.100 kg is fixed at both ends. The tension in the wire is maintained at 20.0 N. What are the frequencies of the first three allowed modes of vibration? If a node is observed at a point 0.400 m from one end, in what mode and with what frequency is it vibrating?

17. Find the fundamental frequency and the next three frequencies that could cause a standing-wave pattern on a string that is 30.0 m long, has a mass per length of 9.00×10^{-3} kg/m, and is stretched to a tension of 20.0 N.

18. A standing wave is established in a 120-cm-long string fixed at both ends. The string vibrates in four segments when driven at 120 Hz. (a) Determine the wavelength. (b) What is the fundamental frequency of the string?

19. A cello A-string vibrates in its first normal mode with a frequency of 220 vibrations/s. The vibrating segment is 70.0 cm long and has a mass of 1.20 g. (a) Find the tension in the string. (b) Determine the frequency of vibration when the string vibrates in three segments.

20. A string of length L, mass per unit length μ, and tension T is vibrating at its fundamental frequency. Describe the effect that each of the following conditions has on the fundamental frequency: (a) The length of the string is doubled, but all other factors are held constant. (b) The mass per unit length is doubled, but all other factors are held constant. (c) The tension is doubled, but all other factors are held constant.

21. A 60.0-cm guitar string under a tension of 50.0 N has a mass per unit length of 0.100 g/cm. What is the highest resonance frequency of the string that can be heard by a person able to hear frequencies of up to 20 000 Hz?

22. A stretched wire vibrates in its first normal mode at a frequency of 400 Hz. What would be the fundamental frequency if the wire were half as long, its diameter were doubled, and its tension were increased four-fold?

23. A violin string has a length of 0.350 m and is tuned to concert G, with $f_G = 392$ Hz. Where must the violinist place her finger to play concert A, with $f_A = 440$ Hz? If this position is to remain correct to one-half the width of a finger (that is, to within 0.600 cm), what is the maximum allowable percentage change in the string's tension?

24. **Review Problem.** A sphere of mass M is supported by a string that passes over a light horizontal rod of length L (Fig. P18.24). Given that the angle is θ and that the fundamental frequency of standing waves in the section of the string above the horizontal rod is f, determine the mass of this section of the string.

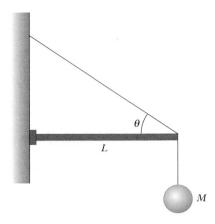

Figure P18.24

25. In the arrangement shown in Figure P18.25, a mass can be hung from a string (with a linear mass density of $\mu = 0.002\ 00$ kg/m) that passes over a light pulley. The string is connected to a vibrator (of constant frequency f), and the length of the string between point P and the pulley is $L = 2.00$ m. When the mass m is either 16.0 kg or 25.0 kg, standing waves are observed; however, no standing waves are observed with any mass between these values. (a) What is the frequency of the vibrator? (*Hint:* The greater the tension in the string, the smaller the number of nodes in the standing wave.) (b) What is the largest mass for which standing waves could be observed?

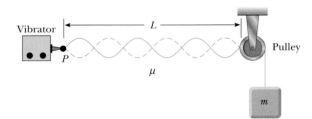

Figure P18.25

26. On a guitar, the fret closest to the bridge is a distance of 21.4 cm from it. The top string, pressed down at this last fret, produces the highest frequency that can be played on the guitar, 2 349 Hz. The next lower note has a frequency of 2 217 Hz. How far away from the last fret should the next fret be?

Section 18.4   Resonance

27. The chains suspending a child's swing are 2.00 m long. At what frequency should a big brother push to make the child swing with greatest amplitude?

28. Standing-wave vibrations are set up in a crystal goblet with four nodes and four antinodes equally spaced

around the 20.0-cm circumference of its rim. If transverse waves move around the glass at 900 m/s, an opera singer would have to produce a high harmonic with what frequency to shatter the glass with a resonant vibration?

29. An earthquake can produce a *seiche* (pronounced "saysh") in a lake, in which the water sloshes back and forth from end to end with a remarkably large amplitude and long period. Consider a seiche produced in a rectangular farm pond, as diagrammed in the cross-sectional view of Figure P18.29 (figure not drawn to scale). Suppose that the pond is 9.15 m long and of uniform depth. You measure that a wave pulse produced at one end reaches the other end in 2.50 s. (a) What is the wave speed? (b) To produce the seiche, you suggest that several people stand on the bank at one end and paddle together with snow shovels, moving them in simple harmonic motion. What must be the frequency of this motion?

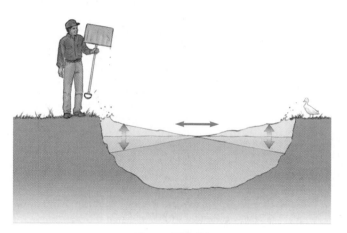

Figure P18.29

30. The Bay of Fundy, Nova Scotia, has the highest tides in the world. Assume that in mid-ocean and at the mouth of the bay, the Moon's gravity gradient and the Earth's rotation make the water surface oscillate with an amplitude of a few centimeters and a period of 12 h 24 min. At the head of the bay, the amplitude is several meters. Argue for or against the proposition that the tide is amplified by standing-wave resonance. Suppose that the bay has a length of 210 km and a depth everywhere of 36.1 m. The speed of long-wavelength water waves is given by $\sqrt{gd}$, where d is the water's depth.

Section 18.5 Standing Waves in Air Columns

Note: In this section, assume that the speed of sound in air is 343 m/s at 20°C and is described by the equation

$$v = (331 \text{ m/s})\sqrt{1 + \frac{T_C}{273°}}$$

at any Celsius temperature T_C.

31. Calculate the length of a pipe that has a fundamental frequency of 240 Hz if the pipe is (a) closed at one end and (b) open at both ends.

32. A glass tube (open at both ends) of length L is positioned near an audio speaker of frequency $f = 0.680$ kHz. For what values of L will the tube resonate with the speaker?

33. The overall length of a piccolo is 32.0 cm. The resonating air column vibrates as a pipe open at both ends. (a) Find the frequency of the lowest note that a piccolo can play, assuming that the speed of sound in air is 340 m/s. (b) Opening holes in the side effectively shortens the length of the resonant column. If the highest note that a piccolo can sound is 4 000 Hz, find the distance between adjacent antinodes for this mode of vibration.

34. The fundamental frequency of an open organ pipe corresponds to middle C (261.6 Hz on the chromatic musical scale). The third resonance of a closed organ pipe has the same frequency. What are the lengths of the two pipes?

35. Estimate the length of your ear canal, from its opening at the external ear to the eardrum. (Do not stick anything into your ear!) If you regard the canal as a tube that is open at one end and closed at the other, at approximately what fundamental frequency would you expect your hearing to be most sensitive? Explain why you can hear especially soft sounds just around this frequency.

36. An open pipe 0.400 m in length is placed vertically in a cylindrical bucket and nearly touches the bottom of the bucket, which has an area of 0.100 m². Water is slowly poured into the bucket until a sounding tuning fork of frequency 440 Hz, held over the pipe, produces resonance. Find the mass of water in the bucket at this moment.

37. A shower stall measures 86.0 cm × 86.0 cm × 210 cm. If you were singing in this shower, which frequencies would sound the richest (because of resonance)? Assume that the stall acts as a pipe closed at both ends, with nodes at opposite sides. Assume that the voices of various singers range from 130 Hz to 2 000 Hz. Let the speed of sound in the hot shower stall be 355 m/s.

38. When a metal pipe is cut into two pieces, the lowest resonance frequency for air in one piece is 256 Hz and that for the other is 440 Hz. (a) What resonant frequency would have been produced by the original length of pipe? (b) How long was the original pipe?

39. As shown in Figure P18.39, water is pumped into a long vertical cylinder at a rate of 18.0 cm³/s. The radius of the cylinder is 4.00 cm, and at the open top of the cylinder is a tuning fork vibrating with a frequency of 200 Hz. As the water rises, how much time elapses between successive resonances?

40. As shown in Figure P18.39, water is pumped into a long vertical cylinder at a volume flow rate R. The radius of

200 Hz

18.0 cm³/s

Figure P18.39 Problems 39 and 40.

the cylinder is r, and at the open top of the cylinder is a tuning fork vibrating with a frequency f. As the water rises, how much time elapses between successive resonances?

41. A tuning fork with a frequency of 512 Hz is placed near the top of the tube shown in Figure 18.15a. The water level is lowered so that the length L slowly increases from an initial value of 20.0 cm. Determine the next two values of L that correspond to resonant modes.

42. A student uses an audio oscillator of adjustable frequency to measure the depth of a water well. Two successive resonances are heard at 51.5 Hz and 60.0 Hz. How deep is the well?

43. A glass tube is open at one end and closed at the other by a movable piston. The tube is filled with air warmer than that at room temperature, and a 384-Hz tuning fork is held at the open end. Resonance is heard when the piston is 22.8 cm from the open end and again when it is 68.3 cm from the open end. (a) What speed of sound is implied by these data? (b) How far from the open end will the piston be when the next resonance is heard?

44. The longest pipe on an organ that has pedal stops is often 4.88 m. What is the fundamental frequency (at 0.00°C) if the nondriven end of the pipe is (a) closed and (b) open? (c) What are the frequencies at 20.0°C?

45. With a particular fingering, a flute sounds a note with a frequency of 880 Hz at 20.0°C. The flute is open at both ends. (a) Find the length of the air column. (b) Find the frequency it produces during the half-time performance at a late-season football game, when the ambient temperature is −5.00°C.

(Optional)
Section 18.6   Standing Waves in Rods and Plates

46. An aluminum rod is clamped one quarter of the way along its length and set into longitudinal vibration by a variable-frequency driving source. The lowest frequency that produces resonance is 4 400 Hz. The speed of sound in aluminum is 5 100 m/s. Determine the length of the rod.

47. An aluminum rod 1.60 m in length is held at its center. It is stroked with a rosin-coated cloth to set up a longitudinal vibration. (a) What is the fundamental frequency of the waves established in the rod? (b) What harmonics are set up in the rod held in this manner? (c) What would be the fundamental frequency if the rod were made of copper?

48. A 60.0-cm metal bar that is clamped at one end is struck with a hammer. If the speed of longitudinal (compressional) waves in the bar is 4 500 m/s, what is the lowest frequency with which the struck bar resonates?

Section 18.7   Beats: Interference in Time

WEB 49. In certain ranges of a piano keyboard, more than one string is tuned to the same note to provide extra loudness. For example, the note at 110 Hz has two strings that vibrate at this frequency. If one string slips from its normal tension of 600 N to 540 N, what beat frequency is heard when the hammer strikes the two strings simultaneously?

50. While attempting to tune the note C at 523 Hz, a piano tuner hears 2 beats/s between a reference oscillator and the string. (a) What are the possible frequencies of the string? (b) When she tightens the string slightly, she hears 3 beats/s. What is the frequency of the string now? (c) By what percentage should the piano tuner now change the tension in the string to bring it into tune?

51. A student holds a tuning fork oscillating at 256 Hz. He walks toward a wall at a constant speed of 1.33 m/s. (a) What beat frequency does he observe between the tuning fork and its echo? (b) How fast must he walk away from the wall to observe a beat frequency of 5.00 Hz?

(Optional)
Section 18.8   Non-Sinusoidal Wave Patterns

52. Suppose that a flutist plays a 523-Hz C note with first harmonic displacement amplitude $A_1 = 100$ nm. From Figure 18.20b, read, by proportion, the displacement amplitudes of harmonics 2 through 7. Take these as the values A_2 through A_7 in the Fourier analysis of the sound, and assume that $B_1 = B_2 = \ldots = B_7 = 0$. Construct a graph of the waveform of the sound. Your waveform will not look exactly like the flute waveform in Figure 18.19b because you simplify by ignoring cosine terms; nevertheless, it produces the same sensation to human hearing.

53. An A-major chord consists of the notes called A, C#, and E. It can be played on a piano by simultaneously striking strings that have fundamental frequencies of 440.00 Hz, 554.37 Hz, and 659.26 Hz. The rich consonance of the chord is associated with the near equality of the frequencies of some of the higher harmonics of the three tones. Consider the first five harmonics of each string and determine which harmonics show near equality.

ADDITIONAL PROBLEMS

54. **Review Problem.** For the arrangement shown in Figure P18.54, $\theta = 30.0°$, the inclined plane and the small pulley are frictionless, the string supports the mass M at the bottom of the plane, and the string has a mass m that is small compared with M. The system is in equilibrium, and the vertical part of the string has a length h. Standing waves are set up in the vertical section of the string. Find (a) the tension in the string, (b) the whole length of the string (ignoring the radius of curvature of the pulley), (c) the mass per unit length of the string, (d) the speed of waves on the string, (e) the lowest-frequency standing wave, (f) the period of the standing wave having three nodes, (g) the wavelength of the standing wave having three nodes, and (h) the frequency of the beats resulting from the interference of the sound wave of lowest frequency generated by the string with another sound wave having a frequency that is 2.00% greater.

Figure P18.54

55. Two loudspeakers are placed on a wall 2.00 m apart. A listener stands 3.00 m from the wall directly in front of one of the speakers. The speakers are being driven by a single oscillator at a frequency of 300 Hz. (a) What is the phase difference between the two waves when they reach the observer? (b) What is the frequency closest to 300 Hz to which the oscillator may be adjusted such that the observer hears minimal sound?

56. On a marimba (Fig. P18.56), the wooden bar that sounds a tone when it is struck vibrates in a transverse standing wave having three antinodes and two nodes. The lowest-frequency note is 87.0 Hz; this note is produced by a bar 40.0 cm long. (a) Find the speed of transverse waves on the bar. (b) The loudness of the emitted sound is enhanced by a resonant pipe suspended vertically below the center of the bar. If the pipe is open at the top end only and the speed of sound in air is 340 m/s, what is the length of the pipe required to resonate with the bar in part (a)?

Figure P18.56 Marimba players in Mexico City. *(Murray Greenberg)*

57. Two train whistles have identical frequencies of 180 Hz. When one train is at rest in the station and is sounding its whistle, a beat frequency of 2.00 Hz is heard from a train moving nearby. What are the two possible speeds and directions that the moving train can have?

58. A speaker at the front of a room and an identical speaker at the rear of the room are being driven by the same oscillator at 456 Hz. A student walks at a uniform rate of 1.50 m/s along the length of the room. How many beats does the student hear per second?

59. While Jane waits on a railroad platform, she observes two trains approaching from the same direction at equal speeds of 8.00 m/s. Both trains are blowing their whistles (which have the same frequency), and one train is some distance behind the other. After the first train passes Jane, but before the second train passes her, she hears beats having a frequency of 4.00 Hz. What is the frequency of the trains' whistles?

60. A string fixed at both ends and having a mass of 4.80 g, a length of 2.00 m, and a tension of 48.0 N vibrates in its second ($n = 2$) natural mode. What is the wavelength in air of the sound emitted by this vibrating string?

61. A string 0.400 m in length has a mass per unit length of 9.00×10^{-3} kg/m. What must be the tension in the string if its second harmonic is to have the same frequency as the second resonance mode of a 1.75-m-long pipe open at one end?

62. In a major chord on the physical pitch musical scale, the frequencies are in the ratios 4:5:6:8. A set of pipes, closed at one end, must be cut so that, when they are sounded in their first normal mode, they produce a major chord. (a) What is the ratio of the lengths of the pipes? (b) What are the lengths of the pipes needed if the lowest frequency of the chord is 256 Hz? (c) What are the frequencies of this chord?

63. Two wires are welded together. The wires are made of the same material, but the diameter of one wire is twice that of the other. They are subjected to a tension of 4.60 N. The thin wire has a length of 40.0 cm and a linear mass density of 2.00 g/m. The combination is fixed at both ends and vibrated in such a way that two antinodes are present, with the node between them being right at the weld. (a) What is the frequency of vibration? (b) How long is the thick wire?

64. Two identical strings, each fixed at both ends, are arranged near each other. If string A starts oscillating in its first normal mode, string B begins vibrating in its third ($n = 3$) natural mode. Determine the ratio of the tension of string B to the tension of string A.

65. A standing wave is set up in a string of variable length and tension by a vibrator of variable frequency. When the vibrator has a frequency f, in a string of length L and under a tension T, n antinodes are set up in the string. (a) If the length of the string is doubled, by what factor should the frequency be changed so that the same number of antinodes is produced? (b) If the frequency and length are held constant, what tension produces $n + 1$ antinodes? (c) If the frequency is tripled and the length of the string is halved, by what factor should the tension be changed so that twice as many antinodes are produced?

66. A 0.010 0-kg, 2.00-m-long wire is fixed at both ends and vibrates in its simplest mode under a tension of 200 N. When a tuning fork is placed near the wire, a beat frequency of 5.00 Hz is heard. (a) What could the frequency of the tuning fork be? (b) What should the tension in the wire be if the beats are to disappear?

WEB 67. If two adjacent natural frequencies of an organ pipe are determined to be 0.550 kHz and 0.650 kHz, calculate the fundamental frequency and length of the pipe. (Use $v = 340$ m/s.)

68. Two waves are described by the equations

$$y_1(x, t) = 5.0 \sin(2.0x - 10t)$$

and

$$y_2(x, t) = 10 \cos(2.0x - 10t)$$

where x is in meters and t is in seconds. Show that the resulting wave is sinusoidal, and determine the amplitude and phase of this sinusoidal wave.

69. The wave function for a standing wave is given in Equation 18.3 as $y = (2A \sin kx) \cos \omega t$. (a) Rewrite this wave function in terms of the wavelength λ and the wave speed v of the wave. (b) Write the wave function of the simplest standing-wave vibration of a stretched string of length L. (c) Write the wave function for the second harmonic. (d) Generalize these results, and write the wave function for the nth resonance vibration.

70. **Review Problem.** A 12.0-kg mass hangs in equilibrium from a string with a total length of $L = 5.00$ m and a linear mass density of $\mu = 0.001\ 00$ kg/m. The string is wrapped around two light, frictionless pulleys that are separated by a distance of $d = 2.00$ m (Fig. P18.70a). (a) Determine the tension in the string. (b) At what frequency must the string between the pulleys vibrate to form the standing-wave pattern shown in Figure P18.70b?

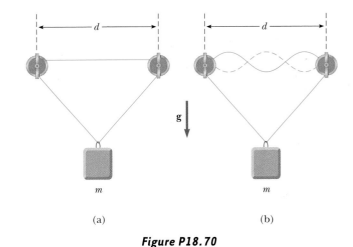

(a)   (b)

Figure P18.70

ANSWERS TO QUICK QUIZZES

18.1 At the antinodes. All particles have the same period $T = 2\pi/\omega$, but a particle at an antinode must travel through the greatest vertical distance in this amount of time and therefore must travel fastest.

18.2 For each natural frequency of the glass, the standing wave must "fit" exactly around the rim. In Figure 18.12a we see three antinodes on the near side of the glass, and thus there must be another three on the far side. This corresponds to three complete waves. In a top view, the wave pattern looks like this (although we have greatly exaggerated the amplitude):

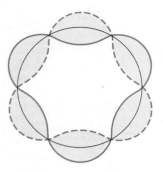

18.3 At highway speeds, a car crosses the ridges on the rumble strip at a rate that matches one of the car's natural frequencies of oscillation. This causes the car to oscillate substantially more than when it is traveling over the randomly spaced bumps of regular pavement. This sudden resonance oscillation alerts the driver that he or she must pay attention.

18.4 (b). With both ends open, the pipe has a fundamental frequency given by Equation 18.11: $f_{open} = v/2L$. With one end closed, the pipe has a fundamental frequency given by Equation 18.12:

$$f_{closed} = \frac{v}{4L} = \frac{1}{2} \frac{v}{2L} = \frac{1}{2} f_{open}$$

part 3

Thermodynamics

We now direct our attention to the study of thermodynamics, which deals with the concepts of heat and temperature. As we shall see, thermodynamics is very successful in explaining the bulk properties of matter and the correlation between these properties and the mechanics of both atoms and molecules.

Historically, the development of thermodynamics paralleled the development of the atomic theory of matter. By the 1820s, chemical experiments had provided solid evidence for the existence of atoms. At that time, scientists recognized that a connection between the theory of heat and temperature and the structure of matter must exist. In 1827, the botanist Robert Brown reported that grains of pollen suspended in a liquid move erratically from one place to another, as if under constant agitation. In 1905, Albert Einstein used kinetic theory to explain the cause of this erratic motion, which today is known as *Brownian motion.* Einstein explained this phenomenon by assuming that the grains are under constant bombardment by "invisible" molecules in the liquid, which themselves move erratically. This explanation gave scientists insight into molecular motion and gave credence to the idea that matter is made up of atoms. A connection was thus forged between the everyday world and the tiny, invisible building blocks that make up this world.

Thermodynamics also addresses more practical questions. Have you ever wondered how a refrigerator is able to cool its contents, what types of transformations occur in a power plant or in the engine of your automobile, or what happens to the kinetic energy of a moving object when the object comes to rest? The laws of thermodynamics and the concepts of heat and temperature provide explanations. In general, thermodynamics is concerned with transformations of matter in all of its states—solid, liquid, and gas.

◀ Richard A. Cooke III/Tony Stone Images.

c h a p t e r

19

Temperature

In our study of mechanics, we carefully defined such concepts as mass, force, and kinetic energy to facilitate our quantitative approach. Likewise, a quantitative description of thermal phenomena requires a careful definition of such important terms as *temperature, heat,* and *internal energy.* This chapter begins with a look at these three entities and with a description of one of the laws of thermodynamics (the poetically named "zeroth law"). We then discuss the three most common temperature scales—Celsius, Fahrenheit, and Kelvin.

Next, we consider why the composition of a body is an important factor when we are dealing with thermal phenomena. For example, gases expand appreciably when heated, whereas liquids and solids expand only slightly. If a gas is not free to expand as it is heated, its pressure increases. Certain substances may melt, boil, burn, or explode when they are heated, depending on their composition and structure.

This chapter concludes with a study of ideal gases on the macroscopic scale. Here, we are concerned with the relationships among such quantities as pressure, volume, and temperature. Later on, in Chapter 21, we shall examine gases on a microscopic scale, using a model that represents the components of a gas as small particles.

Molten lava flowing down a mountain in Kilauea, Hawaii. The temperature of the hot lava flowing from a central crater decreases until the lava is in thermal equilibrium with its surroundings. At that equilibrium temperature, the lava has solidified and formed the mountains. *(Ken Sakomoto/Black Star)*

19.1 TEMPERATURE AND THE ZEROTH LAW OF THERMODYNAMICS

10.3 & 10.4

We often associate the concept of temperature with how hot or cold an object feels when we touch it. Thus, our senses provide us with a qualitative indication of temperature. However, our senses are unreliable and often mislead us. For example, if we remove a metal ice tray and a cardboard box of frozen vegetables from the freezer, the ice tray feels colder than the box even though both are at the same temperature. The two objects feel different because metal is a better thermal conductor than cardboard is. What we need, therefore, is a reliable and reproducible method for establishing the relative hotness or coldness of bodies. Scientists have developed a variety of thermometers for making such quantitative measurements.

We are all familiar with the fact that two objects at different initial temperatures eventually reach some intermediate temperature when placed in contact with each other. For example, when a scoop of ice cream is placed in a room-temperature glass bowl, the ice cream melts and the temperature of the bowl decreases. Likewise, when an ice cube is dropped into a cup of hot coffee, it melts and the coffee's temperature decreases.

To understand the concept of temperature, it is useful to define two often-used phrases: *thermal contact* and *thermal equilibrium.* To grasp the meaning of thermal contact, let us imagine that two objects are placed in an insulated container such that they interact with each other but not with the rest of the world. If the objects are at different temperatures, energy is exchanged between them, even if they are initially not in physical contact with each other. **Heat is the transfer of energy from one object to another object as a result of a difference in temperature between the two.** We shall examine the concept of heat in greater detail in Chapter 20. For purposes of the current discussion, we assume that two objects are in **thermal contact** with each other if energy can be exchanged between them. **Thermal equilibrium** is a situation in which two objects in thermal contact with each other cease to exchange energy by the process of heat.

Let us consider two objects A and B, which are not in thermal contact, and a third object C, which is our thermometer. We wish to determine whether A and B

QuickLab

Fill three cups with tap water: one hot, one cold, and one lukewarm. Dip your left index finger into the hot water and your right index finger into the cold water. Slowly count to 20, then quickly dip both fingers into the lukewarm water. What do you feel?

are in thermal equilibrium with each other. The thermometer (object C) is first placed in thermal contact with object A until thermal equilibrium is reached. From that moment on, the thermometer's reading remains constant, and we record this reading. The thermometer is then removed from object A and placed in thermal contact with object B. The reading is again recorded after thermal equilibrium is reached. If the two readings are the same, then object A and object B are in thermal equilibrium with each other.

We can summarize these results in a statement known as the **zeroth law of thermodynamics** (the law of equilibrium):

Zeroth law of thermodynamics

> If objects A and B are separately in thermal equilibrium with a third object C, then objects A and B are in thermal equilibrium with each other.

This statement can easily be proved experimentally and is very important because it enables us to define temperature. We can think of **temperature** as the property that determines whether an object is in thermal equilibrium with other objects. **Two objects in thermal equilibrium with each other are at the same temperature.** Conversely, if two objects have different temperatures, then they are not in thermal equilibrium with each other.

19.2 THERMOMETERS AND THE CELSIUS TEMPERATURE SCALE

Thermometers are devices that are used to define and measure temperatures. All thermometers are based on the principle that some physical property of a system changes as the system's temperature changes. Some physical properties that change with temperature are (1) the volume of a liquid, (2) the length of a solid, (3) the pressure of a gas at constant volume, (4) the volume of a gas at constant pressure, (5) the electric resistance of a conductor, and (6) the color of an object. For a given substance and a given temperature range, a temperature scale can be established on the basis of any one of these physical properties.

A common thermometer in everyday use consists of a mass of liquid—usually mercury or alcohol—that expands into a glass capillary tube when heated (Fig. 19.1). In this case the physical property is the change in volume of a liquid. Any temperature change can be defined as being proportional to the change in length of the liquid column. The thermometer can be calibrated by placing it in thermal contact with some natural systems that remain at constant temperature. One such system is a mixture of water and ice in thermal equilibrium at atmospheric pressure. On the **Celsius temperature scale,** this mixture is defined to have a temperature of zero degrees Celsius, which is written as 0°C; this temperature is called the *ice point* of water. Another commonly used system is a mixture of water and steam in thermal equilibrium at atmospheric pressure; its temperature is 100°C, which is the *steam point* of water. Once the liquid levels in the thermometer have been established at these two points, the distance between the two points is divided into 100 equal segments to create the Celsius scale. Thus, each segment denotes a change in temperature of one Celsius degree. (This temperature scale used to be called the *centigrade scale* because there are 100 gradations between the ice and steam points of water.)

Thermometers calibrated in this way present problems when extremely accurate readings are needed. For instance, the readings given by an alcohol ther-

Figure 19.1 As a result of thermal expansion, the level of the mercury in the thermometer rises as the mercury is heated by water in the test tube. *(Charles D. Winters)*

mometer calibrated at the ice and steam points of water might agree with those given by a mercury thermometer only at the calibration points. Because mercury and alcohol have different thermal expansion properties, when one thermometer reads a temperature of, for example, 50°C, the other may indicate a slightly different value. The discrepancies between thermometers are especially large when the temperatures to be measured are far from the calibration points.[1]

An additional practical problem of any thermometer is the limited range of temperatures over which it can be used. A mercury thermometer, for example, cannot be used below the freezing point of mercury, which is − 39°C, and an alcohol thermometer is not useful for measuring temperatures above 85°C, the boiling point of alcohol. To surmount this problem, we need a universal thermometer whose readings are independent of the substance used in it. The gas thermometer, discussed in the next section, approaches this requirement.

19.3 THE CONSTANT-VOLUME GAS THERMOMETER AND THE ABSOLUTE TEMPERATURE SCALE

The temperature readings given by a gas thermometer are nearly independent of the substance used in the thermometer. One version is the constant-volume gas thermometer shown in Figure 19.2. The physical change exploited in this device is the variation of pressure of a fixed volume of gas with temperature. When the constant-volume gas thermometer was developed, it was calibrated by using the ice

[1] Two thermometers that use the same liquid may also give different readings. This is due in part to difficulties in constructing uniform-bore glass capillary tubes.

Figure 19.2 A constant-volume gas thermometer measures the pressure of the gas contained in the flask immersed in the bath. The volume of gas in the flask is kept constant by raising or lowering reservoir *B* to keep the mercury level in column *A* constant.

Figure 19.3 A typical graph of pressure versus temperature taken with a constant-volume gas thermometer. The two dots represent known reference temperatures (the ice and steam points of water).

and steam points of water, as follows (a different calibration procedure, which we shall discuss shortly, is now used): The flask was immersed in an ice bath, and mercury reservoir B was raised or lowered until the top of the mercury in column A was at the zero point on the scale. The height h, the difference between the mercury levels in reservoir B and column A, indicated the pressure in the flask at $0°C$.

The flask was then immersed in water at the steam point, and reservoir B was readjusted until the top of the mercury in column A was again at zero on the scale; this ensured that the gas's volume was the same as it was when the flask was in the ice bath (hence, the designation "constant volume"). This adjustment of reservoir B gave a value for the gas pressure at $100°C$. These two pressure and temperature values were then plotted, as shown in Figure 19.3. The line connecting the two points serves as a calibration curve for unknown temperatures. If we wanted to measure the temperature of a substance, we would place the gas flask in thermal contact with the substance and adjust the height of reservoir B until the top of the mercury column in A was at zero on the scale. The height of the mercury column would indicate the pressure of the gas; knowing the pressure, we could find the temperature of the substance using the graph in Figure 19.3.

 Now let us suppose that temperatures are measured with gas thermometers containing different gases at different initial pressures. Experiments show that the **10.3** thermometer readings are nearly independent of the type of gas used, as long as the gas pressure is low and the temperature is well above the point at which the gas liquefies (Fig. 19.4). The agreement among thermometers using various gases improves as the pressure is reduced.

If you extend the curves shown in Figure 19.4 toward negative temperatures, you find, in every case, that the pressure is zero when the temperature is $-273.15°C$. This significant temperature is used as the basis for the **absolute temperature scale,** which sets $-273.15°C$ as its zero point. This temperature is often referred to as **absolute zero.** The size of a degree on the absolute temperature scale is identical to the size of a degree on the Celsius scale. Thus, the conversion between these temperatures is

$$T_C = T - 273.15 \qquad\qquad \textbf{(19.1)}$$

where T_C is the Celsius temperature and T is the absolute temperature.

Because the ice and steam points are experimentally difficult to duplicate, an absolute temperature scale based on a single fixed point was adopted in 1954 by the International Committee on Weights and Measures. From a list of fixed points associated with various substances (Table 19.1), the triple point of water was chosen as the reference temperature for this new scale. The **triple point of water** is the single combination of temperature and pressure at which liquid water, gaseous

Figure 19.4 Pressure versus temperature for three dilute gases. Note that, for all gases, the pressure extrapolates to zero at the temperature $-273.15°C$.

TABLE 19.1 Fixed-Point Temperatures[a]

Fixed Point	Temperature (°C)	Temperature (K)
Triple point of hydrogen	− 259.34	13.81
Boiling point of helium	− 268.93	4.215
Boiling point of hydrogen at 33.36 kPa pressure	− 256.108	17.042
Boiling point of hydrogen	− 252.87	20.28
Triple point of neon	− 246.048	27.102
Triple point of oxygen	− 218.789	54.361
Boiling point of oxygen	− 182.962	90.188
Triple point of water	0.01	273.16
Boiling point of water	100.00	373.15
Freezing point of tin	231.968 1	505.118 1
Freezing point of zinc	419.58	692.73
Freezing point of silver	961.93	1 235.08
Freezing point of gold	1 064.43	1 337.58

[a] All values are from National Bureau of Standards Special Publication 420; U. S. Department of Commerce, May 1975. All values are at standard atmospheric pressure except for triple points and as noted.

water, and ice (solid water) coexist in equilibrium. This triple point occurs at a temperature of approximately 0.01°C and a pressure of 4.58 mm of mercury. On the new scale, which uses the unit *kelvin*, the temperature of water at the triple point was set at 273.16 kelvin, abbreviated 273.16 K. (*Note:* no degree sign "°" is used with the unit kelvin.) This choice was made so that the old absolute temperature scale based on the ice and steam points would agree closely with the new scale based on the triple point. This new absolute temperature scale (also called the **Kelvin scale**) employs the SI unit of absolute temperature, the **kelvin**, which is defined to be **1/273.16 of the difference between absolute zero and the temperature of the triple point of water.**

Figure 19.5 shows the absolute temperature for various physical processes and structures. The temperature of absolute zero (0 K) cannot be achieved, although laboratory experiments incorporating the laser cooling of atoms have come very close.

What would happen to a gas if its temperature could reach 0 K? As Figure 19.4 indicates, the pressure it exerts on the walls of its container would be zero. In Section 19.5 we shall show that the pressure of a gas is proportional to the average kinetic energy of its molecules. Thus, according to classical physics, the kinetic energy of the gas molecules would become zero at absolute zero, and molecular motion would cease; hence, the molecules would settle out on the bottom of the container. Quantum theory modifies this model and shows that some residual energy, called the *zero-point energy*, would remain at this low temperature.

The Celsius, Fahrenheit, and Kelvin Temperature Scales[2]

Equation 19.1 shows that the Celsius temperature T_C is shifted from the absolute (Kelvin) temperature T by 273.15°. Because the size of a degree is the same on the

[2] Named after Anders Celsius (1701–1744), Gabriel Fahrenheit (1686–1736), and William Thomson, Lord Kelvin (1824–1907), respectively.

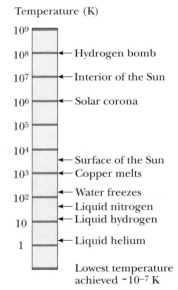

Temperature (K)

- Hydrogen bomb
- Interior of the Sun
- Solar corona
- Surface of the Sun
- Copper melts
- Water freezes
- Liquid nitrogen
- Liquid hydrogen
- Liquid helium

Lowest temperature achieved ~10⁻⁷ K

Figure 19.5 Absolute temperatures at which various physical processes occur. Note that the scale is logarithmic.

two scales, a temperature difference of 5°C is equal to a temperature difference of 5 K. The two scales differ only in the choice of the zero point. Thus, the ice-point temperature on the Kelvin scale, 273.15 K, corresponds to 0.00°C, and the Kelvin-scale steam point, 373.15 K, is equivalent to 100.00°C.

A common temperature scale in everyday use in the United States is the **Fahrenheit scale.** This scale sets the temperature of the ice point at 32°F and the temperature of the steam point at 212°F. The relationship between the Celsius and Fahrenheit temperature scales is

$$T_F = \tfrac{9}{5}T_C + 32°F \qquad\qquad (19.2)$$

Quick Quiz 19.1

What is the physical significance of the factor $\tfrac{9}{5}$ in Equation 19.2? Why is this factor missing in Equation 19.1?

Extending the ideas considered in Quick Quiz 19.1, we use Equation 19.2 to find a relationship between changes in temperature on the Celsius, Kelvin, and Fahrenheit scales:

$$\Delta T_C = \Delta T = \tfrac{5}{9}\Delta T_F \qquad\qquad (19.3)$$

EXAMPLE 19.1   Converting Temperatures

On a day when the temperature reaches 50°F, what is the temperature in degrees Celsius and in kelvins?

Solution   Substituting $T_F = 50°F$ into Equation 19.2, we obtain

$$T_C = \tfrac{5}{9}(T_F - 32) = \tfrac{5}{9}(50 - 32) = \boxed{10°C}$$

From Equation 19.1, we find that

$$T = T_C + 273.15 = 10°C + 273.15 = \boxed{283\ K}$$

A convenient set of weather-related temperature equivalents to keep in mind is that 0°C is (literally) freezing at 32°F, 10°C is cool at 50°F, 30°C is warm at 86°F, and 40°C is a hot day at 104°F.

EXAMPLE 19.2   Heating a Pan of Water

A pan of water is heated from 25°C to 80°C. What is the change in its temperature on the Kelvin scale and on the Fahrenheit scale?

Solution   From Equation 19.3, we see that the change in temperature on the Celsius scale equals the change on the Kelvin scale. Therefore,

$$\Delta T = \Delta T_C = 80°C - 25°C = 55°C = \boxed{55\ K}$$

From Equation 19.3, we also find that

$$\Delta T_F = \tfrac{9}{5}\Delta T_C = \tfrac{9}{5}(55°C) = \boxed{99°F}$$

19.4   THERMAL EXPANSION OF SOLIDS AND LIQUIDS

Our discussion of the liquid thermometer made use of one of the best-known changes in a substance: As its temperature increases, its volume almost always increases. (As we shall see shortly, in some substances the volume decreases when the temperature increases.) This phenomenon, known as **thermal expansion,** has

(a)

(b)

Figure 19.6 (a) Thermal-expansion joints are used to separate sections of roadways on bridges. Without these joints, the surfaces would buckle due to thermal expansion on very hot days or crack due to contraction on very cold days. (b) The long, vertical joint is filled with a soft material that allows the wall to expand and contract as the temperature of the bricks changes. *(a, Frank Siteman, Stock/Boston; b, George Semple)*

an important role in numerous engineering applications. For example, thermal-expansion joints, such as those shown in Figure 19.6, must be included in buildings, concrete highways, railroad tracks, brick walls, and bridges to compensate for dimensional changes that occur as the temperature changes.

Thermal expansion is a consequence of the change in the average separation between the constituent atoms in an object. To understand this, imagine that the atoms are connected by stiff springs, as shown in Figure 19.7. At ordinary temperatures, the atoms in a solid oscillate about their equilibrium positions with an amplitude of approximately 10^{-11} m and a frequency of approximately 10^{13} Hz. The average spacing between the atoms is about 10^{-10} m. As the temperature of the solid increases, the atoms oscillate with greater amplitudes; as a result, the average separation between them increases.[3] Consequently, the object expands.

If thermal expansion is sufficiently small relative to an object's initial dimensions, the change in any dimension is, to a good approximation, proportional to the first power of the temperature change. Suppose that an object has an initial length L_i along some direction at some temperature and that the length increases by an amount ΔL for a change in temperature ΔT. Because it is convenient to consider the fractional change in length per degree of temperature change, we define the **average coefficient of linear expansion** as

$$\alpha \equiv \frac{\Delta L/L_i}{\Delta T}$$

Experiments show that α is constant for small changes in temperature. For purposes of calculation, this equation is usually rewritten as

$$\Delta L = \alpha L_i \, \Delta T \qquad \textbf{(19.4)}$$

or as

$$L_f - L_i = \alpha L_i (T_f - T_i) \qquad \textbf{(19.5)}$$

Figure 19.7 A mechanical model of the atomic configuration in a substance. The atoms (spheres) are imagined to be attached to each other by springs that reflect the elastic nature of the interatomic forces.

Average coefficient of linear expansion

The change in length of an object is proportional to the change in temperature

[3] More precisely, thermal expansion arises from the *asymmetrical* nature of the potential-energy curve for the atoms in a solid. If the oscillators were truly harmonic, the average atomic separations would not change regardless of the amplitude of vibration.

where L_f is the final length, T_i and T_f are the initial and final temperatures, and the proportionality constant α is the average coefficient of linear expansion for a given material and has units of $°C^{-1}$.

It may be helpful to think of thermal expansion as an effective magnification or as a photographic enlargement of an object. For example, as a metal washer is heated (Fig. 19.8), all dimensions, including the radius of the hole, increase according to Equation 19.4.

Table 19.2 lists the average coefficient of linear expansion for various materials. Note that for these materials α is positive, indicating an increase in length with increasing temperature. This is not always the case. Some substances—calcite ($CaCO_3$) is one example—expand along one dimension (positive α) and contract along another (negative α) as their temperatures are increased.

Because the linear dimensions of an object change with temperature, it follows that surface area and volume change as well. The change in volume at constant pressure is proportional to the initial volume V_i and to the change in temperature according to the relationship

$$\Delta V = \beta V_i \Delta T \qquad (19.6)$$

where β is the **average coefficient of volume expansion.** For a solid, the average coefficient of volume expansion is approximately three times the average linear expansion coefficient: $\beta = 3\alpha$. (This assumes that the average coefficient of linear expansion of the solid is the same in all directions.)

To see that $\beta = 3\alpha$ for a solid, consider a box of dimensions ℓ, w, and h. Its volume at some temperature T_i is $V_i = \ell w h$. If the temperature changes to $T_i + \Delta T$, its volume changes to $V_i + \Delta V$, where each dimension changes according to Equation 19.4. Therefore,

$$
\begin{aligned}
V_i + \Delta V &= (\ell + \Delta \ell)(w + \Delta w)(h + \Delta h) \\
&= (\ell + \alpha \ell \, \Delta T)(w + \alpha w \, \Delta T)(h + \alpha h \, \Delta T) \\
&= \ell w h (1 + \alpha \, \Delta T)^3 \\
&= V_i[1 + 3\alpha \, \Delta T + 3(\alpha \, \Delta T)^2 + (\alpha \, \Delta T)^3]
\end{aligned}
$$

Figure 19.8 Thermal expansion of a homogeneous metal washer. As the washer is heated, all dimensions increase. (The expansion is exaggerated in this figure.)

TABLE 19.2 **Average Expansion Coefficients for Some Materials Near Room Temperature**

Material	Average Linear Expansion Coefficient (α) $(°C)^{-1}$	Material	Average Volume Expansion Coefficient (β) $(°C)^{-1}$
Aluminum	24×10^{-6}	Alcohol, ethyl	1.12×10^{-4}
Brass and bronze	19×10^{-6}	Benzene	1.24×10^{-4}
Copper	17×10^{-6}	Acetone	1.5×10^{-4}
Glass (ordinary)	9×10^{-6}	Glycerin	4.85×10^{-4}
Glass (Pyrex)	3.2×10^{-6}	Mercury	1.82×10^{-4}
Lead	29×10^{-6}	Turpentine	9.0×10^{-4}
Steel	11×10^{-6}	Gasoline	9.6×10^{-4}
Invar (Ni–Fe alloy)	0.9×10^{-6}	Air at 0°C	3.67×10^{-3}
Concrete	12×10^{-6}	Helium	3.665×10^{-3}

If we now divide both sides by V_i and then isolate the term $\Delta V/V_i$, we obtain the fractional change in volume:

$$\frac{\Delta V}{V_i} = 3\alpha \, \Delta T + 3(\alpha \, \Delta T)^2 + (\alpha \, \Delta T)^3$$

Because $\alpha \, \Delta T \ll 1$ for typical values of $\Delta T \, (< \sim 100°C)$, we can neglect the terms $3(\alpha \, \Delta T)^2$ and $(\alpha \, \Delta T)^3$. Upon making this approximation, we see that

$$\frac{\Delta V}{V_i} = 3\alpha \, \Delta T$$

$$3\alpha = \frac{1}{V_i}\frac{\Delta V}{\Delta T}$$

Equation 19.6 shows that the right side of this expression is equal to β, and so we have $3\alpha = \beta$, the relationship we set out to prove. In a similar way, you can show that the change in area of a rectangular plate is given by $\Delta A = 2\alpha A_i \, \Delta T$ (see Problem 53).

As Table 19.2 indicates, each substance has its own characteristic average coefficient of expansion. For example, when the temperatures of a brass rod and a steel rod of equal length are raised by the same amount from some common initial value, the brass rod expands more than the steel rod does because brass has a greater average coefficient of expansion than steel does. A simple mechanism called a *bimetallic strip* utilizes this principle and is found in practical devices such as thermostats. It consists of two thin strips of dissimilar metals bonded together. As the temperature of the strip increases, the two metals expand by different amounts and the strip bends, as shown in Figure 19.9.

QuickLab

Tape two plastic straws tightly together along their entire length but with a 2-cm offset. Hold them in a stream of very hot water from a faucet so that water pours through one but not through the other. Quickly hold the straws up and sight along their length. You should be able to see a very slight curvature in the tape caused by the difference in expansion of the two straws. The effect is small, so look closely. Running cold water through the same straw and again sighting along the length will help you see the small change in shape more clearly.

Figure 19.9 (a) A bimetallic strip bends as the temperature changes because the two metals have different expansion coefficients. (b) A bimetallic strip used in a thermostat to break or make electrical contact. (c) The interior of a thermostat, showing the coiled bimetallic strip. Why do you suppose the strip is coiled? *(c, Charles D. Winters)*

If you quickly plunge a room-temperature thermometer into very hot water, the mercury level will go *down* briefly before going up to a final reading. Why?

You are offered a prize for making the most sensitive glass thermometer using the materials in Table 19.2. Which glass and which working liquid would you choose?

EXAMPLE 19.3   Expansion of a Railroad Track

A steel railroad track has a length of 30.000 m when the temperature is 0.0°C. (a) What is its length when the temperature is 40.0°C?

Thermal expansion: The extreme temperature of a July day in Asbury Park, NJ, caused these railroad tracks to buckle and derail the train in the distance. *(AP/Wide World Photos)*

Solution   Making use of Table 19.2 and noting that the change in temperature is 40.0°C, we find that the increase in length is

$$\Delta L = \alpha L_i \Delta T = [11 \times 10^{-6}(°C)^{-1}](30.000 \text{ m})(40.0°C)$$
$$= 0.013 \text{ m}$$

If the track is 30.000 m long at 0.0°C, its length at 40.0°C is

30.013 m.

(b) Suppose that the ends of the rail are rigidly clamped at 0.0°C so that expansion is prevented. What is the thermal stress set up in the rail if its temperature is raised to 40.0°C?

Solution   From the definition of Young's modulus for a solid (see Eq. 12.6), we have

$$\text{Tensile stress} = \frac{F}{A} = Y\frac{\Delta L}{L_i}$$

Because Y for steel is 20×10^{10} N/m² (see Table 12.1), we have

$$\frac{F}{A} = (20 \times 10^{10} \text{ N/m}^2)\left(\frac{0.013 \text{ m}}{30.000 \text{ m}}\right) = \boxed{8.7 \times 10^7 \text{ N/m}^2}$$

Exercise   If the rail has a cross-sectional area of 30.0 cm², what is the force of compression in the rail?

Answer 2.6×10^5 N = 58 000 lb!

The Unusual Behavior of Water

Liquids generally increase in volume with increasing temperature and have average coefficients of volume expansion about ten times greater than those of solids. Water is an exception to this rule, as we can see from its density-versus-temperature curve shown in Figure 19.10. As the temperature increases from 0°C to 4°C, water contracts and thus its density increases. Above 4°C, water expands with increasing temperature, and so its density decreases. In other words, the density of water reaches a maximum value of 1 000 kg/m³ at 4°C.

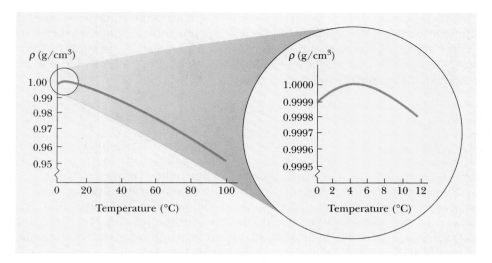

Figure 19.10 How the density of water at atmospheric pressure changes with temperature. The inset at the right shows that the maximum density of water occurs at 4°C.

We can use this unusual thermal-expansion behavior of water to explain why a pond begins freezing at the surface rather than at the bottom. When the atmospheric temperature drops from, for example, 7°C to 6°C, the surface water also cools and consequently decreases in volume. This means that the surface water is denser than the water below it, which has not cooled and decreased in volume. As a result, the surface water sinks, and warmer water from below is forced to the surface to be cooled. When the atmospheric temperature is between 4°C and 0°C, however, the surface water expands as it cools, becoming less dense than the water below it. The mixing process stops, and eventually the surface water freezes. As the water freezes, the ice remains on the surface because ice is less dense than water. The ice continues to build up at the surface, while water near the bottom remains at 4°C. If this were not the case, then fish and other forms of marine life would not survive.

19.5 MACROSCOPIC DESCRIPTION OF AN IDEAL GAS

In this section we examine the properties of a gas of mass m confined to a container of volume V at a pressure P and a temperature T. It is useful to know how these quantities are related. In general, the equation that interrelates these quantities, called the *equation of state*, is very complicated. However, if the gas is maintained at a very low pressure (or low density), the equation of state is quite simple and can be found experimentally. Such a low-density gas is commonly referred to as an *ideal gas.*[4]

---

[4] To be more specific, the assumption here is that the temperature of the gas must not be too low (the gas must not condense into a liquid) or too high, and that the pressure must be low. In reality, an ideal gas does not exist. However, the concept of an ideal gas is very useful in view of the fact that real gases at low pressures behave as ideal gases do. The concept of an ideal gas implies that the gas molecules do not interact except upon collision, and that the molecular volume is negligible compared with the volume of the container.

Figure 19.11 An ideal gas confined to a cylinder whose volume can be varied by means of a movable piston.

It is convenient to express the amount of gas in a given volume in terms of the number of moles n. As we learned in Section 1.3, one mole of any substance is that amount of the substance that contains Avogadro's number $N_A = 6.022 \times 10^{23}$ of constituent particles (atoms or molecules). The number of moles n of a substance is related to its mass m through the expression

$$n = \frac{m}{M} \tag{19.7}$$

where M is the molar mass of the substance (see Section 1.3), which is usually expressed in units of grams per mole (g/mol). For example, the molar mass of oxygen (O_2) is 32.0 g/mol. Therefore, the mass of one mole of oxygen is 32.0 g.

Now suppose that an ideal gas is confined to a cylindrical container whose volume can be varied by means of a movable piston, as shown in Figure 19.11. If we assume that the cylinder does not leak, the mass (or the number of moles) of the gas remains constant. For such a system, experiments provide the following information: First, when the gas is kept at a constant temperature, its pressure is inversely proportional to its volume (Boyle's law). Second, when the pressure of the gas is kept constant, its volume is directly proportional to its temperature (the law of Charles and Gay–Lussac). These observations are summarized by the **equation of state for an ideal gas:**

$$PV = nRT \tag{19.8}$$

In this expression, known as the **ideal gas law,** R is a universal constant that is the same for all gases and T is the absolute temperature in kelvins. Experiments on numerous gases show that as the pressure approaches zero, the quantity PV/nT approaches the same value R for all gases. For this reason, R is called the **universal gas constant.** In SI units, in which pressure is expressed in pascals (1 Pa = 1 N/m^2) and volume in cubic meters, the product PV has units of newton·meters, or joules, and R has the value

The universal gas constant

$$R = 8.315 \text{ J/mol·K} \tag{19.9}$$

If the pressure is expressed in atmospheres and the volume in liters (1 L = 10^3 cm^3 = 10^{-3} m^3), then R has the value

$$R = 0.082 \ 14 \text{ L·atm/mol·K}$$

Using this value of R and Equation 19.8, we find that the volume occupied by 1 mol of any gas at atmospheric pressure and at 0°C (273 K) is 22.4 L.

Now that we have presented the equation of state, we are ready for a formal definition of an ideal gas: **An ideal gas is one for which PV/nT is constant at all pressures.**

The ideal gas law states that if the volume and temperature of a fixed amount of gas do not change, then the pressure also remains constant. Consider the bottle of champagne shown at the beginning of this chapter. Because the temperature of the bottle and its contents remains constant, so does the pressure, as can be shown by replacing the cork with a pressure gauge. Shaking the bottle displaces some carbon dioxide gas from the "head space" to form bubbles within the liquid, and these bubbles become attached to the inside of the bottle. (No new gas is generated by shaking.) When the bottle is opened, the pressure is reduced; this causes the volume of the bubbles to increase suddenly. If the bubbles are attached to the bottle (beneath the liquid surface), their rapid expansion expels liquid from the

QuickLab

Vigorously shake a can of soda pop and then thoroughly tap its bottom and sides to dislodge any bubbles trapped there. You should be able to open the can without spraying its contents all over.

bottle. If the sides and bottom of the bottle are first tapped until no bubbles remain beneath the surface, then when the champagne is opened, the drop in pressure will not force liquid from the bottle. Try the QuickLab, but practice before demonstrating to a friend!

The ideal gas law is often expressed in terms of the total number of molecules N. Because the total number of molecules equals the product of the number of moles n and Avogadro's number N_A, we can write Equation 19.8 as

$$PV = nRT = \frac{N}{N_A} RT$$

$$PV = Nk_B T \qquad \qquad (19.10)$$

where k_B is **Boltzmann's constant,** which has the value

$$k_B = \frac{R}{N_A} = 1.38 \times 10^{-23}\, \text{J/K} \qquad \qquad (19.11)$$

Boltzmann's constant

It is common to call quantities such as P, V, and T the **thermodynamic variables** of an ideal gas. If the equation of state is known, then one of the variables can always be expressed as some function of the other two.

EXAMPLE 19.4   How Many Gas Molecules in a Container?

An ideal gas occupies a volume of 100 cm^3 at 20°C and 100 Pa. Find the number of moles of gas in the container.

$$n = \frac{PV}{RT} = \frac{(100\ \text{Pa})(10^{-4}\ \text{m}^3)}{(8.315\ \text{J/mol}\cdot\text{K})(293\ \text{K})} = 4.10 \times 10^{-6}\ \text{mol}$$

Solution   The quantities given are volume, pressure, and temperature: $V = 100\ \text{cm}^3 = 1.00 \times 10^{-4}\ \text{m}^3$, $P = 100\ \text{Pa}$, and $T = 20°\text{C} = 293\ \text{K}$. Using Equation 19.8, we find that

Exercise   How many molecules are in the container?

Answer 2.47×10^{18} molecules.

EXAMPLE 19.5   Filling a Scuba Tank

A certain scuba tank is designed to hold 66 ft^3 of air when it is at atmospheric pressure at 22°C. When this volume of air is compressed to an absolute pressure of 3 000 lb/in.^{2} and stored in a 10-L (0.35-ft^3) tank, the air becomes so hot that the tank must be allowed to cool before it can be used. If the air does not cool, what is its temperature? (Assume that the air behaves like an ideal gas.)

Solution   If no air escapes from the tank during filling, then the number of moles n remains constant; therefore, using $PV = nRT$, and with n and R being constant, we obtain for the initial and final values:

$$\frac{P_i V_i}{T_i} = \frac{P_f V_f}{T_f}$$

The initial pressure of the air is 14.7 lb/in.^{2}, its final pressure is 3 000 lb/in.^{2}, and the air is compressed from an initial volume of 66 ft^3 to a final volume of 0.35 ft^3. The initial temperature, converted to SI units, is 295 K. Solving for T_f, we obtain

$$T_f = \left(\frac{P_f V_f}{P_i V_i}\right) T_i = \frac{(3\ 000\ \text{lb/in.}^2)(0.35\ \text{ft}^3)}{(14.7\ \text{lb/in.}^2)(66\ \text{ft}^3)}(295\ \text{K})$$

$$= 319\ \text{K}$$

Exercise   What is the air temperature in degrees Celsius and in degrees Fahrenheit?

Answer   45.9°C; 115°F.

Quick Quiz 19.4

In the previous example we used SI units for the temperature in our calculation step but not for the pressures or volumes. When working with the ideal gas law, how do you decide when it is necessary to use SI units and when it is not?

EXAMPLE 19.6   Heating a Spray Can

A spray can containing a propellant gas at twice atmospheric pressure (202 kPa) and having a volume of 125 cm^3 is at 22°C. It is then tossed into an open fire. When the temperature of the gas in the can reaches 195°C, what is the pressure inside the can? Assume any change in the volume of the can is negligible.

Solution   We employ the same approach we used in Example 19.5, starting with the expression

$$\frac{P_i V_i}{T_i} = \frac{P_f V_f}{T_f}$$

Because the initial and final volumes of the gas are assumed to be equal, this expression reduces to

$$\frac{P_i}{T_i} = \frac{P_f}{T_f}$$

Solving for P_f gives

$$P_f = \left(\frac{T_f}{T_i}\right)(P_i) = \left(\frac{468 \text{ K}}{295 \text{ K}}\right)(202 \text{ kPa}) = \boxed{320 \text{ kPa}}$$

Obviously, the higher the temperature, the higher the pressure exerted by the trapped gas. Of course, if the pressure increases high enough, the can will explode. Because of this possibility, you should never dispose of spray cans in a fire.

SUMMARY

Two bodies are in **thermal equilibrium** with each other if they have the same temperature.

The **zeroth law of thermodynamics** states that if objects A and B are separately in thermal equilibrium with a third object C, then objects A and B are in thermal equilibrium with each other.

The SI unit of absolute temperature is the **kelvin,** which is defined to be the fraction 1/273.16 of the temperature of the triple point of water.

When the temperature of an object is changed by an amount ΔT, its length changes by an amount ΔL that is proportional to ΔT and to its initial length L_i:

$$\Delta L = \alpha L_i \Delta T \tag{19.4}$$

where the constant α is the **average coefficient of linear expansion.** The **average volume expansion coefficient** β for a solid is approximately equal to 3α.

An **ideal gas** is one for which PV/nT is constant at all pressures. An ideal gas is described by the **equation of state,**

$$PV = nRT \tag{19.8}$$

where n equals the number of moles of the gas, V is its volume, R is the universal gas constant (8.315 J/mol·K), and T is the absolute temperature. A real gas behaves approximately as an ideal gas if it is far from liquefaction.

QUESTIONS

1. Is it possible for two objects to be in thermal equilibrium if they are not in contact with each other? Explain.

2. A piece of copper is dropped into a beaker of water. If the water's temperature increases, what happens to the temperature of the copper? Under what conditions are the water and copper in thermal equilibrium?

3. In principle, any gas can be used in a constant-volume gas thermometer. Why is it not possible to use oxygen for temperatures as low as 15 K? What gas would you use? (Refer to the data in Table 19.1.)

4. Rubber has a negative average coefficient of linear expansion. What happens to the size of a piece of rubber as it is warmed?

5. Why should the amalgam used in dental fillings have the same average coefficient of expansion as a tooth? What would occur if they were mismatched?

6. Explain why the thermal expansion of a spherical shell made of a homogeneous solid is equivalent to that of a solid sphere of the same material.

7. A steel ring bearing has an inside diameter that is 0.1 mm smaller than the diameter of an axle. How can it be made to fit onto the axle without removing any material?

8. Markings to indicate length are placed on a steel tape in a room that has a temperature of 22°C. Are measurements made with the tape on a day when the temperature is 27°C greater than, less than, or the same length as the object's length? Defend your answer.

9. Determine the number of grams in 1 mol of each of the following gases: (a) hydrogen, (b) helium, and (c) carbon monoxide.

10. An inflated rubber balloon filled with air is immersed in a flask of liquid nitrogen that is at 77 K. Describe what happens to the balloon, assuming that it remains flexible while being cooled.

11. Two identical cylinders at the same temperature each contain the same kind of gas and the same number of moles of gas. If the volume of cylinder A is three times greater than the volume of cylinder B, what can you say about the relative pressures in the cylinders?

12. The pendulum of a certain pendulum clock is made of brass. When the temperature increases, does the clock run too fast, run too slowly, or remain unchanged? Explain.

13. An automobile radiator is filled to the brim with water while the engine is cool. What happens to the water when the engine is running and the water is heated? What do modern automobiles have in their cooling systems to prevent the loss of coolants?

14. Metal lids on glass jars can often be loosened by running them under hot water. How is this possible?

15. When the metal ring and metal sphere shown in Figure Q19.15 are both at room temperature, the sphere can just be passed through the ring. After the sphere is heated, it cannot be passed through the ring. Explain.

Figure Q19.15 (*Courtesy of Central Scientific Company*)

PROBLEMS

1, 2, 3 = straightforward, intermediate, challenging ☐ = full solution available in the *Student Solutions Manual and Study Guide*
WEB = solution posted at **http://www.saunderscollege.com/physics/** 🖥 = Computer useful in solving problem 🖱 = Interactive Physics
☐ = paired numerical/symbolic problems

Section 19.1 Temperature and the Zeroth Law of Thermodynamics
Section 19.2 Thermometers and the Celsius Temperature Scale
Section 19.3 The Constant-Volume Gas Thermometer and the Absolute Temperature Scale

Note: A pressure of 1 atm = 1.013×10^5 Pa = 101.3 kPa.

1. Convert the following to equivalent temperatures on the Celsius and Kelvin scales: (a) the normal human body temperature, 98.6°F; (b) the air temperature on a cold day, − 5.00°F.

2. In a constant-volume gas thermometer, the pressure at 20.0°C is 0.980 atm. (a) What is the pressure at 45.0°C? (b) What is the temperature if the pressure is 0.500 atm?

WEB 3. A constant-volume gas thermometer is calibrated in dry ice (that is, carbon dioxide in the solid state, which has a temperature of − 80.0°C) and in boiling ethyl alcohol (78.0°C). The two pressures are 0.900 atm and

1.635 atm. (a) What Celsius value of absolute zero does the calibration yield? What is the pressure at (b) the freezing point of water and (c) the boiling point of water?

4. There is a temperature whose numerical value is the same on both the Celsius and Fahrenheit scales. What is this temperature?

5. Liquid nitrogen has a boiling point of $-195.81°C$ at atmospheric pressure. Express this temperature in (a) degrees Fahrenheit and (b) kelvins.

6. On a Strange temperature scale, the freezing point of water is $-15.0°S$ and the boiling point is $+60.0°S$. Develop a *linear* conversion equation between this temperature scale and the Celsius scale.

7. The temperature difference between the inside and the outside of an automobile engine is $450°C$. Express this temperature difference on the (a) Fahrenheit scale and (b) Kelvin scale.

8. The melting point of gold is $1\,064°C$, and the boiling point is $2\,660°C$. (a) Express these temperatures in kelvins. (b) Compute the difference between these temperatures in Celsius degrees and in kelvins.

Section 19.4 Thermal Expansion of Solids and Liquids

Note: When solving the problems in this section, use the data in Table 19.2.

9. A copper telephone wire has essentially no sag between poles 35.0 m apart on a winter day when the temperature is $-20.0°C$. How much longer is the wire on a summer day when $T_C = 35.0°C$?

10. The concrete sections of a certain superhighway are designed to have a length of 25.0 m. The sections are poured and cured at $10.0°C$. What minimum spacing should the engineer leave between the sections to eliminate buckling if the concrete is to reach a temperature of $50.0°C$?

11. An aluminum tube is 3.000 0 m long at $20.0°C$. What is its length at (a) $100.0°C$ and (b) $0.0°C$?

12. A brass ring with a diameter of 10.00 cm at $20.0°C$ is heated and slipped over an aluminum rod with a diameter of 10.01 cm at $20.0°C$. Assume that the average coefficients of linear expansion are constant. (a) To what temperature must this combination be cooled to separate them? Is this temperature attainable? (b) If the aluminum rod were 10.02 cm in diameter, what would be the required temperature?

13. A pair of eyeglass frames is made of epoxy plastic. At room temperature ($20.0°C$), the frames have circular lens holes 2.20 cm in radius. To what temperature must the frames be heated if lenses 2.21 cm in radius are to be inserted in them? The average coefficient of linear expansion for epoxy is $1.30 \times 10^{-4}\ (°C)^{-1}$.

14. The New River Gorge bridge in West Virginia is a steel arch bridge 518 m in length. How much does its length change between temperature extremes of $-20.0°C$ and $35.0°C$?

15. A square hole measuring 8.00 cm along each side is cut in a sheet of copper. (a) Calculate the change in the area of this hole if the temperature of the sheet is increased by 50.0 K. (b) Does the result represent an increase or a decrease in the area of the hole?

16. The average coefficient of volume expansion for carbon tetrachloride is $5.81 \times 10^{-4}\ (°C)^{-1}$. If a 50.0-gal steel container is filled completely with carbon tetrachloride when the temperature is $10.0°C$, how much will spill over when the temperature rises to $30.0°C$?

WEB 17. The active element of a certain laser is a glass rod 30.0 cm long by 1.50 cm in diameter. If the temperature of the rod increases by $65.0°C$, what is the increase in (a) its length, (b) its diameter, and (c) its volume? (Assume that $\alpha = 9.00 \times 10^{-6}\ (°C)^{-1}$.)

18. A volumetric glass flask made of Pyrex is calibrated at $20.0°C$. It is filled to the 100-mL mark with $35.0°C$ acetone with which it immediately comes to thermal equilibrium. (a) What is the volume of the acetone when it cools to $20.0°C$? (b) How significant is the change in volume of the flask?

19. A concrete walk is poured on a day when the temperature is $20.0°C$, in such a way that the ends are unable to move. (a) What is the stress in the cement on a hot day of $50.0°C$? (b) Does the concrete fracture? Take Young's modulus for concrete to be $7.00 \times 10^9\ N/m^2$ and the tensile strength to be $2.00 \times 10^9\ N/m^2$.

20. Figure P19.20 shows a circular steel casting with a gap. If the casting is heated, (a) does the width of the gap increase or decrease? (b) The gap width is 1.600 cm when the temperature is $30.0°C$. Determine the gap width when the temperature is $190°C$.

Figure P19.20

21. A steel rod undergoes a stretching force of 500 N. Its cross-sectional area is $2.00\ cm^2$. Find the change in temperature that would elongate the rod by the same amount that the 500-N force does. (*Hint:* Refer to Tables 12.1 and 19.2.)

22. A steel rod 4.00 cm in diameter is heated so that its temperature increases by $70.0°C$. It is then fastened between two rigid supports. The rod is allowed to cool to its original temperature. Assuming that Young's modulus for the steel is $20.6 \times 10^{10}\ N/m^2$ and that its average

coefficient of linear expansion is 11.0×10^{-6} $(°C)^{-1}$, calculate the tension in the rod.

23. A hollow aluminum cylinder 20.0 cm deep has an internal capacity of 2.000 L at 20.0°C. It is completely filled with turpentine and then warmed to 80.0°C. (a) How much turpentine overflows? (b) If the cylinder is then cooled back to 20.0°C, how far below the surface of the cylinder's rim does the turpentine's surface recede?

24. At 20.0°C, an aluminum ring has an inner diameter of 5.000 0 cm and a brass rod has a diameter of 5.050 0 cm. (a) To what temperature must the ring be heated so that it will just slip over the rod? (b) To what common temperature must the two be heated so that the ring just slips over the rod? Would this latter process work?

Section 19.5 Macroscopic Description of an Ideal Gas

25. Gas is contained in an 8.00-L vessel at a temperature of 20.0°C and a pressure of 9.00 atm. (a) Determine the number of moles of gas in the vessel. (b) How many molecules of gas are in the vessel?

26. A tank having a volume of 0.100 m^3 contains helium gas at 150 atm. How many balloons can the tank blow up if each filled balloon is a sphere 0.300 m in diameter at an absolute pressure of 1.20 atm?

27. An auditorium has dimensions 10.0 m $\times$ 20.0 m $\times$ 30.0 m. How many molecules of air fill the auditorium at 20.0°C and a pressure of 101 kPa?

28. Nine grams of water are placed in a 2.00-L pressure cooker and heated to 500°C. What is the pressure inside the container if no gas escapes?

WEB 29. The mass of a hot-air balloon and its cargo (not including the air inside) is 200 kg. The air outside is at 10.0°C and 101 kPa. The volume of the balloon is 400 m^3. To what temperature must the air in the balloon be heated before the balloon will lift off? (Air density at 10.0°C is 1.25 kg/m^3.)

30. One mole of oxygen gas is at a pressure of 6.00 atm and a temperature of 27.0°C. (a) If the gas is heated at constant volume until the pressure triples, what is the final temperature? (b) If the gas is heated until both the pressure and the volume are doubled, what is the final temperature?

31. (a) Find the number of moles in 1.00 m^3 of an ideal gas at 20.0°C and atmospheric pressure. (b) For air, Avogadro's number of molecules has a mass of 28.9 g. Calculate the mass of 1 m^3 of air. Compare the result with the tabulated density of air.

32. A cube 10.0 cm on each edge contains air (with equivalent molar mass 28.9 g/mol) at atmospheric pressure and temperature 300 K. Find (a) the mass of the gas, (b) its weight, and (c) the force it exerts on each face of the cube. (d) Comment on the underlying physical reason why such a small sample can exert such a great force.

33. An automobile tire is inflated with air originally at 10.0°C and normal atmospheric pressure. During the process, the air is compressed to 28.0% of its original volume and its temperature is increased to 40.0°C. (a) What is the tire pressure? (b) After the car is driven at high speed, the tire air temperature rises to 85.0°C and the interior volume of the tire increases by 2.00%. What is the new tire pressure (absolute) in pascals?

34. A spherical weather balloon is designed to expand to a maximum radius of 20.0 m when in flight at its working altitude, where the air pressure is 0.030 0 atm and the temperature is 200 K. If the balloon is filled at atmospheric pressure and 300 K, what is its radius at liftoff?

35. A room of volume 80.0 m^3 contains air having an equivalent molar mass of 28.9 g/mol. If the temperature of the room is raised from 18.0°C to 25.0°C, what mass of air (in kilograms) will leave the room? Assume that the air pressure in the room is maintained at 101 kPa.

36. A room of volume V contains air having equivalent molar mass M (in g/mol). If the temperature of the room is raised from T_1 to T_2, what mass of air will leave the room? Assume that the air pressure in the room is maintained at P_0.

37. At 25.0 m below the surface of the sea (density = 1 025 kg/m^3), where the temperature is 5.00°C, a diver exhales an air bubble having a volume of 1.00 cm^3. If the surface temperature of the sea is 20.0°C, what is the volume of the bubble right before it breaks the surface?

38. Estimate the mass of the air in your bedroom. State the quantities you take as data and the value you measure or estimate for each.

39. The pressure gauge on a tank registers the gauge pressure, which is the difference between the interior and exterior pressures. When the tank is full of oxygen (O_2), it contains 12.0 kg of the gas at a gauge pressure of 40.0 atm. Determine the mass of oxygen that has been withdrawn from the tank when the pressure reading is 25.0 atm. Assume that the temperature of the tank remains constant.

40. In state-of-the-art vacuum systems, pressures as low as 10^{-9} Pa are being attained. Calculate the number of molecules in a 1.00-m^3 vessel at this pressure if the temperature is 27°C.

41. Show that 1 mol of any gas (assumed to be ideal) at atmospheric pressure (101.3 kPa) and standard temperature (273 K) occupies a volume of 22.4 L.

42. A diving bell in the shape of a cylinder with a height of 2.50 m is closed at the upper end and open at the lower end. The bell is lowered from air into sea water (ρ = 1.025 g/cm^3). The air in the bell is initially at 20.0°C. The bell is lowered to a depth (measured to the bottom of the bell) of 45.0 fathoms, or 82.3 m. At this depth, the water temperature is 4.0°C, and the air in the bell is in thermal equilibrium with the water. (a) How high does sea water rise in the bell? (b) To what minimum pressure must the air in the bell be increased for the water that entered to be expelled?

ADDITIONAL PROBLEMS

43. A student measures the length of a brass rod with a steel tape at 20.0°C. The reading is 95.00 cm. What will the tape indicate for the length of the rod when the rod and the tape are at (a) − 15.0°C and (b) 55.0°C?

44. The density of gasoline is 730 kg/m³ at 0°C. Its average coefficient of volume expansion is 9.60×10^{-4}/°C. If 1.00 gal of gasoline occupies 0.003 80 m³, how many extra kilograms of gasoline would you get if you bought 10.0 gal of gasoline at 0°C rather than at 20.0°C from a pump that is not temperature compensated?

45. A steel ball bearing is 4.000 cm in diameter at 20.0°C. A bronze plate has a hole in it that is 3.994 cm in diameter at 20.0°C. What common temperature must they have so that the ball just squeezes through the hole?

46. **Review Problem.** An aluminum pipe 0.655 m long at 20.0°C and open at both ends is used as a flute. The pipe is cooled to a low temperature but is then filled with air at 20.0°C as soon as it is played. By how much does its fundamental frequency change as the temperature of the metal increases from 5.00°C to 20.0°C?

47. A mercury thermometer is constructed as shown in Figure P19.47. The capillary tube has a diameter of 0.004 00 cm, and the bulb has a diameter of 0.250 cm. Neglecting the expansion of the glass, find the change in height of the mercury column that occurs with a temperature change of 30.0°C.

Figure P19.47 Problems 47 and 48.

48. A liquid with a coefficient of volume expansion β just fills a spherical shell of volume V_i at a temperature of T_i (see Fig. P19.47). The shell is made of a material that has an average coefficient of linear expansion α. The liquid is free to expand into an open capillary of area A projecting from the top of the sphere. (a) If the temperature increases by ΔT, show that the liquid rises in the capillary by the amount Δh given by the equation $\Delta h = (V_i/A)(\beta - 3\alpha) \Delta T$. (b) For a typical system, such as a mercury thermometer, why is it a good approximation to neglect the expansion of the shell?

WEB 49. A liquid has a density ρ. (a) Show that the fractional change in density for a change in temperature ΔT is $\Delta\rho/\rho = -\beta \Delta T$. What does the negative sign signify? (b) Fresh water has a maximum density of 1.000 0 g/cm³ at 4.0°C. At 10.0°C, its density is 0.999 7 g/cm³. What is β for water over this temperature interval?

50. A cylinder is closed by a piston connected to a spring of constant 2.00×10^3 N/m (Fig. P19.50). While the spring is relaxed, the cylinder is filled with 5.00 L of gas at a pressure of 1.00 atm and a temperature of 20.0°C. (a) If the piston has a cross-sectional area of 0.010 0 m² and a negligible mass, how high will it rise when the temperature is increased to 250°C? (b) What is the pressure of the gas at 250°C?

Figure P19.50

WEB 51. A vertical cylinder of cross-sectional area A is fitted with a tight-fitting, frictionless piston of mass m (Fig. P19.51). (a) If n moles of an ideal gas are in the cylinder at a temperature of T, what is the height h at which the piston is in equilibrium under its own weight? (b) What is the value for h if $n = 0.200$ mol, $T = 400$ K, $A = 0.008$ 00 m², and $m = 20.0$ kg?

Figure P19.51

52. A bimetallic bar is made of two thin strips of dissimilar metals bonded together. As they are heated, the one with the greater average coefficient of expansion expands more than the other, forcing the bar into an arc, with the outer radius having a greater circumference (Fig. P19.52). (a) Derive an expression for the angle of bending θ as a function of the initial length of the strips, their average coefficients of linear expansion, the change in temperature, and the separation of the centers of the strips ($\Delta r = r_2 - r_1$). (b) Show that the angle of bending decreases to zero when ΔT decreases to zero or when the two average coefficients of expansion become equal. (c) What happens if the bar is cooled?

Figure P19.52

53. The rectangular plate shown in Figure P19.53 has an area A_i equal to ℓw. If the temperature increases by ΔT, show that the increase in area is $\Delta A = 2\alpha A_i \Delta T$, where α is the average coefficient of linear expansion. What approximation does this expression assume? (*Hint:* Note that each dimension increases according to the equation $\Delta L = \alpha L_i \Delta T$.)

Figure P19.53

54. Precise temperature measurements are often made on the basis of the change in electrical resistance of a metal with temperature. The resistance varies approximately according to the expression $R = R_0(1 + AT_C)$, where R_0 and A are constants. A certain element has a resistance of 50.0 ohms (Ω) at 0°C and 71.5 Ω at the freezing point of tin (231.97°C). (a) Determine the constants A and R_0. (b) At what temperature is the resistance equal to 89.0 Ω?

55. **Review Problem.** A clock with a brass pendulum has a period of 1.000 s at 20.0°C. If the temperature increases to 30.0°C, (a) by how much does the period change, and (b) how much time does the clock gain or lose in one week?

56. **Review Problem.** Consider an object with any one of the shapes displayed in Table 10.2. What is the percentage increase in the moment of inertia of the object when it is heated from 0°C to 100°C if it is composed of (a) copper or (b) aluminum? (See Table 19.2. Assume that the average linear expansion coefficients do not vary between 0°C and 100°C.)

57. **Review Problem.** (a) Derive an expression for the buoyant force on a spherical balloon that is submerged in water as a function of the depth below the surface, the volume (V_i) of the balloon at the surface, the pressure (P_0) at the surface, and the density of the water. (Assume that water temperature does not change with depth.) (b) Does the buoyant force increase or decrease as the balloon is submerged? (c) At what depth is the buoyant force one-half the surface value?

58. (a) Show that the density of an ideal gas occupying a volume V is given by $\rho = PM/RT$, where M is the molar mass. (b) Determine the density of oxygen gas at atmospheric pressure and 20.0°C.

59. Starting with Equation 19.10, show that the total pressure P in a container filled with a mixture of several ideal gases is $P = P_1 + P_2 + P_3 + \ldots$, where P_1, $P_2, \ldots$ are the pressures that each gas would exert if it alone filled the container. (These individual pressures are called the *partial pressures* of the respective gases.) This is known as *Dalton's law of partial pressures.*

60. A sample of dry air that has a mass of 100.00 g, collected at sea level, is analyzed and found to consist of the following gases:

$$\text{nitrogen } (N_2) = 75.52 \text{ g}$$
$$\text{oxygen } (O_2) = 23.15 \text{ g}$$
$$\text{argon } (Ar) = 1.28 \text{ g}$$
$$\text{carbon dioxide } (CO_2) = 0.05 \text{ g}$$

as well as trace amounts of neon, helium, methane, and other gases. (a) Calculate the partial pressure (see Problem 59) of each gas when the pressure is 101.3 kPa. (b) Determine the volume occupied by the 100-g sample at a temperature of 15.00°C and a pressure of 1.013×10^5 Pa. What is the density of the air for these conditions? (c) What is the effective molar mass of the air sample?

61. Steel rails for an interurban rapid transit system form a continuous track that is held rigidly in place in concrete. (a) If the track was laid when the temperature was 0°C, what is the stress in the rails on a warm day when the temperature is 25.0°C? (b) What fraction of the yield strength of 52.2×10^7 N/m^2 does this stress represent?

62. (a) Use the equation of state for an ideal gas and the definition of the average coefficient of volume expansion, in the form $\beta = (1/V)\,dV/dT$, to show that the average coefficient of volume expansion for an ideal gas at constant pressure is given by $\beta = 1/T$, where T is the absolute temperature. (b) What value does this expression predict for β at $0°C$? Compare this with the experimental values for helium and air in Table 19.2.

63. Two concrete spans of a 250-m-long bridge are placed end to end so that no room is allowed for expansion (Fig. P19.63a). If a temperature increase of $20.0°C$ occurs, what is the height y to which the spans rise when they buckle (Fig. P19.63b)?

64. Two concrete spans of a bridge of length L are placed end to end so that no room is allowed for expansion (see Fig. P19.63a). If a temperature increase of ΔT occurs, what is the height y to which the spans rise when they buckle (see Fig. P19.63b)?

(a)   (b)

Figure P19.63   Problems 63 and 64.

65. A copper rod and a steel rod are heated. At $0°C$ the copper rod has length L_c, and the steel rod has length L_s. When the rods are being heated or cooled, the difference between their lengths stays constant at 5.00 cm. Determine the values of L_c and L_s.

66. A cylinder that has a 40.0-cm radius and is 50.0 cm deep is filled with air at $20.0°C$ and 1.00 atm (Fig. P19.66a). A 20.0-kg piston is now lowered into the cylinder, compressing the air trapped inside (Fig. P19.66b). Finally, a 75.0-kg man stands on the piston, further compressing the air, which remains at $20°C$ (Fig. P19.66c). (a) How far down (Δh) does the piston move when the man steps onto it? (b) To what temperature should the gas be heated to raise the piston and the man back to h_i?

67. The relationship $L_f = L_i(1 + \alpha\Delta T)$ is an approximation that works when the average coefficient of expansion is small. If α is large, one must integrate the relationship $dL/dT = \alpha L$ to determine the final length. (a) Assuming that the average coefficient of linear expansion is constant as L varies, determine a general expression for the final length. (b) Given a rod of length 1.00 m and a temperature change of $100.0\ °C$, determine the error caused by the approximation when $\alpha = 2.00 \times 10^{-5}\ (°C)^{-1}$ (a typical value for a metal) and

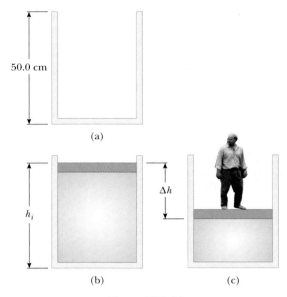

Figure P19.66

when $\alpha = 0.0200\ (°C)^{-1}$ (an unrealistically large value for comparison).

68. A steel wire and a copper wire, each of diameter 2.000 mm, are joined end to end. At $40.0°C$, each has an unstretched length of 2.000 m; they are connected between two fixed supports 4.000 m apart on a tabletop, so that the steel wire extends from $x = -2.000$ m to $x = 0$, the copper wire extends from $x = 0$ to $x = 2.000$ m, and the tension is negligible. The temperature is then lowered to $20.0°C$. At this lower temperature, what are the tension in the wire and the x coordinate of the junction between the wires? (Refer to Tables 12.1 and 19.2.)

69. **Review Problem.**   A steel guitar string with a diameter of 1.00 mm is stretched between supports 80.0 cm apart. The temperature is $0.0°C$. (a) Find the mass per unit length of this string. (Use 7.86×10^3 kg/m³ as the mass density.) (b) The fundamental frequency of transverse oscillations of the string is 200 Hz. What is the tension in the string? (c) If the temperature is raised to $30.0°C$, find the resulting values of the tension and the fundamental frequency. (Assume that both the Young's modulus [Table 12.1] and the average coefficient of linear expansion [Table 19.2] have constant values between $0.0°C$ and $30.0°C$.)

70. A 1.00-km steel railroad rail is fastened securely at both ends when the temperature is $20.0°C$. As the temperature increases, the rail begins to buckle. If its shape is an arc of a vertical circle, find the height h of the center of the buckle when the temperature is $25.0°C$. (You will need to solve a transcendental equation.)

ANSWERS TO QUICK QUIZZES

19.1 The size of a degree on the Fahrenheit scale is $\frac{5}{9}$ the size of a degree on the Celsius scale. This is true because the Fahrenheit range of 32°F to 212°F is equivalent to the Celsius range of 0°C to 100°C. The factor $\frac{9}{5}$ in Equation 19.2 corrects for this difference. Equation 19.1 does not need this correction because the size of a Celsius degree is the same as the size of a kelvin.

19.2 The glass bulb containing most of the mercury warms up first because it is in direct thermal contact with the hot water. It expands slightly, and thus its volume increases. This causes the mercury level in the capillary tube to drop. As the mercury inside the bulb warms up, it also expands. Eventually, its increase in volume is sufficient to raise the mercury level in the capillary tube.

19.3 For the glass, choose Pyrex, which has a lower average coefficient of linear expansion than does ordinary glass. For the working liquid, choose gasoline, which has the largest average coefficient of volume expansion.

19.4 You do not have to convert the units for pressure and volume to SI units as long as the same units appear in the numerator and the denominator. This is not true for ratios of temperature units, as you can see by comparing the ratios 300 K/200 K and 26.85°C/(− 73.15°C). You must always use absolute (kelvin) temperatures when applying the ideal gas law.

c h a p t e r

20

Heat and the First Law of Thermodynamics

Until about 1850, the fields of thermodynamics and mechanics were considered two distinct branches of science, and the law of conservation of energy seemed to describe only certain kinds of mechanical systems. However, mid–19th century experiments performed by the Englishman James Joule and others showed that energy may be added to (or removed from) a system either by heat or by doing work on the system (or having the system do work). Today we know that internal energy, which we formally define in this chapter, can be transformed to mechanical energy. Once the concept of energy was broadened to include internal energy, the law of conservation of energy emerged as a universal law of nature.

This chapter focuses on the concept of internal energy, the processes by which energy is transferred, the first law of thermodynamics, and some of the important applications of the first law. The first law of thermodynamics is the law of conservation of energy. It describes systems in which the only energy change is that of internal energy, which is due to transfers of energy by heat or work. Furthermore, the first law makes no distinction between the results of heat and the results of work. According to the first law, a system's internal energy can be changed either by an energy transfer by heat to or from the system or by work done on or by the system.

James Prescott Joule British physicist (1818–1889)   Joule received some formal education in mathematics, philosophy, and chemistry but was in large part self-educated. His research led to the establishment of the principle of conservation of energy. His study of the quantitative relationship among electrical, mechanical, and chemical effects of heat culminated in his discovery in 1843 of the amount of work required to produce a unit of energy, called the mechanical equivalent of heat. *(By kind permission of the President and Council of the Royal Society)*

20.1 ▷ HEAT AND INTERNAL ENERGY

At the outset, it is important that we make a major distinction between internal energy and heat. **Internal energy is all the energy of a system that is associated with its microscopic components—atoms and molecules—when viewed from a reference frame at rest with respect to the object.** The last part of this sentence ensures that any bulk kinetic energy of the system due to its motion through space is not included in internal energy. Internal energy includes kinetic energy of translation, rotation, and vibration of molecules, potential energy within molecules, and potential energy between molecules. It is useful to relate internal energy to the temperature of an object, but this relationship is limited—we shall find in Section 20.3 that internal energy changes can also occur in the absence of temperature changes.

As we shall see in Chapter 21, the internal energy of a monatomic ideal gas is associated with the translational motion of its atoms. This is the only type of energy available for the microscopic components of this system. In this special case, the internal energy is simply the total kinetic energy of the atoms of the gas; the higher the temperature of the gas, the greater the average kinetic energy of the atoms and the greater the internal energy of the gas. More generally, in solids, liquids, and molecular gases, internal energy includes other forms of molecular energy. For example, a diatomic molecule can have rotational kinetic energy, as well as vibrational kinetic and potential energy.

Heat is defined as the transfer of energy across the boundary of a system due to a temperature difference between the system and its surroundings. When you *heat* a substance, you are transferring energy into it by placing it in contact with surroundings that have a higher temperature. This is the case, for example, when you place a pan of cold water on a stove burner—the burner is at a higher temperature than the water, and so the water gains energy. We shall also use the term *heat* to represent the amount of energy transferred by this method.

Scientists used to think of heat as a fluid called *caloric*, which they believed was transferred between objects; thus, they defined heat in terms of the temperature changes produced in an object during heating. Today we recognize the distinct difference between internal energy and heat. Nevertheless, we refer to quantities

Heat

using names that do not quite correctly define the quantities but which have become entrenched in physics tradition based on these early ideas. Examples of such quantities are *latent heat* and *heat capacity*.

As an analogy to the distinction between heat and internal energy, consider the distinction between work and mechanical energy discussed in Chapter 7. The work done on a system is a measure of the amount of energy transferred to the system from its surroundings, whereas the mechanical energy of the system (kinetic or potential, or both) is a consequence of the motion and relative positions of the members of the system. Thus, when a person does work on a system, energy is transferred from the person to the system. It makes no sense to talk about the work *of* a system—one can refer only to the work done *on* or *by* a system when some process has occurred in which energy has been transferred to or from the system. Likewise, it makes no sense to talk about the heat *of* a system—one can refer to *heat* only when energy has been transferred as a result of a temperature difference. Both heat and work are ways of changing the energy of a system.

It is also important to recognize that the internal energy of a system can be changed even when no energy is transferred by heat. For example, when a gas is compressed by a piston, the gas is warmed and its internal energy increases, but no transfer of energy by heat from the surroundings to the gas has occurred. If the gas then expands rapidly, it cools and its internal energy decreases, but no transfer of energy by heat from it to the surroundings has taken place. The temperature changes in the gas are due not to a difference in temperature between the gas and its surroundings but rather to the compression and the expansion. In each case, energy is transferred to or from the gas by *work,* and the energy change within the system is an increase or decrease of internal energy. The changes in internal energy in these examples are evidenced by corresponding changes in the temperature of the gas.

Units of Heat

As we have mentioned, early studies of heat focused on the resultant increase in temperature of a substance, which was often water. The early notions of heat based on caloric suggested that the flow of this fluid from one body to another caused changes in temperature. From the name of this mythical fluid, we have an energy unit related to thermal processes, the **calorie (cal),** which is defined as **the amount of energy transfer necessary to raise the temperature of 1 g of water from 14.5°C to 15.5°C.**[1] (Note that the "Calorie," written with a capital "C" and used in describing the energy content of foods, is actually a kilocalorie.) The unit of energy in the British system is the **British thermal unit (Btu),** which is defined as **the amount of energy transfer required to raise the temperature of 1 lb of water from 63°F to 64°F.**

Scientists are increasingly using the SI unit of energy, the *joule,* when describing thermal processes. In this textbook, heat and internal energy are usually measured in joules. (Note that both heat and work are measured in energy units. Do not confuse these two means of energy *transfer* with energy itself, which is also measured in joules.)

The calorie

---

[1] Originally, the calorie was defined as the "heat" necessary to raise the temperature of 1 g of water by 1°C. However, careful measurements showed that the amount of energy required to produce a 1°C change depends somewhat on the initial temperature; hence, a more precise definition evolved.

The Mechanical Equivalent of Heat

In Chapters 7 and 8, we found that whenever friction is present in a mechanical system, some mechanical energy is lost—in other words, mechanical energy is not conserved in the presence of nonconservative forces. Various experiments show that this lost mechanical energy does not simply disappear but is transformed into internal energy. We can perform such an experiment at home by simply hammering a nail into a scrap piece of wood. What happens to all the kinetic energy of the hammer once we have finished? Some of it is now in the nail as internal energy, as demonstrated by the fact that the nail is measurably warmer. Although this connection between mechanical and internal energy was first suggested by Benjamin Thompson, it was Joule who established the equivalence of these two forms of energy.

A schematic diagram of Joule's most famous experiment is shown in Figure 20.1. The system of interest is the water in a thermally insulated container. Work is done on the water by a rotating paddle wheel, which is driven by heavy blocks falling at a constant speed. The stirred water is warmed due to the friction between it and the paddles. If the energy lost in the bearings and through the walls is neglected, then the loss in potential energy associated with the blocks equals the work done by the paddle wheel on the water. If the two blocks fall through a distance h, the loss in potential energy is $2mgh$, where m is the mass of one block; it is this energy that causes the temperature of the water to increase. By varying the conditions of the experiment, Joule found that the loss in mechanical energy $2mgh$ is proportional to the increase in water temperature ΔT. The proportionality constant was found to be approximately $4.18 \, \text{J/g} \cdot {}^\circ\text{C}$. Hence, $4.18 \, \text{J}$ of mechanical energy raises the temperature of $1 \, \text{g}$ of water by 1°C. More precise measurements taken later demonstrated the proportionality to be $4.186 \, \text{J/g} \cdot {}^\circ\text{C}$ when the temperature of the water was raised from 14.5°C to 15.5°C. We adopt this "15-degree calorie" value:

$$1 \, \text{cal} \equiv 4.186 \, \text{J} \qquad \textbf{(20.1)}$$

This equality is known, for purely historical reasons, as the **mechanical equivalent of heat.**

Benjamin Thompson (1753–1814). *(North Wind Picture Archives)*

Mechanical equivalent of heat

Thermal insulator

Figure 20.1 Joule's experiment for determining the mechanical equivalent of heat. The falling blocks rotate the paddles, causing the temperature of the water to increase.

EXAMPLE 20.1 ▶ **Losing Weight the Hard Way**

A student eats a dinner rated at 2 000 Calories. He wishes to do an equivalent amount of work in the gymnasium by lifting a 50.0-kg barbell. How many times must he raise the barbell to expend this much energy? Assume that he raises the barbell 2.00 m each time he lifts it and that he regains no energy when he drops the barbell to the floor.

Solution Because 1 Calorie = 1.00×10^3 cal, the work required is 2.00×10^6 cal. Converting this value to joules, we have for the total work required:

$$W = (2.00 \times 10^6 \text{ cal})(4.186 \text{ J/cal}) = 8.37 \times 10^6 \text{ J}$$

The work done in lifting the barbell a distance h is equal to mgh, and the work done in lifting it n times is $nmgh$. We equate this to the total work required:

$$W = nmgh = 8.37 \times 10^6 \text{ J}$$

$$n = \frac{8.37 \times 10^6 \text{ J}}{(50.0 \text{ kg})(9.80 \text{ m/s}^2)(2.00 \text{ m})} = \boxed{8.54 \times 10^3 \text{ times}}$$

If the student is in good shape and lifts the barbell once every 5 s, it will take him about 12 h to perform this feat. Clearly, it is much easier for this student to lose weight by dieting.

20.2 ▶ HEAT CAPACITY AND SPECIFIC HEAT

When energy is added to a substance and no work is done, the temperature of the substance usually rises. (An exception to this statement is the case in which a substance undergoes a change of state—also called a *phase transition*—as discussed in the next section.) The quantity of energy required to raise the temperature of a given mass of a substance by some amount varies from one substance to another. For example, the quantity of energy required to raise the temperature of 1 kg of water by 1°C is 4 186 J, but the quantity of energy required to raise the temperature of 1 kg of copper by 1°C is only 387 J. In the discussion that follows, we shall use heat as our example of energy transfer, but we shall keep in mind that we could change the temperature of our system by doing work on it.

The **heat capacity** C of a particular sample of a substance is defined as the amount of energy needed to raise the temperature of that sample by 1°C. From this definition, we see that if heat Q produces a change ΔT in the temperature of a substance, then

Heat capacity

$$Q = C\Delta T \tag{20.2}$$

The **specific heat** c of a substance is the heat capacity per unit mass. Thus, if energy Q transferred by heat to mass m of a substance changes the temperature of the sample by ΔT, then the specific heat of the substance is

Specific heat

$$c \equiv \frac{Q}{m\Delta T} \tag{20.3}$$

Specific heat is essentially a measure of how thermally insensitive a substance is to the addition of energy. The greater a material's specific heat, the more energy must be added to a given mass of the material to cause a particular temperature change. Table 20.1 lists representative specific heats.

From this definition, we can express the energy Q transferred by heat between a sample of mass m of a material and its surroundings for a temperature change ΔT as

$$Q = mc\Delta T \tag{20.4}$$

For example, the energy required to raise the temperature of 0.500 kg of water by 3.00°C is $(0.500 \text{ kg})(4\ 186 \text{ J/kg} \cdot ^\circ\text{C})(3.00^\circ\text{C}) = 6.28 \times 10^3$ J. Note that when the temperature increases, Q and ΔT are taken to be positive, and energy flows into

TABLE 20.1   Specific Heats of Some Substances at 25°C and Atmospheric Pressure

Substance	Specific Heat c	
	J/kg·°C	cal/g·°C
Elemental Solids		
Aluminum	900	0.215
Beryllium	1 830	0.436
Cadmium	230	0.055
Copper	387	0.092 4
Germanium	322	0.077
Gold	129	0.030 8
Iron	448	0.107
Lead	128	0.030 5
Silicon	703	0.168
Silver	234	0.056
Other Solids		
Brass	380	0.092
Glass	837	0.200
Ice (-5°C)	2 090	0.50
Marble	860	0.21
Wood	1 700	0.41
Liquids		
Alcohol (ethyl)	2 400	0.58
Mercury	140	0.033
Water (15°C)	4 186	1.00
Gas		
Steam (100°C)	2 010	0.48

the system. When the temperature decreases, Q and ΔT are negative, and energy flows out of the system.

Specific heat varies with temperature. However, if temperature intervals are not too great, the temperature variation can be ignored and c can be treated as a constant.[2] For example, the specific heat of water varies by only about 1% from 0°C to 100°C at atmospheric pressure. Unless stated otherwise, we shall neglect such variations.

Measured values of specific heats are found to depend on the conditions of the experiment. In general, measurements made at constant pressure are different from those made at constant volume. For solids and liquids, the difference between the two values is usually no greater than a few percent and is often neglected. Most of the values given in Table 20.1 were measured at atmospheric pressure and room temperature. As we shall see in Chapter 21, the specific heats for

[2] The definition given by Equation 20.3 assumes that the specific heat does not vary with temperature over the interval $\Delta T = T_f - T_i$. In general, if c varies with temperature over the interval, then the correct expression for Q is

$$Q = m \int_{T_i}^{T_f} c \, dT$$

gases measured at constant pressure are quite different from values measured at constant volume.

Imagine you have 1 kg each of iron, glass, and water, and that all three samples are at 10°C. (a) Rank the samples from lowest to highest temperature after 100 J of energy is added to each. (b) Rank them from least to greatest amount of energy transferred by heat if each increases in temperature by 20°C.

QuickLab ▶

In an open area, such as a parking lot, use the flame from a match to pop an air-filled balloon. Now try the same thing with a water-filled balloon. Why doesn't the water-filled balloon pop?

It is interesting to note from Table 20.1 that water has the highest specific heat of common materials. This high specific heat is responsible, in part, for the moderate temperatures found near large bodies of water. As the temperature of a body of water decreases during the winter, energy is transferred from the cooling water to the air by heat, increasing the internal energy of the air. Because of the high specific heat of water, a relatively large amount of energy is transferred to the air for even modest temperature changes of the water. The air carries this internal energy landward when prevailing winds are favorable. For example, the prevailing winds on the West Coast of the United States are toward the land (eastward). Hence, the energy liberated by the Pacific Ocean as it cools keeps coastal areas much warmer than they would otherwise be. This explains why the western coastal states generally have more favorable winter weather than the eastern coastal states, where the prevailing winds do not tend to carry the energy toward land.

A difference in specific heats causes the cheese topping on a slice of pizza to burn you more than a mouthful of crust at the same temperature. Both crust and cheese undergo the same change in temperature, starting at a high straight-from-the-oven value and ending at the temperature of the inside of your mouth, which is about 37°C. Because the cheese is much more likely to burn you, it must release much more energy as it cools than does the crust. If we assume roughly the same mass for both cheese and crust, then Equation 20.3 indicates that the specific heat of the cheese, which is mostly water, is greater than that of the crust, which is mostly air.

Conservation of Energy: Calorimetry

One technique for measuring specific heat involves heating a sample to some known temperature T_x, placing it in a vessel containing water of known mass and temperature $T_w < T_x$, and measuring the temperature of the water after equilibrium has been reached. Because a negligible amount of mechanical work is done in the process, the law of the conservation of energy requires that the amount of energy that leaves the sample (of unknown specific heat) equal the amount of energy that enters the water.[3] This technique is called **calorimetry,** and devices in which this energy transfer occurs are called **calorimeters.**

Conservation of energy allows us to write the equation

$$Q_{cold} = -Q_{hot} \tag{20.5}$$

which simply states that the energy leaving the hot part of the system by heat is equal to that entering the cold part of the system. The negative sign in the equation is necessary to maintain consistency with our sign convention for heat. The

---

[3] For precise measurements, the water container should be included in our calculations because it also exchanges energy with the sample. However, doing so would require a knowledge of its mass and composition. If the mass of the water is much greater than that of the container, we can neglect the effects of the container.

heat Q_{hot} is negative because energy is leaving the hot sample. The negative sign in the equation ensures that the right-hand side is positive and thus consistent with the left-hand side, which is positive because energy is entering the cold water.

Suppose m_x is the mass of a sample of some substance whose specific heat we wish to determine. Let us call its specific heat c_x and its initial temperature T_x. Likewise, let m_w, c_w, and T_w represent corresponding values for the water. If T_f is the final equilibrium temperature after everything is mixed, then from Equation 20.4, we find that the energy transfer for the water is $m_w c_w (T_f - T_w)$, which is positive because $T_f > T_w$, and that the energy transfer for the sample of unknown specific heat is $m_x c_x (T_f - T_x)$, which is negative. Substituting these expressions into Equation 20.5 gives

$$m_w c_w (T_f - T_w) = -m_x c_x (T_f - T_x)$$

Solving for c_x gives

$$c_x = \frac{m_w c_w (T_f - T_w)}{m_x (T_x - T_f)}$$

EXAMPLE 20.2   Cooling a Hot Ingot

A 0.050 0-kg ingot of metal is heated to 200.0°C and then dropped into a beaker containing 0.400 kg of water initially at 20.0°C. If the final equilibrium temperature of the mixed system is 22.4°C, find the specific heat of the metal.

Solution   According to Equation 20.5, we can write

$$m_w c_w (T_f - T_w) = -m_x c_x (T_f - T_x)$$

$$(0.400 \text{ kg})(4\ 186\ \text{J/kg} \cdot °\text{C})(22.4°\text{C} - 20.0°\text{C}) =$$

$$-(0.050\ 0\ \text{kg})(c_x)(22.4°\text{C} - 200.0°\text{C})$$

From this we find that

$$c_x = \boxed{453\ \text{J/kg} \cdot °\text{C}}$$

The ingot is most likely iron, as we can see by comparing this result with the data given in Table 20.1. Note that the temperature of the ingot is initially above the steam point. Thus, some of the water may vaporize when we drop the ingot into the water. We assume that we have a sealed system and thus that this steam cannot escape. Because the final equilibrium temperature is lower than the steam point, any steam that does result recondenses back into water.

Exercise   What is the amount of energy transferred to the water as the ingot is cooled?

Answer   4 020 J.

EXAMPLE 20.3   Fun Time for a Cowboy

A cowboy fires a silver bullet with a mass of 2.00 g and with a muzzle speed of 200 m/s into the pine wall of a saloon. Assume that all the internal energy generated by the impact remains with the bullet. What is the temperature change of the bullet?

Solution   The kinetic energy of the bullet is

$$\tfrac{1}{2}mv^2 = \tfrac{1}{2}(2.00 \times 10^{-3}\ \text{kg})(200\ \text{m/s})^2 = 40.0\ \text{J}$$

Because nothing in the environment is hotter than the bullet, the bullet gains no energy by heat. Its temperature increases because the 40.0 J of kinetic energy becomes 40.0 J of extra internal energy. The temperature change is the same as that which would take place if 40.0 J of energy were transferred by

heat from a stove to the bullet. If we imagine this latter process taking place, we can calculate ΔT from Equation 20.4. Using 234 J/kg · °C as the specific heat of silver (see Table 20.1), we obtain

$$\Delta T = \frac{Q}{mc} = \frac{40.0\ \text{J}}{(2.00 \times 10^{-3}\ \text{kg})(234\ \text{J/kg} \cdot °\text{C})} = \boxed{85.5°\text{C}}$$

Exercise   Suppose that the cowboy runs out of silver bullets and fires a lead bullet of the same mass and at the same speed into the wall. What is the temperature change of the bullet?

Answer   156°C.

20.3 LATENT HEAT

A substance often undergoes a change in temperature when energy is transferred between it and its surroundings. There are situations, however, in which the transfer of energy does not result in a change in temperature. This is the case whenever the physical characteristics of the substance change from one form to another; such a change is commonly referred to as a **phase change.** Two common phase changes are from solid to liquid (melting) and from liquid to gas (boiling); another is a change in the crystalline structure of a solid. All such phase changes involve a change in internal energy but no change in temperature. The increase in internal energy in boiling, for example, is represented by the breaking of bonds between molecules in the liquid state; this bond breaking allows the molecules to move farther apart in the gaseous state, with a corresponding increase in intermolecular potential energy.

As you might expect, different substances respond differently to the addition or removal of energy as they change phase because their internal molecular arrangements vary. Also, the amount of energy transferred during a phase change depends on the amount of substance involved. (It takes less energy to melt an ice cube than it does to thaw a frozen lake.) If a quantity Q of energy transfer is required to change the phase of a mass m of a substance, the ratio $L \equiv Q/m$ characterizes an important thermal property of that substance. Because this added or removed energy does not result in a temperature change, the quantity L is called the **latent heat** (literally, the "hidden" heat) of the substance. The value of L for a substance depends on the nature of the phase change, as well as on the properties of the substance.

From the definition of latent heat, and again choosing heat as our energy transfer mechanism, we find that the energy required to change the phase of a given mass m of a pure substance is

$$Q = mL \qquad\qquad (20.6)$$

Latent heat of fusion L_f is the term used when the phase change is from solid to liquid (*to fuse* means "to combine by melting"), and **latent heat of vaporization**

TABLE 20.2 Latent Heats of Fusion and Vaporization

Substance	Melting Point (°C)	Latent Heat of Fusion (J/kg)	Boiling Point (°C)	Latent Heat of Vaporization (J/kg)
Helium	− 269.65	5.23×10^3	− 268.93	2.09×10^4
Nitrogen	− 209.97	2.55×10^4	− 195.81	2.01×10^5
Oxygen	− 218.79	1.38×10^4	− 182.97	2.13×10^5
Ethyl alcohol	− 114	1.04×10^5	78	8.54×10^5
Water	0.00	3.33×10^5	100.00	2.26×10^6
Sulfur	119	3.81×10^4	444.60	3.26×10^5
Lead	327.3	2.45×10^4	1 750	8.70×10^5
Aluminum	660	3.97×10^5	2 450	1.14×10^7
Silver	960.80	8.82×10^4	2 193	2.33×10^6
Gold	1 063.00	6.44×10^4	2 660	1.58×10^6
Copper	1 083	1.34×10^5	1 187	5.06×10^6

L_v is the term used when the phase change is from liquid to gas (the liquid "vaporizes").[4] The latent heats of various substances vary considerably, as data in Table 20.2 show.

Quick Quiz 20.2

Which is more likely to cause a serious burn, 100°C liquid water or an equal mass of 100°C steam?

To understand the role of latent heat in phase changes, consider the energy required to convert a 1.00-g block of ice at −30.0°C to steam at 120.0°C. Figure 20.2 indicates the experimental results obtained when energy is gradually added to the ice. Let us examine each portion of the red curve.

Part A. On this portion of the curve, the temperature of the ice changes from −30.0°C to 0.0°C. Because the specific heat of ice is 2 090 J/kg·°C, we can calculate the amount of energy added by using Equation 20.4:

$$Q = m_i c_i \Delta T = (1.00 \times 10^{-3}\text{ kg})(2\,090\text{ J/kg·°C})(30.0°\text{C}) = 62.7\text{ J}$$

Part B. When the temperature of the ice reaches 0.0°C, the ice–water mixture remains at this temperature—even though energy is being added—until all the ice melts. The energy required to melt 1.00 g of ice at 0.0°C is, from Equation 20.6,

$$Q = mL_f = (1.00 \times 10^{-3}\text{ kg})(3.33 \times 10^5\text{ J/kg}) = 333\text{ J}$$

Thus, we have moved to the 396 J (= 62.7 J + 333 J) mark on the energy axis.

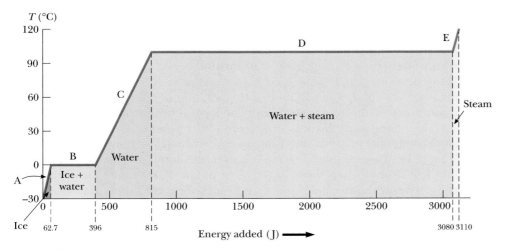

Figure 20.2 A plot of temperature versus energy added when 1.00 g of ice initially at −30.0°C is converted to steam at 120.0°C.

[4] When a gas cools, it eventually *condenses*—that is, it returns to the liquid phase. The energy given up per unit mass is called the *latent heat of condensation* and is numerically equal to the latent heat of vaporization. Likewise, when a liquid cools, it eventually solidifies, and the *latent heat of solidification* is numerically equal to the latent heat of fusion.

Part C. Between 0.0°C and 100.0°C, nothing surprising happens. No phase change occurs, and so all energy added to the water is used to increase its temperature. The amount of energy necessary to increase the temperature from 0.0°C to 100.0°C is

$$Q = m_w c_w \Delta T = (1.00 \times 10^{-3} \text{ kg})(4.19 \times 10^3 \text{ J/kg} \cdot °\text{C})(100.0°\text{C}) = 419 \text{ J}$$

Part D. At 100.0°C, another phase change occurs as the water changes from water at 100.0°C to steam at 100.0°C. Similar to the ice–water mixture in part B, the water–steam mixture remains at 100.0°C—even though energy is being added—until all of the liquid has been converted to steam. The energy required to convert 1.00 g of water to steam at 100.0°C is

$$Q = m L_v = (1.00 \times 10^{-3} \text{ kg})(2.26 \times 10^6 \text{ J/kg}) = 2.26 \times 10^3 \text{ J}$$

Part E. On this portion of the curve, as in parts A and C, no phase change occurs; thus, all energy added is used to increase the temperature of the steam. The energy that must be added to raise the temperature of the steam from 100.0°C to 120.0°C is

$$Q = m_s c_s \Delta T = (1.00 \times 10^{-3} \text{ kg})(2.01 \times 10^3 \text{ J/kg} \cdot °\text{C})(20.0°\text{C}) = 40.2 \text{ J}$$

The total amount of energy that must be added to change 1 g of ice at $-30.0°$C to steam at 120.0°C is the sum of the results from all five parts of the curve, which is 3.11×10^3 J. Conversely, to cool 1 g of steam at 120.0°C to ice at $-30.0°$C, we must remove 3.11×10^3 J of energy.

We can describe phase changes in terms of a rearrangement of molecules when energy is added to or removed from a substance. (For elemental substances in which the atoms do not combine to form molecules, the following discussion should be interpreted in terms of atoms. We use the general term *molecules* to refer to both molecular substances and elemental substances.) Consider first the liquid-to-gas phase change. The molecules in a liquid are close together, and the forces between them are stronger than those between the more widely separated molecules of a gas. Therefore, work must be done on the liquid against these attractive molecular forces if the molecules are to separate. The latent heat of vaporization is the amount of energy per unit mass that must be added to the liquid to accomplish this separation.

Similarly, for a solid, we imagine that the addition of energy causes the amplitude of vibration of the molecules about their equilibrium positions to become greater as the temperature increases. At the melting point of the solid, the amplitude is great enough to break the bonds between molecules and to allow molecules to move to new positions. The molecules in the liquid also are bound to each other, but less strongly than those in the solid phase. The latent heat of fusion is equal to the energy required per unit mass to transform the bonds among all molecules from the solid-type bond to the liquid-type bond.

As you can see from Table 20.2, the latent heat of vaporization for a given substance is usually somewhat higher than the latent heat of fusion. This is not surprising if we consider that the average distance between molecules in the gas phase is much greater than that in either the liquid or the solid phase. In the solid-to-liquid phase change, we transform solid-type bonds between molecules into liquid-type bonds between molecules, which are only slightly less strong. In the liquid-to-gas phase change, however, we break liquid-type bonds and create a situation in which the molecules of the gas essentially are not bonded to each other. Therefore, it is not surprising that more energy is required to vaporize a given mass of substance than is required to melt it.

Quick Quiz 20.3

Calculate the slopes for the A, C, and E portions of Figure 20.2. Rank the slopes from least to greatest and explain what this ordering means.

Problem-Solving Hints

Calorimetry Problems

If you are having difficulty in solving calorimetry problems, be sure to consider the following points:

- Units of measure must be consistent. For instance, if you are using specific heats measured in cal/g · °C, be sure that masses are in grams and temperatures are in Celsius degrees.
- Transfers of energy are given by the equation $Q = mc\Delta T$ only for those processes in which no phase changes occur. Use the equations $Q = mL_f$ and $Q = mL_v$ only when phase changes *are* taking place.
- Often, errors in sign are made when the equation $Q_{cold} = -Q_{hot}$ is used. Make sure that you use the negative sign in the equation, and remember that ΔT is always the final temperature minus the initial temperature.

EXAMPLE 20.4   Cooling the Steam

What mass of steam initially at 130°C is needed to warm 200 g of water in a 100-g glass container from 20.0°C to 50.0°C?

Solution   The steam loses energy in three stages. In the first stage, the steam is cooled to 100°C. The energy transfer in the process is

$$Q_1 = m_s c_s \, \Delta T = m_s (2.01 \times 10^3 \, \text{J/kg} \cdot {}^\circ\text{C}) \, (-30.0^\circ\text{C})$$
$$= -m_s (6.03 \times 10^4 \, \text{J/kg})$$

where m_s is the unknown mass of the steam.

In the second stage, the steam is converted to water. To find the energy transfer during this phase change, we use $Q = -mL_v$, where the negative sign indicates that energy is leaving the steam:

$$Q_2 = -m_s (2.26 \times 10^6 \, \text{J/kg})$$

In the third stage, the temperature of the water created from the steam is reduced to 50.0°C. This change requires an energy transfer of

$$Q_3 = m_s c_w \, \Delta T = m_s (4.19 \times 10^3 \, \text{J/kg} \cdot {}^\circ\text{C}) \, (-50.0^\circ\text{C})$$
$$= -m_s (2.09 \times 10^5 \, \text{J/kg})$$

Adding the energy transfers in these three stages, we obtain

$$Q_{hot} = Q_1 + Q_2 + Q_3$$
$$= -m_s (6.03 \times 10^4 \, \text{J/kg} + 2.26 \times 10^6 \, \text{J/kg}$$
$$+ 2.09 \times 10^5 \, \text{J/kg})$$
$$= -m_s (2.53 \times 10^6 \, \text{J/kg})$$

Now, we turn our attention to the temperature increase of the water and the glass. Using Equation 20.4, we find that

$$Q_{cold} = (0.200 \, \text{kg}) \, (4.19 \times 10^3 \, \text{J/kg} \cdot {}^\circ\text{C}) \, (30.0^\circ\text{C})$$
$$+ (0.100 \, \text{kg}) \, (837 \, \text{J/kg} \cdot {}^\circ\text{C}) \, (30.0^\circ\text{C})$$
$$= 2.77 \times 10^4 \, \text{J}$$

Using Equation 20.5, we can solve for the unknown mass:

$$Q_{cold} = -Q_{hot}$$
$$2.77 \times 10^4 \, \text{J} = -[-m_s (2.53 \times 10^6 \, \text{J/kg})]$$
$$m_s = \boxed{1.09 \times 10^{-2} \, \text{kg} = 10.9 \, \text{g}}$$

EXAMPLE 20.5 Boiling Liquid Helium

Liquid helium has a very low boiling point, 4.2 K, and a very low latent heat of vaporization, 2.09×10^4 J/kg. If energy is transferred to a container of boiling liquid helium from an immersed electric heater at a rate of 10.0 W, how long does it take to boil away 1.00 kg of the liquid?

Solution Because $L_v = 2.09 \times 10^4$ J/kg, we must supply 2.09×10^4 J of energy to boil away 1.00 kg. Because 10.0 W = 10.0 J/s, 10.0 J of energy is transferred to the helium each second. Therefore, the time it takes to transfer 2.09×10^4 J

of energy is

$$t = \frac{2.09 \times 10^4 \text{ J}}{10.0 \text{ J/s}} = 2.09 \times 10^3 \text{ s} \approx \boxed{35 \text{ min}}$$

Exercise If 10.0 W of power is supplied to 1.00 kg of water at 100°C, how long does it take for the water to completely boil away?

Answer 62.8 h.

20.4 WORK AND HEAT IN THERMODYNAMIC PROCESSES

In the macroscopic approach to thermodynamics, we describe the *state* of a system using such variables as pressure, volume, temperature, and internal energy. The number of macroscopic variables needed to characterize a system depends on the nature of the system. For a homogeneous system, such as a gas containing only one type of molecule, usually only two variables are needed. However, it is important to note that a *macroscopic state* of an isolated system can be specified only if the system is in thermal equilibrium internally. In the case of a gas in a container, internal thermal equilibrium requires that every part of the gas be at the same pressure and temperature.

Consider a gas contained in a cylinder fitted with a movable piston (Fig. 20.3). At equilibrium, the gas occupies a volume V and exerts a uniform pressure P on the cylinder's walls and on the piston. If the piston has a cross-sectional area A, the

(a)   (b)

Figure 20.3 Gas contained in a cylinder at a pressure P does work on a moving piston as the system expands from a volume V to a volume $V + dV$.

force exerted by the gas on the piston is $F = PA$. Now let us assume that the gas expands **quasi-statically,** that is, slowly enough to allow the system to remain essentially in thermal equilibrium at all times. As the piston moves up a distance dy, the work done by the gas on the piston is

$$dW = F\, dy = PA\, dy$$

Because $A\, dy$ is the increase in volume of the gas dV, we can express the work done by the gas as

$$dW = P\, dV \qquad\qquad \textbf{(20.7)}$$

Because the gas expands, dV is positive, and so the work done by the gas is positive. If the gas were compressed, dV would be negative, indicating that the work done by the gas (which can be interpreted as work done *on* the gas) was negative.

In the thermodynamics problems that we shall solve, we shall identify the system of interest as a substance that is exchanging energy with the environment. In many problems, this will be a gas contained in a vessel; however, we will also consider problems involving liquids and solids. It is an unfortunate fact that, because of the separate historical development of thermodynamics and mechanics, positive work for a thermodynamic system is commonly defined as the work done *by* the system, rather than that done *on* the system. This is the reverse of the case for our study of work in mechanics. Thus, **in thermodynamics, positive work represents a transfer of energy out of the system.** We will use this convention to be consistent with common treatments of thermodynamics.

The total work done by the gas as its volume changes from V_i to V_f is given by the integral of Equation 20.7:

$$W = \int_{V_i}^{V_f} P\, dV \qquad\qquad \textbf{(20.8)}$$

To evaluate this integral, it is not enough that we know only the initial and final values of the pressure. We must also know the pressure at every instant during the expansion; we would know this if we had a functional dependence of P with respect to V. This important point is true for any process—the expansion we are discussing here, or any other. To fully specify a process, we must know the values of the thermodynamic variables at every state through which the system passes between the initial and final states. In the expansion we are considering here, we can plot the pressure and volume at each instant to create a PV diagram like the one shown in Figure 20.4. The value of the integral in Equation 20.8 is the area bounded by such a curve. Thus, we can say that

the work done by a gas in the expansion from an initial state to a final state is the area under the curve connecting the states in a PV diagram.

As Figure 20.4 shows, the work done in the expansion from the initial state i to the final state f depends on the path taken between these two states, where the *path* on a PV diagram is a description of the thermodynamic process through which the system is taken. To illustrate this important point, consider several paths connecting i and f (Fig. 20.5). In the process depicted in Figure 20.5a, the pressure of the gas is first reduced from P_i to P_f by cooling at constant volume V_i. The gas then expands from V_i to V_f at constant pressure P_f. The value of the work done along this path is equal to the area of the shaded rectangle, which is equal to

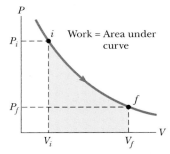

Figure 20.4   A gas expands quasi-statically (slowly) from state i to state f. The work done by the gas equals the area under the PV curve.

Work equals area under the curve in a PV diagram.

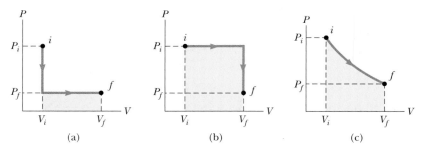

Figure 20.5   The work done by a gas as it is taken from an initial state to a final state depends on the path between these states.

Work done depends on the path between the initial and final states.

$P_f(V_f - V_i)$. In Figure 20.5b, the gas first expands from V_i to V_f at constant pressure P_i. Then, its pressure is reduced to P_f at constant volume V_f. The value of the work done along this path is $P_i(V_f - V_i)$, which is greater than that for the process described in Figure 20.5a. Finally, for the process described in Figure 20.5c, where both P and V change continuously, the work done has some value intermediate between the values obtained in the first two processes. Therefore, we see that **the work done by a system depends on the initial and final states and on the path followed by the system between these states.**

The energy transfer by heat Q into or out of a system also depends on the process. Consider the situations depicted in Figure 20.6. In each case, the gas has the same initial volume, temperature, and pressure and is assumed to be ideal. In Figure 20.6a, the gas is thermally insulated from its surroundings except at the bottom of the gas-filled region, where it is in thermal contact with an energy reservoir. An *energy reservoir* is a source of energy that is considered to be so great that a finite transfer of energy from the reservoir does not change its temperature. The piston is held at its initial position by an external agent—a hand, for instance. When the force with which the piston is held is reduced slightly, the piston rises very slowly to its final position. Because the piston is moving upward, the gas is doing work on

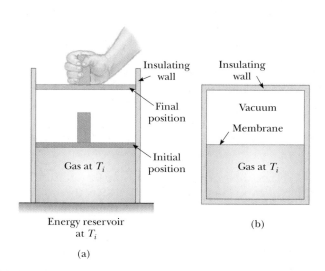

Figure 20.6   (a) A gas at temperature T_i expands slowly while absorbing energy from a reservoir in order to maintain a constant temperature. (b) A gas expands rapidly into an evacuated region after a membrane is broken.

the piston. During this expansion to the final volume V_f, just enough energy is transferred by heat from the reservoir to the gas to maintain a constant temperature T_i.

Now consider the completely thermally insulated system shown in Figure 20.6b. When the membrane is broken, the gas expands rapidly into the vacuum until it occupies a volume V_f and is at a pressure P_f. In this case, the gas does no work because there is no movable piston on which the gas applies a force. Furthermore, no energy is transferred by heat through the insulating wall.

The initial and final states of the ideal gas in Figure 20.6a are identical to the initial and final states in Figure 20.6b, but the paths are different. In the first case, the gas does work on the piston, and energy is transferred slowly to the gas. In the second case, no energy is transferred, and the value of the work done is zero. Therefore, we conclude that **energy transfer by heat, like work done, depends on the initial, final, and intermediate states of the system.** In other words, because heat and work depend on the path, neither quantity is determined solely by the end points of a thermodynamic process.

This device, called *Hero's engine*, was invented around 150 B.C. by Hero in Alexandria. When water is boiled in the flask, which is suspended by a cord, steam exits through two tubes at the sides (in opposite directions), creating a torque that rotates the flask. *(Courtesy of Central Scientific Company)*

20.5 ▷ THE FIRST LAW OF THERMODYNAMICS

When we introduced the law of conservation of mechanical energy in Chapter 8, we stated that the mechanical energy of a system is constant in the absence of nonconservative forces such as friction. That is, we did not include changes in the internal energy of the system in this mechanical model. The first law of thermodynamics is a generalization of the law of conservation of energy that encompasses changes in internal energy. It is a universally valid law that can be applied to many processes and provides a connection between the microscopic and macroscopic worlds.

We have discussed two ways in which energy can be transferred between a system and its surroundings. One is work done by the system, which requires that there be a macroscopic displacement of the point of application of a force (or pressure). The other is heat, which occurs through random collisions between the molecules of the system. Both mechanisms result in a change in the internal energy of the system and therefore usually result in measurable changes in the macroscopic variables of the system, such as the pressure, temperature, and volume of a gas.

To better understand these ideas on a quantitative basis, suppose that a system undergoes a change from an initial state to a final state. During this change, energy transfer by heat Q to the system occurs, and work W is done *by* the system. As an example, suppose that the system is a gas in which the pressure and volume change from P_i and V_i to P_f and V_f. If the quantity $Q - W$ is measured for various paths connecting the initial and final equilibrium states, we find that it is the same for all paths connecting the two states. We conclude that the quantity $Q - W$ is determined completely by the initial and final states of the system, and we call this quantity the **change in the internal energy** of the system. Although Q and W both depend on the path, **the quantity $Q - W$ is independent of the path.** If we use the sumbol E_{int} to represent the internal energy, then the *change* in internal energy ΔE_{int} can be expressed as[5]

$Q - W$ is the change in internal energy

$$\Delta E_{int} = Q - W \qquad \textbf{(20.9)}$$

First-law equation

---

[5] It is an unfortunate accident of history that the traditional symbol for internal energy is U, which is also the traditional symbol for potential energy, as introduced in Chapter 8. To avoid confusion between potential energy and internal energy, we use the symbol E_{int} for internal energy in this book. If you take an advanced course in thermodynamics, however, be prepared to see U used as the symbol for internal energy.

where all quantities must have the same units of measure for energy.[6] Equation 20.9 is known as the **first-law equation** and is a key concept in many applications. As a reminder, we use the convention that Q is positive when energy enters the system and negative when energy leaves the system, and that W is positive when the system does work on the surroundings and negative when work is done on the system.

When a system undergoes an infinitesimal change in state in which a small amount of energy dQ is transferred by heat and a small amount of work dW is done, the internal energy changes by a small amount dE_{int}. Thus, for infinitesimal processes we can express the first-law equation as[7]

First-law equation for infinitesimal changes

$$dE_{int} = dQ - dW$$

The first-law equation is an energy conservation equation specifying that the only type of energy that changes in the system is the internal energy E_{int}. Let us look at some special cases in which this condition exists.

Isolated system

First, let us consider an *isolated system*—that is, one that does not interact with its surroundings. In this case, no energy transfer by heat takes place and the value of the work done by the system is zero; hence, the internal energy remains constant. That is, because $Q = W = 0$, it follows that $\Delta E_{int} = 0$, and thus $E_{int,\,i} = E_{int,\,f}$. We conclude that **the internal energy E_{int} of an isolated system remains constant.**

Next, we consider the case of a system (one not isolated from its surroundings) that is taken through a **cyclic process**—that is, a process that starts and ends at the same state. In this case, the change in the internal energy must again be zero, and therefore the energy Q added to the system must equal the work W done by the system during the cycle. That is, in a cyclic process,

Cyclic process

$$\Delta E_{int} = 0 \quad \text{and} \quad Q = W$$

On a PV diagram, a cyclic process appears as a closed curve. (The processes described in Figure 20.5 are represented by open curves because the initial and final states differ.) It can be shown that **in a cyclic process, the net work done by the system per cycle equals the area enclosed by the path representing the process on a PV diagram.**

If the value of the work done by the system during some process is zero, then the change in internal energy ΔE_{int} equals the energy transfer Q into or out of the system:

$$\Delta E_{int} = Q$$

If energy enters the system, then Q is positive and the internal energy increases. For a gas, we can associate this increase in internal energy with an increase in the kinetic energy of the molecules. Conversely, if no energy transfer occurs during some process but work is done by the system, then the change in internal energy equals the negative value of the work done by the system:

$$\Delta E_{int} = -W$$

---

[6] For the definition of work from our mechanics studies, the first law would be written as $\Delta E_{int} = Q + W$ because energy transfer into the system by either work or heat would increase the internal energy of the system. Because of the reversal of the definition of positive work discussed in Section 20.4, the first law appears as in Equation 20.9, with a minus sign.

[7] Note that dQ and dW are not true differential quantities; however, dE_{int} is. Because dQ and dW are *inexact differentials*, they are often represented by the symbols dQ and dW. For further details on this point, see an advanced text on thermodynamics, such as R. P. Bauman, *Modern Thermodynamics and Statistical Mechanics*, New York, Macmillan Publishing Co., 1992.

For example, if a gas is compressed by a moving piston in an insulated cylinder, no energy is transferred by heat and the work done by the gas is negative; thus, the internal energy increases because kinetic energy is transferred from the moving piston to the gas molecules.

On a microscopic scale, no distinction exists between the result of heat and that of work. Both heat and work can produce a change in the internal energy of a system. Although the macroscopic quantities Q and W are *not* properties of a system, they are related to the change of the internal energy of a system through the first-law equation. Once we define a process, or path, we can either calculate or measure Q and W, and we can find the change in the system's internal energy using the first-law equation.

One of the important consequences of the first law of thermodynamics is that there exists a quantity known as internal energy whose value is determined by the state of the system. The internal energy function is therefore called a *state function*.

20.6 SOME APPLICATIONS OF THE FIRST LAW OF THERMODYNAMICS

Before we apply the first law of thermodynamics to specific systems, it is useful for us to first define some common thermodynamic processes. An **adiabatic process** is one during which no energy enters or leaves the system by heat—that is, $Q = 0$. An adiabatic process can be achieved either by thermally insulating the system from its surroundings (as shown in Fig. 20.6b) or by performing the process rapidly, so that there is little time for energy to transfer by heat. Applying the first law of thermodynamics to an adiabatic process, we see that

> In an adiabatic process, $Q = 0$.

$$\Delta E_{\text{int}} = -W \qquad \text{(adiabatic process)} \qquad \textbf{(20.10)}$$

> First-law equation for an adiabatic process

From this result, we see that if a gas expands adiabatically such that W is positive, then ΔE_{int} is negative and the temperature of the gas decreases. Conversely, the temperature of a gas increases when the gas is compressed adiabatically.

Adiabatic processes are very important in engineering practice. Some common examples are the expansion of hot gases in an internal combustion engine, the liquefaction of gases in a cooling system, and the compression stroke in a diesel engine.

The process described in Figure 20.6b, called an **adiabatic free expansion,** is unique. The process is adiabatic because it takes place in an insulated container. Because the gas expands into a vacuum, it does not apply a force on a piston as was depicted in Figure 20.6a, so no work is done on or by the gas. Thus, in this adiabatic process, both $Q = 0$ and $W = 0$. As a result, $\Delta E_{\text{int}} = 0$ for this process, as we can see from the first law. That is, **the initial and final internal energies of a gas are equal in an adiabatic free expansion.** As we shall see in the next chapter, the internal energy of an ideal gas depends only on its temperature. Thus, we expect no change in temperature during an adiabatic free expansion. This prediction is in accord with the results of experiments performed at low pressures. (Experiments performed at high pressures for real gases show a slight decrease or increase in temperature after the expansion. This change is due to intermolecular interactions, which represent a deviation from the model of an ideal gas.)

> In an adiabatic free expansion, $\Delta E_{\text{int}} = 0$.

A process that occurs at constant pressure is called an **isobaric process.** In such a process, the values of the heat and the work are both usually nonzero. The

> In an isobaric process, P remains constant.

work done by the gas is simply

$$W = P(V_f - V_i) \qquad \text{(isobaric process)} \qquad \textbf{(20.11)}$$

where P is the constant pressure.

A process that takes place at constant volume is called an **isovolumetric process.** In such a process, the value of the work done is clearly zero because the volume does not change. Hence, from the first law we see that in an isovolumetric process, because $W = 0$,

First-law equation for a constant-volume process

$$\Delta E_{int} = Q \qquad \text{(isovolumetric process)} \qquad \textbf{(20.12)}$$

This expression specifies that **if energy is added by heat to a system kept at constant volume, then all of the transferred energy remains in the system as an increase of the internal energy of the system.** For example, when a can of spray paint is thrown into a fire, energy enters the system (the gas in the can) by heat through the metal walls of the can. Consequently, the temperature, and thus the pressure, in the can increases until the can possibly explodes.

A process that occurs at constant temperature is called an **isothermal process.** A plot of P versus V at constant temperature for an ideal gas yields a hyperbolic curve called an *isotherm.* The internal energy of an ideal gas is a function of temperature only. Hence, in an isothermal process involving an ideal gas, $\Delta E_{int} = 0$. For an isothermal process, then, we conclude from the first law that the energy transfer Q must be equal to the work done by the gas—that is, $Q = W$. Any energy that enters the system by heat is transferred out of the system by work; as a result, no change of the internal energy of the system occurs.

In an isothermal process, T remains constant.

Quick Quiz 20.4

In the last three columns of the following table, fill in the boxes with $-$, $+$, or 0. For each situation, the system to be considered is identified.

Situation	System	Q	W	ΔE_{int}
(a) Rapidly pumping up a bicycle tire	Air in the pump			
(b) Pan of room-temperature water sitting on a hot stove	Water in the pan			
(c) Air quickly leaking out of a balloon	Air originally in balloon			

Isothermal Expansion of an Ideal Gas

Suppose that an ideal gas is allowed to expand quasi-statically at constant temperature, as described by the PV diagram shown in Figure 20.7. The curve is a hyperbola (see Appendix B, Eq. B.23), and the equation of state of an ideal gas with T constant indicates that the equation of this curve is $PV = $ constant. The isothermal expansion of the gas can be achieved by placing the gas in thermal contact with an energy reservoir at the same temperature, as shown in Figure 20.6a.

Let us calculate the work done by the gas in the expansion from state i to state f. The work done by the gas is given by Equation 20.8. Because the gas is ideal and the process is quasi-static, we can use the expression $PV = nRT$ for each point on

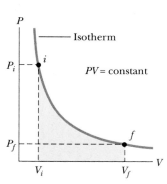

Figure 20.7 The PV diagram for an isothermal expansion of an ideal gas from an initial state to a final state. The curve is a hyperbola.

the path. Therefore, we have

$$W = \int_{V_i}^{V_f} P\,dV = \int_{V_i}^{V_f} \frac{nRT}{V}\,dV$$

Because T is constant in this case, it can be removed from the integral along with n and R:

$$W = nRT \int_{V_i}^{V_f} \frac{dV}{V} = nRT \ln V \Big|_{V_i}^{V_f}$$

To evaluate the integral, we used $\int (dx/x) = \ln x$. Evaluating this at the initial and final volumes, we have

$$W = nRT \ln\left(\frac{V_f}{V_i}\right) \qquad\qquad \textbf{(20.13)}$$

Work done by an ideal gas in an isothermal process

Numerically, this work W equals the shaded area under the PV curve shown in Figure 20.7. Because the gas expands, $V_f > V_i$, and the value for the work done by the gas is positive, as we expect. If the gas is compressed, then $V_f < V_i$, and the work done by the gas is negative.

Quick Quiz 20.5

Characterize the paths in Figure 20.8 as isobaric, isovolumetric, isothermal, or adiabatic. Note that $Q = 0$ for path B.

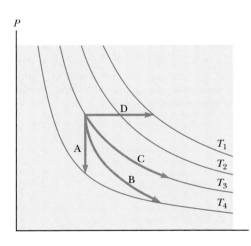

Figure 20.8   Identify the nature of paths A, B, C, and D.

EXAMPLE 20.6   An Isothermal Expansion

A 1.0-mol sample of an ideal gas is kept at 0.0°C during an expansion from 3.0 L to 10.0 L. (a) How much work is done by the gas during the expansion?

Solution   Substituting the values into Equation 20.13, we have

$$W = nRT \ln\left(\frac{V_f}{V_i}\right)$$

$$W = (1.0\ \text{mol})(8.31\ \text{J/mol·K})(273\ \text{K}) \ln\left(\frac{10.0}{3.0}\right)$$

$$= \boxed{2.7 \times 10^3\ \text{J}}$$

(b) How much energy transfer by heat occurs with the surroundings in this process?

Solution From the first law, we find that

$$\Delta E_{int} = Q - W$$
$$0 = Q - W$$

$$Q = W = \boxed{2.7 \times 10^3 \, J}$$

(c) If the gas is returned to the original volume by means of an isobaric process, how much work is done by the gas?

Solution The work done in an isobaric process is given by Equation 20.11. We are not given the pressure, so we need to incorporate the ideal gas law:

$$W = P(V_f - V_i) = \frac{nRT_i}{V_i} (V_f - V_i)$$

$$= \frac{(1.0 \text{ mol}) (8.31 \text{ J/mol} \cdot \text{K}) (273 \text{ K})}{10.0 \times 10^{-3} \text{m}^3}$$

$$\times (3.0 \times 10^{-3} \text{m}^3 - 10.0 \times 10^{-3} \text{m}^3)$$

$$= \boxed{-1.6 \times 10^3 \, J}$$

Notice that we use the initial temperature and volume to determine the value of the constant pressure because we do not know the final temperature. The work done by the gas is negative because the gas is being compressed.

EXAMPLE 20.7 Boiling Water

Suppose 1.00 g of water vaporizes isobarically at atmospheric pressure $(1.013 \times 10^5 \text{ Pa})$. Its volume in the liquid state is $V_i = V_{liquid} = 1.00 \text{ cm}^3$, and its volume in the vapor state is $V_f = V_{vapor} = 1\,671 \text{ cm}^3$. Find the work done in the expansion and the change in internal energy of the system. Ignore any mixing of the steam and the surrounding air—imagine that the steam simply pushes the surrounding air out of the way.

Solution Because the expansion takes place at constant pressure, the work done by the system in pushing away the surrounding air is, from Equation 20.11,

$$W = P(V_f - V_i)$$
$$= (1.013 \times 10^5 \text{ Pa}) (1\,671 \times 10^{-6} \text{ m}^3 - 1.00 \times 10^{-6} \text{ m}^3)$$

$$= \boxed{169 \, J}$$

To determine the change in internal energy, we must know the energy transfer Q needed to vaporize the water. Using Equation 20.6 to calculate the latent heat of vaporization for water, we have

$$Q = mL_v = (1.00 \times 10^{-3} \text{ kg}) (2.26 \times 10^6 \text{ J/kg}) = 2\,260 \, J$$

Hence, from the first law, the change in internal energy is

$$\Delta E_{int} = Q - W = 2\,260 \, J - 169 \, J = \boxed{2.09 \text{ kJ}}$$

The positive value for ΔE_{int} indicates that the internal energy of the system increases. We see that most $(2\,090 \text{ J}/2\,260 \text{ J} = 93\%)$ of the energy transferred to the liquid goes into increasing the internal energy of the system. Only $169 \text{ J}/2\,260 \text{ J} = 7\%$ leaves the system by work done by the steam on the surrounding atmosphere.

EXAMPLE 20.8 Heating a Solid

A 1.0-kg bar of copper is heated at atmospheric pressure. If its temperature increases from 20°C to 50°C, (a) what is the work done by the copper on the surrounding atmosphere?

Solution Because the process is isobaric, we can find the work done by the copper using Equation 20.11, $W = P(V_f - V_i)$. We can calculate the change in volume of the copper using Equation 19.6. Using the average linear expansion coefficient for copper given in Table 19.2, and remembering that $\beta = 3\alpha$, we obtain

$$\Delta V = \beta V_i \Delta T$$
$$= [5.1 \times 10^{-5} (\text{°C})^{-1}] (50°\text{C} - 20°\text{C}) V_i = 1.5 \times 10^{-3} \, V_i$$

The volume V_i is equal to m/ρ, and Table 15.1 indicates that the density of copper is $8.92 \times 10^3 \text{ kg/m}^3$. Hence,

$$\Delta V = (1.5 \times 10^{-3}) \left(\frac{1.0 \text{ kg}}{8.92 \times 10^3 \text{ kg/m}^3} \right) = 1.7 \times 10^{-7} \text{ m}^3$$

The work done is

$$W = P \Delta V = (1.013 \times 10^5 \text{ N/m}^2) (1.7 \times 10^{-7} \text{ m}^3)$$

$$= \boxed{1.7 \times 10^{-2} \, J}$$

(b) What quantity of energy is transferred to the copper by heat?

Solution Taking the specific heat of copper from Table 20.1 and using Equation 20.4, we find that the energy transferred by heat is

$$Q = mc\Delta T = (1.0 \text{ kg})(387 \text{ J/kg} \cdot {}^{\circ}\text{C})(30{}^{\circ}\text{C}) = \boxed{1.2 \times 10^4 \text{ J}}$$

(c) What is the increase in internal energy of the copper?

Solution From the first law of thermodynamics, we have

$$\Delta E_{\text{int}} = Q - W = 1.2 \times 10^4 \text{ J} - 1.7 \times 10^{-2} \text{ J} = \boxed{1.2 \times 10^4 \text{ J}}$$

Note that almost all of the energy transferred into the system by heat goes into increasing the internal energy. The fraction of energy used to do work on the surrounding atmosphere is only about 10^{-6}! Hence, when analyzing the thermal expansion of a solid or a liquid, the small amount of work done by the system is usually ignored.

20.7 ENERGY TRANSFER MECHANISMS

It is important to understand the rate at which energy is transferred between a system and its surroundings and the mechanisms responsible for the transfer. Therefore, let us now look at three common energy transfer mechanisms that can result in a change in internal energy of a system.

Thermal Conduction

The energy transfer process that is most clearly associated with a temperature difference is **thermal conduction.** In this process, the transfer can be represented on an atomic scale as an exchange of kinetic energy between microscopic particles—molecules, atoms, and electrons—in which less energetic particles gain energy in collisions with more energetic particles. For example, if you hold one end of a long metal bar and insert the other end into a flame, you will find that the temperature of the metal in your hand soon increases. The energy reaches your hand by means of conduction. We can understand the process of conduction by examining what is happening to the microscopic particles in the metal. Initially, before the rod is inserted into the flame, the microscopic particles are vibrating about their equilibrium positions. As the flame heats the rod, those particles near the flame begin to vibrate with greater and greater amplitudes. These particles, in turn, collide with their neighbors and transfer some of their energy in the collisions. Slowly, the amplitudes of vibration of metal atoms and electrons farther and farther from the flame increase until, eventually, those in the metal near your hand are affected. This increased vibration represents an increase in the temperature of the metal and of your potentially burned hand.

The rate of thermal conduction depends on the properties of the substance being heated. For example, it is possible to hold a piece of asbestos in a flame indefinitely. This implies that very little energy is conducted through the asbestos. In general, metals are good thermal conductors, and materials such as asbestos, cork, paper, and fiberglass are poor conductors. Gases also are poor conductors because the separation distance between the particles is so great. Metals are good thermal conductors because they contain large numbers of electrons that are relatively free to move through the metal and so can transport energy over large distances. Thus, in a good conductor, such as copper, conduction takes place both by means of the vibration of atoms and by means of the motion of free electrons.

Conduction occurs only if there is a difference in temperature between two parts of the conducting medium. Consider a slab of material of thickness Δx and cross-sectional area A. One face of the slab is at a temperature T_1, and the other face is at a temperature $T_2 > T_1$ (Fig. 20.9). Experimentally, it is found that the

Melted snow pattern on a parking lot surface indicates the presence of underground hot water pipes used to aid snow removal. Energy from the water is conducted from the pipes to the pavement, where it causes the snow to melt. *(Courtesy of Dr. Albert A. Bartlett, University of Colorado, Boulder, CO)*

energy Q transferred in a time Δt flows from the hotter face to the colder one. The rate $Q/\Delta t$ at which this energy flows is found to be proportional to the cross-sectional area and the temperature difference $\Delta T = T_2 - T_1$, and inversely proportional to the thickness:

$$\frac{Q}{\Delta t} \propto A \frac{\Delta T}{\Delta x}$$

It is convenient to use the symbol for power $\mathcal{P}$ to represent the **rate of energy transfer:** $\mathcal{P} = Q/\Delta t$. Note that $\mathcal{P}$ has units of watts when Q is in joules and Δt is in seconds. For a slab of infinitesimal thickness dx and temperature difference dT, we can write the **law of thermal conduction** as

| Law of thermal conduction |

$$\mathcal{P} = kA \left| \frac{dT}{dx} \right| \tag{20.14}$$

where the proportionality constant k is the **thermal conductivity** of the material and $|dT/dx|$ is the **temperature gradient** (the variation of temperature with position).

Suppose that a long, uniform rod of length L is thermally insulated so that energy cannot escape by heat from its surface except at the ends, as shown in Figure 20.10. One end is in thermal contact with an energy reservoir at temperature T_1, and the other end is in thermal contact with a reservoir at temperature $T_2 > T_1$. When a steady state has been reached, the temperature at each point along the rod is constant in time. In this case if we assume that k is not a function of temperature, the temperature gradient is the same everywhere along the rod and is

$$\left| \frac{dT}{dx} \right| = \frac{T_2 - T_1}{L}$$

Thus the rate of energy transfer by conduction through the rod is

$$\mathcal{P} = kA \frac{(T_2 - T_1)}{L} \tag{20.15}$$

Substances that are good thermal conductors have large thermal conductivity values, whereas good thermal insulators have low thermal conductivity values. Table 20.3 lists thermal conductivities for various substances. Note that metals are generally better thermal conductors than nonmetals are.

Figure 20.9   Energy transfer through a conducting slab with a cross-sectional area A and a thickness Δx. The opposite faces are at different temperatures T_1 and T_2.

Figure 20.10   Conduction of energy through a uniform, insulated rod of length L. The opposite ends are in thermal contact with energy reservoirs at different temperatures.

Quick Quiz 20.6

Will an ice cube wrapped in a wool blanket remain frozen for (a) a shorter length of time, (b) the same length of time, or (c) a longer length of time than an identical ice cube exposed to air at room temperature?

For a compound slab containing several materials of thicknesses L_1, L_2, . . . and thermal conductivities k_1, k_2, . . . , the rate of energy transfer through the slab at steady state is

$$\mathcal{P} = \frac{A(T_2 - T_1)}{\sum_i (L_i/k_i)} \tag{20.16}$$

TABLE 20.3	Thermal Conductivities
Substance	**Thermal Conductivity (W/m·°C)**
Metals (at 25°C)	
Aluminum	238
Copper	397
Gold	314
Iron	79.5
Lead	34.7
Silver	427
Nonmetals (approximate values)	
Asbestos	0.08
Concrete	0.8
Diamond	2 300
Glass	0.8
Ice	2
Rubber	0.2
Water	0.6
Wood	0.08
Gases (at 20°C)	
Air	0.023 4
Helium	0.138
Hydrogen	0.172
Nitrogen	0.023 4
Oxygen	0.023 8

where T_1 and T_2 are the temperatures of the outer surfaces (which are held constant) and the summation is over all slabs. The following example shows how this equation results from a consideration of two thicknesses of materials.

EXAMPLE 20.9 Energy Transfer Through Two Slabs

Two slabs of thickness L_1 and L_2 and thermal conductivities k_1 and k_2 are in thermal contact with each other, as shown in Figure 20.11. The temperatures of their outer surfaces are T_1 and T_2, respectively, and $T_2 > T_1$. Determine the temperature at the interface and the rate of energy transfer by conduction through the slabs in the steady-state condition.

Solution If T is the temperature at the interface, then the rate at which energy is transferred through slab 1 is

$$(1) \qquad \mathcal{P}_1 = \frac{k_1 A(T - T_1)}{L_1}$$

The rate at which energy is transferred through slab 2 is

$$(2) \qquad \mathcal{P}_2 = \frac{k_2 A(T_2 - T)}{L_2}$$

When a steady state is reached, these two rates must be equal; hence,

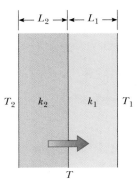

Figure 20.11 Energy transfer by conduction through two slabs in thermal contact with each other. At steady state, the rate of energy transfer through slab 1 equals the rate of energy transfer through slab 2.

$$\frac{k_1A(T-T_1)}{L_1} = \frac{k_2A(T_2-T)}{L_2}$$

Solving for T gives

(3) $$T = \frac{k_1L_2T_1 + k_2L_1T_2}{k_1L_2 + k_2L_1}$$

Substituting (3) into either (1) or (2), we obtain

$$\mathcal{P} = \frac{A(T_2 - T_1)}{(L_1/k_1) + (L_2/k_2)}$$

Extension of this model to several slabs of materials leads to Equation 20.16.

Home Insulation

In engineering practice, the term L/k for a particular substance is referred to as the **R value** of the material. Thus, Equation 20.16 reduces to

$$\mathcal{P} = \frac{A(T_2 - T_1)}{\displaystyle\sum_i R_i} \qquad (20.17)$$

where $R_i = L_i/k_i$. The R values for a few common building materials are given in Table 20.4. In the United States, the insulating properties of materials used in buildings are usually expressed in engineering units, not SI units. Thus, in Table 20.4, measurements of R values are given as a combination of British thermal units, feet, hours, and degrees Fahrenheit.

At any vertical surface open to the air, a very thin stagnant layer of air adheres to the surface. One must consider this layer when determining the R value for a wall. The thickness of this stagnant layer on an outside wall depends on the speed of the wind. Energy loss from a house on a windy day is greater than the loss on a day when the air is calm. A representative R value for this stagnant layer of air is given in Table 20.4.

Energy is conducted from the inside to the exterior more rapidly on the part of the roof where the snow has melted. The dormer appears to have been added and insulated. The main roof does not appear to be well insulated. *(Courtesy of Dr. Albert A. Bartlett, University of Colorado, Boulder, CO)*

TABLE 20.4   **R Values for Some Common Building Materials**

Material	R value (ft² · °F · h/Btu)
Hardwood siding (1 in. thick)	0.91
Wood shingles (lapped)	0.87
Brick (4 in. thick)	4.00
Concrete block (filled cores)	1.93
Fiberglass batting (3.5 in. thick)	10.90
Fiberglass batting (6 in. thick)	18.80
Fiberglass board (1 in. thick)	4.35
Cellulose fiber (1 in. thick)	3.70
Flat glass (0.125 in. thick)	0.89
Insulating glass (0.25-in. space)	1.54
Air space (3.5 in. thick)	1.01
Stagnant air layer	0.17
Drywall (0.5 in. thick)	0.45
Sheathing (0.5 in. thick)	1.32

This thermogram of a home, made during cold weather, shows colors ranging from white and orange (areas of greatest energy loss) to blue and purple (areas of least energy loss). *(Daedalus Enterprises, Inc./Peter Arnold, Inc.)*

EXAMPLE 20.10 ▸ The *R* Value of a Typical Wall

Calculate the total *R* value for a wall constructed as shown in Figure 20.12a. Starting outside the house (toward the front in the figure) and moving inward, the wall consists of 4-in. brick, 0.5-in. sheathing, an air space 3.5 in. thick, and 0.5-in. drywall. Do not forget the stagnant air layers inside and outside the house.

Solution   Referring to Table 20.4, we find that

R_1 (outside stagnant air layer)	$= 0.17 \text{ ft}^2 \cdot {}^\circ\text{F} \cdot \text{h/Btu}$
R_2 (brick)	$= 4.00 \text{ ft}^2 \cdot {}^\circ\text{F} \cdot \text{h/Btu}$
R_3 (sheathing)	$= 1.32 \text{ ft}^2 \cdot {}^\circ\text{F} \cdot \text{h/Btu}$
R_4 (air space)	$= 1.01 \text{ ft}^2 \cdot {}^\circ\text{F} \cdot \text{h/Btu}$
R_5 (drywall)	$= 0.45 \text{ ft}^2 \cdot {}^\circ\text{F} \cdot \text{h/Btu}$
R_6 (inside stagnant air layer)	$= 0.17 \text{ ft}^2 \cdot {}^\circ\text{F} \cdot \text{h/Btu}$
R_{total}	$= \boxed{7.12 \text{ ft}^2 \cdot {}^\circ\text{F} \cdot \text{h/Btu}}$

Exercise   If a layer of fiberglass insulation 3.5 in. thick is placed inside the wall to replace the air space, as shown in Figure 20.12b, what is the new total *R* value? By what factor is the energy loss reduced?

Answer $R = 17 \text{ ft}^2 \cdot {}^\circ\text{F} \cdot \text{h/Btu}$; 2.4.

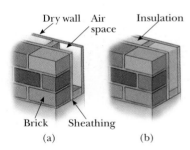

Figure 20.12   An exterior house wall containing (a) an air space and (b) insulation.

Convection

At one time or another, you probably have warmed your hands by holding them over an open flame. In this situation, the air directly above the flame is heated and expands. As a result, the density of this air decreases and the air rises. This warmed mass of air heats your hands as it flows by. **Energy transferred by the movement of a heated substance is said to have been transferred by convection.** When the movement results from differences in density, as with air around a fire, it is referred to as *natural convection*. Air flow at a beach is an example of natural convection, as is the mixing that occurs as surface water in a lake cools and sinks (see

Figure 20.13 Convection currents are set up in a room heated by a radiator.

Chapter 19). When the heated substance is forced to move by a fan or pump, as in some hot-air and hot-water heating systems, the process is called *forced convection*.

If it were not for convection currents, it would be very difficult to boil water. As water is heated in a teakettle, the lower layers are warmed first. The heated water expands and rises to the top because its density is lowered. At the same time, the denser, cool water at the surface sinks to the bottom of the kettle and is heated.

The same process occurs when a room is heated by a radiator. The hot radiator warms the air in the lower regions of the room. The warm air expands and rises to the ceiling because of its lower density. The denser, cooler air from above sinks, and the continuous air current pattern shown in Figure 20.13 is established.

Radiation

The third means of energy transfer that we shall discuss is **radiation.** All objects radiate energy continuously in the form of electromagnetic waves (see Chapter 34) produced by thermal vibrations of the molecules. You are likely familiar with electromagnetic radiation in the form of the orange glow from an electric stove burner, an electric space heater, or the coils of a toaster.

The rate at which an object radiates energy is proportional to the fourth power of its absolute temperature. This is known as **Stefan's law** and is expressed in equation form as

$$\mathscr{P} = \sigma A e T^4 \qquad \textbf{(20.18)}$$

where $\mathscr{P}$ is the power in watts radiated by the object, σ is a constant equal to $5.669\ 6 \times 10^{-8}\ \text{W/m}^2 \cdot \text{K}^4$, A is the surface area of the object in square meters, e is the **emissivity** constant, and T is the surface temperature in kelvins. The value of e can vary between zero and unity, depending on the properties of the surface of the object. The emissivity is equal to the fraction of the incoming radiation that the surface absorbs.

Approximately 1 340 J of electromagnetic radiation from the Sun passes perpendicularly through each 1 m² at the top of the Earth's atmosphere every second. This radiation is primarily visible and infrared light accompanied by a significant amount of ultraviolet radiation. We shall study these types of radiation in detail in Chapter 34. Some of this energy is reflected back into space, and some is absorbed by the atmosphere. However, enough energy arrives at the surface of the Earth each day to supply all our energy needs on this planet hundreds of times over—if only it could be captured and used efficiently. The growth in the number of solar energy–powered houses built in this country reflects the increasing efforts being made to use this abundant energy. Radiant energy from the Sun affects our day-to-day existence in a number of ways. For example, it influences the Earth's average temperature, ocean currents, agriculture, and rain patterns.

What happens to the atmospheric temperature at night is another example of the effects of energy transfer by radiation. If there is a cloud cover above the Earth, the water vapor in the clouds absorbs part of the infrared radiation emitted by the Earth and re-emits it back to the surface. Consequently, temperature levels at the surface remain moderate. In the absence of this cloud cover, there is nothing to prevent this radiation from escaping into space; thus the temperature decreases more on a clear night than on a cloudy one.

As an object radiates energy at a rate given by Equation 20.18, it also absorbs electromagnetic radiation. If the latter process did not occur, an object would eventually radiate all its energy, and its temperature would reach absolute zero. The energy an object absorbs comes from its surroundings, which consist of other objects that radiate energy. If an object is at a temperature T and its surroundings are at a temperature T_0, then the net energy gained or lost each second by the object as a result of radiation is

$$\mathcal{P}_{net} = \sigma Ae(T^4 - T_0{}^4) \qquad \textbf{(20.19)}$$

When an object is in equilibrium with its surroundings, it radiates and absorbs energy at the same rate, and so its temperature remains constant. When an object is hotter than its surroundings, it radiates more energy than it absorbs, and its temperature decreases. An **ideal absorber** is defined as an object that absorbs all the energy incident on it, and for such a body, $e = 1$. Such an object is often referred to as a **black body.** An ideal absorber is also an ideal radiator of energy. In contrast, an object for which $e = 0$ absorbs none of the energy incident on it. Such an object reflects all the incident energy, and thus is an **ideal reflector.**

The Dewar Flask

The *Dewar flask*[8] is a container designed to minimize energy losses by conduction, convection, and radiation. Such a container is used to store either cold or hot liquids for long periods of time. (A Thermos bottle is a common household equivalent of a Dewar flask.) The standard construction (Fig. 20.14) consists of a double-walled Pyrex glass vessel with silvered walls. The space between the walls is evacuated to minimize energy transfer by conduction and convection. The silvered surfaces minimize energy transfer by radiation because silver is a very good reflector and has very low emissivity. A further reduction in energy loss is obtained by reducing the size of the neck. Dewar flasks are commonly used to store liquid nitrogen (boiling point: 77 K) and liquid oxygen (boiling point: 90 K).

To confine liquid helium (boiling point: 4.2 K), which has a very low heat of vaporization, it is often necessary to use a double Dewar system in which the Dewar flask containing the liquid is surrounded by a second Dewar flask. The space between the two flasks is filled with liquid nitrogen.

Newer designs of storage containers use "super insulation" that consists of many layers of reflecting material separated by fiberglass. All of this is in a vacuum, and no liquid nitrogen is needed with this design.

Figure 20.14 A cross-sectional view of a Dewar flask, which is used to store hot or cold substances.

EXAMPLE 20.11   **Who Turned Down the Thermostat?**

A student is trying to decide what to wear. The surroundings (his bedroom) are at 20.0°C. If the skin temperature of the unclothed student is 35°C, what is the net energy loss from his body in 10.0 min by radiation? Assume that the emissivity of skin is 0.900 and that the surface area of the student is 1.50 m².

Solution   Using Equation 20.19, we find that the net rate of energy loss from the skin is

$$\mathcal{P}_{net} = \sigma Ae(T^4 - T_0{}^4)$$
$$= (5.67 \times 10^{-8}\ \text{W/m}^2 \cdot \text{K}^4)(1.50\ \text{m}^2)$$
$$\times\ (0.900)[(308\ \text{K})^4 - (293\ \text{K})^4] = 125\ \text{W}$$

(Why is the temperature given in kelvins?) At this rate, the total energy lost by the skin in 10 min is

$$Q = \mathcal{P}_{net} \times \Delta t = (125\ \text{W})(600\ \text{s}) = \boxed{7.5 \times 10^4\ \text{J}}$$

Note that the energy radiated by the student is roughly equivalent to that produced by two 60-W light bulbs!

---

[8] Invented by Sir James Dewar (1842–1923).

SUMMARY

Internal energy is all of a system's energy that is associated with the system's microscopic components. Internal energy includes kinetic energy of translation, rotation, and vibration of molecules, potential energy within molecules, and potential energy between molecules.

Heat is the transfer of energy across the boundary of a system resulting from a temperature difference between the system and its surroundings. We use the symbol Q for the amount of energy transferred by this process.

The **calorie** is the amount of energy necessary to raise the temperature of 1 g of water from 14.5°C to 15.5°C. The **mechanical equivalent of heat** is 1 cal = 4.186 J.

The **heat capacity** C of any sample is the amount of energy needed to raise the temperature of the sample by 1°C. The energy Q required to change the temperature of a mass m of a substance by an amount ΔT is

$$Q = mc\Delta T \tag{20.4}$$

where c is the **specific heat** of the substance.

The energy required to change the phase of a pure substance of mass m is

$$Q = mL \tag{20.6}$$

where L is the **latent heat** of the substance and depends on the nature of the phase change and the properties of the substance.

The **work done** by a gas as its volume changes from some initial value V_i to some final value V_f is

$$W = \int_{V_i}^{V_f} P \, dV \tag{20.8}$$

where P is the pressure, which may vary during the process. In order to evaluate W, the process must be fully specified—that is, P and V must be known during each step. In other words, the work done depends on the path taken between the initial and final states.

The **first law of thermodynamics** states that when a system undergoes a change from one state to another, the change in its internal energy is

$$\Delta E_{\text{int}} = Q - W \tag{20.9}$$

where Q is the energy transferred into the system by heat and W is the work done by the system. Although Q and W both depend on the path taken from the initial state to the final state, the quantity ΔE_{int} is path-independent. This central equation is a statement of conservation of energy that includes changes in internal energy.

In a **cyclic process** (one that originates and terminates at the same state), $\Delta E_{\text{int}} = 0$ and, therefore, $Q = W$. That is, the energy transferred into the system by heat equals the work done by the system during the process.

In an **adiabatic process,** no energy is transferred by heat between the system and its surroundings ($Q = 0$). In this case, the first law gives $\Delta E_{\text{int}} = -W$. That is, the internal energy changes as a consequence of work being done by the system. In the **adiabatic free expansion** of a gas, $Q = 0$ and $W = 0$; thus, $\Delta E_{\text{int}} = 0$. That is, the internal energy of the gas does not change in such a process.

An **isobaric process** is one that occurs at constant pressure. The work done in such a process is $W = P(V_f - V_i)$.

An **isovolumetric process** is one that occurs at constant volume. No work is done in such a process, so $\Delta E_{\text{int}} = Q$.

An **isothermal process** is one that occurs at constant temperature. The work done by an ideal gas during an isothermal process is

$$W = nRT \ln\left(\frac{V_f}{V_i}\right) \tag{20.13}$$

Energy may be transferred by conduction, convection, or radiation. **Conduction** can be viewed as an exchange of kinetic energy between colliding molecules or electrons. The rate at which energy flows by conduction through a slab of area A is

$$\mathcal{P} = kA\left|\frac{dT}{dx}\right| \tag{20.14}$$

where k is the **thermal conductivity** of the material from which the slab is made and $|dT/dx|$ is the **temperature gradient.** This equation can be used in many situations in which the rate of transfer of energy through materials is important.

In **convection,** a heated substance moves from one place to another.

All bodies emit **radiation** in the form of electromagnetic waves at the rate

$$\mathcal{P} = \sigma A e T^4 \tag{20.18}$$

A body that is hotter than its surroundings radiates more energy than it absorbs, whereas a body that is cooler than its surroundings absorbs more energy than it radiates.

QUESTIONS

1. The specific heat of water is about two times that of ethyl alcohol. Equal masses of alcohol and water are contained in separate beakers and are supplied with the same amount of energy. Compare the temperature increases of the two liquids.

2. Give one reason why coastal regions tend to have a more moderate climate than inland regions do.

3. A small metal crucible is taken from a 200°C oven and immersed in a tub full of water at room temperature (this process is often referred to as *quenching*). What is the approximate final equilibrium temperature?

4. What is the major problem that arises in measuring specific heats if a sample with a temperature greater than 100°C is placed in water?

5. In a daring lecture demonstration, an instructor dips his wetted fingers into molten lead (327°C) and withdraws them quickly, without getting burned. How is this possible? (This is a dangerous experiment that you should *not* attempt.)

6. The pioneers found that placing a large tub of water in a storage cellar would prevent their food from freezing on really cold nights. Explain why.

7. What is wrong with the statement, "Given any two bodies, the one with the higher temperature contains more heat."

8. Why is it possible for you to hold a lighted match, even when it is burned to within a few millimeters of your fingertips?

9. Why is it more comfortable to hold a cup of hot tea by the handle than by wrapping your hands around the cup itself?

10. Figure Q20.10 shows a pattern formed by snow on the roof of a barn. What causes the alternating pattern of snowcover and exposed roof?

Figure Q20.10 Alternating pattern on a snow-covered roof. *(Courtesy of Dr. Albert A. Bartlett, University of Colorado, Boulder, CO)*

11. Why is a person able to remove a piece of dry aluminum foil from a hot oven with bare fingers but burns his or her fingers if there is moisture on the foil?

12. A tile floor in a bathroom may feel uncomfortably cold to your bare feet, but a carpeted floor in an adjoining room at the same temperature feels warm. Why?

13. Why can potatoes be baked more quickly when a metal skewer has been inserted through them?

14. Explain why a Thermos bottle has silvered walls and a vacuum jacket.

15. A piece of paper is wrapped around a rod made half of wood and half of copper. When held over a flame, the paper in contact with the wood burns but the paper in contact with the metal does not. Explain.

16. Why is it necessary to store liquid nitrogen or liquid oxygen in vessels equipped with either polystyrene insulation or a double-evacuated wall?

17. Why do heavy draperies hung over the windows help keep a home warm in the winter and cool in the summer?

18. If you wish to cook a piece of meat thoroughly on an open fire, why should you not use a high flame? (*Note:* Carbon is a good thermal insulator.)

19. When insulating a wood-frame house, is it better to place the insulation against the cooler, outside wall or against the warmer, inside wall? (In either case, an air barrier must be considered.)

20. In an experimental house, polystyrene beads were pumped into the air space between the panes of glass in double-pane windows at night in the winter, and they were pumped out to holding bins during the day. How would this procedure assist in conserving energy in the house?

21. Pioneers stored fruits and vegetables in underground cellars. Discuss the advantages of choosing this location as a storage site.

22. Concrete has a higher specific heat than soil does. Use this fact to explain (partially) why cities have a higher average night-time temperature than the surrounding countryside does. If a city is hotter than the surrounding countryside, would you expect breezes to blow from city to country or from country to city? Explain.

23. When camping in a canyon on a still night, a hiker notices that a breeze begins to stir as soon as the Sun strikes the surrounding peaks. What causes the breeze?

24. Updrafts of air are familiar to all pilots and are used to keep non-motorized gliders aloft. What causes these currents?

25. If water is a poor thermal conductor, why can it be heated quickly when placed over a flame?

26. The United States penny is now made of copper-coated zinc. Can a calorimetric experiment be devised to test for the metal content in a collection of pennies? If so, describe such a procedure.

27. If you hold water in a paper cup over a flame, you can bring the water to a boil without burning the cup. How is this possible?

28. When a sealed Thermos bottle full of hot coffee is shaken, what are the changes, if any, in (a) the temperature of the coffee and (b) the internal energy of the coffee?

29. Using the first law of thermodynamics, explain why the *total* energy of an isolated system is always constant.

30. Is it possible to convert internal energy into mechanical energy? Explain using examples.

31. Suppose that you pour hot coffee for your guests and one of them chooses to drink the coffee after it has been in the cup for several minutes. For the coffee to be warmest, should the person add the cream just after the coffee is poured or just before drinking it? Explain.

32. Suppose that you fill two identical cups both at room temperature with the same amount of hot coffee. One cup contains a metal spoon, while the other does not. If you wait for several minutes, which of the two contains the warmer coffee? Which energy transfer process accounts for this result?

33. A warning sign often seen on highways just before a bridge is "Caution—Bridge Surface Freezes Before Road Surface." Which of the three energy transfer processes is most important in causing a bridge surface to freeze before a road surface on very cold days?

PROBLEMS

1, 2, 3 = straightforward, intermediate, challenging   ☐ = full solution available in the *Student Solutions Manual and Study Guide*
WEB = solution posted at **http://www.saunderscollege.com/physics/**   ▢ = Computer useful in solving problem   ▨ = Interactive Physics
☐ = paired numerical/symbolic problems

Section 20.1   Heat and Internal Energy

1. Water at the top of Niagara Falls has a temperature of 10.0°C. It falls through a distance of 50.0 m. Assuming that all of its potential energy goes into warming of the water, calculate the temperature of the water at the bottom of the Falls.

2. Consider Joule's apparatus described in Figure 20.1. Each of the two masses is 1.50 kg, and the tank is filled with 200 g of water. What is the increase in the temperature of the water after the masses fall through a distance of 3.00 m?

Section 20.2   Heat Capacity and Specific Heat

3. The temperature of a silver bar rises by 10.0°C when it absorbs 1.23 kJ of energy by heat. The mass of the bar is 525 g. Determine the specific heat of silver.

4. A 50.0-g sample of copper is at 25.0°C. If 1 200 J of energy is added to it, what is the final temperature of the copper?

WEB 5. A 1.50-kg iron horseshoe initially at 600°C is dropped into a bucket containing 20.0 kg of water at 25.0°C. What is the final temperature? (Neglect the heat capacity of the container and assume that a negligible amount of water boils away.)

6. An aluminum cup with a mass of 200 g contains 800 g of water in thermal equilibrium at 80.0°C. The combination of cup and water is cooled uniformly so that the temperature decreases at a rate of 1.50°C/min. At what rate is energy being removed by heat? Express your answer in watts.

7. An aluminum calorimeter with a mass of 100 g contains 250 g of water. The calorimeter and water are in thermal equilibrium at 10.0°C. Two metallic blocks are placed into the water. One is a 50.0-g piece of copper at 80.0°C; the other block has a mass of 70.0 g and is originally at a temperature of 100°C. The entire system stabilizes at a final temperature of 20.0°C. (a) Determine the specific heat of the unknown sample. (b) Guess the material of the unknown, using the data given in Table 20.1.

8. Lake Erie contains roughly 4.00×10^{11} m^3 of water. (a) How much energy is required to raise the temperature of this volume of water from 11.0°C to 12.0°C? (b) Approximately how many years would it take to supply this amount of energy with the use of a 1 000-MW wasted energy output of an electric power plant?

9. A 3.00-g copper penny at 25.0°C drops from a height of 50.0 m to the ground. (a) If 60.0% of the change in potential energy goes into increasing the internal energy, what is its final temperature? (b) Does the result you obtained in (a) depend on the mass of the penny? Explain.

10. If a mass m_h of water at T_h is poured into an aluminum cup of mass m_{Al} containing mass m_c of water at T_c, where $T_h > T_c$, what is the equilibrium temperature of the system?

11. A water heater is operated by solar power. If the solar collector has an area of 6.00 m^2 and the power delivered by sunlight is 550 W/m^2, how long does it take to increase the temperature of 1.00 m^3 of water from 20.0°C to 60.0°C?

Section 20.3   Latent Heat

12. How much energy is required to change a 40.0-g ice cube from ice at -10.0°C to steam at 110°C?

13. A 3.00-g lead bullet at 30.0°C is fired at a speed of 240 m/s into a large block of ice at 0°C, in which it becomes embedded. What quantity of ice melts?

14. Steam at 100°C is added to ice at 0°C. (a) Find the amount of ice melted and the final temperature when the mass of steam is 10.0 g and the mass of ice is 50.0 g. (b) Repeat this calculation, taking the mass of steam as 1.00 g and the mass of ice as 50.0 g.

15. A 1.00-kg block of copper at 20.0°C is dropped into a large vessel of liquid nitrogen at 77.3 K. How many kilograms of nitrogen boil away by the time the copper reaches 77.3 K? (The specific heat of copper is 0.092 0 cal/g · °C. The latent heat of vaporization of nitrogen is 48.0 cal/g.)

16. A 50.0-g copper calorimeter contains 250 g of water at 20.0°C. How much steam must be condensed into the water if the final temperature of the system is to reach 50.0°C?

WEB 17. In an insulated vessel, 250 g of ice at 0°C is added to 600 g of water at 18.0°C. (a) What is the final temperature of the system? (b) How much ice remains when the system reaches equilibrium?

18. **Review Problem.**   Two speeding lead bullets, each having a mass of 5.00 g, a temperature of 20.0°C, and a speed of 500 m/s, collide head-on. Assuming a perfectly inelastic collision and no loss of energy to the atmosphere, describe the final state of the two-bullet system.

19. If 90.0 g of molten lead at 327.3°C is poured into a 300-g casting form made of iron and initially at 20.0°C, what is the final temperature of the system? (Assume that no energy loss to the environment occurs.)

Section 20.4   Work and Heat in Thermodynamic Processes

20. Gas in a container is at a pressure of 1.50 atm and a volume of 4.00 m^3. What is the work done by the gas (a) if it expands at constant pressure to twice its initial volume? (b) If it is compressed at constant pressure to one quarter of its initial volume?

WEB 21. A sample of ideal gas is expanded to twice its original volume of 1.00 m^3 in a quasi-static process for which $P = \alpha V^2$, with $\alpha = 5.00$ atm/m^6, as shown in Figure P20.21. How much work is done by the expanding gas?

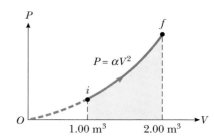

Figure P20.21

22. (a) Determine the work done by a fluid that expands from i to f as indicated in Figure P20.22. (b) How much

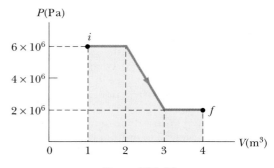

Figure P20.22

work is performed by the fluid if it is compressed from *f* to *i* along the same path?

23. One mole of an ideal gas is heated slowly so that it goes from PV state (P_i, V_i) to $(3P_i, 3V_i)$ in such a way that the pressure of the gas is directly proportional to the volume. (a) How much work is done in the process? (b) How is the temperature of the gas related to its volume during this process?

24. A sample of helium behaves as an ideal gas as energy is added by heat at constant pressure from 273 K to 373 K. If the gas does 20.0 J of work, what is the mass of helium present?

WEB 25. An ideal gas is enclosed in a cylinder with a movable piston on top. The piston has a mass of 8 000 g and an area of 5.00 cm² and is free to slide up and down, keeping the pressure of the gas constant. How much work is done as the temperature of 0.200 mol of the gas is raised from 20.0°C to 300°C?

26. An ideal gas is enclosed in a cylinder that has a movable piston on top. The piston has a mass m and an area A and is free to slide up and down, keeping the pressure of the gas constant. How much work is done as the temperature of n mol of the gas is raised from T_1 to T_2?

27. A gas expands from *I* to *F* along three possible paths, as indicated in Figure P20.27. Calculate the work in joules done by the gas along the paths *IAF, IF,* and *IBF.*

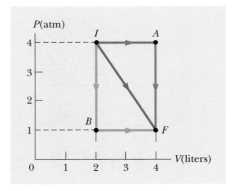

Figure P20.27

Section 20.5   The First Law of Thermodynamics

28. A gas is compressed from 9.00 L to 2.00 L at a constant pressure of 0.800 atm. In the process, 400 J of energy leaves the gas by heat. (a) What is the work done by the gas? (b) What is the change in its internal energy?

29. A thermodynamic system undergoes a process in which its internal energy decreases by 500 J. If, at the same time, 220 J of work is done on the system, what is the energy transferred to or from it by heat?

30. A gas is taken through the cyclic process described in Figure P20.30. (a) Find the net energy transferred to the system by heat during one complete cycle. (b) If the

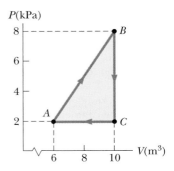

Figure P20.30   Problems 30 and 31.

cycle is reversed—that is, if the process follows the path *ACBA*—what is the net energy input per cycle by heat?

31. Consider the cyclic process depicted in Figure P20.30. If Q is negative for the process *BC*, and if ΔE_{int} is negative for the process *CA*, what are the signs of Q, W, and ΔE_{int} that are associated with each process?

32. A sample of an ideal gas goes through the process shown in Figure P20.32. From *A* to *B*, the process is adiabatic; from *B* to *C*, it is isobaric, with 100 kJ of energy flowing into the system by heat. From *C* to *D*, the process is isothermal; from *D* to *A*, it is isobaric, with 150 kJ of energy flowing out of the system by heat. Determine the difference in internal energy, $E_{int, B} - E_{int, A}$.

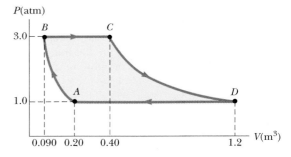

Figure P20.32

Section 20.6   Some Applications of the First Law of Thermodynamics

33. An ideal gas initially at 300 K undergoes an isobaric expansion at 2.50 kPa. If the volume increases from 1.00 m³ to 3.00 m³ and if 12.5 kJ of energy is transferred to the gas by heat, what are (a) the change in its internal energy and (b) its final temperature?

34. One mole of an ideal gas does 3 000 J of work on its surroundings as it expands isothermally to a final pressure of 1.00 atm and a volume of 25.0 L. Determine (a) the initial volume and (b) the temperature of the gas.

35. How much work is done by the steam when 1.00 mol of water at 100°C boils and becomes 1.00 mol of steam at

100°C and at 1.00 atm pressure? Assuming the steam to be an ideal gas, determine the change in internal energy of the steam as it vaporizes.

36. A 1.00-kg block of aluminum is heated at atmospheric pressure such that its temperature increases from 22.0°C to 40.0°C. Find (a) the work done by the aluminum, (b) the energy added to it by heat, and (c) the change in its internal energy.

37. A 2.00-mol sample of helium gas initially at 300 K and 0.400 atm is compressed isothermally to 1.20 atm. Assuming the behavior of helium to be that of an ideal gas, find (a) the final volume of the gas, (b) the work done by the gas, and (c) the energy transferred by heat.

38. One mole of water vapor at a temperature of 373 K cools down to 283 K. The energy given off from the cooling vapor by heat is absorbed by 10.0 mol of an ideal gas, causing it to expand at a constant temperature of 273 K. If the final volume of the ideal gas is 20.0 L, what is the initial volume of the ideal gas?

39. An ideal gas is carried through a thermodynamic cycle consisting of two isobaric and two isothermal processes, as shown in Figure P20.39. Show that the net work done in the entire cycle is given by the equation

$$W_{net} = P_1(V_2 - V_1) \ln \frac{P_2}{P_1}$$

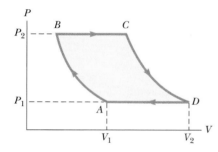

Figure P20.39

40. In Figure P20.40, the change in internal energy of a gas that is taken from A to C is $+800$ J. The work done along the path ABC is $+500$ J. (a) How much energy must be added to the system by heat as it goes from A through B and on to C? (b) If the pressure at point A is five times that at point C, what is the work done by the system in going from C to D? (c) What is the energy exchanged with the surroundings by heat as the cycle goes from C to A? (d) If the change in internal energy in going from point D to point A is $+500$ J, how much energy must be added to the system by heat as it goes from point C to point D?

Section 20.7 Energy Transfer Mechanisms

41. A steam pipe is covered with 1.50-cm-thick insulating material with a thermal conductivity of 0.200 cal/cm · °C · s.

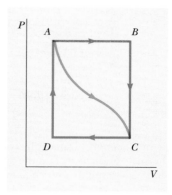

Figure P20.40

How much energy is lost every second by heat when the steam is at 200°C and the surrounding air is at 20.0°C? The pipe has a circumference of 20.0 cm and a length of 50.0 m. Neglect losses through the ends of the pipe.

42. A box with a total surface area of 1.20 m² and a wall thickness of 4.00 cm is made of an insulating material. A 10.0-W electric heater inside the box maintains the inside temperature at 15.0°C above the outside temperature. Find the thermal conductivity k of the insulating material.

43. A glass window pane has an area of 3.00 m² and a thickness of 0.600 cm. If the temperature difference between its surfaces is 25.0°C, what is the rate of energy transfer by conduction through the window?

44. A thermal window with an area of 6.00 m² is constructed of two layers of glass, each 4.00 mm thick and separated from each other by an air space of 5.00 mm. If the inside surface is at 20.0°C and the outside is at −30.0°C, what is the rate of energy transfer by conduction through the window?

45. A bar of gold is in thermal contact with a bar of silver of the same length and area (Fig. P20.45). One end of the compound bar is maintained at 80.0°C, while the opposite end is at 30.0°C. When the rate of energy transfer by conduction reaches steady state, what is the temperature at the junction?

Figure P20.45

46. Two rods of the same length but made of different materials and having different cross-sectional areas are placed side by side, as shown in Figure P20.46. Deter-

Figure P20.46

mine the rate of energy transfer by conduction in terms of the thermal conductivity and the area of each rod. Generalize your result to a system consisting of several rods.

47. Calculate the R value of (a) a window made of a single pane of flat glass $\frac{1}{8}$ in. thick; (b) a thermal window made of two single panes, each $\frac{1}{8}$ in. thick and separated by a $\frac{1}{4}$-in. air space. (c) By what factor is the thermal conduction reduced if the thermal window replaces the single-pane window?

48. The surface of the Sun has a temperature of about 5 800 K. The radius of the Sun is 6.96×10^8 m. Calculate the total energy radiated by the Sun each second. (Assume that $e = 0.965$.)

49. A large, hot pizza floats in outer space. What is the order of magnitude (a) of its rate of energy loss? (b) of its rate of temperature change? List the quantities you estimate and the value you estimate for each.

50. The tungsten filament of a certain 100-W light bulb radiates 2.00 W of light. (The other 98 W is carried away by convection and conduction.) The filament has a surface area of 0.250 mm^2 and an emissivity of 0.950. Find the filament's temperature. (The melting point of tungsten is 3 683 K.)

51. At high noon, the Sun delivers 1 000 W to each square meter of a blacktop road. If the hot asphalt loses energy only by radiation, what is its equilibrium temperature?

52. At our distance from the Sun, the intensity of solar radiation is 1 340 W/m^2. The temperature of the Earth is affected by the so-called "greenhouse effect" of the atmosphere. This effect makes our planet's emissivity for visible light higher than its emissivity for infrared light. For comparison, consider a spherical object with no atmosphere at the same distance from the Sun as the Earth. Assume that its emissivity is the same for all kinds of electromagnetic waves and that its temperature is uniform over its surface. Identify the projected area over which it absorbs sunlight and the surface area over which it radiates. Compute its equilibrium temperature. Chilly, isn't it? Your calculation applies to (a) the average temperature of the Moon, (b) astronauts in mortal danger aboard the crippled *Apollo 13* spacecraft, and (c) global catastrophe on the Earth if widespread fires caused a layer of soot to accumulate throughout the upper atmosphere so that most of the radiation from the Sun was absorbed there rather than at the surface below the atmosphere.

ADDITIONAL PROBLEMS

53. One hundred grams of liquid nitrogen at 77.3 K is stirred into a beaker containing 200 g of water at 5.00°C. If the nitrogen leaves the solution as soon as it turns to gas, how much water freezes? (The latent heat of vaporization of nitrogen is 48.0 cal/g, and the latent heat of fusion of water is 79.6 cal/g.)

54. A 75.0-kg cross-country skier moves across the snow (Fig. P20.54). The coefficient of friction between the skis and the snow is 0.200. Assume that all the snow beneath his skis is at 0°C and that all the internal energy generated by friction is added to the snow, which sticks to his skis until it melts. How far would he have to ski to melt 1.00 kg of snow?

Figure P20.54   A cross-country skier. *(Nathan Bilow/Leo de Wys, Inc.)*

55. An aluminum rod 0.500 m in length and with a cross-sectional area 2.50 cm^2 is inserted into a thermally insulated vessel containing liquid helium at 4.20 K. The rod is initially at 300 K. (a) If one half of the rod is inserted into the helium, how many liters of helium boil off by the time the inserted half cools to 4.20 K? (Assume that the upper half does not yet cool.) (b) If the upper end of the rod is maintained at 300 K, what is the approximate boil-off rate of liquid helium after the lower half has reached 4.20 K? (Aluminum has thermal conductivity of 31.0 J/s · cm · K at 4.2 K; ignore its temperature variation. Aluminum has a specific heat of 0.210 cal/g · °C and density of 2.70 g/cm^3. The density of liquid helium is 0.125 g/cm^3.)

56. On a cold winter day, you buy a hot dog from a street vendor. Into the pocket of your down parka you put the change he gives you: coins consisting of 9.00 g of copper at −12.0°C. Your pocket already contains 14.0 g of silver coins at 30.0°C. A short time later, the temperature of the copper coins is 4.00°C and is increasing at a rate of 0.500°C/s. At this time (a) what is the temperature of the silver coins, and (b) at what rate is it changing? (Neglect energy transferred to the surroundings.)

57. A *flow calorimeter* is an apparatus used to measure the specific heat of a liquid. The technique of flow calorimetry involves measuring the temperature difference between the input and output points of a flowing stream of the liquid while energy is added by heat at a known rate. In one particular experiment, a liquid with a density of 0.780 g/cm^3 flows through the calorimeter at the rate of 4.00 cm^3/s. At steady state, a temperature difference of $4.80°$C is established between the input and output points when energy is supplied by heat at the rate of 30.0 J/s. What is the specific heat of the liquid?

58. A *flow calorimeter* is an apparatus used to measure the specific heat of a liquid. The technique of flow calorimetry involves measuring the temperature difference between the input and output points of a flowing stream of the liquid while energy is added by heat at a known rate. In one particular experiment, a liquid of density ρ flows through the calorimeter with volume flow rate R. At steady state, a temperature difference ΔT is established between the input and output points when energy is supplied at the rate $\mathcal{P}$. What is the specific heat of the liquid?

59. One mole of an ideal gas, initially at 300 K, is cooled at constant volume so that the final pressure is one-fourth the initial pressure. The gas then expands at constant pressure until it reaches the initial temperature. Determine the work done by the gas.

60. One mole of an ideal gas is contained in a cylinder with a movable piston. The initial pressure, temperature, and volume are P_i, V_i, and T_i, respectively. Find the work done by the gas for the following processes and show each process on a PV diagram: (a) An isobaric compression in which the final volume is one-half the initial volume. (b) An isothermal compression in which the final pressure is four times the initial pressure. (c) An isovolumetric process in which the final pressure is triple the initial pressure.

61. An ideal gas initially at P_i, V_i, and T_i is taken through a cycle as shown in Figure P20.61. (a) Find the net work done by the gas per cycle. (b) What is the net energy added by heat to the system per cycle? (c) Obtain a nu-

merical value for the net work done per cycle for 1.00 mol of gas initially at 0°C.

62. Review Problem. An iron plate is held against an iron wheel so that a sliding frictional force of 50.0 N acts between the two pieces of metal. The relative speed at which the two surfaces slide over each other is 40.0 m/s. (a) Calculate the rate at which mechanical energy is converted to internal energy. (b) The plate and the wheel each have a mass of 5.00 kg, and each receives 50.0% of the internal energy. If the system is run as described for 10.0 s and each object is then allowed to reach a uniform internal temperature, what is the resultant temperature increase?

WEB **63.** A "solar cooker" consists of a curved reflecting mirror that focuses sunlight onto the object to be warmed (Fig. P20.63). The solar power per unit area reaching the Earth at the location is 600 W/m^2, and the cooker has a diameter of 0.600 m. Assuming that 40.0% of the incident energy is transferred to the water, how long does it take to completely boil off 0.500 L of water initially at 20.0°C? (Neglect the heat capacity of the container.)

Figure P20.63

64. Water in an electric teakettle is boiling. The power absorbed by the water is 1.00 kW. Assuming that the pressure of the vapor in the kettle equals atmospheric pressure, determine the speed of effusion of vapor from the kettle's spout if the spout has a cross-sectional area of 2.00 cm^2.

65. Liquid water evaporates and even boils at temperatures other than 100°C, depending on the ambient pressure. Suppose that the latent heat of vaporization in Table 20.2 describes the liquid–vapor transition at all temperatures. A chamber contains 1.00 kg of water at 0°C under a piston, which just touches the water's surface. The piston is then raised quickly so that part of the water is vaporized and the other part is frozen (no liquid remains). Assuming that the temperature remains con-

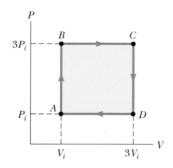

Figure P20.61

stant at 0°C, determine the mass of the ice that forms in the chamber.

66. A cooking vessel on a slow burner contains 10.0 kg of water and an unknown mass of ice in equilibrium at 0°C at time $t = 0$. The temperature of the mixture is measured at various times, and the result is plotted in Figure P20.66. During the first 50.0 min, the mixture remains at 0°C. From 50.0 min to 60.0 min, the temperature increases to 2.00°C. Neglecting the heat capacity of the vessel, determine the initial mass of the ice.

Figure P20.66

67. **Review Problem.** (a) In air at 0°C, a 1.60-kg copper block at 0°C is set sliding at 2.50 m/s over a sheet of ice at 0°C. Friction brings the block to rest. Find the mass of the ice that melts. To describe the process of slowing down, identify the energy input Q, the work output W, the change in internal energy ΔE_{int}, and the change in mechanical energy ΔK for both the block and the ice. (b) A 1.60-kg block of ice at 0°C is set sliding at 2.50 m/s over a sheet of copper at 0°C. Friction brings the block to rest. Find the mass of the ice that melts. Identify Q, W, ΔE_{int}, and ΔK for the block and for the metal sheet during the process. (c) A thin 1.60-kg slab of copper at 20°C is set sliding at 2.50 m/s over an identical stationary slab at the same temperature. Friction quickly stops the motion. If no energy is lost to the environment by heat, find the change in temperature of both objects. Identify Q, W, ΔE_{int}, and ΔK for each object during the process.

68. The average thermal conductivity of the walls (including the windows) and roof of the house depicted in Figure P20.68 is 0.480 W/m · °C, and their average thickness is 21.0 cm. The house is heated with natural gas having a heat of combustion (that is, the energy provided per cubic meter of gas burned) of 9 300 kcal/m³. How many cubic meters of gas must be burned each day to maintain an inside temperature of 25.0°C if the out-

Figure P20.68

side temperature is 0.0°C? Disregard radiation and the energy lost by heat through the ground.

69. A pond of water at 0°C is covered with a layer of ice 4.00 cm thick. If the air temperature stays constant at −10.0°C, how long does it take the ice's thickness to increase to 8.00 cm? (*Hint:* To solve this problem, use Equation 20.14 in the form

$$\frac{dQ}{dt} = kA \frac{\Delta T}{x}$$

and note that the incremental energy dQ extracted from the water through the thickness x of ice is the amount required to freeze a thickness dx of ice. That is, $dQ = L\rho A\,dx$, where ρ is the density of the ice, A is the area, and L is the latent heat of fusion.)

70. The inside of a hollow cylinder is maintained at a temperature T_a while the outside is at a lower temperature T_b (Fig. P20.70). The wall of the cylinder has a thermal conductivity k. Neglecting end effects, show that the rate of energy conduction from the inner to the outer wall in the radial direction is

$$\frac{dQ}{dt} = 2\pi Lk \left[\frac{T_a - T_b}{\ln(b/a)} \right]$$

(*Hint:* The temperature gradient is dT/dr. Note that a radial flow of energy occurs through a concentric cylinder of area $2\pi rL$.)

Figure P20.70

71. The passenger section of a jet airliner has the shape of a cylindrical tube with a length of 35.0 m and an inner radius of 2.50 m. Its walls are lined with an insulating material 6.00 cm in thickness and having a thermal conductivity of 4.00×10^{-5} cal/s · cm · °C. A heater must maintain the interior temperature at 25.0°C while the outside temperature is at -35.0°C. What power must be supplied to the heater if this temperature difference is to be maintained? (Use the result you obtained in Problem 70.)

72. A student obtains the following data in a calorimetry experiment designed to measure the specific heat of aluminum:

Initial temperature of water and calorimeter	70°C
Mass of water	0.400 kg
Mass of calorimeter	0.040 kg
Specific heat of calorimeter	0.63 kJ/kg · °C
Initial temperature of aluminum	27°C
Mass of aluminum	0.200 kg
Final temperature of mixture	66.3°C

Use these data to determine the specific heat of aluminum. Your result should be within 15% of the value listed in Table 20.1.

ANSWERS TO QUICK QUIZZES

20.1 (a) Water, glass, iron. Because water has the highest specific heat (4 186 J/kg · °C), it has the smallest change in temperature. Glass is next (837 J/kg · °C), and iron is last (448 J/kg · °C). (b) Iron, glass, water. For a given temperature increase, the energy transfer by heat is proportional to the specific heat.

20.2 Steam. According to Table 20.2, a kilogram of 100°C steam releases 2.26×10^6 J of energy as it condenses to 100°C water. After it releases this much energy into your skin, it is identical to 100°C water and will continue to burn you.

20.3 C, A, E. The slope is the ratio of the temperature change to the amount of energy input. Thus, the slope is proportional to the reciprocal of the specific heat. Water, which has the highest specific heat, has the least slope.

20.4

Situation	System	Q	W	ΔE_{int}
(a) Rapidly pumping up a bicycle tire	Air in the pump	0	−	+
(b) Pan of room-temperature water sitting on a hot stove	Water in the pan	+	0	+
(c) Air quickly leaking out of a balloon	Air originally in the balloon	0	+	−

(a) Because the pumping is rapid, no energy enters or leaves the system by heat; thus, $Q = 0$. Because work is done *on* the system, this work is negative. Thus, $\Delta E_{int} = Q - W$ must be positive. The air in the pump is warmer. (b) No work is done either by or on the system, but energy flows into the water by heat from the hot burner, making both Q and ΔE_{int} positive. (c) Because the leak is rapid, no energy flows into or out of the system by heat; hence, $Q = 0$. The air molecules escaping from the balloon do work on the surrounding air molecules as they push them out of the way. Thus, W is positive and ΔE_{int} is negative. The decrease in internal energy is evidenced by the fact that the escaping air becomes cooler.

20.5 A is isovolumetric, B is adiabatic, C is isothermal, and D is isobaric.

20.6 c. The blanket acts as a thermal insulator, slowing the transfer of energy by heat from the air into the cube.

c h a p t e r

21

The Kinetic Theory of Gases

Chapter Outline

In Chapter 19 we discussed the properties of an ideal gas, using such macroscopic variables as pressure, volume, and temperature. We shall now show that such large-scale properties can be described on a microscopic scale, where matter is treated as a collection of molecules. Newton's laws of motion applied in a statistical manner to a collection of particles provide a reasonable description of thermodynamic processes. To keep the mathematics relatively simple, we shall consider molecular behavior of gases only, because in gases the interactions between molecules are much weaker than they are in liquids or solids. In the current view of gas behavior, called the *kinetic theory,* gas molecules move about in a random fashion, colliding with the walls of their container and with each other. Perhaps the most important feature of this theory is that it demonstrates that the kinetic energy of molecular motion and the internal energy of a gas system are equivalent. Furthermore, the kinetic theory provides us with a physical basis for our understanding of the concept of temperature.

In the simplest model of a gas, each molecule is considered to be a hard sphere that collides elastically with other molecules and with the container's walls. The hard-sphere model assumes that the molecules do not interact with each other except during collisions and that they are not deformed by collisions. This description is adequate only for monatomic gases, for which the energy is entirely translational kinetic energy. One must modify the theory for more complex molecules, such as oxygen (O_2) and carbon dioxide (CO_2), to include the internal energy associated with rotations and vibrations of the molecules.

21.1 MOLECULAR MODEL OF AN IDEAL GAS

10.5 We begin this chapter by developing a microscopic model of an ideal gas. The model shows that the pressure that a gas exerts on the walls of its container is a consequence of the collisions of the gas molecules with the walls. As we shall see, the model is consistent with the macroscopic description of Chapter 19. In developing this model, we make the following assumptions:

- The number of molecules is large, and the average separation between molecules is great compared with their dimensions. This means that the volume of the molecules is negligible when compared with the volume of the container.
- The molecules obey Newton's laws of motion, but as a whole they move randomly. By "randomly" we mean that any molecule can move in any direction with equal probability. We also assume that the distribution of speeds does not change in time, despite the collisions between molecules. That is, at any given moment, a certain percentage of molecules move at high speeds, a certain percentage move at low speeds, and a certain percentage move at speeds intermediate between high and low.
- The molecules undergo elastic collisions with each other and with the walls of the container. Thus, in the collisions, both kinetic energy and momentum are constant.
- The forces between molecules are negligible except during a collision. The forces between molecules are short-range, so the molecules interact with each other only during collisions.
- The gas under consideration is a pure substance. That is, all of its molecules are identical.

Assumptions of the molecular model of an ideal gas

Although we often picture an ideal gas as consisting of single atoms, we can assume that the behavior of molecular gases approximates that of ideal gases rather

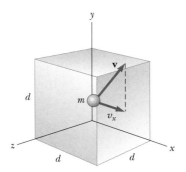

Figure 21.1 A cubical box with sides of length d containing an ideal gas. The molecule shown moves with velocity **v**.

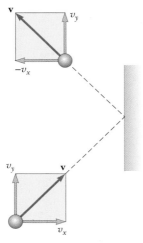

Figure 21.2 A molecule makes an elastic collision with the wall of the container. Its x component of momentum is reversed, while its y component remains unchanged. In this construction, we assume that the molecule moves in the xy plane.

well at low pressures. Molecular rotations or vibrations have no effect, on the average, on the motions that we considered here.

Now let us derive an expression for the pressure of an ideal gas consisting of N molecules in a container of volume V. The container is a cube with edges of length d (Fig. 21.1). Consider the collision of one molecule moving with a velocity **v** toward the right-hand face of the box. The molecule has velocity components v_x, v_y, and v_z. Previously, we used m to represent the mass of a sample, but throughout this chapter we shall use m to represent the mass of one molecule. As the molecule collides with the wall elastically, its x component of velocity is reversed, while its y and z components of velocity remain unaltered (Fig. 21.2). Because the x component of the momentum of the molecule is mv_x before the collision and $-mv_x$ after the collision, the change in momentum of the molecule is

$$\Delta p_x = -mv_x - (mv_x) = -2mv_x$$

Applying the impulse–momentum theorem (Eq. 9.9) to the molecule gives

$$F_1 \,\Delta t = \Delta p_x = -2mv_x$$

where F_1 is the magnitude of the average force exerted by the wall on the molecule in the time Δt. The subscript 1 indicates that we are currently considering only *one* molecule. For the molecule to collide twice with the same wall, it must travel a distance $2d$ in the x direction. Therefore, the time interval between two collisions with the same wall is $\Delta t = 2d/v_x$. Over a time interval that is long compared with Δt, the average force exerted on the molecule for each collision is

$$F_1 = \frac{-2mv_x}{\Delta t} = \frac{-2mv_x}{2d/v_x} = \frac{-mv_x^2}{d} \qquad \textbf{(21.1)}$$

According to Newton's third law, the average force exerted by the molecule on the wall is equal in magnitude and opposite in direction to the force in Equation 21.1:

$$F_{1,\,\text{on wall}} = -F_1 = -\left(\frac{-mv_x^2}{d}\right) = \frac{mv_x^2}{d}$$

Each molecule of the gas exerts a force F_1 on the wall. We find the total force F exerted by all the molecules on the wall by adding the forces exerted by the individual molecules:

$$F = \frac{m}{d}\,(v_{x1}^2 + v_{x2}^2 + \cdots)$$

In this equation, v_{x1} is the x component of velocity of molecule 1, v_{x2} is the x component of velocity of molecule 2, and so on. The summation terminates when we reach N molecules because there are N molecules in the container.

To proceed further, we must note that the average value of the square of the velocity in the x direction for N molecules is

$$\overline{v_x^2} = \frac{v_{x1}^2 + v_{x2}^2 + \cdots + v_{xN}^2}{N}$$

Thus, the total force exerted on the wall can be written

$$F = \frac{Nm}{d}\,\overline{v_x^2}$$

Now let us focus on one molecule in the container whose velocity components are v_x, v_y, and v_z. The Pythagorean theorem relates the square of the speed of this molecule to the squares of these components:

$$v^2 = v_x{}^2 + v_y{}^2 + v_z{}^2$$

Hence, the average value of v^2 for all the molecules in the container is related to the average values of $v_x{}^2$, $v_y{}^2$, and $v_z{}^2$ according to the expression

$$\overline{v^2} = \overline{v_x{}^2} + \overline{v_y{}^2} + \overline{v_z{}^2}$$

Because the motion is completely random, the average values $\overline{v_x{}^2}$, $\overline{v_y{}^2}$, and $\overline{v_z{}^2}$ are equal to each other. Using this fact and the previous equation, we find that

$$\overline{v^2} = 3\overline{v_x{}^2}$$

Thus, the total force exerted on the wall is

$$F = \frac{N}{3}\left(\frac{m\overline{v^2}}{d}\right)$$

Using this expression, we can find the total pressure exerted on the wall:

$$P = \frac{F}{A} = \frac{F}{d^2} = \frac{1}{3}\left(\frac{N}{d^3}\,m\overline{v^2}\right) = \frac{1}{3}\left(\frac{N}{V}\right)m\overline{v^2}$$

$$P = \frac{2}{3}\left(\frac{N}{V}\right)\left(\frac{1}{2}\,m\overline{v^2}\right) \qquad \textbf{(21.2)}$$

Ludwig Boltzmann Austrian theoretical physicist (1844–1906) Boltzmann made many important contributions to the development of the kinetic theory of gases, electromagnetism, and thermodynamics. His pioneering work in the field of kinetic theory led to the branch of physics known as *statistical mechanics*. *(Courtesy of AIP Niels Bohr Library, Lande Collection)*

Relationship between pressure and molecular kinetic energy

This result indicates that **the pressure is proportional to the number of molecules per unit volume and to the average translational kinetic energy of the molecules, $\frac{1}{2}m\overline{v^2}$.** In deriving this simplified model of an ideal gas, we obtain an important result that relates the large-scale quantity of pressure to an atomic quantity—the average value of the square of the molecular speed. Thus, we have established a key link between the atomic world and the large-scale world.

You should note that Equation 21.2 verifies some features of pressure with which you are probably familiar. One way to increase the pressure inside a container is to increase the number of molecules per unit volume in the container. This is what you do when you add air to a tire. The pressure in the tire can also be increased by increasing the average translational kinetic energy of the air molecules in the tire. As we shall soon see, this can be accomplished by increasing the temperature of that air. It is for this reason that the pressure inside a tire increases as the tire warms up during long trips. The continuous flexing of the tire as it moves along the surface of a road results in work done as parts of the tire distort and in an increase in internal energy of the rubber. The increased temperature of the rubber results in the transfer of energy by heat into the air inside the tire. This transfer increases the air's temperature, and this increase in temperature in turn produces an increase in pressure.

Molecular Interpretation of Temperature

We can gain some insight into the meaning of temperature by first writing Equation 21.2 in the more familiar form

$$PV = \frac{2}{3}N\left(\frac{1}{2}m\overline{v^2}\right)$$

Let us now compare this with the equation of state for an ideal gas (Eq. 19.10):

$$PV = Nk_{\mathrm{B}}T$$

Recall that the equation of state is based on experimental facts concerning the macroscopic behavior of gases. Equating the right sides of these expressions, we find that

$$T = \frac{2}{3k_B}\left(\frac{1}{2}m\overline{v^2}\right)$$ (21.3)

Temperature is proportional to average kinetic energy

That is, **temperature is a direct measure of average molecular kinetic energy.**

By rearranging Equation 21.3, we can relate the translational molecular kinetic energy to the temperature:

$$\frac{1}{2}m\overline{v^2} = \frac{3}{2}k_B T$$ (21.4)

Average kinetic energy per molecule

That is, the average translational kinetic energy per molecule is $\frac{3}{2}k_B T$. Because $\overline{v_x^2} = \frac{1}{3}\overline{v^2}$, it follows that

$$\frac{1}{2}m\overline{v_x^2} = \frac{1}{2}k_B T$$ (21.5)

In a similar manner, it follows that the motions in the y and z directions give us

$$\frac{1}{2}m\overline{v_y^2} = \frac{1}{2}k_B T \quad \text{and} \quad \frac{1}{2}m\overline{v_z^2} = \frac{1}{2}k_B T$$

Thus, each translational degree of freedom contributes an equal amount of energy to the gas, namely, $\frac{1}{2}k_B T$. (In general, "degrees of freedom" refers to the number of independent means by which a molecule can possess energy.) A generalization of this result, known as the **theorem of equipartition of energy,** states that

Theorem of equipartition of energy

each degree of freedom contributes $\frac{1}{2}k_B T$ to the energy of a system.

The total translational kinetic energy of N molecules of gas is simply N times the average energy per molecule, which is given by Equation 21.4:

Total translational kinetic energy of N molecules

$$E_{trans} = N\left(\frac{1}{2}m\overline{v^2}\right) = \frac{3}{2}Nk_B T = \frac{3}{2}nRT$$ (21.6)

where we have used $k_B = R/N_A$ for Boltzmann's constant and $n = N/N_A$ for the number of moles of gas. If we consider a gas for which the only type of energy for the molecules is translational kinetic energy, we can use Equation 21.6 to express

TABLE 21.1   Some rms Speeds

Gas	Molar Mass (g/mol)	v_{rms} at 20°C (m/s)
H_2	2.02	1904
He	4.00	1352
H_2O	18.0	637
Ne	20.2	602
N_2 or CO	28.0	511
NO	30.0	494
CO_2	44.0	408
SO_2	64.1	338

the internal energy of the gas. This result implies that the internal energy of an ideal gas depends only on the temperature.

The square root of $\overline{v^2}$ is called the *root-mean-square* (rms) *speed* of the molecules. From Equation 21.4 we obtain, for the rms speed,

$$v_{rms} = \sqrt{\overline{v^2}} = \sqrt{\frac{3k_B T}{m}} = \sqrt{\frac{3RT}{M}} \qquad \textbf{(21.7)}$$

Root-mean-square speed

where M is the molar mass in kilograms per mole. This expression shows that, at a given temperature, lighter molecules move faster, on the average, than do heavier molecules. For example, at a given temperature, hydrogen molecules, whose molar mass is 2×10^{-3} kg/mol, have an average speed four times that of oxygen molecules, whose molar mass is 32×10^{-3} kg/mol. Table 21.1 lists the rms speeds for various molecules at 20°C.

EXAMPLE 21.1 ▶ A Tank of Helium

A tank used for filling helium balloons has a volume of 0.300 m³ and contains 2.00 mol of helium gas at 20.0°C. Assuming that the helium behaves like an ideal gas, (a) what is the total translational kinetic energy of the molecules of the gas?

Solution Using Equation 21.6 with $n = 2.00$ mol and $T = 293$ K, we find that

$$E_{trans} = \tfrac{3}{2}nRT = \tfrac{3}{2}(2.00 \text{ mol})(8.31 \text{ J/mol·K})(293 \text{ K})$$

$$= \boxed{7.30 \times 10^3 \text{ J}}$$

(b) What is the average kinetic energy per molecule?

Solution Using Equation 21.4, we find that the average kinetic energy per molecule is

$$\tfrac{1}{2}m\overline{v^2} = \tfrac{3}{2}k_B T = \tfrac{3}{2}(1.38 \times 10^{-23} \text{ J/K})(293 \text{ K})$$

$$= \boxed{6.07 \times 10^{-21} \text{ J}}$$

Exercise Using the fact that the molar mass of helium is 4.00×10^{-3} kg/mol, determine the rms speed of the atoms at 20.0°C.

Answer 1.35×10^3 m/s.

Quick Quiz 21.1 ▶

At room temperature, the average speed of an air molecule is several hundred meters per second. A molecule traveling at this speed should travel across a room in a small fraction of a second. In view of this, why does it take the odor of perfume (or other smells) several minutes to travel across the room?

21.2 ▶ MOLAR SPECIFIC HEAT OF AN IDEAL GAS

10.5 The energy required to raise the temperature of n moles of gas from T_i to T_f depends on the path taken between the initial and final states. To understand this, let us consider an ideal gas undergoing several processes such that the change in temperature is $\Delta T = T_f - T_i$ for all processes. The temperature change can be achieved by taking a variety of paths from one isotherm to another, as shown in Figure 21.3. Because ΔT is the same for each path, the change in internal energy ΔE_{int} is the same for all paths. However, we know from the first law, $Q = \Delta E_{int} + W$, that the heat Q is different for each path because W (the area under the curves) is different for each path. Thus, the heat associated with a given change in temperature does not have a unique value.

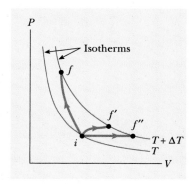

Figure 21.3 An ideal gas is taken from one isotherm at temperature T to another at temperature $T + \Delta T$ along three different paths.

We can address this difficulty by defining specific heats for two processes that frequently occur: changes at constant volume and changes at constant pressure. Because the number of moles is a convenient measure of the amount of gas, we define the **molar specific heats** associated with these processes with the following equations:

$$Q = nC_V \Delta T \qquad \text{(constant volume)} \qquad \textbf{(21.8)}$$

$$Q = nC_P \Delta T \qquad \text{(constant pressure)} \qquad \textbf{(21.9)}$$

where C_V is the **molar specific heat at constant volume** and C_P is the **molar specific heat at constant pressure.** When we heat a gas at constant pressure, not only does the internal energy of the gas increase, but the gas also does work because of the change in volume. Therefore, the heat $Q_{\text{constant } P}$ must account for both the increase in internal energy and the transfer of energy out of the system by work, and so $Q_{\text{constant } P}$ is greater than $Q_{\text{constant } V}$. Thus, C_P is greater than C_V.

In the previous section, we found that the temperature of a gas is a measure of the average translational kinetic energy of the gas molecules. This kinetic energy is associated with the motion of the center of mass of each molecule. It does not include the energy associated with the internal motion of the molecule—namely, vibrations and rotations about the center of mass. This should not be surprising because the simple kinetic theory model assumes a structureless molecule.

In view of this, let us first consider the simplest case of an ideal monatomic gas, that is, a gas containing one atom per molecule, such as helium, neon, or argon. When energy is added to a monatomic gas in a container of fixed volume (by heating, for example), all of the added energy goes into increasing the translational kinetic energy of the atoms. There is no other way to store the energy in a monatomic gas. Therefore, from Equation 21.6, we see that the total internal energy E_{int} of N molecules (or n mol) of an ideal monatomic gas is

> Internal energy of an ideal monatomic gas is proportional to its temperature

$$E_{\text{int}} = \tfrac{3}{2} N k_B T = \tfrac{3}{2} nRT \qquad \textbf{(21.10)}$$

Note that for a monatomic ideal gas, E_{int} is a function of T only, and the functional relationship is given by Equation 21.10. In general, the internal energy of an ideal gas is a function of T only, and the exact relationship depends on the type of gas, as we shall soon explore.

Quick Quiz 21.2

How does the internal energy of a gas change as its pressure is decreased while its volume is increased in such a way that the process follows the isotherm labeled T in Figure 21.4? (a) E_{int} increases. (b) E_{int} decreases. (c) E_{int} stays the same. (d) There is not enough information to determine ΔE_{int}.

If energy is transferred by heat to a system at *constant volume*, then no work is done by the system. That is, $W = \int P \, dV = 0$ for a constant-volume process. Hence, from the first law of thermodynamics, we see that

$$Q = \Delta E_{\text{int}} \qquad \textbf{(21.11)}$$

In other words, all of the energy transferred by heat goes into increasing the internal energy (and temperature) of the system. A constant-volume process from i to f is described in Figure 21.4, where ΔT is the temperature difference between the two isotherms. Substituting the expression for Q given by Equation 21.8 into

Equation 21.11, we obtain

$$\Delta E_{\text{int}} = nC_V\,\Delta T \qquad \textbf{(21.12)}$$

If the molar specific heat is constant, we can express the internal energy of a gas as

$$E_{\text{int}} = nC_V T$$

This equation applies to all ideal gases—to gases having more than one atom per molecule, as well as to monatomic ideal gases.

In the limit of infinitesimal changes, we can use Equation 21.12 to express the molar specific heat at constant volume as

$$C_V = \frac{1}{n}\frac{dE_{\text{int}}}{dT} \qquad \textbf{(21.13)}$$

Let us now apply the results of this discussion to the monatomic gas that we have been studying. Substituting the internal energy from Equation 21.10 into Equation 21.13, we find that

$$C_V = \tfrac{3}{2}R \qquad \textbf{(21.14)}$$

This expression predicts a value of $C_V = \tfrac{3}{2}R = 12.5\ \text{J/mol·K}$ for all monatomic gases. This is in excellent agreement with measured values of molar specific heats for such gases as helium, neon, argon, and xenon over a wide range of temperatures (Table 21.2).

Now suppose that the gas is taken along the constant-pressure path $i \rightarrow f'$ shown in Figure 21.4. Along this path, the temperature again increases by ΔT. The energy that must be transferred by heat to the gas in this process is $Q = nC_P\,\Delta T$. Because the volume increases in this process, the work done by the gas is $W = P\,\Delta V$, where P is the constant pressure at which the process occurs. Applying

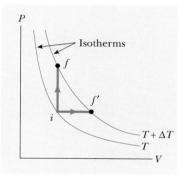

Figure 21.4 Energy is transferred by heat to an ideal gas in two ways. For the constant-volume path $i \rightarrow f$, all the energy goes into increasing the internal energy of the gas because no work is done. Along the constant-pressure path $i \rightarrow f'$, part of the energy transferred in by heat is transferred out by work done by the gas.

TABLE 21.2 **Molar Specific Heats of Various Gases**

Gas	**Molar Specific Heat (J/mol · K)[a]**			
	C_P	C_V	$C_P - C_V$	$\gamma = C_P/C_V$
Monatomic Gases				
He	20.8	12.5	8.33	1.67
Ar	20.8	12.5	8.33	1.67
Ne	20.8	12.7	8.12	1.64
Kr	20.8	12.3	8.49	1.69
Diatomic Gases				
H_2	28.8	20.4	8.33	1.41
N_2	29.1	20.8	8.33	1.40
O_2	29.4	21.1	8.33	1.40
CO	29.3	21.0	8.33	1.40
Cl_2	34.7	25.7	8.96	1.35
Polyatomic Gases				
CO_2	37.0	28.5	8.50	1.30
SO_2	40.4	31.4	9.00	1.29
H_2O	35.4	27.0	8.37	1.30
CH_4	35.5	27.1	8.41	1.31

[a] All values except that for water were obtained at 300 K.

the first law to this process, we have

$$\Delta E_{int} = Q - W = nC_P \Delta T - P\Delta V \tag{21.15}$$

In this case, the energy added to the gas by heat is channeled as follows: Part of it does external work (that is, it goes into moving a piston), and the remainder increases the internal energy of the gas. But the change in internal energy for the process $i \rightarrow f'$ is equal to that for the process $i \rightarrow f$ because E_{int} depends only on temperature for an ideal gas and because ΔT is the same for both processes. In addition, because $PV = nRT$, we note that for a constant-pressure process, $P\Delta V = nR\Delta T$. Substituting this value for $P\Delta V$ into Equation 21.15 with $\Delta E_{int} = nC_V \Delta T$ (Eq. 21.12) gives

$$nC_V \Delta T = nC_P \Delta T - nR\Delta T$$

$$C_P - C_V = R \tag{21.16}$$

This expression applies to *any* ideal gas. It predicts that the molar specific heat of an ideal gas at constant pressure is greater than the molar specific heat at constant volume by an amount R, the universal gas constant (which has the value 8.31 J/mol·K). This expression is applicable to real gases, as the data in Table 21.2 show.

Because $C_V = \frac{3}{2}R$ for a monatomic ideal gas, Equation 21.16 predicts a value $C_P = \frac{5}{2}R = 20.8$ J/mol·K for the molar specific heat of a monatomic gas at constant pressure. The ratio of these heat capacities is a dimensionless quantity γ (Greek letter gamma):

Ratio of molar specific heats for a monatomic ideal gas

$$\gamma = \frac{C_P}{C_V} = \frac{(5/2)R}{(3/2)R} = \frac{5}{3} = 1.67 \tag{21.17}$$

Theoretical values of C_P and γ are in excellent agreement with experimental values obtained for monatomic gases, but they are in serious disagreement with the values for the more complex gases (see Table 21.2). This is not surprising because the value $C_V = \frac{3}{2}R$ was derived for a monatomic ideal gas, and we expect some additional contribution to the molar specific heat from the internal structure of the more complex molecules. In Section 21.4, we describe the effect of molecular structure on the molar specific heat of a gas. We shall find that the internal energy—and, hence, the molar specific heat—of a complex gas must include contributions from the rotational and the vibrational motions of the molecule.

We have seen that the molar specific heats of gases at constant pressure are greater than the molar specific heats at constant volume. This difference is a consequence of the fact that in a constant-volume process, no work is done and all of the energy transferred by heat goes into increasing the internal energy (and temperature) of the gas, whereas in a constant-pressure process, some of the energy transferred by heat is transferred out as work done by the gas as it expands. In the case of solids and liquids heated at constant pressure, very little work is done because the thermal expansion is small. Consequently, C_P and C_V are approximately equal for solids and liquids.

EXAMPLE 21.2   Heating a Cylinder of Helium

A cylinder contains 3.00 mol of helium gas at a temperature of 300 K. (a) If the gas is heated at constant volume, how much energy must be transferred by heat to the gas for its temperature to increase to 500 K?

Solution For the constant-volume process, we have

$$Q_1 = nC_V \Delta T$$

Because $C_V = 12.5$ J/mol·K for helium and $\Delta T = 200$ K, we obtain

$$Q_1 = (3.00 \text{ mol}) (12.5 \text{ J/mol} \cdot \text{K}) (200 \text{ K}) = \boxed{7.50 \times 10^3 \text{ J}}$$

(b) How much energy must be transferred by heat to the gas at constant pressure to raise the temperature to 500 K?

Solution Making use of Table 21.2, we obtain

$$Q_2 = nC_P \Delta T = (3.00 \text{ mol}) (20.8 \text{ J/mol} \cdot \text{K}) (200 \text{ K})$$

$$= \boxed{12.5 \times 10^3 \text{ J}}$$

Exercise What is the work done by the gas in this isobaric process?

Answer $W = Q_2 - Q_1 = 5.00 \times 10^3$ J.

21.3 ADIABATIC PROCESSES FOR AN IDEAL GAS

As we noted in Section 20.6, an adiabatic process is one in which no energy is transferred by heat between a system and its surroundings. For example, if a gas is compressed (or expanded) very rapidly, very little energy is transferred out of (or into) the system by heat, and so the process is nearly adiabatic. (We must remember that the temperature of a system changes in an adiabatic process even though no energy is transferred by heat.) Such processes occur in the cycle of a gasoline engine, which we discuss in detail in the next chapter.

Another example of an adiabatic process is the very slow expansion of a gas that is thermally insulated from its surroundings. In general,

> an **adiabatic process** is one in which no energy is exchanged by heat between a system and its surroundings.

Definition of an adiabatic process

Let us suppose that an ideal gas undergoes an adiabatic expansion. At any time during the process, we assume that the gas is in an equilibrium state, so that the equation of state $PV = nRT$ is valid. As we shall soon see, the pressure and volume at any time during an adiabatic process are related by the expression

$$PV^\gamma = \text{constant} \tag{21.18}$$

Relationship between P and V for an adiabatic process involving an ideal gas

where $\gamma = C_P/C_V$ is assumed to be constant during the process. Thus, we see that all three variables in the ideal gas law—P, V, and T—change during an adiabatic process.

Proof That PV^γ = constant for an Adiabatic Process

When a gas expands adiabatically in a thermally insulated cylinder, no energy is transferred by heat between the gas and its surroundings; thus, $Q = 0$. Let us take the infinitesimal change in volume to be dV and the infinitesimal change in temperature to be dT. The work done by the gas is $P\,dV$. Because the internal energy of an ideal gas depends only on temperature, the change in the internal energy in an adiabatic expansion is the same as that for an isovolumetric process between the same temperatures, $dE_{\text{int}} = nC_V\,dT$ (Eq. 21.12). Hence, the first law of thermodynamics, $\Delta E_{\text{int}} = Q - W$, with $Q = 0$, becomes

$$dE_{\text{int}} = nC_V\,dT = -P\,dV$$

Taking the total differential of the equation of state of an ideal gas, $PV = nRT$, we

Figure 21.5   The *PV* diagram for an adiabatic expansion. Note that $T_f < T_i$ in this process.

Adiabatic process

see that

$$P\,dV + V\,dP = nR\,dT$$

Eliminating dT from these two equations, we find that

$$P\,dV + V\,dP = -\frac{R}{C_V}P\,dV$$

Substituting $R = C_P - C_V$ and dividing by PV, we obtain

$$\frac{dV}{V} + \frac{dP}{P} = -\left(\frac{C_P - C_V}{C_V}\right)\frac{dV}{V} = (1 - \gamma)\frac{dV}{V}$$

$$\frac{dP}{P} + \gamma\frac{dV}{V} = 0$$

Integrating this expression, we have

$$\ln P + \gamma \ln V = \text{constant}$$

which is equivalent to Equation 21.18:

$$PV^\gamma = \text{constant}$$

The *PV* diagram for an adiabatic expansion is shown in Figure 21.5. Because $\gamma > 1$, the *PV* curve is steeper than it would be for an isothermal expansion. By the definition of an adiabatic process, no energy is transferred by heat into or out of the system. Hence, from the first law, we see that ΔE_{int} is negative (the gas does work, so its internal energy decreases) and so ΔT also is negative. Thus, we see that the gas cools ($T_f < T_i$) during an adiabatic expansion. Conversely, the temperature increases if the gas is compressed adiabatically. Applying Equation 21.18 to the initial and final states, we see that

$$P_i V_i^\gamma = P_f V_f^\gamma \tag{21.19}$$

Using the ideal gas law, we can express Equation 21.19 as

$$T_i V_i^{\gamma-1} = T_f V_f^{\gamma-1} \tag{21.20}$$

EXAMPLE 21.3   A Diesel Engine Cylinder

Air at 20.0°C in the cylinder of a diesel engine is compressed from an initial pressure of 1.00 atm and volume of 800.0 cm³ to a volume of 60.0 cm³. Assume that air behaves as an ideal gas with $\gamma = 1.40$ and that the compression is adiabatic. Find the final pressure and temperature of the air.

Solution   Using Equation 21.19, we find that

$$P_f = P_i\left(\frac{V_i}{V_f}\right)^\gamma = (1.00\text{ atm})\left(\frac{800.0\text{ cm}^3}{60.0\text{ cm}^3}\right)^{1.40}$$

$$= \boxed{37.6\text{ atm}}$$

Because $PV = nRT$ is valid during any process and because

no gas escapes from the cylinder,

$$\frac{P_i V_i}{T_i} = \frac{P_f V_f}{T_f}$$

$$T_f = \frac{P_f V_f}{P_i V_i}T_i = \frac{(37.6\text{ atm})(60.0\text{ cm}^3)}{(1.00\text{ atm})(800.0\text{ cm}^3)}(293\text{ K})$$

$$= \boxed{826\text{ K} = 553°\text{C}}$$

The high compression in a diesel engine raises the temperature of the fuel enough to cause its combustion without the use of spark plugs.

21.4 THE EQUIPARTITION OF ENERGY

We have found that model predictions based on molar specific heat agree quite well with the behavior of monatomic gases but not with the behavior of complex gases (see Table 21.2). Furthermore, the value predicted by the model for the quantity $C_P - C_V = R$ is the same for all gases. This is not surprising because this difference is the result of the work done by the gas, which is independent of its molecular structure.

To clarify the variations in C_V and C_P in gases more complex than monatomic gases, let us first explain the origin of molar specific heat. So far, we have assumed that the sole contribution to the internal energy of a gas is the translational kinetic energy of the molecules. However, the internal energy of a gas actually includes contributions from the translational, vibrational, and rotational motion of the molecules. The rotational and vibrational motions of molecules can be activated by collisions and therefore are "coupled" to the translational motion of the molecules. The branch of physics known as *statistical mechanics* has shown that, for a large number of particles obeying the laws of Newtonian mechanics, the available energy is, on the average, shared equally by each independent degree of freedom. Recall from Section 21.1 that the equipartition theorem states that, at equilibrium, each degree of freedom contributes $\frac{1}{2}k_B T$ of energy per molecule.

Let us consider a diatomic gas whose molecules have the shape of a dumbbell (Fig. 21.6). In this model, the center of mass of the molecule can translate in the x, y, and z directions (Fig. 21.6a). In addition, the molecule can rotate about three mutually perpendicular axes (Fig. 21.6b). We can neglect the rotation about the y axis because the moment of inertia I_y and the rotational energy $\frac{1}{2}I_y\omega^2$ about this axis are negligible compared with those associated with the x and z axes. (If the two atoms are taken to be point masses, then I_y is identically zero.) Thus, there are five degrees of freedom: three associated with the translational motion and two associated with the rotational motion. Because each degree of freedom contributes, on the average, $\frac{1}{2}k_B T$ of energy per molecule, the total internal energy for a system of N molecules is

$$E_{\text{int}} = 3N(\tfrac{1}{2}k_B T) + 2N(\tfrac{1}{2}k_B T) = \tfrac{5}{2}Nk_B T = \tfrac{5}{2}nRT$$

We can use this result and Equation 21.13 to find the molar specific heat at constant volume:

$$C_V = \frac{1}{n}\frac{dE_{\text{int}}}{dT} = \frac{1}{n}\frac{d}{dT}\left(\frac{5}{2}nRT\right) = \frac{5}{2}R$$

From Equations 21.16 and 21.17, we find that

$$C_P = C_V + R = \tfrac{7}{2}R$$

$$\gamma = \frac{C_P}{C_V} = \frac{\frac{7}{2}R}{\frac{5}{2}R} = \frac{7}{5} = 1.40$$

These results agree quite well with most of the data for diatomic molecules given in Table 21.2. This is rather surprising because we have not yet accounted for the possible vibrations of the molecule. In the vibratory model, the two atoms are joined by an imaginary spring (see Fig. 21.6c). The vibrational motion adds two more degrees of freedom, which correspond to the kinetic energy and the potential energy associated with vibrations along the length of the molecule. Hence, classical physics and the equipartition theorem predict an internal energy of

$$E_{\text{int}} = 3N(\tfrac{1}{2}k_B T) + 2N(\tfrac{1}{2}k_B T) + 2N(\tfrac{1}{2}k_B T) = \tfrac{7}{2}Nk_B T = \tfrac{7}{2}nRT$$

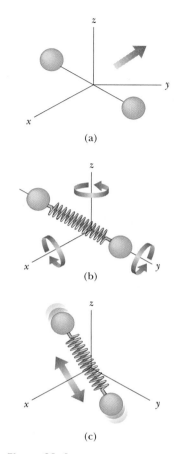

Figure 21.6 Possible motions of a diatomic molecule: (a) translational motion of the center of mass, (b) rotational motion about the various axes, and (c) vibrational motion along the molecular axis.

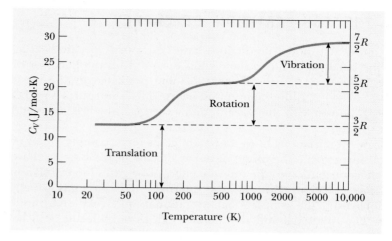

Figure 21.7   The molar specific heat of hydrogen as a function of temperature. The horizontal scale is logarithmic. Note that hydrogen liquefies at 20 K.

and a molar specific heat at constant volume of

$$C_V = \frac{1}{n} \frac{dE_{\text{int}}}{dT} = \frac{1}{n} \frac{d}{dT}\left(\frac{7}{2} nRT\right) = \frac{7}{2} R$$

This value is inconsistent with experimental data for molecules such as H_2 and N_2 (see Table 21.2) and suggests a breakdown of our model based on classical physics.

For molecules consisting of more than two atoms, the number of degrees of freedom is even larger and the vibrations are more complex. This results in an even higher predicted molar specific heat, which is in qualitative agreement with experiment. The more degrees of freedom available to a molecule, the more "ways" it can store internal energy; this results in a higher molar specific heat.

We have seen that the equipartition theorem is successful in explaining some features of the molar specific heat of gas molecules with structure. However, the theorem does not account for the observed temperature variation in molar specific heats. As an example of such a temperature variation, C_V for H_2 is $\frac{5}{2}R$ from about 250 K to 750 K and then increases steadily to about $\frac{7}{2}R$ well above 750 K (Fig. 21.7). This suggests that much more significant vibrations occur at very high temperatures. At temperatures well below 250 K, C_V has a value of about $\frac{3}{2}R$, suggesting that the molecule has only translational energy at low temperatures.

A Hint of Energy Quantization

The failure of the equipartition theorem to explain such phenomena is due to the inadequacy of classical mechanics applied to molecular systems. For a more satisfactory description, it is necessary to use a quantum-mechanical model, in which the energy of an individual molecule is quantized. The energy separation between adjacent vibrational energy levels for a molecule such as H_2 is about ten times greater than the average kinetic energy of the molecule at room temperature. Consequently, collisions between molecules at low temperatures do not provide enough energy to change the vibrational state of the molecule. It is often stated that such degrees of freedom are "frozen out." This explains why the vibrational energy does not contribute to the molar specific heats of molecules at low temperatures.

The rotational energy levels also are quantized, but their spacing at ordinary temperatures is small compared with $k_B T$. Because the spacing between quantized energy levels is small compared with the available energy, the system behaves in accordance with classical mechanics. However, at sufficiently low temperatures (typically less than 50 K), where $k_B T$ is small compared with the spacing between rotational levels, intermolecular collisions may not be sufficiently energetic to alter the rotational states. This explains why C_V reduces to $\frac{3}{2}R$ for H_2 in the range from 20 K to approximately 100 K.

The Molar Specific Heat of Solids

The molar specific heats of solids also demonstrate a marked temperature dependence. Solids have molar specific heats that generally decrease in a nonlinear manner with decreasing temperature and approach zero as the temperature approaches absolute zero. At high temperatures (usually above 300 K), the molar specific heats approach the value of $3R \approx 25\ \text{J/mol}\cdot\text{K}$, a result known as the *DuLong–Petit law*. The typical data shown in Figure 21.8 demonstrate the temperature dependence of the molar specific heats for two semiconducting solids, silicon and germanium.

We can explain the molar specific heat of a solid at high temperatures using the equipartition theorem. For small displacements of an atom from its equilibrium position, each atom executes simple harmonic motion in the x, y, and z directions. The energy associated with vibrational motion in the x direction is

$$E = \tfrac{1}{2}mv_x^2 + \tfrac{1}{2}kx^2$$

The expressions for vibrational motions in the y and z directions are analogous. Therefore, each atom of the solid has six degrees of freedom. According to the equipartition theorem, this corresponds to an average vibrational energy of $6(\tfrac{1}{2}k_B T) = 3k_B T$ per atom. Therefore, the total internal energy of a solid consisting of N atoms is

$$E_{\text{int}} = 3Nk_B T = 3nRT \qquad \textbf{(21.21)}$$

From this result, we find that the molar specific heat of a solid at constant volume is

$$C_V = \frac{1}{n}\frac{dE_{\text{int}}}{dT} = 3R \qquad \textbf{(21.22)}$$

This result is in agreement with the empirical DuLong–Petit law. The discrepancies between this model and the experimental data at low temperatures are again due to the inadequacy of classical physics in describing the microscopic world.

Figure 21.8 Molar specific heat of silicon and germanium. As T approaches zero, the molar specific heat also approaches zero. *(From C. Kittel,* Introduction to Solid State Physics, *New York, Wiley, 1971.)*

> Total internal energy of a solid

> Molar specific heat of a solid at constant volume

21.5 THE BOLTZMANN DISTRIBUTION LAW

Thus far we have neglected the fact that not all molecules in a gas have the same speed and energy. In reality, their motion is extremely chaotic. Any individual molecule is colliding with others at an enormous rate—typically, a billion times per second. Each collision results in a change in the speed and direction of motion of each of the participant molecules. From Equation 21.7, we see that average molecular speeds increase with increasing temperature. What we would like to know now is the relative number of molecules that possess some characteristic, such as a certain percentage of the total energy or speed. The ratio of the number of molecules that have the desired characteristic to the total number of molecules is the probability that a particular molecule has that characteristic.

The Exponential Atmosphere

We begin by considering the distribution of molecules in our atmosphere. Let us determine how the number of molecules per unit volume varies with altitude. Our model assumes that the atmosphere is at a constant temperature T. (This assumption is not entirely correct because the temperature of our atmosphere decreases by about 2°C for every 300-m increase in altitude. However, the model does illustrate the basic features of the distribution.)

According to the ideal gas law, a gas containing N molecules in thermal equilibrium obeys the relationship $PV = Nk_BT$. It is convenient to rewrite this equation in terms of the **number density** $n_V = N/V$, which represents the number of molecules per unit volume of gas. This quantity is important because it can vary from one point to another. In fact, our goal is to determine how n_V changes in our atmosphere. We can express the ideal gas law in terms of n_V as $P = n_V k_B T$. Thus, if the number density n_V is known, we can find the pressure, and vice versa. The pressure in the atmosphere decreases with increasing altitude because a given layer of air must support the weight of all the atmosphere above it—that is, the greater the altitude, the less the weight of the air above that layer, and the lower the pressure.

To determine the variation in pressure with altitude, let us consider an atmospheric layer of thickness dy and cross-sectional area A, as shown in Figure 21.9. Because the air is in static equilibrium, the magnitude PA of the upward force exerted on the bottom of this layer must exceed the magnitude of the downward force on the top of the layer, $(P + dP)A$, by an amount equal to the weight of gas in this thin layer. If the mass of a gas molecule in the layer is m, and if a total of N molecules are in the layer, then the weight of the layer is given by $mgN = mgn_V V = mgn_V A\,dy$. Thus, we see that

$$PA - (P + dP)A = mgn_V A\,dy$$

This expression reduces to

$$dP = -mgn_V dy$$

Because $P = n_V k_B T$ and T is assumed to remain constant, we see that $dP = k_B T\,dn_V$. Substituting this result into the previous expression for dP and rearranging terms, we have

$$\frac{dn_V}{n_V} = -\frac{mg}{k_B T}\,dy$$

Integrating this expression, we find that

Law of atmospheres

$$n_V(y) = n_0 e^{-mgy/k_B T} \tag{21.23}$$

where the constant n_0 is the number density at $y = 0$. This result is known as the **law of atmospheres.**

According to Equation 21.23, the number density decreases exponentially with increasing altitude when the temperature is constant. The number density of our atmosphere at sea level is about $n_0 = 2.69 \times 10^{25}$ molecules/m^3. Because the pressure is $P = n_V k_B T$, we see from Equation 21.23 that the pressure of our atmosphere varies with altitude according to the expression

$$P = P_0 e^{-mgy/k_B T} \tag{21.24}$$

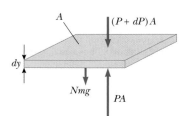

Figure 21.9 An atmospheric layer of gas in equilibrium.

where $P_0 = n_0 k_B T$. A comparison of this model with the actual atmospheric pressure as a function of altitude shows that the exponential form is a reasonable approximation to the Earth's atmosphere.

EXAMPLE 21.4 ▸ High-Flying Molecules

What is the number density of air at an altitude of 11.0 km (the cruising altitude of a commercial jetliner) compared with its number density at sea level? Assume that the air temperature at this height is the same as that at the ground, 20°C.

Solution The number density of our atmosphere decreases exponentially with altitude according to the law of atmospheres, Equation 21.23. We assume an average molecular mass of 28.9 u = 4.80×10^{-26} kg. Taking $y = 11.0$ km, we calculate the power of the exponential in Equation 21.23 to be

$$\frac{mgy}{k_B T} = \frac{(4.80 \times 10^{-26} \text{ kg})(9.80 \text{ m/s}^2)(11\ 000 \text{ m})}{(1.38 \times 10^{-23} \text{ J/K})(293 \text{ K})} = 1.28$$

Thus, Equation 21.23 gives

$$n_V = n_0 e^{-mgy/k_B T} = n_0 e^{-1.28} = \boxed{0.278 n_0}$$

That is, the number density of air at an altitude of 11.0 km is only 27.8% of the number density at sea level, if we assume constant temperature. Because the temperature actually decreases with altitude, the number density of air is less than this in reality.

The pressure at this height is reduced in the same manner. For this reason, high-flying aircraft must have pressurized cabins to ensure passenger comfort and safety.

Computing Average Values

The exponential function $e^{-mgy/k_B T}$ that appears in Equation 21.23 can be interpreted as a probability distribution that gives the relative probability of finding a gas molecule at some height y. Thus, the probability distribution $p(y)$ is proportional to the number density distribution $n_V(y)$. This concept enables us to determine many properties of the atmosphere, such as the fraction of molecules below a certain height or the average potential energy of a molecule.

As an example, let us determine the average height $\bar{y}$ of a molecule in the atmosphere at temperature T. The expression for this average height is

$$\bar{y} = \frac{\int_0^\infty y n_V(y)\ dy}{\int_0^\infty n_V(y)\ dy} = \frac{\int_0^\infty y e^{-mgy/k_B T}\ dy}{\int_0^\infty e^{-mgy/k_B T}\ dy}$$

where the height of a molecule can range from 0 to ∞. The numerator in this expression represents the sum of the heights of the molecules times their number, while the denominator is the sum of the number of molecules. That is, the denominator is the total number of molecules. After performing the indicated integrations, we find that

$$\bar{y} = \frac{(k_B T/mg)^2}{k_B T/mg} = \frac{k_B T}{mg}$$

This expression states that the average height of a molecule increases as T increases, as expected.

We can use a similar procedure to determine the average potential energy of a gas molecule. Because the gravitational potential energy of a molecule at height y is $U = mgy$, the average potential energy is equal to $mg\bar{y}$. Because $\bar{y} = k_B T/mg$, we see that $\overline{U} = mg(k_\mathrm{B}T/mg) = k_\mathrm{B}T$. This important result indicates that the average gravitational potential energy of a molecule depends only on temperature, and not on m or g.

The Boltzmann Distribution

Because the gravitational potential energy of a molecule at height y is $U = mgy$, we can express the law of atmospheres (Eq. 21.23) as

$$n_V = n_0 e^{-U/k_\mathrm{B}T}$$

This means that gas molecules in thermal equilibrium are distributed in space with a probability that depends on gravitational potential energy according to the exponential factor $e^{-U/k_\mathrm{B}T}$.

This exponential expression describing the distribution of molecules in the atmosphere is powerful and applies to any type of energy. In general, the number density of molecules having energy E is

| Boltzmann distribution law |

$$n_V(E) = n_0 e^{-E/k_\mathrm{B}T} \qquad (21.25)$$

This equation is known as the **Boltzmann distribution law** and is important in describing the statistical mechanics of a large number of molecules. It states that **the probability of finding the molecules in a particular energy state varies exponentially as the negative of the energy divided by $k_\mathrm{B}T$.** All the molecules would fall into the lowest energy level if the thermal agitation at a temperature T did not excite the molecules to higher energy levels.

EXAMPLE 21.5   **Thermal Excitation of Atomic Energy Levels**

As we discussed briefly in Section 8.10, atoms can occupy only certain discrete energy levels. Consider a gas at a temperature of 2 500 K whose atoms can occupy only two energy levels separated by 1.50 eV, where 1 eV (electron volt) is an energy unit equal to 1.6×10^{-19} J (Fig. 21.10). Determine the ratio of the number of atoms in the higher energy level to the number in the lower energy level.

Solution   Equation 21.25 gives the relative number of atoms in a given energy level. In this case, the atom has two possible energies, E_1 and E_2, where E_1 is the lower energy level. Hence, the ratio of the number of atoms in the higher energy level to the number in the lower energy level is

$$\frac{n_V(E_2)}{n_V(E_1)} = \frac{n_0 e^{-E_2/k_\mathrm{B}T}}{n_0 e^{-E_1/k_\mathrm{B}T}} = e^{-(E_2-E_1)/k_\mathrm{B}T}$$

In this problem, $E_2 - E_1 = 1.50$ eV, and the denominator of the exponent is

$$k_\mathrm{B}T = (1.38 \times 10^{-23}\,\mathrm{J/K})(2\,500\,\mathrm{K})/1.60 \times 10^{-19}\,\mathrm{J/eV}$$
$$= 0.216\,\mathrm{eV}$$

Therefore, the required ratio is

$$\frac{n(E_2)}{n(E_1)} = e^{-1.50\,\mathrm{eV}/0.216\,\mathrm{eV}} = e^{-6.94} = \boxed{9.64 \times 10^{-4}}$$

This result indicates that at $T = 2\,500$ K, only a small fraction of the atoms are in the higher energy level. In fact, for every atom in the higher energy level, there are about 1 000 atoms in the lower level. The number of atoms in the higher level increases at even higher temperatures, but the distribution law specifies that at equilibrium there are always more atoms in the lower level than in the higher level.

Figure 21.10   Energy level diagram for a gas whose atoms can occupy two energy levels.

21.6 DISTRIBUTION OF MOLECULAR SPEEDS

In 1860 James Clerk Maxwell (1831–1879) derived an expression that describes the distribution of molecular speeds in a very definite manner. His work and subsequent developments by other scientists were highly controversial because direct detection of molecules could not be achieved experimentally at that time. However, about 60 years later, experiments were devised that confirmed Maxwell's predictions.

Let us consider a container of gas whose molecules have some distribution of speeds. Suppose we want to determine how many gas molecules have a speed in the range from, for example, 400 to 410 m/s. Intuitively, we expect that the speed distribution depends on temperature. Furthermore, we expect that the distribution peaks in the vicinity of v_{rms}. That is, few molecules are expected to have speeds much less than or much greater than v_{rms} because these extreme speeds result only from an unlikely chain of collisions.

The observed speed distribution of gas molecules in thermal equilibrium is shown in Figure 21.11. The quantity N_v, called the **Maxwell–Boltzmann distribution function,** is defined as follows: If N is the total number of molecules, then the number of molecules with speeds between v and $v + dv$ is $dN = N_v dv$. This number is also equal to the area of the shaded rectangle in Figure 21.11. Furthermore, the fraction of molecules with speeds between v and $v + dv$ is $N_v dv/N$. This fraction is also equal to the probability that a molecule has a speed in the range v to $v + dv$.

The fundamental expression that describes the distribution of speeds of N gas molecules is

$$N_v = 4\pi N \left(\frac{m}{2\pi k_B T} \right)^{3/2} v^2 e^{-mv^2/2k_B T} \qquad \textbf{(21.26)}$$

where m is the mass of a gas molecule, k_B is Boltzmann's constant, and T is the absolute temperature.[1] Observe the appearance of the Boltzmann factor $e^{-E/k_B T}$ with $E = \frac{1}{2}mv^2$.

As indicated in Figure 21.11, the average speed $\bar{v}$ is somewhat lower than the rms speed. The *most probable speed* v_{mp} is the speed at which the distribution curve reaches a peak. Using Equation 21.26, one finds that

$$v_{rms} = \sqrt{\overline{v^2}} = \sqrt{3k_B T/m} = 1.73\sqrt{k_B T/m} \qquad \textbf{(21.27)}$$

$$\bar{v} = \sqrt{8k_B T/\pi m} = 1.60\sqrt{k_B T/m} \qquad \textbf{(21.28)}$$

$$v_{mp} = \sqrt{2k_B T/m} = 1.41\sqrt{k_B T/m} \qquad \textbf{(21.29)}$$

The details of these calculations are left for the student (see Problems 41 and 62). From these equations, we see that

$$v_{rms} > \bar{v} > v_{mp}$$

Figure 21.12 represents speed distribution curves for N_2. The curves were obtained by using Equation 21.26 to evaluate the distribution function at various speeds and at two temperatures. Note that the peak in the curve shifts to the right

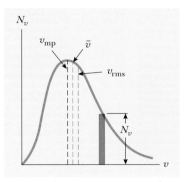

Figure 21.11 The speed distribution of gas molecules at some temperature. The number of molecules having speeds in the range dv is equal to the area of the shaded rectangle, $N_v dv$. The function N_v approaches zero as v approaches infinity.

Maxwell speed distribution function

rms speed

Average speed

Most probable speed

---

[1] For the derivation of this expression, see an advanced textbook on thermodynamics, such as that by R. P. Bauman, *Modern Thermodynamics and Statistical Mechanics,* New York, Macmillan Publishing Co., 1972.

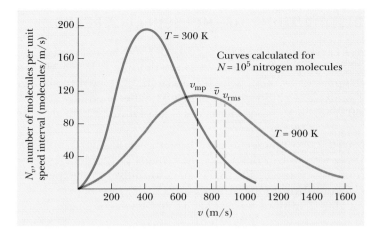

Figure 21.12   The speed distribution function for 10^5 nitrogen molecules at 300 K and 900 K. The total area under either curve is equal to the total number of molecules, which in this case equals 10^5. Note that $v_{rms} > \bar{v} > v_{mp}$.

as T increases, indicating that the average speed increases with increasing temperature, as expected. The asymmetric shape of the curves is due to the fact that the lowest speed possible is zero while the upper classical limit of the speed is infinity.

Quick Quiz 21.3

Consider the two curves in Figure 21.12. What is represented by the area under each of the curves between the 800-m/s and 1 000-m/s marks on the horizontal axis?

Fill one glass with very hot tap water and another with very cold water. Put a single drop of food coloring in each glass. Which drop disperses faster? Why?

Equation 21.26 shows that the distribution of molecular speeds in a gas depends both on mass and on temperature. At a given temperature, the fraction of molecules with speeds exceeding a fixed value increases as the mass decreases. This explains why lighter molecules, such as H_2 and He, escape more readily from the Earth's atmosphere than do heavier molecules, such as N_2 and O_2. (See the discussion of escape speed in Chapter 14. Gas molecules escape even more readily from the Moon's surface than from the Earth's because the escape speed on the Moon is lower than that on the Earth.)

The speed distribution curves for molecules in a liquid are similar to those shown in Figure 21.12. We can understand the phenomenon of evaporation of a liquid from this distribution in speeds, using the fact that some molecules in the liquid are more energetic than others. Some of the faster-moving molecules in the liquid penetrate the surface and leave the liquid even at temperatures well below the boiling point. The molecules that escape the liquid by evaporation are those that have sufficient energy to overcome the attractive forces of the molecules in the liquid phase. Consequently, the molecules left behind in the liquid phase have a lower average kinetic energy; as a result, the temperature of the liquid decreases. Hence, evaporation is a cooling process. For example, an alcohol-soaked cloth often is placed on a feverish head to cool and comfort a patient.

The evaporation process

EXAMPLE 21.6   A System of Nine Particles

Nine particles have speeds of 5.00, 8.00, 12.0, 12.0, 12.0, 14.0, 14.0, 17.0, and 20.0 m/s. (a) Find the particles' average speed.

Solution   The average speed is the sum of the speeds divided by the total number of particles:

$$\bar{v} = \frac{(5.00 + 8.00 + 12.0 + 12.0 + 12.0 + 14.0 + 14.0 + 17.0 + 20.0)\ \text{m/s}}{9}$$

$$= \boxed{12.7\ \text{m/s}}$$

(b) What is the rms speed?

Solution   The average value of the square of the speed is

$$\bar{v^2} = \frac{\begin{pmatrix}5.00^2 + 8.00^2 + 12.0^2 + 12.0^2 + 12.0^2 \\ + 14.0^2 + 14.0^2 + 17.0^2 + 20.0^2\end{pmatrix}\ \text{m}}{9}$$

$$= 178\ \text{m}^2/\text{s}^2$$

Hence, the rms speed is

$$v_{\text{rms}} = \sqrt{\bar{v^2}} = \sqrt{178\ \text{m}^2/\text{s}^2} = \boxed{13.3\ \text{m/s}}$$

(c) What is the most probable speed of the particles?

Solution   Three of the particles have a speed of 12 m/s, two have a speed of 14 m/s, and the remaining have different speeds. Hence, we see that the most probable speed v_{mp} is

$$\boxed{12\ \text{m/s.}}$$

Optional Section

21.7   MEAN FREE PATH

Most of us are familiar with the fact that the strong odor associated with a gas such as ammonia may take a fraction of a minute to diffuse throughout a room. However, because average molecular speeds are typically several hundred meters per second at room temperature, we might expect a diffusion time much less than 1 s. But, as we saw in Quick Quiz 21.1, molecules collide with one other because they are not geometrical points. Therefore, they do not travel from one side of a room to the other in a straight line. Between collisions, the molecules move with constant speed along straight lines. The average distance between collisions is called the **mean free path.** The path of an individual molecule is random and resembles that shown in Figure 21.13. As we would expect from this description, the mean free path is related to the diameter of the molecules and the density of the gas.

We now describe how to estimate the mean free path for a gas molecule. For this calculation, we assume that the molecules are spheres of diameter d. We see from Figure 21.14a that no two molecules collide unless their centers are less than a distance d apart as they approach each other. An equivalent way to describe the

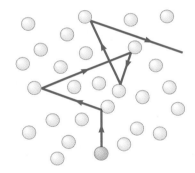

Figure 21.13   A molecule moving through a gas collides with other molecules in a random fashion. This behavior is sometimes referred to as a *random-walk process.* The mean free path increases as the number of molecules per unit volume decreases. Note that the motion is not limited to the plane of the paper.

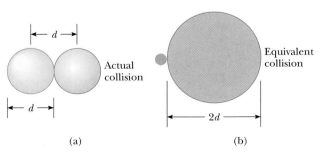

(a)   (b)

Figure 21.14   (a) Two spherical molecules, each of diameter d, collide if their centers are within a distance d of each other. (b) The collision between the two molecules is equivalent to a point molecule's colliding with a molecule having an effective diameter of $2d$.

Figure 21.15 In a time t, a molecule of effective diameter $2d$ sweeps out a cylinder of length $\overline{v}t$, where $\overline{v}$ is its average speed. In this time, it collides with every point molecule within this cylinder.

collisions is to imagine that one of the molecules has a diameter $2d$ and that the rest are geometrical points (Fig. 21.14b). Let us choose the large molecule to be one moving with the average speed $\overline{v}$. In a time t, this molecule travels a distance $\overline{v}t$. In this time interval, the molecule sweeps out a cylinder having a cross-sectional area πd^2 and a length $\overline{v}t$ (Fig. 21.15). Hence, the volume of the cylinder is $\pi d^2 \overline{v}t$. If n_V is the number of molecules per unit volume, then the number of point-size molecules in the cylinder is $(\pi d^2 \overline{v}t)\, n_V$. The molecule of equivalent diameter $2d$ collides with every molecule in this cylinder in the time t. Hence, the number of collisions in the time t is equal to the number of molecules in the cylinder, $(\pi d^2 \overline{v}t)\, n_V$.

The mean free path ℓ equals the average distance $\overline{v}t$ traveled in a time t divided by the number of collisions that occur in that time:

$$\ell = \frac{\overline{v}t}{(\pi d^2 \overline{v}t)\, n_V} = \frac{1}{\pi d^2 n_V}$$

Because the number of collisions in a time t is $(\pi d^2 \overline{v}t)\, n_V$, the number of collisions per unit time, or **collision frequency f,** is

$$f = \pi d^2 \overline{v} n_V$$

The inverse of the collision frequency is the average time between collisions, known as the **mean free time.**

Our analysis has assumed that molecules in the cylinder are stationary. When the motion of these molecules is included in the calculation, the correct results are

| Mean free path | | |

$$\ell = \frac{1}{\sqrt{2}\,\pi d^2 n_V} \tag{21.30}$$

| Collision frequency | | |

$$f = \sqrt{2}\,\pi d^2 \overline{v} n_V = \frac{\overline{v}}{\ell} \tag{21.31}$$

EXAMPLE 21.7 ▶ Bouncing Around in the Air

Approximate the air around you as a collection of nitrogen molecules, each of which has a diameter of 2.00×10^{-10} m. (a) How far does a typical molecule move before it collides with another molecule?

Solution Assuming that the gas is ideal, we can use the equation $PV = Nk_B T$ to obtain the number of molecules per unit volume under typical room conditions:

$$n_V = \frac{N}{V} = \frac{P}{k_B T} = \frac{1.01 \times 10^5 \text{ N/m}^2}{(1.38 \times 10^{-23} \text{ J/K})(293 \text{ K})}$$
$$= 2.50 \times 10^{25} \text{ molecules/m}^3$$

Hence, the mean free path is

$$\ell = \frac{1}{\sqrt{2}\,\pi d^2 n_V}$$

$$= \frac{1}{\sqrt{2}\,\pi (2.00 \times 10^{-10} \text{ m})^2 (2.50 \times 10^{25} \text{ molecules/m}^3)}$$

$$= \boxed{2.25 \times 10^{-7} \text{ m}}$$

This value is about 10^3 times greater than the molecular diameter.

(b) On average, how frequently does one molecule collide with another?

Solution Because the rms speed of a nitrogen molecule at 20.0°C is 511 m/s (see Table 21.1), we know from Equations 21.27 and 21.28 that $\overline{v} = (1.60/1.73)(511 \text{ m/s}) = 473 \text{ m/s}$. Therefore, the collision frequency is

$$f = \frac{\overline{v}}{\ell} = \frac{473 \text{ m/s}}{2.25 \times 10^{-7} \text{ m}} = \boxed{2.10 \times 10^9 \text{/s}}$$

The molecule collides with other molecules at the average rate of about two billion times each second!

The mean free path ℓ is *not* the same as the average separation between particles. In fact, the average separation d between particles is approximately $n_V^{-1/3}$. In this example, the average molecular separation is

$$d = \frac{1}{n_V^{1/3}} = \frac{1}{(2.5 \times 10^{25})^{1/3}} = 3.4 \times 10^{-9} \text{ m}$$

SUMMARY

The pressure of N molecules of an ideal gas contained in a volume V is

$$P = \frac{2}{3} \frac{N}{V} \left(\frac{1}{2} \overline{mv^2} \right) \tag{21.2}$$

The average translational kinetic energy per molecule of a gas, $\frac{1}{2}\overline{mv^2}$, is related to the temperature T of the gas through the expression

$$\frac{1}{2}\overline{mv^2} = \frac{3}{2}k_B T \tag{21.4}$$

where k_B is Boltzmann's constant. Each translational degree of freedom (x, y, or z) has $\frac{1}{2}k_B T$ of energy associated with it.

The **theorem of equipartition of energy** states that the energy of a system in thermal equilibrium is equally divided among all degrees of freedom.

The total energy of N molecules (or n mol) of an ideal monatomic gas is

$$E_{int} = \frac{3}{2}Nk_B T = \frac{3}{2}nRT \tag{21.10}$$

The change in internal energy for n mol of any ideal gas that undergoes a change in temperature ΔT is

$$\Delta E_{int} = nC_V \Delta T \tag{21.12}$$

where C_V is the molar specific heat at constant volume.

The molar specific heat of an ideal monatomic gas at constant volume is $C_V = \frac{3}{2}R$; the molar specific heat at constant pressure is $C_P = \frac{5}{2}R$. The ratio of specific heats is $\gamma = C_P/C_V = \frac{5}{3}$.

If an ideal gas undergoes an adiabatic expansion or compression, the first law of thermodynamics, together with the equation of state, shows that

$$PV^\gamma = \text{constant} \tag{21.18}$$

The **Boltzmann distribution law** describes the distribution of particles among available energy states. The relative number of particles having energy E is

$$n_V(E) = n_0 e^{-E/k_B T} \tag{21.25}$$

The **Maxwell–Boltzmann distribution function** describes the distribution of speeds of molecules in a gas:

$$N_v = 4\pi N \left(\frac{m}{2\pi k_B T} \right)^{3/2} v^2 e^{-mv^2/2k_B T} \tag{21.26}$$

This expression enables us to calculate the **root-mean-square speed,** the **average speed,** and **the most probable speed:**

$$v_{rms} = \sqrt{\overline{v^2}} = \sqrt{3k_B T/m} = 1.73\sqrt{k_B T/m} \tag{21.27}$$

$$\overline{v} = \sqrt{8k_B T/\pi m} = 1.60\sqrt{k_B T/m} \tag{21.28}$$

$$v_{mp} = \sqrt{2k_B T/m} = 1.41\sqrt{k_B T/m} \tag{21.29}$$

QUESTIONS

1. Dalton's law of partial pressures states that the total pressure of a mixture of gases is equal to the sum of the partial pressures of gases making up the mixture. Give a convincing argument for this law on the basis of the kinetic theory of gases.

2. One container is filled with helium gas and another with argon gas. If both containers are at the same temperature, which gas molecules have the higher rms speed? Explain.

3. A gas consists of a mixture of He and N_2 molecules. Do the lighter He molecules travel faster than the N_2 molecules? Explain.

4. Although the average speed of gas molecules in thermal equilibrium at some temperature is greater than zero, the average velocity is zero. Explain why this statement must be true.

5. When alcohol is rubbed on your body, your body temperature decreases. Explain this effect.

6. A liquid partially fills a container. Explain why the temperature of the liquid decreases if the container is then partially evacuated. (Using this technique, one can freeze water at temperatures above 0°C.)

7. A vessel containing a fixed volume of gas is cooled. Does the mean free path of the gas molecules increase, decrease, or remain constant during the cooling process? What about the collision frequency?

8. A gas is compressed at a constant temperature. What happens to the mean free path of the molecules in the process?

9. If a helium-filled balloon initially at room temperature is placed in a freezer, will its volume increase, decrease, or remain the same?

10. What happens to a helium-filled balloon released into the air? Will it expand or contract? Will it stop rising at some height?

11. Which is heavier, dry air or air saturated with water vapor? Explain.

12. Why does a diatomic gas have a greater energy content per mole than a monatomic gas at the same temperature?

13. An ideal gas is contained in a vessel at 300 K. If the temperature is increased to 900 K, (a) by what factor does the rms speed of each molecule change? (b) By what factor does the pressure in the vessel change?

14. A vessel is filled with gas at some equilibrium pressure and temperature. Can all gas molecules in the vessel have the same speed?

15. In our model of the kinetic theory of gases, molecules were viewed as hard spheres colliding elastically with the walls of the container. Is this model realistic?

16. In view of the fact that hot air rises, why does it generally become cooler as you climb a mountain? (Note that air is a poor thermal conductor.)

PROBLEMS

1, 2, 3 = straightforward, intermediate, challenging   ☐ = full solution available in the *Student Solutions Manual and Study Guide*
WEB = solution posted at **http://www.saunderscollege.com/physics/**   💻 = Computer useful in solving problem   📱 = Interactive Physics
☐ = paired numerical/symbolic problems

Section 21.1   Molecular Model of an Ideal Gas

1. Use the definition of Avogadro's number to find the mass of a helium atom.

2. A sealed cubical container 20.0 cm on a side contains three times Avogadro's number of molecules at a temperature of 20.0°C. Find the force exerted by the gas on one of the walls of the container.

3. In a 30.0-s interval, 500 hailstones strike a glass window with an area of 0.600 m^2 at an angle of 45.0° to the window surface. Each hailstone has a mass of 5.00 g and a speed of 8.00 m/s. If the collisions are elastic, what are the average force and pressure on the window?

4. In a time t, N hailstones strike a glass window of area A at an angle θ to the window surface. Each hailstone has a mass m and a speed v. If the collisions are elastic, what are the average force and pressure on the window?

5. In a period of 1.00 s, 5.00×10^{23} nitrogen molecules strike a wall with an area of 8.00 cm^2. If the molecules move with a speed of 300 m/s and strike the wall head-on in perfectly elastic collisions, what is the pressure exerted on the wall? (The mass of one N_2 molecule is 4.68×10^{-26} kg.)

6. A 5.00-L vessel contains 2 mol of oxygen gas at a pressure of 8.00 atm. Find the average translational kinetic energy of an oxygen molecule under these conditions.

7. A spherical balloon with a volume of 4 000 cm^3 contains helium at an (inside) pressure of 1.20×10^5 Pa. How many moles of helium are in the balloon if each helium atom has an average kinetic energy of 3.60×10^{-22} J?

8. The rms speed of a helium atom at a certain temperature is 1 350 m/s. Find by proportion the rms speed of an oxygen molecule at this temperature. (The molar mass of O_2 is 32.0 g/mol, and the molar mass of He is 4.00 g/mol.)

9. (a) How many atoms of helium gas fill a balloon of diameter 30.0 cm at 20.0°C and 1.00 atm? (b) What is the average kinetic energy of the helium atoms? (c) What is the root-mean-square speed of each helium atom?

10. A 5.00-liter vessel contains nitrogen gas at 27.0°C and 3.00 atm. Find (a) the total translational kinetic energy of the gas molecules and (b) the average kinetic energy per molecule.

WEB **11.** A cylinder contains a mixture of helium and argon gas in equilibrium at 150°C. (a) What is the average kinetic energy for each type of gas molecule? (b) What is the root-mean-square speed for each type of molecule?

12. (a) Show that 1 Pa = 1 J/m^3. (b) Show that the density in space of the translational kinetic energy of an ideal gas is $3P/2$.

Section 21.2 Molar Specific Heat of an Ideal Gas

Note: You may use the data given in Table 21.2.

13. Calculate the change in internal energy of 3.00 mol of helium gas when its temperature is increased by 2.00 K.

14. One mole of air ($C_V = 5R/2$) at 300 K and confined in a cylinder under a heavy piston occupies a volume of 5.00 L. Determine the new volume of the gas if 4.40 kJ of energy is transferred to the air by heat.

WEB **15.** One mole of hydrogen gas is heated at constant pressure from 300 K to 420 K. Calculate (a) the energy transferred by heat to the gas, (b) the increase in its internal energy, and (c) the work done by the gas.

16. In a constant-volume process, 209 J of energy is transferred by heat to 1.00 mol of an ideal monatomic gas initially at 300 K. Find (a) the increase in internal energy of the gas, (b) the work it does, and (c) its final temperature.

17. A house has well-insulated walls. It contains a volume of 100 m^3 of air at 300 K. (a) Calculate the energy required to increase the temperature of this air by 1.00°C. (b) If this energy could be used to lift an object of mass m through a height of 2.00 m, what is the value of m?

18. A vertical cylinder with a heavy piston contains air at 300 K. The initial pressure is 200 kPa, and the initial volume is 0.350 m^3. Take the molar mass of air as 28.9 g/mol and assume that $C_V = 5R/2$. (a) Find the specific heat of air at constant volume in units of J/kg · °C. (b) Calculate the mass of the air in the cylinder. (c) Suppose the piston is held fixed. Find the energy input required to raise the temperature of the air to 700 K. (d) Assume again the conditions of the initial state and that the heavy piston is free to move. Find the energy input required to raise the temperature to 700 K.

19. A 1-L Thermos bottle is full of tea at 90°C. You pour out one cup and immediately screw the stopper back on. Make an order-of-magnitude estimate of the change in temperature of the tea remaining in the flask that results from the admission of air at room temperature. State the quantities you take as data and the values you measure or estimate for them.

20. For a diatomic ideal gas, $C_V = 5R/2$. One mole of this gas has pressure P and volume V. When the gas is heated, its pressure triples and its volume doubles. If this heating process includes two steps, the first at con-

stant pressure and the second at constant volume, determine the amount of energy transferred to the gas by heat.

21. One mole of an ideal monatomic gas is at an initial temperature of 300 K. The gas undergoes an isovolumetric process, acquiring 500 J of energy by heat. It then undergoes an isobaric process, losing this same amount of energy by heat. Determine (a) the new temperature of the gas and (b) the work done on the gas.

22. A container has a mixture of two gases: n_1 moles of gas 1, which has a molar specific heat C_1; and n_2 moles of gas 2, which has a molar specific heat C_2. (a) Find the molar specific heat of the mixture. (b) What is the molar specific heat if the mixture has m gases in the amounts $n_1, n_2, n_3, \ldots, n_m$, and molar specific heats $C_1, C_2, C_3, \ldots, C_m$, respectively?

23. **Review Problem.** One mole of an ideal diatomic gas with $C_V = 5R/2$ occupies a volume V_i at a pressure P_i. The gas undergoes a process in which the pressure is proportional to the volume. At the end of the process, it is found that the rms speed of the gas molecules has doubled from its initial value. Determine the amount of energy transferred to the gas by heat.

Section 21.3 Adiabatic Processes for an Ideal Gas

24. During the compression stroke of a certain gasoline engine, the pressure increases from 1.00 atm to 20.0 atm. Assuming that the process is adiabatic and that the gas is ideal, with $\gamma = 1.40$, (a) by what factor does the volume change and (b) by what factor does the temperature change? (c) If the compression starts with 0.016 0 mol of gas at 27.0°C, find the values of Q, W, and ΔE_{int} that characterize the process.

25. Two moles of an ideal gas ($\gamma = 1.40$) expands slowly and adiabatically from a pressure of 5.00 atm and a volume of 12.0 L to a final volume of 30.0 L. (a) What is the final pressure of the gas? (b) What are the initial and final temperatures? (c) Find Q, W, and ΔE_{int}.

26. Air ($\gamma = 1.40$) at 27.0°C and at atmospheric pressure is drawn into a bicycle pump that has a cylinder with an inner diameter of 2.50 cm and a length of 50.0 cm. The down stroke adiabatically compresses the air, which reaches a gauge pressure of 800 kPa before entering the tire. Determine (a) the volume of the compressed air and (b) the temperature of the compressed air. (c) The pump is made of steel and has an inner wall that is 2.00 mm thick. Assume that 4.00 cm of the cylinder's length is allowed to come to thermal equilibrium with the air. What will be the increase in wall temperature?

27. Air in a thundercloud expands as it rises. If its initial temperature was 300 K, and if no energy is lost by thermal conduction on expansion, what is its temperature when the initial volume has doubled?

28. How much work is required to compress 5.00 mol of air at 20.0°C and 1.00 atm to one tenth of the original vol-

ume by (a) an isothermal process and (b) an adiabatic process? (c) What is the final pressure in each of these two cases?

29. Four liters of a diatomic ideal gas ($\gamma = 1.40$) confined to a cylinder is subject to a closed cycle. Initially, the gas is at 1.00 atm and at 300 K. First, its pressure is tripled under constant volume. Then, it expands adiabatically to its original pressure. Finally, the gas is compressed isobarically to its original volume. (a) Draw a *PV* diagram of this cycle. (b) Determine the volume of the gas at the end of the adiabatic expansion. (c) Find the temperature of the gas at the start of the adiabatic expansion. (d) Find the temperature at the end of the cycle. (e) What was the net work done for this cycle?

30. A diatomic ideal gas ($\gamma = 1.40$) confined to a cylinder is subjected to a closed cycle. Initially, the gas is at P_i, V_i, and T_i. First, its pressure is tripled under constant volume. Then, it expands adiabatically to its original pressure. Finally, the gas is compressed isobarically to its original volume. (a) Draw a *PV* diagram of this cycle. (b) Determine the volume of the gas at the end of the adiabatic expansion. (c) Find the temperature of the gas at the start of the adiabatic expansion. (d) Find the temperature at the end of the cycle. (e) What was the net work done for this cycle?

31. During the power stroke in a four-stroke automobile engine, the piston is forced down as the mixture of gas and air undergoes an adiabatic expansion. Assume that (1) the engine is running at 2 500 rpm, (2) the gauge pressure right before the expansion is 20.0 atm, (3) the volumes of the mixture right before and after the expansion are 50.0 and 400 cm³, respectively (Fig.

50.0 cm³

400 cm³

Before

After

Figure P21.31

P21.31), (4) the time involved in the expansion is one-fourth that of the total cycle, and (5) the mixture behaves like an ideal gas, with $\gamma = 1.40$. Find the average power generated during the expansion.

Section 21.4   The Equipartition of Energy

32. A certain molecule has *f* degrees of freedom. Show that a gas consisting of such molecules has the following properties: (1) its total internal energy is $fnRT/2$; (2) its molar specific heat at constant volume is $fR/2$; (3) its molar specific heat at constant pressure is $(f + 2)R/2$; (4) the ratio $\gamma = C_P/C_V = (f + 2)/f$.

WEB 33. Consider 2.00 mol of an ideal diatomic gas. Find the total heat capacity at constant volume and at constant pressure (a) if the molecules rotate but do not vibrate and (b) if the molecules both rotate and vibrate.

34. Inspecting the magnitudes of C_V and C_P for the diatomic and polyatomic gases in Table 21.2, we find that the values increase with increasing molecular mass. Give a qualitative explanation of this observation.

35. In a crude model (Fig. P21.35) of a rotating diatomic molecule of chlorine (Cl_2), the two Cl atoms are 2.00×10^{-10} m apart and rotate about their center of mass with angular speed $\omega = 2.00 \times 10^{12}$ rad/s. What is the rotational kinetic energy of one molecule of Cl_2, which has a molar mass of 70.0 g/mol?

Cl

Cl

Figure P21.35

Section 21.5   The Boltzmann Distribution Law
Section 21.6   Distribution of Molecular Speeds

36. One cubic meter of atomic hydrogen at 0°C contains approximately 2.70×10^{25} atoms at atmospheric pressure. The first excited state of the hydrogen atom has an energy of 10.2 eV above the lowest energy level, which is called the *ground state*. Use the Boltzmann factor to find the number of atoms in the first excited state at 0°C and at 10 000°C.

37. If convection currents (weather) did not keep the Earth's lower atmosphere stirred up, its chemical composition would change somewhat with altitude because the various molecules have different masses. Use the law of the atmospheres to determine how the equilibrium ratio of oxygen to nitrogen molecules changes between sea level and 10.0 km. Assume a uniform temperature of 300 K and take the masses to be 32.0 u for oxygen (O_2) and 28.0 u for nitrogen (N_2).

38. A mixture of two gases diffuses through a filter at rates proportional to the gases' rms speeds. (a) Find the ratio of speeds for the two isotopes of chlorine, ^{35}Cl and ^{37}Cl, as they diffuse through the air. (b) Which isotope moves faster?

39. Fifteen identical particles have various speeds: one has a speed of 2.00 m/s; two have a speed of 3.00 m/s; three have a speed of 5.00 m/s; four have a speed of 7.00 m/s; three have a speed of 9.00 m/s; and two have a speed of 12.0 m/s. Find (a) the average speed, (b) the rms speed, and (c) the most probable speed of these particles.

40. Gaseous helium is in thermal equilibrium with liquid helium at 4.20 K. Even though it is on the point of condensation, model the gas as ideal and determine the most probable speed of a helium atom (mass = 6.64×10^{-27} kg) in it.

41. From the Maxwell–Boltzmann speed distribution, show that the most probable speed of a gas molecule is given by Equation 21.29. Note that the most probable speed corresponds to the point at which the slope of the speed distribution curve, dN_v/dv, is zero.

42. **Review Problem.** At what temperature would the average speed of helium atoms equal (a) the escape speed from Earth, 1.12×10^4 m/s, and (b) the escape speed from the Moon, 2.37×10^3 m/s ? (See Chapter 14 for a discussion of escape speed, and note that the mass of a helium atom is 6.64×10^{-27} kg.)

43. A gas is at 0°C. If we wish to double the rms speed of the gas's molecules, by how much must we raise its temperature?

44. The latent heat of vaporization for water at room temperature is 2 430 J/g. (a) How much kinetic energy does each water molecule that evaporates possess before it evaporates? (b) Find the pre-evaporation rms speed of a water molecule that is evaporating. (c) What is the effective temperature of these molecules (modeled as if they were already in a thin gas)? Why do these molecules not burn you?

(Optional)
Section 21.7 Mean Free Path

45. In an ultrahigh vacuum system, the pressure is measured to be 1.00×10^{-10} torr (where 1 torr = 133 Pa). Assume that the gas molecules have a molecular diameter of 3.00×10^{-10} m and that the temperature is 300 K. Find (a) the number of molecules in a volume of 1.00 m^3, (b) the mean free path of the molecules, and (c) the collision frequency, assuming an average speed of 500 m/s.

46. In deep space it is reported that there is only one particle per cubic meter. Using the average temperature of 3.00 K and assuming that the particle is H$_2$ (with a diameter of 0.200 nm), (a) determine the mean free path of the particle and the average time between collisions.

(b) Repeat part (a), assuming that there is only one particle per cubic centimeter.

47. Show that the mean free path for the molecules of an ideal gas at temperature T and pressure P is

$$\ell = \frac{k_B T}{\sqrt{2}\,\pi d^2 P}$$

where d is the molecular diameter.

48. In a tank full of oxygen, how many molecular diameters d (on average) does an oxygen molecule travel (at 1.00 atm and 20.0°C) before colliding with another O$_2$ molecule? (The diameter of the O$_2$ molecule is approximately 3.60×10^{-10} m.)

49. Argon gas at atmospheric pressure and 20.0°C is confined in a 1.00-m^3 vessel. The effective hard-sphere diameter of the argon atom is 3.10×10^{-10} m. (a) Determine the mean free path ℓ. (b) Find the pressure when the mean free path is $\ell = 1.00$ m. (c) Find the pressure when $\ell = 3.10 \times 10^{-10}$ m.

ADDITIONAL PROBLEMS

50. The dimensions of a room are 4.20 m $\times$ 3.00 m $\times$ 2.50 m. (a) Find the number of molecules of air in it at atmospheric pressure and 20.0°C. (b) Find the mass of this air, assuming that the air consists of diatomic molecules with a molar mass of 28.9 g/mol. (c) Find the average kinetic energy of a molecule. (d) Find the root-mean-square molecular speed. (e) On the assumption that the specific heat is a constant independent of temperature, we have $E_{int} = 5nRT/2$. Find the internal energy in the air. (f) Find the internal energy of the air in the room at 25.0°C.

51. The function $E_{int} = 3.50nRT$ describes the internal energy of a certain ideal gas. A sample comprising 2.00 mol of the gas always starts at pressure 100 kPa and temperature 300 K. For each one of the following processes, determine the final pressure, volume, and temperature; the change in internal energy of the gas; the energy added to the gas by heat; and the work done by the gas: (a) The gas is heated at constant pressure to 400 K. (b) The gas is heated at constant volume to 400 K. (c) The gas is compressed at constant temperature to 120 kPa. (d) The gas is compressed adiabatically to 120 kPa.

52. Twenty particles, each of mass m and confined to a volume V, have various speeds: two have speed v; three have speed $2v$; five have speed $3v$; four have speed $4v$; three have speed $5v$; two have speed $6v$; one has speed $7v$. Find (a) the average speed, (b) the rms speed, (c) the most probable speed, (d) the pressure that the particles exert on the walls of the vessel, and (e) the average kinetic energy per particle.

WEB 53. A cylinder contains n mol of an ideal gas that undergoes an adiabatic process. (a) Starting with the expression

$W = \int P \, dV$ and using the expression $PV^\gamma =$ constant, show that the work done is

$$W = \left(\frac{1}{\gamma - 1}\right)(P_i V_i - P_f V_f)$$

(b) Starting with the first law equation in differential form, prove that the work done also is equal to $nC_V(T_i - T_f)$. Show that this result is consistent with the equation given in part (a).

54. A vessel contains 1.00×10^4 oxygen molecules at 500 K. (a) Make an accurate graph of the Maxwell speed distribution function versus speed with points at speed intervals of 100 m/s. (b) Determine the most probable speed from this graph. (c) Calculate the average and rms speeds for the molecules and label these points on your graph. (d) From the graph, estimate the fraction of molecules having speeds in the range of 300 m/s to 600 m/s.

55. **Review Problem.**   Oxygen at pressures much greater than 1 atm is toxic to lung cells. By weight, what ratio of helium gas (He) to oxygen gas (O_2) must be used by a scuba diver who is to descend to an ocean depth of 50.0 m?

56. A cylinder with a piston contains 1.20 kg of air at 25.0°C and 200 kPa. Energy is transferred into the system by heat as it is allowed to expand, with the pressure rising to 400 kPa. Throughout the expansion, the relationship between pressure and volume is given by

$$P = CV^{1/2}$$

where C is a constant. (a) Find the initial volume. (b) Find the final volume. (c) Find the final temperature. (d) Find the work that the air does. (e) Find the energy transferred by heat. Take $M = 28.9$ g/mol.

WEB 57. The compressibility κ of a substance is defined as the fractional change in volume of that substance for a given change in pressure:

$$\kappa = -\frac{1}{V}\frac{dV}{dP}$$

(a) Explain why the negative sign in this expression ensures that κ is always positive. (b) Show that if an ideal gas is compressed isothermally, its compressibility is given by $\kappa_1 = 1/P$. (c) Show that if an ideal gas is compressed adiabatically, its compressibility is given by $\kappa_2 = 1/\gamma P$. (d) Determine values for κ_1 and κ_2 for a monatomic ideal gas at a pressure of 2.00 atm.

58. **Review Problem.**   (a) Show that the speed of sound in an ideal gas is

$$v = \sqrt{\frac{\gamma R T}{M}}$$

where M is the molar mass. Use the general expression for the speed of sound in a fluid from Section 17.1; the definition of the bulk modulus from Section 12.4; and the result of Problem 57 in this chapter. As a sound

wave passes through a gas, the compressions are either so rapid or so far apart that energy flow by heat is prevented by lack of time or by effective thickness of insulation. The compressions and rarefactions are adiabatic. (b) Compute the theoretical speed of sound in air at 20°C and compare it with the value given in Table 17.1. Take $M = 28.9$ g/mol. (c) Show that the speed of sound in an ideal gas is

$$v = \sqrt{\frac{\gamma k_B T}{m}}$$

where m is the mass of one molecule. Compare your result with the most probable, the average, and the rms molecular speeds.

59. For a Maxwellian gas, use a computer or programmable calculator to find the numerical value of the ratio $N_v(v)/N_v(v_{mp})$ for the following values of v: $v = (v_{mp}/50), (v_{mp}/10), (v_{mp}/2), v_{mp}, 2v_{mp}, 10v_{mp}, 50v_{mp}$. Give your results to three significant figures.

60. A pitcher throws a 0.142-kg baseball at 47.2 m/s (Fig. P21.60). As it travels 19.4 m, the ball slows to 42.5 m/s because of air resistance. Find the change in temperature of the air through which it passes. To find the greatest possible temperature change, you may make the following assumptions: Air has a molar heat capacity of $C_P = 7R/2$ and an equivalent molar mass of 28.9 g/mol. The process is so rapid that the cover of the baseball acts as thermal insulation, and the temperature of the ball itself does not change. A change in temperature happens initially only for the air in a cylinder 19.4 m in length and 3.70 cm in radius. This air is initially at 20.0°C.

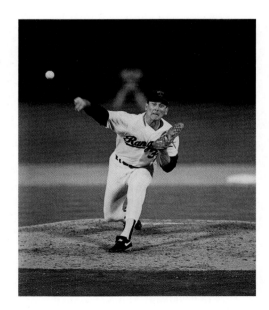

Figure P21.60   Nolan Ryan hurls the baseball for his 5 000th strikeout. *(Joe Patronite/ALLSPORT)*

61. Consider the particles in a *gas centrifuge,* a device that separates particles of different mass by whirling them in a circular path of radius r at angular speed ω. Newton's second law applied to circular motion states that a force of magnitude equal to $m\omega^2 r$ acts on a particle. (a) Discuss how a gas centrifuge can be used to separate particles of different mass. (b) Show that the density of the particles as a function of r is

$$n(r) = n_0 e^{mr^2\omega^2/2k_B T}$$

62. Verify Equations 21.27 and 21.28 for the rms and average speeds of the molecules of a gas at a temperature T. Note that the average value of v^n is

$$\overline{v^n} = \frac{1}{N}\int_0^\infty v^n N_v\, dv$$

and make use of the definite integrals

$$\int_0^\infty x^3 e^{-ax^2}\, dx = \frac{1}{2a^2} \qquad \int_0^\infty x^4 e^{-ax^2}\, dx = \frac{3}{8a^2}\sqrt{\frac{\pi}{a}}$$

63. A sample of a monatomic ideal gas occupies 5.00 L at atmospheric pressure and 300 K (point A in Figure P21.63). It is heated at constant volume to 3.00 atm (point B). Then, it is allowed to expand isothermally to 1.00 atm (point C) and at last is compressed isobarically to its original state. (a) Find the number of moles in the sample. (b) Find the temperatures at points B and C and the volume at point C. (c) Assuming that the specific heat does not depend on temperature, so that $E_{int} = 3nRT/2$, find the internal energy at points A, B,

and C. (d) Tabulate P, V, T, and E_{int} at the states at points A, B, and C. (e) Now consider the processes $A \rightarrow B$, $B \rightarrow C$, and $C \rightarrow A$. Describe just how to carry out each process experimentally. (f) Find Q, W, and ΔE_{int} for each of the processes. (g) For the whole cycle $A \rightarrow B \rightarrow C \rightarrow A$, find Q, W, and ΔE_{int}.

64. *If you can't walk to outer space, can you walk at least half way?* (a) Show that the fraction of particles below an altitude h in the atmosphere is

$$f = 1 - e^{(-mgh/k_B T)}$$

(b) Use this result to show that half the particles are below the altitude $h' = k_B T \ln(2)/mg$. What is the value of h' for the Earth? (Assume a temperature of 270 K, and note that the average molar mass for air is 28.9 g/mol.)

65. This problem will help you to think about the size of molecules. In the city of Beijing, a restaurant keeps a pot of chicken broth simmering continuously. Every morning it is topped off to contain 10.0 L of water, along with a fresh chicken, vegetables, and spices. The soup is thoroughly stirred. The molar mass of water is 18.0 g/mol. (a) Find the number of molecules of water in the pot. (b) During a certain month, 90.0% of the broth was served each day to people who then emigrated immediately. Of the water molecules present in the pot on the first day of the month, when was the last one likely to have been ladled out of the pot? (c) The broth has been simmering for centuries, through wars, earthquakes, and stove repairs. Suppose that the water that was in the pot long ago has thoroughly mixed into the Earth's hydrosphere, of mass 1.32×10^{21} kg. How many of the water molecules originally in the pot are likely to be present in it again today?

66. **Review Problem.** (a) If it has enough kinetic energy, a molecule at the surface of the Earth can escape the Earth's gravitation. Using the principle of conservation of energy, show that the minimum kinetic energy needed for escape is mgR, where m is the mass of the molecule, g is the free-fall acceleration at the surface of the Earth, and R is the radius of the Earth. (b) Calculate the temperature for which the minimum escape kinetic energy is ten times the average kinetic energy of an oxygen molecule.

67. Using multiple laser beams, physicists have been able to cool and trap sodium atoms in a small region. In one experiment, the temperature of the atoms was reduced to 0.240 mK. (a) Determine the rms speed of the sodium atoms at this temperature. The atoms can be trapped for about 1.00 s. The trap has a linear dimension of roughly 1.00 cm. (b) Approximately how long would it take an atom to wander out of the trap region if there were no trapping action?

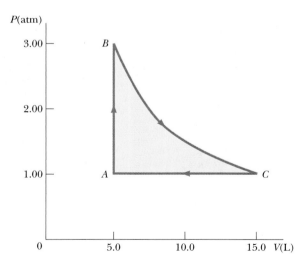

P(atm)

3.00 — B

2.00 —

1.00 — A C

0   5.0   10.0   15.0   *V*(L)

Figure P21.63

Answers to Quick Quizzes

21.1 Although a molecule moves very rapidly, it does not travel far before it collides with another molecule. The collision deflects the molecule from its original path. Eventually, a perfume molecule will make its way from one end of the room to the other, but the path it takes is much longer than the straight-line distance from the perfume bottle to your nose.

21.2 (c) E_{int} stays the same. According to Equation 21.10, E_{int} is a function of temperature only. Along an isotherm, T is constant by definition. Therefore, the internal energy of the gas does not change.

21.3 The area under each curve represents the number of molecules in that particular velocity range. The $T = 900$ K curve has many more molecules moving between 800 m/s and 1000 m/s than does the $T = 300$ K curve.

c h a p t e r

22

Heat Engines, Entropy, and the Second Law of Thermodynamics

The first law of thermodynamics, which we studied in Chapter 20, is a statement of conservation of energy, generalized to include internal energy. This law states that a change in internal energy in a system can occur as a result of energy transfer by heat or by work, or by both. As was stated in Chapter 20, the law makes no distinction between the results of heat and the results of work—either heat or work can cause a change in internal energy. However, an important distinction between the two is not evident from the first law. One manifestation of this distinction is that it is impossible to convert internal energy completely to mechanical energy by taking a substance through a thermodynamic cycle as is done in a *heat engine,* a device we study in this chapter.

Although the first law of thermodynamics is very important, it makes no distinction between processes that occur spontaneously and those that do not. However, we find that only certain types of energy-conversion and energy-transfer processes actually take place. The *second law of thermodynamics,* which we study in this chapter, establishes which processes do and which do not occur in nature. The following are examples of processes that proceed in only one direction, governed by the second law:

- When two objects at different temperatures are placed in thermal contact with each other, energy always flows by heat from the warmer to the cooler, never from the cooler to the warmer.
- A rubber ball dropped to the ground bounces several times and eventually comes to rest, but a ball lying on the ground never begins bouncing on its own.
- An oscillating pendulum eventually comes to rest because of collisions with air molecules and friction at the point of suspension. The mechanical energy of the system is converted to internal energy in the air, the pendulum, and the suspension; the reverse conversion of energy never occurs.

All these processes are *irreversible*—that is, they are processes that occur naturally in one direction only. No irreversible process has ever been observed to run backward—if it were to do so, it would violate the second law of thermodynamics.[1]

From an engineering standpoint, perhaps the most important implication of the second law is the limited efficiency of heat engines. The second law states that a machine capable of continuously converting internal energy completely to other forms of energy in a cyclic process cannot be constructed.

22.1 ▶ HEAT ENGINES AND THE SECOND LAW OF THERMODYNAMICS

A **heat engine** is a device that converts internal energy to mechanical energy. For instance, in a typical process by which a power plant produces electricity, coal or some other fuel is burned, and the high-temperature gases produced are used to convert liquid water to steam. This steam is directed at the blades of a turbine, setting it into rotation. The mechanical energy associated with this rotation is used to drive an electric generator. Another heat engine—the internal combustion engine in an automobile—uses energy from a burning fuel to perform work that results in the motion of the automobile.

---

[1] Although we have never *observed* a process occurring in the time-reversed sense, it is *possible* for it to occur. As we shall see later in the chapter, however, such a process is highly improbable. From this viewpoint, we say that processes occur with a vastly greater probability in one direction than in the opposite direction.

Figure 22.1 This steam-driven locomotive runs from Durango to Silverton, Colorado. It obtains its energy by burning wood or coal. The generated energy vaporizes water into steam, which powers the locomotive. (This locomotive must take on water from tanks located along the route to replace steam lost through the funnel.) Modern locomotives use diesel fuel instead of wood or coal. Whether old-fashioned or modern, such locomotives are heat engines, which extract energy from a burning fuel and convert a fraction of it to mechanical energy. *(© Phil Degginger/Tony Stone Images)*

A heat engine carries some working substance through a cyclic process during which (1) the working substance absorbs energy from a high-temperature energy reservoir, (2) work is done by the engine, and (3) energy is expelled by the engine to a lower-temperature reservoir. As an example, consider the operation of a steam engine (Fig. 22.1), in which the working substance is water. The water in a boiler absorbs energy from burning fuel and evaporates to steam, which then does work by expanding against a piston. After the steam cools and condenses, the liquid water produced returns to the boiler and the cycle repeats.

It is useful to represent a heat engine schematically as in Figure 22.2. The engine absorbs a quantity of energy Q_h from the hot reservoir, does work W, and then gives up a quantity of energy Q_c to the cold reservoir. Because the working substance goes through a cycle, its initial and final internal energies are equal, and so $\Delta E_{int} = 0$. Hence, from the first law of thermodynamics, $\Delta E_{int} = Q - W$, and with no change in internal energy, **the net work W done by a heat engine is equal to the net energy Q_{net} flowing through it.** As we can see from Figure 22.2, $Q_{net} = Q_h - Q_c$; therefore,

$$W = Q_h - Q_c \qquad \textbf{(22.1)}$$

In this expression and in many others throughout this chapter, to be consistent with traditional treatments of heat engines, we take both Q_h and Q_c to be positive quantities, even though Q_c represents energy leaving the engine. In discussions of heat engines, we shall describe energy leaving a system with an explicit minus sign,

Figure 22.2 Schematic representation of a heat engine. The engine absorbs energy Q_h from the hot reservoir, expels energy Q_c to the cold reservoir, and does work W.

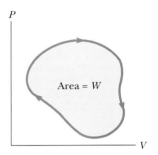

P

Area = W

V

Figure 22.3 *PV* diagram for an arbitrary cyclic process. The value of the net work done equals the area enclosed by the curve.

as in Equation 22.1. Also note that we model the energy input and output for the heat engine as heat, as it often is; however, the energy transfer could occur by another mechanism.

   The net work done in a cyclic process is the area enclosed by the curve representing the process on a *PV* diagram. This is shown for an arbitrary cyclic process in Figure 22.3.

   The **thermal efficiency** *e* of a heat engine is defined as the ratio of the net work done by the engine during one cycle to the energy absorbed at the higher temperature during the cycle:

$$e = \frac{W}{Q_h} = \frac{Q_h - Q_c}{Q_h} = 1 - \frac{Q_c}{Q_h} \tag{22.2}$$

   We can think of the efficiency as the ratio of what you get (mechanical work) to what you give (energy transfer at the higher temperature). In practice, we find that all heat engines expel only a fraction of the absorbed energy as mechanical work and that consequently the efficiency is less than 100%. For example, a good automobile engine has an efficiency of about 20%, and diesel engines have efficiencies ranging from 35% to 40%.

   Equation 22.2 shows that a heat engine has 100% efficiency ($e = 1$) only if $Q_c = 0$—that is, if no energy is expelled to the cold reservoir. In other words, a heat engine with perfect efficiency would have to expel all of the absorbed energy as mechanical work. On the basis of the fact that efficiencies of real engines are well below 100%, the **Kelvin–Planck** form of the **second law of thermodynamics** states the following:

   It is impossible to construct a heat engine that, operating in a cycle, produces no effect other than the absorption of energy from a reservoir and the performance of an equal amount of work.

This statement of the second law means that, during the operation of a heat engine, *W* can never be equal to Q_h, or, alternatively, that some energy Q_c must be

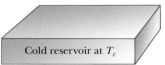

The impossible engine

Figure 22.4 Schematic diagram of a heat engine that absorbs energy Q_h from a hot reservoir and does an equivalent amount of work. It is impossible to construct such a perfect engine.

rejected to the environment. Figure 22.4 is a schematic diagram of the impossible "perfect" heat engine.

The first and second laws of thermodynamics can be summarized as follows: The first law specifies that **we cannot get more energy out of a cyclic process by work than the amount of energy we put in,** and the second law states that **we cannot break even because we must put more energy in, at the higher temperature, than the net amount of energy we get out by work.**

EXAMPLE 22.1 ▶ The Efficiency of an Engine

Find the efficiency of a heat engine that absorbs 2 000 J of energy from a hot reservoir and exhausts 1 500 J to a cold reservoir.

Solution   To calculate the efficiency of the engine, we use

Equation 22.2:

$$e = 1 - \frac{Q_c}{Q_h} = 1 - \frac{1\ 500\ \text{J}}{2\ 000\ \text{J}} = 0.25,\ \text{or}\ \boxed{25\%}$$

Refrigerators and Heat Pumps

Refrigerators and **heat pumps** are heat engines running in reverse. Here, we introduce them briefly for the purposes of developing an alternate statement of the second law; we shall discuss them more fully in Section 22.5.

In a refrigerator or heat pump, the engine absorbs energy Q_c from a cold reservoir and expels energy Q_h to a hot reservoir (Fig. 22.5). This can be accomplished only if work is done *on* the engine. From the first law, we know that the energy given up to the hot reservoir must equal the sum of the work done and the energy absorbed from the cold reservoir. Therefore, the refrigerator or heat pump transfers energy from a colder body (for example, the contents of a kitchen refrigerator or the winter air outside a building) to a hotter body (the air in the kitchen or a room in the building). In practice, it is desirable to carry out this process with a minimum of work. If it could be accomplished without doing any work, then the refrigerator or heat pump would be "perfect" (Fig. 22.6). Again, the existence of

Impossible refrigerator

Figure 22.6   Schematic diagram of an impossible refrigerator or heat pump—that is, one that absorbs energy Q_c from a cold reservoir and expels an equivalent amount of energy to a hot reservoir with $W = 0$.

Refrigerator

Figure 22.5   Schematic diagram of a refrigerator, which absorbs energy Q_c from a cold reservoir and expels energy Q_h to a hot reservoir. Work W is done *on* the refrigerator. A heat pump, which can be used to heat or cool a building, works the same way.

such a device would be in violation of the second law of thermodynamics, which in the form of the **Clausius statement**[2] states:

> It is impossible to construct a cyclical machine whose sole effect is the continuous transfer of energy from one object to another object at a higher temperature without the input of energy by work.

In simpler terms, **energy does not flow spontaneously from a cold object to a hot object.** For example, we cool homes in summer using heat pumps called *air conditioners.* The air conditioner pumps energy from the cool room in the home to the warm air outside. This direction of energy transfer requires an input of energy to the air conditioner, which is supplied by the electric power company.

The Clausius and Kelvin–Planck statements of the second law of thermodynamics appear, at first sight, to be unrelated, but in fact they are equivalent in all respects. Although we do not prove so here, if either statement is false, then so is the other.[3]

22.2 REVERSIBLE AND IRREVERSIBLE PROCESSES

In the next section we discuss a theoretical heat engine that is the most efficient possible. To understand its nature, we must first examine the meaning of reversible and irreversible processes. In a **reversible** process, the system undergoing the process can be returned to its initial conditions along the same path shown on a *PV* diagram, and every point along this path is an equilibrium state. A process that does not satisfy these requirements is **irreversible.**

All natural processes are known to be irreversible. From the endless number of examples that could be selected, let us examine the adiabatic free expansion of a gas, which was already discussed in Section 20.6, and show that it cannot be reversible. The system that we consider is a gas in a thermally insulated container, as shown in Figure 22.7. A membrane separates the gas from a vacuum. When the membrane is punctured, the gas expands freely into the vacuum. As a result of the puncture, the system has changed because it occupies a greater volume after the expansion. Because the gas does not exert a force through a distance on the surroundings, it does no work on the surroundings as it expands. In addition, no energy is transferred to or from the gas by heat because the container is insulated from its surroundings. Thus, in this adiabatic process, the system has changed but the surroundings have not.

For this process to be reversible, we need to be able to return the gas to its original volume and temperature without changing the surroundings. Imagine that we try to reverse the process by compressing the gas to its original volume. To do so, we fit the container with a piston and use an engine to force the piston inward. During this process, the surroundings change because work is being done by an outside agent on the system. In addition, the system changes because the compression increases the temperature of the gas. We can lower the temperature of the gas by allowing it to come into contact with an external energy reservoir. Although this step returns the gas to its original conditions, the surroundings are

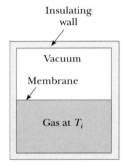

Figure 22.7 Adiabatic free expansion of a gas.

[2] First expressed by Rudolf Clausius (1822–1888).

[3] See, for example, R. P. Bauman, *Modern Thermodynamics and Statistical Mechanics,* New York, Macmillan Publishing Co., 1992.

again affected because energy is being added to the surroundings from the gas. If this energy could somehow be used to drive the engine that we have used to compress the gas, then the net energy transfer to the surroundings would be zero. In this way, the system and its surroundings could be returned to their initial conditions, and we could identify the process as reversible. However, the Kelvin–Planck statement of the second law specifies that the energy removed from the gas to return the temperature to its original value cannot be completely converted to mechanical energy in the form of the work done by the engine in compressing the gas. Thus, we must conclude that the process is irreversible.

We could also argue that the adiabatic free expansion is irreversible by relying on the portion of the definition of a reversible process that refers to equilibrium states. For example, during the expansion, significant variations in pressure occur throughout the gas. Thus, there is no well-defined value of the pressure for the entire system at any time between the initial and final states. In fact, the process cannot even be represented as a path on a *PV* diagram. The *PV* diagram for an adiabatic free expansion would show the initial and final conditions as points, but these points would not be connected by a path. Thus, because the intermediate conditions between the initial and final states are not equilibrium states, the process is irreversible.

Although all real processes are always irreversible, some are almost reversible. If a real process occurs very slowly such that the system is always very nearly in an equilibrium state, then the process can be approximated as reversible. For example, let us imagine that we compress a gas very slowly by dropping some grains of sand onto a frictionless piston, as shown in Figure 22.8. We make the process isothermal by placing the gas in thermal contact with an energy reservoir, and we transfer just enough energy from the gas to the reservoir during the process to keep the temperature constant. The pressure, volume, and temperature of the gas are all well defined during the isothermal compression, so each state during the process is an equilibrium state. Each time we add a grain of sand to the piston, the volume of the gas decreases slightly while the pressure increases slightly. Each grain we add represents a change to a new equilibrium state. We can reverse the process by slowly removing grains from the piston.

A general characteristic of a reversible process is that no dissipative effects (such as turbulence or friction) that convert mechanical energy to internal energy can be present. Such effects can be impossible to eliminate completely. Hence, it is not surprising that real processes in nature are irreversible.

Figure 22.8 A gas in thermal contact with an energy reservoir is compressed slowly as individual grains of sand drop onto the piston. The compression is isothermal and reversible.

22.3 THE CARNOT ENGINE

In 1824 a French engineer named Sadi Carnot described a theoretical engine, now called a **Carnot engine,** that is of great importance from both practical and theoretical viewpoints. He showed that a heat engine operating in an ideal, reversible cycle—called a **Carnot cycle**—between two energy reservoirs is the most efficient engine possible. Such an ideal engine establishes an upper limit on the efficiencies of all other engines. That is, the net work done by a working substance taken through the Carnot cycle is the greatest amount of work possible for a given amount of energy supplied to the substance at the upper temperature. **Carnot's theorem** can be stated as follows:

No real heat engine operating between two energy reservoirs can be more efficient than a Carnot engine operating between the same two reservoirs.

Sadi Carnot French physicist **(1796–1832)** Carnot was the first to show the quantitative relationship between work and heat. In 1824 he published his only work—*Reflections on the Motive Power of Heat*—which reviewed the industrial, political, and economic importance of the steam engine. In it, he defined work as "weight lifted through a height." *(FPG)*

To argue the validity of this theorem, let us imagine two heat engines operating between the *same* energy reservoirs. One is a Carnot engine with efficiency e_C, and the other is an engine with efficiency e, which is greater than e_C. We use the more efficient engine to drive the Carnot engine as a Carnot refrigerator. Thus, the output by work of the more efficient engine is matched to the input by work of the

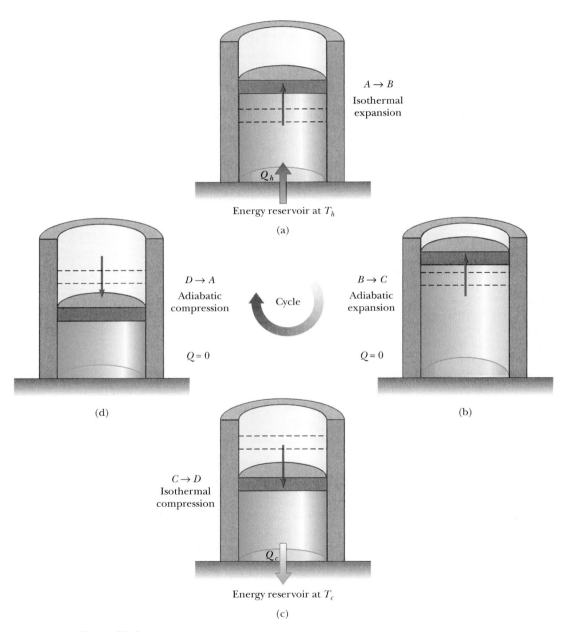

Figure 22.9   The Carnot cycle. In process $A \rightarrow B$, the gas expands isothermally while in contact with a reservoir at T_h. In process $B \rightarrow C$, the gas expands adiabatically ($Q = 0$). In process $C \rightarrow D$, the gas is compressed isothermally while in contact with a reservoir at $T_c < T_h$. In process $D \rightarrow A$, the gas is compressed adiabatically. The upward arrows on the piston indicate that weights are being removed during the expansions, and the downward arrows indicate that weights are being added during the compressions.

Carnot refrigerator. For the *combination* of the engine and refrigerator, then, no exchange by work with the surroundings occurs. Because we have assumed that the engine is more efficient than the refrigerator, the net result of the combination is a transfer of energy from the cold to the hot reservoir without work being done on the combination. According to the Clausius statement of the second law, this is impossible. Hence, the assumption that $e > e_C$ must be false. **All real engines are less efficient than the Carnot engine because they do not operate through a reversible cycle.** The efficiency of a real engine is further reduced by such practical difficulties as friction and energy losses by conduction.

To describe the Carnot cycle taking place between temperatures T_c and T_h, we assume that the working substance is an ideal gas contained in a cylinder fitted with a movable piston at one end. The cylinder's walls and the piston are thermally nonconducting. Four stages of the Carnot cycle are shown in Figure 22.9, and the *PV* diagram for the cycle is shown in Figure 22.10. The Carnot cycle consists of two adiabatic processes and two isothermal processes, all reversible:

1. Process $A \rightarrow B$ (Fig. 22.9a) is an isothermal expansion at temperature T_h. The gas is placed in thermal contact with an energy reservoir at temperature T_h. During the expansion, the gas absorbs energy Q_h from the reservoir through the base of the cylinder and does work W_{AB} in raising the piston.
2. In process $B \rightarrow C$ (Fig. 22.9b), the base of the cylinder is replaced by a thermally nonconducting wall, and the gas expands adiabatically—that is, no energy enters or leaves the system. During the expansion, the temperature of the gas decreases from T_h to T_c and the gas does work W_{BC} in raising the piston.
3. In process $C \rightarrow D$ (Fig. 22.9c), the gas is placed in thermal contact with an energy reservoir at temperature T_c and is compressed isothermally at temperature T_c. During this time, the gas expels energy Q_c to the reservoir, and the work done by the piston on the gas is W_{CD}.
4. In the final process $D \rightarrow A$ (Fig. 22.9d), the base of the cylinder is replaced by a nonconducting wall, and the gas is compressed adiabatically. The temperature of the gas increases to T_h, and the work done by the piston on the gas is W_{DA}.

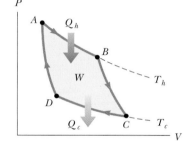

Figure 22.10   *PV* diagram for the Carnot cycle. The net work done, *W*, equals the net energy received in one cycle, $Q_h - Q_c$. Note that $\Delta E_{int} = 0$ for the cycle.

The net work done in this reversible, cyclic process is equal to the area enclosed by the path *ABCDA* in Figure 22.10. As we demonstrated in Section 22.1, because the change in internal energy is zero, the net work *W* done in one cycle equals the net energy transferred into the system, $Q_h - Q_c$. The thermal efficiency of the engine is given by Equation 22.2:

$$e = \frac{W}{Q_h} = \frac{Q_h - Q_c}{Q_h} = 1 - \frac{Q_c}{Q_h}$$

In Example 22.2, we show that for a Carnot cycle

$$\frac{Q_c}{Q_h} = \frac{T_c}{T_h} \tag{22.3}$$

Ratio of energies for a Carnot cycle

Hence, the thermal efficiency of a Carnot engine is

$$e_C = 1 - \frac{T_c}{T_h} \tag{22.4}$$

Efficiency of a Carnot engine

This result indicates that **all Carnot engines operating between the same two temperatures have the same efficiency.**

Equation 22.4 can be applied to any working substance operating in a Carnot cycle between two energy reservoirs. According to this equation, the efficiency is zero if $T_c = T_h$, as one would expect. The efficiency increases as T_c is lowered and as T_h is raised. However, the efficiency can be unity (100%) only if $T_c = 0$ K. Such reservoirs are not available; thus, the maximum efficiency is always less than 100%. In most practical cases, T_c is near room temperature, which is about 300 K. Therefore, one usually strives to increase the efficiency by raising T_h.

EXAMPLE 22.2   Efficiency of the Carnot Engine

Show that the efficiency of a heat engine operating in a Carnot cycle using an ideal gas is given by Equation 22.4.

Solution   During the isothermal expansion (process $A \rightarrow B$ in Figure 22.9), the temperature does not change. Thus, the internal energy remains constant. The work done by a gas during an isothermal expansion is given by Equation 20.13. According to the first law, this work is equal to Q_h, the energy absorbed, so that

$$Q_h = W_{AB} = nRT_h \ln \frac{V_B}{V_A}$$

In a similar manner, the energy transferred to the cold reservoir during the isothermal compression $C \rightarrow D$ is

$$Q_c = |W_{CD}| = nRT_c \ln \frac{V_C}{V_D}$$

We take the absolute value of the work because we are defining all values of Q for a heat engine as positive, as mentioned earlier. Dividing the second expression by the first, we find that

$$(1) \qquad \frac{Q_c}{Q_h} = \frac{T_c}{T_h} \frac{\ln(V_C/V_D)}{\ln(V_B/V_A)}$$

We now show that the ratio of the logarithmic quantities is unity by establishing a relationship between the ratio of volumes. For any quasi-static, adiabatic process, the pressure and volume are related by Equation 21.18:

$$(2) \qquad PV^\gamma = \text{constant}$$

During any reversible, quasi-static process, the ideal gas must also obey the equation of state, $PV = nRT$. Solving this expression for P and substituting into (2), we obtain

$$\frac{nRT}{V} V^\gamma = \text{constant}$$

which we can write as

$$TV^{\gamma-1} = \text{constant}$$

where we have absorbed nR into the constant right-hand side. Applying this result to the adiabatic processes $B \rightarrow C$ and $D \rightarrow A$, we obtain

$$T_h V_B^{\gamma-1} = T_c V_C^{\gamma-1}$$

$$T_h V_A^{\gamma-1} = T_c V_D^{\gamma-1}$$

Dividing the first equation by the second, we obtain

$$(V_B/V_A)^{\gamma-1} = (V_C/V_D)^{\gamma-1}$$

$$(3) \qquad \frac{V_B}{V_A} = \frac{V_C}{V_D}$$

Substituting (3) into (1), we find that the logarithmic terms cancel, and we obtain the relationship

$$\frac{Q_c}{Q_h} = \frac{T_c}{T_h}$$

Using this result and Equation 22.2, we see that the thermal efficiency of the Carnot engine is

$$e_C = 1 - \frac{Q_c}{Q_h} = 1 - \frac{T_c}{T_h}$$

which is Equation 22.4, the one we set out to prove.

EXAMPLE 22.3   The Steam Engine

A steam engine has a boiler that operates at 500 K. The energy from the burning fuel changes water to steam, and this steam then drives a piston. The cold reservoir's temperature is that of the outside air, approximately 300 K. What is the maximum thermal efficiency of this steam engine?

Solution   Using Equation 22.4, we find that the maximum thermal efficiency for any engine operating between these temperatures is

$$e_C = 1 - \frac{T_c}{T_h} = 1 - \frac{300 \text{ K}}{500 \text{ K}} = 0.4, \text{ or } \boxed{40\%}$$

You should note that this is the highest *theoretical* efficiency of the engine. In practice, the efficiency is considerably lower.

Exercise Determine the maximum work that the engine can perform in each cycle if it absorbs 200 J of energy from the hot reservoir during each cycle.

Answer 80 J.

EXAMPLE 22.4 ▶ **The Carnot Efficiency**

The highest theoretical efficiency of a certain engine is 30%. If this engine uses the atmosphere, which has a temperature of 300 K, as its cold reservoir, what is the temperature of its hot reservoir?

Solution We use the Carnot efficiency to find T_h:

$$e_C = 1 - \frac{T_c}{T_h}$$

$$T_h = \frac{T_c}{1 - e_C} = \frac{300 \text{ K}}{1 - 0.30} = \boxed{430 \text{ K}}$$

22.4 ▶ **GASOLINE AND DIESEL ENGINES**

In a gasoline engine, six processes occur in each cycle; five of these are illustrated in Figure 22.11. In this discussion, we consider the interior of the cylinder above the piston to be the system that is taken through repeated cycles in the operation of the engine. For a given cycle, the piston moves up and down twice. This represents a four-stroke cycle consisting of two upstrokes and two downstrokes. The processes in the cycle can be approximated by the **Otto cycle,** a *PV* diagram of which is illustrated in Figure 22.12:

1. During the *intake stroke* $O \rightarrow A$ (Fig. 22.11a), the piston moves downward, and a gaseous mixture of air and fuel is drawn into the cylinder at atmospheric pressure. In this process, the volume increases from V_2 to V_1. This is the energy input part of the cycle, as energy enters the system (the interior of the cylinder) as internal energy stored in the fuel. This is energy transfer by *mass transfer—* that is, the energy is carried with a substance. It is similar to convection, which we studied in Chapter 20.

2. During the *compression stroke* $A \rightarrow B$ (Fig. 22.11b), the piston moves upward, the air–fuel mixture is compressed adiabatically from volume V_1 to volume V_2, and the temperature increases from T_A to T_B. The work done by the gas is negative, and its value is equal to the area under the curve AB in Figure 22.12.

3. In process $B \rightarrow C$, combustion occurs when the spark plug fires (Fig. 22.11c). This is not one of the strokes of the cycle because it occurs in a very short period of time while the piston is at its highest position. The combustion represents a rapid transformation from internal energy stored in chemical bonds in the fuel to internal energy associated with molecular motion, which is related to temperature. During this time, the pressure and temperature in the cylinder increase rapidly, with the temperature rising from T_B to T_C. The volume, however, remains approximately constant because of the short time interval. As a result, approximately no work is done by the gas. We can model this process in the *PV* diagram (Fig. 22.12) as that process in which the energy Q_h enters the system. However, in reality this process is a transformation of energy already in the cylinder (from process $O \rightarrow A$) rather than a transfer.

4. In the *power stroke* $C \rightarrow D$ (Fig. 22.11d), the gas expands adiabatically from V_2 to

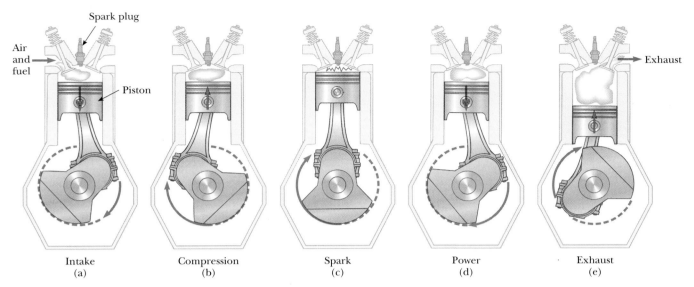

Spark plug

Air
and →
fuel

Piston

Exhaust

| Intake | Compression | Spark | Power | Exhaust |
| (a) | (b) | (c) | (d) | (e) |

Figure 22.11   The four-stroke cycle of a conventional gasoline engine. (a) In the intake stroke, air is mixed with fuel. (b) The intake valve is then closed, and the air–fuel mixture is compressed by the piston. (c) The mixture is ignited by the spark plug, with the result that the temperature of the mixture increases. (d) In the power stroke, the gas expands against the piston. (e) Finally, the residual gases are expelled, and the cycle repeats.

Figure 22.12   *PV* diagram for the Otto cycle, which approximately represents the processes occurring in an internal combustion engine.

Efficiency of the Otto cycle

V_1. This expansion causes the temperature to drop from T_C to T_D. Work is done by the gas in pushing the piston downward, and the value of this work is equal to the area under the curve *CD*.

5. In the process $D \rightarrow A$ (not shown in Fig. 22.11), an exhaust valve is opened as the piston reaches the bottom of its travel, and the pressure suddenly drops for a short time interval. During this interval, the piston is almost stationary and the volume is approximately constant. Energy is expelled from the interior of the cylinder and continues to be expelled during the next process.

6. In the final process, the *exhaust stroke* $A \rightarrow O$ (Fig. 22.11e), the piston moves upward while the exhaust valve remains open. Residual gases are exhausted at atmospheric pressure, and the volume decreases from V_1 to V_2. The cycle then repeats.

If the air–fuel mixture is assumed to be an ideal gas, then the efficiency of the Otto cycle is

$$e = 1 - \frac{1}{(V_1/V_2)^{\gamma-1}} \qquad (22.5)$$

where γ is the ratio of the molar specific heats C_P/C_V for the fuel–air mixture and V_1/V_2 is the **compression ratio.** Equation 22.5, which we derive in Example 22.5, shows that the efficiency increases as the compression ratio increases. For a typical compression ratio of 8 and with $\gamma = 1.4$, we predict a theoretical efficiency of 56% for an engine operating in the idealized Otto cycle. This value is much greater than that achieved in real engines (15% to 20%) because of such effects as friction, energy transfer by conduction through the cylinder walls, and incomplete combustion of the air–fuel mixture.

Diesel engines operate on a cycle similar to the Otto cycle but do not employ a spark plug. The compression ratio for a diesel engine is much greater than that for a gasoline engine. Air in the cylinder is compressed to a very small volume, and, as a consequence, the cylinder temperature at the end of the compression stroke is very high. At this point, fuel is injected into the cylinder. The temperature is high enough for the fuel–air mixture to ignite without the assistance of a spark plug. Diesel engines are more efficient than gasoline engines because of their greater compression ratios and resulting higher combustion temperatures.

EXAMPLE 22.5   Efficiency of the Otto Cycle

Show that the thermal efficiency of an engine operating in an idealized Otto cycle (see Figs. 22.11 and 22.12) is given by Equation 22.5. Treat the working substance as an ideal gas.

Solution   First, let us calculate the work done by the gas during each cycle. No work is done during processes $B \rightarrow C$ and $D \rightarrow A$. The work done by the gas during the adiabatic compression $A \rightarrow B$ is negative, and the work done by the gas during the adiabatic expansion $C \rightarrow D$ is positive. The value of the net work done equals the area of the shaded region bounded by the closed curve in Figure 22.12. Because the change in internal energy for one cycle is zero, we see from the first law that the net work done during one cycle equals the net energy flow through the system:

$$W = Q_h - Q_c$$

Because processes $B \rightarrow C$ and $D \rightarrow A$ take place at constant volume, and because the gas is ideal, we find from the definition of molar specific heat (Eq. 21.8) that

$$Q_h = nC_V(T_C - T_B) \quad \text{and} \quad Q_c = nC_V(T_D - T_A)$$

Using these expressions together with Equation 22.2, we obtain for the thermal efficiency

$$(1) \qquad e = \frac{W}{Q_h} = 1 - \frac{Q_c}{Q_h} = 1 - \frac{T_D - T_A}{T_C - T_B}$$

We can simplify this expression by noting that processes $A \rightarrow B$ and $C \rightarrow D$ are adiabatic and hence obey the relationship $TV^{\gamma-1} = $ constant, which we obtained in Example 22.2. For the two adiabatic processes, then,

$$A \rightarrow B: \qquad T_A V_A{}^{\gamma-1} = T_B V_B{}^{\gamma-1}$$

$$C \rightarrow D: \qquad T_C V_C{}^{\gamma-1} = T_D V_D{}^{\gamma-1}$$

Using these equations and relying on the fact that $V_A = V_D = V_1$ and $V_B = V_C = V_2$, we find that

$$T_A V_1{}^{\gamma-1} = T_B V_2{}^{\gamma-1}$$

$$(2) \qquad T_A = T_B\left(\frac{V_2}{V_1}\right)^{\gamma-1}$$

$$T_D V_1{}^{\gamma-1} = T_C V_2{}^{\gamma-1}$$

$$(3) \qquad T_D = T_C\left(\frac{V_2}{V_1}\right)^{\gamma-1}$$

Subtracting (2) from (3) and rearranging, we find that

$$(4) \qquad \frac{T_D - T_A}{T_C - T_B} = \left(\frac{V_2}{V_1}\right)^{\gamma-1}$$

Substituting (4) into (1), we obtain for the thermal efficiency

$$(5) \qquad e = 1 - \frac{1}{(V_1/V_2)^{\gamma-1}}$$

which is Equation 22.5.

We can also express this efficiency in terms of temperatures by noting from (2) and (3) that

$$\left(\frac{V_2}{V_1}\right)^{\gamma-1} = \frac{T_A}{T_B} = \frac{T_D}{T_C}$$

Therefore, (5) becomes

$$(6) \qquad e = 1 - \frac{T_A}{T_B} = 1 - \frac{T_D}{T_C}$$

During the Otto cycle, the lowest temperature is T_A and the highest temperature is T_C. Therefore, the efficiency of a Carnot engine operating between reservoirs at these two temperatures, which is given by the expression $e_C = 1 - (T_A/T_C)$, is *greater* than the efficiency of the Otto cycle given by (6), as expected.

APPLICATION   Models of Gasoline and Diesel Engines

We can use the thermodynamic principles discussed in this and earlier chapters to model the performance of gasoline and diesel engines. In both types of engine, a gas is first compressed in the cylinders of the engine and then the fuel–air mixture is ignited. Work is done on the gas during compression, but significantly more work is done on the piston by the mixture as the products of combustion expand in the cylinder. The power of the engine is transferred from the piston to the crankshaft by the connecting rod.

Two important quantities of either engine are the **displacement volume,** which is the volume displaced by the piston as it moves from the bottom to the top of the cylinder, and the com-

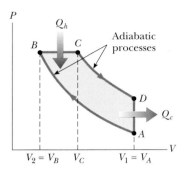

Figure 22.13   *PV diagram for an ideal diesel engine.*

pression ratio r, which is the ratio of the maximum and minimum volumes of the cylinder (see p. 680). In our notation, $r = V_A/V_B$, or V_1/V_2 in Eq. 22.5. Most gasoline and diesel engines operate with a four-cycle process (intake, compression, power, exhaust), in which the net work of the intake and exhaust cycles can be considered negligible. Therefore, power is developed only once for every two revolutions of the crankshaft.

In a diesel engine, only air (and no fuel) is present in the cylinder at the beginning of the compression. In the idealized diesel cycle of Figure 22.13, air in the cylinder undergoes an adiabatic compression from A to B. Starting at B, fuel is injected into the cylinder in such a way that the fuel–air mixture undergoes a constant-pressure expansion to an intermediate volume $V_C(B \rightarrow C)$. The high temperature of the mixture causes combustion, and the power stroke is an adiabatic expansion back to $V_D = V_A(C \rightarrow D)$. The exhaust valve is opened, and a constant-volume output of energy occurs $(D \rightarrow A)$ as the cylinder empties.

To simplify our calculations, we assume that the mixture in the cylinder is air modeled as an ideal gas. We use specific heats c instead of molar specific heats C and assume constant values for air at 300 K. We express the specific heats and the universal gas constant in terms of unit masses rather than moles. Thus, $c_V = 0.718$ kJ/kg·K, $c_P = 1.005$ kJ/kg·K, $\gamma = c_P/c_V = 1.40$, and $R = c_P - c_V = 0.287$ kJ/kg·K = 0.287 kPa·m³/kg·K.

A 3.00-L Gasoline Engine

Let us calculate the power delivered by a six-cylinder gasoline engine that has a displacement volume of 3.00 L operating at 4 000 rpm and having a compression ratio of $r = 9.50$. The air–fuel mixture enters a cylinder at atmospheric pressure and an ambient temperature of 27°C. During combustion, the mixture reaches a temperature of 1 350°C.

First, let us calculate the work done by an individual cylinder. Using the initial pressure $P_A = 100$ kPa and the initial temperature $T_A = 300$ K, we calculate the initial volume and the mass of the air–fuel mixture. We know that the ratio of the initial and final volumes is the compression ratio,

$$\frac{V_A}{V_B} = r = 9.50$$

We also know that the difference in volumes is the displacement volume. The 3.00-L rating of the engine is the total displacement volume for all six cylinders. Thus, for one cylinder,

$$V_A - V_B = \frac{3.00 \text{ L}}{6} = \frac{3.00 \times 10^{-3}\, \text{m}^3}{6} = 0.500 \times 10^{-3}\, \text{m}^3$$

Solving these two equations simultaneously, we find the initial and final volumes:

$$V_A = 0.559 \times 10^{-3}\, \text{m}^3 \qquad V_B = 0.588 \times 10^{-4}\, \text{m}^3$$

Using the ideal gas law (in the form $PV = mRT$, because we are using the universal gas constant in terms of mass rather than moles), we can find the mass of the air–fuel mixture:

$$m = \frac{P_A V_A}{R T_A} = \frac{(100 \text{ kPa})(0.559 \times 10^{-3}\, \text{m}^3)}{(0.287 \text{ kPa·m}^3/\text{kg·K})(300 \text{ K})}$$

$$= 6.49 \times 10^{-4}\, \text{kg}$$

Process $A \rightarrow B$ (see Fig. 22.12) is an adiabatic compression, and this means that $PV^\gamma = $ constant; hence,

$$P_B V_B{}^\gamma = P_A V_A{}^\gamma$$

$$P_B = P_A \left(\frac{V_A}{V_B}\right)^\gamma = P_A(r)^\gamma = (100 \text{ kPa})(9.50)^{1.40}$$

$$= 2.34 \times 10^3 \text{ kPa}$$

Using the ideal gas law, we find that the temperature after the compression is

$$T_B = \frac{P_B V_B}{mR} = \frac{(2.34 \times 10^3 \text{ kPa})(0.588 \times 10^{-4}\, \text{m}^3)}{(6.49 \times 10^{-4}\, \text{kg})(0.287 \text{ kPa·m}^3/\text{kg·K})}$$

$$= 739 \text{ K}$$

In process $B \rightarrow C$, the combustion that transforms the internal energy in chemical bonds into internal energy of molecular motion occurs at constant volume; thus, $V_C = V_B$. Combustion causes the temperature to increase to $T_C = 1\,350°C = 1\,623$ K. Using this value and the ideal gas law, we can calculate P_C:

$$P_C = \frac{mRT_C}{V_C}$$

$$= \frac{(6.49 \times 10^{-4}\, \text{kg})(0.287 \text{ kPa·m}^3/\text{kg·K})(1\,623 \text{ K})}{(0.588 \times 10^{-4}\, \text{m}^3)}$$

$$= 5.14 \times 10^3 \text{ kPa}$$

Process $C \rightarrow D$ is an adiabatic expansion; the pressure after the expansion is

$$P_D = P_C \left(\frac{V_C}{V_D}\right)^\gamma = P_C \left(\frac{V_B}{V_A}\right)^\gamma = P_C \left(\frac{1}{r}\right)^\gamma$$

$$= (5.14 \times 10^3 \text{ kPa})\left(\frac{1}{9.50}\right)^{1.40} = 220 \text{ kPa}$$

Using the ideal gas law again, we find the final temperature:

$$T_D = \frac{P_D V_D}{mR} = \frac{(220 \text{ kPa})(0.559 \times 10^{-3} \text{ m}^3)}{(6.49 \times 10^{-4} \text{ kg})(0.287 \text{ kPa} \cdot \text{m}^3/\text{kg} \cdot \text{K})}$$

$$= 660 \text{ K}$$

Now that we have the temperatures at the beginning and end of each process of the cycle, we can calculate the net energy transfer and net work done by each cylinder every two cycles. From Equation 21.8, we can state

$$Q_h = Q_{in} = mc_V(T_C - T_B)$$
$$= (6.49 \times 10^{-4} \text{ kg})(0.718 \text{ kJ/kg} \cdot \text{K})(1\,623 \text{ K} - 739 \text{ K})$$
$$= 0.412 \text{ kJ}$$

$$Q_c = Q_{out} = mc_V(T_D - T_A)$$
$$= (6.49 \times 10^{-4} \text{ kg})(0.718 \text{ kJ/kg} \cdot \text{K})(660 \text{ K} - 300 \text{ K})$$
$$= 0.168 \text{ kJ}$$

$$W_{net} = Q_{in} - Q_{out} = 0.244 \text{ kJ}$$

From Equation 22.2, the efficiency is $e = W_{net}/Q_{in} = 59\%$. (We can also use Equation 22.5 to calculate the efficiency directly from the compression ratio.)

Recalling that power is delivered every other revolution of the crankshaft, we find that the net power for the six-cylinder engine operating at 4 000 rpm is

$$\mathcal{P}_{net} = 6\left(\frac{1}{2 \text{ rev}}\right)(4\,000 \text{ rev/min})(1 \text{ min}/60 \text{ s})(0.244 \text{ kJ})$$

$$= 49 \text{ kW} = 66 \text{ hp}$$

A 2.00-L Diesel Engine

Let us calculate the power delivered by a four-cylinder diesel engine that has a displacement volume of 2.00 L and is operating at 3 000 rpm. The compression ratio is $r = V_A/V_B = 22.0$, and the **cutoff ratio**, which is the ratio of the volume change during the constant-pressure process $B \rightarrow C$ in Figure 22.13, is $r_c = V_C/V_B = 2.00$. The air enters each cylinder at the beginning of the compression cycle at atmospheric pressure and at an ambient temperature of 27°C.

Our model of the diesel engine is similar to our model of the gasoline engine except that now the fuel is injected at point B and the mixture self-ignites near the end of the compression cycle $A \rightarrow B$, when the temperature reaches the ignition temperature. We assume that the energy input occurs in the constant-pressure process $B \rightarrow C$, and that the expansion process continues from C to D with no further energy transfer by heat.

Let us calculate the work done by an individual cylinder that has an initial volume of $V_A = (2.00 \times 10^{-3} \text{ m}^3)/4 = 0.500 \times 10^{-3} \text{ m}^3$. Because the compression ratio is quite high, we approximate the maximum cylinder volume to be the displacement volume. Using the initial pressure $P_A = 100 \text{ kPa}$ and initial temperature $T_A = 300 \text{ K}$, we can calculate the mass of the air in the cylinder using the ideal gas law:

$$m = \frac{P_A V_A}{RT_A} = \frac{(100 \text{ kPa})(0.500 \times 10^{-3} \text{ m}^3)}{(0.287 \text{ kPa} \cdot \text{m}^3/\text{kg} \cdot \text{K})(300 \text{ K})} = 5.81 \times 10^{-4} \text{ kg}$$

Process $A \rightarrow B$ is an adiabatic compression, so $PV^\gamma = $ constant; thus,

$$P_B V_B{}^\gamma = P_A V_A{}^\gamma$$
$$P_B = P_A\left(\frac{V_A}{V_B}\right)^\gamma = (100 \text{ kPa})(22.0)^{1.40} = 7.57 \times 10^3 \text{ kPa}$$

Using the ideal gas law, we find that the temperature of the air after the compression is

$$T_B = \frac{P_B V_B}{mR} = \frac{(7.57 \times 10^3 \text{ kPa})(0.500 \times 10^{-3} \text{ m}^3)\left(\dfrac{1}{22.0}\right)}{(5.81 \times 10^{-4} \text{ kg})(0.287 \text{ kPa} \cdot \text{m}^3/\text{kg} \cdot \text{K})}$$

$$= 1.03 \times 10^3 \text{ K}$$

Process $B \rightarrow C$ is a constant-pressure expansion; thus, $P_C = P_B$. We know from the cutoff ratio of 2.00 that the volume doubles in this process. According to the ideal gas law, a doubling of volume in an isobaric process results in a doubling of the temperature, so

$$T_C = 2T_B = 2.06 \times 10^3 \text{ K}$$

Process $C \rightarrow D$ is an adiabatic expansion; therefore,

$$P_D = P_C\left(\frac{V_C}{V_D}\right)^\gamma = P_C\left(\frac{V_C}{V_B}\frac{V_B}{V_D}\right)^\gamma = P_C\left(r_c\frac{1}{r}\right)^\gamma$$

$$= (7.57 \times 10^3 \text{ kPa})\left(\frac{2.00}{22.0}\right)^{1.40} = 264 \text{ kPa}$$

We find the temperature at D from the ideal gas law:

$$T_D = \frac{P_D V_D}{mR} = \frac{(264 \text{ kPa})(0.500 \times 10^{-3} \text{ m}^3)}{(5.81 \times 10^{-4} \text{ kg})(0.287 \text{ kPa} \cdot \text{m}^3/\text{kg} \cdot \text{K})}$$

$$= 792 \text{ K}$$

Now that we have the temperatures at the beginning and end of each process, we can calculate the net energy transfer by heat and the net work done by each cylinder every two cycles:

$$Q_h = Q_{in} = mc_P(T_C - T_B) = 0.601 \text{ kJ}$$
$$Q_c = Q_{out} = mc_V(T_D - T_A) = 0.205 \text{ kJ}$$
$$W_{net} = Q_{in} - Q_{out} = 0.396 \text{ kJ}$$

The efficiency is $e = W_{net}/Q_{in} = 66\%$.

The net power for the four-cylinder engine operating at 3 000 rpm is

$$\mathcal{P}_{net} = 4\left(\frac{1}{2 \text{ rev}}\right)(3\,000 \text{ rev/min})(1 \text{ min}/60 \text{ s})(0.396 \text{ kJ})$$

$$= 39.6 \text{ kW} = 53 \text{ hp}$$

Of course, modern engine design goes beyond this simple thermodynamic treatment, which uses idealized cycles.

22.5 ▶ HEAT PUMPS AND REFRIGERATORS

In Section 22.1 we introduced a heat pump as a mechanical device that moves energy from a region at lower temperature to a region at higher temperature. Heat pumps have long been used for cooling homes and buildings, and they are now becoming increasingly popular for heating them as well. The heat pump contains two sets of metal coils that can exchange energy by heat with the surroundings: one set on the outside of the building, in contact with the air or buried in the ground; and the other set in the interior of the building. In the heating mode, a circulating fluid flowing through the coils absorbs energy from the outside and releases it to the interior of the building from the interior coils. The fluid is cold and at low pressure when it is in the external coils, where it absorbs energy by heat from either the air or the ground. The resulting warm fluid is then compressed and enters the interior coils as a hot, high-pressure fluid, where it releases its stored energy to the interior air.

An air conditioner is simply a heat pump operating in the cooling mode, with its exterior and interior coils interchanged. Energy is absorbed into the circulating fluid in the interior coils; then, after the fluid is compressed, energy leaves the fluid through the external coils. The air conditioner must have a way to release energy to the outside. Otherwise, the work done on the air conditioner would represent energy added to the air inside the house, and the temperature would increase. In the same manner, a refrigerator cannot cool the kitchen if the refrigerator door is left open. The amount of energy leaving the external coils (Fig. 22.14) behind or underneath the refrigerator is greater than the amount of energy removed from the food or from the air in the kitchen if the door is left open. The difference between the energy out and the energy in is the work done by the electricity supplied to the refrigerator.

Figure 22.15 is a schematic representation of a heat pump. The cold temperature is T_c, the hot temperature is T_h, and the energy absorbed by the circulating fluid is Q_c. The heat pump does work W on the fluid, and the energy transferred from the pump to the building in the heating mode is Q_h.

The effectiveness of a heat pump is described in terms of a number called the **coefficient of performance** (COP). In the heating mode, the COP is defined as the ratio of the energy transferred to the hot reservoir to the work required to transfer that energy:

$$\text{COP (heating mode)} \equiv \frac{\text{Energy transferred at high temperature}}{\text{Work done by pump}} = \frac{Q_h}{W} \qquad \textbf{(22.6)}$$

Note that the COP is similar to the thermal efficiency for a heat engine in that it is a ratio of what you get (energy delivered to the interior of the building) to what you give (work input). Because Q_h is generally greater than W, typical values for the COP are greater than unity. It is desirable for the COP to be as high as possible, just as it is desirable for the thermal efficiency of an engine to be as high as possible.

If the outside temperature is 25°F or higher, then the COP for a heat pump is about 4. That is, the amount of energy transferred to the building is about four times greater than the work done by the motor in the heat pump. However, as the outside temperature decreases, it becomes more difficult for the heat pump to extract sufficient energy from the air, and so the COP decreases. In fact, the COP can fall below unity for temperatures below the midteens. Thus, the use of heat pumps that extract energy from the air, while satisfactory in moderate climates, is not appropriate in areas where winter temperatures are very low. It is possible to

Figure 22.14 The coils on the back of a refrigerator transfer energy by heat to the air. The second law of thermodynamics states that this amount of energy must be greater than the amount of energy removed from the contents of the refrigerator (or from the air in the kitchen, if the refrigerator door is left open). (*Charles D. Winters*)

use heat pumps in colder areas by burying the external coils deep in the ground. In this case, the energy is extracted from the ground, which tends to be warmer than the air in the winter.

In an electric heater, electrical energy can be converted to internal energy with an efficiency of 100%. By what percentage does the cost of heating your home change when you replace your electric heating system with a heat pump that has a COP of 4? Assume that the motor running the heat pump is 100% efficient.

Figure 22.15 Schematic diagram of a heat pump, which absorbs energy Q_c from a cold reservoir and expels energy Q_h to a hot reservoir. Note that this diagram is the same as that for the refrigerator shown in Figure 22.5.

Theoretically, a Carnot-cycle heat engine run in reverse constitutes the most effective heat pump possible, and it determines the maximum COP for a given combination of hot and cold reservoir temperatures. Using Equations 22.1 and 22.3, we see that the maximum COP for a heat pump in its heating mode is

$$\text{COP}_\text{C}(\text{heating mode}) = \frac{Q_h}{W}$$

$$= \frac{Q_h}{Q_h - Q_c} = \frac{1}{1 - \dfrac{Q_c}{Q_h}} = \frac{1}{1 - \dfrac{T_c}{T_h}} = \frac{T_h}{T_h - T_c}$$

For a heat pump operating in the cooling mode, "what you get" is energy removed from the cold reservoir. The most effective refrigerator or air conditioner is one that removes the greatest amount of energy from the cold reservoir in exchange for the least amount of work. Thus, for these devices we define the COP in terms of Q_c:

$$\text{COP (cooling mode)} = \frac{Q_c}{W} \tag{22.7}$$

A good refrigerator should have a high COP, typically 5 or 6.

The greatest possible COP for a heat pump in the cooling mode is that of a heat pump whose working substance is carried through a Carnot cycle in reverse:

$$\text{COP}_\text{C (cooling mode)} = \frac{T_c}{T_h - T_c}$$

As the difference between the temperatures of the two reservoirs approaches zero in this expression, the theoretical COP approaches infinity. In practice, the low temperature of the cooling coils and the high temperature at the compressor limit the COP to values below 10.

22.6 > ENTROPY

10.10
&
10.11

The zeroth law of thermodynamics involves the concept of temperature, and the first law involves the concept of internal energy. Temperature and internal energy are both state functions—that is, they can be used to describe the thermodynamic state of a system. Another state function—this one related to the second law of thermodynamics—is **entropy** S. In this section we define entropy on a macroscopic scale as it was first expressed by Clausius in 1865.

Consider any infinitesimal process in which a system changes from one equilibrium state to another. If dQ_r is the amount of energy transferred by heat when the system follows a reversible path between the states, then the change in entropy dS is equal to this amount of energy for the reversible process divided by the absolute temperature of the system:

Clausius definition of change in entropy

$$dS = \frac{dQ_r}{T} \qquad (22.8)$$

We have assumed that the temperature is constant because the process is infinitesimal. Since we have claimed that entropy is a state function, **the change in entropy during a process depends only on the end points and therefore is independent of the actual path followed.**

The subscript r on the quantity dQ_r is a reminder that the transferred energy is to be measured along a reversible path, even though the system may actually have followed some irreversible path. When energy is absorbed by the system, dQ_r is positive and the entropy of the system increases. When energy is expelled by the system, dQ_r is negative and the entropy of the system decreases. Note that Equation 22.8 defines not entropy but rather the *change* in entropy. Hence, the meaningful quantity in describing a process is the *change* in entropy.

Entropy was originally formulated as a useful concept in thermodynamics; however, its importance grew tremendously as the field of statistical mechanics developed because the analytical techniques of statistical mechanics provide an alternative means of interpreting entropy. In statistical mechanics, the behavior of a substance is described in terms of the statistical behavior of its atoms and molecules. One of the main results of this treatment is that **isolated systems tend toward disorder and that entropy is a measure of this disorder.** For example, consider the molecules of a gas in the air in your room. If half of the gas molecules had velocity vectors of equal magnitude directed toward the left and the other half had velocity vectors of the same magnitude directed toward the right, the situation would be very ordered. However, such a situation is extremely unlikely. If you could actually view the molecules, you would see that they move haphazardly in all directions, bumping into one another, changing speed upon collision, some going fast and others going slowly. This situation is highly disordered.

The cause of the tendency of an isolated system toward disorder is easily explained. To do so, we distinguish between *microstates* and *macrostates* of a system. A **microstate** is a particular description of the properties of the individual molecules of the system. For example, the description we just gave of the velocity vectors' of the air molecules in your room being very ordered refers to a particular microstate, and the more likely likely haphazard motion is another microstate—one that represents disorder. A **macrostate** is a description of the conditions of the system from a macroscopic point of view and makes use of macroscopic variables such as pressure, density, and temperature. For example, in both of the microstates described for the air molecules in your room, the air molecules are distributed uniformly throughout the volume of the room; this uniform density distribution is a macrostate. We could not distinguish between our two microstates by making a macroscopic measurement—both microstates would appear to be the same macroscopically, and the two macrostates corresponding to these microstates are equivalent.

For any given macrostate of the system, a number of microstates are possible, or *accessible*. Among these microstates, it is assumed that all are equally probable. However, when all possible microstates are examined, it is found that far more of them are disordered than are ordered. Because all of the microstates are equally probable, it is highly likely that the actual macrostate is one resulting from one of the highly disordered microstates, simply because there are many more of them. Similarly, the probability of a macrostate's forming from disordered microstates is greater than the probability of a macrostate's forming from ordered microstates.

All physical processes that take place in a system tend to cause the system and its surroundings to move toward more probable macrostates. The more probable macrostate is always one of greater disorder. If we consider a system and its surroundings to include the entire Universe, then the Universe is always moving toward a macrostate corresponding to greater disorder. Because entropy is a measure of disorder, an alternative way of stating this is **the entropy of the Universe increases in all real processes.** This is yet another statement of the second law of thermodynamics that can be shown to be equivalent to the Kelvin–Planck and Clausius statements.

> In real processes, the disorder of the Universe increases

To calculate the change in entropy for a finite process, we must recognize that T is generally not constant. If dQ_r is the energy transferred by heat when the system is at a temperature T, then the change in entropy in an arbitrary reversible process between an initial state and a final state is

$$\Delta S = \int_i^f dS = \int_i^f \frac{dQ_r}{T} \qquad \text{(reversible path)} \qquad \textbf{(22.9)}$$

> Change in entropy for a finite process

As with an infinitesimal process, the change in entropy ΔS of a system going from one state to another has the same value for *all* paths connecting the two states. That is, the finite change in entropy ΔS of a system depends only on the properties of the initial and final equilibrium states. Thus, we are free to choose a particular reversible path over which to evaluate the entropy in place of the actual path, as long as the initial and final states are the same for both paths.

Quick Quiz 22.2

Which of the following is true for the entropy change of a system that undergoes a reversible, adiabatic process? (a) $\Delta S < 0$. (b) $\Delta S = 0$. (c) $\Delta S > 0$.

Let us consider the changes in entropy that occur in a Carnot heat engine operating between the temperatures T_c and T_h. In one cycle, the engine absorbs energy Q_h from the hot reservoir and expels energy Q_c to the cold reservoir. These energy transfers occur only during the isothermal portions of the Carnot cycle; thus, the constant temperature can be brought out in front of the integral sign in Equation 22.9. The integral then simply has the value of the total amount of energy transferred by heat. Thus, the total change in entropy for one cycle is

$$\Delta S = \frac{Q_h}{T_h} - \frac{Q_c}{T_c}$$

where the negative sign represents the fact that energy Q_c is expelled by the system, since we continue to define Q_c as a positive quantity when referring to heat engines. In Example 22.2 we showed that, for a Carnot engine,

$$\frac{Q_c}{Q_h} = \frac{T_c}{T_h}$$

Using this result in the previous expression for ΔS, we find that the total change in entropy for a Carnot engine operating in a cycle is *zero*:

$$\Delta S = 0$$

> The change in entropy for a Carnot cycle is zero

Now let us consider a system taken through an arbitrary (non-Carnot) cycle. Because entropy is a state function—and hence depends only on the properties of a given equilibrium state—we conclude that $\Delta S = 0$ for *any* cycle. In general, we can write this condition in the mathematical form

> $\Delta S = 0$ for any reversible cycle

$$\oint \frac{dQ_r}{T} = 0 \qquad \text{(22.10)}$$

where the symbol $\oint$ indicates that the integration is over a closed path.

Quasi-Static, Reversible Process for an Ideal Gas

Let us suppose that an ideal gas undergoes a quasi-static, reversible process from an initial state having temperature T_i and volume V_i to a final state described by T_f and V_f. Let us calculate the change in entropy of the gas for this process.

Writing the first law of thermodynamics in differential form and rearranging the terms, we have $dQ_r = dE_{int} + dW$, where $dW = P\,dV$. For an ideal gas, recall that $dE_{int} = nC_V\,dT$ (Eq. 21.12), and from the ideal gas law, we have $P = nRT/V$. Therefore, we can express the energy transferred by heat in the process as

$$dQ_r = dE_{int} + P\,dV = nC_V\,dT + nRT\,\frac{dV}{V}$$

We cannot integrate this expression as it stands because the last term contains two variables, T and V. However, if we divide all terms by T, each of the terms on the right-hand side depends on only one variable:

$$\frac{dQ_r}{T} = nC_V\,\frac{dT}{T} + nR\,\frac{dV}{V} \qquad \text{(22.11)}$$

Assuming that C_V is constant over the interval in question, and integrating Equation 22.11 from the initial state to the final state, we obtain

$$\Delta S = \int_i^f \frac{dQ_r}{T} = nC_V \ln \frac{T_f}{T_i} + nR \ln \frac{V_f}{V_i} \qquad \text{(22.12)}$$

This expression demonstrates mathematically what we argued earlier—that ΔS depends only on the initial and final states and is independent of the path between the states. Also, note in Equation 22.12 that ΔS can be positive or negative, depending on the values of the initial and final volumes and temperatures. Finally, for a cyclic process ($T_i = T_f$ and $V_i = V_f$), we see from Equation 22.12 that $\Delta S = 0$. This is evidence that entropy is a state function.

EXAMPLE 22.6   Change in Entropy—Melting

A solid that has a latent heat of fusion L_f melts at a temperature T_m. (a) Calculate the change in entropy of this substance when a mass m of the substance melts.

Solution   Let us assume that the melting occurs so slowly that it can be considered a reversible process. In this case the temperature can be regarded as constant and equal to T_m.

Making use of Equations 22.9 and that for the latent heat of fusion $Q = mL_f$ (Eq. 20.6), we find that

$$\Delta S = \int \frac{dQ_r}{T} = \frac{1}{T_m} \int dQ = \frac{Q}{T_m} = \boxed{\frac{mL_f}{T_m}}$$

Note that we are able to remove T_m from the integral because the process is isothermal. Note also that ΔS is positive. This means that when a solid melts, its entropy increases because the molecules are much more disordered in the liquid state than they are in the solid state. The positive value for ΔS also means that the substance in its liquid state does not spontaneously transfer energy from itself to the surroundings and freeze because to do so would involve a spontaneous decrease in entropy.

(b) Estimate the value of the change in entropy of an ice cube when it melts.

Solution Let us assume an ice tray makes cubes that are about 3 cm on a side. The volume per cube is then (very roughly) 30 cm^3. This much liquid water has a mass of 30 g. From Table 20.2 we find that the latent heat of fusion of ice is 3.33×10^5 J/kg. Substituting these values into our answer for part (a), we find that

$$\Delta S = \frac{mL_f}{T_m} = \frac{(0.03 \text{ kg})(3.33 \times 10^5 \text{ J/kg})}{273 \text{ K}} = \boxed{4 \times 10^1 \text{ J/K}}$$

We retain only one significant figure, in keeping with the nature of our estimations.

22.7 ENTROPY CHANGES IN IRREVERSIBLE PROCESSES

By definition, calculation of the change in entropy requires information about a reversible path connecting the initial and final equilibrium states. To calculate changes in entropy for real (irreversible) processes, we must remember that entropy (like internal energy) depends only on the *state* of the system. That is, entropy is a state function. Hence, the change in entropy when a system moves between any two equilibrium states depends only on the initial and final states. We can show that if this were not the case, the second law of thermodynamics would be violated.

We now calculate the entropy change in some irreversible process between two equilibrium states by devising a reversible process (or series of reversible processes) between the same two states and computing $\Delta S = \int dQ_r / T$ for the reversible process. In irreversible processes, it is critically important that we distinguish between Q, the actual energy transfer in the process, and Q_r, the energy that would have been transferred by heat along a reversible path. Only Q_r is the correct value to be used in calculating the entropy change.

As we shall see in the following examples, the change in entropy for a system and its surroundings is always positive for an irreversible process. In general, the total entropy—and therefore the disorder—always increase in an irreversible process. Keeping these considerations in mind, we can state the second law of thermodynamics as follows:

The total entropy of an isolated system that undergoes a change can never decrease.

Furthermore, **if the process is irreversible, then the total entropy of an isolated system always increases. In a reversible process, the total entropy of an isolated system remains constant.**

When dealing with a system that is not isolated from its surroundings, remember that the increase in entropy described in the second law is that of the system *and* its surroundings. When a system and its surroundings interact in an irreversible process, the increase in entropy of one is greater than the decrease in entropy of the other. Hence, we conclude that **the change in entropy of the Universe must be greater than zero for an irreversible process and equal to zero for a reversible process.** Ultimately, the entropy of the Universe should reach a maximum value. At this value, the Universe will be in a state of uniform temperature and density. All physical, chemical, and biological processes will cease because a state of perfect disorder implies that no energy is available for doing work. This gloomy state of affairs is sometimes referred to as the heat death of the Universe.

Quick Quiz 22.3

In the presence of sunlight, a tree rearranges an unorganized collection of carbon dioxide and water molecules into the highly ordered collection of molecules we see as leaves and branches. True or false: This reduction of entropy in the tree is a violation of the second law of thermodynamics. Explain your response.

Entropy Change in Thermal Conduction

Let us now consider a system consisting of a hot reservoir and a cold reservoir in thermal contact with each other and isolated from the rest of the Universe. A process occurs during which energy Q is transferred by heat from the hot reservoir at temperature T_h to the cold reservoir at temperature T_c. Because the cold reservoir absorbs energy Q, its entropy increases by Q/T_c. At the same time, the hot reservoir loses energy Q, and so its entropy change is $-Q/T_h$. Because $T_h > T_c$, the increase in entropy of the cold reservoir is greater than the decrease in entropy of the hot reservoir. Therefore, the change in entropy of the system (and of the Universe) is greater than zero:

$$\Delta S_U = \frac{Q}{T_c} + \frac{-Q}{T_h} > 0$$

EXAMPLE 22.7   **Which Way Does the Energy Flow?**

A large, cold object is at 273 K, and a large, hot object is at 373 K. Show that it is impossible for a small amount of energy—for example, 8.00 J—to be transferred spontaneously from the cold object to the hot one without a decrease in the entropy of the Universe and therefore a violation of the second law.

Solution We assume that, during the energy transfer, the two objects do not undergo a temperature change. This is not a necessary assumption; we make it only to avoid using integral calculus in our calculations. The process as described is irreversible, and so we must find an equivalent reversible process. It is sufficient to assume that the objects are connected by a poor thermal conductor whose temperature spans the range from 273 K to 373 K. This conductor transfers energy slowly, and its state does not change during the process. Under this assumption, the energy transfer to or from each object is reversible, and we may set $Q = Q_r$. The entropy change of the hot object is

$$\Delta S_h = \frac{Q_r}{T_h} = \frac{8.00\ \text{J}}{373\ \text{K}} = 0.021\ 4\ \text{J/K}$$

The cold object loses energy, and its entropy change is

$$\Delta S_c = \frac{Q_r}{T_c} = \frac{-8.00\ \text{J}}{273\ \text{K}} = -0.029\ 3\ \text{J/K}$$

We consider the two objects to be isolated from the rest of the Universe. Thus, the entropy change of the Universe is just

that of our two-object system, which is

$$\Delta S_U = \Delta S_c + \Delta S_h = -0.007\ 9\ \text{J/K}$$

This decrease in entropy of the Universe is in violation of the second law. That is, **the spontaneous transfer of energy from a cold to a hot object cannot occur.**

In terms of disorder, let us consider the violation of the second law if energy were to continue to transfer spontaneously from a cold object to a hot object. Before the transfer, a certain degree of order is associated with the different temperatures of the objects. The hot object's molecules have a higher average energy than the cold object's molecules. If energy spontaneously flows from the cold object to the hot object, then, over a period of time, the cold object will become colder and the hot object will become hotter. The difference in average molecular energy will become even greater; this would represent an increase in order for the system and a violation of the second law.

In comparison, the process that does occur naturally is the flow of energy from the hot object to the cold object. In this process, the difference in average molecular energy decreases; this represents a more random distribution of energy and an increase in disorder.

Exercise Suppose that 8.00 J of energy is transferred from a hot object to a cold one. What is the net entropy change of the Universe?

Answer $+0.007\ 9\ \text{J/K}$.

Entropy Change in a Free Expansion

Let us again consider the adiabatic free expansion of a gas occupying an initial volume V_i (Fig. 22.16). A membrane separating the gas from an evacuated region is broken, and the gas expands (irreversibly) to a volume V_f. Let us find the changes in entropy of the gas and of the Universe during this process.

The process is clearly neither reversible nor quasi-static. The work done by the gas against the vacuum is zero, and because the walls are insulating, no energy is transferred by heat during the expansion. That is, $W = 0$ and $Q = 0$. Using the first law, we see that the change in internal energy is zero. Because the gas is ideal, E_{int} depends on temperature only, and we conclude that $\Delta T = 0$ or $T_i = T_f$.

To apply Equation 22.9, we cannot use $Q = 0$, the value for the irreversible process, but must instead find Q_r; that is, we must find an equivalent reversible path that shares the same initial and final states. A simple choice is an isothermal, reversible expansion in which the gas pushes slowly against a piston while energy enters the gas by heat from a reservoir to hold the temperature constant. Because T is constant in this process, Equation 22.9 gives

$$\Delta S = \int_i^f \frac{dQ_r}{T} = \frac{1}{T}\int_i^f dQ_r$$

For an isothermal process, the first law of thermodynamics specifies that $\int_i^f dQ_r$ is equal to the work done by the gas during the expansion from V_i to V_f, which is given by Equation 20.13. Using this result, we find that the entropy change for the gas is

$$\Delta S = nR \ln \frac{V_f}{V_i} \tag{22.13}$$

Because $V_f > V_i$, we conclude that ΔS is positive. This positive result indicates that both the entropy and the disorder of the gas increase as a result of the irreversible, adiabatic expansion.

Because the free expansion takes place in an insulated container, no energy is transferred by heat from the surroundings. (Remember that the isothermal, reversible expansion is only a *replacement* process that we use to calculate the entropy change for the gas; it is not the *actual* process.) Thus, the free expansion has no effect on the surroundings, and the entropy change of the surroundings is zero. Thus, the entropy change for the Universe is positive; this is consistent with the second law.

Figure 22.16 Adiabatic free expansion of a gas. When the membrane separating the gas from the evacuated region is ruptured, the gas expands freely and irreversibly. As a result, it occupies a greater final volume. The container is thermally insulated from its surroundings; thus, $Q = 0$.

EXAMPLE 22.8 ▶ **Free Expansion of a Gas**

Calculate the change in entropy for a process in which 2.00 mol of an ideal gas undergoes a free expansion to three times its initial volume.

Solution Using Equation 22.13 with $n = 2.00$ mol and $V_f = 3V_i$, we find that

$$\Delta S = nR \ln \frac{V_f}{V_i} = (2.00 \text{ mol})(8.31 \text{ J/mol} \cdot \text{K}) \, (\ln 3)$$

$$= 18.3 \text{ J/K}$$

It is easy to see that the gas is more disordered after the expansion. Instead of being concentrated in a relatively small space, the molecules are scattered over a larger region.

Entropy Change in Calorimetric Processes

A substance of mass m_1, specific heat c_1, and initial temperature T_1 is placed in thermal contact with a second substance of mass m_2, specific heat c_2, and initial temperature $T_2 > T_1$. The two substances are contained in a calorimeter so that no energy is lost to the surroundings. The system of the two substances is allowed to reach thermal equilibrium. What is the total entropy change for the system?

First, let us calculate the final equilibrium temperature T_f. Using the techniques of Section 20.2—namely, Equation 20.5, $Q_{cold} = -Q_{hot}$, and Equation 20.4, $Q = mc\,\Delta T$, we obtain

$$m_1 c_1 \Delta T_1 = -m_2 c_2 \Delta T_2$$

$$m_1 c_1 (T_f - T_1) = -m_2 c_2 (T_f - T_2)$$

Solving for T_f, we have

$$T_f = \frac{m_1 c_1 T_1 + m_2 c_2 T_2}{m_1 c_1 + m_2 c_2} \qquad \textbf{(22.14)}$$

The process is irreversible because the system goes through a series of non-equilibrium states. During such a transformation, the temperature of the system at any time is not well defined because different parts of the system have different temperatures. However, we can imagine that the hot substance at the initial temperature T_2 is slowly cooled to the temperature T_f as it comes into contact with a series of reservoirs differing infinitesimally in temperature, the first reservoir being at T_2 and the last being at T_f. Such a series of very small changes in temperature would approximate a reversible process. We imagine doing the same thing for the cold substance. Applying Equation 22.9 and noting that $dQ = mc\,dT$ for an infinitesimal change, we have

$$\Delta S = \int_1 \frac{dQ_{cold}}{T} + \int_2 \frac{dQ_{hot}}{T} = m_1 c_1 \int_{T_1}^{T_f} \frac{dT}{T} + m_2 c_2 \int_{T_2}^{T_f} \frac{dT}{T}$$

where we have assumed that the specific heats remain constant. Integrating, we find that

Change in entropy for a calorimetric process

$$\Delta S = m_1 c_1 \ln \frac{T_f}{T_1} + m_2 c_2 \ln \frac{T_f}{T_2} \qquad \textbf{(22.15)}$$

where T_f is given by Equation 22.14. If Equation 22.14 is substituted into Equation 22.15, we can show that one of the terms in Equation 22.15 is always positive and the other is always negative. (You may want to verify this for yourself.) The positive term is always greater than the negative term, and this results in a positive value for ΔS. Thus, we conclude that the entropy of the Universe increases in this irreversible process.

Finally, you should note that Equation 22.15 is valid only when no mixing of different substances occurs, because a further entropy increase is associated with the increase in disorder during the mixing. If the substances are liquids or gases and mixing occurs, the result applies only if the two fluids are identical, as in the following example.

EXAMPLE 22.9   Calculating ΔS for a Calorimetric Process

Suppose that 1.00 kg of water at 0.00°C is mixed with an equal mass of water at 100°C. After equilibrium is reached, the mixture has a uniform temperature of 50.0°C. What is the change in entropy of the system?

Solution   We can calculate the change in entropy from Equation 22.15 using the values $m_1 = m_2 = 1.00$ kg, $c_1 = c_2 = 4\,186$ J/kg·K, $T_1 = 273$ K, $T_2 = 373$ K, and $T_f = 323$ K:

$$\Delta S = m_1 c_1 \ln \frac{T_f}{T_1} + m_2 c_2 \ln \frac{T_f}{T_2}$$

$$= (1.00 \text{ kg})(4\,186 \text{ J/kg·K}) \ln\left(\frac{323 \text{ K}}{273 \text{ K}}\right)$$

$$+ (1.00 \text{ kg})(4\,186 \text{ J/kg·K}) \ln\left(\frac{323 \text{ K}}{373 \text{ K}}\right)$$

$$= 704 \text{ J/K} - 602 \text{ J/K} = \boxed{102 \text{ J/K}}$$

That is, as a result of this irreversible process, the increase in entropy of the cold water is greater than the decrease in entropy of the warm water. Consequently, the increase in entropy of the system is 102 J/K.

Optional Section

22.8 ENTROPY ON A MICROSCOPIC SCALE[4]

As we have seen, we can approach entropy by relying on macroscopic concepts and using parameters such as pressure and temperature. We can also treat entropy from a microscopic viewpoint through statistical analysis of molecular motions. We now use a microscopic model to investigate once again the free expansion of an ideal gas, which was discussed from a macroscopic point of view in the preceding section.

In the kinetic theory of gases, gas molecules are represented as particles moving randomly. Let us suppose that the gas is initially confined to a volume V_i, as shown in Figure 22.17a. When the partition separating V_i from a larger container is removed, the molecules eventually are distributed throughout the greater volume V_f (Fig. 22.17b). For a given uniform distribution of gas in the volume, there are a large number of equivalent microstates, and we can relate the entropy of the gas to the number of microstates corresponding to a given macrostate.

We count the number of microstates by considering the variety of molecular locations involved in the free expansion. The instant after the partition is removed (and before the molecules have had a chance to rush into the other half of the container), all the molecules are in the initial volume. We assume that each molecule occupies some microscopic volume V_m. The total number of possible locations of a single molecule in a macroscopic initial volume V_i is the ratio $w_i = V_i/V_m$, which is a huge number. We use w_i here to represent the number of *ways* that the molecule can be placed in the volume, or the number of microstates, which is equivalent to the number of available locations. We assume that the molecule's occupying each of these locations is equally probable.

As more molecules are added to the system, the number of possible ways that the molecules can be positioned in the volume multiplies. For example, in considering two molecules, for every possible placement of the first, all possible placements of the second are available. Thus, there are w_1 ways of locating the first molecule, and for each of these, there are w_2 ways of locating the second molecule. The total number of ways of locating the two molecules is $w_1 w_2$.

Neglecting the very small probability of having two molecules occupy the same location, each molecule may go into any of the V_i/V_m locations, and so the number of ways of locating N molecules in the volume becomes $W_i = w_i^N = (V_i/V_m)^N$. ($W_i$ is not to be confused with work.) Similarly, when the volume is increased to V_f, the number of ways of locating N molecules increases to $W_f = w_f^N = (V_f/V_m)^N$. The ratio of the number of ways of placing the molecules in the volume for the

(a)

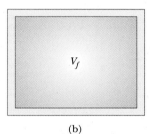

(b)

Figure 22.17 In a free expansion, the gas is allowed to expand into a region that was previously a vacuum.

[4] This section was adapted from A. Hudson and R. Nelson, *University Physics*, Philadelphia, Saunders College Publishing, 1990.

initial and final configurations is

$$\frac{W_f}{W_i} = \frac{(V_f/V_m)^N}{(V_i/V_m)^N} = \left(\frac{V_f}{V_i}\right)^N$$

If we now take the natural logarithm of this equation and multiply by Boltzmann's constant, we find that

$$k_B \ln\left(\frac{W_f}{W_i}\right) = nN_A k_B \ln\left(\frac{V_f}{V_i}\right)$$

where we have used the equality $N = nN_A$. We know from Equation 19.11 that $N_A k_B$ is the universal gas constant R; thus, we can write this equation as

$$k_B \ln W_f - k_B \ln W_i = nR \ln\left(\frac{V_f}{V_i}\right) \tag{22.16}$$

From Equation 22.13 we know that when n mol of a gas undergoes a free expansion from V_i to V_f, the change in entropy is

$$S_f - S_i = nR \ln\left(\frac{V_f}{V_i}\right) \tag{22.17}$$

Note that the right-hand sides of Equations 22.16 and 22.17 are identical. Thus, we make the following important connection between entropy and the number of microstates for a given macrostate:

Entropy (microscopic definition)

$$S \equiv k_B \ln W \tag{22.18}$$

The more microstates there are that correspond to a given macrostate, the greater is the entropy of that macrostate. As we have discussed previously, there are many more disordered microstates than ordered microstates. Thus, Equation 22.18 indicates mathematically that **entropy is a measure of microscopic disorder.** Although in our discussion we used the specific example of the free expansion of an ideal gas, a more rigorous development of the statistical interpretation of entropy would lead us to the same conclusion.

Imagine the container of gas depicted in Figure 22.18a as having all of its molecules traveling at speeds greater than the mean value on the left side and all of its molecules traveling at speeds less than the mean value on the right side (an ordered microstate). Compare this with the uniform mixture of fast- and slow-mov-

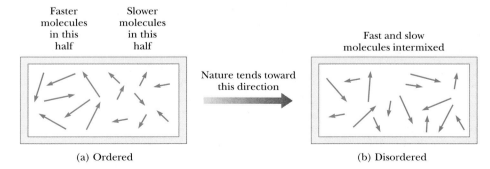

Figure 22.18 A container of gas in two equally probable states of molecular motion. (a) An ordered arrangement, which is one of a few and therefore a collectively unlikely set. (b) A disordered arrangement, which is one of many and therefore a collectively likely set.

Figure 22.19 By tossing a coin into a jar, the carnival-goer can win the fish in the jar. It is more likely that the coin will land in a jar containing a goldfish than in the one containing the black fish.

ing molecules in Figure 22.18b (a disordered microstate). You might expect the ordered microstate to be very unlikely because random motions tend to mix the slow- and fast-moving molecules uniformly. Yet *individually* each of these microstates is equally probable. However, there are far more disordered microstates than ordered microstates, and so a macrostate corresponding to a large number of equivalent disordered microstates is much more probable than a macrostate corresponding to a small number of equivalent ordered microstates.

Figure 22.19 shows a real-world example of this concept. There are two possible macrostates for the carnival game—winning a goldfish and winning a black fish. Because only one jar in the array of jars contains a black fish, only one possible microstate corresponds to the macrostate of winning a black fish. A large number of microstates are described by the coin's falling into a jar containing a goldfish. Thus, for the macrostate of winning of a goldfish, there are many equivalent microstates. As a result, the probability of winning a goldfish is much greater than the probability of winning a black fish. If there are 24 goldfish and 1 black fish, the probability of winning the black fish is 1 in 25. This assumes that all microstates have the same probability, a situation that may not be quite true for the situation shown in Figure 22.19. If you are an accurate coin tosser and you are aiming for the edge of the array of jars, then the probability of the coin's landing in a jar near the edge is likely to be greater than the probability of its landing in a jar near the center.

Let us consider a similar type of probability problem for 100 molecules in a container. At any given moment, the probability of one molecule's being in the left part of the container shown in Figure 22.20a as a result of random motion is $\frac{1}{2}$. If there are two molecules, as shown in Figure 22.20b, the probability of both being in the left part is $(\frac{1}{2})^2$ or 1 in 4. If there are three molecules (Fig. 22.20c), the probability of all of them being in the left portion at the same moment is $(\frac{1}{2})^3$, or 1 in 8. For 100 independently moving molecules, the probability that the 50 fastest ones will be found in the left part at any moment is $(\frac{1}{2})^{50}$. Likewise, the probability that the remaining 50 slower molecules will be found in the right part at any moment is $(\frac{1}{2})^{50}$. Therefore, the probability of finding this fast-slow separation as a result of random motion is the product $(\frac{1}{2})^{50}(\frac{1}{2})^{50} = (\frac{1}{2})^{100}$, which corresponds to about 1 in 10^{30}. When this calculation is extrapolated from 100 molecules to the number in 1 mol of gas (6.02×10^{23}), the ordered arrangement is found to be *extremely* improbable!

QuickLab

Roll a pair of dice 100 times and record the total number of spots appearing on the dice for each throw. Which total comes up most frequently? Is this expected?

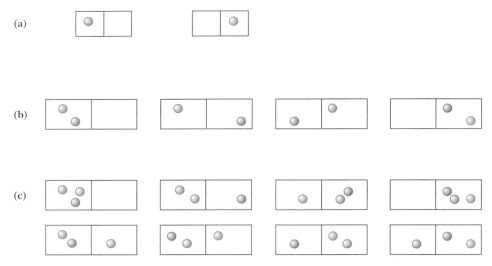

Figure 22.20 (a) One molecule in a two-sided container has a 1-in-2 chance of being on the left side. (b) Two molecules have a 1-in-4 chance of being on the left side at the same time. (c) Three molecules have a 1-in-8 chance of being on the left side at the same time.

EXAMPLE 22.10   Adiabatic Free Expansion—One Last Time

Let us verify that the macroscopic and microscopic approaches to the calculation of entropy lead to the same conclusion for the adiabatic free expansion of an ideal gas. Suppose that 1 mol of gas expands to four times its initial volume. As we have seen for this process, the initial and final temperatures are the same. (a) Using a macroscopic approach, calculate the entropy change for the gas. (b) Using statistical considerations, calculate the change in entropy for the gas and show that it agrees with the answer you obtained in part (a).

Solution (a) Using Equation 22.13, we have

$$\Delta S = nR \ln\left(\frac{V_f}{V_i}\right) = (1)R \ln\left(\frac{4V_i}{V_i}\right) = \boxed{R \ln 4}$$

(b) The number of microstates available to a single molecule in the initial volume V_i is $w_i = V_i/V_m$. For 1 mol (N_A molecules), the number of available microstates is

$$W_i = w_i{}^{N_A} = \left(\frac{V_i}{V_m}\right)^{N_A}$$

The number of microstates for all N_A molecules in the final volume $V_f = 4V_i$ is

$$W_f = \left(\frac{V_f}{V_m}\right)^{N_A} = \left(\frac{4V_i}{V_m}\right)^{N_A}$$

Thus, the ratio of the number of final microstates to initial microstates is

$$\frac{W_f}{W_i} = 4^{N_A}$$

Using Equation 22.18, we obtain

$$\Delta S = k_B \ln W_f - k_B \ln W_i = k_B \ln\left(\frac{W_f}{W_i}\right)$$

$$= k_B \ln(4^{N_A}) = N_A k_B \ln 4 = \boxed{R \ln 4}$$

The answer is the same as that for part (a), which dealt with macroscopic parameters.

CONCEPTUAL EXAMPLE 22.11   Let's Play Marbles!

Suppose you have a bag of 100 marbles. Fifty of the marbles are red, and 50 are green. You are allowed to draw four marbles from the bag according to the following rules: Draw one marble, record its color, and return it to the bag. Then draw another marble. Continue this process until you have drawn and returned four marbles. What are the possible

macrostates for this set of events? What is the most likely macrostate? What is the least likely macrostate?

Solution Because each marble is returned to the bag before the next one is drawn, the probability of drawing a red marble is always the same as the probability of drawing a green one. All the possible microstates and macrostates are shown in Table 22.1. As this table indicates, there is only one way to draw four red marbles, and so there is only one microstate. However, there are four possible microstates that correspond to the macrostate of one green marble and three red marbles; six microstates that correspond to two green marbles and two red marbles; four microstates that correspond to three green marbles and one red marble; and one microstate that corresponds to four green marbles. The most likely macrostate—two red marbles and two green marbles—corresponds to the most disordered microstates. The least likely macrostates—four red marbles or four green marbles—correspond to the most ordered microstates.

TABLE 22.1   Possible Results of Drawing Four Marbles from a Bag

Macrostate	Possible Microstates	Total Number of Microstates
All R	RRRR	1
1G, 3R	RRRG, RRGR, RGRR, GRRR	4
2G, 2R	RRGG, RGRG, GRRG, RGGR, GRGR, GGRR	6
3G, 1R	GGGR, GGRG, GRGG, RGGG	4
All G	GGGG	1

SUMMARY

A **heat engine** is a device that converts internal energy to other useful forms of energy. The net work done by a heat engine in carrying a working substance through a cyclic process ($\Delta E_{int} = 0$) is

$$W = Q_h - Q_c \qquad \text{(22.1)}$$

where Q_h is the energy absorbed from a hot reservoir and Q_c is the energy expelled to a cold reservoir.

The **thermal efficiency** e of a heat engine is

$$e = \frac{W}{Q_h} = 1 - \frac{Q_c}{Q_h} \qquad \text{(22.2)}$$

The **second law of thermodynamics** can be stated in the following two ways:

- It is impossible to construct a heat engine that, operating in a cycle, produces no effect other than the absorption of energy from a reservoir and the performance of an equal amount of work (the Kelvin–Planck statement).
- It is impossible to construct a cyclic machine whose sole effect is the continuous transfer of energy from one object to another object at a higher temperature without the input of energy by work (the Clausius statement).

In a **reversible** process, the system can be returned to its initial conditions along the same path shown on a PV diagram, and every point along this path is an equilibrium state. A process that does not satisfy these requirements is **irreversible**. **Carnot's theorem** states that no real heat engine operating (irreversibly) between the temperatures T_c and T_h can be more efficient than an engine operating reversibly in a Carnot cycle between the same two temperatures.

The **thermal efficiency** of a heat engine operating in the Carnot cycle is

$$e_C = 1 - \frac{T_c}{T_h} \qquad \text{(22.4)}$$

You should be able to use this equation (or an equivalent form involving a ratio of heats) to determine the maximum possible efficiency of any heat engine.

The second law of thermodynamics states that when real (irreversible) processes occur, the degree of disorder in the system plus the surroundings increases. When a process occurs in an isolated system, the state of the system becomes more disordered. The measure of disorder in a system is called **entropy** *S*. Thus, another way in which the second law can be stated is

- The entropy of the Universe increases in all real processes.

The **change in entropy** *dS* of a system during a process between two infinitesimally separated equilibrium states is

$$dS = \frac{dQ_r}{T} \tag{22.8}$$

where dQ_r is the energy transfer by heat for a reversible process that connects the initial and final states. The change in entropy of a system during an arbitrary process between an initial state and a final state is

$$\Delta S = \int_i^f \frac{dQ_r}{T} \tag{22.9}$$

The value of ΔS for the system is the same for all paths connecting the initial and final states. The change in entropy for a system undergoing any reversible, cyclic process is zero, and when such a process occurs, the entropy of the Universe remains constant.

From a microscopic viewpoint, entropy is defined as

$$S \equiv k_B \ln W \tag{22.18}$$

where k_B is Boltzmann's constant and *W* is the number of microstates available to the system for the existing macrostate. Because of the statistical tendency of systems to proceed toward states of greater probability and greater disorder, all natural processes are irreversible, and entropy increases. Thus, entropy is a measure of microscopic disorder.

QUESTIONS

1. Is it possible to convert internal energy to mechanical energy? Describe a process in which such a conversion occurs.

2. What are some factors that affect the efficiency of automobile engines?

3. In practical heat engines, which are we able to control more: the temperature of the hot reservoir, or the temperature of the cold reservoir? Explain.

4. A steam-driven turbine is one major component of an electric power plant. Why is it advantageous to have the temperature of the steam as high as possible?

5. Is it possible to construct a heat engine that creates no thermal pollution? What does this tell us about environmental considerations for an industrialized society?

6. Discuss three common examples of natural processes that involve an increase in entropy. Be sure to account for all parts of each system under consideration.

7. Discuss the change in entropy of a gas that expands (a) at constant temperature and (b) adiabatically.

8. In solar ponds constructed in Israel, the Sun's energy is concentrated near the bottom of a salty pond. With the proper layering of salt in the water, convection is prevented, and temperatures of 100°C may be reached. Can you estimate the maximum efficiency with which useful energy can be extracted from the pond?

9. The vortex tube (Fig. Q22.9) is a T-shaped device that takes in compressed air at 20 atm and 20°C and gives off air at −20°C from one flared end and air at 60°C from the other flared end. Does the operation of this device vi-

Compressed air in

Cold air −20°C   Hot air + 60°C

Ranque-Hilsch vortex tube

Figure Q22.9

olate the second law of thermodynamics? If not, explain why not.

10. Why does your automobile burn more gas in winter than in summer?

11. Can a heat pump have a coefficient of performance (COP) less than unity? Explain.

12. Give some examples of irreversible processes that occur in nature.

13. Give an example of a process in nature that is nearly reversible.

14. A thermodynamic process occurs in which the entropy of a system changes by − 8.0 J/K. According to the second law of thermodynamics, what can you conclude about the entropy change of the environment?

15. If a supersaturated sugar solution is allowed to evaporate slowly, sugar crystals form in the container. Hence, sugar molecules go from a disordered form (in solution) to a highly ordered crystalline form. Does this process violate the second law of thermodynamics? Explain.

16. How could you increase the entropy of 1 mol of a metal that is at room temperature? How could you decrease its entropy?

17. A heat pump is to be installed in a region where the average outdoor temperature in the winter months is − 20°C. In view of this, why would it be advisable to place the outdoor compressor unit deep in the ground? Why are heat pumps not commonly used for heating in cold climates?

18. Suppose your roommate is "Mr. Clean" and tidies up your messy room after a big party. That is, your roommate is increasing order in the room. Does this represent a violation of the second law of thermodynamics?

19. Discuss the entropy changes that occur when you (a) bake a loaf of bread and (b) consume the bread.

20. The device shown in Figure Q22.20, which is called a thermoelectric converter, uses a series of semiconductor cells to convert internal energy to electrical energy. In the photograph on the left, both legs of the device are at the same temperature and no electrical energy is produced. However, when one leg is at a higher temperature than the other, as shown in the photograph on the right, electrical energy is produced as the device extracts energy from the hot reservoir and drives a small electric motor. (a) Why does the temperature differential produce electrical energy in this demonstration? (b) In what sense does this intriguing experiment demonstrate the second law of thermodynamics?

21. A classmate tells you that it is just as likely for all the air molecules in the room you are both in to be concentrated in one corner (with the rest of the room being a vacuum) as it is for the air molecules to be distributed uniformly about the room in their current state. Is this true? Why doesn't the situation he describes actually happen?

Figure Q22.20 *(Courtesy of PASCO Scientific Company)*

PROBLEMS

1, 2, 3 = straightforward, intermediate, challenging ☐ = full solution available in the *Student Solutions Manual and Study Guide*
WEB = solution posted at **http://www.saunderscollege.com/physics/** 💻 = Computer useful in solving problem ▦ = Interactive Physics
☐ = paired numerical/symbolic problems

Section 22.1   Heat Engines and the Second Law of Thermodynamics
Section 22.2   Reversible and Irreversible Processes

1. A heat engine absorbs 360 J of energy and performs 25.0 J of work in each cycle. Find (a) the efficiency of the engine and (b) the energy expelled to the cold reservoir in each cycle.

2. The energy absorbed by an engine is three times greater than the work it performs. (a) What is its thermal efficiency? (b) What fraction of the energy absorbed is expelled to the cold reservoir?

3. A particular engine has a power output of 5.00 kW and an efficiency of 25.0%. Assuming that the engine expels 8 000 J of energy in each cycle, find (a) the energy absorbed in each cycle and (b) the time for each cycle.

4. A heat engine performs 200 J of work in each cycle and has an efficiency of 30.0%. For each cycle, how much energy is (a) absorbed and (b) expelled?

5. An ideal gas is compressed to half its original volume while its temperature is held constant. (a) If 1 000 J of energy is removed from the gas during the compression, how much work is done on the gas? (b) What is the change in the internal energy of the gas during the compression?

6. Suppose that a heat engine is connected to two energy reservoirs, one a pool of molten aluminum (660°C) and the other a block of solid mercury (− 38.9°C). The engine runs by freezing 1.00 g of aluminum and melting 15.0 g of mercury during each cycle. The heat of fusion of aluminum is 3.97×10^5 J/kg; the heat of fusion of mercury is 1.18×10^4 J/kg. What is the efficiency of this engine?

Section 22.3   The Carnot Engine

7. One of the most efficient engines ever built (actual efficiency 42.0%) operates between 430°C and 1 870°C. (a) What is its maximum theoretical efficiency? (b) How much power does the engine deliver if it absorbs 1.40×10^5 J of energy each second from the hot reservoir?

8. A heat engine operating between 80.0°C and 200°C achieves 20.0% of the maximum possible efficiency. What energy input will enable the engine to perform 10.0 kJ of work?

9. A Carnot engine has a power output of 150 kW. The engine operates between two reservoirs at 20.0°C and 500°C. (a) How much energy does it absorb per hour? (b) How much energy is lost per hour in its exhaust?

10. A steam engine is operated in a cold climate where the exhaust temperature is 0°C. (a) Calculate the theoretical maximum efficiency of the engine, using an intake steam temperature of 100°C. (b) If superheated steam at 200°C were used instead, what would be the maximum possible efficiency?

11. An ideal gas is taken through a Carnot cycle. The isothermal expansion occurs at 250°C, and the isothermal compression takes place at 50.0°C. Assuming that the gas absorbs 1 200 J of energy from the hot reservoir during the isothermal expansion, find (a) the energy expelled to the cold reservoir in each cycle and (b) the net work done by the gas in each cycle.

12. The exhaust temperature of a Carnot heat engine is 300°C. What is the intake temperature if the efficiency of the engine is 30.0%?

13. A power plant operates at 32.0% efficiency during the summer when the sea water for cooling is at 20.0°C. The plant uses 350°C steam to drive turbines. Assuming that the plant's efficiency changes in the same proportion as the ideal efficiency, what would be the plant's efficiency in the winter, when the sea water is at 10.0°C?

14. Argon enters a turbine at a rate of 80.0 kg/min, a temperature of 800°C, and a pressure of 1.50 MPa. It expands adiabatically as it pushes on the turbine blades and exits at a pressure of 300 kPa. (a) Calculate its temperature at the time of exit. (b) Calculate the (maximum) power output of the turning turbine. (c) The turbine is one component of a model closed-cycle gas turbine engine. Calculate the maximum efficiency of the engine.

15. A power plant that would make use of the temperature gradient in the ocean has been proposed. The system is to operate between 5.00°C (water temperature at a depth of about 1 km) and 20.0°C (surface water temperature). (a) What is the maximum efficiency of such a system? (b) If the power output of the plant is 75.0 MW, how much energy is absorbed per hour? (c) In view of your answer to part (a), do you think such a system is worthwhile (considering that there is no charge for fuel)?

16. A 20.0%-efficient real engine is used to speed up a train from rest to 5.00 m/s. It is known that an ideal (Carnot) engine having the same cold and hot reservoirs would accelerate the same train from rest to a speed of 6.50 m/s using the same amount of fuel. Assuming that the engines use air at 300 K as a cold reservoir, find the temperature of the steam serving as the hot reservoir.

17. A firebox is at 750 K, and the ambient temperature is 300 K. The efficiency of a Carnot engine doing 150 J of work as it transports energy between these constant-temperature baths is 60.0%. The Carnot engine must absorb energy 150 J/0.600 = 250 J from the hot reser-

voir and release 100 J of energy into the environment. To follow Carnot's reasoning, suppose that some other heat engine S could have an efficiency of 70.0%. (a) Find the energy input and energy output of engine S as it does 150 J of work. (b) Let engine S operate as in part (a) and run the Carnot engine in reverse. Find the total energy the firebox puts out as both engines operate together and the total energy absorbed by the environment. Show that the Clausius statement of the second law of thermodynamics is violated. (c) Find the energy input and work output of engine S as it exhausts 100 J of energy. (d) Let engine S operate as in (c) and contribute 150 J of its work output to running the Carnot engine in reverse. Find the total energy that the firebox puts out as both engines operate together, the total work output, and the total energy absorbed by the environment. Show that the Kelvin–Planck statement of the second law is violated. Thus, our assumption about the efficiency of engine S must be false. (e) Let the engines operate together through one cycle as in part (d). Find the change in entropy of the Universe. Show that the entropy statement of the second law is violated.

18. At point A in a Carnot cycle, 2.34 mol of a monatomic ideal gas has a pressure of 1 400 kPa, a volume of 10.0 L, and a temperature of 720 K. It expands isothermally to point B, and then expands adiabatically to point C, where its volume is 24.0 L. An isothermal compression brings it to point D, where its new volume is 15.0 L. An adiabatic process returns the gas to point A. (a) Determine all the unknown pressures, volumes, and temperatures as you fill in the following table:

	P	V	T
A	1 400 kPa	10.0 L	720 K
B			
C		24.0 L	
D		15.0 L	

(b) Find the energy added by heat, the work done, and the change in internal energy for each of the following steps: $A \rightarrow B$, $B \rightarrow C$, $C \rightarrow D$, and $D \rightarrow A$. (c) Show that $W_{net}/Q_{in} = 1 - T_C/T_A$, the Carnot efficiency.

Section 22.4 Gasoline and Diesel Engines

19. In a cylinder of an automobile engine just after combustion, the gas is confined to a volume of 50.0 cm³ and has an initial pressure of 3.00×10^6 Pa. The piston moves outward to a final volume of 300 cm³, and the gas expands without energy loss by heat. (a) If $\gamma = 1.40$ for the gas, what is the final pressure? (b) How much work is done by the gas in expanding?

20. A gasoline engine has a compression ratio of 6.00 and uses a gas for which $\gamma = 1.40$. (a) What is the efficiency of the engine if it operates in an idealized Otto cycle?

(b) If the actual efficiency is 15.0%, what fraction of the fuel is wasted as a result of friction and energy losses by heat that could by avoided in a reversible engine? (Assume complete combustion of the air–fuel mixture.)

21. A 1.60-L gasoline engine with a compression ratio of 6.20 has a power output of 102 hp. Assuming that the engine operates in an idealized Otto cycle, find the energy absorbed and exhausted each second. Assume that the fuel–air mixture behaves like an ideal gas, with $\gamma = 1.40$.

22. The compression ratio of an Otto cycle, as shown in Figure 22.12, is $V_A/V_B = 8.00$. At the beginning A of the compression process, 500 cm³ of gas is at 100 kPa and 20.0°C. At the beginning of the adiabatic expansion, the temperature is $T_C = 750$°C. Model the working fluid as an ideal gas, with $E_{int} = nC_VT = 2.50nRT$ and $\gamma = 1.40$. (a) Fill in the following table to track the states of the gas:

	T (K)	P (kPa)	V (cm³)	E_{int}
A	293	100	500	
B				
C	1 023			
D				
A				

(b) Fill in the following table to track the processes:

	Q	W	ΔE_{int}
$A \rightarrow B$			
$B \rightarrow C$			
$C \rightarrow D$			
$D \rightarrow A$			
$ABCDA$			

(c) Identify the energy input Q_h, the energy exhaust Q_c, and the net output work W. (d) Calculate the thermal efficiency. (e) Find the number of revolutions per minute that the crankshaft must complete for a one-cylinder engine to have an output power of 1.00 kW = 1.34 hp. (*Hint:* The thermodynamic cycle involves four piston strokes.)

Section 22.5 Heat Pumps and Refrigerators

23. What is the coefficient of performance of a refrigerator that operates with Carnot efficiency between the temperatures -3.00°C and $+27.0$°C?

24. What is the maximum possible coefficient of performance of a heat pump that brings energy from outdoors at -3.00°C into a 22.0°C house? (*Hint:* The heat pump does work W, which is also available to warm up the house.)

25. An ideal refrigerator or ideal heat pump is equivalent to a Carnot engine running in reverse. That is, energy Q_c is absorbed from a cold reservoir, and energy Q_h is rejected to a hot reservoir. (a) Show that the work that must be supplied to run the refrigerator or heat pump is

$$W = \frac{T_h - T_c}{T_c} Q_c$$

(b) Show that the coefficient of performance (COP) of the ideal refrigerator is

$$\text{COP} = \frac{T_c}{T_h - T_c}$$

26. A heat pump (Fig. P22.26) is essentially a heat engine run backward. It extracts energy from colder air outside and deposits it in a warmer room. Suppose that the ratio of the actual energy entering the room to the work done by the device's motor is 10.0% of the theoretical maximum ratio. Determine the energy entering the room per joule of work done by the motor when the inside temperature is 20.0°C and the outside temperature is −5.00°C.

Figure P22.26

WEB 27. How much work does an ideal Carnot refrigerator require to remove 1.00 J of energy from helium at 4.00 K and reject this energy to a room-temperature (293-K) environment?

28. How much work does an ideal Carnot refrigerator require to remove energy Q from helium at T_c and reject this energy to a room-temperature environment at T_h?

29. A refrigerator has a coefficient of performance equal to 5.00. Assuming that the refrigerator absorbs 120 J of energy from a cold reservoir in each cycle, find (a) the work required in each cycle and (b) the energy expelled to the hot reservoir.

30. A refrigerator maintains a temperature of 0°C in the cold compartment with a room temperature of 25.0°C. It removes energy from the cold compartment at the rate 8 000 kJ/h. (a) What minimum power is required

to operate the refrigerator? (b) At what rate does the refrigerator exhaust energy into the room?

Section 22.6   Entropy

31. An ice tray contains 500 g of water at 0°C. Calculate the change in entropy of the water as it freezes slowly and completely at 0°C.

32. At a pressure of 1 atm, liquid helium boils at 4.20 K. The latent heat of vaporization is 20.5 kJ/kg. Determine the entropy change (per kilogram) of the helium resulting from vaporization.

33. Calculate the change in entropy of 250 g of water heated slowly from 20.0°C to 80.0°C. (*Hint:* Note that $dQ = mc\,dT$.)

34. An airtight freezer holds 2.50 mol of air at 25.0°C and 1.00 atm. The air is then cooled to − 18.0°C. (a) What is the change in entropy of the air if the volume is held constant? (b) What would the change be if the pressure were maintained at 1 atm during the cooling?

Section 22.7   Entropy Changes in Irreversible Processes

35. The temperature at the surface of the Sun is approximately 5 700 K, and the temperature at the surface of the Earth is approximately 290 K. What entropy change occurs when 1 000 J of energy is transferred by radiation from the Sun to the Earth?

36. A 1.00-kg iron horseshoe is taken from a furnace at 900°C and dropped into 4.00 kg of water at 10.0°C. Assuming that no energy is lost by heat to the surroundings, determine the total entropy change of the system (horseshoe and water).

WEB 37. A 1 500-kg car is moving at 20.0 m/s. The driver brakes to a stop. The brakes cool off to the temperature of the surrounding air, which is nearly constant at 20.0°C. What is the total entropy change?

38. How fast are you personally making the entropy of the Universe increase right now? Make an order-of-magnitude estimate, stating what quantities you take as data and the values you measure or estimate for them.

39. One mole of H_2 gas is contained in the left-hand side of the container shown in Figure P22.39, which has equal volumes left and right. The right-hand side is evacuated. When the valve is opened, the gas streams into the right-hand side. What is the final entropy change of the gas? Does the temperature of the gas change?

Figure P22.39

40. A rigid tank of small mass contains 40.0 g of argon, initially at 200°C and 100 kPa. The tank is placed into a reservoir at 0°C and is allowed to cool to thermal equilibrium. Calculate (a) the volume of the tank, (b) the change in internal energy of the argon, (c) the energy transferred by heat, (d) the change in entropy of the argon, and (e) the change in entropy of the constant-temperature bath.

41. A 2.00-L container has a center partition that divides it into two equal parts, as shown in Figure P22.41. The left-hand side contains H_2 gas, and the right-hand side contains O_2 gas. Both gases are at room temperature and at atmospheric pressure. The partition is removed, and the gases are allowed to mix. What is the entropy increase of the system?

| 0.044 mol H_2 | 0.044 mol O_2 |

Figure P22.41

42. A 100 000-kg iceberg at $-5.00°C$ breaks away from the polar ice shelf and floats away into the ocean, at $5.00°C$. What is the final change in the entropy of the system after the iceberg has completely melted? (The specific heat of ice is 2010 J/kg · °C.)

43. One mole of an ideal monatomic gas, initially at a pressure of 1.00 atm and a volume of 0.025 0 m^3, is heated to a final state with a pressure of 2.00 atm and a volume of 0.040 0 m^3. Determine the change in entropy of the gas for this process.

44. One mole of a diatomic ideal gas, initially having pressure P and volume V, expands so as to have pressure $2P$ and volume $2V$. Determine the entropy change of the gas in the process.

(Optional)
Section 22.8   Entropy on a Microscopic Scale

45. If you toss two dice, what is the total number of ways in which you can obtain (a) a 12 and (b) a 7?

46. Prepare a table like Table 22.1 for the following occurrence. You toss four coins into the air simultaneously and then record the results of your tosses in terms of the numbers of heads and tails that result. For example, HHTH and HTHH are two possible ways in which three heads and one tail can be achieved. (a) On the basis of your table, what is the most probable result of a toss? In terms of entropy, (b) what is the most ordered state, and (c) what is the most disordered?

47. Repeat the procedure used to construct Table 22.1 (a) for the case in which you draw three marbles from your bag rather than four and (b) for the case in which you draw five rather than four.

ADDITIONAL PROBLEMS

48. Every second at Niagara Falls, some 5 000 m^3 of water falls a distance of 50.0 m (Fig. P22.48). What is the increase in entropy per second due to the falling water? (Assume that the mass of the surroundings is so great that its temperature and that of the water stay nearly constant at 20.0°C. Suppose that a negligible amount of water evaporates.)

Figure P22.48   Niagara Falls. *(Jan Kopec/Tony Stone Images)*

49. If a 35.0%-efficient Carnot heat engine is run in reverse so that it functions as a refrigerator, what would be the engine's (that is, the refrigerator's) coefficient of performance (COP)?

50. How much work does an ideal Carnot refrigerator use to change 0.500 kg of tap water at 10.0°C into ice at $-20.0°C$? Assume that the freezer compartment is held at $-20.0°C$ and that the refrigerator exhausts energy into a room at 20.0°C.

51. A house loses energy through the exterior walls and roof at a rate of 5 000 J/s = 5.00 kW when the interior temperature is 22.0°C and the outside temperature is $-5.00°C$. Calculate the electric power required to maintain the interior temperature at 22.0°C for the following two cases: (a) The electric power is used in electric resistance heaters (which convert all of the electricity supplied into internal energy). (b) The electric power is used to drive an electric motor that operates the compressor of a heat pump (which has a coefficient of performance [COP] equal to 60.0% of the Carnot-cycle value).

52. A heat engine operates between two reservoirs at $T_2 = 600$ K and $T_1 = 350$ K. It absorbs 1 000 J of energy from the higher-temperature reservoir and performs 250 J of work. Find (a) the entropy change of the Universe ΔS_U for this process and (b) the work W that could have been done by an ideal Carnot engine operating between these two reservoirs. (c) Show that the difference between the work done in parts (a) and (b) is $T_1 \Delta S_U$.

53. Figure P22.53 represents n mol of an ideal monatomic gas being taken through a cycle that consists of two isothermal processes at temperatures $3T_i$ and T_i and two constant-volume processes. For each cycle, determine,

Figure P22.53

in terms of n, R, and T_i, (a) the net energy transferred by heat to the gas and (b) the efficiency of an engine operating in this cycle.

54. A refrigerator has a coefficient of performance (COP) of 3.00. The ice tray compartment is at $-20.0°C$, and the room temperature is 22.0°C. The refrigerator can convert 30.0 g of water at 22.0°C to 30.0 g of ice at $-20.0°C$ each minute. What input power is required? Give your answer in watts.

55. An ideal (Carnot) freezer in a kitchen has a constant temperature of 260 K, while the air in the kitchen has a constant temperature of 300 K. Suppose that the insulation for the freezer is not perfect, such that some energy flows into the freezer at a rate of 0.150 W. Determine the average power that the freezer's motor needs to maintain the constant temperature in the freezer.

56. An electric power plant has an overall efficiency of 15.0%. The plant is to deliver 150 MW of power to a city, and its turbines use coal as the fuel. The burning coal produces steam, which drives the turbines. The steam is then condensed to water at 25.0°C as it passes through cooling coils in contact with river water. (a) How many metric tons of coal does the plant consume each day (1 metric ton = 10^3 kg)? (b) What is the total cost of the fuel per year if the delivered price is $8.00/metric ton? (c) If the river water is delivered at 20.0°C, at what minimum rate must it flow over the cooling coils in order that its temperature not exceed 25.0°C? (*Note:* The heat of combustion of coal is 33.0 kJ/g.)

57. A power plant, having a Carnot efficiency, produces 1 000 MW of electrical power from turbines that take in steam at 500 K and reject water at 300 K into a flowing river. Assuming that the water downstream is 6.00 K warmer due to the output of the power plant, determine the flow rate of the river.

58. A power plant, having a Carnot efficiency, produces electric power $\mathcal{P}$ from turbines that take in energy from steam at temperature T_h and discharge energy at temperature T_c through a heat exchanger into a flowing river. Assuming that the water downstream is warmer by ΔT due to the output of the power plant, determine the flow rate of the river.

59. An athlete whose mass is 70.0 kg drinks 16 oz (453.6 g) of refrigerated water. The water is at a temperature of 35.0°F. (a) Neglecting the temperature change of her body that results from the water intake (that is, the body is regarded as a reservoir that is always at 98.6°F), find the entropy increase of the entire system. (b) Assume that the entire body is cooled by the drink and that the average specific heat of a human is equal to the specific heat of liquid water. Neglecting any other energy transfers by heat and any metabolic energy release, find the athlete's temperature after she drinks the cold water, given an initial body temperature of 98.6°F. Under *these* assumptions, what is the entropy increase of the entire system? Compare this result with the one you obtained in part (a).

60. One mole of an ideal monatomic gas is taken through the cycle shown in Figure P22.60. The process $A \rightarrow B$ is a reversible isothermal expansion. Calculate (a) the net work done by the gas, (b) the energy added to the gas, (c) the energy expelled by the gas, and (d) the efficiency of the cycle.

Figure P22.60

61. Calculate the increase in entropy of the Universe when you add 20.0 g of 5.00°C cream to 200 g of 60.0°C coffee. Assume that the specific heats of cream and coffee are both 4.20 J/g·°C.

62. In 1993 the federal government instituted a requirement that all room air conditioners sold in the United States must have an energy efficiency ratio (EER) of 10 or higher. The EER is defined as the ratio of the cooling capacity of the air conditioner, measured in Btu/h, to its electrical power requirement in watts. (a) Convert the EER of 10.0 to dimensionless form, using the conversion 1 Btu = 1 055 J. (b) What is the appropriate name for this dimensionless quantity? (c) In the 1970s it was common to find room air conditioners with EERs of 5 or lower. Compare the operating costs for 10 000-Btu/h air conditioners with EERs of 5.00 and 10.0 if each air conditioner were to operate for 1 500 h during the summer in a city where electricity costs 10.0¢ per kilowatt-hour.

63. One mole of a monatomic ideal gas is taken through the cycle shown in Figure P22.63. At point A, the pressure, volume, and temperature are P_i, V_i, and T_i, respectively. In terms of R and T_i, find (a) the total energy entering the system by heat per cycle, (b) the total energy leaving the system by heat per cycle, (c) the efficiency of an engine operating in this cycle, and (d) the efficiency of an engine operating in a Carnot cycle between the same temperature extremes.

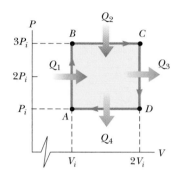

Figure P22.63

64. One mole of an ideal gas expands isothermally. (a) If the gas doubles its volume, show that the work of expansion is $W = RT \ln 2$. (b) Because the internal energy E_{int} of an ideal gas depends solely on its temperature, no change in E_{int} occurs during the expansion. It follows from the first law that the heat input to the gas during the expansion is equal to the energy output by work. Why does this conversion *not* violate the second law?

65. A system consisting of n mol of an ideal gas undergoes a reversible, *isobaric* process from a volume V_i to a volume $3V_i$. Calculate the change in entropy of the gas. (*Hint:* Imagine that the system goes from the initial state to the final state first along an isotherm and then along an adiabatic path—no change in entropy occurs along the adiabatic path.)

66. Suppose you are working in a patent office, and an inventor comes to you with the claim that her heat engine, which employs water as a working substance, has a thermodynamic efficiency of 0.61. She explains that it operates between energy reservoirs at 4°C and 0°C. It is a very complicated device, with many pistons, gears, and pulleys, and the cycle involves freezing and melting. Does her claim that $e = 0.61$ warrant serious consideration? Explain.

67. An idealized diesel engine operates in a cycle known as the *air-standard diesel cycle*, as shown in Figure 22.13. Fuel is sprayed into the cylinder at the point of maximum compression B. Combustion occurs during the expansion $B \rightarrow C$, which is approximated as an isobaric process. Show that the efficiency of an engine operating in this idealized diesel cycle is

$$e = 1 - \frac{1}{\gamma}\left(\frac{T_D - T_A}{T_C - T_B}\right)$$

68. One mole of an ideal gas ($\gamma = 1.40$) is carried through the Carnot cycle described in Figure 22.10. At point A, the pressure is 25.0 atm and the temperature is 600 K. At point C, the pressure is 1.00 atm and the temperature is 400 K. (a) Determine the pressures and volumes at points A, B, C, and D. (b) Calculate the net work done per cycle. (c) Determine the efficiency of an engine operating in this cycle.

69. A typical human has a mass of 70.0 kg and produces about 2 000 kcal (2.00×10^6 cal) of metabolic energy per day. (a) Find the rate of metabolic energy production in watts and in calories per hour. (b) If none of the metabolic energy were transferred out of the body, and the specific heat of the human body is 1.00 cal/g·°C, what is the rate at which body temperature would rise? Give your answer in degrees Celsius per hour and in degrees Fahrenheit per hour.

70. Suppose that 1.00 kg of water at 10.0°C is mixed with 1.00 kg of water at 30.0°C at constant pressure. When the mixture has reached equilibrium, (a) what is the final temperature? (b) Take $c_P = 4.19$ kJ/kg·K for water. Show that the entropy of the system increases by

$$\Delta S = 4.19 \ln\left[\left(\frac{293}{283}\right)\left(\frac{293}{303}\right)\right] \text{ kJ/K}$$

(c) Verify numerically that $\Delta S > 0$. (d) Is the mixing an irreversible process?

Answers to Quick Quizzes

22.1 The cost of heating your home decreases to 25% of the original cost. With electric heating, you receive the same amount of energy for heating your home as enters it by electricity. The COP of 4 for the heat pump means that you are receiving four times as much energy as the energy entering by electricity. With four times as much energy per unit of energy from electricity, you need only one-fourth as much electricity.

22.2 (b) Because the process is reversible and adiabatic, $Q_r = 0$; therefore, $\Delta S = 0$.

22.3 False. The second law states that the entropy of the *Universe* increases in real processes. Although the organization of molecules into ordered leaves and branches represents a decrease in entropy *of the tree*, this organization takes place because of a number of processes in which the tree interacts with its surroundings. If we include the entropy changes associated with all these processes, the entropy change of the Universe during the growth of a tree is still positive.

TABLE A.1 Conversion Factors

Length

	m	cm	km	in.	ft	mi
1 meter	1	10^2	10^{-3}	39.37	3.281	6.214×10^{-4}
1 centimeter	10^{-2}	1	10^{-5}	0.393 7	3.281×10^{-2}	6.214×10^{-6}
1 kilometer	10^3	10^5	1	3.937×10^4	3.281×10^3	0.621 4
1 inch	2.540×10^{-2}	2.540	2.540×10^{-5}	1	8.333×10^{-2}	1.578×10^{-5}
1 foot	0.304 8	30.48	3.048×10^{-4}	12	1	1.894×10^{-4}
1 mile	1 609	1.609×10^5	1.609	6.336×10^4	5 280	1

Mass

	kg	g	slug	u
1 kilogram	1	10^3	6.852×10^{-2}	6.024×10^{26}
1 gram	10^{-3}	1	6.852×10^{-5}	6.024×10^{23}
1 slug	14.59	1.459×10^4	1	8.789×10^{27}
1 atomic mass unit	1.660×10^{-27}	1.660×10^{-24}	1.137×10^{-28}	1

Note: 1 metric ton = 1 000 kg.

Time

	s	min	h	day	yr
1 second	1	1.667×10^{-2}	2.778×10^{-4}	1.157×10^{-5}	3.169×10^{-8}
1 minute	60	1	1.667×10^{-2}	6.994×10^{-4}	1.901×10^{-6}
1 hour	3 600	60	1	4.167×10^{-2}	1.141×10^{-4}
1 day	8.640×10^4	1 440	24	1	2.738×10^{-5}
1 year	3.156×10^7	5.259×10^5	8.766×10^3	365.2	1

Speed

	m/s	cm/s	ft/s	mi/h
1 meter per second	1	10^2	3.281	2.237
1 centimeter per second	10^{-2}	1	3.281×10^{-2}	2.237×10^{-2}
1 foot per second	0.304 8	30.48	1	0.681 8
1 mile per hour	0.447 0	44.70	1.467	1

Note: 1 mi/min = 60 mi/h = 88 ft/s.

continued

TABLE A.1   *Continued*

Force

	N	dyn	lb
1 newton	1	10^5	0.224 8
1 pound	4.448	4.448×10^5	1

Work, Energy, Heat

	J	ft·lb
1 joule	1	0.737 6
1 ft·lb	1.356	1
1 eV	1.602×10^{-19}	1.182×10^{-19}
1 cal	4.186	3.087
1 Btu	1.055×10^3	7.779×10^2
1 kWh	3.600×10^6	2.655×10^6

	eV	cal	Btu	kWh
1 joule	6.242×10^{18}	0.238 9	9.481×10^{-4}	2.778×10^{-7}
1 erg	6.242×10^{11}	2.389×10^{-8}	9.481×10^{-11}	2.778×10^{-14}
1 ft·lb	8.464×10^{18}	0.323 9	1.285×10^{-3}	3.766×10^{-7}
1 eV	1	3.827×10^{-20}	1.519×10^{-22}	4.450×10^{-26}
1 cal	2.613×10^{19}	1	3.968×10^{-3}	1.163×10^{-6}
1 Btu	6.585×10^{21}	2.520×10^2	1	2.930×10^{-4}
1 kWh	2.247×10^{25}	8.601×10^5	3.413×10^2	1

Pressure

	Pa	atm
1 pascal	1	9.869×10^{-6}
1 atmosphere	1.013×10^5	1
1 centimeter mercury[a]	1.333×10^3	1.316×10^{-2}
1 pound per inch^2	6.895×10^3	6.805×10^{-2}
1 pound per foot^2	47.88	4.725×10^{-4}

	cm Hg	lb/in.^{2}	lb/ft^2
1 newton per meter^2	7.501×10^{-4}	1.450×10^{-4}	2.089×10^{-2}
1 atmosphere	76	14.70	2.116×10^3
1 centimeter mercury[a]	1	0.194 3	27.85
1 pound per inch^2	5.171	1	144
1 pound per foot^2	3.591×10^{-2}	6.944×10^{-3}	1

[a] At 0°C and at a location where the acceleration due to gravity has its "standard" value, 9.806 65 m/s^2.

TABLE A.2　Symbols, Dimensions, and Units of Physical Quantities

Quantity	Common Symbol	Unit[a]	Dimensions[b]	Unit in Terms of Base SI Units
Acceleration	$\mathbf{a}$	m/s^2	L/T^2	m/s^2
Amount of substance	n	mole		mol
Angle	θ, ϕ	radian (rad)	1	
Angular acceleration	$\boldsymbol{\alpha}$	rad/s^2	T^{-2}	s^{-2}
Angular frequency	ω	rad/s	T^{-1}	s^{-1}
Angular momentum	$\mathbf{L}$	$kg \cdot m^2/s$	ML^2/T	$kg \cdot m^2/s$
Angular velocity	$\boldsymbol{\omega}$	rad/s	T^{-1}	s^{-1}
Area	A	m^2	L^2	m^2
Atomic number	Z			
Capacitance	C	farad (F)	Q^2T^2/ML^2	$A^2 \cdot s^4/kg \cdot m^2$
Charge	q, Q, e	coulomb (C)	Q	$A \cdot s$
Charge density				
Line	λ	C/m	Q/L	$A \cdot s/m$
Surface	σ	C/m^2	Q/L^2	$A \cdot s/m^2$
Volume	ρ	C/m^3	Q/L^3	$A \cdot s/m^3$
Conductivity	σ	$1/\Omega \cdot m$	Q^2T/ML^3	$A^2 \cdot s^3/kg \cdot m^3$
Current	I	AMPERE	Q/T	A
Current density	$\mathbf{J}$	A/m^2	Q/T^2	A/m^2
Density	ρ	kg/m^3	M/L^3	kg/m^3
Dielectric constant	κ			
Displacement	$\mathbf{r, s}$	METER	L	m
Distance	d, h			
Length	ℓ, L			
Electric dipole moment	$\mathbf{p}$	$C \cdot m$	QL	$A \cdot s \cdot m$
Electric field	$\mathbf{E}$	V/m	ML/QT^2	$kg \cdot m/A \cdot s^3$
Electric flux	Φ_E	$V \cdot m$	ML^3/QT^2	$kg \cdot m^3/A \cdot s^3$
Electromotive force	$\boldsymbol{\varepsilon}$	volt (V)	ML^2/QT^2	$kg \cdot m^2/A \cdot s^3$
Energy	E, U, K	joule (J)	ML^2/T^2	$kg \cdot m^2/s^2$
Entropy	S	J/K	$ML^2/T^2 \cdot K$	$kg \cdot m^2/s^2 \cdot K$
Force	$\mathbf{F}$	newton (N)	ML/T^2	$kg \cdot m/s^2$
Frequency	f	hertz (Hz)	T^{-1}	s^{-1}
Heat	Q	joule (J)	ML^2/T^2	$kg \cdot m^2/s^2$
Inductance	L	henry (H)	ML^2/Q^2	$kg \cdot m^2/A^2 \cdot s^2$
Magnetic dipole moment	$\boldsymbol{\mu}$	$N \cdot m/T$	QL^2/T	$A \cdot m^2$
Magnetic field	$\mathbf{B}$	tesla (T) $(= Wb/m^2)$	M/QT	$kg/A \cdot s^2$
Magnetic flux	Φ_B	weber (Wb)	ML^2/QT	$kg \cdot m^2/A \cdot s^2$
Mass	m, M	KILOGRAM	M	kg
Molar specific heat	C	$J/mol \cdot K$		$kg \cdot m^2/s^2 \cdot mol \cdot K$
Moment of inertia	I	$kg \cdot m^2$	ML^2	$kg \cdot m^2$
Momentum	$\mathbf{p}$	$kg \cdot m/s$	ML/T	$kg \cdot m/s$
Period	T	s	T	s
Permeability of space	μ_0	$N/A^2 (= H/m)$	ML/Q^2T	$kg \cdot m/A^2 \cdot s^2$
Permittivity of space	ϵ_0	$C^2/N \cdot m^2 (= F/m)$	Q^2T^2/ML^3	$A^2 \cdot s^4/kg \cdot m^3$
Potential	V	volt (V) $(= J/C)$	ML^2/QT^2	$kg \cdot m^2/A \cdot s^3$
Power	$\mathcal{P}$	watt (W) $(= J/s)$	ML^2/T^3	$kg \cdot m^2/s^3$

continued

TABLE A.2   Continued

Quantity	Common Symbol	Unit[a]	Dimensions[b]	Unit in Terms of Base SI Units
Pressure	P	pascal (Pa) $= (N/m^2)$	M/LT^2	$kg/m \cdot s^2$
Resistance	R	ohm $(\Omega)(=V/A)$	ML^2/Q^2T	$kg \cdot m^2/A^2 \cdot s^3$
Specific heat	c	$J/kg \cdot K$	$L^2/T^2 \cdot K$	$m^2/s^2 \cdot K$
Speed	v	m/s	L/T	m/s
Temperature	T	KELVIN	K	K
Time	t	SECOND	T	s
Torque	τ	$N \cdot m$	ML^2/T^2	$kg \cdot m^2/s^2$
Volume	V	m^3	L^3	m^3
Wavelength	λ	m	L	m
Work	W	joule $(J)(=N \cdot m)$	ML^2/T^2	$kg \cdot m^2/s^2$

[a] The base SI units are given in uppercase letters.

[b] The symbols M, L, T, and Q denote mass, length, time, and charge, respectively.

TABLE A.3   Table of Atomic Masses[a]

Atomic Number Z	Element	Symbol	Chemical Atomic Mass (u)	Mass Number (* Indicates Radioactive) A	Atomic Mass (u)	Percent Abundance	Half-Life (If Radioactive) $T_{1/2}$
0	(Neutron)	n		1*	1.008 665		10.4 min
1	Hydrogen	H	1.007 9	1	1.007 825	99.985	
	Deuterium	D		2	2.014 102	0.015	
	Tritium	T		3*	3.016 049		12.33 yr
2	Helium	He	4.002 60	3	3.016 029	0.000 14	
				4	4.002 602	99.999 86	
				6*	6.018 886		0.81 s
3	Lithium	Li	6.941	6	6.015 121	7.5	
				7	7.016 003	9.25	
				8*	8.022 486		0.84 s
4	Beryllium	Be	9.012 2	7*	7.016 928		53.3 days
				9	9.012 174	100	
				10*	10.013 534		1.5×10^6 yr
5	Boron	B	10.81	10	10.012 936	19.9	
				11	11.009 305	80.1	
				12*	12.014 352		0.020 2 s
6	Carbon	C	12.011	10*	10.016 854		19.3 s
				11*	11.011 433		20.4 min
				12	12.000 000	98.90	
				13	13.003 355	1.10	
				14*	14.003 242		5 730 yr
				15*	15.010 599		2.45 s
7	Nitrogen	N	14.006 7	12*	12.018 613		0.011 0 s
				13*	13.005 738		9.96 min
				14	14.003 074	99.63	
				15	15.000 108	0.37	
				16*	16.006 100		7.13 s
				17*	17.008 450		4.17 s

TABLE A.3   Continued

Atomic Number Z	Element	Symbol	Chemical Atomic Mass (u)	Mass Number (* Indicates Radioactive) A	Atomic Mass (u)	Percent Abundance	Half-Life (If Radioactive) $T_{1/2}$
8	Oxygen	O	15.999 4	14*	14.008 595		70.6 s
				15*	15.003 065		122 s
				16	15.994 915	99.761	
				17	16.999 132	0.039	
				18	17.999 160	0.20	
				19*	19.003 577		26.9 s
9	Fluorine	F	18.998 40	17*	17.002 094		64.5 s
				18*	18.000 937		109.8 min
				19	18.998 404	100	
				20*	19.999 982		11.0 s
				21*	20.999 950		4.2 s
10	Neon	Ne	20,180	18*	18.005 710		1.67 s
				19*	19.001 880		17.2 s
				20	19.992 435	90.48	
				21	20.993 841	0.27	
				22	21.991 383	9.25	
				23*	22.994 465		37.2 s
11	Sodium	Na	22.989 87	21*	20.997 650		22.5 s
				22*	21.994 434		2.61 yr
				23	22.989 770	100	
				24*	23.990 961		14.96 h
12	Magnesium	Mg	24.305	23*	22.994 124		11.3 s
				24	23.985 042	78.99	
				25	24.985 838	10.00	
				26	25.982 594	11.01	
				27*	26.984 341		9.46 min
13	Aluminum	Al	26.981 54	26*	25.986 892		7.4×10^5 yr
				27	26.981 538	100	
				28*	27.981 910		2.24 min
14	Silicon	Si	28.086	28	27.976 927	92.23	
				29	28.976 495	4.67	
				30	29.973 770	3.10	
				31*	30.975 362		2.62 h
				32*	31.974 148		172 yr
15	Phosphorus	P	30.973 76	30*	29.978 307		2.50 min
				31	30.973 762	100	
				32*	31.973 908		14.26 days
				33*	32.971 725		25.3 days
16	Sulfur	S	32.066	32	31.972 071	95.02	
				33	32.971 459	0.75	
				34	33.967 867	4.21	
				35*	34.969 033		87.5 days
				36	35.967 081	0.02	
17	Chlorine	Cl	35.453	35	34.968 853	75.77	
				36*	35.968 307		3.0×10^5 yr
				37	36.965 903	24.23	

continued

TABLE A.3 Continued

Atomic Number Z	Element	Symbol	Chemical Atomic Mass (u)	Mass Number (* Indicates Radioactive) A	Atomic Mass (u)	Percent Abundance	Half-Life (If Radioactive) $T_{1/2}$
18	Argon	Ar	39.948	36	35.967 547	0.337	
				37*	36.966 776		35.04 days
				38	37.962 732	0.063	
				39*	38.964 314		269 yr
				40	39.962 384	99.600	
				42*	41.963 049		33 yr
19	Potassium	K	39.098 3	39	38.963 708	93.258 1	
				40*	39.964 000	0.011 7	1.28×10^9 yr
				41	40.961 827	6.730 2	
20	Calcium	Ca	40.08	40	39.962 591	96.941	
				41*	40.962 279		1.0×10^5 yr
				42	41.958 618	0.647	
				43	42.958 767	0.135	
				44	43.955 481	2.086	
				46	45.953 687	0.004	
				48	47.952 534	0.187	
21	Scandium	Sc	44.955 9	41*	40.969 250		0.596 s
				45	44.955 911	100	
22	Titanium	Ti	47.88	44*	43.959 691		49 yr
				46	45.952 630	8.0	
				47	46.951 765	7.3	
				48	47.947 947	73.8	
				49	48.947 871	5.5	
				50	49.944 792	5.4	
23	Vanadium	V	50.941 5	48*	47.952 255		15.97 days
				50*	49.947 161	0.25	1.5×10^{17} yr
				51	50.943 962	99.75	
24	Chromium	Cr	51.996	48*	47.954 033		21.6 h
				50	49.946 047	4.345	
				52	51.940 511	83.79	
				53	52.940 652	9.50	
				54	53.938 883	2.365	
25	Manganese	Mn	54.938 05	54*	53.940 361		312.1 days
				55	54.938 048	100	
26	Iron	Fe	55.847	54	53.939 613	5.9	
				55*	54.938 297		2.7 yr
				56	55.934 940	91.72	
				57	56.935 396	2.1	
				58	57.933 278	0.28	
				60*	59.934 078		1.5×10^6 yr
27	Cobalt	Co	58.933 20	59	58.933 198	100	
				60*	59.933 820		5.27 yr
28	Nickel	Ni	58.693	58	57.935 346	68.077	
				59*	58.934 350		7.5×10^4 yr
				60	59.930 789	26.223	
				61	60.931 058	1.140	
				62	61.928 346	3.634	
				63*	62.929 670		100 yr
				64	63.927 967	0.926	

TABLE A.3 *Continued*

Atomic Number Z	Element	Symbol	Chemical Atomic Mass (u)	Mass Number (* Indicates Radioactive) A	Atomic Mass (u)	Percent Abundance	Half-Life (If Radioactive) $T_{1/2}$
29	Copper	Cu	63.54	63	62.929 599	69.17	
				65	64.927 791	30.83	
30	Zinc	Zn	65.39	64	63.929 144	48.6	
				66	65.926 035	27.9	
				67	66.927 129	4.1	
				68	67.924 845	18.8	
				70	69.925 323	0.6	
31	Gallium	Ga	69.723	69	68.925 580	60.108	
				71	70.924 703	39.892	
32	Germanium	Ge	72.61	70	69.924 250	21.23	
				72	71.922 079	27.66	
				73	72.923 462	7.73	
				74	73.921 177	35.94	
				76	75.921 402	7.44	
33	Arsenic	As	74.921 6	75	74.921 594	100	
34	Selenium	Se	78.96	74	73.922 474	0.89	
				76	75.919 212	9.36	
				77	76.919 913	7.63	
				78	77.917 307	23.78	
				79*	78.918 497		$\leqslant 6.5 \times 10^4$ yr
				80	79.916 519	49.61	
				82*	81.916 697	8.73	1.4×10^{20} yr
35	Bromine	Br	79.904	79	78.918 336	50.69	
				81	80.916 287	49.31	
36	Krypton	Kr	83.80	78	77.920 400	0.35	
				80	79.916 377	2.25	
				81*	80.916 589		2.1×10^5 yr
				82	81.913 481	11.6	
				83	82.914 136	11.5	
				84	83.911 508	57.0	
				85*	84.912 531		10.76 yr
				86	85.910 615	17.3	
37	Rubidium	Rb	85.468	85	84.911 793	72.17	
				87*	86.909 186	27.83	4.75×10^{10} yr
38	Strontium	Sr	87.62	84	83.913 428	0.56	
				86	85.909 266	9.86	
				87	86.908 883	7.00	
				88	87.905 618	82.58	
				90*	89.907 737		29.1 yr
39	Yttrium	Y	88.905 8	89	88.905 847	100	
40	Zirconium	Zr	91.224	90	89.904 702	51.45	
				91	90.905 643	11.22	
				92	91.905 038	17.15	
				93*	92.906 473		1.5×10^6 yr
				94	93.906 314	17.38	
				96	95.908 274	2.80	

continued

TABLE A.3 *Continued*

Atomic Number Z	Element	Symbol	Chemical Atomic Mass (u)	Mass Number (* Indicates Radioactive) A	Atomic Mass (u)	Percent Abundance	Half-Life (If Radioactive) $T_{1/2}$
41	Niobium	Nb	92.906 4	91*	90.906 988		6.8×10^2 yr
				92*	91.907 191		3.5×10^7 yr
				93	92.906 376	100	
				94*	93.907 280		2×10^4 yr
42	Molybdenum	Mo	95.94	92	91.906 807	14.84	
				93*	92.906 811		3.5×10^3 yr
				94	93.905 085	9.25	
				95	94.905 841	15.92	
				96	95.904 678	16.68	
				97	96.906 020	9.55	
				98	97.905 407	24.13	
				100	99.907 476	9.63	
43	Technetium	Tc		97*	96.906 363		2.6×10^6 yr
				98*	97.907 215		4.2×10^6 yr
				99*	98.906 254		2.1×10^5 yr
44	Ruthenium	Ru	101.07	96	95.907 597	5.54	
				98	97.905 287	1.86	
				99	98.905 939	12.7	
				100	99.904 219	12.6	
				101	100.905 558	17.1	
				102	101.904 348	31.6	
				104	103.905 428	18.6	
45	Rhodium	Rh	102.905 5	103	102.905 502	100	
46	Palladium	Pd	106.42	102	101.905 616	1.02	
				104	103.904 033	11.14	
				105	104.905 082	22.33	
				106	105.903 481	27.33	
				107*	106.905 126		6.5×10^6 yr
				108	107.903 893	26.46	
				110	109.905 158	11.72	
47	Silver	Ag	107.868	107	106.905 091	51.84	
				109	108.904 754	48.16	
48	Cadmium	Cd	112.41	106	105.906 457	1.25	
				108	107.904 183	0.89	
				109*	108.904 984		462 days
				110	109.903 004	12.49	
				111	110.904 182	12.80	
				112	111.902 760	24.13	
				113*	112.904 401	12.22	9.3×10^{15} yr
				114	113.903 359	28.73	
				116	115.904 755	7.49	
49	Indium	In	114.82	113	112.904 060	4.3	
				115*	114.903 876	95.7	4.4×10^{14} yr
50	Tin	Sn	118.71	112	111.904 822	0.97	
				114	113.902 780	0.65	
				115	114.903 345	0.36	
				116	115.901 743	14.53	
				117	116.902 953	7.68	

TABLE A.3	*Continued*						

Atomic Number Z	Element	Symbol	Chemical Atomic Mass (u)	Mass Number (* Indicates Radioactive) A	Atomic Mass (u)	Percent Abundance	Half-Life (If Radioactive) $T_{1/2}$
(50)	(Tin)			118	117.901 605	24.22	
				119	118.903 308	8.58	
				120	119.902 197	32.59	
				121*	120.904 237		55 yr
				122	121.903 439	4.63	
				124	123.905 274	5.79	
51	Antimony	Sb	121.76	121	120.903 820	57.36	
				123	122.904 215	42.64	
				125*	124.905 251		2.7 yr
52	Tellurium	Te	127.60	120	119.904 040	0.095	
				122	121.903 052	2.59	
				123*	122.904 271	0.905	1.3×10^{13} yr
				124	123.902 817	4.79	
				125	124.904 429	7.12	
				126	125.903 309	18.93	
				128*	127.904 463	31.70	$> 8 \times 10^{24}$ yr
				130*	129.906 228	33.87	$\leq 1.25 \times 10^{21}$ yr
53	Iodine	I	126.904 5	127	126.904 474	100	
				129*	128.904 984		1.6×10^{7} yr
54	Xenon	Xe	131.29	124	123.905 894	0.10	
				126	125.904 268	0.09	
				128	127.903 531	1.91	
				129	128.904 779	26.4	
				130	129.903 509	4.1	
				131	130.905 069	21.2	
				132	131.904 141	26.9	
				134	133.905 394	10.4	
				136*	135.907 215	8.9	$\geq 2.36 \times 10^{21}$ yr
55	Cesium	Cs	132.905 4	133	132.905 436	100	
				134*	133.906 703		2.1 yr
				135*	134.905 891		2×10^{6} yr
				137*	136.907 078		30 yr
56	Barium	Ba	137.33	130	129.906 289	0.106	
				132	131.905 048	0.101	
				133*	132.905 990		10.5 yr
				134	133.904 492	2.42	
				135	134.905 671	6.593	
				136	135.904 559	7.85	
				137	136.905 816	11.23	
				138	137.905 236	71.70	
57	Lanthanum	La	138.905	137*	136.906 462		6×10^{4} yr
				138*	137.907 105	0.090 2	1.05×10^{11} yr
				139	138.906 346	99.909 8	
58	Cerium	Ce	140.12	136	135.907 139	0.19	
				138	137.905 986	0.25	
				140	139.905 434	88.43	
				142*	141.909 241	11.13	$> 5 \times 10^{16}$ yr
59	Praseodymium	Pr	140.907 6	141	140.907 647	100	

continued

TABLE A.3   Continued

Atomic Number Z	Element	Symbol	Chemical Atomic Mass (u)	Mass Number (* Indicates Radioactive) A	Atomic Mass (u)	Percent Abundance	Half-Life (If Radioactive) $T_{1/2}$
60	Neodymium	Nd	144.24	142	141.907 718	27.13	
				143	142.909 809	12.18	
				144*	143.910 082	23.80	2.3×10^{15} yr
				145	144.912 568	8.30	
				146	145.913 113	17.19	
				148	147.916 888	5.76	
				150*	149.920 887	5.64	$> 1 \times 10^{18}$ yr
61	Promethium	Pm		143*	142.910 928		265 days
				145*	144.912 745		17.7 yr
				146*	145.914 698		5.5 yr
				147*	146.915 134		2.623 yr
62	Samarium	Sm	150.36	144	143.911 996	3.1	
				146*	145.913 043		1.0×10^{8} yr
				147*	146.914 894	15.0	1.06×10^{11} yr
				148*	147.914 819	11.3	7×10^{15} yr
				149*	148.917 180	13.8	$> 2 \times 10^{15}$ yr
				150	149.917 273	7.4	
				151*	150.919 928		90 yr
				152	151.919 728	26.7	
				154	153.922 206	22.7	
63	Europium	Eu	151.96	151	150.919 846	47.8	
				152*	151.921 740		13.5 yr
				153	152.921 226	52.2	
				154*	153.922 975		8.59 yr
				155*	154.922 888		4.7 yr
64	Gadolinium	Gd	157.25	148*	147.918 112		75 yr
				150*	149.918 657		1.8×10^{6} yr
				152*	151.919 787	0.20	1.1×10^{14} yr
				154	153.920 862	2.18	
				155	154.922 618	14.80	
				156	155.922 119	20.47	
				157	156.923 957	15.65	
				158	157.924 099	24.84	
				160	159.927 050	21.86	
65	Terbium	Tb	158.925 3	159	158.925 345	100	
66	Dysprosium	Dy	162.50	156	155.924 277	0.06	
				158	157.924 403	0.10	
				160	159.925 193	2.34	
				161	160.926 930	18.9	
				162	161.926 796	25.5	
				163	162.928 729	24.9	
				164	163.929 172	28.2	
67	Holmium	Ho	164.930 3	165	164.930 316	100	
				166*	165.932 282		1.2×10^{3} yr
68	Erbium	Er	167.26	162	161.928 775	0.14	
				164	163.929 198	1.61	
				166	165.930 292	33.6	

TABLE A.3 *Continued*

Atomic Number Z	Element	Symbol	Chemical Atomic Mass (u)	Mass Number (* Indicates Radioactive) A	Atomic Mass (u)	Percent Abundance	Half-Life (If Radioactive) $T_{1/2}$
(68)	(Erbium)			167	166.932 047	22.95	
				168	167.932 369	27.8	
				170	169.935 462	14.9	
69	Thulium	Tm	168.934 2	169	168.934 213	100	
				171*	170.936 428		1.92 yr
70	Ytterbium	Yb	173.04	168	167.933 897	0.13	
				170	169.934 761	3.05	
				171	170.936 324	14.3	
				172	171.936 380	21.9	
				173	172.938 209	16.12	
				174	173.938 861	31.8	
				176	175.942 564	12.7	
71	Lutecium	Lu	174.967	173*	172.938 930		1.37 yr
				175	174.940 772	97.41	
				176*	175.942 679	2.59	3.78×10^{10} yr
72	Hafnium	Hf	178.49	174*	173.940 042	0.162	2.0×10^{15} yr
				176	175.941 404	5.206	
				177	176.943 218	18.606	
				178	177.943 697	27.297	
				179	178.945 813	13.629	
				180	179.946 547	35.100	
73	Tantalum	Ta	180.947 9	180	179.947 542	0.012	
				181	180.947 993	99.988	
74	Tungsten (Wolfram)	W	183.85	180	179.946 702	0.12	
				182	181.948 202	26.3	
				183	182.950 221	14.28	
				184	183.950 929	30.7	
				186	185.954 358	28.6	
75	Rhenium	Re	186.207	185	184.952 951	37.40	
				187*	186.955 746	62.60	4.4×10^{10} yr
76	Osmium	Os	190.2	184	183.952 486	0.02	
				186*	185.953 834	1.58	2.0×10^{15} yr
				187	186.955 744	1.6	
				188	187.955 832	13.3	
				189	188.958 139	16.1	
				190	189.958 439	26.4	
				192	191.961 468	41.0	
				194*	193.965 172		6.0 yr
77	Iridium	Ir	192.2	191	190.960 585	37.3	
				193	192.962 916	62.7	
78	Platinum	Pt	195.08	190*	189.959 926	0.01	6.5×10^{11} yr
				192	191.961 027	0.79	
				194	193.962 655	32.9	
				195	194.964 765	33.8	
				196	195.964 926	25.3	
				198	197.967 867	7.2	
79	Gold	Au	196.966 5	197	196.966 543	100	

continued

TABLE A.3   *Continued*

Atomic Number Z	Element	Symbol	Chemical Atomic Mass (u)	Mass Number (* Indicates Radioactive) A	Atomic Mass (u)	Percent Abundance	Half-Life (If Radioactive) $T_{1/2}$
80	Mercury	Hg	200.59	196	195.965 806	0.15	
				198	197.966 743	9.97	
				199	198.968 253	16.87	
				200	199.968 299	23.10	
				201	200.970 276	13.10	
				202	201.970 617	29.86	
				204	203.973 466	6.87	
81	Thallium	Tl	204.383	203	202.972 320	29.524	
				204*	203.973 839		3.78 yr
				205	204.974 400	70.476	
		(Ra E″)		206*	205.976 084		4.2 min
		(Ac C″)		207*	206.977 403		4.77 min
		(Th C″)		208*	207.981 992		3.053 min
		(Ra C″)		210*	209.990 057		1.30 min
82	Lead	Pb	207.2	202*	201.972 134		5×10^4 yr
				204*	203.973 020	1.4	$\geq 1.4 \times 10^{17}$ yr
				205*	204.974 457		1.5×10^7 yr
				206	205.974 440	24.1	
				207	206.975 871	22.1	
				208	207.976 627	52.4	
		(Ra D)		210*	209.984 163		22.3 yr
		(Ac B)		211*	210.988 734		36.1 min
		(Th B)		212*	211.991 872		10.64 h
		(Ra B)		214*	213.999 798		26.8 min
83	Bismuth	Bi	208.980 3	207*	206.978 444		32.2 yr
				208*	207.979 717		3.7×10^5 yr
				209	208.980 374	100	
		(Ra E)		210*	209.984 096		5.01 days
		(Th C)		211*	210.987 254		2.14 min
				212*	211.991 259		60.6 min
		(Ra C)		214*	213.998 692		19.9 min
				215*	215.001 836		7.4 min
84	Polonium	Po		209*	208.982 405		102 yr
		(Ra F)		210*	209.982 848		138.38 days
		(Ac C′)		211*	210.986 627		0.52 s
		(Th C′)		212*	211.988 842		0.30 μs
		(Ra C′)		214*	213.995 177		164 μs
		(Ac A)		215*	214.999 418		0.001 8 s
		(Th A)		216*	216.001 889		0.145 s
		(Ra A)		218*	218.008 965		3.10 min
85	Astatine	At		215*	214.998 638		≈ 100 μs
				218*	218.008 685		1.6 s
				219*	219.011 294		0.9 min
86	Radon	Rn					
		(An)		219*	219.009 477		3.96 s
		(Tn)		220*	220.011 369		55.6 s
		(Rn)		222*	222.017 571		3.823 days
87	Francium	Fr					
		(Ac K)		223*	223.019 733		22 min

TABLE A.3   Continued

Atomic Number Z	Element	Symbol	Chemical Atomic Mass (u)	Mass Number (* Indicates Radioactive) A	Atomic Mass (u)	Percent Abundance	Half-Life (If Radioactive) $T_{1/2}$
88	Radium	Ra					
		(Ac X)		223*	223.018 499		11.43 days
		(Th X)		224*	224.020 187		3.66 days
		(Ra)		226*	226.025 402		1 600 yr
		(Ms Th$_1$)		228*	228.031 064		5.75 yr
89	Actinium	Ac		227*	227.027 749		21.77 yr
		(Ms Th$_2$)		228*	228.031 015		6.15 h
90	Thorium	Th	232.038 1				
		(Rd Ac)		227*	227.027 701		18.72 days
		(Rd Th)		228*	228.028 716		1.913 yr
				229*	229.031 757		7 300 yr
		(Io)		230*	230.033 127		75.000 yr
		(UY)		231*	231.036 299		25.52 h
		(Th)		232*	232.038 051	100	1.40×10^{10} yr
		(UX$_1$)		234*	234.043 593		24.1 days
91	Protactinium	Pa		231*	231.035 880		32.760 yr
		(Uz)		234*	234.043 300		6.7 h
92	Uranium	U	238.028 9	232*	232.037 131		69 yr
				233*	233.039 630		1.59×10^5 yr
				234*	234.040 946	0.005 5	2.45×10^5 yr
		(Ac U)		235*	235.043 924	0.720	7.04×10^8 yr
				236*	236.045 562		2.34×10^7 yr
		(UI)		238*	238.050 784	99.274 5	4.47×10^9 yr
93	Neptunium	Np		235*	235.044 057		396 days
				236*	236.046 560		1.15×10^5 yr
				237*	237.048 168		2.14×10^6 yr
94	Plutonium	Pu		236*	236.046 033		2.87 yr
				238*	238.049 555		87.7 yr
				239*	239.052 157		2.412×10^4 yr
				240*	240.053 808		6 560 yr
				241*	241.056 846		14.4 yr
				242*	242.058 737		3.73×10^6 yr
				244*	244.064 200		8.1×10^7 yr

[a] The masses in the sixth column are atomic masses, which include the mass of Z electrons. Data are from the National Nuclear Data Center, Brookhaven National Laboratory, prepared by Jagdish K. Tuli, July 1990. The data are based on experimental results reported in *Nuclear Data Sheets* and *Nuclear Physics* and also from *Chart of the Nuclides,* 14th ed. Atomic masses are based on those by A. H. Wapstra, G. Audi, and R. Hoekstra. Isotopic abundances are based on those by N. E. Holden.

These appendices in mathematics are intended as a brief review of operations and methods. Early in this course, you should be totally familiar with basic algebraic techniques, analytic geometry, and trigonometry. The appendices on differential and integral calculus are more detailed and are intended for those students who have difficulty applying calculus concepts to physical situations.

B.1 ► SCIENTIFIC NOTATION

Many quantities that scientists deal with often have very large or very small values. For example, the speed of light is about 300 000 000 m/s, and the ink required to make the dot over an i in this textbook has a mass of about 0.000 000 001 kg. Obviously, it is very cumbersome to read, write, and keep track of numbers such as these. We avoid this problem by using a method dealing with powers of the number 10:

$$10^0 = 1$$

$$10^1 = 10$$

$$10^2 = 10 \times 10 = 100$$

$$10^3 = 10 \times 10 \times 10 = 1000$$

$$10^4 = 10 \times 10 \times 10 \times 10 = 10\ 000$$

$$10^5 = 10 \times 10 \times 10 \times 10 \times 10 = 100\ 000$$

and so on. The number of zeros corresponds to the power to which 10 is raised, called the **exponent** of 10. For example, the speed of light, 300 000 000 m/s, can be expressed as 3×10^8 m/s.

In this method, some representative numbers smaller than unity are

$$10^{-1} = \frac{1}{10} = 0.1$$

$$10^{-2} = \frac{1}{10 \times 10} = 0.01$$

$$10^{-3} = \frac{1}{10 \times 10 \times 10} = 0.001$$

$$10^{-4} = \frac{1}{10 \times 10 \times 10 \times 10} = 0.000\ 1$$

$$10^{-5} = \frac{1}{10 \times 10 \times 10 \times 10 \times 10} = 0.000\ 01$$

In these cases, the number of places the decimal point is to the left of the digit 1 equals the value of the (negative) exponent. Numbers expressed as some power of 10 multiplied by another number between 1 and 10 are said to be in **scientific notation.** For example, the scientific notation for 5 943 000 000 is 5.943×10^9 and that for 0.000 083 2 is 8.32×10^{-5}.

When numbers expressed in scientific notation are being multiplied, the following general rule is very useful:

$$10^n \times 10^m = 10^{n+m} \qquad \text{(B.1)}$$

where n and m can be *any* numbers (not necessarily integers). For example, $10^2 \times 10^5 = 10^7$. The rule also applies if one of the exponents is negative: $10^3 \times 10^{-8} = 10^{-5}$.

When dividing numbers expressed in scientific notation, note that

$$\frac{10^n}{10^m} = 10^n \times 10^{-m} = 10^{n-m} \qquad \text{(B.2)}$$

EXERCISES

With help from the above rules, verify the answers to the following:

1. $86\ 400 = 8.64 \times 10^4$
2. $9\ 816\ 762.5 = 9.816\ 762\ 5 \times 10^6$
3. $0.000\ 000\ 039\ 8 = 3.98 \times 10^{-8}$
4. $(4 \times 10^8)(9 \times 10^9) = 3.6 \times 10^{18}$
5. $(3 \times 10^7)(6 \times 10^{-12}) = 1.8 \times 10^{-4}$
6. $\dfrac{75 \times 10^{-11}}{5 \times 10^{-3}} = 1.5 \times 10^{-7}$
7. $\dfrac{(3 \times 10^6)(8 \times 10^{-2})}{(2 \times 10^{17})(6 \times 10^5)} = 2 \times 10^{-18}$

B.2 ▶ ALGEBRA

Some Basic Rules

When algebraic operations are performed, the laws of arithmetic apply. Symbols such as x, y, and z are usually used to represent quantities that are not specified, what are called the **unknowns.**

First, consider the equation

$$8x = 32$$

If we wish to solve for x, we can divide (or multiply) each side of the equation by the same factor without destroying the equality. In this case, if we divide both sides by 8, we have

$$\frac{8x}{8} = \frac{32}{8}$$

$$x = 4$$

Next consider the equation

$$x + 2 = 8$$

In this type of expression, we can add or subtract the same quantity from each side. If we subtract 2 from each side, we get

$$x + 2 - 2 = 8 - 2$$
$$x = 6$$

In general, if $x + a = b$, then $x = b - a$.

Now consider the equation

$$\frac{x}{5} = 9$$

If we multiply each side by 5, we are left with x on the left by itself and 45 on the right:

$$\left(\frac{x}{5}\right)(5) = 9 \times 5$$
$$x = 45$$

In all cases, *whatever operation is performed on the left side of the equality must also be performed on the right side.*

The following rules for multiplying, dividing, adding, and subtracting fractions should be recalled, where a, b, and c are three numbers:

	Rule	**Example**
Multiplying	$\left(\dfrac{a}{b}\right)\left(\dfrac{c}{d}\right) = \dfrac{ac}{bd}$	$\left(\dfrac{2}{3}\right)\left(\dfrac{4}{5}\right) = \dfrac{8}{15}$
Dividing	$\dfrac{(a/b)}{(c/d)} = \dfrac{ad}{bc}$	$\dfrac{2/3}{4/5} = \dfrac{(2)(5)}{(4)(3)} = \dfrac{10}{12}$
Adding	$\dfrac{a}{b} \pm \dfrac{c}{d} = \dfrac{ad \pm bc}{bd}$	$\dfrac{2}{3} - \dfrac{4}{5} = \dfrac{(2)(5) - (4)(3)}{(3)(5)} = -\dfrac{2}{15}$

EXERCISES

In the following exercises, solve for x:

Answers

1. $a = \dfrac{1}{1 + x}$ $x = \dfrac{1 - a}{a}$

2. $3x - 5 = 13$ $x = 6$

3. $ax - 5 = bx + 2$ $x = \dfrac{7}{a - b}$

4. $\dfrac{5}{2x + 6} = \dfrac{3}{4x + 8}$ $x = -\dfrac{11}{7}$

Powers

When powers of a given quantity x are multiplied, the following rule applies:

$$x^n x^m = x^{n+m} \tag{B.3}$$

For example, $x^2x^4 = x^{2+4} = x^6$.

When dividing the powers of a given quantity, the rule is

$$\frac{x^n}{x^m} = x^{n-m} \tag{B.4}$$

For example, $x^8/x^2 = x^{8-2} = x^6$.

A power that is a fraction, such as $\frac{1}{3}$, corresponds to a root as follows:

$$x^{1/n} = \sqrt[n]{x} \tag{B.5}$$

For example, $4^{1/3} = \sqrt[3]{4} = 1.5874$. (A scientific calculator is useful for such calculations.)

Finally, any quantity x^n raised to the mth power is

$$(x^n)^m = x^{nm} \tag{B.6}$$

Table B.1 summarizes the rules of exponents.

TABLE B.1
Rules of Exponents

$$x^0 = 1$$
$$x^1 = x$$
$$x^n x^m = x^{n+m}$$
$$x^n/x^m = x^{n-m}$$
$$x^{1/n} = \sqrt[n]{x}$$
$$(x^n)^m = x^{nm}$$

EXERCISES

Verify the following:

1. $3^2 \times 3^3 = 243$
2. $x^5 x^{-8} = x^{-3}$
3. $x^{10}/x^{-5} = x^{15}$
4. $5^{1/3} = 1.709\ 975$ \qquad (Use your calculator.)
5. $60^{1/4} = 2.783\ 158$ \qquad (Use your calculator.)
6. $(x^4)^3 = x^{12}$

Factoring

Some useful formulas for factoring an equation are

$$ax + ay + az = a(x + y + x) \qquad \text{common factor}$$
$$a^2 + 2ab + b^2 = (a + b)^2 \qquad \text{perfect square}$$
$$a^2 - b^2 = (a + b)(a - b) \qquad \text{differences of squares}$$

Quadratic Equations

The general form of a quadratic equation is

$$ax^2 + bx + c = 0 \tag{B.7}$$

where x is the unknown quantity and a, b, and c are numerical factors referred to as **coefficients** of the equation. This equation has two roots, given by

$$x = \frac{-b \pm \sqrt{b^2 - 4ac}}{2a} \tag{B.8}$$

If $b^2 \geq 4ac$, the roots are real.

EXAMPLE 1

The equation $x^2 + 5x + 4 = 0$ has the following roots corresponding to the two signs of the square-root term:

$$x = \frac{-5 \pm \sqrt{5^2 - (4)(1)(4)}}{2(1)} = \frac{-5 \pm \sqrt{9}}{2} = \frac{-5 \pm 3}{2}$$

$$x_+ = \frac{-5 + 3}{2} = \boxed{-1} \quad x_- = \frac{-5 - 3}{2} = \boxed{-4}$$

where x_+ refers to the root corresponding to the positive sign and x_- refers to the root corresponding to the negative sign.

EXERCISES

Solve the following quadratic equations:

Answers

1. $x^2 + 2x - 3 = 0$ $x_+ = 1$ $x_- = -3$
2. $2x^2 - 5x + 2 = 0$ $x_+ = 2$ $x_- = \frac{1}{2}$
3. $2x^2 - 4x - 9 = 0$ $x_+ = 1 + \sqrt{22}/2$ $x_- = 1 - \sqrt{22}/2$

Linear Equations

A linear equation has the general form

$$y = mx + b \tag{B.9}$$

where m and b are constants. This equation is referred to as being linear because the graph of y versus x is a straight line, as shown in Figure B.1. The constant b, called the **y-intercept,** represents the value of y at which the straight line intersects the y axis. The constant m is equal to the **slope** of the straight line and is also equal to the tangent of the angle that the line makes with the x axis. If any two points on the straight line are specified by the coordinates (x_1, y_1) and (x_2, y_2), as in Figure B.1, then the slope of the straight line can be expressed as

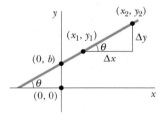

$$\text{Slope} = \frac{y_2 - y_1}{x_2 - x_1} = \frac{\Delta y}{\Delta x} = \tan \theta \tag{B.10}$$

Figure B.1

Note that m and b can have either positive or negative values. If $m > 0$, the straight line has a *positive* slope, as in Figure B1. If $m < 0$, the straight line has a *negative* slope. In Figure B.1, both m and b are positive. Three other possible situations are shown in Figure B.2.

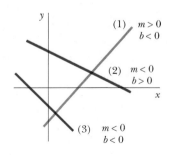

EXERCISES

1. Draw graphs of the following straight lines:
   (a) $y = 5x + 3$ (b) $y = -2x + 4$ (c) $y = -3x - 6$
2. Find the slopes of the straight lines described in Exercise 1.

Figure B.2

Answers   (a) 5   (b) -2 (c) -3

3. Find the slopes of the straight lines that pass through the following sets of points:
(a) $(0, -4)$ and $(4, 2)$, (b) $(0, 0)$ and $(2, -5)$, and (c) $(-5, 2)$ and $(4, -2)$

Answers (a) $3/2$ (b) $-5/2$ (c) $-4/9$

Solving Simultaneous Linear Equations

Consider the equation $3x + 5y = 15$, which has two unknowns, x and y. Such an equation does not have a unique solution. For example, note that $(x = 0, y = 3)$, $(x = 5, y = 0)$, and $(x = 2, y = 9/5)$ are all solutions to this equation.

If a problem has two unknowns, a unique solution is possible only if we have *two* equations. In general, if a problem has n unknowns, its solution requires n equations. In order to solve two simultaneous equations involving two unknowns, x and y, we solve one of the equations for x in terms of y and substitute this expression into the other equation.

EXAMPLE 2

Solve the following two simultaneous equations:

$$(1) \quad 5x + y = -8$$

$$(2) \quad 2x - 2y = 4$$

Solution From (2), $x = y + 2$. Substitution of this into (1) gives

$$5(y + 2) + y = -8$$

$$6y = -18$$

$$y = -3$$

$$x = y + 2 = \boxed{-1}$$

Alternate Solution Multiply each term in (1) by the factor 2 and add the result to (2):

$$10x + 2y = -16$$

$$\underline{2x - 2y = 4}$$

$$12x = -12$$

$$x = -1$$

$$y = x - 2 = \boxed{-3}$$

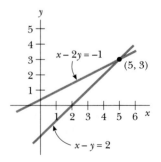

Figure B.3

Two linear equations containing two unknowns can also be solved by a graphical method. If the straight lines corresponding to the two equations are plotted in a conventional coordinate system, the intersection of the two lines represents the solution. For example, consider the two equations

$$x - y = 2$$

$$x - 2y = -1$$

These are plotted in Figure B.3. The intersection of the two lines has the coordinates $x = 5$, $y = 3$. This represents the solution to the equations. You should check this solution by the analytical technique discussed above.

EXERCISES

Solve the following pairs of simultaneous equations involving two unknowns:

Answers

1. $x + y = 8$ $x = 5, y = 3$
$\quad\; x - y = 2$

2. $98 - T = 10a$ $T = 65, a = 3.27$
   $T - 49 = 5a$
3. $6x + 2y = 6$ $x = 2, y = -3$
   $8x - 4y = 28$

Logarithms

Suppose that a quantity x is expressed as a power of some quantity a:

$$x = a^y \tag{B.11}$$

The number a is called the **base** number. The **logarithm** of x with respect to the base a is equal to the exponent to which the base must be raised in order to satisfy the expression $x = a^y$:

$$y = \log_a x \tag{B.12}$$

Conversely, the **antilogarithm** of y is the number x:

$$x = \mathrm{antilog}_a y \tag{B.13}$$

In practice, the two bases most often used are base 10, called the *common* logarithm base, and base $e = 2.718 \ldots$, called Euler's constant or the *natural* logarithm base. When common logarithms are used,

$$y = \log_{10} x \qquad (\text{or } x = 10^y) \tag{B.14}$$

When natural logarithms are used,

$$y = \ln_e x \qquad (\text{or } x = e^y) \tag{B.15}$$

For example, $\log_{10} 52 = 1.716$, so that $\mathrm{antilog}_{10} 1.716 = 10^{1.716} = 52$. Likewise, $\ln_e 52 = 3.951$, so $\mathrm{antiln}_e 3.951 = e^{3.951} = 52$.

In general, note that you can convert between base 10 and base e with the equality

$$\ln_e x = (2.302\ 585) \log_{10} x \tag{B.16}$$

Finally, some useful properties of logarithms are

$$\log(ab) = \log a + \log b$$
$$\log(a/b) = \log a - \log b$$
$$\log(a^n) = n \log a$$
$$\ln e = 1$$
$$\ln e^a = a$$
$$\ln\left(\frac{1}{a}\right) = -\ln a$$

B.3 GEOMETRY

The **distance** d between two points having coordinates (x_1, y_1) and (x_2, y_2) is

$$d = \sqrt{(x_2 - x_1)^2 + (y_2 - y_1)^2} \tag{B.17}$$

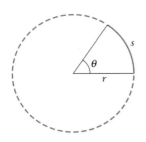

Figure B.4

Radian measure: The arc length s of a circular arc (Fig. B.4) is proportional to the radius r for a fixed value of θ (in radians):

$$s = r\theta$$

$$\theta = \frac{s}{r}$$

(B.18)

Table B.2 gives the areas and volumes for several geometric shapes used throughout this text:

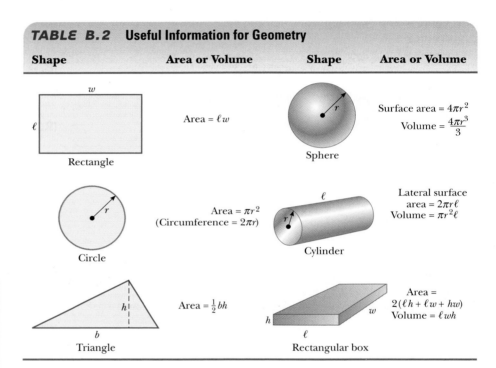

TABLE B.2   Useful Information for Geometry

Shape	Area or Volume	Shape	Area or Volume
Rectangle	Area = ℓw	Sphere	Surface area = $4\pi r^2$ Volume = $\frac{4\pi r^3}{3}$
Circle	Area = πr^2 (Circumference = $2\pi r$)	Cylinder	Lateral surface area = $2\pi r\ell$ Volume = $\pi r^2 \ell$
Triangle	Area = $\frac{1}{2}bh$	Rectangular box	Area = $2(\ell h + \ell w + hw)$ Volume = $\ell w h$

Figure B.5

The equation of a **straight line** (Fig. B.5) is

$$y = mx + b$$

(B.19)

where b is the y-intercept and m is the slope of the line.

The equation of a **circle** of radius R centered at the origin is

$$x^2 + y^2 = R^2$$

(B.20)

The equation of an **ellipse** having the origin at its center (Fig. B.6) is

$$\frac{x^2}{a^2} + \frac{y^2}{b^2} = 1$$

(B.21)

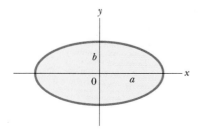

Figure B.6

where a is the length of the semi-major axis (the longer one) and b is the length of the semi-minor axis (the shorter one).

The equation of a **parabola** the vertex of which is at $y = b$ (Fig. B.7) is

$$y = ax^2 + b \tag{B.22}$$

The equation of a **rectangular hyperbola** (Fig. B.8) is

$$xy = \text{constant} \tag{B.23}$$

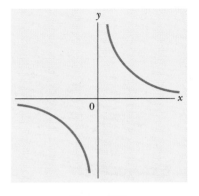

Figure B.7

B.4 ▸ TRIGONOMETRY

That portion of mathematics based on the special properties of the right triangle is called trigonometry. By definition, a right triangle is one containing a 90° angle. Consider the right triangle shown in Figure B.9, where side a is opposite the angle θ, side b is adjacent to the angle θ, and side c is the hypotenuse of the triangle. The three basic trigonometric functions defined by such a triangle are the sine (sin), cosine (cos), and tangent (tan) functions. In terms of the angle θ, these functions are defined by

$$\sin \theta \equiv \frac{\text{side opposite } \theta}{\text{hypotenuse}} = \frac{a}{c} \tag{B.24}$$

$$\cos \theta \equiv \frac{\text{side adjacent to } \theta}{\text{hypotenuse}} = \frac{b}{c} \tag{B.25}$$

$$\tan \theta \equiv \frac{\text{side opposite } \theta}{\text{side adjacent to } \theta} = \frac{a}{b} \tag{B.26}$$

The Pythagorean theorem provides the following relationship between the sides of a right triangle:

$$c^2 = a^2 + b^2 \tag{B.27}$$

From the above definitions and the Pythagorean theorem, it follows that

$$\sin^2 \theta + \cos^2 \theta = 1$$

$$\tan \theta = \frac{\sin \theta}{\cos \theta}$$

Figure B.8

The cosecant, secant, and cotangent functions are defined by

$$\csc \theta \equiv \frac{1}{\sin \theta} \qquad \sec \theta \equiv \frac{1}{\cos \theta} \qquad \cot \theta \equiv \frac{1}{\tan \theta}$$

The relationships below follow directly from the right triangle shown in Figure B.9:

$$\sin \theta = \cos(90° - \theta)$$

$$\cos \theta = \sin(90° - \theta)$$

$$\cot \theta = \tan(90° - \theta)$$

Some properties of trigonometric functions are

$$\sin(-\theta) = -\sin \theta$$

$$\cos(-\theta) = \cos \theta$$

$$\tan(-\theta) = -\tan \theta$$

The following relationships apply to *any* triangle, as shown in Figure B.10:

$$\alpha + \beta + \gamma = 180°$$

a = opposite side
b = adjacent side
c = hypotenuse

Figure B.9

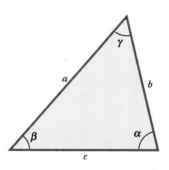

Figure B.10

$$\text{Law of cosines} \quad \begin{aligned} a^2 &= b^2 + c^2 - 2bc \cos \alpha \\ b^2 &= a^2 + c^2 - 2ac \cos \beta \\ c^2 &= a^2 + b^2 - 2ab \cos \gamma \end{aligned}$$

$$\text{Law of sines} \quad \frac{a}{\sin \alpha} = \frac{b}{\sin \beta} = \frac{c}{\sin \gamma}$$

Table B.3 lists a number of useful trigonometric identities.

TABLE B.3 Some Trigonometric Identities

$\sin^2 \theta + \cos^2 \theta = 1$	$\csc^2 \theta = 1 + \cot^2 \theta$
$\sec^2 \theta = 1 + \tan^2 \theta$	$\sin^2 \dfrac{\theta}{2} = \frac{1}{2}(1 - \cos \theta)$
$\sin 2\theta = 2 \sin \theta \cos \theta$	$\cos^2 \dfrac{\theta}{2} = \frac{1}{2}(1 + \cos \theta)$
$\cos 2\theta = \cos^2 \theta - \sin^2 \theta$	$1 - \cos \theta = 2 \sin^2 \dfrac{\theta}{2}$
$\tan 2\theta = \dfrac{2 \tan \theta}{1 - \tan^2 \theta}$	$\tan \dfrac{\theta}{2} = \sqrt{\dfrac{1 - \cos \theta}{1 + \cos \theta}}$

$$\sin(A \pm B) = \sin A \cos B \pm \cos A \sin B$$
$$\cos(A \pm B) = \cos A \cos B \mp \sin A \sin B$$
$$\sin A \pm \sin B = 2 \sin[\tfrac{1}{2}(A \pm B)]\cos[\tfrac{1}{2}(A \mp B)]$$
$$\cos A + \cos B = 2 \cos[\tfrac{1}{2}(A + B)]\cos[\tfrac{1}{2}(A - B)]$$
$$\cos A - \cos B = 2 \sin[\tfrac{1}{2}(A + B)]\sin[\tfrac{1}{2}(B - A)]$$

EXAMPLE 3

Consider the right triangle in Figure B.11, in which $a = 2$, $b = 5$, and c is unknown. From the Pythagorean theorem, we have

$$c^2 = a^2 + b^2 = 2^2 + 5^2 = 4 + 25 = 29$$

$$c = \sqrt{29} = \boxed{5.39}$$

To find the angle θ, note that

$$\tan \theta = \frac{a}{b} = \frac{2}{5} = 0.400$$

From a table of functions or from a calculator, we have

$$\theta = \tan^{-1}(0.400) = \boxed{21.8°}$$

where $\tan^{-1}(0.400)$ is the notation for "angle whose tangent is 0.400," sometimes written as arctan (0.400).

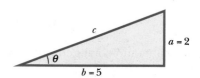

Figure B.11

EXERCISES

1. In Figure B.12, identify (a) the side opposite θ and (b) the side adjacent to ϕ and then find (c) $\cos \theta$, (d) $\sin \phi$, and (e) $\tan \phi$.

   Answers   (a) 3, (b) 3, (c) $\frac{4}{5}$, (d) $\frac{4}{5}$, and (e) $\frac{4}{3}$

2. In a certain right triangle, the two sides that are perpendicular to each other are 5 m and 7 m long. What is the length of the third side?

   Answer   8.60 m

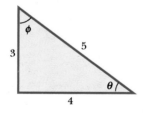

Figure B.12

3. A right triangle has a hypotenuse of length 3 m, and one of its angles is 30°. What is the length of (a) the side opposite the 30° angle and (b) the side adjacent to the 30° angle?

Answers (a) 1.5 m, (b) 2.60 m

B.5 ▶ SERIES EXPANSIONS

$$(a + b)^n = a^n + \frac{n}{1!} a^{n-1}b + \frac{n(n - 1)}{2!} a^{n-2}b^2 + \cdots$$

$$(1 + x)^n = 1 + nx + \frac{n(n - 1)}{2!} x^2 + \cdots$$

$$e^x = 1 + x + \frac{x^2}{2!} + \frac{x^3}{3!} + \cdots$$

$$\ln(1 \pm x) = \pm x - \tfrac{1}{2}x^2 \pm \tfrac{1}{3}x^3 - \cdots$$

$$\sin x = x - \frac{x^3}{3!} + \frac{x^5}{5!} - \cdots$$

$$\cos x = 1 - \frac{x^2}{2!} + \frac{x^4}{4!} - \cdots$$

$$\tan x = x + \frac{x^3}{3} + \frac{2x^5}{15} + \cdots \qquad |x| < \pi/2$$

x in radians

For $x \ll 1$, the following approximations can be used[1]:

$$(1 + x)^n \approx 1 + nx \qquad \sin x \approx x$$

$$e^x \approx 1 + x \qquad \cos x \approx 1$$

$$\ln(1 \pm x) \approx \pm x \qquad \tan x \approx x$$

B.6 ▶ DIFFERENTIAL CALCULUS

In various branches of science, it is sometimes necessary to use the basic tools of calculus, invented by Newton, to describe physical phenomena. The use of calculus is fundamental in the treatment of various problems in Newtonian mechanics, electricity, and magnetism. In this section, we simply state some basic properties and "rules of thumb" that should be a useful review to the student.

First, a **function** must be specified that relates one variable to another (such as a coordinate as a function of time). Suppose one of the variables is called y (the dependent variable), the other x (the independent variable). We might have a function relationship such as

$$y(x) = ax^3 + bx^2 + cx + d$$

If a, b, c, and d are specified constants, then y can be calculated for any value of x. We usually deal with continuous functions, that is, those for which y varies "smoothly" with x.

---

[1] The approximations for the functions $\sin x$, $\cos x$, and $\tan x$ are for $x \leq 0.1$ rad.

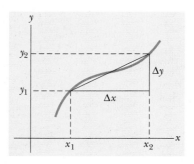

Figure B.13

The **derivative** of y with respect to x is defined as the limit, as Δx approaches zero, of the slopes of chords drawn between two points on the y versus x curve. Mathematically, we write this definition as

$$\frac{dy}{dx} = \lim_{\Delta x \to 0} \frac{\Delta y}{\Delta x} = \lim_{\Delta x \to 0} \frac{y(x + \Delta x) - y(x)}{\Delta x} \qquad \textbf{(B.28)}$$

where Δy and Δx are defined as $\Delta x = x_2 - x_1$ and $\Delta y = y_2 - y_1$ (Fig. B.13). It is important to note that dy/dx *does not* mean dy divided by dx, but is simply a notation of the limiting process of the derivative as defined by Equation B.28.

A useful expression to remember when $y(x) = ax^n$, where a is a *constant* and n is *any* positive or negative number (integer or fraction), is

$$\frac{dy}{dx} = nax^{n-1} \qquad \textbf{(B.29)}$$

If $y(x)$ is a polynomial or algebraic function of x, we apply Equation B.29 to *each* term in the polynomial and take $d[\text{constant}]/dx = 0$. In Examples 4 through 7, we evaluate the derivatives of several functions.

EXAMPLE 4

Suppose $y(x)$ (that is, y as a function of x) is given by

$$y(x) = ax^3 + bx + c$$

where a and b are constants. Then it follows that

$$y(x + \Delta x) = a(x + \Delta x)^3$$
$$+ \, b(x + \Delta x) + c$$
$$y(x + \Delta x) = a(x^3 + 3x^2\Delta x + 3x\Delta x^2 + \Delta x^3)$$
$$+ \, b(x + \Delta x) + c$$

so

$$\Delta y = y(x + \Delta x) - y(x) = a(3x^2\Delta x + 3x\Delta x^2 + \Delta x^3)$$
$$+ \, b\Delta x$$

Substituting this into Equation B.28 gives

$$\frac{dy}{dx} = \lim_{\Delta x \to 0} \frac{\Delta y}{\Delta x} = \lim_{\Delta x \to 0} [3ax^2 + 3x\Delta x + \Delta x^2] + b$$

$$\frac{dy}{dx} = 3ax^2 + b$$

EXAMPLE 5

$$y(x) = 8x^5 + 4x^3 + 2x + 7$$

Solution Applying Equation B.29 to each term independently, and remembering that d/dx (constant) $= 0$, we have

$$\frac{dy}{dx} = 8(5)x^4 + 4(3)x^2 + 2(1)x^0 + 0$$

$$\frac{dy}{dx} = 40x^4 + 12x^2 + 2$$

Special Properties of the Derivative

A. Derivative of the product of two functions If a function $f(x)$ is given by the product of two functions, say, $g(x)$ and $h(x)$, then the derivative of $f(x)$ is defined as

$$\frac{d}{dx} f(x) = \frac{d}{dx}[g(x)h(x)] = g\frac{dh}{dx} + h\frac{dg}{dx} \qquad \textbf{(B.30)}$$

B. Derivative of the sum of two functions   If a function $f(x)$ is equal to the sum of two functions, then the derivative of the sum is equal to the sum of the derivatives:

$$\frac{d}{dx} f(x) = \frac{d}{dx}[g(x) + h(x)] = \frac{dg}{dx} + \frac{dh}{dx} \qquad \textbf{(B.31)}$$

C. Chain rule of differential calculus   If $y = f(x)$ and $x = f(z)$, then dy/dx can be written as the product of two derivatives:

$$\frac{dy}{dx} = \frac{dy}{dz}\frac{dz}{dx} \qquad \textbf{(B.32)}$$

D. The second derivative   The second derivative of y with respect to x is defined as the derivative of the function dy/dx (the derivative of the derivative). It is usually written

$$\frac{d^2y}{dx^2} = \frac{d}{dx}\left(\frac{dy}{dx}\right) \qquad \textbf{(B.33)}$$

EXAMPLE 6

Find the derivative of $y(x) = x^3/(x + 1)^2$ with respect to x.

Solution   We can rewrite this function as $y(x) = x^3(x + 1)^{-2}$ and apply Equation B.30:

$$\frac{dy}{dx} = (x + 1)^{-2}\frac{d}{dx}(x^3) + x^3\frac{d}{dx}(x + 1)^{-2}$$

$$= (x + 1)^{-2}3x^2 + x^3(-2)(x + 1)^{-3}$$

$$\frac{dy}{dx} = \frac{3x^2}{(x + 1)^2} - \frac{2x^3}{(x + 1)^3}$$

EXAMPLE 7

A useful formula that follows from Equation B.30 is the derivative of the quotient of two functions. Show that

$$\frac{d}{dx}\left[\frac{g(x)}{h(x)}\right] = \frac{h\dfrac{dg}{dx} - g\dfrac{dh}{dx}}{h^2}$$

Solution   We can write the quotient as gh^{-1} and then apply Equations B.29 and B.30:

$$\frac{d}{dx}\left(\frac{g}{h}\right) = \frac{d}{dx}(gh^{-1}) = g\frac{d}{dx}(h^{-1}) + h^{-1}\frac{d}{dx}(g)$$

$$= -gh^{-2}\frac{dh}{dx} + h^{-1}\frac{dg}{dx}$$

$$= \frac{h\dfrac{dg}{dx} - g\dfrac{dh}{dx}}{h^2}$$

Some of the more commonly used derivatives of functions are listed in Table B.4.

B.7   INTEGRAL CALCULUS

We think of integration as the inverse of differentiation. As an example, consider the expression

$$f(x) = \frac{dy}{dx} = 3ax^2 + b \qquad \textbf{(B.34)}$$

which was the result of differentiating the function

$$y(x) = ax^3 + bx + c$$

TABLE B.4
Derivatives for Several Functions

$$\frac{d}{dx}(a) = 0$$

$$\frac{d}{dx}(ax^n) = nax^{n-1}$$

$$\frac{d}{dx}(e^{ax}) = ae^{ax}$$

$$\frac{d}{dx}(\sin ax) = a\cos ax$$

$$\frac{d}{dx}(\cos ax) = -a\sin ax$$

$$\frac{d}{dx}(\tan ax) = a\sec^2 ax$$

$$\frac{d}{dx}(\cot ax) = -a\csc^2 ax$$

$$\frac{d}{dx}(\sec x) = \tan x \sec x$$

$$\frac{d}{dx}(\csc x) = -\cot x \csc x$$

$$\frac{d}{dx}(\ln ax) = \frac{1}{x}$$

Note: The letters a and n are constants.

$$\frac{d}{dx}\cos(ax+b) = -a\sin(ax+b)$$

in Example 4. We can write Equation B.34 as $dy = f(x)\,dx = (3ax^2 + b)\,dx$ and obtain $y(x)$ by "summing" over all values of x. Mathematically, we write this inverse operation

$$y(x) = \int f(x)\,dx$$

For the function $f(x)$ given by Equation B.34, we have

$$y(x) = \int (3ax^2 + b)\,dx = ax^3 + bx + c$$

where c is a constant of the integration. This type of integral is called an *indefinite integral* because its value depends on the choice of c.

A general **indefinite integral** $I(x)$ is defined as

$$I(x) = \int f(x)\,dx \qquad \text{(B.35)}$$

where $f(x)$ is called the *integrand* and $f(x) = \dfrac{dI(x)}{dx}$.

For a *general continuous* function $f(x)$, the integral can be described as the area under the curve bounded by $f(x)$ and the x axis, between two specified values of x, say, x_1 and x_2, as in Figure B.14.

The area of the blue element is approximately $f(x_i)\Delta x_i$. If we sum all these area elements from x_1 and x_2 and take the limit of this sum as $\Delta x_i \to 0$, we obtain the *true* area under the curve bounded by $f(x)$ and x, between the limits x_1 and x_2:

$$\text{Area} = \lim_{\Delta x_i \to 0} \sum_i f(x_i)\,\Delta x_i = \int_{x_1}^{x_2} f(x)\,dx \qquad \text{(B.36)}$$

Integrals of the type defined by Equation B.36 are called **definite integrals.**

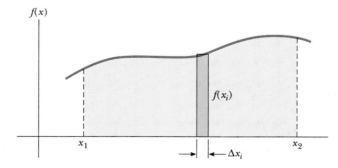

Figure B.14

One common integral that arises in practical situations has the form

$$\int x^n\,dx = \frac{x^{n+1}}{n+1} + c \qquad (n \neq -1) \qquad \text{(B.37)}$$

This result is obvious, being that differentiation of the right-hand side with respect to x gives $f(x) = x^n$ directly. If the limits of the integration are known, this integral becomes a *definite integral* and is written

$$\int_{x_1}^{x_2} x^n\,dx = \frac{x_2^{\,n+1} - x_1^{\,n+1}}{n+1} \qquad (n \neq -1) \qquad \text{(B.38)}$$

EXAMPLES

1. $\int_0^a x^2 \, dx = \dfrac{x^3}{3}\Big]_0^a = \dfrac{a^3}{3}$

2. $\int_0^b x^{3/2} \, dx = \dfrac{x^{5/2}}{5/2}\Big]_0^b = \dfrac{2}{5} b^{5/2}$

3. $\int_3^5 x \, dx = \dfrac{x^2}{2}\Big]_3^5 = \dfrac{5^2 - 3^2}{2} = 8$

Partial Integration

Sometimes it is useful to apply the method of *partial integration* (also called "integrating by parts") to evaluate certain integrals. The method uses the property that

$$\int u \, dv = uv - \int v \, du \tag{B.39}$$

where u and v are *carefully* chosen so as to reduce a complex integral to a simpler one. In many cases, several reductions have to be made. Consider the function

$$I(x) = \int x^2 e^x \, dx$$

This can be evaluated by integrating by parts twice. First, if we choose $u = x^2$, $v = e^x$, we get

$$\int x^2 e^x \, dx = \int x^2 \, d(e^x) = x^2 e^x - 2 \int e^x x \, dx + c_1$$

Now, in the second term, choose $u = x$, $v = e^x$, which gives

$$\int x^2 e^x \, dx = x^2 e^x - 2xe^x + 2 \int e^x \, dx + c_1$$

or

$$\int x^2 e^x \, dx = x^2 e^x - 2xe^x + 2e^x + c_2$$

The Perfect Differential

Another useful method to remember is the use of the *perfect differential,* in which we look for a change of variable such that the differential of the function is the differential of the independent variable appearing in the integrand. For example, consider the integral

$$I(x) = \int \cos^2 x \sin x \, dx$$

This becomes easy to evaluate if we rewrite the differential as $d(\cos x) = -\sin x \, dx$. The integral then becomes

$$\int \cos^2 x \sin x \, dx = -\int \cos^2 x \, d(\cos x)$$

If we now change variables, letting $y = \cos x$, we obtain

$$\int \cos^2 x \sin x \, dx = -\int y^2 dy = -\dfrac{y^3}{3} + c = -\dfrac{\cos^3 x}{3} + c$$

Table B.5 lists some useful indefinite integrals. Table B.6 gives Gauss's probability integral and other definite integrals. A more complete list can be found in various handbooks, such as *The Handbook of Chemistry and Physics,* CRC Press.

TABLE B.5 Some Indefinite Integrals (An arbitrary constant should be added to each of these integrals.)

$$\int x^n \, dx = \frac{x^{n+1}}{n+1} \qquad (\text{provided } n \neq -1)$$

$$\int \frac{dx}{x} = \int x^{-1} \, dx = \ln x$$

$$\int \frac{dx}{a+bx} = \frac{1}{b} \ln(a+bx)$$

$$\int \frac{x \, dx}{a+bx} = \frac{x}{b} - \frac{a}{b^2} \ln(a+bx)$$

$$\int \frac{dx}{x(x+a)} = -\frac{1}{a} \ln \frac{x+a}{x}$$

$$\int \frac{dx}{(a+bx)^2} = -\frac{1}{b(a+bx)}$$

$$\int \frac{dx}{a^2+x^2} = \frac{1}{a} \tan^{-1} \frac{x}{a}$$

$$\int \frac{dx}{a^2-x^2} = \frac{1}{2a} \ln \frac{a+x}{a-x} \qquad (a^2-x^2 > 0)$$

$$\int \frac{dx}{x^2-a^2} = \frac{1}{2a} \ln \frac{x-a}{x+a} \qquad (x^2-a^2 > 0)$$

$$\int \frac{x \, dx}{a^2 \pm x^2} = \pm \tfrac{1}{2} \ln(a^2 \pm x^2)$$

$$\int \frac{dx}{\sqrt{a^2-x^2}} = \sin^{-1} \frac{x}{a} = -\cos^{-1} \frac{x}{a} \qquad (a^2-x^2 > 0)$$

$$\int \frac{dx}{\sqrt{x^2 \pm a^2}} = \ln(x + \sqrt{x^2 \pm a^2})$$

$$\int \frac{x \, dx}{\sqrt{a^2-x^2}} = -\sqrt{a^2-x^2}$$

$$\int \frac{x \, dx}{\sqrt{x^2 \pm a^2}} = \sqrt{x^2 \pm a^2}$$

$$\int \sqrt{a^2-x^2} \, dx = \tfrac{1}{2}\left(x\sqrt{a^2-x^2} + a^2 \sin^{-1} \frac{x}{a} \right)$$

$$\int x\sqrt{a^2-x^2} \, dx = -\tfrac{1}{3}(a^2-x^2)^{3/2}$$

$$\int \sqrt{x^2 \pm a^2} \, dx = \tfrac{1}{2}[x\sqrt{x^2 \pm a^2} \pm a^2 \ln(x + \sqrt{x^2 \pm a^2})]$$

$$\int x(\sqrt{x^2 \pm a^2}) \, dx = \tfrac{1}{3}(x^2 \pm a^2)^{3/2}$$

$$\int e^{ax} \, dx = \frac{1}{a} e^{ax}$$

$$\int \ln ax \, dx = (x \ln ax) - x$$

$$\int xe^{ax} \, dx = \frac{e^{ax}}{a^2}(ax-1)$$

$$\int \frac{dx}{a+be^{cx}} = \frac{x}{a} - \frac{1}{ac} \ln(a+be^{cx})$$

$$\int \sin ax \, dx = -\frac{1}{a} \cos ax$$

$$\int \cos ax \, dx = \frac{1}{a} \sin ax$$

$$\int \tan ax \, dx = \frac{1}{a} \ln(\cos ax) = \frac{1}{a} \ln(\sec ax)$$

$$\int \cot ax \, dx = \frac{1}{a} \ln(\sin ax)$$

$$\int \sec ax \, dx = \frac{1}{a} \ln(\sec ax + \tan ax) = \frac{1}{a} \ln\left[\tan\left(\frac{ax}{2} + \frac{\pi}{4}\right)\right]$$

$$\int \csc ax \, dx = \frac{1}{a} \ln(\csc ax - \cot ax) = \frac{1}{a} \ln\left(\tan \frac{ax}{2}\right)$$

$$\int \sin^2 ax \, dx = \frac{x}{2} - \frac{\sin 2ax}{4a}$$

$$\int \cos^2 ax \, dx = \frac{x}{2} + \frac{\sin 2ax}{4a}$$

$$\int \frac{dx}{\sin^2 ax} = -\frac{1}{a} \cot ax$$

$$\int \frac{dx}{\cos^2 ax} = \frac{1}{a} \tan ax$$

$$\int \tan^2 ax \, dx = \frac{1}{a}(\tan ax) - x$$

$$\int \cot^2 ax \, dx = -\frac{1}{a}(\cot ax) - x$$

$$\int \sin^{-1} ax \, dx = x(\sin^{-1} ax) + \frac{\sqrt{1-a^2x^2}}{a}$$

$$\int \cos^{-1} ax \, dx = x(\cos^{-1} ax) - \frac{\sqrt{1-a^2x^2}}{a}$$

$$\int \frac{dx}{(x^2+a^2)^{3/2}} = \frac{x}{a^2\sqrt{x^2+a^2}}$$

$$\int \frac{x \, dx}{(x^2+a^2)^{3/2}} = -\frac{1}{\sqrt{x^2+a^2}}$$

TABLE B.6   Gauss's Probability Integral and Other Definite Integrals

$$\int_0^\infty x^n e^{-ax} \, dx = \frac{n!}{a^{n+1}}$$

$$I_0 = \int_0^\infty e^{-ax^2} \, dx = \frac{1}{2}\sqrt{\frac{\pi}{a}} \qquad \text{(Gauss's probability integral)}$$

$$I_1 = \int_0^\infty x e^{-ax^2} \, dx = \frac{1}{2a}$$

$$I_2 = \int_0^\infty x^2 e^{-ax^2} \, dx = -\frac{dI_0}{da} = \frac{1}{4}\sqrt{\frac{\pi}{a^3}}$$

$$I_3 = \int_0^\infty x^3 e^{-ax^2} \, dx = -\frac{dI_1}{da} = \frac{1}{2a^2}$$

$$I_4 = \int_0^\infty x^4 e^{-ax^2} \, dx = \frac{d^2 I_0}{da^2} = \frac{3}{8}\sqrt{\frac{\pi}{a^5}}$$

$$I_5 = \int_0^\infty x^5 e^{-ax^2} \, dx = \frac{d^2 I_1}{da^2} = \frac{1}{a^3}$$

.

.

.

$$I_{2n} = (-1)^n \frac{d^n}{da^n} I_0$$

$$I_{2n+1} = (-1)^n \frac{d^n}{da^n} I_1$$

Group I   Group II   Transition elements

Group I	Group II							
H 1 1.008 0 $1s^1$								
Li 3 6.94 $2s^1$	**Be** 4 9.012 $2s^2$							
Na 11 22.99 $3s^1$	**Mg** 12 24.31 $3s^2$							
K 19 39.102 $4s^1$	**Ca** 20 40.08 $4s^2$	**Sc** 21 44.96 $3d^1 4s^2$	**Ti** 22 47.90 $3d^2 4s^2$	**V** 23 50.94 $3d^3 4s^2$	**Cr** 24 51.996 $3d^5 4s^1$	**Mn** 25 54.94 $3d^5 4s^2$	**Fe** 26 55.85 $3d^6 4s^2$	**Co** 27 58.93 $3d^7 4s^2$
Rb 37 85.47 $5s^1$	**Sr** 38 87.62 $5s^2$	**Y** 39 88.906 $4d^1 5s^2$	**Zr** 40 91.22 $4d^2 5s^2$	**Nb** 41 92.91 $4d^4 5s^1$	**Mo** 42 95.94 $4d^5 5s^1$	**Tc** 43 (99) $4d^5 5s^2$	**Ru** 44 101.1 $4d^7 5s^1$	**Rh** 45 102.91 $4d^8 5s^1$
Cs 55 132.91 $6s^1$	**Ba** 56 137.34 $6s^2$	57-71*	**Hf** 72 178.49 $5d^2 6s^2$	**Ta** 73 180.95 $5d^3 6s^2$	**W** 74 183.85 $5d^4 6s^2$	**Re** 75 186.2 $5d^5 6s^2$	**Os** 76 190.2 $5d^6 6s^2$	**Ir** 77 192.2 $5d^7 6s^2$
Fr 87 (223) $7s^1$	**Ra** 88 (226) $7s^2$	89-103**	**Rf** 104 (261) $6d^2 7s^2$	**Db** 105 (262) $6d^3 7s^2$	**Sg** 106 (263)	**Bh** 107 (262)	**Hs** 108 (265)	**Mt** 109 (266)

Legend:
Symbol — **Ca** — Atomic number 20
Atomic mass † — 40.08
$4s^2$ — Electron configuration

*Lanthanide series

La 57 138.91 $5d^1 6s^2$	**Ce** 58 140.12 $5d^1 4f^1 6s^2$	**Pr** 59 140.91 $4f^3 6s^2$	**Nd** 60 144.24 $4f^4 6s^2$	**Pm** 61 (147) $4f^5 6s^2$	**Sm** 62 150.4 $4f^6 6s^2$

**Actinide series

Ac 89 (227) $6d^1 7s^2$	**Th** 90 (232) $6d^2 7s^2$	**Pa** 91 (231) $5f^2 6d^1 7s^2$	**U** 92 (238) $5f^3 6d^1 7s^2$	**Np** 93 (239) $5f^4 6d^1 7s^2$	**Pu** 94 (239) $5f^6 6d^0 7s^2$

▫ Atomic mass values given are averaged over isotopes in the percentages in which they exist in nature.
† For an unstable element, mass number of the most stable known isotope is given in parentheses.
†† Elements 110, 111, 112, and 114 have not yet been named.
††† For a description of the atomic data, visit **physics.nist.gov/atomic**

		Group III	Group IV	Group V	Group VI	Group VII	Group 0
						H 1 1.008 0 $1s^1$	**He** 2 4.002 6 $1s^2$
		B 5 10.81 $2p^1$	**C** 6 12.011 $2p^2$	**N** 7 14.007 $2p^3$	**O** 8 15.999 $2p^4$	**F** 9 18.998 $2p^5$	**Ne** 10 20.18 $2p^6$
		Al 13 26.98 $3p^1$	**Si** 14 28.09 $3p^2$	**P** 15 30.97 $3p^3$	**S** 16 32.06 $3p^4$	**Cl** 17 35.453 $3p^5$	**Ar** 18 39.948 $3p^6$

Ni 28 58.71 $3d^8 4s^2$	**Cu** 29 63.54 $3d^{10}4s^1$	**Zn** 30 65.37 $3d^{10}4s^2$	**Ga** 31 69.72 $4p^1$	**Ge** 32 72.59 $4p^2$	**As** 33 74.92 $4p^3$	**Se** 34 78.96 $4p^4$	**Br** 35 79.91 $4p^5$	**Kr** 36 83.80 $4p^6$
Pd 46 106.4 $4d^{10}$	**Ag** 47 107.87 $4d^{10}5s^1$	**Cd** 48 112.40 $4d^{10}5s^2$	**In** 49 114.82 $5p^1$	**Sn** 50 118.69 $5p^2$	**Sb** 51 121.75 $5p^3$	**Te** 52 127.60 $5p^4$	**I** 53 126.90 $5p^5$	**Xe** 54 131.30 $5p^6$
Pt 78 195.09 $5d^9 6s^1$	**Au** 79 196.97 $5d^{10}6s^1$	**Hg** 80 200.59 $5d^{10}6s^2$	**Tl** 81 204.37 $6p^1$	**Pb** 82 207.2 $6p^2$	**Bi** 83 208.98 $6p^3$	**Po** 84 (210) $6p^4$	**At** 85 (218) $6p^5$	**Rn** 86 (222) $6p^6$
110†† (269)	111†† (272)	112†† (277)		114†† (289)				

Eu 63 152.0 $4f^7 6s^2$	**Gd** 64 157.25 $5d^1 4f^7 6s^2$	**Tb** 65 158.92 $5d^1 4f^8 6s^2$	**Dy** 66 162.50 $4f^{10}6s^2$	**Ho** 67 164.93 $4f^{11}6s^2$	**Er** 68 167.26 $4f^{12}6s^2$	**Tm** 69 168.93 $4f^{13}6s^2$	**Yb** 70 173.04 $4f^{14}6s^2$	**Lu** 71 174.97 $5d^1 4f^{14}6s^2$
Am 95 (243) $5f^7 6d^0 7s^2$	**Cm** 96 (245) $5f^7 6d^1 7s^2$	**Bk** 97 (247) $5f^8 6d^1 7s^2$	**Cf** 98 (249) $5f^{10}6d^0 7s^2$	**Es** 99 (254) $5f^{11}6d^0 7s^2$	**Fm** 100 (253) $5f^{12}6d^0 7s^2$	**Md** 101 (255) $5f^{13}6d^0 7s^2$	**No** 102 (255) $6d^0 7s^2$	**Lr** 103 (257) $6d^1 7s^2$

APPENDIX D · *SI Units*

TABLE D.1 SI Units

Base Quantity	SI Base Unit Name	Symbol
Length	Meter	m
Mass	Kilogram	kg
Time	Second	s
Electric current	Ampere	A
Temperature	Kelvin	K
Amount of substance	Mole	mol
Luminous intensity	Candela	cd

TABLE D.2 Some Derived SI Units

Quantity	Name	Symbol	Expression in Terms of Base Units	Expression in Terms of Other SI Units
Plane angle	radian	rad	m/m	
Frequency	hertz	Hz	s^{-1}	
Force	newton	N	$kg \cdot m/s^2$	J/m
Pressure	pascal	Pa	$kg/m \cdot s^2$	N/m^2
Energy; work	joule	J	$kg \cdot m^2/s^2$	$N \cdot m$
Power	watt	W	$kg \cdot m^2/s^3$	J/s
Electric charge	coulomb	C	$A \cdot s$	
Electric potential	volt	V	$kg \cdot m^2/A \cdot s^3$	W/A
Capacitance	farad	F	$A^2 \cdot s^4/kg \cdot m^2$	C/V
Electric resistance	ohm	Ω	$kg \cdot m^2/A^2 \cdot s^3$	V/A
Magnetic flux	weber	Wb	$kg \cdot m^2/A \cdot s^2$	$V \cdot s$
Magnetic field intensity	tesla	T	$kg/A \cdot s^2$	
Inductance	henry	H	$kg \cdot m^2/A^2 \cdot s^2$	$T \cdot m^2/A$

All Nobel Prizes in physics are listed (and marked with a P), as well as relevant Nobel Prizes in Chemistry (C). The key dates for some of the scientific work are supplied; they often antedate the prize considerably.

1901 (P) *Wilhelm Roentgen* for discovering x-rays (1895).

1902 (P) *Hendrik A. Lorentz* for predicting the Zeeman effect and *Pieter Zeeman* for discovering the Zeeman effect, the splitting of spectral lines in magnetic fields.

1903 (P) *Antoine-Henri Becquerel* for discovering radioactivity (1896) and *Pierre and Marie Curie* for studying radioactivity.

1904 (P) *Lord Rayleigh* for studying the density of gases and discovering argon.
(C) *William Ramsay* for discovering the inert gas elements helium, neon, xenon, and krypton, and placing them in the periodic table.

1905 (P) *Philipp Lenard* for studying cathode rays, electrons (1898–1899).

1906 (P) *J. J. Thomson* for studying electrical discharge through gases and discovering the electron (1897).

1907 (P) *Albert A. Michelson* for inventing optical instruments and measuring the speed of light (1880s).

1908 (P) *Gabriel Lippmann* for making the first color photographic plate, using interference methods (1891).
(C) *Ernest Rutherford* for discovering that atoms can be broken apart by alpha rays and for studying radioactivity.

1909 (P) *Guglielmo Marconi* and *Carl Ferdinand Braun* for developing wireless telegraphy.

1910 (P) *Johannes D. van der Waals* for studying the equation of state for gases and liquids (1881).

1911 (P) *Wilhelm Wien* for discovering Wien's law giving the peak of a blackbody spectrum (1893).
(C) *Marie Curie* for discovering radium and polonium (1898) and isolating radium.

1912 (P) *Nils Dalén* for inventing automatic gas regulators for lighthouses.

1913 (P) *Heike Kamerlingh Onnes* for the discovery of superconductivity and liquefying helium (1908).

1914 (P) *Max T. F. von Laue* for studying x-rays from their diffraction by crystals, showing that x-rays are electromagnetic waves (1912).
(C) *Theodore W. Richards* for determining the atomic weights of sixty elements, indicating the existence of isotopes.

1915 (P) *William Henry Bragg* and *William Lawrence Bragg*, his son, for studying the diffraction of x-rays in crystals.

1917 (P) *Charles Barkla* for studying atoms by x-ray scattering (1906).

1918 (P) *Max Planck* for discovering energy quanta (1900).

1919 (P) *Johannes Stark,* for discovering the Stark effect, the splitting of spectral lines in electric fields (1913).

1920 (P) *Charles-Édouard Guillaume* for discovering invar, a nickel-steel alloy with low coefficient of expansion.
(C) *Walther Nernst* for studying heat changes in chemical reactions and formulating the third law of thermodynamics (1918).

1921 (P) *Albert Einstein* for explaining the photoelectric effect and for his services to theoretical physics (1905).
(C) *Frederick Soddy* for studying the chemistry of radioactive substances and discovering isotopes (1912).

1922 (P) *Niels Bohr* for his model of the atom and its radiation (1913).
(C) *Francis W. Aston* for using the mass spectrograph to study atomic weights, thus discovering 212 of the 287 naturally occurring isotopes.

1923 (P) *Robert A. Millikan* for measuring the charge on an electron (1911) and for studying the photoelectric effect experimentally (1914).

1924 (P) *Karl M. G. Siegbahn* for his work in x-ray spectroscopy.

1925 (P) *James Franck* and *Gustav Hertz* for discovering the Franck-Hertz effect in electron-atom collisions.

1926 (P) *Jean-Baptiste Perrin* for studying Brownian motion to validate the discontinuous structure of matter and measure the size of atoms.

1927 (P) *Arthur Holly Compton* for discovering the Compton effect on x-rays, their change in wavelength when they collide with matter (1922), and *Charles T. R. Wilson* for inventing the cloud chamber, used to study charged particles (1906).

1928 (P) *Owen W. Richardson* for studying the thermionic effect and electrons emitted by hot metals (1911).

1929 (P) *Louis Victor de Broglie* for discovering the wave nature of electrons (1923).

1930 (P) *Chandrasekhara Venkata Raman* for studying Raman scattering, the scattering of light by atoms and molecules with a change in wavelength (1928).

1932 (P) *Werner Heisenberg* for creating quantum mechanics (1925).

1933 (P) *Erwin Schrödinger* and *Paul A. M. Dirac* for developing wave mechanics (1925) and relativistic quantum mechanics (1927).
(C) *Harold Urey* for discovering heavy hydrogen, deuterium (1931).

1935 (P) *James Chadwick* for discovering the neutron (1932).
(C) *Irène* and *Frédéric Joliot-Curie* for synthesizing new radioactive elements.

1936 (P) *Carl D. Anderson* for discovering the positron in particular and antimatter in general (1932) and *Victor F. Hess* for discovering cosmic rays.
(C) *Peter J. W. Debye* for studying dipole moments and diffraction of x-rays and electrons in gases.

1937 (P) *Clinton Davisson* and *George Thomson* for discovering the diffraction of electrons by crystals, confirming de Broglie's hypothesis (1927).

1938 (P) *Enrico Fermi* for producing the transuranic radioactive elements by neutron irradiation (1934–1937).

1939 (P) *Ernest O. Lawrence* for inventing the cyclotron.

1943 (P) *Otto Stern* for developing molecular-beam studies (1923), and using them to discover the magnetic moment of the proton (1933).

1944 (P) *Isidor I. Rabi* for discovering nuclear magnetic resonance in atomic and molecular beams.
(C) *Otto Hahn* for discovering nuclear fission (1938).

1945 (P) *Wolfgang Pauli* for discovering the exclusion principle (1924).

1946 (P) *Percy W. Bridgman* for studying physics at high pressures.

1947 (P) *Edward V. Appleton* for studying the ionosphere.

1948 (P) *Patrick M. S. Blackett* for studying nuclear physics with cloud-chamber photographs of cosmic-ray interactions.

1949 (P) *Hideki Yukawa* for predicting the existence of mesons (1935).

1950 (P) *Cecil F. Powell* for developing the method of studying cosmic rays with photographic emulsions and discovering new mesons.

1951 (P) *John D. Cockcroft* and *Ernest T. S. Walton* for transmuting nuclei in an accelerator (1932).
(C) *Edwin M. McMillan* for producing neptunium (1940) and *Glenn T. Seaborg* for producing plutonium (1941) and further transuranic elements.

1952 (P) *Felix Bloch* and *Edward Mills Purcell* for discovering nuclear magnetic resonance in liquids and gases (1946).

1953 (P) *Frits Zernike* for inventing the phase-contrast microscope, which uses interference to provide high contrast.

1954 (P) *Max Born* for interpreting the wave function as a probability (1926) and other quantum-mechanical discoveries and *Walther Bothe* for developing the coincidence method to study subatomic particles (1930–1931), producing, in particular, the particle interpreted by Chadwick as the neutron.

1955 (P) *Willis E. Lamb, Jr.,* for discovering the Lamb shift in the hydrogen spectrum (1947) and *Polykarp Kusch* for determining the magnetic moment of the electron (1947).

1956 (P) *John Bardeen, Walter H. Brattain,* and *William Shockley* for inventing the transistor (1956).

1957 (P) *T.-D. Lee* and *C.-N. Yang* for predicting that parity is not conserved in beta decay (1956).

1958 (P) *Pavel A. Čerenkov* for discovering Čerenkov radiation (1935) and *Ilya M. Frank* and *Igor Tamm* for interpreting it (1937).

1959 (P) *Emilio G. Segrè* and *Owen Chamberlain* for discovering the antiproton (1955).

1960 (P) *Donald A. Glaser* for inventing the bubble chamber to study elementary particles (1952).
(C) *Willard Libby* for developing radiocarbon dating (1947).

1961 (P) *Robert Hofstadter* for discovering internal structure in protons and neutrons and *Rudolf L. Mössbauer* for discovering the Mössbauer effect of recoilless gamma-ray emission (1957).

1962 (P) *Lev Davidovich Landau* for studying liquid helium and other condensed matter theoretically.

1963 (P) *Eugene P. Wigner* for applying symmetry principles to elementary-particle theory and *Maria Goeppert Mayer* and *J. Hans D. Jensen* for studying the shell model of nuclei (1947).

1964 (P) *Charles H. Townes, Nikolai G. Basov,* and *Alexandr M. Prokhorov* for developing masers (1951–1952) and lasers.

1965 (P) *Sin-itiro Tomonaga, Julian S. Schwinger,* and *Richard P. Feynman* for developing quantum electrodynamics (1948).

1966 (P) *Alfred Kastler* for his optical methods of studying atomic energy levels.

1967 (P) *Hans Albrecht Bethe* for discovering the routes of energy production in stars (1939).

1968 (P) *Luis W. Alvarez* for discovering resonance states of elementary particles.

1969 (P) *Murray Gell-Mann* for classifying elementary particles (1963).

1970 (P) *Hannes Alfvén* for developing magnetohydrodynamic theory and *Louis Eugène Félix Néel* for discovering antiferromagnetism and ferrimagnetism (1930s).

1971 (P) *Dennis Gabor* for developing holography (1947).

(C) *Gerhard Herzberg* for studying the structure of molecules spectroscopically.

1972 (P) *John Bardeen, Leon N. Cooper,* and *John Robert Schrieffer* for explaining superconductivity (1957).

1973 (P) *Leo Esaki* for discovering tunneling in semiconductors, *Ivar Giaever* for discovering tunneling in superconductors, and *Brian D. Josephson* for predicting the Josephson effect, which involves tunneling of paired electrons (1958–1962).

1974 (P) *Anthony Hewish* for discovering pulsars and *Martin Ryle* for developing radio interferometry.

1975 (P) *Aage N. Bohr, Ben R. Mottelson,* and *James Rainwater* for discovering why some nuclei take asymmetric shapes.

1976 (P) *Burton Richter* and *Samuel C. C. Ting* for discovering the J/psi particle, the first charmed particle (1974).

1977 (P) *John H. Van Vleck, Nevill F. Mott,* and *Philip W. Anderson* for studying solids quantum-mechanically.

(C) *Ilya Prigogine* for extending thermodynamics to show how life could arise in the face of the second law.

1978 (P) *Arno A. Penzias* and *Robert W. Wilson* for discovering the cosmic background radiation (1965) and *Pyotr Kapitsa* for his studies of liquid helium.

1979 (P) *Sheldon L. Glashow, Abdus Salam,* and *Steven Weinberg* for developing the theory that unified the weak and electromagnetic forces (1958–1971).

1980 (P) *Val Fitch* and *James W. Cronin* for discovering CP (charge-parity) violation (1964), which possibly explains the cosmological dominance of matter over antimatter.

1981 (P) *Nicolaas Bloembergen* and *Arthur L. Schawlow* for developing laser spectroscopy and *Kai M. Siegbahn* for developing high-resolution electron spectroscopy (1958).

1982 (P) *Kenneth G. Wilson* for developing a method of constructing theories of phase transitions to analyze critical phenomena.

1983 (P) *William A. Fowler* for theoretical studies of astrophysical nucleosynthesis and *Subramanyan Chandrasekhar* for studying physical processes of importance to stellar structure and evolution, including the prediction of white dwarf stars (1930).

1984 (P) *Carlo Rubbia* for discovering the W and Z particles, verifying the electroweak unification, and *Simon van der Meer,* for developing the method of stochastic cooling of the CERN beam that allowed the discovery (1982–1983).

1985 (P) *Klaus von Klitzing* for the quantized Hall effect, relating to conductivity in the presence of a magnetic field (1980).

1986 (P) *Ernst Ruska* for inventing the electron microscope (1931), and *Gerd Binnig* and *Heinrich Rohrer* for inventing the scanning-tunneling electron microscope (1981).

1987 (P) *J. Georg Bednorz* and *Karl Alex Müller* for the discovery of high temperature superconductivity (1986).

1988 (P) *Leon M. Lederman, Melvin Schwartz,* and *Jack Steinberger* for a collaborative experiment that led to the development of a new tool for studying the weak nuclear force, which affects the radioactive decay of atoms.

1989 (P) *Norman Ramsay* (U.S.) for various techniques in atomic physics; and *Hans Dehmelt* (U.S.) and *Wolfgang Paul* (Germany) for the development of techniques for trapping single charge particles.

1990 (P) *Jerome Friedman, Henry Kendall* (both U.S.), and *Richard Taylor* (Canada) for experiments important to the development of the quark model.

1991 (P) *Pierre-Gilles de Gennes* for discovering that methods developed for studying order phenomena in simple systems can be generalized to more complex forms of matter, in particular to liquid crystals and polymers.

1992 (P) *George Charpak* for developing detectors that trace the paths of evanescent subatomic particles produced in particle accelerators.

1993 (P) *Russell Hulse* and *Joseph Taylor* for discovering evidence of gravitational waves.

1994 (P) *Bertram N. Brockhouse* and *Clifford G. Shull* for pioneering work in neutron scattering.

1995 (P) *Martin L. Perl* and *Frederick Reines* for discovering the tau particle and the neutrino, respectively.

1996 (P) *David M. Lee, Douglas C. Osheroff,* and *Robert C. Richardson* for developing a superfluid using helium-3.

1997 (P) *Steven Chu, Claude Cohen-Tannoudji,* and *William D. Phillips* for developing methods to cool and trap atoms with laser light.

1998 (P) *Robert B. Laughlin, Horst L. Störmer,* and *Daniel C. Tsui* for discovering a new form of quantum fluid with fractionally charged excitations.

Answers to Odd-Numbered Problems

Chapter 1

1. $2.15 \times 10^4 \ \text{kg/m}^3$
3. 184 g
5. (a) $7.10 \ \text{cm}^3$ (b) $1.18 \times 10^{-29} \ \text{m}^3$ (c) 0.228 nm
   (d) $12.7 \ \text{cm}^3$, $2.11 \times 10^{-29} \ \text{m}^3$, 0.277 nm
7. (a) $4.00 \ \text{u} = 6.64 \times 10^{-24} \ \text{g}$ (b) $55.9 \ \text{u} = 9.29 \times 10^{-23} \ \text{g}$ (c) $207 \ \text{u} = 3.44 \times 10^{-22} \ \text{g}$
9. (a) $9.83 \times 10^{-16} \ \text{g}$ (b) 1.06×10^7 atoms
11. (a) 4.01×10^{25} molecules (b) 3.65×10^4 molecules
13. no
15. (b) only
17. $0.579t \ \text{ft}^3/\text{s} + 1.19 \times 10^{-9}t^2 \ \text{ft}^3/\text{s}^2$
19. $1.39 \times 10^3 \ \text{m}^2$
21. (a) 0.071 4 gal/s (b) $2.70 \times 10^{-4} \ \text{m}^3/\text{s}$ (c) 1.03 h
23. $4.05 \times 10^3 \ \text{m}^2$
25. $11.4 \times 10^3 \ \text{kg/m}^3$
27. 1.19×10^{57} atoms
29. (a) 190 y (b) 2.32×10^4 times
31. 151 μm
33. 1.00×10^{10} lb
35. $3.08 \times 10^4 \ \text{m}^3$
37. 5.0 m
39. 2.86 cm
41. $\sim 10^6$ balls
43. $\sim 10^7$ or 10^8 rev
45. $\sim 10^7$ or 10^8 blades
47. $\sim 10^2$ kg; $\sim 10^3$ kg
49. $\sim 10^2$ tuners
51. (a) $(346 \pm 13) \ \text{m}^2$ (b) (66.0 ± 1.3) m
53. $(1.61 \pm 0.17) \times 10^3 \ \text{kg/m}^3$
55. 115.9 m
57. 316 m
59. $4.50 \ \text{m}^2$
61. 3.41 m
63. 0.449%
65. (a) 0.529 cm/s (b) 11.5 cm/s
67. 1×10^{10} gal/y
69. $\sim 10^{11}$ stars

Chapter 2

1. (a) 2.30 m/s (b) 16.1 m/s (c) 11.5 m/s
3. (a) 5 m/s (b) 1.2 m/s (c) -2.5 m/s (d) -3.3 m/s
   (e) 0
5. (a) 3.75 m/s (b) 0

7. (a)

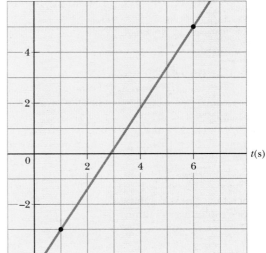

(b) 1.60 m/s
9. (a) -2.4 m/s (b) -3.2 m/s (c) 4.0 s
11. (a) 5.0 m/s (b) -2.5 m/s (c) 0 (d) 5.0 m/s
13. $1.34 \times 10^4 \ \text{m/s}^2$
15. (a)

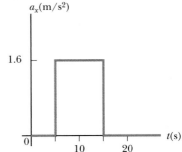

(b) $1.6 \ \text{m/s}^2$ and $0.80 \ \text{m/s}^2$
17. (a) 2.00 m (b) -3.00 m/s (c) $-2.00 \ \text{m/s}^2$
19. (a) $1.3 \ \text{m/s}^2$ (b) $2.0 \ \text{m/s}^2$ at 3 s (c) at $t = 6$ s and for $t > 10$ s (d) $-1.5 \ \text{m/s}^2$ at 8 s
21. $2.74 \times 10^5 \ \text{m/s}^2$, which is $2.79 \times 10^4 \ g$
23. (a) 6.61 m/s (b) $-0.448 \ \text{m/s}^2$
25. $-16.0 \ \text{cm/s}^2$
27. (a) 2.56 m (b) -3.00 m/s
29. (a) 8.94 s (b) 89.4 m/s
31. (a) 20.0 s (b) no
33. $x_f - x_i = v_{xf} t - a_x t^2/2$; $v_{xf} = 3.10$ m/s

35. (a) 35.0 s   (b) 15.7 m/s

37. (a) -202 m/s^2 (b) 198 m

39. (a) 3.00 m/s   (b) 6.00 s   (c) -0.300 m/s^2
   (d) 2.05 m/s

41. (a) -4.90 m, -19.6 m, -44.1 m   (b) -9.80 m/s,
   -19.6 m/s, -29.4 m/s

43. (a) 10.0 m/s up   (b) 4.68 m/s down

45. No. In 0.2 s the bill falls out from between David's fingers.

47. (a) 29.4 m/s   (b) 44.1 m

49. (a) 7.82 m   (b) 0.782 s

51. (a) 1.53 s   (b) 11.5 m   (c) -4.60 m/s, -9.80 m/s^2

53. (a) $a_x = a_{xi} + Jt$, $v_x = v_{xi} + a_{xi}t + \frac{1}{2}Jt^2$,
   $x = x_i + v_{xi}t + \frac{1}{2}a_{xi}t^2 + \frac{1}{6}Jt^3$

55. 0.222 s

57. 0.509 s

59. (a) 41.0 s   (b) 1.73 km   (c) -184 m/s

61. $v_{xi}t + at^2/2$, in agreement with Equation 2.11

63. (a) 5.43 m/s^2 and 3.83 m/s^2 (b) 10.9 m/s and 11.5 m/s
   (c) Maggie by 2.62 m

65. (a) 45.7 s   (b) 574 m   (c) 12.6 m/s   (d) 765 s

67. (a) 2.99 s   (b) -15.4 m/s   (c) 31.3 m/s down and
   34.9 m/s down

69. (a) 5.45 s   (b) 73.0 m   (c) $v_{Stan} = 22.6$ m/s, $v_{Kathy} = 26.7$ m/s

71. (a) See top of next column.
   (b) See top of next column.

73. $0.577v$

Chapter 3

1. $(-2.75, -4.76)$ m

3. 1.15; 2.31

5. (a) 2.24 m   (b) 2.24 m at 26.6° from the positive x axis.

7. (a) 484 m   (b) 18.1° north of west

9. 70.0 m

11. (a) approximately 6.1 units at 112°   (b) approximately
   14.8 units at 22°

13. (a) 10.0 m   (b) 15.7 m   (c) 0

15. (a) 5.2 m at 60°   (b) 3.0 m at 330°   (c) 3.0 m at 150°
   (d) 5.2 m at 300°

17. approximately 420 ft at $-3°$

19. 5.83 m at 59.0° to the right of his initial direction

21. 1.31 km north and 2.81 km east

23. (a) 10.4 cm   (b) 35.5°

25. 47.2 units at 122° from the positive x axis.

27. $(-25.0\mathbf{i})$ m $+ (43.3\mathbf{j})$ m

29. 7.21 m at 56.3° from the positive x axis.

31. (a) $2.00\mathbf{i} - 6.00\mathbf{j}$ (b) $4.00\mathbf{i} + 2.00\mathbf{j}$ (c) 6.32   (d) 4.47
   (e) 288°; 26.6° from the positive x axis.

33. (a) $(-11.1\mathbf{i} + 6.40\mathbf{j})$ m   (b) $(1.65\mathbf{i} + 2.86\mathbf{j})$ cm
   (c) $(-18.0\mathbf{i} - 12.6\mathbf{j})$ in.

35. 9.48 m at 166°

37. (a) 185 N at 77.8° from the positive x axis
   (b) $(-39.3\mathbf{i} - 181\mathbf{j})$ N

39. $\mathbf{A} + \mathbf{B} = (2.60\mathbf{i} + 4.50\mathbf{j})$ m

Chapter 2, Problem 71(a)

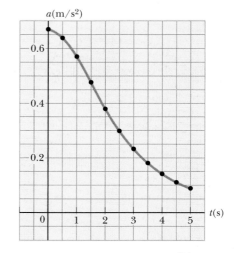

Chapter 2, Problem 71(b)

41. 196 cm at $-14.7°$ from the positive x axis.

43. (a) $8.00\mathbf{i} + 12.0\mathbf{j} - 4.00\mathbf{k}$ (b) $2.00\mathbf{i} + 3.00\mathbf{j} - 1.00\mathbf{k}$
   (c) $-24.0\mathbf{i} - 36.0\mathbf{j} + 12.0\mathbf{k}$

45. (a) 5.92 m is the magnitude of $(5.00\mathbf{i} - 1.00\mathbf{j} - 3.00\mathbf{k})$ m
   (b) 19.0 m is the magnitude of $(4.00\mathbf{i} - 11.0\mathbf{j} + 15.0\mathbf{k})$ m

47. 157 km

49. (a) $-3.00\mathbf{i} + 2.00\mathbf{j}$ (b) 3.61 at 146° from the positive
   x axis.   (c) $3.00\mathbf{i} - 6.00\mathbf{j}$

51. (a) $49.5\mathbf{i} + 27.1\mathbf{j}$ (b) 56.4 units at 28.7° from the positive x axis.

53. 1.15°

55. (a) 2.00, 1.00, 3.00   (b) 3.74   (c) $\theta_x = 57.7°$, $\theta_y = 74.5°$,
   $\theta_z = 36.7°$

57. 240 m at 237°

59. 390 mi/h at 7.37° north of east

61. $\mathbf{R}_1 = a\mathbf{i} + b\mathbf{j}$; $R_1 = \sqrt{a^2 + b^2}$ (b) $\mathbf{R}_2 = a\mathbf{i} + b\mathbf{j} + c\mathbf{k}$

Chapter 4

1. (a) 4.87 km at 209° from east (b) 23.3 m/s
   (c) 13.5 m/s at 209°
3. (a) $(18.0t)\mathbf{i} + (4.00t - 4.90t^2)\mathbf{j}$
   (b) $18.0\mathbf{i} + (4.00 - 9.80t)\mathbf{j}$ (c) $-9.80\mathbf{j}$
   (d) $(54.0\mathbf{i} - 32.1\mathbf{j})$ m
   (e) $(18.0\mathbf{i} - 25.4\mathbf{j})$ m/s (f) $(-9.80\mathbf{j})$ m/s^2
5. (a) $(2.00\mathbf{i} + 3.00\mathbf{j})$ m/s^2
   (b) $(3.00t + t^2)\mathbf{i}$ m, $(1.50\,t^2 - 2.00t)\mathbf{j}$ m
7. (a) $(0.800\mathbf{i} - 0.300\mathbf{j})$ m/s^2 (b) 339°
   (c) $(360\mathbf{i} - 72.7\mathbf{j})$ m, $-15.2°$
9. (a) $(3.34\mathbf{i})$ m/s (b) $-50.9°$
11. (a) 20.0° (b) 3.05 s
13. $x = 7.23$ km $y = 1.68$ km
15. 53.1°
17. 22.4° or 89.4°
19. (a) The ball clears by 0.889 m (b) while descending
21. $d \tan \theta_i - gd^2/(2v_i^2 \cos^2\theta_i)$
23. (a) 0.852 s (b) 3.29 m/s (c) 4.03 m/s (d) 50.8°
    (e) 1.12 s
25. 377 m/s^2
27. 10.5 m/s, 219 m/s^2
29. (a) 6.00 rev/s (b) 1.52 km/s^2 (c) 1.28 km/s^2
31. 1.48 m/s^2 inward at 29.9° behind the radius
33. (a) 13.0 m/s^2 (b) 5.70 m/s (c) 7.50 m/s^2
35. (a)

   (b) 29.7 m/s^2 (c) 6.67 m/s at 36.9° above the
   horizontal
37. 2.02×10^3 s; 21.0% longer
39. 153 km/h at 11.3° north of west
41. (a) 36.9° (b) 41.6° (c) 3.00 min
43. 15.3 m
45. $2\,v_i t \cos \theta_i$
47. (b) $45° + \phi/2$; $v_i^2(1 - \sin \phi)/g \cos^2\phi$
49. (a) 41.7 m/s (b) 3.81 s (c) $(34.1\mathbf{i} - 13.4\mathbf{j})$ m/s; 36.6 m/s
51. (a) 25.0 m/s^2 (radial); 9.80 m/s^2 (tangential)
    (b)

(c) 26.8 m/s^2 inward at 21.4° below the horizontal
53. 8.94 m/s at $-63.4°$ relative to the positive x axis.
55. 20.0 m
57. (a) 0.600 m (b) 0.402 m (c) 1.87 m/s^2 toward center
    (d) 9.80 m/s^2 down
59. (a) 6.80 km (b) 3.00 km vertically above the impact
    point (c) 66.2°
61. (a) 46.5 m/s (b) $-77.6°$ (c) 6.34 s
63. (a) 1.53 km (b) 36.2 s (c) 4.04 km
65. (a) 20.0 m/s, 5.00 s (b) $(16.0\mathbf{i} - 27.1\mathbf{j})$ m/s (c) 6.54 s
    (d) $(24.6\mathbf{i})$ m
67. (a) 43.2 m (b) $(9.66\mathbf{i} - 25.5\mathbf{j})$ m/s
69. Imagine you are shaking down the mercury in a fever
    thermometer. Starting with your hand at the level of your
    shoulder, move your hand down as fast as you can and
    snap it around an arc at the bottom. ~ 100 m/s$^2 \approx 10\,g$

Chapter 5

1. (a) 1/3 (b) 0.750 m/s^2
3. $(6.00\mathbf{i} + 15.0\mathbf{j})$ N; 16.2 N
5. 312 N
7. (a) $x = vt/2$ (b) $F_g v\mathbf{i}/gt + F_g\mathbf{j}$
9. (a) $(2.50\mathbf{i} + 5.00\mathbf{j})$ N (b) 5.59 N
11. (a) 3.64×10^{-18} N (b) 8.93×10^{-30} N is 408 billion
    times smaller.
13. 2.38 kN
15. (a) 5.00 m/s^2 at 36.9° (b) 6.08 m/s^2 at 25.3°
17. (a) $\sim 10^{-22}$ m/s^2 (b) $\sim 10^{-23}$ m
19. (a) 0.200 m/s^2 forward (b) 10.0 m (c) 2.00 m/s
21. (a) 15.0 lb up (b) 5.00 lb up (c) 0
23. 613 N

27. (a) 49.0 N (b) 98.0 N (c) 24.5 N
29. 8.66 N east
31. 100 N and 204 N
33. 3.73 m
35. $a = F/(m_1 + m_2)$; $T = F m_1/(m_1 + m_2)$
37. (a) $F_x > 19.6$ N (b) $F_x \le -78.4$ N
    (c) See top of next page.
39. (a) 706 N (b) 814 N (c) 706 N (d) 648 N
41. $\mu_s = 0.306$; $\mu_k = 0.245$
43. (a) 256 m (b) 42.7 m
45. (a) 1.78 m/s^2 (b) 0.368 (c) 9.37 N (d) 2.67 m/s
47. (a) 0.161 (b) 1.01 m/s^2
49. 37.8 N

Chapter 5, Problem 37(c)

Chapter 5, Problem 55(a)

51. (a)

Chapter 5, Problem 67(a)

71. (a)

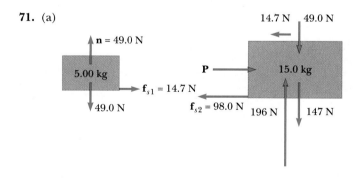

(b) 27.2 N, 1.29 m/s^2
53. Any value between 31.7 N and 48.6 N
55. (a) See top of next column.
  (b) 0.408 m/s^2 (c) 83.3 N
57. 1.18 kN
59. (a) $Mg/2$, $Mg/2$, $Mg/2$, $3Mg/2$, Mg (b) $Mg/2$

61. (b)

θ	0	15.0°	30.0°	45.0°	60.0°
P(N)	40.0	46.4	60.1	94.3	260

63. (a) 19.3°   (b) 4.21 N
65. (a) 2.13 s   (b) 1.67 m
67. (a) See next column.
  Static friction between the two blocks accelerates the upper block.   (b) 34.7 N   (c) 0.306
69. $(M + m_1 + m_2)(m_2 g/m_1)$

(b) 113 N   (c) 0.980 m/s^2 and 1.96 m/s^2
73. (a) 0.087 1   (b) 27.4 N
75. (a) 30.7°   (b) 0.843 N
77. (a) 3.34   (b) Either the car would flip over backwards, or
  the wheels would skid, spinning in place, and the time
  would increase.

Chapter 6

1. (a) 8.00 m/s   (b) 3.02 N
3. Any speed up to 8.08 m/s

5. 6.22×10^{-12} N

7. (a) 1.52 m/s^2 (b) 1.66 km/s (c) $6\,820$ s

9. (a) static friction (b) $0.085\,0$

11. $v \le 14.3$ m/s

13. (a) 68.6 N toward the center of the circle and 784 N up
(b) 0.857 m/s^2

15. No. The jungle lord needs a vine of tensile strength 1.38 kN.

17. (a) 4.81 m/s (b) 700 N up

19. 3.13 m/s

21. (a) 2.49×10^4 N up (b) 12.1 m/s

23. (a) 0.822 m/s^2 (b) 37.0 N (c) 0.0839

25. (a) $17.0°$ (b) 5.12 N

27. (a) 491 N (b) 50.1 kg (c) 2.00 m/s^2

29. $0.0927°$

31. (a) 32.7 s^{-1} (b) 9.80 m/s^2 (c) 4.90 m/s^2

33. 3.01 N

35. (a) 1.47 N·s/m (b) 2.04×10^{-3} s (c) 2.94×10^{-2} N

37. (a) 0.0347 s^{-1} (b) 2.50 m/s (c) $a = -cv$

39. $\sim 10^1$ N

41. (a) 13.7 m/s down

(b)

t (s)	x (m)	v (m/s)
0	0	0
0.2	0	-1.96
0.4	-0.392	-3.88
. . .	. . .	. . .
1.0	-3.77	-8.71
. . . 2.0	-14.4	-12.56
. . . 4.0	-41.0	-13.67

43. (a) 49.5 m/s and 4.95 m/s

(b)

t (s)	y (m)	v (m/s)
0	$1\,000$	0
. . . 1	995	-9.7
. . . 2	980	-18.6
. . . 10	674	-47.7
. . . 10.1	671	-16.7
. . . 12	659	-4.95
. . . 145	0	-4.95

45. (a) 2.33×10^{-4} kg/m (b) 53 m/s (c) 42 m/s. The second trajectory is higher and shorter. In both, the ball attains maximum height when it has covered 57% of its horizontal range, and it attains minimum speed somewhat later. The impact speeds also are both about 30 m/s.

47. (a) $mg - mv^2/R$ (b) $\sqrt{gR}$

49. (a) 2.63 m/s^2 (b) 201 m (c) 17.7 m/s

51. (a) 9.80 N (b) 9.80 N (c) 6.26 m/s

53. (b) 732 N down at the equator and 735 N down at the poles

59. (a) 1.58 m/s^2 (b) 455 N (c) 329 N (d) 397 N upward and $9.15°$ inward

61. (a) 5.19 m/s (b) Child + seat:

$T = 555$ N

63. (b) 2.54 s; 23.6 rev/min

65. 215 N horizontally inward

67. (a) either $70.4°$ or $0°$ (b) $0°$

69. 12.8 N

71. (a)

t (s)	d (m)
0	0
1	4.9
2	18.9
. . . 5	112.6
. . . 10	347.0
11	399.1
. . . 15	611.3
. . . 20	876.5

(b)

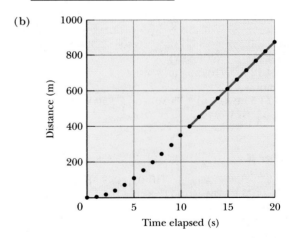

(c) The graph is straight for 11 s $< t <$ 20 s, with slope 53.0 m/s.

Chapter 7

1. 15.0 MJ

3. (a) 32.8 mJ (b) -32.8 mJ

5. (a) 31.9 J (b) 0 (c) 0 (d) 31.9 J

7. 4.70 kJ

9. 14.0

11. (a) 16.0 J (b) $36.9°$

13. (a) $11.3°$ (b) $156°$ (c) $82.3°$

15. (a) 24.0 J   (b) − 3.00 J   (c) 21.0 J
17. (a) 7.50 J   (b) 15.0 J   (c) 7.50 J   (d) 30.0 J
19. (a) 0.938 cm   (b) 1.25 J
21. 0.299 m/s
23. 12.0 J
25. (b) mgR
27. (a) 1.20 J   (b) 5.00 m/s   (c) 6.30 J
29. (a) 60.0 J   (b) 60.0 J
31. (a) $\sqrt{2W/m}$ (b) W/d
33. (a) 650 J   (b) − 588 J   (c) 0   (d) 0   (e) 62.0 J
     (f) 1.76 m/s
35. (a) − 168 J   (b) − 184 J   (c) 500 J   (d) 148 J
     (e) 5.64 m/s
37. 2.04 m
39. (a) 22 500 N   (b) 1.33×10^{-4} s
41. (a) 0.791 m/s   (b) 0.531 m/s
43. 875 W
45. 830 N
47. (a) 5 910 W   (b) It is 53.0% of 11 100 W
49. (a) 0.013 5 gal   (b) 73.8   (c) 8.08 kW
51. 5.90 km/L
53. (a) 5.37×10^{-11} J   (b) 1.33×10^{-9} J
55. 90.0 J
59. (a) $(2 + 24t^2 + 72t^4)$ J   (b) $12t$ m/s^2; $48t$ N
     (c) $(48t + 288t^3)$ W   (d) 1 250 J
61. − 0.047 5 J
63. 878 kN
65. (b) 240 W
67. (a) $\mathbf{F}_1 = (20.5\mathbf{i} + 14.3\mathbf{j})$ N; $\mathbf{F}_2 = (− 36.4\mathbf{i} + 21.0\mathbf{j})$ N
     (b) $(− 15.9\mathbf{i} + 35.3\mathbf{j})$ N   (c) $(− 3.18\mathbf{i} + 7.07\mathbf{j})$ m/s^2
     (d) $(− 5.54\mathbf{i} + 23.7\mathbf{j})$ m/s   (e) $(− 2.30\mathbf{i} + 39.3\mathbf{j})$ m
     (f) 1 480 J   (g) 1 480 J
69. (a) 4.12 m   (b) 3.35 m
71. 1.68 m/s
73. (a) 14.5 m/s   (b) 1.75 kg   (c) 0.350 kg
75. 0.799 J

Chapter 8

1. (a) 259 kJ, 0, − 259 kJ   (b) 0, − 259 kJ, − 259 kJ
3. (a) − 196 J   (b) − 196 J   (c) − 196 J. The force is con-
     servative.
5. (a) 125 J   (b) 50.0 J   (c) 66.7 J   (d) Nonconservative.
     The results differ.
7. (a) 40.0 J   (b) − 40.0 J   (c) 62.5 J
9. (a) $Ax^2/2 − Bx^3/3$ (b) $\Delta U = 5A/2 − 19B/3$;
     $\Delta K = − 5A/2 + 19B/3$
11. 0.344 m
13. (a) $v_B = 5.94$ m/s; $v_C = 7.67$ m/s   (b) 147 J
15. $v = (3gR)^{1/2}$, 0.098 0 N down
17. 10.2 m
19. (a) 19.8 m/s   (b) 78.4 J   (c) 1.00
21. (a) 4.43 m/s   (b) 5.00 m
23. (a) 18.5 km, 51.0 km   (b) 10.0 MJ
25. (b) 60.0°
27. 5.49 m/s

29. 2.00 m/s, 2.79 m/s, 3.19 m/s
31. 3.74 m/s
33. (a) − 160 J   (b) 73.5 J   (c) 28.8 N   (d) 0.679
35. 489 kJ
37. (a) 1.40 m/s   (b) 4.60 cm after release   (c) 1.79 m/s
39. 1.96 m
41. (A/r^2) away from the other particle
43. (a) $r = 1.5$ mm, stable; 2.3 mm, unstable; 3.2 mm, stable;
     $r \rightarrow \infty$ mm neutral   (b) − 5.6 J < E < 1 J
     (c) 0.6 mm < r < 3.6 mm   (d) 2.6 J   (e) 1.5 mm
     (f) 4 J
45. (a) + at Ⓑ, − at Ⓓ, 0 at Ⓐ, Ⓒ, and Ⓔ   (b) Ⓒ stable;
     Ⓐ and Ⓔ unstable
     (c)

47. (b)

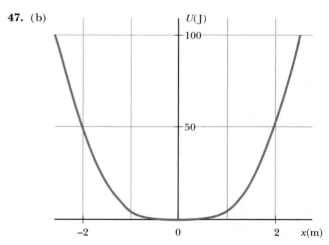

     Equilibrium at $x = 0$ (c) $v = \sqrt{0.800 \text{J}/m}$
49. (a) 1.50×10^{-10} J   (b) 1.07×10^{-9} J   (c) 9.15×10^{-10} J
51. 48.2° Note that the answer is independent of the pump-
     kin's mass and of the radius of the dome.
53. (a) 0.225 J   (b) $\Delta E_f = − 0.363$ J   (c) No; the normal
     force changes in a complicated way.
55. $\sim 10^2$ W sustainable power
57. 0.327
59. (a) 23.6 cm   (b) 5.90 m/s^2 up the incline; no.
     (c) Gravitational potential energy turns into kinetic en-
     ergy plus elastic potential energy and then entirely into
     elastic potential energy.
61. 1.25 m/s

63. (a) 0.400 m   (b) 4.10 m/s   (c) The block stays on the track.
65. (b) 2.06 m/s
67. (b) 1.44 m   (c) 0.400 m   (d) No. A very strong wind pulls the string out horizontally (parallel to the ground). The largest possible equilibrium height is equal to L.
71. (a) 6.15 m/s   (b) 9.87 m/s
73. 0.923 m/s

Chapter 9

1. (a) $(9.00\mathbf{i} - 12.0\mathbf{j})$ kg·m/s   (b) 15.0 kg·m/s at 307°
3. 6.25 cm/s west
5. $\sim 10^{-23}$ m/s
7. (b) $p = \sqrt{2mK}$
9. (a) 13.5 N·s   (b) 9.00 kN   (c) 18.0 kN
11. 260 N normal to the wall
13. 15.0 N in the direction of the initial velocity of the exiting water stream
15. 65.2 m/s
17. 301 m/s
19. (a) $v_{gx} = 1.15$ m/s   (b) $v_{px} = -0.346$ m/s
21. (a) 20.9 m/s east   (b) 8.68 kJ into internal energy
23. (a) 2.50 m/s   (b) 37.5 kJ   (c) Each process is the time-reversal of the other. The same momentum conservation equation describes both.
25. (a) 0.284   (b) 115 fJ and 45.4 fJ
27. 91.2 m/s
29. (a) 2.88 m/s at 32.3° north of east   (b) 783 J into internal energy
31. No; his speed was 41.5 mi/h.
33. 2.50 m/s at $-60.0°$
35. $(3.00\mathbf{i} - 1.20\mathbf{j})$ m/s
37. Orange: $v_i \cos \theta$; yellow: $v_i \sin \theta$
39. (a) $(-9.33\mathbf{i} - 8.33\mathbf{j})$ Mm/s   (b) 439 fJ
41. $\mathbf{r}_{CM} = (11.7\mathbf{i} + 13.3\mathbf{j})$ cm
43. 0.006 73 nm from the oxygen nucleus along the bisector of the angle
45. (a) 15.9 g   (b) 0.153 m
47. 0.700 m
49. (a) $(1.40\mathbf{i} + 2.40\mathbf{j})$ m/s   (b) $(7.00\mathbf{i} + 12.0\mathbf{j})$ kg·m/s
51. (a) 39.0 MN up   (b) 3.20 m/s² up
53. (a) 442 metric tons   (b) 19.2 metric tons
55. (a) $(1.33\mathbf{i})$ m/s   (b) $(-235\mathbf{i})$ N   (c) 0.680 s
(d) $(-160\mathbf{i})$ N·s and $(+160\mathbf{i})$ N·s   (e) 1.81 m
(f) 0.454 m   (g) -427 J   (h) $+107$ J
(i) Equal friction forces act through different distances on person and cart to do different amounts of work on them. The total work on both together, -320 J, becomes $+320$ J of internal energy in this perfectly inelastic collision.
57. 1.39 km/s
59. 240 s
61. 0.980 m
63. (a) 6.81 m/s   (b) 1.00 m
65. $(3Mgx/L)\mathbf{j}$

67. (a) 3.75 kg·m/s²   (b) 3.75 N   (c) 3.75 N   (d) 2.81 J
(e) 1.41 J   (f) Friction between sand and belt converts half of the input work into internal energy.
69. (a) As the child walks to the right, the boat moves to the left and the center of mass remains fixed.   (b) 5.55 m from the pier (c) No, since 6.55 m is less than 7.00 m.
71. (a) 100 m/s   (b) 374 J
73. (a) $\sqrt{2}\, v_i$ for m and $\sqrt{2/3}\, v_i$ for $3m$ (b) 35.3°
75. (a) 3.73 km/s   (b) 153 km

Chapter 10

1. (a) 4.00 rad/s²   (b) 18.0 rad
3. (a) 1 200 rad/s   (b) 25.0 s
5. (a) 5.24 s   (b) 27.4 rad
7. (a) 5.00 rad, 10.0 rad/s, 4.00 rad/s²   (b) 53.0 rad, 22.0 rad/s, 4.00 rad/s²
9. 13.7 rad/s²
11. $\sim 10^7$ rev/y
13. (a) 0.180 rad/s   (b) 8.10 m/s² toward the center of the track
15. (a) 8.00 rad/s   (b) 8.00 m/s, $a_r = -64.0$ m/s², $a_t = 4.00$ m/s²   (c) 9.00 rad
17. (a) 54.3 rev   (b) 12.1 rev/s
19. (a) 126 rad/s   (b) 3.78 m/s   (c) 1.27 km/s²
(d) 20.2 m
21. (a) $-2.73\mathbf{i}$ m $+ 1.24\mathbf{j}$ m   (b) second quadrant, 156°
(c) $-1.85\mathbf{i}$ m/s $- 4.10\mathbf{j}$ m/s
(d) into the third quadrant at 246°

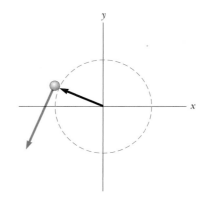

(e) $6.15\mathbf{i}$ m/s² $- 2.78\mathbf{j}$ m/s²
(f) $24.6\mathbf{i}$ N $- 11.1\mathbf{j}$ N
23. (a) 92.0 kg·m², 184 J   (b) 6.00 m/s, 4.00 m/s, 8.00 m/s, 184 J
25. (a) 143 kg·m²   (b) 2.57 kJ
29. 1.28 kg·m²
31. $\sim 10^0 = 1$ kg·m²
33. -3.55 N·m
35. 882 N·m
37. (a) 24.0 N·m   (b) 0.035 6 rad/s²   (c) 1.07 m/s²
39. (a) 0.309 m/s²   (b) 7.67 N and 9.22 N
41. (a) 872 N   (b) 1.40 kN

43. 2.36 m/s
45. (a) 11.4 N, 7.57 m/s², 9.53 m/s down   (b) 9.53 m/s
49. (a) $2(Rg/3)^{1/2}$ (b) $4(Rg/3)^{1/2}$ (c) $(Rg)^{1/2}$
51. $\frac{1}{3}\ell$
53. (a) 1.03 s   (b) 10.3 rev
55. (a) 4.00 J   (b) 1.60 s   (c) yes
57. (a) 12.5 rad/s   (b) 128 rad
59. (a) $(3g/L)^{1/2}$ (b) $3g/2L$ (c) $-\frac{3}{2}g\mathbf{i} - \frac{3}{4}g\mathbf{j}$
   (d) $-\frac{3}{2}Mg\mathbf{i} + \frac{1}{4}Mg\mathbf{j}$
61. $\alpha = g(h_2 - h_1)/2\pi R^2$
63. (b) $2gM(\sin\theta - \mu\cos\theta)(m + 2M)^{-1}$
65. 139 m/s
67. 5.80 kg·m²; the height makes no difference.
69. (a) 2 160 N·m   (b) 439 W
71. (a) 118 N and 156 N   (b) 1.19 kg·m²
73. (a) $\alpha = -0.176$ rad/s²   (b) 1.29 rev   (c) 9.26 rev

Chapter 11

1. (a) 500 J   (b) 250 J   (c) 750 J
3. $\frac{7}{10}Mv^2$
5. (a) $\frac{2}{3}g\sin\theta$ for the disk, larger than $\frac{1}{2}g\sin\theta$ for the hoop
   (b) $\frac{1}{3}\tan\theta$
7. 1.21×10^{-4} kg·m². The height is unnecessary.
9. $-7.00\mathbf{i} + 16.0\mathbf{j} - 10.0\mathbf{k}$
11. (a) $-17.0\mathbf{k}$ (b) 70.5°
13. (a) 2.00 N·m   (b) $\mathbf{k}$
15. (a) negative z direction   (b) positive z direction
17. 45.0°
19. $(17.5\mathbf{k})$ kg·m²/s
21. $(60.0\mathbf{k})$ kg·m²/s
23. $mvR[\cos(vt/R) + 1]\mathbf{k}$
25. (a) zero   (b) $(-mv_i^3 \sin^2\theta\cos\theta/2g)\mathbf{k}$
   (c) $(-2mv_i^3 \sin^2\theta\cos\theta/g)\mathbf{k}$ (d) The downward force
   of gravity exerts a torque in the $-z$ direction.
27. $-m\ell \ gt\cos\theta\,\mathbf{k}$
29. 4.50 kg·m²/s up
31. (a) 0.433 kg·m²/s   (b) 1.73 kg·m²/s
33. (a) $\omega_f = \omega_i I_1/(I_1 + I_2)$ (b) $I_1/(I_1 + I_2)$
35. (a) 1.91 rad/s   (b) 2.53 J, 6.44 J
37. (a) 0.360 rad/s counterclockwise   (b) 99.9 J
39. (a) $mv\ell$ down   (b) $M/(M + m)$
41. (a) $\omega = 2mv_i d/(M + 2m)R^2$ (b) No; some mechanical
   energy changes into internal energy.
43. (a) 2.19×10^6 m/s   (b) 2.18×10^{-18} J
   (c) 4.13×10^{16} rad/s
45. $[10Rg(1 - \cos\theta)/7r^2]^{1/2}$
51. (a) $2.70R$ (b) $F_x = -\frac{20}{7}mg, F_y = -mg$
53. 0.632
55. (a) $v_i r_i/r$ (b) $T = (mv_i^2 r_i^2)r^{-3}$ (c) $\frac{1}{2}mv_i^2(r_i^2/r^2 - 1)$
   (d) 4.50 m/s, 10.1 N, 0.450 J
57. 54.0°
59. (a) 3 750 kg·m²/s   (b) 1.88 kJ   (c) 3 750 kg·m²/s
   (d) 10.0 m/s   (e) 7.50 kJ   (f) 5.62 kJ
61. $(M/m)[3ga(\sqrt{2} - 1)]^{1/2}$
63. (c) $(8Fd/3M)^{1/2}$

67. (a) 0.800 m/s², 0.400 m/s²   (b) 0.600 N backward on
   the plank and forward on the roller, at the top of each
   roller; 0.200 N forward on each roller and backward on
   the floor, at the bottom of each roller.

Chapter 12

1. 10.0 N up; 6.00 N·m counterclockwise
3. $[(m_1 + m_b)d + m_1\ell/2]/m_2$
5. -0.429 m
7. (3.85 cm, 6.85 cm)
9. $(-1.50$ m, -1.50 m)
11. (a) 859 N   (b) 1 040 N left and upward at 36.9°
13. (a) $f_s = 268$ N, $n = 1\ 300$ N   (b) 0.324
15. (a) 1.04 kN at 60.0°   (b) $(370\mathbf{i} + 900\mathbf{j})$ N
17. 2.94 kN on each rear wheel and 4.41 kN on each front
   wheel
19. (a) 29.9 N   (b) 22.2 N
21. (a) 35.5 kN   (b) 11.5 kN   (c) -4.19 kN
23. 88.2 N and 58.8 N
25. 4.90 mm
27. 0.023 8 mm
29. 0.912 mm
31. $\dfrac{8m_1 m_2 g L_i}{\pi d^2 Y(m_1 + m_2)}$

33. (a) 3.14×10^4 N   (b) 6.28×10^4 N
35. 1.80×10^8 N/m²
37. $n_A = 5.98 \times 10^5$ N, $n_B = 4.80 \times 10^5$ N
39. (a) 0.400 mm   (b) 40.0 kN   (c) 2.00 mm   (d) 2.40 mm
   (e) 48.0 kN
41. (a)

   (b) 69.8 N   (c) 0.877L
43. (a) 160 N right   (b) 13.2 N right   (c) 292 N up
   (d) 192 N
45. (a) $T = F_g(L + d)/\sin\theta(2L + d)$
   (b) $R_x = F_g(L + d)\cot\theta/(2L + d); R_y = F_g L/(2L + d)$
47. 0.789 L
49. 5.08 kN, $R_x = 4.77$ kN, $R_y = 8.26$ kN
51. $T = 2.71$ kN, $R_x = 2.65$ kN
53. (a) $\mu_k = 0.571$; the normal force acts 20.1 cm to the left
   of the front edge of the sliding cabinet.   (b) 0.501 m

55. (b) $60.0°$
57. (a) $M = (m/2)(2\mu_s \sin\theta - \cos\theta)(\cos\theta - \mu_s \sin\theta)^{-1}$
  (b) $R = (m + M)g(1 + \mu_s^2)^{1/2}$,
  $F = g[M^2 + \mu_s^2(m + M)^2]^{1/2}$
59. (a) 133 N (b) $n_A = 429$ N and $n_B = 257$ N
  (c) $R_x = 133$ N and $R_y = -257$ N
61. 66.7 N
65. 1.09 m
67. (a) 4 500 N (b) 4.50×10^6 N/m² (c) yes.
69. (a) $P_y = (F_g/L)(d - ah/g)$ (b) 0.306 m
  (c) $\mathbf{P} = (-306\mathbf{i} + 553\mathbf{j})$ N
71. $n_A = n_E = 6.66$ kN; $F_{AB} = 10.4$ kN $= F_{BC} = F_{DC} = F_{DE}$;
  $F_{AC} = 7.94$ kN $= F_{CE}$; $F_{BD} = 15.9$ kN

Chapter 13

1. (a) 1.50 Hz, 0.667 s (b) 4.00 m (c) π rad (d) 2.83 m
3. (a) 20.0 cm (b) 94.2 cm/s as the particle passes through equilibrium (c) 17.8 m/s² at the maximum displacement from equilibrium
5. (b) 18.8 cm/s, 0.333 s (c) 178 cm/s², 0.500 s
  (d) 12.0 cm
7. 0.627 s
9. (a) 40.0 cm/s, 160 cm/s² (b) 32.0 cm/s, -96.0 cm/s²
  (c) 0.232 s
11. 40.9 N/m
13. (a) 0.750 m (b) $x = -(0.750$ m$)\sin(2.00t/s)$
15. 0.628 m/s
17. 2.23 m/s
19. (a) 28.0 mJ (b) 1.02 m/s (c) 12.2 mJ (d) 15.8 mJ
21. (a) 2.61 m/s (b) 2.38 m/s
23. 2.60 cm and -2.60 cm

Chapter 13, Problem 57(a)

25. (a) 35.7 m (b) 29.1 s
27. $\sim 10^0$ s
29. (a) 0.817 m/s (b) 2.54 rad/s² (c) 0.634 N
33. 0.944 kg·m²
37. (a) 5.00×10^{-7} kg·m² (b) 3.16×10^{-4} N·m/rad
39. The x coordinate of the crank pin is $A \cos\omega t$.
41. 1.00×10^{-3} s^{-1}
43. (a) 2.95 Hz (b) 2.85 cm
47. Either 1.31 Hz or 0.641 Hz
49. 6.58 kN/m
51. (a) $2Mg$; $Mg(1 + y/L)$ (b) $T = (4\pi/3)(2L/g)^{1/2}$; 2.68 s
53. 6.62 cm
55. 9.19×10^{13} Hz
57. (a) See bottom of preceding column.
  (b) $\dfrac{dT}{dt} = \dfrac{\pi(dM/dt)}{2\rho a^2 g^{1/2}[L_i + (dM/dt)\,t/2\rho a^2]^{1/2}}$
  (c) $T = 2\pi g^{-1/2}[L_i + (dM/dt)\,t/2\rho a^2]^{1/2}$
59. $f = (2\pi L)^{-1}(gL + kh^2/M)^{1/2}$
61. (a) 3.56 Hz (b) 2.79 Hz (c) 2.10 Hz
63. (a) 3.00 s (b) 14.3 J (c) 25.5°
65. 0.224 rad/s

Chapter 14

1. $\sim 10^{-7}$ N toward you
3. $\mathbf{g} = (Gm/\ell^2)(\frac{1}{2} + \sqrt{2})$ toward the opposite corner
5. $(-100\mathbf{i} + 59.3\mathbf{j})$ pN
7. (a) 4.39×10^{20} N (b) 1.99×10^{20} N (c) 3.55×10^{22} N
9. 0.613 m/s² toward the Earth
11.

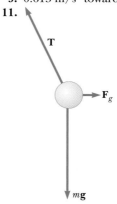

1.000 m $-$ 61.3 nm
15. 12.6×10^{31} kg
17. 1.27
19. 1.90×10^{27} kg
21. 8.92×10^7 m
25. $g = 2MGr(r^2 + a^2)^{-3/2}$ toward the center of mass
27. (a) -4.77×10^9 J (b) 569 N (c) 569 N up
29. (a) 1.84×10^9 kg/m³ (b) 3.27×10^6 m/s²
  (c) -2.08×10^{13} J
31. (a) -1.67×10^{-14} J (b) At the center
33. 1.58×10^{10} J
35. (a) 1.48 h (b) 7.79 km/s (c) 6.43×10^9 J

37. 1.66×10^4 m/s

41. 15.6 km/s

43. $GM_E m/12 R_E$

45. $2 GmM/\pi R^2$ straight up in the picture

47. (a) 7.41×10^{-10} N   (b) 1.04×10^{-8} N
     (c) 5.21×10^{-9} N

49. 2.26×10^{-7}

51. (b) 1.10×10^{32} kg

53. (b) $GMm/2R$

55. 7.79×10^{14} kg

57. 7.41×10^{-10} N

59. $v_{\text{esc}} = (8\pi G\rho/3)^{1/2} R$

61. (a) $v_1 = m_2 (2G/d)^{1/2} (m_1 + m_2)^{-1/2}$
     $v_2 = m_1 (2G/d)^{1/2} (m_1 + m_2)^{-1/2}$
     $v_{\text{rel}} = (2G/d)^{1/2} (m_1 + m_2)^{1/2}$
     (b) $K_1 = 1.07 \times 10^{32}$ J, $K_2 = 2.67 \times 10^{31}$ J

63. (a) $A = M/\pi R^4$ (b) $F = GmM/r^2$ toward the center
     (c) $F = GmMr^2/R^4$ toward the center

65. 119 km

67. (a) -36.7 MJ   (b) 9.24×10^{10} kg·m²/s
     (c) 5.58 km/s, 10.4 Mm   (d) 8.69 Mm   (e) 134 min

71.

t (s)	x (m)	y (m)	v_x (m/s)	v_y (m/s)
0	0	12 740 000	5 000	0
10	50 000	12 740 000	4 999.9	-24.6
20	99 999	12 739 754	4 999.7	-49.1
30	149 996	12 739 263	4 999.4	-73.7 . . .

The object does not hit the Earth; its minimum radius is
$1.33 R_E$. Its period is 1.09×10^4 s. A circular orbit would
require speed 5.60 km/s.

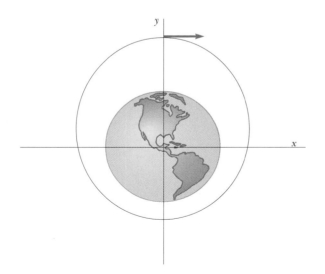

Chapter 15

1. 0.111 kg

3. 6.24 MPa

5. 5.27×10^{18} kg

7. 1.62 m

9. 7.74×10^{-3} m²

11. 271 kN horizontally backward

13. $P_0 + (\rho d/2)(g^2 + a^2)^{1/2}$

15. 0.722 mm

17. 10.5 m; no, some alcohol and water evaporate.

19. 12.6 cm

21. 1.07 m²

23. (a) 9.80 N   (b) 6.17 N

25. (a) 7.00 cm   (b) 2.80 kg

27. $\rho_{\text{oil}} = 1\,250$ kg/m³; $\rho_{\text{sphere}} = 500$ kg/m³

29. 1 430 m³

31. 2.67×10^3 kg

33. (a) 1.06 m/s   (b) 4.24 m/s

35. (a) 17.7 m/s   (b) 1.73 mm

37. 31.6 m/s

39. 68.0 kPa

41. 103 m/s

43. (a) 4.43 m/s   (b) The siphon can be no higher than
     10.3 m.

45. $2\sqrt{h(h_0 - h)}$

47. 0.258 N

49. 1.91 m

53. 709 kg/m³

55. top scale 17.3 N; bottom scale 31.7 N

59. 90.04%

61. 4.43 m/s

63. (a) 10.3 m   (b) 0

65. (a) 18.3 mm   (b) 14.3 mm   (c) 8.56 mm

67. (a) 2.65 m/s   (b) 2.31×10^4 Pa

69. (a) 1.25 cm   (b) 13.8 m/s

Chapter 16

1. $y = 6 [(x - 4.5t)^2 + 3]^{-1}$

3. (a) left   (b) 5.00 m/s

5. (a) longitudinal   (b) 665 s

7. (a) 156°   (b) 0.058 4 cm

9. (a) y_1 in $+x$ direction, y_2 in $-x$ direction   (b) 0.750 s
     (c) 1.00 m

11. 30.0 N

13. 1.64 m/s²

15. 13.5 N

17. 586 m/s

19. 32.9 ms

21. 0.329 s

23. (a) See top of next page   (b) 0.125 s

25. 0.319 m

27. 2.40 m/s

29. (a) 0.250 m   (b) 40.0 rad/s   (c) 0.300 rad/m
     (d) 20.9 m   (e) 133 m/s   (f) $+x$

31. (a) $y = (8.00$ cm$) \sin(7.85x + 6\pi t)$
     (b) $y = (8.00$ cm$) \sin(7.85x + 6\pi t - 0.785)$

33. (a) 0.500 Hz, 3.14 rad/s   (b) 3.14 rad/m
     (c) $(0.100$ m$) \sin(3.14x/$m $- 3.14t/$s$)$

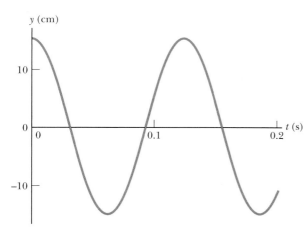

Chapter 16, Problem 23(a)

   (d) $(0.100 \text{ m}) \sin(-3.14t/s)$
   (e) $(0.100 \text{ m}) \sin(4.71 \text{ rad} - 3.14t/s)$ (f) 0.314 m/s
35. 2.00 cm, 2.98 m, 0.576 Hz, 1.72 m/s
37. (b) 3.18 Hz
41. 55.1 Hz
43. (a) 62.5 m/s   (b) 7.85 m   (c) 7.96 Hz   (d) 21.1 W
45. (a) $A = 40.0$ (b) $A = 7.00$, $B = 0$, $C = 3.00$. One can
   take the dot product of the given equation with each one
   of $\mathbf{i}$, $\mathbf{j}$, and $\mathbf{k}$.   (c) By inspection, $A = 0$, $B = 7.00$ mm,
   $C = 3.00/\text{m}$, $D = 4.00/\text{s}$, $E = 2.00$. Consider the average
   value of both sides of the given equation to find A. Then
   consider the maximum value of both sides to find B. You
   can evaluate the partial derivative of both sides of the
   given equation with respect to x and separately with respect to t to obtain equations yielding C and D upon chosen substitutions for x and t. Then substitute $x = 0$
   and $t = 0$ to obtain E.
47. It is if $v = (T/\mu)^{1/2}$
49. ~1 min
51. (a) $3.33\mathbf{i}$ m/s   (b) -5.48 cm   (c) 0.667 m, 5.00 Hz
   (d) 11.0 m/s
53. $(Lm/Mg \sin \theta)^{1/2}$
55. (a) 39.2 N   (b) 0.892 m   (c) 83.6 m/s
57. 14.7 kg
61. (a) $(0.707)2(L/g)^{1/2}$ (b) $L/4$
63. 3.86×10^{-4}
65. (a) $v = (2T_0/\mu_0)^{1/2} = v_0 2^{1/2}$
   $v' = (2T_0/3\mu_0)^{1/2} = v_0 (2/3)^{1/2}$
   (b) $0.966t_0$
67. 130 m/s, 1.73 km

Chapter 17
1. 5.56 km
3. 7.82 m
5. (a) 27.2 s   (b) 25.7 s; the interval in (a) is longer
7. (a) 153 m/s   (b) 614 m
9. (a) amplitude 2.00 μm, wavelength 40.0 cm,
   speed 54.6 m/s   (b) -0.433 μm   (c) 1.72 mm/s

11. $\Delta P = (0.2 \text{ Pa}) \sin(62.8x/\text{m} - 2.16 \times 10^4 \text{ } t/s)$
13. (a) 6.52 mm   (b) 20.5 m/s
15. 5.81 m
17. 66.0 dB
19. (a) 3.75 W/m^2 (b) 0.600 W/m^2
21. (a) 1.32×10^{-4} W/m^2 (b) 81.2 dB
23. 65.6 dB
25. (a) 65.0 dB   (b) 67.8 dB   (c) 69.6 dB
27. 1.13 μW
29. (a) 30.0 m   (b) 9.49×10^5 m
31. (a) 332 J   (b) 46.4 dB
33. (a) 75.7-Hz drop   (b) 0.948 m
35. 26.4 m/s
37. 19.3 m
39. (a) 338 Hz   (b) 483 Hz
41. 56.4°
43. (a) 56.3 s   (b) 56.6 km farther along
45. 400 m; 27.5%
47. (a) 23.2 cm   (b) 8.41×10^{-8} m   (c) 1.38 cm
49. (a) 0.515/min   (b) 0.614/min
51. 7.94 km
53. (a) 55.8 m/s   (b) 2 500 Hz
55. Bat is gaining on the insect at the rate of 1.69 m/s.
57. (a)

   (b) 0.343 m   (c) 0.303 m   (d) 0.383 m
   (e) 1.03 kHz
59. (a) 0.691 m   (b) 691 km
61. 1204.2 Hz
63. (a) 0.948°   (b) 4.40°
65. 1.34×10^4 N
67. 95.5 s
69. (b) 531 Hz
71. (a) 6.45   (b) 0
73. ~10^{11} Hz

Chapter 18
1. (a) 9.24 m   (b) 600 Hz
3. 5.66 cm
5. 91.3°
7. (a) 2   (b) 9.28 m and 1.99 m
9. 15.7 m, 31.8 Hz, 500 m/s
11. At 0.089 1 m, 0.303 m, 0.518 m, 0.732 m, 0.947 m, and
   1.16 m from one speaker
13. (a) 4.24 cm   (b) 6.00 cm   (c) 6.00 cm   (d) 0.500 cm,
   1.50 cm, and 2.50 cm
17. 0.786 Hz, 1.57 Hz, 2.36 Hz, and 3.14 Hz
19. (a) 163 N   (b) 660 Hz
21. 19.976 kHz

23. 31.2 cm from the bridge; 3.84%
25. (a) 350 Hz   (b) 400 kg
27. 0.352 Hz
29. (a) 3.66 m/s   (b) 0.200 Hz
31. (a) 0.357 m   (b) 0.715 m
33. (a) 531 Hz   (b) 42.5 mm
35. around 3 kHz
37. n(206 Hz) for $n = 1$ to 9, and n(84.5 Hz) for $n = 2$ to 23
39. 239 s
41. 0.502 m and 0.837 m
43. (a) 350 m/s   (b) 1.14 m
45. (a) 19.5 cm   (b) 841 Hz
47. (a) 1.59 kHz   (b) odd-numbered harmonics
     (c) 1.11 kHz
49. 5.64 beats/s
51. (a) 1.99 beats/s   (b) 3.38 m/s
53. The second harmonic of E is close to the third harmonic
of A, and the fourth harmonic of C# is close to the fifth
harmonic of A.
55. (a) 3.33 rad   (b) 283 Hz
57. 3.85 m/s away from the station or 3.77 m/s toward the
station
59. 85.7 Hz
61. 31.1 N
63. (a) 59.9 Hz   (b) 20.0 cm
65. (a) 1/2   (b) $[n/(n + 1)]^2 \, T$ (c) 9/16
67. 50.0 Hz, 1.70 m
69. (a) $2A \sin(2\pi x/\lambda) \cos(2\pi vt/\lambda)$
     (b) $2A \sin(\pi x/L) \cos(\pi vt/L)$
     (c) $2A \sin(2\pi x/L) \cos(2\pi vt/L)$
     (d) $2A \sin(n\pi x/L) \cos(n\pi vt/L)$

Chapter 19

1. (a) 37.0°C = 310 K   (b) -20.6°C = 253 K
3. (a) -274°C   (b) 1.27 atm   (c) 1.74 atm
5. (a) -320°F   (b) 77.3 K
7. (a) 810°F   (b) 450 K
9. 3.27 cm
11. (a) 3.005 8 m   (b) 2.998 6 m
13. 55.0°C
15. (a) 0.109 cm²   (b) increase
17. (a) 0.176 mm   (b) 8.78 μm   (c) 0.093 0 cm³
19. (a) 2.52 MN/m²   (b) It will not break.
21. 1.14°C
23. (a) 99.4 cm³   (b) 0.943 cm
25. (a) 3.00 mol   (b) 1.80×10^{24} molecules
27. 1.50×10^{29} molecules
29. 472 K
31. (a) 41.6 mol   (b) 1.20 kg, in agreement with the
tabulated density
33. (a) 400 kPa   (b) 449 kPa
35. 2.27 kg
37. 3.67 cm³
39. 4.39 kg
43. (a) 94.97 cm   (b) 95.03 cm

45. 208°C
47. 3.55 cm
49. (a) Expansion makes density drop.   (b) 5×10^{-5} (°C)$^{-1}$
51. (a) $h = nRT/(mg + P_0A)$ (b) 0.661 m
53. $\alpha \, \Delta T$ is much less than 1.
55. (a) 9.49×10^{-5} s   (b) 57.4 s lost
57. (a) $\rho g P_0 V_i (P_0 + \rho g d)^{-1}$ (b) decrease   (c) 10.3 m
61. (a) 5.00 MPa   (b) 9.58×10^{-3}
63. 2.74 m
65. $L_c = 9.17$ cm, $L_s = 14.2$ cm
67. (a) $L_f = L_i e^{\alpha \Delta T}$ (b) 2.00×10^{-4}%; 59.4%
69. (a) 6.17×10^{-3} kg/m   (b) 632 N   (c) 580 N; 192 Hz

Chapter 20

1. $(10.0 + 0.117)$°C
3. 0.234 kJ/kg · °C
5. 29.6°C
7. (a) 0.435 cal/g · °C   (b) beryllium
9. (a) 25.8°C   (b) No
11. 50.7 ks
13. 0.294 g
15. 0.414 kg
17. (a) 0°C   (b) 114 g
19. 59.4°C
21. 1.18 MJ
23. (a) $4P_iV_i$ (b) $T = (P_i/nRV_i)\, V^2$
25. 466 J
27. 810 J, 506 J, 203 J
29. $Q = -720$ J
31.

	Q	W	ΔE_{int}
BC	$-$	0	$-$
CA	$-$	$-$	$-$
AB	$+$	$+$	$+$

33. (a) 7.50 kJ   (b) 900 K
35. 3.10 kJ; 37.6 kJ
37. (a) 0.041 0 m³   (b) -5.48 kJ   (c) -5.48 kJ
41. 2.40×10^6 cal/s
43. 10.0 kW
45. 51.2°C
47. (a) 0.89 ft² · °F · h/Btu   (b) 1.85 ft² · °F · h/Btu   (c) 2.08
49. (a) $\sim 10^3$ W   (b) decreasing at $\sim 10^{-1}$ K/s
51. 364 K
53. 47.7 g
55. (a) 16.8 L   (b) 0.351 L/s
57. 2.00 kJ/kg · °C
59. 1.87 kJ
61. (a) $4P_iV_i$ (b) $4P_iV_i$ (c) 9.08 kJ
63. 5.31 h
65. 872 g
67. (a) 15.0 mg. Block: $Q = 0$, $W = +5.00$ J, $\Delta E_{int} = 0$,
     $\Delta K = -5.00$ J; Ice: $Q = 0$, $W = -5.00$ J, $\Delta E_{int} = 5.00$ J,
     $\Delta K = 0$.

(b) 15.0 mg. Block: $Q = 0$, $W = 0$, $\Delta E_{int} = 5.00$ J,
$\Delta K = -5.00$ J; Metal: $Q = 0$, $W = 0$, $\Delta E_{int} = 0$, $\Delta K = 0$.
(c) 0.004 04°C. Moving slab: $Q = 0$, $W = +2.50$ J,
$\Delta E_{int} = 2.50$ J, $\Delta K = -5.00$ J; Stationary slab: $Q = 0$,
$W = -2.50$ J, $\Delta E_{int} = 2.50$ J, $\Delta K = 0$

69. 10.2 h
71. 9.32 kW

Chapter 21

1. 6.64×10^{-27} kg
3. 0.943 N; 1.57 Pa
5. 17.6 kPa
7. 3.32 mol
9. (a) 3.53×10^{23} atoms   (b) 6.07×10^{-21} J
(c) 1.35 km/s
11. (a) 8.76×10^{-21} J for both   (b) 1.62 km/s for helium;
514 m/s for argon
13. 75.0 J
15. (a) 3.46 kJ   (b) 2.45 kJ   (c) 1.01 kJ
17. (a) 118 kJ   (b) 6.03×10^3 kg
19. Between 10^{-2}°C and 10^{-3}°C
21. (a) 316 K   (b) 200 J
23. $9\,P_iV_i$
25. (a) 1.39 atm   (b) 366 K, 253 K   (c) 0, 4.66 kJ, -4.66 kJ
27. 227 K
29. (a)

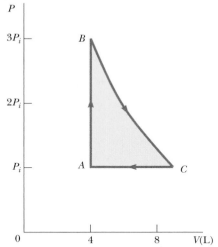

(b) 8.79 L   (c) 900 K   (d) 300 K   (e) 336 J
31. 25.0 kW
33. (a) 9.95 cal/K, 13.9 cal/K   (b) 13.9 cal/K, 17.9 cal/K
35. 2.33×10^{-21} J
37. The ratio of oxygen to nitrogen molecules decreases to
85.5% of its sea-level value.
39. (a) 6.80 m/s   (b) 7.41 m/s   (c) 7.00 m/s
43. 819°C
45. (a) 3.21×10^{12} molecules   (b) 778 km
(c) 6.42×10^{-4} s^{-1}
49. (a) 9.36×10^{-8} m   (b) 9.36×10^{-8} atm   (c) 302 atm
51. (a) 100 kPa, 66.5 L, 400 K, 5.82 kJ, 7.48 kJ, 1.66 kJ

(b) 133 kPa, 49.9 L, 400 K, 5.82 kJ, 5.82 kJ, 0
(c) 120 kPa, 41.6 L, 300 K, 0, -910 J, -910 J
(d) 120 kPa, 43.3 L, 312 K, 722 J, 0, -722 J
55. 0.625
57. (a) Pressure increases as volume decreases.
(d) 0.500 atm^{-1}, 0.300 atm^{-1}
59. 1.09×10^{-3}; 2.69×10^{-2}; 0.529; 1.00; 0.199;
1.01×10^{-41}; 1.25×10^{-1082}
61. (a) Larger-mass molecules settle to the outside.
63. (a) 0.203 mol   (b) $T_B = T_C = 900$ K; $V_C = 15.0$ L

(c, d)	*P* (atm)	*V* (L)	*T* (K)	E_{int} (kJ)
A	1	5	300	0.760
B	3	5	900	2.28
C	1	15	900	2.28

(e) For $A \rightarrow B$, lock the piston in place and put the cylinder into an oven at 900 K. For $B \rightarrow C$, keep the gas in the
oven while gradually letting the gas expand to lift a load
on the piston as far as it can. For $C \rightarrow A$, move the cylinder from the oven back to the 300-K room and let the gas
cool and contract.

(f, g)	*Q* (kJ)	*W* (kJ)	ΔE_{int} (kJ)
$A \rightarrow B$	1.52	0	1.52
$B \rightarrow C$	1.67	1.67	0
$C \rightarrow A$	-2.53	-1.01	-1.52
ABCA	0.656	0.656	0

65. (a) 3.34×10^{26} molecules   (b) during the 27th day
(c) 2.53×10^6
67. (a) 0.510 m/s   (b) 20 ms

Chapter 22

1. (a) 6.94%   (b) 335 J
3. (a) 10.7 kJ   (b) 0.533 s
5. (a) 1.00 kJ   (b) 0
7. (a) 67.2%   (b) 58.8 kW
9. (a) 869 MJ   (b) 330 MJ
11. (a) 741 J   (b) 459 J
13. 0.330 or 33.0%
15. (a) 5.12%   (b) 5.27 TJ/h   (c) As conventional energy
sources become more expensive, or as their true costs are
recognized, alternative sources become economically vi-
able.
17. (a) 214 J, 64.3 J
(b) -35.7 J, -35.7 J. The net effect is the transport of
energy from the cold to the hot reservoir without expen-
diture of external work.
(c) 333 J, 233 J
(d) 83.3 J, 83.3 J, 0. The net effect is the expulsion of the
energy entering the system by heat, entirely by work, in a
cyclic process.
(e) -0.111 J/K. The entropy of the Universe has
decreased.

19. (a) 244 kPa   (b) 192 J
21. 146 kW, 70.8 kW
23. 9.00
27. 72.2 J
29. (a) 24.0 J   (b) 144 J
31. -610 J/K
33. 195 J/K
35. 3.27 J/K
37. 1.02 kJ/K
39. 5.76 J/K. Temperature is constant if the gas is ideal.
41. 0.507 J/K
43. 18.4 J/K
45. (a) 1   (b) 6
47. (a)

Macrostate	Possible Microstates	Total Number of Microstates
All R	RRR	1
2R, 1G	RRG, RGR, GRR	3
1R, 2G	GRR, GRG, RGG	3
All G	GGG	1

(b)

Macrostate	Possible Microstates	Total Number of Microstates
All R	RRRRR	1
4R, 1G	RRRRG, RRRGR, RRGRR, RGRRR, GRRRR	5
3R, 2G	RRRGG, RRGRG, RGRRG, GRRRG, RRGGR, RGRGR, GRRGR, RGGRR, GRGRR, GGRRR	10
2R, 3G	GGGRR, GGRGR, GRGGR, RGGGR, GGRRG, GRGRG, RGGRG, GRRGG, RGRGG, RRGGG	10
1R, 4G	GGGGR, GGGRG, GGRGG, GRGGG, RGGGG	5
All G	GGGGG	1

49. 1.86
51. (a) 5.00 kW   (b) 763 W
53. (a) $2nRT_i \ln 2$ (b) 0.273
55. 23.1 mW
57. 5.97×10^4 kg/s
59. (a) 3.19 cal/K   (b) 98.19°F, 2.59 cal/K
61. 1.18 J/K
63. (a) $10.5nRT_i$ (b) $8.50nRT_i$ (c) 0.190   (d) 0.833
65. $nC_P \ln 3$
69. (a) 96.9 W $= 8.33 \times 10^4$ cal/hr
     (b) 1.19°C/h $= 2.14$°F/h

Page numbers in *italics* indicate illustrations; page numbers followed by an "n" indicate footnotes, page numbers followed by "t" indicate tables.

Standard Abbreviations and Symbols for Units

Symbol	Unit	Symbol	Unit
A	ampere	K	kelvin
Å	angstrom	kcal	kilocalorie
u	atomic mass unit	kg	kilogram
atm	atmosphere	kmol	kilomole
Btu	British thermal unit	L	liter
C	coulomb	lb	pound
°C	degree Celsius	m	meter
cal	calorie	min	minute
eV	electron volt	mol	mole
°F	degree Fahrenheit	N	newton
F	farad	Pa	pascal
ft	foot	rad	radian
G	gauss	rev	revolution
g	gram	s	second
H	henry	T	tesla
h	hour	V	volt
hp	horsepower	W	watt
Hz	hertz	Wb	weber
in.	inch	Ω	ohm
J	joule		

Mathematical Symbols Used in the Text and Their Meaning

Symbol	Meaning
$=$	is equal to
$\equiv$	is defined as
$\neq$	is not equal to
$\propto$	is proportional to
$\sim$	is on the order of
$>$	is greater than
$<$	is less than
$>> (<<)$	is much greater (less) than
$\approx$	is approximately equal to
Δx	the change in x
$\sum_{i=1}^{N} x_i$	the sum of all quantities x_i from $i = 1$ to $i = N$
$\lvert x \rvert$	the magnitude of x (always a nonnegative quantity)
$\Delta x \to 0$	Δx approaches zero
$\dfrac{dx}{dt}$	the derivative of x with respect to t
$\dfrac{\partial x}{\partial t}$	the partial derivative of x with respect to t
$\displaystyle\int$	integral